D0858263

Author's Handbook of
STYLES
for Life Science Journals

rec'd 2/12/96 Beverly Bks $89.51

Michel C. Atlas
Kornhauser Health Sciences Library
University of Louisville

CRC Press
Boca Raton   New York   London   Tokyo

Library of Congress Cataloging-in-Publication Data

Atlas, Michel C.
    Author's handbook of styles for life science journals / Michel C. Atlas.
      p.   cm.
    Includes index.
    ISBN 0-8493-2503-X
    1. Medical writing--Handbooks, manuals, etc. 2. Authorship--Style manuals. 3. Medical literature--
Handbooks, manuals, etc. I. Title.
    R119.A85 1995
    808'.06661--dc20

                  95-38010
                      CIP

This book contains information obtained from authentic and highly regarded sources. Reprinted material is quoted with permission, and sources are indicated. A wide variety of references are listed. Reasonable efforts have been made to publish reliable data and information, but the author and the publisher cannot assume responsibility for the validity of all materials or for the consequences of their use.

Neither this book nor any part may be reproduced or transmitted in any form or by any means, electronic or mechanical, including photocopying, microfilming, and recording, or by any information storage or retrieval system, without prior permission in writing from the publisher.

CRC Press, Inc.'s consent does not extend to copying for general distribution, for promotion, for creating new works, or for resale. Specific permission must be obtained in writing from CRC Press for such copying.

Direct all inquiries to CRC Press, Inc., 2000 Corporate Blvd., N.W., Boca Raton, Florida 33431.

© 1996 by CRC Press, Inc.

No claim to original U.S. Government works
International Standard Book Number 0-8493-2503-X
Library of Congress Card Number 95-38010
Printed in the United States of America 1 2 3 4 5 6 7 8 9 0
Printed on acid-free paper

PREFACE

Over 2 million articles are published in the biomedical sciences alone each year. Each of the manuscripts for these articles, not to mention those that never get published, must be prepared according to the specific styles of the journal to which they will be submitted. This requires consultation with the instructions to the authors for each of the 20,000 biomedical journals currently estimated as being published in the world today. Each journal requires that the articles submitted to it for consideration for publication be written, arranged, formatted, copied, and submitted in accordance with its own idiosyncratic prerequisites and style. The mode and manner of submission, qualifications for submission, and necessary items to include with the submission may vary slightly or extensively from journal to journal even within a specific specialty, from the same sponsoring agency or commercial publisher.

Authors need to know whether the journal to which they are considering submitting a manuscript is the appropriate outlet for their manuscript. They need to know whether their submission falls within the scope of that journal's publication efforts. They must know what specific information in the way of scientific evidence, ethical considerations (copyright, multiple authorship, multiple publication, conflict of interest, ethical treatment of experimental subjects, etc.), statistical analyses, and other requirements (forms to be signed, fees to be paid, permissions to be obtained and submitted, etc.) must be included in or submitted with their manuscript before their paper can be considered complete for submission.

A manuscript must be submitted in the proper manner before a journal's editor will even begin to consider entering it into the publication process. Journal articles may be returned unconsidered to their authors for non-scientific and seemingly trivial reasons like not sending the proper number of copies, manuscripts being too long or being printed in unacceptable fonts or on unacceptable paper, or the reference list being arranged in a format inappropriate for that journal. Many things, in addition to the preparation of scientific evidence, must be planned in advance of the submission of an article for publication in a life science journal. Finding the information about manuscript preparation for a specific journal is often difficult. Instructions to authors appear in only some issues of most journals. If these are not available time is wasted while trying to find the necessary style information. Authors cannot afford such delays.

Attempts to standardize the format of journals by a group, now known as the International Committee of Medical Journal Editors, has been only marginally successful. This group has been working for almost 30 years to bring some level of cohesion and uniformity in the requirements for publication in biomedical journals. The fourth edition of the "Uniform Requirements for Manuscripts Submitted to Biomedical Journals"was published in January 1993. Over 250 journals from around the world have agreed to "consider for publication manuscripts prepared in accordance" with the requirements. The requirements developed by this group, which include ethical considerations for conducting experiments, is included on pages 681-691 of the *Author's Handbook of Styles for Life Science Journals*.

The *Author's Handbook of Styles for Life Science Journals* is designed to assist scientists and their editorial and/or secretarial assistants in the correct and complete compilation of a manuscript for submission to their journal of choice. The 440 journals included are all English-language journals that accept unsolicited articles for consideration for publication. Titles were selected for inclusion because of their position on the "Journal Rankings by Times Cited" in the 1991 *Science Citation Index Journal Citation Report*. Journals from outside the fields of the life sciences are included on this list, but journals from these other fields, such as engineering, physics, and chemistry, are not included here. Inclusion in the *List of Serials Indexed for Online Users* of the National Library of Medicine's *Medline* electronic database of 4000 biomedical journals was used as the criterion for whether a doubtful title should be listed. Non-English language, non-life science, and journals that print only invited articles are not included. Additional titles were added to give the book a wider scope of coverag, as the *SCI* listing and *Medline* are heavily weighted towards the biomedical sciences, so other life science journals were added based on their general level of prestige and reputation. Some new titles that promise to be important to their fields, like *Nature Medicine* and *Emerging Infectious Diseases* were also included. Just as instructions for authors always include pleas for conciseness and brevity, so too were efforts made to keep this volume short, and so not all titles each user might wish to see could be included.

Journal titles in the *Author's Handbook of Styles for Life Science Journals* are listed alphabetically; their sponsoring societies or other agencies and/or publishers are also listed. The entries for each journal are arranged as Aim and Scope, Manuscript Submission, Manuscript Format, Other Formats, Other, and Source. The Aim and Scope of a journal was copied, generally verbatim, directly from those journals that printed theirs regularly and prominently. Manuscript Submission includes requirements for submission such as unique and single submission, requirements for authorship, and transfer of copyright. The term 'corresponding author' is used generically to refer to the person assigned to examine proofs, order reprints, pay page charges, give final approval, receive and dispense reprint requests, and/or generally speak for all authors on a paper. Manuscript Format gives instructions on how to arrange the manuscript of a general research article, such as how or whether to divide the paper, how to construct the abstract, what kind of paper and fonts should be used, and how to compose illustrations, tables, and references. Other Formats tells what other sections the journal has in addition to the regular original research article and how submissions for those sections are to be prepared. Items included under Other are miscellaneous specific instructions for such things as the use of abbreviations, data deposition, page charges, nomenclature, and sources of information for further assistance. Source cites the location of the information presented; front matter refers to the unnumbered pages at the beginning of a journal issue; back matter is the unnumbered pages at the end.

The information given in the instructions to authors within many journals is often contradictory, incomplete, repetitious, confusing, inconsistent, and/or poorly printed. Journal instructions are usually expressed conditionally, with phrases like "when appropriate" and "if available." Most such phrasing has been deleted as the compiler assumes that authors will not worry about supplying such things as e-mail addresses if they do not have them and copyright releases when they have not used any previously published information. Page and size allotments are maximums. When dimensions were given in both inches and centimeters, only inches were included. Many instructions to authors contain extensive lists of abbreviations and symbols; only short lists are included here. The location of lengthier lists is cited; as are detailed explanations of such things as guidelines for the care of experimental subjects. For the sake of saving space, most examples, except for references, were removed. The information presented in *Author's Handbook of Styles for Life Science Journals* improves those instructions, presenting them in a more uniform, clear, readable manner that will be relatively easy to follow. It should be an invaluable resource to secretaries who have to type the manuscript, especially the citations, in a specific style. Every author should find this book useful.

I wish to thank Harvey Kane, Barbara Turgeon, and Larry Parks for their assistance in preparing this manuscript.

Michel C. Atlas, M.L.S.
Kornhauser Health Sciences Library
University of Louisville

TABLE OF CONTENTS

Acta Endocrinologica
see European Journal of Endocrinology

Acta Medica Scandinavica
see Journal of Internal Medicine

Acta Neurologica Scandinavica
Munksgaard International Publishers

Mogens Dam
University Clinic of Neurology
Hvidovre Hospital, DK-2650 Hvidovre, Denmark
Ronald J. Polinsky
5 Homewood Drive
Morris Township, NJ 07960
　　　　From N. America, S. America, Japan

AIM AND SCOPE

Acta Neurologica Scandinavica publishes manuscripts in English of a high scientific quality representing original clinical, diagnostic or experimental work in neurology and neurosurgery, especially those which bring new knowledge and observations from the application of therapies or techniques in the combating of a broad spectrum of neurological disease and neurodegenerative disorders. Relevant articles on the basic neurosciences will be published where they extend present understanding of such disorders.

MANUSCRIPT SUBMISSION

Consult a current issue for style and format. Do not exceed 10 printed pages; longer articles will be subject to page charges. Submit manuscripts in triplicate, typewritten, double-spaced and with broad margins. If possible, also supply a disc copy, stating type of system used.

Authors submitting a paper understand that the work has not been published before, is not being considered for publication elsewhere and has been read and approved by all authors. Submission means that the authors automatically agree to assign exclusive copyright to Munksgaard International Publishers upon acceptance for publication.

MANUSCRIPT FORMAT

Divide papers into: abstract (120 words), 3-8 key words, introduction, material and methods, results, discussion, acknowledgments, and references. Use National Library of Medicine's *Medical Subject Headings* as a guide for key words. Title page should contain a concise title, authors' names and initials, names in English of departments and institutions to be attributed, and their city and country of location. If the title exceeds 40 characters (letters and spaces) include a brief running head. Give name and full postal address, including postcodes, of corresponding and reprint request author.

References: Keep to pertinent minimum. Number consecutively in order of appearance in text, in accordance with Vancouver system. Identify references in text, tables, and legends by Arabic numerals (in parentheses). Number references cited only in tables or figure legends in accordance with a sequence established by the first identification of that figure or table in text. Base style on *Index Medicus*.

Avoid using abstracts as references. Include manuscripts accepted, but not published; designate abbreviated title of journal followed by (in press). Cite information from manuscripts not yet accepted as personal communication. Verify references against original documents. Abbreviate titles in accordance with *Index Medicus*.

STANDARD JOURNAL ARTICLE

List all authors when 6 or less; when 7 or more, list only the first 3 and add et al.

Gajdos PH, Outin H, Eikharrat D et al. High-dose intravenous gamma-globulin for myasthenia gravis. Lancet 1984: i: 406-7.

CORPORATE AUTHOR

International Committee of Medical Journal Editors. Uniform requirements for manuscripts submitted to biomedical journals. Ann Intern Med 1988: 108: 258–65.

CHAPTER IN A BOOK

Fieschi C. Strategies for the treatment of brain infarction. In: Battistini L et al, eds. Acute brain ischemia—medical and surgical therapy. New York: Raven Press, 1986: 1–11.

PERSONAL AUTHORS

Hynd GW, Grant Willis W. Pediatric neuropsychology, Tokyo: Grune & Stratton, 1988.

SYMPOSIAL WORKS

Rausch C, Murgatroyd PW, Schnickelheim Z, eds. Cerebral glucose utilization testing. New York: Medical Press, 1989: 141-9.

DISSERTATION OR THESIS

Matlin RK. Percutaneous retrogasserian glycerol rhizotomy. Thesis. Berkeley: University of California, 1988.

Illustrations: Figures should clarify text. Keep to a minimum. Details must be large enough to retain clarity after reduction in size to fit single-column width (81 mm). In exceptional cases two-column (120 mm) or full-page width (168 mm) will be accepted. Illustrations must fit proportions of printed page. Submit two copies of each illustration, identify with a label on back indicating number, author's name and top. Have line drawings professionally drawn; half-tones should exhibit high contrast. Type figure legends on a separate page at end of manuscript. Color illustrations must be paid for by author. Submit original color transparencies and two sets of color prints.

Tables: Type on separate sheets numbered with Arabic numerals and a self-explanatory heading.

OTHER FORMATS

CASE HISTORIES—Two printed pages.
LETTERS TO THE EDITOR—Debate within the specialty.
SHORT COMMUNICATIONS—Papers requiring rapid publication because of their significance and timeliness. Do not exceed two printed pages.

OTHER

Abbreviations and Symbols: Use: *Units, Symbols and Abbreviations. A Guide for Biological and Medical Editors and Authors*, (1988, The Royal Society of Medicine,

London). Give unfamiliar terms or abbreviations in full at first mention. Use metric units. Use no Roman numerals; in decimals use decimal point, not comma.

Page Charges: Articles in excess of 10 printed pages will incur author fees of 650 Danish kroner per page.

SOURCE

January 1994, 89(1): inside back cover.

Acta Neuropathologica
Springer International

K. Jellinger
L.B. Instut fur Kinische Neurobiolgie
Krankenhaus Lainz, Wolkersbergenstrasse 1
A-113 Wien Austria
Telephone: 02 22/8 01 10/3431; Fax: 02 22/8 04 54 01

AIM AND SCOPE

Acta Neuropathologica publishes manuscripts dealing with diseased and experimentally altered nervous tissue. The journal's scope includes histology, cytology, fine structure, and associated biochemical, histochemical, cytochemical, immunological and physiological findings. The Editors are especially interested in interdisciplinary studies that correlate abnormalities in structure and function and that investigate the pathogenesis of neurological disease.

MANUSCRIPT SUBMISSION

Submit one original manuscript (typed double-spaced with wide margins on one side only) plus two copies (photocopied on both sides) of entire text and two sets of figures mounted on normal bond paper. Limit abbreviations to those in common use. Spell out abbreviations the first time they occur. Indicate where figures are to be inserted in left margin. Materials and methods section, formulas, footnotes, and tables are published in small print. Mark other sections to be published in small print by a vertical line and letter "P" in left margin. Regular papers should not exceed 20 typewritten pages; five pages maximum for Case reports.

MANUSCRIPT FORMAT

Regular papers and Case reports should include: Title page, Summary, Key words, Introduction, Materials and methods, Results, Discussion, Acknowledgments, References, tables and figure legends. In shorter Regular papers, Results and Discussion sections may be combined.

Title Page: Include title; name(s) of author(s); name of institution where work was done; footnotes, if any, to title; name and address of corresponding author; telex or fax number. Capitalize only first word and words that are always capitalized in title, subtitle, and all subheadings.

Abstract and Key Words: Brief (250 words) resume of most important results and conclusions. List up to 5 key words for subject indexing taken from National Library of Medicine's *Medical Subject Headings*.

Footnotes: Footnotes to title appear without symbols;, number those to text consecutively; indicate footnotes to tables by superscript lower-case letters.

References: Authors are responsible for correct references and text citations. List all references cited in text and tables under first author's name in alphabetical order. For journal articles give name(s) and initial(s) of all author(s), year of publication (in parentheses), title, journal name as abbreviated in *Index Medicus*, volume number, opening and closing page numbers; for books, both authored and edited give name(s) and initial(s) of all author(s), year of publication (in parentheses), title, editors (if any), edition, volume number, publisher, city, first and last page numbers. If entire book is referenced, no page numbers are required.

JOURNAL

Herman MM, Rubinstein LJ (1984) Divergent glial and neuronal differentiation in a cerebellar neuroblastoma in an organ culture system: in vitro occurrence of synaptic ribbons. Acta Neuropathol (Berl) 65:10–24.

CHAPTER IN BOOK

Low P (1984) Endoneurial fluid pressure and microenvironment of nerve. In: Dyck PT, Thomas PK, Lambert EH, Bunge R (eds) Peripheral neuropathy, 2nd edn. Saunders, Philadelphia, pp 599–617.

BOOK

Sternberger LA (1979) Immunocytochemistry, 2nd edn. Wiley, New York.

Reference single publications cited in text as numbers in square brackets referring to an alphabetically ordered and numbered list of references. In reference list include only published papers and those in press. Cite unpublished data and personal communications in parentheses.

Illustrations: Restrict figures to those needed for clarity to a maximum of 8. Authors will pay for additional figures. Do not submit previously published illustrations except in review articles. Mount individual illustrations and those that are to appear together as one figure on regular bond paper. Do not exceed page print area of 176 mm × 236 mm or column width of 86 mm, including space for legends. Mount all figures separately and number consecutively. Label lightly on back with a soft pencil to indicate figure top, figure number, author's name, and manuscript title.

Line Drawings—Submit mounted good-quality prints of correct size for direct 1:1 reproduction. Computer drawings are acceptable if they are of comparable quality to line drawings. Lines and curves must be smooth. Letters, numbers of symbols used as labels clearly legible. Use letters 2 mm high.

Halftone Illustrations—Submit mounted photographic prints of good technical quality and contrast that are desired size for reproduction. Letters or numbers used as labels about 23 mm high.

Legends: Submit concise and informative legends on a separate page with magnifications for micrographs.

OTHER

Color Charges: Authors will be charged DM 1200 for the first page and DM 600 for each additional page towards costs for reproduction and printing of color illustrations.

SOURCE

1994, 87(1): inside back cover.

Acta Oncologica
Scandinavian University Press

Acta Oncologica
P.O. Box 3255
S-103 65 Stockholm, Sweden

AIM AND SCOPE

Acta Oncologica is a journal for the clinical oncologist and accepts articles within all fields of clinical cancer research. Articles on tumour pathology, experimental oncology and radiobiology, cancer epidemiology and medical radiophysics are published, especially if they have clinical interest. Scientific articles on cancer nursing and psychologic or social aspects of cancer are also welcomed.

MANUSCRIPT SUBMISSION

Submit 3 good quality photocopies of all written material, figures and tables and a cover letter, signed by all authors, verifying that no substantial part of the paper has been or may be published elsewhere.

Type on one side of paper only with double-spacing (at least 1 cm between lines) and margins of at least 4 cm on left side and at top.

Follow Vancouver style for technical requirements for manuscripts (*Br Med J* 1988; 296:401–5 or *Nordisk Medicin* 1988; 103: 93-6).

MANUSCRIPT FORMAT

Title Page: Include title of article, first name, middle initial, and last name of each author with institutional affiliation, name and address of corresponding author, source(s) of support in the form of grants, etc., and a short running title.

Abstract: Do not exceed 150 words on a separate sheet.

Text: Divide into Introduction, Material and Methods, Results, and Discussion. Abbreviations, which are not generally accepted, should be spelled out when first used in text. Avoid uncommon abbreviations and clinical jargon. Footnotes are not accepted.

Tables: Type each table double-spaced on a separate sheet. Do not submit tables as photographs. Supply a brief title for each table. Give each column a short or abbreviated heading. Place explanatory matter in notes under table, not in heading. Identify statistical measures of variations, such as standard deviation and standard error of the mean.

Illustrations: Figures should be professionally drawn and photographed; freehand or typewritten lettering is not accepted. Computer-drawn figures can be accepted if of good quality. Make letters and numerals large enough to allow reduction of a diagram to one column width (8 cm).

Submit figures unmounted as glossy black-and-white photographs usually about 5×7 in. Put titles and detailed explanations in legends, not in illustrations. Costs of color drawings or photographs are defrayed by author(s).

References: Number references consecutively in order first mentioned in text. Identify references in text, tables, and legends by Arabic numerals in parenthesis. Number references cited only in tables or in legends to figures in accordance with first mention in text of particular table or illustration. Abbreviate journal titles according to *Index Medicus*. Type reference list double-spaced; include name(s) of author(s) followed by initial(s), full title of article, name of journal, year of publication, volume number and number of first and last page. Reference to books should indicate author(s), title of chapter, book title, name of editor(s), place, publisher, year of publication, and number of first and last page. List all authors when six or less; when seven or more, list only first three and add et al.

OTHER FORMATS

CORRESPONDENCE AND SHORT COMMUNICATIONS— Present discussion concerning published articles, preliminary original reports, case reports, technical notes and similar material. Contributions can be written either as Letters to the Editor or arranged according to instructions for full-length reports. In latter case strictly limit Introduction, Discussion and References. Tables and illustrations, if necessary, should be kept at a minimum and be possible to print within one single-column width.

OTHER

Units: International System of Units (SI)

SOURCE

1994, 33(1): inside back cover.

Acta Otolaryngologica
Scandinavian University Press

Anne Bindslev, Managing Editor
Scandinavian University Press
Box 3255
S-103 65 Stockholm, Sweden

AIM AND SCOPE

The journal publishes original articles on basic research as well as clinical studies in the field of otolaryngology and related subdisciplines. Review articles presenting the present state of the art and containing an internationally representative bibliography are published. Short communications are preliminary reports and other short contributions. Letters to the editor provide a forum for comments on published articles. Case reports are not accepted.

MANUSCRIPT SUBMISSION

Follow rules stated in "Uniform requirements for manuscripts submitted to biomedical journals" (see *Br Med J 1988*; 296: 401-5, or *Nord Med* 1988; 103: 93-96). Manuscripts estimated to exceed 8 printed pages, including figures and tables, will be returned.

Submit two sets of all material, including illustrations and tables. Double-space entire manuscript. Leave wide margins.

MANUSCRIPT FORMAT

Organize paper: title page, abstract page, introduction, material and methods, results, discussion, acknowledgments, list of references and, after the last reference, title, name, and full address of corresponding author. Begin each section, and figure and table legends, on separate sheets. Type page number in upper right corner of each page.

Title Page: Include: title of manuscript, name of author(s), name and address of department(s) and institution(s), and a 'running title' of no more than 50 characters.

Abstract: Give name of author(s), title of manuscript, name of journal, abstract text (no more than 20 typewritten lines), and key words (avoid words already in title).

References: Use Vancouver system. Use no more than 25 references per manuscript. Number consecutively in order first mentioned in text. Identify references in text by Arabic numerals (in parentheses). Abbreviate journal titles according to *Index Medicus*.

1. Mercke U, Hakansson CH, Toremalm NG. A method for standardized studies of mucociliary activity. Acta Otolaryngol (Stockh) 1974; 78: 1 18-23.

2. Widdicombe JG. Nervous receptors in the respiratory tract and lungs. In: Hornbein T, ed. Lung biology in health and disease. Vol 17, Part 1. Regulation of breathing. New York: Dekker, 1981: 429-72.

Cite information from papers not yet accepted for publication in text as "unpublished observation(s)" or "personal communication" (in parentheses).

Illustrations: Submit unmounted illustrations in duplicate, professionally drawn or photographed and delivered as glossy black-and-white photographic prints. Computerdrawn figures are accepted provided they are of high quality. Photomicrographs must have internal scale markers, and symbols, arrows or letters should contrast with background. Type figure number, name of first author, and arrow indicating 'top' on a gummed label affixed to back of each illustration.

OTHER

Page Charges: First 4 pages are free. Additional pages 1400 SEK per page. Expense for color photos must be borne by the author(s).

SOURCE

January 1994, 114(1): inside back cover.

Acta Paediatrica
Scandinavian University Press

Editorial Office, *Acta Paediatrica*
Karolinska Hospital
Department of Paediatrics
Box 60500
S-10401 Stockholm, Sweden
Telephone: +46 8 729 24 87; Fax: +46 8 729 40 34

AIMS AND SCOPE

Acta Paediatrica publishes papers in English, covering both clinical and experimental research, in all fields of paediatrics including developmental physiology. *Acta Paediatrica* accepts review articles, original articles, short communications, therapeutic notes and case reports. Review articles should give the present state-of-the-art on topics of clinical importance and include an internationally relevant bibliography. Case reports should contain relevant clinical data, new diagnostic methods or laboratory findings with clinical implications. Short communications and therapeutic notes are intended as preliminary reports. A Letters to the Editor section constitutes a forum for comments and short discussions.

MANUSCRIPT SUBMISSION

Prepare manuscripts in English in accordance with the *"Uniform Requirements for Manuscripts Submitted to Biomedical Journal"* (*Br Med J* 1991;302:338–41).

Make a full statement to Editor-in-Chief about any prior report that could be regarded as duplication or as very similar to work submitted. Submit approval of paper for publication, signed by all authors, to Editorial Office. State clearly in paper that study has been approved by institutional ethics committee.

Be concise. Papers exceeding five printed pages (including illustrations, tables and references) will incur a page charge for each extra page. Short communications may not exceed two printed pages and Case Reports three printed pages.

Submit manuscripts in triplicate, including tables and illustrations. Double space entire manuscript.

MANUSCRIPT FORMAT

Start each part on a new page: title page, with author's name and affiliation, address of corresponding author and a short title, (abstract including keywords, text, references, tables, and figure legends.

Title Page: Leave 7-8 cm space at top of page. Avoid subtitles. This example shows content, capitalization, underlining (for italics) and spacing.

Mechanics of breathing in the newborn

L Andersson and K. Pettersson

Department of Paediatrics, University Hospital, Lund, Sweden

Short title: Studies in neonatal hypoglycaemia

Corresponding author: K. Pettersson, Department of Paediatrics, University Hospital, S-221 85 Lund, Sweden. Telephone: +00 0 000 00 00; Fax: +00 0 000 00 00.

Abstract: Do not exceed 150 words, including key words listed alphabetically.

Kohler L, Holst K. Dental health of four-year-old children. Acta Paediatr 00, Stockholm. ISSN 0803-5253

An unselected population of 1567 four-year-old children. . . . compounds have also been detected in human milk. *Caries, cow's milk, mother's milk, pesticides.*

Text: Leave a left margin of about 4 cm. Number pages in top right corner, beginning with title page. Headings (left margin): Patients and methods, Results, Discussion, Acknowledgments, References.

References: Number references consecutively in order first mentioned in text. Identify references in text, tables and legends by Arabic numerals (in parentheses). Abbreviate journal titles according to *Index Medicus*. Observe punctuation carefully.

1. Kuhl C, Andersen GE, Hertel J, Molsted-Pedersen L. Metabolic events in infants of diabetic mothers during the first 24 hours after birth. Acta Paediatr Scand 1982;71:19–25

(List all authors when six or less; when seven or more list first six and add et al.).

2. Feigin RD. Bacterial meningitis beyond the neonatal period. In: Feigin RD, Cherry ID, eds. Textbook of pediatric infectious diseases, 2nd ed. Philadelphia: WB Saunders, 1987

3. Jones G. Textbook of paediatrics. Uppsala: Almqvist & Wiksell, 1974: 193-9

4. D'Hondt E, Berge E, Colinet G. Production and quality control of the Oka strain live varicella vaccine. Postgrad Med J 1985;61(suppl 4):53–6

Notes/Footnotes: Incorporate notes/footnotes in text, within parentheses.

Illustrations: Submit glossy prints rather than original illustrations. Identify them on back by name and number. Make camera-ready with an allowance of 25-50%, for reduction. If necessary, lightly indicate top of figure on reverse. Do not mount photographs.

OTHER

Abbreviations: Do not use in title or abstract. In text use only standard abbreviations, i.e., those listed in latest editions of any recognized medical dictionaries (e.g., *Dorland's, Butterworth's*). The full term for which an abbreviation stands must precede its first use in text, unless it is a standard unit of measurement. Redefine abbreviations used in figure legends.

Units: SI system.

SOURCE
January 1994, 83(1): inside back cover.

Acta Physiologica Scandinavica
Scandinavian Physiological Society
Blackwell Scientific Publications

Professor Boris Uvnas, Chief Editor
Department of Pharmacology, Karolinska Institutet
S-171 77 Stockholm, Sweden

AIM AND SCOPE
Publishes original contributions to physiology and related sciences.

MANUSCRIPT SUBMISSION
Submit three copies. Be concise. Papers estimated to exceed 12 printed pages (including illustrations, tables and references) will usually not be accepted or, authors may be charged for excess pages. Authors may also be charged for costs of illustrations and tables considered too numerous, as well as for extra costs of redrawings and corrections.

MANUSCRIPT FORMAT
Type manuscripts double-spaced throughout on one side of standard A4 paper with a 5-cm-wide margin. Number all pages of text at top right corner, beginning with title page. Use capitals for main headings; underline subheadings. Organize manuscript as follows. Begin each section with a new page.

Title Page: Author's name, institution, university and country and, if title is longer than 40 letters and spaces, a short title not exceeding this limit for running heads. Include name and full postal address of corresponding and reprint request author.

Title should be as informative as possible and must not include abbreviations and symbols. Animal species should normally appear in title. Example: *Acta Physiol Scand* 1993, 000, 000–000

Trophic effects of hypophyseal hormones on resistance vessels. (Initial capitals only)

B. FOLKOW[1], O. G. P. ISAKSSON[1], G. KARLSTROM, A. F. LEVER[2] and N. NORDLANDER[1] (Authors' names and initials with capitals.)

[1]Department of Physiology, University of Goteborg, Sweden and [2]Medical Research Council Blood Pressure Unit, Western Infirmary, Glasgow, UK. (Addresses - department, university and country only.)

Short title. *Pituitary function and cardiovascular adaptation* (No more than 40 letters and spaces.)

Professor B. Folkow, Department of Physiology, University of Goteborg, Medicinaregatan 11, S-413 90 Goteborg, Sweden (Full postal address of the author to whom all correspondence is to be sent. Supply fax number if possible.)

Abstract: Do not exceed 250 words. Do not use abbreviations. Add up to 8 key words in alphabetical order.

FOLKOW, B., ISAKSSON, O.G.P., KARLSTROM, G., LEVER, A.F. & NORDLANDER, M. 1993. Trophic effects of hypophyseal hormones on resistance vessels and the heart in normotensive and renal hypertensive rats. *Acta Physiol Scand* 000, 000—000. Received .... accepted .... ISSN 0001-6772. Department of Physiology, University of Goteborg, Sweden, and MRC Blood Pressure Unit, Western Infirmary, Glasgow, UK.

Experiments were performed on isolated...

References: Restricted in number as much as possible. Abbreviate journal titles according to *Index Medicus*. Book references should include editor(s), publisher and place of publication. In text, references follow Harvard system in style: Folkow & Karlstrom (1984) or Folkow et al. (1988) if there are more than two authors. Check carefully that the references in text and in reference list agree.

List references, double-spaced, at end of paper in alphabetical order of (first) authors as illustrated:

FOLROW B. 1956. Structural, myogenic, humoral and nervous factors controlling peripheral resistance. In: M.

Harington (ed.) *Hypotensive Drugs*, pp. 163-173. Pergamon Press, London.

FOLROW, B, GRIMBY, G. & THULESIUS, O. 1958. Adaptive structural changes of the vascular walls in hypertension and their relation to the control of peripheral resistance. *Acta Physiol Scand* 44, 255-272.

CHRISTENSON, N.J. 1984. Studies in the bioconversion of arachidonic acid in human pregnant reproductive tissue./Thesis. Karolinska Institute, Stockholm, Sweden, ISBN 91-7222-7389.

ERXLEBEN, C. 1990. Adaptation in the crayfish stretch receptor neuron is in part mediated by Ca^{2+}-influx through stretch-activated channels. In: N. Elsner & G. Roth (eds) *Proc 48th Gottingen Neurobiol Conf.* pp. 00-00. Publisher, place.

Figures: Submit three copies of each. Indicate in margin where figure is to appear. Unless otherwise indicated, tables and figures will be placed near their first mention in text. Line drawings should either be drawn by hand or in black ink produced on a good-quality printer. Photographs should be glossy prints of good contrast. Number figures in sequence with Arabic numerals, identify on back by number, name of (first) author and first words of title. If necessary, mark top of figure with↑. Deliver prints in a special envelope marked with author's name and title of manuscript.

Submit figures of a size suitable to withstand 50% reduction for either single column (67 mm) or double column (138 mm). Letters, numbers and figures should be 1.3–1.6 mm high when printed.

Each figure should have a legend containing sufficient information to make figure intelligible without reference to text. Type all legends together, double-spaced, on a separate sheet(s). Do not mount photographs.

Put headings and identifications of symbols and histograms in figure legend and not in figure. Label figure sections in lower case in parentheses, (a), (b), etc.

Label axes centrally with initial capital letter only. Do not use abbreviations. Vertically label vertical axis. Put axis units after label in parentheses. Curve points should not be so large as to merge on a small reduction.

For histograms use:

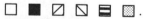

Dots must be sufficiently large that they will not disappear on reduction.

Tables: Limit number of tables. Indicate in margin where in manuscript table is to appear. Type double-spaced on separate sheets with self-explanatory heading and three dividing lines to separate titles and data. Do not use vertical rules.

Footnotes: Avoid use. If essential, number consecutively with Arabic numerals and type at foot of appropriate page.

OTHER

Electronic Submission: Manuscripts may be submitted on disk. Also submit three copies of the typescript. 5.25, 3.5, and 3 in. disks of any density can be accepted. Any software can be accessed; specify type of computer/word processor used and software package used. Supply in both word processor and ASCII formats if possible. Define keyboard characters used to represent a character not on keyboard (e.g., Greek). Create separate files for main text, reference list, figure legends, and tables.

Experimentation, Animal: Conform with the legal requirements in the country in question.

Experimentation, Human: Human experiments involving procedures which are not therapeutic and carry a significant risk of harm must include a statement that experiments were performed with understanding of each subject. Include a statement of approval by Ethics Committee of University.

Spelling: Use British spellings for haemoglobin, anaemia, etc., and 'z' spellings for characterize, categorize, etc., but not for dialyse, analyse, etc. For derived units use negative indices t: mmol 1^{-1}, counts mg^{-1} (not mmol/l, counts/mg, etc.).

Statistical Analyses: Use appropriate statistical methods to establish significance. When data deviate from a normal distribution, use non-parametric tests or perform transformation of data. When more than two simultaneous comparisons are made, use multiple t-test or similar techniques (not Student's t-test).

SOURCE

May 1994, 151(1): 131-134.

Acta Psychiatrica Scandinavica
Munksgaard International Publishers, Ltd.

Professor Jan Otto Ottoson
Department of Psychiatry
Sahlgrenska Hospital
S-413 45 Gothenburg Sweden

AIM AND SCOPE

Publishes original papers on psychiatry and adjacent fields.

MANUSCRIPT SUBMISSION

The work cannot have been published before, cannot be considered for publication elsewhere and must have been read and approved by all authors. Submission means that authors agree to assign exclusive copyright to Munksgaard International Publishers upon acceptance for publication.

Articles may not exceed 10 printed pages; longer articles will be subject to extra-page charges. Provide manuscript on a floppy disk if possible; along with 3 printed, double-spaced copies with broad margins. Use newly formatted disks and clearly label journal, author and title, file content, format, software and program version number.

MANUSCRIPT FORMAT

Divide papers into: abstract (120 words), 3-8 key words, introduction, material and methods, results, discussion, acknowledgments (if any) and references. Use National Library of Medicine's *Medical Subject Heading* as a guide to key words.

Title Page: Concise, informative title, authors' names and initials, names in English of departments and institutions to be attributed, and their city and country of location. If title exceeds 40 characters (letters and spaces) include a brief running head. State name, telephone and fax number and full postal address, including any postcodes of corresponding and reprint request author.

References: Keep to pertinent minimum and number consecutively in order of appearance in text, in accordance with Vancouver system. Identify references in text, tables, and legends by Arabic numerals (in parentheses). References cited only in tables or figure legends are numbered in accordance with a sequence established by first identification of that figure or table in text. Avoid using abstracts as references. Include manuscripts accepted, but not published; designate abbreviated title of journal followed by (in press). Cite information from manuscripts not yet accepted in text as personal communication. Verify references against original documents. Abbreviate titles in accordance with *Index Medicus*. Examples of style are based on *Index Medicus*.

STANDARD JOURNAL ARTICLE

List all authors when 6 or less; when 7 or more, list only first 3 and add et al.

GAJDOS PH, OUTIN H, EIKHARRAT D et al. High-dose intravenous gamma-globulin for myasthenia gravis. Lancet 1984: i: 406-7.

CORPORATE AUTHOR

International Committee of Medical Journal Editors. Uniform requirements for manuscripts submitted to biomedical journals. Ann Intern Med 1988: 108: 258–65.

CHAPTER IN BOOK

FIESCHI C. Strategies for the treatment of brain infarction. In: BATTISTINI, L. et al., ed. Acute brain ischemia—medical and surgical therapy. New York: Raven Press, 1986: 1-11.

Illustrations: Figures should clarify text and be kept to a minimum. Make details large enough to retain clarity after reduction in size. Plan to fit proportions of printed page. Submit two copies of each, identify with a label on back indicating number, author's name and top. Line drawings should be professionally drawn; half-tones should exhibit high contrast. Type figure legends on a separate page at end of manuscript. Color illustrations must be paid for by author. Submit original color transparencies and two sets of color prints.

Tables: Type on separate sheets numbered with Arabic numerals, and self-explanatory headings.

OTHER FORMATS

SHORT COMMUNICATIONS—Cover clearly defined issues relevant to psychiatric practice or report clinical observations that could form basis of systematic studies. Should not exceed 2 printed pages (about 900 words).

SPECIAL CONTRIBUTIONS—Review articles cover topics of interest to clinical psychiatrists and, if possible, practical implications. Do not exceed 8 printed pages including references. Include an abstract and key words.

OTHER

Abbreviations and Symbols: Use: *Units, symbols and abbreviations. A guide for biological and medical editors and authors* (1988, London: The Royal Society of Medicine). Give unfamiliar terms or abbreviations in full at first mention: All units should be metric. Use no Roman numerals; in decimals use decimal point, not comma.

Page Charges: Pages in excess of 10 printed pages are 650 Danish kroner per page.

SOURCE

January 1994, 89(1): inside back cover.

Agents and Actions

All contributions may be sent to any relevant section editor:

Managing Editor
Dr. M. J. Parnham
Parnham Advisory Services
Hankelstr. 43, D-5300 Bonn 1
Telephone: ++49/228/25 91 29; Fax: ++49/228/25 66 63

Dr. M. A. Bray
Ciba-Geigy AG
R 1056.215, CH–4002 Basel
Telephone: ++41/61/697-40 76; Fax: ++41/61/697-84 03
News, Views and Meeting Reports

Prof. Dr. W. Lorenz
Klinikum der Philipps-Universitat
Zentrum fur Operative Medizin I
Inst. f. Theoretische Chirurgie
Baldingerstr./Lahnberge
D-3350 Marburg
Telephone: ++49/6421/28 22 20; Fax: ++49/6421/28 22 25
Allergy and Histamine

Dr. A. J. Lewis
Signal Pharmaceuticals, Inc.
5555 Oberlin Drive
San Diego, CA 92121
Telephone: ++1/619/558 75 00; Fax: ++1/619/558 75 13
Allergy and Histamine

Dr. I. Ahnfelt-Ronne
Growth and Vascular Biology
Novo Nordisk A/S
Niels Steensensvej 1, DK-2820 Gentofte
Telephone: ++45/44/44 88 88; Fax: ++45/31/65 45 74
Inflammation

Dr. G. Carter
Abbott Laboratories
D–462 AP9
Abbott Park, IL 60064-3500
Telephone: ++1/708/937 98 41; Fax: ++1/708/938 50 34
Inflammation

Dr. R. Pettipher
Pfizer Inc.
Central Research Division
Eastern Point Road
Groton, CT 06340
Telephone: ++1/203/441-44 90; Fax: ++1/203/441-57 19
Molecular Immunopathology

Dr. G. A. Higgs
Research Division
Celltech Ltd.
216 Bath Road
Berks. SL1 4EN/UK
Telephone: ++44/753/53 46 55; Fax: ++44/753/53 66 32
    Molecular Immunopathology
Dr. R. O. Day
Dept. of Clinical Pharmacol. and Toxicol.
St. Vincent's Hospital Sidney, Victoria Street
Darlinghurst Sidney N.S.W. 2010/Australia
Telephone: ++61/2/339 11 11; Fax: ++61/2/332 41 42
    Clinical Pharmacology
Prof. Dr. K. Brune
Inst. of Pharmacol. and Toxicol.
University of Erlangen-Nurnberg
Universitatsstr. 22, D-8520 Erlangen
Telephone: ++49/9131/85 22 92; Fax:++49/9131/20 61 19
    Pain
Prof. Dr. W. B. van den Berg
Dept. of Rheumatology
University Hospital St. Radboud
Geert Grooteplein 8
NL–6525 GA Nijmegen
Telephone: ++31/80/61 65 40; Fax: ++31/80/54 14 33
    Bone and Cartilage
Dr. R. J. Smith
Upjohn Company
Hypersensitivity Diseases Research
7000 Portage Road
Kalamazoo, MI 49001-0199/USA
Telephone: ++1/616/385 69 64; Fax: ++1/616/384 93 24
    Bone and Cartilage

AIM AND SCOPE

Agents and Actions publishes papers from the following fields of research:

ALLERGY AND HISTAMINE—Mediators of allergy and immediate hypersensitivity and effects of drugs, including the general actions of histamine and their inhibition.

INFLAMMATION—Inflammatory processes, models, mediators and the effects of drugs.

MOLECULAR IMMUNOPATHOLOGY—Regulation of inflammatory processes by the immune system, cytokines and the effects of drugs.

PAIN—Chemical mediation of pain and the effects of analgesic drugs.

BONE AND CARTILAGE—Metabolic processes and the effects of mediators and drugs.

NEWS AND VIEWS—Meeting reports, letters, opinions, people, policies and general information relating to the above fields of research.

MANUSCRIPT SUBMISSION

Base contributions on new results obtained experimentally and should constitute a significant contribution to one of the above fields. Papers reassessing concepts based on earlier results will also be considered, but not data of a preliminary or screening nature. Contributions that have already been published or accepted for publication with essentially the same content will not be considered, except when results have been published only in an abbreviated form (e.g., abstract).

MANUSCRIPT FORMAT

Manuscripts should be typed double-spaced on one side of 8.5 × 11 in. paper with a 1.5 in. margin on left. Submit original and 2 copies. Number all pages including references, tables and figure legends.

Title Page: List names and addresses of authors on separate title page. If an author's present address differs from that at which work was carried out, give as footnote on title page and indicate at appropriate place in author list by superscript numbers. If correspondence is not to be addressed to first author, so indicate in a footnote with an additional superscript number. Specify a maximum of 5 keywords and provide a running head.

Abstract: Type a self-explanatory 100–150 word abstract on a separate sheet. Present problems, suggest scope and plan of experiments, indicate significant findings and conclusions.

Introduction: Give a clear, concise account of background of problem and rationale of investigation. Cite only work with a direct bearing on present investigation. Avoid exhaustive reviews of literature. Do not indicate results obtained.

Methods and Materials: Procedures must be given in sufficient detail to permit repetition of work. Summarize published procedures, detailing modifications. Give sources of supply for drugs, chemicals etc. Abbreviate cumbersome chemical names for later reference. Mark registered trade names with ®. Give drug doses as unit weight/body weight, e.g., mmol/kg or mg/kg; give concentrations in terms of molarity (e.g., nmol/l or, µmol/l), or as unit weight per unit volume (e.g., µg/ml). Molarity should refer to the active component, whether salt, acid or base. Insert sub-sections wherever possible.

Results: Describe experimental results in sufficient detail to allow repeat of experiment. Results on new drugs and comparative drug studies should include dose-response curves. Provide absolute control values where percentage changes are given. Wherever possible, give mean results with confidence limits or with standard errors of the means and number of observations. Do not use standard errors of the means for values which are not normally distributed (e.g., percentages); use non-parametric statistical analyses. Give significance as values of probability. Name statistical tests used. Do not repeat data in text, tables and figures. Do not include conclusions and theoretical considerations in this section. Divide data into sub-sections to clarify results.

Discussion: Provide an interpretation of results obtained against background of existing knowledge. Do not recapitulate or summarize results. Restrict citations to those which relate to findings.

Acknowledgment: Be brief. Acknowledge sources of drugs not freely available.

References: In text give as consecutive numbers in square brackets. Avoid excessive citation. Type reference list on a separate sheet(s) at end of manuscript, numbered according to consecutive citation in text:

K. E. Barrett, J. R. Minor and D. D. Metcalfe, *Histamine secretion induced by N-acetyl cysteine.* Agents Actions 16, 114–151 (1985).

I. F. Skidmore and C. J. Vardy, *The mediators of bronchial asthma and the mechanism of their release.* In *Pathophysiology and Treatment of Asthma and Arthritis.* Agents and Actions Suppl. vol. 14. (Ed. P. R. Saxena and G. R. Elliot) pp. 33–48, Birkhauser, Basel 1984.

S. Siegel, Nonparametric Statistics for the Behavioral Sciences. pp. 116–127. McGraw-Hill Kogakusha, Tokyo 1956.

Journal abbreviations should correspond to those in *World List of Scientific Periodicals.*

Cite a paper which has been accepted for publication but which has not appeared in the reference list with words "in press" after abbreviated name of journal. Mention unpublished experiments, papers in preparation and personal communications in text only; do not include in reference list. Use personal communications only with written authorization from communicator.

Illustrations: Make illustrations and legends self explanatory without reference to text, number unobtrusively on back of all submitted figures.

Tables: Use separate sheets and prepare for use in single column (7 cm) or page width (15cm). Number consecutively with Arabic numerals in italics. Give a brief descriptive caption. Separate caption, headings and footnotes from each other and from body of table by horizontally ruled lines. Do not use more than 85 characters in a line. Never use vertical lines. Make reference to statistical significance with superscript asterisks. Otherwise, and where statistical significance is indicated together with footnotes, make references using superscript lower case letters.

Figures: Make suitable for reduction to fit a single (7 cm) or double column (maximum 15 cm). Submit original drawings in black ink or good photographic copies. Lettering should be not less than 4 mm high and lines not less than 0.4 mm thick on original drawings. Symbols should be not less than 2.5 mm in diameter; choose from: ●, ■, ▲, O, ▢, △

Photographs and Micrographs: Print on glossy paper. The size should be larger than, but not more than twice as large as final size in journal (i.e., 7 or 15 cm). Letter directly on micrographs.

Legends: Type on a separate sheet(s) of paper paginated as part of paper.

OTHER FORMATS

COMMENTARIES—Forum for personal views on recent developments arising from authors' own or other research. Take form of a mini-review, pointing out relevance of new findings to recent literature and indicating ways in which new findings will affect development of research in that field. Submit in triplicate. Do not exceed 2000 words plus two figures or tables. Provide a short summary of a single paragraph.

MEETING REPORTS, NEWS AND VIEWS—To facilitate rapid dissemination of information, reports of meetings falling within scope of journal. Do not exceed 2000 words; do not include a summary. Submit in duplicate. Include name, dates and site of meeting together with name and address of reporting author on title page. Begin text with a statement explaining aim of meeting. Further subdivisions should reflect major themes of meeting. When citing a specific communication, give city and country from which cited person comes in brackets.

REVIEWS—Reviews on any topic related to themes of the journal. Submit in triplicate with an abstract, introduction and conclusion, but otherwise subdivide according to authors' discretion. Point out weak or critical areas in field and indicate direction the field may take in the future. Do not exceed 10 printed pages.

SHORT COMMUNICATIONS—Intended for rapid publication of new results of particular importance to other workers. Submit in triplicate. Combination of "Results" and "Discussion" sections may be permitted. Length, including title, summary and references, must not exceed 1200 words plus one figure or one table. Provide a short summary, one paragraph.

OTHER

Abbreviations: Excessive use of abbreviations is strongly discouraged. Define all abbreviations when first used by placing them in brackets after the full term.

Experimentation, Animal: Experiments causing pain or discomfort to animals must be performed according to the guidelines of the International Association for the Study of Pain. See *Pain* 16:109-110 (1983).

Experimentation, Human: Clinical studies must be performed in accordance with the "Declaration of Helsinki" and its amendment in "Tokyo and Venice."

Units: Systéme Internationale (S.I.). Examples of these and of chemical and biochemical abbreviations and nomenclature may be found in *Biochem. J.* (1975)145, 1-20 or in *Br. J. Pharmac.* (1984) 81, 3-10.

SOURCE

March 1994, 41(1/2): back matter.

AIDS
Current Science Ltd.

AIDS Editorial Office, Current Science Ltd.
34-42 Cleveland Street
London W1P 4FB UK
Telephone: +44(0)71 323 0323; Fax: +44(0) 71 580 1938
Current Science
20 North Third Street
Philadelphia, PA 19106
Fax: (215) 574-2270

AIM AND SCOPE

AIDS publishes papers reporting original clinical, scientific, epidemiological and social research which are of a high standard and which contribute to the advancement of knowledge in the field of the acquired immune deficiency syndrome.

MANUSCRIPT SUBMISSION

Manuscripts must not have been submitted simultaneously to another journal or have been published elsewhere. Disclose any affiliations, including financial, consultant, or institutional associations, that might lead to bias or conflict of interest.

Submit papers to a journal section: Basic Science, Clinical, Epidemiology and Social.

These instructions comply with those of the International Committee of Medical Journal Editors. See: Uniform requirements for manuscripts submitted to biomedical journals. *N Engl J Med* 1991, **324**:424–428.

MANUSCRIPT FORMAT

Submit three copies (two photocopies). Use white bond paper and 3 cm margin. Double-space throughout, including: title page, abstract and keywords, text, acknowledgments, references, tables and captions. Begin each section on a separate sheet. Number pages consecutively, beginning with title page. Put page number in top right hand corner of each page.

Title Page: Full title and short title 40 characters including spaces (identify as 'running head'). Give first name, middle initial and last name of each author. If work is attributed to a department or institution, give full name. Include any disclaimers, name and address of corresponding author, and name and address of reprint request author. If no reprints, so state. List sources of support (grants, equipment, drugs, or any combination).

Structured Abstract: 250 words on second page.

For articles containing original data concerning course, cause, diagnosis, treatment, prevention or economic analysis of a clinical disorder or an intervention to improve the quality of health care:

Objective—State main question or objective of study and major hypothesis tested, if any.

Design—Describe study design. Indicate use of randomization, blinding, criterion standards for diagnostic tests, temporal direction (retrospective or prospective), etc.

Setting—Indicate study setting, including level of clinical care (e.g., primary or tertiary; private practice or institutional).

Patients, Participants—State selection procedures, entry criteria and numbers of participants entering and finishing study.

Interventions—Describe essential features of interventions including method and duration of administration.

Main Outcome Measure(s)—Indicate primary study outcome measures as planned before data collection began. State if hypothesis was formulated during or after data collection.

Results—Describe measurements not evident from nature of main results. Indicate blinding and absolute values when risk changes or effect sizes are given.

Conclusions—State only conclusions directly supported by data, with their clinical application (avoiding overgeneralization). Give equal emphasis to positive and negative findings of equal scientific merit.

For articles concerning original scientific research:

Objective(s)—State primary objective (if appropriate).

Design—State principal reasoning for procedures adopted.

Methods—State procedures used.

Results—State main results. Keep numerical data to a minimum.

Conclusions—State conclusions drawn from data given.

Key Words: List 3–10 keywords or short phrases. When possible, take terms from *Medical Subject Headings* list of *Index Medicus*.

Text: Divide full papers of an experimental or observational nature into sections headed Introduction, Methods, Results and Discussion. Editorial reviews may require a different format.

Acknowledgments: Acknowledge only those who have made a substantial contribution to study. Obtain written permission from people acknowledged by name.

References: Number consecutively in order of first appearance in text. Assign Arabic numerals in brackets. Include names of all authors when six or less; when seven or more, list only first three and add *et al*. Include full title and source information. Abbreviate journal names as in *Index Medicus*.

JOURNAL

Curran JW, Morgan WM, Hardy AM, Jaffe HW, Darrow WW, Dowdle WR: **The epidemiology of AIDS: current status and future prospects.** *Science* 1985, **229**:1352–1357.

BOOK

Rawls WE, Camione-Paccardo J: **Epidemiology of herpes simplex virus type 1 and type 2 infections.** In *The Human Herpesviruses: An Interdisciplinary Perspective*. Edited by Nahmias AJ, Dowdle WR, Schinazi RF. New York: Elsevier; 1981:137-152.

MEETING ABSTRACT

Booth R, Watters J: **Factors associated with safer sex and needle hygiene among drug users in eight USA cities.** *IX International conference on AIDS/IV STD World Congress.* Berlin, June 1993 [abstract PO-C15-2943].

Do not include personal communications and unpublished work in reference list. Put in parentheses in text. Cite submitted but not yet accepted work in parentheses in text. Include unpublished work accepted for publication but not yet released in reference list with 'in press' in parentheses beside name of journal. Verify references against original documents.

Illustrations: Refer to figures and tables in order of appearance in text by Arabic numerals in parentheses, e.g., (Fig. 2). Have figures professionally drawn and photographed; freehand or typewritten lettering is unacceptable. Paste a label on back with figure number, title of paper, author's name and indicate top. Do not mount illustrations. Present half-tone illustrations as sharp, glossy, black and white prints.

Photomicrographs must have internal scale markers. Identities of people in photographs must be obscured or submit written permission to use photograph. If a figure has been published before, acknowledge original source and submit written permission from copyright holder. Permission is required regardless of authorship or publisher, except for documents in public domain.

Tables: Type each double-spaced on a separate sheet. Do not submit as photographs. Give each an Arabic numeral and a brief title. Do not use vertical rule.

Legends: Type double-spaced, beginning on a separate page. Explain internal scales. Identify staining methods for photomicrographs.

OTHER FORMATS

CORRESPONDENCE—750 words and one figure or table.

SHORT COMMUNICATIONS—2000 words and up to three figures or tables of original research. Identify submissions for short communications in cover letter.

OTHER

Abbreviations: Define on first appearance in text. Avoid those not accepted by international bodies.

Units: Système Internationale (SI) units.

SOURCE

January 1994, 8(1): end of issue.

AJNR
see American Journal of Neuroradiology

AJR: American Journal of Roentgenology
American Roentgen Ray Society

AJR Editorial Office
2223 Avenida de la Playa, Suite 103
La Jolla, CA 92037-3218
Telephone: (619) 459-2229; Fax: (619) 459-8814

AIM AND SCOPE

AJR publishes original contributions to the advancement of medical diagnosis and treatment.

MANUSCRIPT SUBMISSION

Submitted manuscripts should not contain previously published material and should not be under consideration for publication elsewhere.

Manuscript guidelines are based on Uniform Requirements for Manuscripts Submitted to Biomedical Journals (*Ann Intern Med* 1988;108:258–265).

Submit two copies of manuscript (original and a photocopy) and two complete sets of figures. Type manuscript, including references, figure legends, and tables, double-spaced on 8.5 × 11 in. on nonerasable paper. Do not justify right margin. Number all pages consecutively beginning with abstract. Do not put authors' names on manuscript pages.

MANUSCRIPT FORMAT

Organize: title page, blind title page (title only), abstract, introduction, methods, results, discussion, acknowledgments, references, tables, figure legends, and figures.

Title Page: Include title; names and complete addresses (include zip code) of all authors; current addresses of authors who have moved; acknowledgment of grant or other assistance. Clearly identify corresponding author with a current address, telephone number, and fax number. A blind title page gives only the title for use in review process.

Abstract: Clearly state (250 words) purpose, methods, results, and conclusions of study. Include actual data. Use no abbreviations or reference citations. Organize into paragraphs:

Objective—In one or two sentences, indicate article's specific goal or purpose and indicate why it is worthy of attention. Explain hypothesis tested, dilemma resolved, or deficiency remedied. Objective must be identical to one given in title and introduction.

Subjects (or Materials) and Methods—Describe succinctly methods used to achieve objective, state what was done, how it was done, how bias was controlled, what data were collected, and how data were analyzed.

Results—Present findings of procedures. Results should flow logically from methods described and not stray from paper's specific objective. Include as many specific data as possible within one typewritten page.

Conclusion—In one or two sentences, present message to be remembered. Describe conclusion of study, based solely on data provided in body of abstract. Conclusions must relate directly to objective as defined in title and first paragraph of abstract.

Introduction: Briefly describe purpose of investigation; explain why it is important.

Methods: Describe research plan, materials (or subjects), and methods used, in that order. Explain in detail how disease was confirmed and how subjectivity in observations was controlled.

Results: Present results in a clear, logical sequence. If tables are used, do not duplicate tabular data in text. Describe important trends and points.

Discussion: Describe limitations of research plan, materials (or subjects), and methods. When results differ from those of previous investigators, explain discrepancy.

References: 35 maximum. Type double-spaced starting on a separate page. Number consecutively in order of appearance in text. Cite all references in text and enclose in brackets and type on line with text (not superscript).

Do not cite unpublished data in reference list. Cite parenthetically in text, for example, (Smith DJ, personal communication), (Smith DJ, unpublished data). This includes papers submitted, but not yet accepted, for publication.

Do not cite papers presented at meetings in reference list. Cite parenthetically in text (e.g., Smith DJ et al., presented at the American Roentgen Ray Society meeting, May 1990). After first mention, use (Smith DJ et al., ARRS meeting, May 1990).

Give inclusive page numbers (e.g., 333-335). Abbreviate journal names according to *Index Medicus*. List all authors when six or fewer; when seven or more authors, list first three, followed by "et al."

JOURNAL ARTICLE

1. Long RS, Roe EW, Wu EU, et al. Membrane oxygenation: radiographic appearance. *AJR* **1986**;146:1257–1260

BOOK

2. Smith LW, Cohen AR. *Pathology of tumors,* 6th ed. Baltimore: Williams & Wilkins, **1977**:100-109

CHAPTER IN A BOOK

3. Breon AJ. Serum monitors of bone metastasis. In: Clark SA, ed. *Bone metastases.* Baltimore; Williams & Wilkins, **1983**:165-180

Tables: Type each double-spaced on a separate page without vertical or horizontal rules. Give short, descriptive titles. Make self-explanatory. Do not duplicate data in text or figures. Do not exceed two pages in length. Each must contain at least four lines of data. Number in order cited in text. Define abbreviations in explanatory note below table.

Figures and Legends: Submit two complete sets of original unmounted figures in labeled envelopes. Submit clean, unscratched, glossy prints, uniform in size and magnification, with white borders. Submit a separate print (not larger than 5 × 7 in.) for each figure part. Label each on back with figure number and arrow indicating top. Use gummed labels on back of black-and-white prints. Never use labels on color figures, write figure number on back lightly in pencil. Never use ink on front or back of any figures. Do not write authors' names on backs of figures.

Use only removable (rub-on) arrows and letters. Use symbols uniform in size and style and not broken or cracked.

Do line drawings in black ink on a white background. They must be professional in quality. Use the same size type. Submit only glossy prints. Obtain written permission for use of previously published illustrations. Include copies of permission letters and appropriate credit line in legends.

Type legends double-spaced. Make sure figure numbers correspond with order in which figures are cited in text. Write legends in accordance with *AJR* style; refer to any issue.

OTHER FORMATS

CASE REPORTS—Brief description of a special case that provides a message that transcends individual patient. No abstract. Introduction is a short paragraph giving general background and specific interest of case. Describe one case in detail (similar ones can be mentioned briefly in discussion). Emphasize the radiologic aspects; limit clinical information to that necessary to provide a background for radiology. Discussion should be succinct and focus on specific message and relevance of radiologic methods. Literature review is not appropriate.

Maximum of five double-spaced, typewritten pages, including references but not title page or figure legends. Maximum of eight references. Maximum of four figures unless text is shortened accordingly. Legends must not repeat text. Tables and acknowledgments are not appropriate.

COMPUTER PAGE ARTICLES —Practical computer applications to radiology. Include a title page and an abstract. Maximum of eight double-spaced, typed pages. Maximum of five references. Maximum of five figures and tables. Computer printouts are not acceptable. Figures must be submitted as 5 × 7 in. glossy prints.

LETTERS TO THE EDITOR AND REPLIES—Objective and constructive criticism of published articles. Letters may also discuss matters of general interest to radiologists. Disclose financial associations or other possible conflicts of interest.

Type double-spaced on nonletterhead paper, with no greeting or salutation. Titles should be short and pertinent. Title for a reply is simply "Reply." Do not use abbreviations in title or text. Authors names and affiliations should appear at end of letter. Do not end a letter with a handwritten signature. Maximum of four authors. Maximum of two double-spaced, typewritten pages, including references. Maximum of four references. Maximum of two figures or one figure with two parts. Tables and acknowledgments are not appropriate.

OPINIONS, COMMENTARIES, AND PERSPECTIVES—Special articles dealing with controversial topics or issues of special concern to radiologists. Include a title page but no abstract. Use headings to break up the text.

Maximum of five double-spaced, typed pages. Maximum of five references. Maximum of four tables and figures.

PICTORIAL ESSAYS—Message is conveyed through illustrations and legends, serving primarily as teaching tools. Introduction serves as abstract, defining scope of paper and explaining its importance on half of a double-spaced typed page.

Maximum of four double-spaced, typed pages, including references but not title page or figure legends. Maximum of four references. Maximum of 30 figure parts. Number should be as few as necessary to convey message of paper. Tables and acknowledgments are not appropriate.

TECHNICAL NOTES—Brief description of a specific technique or procedure, modification of a technique, or equipment.

No abstract, headings or subheadings are required. If headings are used, they should be a combination of "Case Report," "Materials and Methods," "Results," and "Discussion." Give a brief one-paragraph introduction for general background. Limit discussion to specific message, including uses of technique or equipment. Literature reviews and lengthy case reports are not appropriate. Maximum of five double-spaced, typewritten pages, including references but not title page or figure legends. Maximum of eight references. Maximum of two figures, unless text is shortened accordingly. Tables and acknowledgments are not appropriate.

OTHER

Abbreviations: Keep use of unfamiliar acronyms and abbreviations to a minimum. Define at first mention, followed by abbreviation in parentheses.

Experimentation, Animal: State compliance with NIH guidelines for use of laboratory animals.

Experimentation, Human: Informed consent must be obtained from patients who participated in clinical investigations.

Forms: Submit a completed copy of Author's Checklist and signed copyright agreement from journal issue.

Supplies: Give names and locations (city and state only) of manufacturers of equipment and nongeneric drugs.

Units: Use metric measurements or give the metric equivalent in parentheses.

SOURCE

January 1994, 162(1): A9–A11.

Alcohol

Pergamon Press

R. D. Meyers
Department of Pharmacology, School of Medicine
East Carolina University
Greenville, NC 27858

AIMS AND SCOPE

Alcohol will publish original reports in English of new and systematic studies in the various fields of alcohol research. All aspects of alcohol's biological action will be covered including anatomy, biochemistry, cell biology, physiology, pharmacology, toxicology, behavior, social anthropology, and clinical problems. Brief communications will be included which describe a new method, technique or apparatus or which present results of experiments that can be reported briefly with a minimal number of figures and tables.

MANUSCRIPT SUBMISSION

Submit doubled-spaced typed manuscripts with wide margins in triplicate on good quality paper. If a word processor is used, a letter quality printer must be used and computer generated illustrations must be of the same quality as professional line drawings.

MANUSCRIPT FORMAT

Title page: title of paper; author(s); laboratory or institution of origin with city, state, zip code, and country; complete address for mailing proofs, a running head (40 characters including spaces). Type references, footnotes, and legends for illustrations on separate sheets, double-spaced. Identify illustrations (unmounted photographs) on reverse with figure number and author(s) name, when necessary mark top clearly. Type each table double-spaced on a separate sheet. Do not use italics for emphasis.

Title: Not longer than 85 characters, including spaces.

Abstract: Not to exceed 170 words; suitable for use by abstracting journals; and prepared as follows:

MYERS, R. D., C. MELCHIOR AND C. GISOLFI. *Feeding and body temperature: Changes produced by access calcium ions* ... ALCOHOL. Marked differences in extent of diffusion have been ...

Include 3-12 (or more) words or short phrases suitable for indexing terms at bottom of abstract page.

Footnotes: Number title page footnotes consecutively. If senior author is not to receive reprint requests, designate in footnote to whom requests should be sent. Text footnotes should not be used; incorporate material into text.

References: Prepare literature cited according to Numbered/Alphabetized style of Council of Biology Editors. Cite references by number, in parentheses, within text (only one reference to a number), and list in alphabetical order (double-spaced) on a separate sheet at end of manuscript. Do not recite names of authors within text. Journal citations in reference list contain: surnames and initials of all authors (surname precedes initials); title of article; journal title abbreviated as listed in the *List of Journals Indexed in Index Medicus;* volume, inclusive pages, and year.

1. Banks, W. A.; Kastin, A. J. Peptides and the blood-brain barrier: Lipophilicity as a predictor of permeability. Brain Res. Bull. 15: 287-292; 1985.

Book references: author, title, city of publication, publisher, year, and pages.

1. Mello, N. K. Behavioral studies of alcoholism. In: Kissin, B.; Begleiter, H., eds. The biology of alcoholism, vol. 2. Physiology and behavior. New York: Plenum Press; 1972:219–291.

2. Myers, R. D. Handbook of drug and chemical stimulation of the brain. New York: Van Nostrand Reinhold Company; 1974.

Illustrations: Prepare for use in a single column width if possible. Draw and letter all drawings to the same scale for reduction to a given size. Refer to all illustrations as figures and number in Arabic numerals. Letter in India ink or other suitable material in a size proportionate to illustrations. Size lettering so that its smallest elements (subscripts or superscripts) will be readable when reduced. When possible place all lettering within framework of illustration. Put symbol key on face of chart. Use these standard symbols: ○ ● △ ▲ □ ■ +. Give actual magnification of all photomicrographs. Indicate dimension scale. Submit only sharply contrasting unmounted photographs of figures on glossy paper in black and white unless color reproduction is requested. Submit color prints in actual size. Authors are responsible for additional costs.

Tables: Give each a brief heading; put explanatory matter in footnotes, not in title. Indicate table footnotes in table body in order of appearance with symbols: *, †, ‡, §, ¶, #, **, etc. Do not duplicate material in text or illustrations. Omit vertical rules. Use short or abbreviated column heads. Identify statistical measures of variation, SD, SE, etc. Do not submit analysis of variance tables, but incorporate significant F's where appropriate within text. Form for reporting F value is: $F(11, 20) = 3.05, p < 0.01$.

Formulas and Equations: Keep structural chemical formulas, process flow-diagrams, and complicated mathematical expressions to a minimum. Draw chemical formulas and flow-diagrams in India ink for reproduction as line cuts.

OTHER

Anesthesia: In describing surgical procedures on animals, specify type and dosage of anesthetic agent. Curarizing agents are not anesthetics; if used, provide evidence that anesthesia of suitable grade and duration was employed.

Drugs: Capitalize proprietary (trademarked) names. Chemical name should precede trade, popular name, or abbreviation of a drug at first use.

Nomenclature: Follow: Royal Society Conference of Editors. Metrication in Scientific Journals, *Am. Sci.* 56:159-164;1968.

Units: Use metric system.

SOURCE

January-February 1994, 11(1): inside back cover.

American Heart Journal
Mosby-Year Book, Inc.

Dean T. Mason, MD
Editorial Offices
Western Heart Institute
St. Mary's Hospital & Medical Center
450 Stanyan Street
San Francisco, California 94117
Telephone: (415) 750-5507

AIM AND SCOPE

The *Journal* will consider for publication suitable original manuscripts on topics pertaining to the broad discipline of cardiovascular diseases.

MANUSCRIPT SUBMISSION

Submit original copy of manuscript and all supporting material plus two photocopies (not carbon copies). Type on one side only, double-spaced, on 8.5 × 11 in. paper with adequate margins. Use appropriate subheadings throughout body of text (Methods, Results, and Discussion sections).

In accordance with the Copyright Act of 1976, which became effective January 1, 1978, submit following statement signed by senior or corresponding author:

"The undersigned author transfers all copyright ownership of the manuscript entitled [title of article] to Mosby Year Book, Inc., in the event the work is published. The undersigned author warrants that the article is original, is not under consideration by another journal, has not been previously published, and the data in the manuscript have been reviewed by all authors, who agree with the analysis of the data and the conclusions reached in the manuscript. I sign for and accept responsibility for releasing this material on behalf of any and all coauthors."

This statement should also include any financial or other relation that might pose a potential conflict of interest. Authors will be consulted, when possible, regarding republication of their material.

MANUSCRIPT FORMAT

Original, in-depth clinical investigations and clinically related experimental animal research articles are published as CLINICAL INVESTIGATIONS. Number title page as page 1, summary page as page 2, and continue throughout references, figure legends, and tables. Place page numbers in upper right corner of each page.

Title Page: Include title, authors' full names, highest earned academic degrees, and institutional affiliations and locations. If title is long, include a short title (42 characters) for running head. Designate one author (provide address and business and home telephone numbers) to receive correspondence, galley proofs, and reprint requests.

Abstract/Summary: 150 words, typed double-spaced on a separate sheet.

References: Number reference list in order of appearance in text. Type double-spaced on a separate sheet. Use format in "Uniform Requirements for Manuscripts Submitted to Biomedical Journals" (*N Engl J Med* 1991;324:424-8). Reference citations to periodicals include, in order: names of all authors, title, journal, year, volume, and pages; for example, Hope RR, Scherlag BJ, Lazzara R. Excitation of ischemic myocardium: altered properties of conduction, refractoriness, and excitability. Am Heart J 1980;99:753-65. Follow journal abbreviation used in *Cumulated Index Medicus*. Book references include, in order: names of all authors, chapter title, editor(s), book title, volume (if any), edition (if any), city, publisher, year, and inclusive pages of citation (if any); for example, Sherry S. Detection of thrombi. In: Strauss HE, Pitt B, James AE, eds. Cardiovascular nuclear medicine. St Louis: MosbyYear Book, Inc, 1974:273.

Unpublished data and personal communications may be cited in text; do not list as references. Authors are responsible for accuracy of references.

Illustrations and Legends: Submit three complete sets of glossy prints. Number illustrations in order of mention in text. Mark lightly on back with first author's last name and an arrow to indicate top. A reasonable number of black-and-white illustrations will be reproduced free of cost; special arrangements must be made for color plates, elaborate tables, or extra illustrations. Use black India ink and a black ribbon. All lettering must be professional. Do not send original artwork, x-ray films, or ECG tracings. Glossy photographs are preferred; good black-and-white contrast is essential. Preferred size is 5 × 7 in. Type suitable figure legends, double-spaced, on a separate sheet at end of manuscript. If figure has been taken from previously copyrighted material, give full credit to the original source in legends.

Tables: Make self-explanatory. Number in Roman numerals according to mention in text. Provide brief titles for each. Type each on a separate page. Define abbreviations in a footnote at end of table. If any material has been taken from previously copyrighted material, give full credit to the original source in a footnote.

OTHER FORMATS

BRIEF COMMUNICATIONS—500 words (without an abstract), one or two illustrations, and six references. Brief commentary and/or selected patient reports of special interest.

CURRICULUM IN CARDIOLOGY—Concise, comprehensive, contemporary articles presenting core, updated information necessary for continuing medical education of the modern clinical cardiologists.

EDITORIALS—Brief, substantiated commentary on special subjects.

FORUM: LETTERS TO THE EDITOR—Brief letters or notes regarding published material or information of timely interest.

PROGRESS IN CARDIOLOGY—Concise, comprehensive articles devoted to innovative new areas of development in cardiovascular diseases important to clinical cardiologists.

OTHER

Abbreviations: Use standard abbreviations consistently. Spell out unusual or coined abbreviations at first use in text, followed in parentheses by abbreviation.

Experimentation, Human: State that informed consent was obtained when applicable. Note that study was approved by institutional committee on human research.

Permissions: Direct quotations, tables, or illustrations that have appeared in copyrighted material must be accompanied by written permission for use from copyright owner and original author along with complete information as to source. Photographs of identifiable persons must be accompanied by signed releases showing informed consent.

Style: Consult latest editions of the Council of *Biology Editors Style Manual, The Chicago Manual of Style*, or the AMA's *Manual of Style,* 8th edition for current usage.

SOURCE

January 1994, 127(1): 11A–12A.

American Journal of Botany
Botanical Society of America, Inc.

Dr. Karl J. Niklas
Section of Plant Botany
Cornell University
Ithaca, NY 14853-5908
Telephone: (607) 254-4708; Fax: (607) 254-4695;
E-mail kjn2@cornell.edu.

MANUSCRIPT SUBMISSION

At least one author must be a member when manuscript is submitted, and during year of publication (except for Special Papers). Members are entitled to eight free printed pages per volume.

Submit three copies of manuscript and three copies of illustrations. Do not submit original illustrations until acceptance. In cover letter, include in one paragraph: author(s) name(s), title, number of pages, figures, and tables, any special instructions, any pending address change, and phone and fax numbers for corresponding author. Names, addresses, and fax numbers of possible objective reviewers. Submit stamped, self-addressed large envelope (9 × 12 or 10 × 13 in.) for return of reviews and manuscripts (for U.S. authors only).

Submit checklist with completed items marked. Photocopy "Checklist for Preparation of Manuscripts and Illustrations" printed in each issue. Check items and submit with manuscript.

Assemble manuscript in order: Title page, Footnote page, Abstract page, Text, Literature Cited, Tables, Figure legends, Figures. Do not staple manuscript.

MANUSCRIPT FORMAT

Consult current issues for additional guidance on format. Type manuscripts on 8.5 × 11 in. paper (do not use dot-matrix printer). Double-space throughout, including tables, figure legends, and literature cited. Leave a 1 in. margin on all sides. Avoid hyphens or dashes at ends of lines; do not divide a word at ends of lines. Do not justify right margin.

Do not use italic type. Underline words to be italicized. Underline items to be set boldface with a wavy line. Do not italicize or underline common Latin words or phrases (e.g., et al., i.e., sensu, etc.). Do not use footnotes in text.

Type last name(s) of author(s) and page number in upper right corner of page 2 and all pages following.

Cite each figure and table in text. Organize text so that they are cited in numerical order.

Title: (Page 1) Do not number. Give a running head 8 lines below top of page. Include each author's surname (first author, then et al. for three or more) and a short title (65 characters including spaces for names and title).

Center title near middle of page. Follow last word by superscript 1 (for footnote #1 to appear on footnote page).

Do not add authority to species name in title; include family name in parentheses after the species name.

Below title, include each author's first name, middle initial, surname, affiliation, and unabbreviated complete address. Include superscript number after an author's name to indicate a different address in a footnote. In multiauthored works, include another footnote superscript number to indicate author responsible for correspondence and requests for reprints.

Footnotes: (Page 2) Every article must include this footnote:

[1]Manuscript received _____; revision accepted _____(Editor will fill in dates.)

Brief acknowledgments, if appropriate, are part of this footnote as a separate, unnumbered paragraph.

Other footnotes with author information are permitted; match footnote superscript numbers with those on title page.

Abstract: (Page 3) Single, concise paragraph (200 words), that includes paper's intent, materials and methods, results, and significance. If references must be cited, include journal name, volume number, pages, and year, all in parentheses.

Text: (Page 4 etc.) Main headings are all caps and centered on one line.

Typical main headings are: MATERIALS and METHODS, RESULTS, and DISCUSSION.

In MATERIALS and METHODS include location of manufacturer or supplier with brand name items.

Incorporate summary or conclusions in DISCUSSION.

Undeline second level headings with normal indentation: they will be set boldface italic. Capitalize first letter of first word. Headings are followed by two hyphens, with text following immediately.

Third level headings, if needed, will be typeset italic only. Do not underline.

Literature Citations in Text: Cite one author as Jones (1990) or (Jones, 1990); two authors as Jones and Jackson (1990) or (Jones and Jackson, 1990); three authors as Jones, Jackson, and Smith (1990) or (Jones, Jackson, and Smith, 1990; more than three authors as Jones et al. (1990) or (Jones et al., 1990). Include all author names in literature cited. Cite manuscripts accepted for publication but not yet published as Jones and Smith (in press) or (Jones and Smith, in press). For unpublished data, include initial, surname, and institution. Within parentheses, use a semicolon to separate different types of citations (Fig. 4; Table 2) (Jones and Smith, 1988; Felix and Anderson, 1989). Cite several references within parentheses by year, with oldest one first.

List each reference cited in text in Literature Cited section, and vice versa. Double check for spelling and details of publication. Cite multiple references in chronological order in text.

Literature Cited: (Continue page numbering) Verify all entries against original sources, especially those in languages other than English. Capitalize all nouns in German. Check diacritical marks. Put citations in alphabetical order by first author's surname. Single author titles precede multiauthored titles by same senior author, regardless of date.

List works by the same author(s) chronologically, beginning with earliest date of publication. Use a long dash when author(s) is the same as in immediately preceding citation.

"In press" citations must have been accepted for publication; include name of journal or publisher.

Insert a period and space after each initial of an author's name. Leave a space between volume and page numbers. Write out journal names in full and underline. Capitalize only first word and proper nouns in book titles.

BOOK

JONES, A. B. 1986. Leaf anatomy. Wiley, New York, NY.

CHAPTER OR ARTICLE IN LARGER WORK

_____, 1990. Leaf anatomy. *In* J. Sanders and R. Richards [eds.], Botanical studies. vol. 2, chapt. 3, 200–234. Wiley, New York, NY.

JOURNAL ARTICLE

_____, AND T. THOMPSON. 1984. Ovule development. *American Journal of Botany* 80: 234–240.

PART OF YEARLY ABSTRACT ISSUE

_____, AND _____. 1990. Ovule development. *American Journal of Botany* 80: 221 (Abstract)

JOURNAL ARTICLE IN PRESS

_____, _____, AND F. ROBERTS. In press. Ovule development. *American Journal of Botany*.

Tables: (Continue page numbering) Double-space, using legal size paper if necessary to allow adequate margins. Start each on a separate sheet. Number in Arabic followed by a period. Capitalize first word of title, all others are lowercase unless a proper noun. Place a period at end of title. Do not reduce type size of tables: use same size type as in text. Capitalize each entry in each column. Indicate footnotes by lowercase superscript letters (a. b. c, etc.). Do not use vertical lines.

Figure Legends: (Continue page numbering) Double-space legends and group according to figure arrangements. Quadruple space between groups. Do not use a separate page for each group. Type legends in paragraph form starting with statement of inclusive numbers:

Figs. 3-5. Seeds and seedlings of orchids. 3. At germination. 4. 2 wk after germination. 5. Seedlings.

If many symbols occur in figures, include an explanatory section just above first figure group and separated from it by a quadruple space.

Do not include "exotic symbols" (lines, dots, triangles, etc.) in figure legends; either label in figure or refer to by name in legend.

Illustrations: Include black-and-white halftones (photographs), drawings, or graphs. Consult Editor about color. Reproduction is virtually identical to what is submitted; prepare illustrations using professional standards. Flaws will not be corrected. Consult issues for examples.

Prepare original photographic figures and plates in dimensions: any length up to a maximum of 9.33 in.; width either 3.5 in. for one column or 7.33 in. for full page width. Submit original black-and-white illustrations or other types of high quality reproductions (photomechanical transfers or other diffusion transfer type processes).

Group several photos into one or more plates. Butt photos together with no space between. Printer will add white separator lines. Affix photos with dry mount paper or equivalent on white posterboard. Leave 1 in. margin at top and bottom; 0.5 in. on sides.

Group several drawings to form a plate of drawings. Grouped photos and/or drawings must comprise a symmetrical rectangle or square. Number each figure in a plate; ungrouped single figures need not be numbered. Number all figures in Arabic numerals; do not use letters.

Illustrations magnified more than ten times require a bar scale; include numerical magnification in caption.

Use only press-on symbols and letters in figures: handwritten or typed symbols are unacceptable. Avoid oversized symbols (use capital letter size in text or upper size limit of symbols).

Reviewer's copies of halftone figures and plates must be photographic reproductions or special machine-reproduced copies that approach quality of originals; ordinary xerox-type copies are unacceptable.

Write author(s) name(s) and figure number(s) on back of each figure or plate, on both originals and review copies.

OTHER FORMATS

SPECIAL PAPERS—Mostly reviews of limited scope on timely subjects; usually solicited by Special Papers Committee (Chair: Dr. Darleen A. DeMason, Dept. of Botany and Plant Science, University of California, Riverside, CA 92521; telephone (714) 787-3580). Discuss ideas for possible contributions with Dr. DeMason. Manuscripts are subject to usual review. Membership requirement and page charges are waived.

OTHER

Abbreviations: Use "Figure" only to start a sentence; otherwise "Fig." if singular, "Figs." if plural. Use these abbreviations without spelling out: hr, min, sec, yr, mo, wk, d, diam, cm, mm, DNA, RNA; designate temperature as 30°C.

Use these statistical symbols: SE, SD, df, *N, F, P*, cv, *r*.

Write out other abbreviations first time used in text: abbreviate thereafter: "Transmission electron microscope (TEM) was used...."

Numbers: Write out one to ten unless a measurement (e.g., four petals, 3 mm, 35 sites, 6 yr). Use 1,000 instead of 1000; 0.13 instead of .13; % instead of percent.

Page Charges: $135 per page for each page over eight per author per volume. Authors are expected to pay publication costs if they have funds.

SOURCE

January 1995, 82(1): front matter.

American Journal of Cardiology
Reed Elsevier Medical Publication

William C. Roberts, MD
Baylor Cardiovascular Institute
Baylor University Medical Center
Wadley Tower No. 457, 3600 Gaston Avenue
Dallas, Texas 75246

MANUSCRIPT SUBMISSION

Manuscripts must be submitted solely to the *American Journal of Cardiology*. Manuscripts become property of publisher. Work must not have been previously published. Data must have been reviewed by all authors, who agree with analysis of data and conclusions reached.

State in cover letter the significance and uniqueness of work. State whether blinded or nonblinded reviews are preferred.

Submit 3 copies (1 original and 2 copies) of entire manuscript including text, references, figures, legends and tables. Arrange paper as: title page; second title page, title only (page not numbered); abstract; text; acknowledgments; references; figure legends; miniabstract for Table of Contents (page not numbered); and tables. Number title page as 1, abstract page as 2, and so forth. Use 8.5 × 11 in. paper, on one side only, double-spaced (including references) with 1 in. margins.

MANUSCRIPT FORMAT

Title Page, Abstract and Miniabstract: For complete title page, include title, full first or middle and last names of all authors and their academic degrees. List institution and address (including city and state or city and country) from which work originated and provide information about grant support if necessary. State if work was supported by a grant from a pharmaceutical company. Add at bottom: "Address for reprints:" with full name and address with zip code. Add a 2- to 6-word running head. Limit abstract to 250 words. Limit Table of Contents miniabstract to 150 words and include full title and names (without academic degrees) of authors. List 2 to 6 key words for subject indexing at end of the Abstract or Brief Report.

References: List all authors, year, volume and inclusive pages for all journal references, and specific page numbers for all book references. Do not use abstracts as references. Do not use periods after authors' initials or after abbreviations for titles of journals. Check *Index Medicus* or *Annals of Internal Medicine* (June 1982) for journal titles and abbreviations. Personal communications and unpublished observations do not constitute references, but may be mentioned in text.

JOURNAL

Harvey W, Heberden W, Withering W, Stokes W, Murrell W, Einthoven W, Osler W. Anomalies and curiosities of cardiology and of cardiologists. Reflections of famous medical Williams. *Am J Cardiol* 1984;53:900–915.

CHAPTER IN BOOK

Cabot RC, White PD, Taussig, HB, Levine SA, Wood P, Friedberg CK, Nadas AS, Hurst JW, Braunwald E. How to write cardiologic textbooks. In: Hope JA, ed. A Treatise on Disease of the Heart and Great Vessels. London: Yorke Medical Books, 1984:175–200.

BOOK

Carrel A, Cutler EC, Gross RE, Blalock A, Crafoord C, Brock RC, Bailey CP, DeBakey ME. The Closing of Holes, Replacing of Valves and Inserting of Pipes, or How Cardiovascular Surgeons Deal with Knives, Knaves and Knots. New York: Yorke University Press, 1984:903.

Figures: Submit 3 glossy unmounted prints of each photograph and drawing. Use black ink for all line drawings. Use arrows to designate special features. Crop photographs to show only essential fields. Identify figures on back by article title. Number figures in order mentioned in text. Indicate figure tops. Submit written permission from publisher and author to reproduce any previously published figures. Limit figures to number necessary to present message clearly. Type figure legends on separate page, double-spaced. Identify at end of each legend and in alphabetical order all abbreviations. Cost of color reproduction is paid by author.

Tables: Type on a separate page, double-spaced. Number in Roman numerals and title each table. Identify in alphabetical order at bottom all abbreviations used. When tabulating numbers of patients, use no more than 2 lines, preferably only 1 line, per patient. Use a plus sign (+) to indicate "positive" or "present," a zero (0) for "negative" or "absent," and dash (-) for "no information available" or "not done." Do not use "yes" or "no" or "none."

OTHER FORMATS

BRIEF REPORTS AND CASE REPORTS—4 text pages (including title page) and 6 references. Neither an abstract nor a Table of Contents miniabstract is required. Subheadings are not used. Provide a summary sentence at end.

READERS' COMMENTS—Provide a cover letter on own stationery to Editor in Chief stating why letter should be published. Type letter with references, double-spaced, on a separate sheet of plain white paper and limit to 2 pages. Submit original and 2 copies. Provide title for letter at top of page. At end of letter, include writer's full name and city and state (or country if outside U.S.). Do not include author's title and institution. Letters must be received (from authors within the U.S.) within 2 months of article's publication.

OTHER

Abbreviations: Use no more than 3 per manuscript. Use on every page of manuscript after initially spelled out (followed by abbreviation) in both abstract and introduction. Abbreviations are usually limited to terms in title.

Reviewers: Provide names and addresses of several non-local experts. Names of investigators considered unlikely to give nonbiased reviews also may be submitted. This request is honored.

Style: Use generic names of drugs. Do not spell out any number, including those less than 10, except when opening a sentence. Try not to begin sentences with numbers. Use symbols for less than ($<$), greater than ($>$) and percent (%). Indent for paragraphs except first one in both abstract and introduction. Consult Uniform Requirements for Manuscripts Submitted to Biomedical Journals (*The Annals of Internal Medicine* June 1982;96: 766-771), and also the *Stylebook/Editorial Manual of the AMA.*

SOURCE

July 1, 1994, 74(1): 104.

American Journal of Clinical Nutrition
American Society for Clinical Nutrition, Inc.

Editor's Office
American Journal of Clinical Nutrition
PO Box 12157
Berkeley, CA 94701-3157

Dept. of Nutritional Sciences
Morgan Hall, Room 119
University of California
Berkeley, CA 94701-3157
Telephone: (510) 642-2552; Fax: (510) 642-5867

AIM AND SCOPE

The primary purpose of *The American Journal of Clinical Nutrition (AJCN)* is to publish original basic and clinical studies relevant to human nutrition. Epidemiological studies relevant to basic and clinical nutrition will also be con-

sidered. Sections of the journal are: Editorials; Perspectives; Review Articles; Special Articles; Commentaries; and Original Research Communications: Nutritional status and body composition; Energy metabolism; Obesity and eating disorders; Growth and development; Aging; Hormonal interactions; Pregnancy and birth outcome; Lactation and infant feeding; Lipids and fatty acids; Proteins and amino acids; Carbohydrates; Vitamins; Minerals; Bone; Parenteral and enteral nutrition; Nutrition and disease; Drug-nutrient interactions; Immunologic responses; Nutrition and human performance; Letters to the Editor; Book Reviews; and Opinions.

MANUSCRIPT SUBMISSION

Manuscripts must not have been published, simultaneously submitted, or already accepted for publication elsewhere. Manuscripts that are unnecessarily complex, poorly written, extremely long, or with an excessive number of tables or illustrations are not acceptable.

Submit original and three copies (total = 4) of manuscript with one set of original figures and four photocopies of figures (total = 5). Identify author responsible for correspondence. Include copies of permissions to reproduce material or to use illustrations of identifiable subjects as well as a completed copyright transfer and ethics statement agreement form (*Am J Clin Nutr* 61(1):171, January 1995).

MANUSCRIPT FORMAT

Type white bond paper or ISO A4, with margins of at least 25 mm. Do not justify right margin. Use a good quality printer (not dot matrix). Double space throughout, including title page, abstract, text, acknowledgments, references, tables, and figure legends. Insert line numbers in left margin. Begin each section on separate page: title page, abstract and key words, text, acknowledgments, references, individual tables, and legends. Number pages consecutively beginning with title page. Type page number in upper-right corner of each page.

Title Page: Include title of article, beginning with a key word if possible, capitalize only first letter of first word; first name, middle initial, and last name of each author; names of departments and institutions to which work should be attributed; disclaimers, if any; name, mailing address, and telephone and fax numbers of corresponding and reprint request authors or a statement that reprints will not be available; sources of support (grants, equipment, drugs, or all of these); and a short running head (40 characters counting letters and spaces) identified at foot of title page.

Abstract and Key Words: Abstract (page 2) should contain $\leq$150 words. State purposes of study or investigation, basic procedures (study subjects or experimental animals and observational and analytical methods), main findings (give specific data and their statistical significance, if possible), and principal conclusions. Emphasize new and important aspects of study or observations.

Below abstract, provide and identify 3-10 key words and/or short phrases (20 words total) for indexing. Use terms from National Library of Medicine's *Medical Subject Headings.*

Text: Divide text of observational and experimental articles into sections: introduction, subjects (or materials) and methods, results, and discussion. Long articles may need subheadings. Consult recent issues for guidance on other types of articles, such as book reviews and editorials.

Introduction: Clearly state purpose of article. Summarize rationale. Give only strictly pertinent references. Do not review subject intensively. Do not include methods, data, or conclusions.

Materials and Methods: Describe selection of observational or experimental subjects clearly, including controls. Identify methods, apparatus (manufacturer's name and address in parentheses), and procedures in sufficient detail to allow other workers to reproduce results. Do not use trademark names as generic terms; if a trademark name must be used, specify manufacturer's name and address. Give references to established methods, including statistical methods; provide references and brief descriptions of methods that have been published but are not well known; describe new or substantially modified methods, giving reasons for using them and evaluating their limitations. Identify all drugs and chemicals used, including generic names, dosages, and routes of administration.

Results: Present in logical sequence in text, tables and illustrations. Do not repeat data from tables and/or illustrations; emphasize or summarize only important observations.

Discussion: Emphasize new and important aspects of study and conclusions that follow. Do not repeat data or other material in Introduction or Results. Include implications of findings and their limitations and relate observations to other relevant studies. Link conclusions with goals of study; avoid unqualified statements and conclusions. State new hypotheses when warranted but clearly label as such. Recommendations, when appropriate, may be included.

Acknowledgments: Acknowledge only persons who have made substantive contributions; do not acknowledge editorial or secretarial help. Obtain permission from everyone acknowledged by name.

References: Number consecutively in order first mentioned in text. Identify references in text, tables, and legends by Arabic numerals in parentheses. Number references cited in tables or figure legends according to first citation in text, table, or figure. Appendixes should have a separate reference section.

Avoid using abstracts as references (if an abstract is used include "abstr" in parentheses at end of citation). Doctoral theses may be used. Do not use unpublished observations and personal communications as references; references to written, not oral, communications may be inserted (in parentheses, with date) in text. Do not use unpublished (albeit printed) abstracts from scientific meetings as references. Include among references manuscripts accepted but not yet published; designate journal, followed by "in press" in parentheses. Cite information from manuscripts submitted but not yet accepted in text as "unpublished observations" in parentheses. Report titles in original language and provide English translation in parentheses.

Verify references against original documents.

JOURNALS

Standard journal article (List all authors when six or fewer, when seven or more, list only first three and add et al.) Abbreviate journal titles according to *Index Medicus* style.

Soter NA, Wasserman SI. Austen KF. Cold urticaria: release into the circulation of histamine and eosinophil chemotactic factor of anaphylaxis during cold challenge. N Engl J Med 1976,294:687-90.

Committee on Enzymes of the Scandinavian Society for Clinical Chemistry and Clinical Physiology. Recommended method for the determination of gammaglutamyltransferase in blood. Scand J Clin Lab Invest 1976;36:119-25.

Anonymous
Epidemiology for primary health care. Int J Epidemiol 1976;5:224-5.

BOOKS AND OTHER MONOGRAPHS

Osler AG. Complement: mechanisms and functions. Englewood Cliffs, NJ: Prentice-Hall. 1976.

National Research Council Recommended dietary allowances. 10th ed. Washington, DC: National Academy Press, 1989.

Rhodes AJ, Van Rooyen CE, comps. Textbook of virology: for students and practitioners of medicine and the other health sciences. 5th ed. Baltimore: Williams & Wilkins, 1968.

CHAPTER IN BOOK

Weinstein L, Swartz MN. Pathogenic properties of invading microorganisms. In: Sodeman WA Jr, Sodeman WA, eds. Pathologic physiology: mechanisms of disease. Philadelphia: WB Saunders, 1974:457-72.

Agency Publication

National Center for Health Statistics. Centers for Disease Control. NCHS growth curves for children, birth–18 years. Washington, DC: US Government Printing Office, 1978. [Series 11. 165. DHEW publication (PHS) 78 1650.]

Tables: Type on separate sheets; use double spacing. Number consecutively with Arabic numerals. Supply brief titles. Give each column a short or abbreviated heading. Place explanatory matter in footnotes, not in heading. Each table should contain enough detail (including statistics) so it is intelligible without reference to text. Explain in footnotes all nonstandard abbreviations. Explanations are not needed for LDL, VLDL, HDL, BMI, or ANOVA. For footnotes, use symbols in sequence: *,†, ‡, §, ‖, ¶, **, ††.... The first appearance in a horizontal row determines order of footnotes. Identify statistical measures of variations such as SD and SEM. Omit internal horizontal and vertical rules. Cite each table in text in consecutive order. If data from another published or unpublished source are used, obtain permission and acknowledge fully.

Illustrations: Cite each figure in text in consecutive order. If a figure has been published, acknowledge original source and submit written permission from copyright holder. Permission is required, regardless of authorship or publisher, except for public domain documents. Paste a label on back of each figure indicating figure number, author names, and figure top. Do not write on back of figures (except photocopies), mount on cardboard, or scratch or mar using paper clips. Do not bend figures.

Line Drawings—Submit in a protective envelope one complete set of glossy black-and-white photographic prints, usually 127×173 mm, but no larger than 203×254 mm. High-quality laser-printed line graphs are acceptable. Letters, numbers, and symbols should be clear and of sufficient size to withstand reduction. Attach to each copy of manuscript (including original) a high-quality photocopy of each line drawing. Identify figure number and first author on front of each photocopy.

Black-and-White Photographs (Half-Tones)—Submit five copies of each figure. Submit in a protective envelope one complete set of glossy black-and-white photographic prints, usually 127×173 mm, but no larger than 203×254 mm; attach other four copies to each manuscript copy. Attach an appropriate label to back of each print. Photomicrographs must have internal scale markers. Make symbols, arrows, or letters contrast with background. If photographs of people are used, either subjects must not be identifiable or pictures must be accompanied by written permission for use.

Color Photographs—Submit five color prints and one color negative or positive transparency. Submit in a protective envelope one complete set of prints and the negative or positive transparency. Attach other prints to manuscript copies. Attach an appropriate label to the back of each print. Authors must pay cost of color illustrations.

Legends: Type legends for all figures, double-spaced, on a separate page (not on figures). Give each enough detail (including statistics) to make figure intelligible without reference to text. Identify and explain symbols, arrows, numbers, or letters used to identify parts of illustrations, in legend. Explain internal scale. Identify method of staining in photomicrographs.

OTHER FORMATS

LETTERS TO THE EDITOR—1000 words, 10 references. Letters on subjects of nutritional importance are encouraged. Letters that refer to a recent journal article should be received within 12 weeks of article's publication.

OTHER

Abbreviations: Use only standard abbreviations. Consult: *CBE Style Manual* and *ASTM Standard for Metric Practice* (E380-89a. American Society for Testing and Materials. Philadelphia: ASTM, 1989). Avoid abbreviations in title. Complete form should precede first use of an abbreviation in text unless it is a standard unit of measurement.

Data Submission: Additional detailed tables, appendixes, mathematical derivations, extra figures, and other supplementary matter too costly to be included may be deposited with the National Auxiliary Publications Service of the American Society for Information Sciences, at no charge to author and with author's permission. Alternatively, material may be made available by author. If articles are deposited, a footnote is added stating that photostat or microfilm copies are available and noting deposit number and cost.

Experimentation, Animal: Indicate whether institution's or National Research Council's guide for the care and use of laboratory animals was followed.

Experimentation, Human: Indicate whether procedures followed were in accord with ethical standards of responsible committee on human experimentation (institutional or regional) or in accord with the Helsinki Declaration of 1975 as revised in 1983. Do not use patients' names, initials, or hospital numbers.

Nomenclature: Follow vitamin nomenclature policy of the American Institute of Nutrition (see *Am J Clin Nutr* 1989;50:206-15).

Page Charges: $40.00 per printed journal page; three double-spaced manuscript pages equal approximately one journal page.

Reviewers: Provide names, fields of interest, addresses, and telephone and fax numbers of 4-6 potential unbiased reviewers from outside authors' institution.

Statistical Analyses: Describe statistical methods with enough detail to enable a knowledgeable reader with access to original data to verify reported results. Quantify findings and present with appropriate indicators of measurement error or uncertainty (such as confidence intervals, SDs, or SEMs). Discuss eligibility of experimental subjects. Give details about randomization. Describe methods for, and success of, any blinding of observations. Report treatment complications. Give numbers of observations. Report losses to observation (such as dropouts from a clinical trial). References for study design and statistical methods should be to standard works when possible, rather than to papers where designs or methods were originally reported. Specify any general use computer programs used. Put general descriptions of statistical methods in Methods section. When data are summarized in Results section, specify statistical methods used for analysis. Avoid nontechnical uses of technical statistical terms. Avoid value judgments about results of statistical analyses with phrases like "nearly reached significance." Detailed statistical analyses, mathematical derivations, etc. may sometimes be suitably presented as an appendix.

Style: Use active voice. Use past tense when describing and discussing experimental work. Use present tense for reference to existing knowledge or prevailing concepts and for stating conclusions from experimental work. Do not use "level" when referring to a concentration. Use "energy" instead of "calorie" or "caloric." Clearly differentiate previous knowledge and new contributions.

Follow "Uniform Requirements for Manuscripts Submitted to Biomedical Journals" (*Br Med J* 1988;296:401-5) and the *CBE Style Manual* (*Council of Biology Editors style manual: a guide for authors, editors, and publishers in the biological sciences.* 5th ed. Bethesda, MD: Council of Biology Editors, 1983).

Units: SI units. See *Annals of Internal Medicine* 1987;106:114-29 for frequently used SI conversion factors. If molecular weight of a substance is known, express measurement in moles. Express dosage forms and dietary ingredients as gram or mole quantities. Express energy in joules. Give temperatures in degrees Celsius; blood pressures in millimeters of mercury. Use of katals to report enzyme activity is optional.

Retain non-SI units in parentheses in text after first expressing values in SI units. All table values must be in SI units; conversion factors to non-SI units should be provided in footnotes to tables that use following measurements:

amino acids(blood)	mg/dL
bile acids (blood)	μg/mL
bilirubin (blood)	mg/dL
β-carotene (blood)	μg/dL
cholesterol (blood)	mg/dL
creatinine (blood)	mg/dL
(urine)	g/24h
fatty acids, nonesterified (blood)	mg/dL
monosaccharides (blood: glucose, fructose)	mg/dL
triglycerides (blood)	mg/dL
urea nitrogen (blood)	mg/dL
(urine)	g/24h
minerals (blood)	μg or mg/dL, as appropriate
vitamins (blood)	ng, pg, μg, or mg/dL, as appropriate
energy	kcal
radiation	Ci

SOURCE

January 1994, 59 (1): 137–142.
January issues contain Information for Authors.

American Journal of Clinical Oncology: Cancer Clinical Trials

Raven Press

Luther W. Brady, M.D.
The Hahnemann University
Broad & Vine Streets
Philadelphia, PA 19102
Telephone: (215) 762-4453; Fax (215) 762-8523.

MANUSCRIPT SUBMISSION

Submit an original and two copies of each manuscript, double-spaced and typewritten on 8.5 × 11 in. bond paper, with two sets of illustrations. Submit only previously unpublished work. In a cover letter give address and telephone number of responsible author. Indicate name and address of author to whom page proofs and reprint requests are to be sent. Include a statement expressly transferring copyright to Raven Press in the event the paper is published in the Journal. All authors must sign this statement.

MANUSCRIPT FORMAT

Title Page: Make titles brief and specific. Indicate author(s), affiliation, and source of support.

Abstract: Type, double-spaced on a separate page. Do not exceed 200 words. Make factual and comprehensive.

Key Words: List on a separate sheet.

Text: Number each page after first consecutively, including legends, tables and references, with last name of senior author at page top. Provide a short version of title (40 characters) for running heads. Follow NCI Methodologic Guidelines for Reports of Clinical Trials published in the June 1986 issue of this journal. Quotations must be accurate. Give full credit to source. Obtain and submit permission to quote and reproduce material previously published with manuscript.

Tables: Type double-spaced, each on a separate page. Number and give brief specific titles. Data must be logically and clearly organized; make self-explanatory to supplement, not duplicate text.

Illustrations: Crop and leave unmounted halftones (photographs, photomicrographs, electron micrographs, roentgenograms). Single photographs or composites should fit within print area (6 × 7 in.). Letter halftones grouped under one figure number "a," "b," "c," etc. on lower right corner. On figure back, lightly write in pencil figure number, indicate top edge, and last name of senior author.

Legends: Type double-spaced in consecutive order on a separate sheet of 8.5 × 11 in. white bond paper; be brief and specific. Do not indicate magnification for light photomicrographs unless visual inspection is inadequate for judging relative or absolute size. Use internal scale markers for electron micrographs. Indicate type of stain used in legend only if it is other than hematoxylin and eosin.

References: Arrange in order of appearance, number consecutively. Type double-spaced. Abbreviate journal titles, without periods, as in *Index Medicus*. Give journal references in the following order: author, article title (with subtitle, if any), journal abbreviation, year of publication, volume number in Arabic numerals, and inclusive pages. Order for book references: author, title, volume (if more than one), edition number (if other than first), city of publication, publisher, and year. Chapter in a book, order changes: author of chapter, title of chapter, "In," editors, book title, volume, edition, city, publisher, year, inclusive pages of chapter. Verify bibliographic accuracy and check every reference in manuscript and proofread again in page proofs.

OTHER

Electronic Submission: Submit in conjunction with paper version. Preferred storage medium is MS-DOS compatible format. Macintosh compatible format is also accepted. Use either Microsoft Word, WordPerfect, WordStar, or XYWrite. Label disk with author name, item title, journal title, type of equipment used to generate disk, word processing program (including version number), and filenames. File must agree with final submitted paper manuscript. Do not include any extraneous formatting instructions.

SOURCE

February 1994, 17(1): front matter.

American Journal of Clinical Pathology
American Society of Clinical Pathologists
J. B. Lippincott

American Journal of Clinical Pathology
2100 West Harrison Street
Chicago, Illinois 60612-3798

M. Desmond Burke, MD
Director of Clinical Laboratories
State University of New York at Stony Brook
University Hospital, Level 3-532
Stony Brook, NY 11794-7300
            Review articles on Clinical Pathology

Stacey E. Mills, MD
Department of Pathology - Box 214
University of Virginia Health Sciences Center
Charlottesville, VA 22908
            Review articles on Anatomic Pathology

AIM AND SCOPE

The *American Journal of Clinical Pathology* is devoted to prompt publication of original studies and observations in clinical and anatomic pathology. Original papers relating to laboratory use, management, and information science will be given consideration.

MANUSCRIPT SUBMISSION

Articles must be submitted solely to *American Journal of Clinical Pathology*. No substantial portion may be submitted elsewhere. Manuscripts based primarily on data published previously are not acceptable. Send original typewritten manuscript and two photocopies with three sets of illustrations.

Cover Letter: State that all authors have read and approve final manuscript; (if applicable) that authors will assume cost of reproducing color illustrations; and that permission has been obtained from individuals (who have contributed significantly to study) acknowledged by name. In conformance with Copyright Revision Act of 1976, have all authors sign (except federal employees):

"In consideration of the *American Journal of Clinical Pathology* taking action in reviewing and editing the above named Article, the Author(s) hereby assign(s) to the American Society of Clinical Pathologists its legal representatives, successors and assigns all publishing rights and all other rights to the Article throughout the world, in any language and in any medium, including the right to procure the copyright thereon and the right to secure any renewals, reissues and extensions of any such copyrights in the United States or in any foreign country."

Author(s) may request permission to reuse manuscript or portions thereof. Make requests in writing. Permission shall not be unreasonably withheld.

State that Article is original work of Author(s) except for material in public domain and such excerpts from other works for which written permission of copyright owners has been obtained; and that Article has not previously been published, submitted, or accepted for publication elsewhere; that each author has contributed significantly to this paper, understand it and endorses it; and that authors have no relationship, financial or otherwise, with any manufacturers or distributors of products evaluated in paper, or alternatively, disclose any such relationship in a footnote to paper, or, requirement is irrelevant to this paper.

MANUSCRIPT FORMAT

Type manuscripts double-spaced throughout, with 1 in. margins, on bond paper, 8.5 × 11 in. Begin a new page for each section. Number pages consecutively.

Title Page: Include: concise title; first name, middle initial, last name of each author plus his/her highest degree; institutional affiliation of each author; name and address of author to whom reprint requests should be addressed: acknowledgment of source(s) of support; a brief title (40 characters total, including spaces); and disclaimers, if any.

Abstract: 150 words. State study's purposes, procedures, and significant findings with emphasis on new observations.

Key Words: 3 to 10 below abstract.

Text: Introduction: Clearly state purpose and rationale of study. Materials and Methods: Reference established techniques. Describe new or modified methods in sufficient detail to allow duplication. Results: Describe concisely and in logical fashion. Discussion: Emphasize novel or significant aspects of the study; do not base conclusions on unpublished observations or data derived solely from previous literature.

References: Include only articles cited in text. List in order of citation. When four or fewer authors, list all. When six or more, list first three and add "et al." Abbreviate journal titles according to *Index Medicus*. Give inclusive page references.

1. Rosen PP, Kimmel M. Juvenile papillomatosis of the breast: A follow up study of 42 patients biopsied before 1979. Am J Clin Pathol 1990;93:599-603.

Illustrations: Cite figures consecutively in text. Submit as sharp, glossy, black and white, unmounted photographs. All illustrations must fit journal columns (3.5 × ≤9 in. [single column]; 7 × ≤9 in, [double column]). Figures outside this range will be cropped or reduced. Art work and graphics must be of professional quality and legible. Indicate figure number and first author on top-back of all illustrations, by gummed labeling or writing lightly in pencil. Submit figure legends on a separate page, double-spaced. Include stains and magnifications. Photographs of persons should render them unidentifiable or include their written permission. Authors assume cost of reproducing color photographs.

Tables: Cite tables in text in consecutive order. Type double-spaced on separate pages. Give each column a short heading. Place explanatory material (with nonstandard abbreviations if not explained in text) in footnotes. Do not use internal horizontal and vertical lines.

Footnotes: Limit use. Use symbols in order *, †, ‡, §, ||, ¶, **, ††.

OTHER FORMATS

CASE REPORTS—Four printed pages (about 12 double-spaced manuscript pages).

OTHER

Abbreviations: Use standard abbreviations (see *Council for Biology Editors Style Manual*, Council for Biology Editors, 4th Edition, Arlington, VA, 1978). The full term, followed by abbreviation in parenthesis, should precede first use of abbreviation in text, except for standard units of measurement.

Experimentation, Animal: Include a statement indicating that institution's or National Research Council's guide for care and use of laboratory animals was followed.

Experimentation, Human: Include a statement that procedures followed were in accord with ethical standards established by institution in which experiments were performed or are in accord with Helsinki Declaration of 1975.

Nomenclature: Use current editions of *Dorland's Medical Dictionary* and *Webster's International Dictionary* as references. Use *Bergey's Manual of Determinative Bacteriology* for nomenclature of bacteria. Identify drugs and chemicals by generic names, followed, in parentheses, by chemical formulas when appropriate. Follow TNM method for staging tumors (American Joint Committee on Cancer, *Manual for Staging of Cancer,* 3rd edition, J. B. Lippincott Co., Philadelphia, 1988).

Units: Use SI units, followed in parentheses, by conventional metric units, Report pH, gas pressure measurements (pO_2 and pCO_2), and osmolality in conventional units only. Express temperature in degrees Celsius and enzyme activity in International Units per liter (U/L). Base all SI concentration units on a volume of 1 liter (symbol L). Express as amount of substance (mole) or mass (gram) units using appropriate prefixes, such as milli- or micro-. Make conversions using factors given in *Am J Clin Pathol* 1987; 87: 140-151. In describing reagent preparation, give weights and volumes in conventional metric units only.

SOURCE

January 1994, 101(1): 120.

American Journal of Diseases of Children
see Archives of Pediatric & Adolescent Medicine

American Journal of Gastroenterology
American College of Gastroenterology
Williams & Wilkins

Rowen K. Zetterman, M.D. F.A.C.G.
Department of Internal Medicine
University of Nebraska Medical Center
600 S. 42nd Street, Omaha, NE 68198-2035.

AIM AND SCOPE

The *American Journal of Gastroenterology* publishes original manuscripts, case reports and reviews of clinical topics in gastroenterology.

MANUSCRIPT SUBMISSION

Articles are accepted for publication on condition that they are contributed solely to *The American Journal of Gastroenterology*. Submit an original and three copies (four complete sets including original illustrations) of all manuscripts and reviews and an original and two duplicate copies (three complete sets) of all case reports, editorials, or letters to the editor. In view of the Copyright Revision Act of 1976, include in cover letter:

In consideration of the American College of Gastroenterology taking action in reviewing and editing my submission, the author(s) undersigned hereby transfer(s), assign(s) or otherwise convey(s) all copyright ownership to the American College of Gastroenterology in the event that such work is published by the American College of Gastroenterology." The letter must be signed by all authors.

MANUSCRIPT FORMAT

Use 8.5 × 11 in. white paper. Double-space throughout, including references, footnotes, tables, and legends. Leave sufficient margins at top, bottom, and sides of each page. Number all pages consecutively beginning with title page with the senior author's name typed in the upper right corner of each page. Arrange manuscript as title page, abstract, introduction, materials and methods, results, discussion, references, tables, figure legends, and figures.

Title Page: On a separate page include manuscript title, author(s) name(s), affiliations of each author, institution and department where work was done, and acknowledgment of any financial support. Type name, full address, and telephone number of corresponding and reprint request author in lower right corner.

Abstract: 250 words, divided into Objectives, Methods, Results, and Conclusions for original contributions. Do not use abbreviations or references to figures or tables.

References: Number all references consecutively and list in order of appearance in text. Type double-spaced. Use journal abbreviations found in *Index Medicus.*

ARTICLE WITH 3 OR LESS AUTHORS

1. Thiese ND, Rotterdam H, Dieterich D. Cytomegalovirus esophagitis in AIDS: Diagnosis by endoscopic biopsy. Am J Gastroenterol 1991:86:1123-6.

ARTICLE WITH MORE THAN 3 AUTHORS

2. Schindlbeck NE, Ippisch H, Klauser AG, et al. Which pH threshold is best in esophageal pH monitoring? Am J Gastroenterol 1991:86:1138-41.

BOOK

3. Reuter SR, Redman HC. Gastrointestinal angiography, 2nd ed. Philadelphia: WB Saunders. 1977.

CHAPTER IN BOOK

4. Ansel HJ. Normal pancreatic duct. In: Stewart ET, Vennes JA. Geenen JE, eds. Atlas of endoscopic retrograde cholangiopancreatography. St. Louis: CV Mosby, 1977:43-7.

Tables: Give each an appropriate title. Make each self-explanatory. Do not duplicate text. Data should be logical and well organized so that it can be used to compare or classify related items.

Type double-spaced, including headings, on 8.5 × 11 in. paper. Do not use larger sheets. If a table must be continued, use a second sheet and repeat all column and line heading. Arrange tables such that when they are printed, they will not be wider than width of one typed page (6.5 in.). Number tables consecutively in Arabic numerals beginning with 1.

Illustrations: Submit four sets of original illustrations. Place each set in a separate envelope labeled with senior author's name and manuscript title. Number all illustrations of graphs, artwork, and photographs in consecutive Arabic numerals. Submit as high-contrast, glossy prints, 5 × 7 in. Make lettering sufficiently large to be legible upon reduction. Do not submit original artwork unless requested. Submit x-ray photographs as high-contrast, glossy prints, with light and dark areas the same as original x-ray. Affix gummed label to back of each illustration with name of senior author, manuscript title, figure number, and arrow indicating top.

Authors will bear full costs of color illustrations.

Legends: Type double-spaced on a separate sheet. When appropriate, place arrows on photographs and drawings to indicate area to which reference is made. For photomicrographs, indicate magnification and stain utilized.

OTHER FORMATS

CASE REPORTS—Include a detailed analysis of case and a review of the available literature. If not, submit as letter to the editor. Include brief abstract describing case(s) and literature review.

LETTERS TO THE EDITOR—Comment on articles recently published in *The American Journal of Gastroenterology.* Brief case reports or new findings may also be submitted and include a single supporting illustration. Double-spaced and limit to 1.5 pages.

REVIEWS—Substantive reviews of clinical topics in gastroenterology and liver disease.

OTHER

Experimentation, Animal and Human: Include evidence of peer review for all clinical research papers submitted from U.S. Include date of project approval, when available.

Units: Metric units.

SOURCE

January 1994, 89(1): front matter.

American Journal of Epidemiology

Society for Epidemiologic Research
The Johns Hopkins University School of Hygiene and Public Health

American Journal of Epidemiology
Candler Building, Suite 840
111 Market Place
Baltimore, MD 21202

MANUSCRIPT SUBMISSION

Submit cover letter in duplicate; include name, address, and telephone and fax numbers of corresponding author. Do not submit articles or letters by fax.

All authors must sign a statement indicating that they have read and approve of contents of submitted manuscript.

Manuscripts must not have been previously published, must not essentially duplicate already published material, and may not be simultaneously considered for publication elsewhere. If manuscript is based on findings of an article in press or such an article is cited in support of current findings, submit copies.

Follow code of ethics for authorship discussed in articles in *Annals of Internal Medicine* (e.g., see *Ann Intern Med* 1984;100: 592-4, 1986;104:257-62, 266-7, etc.).

Authors of accepted manuscripts must sign a Publication Agreement for assignment of copyright to The Johns Hopkins University School of Hygiene and Public Health. Published manuscript becomes sole property of the *Journal* and may not be submitted elsewhere for later publication without written permission.

If manuscript contains tables or figures fully reproduced or closely adapted from a previously published material (including author's own published work), supply written permission from original copyright holder and author.

MANUSCRIPT FORMAT

Reports of original research are published as Original Contributions (4000 words) and Brief Original Contributions (1500 words). Reports comprise field and laboratory studies on infectious diseases and noninfectious acute and chronic diseases or studies on statistical or methodological issues. Length guidelines are exclusive of tables, figures, references, and abstract.

Type manuscript double-spaced, with 1.25 in. margins at top, bottom, and sides of page, on one side of 8.5 × 11 in. paper. Put author's name and page number at top right of each page. Print should be letter quality.

Submit original typed manuscript, with duplicate 4 × 6 in. copies of figures, and three complete nonreturnable copies. Each copy should include a photocopy of each figure and table.

Cover Page: Provide title and authors only (without degrees or affiliations). Use superior footnote numbers to indicate affiliations, which go on a separate page.

Footnotes Page: A separate page should give (in order) an alphabetical list of approved abbreviations, author affiliations, and complete address for reprints. Use U.S. Postal Service abbreviations for states. Provide a running head below, a short title (50 letters and spaces).

Grants and Acknowledgments: List grants or agencies supporting work and acknowledgments in order after text and before references. Do not use degrees; "Dr." should precede name of each person with a medical and/or doctoral degree. Do not thank reviewers or study subjects.

Abstract: 200 words for an Original Contribution or 100 words for a Brief Original Contribution.Type as a single, double-spaced paragraph. Include study year(s), location, and population studied. Use impersonal "the author(s)."

Key Words: List eight keywords in alphabetical order, below abstract selected from a current copy of *Medical Subject Headings* of *Index Medicus* (National Library of

Medicine). In submissions without abstract (including letters to the editor), list on a separate page.

Headings: Do not number sections. In Original Contributions and Brief Original Contributions use MATERIALS AND METHODS, RESULTS, and DISCUSSION. Headings may not apply to statistical papers. Place flush left on page in capital letters; do not underline or use bold lettering. Do not use "Introduction" as a heading.

Second-level headings are flush left; capitalize only first letter of first words. Do not underline.

Third-level headings are underlined and begin paragraph (capitalize only first letter of first word), followed by a period.

Disclaimers: *Journal* does not print disclaimers, such as a statement that a government agency does not support findings or does not endorse use of products mentioned in paper.

Footnotes: Avoid footnotes to text. Incorporate material into text (e.g., in parentheses).

Tables: Do not include lengthy, encyclopedic tables; incorporate creative synthesis. Type each on a separate page. Submit an original typed version of each table rather than a reduced photocopy. Number (Arabic numerals) in sequence mentioned in text. Make concise and self-explanatory, with adequate headings and footnotes. Use single top rule, single rule below boxhead, and single bottom rule. Do not use rules within body. Clearly delineate column headings, with straddle rules over pertinent columns to indicate subcategories. Table titles give details on place of study, time of study, and study population. Type TABLE 1 flush left, followed by a period and title; capitalize only first word and proper nouns. In text, use lowercase beginning letter for words table and figure. Leave blank spaces for no entry. For footnotes, use symbols in order: (*, †, ‡, §, ||, ¶, #). Use asterisks (*, **, ***) for p values. If p values repeat in successive tables, define in a footnote.

Figures: Submit figures submitted in duplicate as 4 × 6 in. laserjet or glossy prints, with name of author, figure number, and TOP marked lightly on back of each print (use a label, or mark in pencil; do not use ink). Make letters, numbers, decimal points, and symbols large and sharp enough to be readable when figures are reduced. Fgures are reduced either to fit in one- or two-column width of a journal page. In maps, add scale (in kilometers or meters) and direction north.

Do not include figure titles and legends on prints; type double-spaced together on one page. Each legend should be a separate paragraph and should include details on place of study, time of study, and study population. Define acronyms used in figure in legend. Include photocopies of figures with each manuscript copy.

Have graphs professionally designed; make clear, and (together with caption) self-explanatory. Draft so that they will be legible if reduced for printing. Photographs should be high-quality prints, suitable for reduction to page size.

References: Number consecutively in order of mention in text. Reference numbers in text are full-sized Arabic numerals placed in parentheses within sentence. For three or more consecutive references cited at once, give as (1–4).

Give other references as (4, 5, 12) with spaces between reference numbers.

When directly quoting material in text, give reference number followed by page number(s) of quotation, e.g., (24, p. 65).

Reference all statements of scientific fact. Failure to do so may cause considerable delay in processing the manuscript and may necessitate renumbering references.

Insert references to personal, written communications in parentheses in text, not in reference list. Give name, institutional affiliation, "personal communication," and year (in order). Do not cite verbal communications as supporting documentation.

Reference information from a manuscript submitted for publication but not yet accepted in parentheses in text. Give name, institutional affiliation, and "unpublished manuscript." Cite unpublished data in text. Do not refer to "forthcoming" papers or promise future publication of results.

Verify references against original documents; be exact as to authors' names, initials, and article title. Supply inclusive pages; if only one page number, indicate whether reference is a letter or an abstract in parentheses after title. Include in reference list manuscripts accepted but not yet published; designate journal followed by "(in press)." For references to papers presented at conferences, give location, month, and year of conference.

For articles printed in a language other than English, indicate language in parentheses after article title.

Type double-spaced. Abbreviate titles of journals according to *List of Journals Indexed in Index Medicus* (National Library of Medicine). If more than three authors, list only three and add "et al." Telescope page numbers, e.g., 132-6.

STANDARD JOURNAL ARTICLE

Black RE, Brown KH, Becker S, et al. Longitudinal studies of infectious diseases and physical growth of children in rural Bangladesh. II. Incidence of diarrhea and association with known pathogens. Am J Epidemiol 1982;115:315–24.

COMMITTEE OR CORPORATE AUTHOR

WHO MONICA Project Principal Investigators. The World Health Organization MONICA Project (Monitoring Trends and Determinants in Cardiovascular Disease): a major international collaboration. J Clin Epidemiol 1988;41:105–14.

BOOK

Last JM, ed. Public health and preventive medicine. 12th ed. Norwalk, CT: Appleton-Century-Crofts, 1986.

CHAPTER IN BOOK

Vine MF, McFarland LT. Markers of susceptibility. In: Hulka BS, Wilcosky TC, Griffith JD, eds. Biological markers in epidemiology. New York: Oxford University Press, 1990:196–213.

AGENCY PUBLICATION

Roberts J, Rowland M. Hypertension in adults 25-74 years of age, United States, 1971–1975. Hyattsville, MD: National Center for Health Statistics, 1981. (Vital and health statistics, Series 11: Data from the National Health Survey, no. 221) (DHHS publication no. (PHS) 81–1671).

CONFERENCE PRESENTATION

Pourier MC. DNA adduct determination in human tissues and blood cell DNA samples. Presented at the Mary Lasker Conference of the American Cancer Society, Sarasota, Florida, April 1991.

OTHER FORMATS

Reviews and Commentary–4000 word signed editorials, reviews, and comments on various aspects of epidemiologic research.

Letters to the Editor–500-600 words referring to articles previously published in *Journal*. Convey views as concisely as possible; include a reasonable number of references. Letter is usually sent to first author of original paper for response in same issue in which letter is published. Responding author(s) should reference critique before citing other references.

OTHER

Acronyms: To improve clarity and readability, strictly limit use. Only acronyms and abbreviations needed to stand for long, involved terms, such as hepatitis B surface antigen (HBsAg), are allowed. Use acronyms CI, RR, OR, and SMR in parentheses after being defined at first use. Use acronyms and abbreviations in tables and figures if defined in table legends or footnotes and figure legends. Define acronyms in parentheses in the text at first appearance, and in abbreviations footnote.

Equations: Write mathematical equations for typesetting on a single line, as $a/(a + b)$ and $\exp(-x)$. With proper use of braces, brackets, parentheses, and exponents, even fairly complicated expressions can be put in this form. To refer to an equation in text, use "equation 6" or "expression 6."

Electronic Submission: Submit a DOS diskette either 5.25 or 3.5 in. with final revision. Specify software used. If using WordPerfect, do not use End Notes feature for references. Type references in journal style at end of document.

Numbers: Write out numbers under 10, except for decimals, percentages, measurements, and units of time; express numbers 10 and over in Arabic numerals. Use numerals for case 1, subject 2, etc. Insert commas in numbers with four or more digits.

For decimal fractions less than 1.00, use zero in whole-number position, e.g., 0.001. Use spaces before and after any mathematical symbol.

Page Charges: No charge for papers of four printed pages or less. For longer papers: Original Contributions, $75 per page for fifth and sixth printed pages and $90 per page for seventh and succeeding pages (multi-part papers in a single issue are considered a single contribution). Reviews and Commentaries $60 per page for all pages over four.

Revisions: If manuscript is revised, send original and three complete copies of revision. Address each point raised by reviewers. On two copies of revised manuscript, indicate where changes were made (use a highlighter). In letter with revised manuscript, indicate reasons for not accepting particular suggestions by reviewers and editor.

Spelling: Use American English spelling. Follow *Webster's Third New International Dictionary* or *Merriam Webster's Collegiate Dictionary* Tenth Edition for spelling and word division. Follow *Merck Index* and *Stedman's Medical Dictionary* for spelling chemical names and medical terms.

Statistical Analyses: Style for probability is $p < 0.01$, with a lowercase p.

Express confidence intervals without an equal sign and with a dash (not a comma) between values (e.g., 95 percent confidence interval 1.20-1.90). When a confidence level includes a minus value, use "to" instead of a dash, i.e., 95 percent confidence interval -0.01 to 0.05.

Regression coefficients are not suitable for final results of regression analyses; convert into more generally meaningful terms (e.g., rates, means, relative risks, odds ratios). Regression coefficients are unit dependent for continuous variables and category dependent for discrete or ordinal variables, specify units or categories. i.e., as parenthetical statements in text or in table footnotes or figure legends.

Style: A current copy of *Journal* will show correct format and style. Use in conjunction with the *CBE Style Manual*, (Council of Biology Editors, Inc.,Chicago, IL).

Trade Names: For products used in experiments (particularly those referred to by a trade name), give manufacturer's full name and location (in parentheses). When possible, use generic names of drugs.

Units: Use metric system for all measurements. Where U.S. measurements must be used, give metric equivalents in parentheses.

SOURCE

July 1, 1994, 140(1): back matter.

American Journal of Human Genetics
American Source Society of Human Genetics
University of Chicago Press

Paul H. Byers M.D.
Department of Pathology, SM-30, Hitchcock 202
University of Washington
Seattle, WA 98195

AIM AND SCOPE

The American Journal of Human Genetics is a record of research and review relating to heredity in humans; to the applications of genetic principles in medicine, psychology, anthropology, and social services; and to areas of molecular and cell biology relevant to human genetics.

Original articles, review articles (3000 words) and minireviews (1500 words) on timely subjects concerning all aspects of human genetics and related areas are considered, as are brief commentaries on previously published material, editorials, and book reviews. Descriptions of single mutations unless the mutation is of highly unusual significance will not be published.

MANUSCRIPT SUBMISSION

Only papers submitted solely to the *Journal* will be considered. Authors of accepted manuscripts must sign an agreement transferring copyright to The American Society of Human Genetics.

MANUSCRIPT FORMAT

Send three copies (including three sets of original figures, tables, and references). Include a letter from corresponding author, containing helpful information such as type of article that manuscript will represent in *Journal*, and information on prior or concurrent submissions to other publications. Include telephone number(s), fax number, and direct mailing address.

Type entire manuscript including references, tables, and figure legends, double-spaced, on heavyweight bond paper. Make page margins 1.5 in. on top, bottom, and left and right sides. Use only letter quality word-processing output. Identify handwritten items (e.g., Greek letters) in margin. Begin each manuscript component on a separate page. Number pages consecutively, beginning with title page and continuing through summary, text, references, tables, and figure legends.

Title Page: Give a concise title, names and current affiliations of all authors, and name and complete mailing address of contact author. Include a running head (40 characters including spaces).

Summary: No more than one manuscript page, stating purpose of study, basic procedures and analytical methods, description of observations and findings, and principal conclusion.

Text: Be concise. Include the following sections: Introduction, Material and Methods, Results, Discussion, and Acknowledgments. Long articles, case studies, and editorials may require subheadings within some sections or different formats.

Introduction summarizes rationale for study without reviewing subject extensively. Clearly describe (use past tense) selection of observational or experimental subjects and identify methods, apparatus, and procedures in detail. Provide references for established methods, with brief descriptions for published but not well-known methods; describe new or modified methods, giving reasons for their use and evaluating their limitations.

Present results in logical sequence in text, tables, and illustrations. Do not repeat in text all data in tables and illustrations. Present detailed statistical analyses in an appendix.

Acknowledgments: Acknowledge those who have made substantive contributions. Obtain permission from those acknowledged by name.

References: Place in text in parentheses with author name(s) and year of publication. Put text citations of two or more works at the same time in chronological order. When citing a paper by three or more authors, name first author with "et al." Arrange references section alphabetically by author name(s) and then chronologically. Include written permission from scientist(s) named when using "personal communication" as a reference citation.

List names of first seven authors in reference list. If more than seven add "et al."

JOURNAL

Barker D, Holm T, White R (1984) A locus on chromosome 11p with multiple restriction site polymorphisms. Am J Hum Genet 36:1159-1171

CHAPTER IN A BOOK

Goldstein JL, Brown MS (1983) Familial hypercholesterolemia. In: Stanbury JB, Wyngaarden JB, Frederickson DS, Goldstein JL, Brown MS (eds) The metabolic basis of inherited disease, 5th ed. McGraw-Hill, New York, pp 672-712

ENTIRE VOLUME

Dayhoff MV (ed) (1978) Atlas of protein sequence and structure. Vol 5, suppl 3. National Biomedical Research Foundation, Washington, DC

MEETING PAPER

Novacek MJ, Wyss AR (1985) Morphology, molecules, and eutherian phylogeny: the search for congruence. Paper presented at the Fourth International Theriological Congress, Edmonton, Alberta, August 13-20

REPORT

Hasstedt SJ, Cartwright PE (1981) PAP: pedigree analysis package. Tech rep no 13, Department of Biophysics and Computing, University of Utah, Salt Lake City

THESIS/ DISSERTATION

Khoury MJ (1985) A genealogic study of inbreeding and prereproductive morality in the Old Order of Amish. PhD thesis, Johns Hopkins University, Baltimore

WORK "IN PRESS"

King CE. Age-specific selection. II. The interaction of growth factors. J Exp Biol (in press)

WORK "SUBMITTED" (I.E., NOT YET ACCEPTED)

Feuchter A, Mager D. Human repetitive LTR-like sequences can promote and enhance heterologous gene expression (submitted)

Tables: Type double-spaced on separate sheets. Construct exceptionally large tables and pedigree diagrams in sections. Number consecutively. Give each table a brief title and each column a short heading. Use footnotes denoted by letter to explain all nonstandard abbreviations.

Illustrations: Submit professional quality figures in triplicate. Submit half-tones and black-and-white line drawings as glossy photographs. Figures are usually set as single column (80 mm), sometimes double column (approx. 165 mm), or rarely broadside (approx. 225 mm). Make figure size commensurate with information contained. Prepare figures for publication size (5×7 in.) or other convenient size. Lettering must be such that when reduced the smallest letter will be no less than 2.0 mm in height. Indicate magnification by micron bar or in legend. Use same typeface for all lettering, use similar type sizes, use only sans-serif typefaces. Include symbol explanations in figure. Crop and insert letters for panel/section designations.

Label all illustrations on back to indicate figure number, names of authors, and top. A signed consent form must accompany any photograph of a live-born and/or recognizable human subject.

Legends for Illustrations: Type double-spaced on a separate sheet; insert at end of manuscript after tables. Describe illustrations so they can be understood apart from text. Explain any labels or symbols in figures.

OTHER

Color Charges: $1,000 minimum.

Equations: Type carefully with spacing in final format.

Nomenclature, Abbreviations, and Symbols: Use genetic notation and symbols for human genes approved by Human Gene Mapping (HGM). See Shows et al. "Guidelines for Human Gene Nomenclature" (*Cytogenet. Cell Genet.* 46:11–28, 1987), McAlpine et al. "The 1988 Catalog of Mapped Genes and Report of the Nomenclature Committee" (*Cytogenet. Cell Genet.* 49:4-38,1988), and Kidd et al. "Report of the Committee on Human Gene Mapping by Recombinant DNA Techniques" (*Cytogenet. Cell Genet.* 49:132-218,1988). Symbols for human genes not included in above may be obtained from P. J. McAlpine (cochair, HGM Nomenclature Committee), Department of Human Genetics, University of Manitoba, T250-770 Bannatyne Avenue, Winnipeg, Manitoba R3E 0W3, Canada; Telephone: (204) 788-6393; Fax: (204) 786-8712; E-mail: GENMAP@UOFMCC.

Use following standard abbreviations (for others, follow CBE *Style Manual*):

AMP	adenosine monophosphate
ATP	adenosine triphosphate
BrdUrd	5-bromodeoxyuridine
BSA	bovine serum albumin
df	degree(s) of freedom
DEAE	diethylaminoethyl
DTT	dithiothreitol
DZ	dizygotic
EDTA	ethylenediaminetetraacetate
FCS	fetal calf serum
FSH	follicle-stimulating hormone
mRNA	messenger RNA
mtDNA	mitochondrial DNA
MZ	monozygotic
NAD (NADH)	nicotinamide adenine dinucleotide
NAD (NADP)	nicotinamide adenine dinucleotide phosphate
PBS	phosphate-buffered saline
PAGE	polyacrylamide-gel electrophoresis
PCR	polymerase chain reaction
PIC	polymorphic information content
RFLP	restriction-fragment-length polymorphism
rRNA	ribosomal RNA
SSC	saline sodium citrate
SDS	sodium dodecyl sulfate
SD	standard deviation
SEM	standard error of the mean
TSH	thyroid-stimulating hormone
tRNA	transfer RNA
UV	ultraviolet
VNTR	variable number of tandem repeats

Page Charges: $20.00 per printed page. (Three double-spaced, typed manuscript pages equal approximately one printed page.) If the authors are unable to pay, payment may be waived.

Reviewers: Suggest, in cover letter.

Terminology and Abbreviations: Use CBE *Style Manual: A Guide for Authors, Editors, and Publishers in the Biological Sciences* (5th ed., Council of Biology Editors, Inc., Bethesda, MD) and *The Chicago Manual of Style.*

Units: Metric units or other internationally accepted units.

SOURCE

July 1994, 55(1): back matter.

American Journal of Medical Genetics
Wiley-Liss, Inc.

John M. Opitz, MD
Shodair Hospital
PO Box 5539, 840 Helena Avenue
Helena, MT 59604
Telephone: (406) 444-7522; Fax: (406) 444-7536
Dr. Giovanni Neri
Instituto di Genetic Umana
Universita Cattolica
Largo Francesco,
V'lto, 00168, Rome, Italy
                    In Europe
Dr. Enid F. Gilbert-Barness, MD
Department of Pathology
University of South Florida, Tampa General Hospital
Tampa , FL 33603
                    Clinicopathological conferences
Dr. David J. Prieur.
Department of Veterinary Microbiology and Pathology
Washington State University
Pullman, WA 99164-7040
                    Animal models
Philip Reilly, J.D., M.D.
Medical Director
Eunice Kennedy Shriver Center for Mental Retardation
200 Trapelo Road
Waltham, MA 02254
                    Legal/ethical matters

AIM AND SCOPE

The Journal's primary purpose is to report original research in: Clinical genetics, including description of new syndromes, new causal and pathogenetic insights into known syndromes, advances in carrier detection, and genetic counseling; Nosology (systematic, analytical) in clinical genetics; Clinical cytogenetics and delineation of syndromes due to chromosome aberration; Prenatal diagnosis, reproductive genetics, and the genetics of prenatal and perinatal death in humans; Medical/genetical anthropometry and anthropology; Legal, social, and ethical issues and historical aspects of medical genetics; Formal human genetics pertinent to genetic counseling and the analysis of genetic data; Genetically pertinent research in other medical disciplines, such as orthopedics, neurology, hematology, cardiology, ophthalmology, endocrinology, and oncology; and Molecular and biochemical genetics.

MANUSCRIPT SUBMISSION

Manuscripts must be submitted solely to this journal. Manuscripts must be approved for publication by all coauthors and institutions at which work was performed. Manuscripts must not have been published, either in whole or in part, in another publication of any type, professional or lay. Upon acceptance, copyright will be transferred to publisher.

MANUSCRIPT FORMAT

Submit all material in triplicate typed (double-spaced) on 8.5 × 11 in. paper with 1 in. top, bottom, left and right margins. Number all pages sequentially beginning with title page.

Title Page: Include a concise title, authors' names (without degrees or titles), identification of each author's specific affiliation, a running title (45 characters), and name, complete address, and telephone and fax numbers of contact person.

Abstract: Page 2: succinct condensation of contents of article (250 words). Present essential points, purpose of research, a coherent summary of findings, and a concise presentation of conclusions.

Key Words: Used for indexing and should be specific. Type words that reflect central topic of article on abstract page.

Running Heads: If less than four authors, use surnames of all authors as left page running head. If four or more, use "Doe et al." Provide a shorter version of title to be used as a right page running head. Indicate these running heads (author and short title) on title page.

Text: Be concise. Follow format: Introduction, Materials and Methods, Results, and Discussion run together consecutively. Use subheads and paragraph titles. List acknowledgments before the references.

References: Do not include discursive comments of a subsidiary nature in reference list. Put in footnotes at page bottom, identified by an asterisk, dagger, etc. Make reference to original publications and not to reviews, unless a general topic is referred to. Verify references against original publications. Submit signed permissions from persons acknowledged or cited under "personal communications" for use of their name and/or data.

In extensive reviews only, use number system if there are many references. Arrange reference list in alphabetical order. Make citations in text by number in brackets.

In all other articles, cite references by author and year of publication in text and in alphabetical order in bibliography without numbering. Double space references.

Abbreviations should conform to *Index Medicus*.

JOURNALS

Chitayat D, Hodgkinson KA, Ginsburg O, Dimmick J, Watters GV (1992): King syndrome: A genetically heterogenous phenotype due to congenital myopathies. Am J Med Genet 43:954-956.

BOOKS

Wilkins AS (1993): "Genetic Analysis of Animal Development." New York: Wiley-Liss, pp. 83-160.

CHAPTERS IN BOOKS

Whyte MP (1993): Osteopetrosis and the heritable forms of rickets. In Royce PM, Steinmann B (eds): "Connective Tissue and Its Heritable Disorders," New York: Wiley-Liss, pp. 563-589.

For references to government publications include department, bureau or office, title, location of publisher, year, pages cited, and publication series, report, or monograph.

Legends: Type double-spaced on a separate sheet at end of MS. Legends need not be complete sentences but should enable the reader to identify figure components without referring to text. Explain abbreviations on figure. Identify all elements of every figure using descriptive phrases, especially elements of pedigrees.

If figure presents labelling alphabetics (a, b, c), type legend to reflect exact presentation of those alphabetics. Type table heads on tables and not on a separate page.

Illustrations: Provide three sets of illustrations (3 prints of each original negative); no Xerox copies. Halftone photographs with insufficient contrast will not reproduce well. Submit glossy black and white photographs, 12 × 18 cm (5 × 7 in.) in size. Submit one original chromosome composite and 2 photographic copies). Do not submit original records, graphs, radiographic plates, or artwork. All lettering, graphics, and artwork must be professional quality and must be legible and clear after reduction in size. Typewritten- or hand-lettering is unacceptable. Original prints of individual components of original halftone composite illustrations (rather than a copy of the entire composite figure) must be submitted together with instructions on how illustration is to be composed. Number figures with Arabic numerals.

Identify each illustration with a gummed label on back listing illustration number, name of lead author, title of manuscript, and an arrow indicating top.

Tables: Must be self-explanatory and not duplicate information in text. Give each table a title and number with Roman numerals in order of appearance in text.

OTHER FORMATS

ANIMAL MODELS—Submit to: Dr. David J. Prieur, Department of Veterinary Microbiology and Pathology, Washington State University, Pullman Washington 99164-7040.

BOOK REVIEWS AND BOOK REVIEW ESSAYS—See Szbalski on Chargaff's "Heraclitean Fire," *Am J Med Gen* 4:205209, 1979.

BRIEF CLINICAL REPORTS—Include abstract and key words.

CLINICOPATHOLOGICAL CONFERENCES—See CPC of the Zellweger syndrome in this Journal 7:171,1980. CPC's do not need an abstract or key words; circle words on proofs for indexing purposes.

EDITORIALS (ON GENERAL MATTERS) AND (INVITED) EDITORIAL COMMENTS—Usually on a matter published in same issue of journal. Include a list of key words.

HISTORICAL/BIOGRAPHICAL ESSAYS—Include historical notes on certain subjects, biographical essays, and invited autobiographical material.

LETTERS TO THE EDITOR—Keep comments on a previous work published in journal brief and restricted to one comment (with or without author's reply). On proofs, circle words for indexing. On new matters, letters should be brief but may be illustrated and have 1 or 2 tables; supply key words.

NEW SYNDROME—Noncytogenetic case reports of possibly new syndromes observed in sporadic and familial cases. Use telegraphic style citing primarily abnormal findings (quantified as accurately as possible) but also important normal findings. List findings by region from head to toes. In case of prenatal, birth and family histories and references to the literature phrase such as "unremarkable" or "non- contributory" should be based on the assumption that a careful pertinent search has been made.

Before submission, case reports must be seen by at least two well-known syndromologists to see if other published or unpublished cases are known. Their opinion must accompany report. Submit only in Montreal and Helena offices.

Keep high quality illustrations to a minimum. Include face and submit a photo-permit. Roentgenological illustrations are accepted if they show something unusual or distinctive.

Keep number of authors on case reports to a minimum, preferably no more than two.

Minimum of pertinent discussion of diagnosis, nosology, cause, and selected references. No duplication of text and legends.

Readers are invited to respond in letters to the editor(s) in Helena. If these lead to substantial clarification or identification of a new syndrome they will be published in this section in future issues.

SECTION ON LEGAL/ETHICAL MATTERS—Submit to: Philip Reilly, JD, MD, Medical Director Enuice Kennedy Shriver Center for Mental Retardation, 200 Trapelo Road, Waltham, MA 02254

OTHER

Abbreviations: Provide initial definitions for unusual abbreviations. Abbreviate units of measurement according to *Style Manual for Biological Journals,* American Institute for Biological Sciences.

BW (birth weight), BL (birth length), TBL (total body length until height is used on growth chart), OFC (occipito-frontal head circumference), TRC (total ridge count) and other standard dermatoglyphic symbols (q.v. Penrose LS: Memorandum of dermatoglyphic nomenclature, BD:OAS IV/3, June 1968), CHD (specified congenital heart defect).

Experimentation, Human: Submit a statement of compliance with Code of Ethics of World Medical Association (Declaration of Helsinki) and standards established by author's Institutional Review Board and granting agency. If photographs or any other identifiable data are included, submit a copy of the signed consent.

Pedigree: Pedigrees for animals should not include indications of consanguinity of parents, lack of marriage of parents, indications of order of birth, designations of half sibs, nor indications of zygosity of littermates. When animals of different litters of same sire and dam are combined, group animals by litter and separate litters by diagonal slashes on offspring line. Perform consanguinity analysis on standard inbreeding diagram of Sew-all Wright.

Style: See Steve Aronson [*Current Contents* 20/2:6-15, 1977]. Use standard dictionaries and grammar texts and E. B. White's edition of Strunk's *Elements of Style* (3rd edition) [NYC: Macmillan Company, 1979].

Symbols: Use approved list of pedigree symbols *Am J Med Gen* 49(1): 148. Exercise caution in using unmarried/illegitimate dashed lines and in identifying such offspring in pedigrees.

Units: Metric or other internationally accepted units.

American Journal of Medical Genetics: Neuropsychiatric Genetics

Ming T. Tsuang, M.D., Ph.D.
Harvard University
1280-A Belmont St. Suite 192
Brockton, MA 02401
Telephone: (508) 583-4500 ext. 729
Fax: (508) 586-6791

MANUSCRIPT FORMAT

Instructions are the same except: Limit keywords to five. Put corresponding author's name on each page.

OTHER FORMATS

BRIEF CLINICAL REPORTS—Include an abstract and key words.

BRIEF RESEARCH COMMUNICATIONS—2,500 words reporting preliminary data or ongoing work that does not warrant a full-length article. Include Abstract; do not use section headings except for Acknowledgments and References, which should conform to journal style.

EDITORIALS (ON GENERAL MATTERS) AND (INVITED) EDITORIAL COMMENTS—Speculative discussions, and proposals of hypotheses amenable for testing may be submitted. Include a list of key words. Speculative discussions, and proposals of hypotheses amenable for testing are considered.

LETTERS TO THE EDITOR—Are welcomed and may be altered for space and clarity.

RESEARCH ARTICLES—Reports of novel research on genetic mechanisms underlying psychiatric and neurological disorders. Include Abstract, Materials, Methods, and Results sections. No restrictions on number of pages or figures.

REVIEW ARTICLES—Review of a specific field through an appropriate literature survey. Usually invited. Proposals and suggestions for topics are welcomed. Include an Abstract. Materials and Methods and Results sections are not required. Be as concise as possible.

SOURCE

January 1, 1994, 49: 140–148.

American Journal of Medicine
Association of Professors of Medicine

J. Claude Bennett, M.D.
Department of Medicine
University of Alabama at Birmingham
Diabetes Building, Room 405, University Station
Birmingham, AL 35294-0012

AIM AND SCOPE

The editors of the *American Journal of Medicine* welcome concise articles devoted to internal medicine and its specialties. Case reports will not be accepted.

MANUSCRIPT SUBMISSION

Articles must be contributed solely to *The American Journal of Medicine* and become property of publisher. The publisher reserves copyright and renewal on all published material.

Disclose any affiliation with any organization with a financial interest, direct or indirect in subject matter or materials discussed in manuscript that may affect conduct or reporting of work. Information about potential conflict of interest will be made available to reviewers and will be published with manuscript at discretion of editors.

List research or project support in an acknowledgment. If work was done while author was an employee of U.S. Federal Government, indicate in cover letter because publication of this work is not protected by the Copyright Act and there is no copyright to be transferred.

Indicate in cover letter category of manuscript (Clinical Study, Review Special Article, Brief Clinical Observation, Letter).

Submit three copies of all element: text, references, legends, tables, and figures.

MANUSCRIPT FORMAT

Arrange paper: title page; abstract; text; references; legends; tables; and figures.

Number all pages in sequence, beginning with title page as 1, abstract as 2, etc.

Type on 8.5×11 in. opaque white bond paper, on one side of each sheet only,) double-spaced, with wide margins on all four sides.

Follow "Uniform Requirements for Manuscripts Submitted to Biomedical Journals" of the International Committee of Medical Journal Editors (*Ann Intern Med* 1988; 108: 258-65).

Title Page and Abstract: Include first and last names of all authors, highest academic degrees, name of department(s) and institution(s) from which work originated, information about grants, and name and address and telephone and fax numbers of reprint requests author.

Provide a running title of three to six words and key words.

For clinical studies, submit a structured abstract, containing these labeled elements: Purpose, Patients and Methods (or Materials and Methods), Result, and Conclusion.

Provide an abstract with reviews and special articles.

Text: Limit use of acronyms or abbreviations to units of measurement. Cite every reference, figure, and table in numerical order. (Order of mention in text determines number given to each.)

Place acknowledgments at end of text, before references.

References: Number in order of mention. Provide concluding page numbers for journal references and specific page numbers for book references. Indicate all abstracts and letters.

Follow general arrangement, abbreviations, and punctuation as given in "Uniform Requirements for Manuscripts Submitted to Biomedical Journals." For periodicals, follow *Index Medicus*, listing all authors when six or fewer; when seven or more, list only first three and add *et al*. Periods are not used after authors' initials.

Legends: Start at top of new page, number pages in sequence after last page of references. Identify all abbreviations used in figure at end of each legend. Indicate staining method and magnification for all photomicrographs.

Tables: Start each table at top of new page. Title each table and number (in Roman numerals) in order mentioned in text. Identify abbreviations used in a footnote to each table.

Submit only typed originals of tables; if more than one standard 8.5×11 in. sheet is required, tape sheets together carefully. Do not reduce size of table by photocopying.

Figures: Submit three glossy prints (not originals) of each photo and drawing. Use black ink for all charts (line drawings). Make decimals, broken lines, etc., strong enough for reproduction. Make all lettering approximately equal size. Use arrows to designate special features.

Submit high-contrast prints for roentgenographic photographs; some detail is lost in reproduction. Crop photomicrographs to show only essential field.

Include titles in figure legends, not in figures. Identify figures on back by number and first author's name; indicate top. Number figures in order in which they are mentioned in the text, using Arabic numerals (Figure 1, 2, etc.).

Submit written permission from publisher and author to reproduce any previously published figures.

OTHER FORMATS

BRIEF CLINICAL OBSERVATIONS—3 typed double-spaced pages including title page. Review articles must not exceed 30 typed double-spaced pages, including title page, and limit the number of references.

RAPID PUBLICATION—Unique manuscripts involving new therapeutics and/or the utilization of new technologies may be considered for rapid publication. Request that manuscripts be considered for rapid publication in cover letter.

OTHER

Units: Report length, height, weight, and volume in metric units (meter, kilogram, liter) or their decimal multiples. Give temperatures in degrees Celsius and blood pressures in millimeters of mercury.

SOURCE

July 1994, 97 (1): A4.

American Journal of Neuroradiology

American Society of Neuroradiology
American Society of Head and Neck Radiology
American Society of Interventional and
    Therapeutic Neuroradiology

Michael S. Huckman, M.D.
Department of Diagnostic Radiology
Rush-Presbyterian-St. Luke's Medical Center
1653 W. Congress Parkway
Chicago, IL 60612,
Telephone: (312) 942-2249; Fax: (312) 942-7769

AIM AND SCOPE

AJNR publishes original contributions to the advancement of radiologic knowledge (brain, head, neck, spine, organs of special sense, and related sciences).

MANUSCRIPT SUBMISSION

Submitted manuscripts should not contain previously published material and should not be under consideration for publication elsewhere.

Submit three copies of manuscript (original and two photocopies); three complete sets of figures and copyright form (*AJNR* 15(1): 122) signed by all authors.

Manuscripts may not contain statements that would provide reviewers with authors' identities or name of institution at which work was done. Refer to their previous work with "Previous studies have shown . . ." with appropriate references in bibliography.

MANUSCRIPT FORMAT

Type double-spaced, including references, figure legends, and tables, on 8.5 ×11 in. nonerasable paper. Do not justify right margins. Number all manuscript pages consecutively beginning with abstract. Authors' names do not appear on manuscript pages.

Organize manuscript: title page, blind title page (title only), abstract, introduction, methods, results, discussion, acknowledgments, references, tables, figure legends, and figures. Give acknowledgments on a separate page.

Title Page: Title of article; names and complete addresses (including zip code) of all authors; current addresses of authors who have moved; acknowledgment of grant or other assistance. Identify corresponding author with a a current address, telephone and fax numbers.

Include two copies of title page without authors' names.

Abstract: 200 words concisely stating Purpose, Methods, Results, and Conclusion of study, each in a separately headed paragraph. Include actual data. Do not use abbreviations or reference citations.

Introduction: Briefly describe purpose of investigation, including relevant background information.

Methods: Describe research plan, materials (or subjects), and methods used, in that order. Explain in detail how disease was confirmed and how subjectivity in observations was controlled.

Results: Present in logical sequence. If tables are used, do not duplicate data in text.

Discussion: Describe limitations of research plan, materials (or subjects), and methods, considering both purpose and outcome of study. When results differ from those of previous investigators, explain discrepancies.

References: Type double-spaced starting on a separate page. Number consecutively in order of appearance in text. Cite all references in text enclosed in brackets and typed on text line (not superscript).

Cite unpublished data, including meeting abstracts, parenthetically in text; (Smith DJ, personal communication), (Smith DJ, unpublished data), (Smith DJ, paper presented at a meeting). This includes papers submitted, but not yet accepted. Obtain permission to cite such information in writing from at least one author.

Give inclusive page numbers. Abbreviate journal names per *Index Medicus.* List all authors when six or less; when seven or more, list first three, followed by "et al."

JOURNALS

1. Furuya Y, Ryu H, Uemura K, et al. MRI of intracranial Neurovascular Compression. *J Comput Assist Tomogr* 1992:16:503-50

BOOKS

2. Smith LW. Cohen AR. *Pathology of brain tumors,* 6th ed. Baltimore: Williams & Wilkins, 1977: 100 109

CHAPTER IN BOOK

3. Breon AJ. Serum monitors of bone metastasis. In: Clark SA. ed. *Bone metastases.* Baltimore: Williams & Wilkins. 1983:16–180

Tables: Make self-explanatory. Do not duplicate data in text or figures. Check all arithmetic for accuracy. Ensure that tabular data agree with data in text.

Type each double-spaced on a separate page without vertical or horizontal rules with a short, descriptive title, and number in order cited in text. Define abbreviations in an explanatory note below each table.

Figures: Submit three complete sets of original figures unmounted in three separate envelopes. Submit as clean, unscratched, 5 × 7 in. glossy prints with white borders. Submit separate prints for each figure part. Give all figure parts relating to one patient the same figure number.

Label on back with figure number and an arrow indicating TOP. For black-and-white figures, use a gummed label affixed to back. Never use labels on color figures, write figure number on back lightly in pencil. Never use ink on front or back of any figures.

Use only removable (rub-on) arrows and letters. Use symbols uniform in size and style and not broken or cracked. Images must be uniform in size and magnification.

Present axial CT and MR scans as if viewed from below, coronal scans or radiographs as if patient is facing reader, and sagittal projections or lateral images with patient facing reader's left. Illustrations should not have marks, circles, or numbers around image.

Do line drawings in black ink on a white background. They must be professional in quality, and all use same size type. Glossy prints are preferable.

Obtain written permission for use of all previously published illustrations (include copies of permission letters), and give credit line in legends.

Indicate willingness to cover cost of color illustrations.

Legends: Typed double-spaced. Figure numbers correspond with order in which cited in text.

OTHER FORMATS

ABBREVIATED CASE REPORTS—One journal page (two and one half double-spaced typewritten pages) with a 8 references and four figure parts. Brief description of one or two cases or a technique.

CASE REPORTS—Concise case reports present medically important and educational unusual experiences. A 50-word summary describing gist of work is required opening paragraph.

LETTERS TO THE EDITOR AND REPLIES—Offer objective and constructive criticism of published articles. Matters of general interest to neuroradiologists may also be discussed. Type double-spaced, no more than two pages, on nonletterhead paper, with no heading or salutation. Put signature and affiliation at end of letter.

TECHNICAL NOTES—Brief descriptions of new techniques or significant modifications of techniques that are directly applicable to clinical practice or research (models). A 50-word summary describing gist of work is required in opening paragraph.

OTHER

Abbreviations: Keep use of unfamiliar acronyms and abbreviations to a minimum. When abbreviations are used, define at first mention, followed by abbreviation in parentheses.

Drugs: Give names and locations (city and state) of manufacturers for equipment and nongeneric drugs.

Experimentation, Animal: Experiments must comply with NIH guidelines.

Experimentation, Human: Informed consent must be obtained from patients who participated in clinical investigations.

Units: Use metric measurements or give metric equivalent in parentheses.

SOURCE

January 1994, 15(1): A7-A8.

American Journal of Obstetrics and Gynecology

American Gynecological and Obstetrical Society

Central Association of Obstetricians and Gynecologists

Pacific Coast Obstetrical and Gynecological Society

American Board of Obstetrics and Gynecology

Society of Perinatal Obstetricians

Society of Gynecological Surgeons

Mosby-Year Book, Inc.

Dr. Frederick P. Zuspan
Ohio State University
1395 Granview Ave.
Columbus, Ohio 43212
Telephone: (614) 486-4044; Fax: (614) 486-6176

Manuscripts from the eastern and southeastern United States, and England, Israel, and Italy, manuscripts written for Clinical Opinion and Current Development, Letters to the Editors, and all manuscripts presented before one of the official sponsoring societies, except the Society for Gynecologic Investigation.

Dr. J Peter Van Dorsten, Department of Obstetrics and Gynecology, Medical University of South Carolina, 171 Ashley Ave., Charleston, SC 29425

Manuscripts from The Society of Perinatal Obstetricians

Dr. E. J. Quilligan
University of California, Irvine Medical Center
Department of Obstetrics and Gynecology
101 The City Drive, South, Building 26
Orange, CA 92668
Telephone: (714) 456-6572; Fax: (714) 456-6073)

Manuscripts from states west of Pennsylvania, West Virginia, Kentucky, Tennessee, Mississippi, Hawaii, Alaska, Canada, and abroad (except England, Israel, and Italy).

Dr. Roger A Lobo, Women's Hospital 1240 North Mission Road, Room I M2, Los Angeles, CA 90033

Manuscripts from The Society for Gynecologic Investigation

MANUSCRIPT SUBMISSION

Complete and submit copyright statement that follows Information for Authors in each issue, signed by all authors.

If a report by same author(s) or some of same authors has been previously published in any medium that deals in any respect whatsoever with same patients, same animals, same laboratory experiments, or same data, in part or in full, submit two reprints of article(s) or two copies of manuscript (full-length report or abstract). Submit two copies of new manuscripts. Inform Editor of circumstances, similarities, and differences of reports. This also applies to submission of a manuscript in which a few different patients, animals, laboratory experiments, or data were added to those reported in a previous publication or in a submitted or accepted manuscript.

For multi-authored manuscripts, each author must qualify by having participated actively and sufficiently in study. Inclusion is based only on substantial contributions to concept and design, or analysis and interpretation of data and drafting manuscript or revising it critically for important intellectual content; and on final approval by each author of version of manuscript. Recognize others contributing to work in an Acknowledgment. Confirm in cover letter that all authors fulfilled conditions.

State any commercial association that might pose a conflict of interest, such as ownership, stock holdings, equity interests and consultant activities, or patent-licensing situations. If manuscript is accepted, author and Editor will determine how best to release information. Listing of sources of support and institutional affiliations on title page does not imply a conflict of interest; only where there is a possible conflict of interest is the author(s) expected to inform the Editor.

MANUSCRIPT FORMAT

Submit an original and two good-quality photocopies of manuscript and three sets of black-and-white glossy prints of illustrations. Type double-spaced on one side only of 8.5 × 11 in. white bond paper with 1 in. margins at top, bottom, and sides. Number pages consecutively in upper right corner in order: title page, condensation, abstract, body of text, acknowledgments, references, figure legends, and tables.

Follow "Uniform Requirements for Manuscripts Submitted to Biomedical Journals" of the International Committee of Medical Journal Editors (*N Engl J Med* 1991;324:424-8).

Title Page: On page 1 list in sequence title (concise and suitable for indexing purposes); author line with first name, middle initial, and last name of each author and each author's highest academic degree (no honors degrees); city(ies), state(s) in which study was conducted; divisional, or departmental, and institutional affiliations when study was performed; source(s) of financial support; presented line, if applicable; disclaimers, if any; name, address, business and home telephone numbers, and fax number of corresponding and reprint request authors (if reprints will not be available, so state).

Condensation: On page 2 provide a brief, concise condensation, in a single 25 word sentence to appear on Contents page delineating essential point(s).

Abstract: On page 3, type abstract headed by article title and name(s) of author(s). List 3 to 5 key words or short phrases for indexing purposes.

Abstracts may be structured or standard depending on type of article. Structured abstracts (150 words) are used for regular articles and society articles and contain: Objective(s); Study Design: Results; and Conclusion(s). Objective(s) reflects purpose of study, hypothesis being tested. Study Design includes study setting, subjects (number and type), treatment or intervention, and type of statistical analysis. Results include study outcome and statistical significance if appropriate. Conclusion(s) states significance of results.

Text: Use first person and active voice if appropriate. Passive voice is generally more effective for describing techniques or observations.

Organize regular articles into: Introduction; Material and Methods, Results, and Comment. Summarize findings in a short paragraph at the end of Comment section.

Introduction: State purpose and rationale for study. Cite only most pertinent references as background.

Material and Methods: Describe, in sufficient detail to permit other workers to repeat study, the plan, patients, experimental animals or other species, material, and controls, methods and procedures utilized, and statistical method(s) employed.

Results: Present detailed findings. Mention all tables and/or figures. Figures and tables should supplement, not duplicate, text. Emphasize only important observations; do not compare observations with those of others. Put comparisons and comments in Comment section.

Comments: State importance and significance of findings. Do not repeat details in Results section. Limit opinions to those strictly indicated by facts. Compare findings with those of others. Do not present new data.

Acknowledgments: Acknowledge only persons who have made substantive contributions.

References: 25 are allowed. Case Reports and Brief Communications are limited to 2. Manuscripts for Current Development section have no limit. Number references consecutively in order mentioned in text. Use format of "Uniform Requirements for Manuscripts Submitted to Biomedical Journals" (Vancouver style) (*N Engl J Med* 1991;324:424-8). Use journal titles abbreviations per *Index Medicus*.

If six or fewer authors, list all; if seven or more authors, list three then et al.

JOURNAL

Flamm BL, Dunnett C, Fischermann E, Quilligan EJ. Vaginal delivery following cesarean section: use of oxytocin augmentation and epidural anesthesia with internal tocodynamic and internal fetal monitoring. AM J OBSTET GYNECOL 1984;148:759-63.

BOOKS

Ledger WJ. Dystocia and prolonged labor. In: Wilson JR, Carrington ER, eds. Obstetrics and gynecology. St. Louis: CV Mosby, 1987 vol 8:474-93.

Personal communications and unpublished data, if essential, may be referred to, within parentheses, at appropriate locations in text. Obtain and submit written and signed permission for use from individual being quoted. Published abstracts can be used as numbered references; however, reference to complete published articles is preferred.

Figures: Includes all types of illustrations such as graphs, diagrams, photographs, flow charts, and line drawings. A reasonable number of figures will be reproduced without charge. Special arrangements must be made for color figures at additional charge.

Cite figures consecutively in text in Arabic numerals. Identify with a gummed label on back with author(s) name(s), title of article, figure number, and TOP marked clearly. Be consistent in size. Do not use paper clips or mar surface of figures.

Black-and-White Figures—Submit three sets of 3 × 4 in. (minimum) to 5 × 7 in. (maximum) unmounted, glossy prints. Use commercially available paste-on letters (or numbers) or a professional artist; typed or freehand lettering is not acceptable. Keep lettering in proportion to figure. Submit original drawings, appropriately done in black India ink, roentgenograms, and other material as glossy prints with good black-and-white contrast.

Color Figures—Submit original transparencies and two sets of unmounted prints on glossy (smooth surface) paper.

Polaroid prints are not acceptable. Color transparencies must have a color balance (consistency in lighting and film speed); 35 mm transparencies are enlarged to twice original size. Indicate top, first author's last name, and figure number on front of each transparency and back of each print.

Computer-Generated Figures—Must be legible and clearly printed in jet black ink on heavy coated paper. Patterns or shadings must be dark enough for reproduction and must be distinguishable from each other. Lines, symbols, and letters should be smooth and complete. Legend should not appear. Indicate on back name of first author, figure number and top. Submit three unmounted original individual laser or plotter prints. Laser prints should be full size at 300 dots per in. (DPI) or greater full-page resolution; multiple illustrations on a page are not accepted. Dot matrix prints and photographic halftones are not accepted. Color figures are accepted, but special arrangements must be made with Editors. Colors must be dark enough and of sufficient contrast for reproduction.

Legends: Type together in numeric order double-spaced separate from manuscript text. Number page in sequence after references. Provide original magnifications. If figure was taken from copyrighted material, give full credit to original source in legend.

Tables: Type on separate sheets, one table to a page. Put at end of text. Number in Roman numerals. Cite each in sequence at appropriate points in text. Make titles brief and indicate: purpose or content. Define each column by headings. Explain abbreviations and special designations in footnotes. If a table or any part thereof was taken from copyrighted material, give full credit to original source in a footnote. Make arrangements with Editors for elaborate tables.

OTHER FORMATS

CASE REPORTS AND BRIEF CLINICAL AND BASIC SCIENCE COMMUNICATIONS—700 words, 2 references. Include standard abstract (50 words), 3 to 5 key words/phrases for indexing, and short title. If tables and/or figures are used, deduct an equivalent number of words from total.

CLINICAL OPINION—3000 words. Include standard abstract (50 to 150 words), 3 to 5 key words/ phrases, and short title. Submit to Dr. Zuspan.

CURRENT DEVELOPMENT—6000 words. Include standard abstract (50 to 150 words), 3 to 5 key words/phrases, and short title. Submit to Dr. Zuspan.

LETTERS TO THE EDITORS—Submit in duplicate, typed double-spaced, and 400 words (excluding references, name[s] and addresses] of the signer[s], and the phrase "To the Editors"). Include a signed copyright statement. If two or more signers, one author may sign copyright statement but must add: "I sign for and accept responsibility for releasing this letter on behalf of any and all cosigners." Send to Dr. Zuspan.

Two types of letters are considered: Comments on articles that have appeared in the *Journal* should be brief and directly related to the published article. Letters may be published together with a reply from the original author. Brief case presentations or short reports of a pertinent observation in the form of a Letter to the Editors.

OTHER

Abbreviations: Use only standard abbreviations. Consult *Council of Biology Editors Style Manual* or *AMA's Manual of Style*. Do not use abbreviations in title; avoid use in abstract, and keep to a minimum in text. The full term for which an abbreviation stands precedes first use in text.

Checklist: Complete and submit checklist that appears with copyright statement in each *Journal* issue.

Experimentation, Human: Manuscripts must have approval of requisite institutional authority. Local institutional approval must have been acquired before experiment was started. Indicate approval in Material and Methods section.

Experimentation, Animal: State in Material and Methods section that guidelines for the care and use of animals approved by the local institution were followed. Name type of nonhuman animals or other species used in title, abstract, key words, and Material and Methods section.

Nomenclature: Use either generic, chemical, or proprietary names of drugs. If generic or chemical name is used, place proprietary name in parentheses after first mention, with manufacturer's name, city and state.

Permissions: Direct quotations, tables, or figures from copyrighted material must be accompanied by written permission for use from copyright owner and original author along with complete reference information. Photographs of identifiable persons must be accompanied by signed releases. If not, mask all recognizable features.

Reviewers: Suggest at least three. Give addresses.

SOURCE

January 1994,170(1): 19A–24A.
Information for Authors is published in each issue.

American Journal of Ophthalmology
Ophthalmic Publishing Company

Bradley R. Straatsma, MD
American Journal of Ophthalmology
Suite 660, 77 West Wacker Dr.
Chicago, IL 60601-1632
Telephone: (312) 629-1690; Fax: (312) 629-1744

AIM AND SCOPE

The *American Journal of Ophthalmology* welcomes the submission of manuscripts describing clinical investigations, clinical observations, and clinically relevant laboratory studies related to ophthalmology.

MANUSCRIPT SUBMISSION

Designate one author as corresponding author. Submit a signed copy of disclosure statement and copyright transfer published in *Journal* each month. Copyright transfer confirms that work is original and does not contain data published previously or submitted for publication elsewhere. If data were presented at a scientific meeting, state place, date of presentation, and auspices of meeting on title page. Disclosure statement requires disclosure of any proprietary interests related to manuscript, including financial interests, such as commercial support, compensation or reimbursement, and patent rights or equity ownership. Government employees must include this statement: "The manuscript, including figures and tables, referred to in this Disclosure Statement and Copyright Transfer is a work of authorship prepared as part of the undersigned author(s)' official duties as an officer or employee of the United States Government and is therefore in the public domain. Should the manuscript ever be determined to be subject to copyright, I (we) hereby transfer, assign, or otherwise convey all copyright ownership in the above named manuscript to the American Journal of Ophthalmology."

MANUSCRIPT FORMAT

Submit original and two duplicate copies of typescript and illustrations. Use 8.5 × 11 in. heavy, white, bond paper. Use 1.5-in. margins on all four sides. Indent paragraphs .5 in. Do not justify right margins. Double space entire typescript, including title page, quotations, footnotes, acknowledgment, references, legends, and tables. Use letter quality type, not smaller than 12 pitch or 11 point. Do not underline or use italics, cursive, condensed, boldface, expanded, or reduced type. Number each page consecutively beginning with title page in upper right corner. List first author's name and short running head (60 characters and spaces) under page number. Use only abbreviations that are widely used and understood. Spell out abbreviation in parenthesis at first time use. Abbreviate standard measurements used with numeric quantities. Numeric equivalents must precede percentages. When using P values, state actual value rather than an inequality (e.g., P = .032, not P < .05). Report basic summary statistics, such as mean, standard error and confidence limits. Specify models such as analysis of variance, covariance, and multiple regressions.

Arrange manuscript: Title page; Summary; Introductory text; Material and Methods, Subjects and Methods, or Case Reports; Results; Discussion; References; Legends for illustrations; and Tables.

Title Page: Page 1: list title (86 characters and spaces, applicable for indexing), each author's name with highest academic degree and departmental affiliation. Indicate department and institution where study was performed. Acknowledge sponsoring organizations and grant support. Include recipient's name and location of sponsoring organization. Give name and address for reprint requests. Disclose proprietary interests, revealing nature and extent. Do not acknowledge consultants, editorial assistants, photographers, artists, laboratory assistants, secretaries, and others who assist in preparation. Acknowledge statistical assistance at end of article before references.

Abstract: 250 word structured abstract on a separate page with main purpose, data, and manuscript message. Headings: Purpose (principal question, objective or hypothesis); Methods (study design with randomization, masking, criteria and temporal direction-study setting-note selection procedures, entry criteria, and numbers for subjects); Results (outcome and measurements, confidence intervals, level of statistical significance, and limitations of data); and Conclusions (only those supported directly by data, describe clinical applications).

Text: Follow style used by *Journal.*

Introduction: Describe purpose of study, research rationale and hypothesis tested.

Methods: Include study design, randomization, masking, criteria standards, and temporal aspects (retrospective or prospective). Specify study setting (multicenter, institutional, primary care or referral practice). Identify participants and selection procedure, criteria and numbers. State Interventional procedures, outcome measures, and statistical analyses procedures.

Results: Describe outcomes and measurements. Cite tables and figures in numerical order. Include confidence intervals and exact P values or other indications of statistical significance.

Discussion: Elucidate results, identify statistically or clinically significant limitations or qualifications of study. State conclusions directly supported by data. Do not generalize or speculate. State whether additional study is required, give equal emphasis to positive and negative findings. Conclude with clinical applications or implications.

References: Number references consecutively, both in text and in reference list. Type numbers in text as superscripts. Corresponding author is responsible for complete and accurate references, including proper capitalization and accent marks.

Parenthetically incorporate into text personal communications, posters, program abstracts, unpublished data, manufacturers' manuals, and oral discussions that reader cannot retrieve. Cite personal communication in the text and provide authorization for use. References to studies that have been accepted but have not yet been published must indicate publication in which they will appear.

References to journals include: author(s) (if more than six, list first six and add "et al"); title; journal name as abbreviated in *Index Medicus*; year; volume number; and inclusive page numbers.

References to books include: author(s) or editor(s); book title; edition; city of publication; publisher; copyright year; inclusive pages of section cited. Include chapter title in references to book chapters.

JOURNALS

Arnold AC, Hepler RS. Fluorescein angiography in acute nonarteritic ischemic optic neuropathy. Am J Ophthalmol 1994;17:222-30.

BOOKS

Gass JDM. Stereoscopic atlas of macular diseases: diagnosis and treatment, 3rd ed. St. Louis: C. V. Mosby, 1987:230-1.

CHAPTERS IN BOOKS

Mannis MJ. Bacterial conjunctivitis. In: Kaufman HE, Barron BA, McDonald MB, Waltman SR, editors The cornea. New York: Churchill Livingstone, 1988:189-9.

Illustrations: Submit original and two copies of each. Illustrations should be high quality 5 × 7 in. prints and cannot exceed 8.5 × 11 in. Do not mount graphs, diagrams, or photographs. Number in order of appearance in text. Indicate illustration number, arrow indicating top, and first author's name on a label on back. Do not put any text on illustration face.

Line drawings must be suitable for reproduction and prepared either by a high-quality graphic laser printer or a graphic artist. Photographs must have a glossy finish and sharp contrast. Lettering, arrows, and the like must be large enough for reduction. Crop photographs on a tissue-paper overlay. If patient is identifiable, supply evidence of permission to publish

Legends: Give each a descriptive legend. Type legends in sequence on a separate sheet as follows:

Fig. 1 (Jones, Smith, and Brown). Histologic section of the left eye of Patient 1 shows infiltrating histiocytes (hematoxylin and eosin, $\times70$).

If more than three authors, use name of first author followed by "and associates."

Label multiple part figures from left to right and top to bottom as follows:

Fig. 1 (Smith and associates). Case 3. Top left, The patient preoperatively. Top right, three days postoperatively. Bottom left, Four months postoperatively. Bottom right, One year postoperatively.

Visual field charts and charts for plotting Farnsworth-Munsell color testing used in *Journal* will be supplied to authors on request without charge. Write to Senior Production Associate, at above address or telephone number.

Tables: Number consecutively in Arabic numerals in order of citation in text. Give each a brief title. Put table number in first line of table and table title on second line. Double-space. Do not underline. Do not use vertical lines, display type, or modified type. Give each column a short heading. Use following symbols for footnotes in order: *, †, ‡, §, ||, ¶, and #.

OTHER FORMATS

BRIEF REPORTS—500 words; 5 references on a separate page; two illustrations with legends on separate page. Submit original and two copies. Succinct manuscripts describing clinical investigations, clinical observations, new instruments or techniques or laboratory investigations relevant to clinical ophthalmology. Do not duplicate previously published or submitted data. Include signed Disclosure Statement and Copyright Transfer. Include: Title page; Abstract Page; Text; References; and Legends. Title page follows format of original article. Put structured abstract on a separate page stating report message in an independent 75 word statement under: Purpose/Methods and Results/Conclusions.

LETTERS—Submit error corrections, support or different points of view and additional information about recent articles within six weeks of publication.

NEWS—75 words. Type items regarding meetings, postgraduate courses, lectures, honors, elective offices, and appointments double-spaced on white paper with 1 in. margins.

OTHER

Color Charges: $500 per page. Submit three high-quality prints.

Style: *The Chicago Manual of Style*, 14th ed. Chicago: University of Chicago Press, 1993.

Huth E, editor. *Scientific Style and Format. The CBE manual for Authors, Editors, and Publishers*, 6th ed. Cambridge: Cambridge University Press, 1994.

Strunk W. Jr., White EB. *The Elements of Style*, 3rd ed. New York: Macmillan Publishing Co., 1979.

SOURCE

July 1994, 118(1): 29-33.

American Journal of Pathology
American Society for Investigative Pathology

Nelson Fausto, M.D.
9650 Rockville Pike, Bethesda, MD 20814-3993
Telephone: (301) 571-0107; Fax: (301) 571-0108

AIM AND SCOPE

The *American Journal of Pathology* seeks to publish high-quality original papers on mechanisms of disease. We accept manuscripts that report important findings on disease pathogenesis or basic biological mechanisms that relate to disease, without preference for a specific method of analysis. High priority is given to studies on human disease and relevant experimental models that use cellular, molecular, biochemical, and immunological approaches in conjunction with morphology, and to manuscripts that report the development of new molecular probes and their application in studies of disease pathogenesis and diagnosis.

MANUSCRIPT SUBMISSION

Submit five copies of manuscript and four sets of figures along with a cover letter giving full name, address, telephone and fax numbers of corresponding author.

MANUSCRIPT FORMAT

Type papers double-spaced throughout (including references, tables, and figure legends. Print on a letter-quality printer. Do not use dot matrix printers. Number all pages in top right corner starting with title page.

Title Page: Concise title; names of all authors; department, institution, and address where research was performed; number of text pages, tables, figures; a short running head (40 characters); grant numbers and source(s) of support; names and addresses of corresponding and reprint request authors.

Abstract and Key Words: 150 words or less on a separate sheet intelligible to general reader without reference to text. Avoid abbreviations.

Text: Sections may include Introduction, Materials and Methods, Results, Discussion, and Acknowledgments and need not begin on new pages.

References: Begin on a new page, double-spaced, and numbered in order of citation in text.

JOURNALS

Meyrick B, Reid L: Hypoxia-induced structural changes in the media and adventitia of the rat hilar pulmonary artery and their regression. Am J Pathol 1980,100:151-178

BOOKS

Fishman AP: Pulmonary Hypertension and Cor Pulmonale. Pulmonary Diseases and Disorders. Edited by Fishman AP. New York, McGraw-Hill, 1988, pp. 999-1048

IN PRESS

For papers accepted for publication. Cite as for journal with (in press) in place of volume and page numbers.

SUBMITTED PAPERS AND UNPUBLISHED DATA

Cite in text only.

Tables: Type double-spaced on separate pages. Number with Arabic numerals.

Figures: Submit four sets. If multiple parts, affix labels identifying each part on figure front. Mark back with a top directional arrow, first author's name and figure number. Identify one set and mount other to show preferred layout or placing of insets.

Size figures to fit one (7.4 cm) or two (15.5cm) column width with a maximum plate size of 15.5 × 22.5 cm

Color Figures: Submit original prints or negatives. Do not mount. $500 per figure; $750 second figure; $1000 all subsequent figures. $125 flat charge for each additional image above four in a single figure. A figure is an image or set of images, up to 4, identified by a single legend.

Legends: State staining methods and degree of magnification Use scale bars on photograph and specify in legend.

OTHER FORMATS

METHODS AND MODELS FOR THE ANALYSIS OF DISEASE MECHANISMS—Includes Technical Advances and Animal Models. Submit cover letter and abstract to editor for consideration. Reviews and Commentaries are also solicited.

SHORT COMMUNICATIONS—12 double-spaced typed pages and three figures. Provides a mechanism for the rapid publication of timely and significant findings. Manuscripts should be concise but definitive. Follow guidelines for regular articles. Condense Materials and Methods section but present it in sufficient detail to permit evaluation by reviewers and reproduction of experiments by other scientists.

OTHER

Page Charges: $50 manuscript processing fee. $40 per printed page.

Style: Follow The Uniform. Requirements for Manuscripts Submitted to Biomedical Journals (*N Engl J Med*, 1991, 324:424-428). For standard abbreviations see *CBE Style Manual* (5th ed., 1983).

SOURCE

July 1994, 145(1): 237-238.

American Journal of Physiology
American Physiological Society

Editor, *American Journal of Physiology*
American Physiological Society
9650 Rockville Pike
Bethesda, MD 20814-3991

AIM AND SCOPE

American Journal of Physiology: Cell Physiology is dedicated to promoting contemporary and innovative approaches to the study of cell and general physiology. It publishes original papers dealing with normal and abnormal cell function, including the structure and function of cell membranes, contractile systems, and cellular organelles, as well as mechanisms of development, cell-to-cell interaction, gene expression, and neural, endocrine, and metabolic control. Reports of research utilizing biochemistry, biophysics, molecular biology, morphology, and immunology and contributing to the knowledge of cell physiology are especially welcome. Theoretical as well as experimental studies are sought.

American Journal of Physiology: Endocrinology and Metabolism publishes the results of original studies about endocrine and metabolic systems on any level of organization. Results of molecular, subcellular, and cellular studies in whole animals or humans will be considered. Specific themes include mechanisms of hormone and growth factor action; hormonal or metabolite control of organic and inorganic metabolism; paracrine and autocrine control of endocrine cell performance; activation of gene expression; function and activation of hormone receptors; endocrine or metabolic control of ion channels and membrane function; differentiation of endocrine and reproductive cell function; temporal analysis of hormone secretion and metabolism; and mathematical modeling and kinetic analysis of hormone action or metabolism. Novel molecular, immunological, magnetic resonance, or electrophysiological studies of hormone action or receptor activation are also welcome.

American Journal of Physiology: Gastrointestinal and Liver Physiology publishes original papers dealing with normal or abnormal function of the alimentary canal and its accessory organs, including the salivary glands, pancreas, gallbladder, and liver. Authors are encouraged to submit manuscripts dealing with digestion, secretion, absorption, metabolism, and motility relevant to these organs as well as those dealing with neural, endocrine, and circulatory control mechanisms and with inflammatory processes. Reports of research utilizing techniques, such as biochemistry, molecular biology, cell biology, morphology, and immunology, that contribute to the knowledge of physiology or pathophysiology are especially welcome. Manuscripts reporting studies at any level of organization from molecular events to the whole animal are appropriate.

American Journal of Physiology: Lung Cellular and Molecular Physiology publishes original research dealing with molecular, cellular, and morphological aspects of normal and abnormal function of cells and components of the respiratory system, including the nose and sinuses, the conducting airways, lung parenchyma and pleura, neural cells involved in control of breathing, neuroendrocrine and immunological cells in the lung, and cells of the diaphragm and thoracic muscles. Areas of interest including gas-exchange and metabolic control at a cellular level, regulatory and informational molecules, gene expression, macromolecules and their turnover, cell-to-cell and cell-matrix interactions, cell motility, secretory mechanisms, membrane function, surfactant, matrix components, mucus and lining materials, lung defenses, macrophage function, electrolyte and water transport, development and differentiation of the respiratory system, and response to the environment. Reports of research using innovative approaches of molecular and cell biology, cell physiology, molecular genetics, biochemistry, biophysics, and morphology are especially welcome.

American Journal of Physiology: Heart and Circulatory Physiology publishes original investigations on the physiology of the heart, blood vessels, and lymphatics, including experimental and theoretical studies of cardiovascular function at all levels of organization ranging from the intact animal to the cellular, subcellular, and molecular levels. It embraces new descriptions of these functions and of their control systems, as well as their bases in biochemistry, biophysics, genetics, and cell biology. Preference is given to research that provides significant new insights into the mechanisms that determine the performance of the normal and abnormal heart and circulation.

American Journal of Physiology: Regulatory, Integrative and Comparative Physiology publishes original articles that illuminate physiological processes at all levels of biological organization. The editors wish to attract papers from physiologists who are united by a broad interest in regulation, integration, and homeostasis.

American Journal of Physiology: Renal, Fluid and Electrolyte Physiology publishes original manuscripts on a broad range of subjects relating to the kidney, urinary tract, and their respective cells and vasculature, as well as to the control of body fluid volume and composition. Investigations may involve human or animal models, individual cell types, and isolated membrane systems. Authors are encouraged to submit reports on research using a wide range of approaches to the study of function in these systems, such as biochemistry, immunology, genetics, mathematical modeling, molecular biology, and physiological methodologies. In addition, papers on the pathophysiological basis of disease processes of the kidney, urinary tract, and the regulation of body fluids are encouraged.

The American Physiological Society also publishes the following research journals: *American Journal of Physiology* (consolidated); *Advances in Physiology Education; Journal of Applied Physiology;* and *Journal of Neurophysiology.* Acceptance of manuscripts is based on scientific content and presentation of the material; membership in the Society is not a prerequisite for publication.

MANUSCRIPT SUBMISSION

In cover letter state that research reported is original and will not be submitted for publication elsewhere until a publication decision has been made by APS1. Except in reviews and invited editorials, APS will not accept manuscripts in which, other than in abstracts, a significant portion of the data, as figures and tables has been published elsewhere. If letter is signed only by corresponding author, he/she is acting as the agent for all authors. Each author acknowledges: that he/she has made an important scientific contribution to study and is thoroughly familiar with primary data; that manuscript is a truthful, original piece of work (use of other investigators' data or ideas is acceptable if carefully documented and, when appropriate, permission of other investigator(s) is given, provide appropriate reprints or preprints; that s/he has read complete manuscript and takes responsibility for its content and completeness; and understands that if a paper or part of a paper is found to be faulty or fraudulent, all authors share responsibility.

APS journals seek definitive papers that present the entire contents of a research project. Present all data from a group of subjects, animals, or samples together in a single paper. If not, then cross-reference manuscript. Use identical subject, animal, and sample numbers in the different manuscripts to identify their commonality. If a paper depends critically on another unpublished paper, include three copies for reviewers.

A copyright transfer form will be sent to submitting author. Transfer form must be completed and returned before work is typeset.

MANUSCRIPT FORMAT

Type manuscript double-spaced with wide margins on 8.5 × 11 in. paper. Submit four copies, including figures. Send glossies for photomicrographs, gels, and other halftones. Photocopies of line drawings are acceptable at submission as long as two sets of glossies are provided when manuscript is accepted; for computer-generated laser prints, use paper recommended for camera-ready copy.

Number pages in upper right corner (beginning with first text page). Arrange in order: title page, abstract and index terms, text, text footnotes, acknowledgments, references, figure legends, tables, illustrations.

Title Page: Title of article; author(s); department and institution where work was done, city, state or country, and zip code; abbreviated title for running head (55 characters including spaces); name and address for mailing proofs; and contact telephone and fax numbers. Put abstract and index terms on a separate sheet, double-spaced.

Text, footnotes, acknowledgments, references, and figure legends begin on separate sheets, all lines double-spaced. Type each table on a separate sheet, double-spaced. Make text clear and concise; conform to accepted standards of English style and usage. Define unfamiliar or new terms when first used. Do not use jargon, cliches, and laboratory slang.

Identify illustrations on reverse (lightly with a soft pencil) with figure number and name of first author; when necessary, mark TOP.

Title: 110 characters including spaces. Make informative. Use no unnecessary words like "Studies in . . . ,"

Abstract: 170 word informative one-paragraph abstract. State concisely what was done and why (including species and state of anesthesia), what was found (in terms of data), and what was concluded.

Key Words: Append 3-5 words or short phrases not included in title to abstract.

Promissory Notes: Do not include either implicit or explicit promises that future work will be published.

Footnotes: Number text footnotes consecutively throughout. Assemble, double-spaced, on one sheet.

References: Limit to 30 directly pertinent published works or papers that have been accepted for publication. Cite abstracts, properly identified (Abstract), only if sole source. Verify accuracy. Type separately, double-spaced (do not single-space any line), alphabetically by author; numbered serially, with only one reference per number. Include number appropriate to each reference in parentheses at proper point in text.

JOURNAL ARTICLES

Last name of first author, followed by initials, initials and last names of each coauthor; title of article (first word only capitalized); name of journal abbreviated as in *Serial Sources for the BIOSIS Data Base* (BioSciences Information Service), volume, inclusive pages, and year.

1. Chisholm, D. J., J. D. Young, and L. Lazarus. The gastrointestinal stimulus to insulin release. *J. Clin. Invest.* 48:1453–1460,1969.

BOOK REFERENCES

Author(s) as above; title of book (main words capitalized); city of publication; publisher; year and pages.

Include references to government technical documents only when their availability is assured. For style of citation of these documents, congress proceedings, chapters in books, etc., consult recent Journal issues.

Do not include citations such as "unpublished observations" or "personal communication" in reference list; add in parentheses in text. Secure permission of person cited for "personal communications."

Illustrations: Submit sharp, unmounted glossy photographic prints (or computer-generated laser prints on camera-ready paper) not larger than 8.5 × 11 in. Number illustrations consecutively with Arabic numerals; refer to as figures. Prepare figures for single-column width (3.5 in); otherwise, for double-column width (7 in.). Draw and letter all drawings for reduction to a given size to same scale.

Prepare graphs such as electrocardiograms, kymograms, and oscillograms so that crosshatched background is eliminated. Use blue-ruled instead of black-ruled recording paper for original records.

Give actual magnification of photomicrographs. Editorial Office will make corrections for reduction. A length scale on print is preferable.

Designate special features on photomicrographs by letters, numerals, arrows, and other symbols that contrast with background.

Have lettering done in India ink by a draftsman with graphic arts transfer lettering, or with a graphic arts lettering system. If transfer lettering is used, photograph figures for final print. Freehand, typewritten, or computer-generated dot-matrix lettering is not acceptable. Lettering and symbols must be proportionate to size of illustration to be legible after reduction. Size lettering so that smallest elements will be not less than 2 mm high after reduction.

When possible, place all lettering within framework of the illustration; insert key to symbols on face of figure. When figure is so filled that symbols must be explained in legend, use only standard characters: ❏ ■ ○ ● ◑ △ ▲ ×.

Use photographs of equipment sparingly; good line drawings are more informative. Photographs of animals are not acceptable; good line drawings are more effective.

Figures in color are accepted if author assumes all printing costs. Supply three positive glossy color prints.

Indicate approximate position of each figure in margin of manuscript.

Give each figure a legend. Group legends in numerical order and type double-spaced on one or more sheets.

Tables: Submit illustrations rather than tables. Do not duplicate material in text or illustrations. Indicate approximate position in margin of manuscript.

Submit statistical summary tables when possible rather than tables with many lines of individual values. On Editor's recommendation and with author's approval, Editorial Office will deposit lengthy tables of data.

Number tables consecutively with Arabic numerals; prepare with size of Journal page in mind: 3.5 in. wide, single column; 7 in. wide, double column. Type double-spaced on a separate sheet. Give each a brief title; place explanatory matter in footnotes, not in title. Omit horizontal and vertical rules and nonsignificant decimal places in tabular data. Use short or abbreviated column heads; explain if necessary in footnotes. Identify statistical measures of variations, SD, SE, etc. List table footnotes in order of appearance; identify by standard symbols *, †, ‡, § for four or fewer; for five or more, use consecutive superior letters throughout.

OTHER FORMATS

HISTORY ARTICLES—Length should be comparable to a research report in Journals.

LETTERS TO THE EDITOR—500 words. Type letters, including an informative title, double-spaced. Submit three copies. If a letter is acceptable, a copy is sent to original author, if applicable; author will have an opportunity to provide a rebuttal with new material that will be considered for publication with letter.

MODELING IN PHYSIOLOGY—Submitted via APS Publications Office directly to Editor of Modeling in Physiology (MIP). Application of models in physiology or related areas are emphasized. Submit original research contributions, critiques, reviews, survey papers, or tutorials.

Submit letters to the editor, highlight controversies, ambiguities, or misapplication of theory or method. Mathematics and technical jargon are welcome, but must be relevant and clearly explained and presented. Make articles reasonably self-contained, not dependent on a series of highly technical previous publications, unless well known.

Modeling developments should conform to standard modeling practice. Specify appropriate measures of variability for quantitative results based all or in part on a model and experimental data. For example, if a model is used to estimate model parameter values from data, report variability measures for these estimates (e.g., confidence limits, standard deviations, coefficients of variation) as well as (point) estimates themselves. Indicate how such values were estimated (what algorithms, programs, etc.), including method for calculating variability estimates. Discuss meaning (or lack thereof) of reported parameter values, in context of purpose of modeling effort (e.g., physiological significance). If a model includes parameter (or variable) values estimated from data, and model is used to predict or explain something (e.g., a physiological implication), include an analysis/discussion of how variability in estimated values affects predictions, explanations, or conclusion. If a computer simulation, or simulation model is used, and numerical values are used to generate simulated solutions pertinent to reported results, evaluate sensitivity of such solutions/results to these parameters for numerical values used in simulations. Simulations are usually used to explain or predict real system behavior, which normally depends on numerical values of parameters used in simulation. Therefore, some form of parameter sensitivity analysis is needed to support results based on model.

RAPID COMMUNICATION—Four journal pages, including figures, tables, and references (one printed page equals four double-spaced typewritten pages or three figures or tables). Submit short manuscripts containing results of unusual interest identified as such.

SPECIAL COMMUNICATIONS—Manuscripts that describe new methods, new apparatus, techniques with physiological applicability, and critiques of methods and techniques.

SUPPLEMENTARY MATERIAL—Extensive tables of data, appendixes, mathematical derivations, extra figures, computer printouts, and other supplementary material too costly to include may be submitted for deposition (without charge) with National Auxiliary Publications Service (NAPS), c/o Microfiche Publications, P. O. Box 3513, Grand Central Station, New York, NY 10017. Submit material with manuscript for review. On acceptance, it will be deposited by Editorial office with NAPS. A footnote will be added noting availability of material on microfiche and giving NAPS Document Number.

OTHER

Abbreviations, Symbols, and Terminology: Include a list of new or special abbreviations used, with spelled-out form or definition. Internationally accepted biochemical abbreviations, such as ADP, NADH, and P_i, do not need to be defined; define other frequently used abbreviations only at first mention. A list of accepted abbreviations appears following contents of January and July issues of APS Journals. For commonly accepted abbreviations, word usage, symbols, etc., see *CBE Style Manual* (5th ed., 1983). Use chemical and biochemical terms and abbreviations in accordance with recommendations of IUPAC-IUB Combined Commission on Biochemical Nomenclature. Isotope specification should conform to IUPAC system. For style in specialized fields see: "Glossary on respiration and gas exchange" (*J. Appl. Physiol.* 34: 549-558, 1973); "Glossary of terms for thermal physiology" (*J. Appl. Physiol.* 35: 941-961, 1973).

Drugs: Capitalize proprietary (trademarked) names; check spelling carefully. Chemical or generic name precedes trade name or abbreviation of a drug at first use.

Capitalize and check spelling of trade names of chemicals or equipment.

Electronic Submission: $100 charge for accepted manuscripts for which a disk is not provided for final revised, accepted version of the paper. Any popular word-processing software can be used on 3.5 or 5.25 in. low- or high-density diskettes.

Experimentation, Animal and Human: The Society endorses principles embodied in Declaration of Helsinki (last page of instructions to authors). Conduct investigations involving humans in conformity with these principles. Conduct animal experimentation in conformity with "Guiding Principles for Research Involving Animals and Human Beings" (last page of instructions to authors). In describing surgical procedures on animals, specify type and dosage of anesthetic agent. Curarizing agents are not anesthetics; if used, provide evidence that anesthesia was of suitable grade and duration. Papers in which evidence of adherence to these principles is not apparent will be refused.

Materials Sharing: Work must necessarily be independently verifiable. Authors describing results derived from the use of antibodies, recombinant plasmids and cloned DNAs, mutant cell lines or viruses, and other similarly unique materials are expected to make such materials available to qualified investigators on request.

Submit published nucleic acid/amino acid sequences to a widely accessible data bank. Sequence data submission forms for the National Biomedical Research Foundation-Protein Identification Resource Database (NBRF-PIR) are available from APS Publications Office.

Mathematical Formulas and Equations: Address mathematical aspects to readers who are not mathematicians. Present mathematical strategy, assumptions on which mathematics are based, and a summary of meaning of final mathematical statement and limitations. Lengthy or complex mathematical developments central to an article are often put in an appendix. Submit note for referees filling in details of mathematics not explicitly stated and not needed in article proper or in appendix.

Simplify structural chemical formulas and complicated mathematical equations; carefully check. Clearly identify all subscripts, superscripts, Greek letters, and other unusual characters in penciled notes in margin where they first appear. Distinguish between 1 (one) and letter l (el), 0 (zero) and letter O, $\times$ (multiplication sign) and letter x. Use slant line (/) for simple fractions $(a+b)/(x+y)$ in text rather than built-up fraction $\frac{a+b}{x+y}$ which should be used if equation is offset from text.

Use subscripts or superscripts wherever feasible and appropriate, because they simplify equations by eliminating extraneous operations [$R_A R_D$ instead of RA·RD or (RA)(RD)]. Use circles for pools in compartmental or flow-type models and whole arrows $\longrightarrow$ for interconnections or flows (not arrows with half-heads $\rightharpoonup$, as in reversible chemical equations). Do not use nonstandard mathematical notations; e.g., do not use computer symbols in equations (* for multiplication or ** for exponentiation). Use lowercase letters for time-varying symbols in compartmental model equations, preferably $q(t)$ for masses, $c(t)$ for concentrations, with subscripts as needed. Convention for numerical subscripts for rate constants (k_{21}) is as in most life sciences but opposite to that in pharmacokinetics; i.e., k_{ij} is fractional rate of transfer from compartment j to compartment i (or to compartment i from compartment j). Notation is consistent with standard nomenclature in applied mathematics for matrices and matrix manipulation algorithms in commercial software packages for scientific/mathematical computations involving matrices. In addition to defining symbols as they appear in text, include a table of nomenclature in articles that utilize several or more different symbols, specifying units (dimensions) as well as each definition.

Page Charges: $50 per printed page. Editorial consideration is not related to acceptance of page charge; it is expected that charge will be paid by author's research funds or institution that supported research.

Style: Follow *Webster's Third New International Dictionary* for spelling, compounding, and word division.

SOURCE

December 1994, 267(6 pt 3): back matter.

American Journal of Psychiatry
American Psychiatric Association

Nancy C. Andreasen, MD, PhD
1400 K St., N.W., Washington, D.C. 20005
Telephone: (202) 682-6020; Fax: (202) 682-6016

MANUSCRIPT SUBMISSION

Submit original manuscript and four copies. In a cover letter indicate that paper is intended for publication, state number of figures and words, and specify for which Journal section it is being submitted.

All authors must qualify for authorship. Each author should have participated sufficiently in the work to take public responsibility for its content. Base authorship credit only on substantial contributions to conception and design, or analysis and interpretation of data, and drafting the article or revising it critically for important intellectual content and on final approval of version to be published. All conditions must be met. Acquisition of funding or collection of data does not justify authorship, nor does general supervision of research group.

Acknowledge all forms of support, including drug company support, in author's footnote. Disclose in cover letter any commercial or financial involvements that might present an appearance of a conflict of interest, including but not limited to institutional or corporate affiliations not already specified, paid consultancies, stock ownership or other equity interests, and patent ownership. Information may be shared with reviewers. Such involvements will not be grounds for automatic rejection. If accepted for publication, Editor and authors will consult on whether, and to what extent, to include this information.

Manuscripts are accepted for consideration with the understanding that they represent original material, have not been published previously, are not being considered for publication elsewhere, and have been approved by each author. Any form of publication other than an abstract (400 words) constitutes prior publication. If manuscript contains data or clinical observations already used in published or in press papers, submitted for publication, or to be submitted shortly, provide this information and copies of those papers. Include an explanation of differences between papers.

The Journal requires approval of manuscript submission by all authors in addition to transfer of copyright to the American Psychiatric Association. Obtain letters of permission from publishers for use of extensive quotations (more than 500 words). Tables or figures that have been previously published or submitted elsewhere will not be published.

MANUSCRIPT FORMAT

Double space all parts, including case reports, quotations, references, and tables. Type manuscripts in upper and lowercase on one side only of 8.5×11 in. nonerasable bond paper. Make all four margins 1.5 in. Begin each section on a new page. Arrange as: title page, abstract, text, references, and tables and/or figures. Number all pages.

Title Page: Give a brief, informative title. Avoid two-part titles. Limit authors listed in byline to principal researchers and/or writers. Acknowledge collaborators in a footnote. Use authors' first names not initials. Include degrees. Note number of words (including abstract, text, references, tables and figures) in upper right corner. If paper was presented a meeting, give name, location and inclusive dates. Provide department, institution, city, and state where work was done. Give full address for reprint requests. Acknowledge grant support in a separate paragraph. Include full name of granting agency and grant number. Do not acknowledge persons involved with preparation or typing of manuscripts. Acknowledgments should not exceed four typed lines. Drug company support of any kind must be acknowledged.

Abstract: For special articles—250 words: Objective (primary purpose of the review article); Method (data sources, study selection-number of studies selected for review and how selected, data extraction-rules for abstracting data and how applied); Results (methods of data synthesis, key findings); and Conclusions (potential applications and research needs). For regular articles—250 words, Objective (questions addressed); Method (design of study, setting, location and level of clinical care, patients or participants manner of selection and number who entered and completed study, interventions (if any), main outcome measures); Results (key findings); and Conclusions (including direct clinical applications). For brief reports and other types of articles include unstructured abstracts (100 words).

Text: Clearly state hypothesis, names of statistical tests used, whether tests were one- or two-tailed, and what test was used for each set of data. Use standard deviations, rather than standard errors of the mean. Reference statistical tests that are not well known. All significant and important nonsignificant results must include test value, degree(s) of freedom, and probability.

References: Number and list by order of appearance in text; text citation is followed by appropriate reference number in parentheses. References in tables and figures are numbered as though tables and figures were part of text.

Restrict references to closely pertinent material. Verify accuracy of citation. References should conform exactly to original spelling, accents, punctuation, etc. Be sure all references listed have been cited in text.

Do not include personal communications, unpublished manuscripts, manuscripts submitted but not yet accepted, and similar unpublished items in reference list, note in text. Obtain permission to refer to another individual's unpublished observations. Cite manuscripts "in press" as such in reference list; include name of journal or publisher and location.

List all authors; do not use "et al." Abbreviate journal names as in *Index Medicus;* journals not indexed there should not be abbreviated.

1. Noyes R Jr, DuPont RL Jr, Pecknold JC, Rifkin A, Rubin RT, Swinson RP, Ballenger JC, Burrows GD: Alprazolam in panic disorder and agoraphobia, results from a multicenter trial, II: patient acceptance, side effects, and safety. Arch Gen Psychiatry 1988; 45:423–428

2. Kaplan HI, Sadock BJ (eds): Comprehensive Textbook of Psychiatry 4th ed, vol 2. Baltimore, Williams & Wilkins, 1985

3. Fyer AJ, Manuzza S, Endicott J: Differential diagnosis and assessments of anxiety: recent developments, in Psychopharmacology: The Third Generation of Progress. Edited by Meltzer HY. New York, Raven Press, 1987

Illustrations: Do not submit tables or figures that have been submitted elsewhere or previously published or that duplicate material contained in text or each other. Do not use tables and figures to convey data which could be given succinctly in text. A double-spaced table or a figure that fills one-half of a vertical manuscript page equals 100 words of text; one that fills one-half of a horizontal page equals 150 words.

Tables: Double-space and make no wider than 120 typewriter characters, including spaces, and no longer than 70 lines. Values expressed in the same unit of measurement should read down, not across; when percentages are given, also give appropriate numbers.

Figures: Have figures professionally prepared. Submit glossy or other camera-ready prints. Computer-generated figures that do not meet quality printing standards will be returned. Type figure titles and footnotes on a separate page. Figures are visual expressions of data trends or relationships. Convert figures that represent numerical data which could be expressed more succinctly or clearly in tabular form to tables. Line graphs should show change in continuous variables; present comparisons of like values in different groups as bar graphs.

Use 7 point or larger (after reduction) sans serif for figure type. When space on horizontal axis is insufficient, place headings diagonally on axis. Do not use idiosyncratic abbreviations.

Do not use solid black shading; rather, include outlined white among shadings. Run heading for vertical axis of a graph vertically along axis, not horizontally at top or bottom. Place headings for horizontal axis below that axis, not at top of graph. Do not use error bars. Do not extend vertical or horizontal axis of a graph beyond point needed for data shown. In a graph comparing different groups of subjects, show number of subjects in each group with name of group—in key, in headings below horizontal axis, or in title. Each graph should contain only one vertical and one horizontal axis (except when variables are displayed on two vertical axes). Key should appear within or above figure but should not widen figure. Avoid placing other type (e.g., number of subjects, statistical values) within a graph. Cite footnotes (including p values) with superscript letters in title or in axis labels and. List in order cited in figure. Prepare multiple figures for the same article as a set, and the type should be approximately the same size after reduction.

OTHER FORMATS

BRIEF REPORTS—1500 words, including a 100 word abstract, references, tables, and figures. Present data from preliminary studies with suggestive findings warranting further, more definitive investigation, worthwhile replication studies, and negative studies of important topics.

LETTERS TO THE EDITOR—500 words, including references; no tables or figures. Include notation "for publication." Send three copies. All authors must sign. Submit letters critical of an article published in the *Journal* within 6 weeks of publication. Include title and author of article and month and year of publication.

REGULAR ARTICLES—3800 words, including a 250 word abstract, references, tables, and figures. Original communications of scientific excellence in psychiatric medicine and advances in clinical research, containing new data derived from a sizable series of patients.

SPECIAL ARTICLES—Maximum 7500 words, including an abstract (250 words), references, tables, and figures. Overview articles that bring together important information on a topic of general interest to psychiatry. Check with Editorial Office.

OTHER

Abbreviations: Spell out all abbreviations (other than units of measure) at first use. Idiosyncratic abbreviations should not be used.

Color Illustrations: A cost estimate will be provided at the time of first decision.

Drugs: Use generic not trade names. Use trade or manufacturers' names only if drug or equipment is experimental or unavailable in this country or if such information is crucial to evaluation of results or replication of study.

Experimentation, Human: Include a statement that informed consent was obtained after procedure(s) had been fully explained. In the case of children, include information about whether child's assent was obtained. Protect patient anonymity. Avoid identifying information such as names, initials, hospital numbers, and dates. Disguise identifying information when discussing patients' characteristics and personal history.

Style: Follow International Committee of Medical Journal Editors. Uniform Requirements for Manuscripts Submitted to Biomedical Journals, *N Engl J Med* 1991; 324:42–428.

SOURCE

January 1994, 151(1): A31–A34.

American Journal of Public Health
American Public Health Association

Mervyn Susser, Editor
American Journal of Public Health
1015 15th Street NW
Washington, D.C. 20005
Hugh H. Tilson, MD
Director, ESP Division
Burroughs Wellcome Co, 3030 Cornwallis Road
Research Triangle Park, NC 27709
        Notes from the Field
Elizabeth Fee, PhD
The Johns Hopkins University
School of Hygiene and Public Health
624 N Broadway, Baltimore, MD 21205
        Public Health Then and Now
Wendy Mariner, JD, LLM, MPH
Professor of Health Law
Boston University School of Public Health
80 E Concord Street
Boston, MA 02118
        Health Law and Ethics
Bruce C. Vladeck, PhD
United Hospital Fund of New York
55 Fifth Avenue
New York, NY 10003
        Policy Forum

AIM AND SCOPE

Relevance to public health sets the bounds to the broad interests of the *American Journal of Public Health*. We invite contributions in: Original unpublished work in research, research methods, and program evaluation; analytic reviews or commentaries, including health policy analysis; and reports for special departments.

MANUSCRIPT SUBMISSION

Submit a cover letter signed by all authors. Name one author correspondent (with address, telephone and fax numbers). State that all authors have participated in three activities (conception/design or analysis/interpretation; writing; approval of final version) and can take public responsibility for paper's content and that material has neither been published nor is being considered elsewhere. Disclose all possible conflicts of interest, e.g., financial arrangements in relation to products studied. Include a brief indication of your sense of the main interest of the paper to the readers of *AJPH*. Mention supplementary unpublished material enclosed to support your results as well as overlapping material already published. Enclose two copies of work important to the paper, published or unpublished.

Send five copies of paper, complete with all components intended for publication.

Be concise. Articles—4500 words; Commentaries—2500 words; Briefs—1000 words (excluding references, tables, and figures); and Letters to the Editor—400 words.

More than six authors will need justification. Follow the criteria suggested by the International Committee of Medical Journal Editors *(N Engl J Med* 1991;324:424-428).

MANUSCRIPT FORMAT

Type everything double-spaced on one side of a sheet of 8.5×11 in. paper, with 1 in. margins on all sides.

First Title Page: Include: main title; "running head" (45 characters); names of all authors (full first names), degrees and institutional affiliations at time of work; name and address for correspondence and reprint requests; number of words in text and number of tables, figures, if any; and key words best chosen from National Library of Medicine. Medical Subject Headings Annotated Alphabetical List.

Second Title Page: Remove anything that obviously indicates identity from paper.

Abstract: Articles—180 words under four heads: Objectives (hypotheses etc.); Methods (design, population, analysis); Results; Conclusions. Briefs—100 word unstructured abstract.

References: Follow "Vancouver style" (see *AMA Style Manual* or *N Engl J Med* 1991;324:424-428). List up to six authors; for more, list first three and add et al. Number references in order cited in text, tables, and figures. Place numbers at end of sentence or, when two or more refer to different points, after relevant point.

Verify references in original or cite secondary source. For these and for direct quotations give page numbers. List only references accessible to readers. Cite personal communications with source and date in text only.

For *Public Health Then and Now* use endnote reference style of *The Chicago Manual of Style* (13th ed. Chicago, IL: The University of Chicago Press; 1982: 399-510).

Acronyms: Use only those acronyms in universal use.

Footnotes: These are discouraged except in tables.

Tables and Figures: Arrange each on a separate sheet with a title as a self-contained unit. Figures should be professionally drawn or prepared on a computer and laser printed, with a separate sheet giving legends.

OTHER FORMATS

HEALTH LAW AND ETHICS—Common ground as well as conflicts between public health, law, and ethics.

NOTES FROM THE FIELD—Field and teaching experiences of more than local interest.

POLICY FORUM—Topics of current or future import for public health.

PUBLIC HEALTH THEN AND NOW—History that bears on contemporary public health.

OTHER

Reviewers: Suggest 3 in cover letter.

Style: Follow guidelines of the International Committee (*N Engl J Med* 1991;324:424-428) and the *American Medical Association Manual of Style* (8th ed. Baltimore, MD: Williams & Wilkins; 1989).

SOURCE

January 1994, 84(1): 6.
"What *AJPH* Authors Should Know" is printed in each issue.

American Journal of Respiratory and Critical Care Medicine
American Thoracic Society

Editorial Office
American Thoracic Society
1740 Broadway
New York, NY 10019-4374

AIM AND SCOPE

The *American Journal of Respiratory and Critical Care Medicine* publishes original papers on laboratory and clinical research and clinical observations that are pertinent to respiratory biology and medicine and critical care.

MANUSCRIPT SUBMISSION

Submit four manuscript copies and four photographic prints of each illustration.

In a cover letter list name, address, and telephone and fax numbers of corresponding author. State that submitted material has not been published and is not being considered for publication elsewhere. Submission indicates tacit acknowledgment that all authors have made significant contributions to study and have read and approved manuscript.

Disclose any direct commercial association that might lead to a conflict of interest. This will be withheld from reviewers and will not affect acceptance. At publication, information will be disclosed in a footnote to manuscript. Also disclose less direct associations, such as consultancies, stock ownerships in an authors' or relatives' name, patents, etc. Extent and means of possible disclosure will be determined by discussion with Editor following acceptance.

Contributions may not contain a significant portion of material published or submitted for publication elsewhere, except abstracts (400 words). Editors will determine what constitutes significant duplicate publication. If any material has been published or submitted elsewhere, send four copies of a reprint or preprint with submission.

MANUSCRIPT FORMAT

Type manuscripts on white bond paper 8.5 × 11 in. with margins at least 1 in. Double space throughout. Organize as: title page, abstract and key words, text, acknowledgments, references, figure legends, footnotes, tables, and figures. Begin each on a separate page. Number pages consecutively, beginning with title page.

Title Page: List title (85 characters); first name, middle initial, and last name of each author; name of department(s) and institution(s) to which work should be attributed; name and address of reprint request author, if not senior author or department of origin; all source(s) of support in the form of grants, gifts, equipment, and/or drugs; and a short running head (35 characters counting letters and spaces).

Abstract and Key Words: 200 words, written for both clinicians and basic investigators, state hypothesis or central question, study subjects or experimental animals, observational and analytical methods, main findings, and principal conclusions. Use only approved abbreviations. Provide and identify 3 to 5 key words or short phrases for indexing. Use terms from National Library of Medicine's *Medical Subject Headings* list.

Text: Divide into: Introduction, Methods, Results, and Discussion, where feasible. Long articles may require subheadings. Be concise. Excessive length reduces likelihood of acceptance. In introduction clearly state hypothesis or central question, any background material and supporting evidence, and explain experimental approach.

Do not make statements referring to work in progress or in prospect that imply future publication. Do not cite unpublished works in References, cite them fully parenthetically within text. Submit written permission from author for citation of unpublished work.

Acknowledgments: Group into one paragraph after Discussion.

References: Limit to 35, typed double-spaced. Begin on separate sheet. Number in order of appearance in text. Include all authors' names (do not use "et al."), complete article titles, and articles in press. Supply inclusive page numbers. Do not include submitted manuscripts which have not been accepted for publication. For articles cited in References as in press, submit four copies. Use abbreviations for journal names in *Index Medicus*. Spell out names of journals not listed. Cite a reference for statistical methods used.

JOURNAL ARTICLES

1. Jones DA, Howell S, Roussos C, Edwards RHT. Low-frequency fatigue in isolated skeletal muscles and the effects of methylxanthines. Clin Sci 1982; 63:161-7.

IN PRESS

2. Lakatos E, DeMets DL, Kannel WB, Sorlie P, MacNamara P. Influence of cigarette smoking on lung function and COPD incidence. The Framingham study. J Chronic Dis (In Press).

BOOKS

3. Snedecor GW, Cochran WG. Statistical methods. 6th ed. Ames: Iowa State University Press, 1967; 258–96.

ARTICLES IN BOOKS

4. Rall TW. Central nervous system stimulants (continued): the xanthines. In: Gilman AG, Goodman LS, Gilman A, eds. The pharmacological basis of therapeutics. 6th ed. New York: Macmillan, 1980; 595–607.

GOVERNMENT OR ASSOCIATION REPORT

5. U.S. Public Health Service. Smoking and health. A report on the Surgeon General. Washington, DC: U.S. Government Printing Office, 1979. DHEW Publication No. (PHS)79–50066.

Tables: Design tables to fit vertically a width of 3.5 in. for single column or up to 7.25 in. for double column. Tables that do not fit will be returned. Type double-spaced on separate sheets. Do not submit as photographs. Number consecutively, give each a brief title, and cite each in text. Avoid arbitrary labels or classifications, such as groups A and B. Use specific descriptors, such as "control" and "hypoxia." Explain all non-standard abbreviations in footnotes; use symbols in sequence: *, †, ‡, §, ||, ¶,**, ††, etc.

Illustrations: Submit four photographic prints of each. Submit good quality, unmounted glossy prints, sized to be reduced to 3.5 in. (single column) width, and not exceeding 7.5 in. (double column). Size symbols and lettering in scale with figure. Label abscissa and ordinate of each graph clearly. Make all figures same point size. Submit multipart figures as single composites, label each panel (e.g., A, B). Computer graphics are discouraged.

Color prints are preferred to transparencies. Cost will be borne by authors.

Mark back of each illustration with its number, first author's name, and top indicated. Mark lightly or use a label; do not use paperclips. Place title in figure legend, not on figure.

Legends: Should convey findings. Type double-spaced, on separate pages with Arabic numerals corresponding to illustrations. Explain and identify symbols, arrows, numbers, or letters Explain internal scale and identify staining method in photomicrographs.

OTHER FORMATS

BRIEF COMMUNICATIONS—10 double-spaced typewritten pages including references. Concise articles with significant new observations, whether experimental or clinical. Follow format described for major articles, including illustrations and tables.

CASE REPORTS—10 typewritten pages including references. Provide new information concerning etiology, mechanism, or management of a disease process. Collections of several cases are more desirable than single case reports. New information must be substantiated by scientific, not circumstantial, evidence. Reports of coexistence of two diseases or conditions without proof of causal relation are discouraged.

LETTERS TO THE EDITOR—500 words, signed, discussing previously published material or controversies. Submit presentations of unpublished investigations as Brief Communications. Letters which confirm previously published material without adding significant new information are less likely to be published. Do not use illustrations or tables. Include references parenthetically in body unless several are cited, then present as footnotes. Otherwise, use no footnotes.

SPECIAL FEATURES—Include Editorial, State of the Art, Pulmonary Perspective, Clinical Commentary, Workshop Summary, etc. Contact Editor before submission.

OTHER

Abbreviations: Use abbreviations for laboratory or chemical terms or disease process only after it has been written in full with abbreviation in parentheses immediately after. Some common terms may be abbreviated (Po_2, PCo_2, N_2, CO, $Paco_2$, $Pvco_2$, Sao_2, $AaPo_2$, DLCO, FVC, FEV_1, etc). Abbreviations should not begin a sentence. Avoid specialized jargon.

Conventions: Classify all cases of tuberculosis and all designators of mycobacteria according to 1990 edition of *Diagnostic Standards and Classification of Tuberculosis* (American Lung Association).

Drugs: Use generic names. Provide location (city, state, country) after first reference to manufacturer.

Experimentation, Animal: Conform to NIH guidelines (*Guide for the Care and Use of Laboratory Animals*, NIH Publication No. 86-23, Revised 1985, U.S. Government Printing Office, Washington, D.C.). In descriptions of surgical procedures include name, dose, and route of administration of anesthetic agent. Paralyzing agents are not an acceptable alternative to anesthesia and should be used only in conjunction with suitable anesthetic agents.

Experimentation, Human: Follow recommendations in the Declaration of Helsinki. Editors will reject any manuscript containing studies that do not conform. Include a statement in text that protocols were approved by institutional review board for human studies and that informed written consent was obtained from subjects or their surrogates if required by the institutional review board.

Page Charges: $40 per printed page. Authors may apply for a waiver by writing to Editor.

Statistical Analyses: Describe statistical methods. Distinguish standard error of the mean (± SEM) from standard deviation (± SD).

Style: Follow Uniform Requirements for Submission to Biomedical Journals.

Terms and Symbols: Follow "ATS Committee Report on Pulmonary Physiology Symbols" for terms and symbols pertaining to pulmonary physiology.

SOURCE

July 1994, 150(1): 291-294.
Instructions to Authors appear in January and July issues.

American Journal of Surgery
Society for Surgery of the Alimentary Tract
North Pacific Surgical Association
Society for Clinical Vascular Surgery
Society of Head and Neck Surgeons
Southwestern Surgical Congress
Association for Surgical Education

Hiram C. Polk, Jr., MD
Department of Surgery
University of Louisville
Louisville, KY 40292

AIM AND SCOPE

The Editors and Publisher of *The American Journal of Surgery* invite concise original articles in the fields of clinical and experimental surgery as well as: descriptions of modern operative techniques ("How I Do It"); review articles; surgical pharmacology articles; brief reports. No case reports will be accepted.

MANUSCRIPT SUBMISSION

Send a cover letter from corresponding author containing a statement that the manuscript has been seen and approved by all authors and material is previously unpublished. Acknowledge financial support or potential conflicts of interest. Submit original and three copies of all elements.

MANUSCRIPT FORMAT

Typed double-space on 8.5 × 11 in. white bond paper with 1 in. margins. Do not exceed 15 pages, excluding simple tables, and no more than 40 references. Number all pages beginning with title page. Readability is important. Strive to enlighten, not impress. Avoid medical jargon, wordiness, and overlong sentences. Review and surgical pharmacology articles must not exceed 25 pages or contain more than 100 references.

Affiliated organizations should limit their manuscript pages to 12, excluding simple tables and figures, to avoid page charges. Submit not more than 5 tables, figures, or combination thereof. Use no more than 40 references.

Title Page: Include first and last name of all authors, highest academic degrees, name of department and institution from which work originated, name and address of reprint request author, and telephone and fax numbers for corresponding author.

Abstract: 150 words. For clinical, experimental, and brief reports include four paragraphs labeled Background, Methods, Results, Conclusions. Label review article paragraphs Background, Data Sources, Conclusions. Use unstructured abstracts for "How I Do It" and surgical pharmacology articles.

List three to six key words for indexing on a separate page. Write a 2- to 3-sentence summary to appear in table of contents.

Figures: Illustrations must increase understanding of text. Mention figures in text in consecutive order. Provide three sets of 5 × 7 in. glossy photographs in sharp focus, not mounted. All printing must be professional. Reproduce pen and ink and black and white line art with no gray tones by photomechanical transfer (PMT). Make decimals, broken lines, and lettering strong enough for reproduction. Only high-quality computer-generated figures will be considered. Radiographs must be of high quality and should have arrows, etc. Color cost must be paid by author.

Attach gummed label to figure back indicating figure number, author's name, paper title, and top of figure. Mention figures in the text in consecutive order. Number of illustrations may be limited.

Legends: Type double-spaced on separate page. Indicate magnification and method of staining for photomicrographs.

Tables: Should supplement rather than duplicate text. Do not repeat information in text. Give each table a title and type double-space without horizontal or vertical lines on 8.5 × 11 in. paper. Number consecutively with Roman numerals in order of first mention in text. Define all abbreviations used in a footnote.

References: Follow: International Committee of Medical Editors. Uniform requirements for manuscripts submitted to biomedical journals. *Ann Intern Med* 1988; 108: 258-65 or *Br Med J* 1988; 296 (6619): 401-5. Type double-space on a separate sheet. Number i and list in order mentioned in text. Insert appropriate numbers in brackets in text. Use *Index Medicus* style for abbreviating journal titles. List all authors if four or fewer; for five or more, list first three and add *et al.*

JOURNAL ARTICLES

Cullen JJ, Eagon JC, Dozois EJ, Kelly KA. Treatment of acute postoperative ileus with octreotide. *Am J Surg* 1993;165:113-120.

BOOKS

Fry DE. *Multiple System Organ Failure*. St. Louis: Mosby-Year Book, Inc.; 1992.

CHAPTERS IN BOOKS

Zucker. KA, Bailey RW. Diagnostic and therapeutic laparoscopy for the general surgeon. In: Polk HC Jr., Gardner B, Stone HH, eds. Basic Surgery. 4th ed. St. Louis: Quality Medical Publishing; 1993;95-121.

For articles in press, supply journal name and, if available, volume number and year. Include references to unpublished material, including written (not verbal) personal communications, parenthetically in text.

Acknowledgments: Footnotes of acknowledgment to individuals for help received are not published.

OTHER FORMATS

BRIEF REPORTS—7 typed double-space pages including title page, structured abstract, and references, not including a short 2- to 4-sentence summary and list of key words on separate pages. No more than 3 tables, illustrations, or combination thereof; 10 references maximum. Do not use subheadings within text.

OTHER

Acronyms and Abbreviations: Do not use within text, other than for units of measure.

Drugs: Use generic names; brand name may be included in parentheses at first mention.

Informed Consent: State in Methods section that informed consent was obtained from patients and approval from designated institutional review board. Submit permission to reproduce photographs of patients whose identity is not disguised. Obtain permission of author and publisher and send a copy of permission with manuscript for previously published illustrations.

Statistical Analyses: Reference methods used. See Yancey JM. Ten rules for reading clinical research papers. *Am J Surg* 1990; 159:533–539.

Style: Strunk and Whites' *Elements of Style*. Round off tenths of percents to whole numbers. Do not reference individual patients in text or tables. Round off tenths of percentages to whole numbers.

SOURCE

July 1994, 168: 73.

American Journal of Surgical Pathology
Raven Press

Stephen S. Sternberg, M.D.
Memorial Sloan-Kettering Cancer Center
1275 York Avenue, New York, NY 10021
Telephone: (212) 639-5908; Fax: (212) 794-6223

MANUSCRIPT SUBMISSION

Include a cover letter with name, address, and telephone number of corresponding author. Submit in triplicate, one original and two copies, on standard 8.5 × 11 in. paper with 1-in. margins. Do not use "erasable" bond. If possible include a disc. Type all copy, including captions, footnotes, and references double-spaced and on one side of sheet only. Print material prepared on a word processor with a letter quality printer.

Only previously unpublished material will be considered for publication. Material submitted to the Journal must not be under consideration for publication elsewhere.

A signed copyright transfer agreement is required with each manuscript submission. Agreement is printed in journal issues.

MANUSCRIPT FORMAT

Divide paper into sections with appropriate headings: title page, abstract/key word page, introduction, methods, results, discussion, acknowledgments, references, tables, figure captions. Number all pages sequentially, title page is page 1.

Title Page: Title, list of authors with affiliations, sources of support, and full address for corresponding and reprint request authors.

Abstract and Key Words: On next page, 200 words. Be factual and comprehensive. Avoid abbreviations and general statements. Provide a list of suggested key words or phrases below.

References: Place all references at end of text in alphabetical order, numbered consecutively. Type double-spaced.

Epstein JI. Evaluation of radical prostatectomy capsular margins of resection. *Am J Surg Pathol* 1990;14:626–32.

Nash A. *Soft tissue sarcomas: histological diagnosis.* New York: Raven Press, 1989.

Santa Cruz DJ, Leyva WH. Neoplasms of skin. In: Sternberg SS, ed. *Diagnostic surgical pathology.* New York: Raven Press, 1989:59–102.

For journal abbreviations follow *Index Medicus.* Provide all author names where seven or fewer. When more than seven, list first three and use et al. Verify bibliographic accuracy of all references. Indicate "personal communica-tions" and "unpublished observations" within text but exclude from reference list. Use only with permission.

Tables: Cite sequentially in text, number, supply with suitable explanatory legends and column headings, type on separate sheets, and place at end of manuscript following references. Make self-explanatory to supplement, not duplicate, material in text.

Figures: Number all figures, cite in numerical order, and supply with legends. Submit all artwork in duplicate in camera ready form. Mark (lightly in pencil) on back with lead author's name, figure number, and orientation of image.

Submit line art as glossy photoprints with all linework and lettering in black and lineweight suitable for reduction. Lettering should be done mechanically so that it will remain legible after reduction.

Submit photomicrographs and other photographic images in duplicate as unscreened, unmounted original black and white prints for reproduction at 100%. Size prints to fit within one column (4–8 cm), or across two columns (between 10–16.9 cm).

Legends: Type all together on a separate sheet following tables and references. Make brief and specific. Do not indicate magnification for light photomicrographs unless visual inspection is inadequate for judging relative or absolute size. Use scale markers in electron micrographs. Indicate type of stain used only if it is other than hematoxylin or eosin.

OTHER

Page Charges: Color illustrations only.

Symbols and Equations: Type, where possible, all mathematical and chemical symbols, equations, and formulas. Identify in margin all unusual symbols when first used.

SOURCE

January 1994, 18(1): front matter and inside back cover.

American Journal of the Medical Sciences
Southern Society for Clinical Investigation
J.B. Lippincott Co.

Managing Editor,
The American Journal of the Medical Sciences
Atlanta Dept. of Veterans Affairs
Medical Center Room 2B 117, Medical Service (111)
1670 Clairmont Road
Decatur, GA 30033

AIM AND SCOPE

The American Journal of the Medical Sciences considers for publication manuscripts reporting original clinical or laboratory investigations, Brief Reviews, Rapid Communications, and Case Reports.

MANUSCRIPT SUBMISSION

Submit an accompanying statement indicating that material has not been previously published elsewhere and is being considered solely by *AJMS*. Indicate that authors "transfer, assign or otherwise convey all copyright ownership of the article to The Southern Society for Clinical Investigation if the article is published."

MANUSCRIPT FORMAT

Submit manuscripts in triplicate, with triplicate figures and tables. Type double-spaced on one side of 8 × 11 in. bond paper with 1 in. margins on all sides. Do not staple. Format: title page, abstract, introduction, methods, results, discussion, acknowledgments, references, tables, figure legends, and figures. Indicate each section by centered headings. Use Arabic numbering system. Define abbreviations at first appearance.

Title Page: On a separate page include: title of manuscript (90 characters), author(s) name(s) and degrees, laboratory or institution of origin with city and state, address(es) for mailing proofs and reprint requests, and an abbreviated title (four words) for a running head. Begin page numbering. Credit preliminary reports or abstracts in a footnote to title.

Abstract: 150 word summary of relevant information. List five key indexing terms.

References: Cite consecutively in text; type double-spaced at end of article. Abbreviate journal names according to *Index Medicus*. Cite articles "In Press" in references. Do not include citations such as "In preparation," "Submitted for publication," "Unpublished data," and "Personal communication" in references; insert in text in parentheses. List all authors for each citation (up to six, use et al. for more); give inclusive pagination. Cite journal titles as they existed at publication; do not abbreviate unlisted journals. Verify bibliographic accuracy. Limit to 30 citations.

ARTICLE

Iber FL, McGonagle T, Serebro HA, Luebbers E, Bayless TM, Hendrix TR. Unidirectional sodium flux in small intestine in experimental canine cholera. Am J Med Sci. 1969;258:310–50.

BOOKS

Finneson BE. Low Back Pain. Philadelphia: J.B. Lippincott; 1984:164–9.

SECONDARY CITATIONS

Greenberg MS. Ulcerative, vesicular, and bullous lesions. In: Lynch MA; eds. Burket's Oral Medicine. 4th ed. Philadelphia: J.B. Lippincott; 1984:164–9.

Illustrations: Enclose three sets of unmounted glossy prints, preferably 5 × 7, in a separate envelope backed by cardboard. On back of each figure mark figure number and name of author, and TOP lightly in pencil. List figure legends on a separate sheet .

Tables: Give Arabic numbers and brief titles; double-space on separate pages. Do not use vertical lines or ditto marks. Center table number at top of page with table title beneath it. Designate footnotes within table. Explain abbreviations at bottom of table after footnotes.

Footnotes: Cite footnotes to text and tables as necessary; type at foot of appropriate page, separated from text or table by a horizontal line. Designate footnotes by symbols * † ‡ § ¶.

OTHER FORMATS

BRIEF REVIEWS—25 double-spaced, typed pages, including figures, tables and references, summarizing present state of knowledge concerning a particular aspect of research.

RAPID COMMUNICATIONS—Six double-spaced pages, including figures, tables and references. New and important research that justifies their accelerated appearance.

OTHER

Abbreviations and Nomenclature: Use *Style Manual for Biological Journal* (ed. 2,1964, American Institute of Biological Sciences, Washington, D.C.) for abbreviations. Confine to tables; keep to a minimum in text. Do not use periods after abbreviations. Use generic names for drugs.

Experimentation, Human: Conduct clinical research in accordance with the guiding principles for human experimentation summarized in the Declaration of Helsinki.

Page Charges: $40 per printed page. Authors are charged for color illustrations (photos, slides, graphs, etc.). There is no charge for black-and-white illustrations.

SOURCE

January 1995, 309(1): 66.

American Journal of Tropical Medicine and Hygiene
American Society of Tropical Medicine and Hygiene

Allen Press
McWilson Warren
3088 Briarcliff Road, Suite A1
Atlanta., Georgia 30329

MANUSCRIPT SUBMISSION

Type manuscripts, letter quality in 10 pitch type, on one side of paper. Submit in quadruplicate on paper 8.5 × 11 in., numbered and double-spaced throughout, including references, tables, footnotes, and figure legends.

Include a signed letter from corresponding author indicating that all authors concur with submission and that material has not been, and will not be, submitted elsewhere for publication as long as it is under consideration.

MANUSCRIPT FORMAT

Supply: A short title (50 letters) as a running head; full title; authors' names and affiliations; address, telephone and fax numbers of corresponding author; a table with number of text pages, references, tables, and figures; a concise abstract; an introductory paragraph; separate sections for Materials and Methods, Results and Discussion; separate paragraphs for acknowledgments, listing of financial support; addresses for authors, and address for reprint requests, if ordered; list of references cited; and legends for illustrations.

Tables: Type each on a separate sheet and number serially in Arabic numerals. Design tables for printing in 1-column width if possible, two at most. Tables cannot require more than full-page width. Cite all in text.

Illustrations: Number and cite in text all graphs, drawings and photographs illustrating specific points. Submit only glossy photographic prints or drawings in black India ink on white drawing paper or tracing linen or computer-generated illustrations printed at high resolution (300 ANSI). Insert scale bar in micrographs, indicating length in legend. Prepare illustrations for single column (2 5/8 in.) or page width (5.5 in.). Color illustrations must be paid for by author. Place figure number, name of senior author, and arrow indicating top on back of each illustration. Do not mount one of the four sets of illustrations; label these with label affixed to back.

Legends should be separate, not attached to or written on illustrations.

References: Cite by consecutive numbers in text. Number in superscript. Abbreviate journal names in style of the National Library of Medicine. Do not cite theses, abstracts, proceedings, work in progress, or manuscripts submitted for publication. If necessary, cite information in text as (J. Doe, Univ. of North Carolina, unpublished data). List numerically in order of appearance. List all authors.

JOURNAL ARTICLES

1. Barreto ML, 1993. Use of risk factors obtained by questionnaires in the screening for *Schistosoma mansoni* infection. *Am J Trop Med Hyg 48*: 742-747.

BOOKS

2. Peters CJ, LeDuc JW, 1984. Bunyaviruses, phleboviruses, and related viruses. Belshe RB, ed. *Textbook of human virology*. Littleton, MA: PSG Publishing Co., 572-576.

OTHER FORMATS

LETTERS TO THE EDITOR—On topics of general interest to Society members, such as those dealing with tropical medicine, public health, and hygiene or recently published Journal articles. Submit in duplicate, typewritten, double-spaced and on disk, if possible. Have all authors sign. Should not contain unpublished data.

SHORT REPORT—One printed page (500 words) or two. For important preliminary observations, technique modifications or data that does not warrant publication as a full paper. Limit number of authors, figures, tables, references. Include same items as regular manuscripts, including a very brief abstract; delineations within text are not necessary.

OTHER

Abbreviations and Symbols: Use *CBE Style Manual;* spell out first time appear in both abstract and text.

Drugs: Use proprietary names of drugs or chemicals in conjunction with generic name when substance is first mentioned in abstract and text. Use generic name elsewhere.

Page Charges: $100.00 per printed page. Request Authors waiver of fee upon submission.

SOURCE

January 1994, 50(1): v-vi.

American Journal of Veterinary Research

American Veterinary Medicine Association
American Journal of Veterinary Research

American Veterinary Medicine Association
1931 N. Meacham Rd., Suite 100
Schaumburg, IL 60173-4360
Telephone: 1-800-248-2862

MANUSCRIPT SUBMISSION

Manuscripts must be approved by each author. Neither the article nor any of its parts may be under concurrent consideration by any other publication. Each author should have generated a part of the intellectual content and should agree with all interpretations and conclusions. By-line should not include more than ten authors. Designate corresponding author and provide complete address and telephone number.

MANUSCRIPT FORMAT

Type manuscripts double-space typed (including footnotes, references, tables, and figure legends) on 8.5 × 11 in. paper, with not less than 1 in. margins (top, bottom, sides). Number each page. Computer printed manuscripts must be of letter quality. Submit three copies of manuscript and figures.

Summary: Provide a stand alone synopsis, telling briefly what was done and learned.

Introduction: A brief meaningful statement of purpose.

Materials and Methods: Include method(s) of statistical analysis; describe work so that others can reproduce experiments, but exclude readily available detailed, published descriptions; cite by appropriate references.

Results: Concise, logical presentation of data. Do not reference other reports.

Discussion: Discuss, evaluate, or interpret results. Do not introduce new data or reference work in progress.

References: Limit to what is relevant and necessary. Refer to by superscript numbers. Refer to abstracts, personal communications, and theses by footnotes; cite footnotes by alphabetical, superscripted, lowercase letters.

Type double-space. List in order cited in text. For manuscripts in press (i.e., accepted and in final form), send copies.

Tables: Cite in text in numerical order. Make self explanatory, easily understood, simply constructed, and supplement text. Type double-spaced on a separate sheet (not a photograph) in style shown in recent issues. Give each vertical column including stub column, a column heading; omit vertical rules. No blank spaces in data columns; instead, insert an ellipsis (3 dots or dashes) or indicate NA for "not applicable" or ND for "not determined."

Figures: Submit three complete sets, including glossy print photographs, drawings, graphs, charts, and tracings. Identify each on back with first author's name, figure number, and an arrow indicating top. Use wax pencil to show crop marks on borders. If illustration cannot be cropped because of loss of important features or loss of perspective, indicate whether it can be reduced to 1 or 2 journal column widths. Insert internal scale marker on photomicrographs and electron micrographs. To express magnification by use of an internal scale marker, divide marker length by original magnification.

Drawings, Graphs, and Charts—Make clear and accurate with India ink on white paper. Submit 3 glossy prints. Make numbers or letters large enough to allow for reduction.

Legends: Double-space on separate sheet. Make brief and adequate, include stain where applicable.

OTHER

Experimentation, Animal: Reports suggesting that animals had been subjected to adverse, stressful, or harsh conditions or treatment will not be processed unless it is demonstrated convincingly that the knowledge gained was of sufficient value to justify adverse conditions or treatment.

Page Charges: $50 per manuscript nonrefundable submission fee. Manuscript processing fee = $70 per printed page or fraction. Cost of color Illustrations billed to author.

Trade Names: Identify products and equipment by chemical or generic names or descriptions. Trade names may be included in a lettered footnote with name and location (city and state) of manufacturer when identification of product or equipment is essential.

SOURCE

January 1994, 55(1): 179 [updated August 1992] .

American Review of Respiratory Disease
see American Journal of Respiratory and Critical Care Medicine

Anaesthesia
Association of Anaesthetists of Great Britain and Ireland
W.B. Saunders

Dr. M. Morgan
Department of Anaesthetics
Royal Postgraduate Medical School
Hammersmith Hospital
London W12 OHS U.K.

MANUSCRIPT SUBMISSION

Submit manuscript to only one journal at a time. Manuscript may not have been published, simultaneously submitted, or accepted for publication elsewhere. Submit two copies of manuscripts. Have cover letter signed personally by all authors.

MANUSCRIPT FORMAT

Type manuscripts on white bond paper 8 × 10.5 in. or 8.5 × 11 in. or ISO A4 with 1 in. margins. Use double, and preferably triple, spacing throughout, including references. Do not use a dot matrix printer. Unseparated, fanfolded manuscripts may be returned. Use following sections in order each beginning on a new page: title page, summary and key words, text, acknowledgments, references, individual tables, and legends for figures. Number pages consecutively, beginning with title page.

Title Page: Place corresponding author name and address in top left corner.

Make main title as short as possible. Type in capitals across center of title page.

Type a subsidiary title or not more than 12 words, if necessary, in lower case beneath main title.

Type names of authors in capitals across title page beneath titles without degrees or designations. Initials should precede surnames. Place "AND" before name of last author.

Type a line across title page below author(s) name(s) in capitals.

Type author(s) name(s), degrees and designations in lower case below line. Initials should precede name of each author and his (her) degrees, without full stops between letters, and appointment (e.g., Consultant, Registrar, etc.) should follow. A full postal institutional address should follow.

Summary and Key Words: 150 words, stating study purpose, basic procedures, main findings and their statistical significance, and principal conclusions.

Provide and identify as such, three to 10 key words or short phrases for indexing. Use terms from *Medical Subject Headings* list from *Index Medicus*.

Text: Divide into sections headed: Introduction, Methods, Results and Discussion. Long articles will need subheadings within some sections. Use three heading levels: (i) CAPITALS across center of page; (ii) underlined words (typed at left side of page above paragraph which they precede); (iii) underlined words at beginning of a paragraph.

Acknowledgments: Obtain written permission for publication of reproduced figures and tables from authors and publishers and from everyone acknowledged by name.

References: Number references consecutively in order mentioned in text. Identify references in text, tables and legends by Arabic numerals. Number references cited only in tables or in legends to figures in accordance with a sequence established by first identification in text of table or illustration.

Use form of reference of U.S. National Library of Medicine in *Index Medicus* as published by *Key Words in Anesthesiology*, N. M. Greene. Include first and last pages in all references.

Give journal titles in full. Avoid using abstracts as references. Do not use 'unpublished observations' and 'personal communications'. Insert references to written, not verbal, communications (in parentheses) in text. Include manuscripts accepted but not yet published; designate journal followed by (in press) in parentheses. Cite information from manuscripts submitted but not yet accepted in text as (unpublished observations) in parentheses.

Verify references against original documents.

JOURNALS (LIST ALL AUTHORS)

SOTER NA, WASSERMAN SI, AUSTEN KF. Cold urticaria: release into the circulation of histamine and eosinophil chemotactic factor of anaphylaxis during cold challenge. *New England Journal of Medicine* 1976; 294:687–90.

The Committee on Enzymes of the Scandinavian Society for Clinical Chemistry and Clinical Physiology. Recommended method for the determination of gamma-glutamyltransferase in blood. *Scandinavian Journal of Clinical Laboratory Investigation* 1976; 36: 119–25.

Anonymous. Epidemiology for primary health care. *International Journal of Epidemiology* 1976; 5:224–5.

BOOKS AND OTHER MONOGRAPHS

OSLER AG. Complement: mechanisms and functions. New York: Prentice-Hall, 1976.

American Medical Association Department of Drugs. AMA drug evaluations, 3rd edn. New York: Publishing Sciences Group, 1977.

RHODES AJ, VAN ROOVEN CE, *comps. Textbook of virology: for students and practitioners of medicine and other health sciences*, 5th edn. Baltimore: The Williams & Wilkins Co., 1968.

CHAPTER IN BOOK

WEINSTEIN L, SWARTZ MN. Pathogenic properties of invading micro-organisms. In: SODEMAN WA JR, SODEMAN WA, eds. *Pathologic physiology: mechanisms of disease.* Philadelphia: W. B. Saunders. 1974: 457–72

AGENCY PUBLICATION

National Center for Health Statistics. Acute conditions: incidence and associated disability, United States, July 1968—June 1969. Rockville, MD: National Center for Health Statistics, 1972. (*Vital and health statistics*, Series 10: Data from the National Health Survey. No 69) [DHEW publication No. (HSM) 72-1036].

NEWSPAPER ARTICLE

SHAFFER RA. Advances in chemistry are starting to unlock mysteries of the brain: discoveries could help to cure alcoholism and insomnia, explain mental illness. How the messengers work. *Wall Street Journal* 1977 Aug 12: 1 (col 1), 10 (col 1).

MAGAZINE ARTICLE

ROUECHE B. Annals of medicine: the Santa Claus culture. *The New Yorker* 1971 Sept 4: 66-81.

Tables: Do not include in text. Do not submit as photographs. Cite each table in text in consecutive order. Indicate approximate position of each table to text in left margin of appropriate manuscript page. Start a new sheet for each. Space material adequately. Number consecutively with Arabic numerals. Place author(s) names(s) in top right corner. Supply a brief title. Give each column a short heading. Place explanatory matter in footnotes. Explain all non-standard abbreviations; use the following symbols in sequence: *, †, ‡, §, ||, ¶,**, ††, etc. Identify statistical measures of variations such as SD and SEM. Place legends on the face of table.

Figures: Make each one separate. Do not include in text; collect together in an envelope. Indicate approximate position of each to text in left margin of appropriate page. Submit original figures as India ink drawings on good quality plain paper or tracing paper, or photographs (glossy prints with good contrast). Do not mount. Make about twice size of published version. Identify each by its sequential number in pencil on reverse: include title and authors' names. Mark top on reverse by an arrow and word TOP. Put keys and other explanations in legend. Remove all sources of identification of patients.

Legends: Type double-spaced with Arabic numerals. Explain symbols, arrows, numbers or letters clearly.

OTHER FORMATS

LETTERS FOR PUBLICATION—Type on one side of paper only, double-space with wide margins. Prepare copy in style and format of Correspondence section. Follow advice about references and other matters above. Give degrees and diplomas of each author in a cover letter signed personally by all authors.

OTHER

Abbreviations: Do not use except for units of measurement (e.g., mg., cm, etc.).

Conventions: Give statistics and measurements in figures except that numerals one to nine should be in words if not followed by a measurement symbol (e.g., "two patients" but 2.0 mg). Use 24 hour clock.

Experimentation, Human: Investigations must conform to appropriate ethical standards including voluntary, informed consent and acceptance by an ethics committee.

Statistical Analyses: State tests used. Results should include 95% confidence limits for main findings and probability estimates.

Style: Follow "Uniform requirements for manuscripts submitted to biomedical journals" (British Medical Journal 1979;1: 532-5) except give titles of journals in references in full.

Units: Use SI Conversion Tables in January 1978 issue of *Anaesthesia. Units, Symbols and Abbreviations*: *A Guide for Biological and Medical Editors and Authors* (Royal Society of Medicine, London), except record vascular pressures in mmHg and cmH_2O. Do not use imperial measurements except in an historic context. Follow *Uniform requirements for manuscripts submitted to biomedical journals (British Medical Journal* 1979; 1: 532-5).

SOURCE

January 1994, 49(1):back matter.

Analyst
Royal Society of Chemistry

Editor, *The Analyst*
The Royal Society of Chemistry
Thomas Graham House, Science Park
Milton Road, Cambridge CB4 4WF
Telephone: +44(0)223 420066; Fax: +44 (0)223 420247

Dr. J. F. Tyson
Department of Chemistry
University of Massachusetts
Amherst, MA 01003

AIM AND SCOPE

The Analyst publishes original research papers on all aspects of the theory and practice of analytical chemistry, fundamental and applied, inorganic and organic, including chemical, physical, biochemical, biomedical, clinical, pharmaceutical, biological, automatic and computer-based methods. Papers on new techniques and instrumentation, detectors and sensors, and new areas of application with due attention to overcoming limitations and to underlying principles are all equally welcome. All contributions should represent a significant development in the particular field of analysis and are judged on the criteria of (i) originality and quality of scientific content and (ii) appropriateness of the length to content of new science. Thus, papers reporting results which would be routinely predicted or result from application of standard procedures or techniques are unlikely to prove acceptable in the absence of other attributes which themselves make publication desirable. Applications papers (particularly, those dealing with spectrophotometry or chelating reagents for high-performance liquid chromatography) must contain a comparison with existing methods and demonstrate advantages over accepted methods before publication can be considered.

MANUSCRIPT SUBMISSION

Type double-spaced on one side of paper. Have available copies of any related, relevant, unpublished material and raw data. Send three copies of text and illustrations.
Same material may not be under consideration for publication by another journal in any language.

MANUSCRIPT FORMAT

Follow style and usage in recent copies of *The Analyst*. Be concise. Adopt a logical order of presentation, with suitable paragraph or section headings. Include at least one forename with each author's family name. Indicate corresponding author.

Title: Be brief, but adequately indicate original features of work. Include analyte being determined or identified, matrix and analytical method used.

Summary: 250 words, give salient features and novel aspects. Include relevant quantitative information such as detection limits, precision and accuracy data. Compare with existing methods and demonstrate advantages over accepted methods.

Key Words: Five keywords or key phrases, indicating topics of importance, after summary.

Aim of Investigation: A concise introductory statement indicating study aims and need for work. Support by judicious citation of relevant published literature. Explain usefulness of work, again with reference to appropriate literature. Conclude with an unambiguous statement of work's novel features.

Description of the Experimental Procedures: Include information about all experimental work conducted including, if appropriate, a brief description of relevant preliminary work. Justify choice of any optimization procedure (in accordance with some accepted protocol), and state figure of merit clearly. If appropriate, include information about experiments in which relevant parameters were varied (and about any experimental design used). Include information on how any new method was validated, and describe statistical procedures used. Experimental procedures should be concise; do not give detailed descriptions of well-known operations. Mention suppliers of equipment and materials and their locations.

Support descriptions of methods by experimental results showing accuracy, precision and selectivity.

Discuss optimization and/or experimental design procedures used and statistical procedures applied for evaluation of results.

Results and Discussion: Do not describe experimental work. Present results in tabular or diagrammatic form, followed by appropriate statistical evaluation. Avoid subjective descriptions of results. Comment on scope of method and its validity, followed by statement of conclusions drawn from work.

Acknowledgments: Include contributors other than co-authors, companies or sponsors in a separate paragraph at the end of the paper. Give titles, but not degrees.

References: Number serially in text with superscript figures. e.g., Foote and Delves,[1] Burns *et al.*[2] or . . . in a recent paper[3] . . . Collect in numerical order under 'References' at end of paper. List all authors' names and initials. Double-space.

JOURNALS

Abbreviate titles according to *Chemical Abstracts Service Source Index (CASSI)*.

1 Ebdon, J. R., Lucas, D. M., Soutar, I., and Swanson, L., *Anal. Proc.*, 1993, **30**, 431.

2 Dawson, J. B., Snook, R. D., and Price, W. J., *J. Anal. At. Spectrom.*, 1993, **8**, 331R. 3 Analytical Methods Committee, *Analyst*, 1993,**118**, 1217.

4 Economou, A., and Fielden, P. R., *Analyst,* 1993,**118**, 47; and references cited therein.

5 James, D., Thiel, D. V., Bushell, G. R., and MacKay-Sim, A., unpublished work.

6 Platteau, O., *Analyst,* in the press.

7 Vousden, L. P., personal communication.

8 Appelqvist, R., Ph D Thesis, University of Lund, Sweden, 1993.

BOOKS

Give edition (if not first), publisher, place and date of publication, followed by page number.

1 Hanai, T., in *Liquid Chromatography in Analysis,* ed. Hanai, T., Elsevier, Amsterdam, 1991, pp. 21-46.

2 Lajunen, L. H. J., *Spectrochemical Analysis by Atomic Absorption and Emission,* The Royal Society of Chemistry, Cambridge, 1992.

3 *British Pharmacopoeia 1992*, HM Stationery Office, London, 1992,vol. 1,p.40.

4 Marmion, D. M., *Handbook of US Colorants, Foods, Drugs, Cosmetics and Medical Devices*, Wiley, New York, 3rd edn., 1991.

5 Szepesi, G., *How to Use Reverse-Phase HPLC,* VCH, Weinheim, 1992, ch. 2.

6 *Food and Cancer Prevention: Chemical and Biological Aspects,* Proceedings of an International Conference sponsored by the Food Chemistry Group of the Royal Society of Chemistry, Norwich, September 13-16, 1992, eds. Waldron, K. W., Johnson, I. T., and Fenwick, G. R., The Royal Society of Chemistry, Cambridge, 1993.

7 CRC *Handbook of Chemistry and Physics,* ed. Weast, R. C., CRC Press, Boca Raton, FL, 72nd edn., 1992, sect. D-100.

Check references against original papers. Do not use references to conference abstracts which have not been published in open literature. Keep number of references to a minimum.

Tables and Diagrams: Put each table and illustration on a separate sheet at end of text. Use either tables or graphs but not both for same set of results, unless giving important additional information Give table columns brief headings. Title tables to be understandable without reference to text.

Arrange two column tables horizontally. Convey information given by a straight-line calibration graph as an equation or statement in text.

Use SI units for column headings and graph axis labels. Express numerical values of physical quantities without dimensions, i.e., the quotient of the symbol for the physical quantity and the symbol for the unit used, e.g., $c/\text{mol dm}^3$, or some mathematical function of a number, e.g., $\ln(c/\text{mol dm}^3)$. For dimensionless units, i.e., ratios such as % or ppm, indicate type of ratio in parentheses, e.g., c (%) or c (ppm). Do not use diagonal line (solidus) to represent 'per'. In accordance with SI system, units such as grams per milliliter are already expressed in the form g ml^{-1}. 'Combined' unit, g ml^{-1}, must not have any 'intrusive' numbers. To express concentration in grams per 100 milliliters, use 'per': Concentration/g per 100 ml. Express concentrations in grams per liter (g l^{-1}) rather than grams per 100 ml.

Letter diagrams uniformly in line thicknesses and lettering size and style. Draw all diagrams carefully and clearly on good quality paper. Letter carefully and clearly. If possible, supply chromatograms and spectra, complicated flow charts, circuit diagrams, etc., as artwork for direct reproduction. Do not letter clearest copy.

Send three complete sets of illustrations. Two sets may be made by any convenient copying process.

Legends: Type captions separately. Wherever possible, place extensive identifying lettering in caption rather than on lines on graphs, etc.

Photographs: Submit if they convey essential information that cannot be shown in any other way. Submit as glossy or matte prints made to give maximum detail. Submit color photographs only when a black-and-white photograph fails to show some vital feature and can be supplied either as prints or transparencies.

OTHER FORMATS

COMMUNICATIONS—Brief descriptions of work that has progressed to a stage at which it is likely to be valuable to workers faced with similar problems. Rapid publication is enhanced if diagrams are omitted, tables and formula can be included.

REVIEWS—Critical evaluation of existing state of knowledge on a particular facet of analytical chemistry. Original work may be included. Do not submit simple literature surveys. Contact Editor before beginning work.

OTHER

Abbreviations: Omit abbreviational full stops after common contractions of metric units (e.g., ml g. μg. mm) and other units represented by symbols. Avoid abbreviations in text other than those of recognized units except after definition. Use upper case letters without points for abbreviations for techniques and associated terms, subsequent to definition, e.g., HPLC, AAS, XRF, UV, NMR, SCE. Me, Et, Pr^n, Bu^n, Bu^i, Bu^s, Bu^t, Ph, Ac, Alk, Ar and Hal can be used; define others. Write carboxy groups CO_2R, not CO-OR. Indicate substituents by R (one) or by R^1, R^2, R^3 . . . (more than one).

State percentage concentrations of solutions in internationally recognized terms. Use symbols 'm' instead of 'w' for mass and 'v' for volume. Examples of percentages with acceptable alternative in parentheses: % m/m (g per 100 g); % m/v (g per 100 ml); % v/v. Implications of use of term 'mass' are that 'relative atomic mass' of an element (A_r) replaces atomic weight, and 'relative molecular mass' of a substance (M_r) replaces molecular weight.

Give concentrations of solutions of common acids as dilutions of concentrated acids, such as 'dilute hydrochloric acid (1 + 4)', which signifies 1 volume of concentrated acid mixed with 4 volumes of water. Express molarity as a decimal fraction (e.g., 0.375 mol dm^{-3}).

Nomenclature: Use current internationally recognized (IUPAC) chemical nomenclature. Use common trivial names, defined in terms of IUPAC nomenclature.

Spelling: Use *Oxford English Dictionary.*

Style: Give dimensions in meters (m) or millimeters (mm); express temperatures in K or °C (not °F); express wavelengths in nanometers (nm) (not mμ); express frequency in Hz (or kHz, etc.), not in c/s or c.p.s; denote rotational frequency by use of s^{-1}; in mass spectrometry, express signal intensity in counts s^{-1} and not in Hz; express radionuclide activity in becquerels (Bq); do not use micron (μ); 10^{-6} m will be 1 μm.

Symbols and Units: SI system. See 'Green Book': *Quantities, Units and Symbols* and *Physical Chemistry* (Blackwell, Oxford, 2nd edition, 1993). Explain non-SI units unless definition is obvious. Indicate derivation of derived non-SI units. (see SI appendix *Analyst* 119:157-158)

SOURCE

January 1994, 119(1):155-158.

Analytical Biochemistry: Methods in the Biological Sciences
Academic Press

Analytical Biochemistry, Editorial Office
525 B Street, Suite 1900

San Diego, California 92101-4495
Telephone: (619) 699-6469; Fax: (619) 699-6859

AIM AND SCOPE

Analytical Biochemistry is an international journal that publishes original material on methods and methodology of interest to the biological sciences and all fields that impinge on biochemical investigation. In addition to the expected techniques that apply to biochemical preparations and analysis, the following are within the scope of the journal: cell biology and cell, tissue, and organ culture methods that are of general application; membranes and membrane proteins; procedures of interest in the field of molecular genetics including those necessary for cloning, sequencing, and synthesis of nucleic acids and for mutagenesis; purification of enzymes and other proteins, but only if the methods are both novel and applicable generally; immunological techniques both analytical and preparative; immunoassays only if a unique approach is introduced rather than the determination of a substance not previously assayable; and pharmacological and toxicological research techniques that offer a biochemical approach.

MANUSCRIPT SUBMISSION

Manuscripts are accepted for review with the understanding that work has not been, will not be nor is presently submitted elsewhere, and that its submission has been approved by all authors and by institution where work was carried out; and that any person cited as a source of personal communications has approved such citation.

Submit manuscripts in quadruplicate. Include all figures (originals or glossy prints) and tables. Original copy must be typewritten, double-spaced, on one side of white bond paper (8.5 × 11 in.), with 1 in. margins on all sides. Double space everything everywhere.

MANUSCRIPT FORMAT

Be concise and consistent in style, spelling, and use of abbreviations. On a separate title page note article title, authors' names (without degrees), complete affiliations, a running title (50 letters and spaces), an abstract (200 words), and corresponding author's address, telephone number and fax or telex number. Omit words such as novel, rapid, improved, simple, sensitive, efficient, convenient, new from title. Do not use abbreviations in title or running title. Number all pages. Indicate appropriate subject category on title page.

Text: Suggested organization is: abstract; introductory statement; Materials and Methods; Results; Discussion; Acknowledgments; References. Sections may be combined for clarity. Variations are acceptable, but abstracts are required.

Figures: Submit four sets. Number consecutively with Arabic numerals in order of mention in text: give each a descriptive legend. Type legends together on a separate sheet, double-spaced. A convenient size for drawings is 8.5 × 11 in. Plan to fit proportions of printed page (7.13 × 9 in.; column width, 3.5 in.). Lettering should be large enough that in printed version it is 6-8 points. Except for single-letter part designations, lettering should not be larger than 10 points. Make a drawing with India ink on tracing linen, smooth surface white paper, or Bristol board. High-quality computer graphics are acceptable. Use blue ruled graph paper. Ink grid lines to be shown in black. Properly prepared glossy prints are acceptable. Keep photographs to a minimum; submit as single, unmounted glossy prints with strong contrast; indicate magnification by a scale where possible. Avoid simple histograms; use a table or a paragraph in text. Authors must defray cost of color illustrations.

Tables: Type on separate pages, number consecutively with Arabic numerals in order of mention in text. Give short explanatory titles.

Submit complex tables or sequence information as camera-ready copy, typed single-spaced and preferably 5 × 7-in glossy prints.

Footnotes: Designate in text by superscript numbers and type on a separate sheet following References. Identify footnotes to tables with lowercase italic superscript letters and place at bottom of table.

References: Cite in text by Arabic numerals in parentheses; list at end of paper in consecutive order. Submit copies of manuscripts listed as "in press." Use abbreviations of journal titles from *Chemical Abstracts Service Source Index, 1985*.

1. Gersten, D. M., and Gabriel, O. (1992) *Anal. Biochem.* **203**, 181–186.

2. Birren, B., and Lai, E. (1994) Pulsed Field Gel Electrophoresis: A Practical Guide, Academic Press, San Diego.

3. Compton T. (1990) *in* PCR Protocols: A Guide to Methods and Applications (Innis, M. A., Gelfand, D. H., Sninsky, J. J., and White, T. J., Eds.), pp. 39-45, Academic Press, San Diego.

OTHER FORMATS

NOTES & TIPS—Two printed pages including all tables and figures, designed to accommodate methods that can be summarized in a shorter format allowing more rapid publication, as well as to provide for helpful "kitchen tricks." No formal organization is required; abstracts are not used; state problem and means of resolution in opening paragraph; conclude with a brief summary statement.

OTHER

Names of Chemical or Organic Substances: Follow IUPAC-IUB Joint Combined Commission on Biochemical Nomenclature (JCBN) published in extended Instructions for Authors (July issue of journal in odd-numbered years). Draw attention to any particular chemical or biological hazards that may be involved in carrying out described experiments. Describe relevant safety precautions; reference relevant standards.

SOURCE

July 1994, 220(1): back matter.
An extended Instructions to Authors is published in the July issue during odd-numbered years.

Anatomical Record

American Association of Anatomists
Wiley-Liss, Inc.

Dr. Aaron J. Ladman
Department of Anatomy
Mail Stop 408
Hahnemann University
Broad & Vine Streets
Philadelphia PA 19102

AIM AND SCOPE

The *Anatomical Record* publishes original research primarily on cellular and supracellular (tissue and organ) events as they help to define mechanisms of biologic function, descriptive studies particularly emphasizing mammalian systems if there is an experimental component to the design. Subject matter includes: Theoretical Biology; Molecular and Cellular Biology; General Histology and Cytology; Connective Tissue Biology; Myology; Urinary Biology; Reproductive Biology; Gastroenterology; Respiratory Biology: Cardiovascular Biology; Endocrinology; Immunobiology: Developmental Biology; Neurobiology: Special Senses; Gross Anatomy; Methods; and Education.

MANUSCRIPT SUBMISSION

Type double-spaced throughout on one side of bond or heavy-bodied paper 8.5 × 11 in. with a 1 in. margin on all sides. Number each line of each page consecutively, starting with number one. Submit original and three copies of all materials.

In cover letter include: names, telephone and fax numbers of authors; paper title and statement of specific results and significance; statement that material has not been published and is not under consideration elsewhere; and names, addresses, telephone numbers and fields of interest of three to five qualified referees. Submit any paper in press or under consideration elsewhere that includes information helpful for evaluation. Include written permission from author's whose work is cited as personal communication, unpublished work, or work in press.

MANUSCRIPT FORMAT

Use subdivisions in sequence: Title Page, Abstract, Text, Acknowledgments, Literature Cited, Footnotes, and Tables,

Figure Legends: Start each on new page.

Title Page: Include: title of paper; full name of author(s); institutional affiliation and complete address; telephone and fax numbers; running title (45 letters and spaces); and name and address of corresponding author.

Abstract: 250 words that serves in lieu of a concluding summary. Write in complete sentences, and format into background, methods, results, and conclusions. Concisely state significant findings without reference to rest of paper. Add three to eight key words for indexing.

Text: Divide into Introduction; Materials and Methods; Results or Observations and Discussion.

Literature Cited: In text make references to literature chronologically by author's name followed by year of publication, e.g, . . . studies by Briggs (1975) reveal. . . When references are made to more than one paper by the same author in the same year, designate in text as (Kelley, 1970a, b) and in literature list as:

Kelley, R.O. 1970a An electron microscopic study of mesenchyme during development of interdigital spaces in man. Anat. Rec., *168*:43-54.

Kelley, R.O. 1970b Fine structure of apical, digital and interdigital cells during limb morphogenesis in man. In: Proceedings of the VIIth International Congress of Electron Microscopy, Vol. III:381–382.

Arrange list alphabetically as: Author's name(s), publication year, complete title, volume, and inclusive pages.

Hather, B.M. and R.S. Hikida 1988 Properties of standard avian slow muscle grafts following long-term regeneration. J. Exp. Zool., *246*:115-123.

Jessell, T.M., P. Bovolenta, M. Plaszek, M. Tessier-Lavigne, and J. Dodd 1989 Polarity and patterning in the neural tube: the origin and function of the floor plate. In: Cellular Basis of Morphogenesis, Ciba Foundation Symposium. Wiley, Chichester, pp. 257-282.

Schoenwolf, G.C., H. Bortier, and L. Vakaet 1989 Fate mapping the avian neural plate with quail/chick chimeras: origin of prospective median wedge cells. J. Exp. Zool. *249*:271-278.

Schoenwolf, G.C., V. Garcia-Martinez, and M.S. Dias 1992 Mesoderm movement and fate during avian gastrulation and neurulation. Dev. Dyn. *193*:235-248.

Sternberger, L.A. 1986 Immunocytochemistry, 3rd ed. John Wiley & Sons, New York

Trotter, J.A. 1990 Interfiber tension transmission in series-fibered muscles of the cat hindlimb. J. Morphol. *206*:351-361.

Trotter, J.A., K. Corbett, and B.P. Avner 1981 Structure and function of the murine muscle-tendon junction. Anat. Rec. *201*:293-302.

Trotter, J.A., J.D. Salgado, R. Ozbaysal, and A.S.   Gaunt 1992 The composite structure of quail pectorlis muscle. J. Morphol. *212*:27-35.

Footnotes: Number consecutively. Clearly indicate corresponding numbers in text. Number additional references to same footnote with next consecutive number:
[1]Material used for this experiment was
[2]provided by . . .
[2]See footnote 2, page . . .

Tables: Cite all tables in text. Make titles complete but brief. Present information other than that defining the data as footnotes. Make tables simple and uncomplicated, with as few vertical and horizontal rules as possible. Type footnotes directly beneath and number 1, 2, 3, etc., not in sequence with text footnotes.

Figures: Cite all figures in text. Give figures legends. Number figures, including charts and graphs, consecutively throughout text. In text refer to figures by figure numbers. If possible integrate figures into text. Group figures to fit a single page with their appropriate legend. Use references to relevant text to reduce legend length and avoid redundancy. List abbreviations used alphabetically and place before first figure in which mentioned, e.g.,

AchE	acetylcholinesterase
CP	cortical plate
Smc	primary somatosensory cortex
V	ventral

Illustrations: Limit number of figures to that which adequately presents findings. Submit original drawings or high-quality photographic prints. Photographic prints must have good contrast and be of uniform tone.

Submit original illustrations, and three sets of good-contrast photographic copies. If original drawings are too large for shipment, submit photographic prints. Indicate on reverse: Author's name; Figure number; Top of illustration; Reduction requested; and "Review copy" on those intended for reviewers. Do not fasten with paper clips, staples, etc. Ship flat and protected.

Mount photomicrographs and illustrations: trimmed straight on all sides and "squared," on hard, strong bristol board (0.4 mm thick) with at least a 1 in. margin; attach using appropriate dry mounting materials, or white or colorless cement or glue; when two or more figures are assembled, mount close together and separate by no more than 1/8 in.; mount color figures on a lighter weight flexible bristol paper (0.2 mm thick).

Lettering and Labels—Letter and number with printed paste-on or transfer labels, sturdy and large for handling and reduction. Spray with clear adhesive. Label directly on drawing or photographic print, not on overlay. Place all labeling at least 1/8 in. from edges. Place white labels over dark backgrounds and black labels over light backgrounds, or shadow labels with an appropriately light or dark highlight.

Reduction to Printed Size—Indicate desired reduction desired. Illustrations cannot be reduced to less than 20% of original size; original line drawings cannot exceed 11 × 14 in.; lettering; labels must be readable after reduction. Reduced minimum height of a capital letter should not be less than 2.5 mm for a photomicrograph and 1 mm for a graph or chart. Printed individual figures or group of figures should not exceed 6 13/16 in. × 8 15/16 in. or 3 5/16 in. × 8 15/16 in. (single-column placement). Figures may take an entire page; mark "print as bleed"; mark 1/8 in. on all four sides to be eliminated.

Black-and-White Prints—Use white nonmatte paper.

Line Drawings—Draw with black ink on medium-weight white paper or light-weight artboard. Photographic prints may be submitted. Use stippling and hatching to achieve tonal quality. Avoid use of shading (pencil, wash, or airbrush) for tonal effect unless drawing is to be reproduced as a halftone. Use blue-ruled paper for original graphs.

Color Prints—Submit as prints or transparencies. Mark critical image area on transparency frame.

Stereo Illustrations—Carefully crop mounted pairs so precisely the same territory is covered by each photograph. Position details on pairs 65-70 mm apart. Use a cartographer's stereo view to mount pairs for viewing. Mount stereo pairs of transmission electron microscope images of sectioned materials for either cross-eyed or stereo viewer imaging.

Scanning electron microscope images, freeze-fracture images and photographic images of gross specimens will suffer a false-stereo confusion unless mounted for viewing. Care should be taken that the axis of tilt employed in production of stereo pairs is vertical as viewed in the final mounted prints. Perception of depth in freeze-fracture images will be enhanced if the direction of metal deposition on the replica is parallel to the axis or tilt.

Add labels to one member of a stereo pair.

OTHER

Abbreviations: *CBE Style Manual* (5th edition, 1983, Council of Biology Editors, Inc., Chicago)

Experimentation, Animal: Conduct in a humane manner and in accordance with all local, state, and federal guidelines for care and utilization of laboratory animals. Follow "Guide for the Care and Use of Laboratory Animals" [DHEW Publication No. (NIH) 73-23]. Include details of food and water regimen; light cycles; appropriate tranquilizers, analgesics, anesthetics and care performed in surgical procedures; and manner of euthanization.

Nomenclature: Follow Nomina Anatomica, Nomina Embryologica, Nomina Anatomica Veterinaria, and Nomina Anatomica Avium.

Numbers: Always spell out numbers when first word in a sentence; abbreviations cannot follow such numbers. Use Arabic numbers to indicate time, weight and measurements when followed by abbreviations (e.g., 2 mm; 1 sec; 3 ml). Write out numbers one to ten in text. Use numerals for higher numbers.

Page Charges: $90 per page for tables and illustrations. One page of color is $950, second and subsequent pages, up to four, $500 each.

Spelling: *Webster's International Dictionary.*

Style: Do not hyphenate at end of lines. Do not begin sentences with abbreviations. Spell out word Figure in text, except when in parentheses.

Units: Metric.

SOURCE

May 1994, 239: 113-117.
Instructions to Authors appear in first issue of each year of Journal.

Anesthesia & Analgesia

International Anesthesia Research Society
Society of Cardiovascular Anesthesiologists
Williams & Wilkins

King C. Kryger, PhD, Managing Editor
Anesthesia & Analgesia
The Hearst Building, 5 Third Street, Suite 1216
San Francisco, CA 94103

AIM AND SCOPE

Anesthesia & Analgesia publishes original articles, case reports, technical communications, review articles, special articles, medical intelligence articles, editorials, book reviews, and letters to the editor.

MANUSCRIPT SUBMISSION

Include a cover letter that states that none of the material in the manuscript has been published previously nor is currently under consideration for publication elsewhere. Refer to all submissions and previous reports that might be regarded as prior or duplicate publication of the same, or very similar work, including title page and abstract. Disclose any potential conflicts of interest. Letter must be signed by all authors. Give a word count of manuscript.

Each author must have participated in work to the extent that he or she could publicly defend its contents; and be prepared to sign a statement that he or she has read the manuscript and agrees with its publication.

Follow "Uniform Requirements for Manuscripts Submitted to Biomedical Journals," reprinted in *The New England Journal of Medicine* 1991; 324:424-8.

MANUSCRIPT FORMAT

Use white bond paper, 8.5 by 11 in. or ISO A4, with 1 in. margins. Use double- or triple-spacing throughout, including references and table and figure legends.

Begin each section on a separate page; title page, abstract and key words, text, acknowledgments, references, tables (each, with title and footnotes, on a separate page), and legends. Type on one side of paper. Number pages consecutively, beginning with title page, in upper right corner.

Submit one original plus three copies of manuscript and four sets of figures. Include letters granting permission to reproduce previously published materials or illustrations that may identify subjects.

Title Page: Concise but informative title; short running head (40 letters and spaces) identified and placed at bottom of page. First name, middle initial, and last name of each author, with highest academic degree(s) including fellowship and board affiliations. Name of department(s) and institution(s) to which work should be attributed. Disclaimers, if applicable. Name, address, telephone, and fax numbers of corresponding and reprint request authors, or a statement that reprints will not be available. Financial support from foundations, institutions, pharmaceutical, and other private companies.

Abstract: 200 words. Only for general articles. State purposes of study, basic procedure (study subjects or experimental animals; observational and analytic methods), main findings (specific data and statistical significance), and principal conclusions. Emphasize new and important aspects.

Key Words: 3 to 10 key words or short phrases for indexing below abstract.

Introduction: Purpose of article, rationale for study or observation. Give only strictly pertinent references. Do not review subject extensively.

Methods: Describe selection of observational or experimental subjects, including controls. Identify methods, apparatus (manufacturer's name and address in parentheses), and procedures in sufficient detail to allow reproduction of results. Give references to established methods, including statistical methods; provide references and brief descriptions for published but not well-known methods; describe new or substantially modified methods, give reasons for using them, and evaluate limitations.

Identify all drugs and chemicals used, including generic name(s), dosage(s), and route(s) of administration. Cite drug or chemical name followed by generic name in parentheses. Use generic names.

Results: Present in logical sequence in text, tables, and illustrations. Do not repeat data in tables and/or illustrations; emphasize or summarize only important observations.

Discussion: Emphasize new and important aspects of study and conclusions. Do not repeat data in Results. Include implications and limitations of findings. Relate observations to other relevant studies. Link conclusions with goals of study. Avoid unqualified statements and conclusions.

References: No more than 25; 40 for review articles. All must be available to readers. Cite only references to books and articles or abstracts published in peer-reviewed *Index Medicus* journals. Do not use abstracts more than five years old or appearing only in meeting program.

Number consecutively in order mentioned. Identify in text, tables, and legends by Arabic numerals (in parentheses, on line). Use format of *Index Medicus*. Abbreviate journal titles according to *Index Medicus*. Verify references against original documents. Check for nonduplication.

JOURNAL

List all authors when four or less; when five or more, list first three and add et al.

Rigler ML, Drasner K, Krejcie TC, et al. Cauda equina syndrome after continuous spinal anesthesia. Anesth Analg 1991;72:275–81.

BOOKS

Eisen HM. Immunology: an introduction to molecular and cellular principles of the immune response. 5th ed. New York: Harper and Row, 1974.

CHAPTER IN A BOOK

Weinstein L, Swartz NM. Pathogenic properties of invading microorganisms. In: Sodeman WA Jr, Sodeman WA, eds. Pathologic physiology: mechanisms of disease. Philadelphia: WB Saunders, 1974:457–72.

PUBLISHED PROCEEDINGS PAPER

DuPont B. Bone marrow transplantation in severe combined immunodeficiency with an unrelated MLC compatible donor. In: White HJ, Smith R, eds. Proceedings of the third annual meeting of the International Society for Experimental Hematology. Houston: International Society for Experimental Hematology, 1974: 44–6.

Tables: No more than three or a combination of 6 total with figures. Type on separate sheets. Do not submit as photographs. Number consecutively. Supply brief titles. Give each column a short or abbreviated heading. Place explanatory matter in footnotes, Define all abbreviations, use lowercase italicized letters in alphabetical order. Do not use internal horizontal or vertical rules. Cite in text in consecutive order.

If data are used from another published (or unpublished) source, submit written permission from both author and publisher and acknowledge fully.

Illustrations: Submit 4 complete sets of sharp, unmounted black-and-white figures. All must be glossy illustrations. Figures must be professionally drawn and photographed. Submit photographs 127 by 173 mm but no larger than 203 by 254 mm.

Many computer-generated figures are unsatisfactory and should be professionally redrawn. Lettering should be adequate to retain clarity after reduction (final size is 1.5 mm high). Make symbols, cross-hatching, and stippling sharp and distinct to retain uniqueness after reduction.

Paste label on back indicating figure number, names of authors, and figure top. Do not write on back of figures.

Put internal scale markers on photographs. Use contrasting symbols, arrows, and letters in photomicrographs.

Cite in text in consecutive order. If previously published, acknowledge original source and submit written permission from both author and publisher. Permission is required, regardless of authorship or publisher, except for documents in public domain.

If a human subject is identifiable, written consent must be obtained from patient or legal guardian.

Legends: Type double-spaced on separate page, with corresponding Arabic numerals. Identify and explain symbols, arrows, numbers, or letters. Define all abbreviations.

OTHER FORMATS

BOOK REVIEWS—750 words, report on current literature in anesthesiology.

BRIEF REPORTS—1000 words, describe clinical or laboratory investigations that do not require the breadth of experimentation or support of a general investigative article.

CASE REPORTS—800 words, describe either new and instructive cases, anesthetic techniques and equipment of demonstrable originality, usefulness, and safety, or new information on diseases of importance to anesthesia.

EDITORIALS—Solicited by Editorial Board.

LETTERS TO THE EDITOR—200 words or less, brief constructive comments concerning previously published articles or brief notes of general interest. Double-space. Submit a title and three copies.

MEDICAL INTELLIGENCE ARTICLES—3000 words, collate, describe, and evaluate previously published material to aid in evaluating new concepts or updating old concepts or topics germane to anesthesiology.

ORIGINAL ARTICLES—3000 words, describe clinical or laboratory investigations.

REVIEW ARTICLES—4000 words, collate, describe, and evaluate previously published material to aid in evaluating new concepts.

SPECIAL ARTICLES—2000 words, describe literature, education, societies, and other topical interests of a historical or current trend in anesthesiology.

TECHNICAL COMMUNICATIONS—1500 words, describe instrumentation and analytic techniques.

OTHER

Abbreviations: At first mention in text, spell out in full, followed by abbreviation in parentheses. Do not synthesize new or unusual abbreviations. When many are used, include all in a box at start of article. Define all except those approved by International System of Units.

Consult the following sources: CBE Style Manual Committee. *Council of Biology Editors Style Manual: A Guide for Authors, Editors, and Publishers in the Biological Sciences.* 5th ed. Bethesda, Maryland: Council of Biology Editors, 1983; American Medical Association. *Manual of Style.* 8th ed. Baltimore, Maryland: Williams & Wilkins, 1989.

Experimentation, Animal: State that study was approved by authors' institutional animal investigation committee.

Experimentation, Human: State in text that study was approved by authors' institutional human investigation committee and written informed consent was obtained from all subjects or their parents. Human subjects should not be identifiable. Do not use patients' names, initials, or hospital numbers.

Units: Use metric units for distance/length and weight. Use SI units for clinical laboratory and hematologic data with, if desired, conventional metric units in parentheses.

SOURCE

August 1994, 79(2): front matter.

Anesthesiology
American Society of Anesthesiologists, Inc.
J. B. Lippincott

Lawrence J. Saidman, M.D.
University of California, San Diego
9500 Gilman Drive, La Jolla, California 92093-0815
Telephone: (619) 543-5710; Fax: (619) 543-6162

MANUSCRIPT SUBMISSION

Give name and address of corresponding author. State that material has not been submitted for publication or published in whole or in part elsewhere. State that all authors attest to validity and legitimacy of data. Do not submit preliminary communications or fragments of material in several articles.

Include statement: "In consideration of *Anesthesiology* taking action in reviewing and editing my (our) submission, the author(s) undersigned hereby transfers, assigns, or otherwise conveys all copyright ownership to the American Society of Anesthesiologists, Inc. in the event that such work is published in *Anesthesiology*." Letter must be signed by all authors. Disclose commercial associations that might pose a conflict of interest. Disclose other associations (e.g., consultancies, other equity interests, or patent licensing arrangements) in cover letter. If manuscript is accepted, editor will discuss how best to disclose relevant information.

MANUSCRIPT FORMAT

Submit original and four copies. Type double-spaced (including title, abstract, text, tables, legends, and references) on 8.5 × 11 in. white bond paper. Do not single space anywhere. Keep margins of at least 2.5 cm on all sides of all pages. Number pages consecutively, beginning with title page in upper right corner of each page. Arrange as:

Title Page: Page 1. Make title brief and useful for indexing. List all authors' names with highest academic ranks and affiliation. Restrict authorship to direct participants. Specify in a separate paragraph where work was done, if supported by a grant or otherwise, and meeting, if any, at which paper has been presented. Type an abbreviated title, stating article's essence (50 characters) at bottom of page.

Abstract: Page 2. 250 words, in labeled paragraphs Background, Methods, Results, and Conclusions, briefly describing problem addressed in study, how study was performed, salient results, and what authors conclude from results (see *N Engl J Med* 323:56, 1990). List key words and terms in alphabetical order for indexing.

Text: Page 3. Arrange as: Introduction; Methods and Materials; Results; and Discussion. Start each on new page. Describe statistical methodology used in Methods and Materials section.

References: Start on a new page with page numbers continuing text page numbering. Double space. Number in sequence of appearance in text. Restrict to those with direct bearing on work. Include names and initials of all authors, title, abbreviated titles of medical journals as appear in *Index Medicus* and *Science Citation Index*, volume, inclusive page numbers, and year. Type reference numbers in text as superscripts. Do not include articles published without peer review, including material appearing in programs of meetings or in organizational publications. Include only references accessible to all readers. Abstracts are acceptable only if published within 3 years in an indexed journal. Identify abstracts, editorials, and letters as such parenthetically following title. Manuscripts in preparation or submitted for publication are not acceptable. For manuscripts "in press," submit copies of manuscripts.

JOURNAL

Benhamou D, Ecoffey C, Rouby JJ, Fusciardi J, Viars P: Impact of changes in operating pressure during high frequency jet ventilation. Anesth Analg 63:19-24, 1984

BOOK

Cousins MJ, Bridenbaugh PO: Neural Blockade in Clinical Anesthesia and Management of Pain. Philadelphia, JB Lippincott, 1980, pp 526-530

CHAPTER IN A BOOK

Bresler DE, Katz RL: Chronic pain and alternatives to neural blockade, Neural Blockade in Clinical Anesthesia and Management of Pain. Edited by Cousins MJ, Bridenbaugh PO. Philadelphia, JB Lippincott, 1980, pp 651-678

Illustrations: For previously published tables or illustrations submit a statement that permission for reproduction has been obtained from author and publisher. Submit two (21.6 × 27.9 cm maximum) unmounted, untrimmed glossy prints or high-quality laser prints of each line drawing graph and four glossy prints of electron micrographs, x-rays, or echocardiograms. Make contents and lettering proportional. Paste a label on back indicating number in order of appearance, author's name, and top edge. Have line drawings, graphs, charts, and lettering done professionally.

Legends: Must accompany each illustration. List on a single, unnumbered page at end of manuscript

Tables: Number with Arabic numerals consecutively in order of appearance. Type on separate sheets with captions above tabular material. Double-space throughout. Do not number page. Do not submit as photographs. Deposit excess data in a repository such as National Auxiliary Publication Service (NAPS).

Acknowledgments: In a footnote, list all funding sources.

OTHER FORMATS

CASE REPORTS—500-1500 word educational Reports that describe a single case or a small series of cases and draw attention to important clinical situations, unusual clinical phenomena, or a new treatment or complication are most appropriate. Abstract not required.

CLINICAL INVESTIGATIONS—1500-4000 words presenting results of original important clinical research.

CONTEMPORARY ISSUES FORUM—500-1500 words, no abstract. New, conflicting or controversial information or points of view relevant to specialty.

LABORATORY INVESTIGATIONS—1500-4000 words describing results of original, important laboratory research relevant to anesthesiology.

LABORATORY REPORTS—500-1550 words, briefly describing results of laboratory research, or of new research equipment or techniques. Include abstract and key words.

LETTERS TO THE EDITOR—250-500 word objective, constructive, and educational Letters criticizing published material or discussing matters of general interest to anesthesiologist. A few references, a small table, or a pertinent illustration may be used. Double-space. Submit original and two copies.

MEDICAL INTELLIGENCE ARTICLES—2000-5000 word, in-depth reviews of matters of topical interest in clinical practice or research, shorter and more circumscribed in scope than Reviews. Abstracts not required.

REPORTS OF SCIENTIFIC MEETINGS—500-1500 word summaries of meetings that contain information of interest to specialty, but not attended by large numbers of anesthesiologists. Include reports of subspecialty meetings, basic science meetings, or clinical meetings of other specialties. Intended to disseminate significant new information, not comprehensive summaries of complete meeting.

REVIEWS—300 to 10000 word comprehensive surveys, that synthesize older ideas and suggest new ones. Cover broad areas, and may be clinical, investigational, or basic science in nature. Provide critiques of literature and are more than annotated bibliographies. References may be alphabetized or listed numerically in order of appearance. Abstract not required. Concluding paragraph or summary is desired.

SPECIAL ARTICLES—Articles that do not readily fall into above categories may be published as Special Articles (e.g., articles on history, education, demography, etc.)

OTHER

Abbreviations: Define all abbreviations except those approved by International System of Units for length, mass, time, electric current, temperature, luminous intensity, and amount of substance. Provide a footnote or box at beginning to define abbreviations when great numbers of abbreviations are used. Do not create abbreviations for drugs, procedures, or substrates.

Drugs: Use generic names and manufacturer (with location), lot number, and formulation. If brand name is used, put in parentheses after generic name.

Experimentation, Animal: Conform to Guiding Principles in the Care and Use of Animals as approved by the Council of the American Physiologic Society and published in their Guide for Authors. Obtain approval of appropriate institutional Animal Care Committee. Provide explicit justification for omission of anesthetics in scientific protocol.

Experimentation, Human: Avoid use of patients' names, initials, or hospital numbers. Patients must not be recognizable in photographs unless written consent has been obtained. Experiments must conform to ethical standards, including, where appropriate, a statement as to informed consent and institutional approval of investigation.

Units: Report data as multiples or submultiples of International System of Units (SI). Standard units for length, mass, and volume are meter, kilogram, and liter. Report concentrations or amounts of substances in molar units (moles per liter) rather than in mass units. Preferred units for pressures are either mmHg or cmH$_2$0. When more than two items are present, use negative exponents (ml/kg/min should be ml·kg^{-1}·min^{-1}). SI Unit for pressure (kPa) may be used, with pressure units of mmHg or cmH$_2$0 in parentheses. See Normal Reference Laboratory Values, *N Engl J Med* 298:34-45,1978.

SOURCE

July 1994, 81(1): 43A-44A. Revised January 1994.

Annals of Internal Medicine
American College of Physicians

The Editors, *Annals of Internal Medicine*
Independence Mall West, Sixth Street at Race
Philadelphia, PA 19106-1572
Telephone: (215) 351-2400; Fax: (213) 351-2644

MANUSCRIPT SUBMISSION

List only those who contributed to intellectual content of paper as authors. Authors must: conceive and plan work, interpret evidence, or both; write paper or review successive versions and take part in revisions; approve final version.

Reveal financial support, including equipment and drugs, on title page. Describe in cover letter any financial interests, direct or indirect, that might affect work. If uncertain as to what might be a potential conflict of interest, err on the side of full disclosure. Information may be made available to reviewers and may be published, at editor's discretion.

In cover letter, give full details on previous publication of any content of paper; e.g., reworked data; patients already described; content in another format such as proceedings of a meeting, a chapter in a book, or a letter to the editor; content already published in another language; content reported as a result of press coverage of oral presentation with scientific detail sufficient to support the main conclusion. Previous publication of some content of a paper does not necessarily preclude publication, but failure of a full disclosure is a breach of scientific ethics.

Indicate: *Annals* section; that contents have not been published elsewhere; that paper is not being submitted elsewhere (or provide information on possible prior publication); potential conflicts of interest, especially possible overlap with prior publications; and name, mailing address, telephone number, and fax number of corresponding author.

Obtain and submit permission to use work (text, figures, or tables) from other publications. Include bibliographic information for original source, intended use of work, and name of publication in which it will be used in permission requests.

Submit four copies of manuscript and three original glossy prints of all figures; two copies of cover letter; and one copy of completed transfer-of-copyright form (first 20 pages of each issue).

MANUSCRIPT FORMAT

Follow "Uniform Requirements for Manuscripts Submitted to Biomedical Journals of the International Committee of Medical Journal Editors" (*JAMA* 1993;269:2282-6). Type all parts of manuscript, including tables and figure legends, double-spaced on white bond paper 8.5 × 11 in. or ISO A4. Arrange as: title page, abstract, text, references, tables in numerical sequence, and figure legends. Begin each on a separate page. Number consecutively, starting with title page.

Title Page: Main title and subtitle (if any). Authors in order of appearance on title page (full first names preferred). Abbreviate academic degrees or academic certification; do not include memberships. Institutional affiliation of authors during study. Financial support information, including grant number and granting agency, other support, such as for equipment and drugs. Short title (50 characters). Name, address, telephone and fax numbers of corresponding and reprint request authors; number of words, exclusive of abstract, references, tables, figures, and figure legends.

Abstract: 250 words. For original research and review articles do structured abstracts (Haynes RB, Mulrow CD, Huth EJ, Altman DG, Gardner MJ. More informative abstracts revisited. *Ann Intern Med.* 1990;113:69–76). For original research organize as: Objective; Design; Setting; Patients; Intervention (if any); Measurements; Results; and Conclusions. For Reviews: Purpose; Data Sources; Study Selection; Data Extraction; Results; and Conclusions.

Headings in Text: Position flush with left margin. Use only three levels. First level: Initial capital letters, boldface. Second level: Initial capital letters, regular type. Third level: Initial capital letters, italics. Keep short (three or four words); do not use abbreviations.

References: Number in order cited. Use Arabic numerals within parentheses. Use National Library of Medicine reference style, including journal abbreviations. Provide complete data. Include "available from" note for documents that may not be readily accessible. Cite symposium papers only from published proceedings. When citing an article or book accepted for publication but not yet published, include journal title (or publisher) and year of expected publication. Include references to unpublished material in text, not in references and submit a letter of permission from cited persons to cite such communications. Do not use *ibid.* or *op cit.*

JOURNALS

List all authors when six or fewer; when seven or more, list only first six and add *et al.*

Standard Article

Bernstein H, Gold H. Sodium diphenylhydantoin in the treatment of arrhythmias. JAMA. 1965;191:695-9.

Corporate Author

The Royal Marsden Hospital Bone Marrow Transplantation Team. Failure of syngeneic bone marrow graft without preconditioning in post-hepatitis marrow aplasia. Lancet. 1977; 2:242–4.

Supplement

Mastri AR. Neuropathy of diabetic neurogenic bladder. Ann Intern Med. 1980;92 (2 Pt 2):316–8.

Special Format (also applies to abstracts)

Cahal DA. Methyldopa and haemolytic anemia [Letter]. Lancet. 1975;1:201.

BOOKS

List all authors or editors when six or fewer; when seven or more, list only the first six and add *et al.*

Author

Eisen HN. Immunology: An Introduction to Molecular and Cellular Principles of the Immune Response. 5th ed. New York: Harper and Row, 1974:406.

Editors

Dausset J, Colombani J; eds. Histocompatibility Testing 1972. Copenhagen: Munksgaard; 1973:12–8.

Chapter in a Book

Hellstrom I, Hellstrom KE. Lymphocyte-mediated cytotoxic reactions and blocking serum factors in tumor-bearing individuals. In: Brent L, Holbrow J; eds. Progress in Immunology II. v. 5. New York: American Elsevier; 1974: 147-57.

Published Proceedings Paper (with ordering information)

DuPont B. Bone marrow transplantation in combined immunodeficiency. In: White HJ, Smith R; eds. Proceedings of the Third Annual Meeting of the International Society for Experimental Hematology. Houston: International So-

ciety for Experimental Hematology; 1974:44-6. Available from the American Cancer Society.

OTHER CITATIONS IN REFERENCE LIST

In Press (must have journal title)

Dienstag JL. Experimental infection in chimpanzees with hepatitis A virus. J Infect Dis. 1975. In press.

Magazine Article

Roueche B. Annals of medicine: the Santa Claus culture. The New Yorker. 1971. Sep 4:66-81.

IN-TEXT CITATIONS OF UNPUBLISHED MATERIAL (TO BE PLACED WITHIN PARENTHESES)

Personal communication

(Strott CA, Nugent CA. Personal communication).

Unpublished Papers

(Lerner RA, Dixon FJ. The induction of acute glomerulonephritis in rats. In preparation.)

(Smith J. New agents for cancer chemotherapy. Presented at the Third Annual Meeting of the American Cancer Society, June 13, 1983, New York.)

Acknowledgments: Acknowledge only contributors to scientific content or providers of technical support. Submit written permission from persons acknowledged for other than financial or technical support.

Current Mailing Addresses: After acknowledgments list current mailing addresses of all authors.

Footnotes: Use only on title page and in tables, not within text. Use symbols in order: *, †, ‡, §, ||, ¶,**, ††, ‡‡. Do not use numbers or letters.

Illustrations: Total number (sum of tables and figures) may not exceed half number of typed text pages minus one.

Tables: Double space. Do not submit as glossy prints. Number with Arabic numerals, in order cited in text. Title should concisely describe table content so it can be understood without text. Give units of measure for all numerical data within a column or row under a column heading or at the end of a side heading only if all numerical data within column or row are in those units. Abbreviations not permitted in text may be used, but explain in footnotes.

Figures: Have professionally drawn or prepared using a computer and high-resolution printer. Make lettering uniform in style; use initial capital, then lower case. Number in order cited in text. Use scale markers in photomicrographs to show degree of magnification. Submit three glossy prints of each. Indicate on a label name of first author, figure number, and top of figure; paste label on back. Do not mount figures.

Color figures published if essential. Send one transparency and three color prints. Indicate cropping, if needed, with transparent overlay.

Legends: Begin with a short title using partial sentences. Explain all abbreviations and symbols, even if explained in text. Give stain and magnification data at end of legend for each part of figure. If no scale marker, give original magnification. Acknowledge original sources of borrowed material using wording specified by original publisher. If none specified, cite authors, reference number, and publisher. Submit letters of permission from copyright holder.

OTHER FORMATS

ABROAD—2500 word (10 references, 1 figure or table) Reports on health care in countries other than the U.S.

ACADEMIA AND CLINIC—4000 word descriptions and evaluations of important innovations in medical education.

AD LIBITUM—Prose (900 words) and poetry (80 lines) conveying insight rather than opinion.

ARTICLES—4000 words describing original research on causes, mechanisms, diagnosis, course, treatment, and prevention of disease.

BASIC SCIENCE REVIEWS—5000 words, detailed surveys of relevant scientific methods.

BRIEF COMMUNICATIONS—1500 words, <20 references, 2 figures or tables; preliminary or limited results of original research on causes, mechanisms, diagnosis, course, treatment, and prevention of disease.

BRIEF REPORTS—750 words maximum; 10 references; 1 figure or table, 150 word abstract. Clinical association, case reports, and reports of adverse effects.

CLINICAL CONFERENCES—5000 word edited transcripts of clinical staff conferences on basic and clinical topics.

HISTORY OF MEDICINE—4000 word essays, reports, or biographies on evolution of medicine.

IN THE BALANCE—2500 words; dissenting views on diagnosis and treatment especially in cases of common, important problems. Can have two or more selections and an editorial.

LETTERS—300 words, three authors, and five references. Submit original and two copies, typed double-spaced. Must be received within 6 weeks of an article's publication. Tables and figures are included only selectively. Opinions on papers published in *Annals* and on other current topics and short reports of clinical interest.

LITERATURE OF MEDICINE—Essays: on the medical literature, information retrieval, writing, or bibliographic or other research tools. Reviews: Short, critical reviews of books, software, and audiovisual material.

MEDICINE AND PUBLIC ISSUES—2500 words on the economic, ethical, sociologic, and political environment of medicine.

ON BEING A DOCTOR—2500 word essays on the human experience of being a doctor.

PERSPECTIVE—2500 word essays with references, expressing opinions, presenting hypotheses, or considering controversial issues.

POSITION PAPERS—Summaries of official positions on pertinent medical practice issues. Reference list or bibliography of sources used in formulating the position may be included.

QUOTATIONS—100 words; inspirational pieces on the study and practice of medicine or health. Include citation.

REVIEWS—5000 word detailed critical surveys and meta-analyses of published research relevant to clinical problems.

OTHER

Abbreviations: Do not use unless absolutely necessary; abbreviate names of symptoms or disease or anatomic and histologic characteristics; explain abbreviations for units of measurement or standard scientific symbols. Do abbreviate: long names of chemical substances, terms for therapeutic combination, names of tests and procedures better known by their abbreviations, units of measurement with numerals; in figures and tables to save space. Explain all abbreviations used in figure legends or table footnotes.

Experimentation, Human: Authors' institutional review board must assure that informed consent forms have been obtained. State in methods section, when appropriate, ethical guidelines followed. If patients are recognizable in illustrations, submit signed release forms (or copies).

Proprietary and Generic Names: Use generic names for all drugs. Include proprietary name: if more commonly known; to differentiate among drug forms; if a specific trade preparation was used in a study or involved in an adverse effect. Instruments may be referred to by proprietary name; give name and location of manufacturers in text, within parentheses.

Style: CBE Style Manual Committee. *CBE Style Manual: A Guide for Authors, Editors, and Publishers in the Biological Sciences.* 5th ed. Bethesda, Maryland: Council of Biology Editors; 1983 and Huth EJ. *Medical Style and Format.* Philadelphia: ISI Press; 1986.

Units: Use SI units. See Young D. Implementation of SI units for clinical laboratory data: style specifications and conversion tables. *Ann Intern Med.* 1987;106:114–29. When other units are widely used, indicate in parentheses after SI unit.

SOURCE

July 1, 1994, 121(1): I8–I11.

Annals of Neurology
American Neurological Association
Child Neurology Society

R. A. Fishman, Editor
Annals of Neurology
University of California, San Francisco
505 Parnasus Avenue - Box 0114
San Francisco, CA 94143-0114
Telephone: (415) 476-8332; Fax: (415) 476-8776

AIM AND SCOPE

The *Annals of Neurology* publishes articles on all aspects of the human nervous system, both normal and abnormal, and studies of other species that bear upon the mechanisms or treatment of human neurological disorders.

MANUSCRIPT SUBMISSION

Submissions must not be under consideration by any other journal. No part may have appeared previously in any but abstract form. Transfer of copyright to the American Neurological Association is a condition of publication.

Disclose any potential inherent financial interest in cover letter. Cite funding sources. Disclose commercial considerations, such as an equity interest, patent rights, or corporate affiliations including consultantships, for any product or process mentioned. Information will be acknowledged at publication in a format agreeable to Chief Editor and responsible author. Failure to disclose may result in rejection or retraction if undisclosed financial interests are found after publication.

Manuscripts deriving from abstracts from the annual American Neurological Association meeting shall be offered first to the *Annals*. If authors wish to submit elsewhere, obtain permission from Chief Editor.

In cover letter specify: type of submission (Original Article, etc.); responsible author with complete mailing address, telephone and fax numbers; statement certifying that all coauthors have seen and agree with manuscript's contents; disclosure of any financial interest; names of potential reviewers and any request that a specific individual not be invited to review (optional).

MANUSCRIPT FORMAT

Submit three high-quality copies. Type double-spaced on standard-sized, heavy-duty bond, using ample margins, including reference lists, tables, and figure legends. Number pages consecutively: title page, abstract, text, acknowledgments and support lines, references, tables, legends. Place first author's name and page number in upper right corner of each page (including references, tables, and figure legends). Use letter quality print.

Title Page: Provide a title (80 spaces). List author(s)'s name, highest degree, institutional affiliation, and address. Indicate reprint request and communicating authors. Provide a running-head title (30 spaces).

Abstract: 10-18 lines (200 words). Summarize problem, method of study, results and scientific conclusion.

References: Type double-spaced throughout. Arrange in order of citation, and cite by number. Check for errors of citation or attribution.

Use *Index Medicus* style, give complete publication data. For journal articles: author(s) (up to four; if more, cite first three then et al.); title; journal name (use *Index Medicus* abbreviation); year; volume number; opening and closing page numbers. For books: author(s); chapter title; editor(s); book title; edition number, if not first; city of publication; publisher; year; and specific pages. Optional: series title, chapter number, name of translator, etc. Punctuate only between authors' names—e.g., Smith AB, Jones CD.

Submit a copy of any article cited as "in press." Give unpublished data (including articles submitted but not in press) and personal communications in parentheses in text, not as references.

Tables: Type double-spaced on pages separate from text. Number pages consecutively with text and provide table numbers and titles. Hold length to one standard-sized manuscript page if possible. If a table continues past one page, repeat all heads and stub (left) column. Number in order of citation in text. Do not use photocopy reduction. Data should not duplicate material in text or illustrations.

Figures: Submit one set of high-contrast, glossy, black-and-white illustrations. Two additional sets may be prints or photocopies, depending on material. Label on back indicating top, figure number, and first author. Do not mount. Do not use paper clips.

Illustrations: Have graphs, diagrams, line drawings, etc., done professionally. Plan for publication at 8-cm column width. Labeling should allow for legibility following reduction. Crop photomicrographs to column width unless full-page width (17 cm) is required.

Submit a signed release from persons recognizable in photograph. Masking eyes to hide identity is not sufficient.

Color figures may be accepted for publication. Submit positive color transparencies 35 mm, with color prints, with necessary internal labeling (e.g., arrows, bars).

Legends: Type double-spaced on pages separate from text. Do not paste or fasten legends and figures together. Provide one legend for each figure, and number in sequence. Number pages consecutively with text. For photomicrographs indicate stain and use an internal scale marker.

OTHER FORMATS

BRIEF COMMUNICATIONS—1000 words (110 lines), 15 references or fewer, no more than a total of three tables or illustrations, 5-6 line abstract.

LETTERS TO THE EDITOR AND REPLIES—400 words (40 lines), may include one table or figure if essential to point, and 5 or fewer references. Tabular or illustrative material must fit column width (8 cm).

OTHER

Color Charge: $1,000 for first illustration and $500 for each additional piece, per page.

Experimentation, Human: Specify, and verify if requested, that studies received prior approval by appropriate bodies and that informed consent was obtained from each patient or volunteer.

Permissions: Submit written permission to use nonoriginal material (quotations exceeding 100 words, any table or illustration) from both the original author and the publisher. Credit the source in a text or table footnote or in a legend.

Statistical Analyses: Identify in Methods section statistical procedures used. State rationale for choosing particular statistical methods.

SOURCE

July 1994 36(1): A6.

Annals of Surgery
J. B. Lippincott, Co.

Chair, Editorial Board of *Annals of Surgery*
David C. Sabiston, Jr., M.D.
Department of Surgery
Duke University Medical Center
Durham, North Carolina 27710

AIM AND SCOPE

Annals of Surgery is devoted to the surgical sciences and considers only papers judged to offer significant contributions to the advancement of surgical knowledge.

MANUSCRIPT SUBMISSION

Submit four copies of manuscript and illustrations.

Manuscripts must not have been published, simultaneously submitted, or already accepted for publication elsewhere. Submit copies of any possibly duplicative published material.

In a cover letter from corresponding author, state that manuscript has been seen and approved by all authors. Give any additional information that may be helpful to editor, e.g., type of article, information on publication of any part of manuscript, and whether author(s) will pay to reproduce color illustrations. Include copies of permissions to reproduce published material or to use illustrations of identifiable subjects.

Follow "Uniform Requirements for Manuscripts Submitted to Biomedical Journals."

MANUSCRIPT FORMAT

Use white bond paper, 8×10.5 in. or 8.5×11 in. or ISO A4 with at least 1 in. margins. Double space throughout. Begin each section on separate pages: title page, mini abstract, abstract, text, acknowledgments, references, individual tables, and legends. Number pages consecutively, beginning with title page. Type page number in upper right corner of each page.

Title Page: Include: a concise, informative title; a short running head or footline (40 characters counting letters and spaces) at foot of page and identified; first name, middle initial, and last name of each author, with highest academic degree(s); name of department(s) and institution(s) to which work is attributed; disclaimers, if any; name and address of corresponding and reprint request authors or state that reprints will not be available; source(s) of support, e.g., grants, equipment, drugs, etc.

Mini Abstract: Short description (three sentences, 50 words) to appear on table of contents.

Abstract: 150 words, stating purposes of study or investigation, basic procedures (study subjects or experimental animals and observational and analytic methods), main findings (specific data and their statistical significance, if possible), and principal conclusions. Emphasize new and important aspects of study or observations. Use only approved abbreviations.

Text: Text of observational and experimental articles is usually divided into: Introduction, Methods, Results, and Discussion. Long articles may need subheadings within sections. For other types of articles, consult journal for guidance.

Introduction: Clearly state article's purpose. Summarize rationale for study or observation. Give only strictly pertinent references. Do not review subject extensively.

Methods: Describe selection of observational or experimental subjects (patients or experimental animals, including controls). Identify methods, apparatus (manufacturer's name and address in parentheses), and procedures sufficiently to allow others to reproduce results. Reference established methods, including statistical methods; provide references and brief descriptions of methods that have been published but are not well known; describe new or substantially modified methods, give reasons for use, and evaluate limitations.

Identify all drugs and chemicals used. Include generic name(s), dosage(s), and route(s) of administration. Do not use patients' names, initials, or hospital numbers.

Include numbers of observations and statistical significance of findings when appropriate. Detailed statistical analyses, mathematical derivations, etc. may be presented in an appendix.

Results: Present in logical sequence in text, tables, and illustrations. Do not repeat in text data in tables and/or illustrations; emphasize or summarize only important observations.

Discussion: Emphasize new and important aspects of study and conclusions that follow. Do not repeat in detail data in Results. Include implications and limitations of findings. Relate observations to other relevant studies. Link conclusions with study goals but avoid unqualified statements and conclusions not completely supported by data. State new hypotheses when warranted, clearly labeled as such. Recommendations, when appropriate, may be included.

Acknowledgments: Acknowledge only persons who have made substantive contributions. Obtain written permission from everyone acknowledged by name.

References: Number consecutively as mentioned in text. Identify references in text, tables, and legends by Arabic numerals (in parenthesis). Number references cited only in tables or in figure legends in sequence established by first identification in text of table or illustration.

Use reference form of U.S. National Library of Medicine's *Index Medicus*. Abbreviate journal titles following *Index Medicus*. Avoid using abstracts as references; do not use unpublished observations and personal communications. Insert references to written, not verbal, communications (in parentheses) in text. Include manuscripts accepted but not yet published; designated as "in press" (in parentheses). Cite information from manuscripts submitted but not yet accepted in text as "unpublished observations" (in parentheses). Verify references against original documents.

JOURNAL

Standard Journal Article. (List all authors when six or less; when seven or more, list first three and add et al.)

Soter NA, Wasserman SI, Austen KF. Cold urticaria: release into the circulation of histamine and eosinophil chemotactic factor of anaphylaxis during cold challenge. N Engl J Med 1976; 294:687–690.

Corporate Author

The Committee on Enzymes of the Scandinavian Society for Clinical Chemistry and Clinical Physiology. Recommended method for the determination of gammaglutamyl-transferase in blood. Scand J Clin Lab Invest 1976; 36:119–125.

Anonymous

Epidemiology for primary health care. Int J Epidemiol 1976; 5:224-225.

BOOKS AND OTHER MONOGRAPHS

Personal Author(s)

Osler AG. Complement: mechanisms and functions. Englewood Cliffs: Prentice-Hall. 1976.

Corporate Author

American Medical Association Department of Drugs. AMA drug evaluations. 3rd ed. Littleton: Publishing Sciences Group. 1977.

Editor, Compiler, Chairman as Author

Rhodes AJ, Van Rooyen CE, comps. Textbook of virology: for students and practitioners of medicine and the other health sciences. 5th ed. Baltimore: Williams & Wilkins, 1968.

Chapter in Book

Weinstein L, Swanz M N. Pathogenic properties of invading microorganisms. In: Sodeman WA Jr, Sodeman WA, eds. Pathologic physiology: mechanisms of disease. Philadelphia: WB Saunders, 1974:457-471.

Agency Publication

National Center for Health Statistics. Acute conditions: incidence and associated disability, United States July 1968-June 1969. Rockville, MD.: National Center for Health Statistics, 1972. (Vital and health statistics. Series 10: Data from the National Health Survey, no. 69) (DHEW publication no. (HSM)72-1036).

OTHER ARTICLES

Newspaper Article

Shaffer RA. Advances in chemistry are starting to unlock mysteries of the brain: discoveries could help cure alcoholism and insomnia, explain mental illness. How the messengers work. Wall Street Journal 1977 Aug 12:1(col. 1), 10(col. l).

Magazine Article

Roueché B. Annals of medicine: the Santa Claus culture. The New Yorker 1971 Sep 4:66-81.

Tables: Type on separate sheets; double-space. Omit internal horizontal and vertical rules. If data is from another published or unpublished source, obtain permission and acknowledge fully. Identify statistical measures of variations such as SD and SEM.

Do not submit as photographs. Number consecutively and supply brief titles. Give each column short headings. Place explanatory matter in footnotes. Explain in footnotes all nonstandard abbreviations. For footnotes, use symbols in sequence: *, †, ‡, §, ||, ¶, **, ††. . . .

Illustrations: Black-and-white illustrations and tabular material are published free in moderate numbers. Excess illustrations, excess tabular material, and all color illustrations are charged to author. Have figures professionally drawn and photographed; freehand or typewritten letter-

ing is unacceptable. Instead of original drawings, roentgenograms, and other material, send sharp, glossy black-and-white photographic prints, usually 5×7 in., but no larger than 8 by 10 in. Make letters, numbers, and symbols clear and even throughout, and of sufficient size for reduction. Put titles and detailed explanations in legends, not on illustrations themselves.

Paste a label on figure back indicating figure number, names of authors, and top of figure. Do not write on figure back, mount on cardboard, or scratch or mar with paper clips. Put internal scale markers on photomicrographs. Make symbols, arrows, or letters contrast with background.

If photographs of identifiable persons are used, submit written permission for use.

Cite each figure in text in consecutive order. If a figure has been published, acknowledge original source and submit written permission from copyright holder. Permission is required, regardless of authorship or publisher, except for documents in public domain.

For color illustrations, supply four positive color prints.

Legends: Type double-spaced, starting on a separate page with Arabic numerals corresponding to illustrations. Explain and identify symbols, arrows, numbers, or letters used to identify parts of illustrations. Explain internal scale and identify method of staining in photomicrographs.

OTHER

Abbreviations: Use only standard abbreviations. Avoid use in title. Full term should precede first use of abbreviation in text unless a standard unit of measurement. Consult following sources for standard abbreviations: Iverson, Cheryl et al., eds. *American Medical Association Manual of Style*, 8th ed. Baltimore: Williams & Wilkins, 1989. CBE Style Manual Committee. *Council of Biology Editors Style Manual: A Guide for Authors, Editors, and Publishers in the Biological Sciences*, 4th ed. Arlington: Council of Biology Editors, 1978. O'Connor M, Woodford FP. *Writing Scientific Papers in English: an ELSE-Ciba Foundation Guide for Authors.* Amsterdam, Oxford, New York

Data Submission: Editor may recommend that additional tables containing important backup data too extensive to be published may be deposited with the National Auxiliary Publications Service or made available by the author(s). If so, an appropriate statement will be added to text. Submit such tables with manuscript.

Experimentation, Animal: Indicate whether institution's or National Research Council's guide for the care and use of laboratory animals was followed.

Experimentation, Human: Indicate whether procedures followed were in accord with the ethical standards of the Committee on Human Experimentation of the institution in which experiments were done or in accord with the Helsinki Declaration of 1975.

Units: Report measurements in units in which they were made.

SOURCE
July 1994, 220(1): back matter.

Annals of the Rheumatic Diseases

Arthritis and Rheumatism Council for Research in Great Britain and the Commonwealth
BMJ Publishing Group

Dr. Michael Doherty
Rheumatology Unit, City Hospital
Nottingham NG5 1PB
Telephone: 0602 857112; Fax: 0602 857104

AIM AND SCOPE

The *Annals* publishes original work on all aspects of rheumatology and disorders of connective tissue. Laboratory and clinical studies are equally welcome.

MANUSCRIPT SUBMISSION

Submissions must have not been and will not be published elsewhere. Each author must sign cover letter as evidence of consent to publication. All authors must transfer copyright to journal before publication.

Type on one side of paper only. Using double spacing and ample margins. Submit three copies of articles and letters, with three copies of all tables or figures.

MANUSCRIPT FORMAT

Title Sheet: State title, authors, their department(s) and institution(s) and name, postal address (± telephone/fax no) of corresponding author.

Abstract (Structured): 250 words, summarizing problem being considered, how study was performed, salient results, and principal conclusions, under headings Objectives, Methods, Results, and Conclusions.

Introduction: Brief description of what led to study. Do not include current results and conclusions.

Patients/Methods: Details of conduct of study. Explain statistical methods at end of section.

Results: Avoid repetition in text and tables. Comment on validity and significance of results. Restrict broader discussion of implications to Discussion. Use subheadings to aid clarity of presentation here and in Methods.

Discussion: Place nature and findings of study in context of other relevant published data. Avoid undue extrapolation. Include acknowledgments, including financial support and industry affiliations, at section end.

References: Follow Vancouver style. Cite numerically and list in order cited in text. In text, place reference number between parentheses on line, not superscript. List all authors. Abbreviate journal titles as in *Index Medicus*. Author is responsible for accuracy and completeness of references.

Jones A, Doherty M. The time has come the walrus said. *Ann Rheum Dis* 1992; **51:** 434–5.

Muir H. Current and future trends in articular cartilage research in osteoarthritis. In: Kuettner K, ed. *Articular cartilage biochemistry*. New York: Raven, 1986: 423–41.

Tables: Type each on a separate sheet, with a heading, and no vertical rules.

Figures: List legends to figures on a separate sheet. For photomicrographs include stain used; indicate magnification by a bar marker on figure. Label illustrations on back with first author's name, number in order of appearance in text, and indicate top. Submit radiographs as prints. If appropriate, colored illustrations may be published, at author's cost.

OTHER FORMATS

CONCISE REPORT—1500 words, 15 references, one table, and two figures. Present laboratory or clinical work, collected case reports or, exceptionally, single case reports. Format is identical to a full paper. For cases, 'substitute Case reports' for Methods and Results.

HYPOTHESIS ARTICLE—Present interesting theory, discussed in relation to published data. Discuss with editor.

LETTER—600 words, 10 references; one table and one figure. Instructions for references, tables, and figures are same as for full length articles. Comments on recent articles published in the *Annals*. Original observations relating to short clinical or laboratory studies or single case reports may also be presented.

REVIEW ARTICLE—Usually commissioned, discuss directly with editor possible topics.

OTHER

Abbreviations: Must be defined.

Units: SI units.

SOURCE
January 1994, 53(1): inside back cover.

Annals of Thoracic Surgery

Society of Thoracic Surgeons
Southern Thoracic Surgical Association
Elsevier Science Inc.

Thomas B. Ferguson, MD
3108 Queeny Tower - Barnes Hospital Plaza
St. Louis, MO 63110-1041
Telephone: (314) 361-6084; Fax: (314) 367-0585

AIM AND SCOPE

The Annals of Thoracic Surgery publishes original papers on topics in thoracic and cardiovascular surgery. Also featured are case reports, "how to do it" articles, articles on our surgical heritage, collective and current reviews, correspondence, and book reviews.

MANUSCRIPT SUBMISSION

All authors must agree to statements regarding conflict of interest, scientific responsibility, exclusive publication, and assignment of copyright.

Submit original and two duplicate manuscripts with three clearly separated and labeled sets of illustrations.

Disclose any commercial association that might pose a conflict of interest. Acknowledge all funding sources in a footnote. Bring institutional or corporate affiliations that might constitute a conflict to attention of the Editor.

Before publication, each author will be required to certify that he or she has participated sufficiently in the work to take responsibility for a meaningful share of the content of the manuscript, and that this participation included: conception or design of the experiment(s), or collection and analysis or interpretation of data; drafting manuscript or revising its intellectual content; and approval of final version to be published.

The following statement must be signed: "I certify that none of the material in this manuscript has been published previously, and that none of this material is currently under consideration for publication elsewhere." This includes symposia, transactions, books, articles published by invitation, and preliminary publications of any kind except an abstract of 400 words or fewer.

MANUSCRIPT FORMAT

Type double-spaced throughout (including title page, abstract, text, references, tables, and legends) on one side of 8.5 × 11 in. opaque bond paper with 1.25 in. margins all around. Do not exceed 26 pages. Type last name of first author in upper right corner of each page, including title page.

Arrange manuscript as: title page, abstract, text, references, tables, and legends. Number pages consecutively, beginning with title page as page 1 and ending with legend page.

Text for case reports should be no more than 4 pages (1200 words). If tables or illustrations are included, reduce text by half page or 150 words. Keep text for a "how to do it" article even briefer, but with detailed illustrative material.

Title Page: Use as short a title as possible (95 letters and spaces, 85 for case reports). List a short title (40 characters) for a running head. Include only names of authors directly affiliated with work, state single highest academic degree. Include name and location of no more than two institutional affiliations.

If paper has been or is to be presented at annual meetings of The Society of Thoracic Surgeons, the Southern Thoracic Surgical Association, or another scientific organization, in a footnote give meeting name, location, and dates. At page bottom, type "Address reprint requests to Dr . . ." followed by last name, exact postal address with zip code, telephone number, and fax number of corresponding and reprint request author.

Abstracts: 175 words for an original article; 50-75 words for case reports and "how to do it" articles. No abstract for letters to editor. Include 3 to 5 keywords to assist in cross-indexing.

Text: Organize as: Introduction, Material and Methods, Results, and Comment. Cite references, illustrations, and tables in numerical order in text. Order of mention in text determines number given to each.

Acknowledgments: Include complete grant or subsidy information. Place at end of text before references.

References: Be selective. Cited work must pertain directly to reported work. Original articles should have no more than 20 references; case reports, 8 references; and "how to do it" articles, 5 references.

Identify references in text using Arabic numerals in parentheses on line. Do not cite personal communications, manuscripts in preparation, and other unpublished data in reference list; mention in text in parentheses.

Type double-spaced on a separate sheet. Number consecutively in order mentioned in text.

Provide inclusive page numbers for journal articles; cite specific page numbers from books.

Double-check for accuracy, completeness, and nonduplication.

Use abbreviations of journals from *Index Medicus*:

JOURNAL ARTICLE

8. Grillo HC, Suen HC, Mathisen DJ, Wain JC. Resectional management of thyroid carcinoma invading the airway. Ann Thorac Surg 1992;54:3–10.

(List all authors if 6 or fewer; otherwise list first 3 and add "et al.")

CHAPTER IN A BOOK

12. Vouhe PR. Transplantation of thoracic organs in children. In: Fallis JC, Filler RM, Lemoine G, eds. Pediatric thoracic surgery. New York: Elsevier, 1991:319–29.

BOOK (PERSONAL AUTHORS OR EDITORS)

18. Shields TW, ed. Mediastinal surgery. Philadelphia: Lea & Febiger, 1991:1–10.

Tables: Type double-spaced on separate sheets, each with a Arabic number and title above table and explanatory notes and legends below. Make self-explanatory. Data should not be duplicated in text or illustrations. Provide an alphabetical key to each table to identify abbreviations; place key below explanatory notes.

Include written permission from both author and publisher to reproduce previously published table(s).

Legends: Type double-spaced on a single sheet. Use Arabic numbers corresponding to order of occurrence in text. Alphabetically identify all abbreviations in illustrations at end of each legend. Give type of stain and magnification power for all photomicrographs.

Include written permission from both author and publisher to reproduce previously published illustration(s).

Enclose signed releases for recognizable (unmasked) photographs of human beings.

Illustrations: Submit triplicate sets, each in its own envelope, of unmounted and untrimmed black and white professionally prepared glossy prints. Submit components of the same illustration (e.g., parts A and B) separately. Provide photographs, rather than original art or photocopies. Prepare drawings and graphs with black India ink on white background. Do not use type or computer print. Keep symbols and shading simple. Use widely spaced vertical, horizontal and diagonal lines. Use American spellings. Make lettering of adequate size to retain clarity after reduction. Illustrations will be (3.25 in.) or two columns (6.75 in.) wide, or an intermediate width ranging from 4 to 4.75 in.

Roentgenogram reproductions should be no smaller than 5 × 7 in.

For color illustrations, submit both positive 35-mm transparencies and color prints.

Place first author's last name, figure number, and arrow indicating top on a gummed label on back.

Footnotes: Place at the bottom of manuscript page on which cited. Credit suppliers of drugs, equipment, and other material described at length in footnotes, giving company name and location.

OTHER

Abbreviations: Use *American Medical Association Manual of Style*, 8th ed. Define at first appearance. Avoid use in title and abstract. Spell out, or define in parentheses, all abbreviations, even if commonly employed.

Color Charges: $1150 for first color illustration on each page and $250 for every additional color illustration on same page.

Experimentation, Human: Include date of approval by local institutional human research committee or ethical guidelines followed by investigators in Material and Methods sections.

Experimentation, Animal: State assurances in Material and Methods section that all animals have received humane care in compliance with "Guide for the Care and Use of Laboratory Animals" published by National Institutes of Health (NIH Publication No. 85-23, revised 1985).

Style: For statistical nomenclature and data analysis, follow "Guidelines for Data Reporting and Nomenclature" *The Annals of Thoracic Surgery,* 1988;46:260-1 and found in the program book for the annual meeting of The Society of Thoracic Surgeons.

Units: Standard metric or Systeme International units.

SOURCE

July 1994, 58(1): A29-A30.

Antimicrobial Agents and Chemotherapy
American Society for Microbiology

Journals Division, American Society for Microbiology
1325 Massachusetts Ave., N.W.
Washington, D.C. 20005-4171

AIM AND SCOPE

AAC is an interdisciplinary journal devoted to the dissemination of knowledge relating to all aspects of antimicrobial, antiparasitic, and anticancer agents and chemotherapy. Reports involving studies on or with antimicrobial, antiviral (including antiretroviral), antiparasitic, or anticancer agents are within its purview. Studies involving animal models, pharmacologic characterization, and clinical trials are appropriate. ASM publishes a number of journals covering various aspects of microbiology. *Journal of Bacteriology* papers describe the use of antimicrobial or anticancer agents as tools for elucidating the basic biological processes of microorganisms. *Journal of Clinical Microbiology* papers describe the use of antimicrobial, antiparasitic, or anticancer agents as tools in the isolation, identification, or epidemiology of microorganisms associated with disease, are concerned with quality control procedures for diffusion, elution, or dilution tests for determining susceptibilities to antimicrobial agents in clinical laboratories; or deal with applications of commercially prepared tests or kits to assays performed in clinical laboratories to measure the activities of established antimicrobial agents or their concentrations in body fluids. Manuscripts on the development or modification of assay methods and validation of their sensitivity and specificity are appropriate for AAC. *Applied and Environmental Microbiology* or the *Journal of Clinical Microbiology* papers describe new or novel methods or improvements in media and culture conditions. If these methods are applied to the study of problems related to the production or activity of antimicrobial agents, they may be considered for *AAC*. Submit papers that include extensive taxonomic material (e.g., descriptions of new taxa) to the *International Journal of Systematic Bacteriology*. If the main thrust of the manuscript is not taxonomy, divide the manuscript and submit the taxonomic portion to IJSB. If such division would weaken the main thrust, submit the manuscript to the journal of choice.

Manuscripts rejected by one ASM journal on scientific grounds or on the basis of its general suitability for publication are considered rejected by all other ASM journals.

MANUSCRIPT SUBMISSION

In cover letter state: journal to which manuscript is being submitted, most appropriate journal section, complete mailing address (including street), telephone and fax numbers of corresponding author, BITNET or other electronic mail address if available. Include written assurance of permission to cite personal communications and preprints.

Submit three complete copies, including figures and tables. Type every portion double-spaced (6 mm between lines), including figure legends, table footnotes, and References. Number all pages in sequence, including abstract, figure legends, and tables. Place last two items after References section. Keep 1 in. margins on all four sides. Use line numbers if possible. Make these characters easily distinguishable: numeral zero (0) and letter "oh" (O); numeral one (1), letter "el" (l), and letter "eye" (I); and multiplication sign (×) and letter "ex" (x). If cannot, mark these items at first occurrence for cell lines, strain and genetic designations, viruses, etc., on modified manuscript to be identified by copy editor.

Enclose three copies of each "in press" and "submitted" manuscript.

Present only reports of original research. All authors must agree to submission and are responsible for content, including appropriate citations and acknowledgments, and must agree to the decisions of corresponding author. Submission is a guarantee that manuscript, or one substantially the same, was not published previously, is not being considered or published elsewhere, and was not rejected on scientific grounds by another ASM journal.

An author is one who made a substantial contribution to the "overall design and execution of the experiments";.

Acknowledge individuals who provided assistance, e.g., supplied strains or reagents or critiqued the paper, in the Acknowledgment section.

Follow definition of primary publication (*How to Write and Publish a Scientific Paper*, 3rd ed., Robert A. Day): "the first publication of original research results, in a form whereby peers of the author can repeat the experiments and test the conclusion, and in a journal or other source document readily available within the scientific community."

Do not submit work published in conference reports, symposium proceedings, technical bulletins, or any other retrievable source. Preliminary disclosures of research findings published in abstract form as an adjunct to a meeting is not considered prior publication.

Acknowledge prior publication of data in manuscript even though author(s) do not consider it in violation of ASM policy. Submit copies of relevant work.

Obtain permissions from both original publisher and original author [i.e., copyright owner(s)] to reproduce figures, tables, or text (in whole or in part) from previous publications. Submit signed permissions, identified as to relevant item in manuscript (e.g., "permissions for Fig. 1 in AAC 123-93").

MANUSCRIPT FORMAT

Title, Running Title, and Byline: Numbered series titles are not permitted. Avoid main title/subtitle arrangement, complete sentences, and unnecessary articles. On title page, include title, running title (54 characters and spaces), name of each author, address(es) of institution(s) at which work was performed, each author's affiliation, and footnote with present address of any author no longer where work was performed. Place an asterisk after name of author to whom inquiries should be directed, and give telephone and fax numbers. Complete mailing address, telephone and fax numbers and e-mail address of corresponding author is printed as a footnote if desired. Place in lower left of title page and label "Correspondent Footnote."

Abstract: 250 words. Concisely summarize basic content of paper without extensive experimental details. Avoid abbreviations. Do not include diagrams. When a reference is essential, use References citation but omit article title. Conclude with a summary statement. Make complete and understandable without reference to text.

Introduction: Supply sufficient background information to allow readers to understand and evaluate results without referring to previous publications. Provide rationale for study. Chose references carefully to provide the most salient background, not an exhaustive review.

Materials and Methods: Include sufficient technical information to allow experiments to be repeated. When centrifugation conditions are critical, give enough information to enable repetition of procedure: make of centrifuge, model of rotor, temperature, time at maximum speed, and centrifugal force (x g rather than revolutions per minute). For commonly used materials and methods (e.g., media and protein determinations), a reference is sufficient. If several alternative methods are commonly used, identify method and cite reference. For example, state "cells were broken by ultrasonic treatment as previously described

(9)" not "cells were broken as previously described (9)." Allow reader to assess method without constant reference to previous publications. Describe new methods completely. Give sources of unusual chemicals, equipment, or microbial strains. When large numbers of microbial strains or mutants are used, include tables identifying sources and properties of strains, mutants, bacteriophages, plasmids, etc.

Describe a method, strain, etc., used in only one of several experiments in Results or very briefly (one or two sentences) in a table footnote or figure legend.

Results: Include rationale or design of experiments; reserve extensive interpretation of results for Discussion. Present results as concisely as possible as text, table(s), or figure(s). Avoid extensive use of graphs to present data more concisely or more quantitatively presented in text or tables. Limit photographs (particularly photomicrographs and electron micrographs) to those absolutely necessary to show experimental findings. Number figures and tables in order cited in text. Be sure all figures and tables are cited.

Discussion: Provide an interpretation of results in relation to previously published work and to experimental system at hand. Do not repeat Results or reiterate introduction. In short papers, combine Results and Discussion sections.

Acknowledgments: Acknowledge financial assistance and personal assistance in separate paragraphs. For grant support state: "This work was supported in part by Public Health Service grant CA-01234 from the National Cancer Institute." Absence of such an acknowledgment assumes that no support was received.

Appendixes: Appendixes are permitted. Do not use titles, authors, and References sections that are distinct from those of primary article. Label equations, tables, and figures with letter "A" preceding numeral to distinguish from those cited in main body of text.

References: Include all relevant sources. All listed references must be cited in text. Arrange in alphabetical order by first author. Number consecutively. Abbreviate journal names according to *Serial Sources for the BIOSIS Data Base* (BioSciences Information Service, 1992).

1. **Andrews, F. A., W. G. Beggs, and G. A. Sarosi.** 1977. Influence of antioxidants on the bioactivity of amphotericin B. Antimicrob. Agents Chemother. 11:615–619.

2. **Berry, L. J., R. N. Moore, K. J. Goodrum, and R. E. Couch, Jr.** 1977. Cellular requirements for enzyme inhibition by endotoxin in mice, p. 321–325. In D. Schlessinger (ed.), Microbiology—1977. American Society for Microbiology, Washington, D. C.

3. **Cox, C. S., B. R. Brown, and J. C. Smith.** J. Gen. Genet., in press.*

4. **Dhople, A., I. Ortega, and C. Berauer.** 1989. Effect of oxygen on in vitro growth of *Mycobacterium leprae*, abst. U-82, p. 168. Abstr. 89th Annu. Meet. Am. Soc. Microbiol. 1989.

5. **Finegold, S. M., W. E. Shepherd, and E. H. Spaulding.** 1977. Cumitech 5, Practical anaerobic bacteriol-

ogy. Coordinating ed., W. E. Shepherd. American Society for Microbiology, Washington, D. C.

6. **Fitzgerald, G., and D. Shaw.** *In* A. E. Waters (ed.), Clinical microbiology, in press. EFH Publishing Co., Boston.

7. **Gill, T. J., III.** 1976. Principles of radioimmunoassay, p. 169–171. *In* N. R. Rose and H. Friedman (ed.), Manual of clinical immunology. American Society for Microbiology, Washington, D. C.

8. **Gustlethwaite, F. P.** 1985. Letter. Lancet II:327.

9. **Jacoby, J., R. Grimm, J. Bostic, V. Dean, and G. Starke.** Submitted for publication.

10. **Jensen, C., and D. S. Schumacher.** Unpublished data.

11. **Jones, A.** (Yale University). 1990. Personal communication.

12. **Leadbetter, E. R.** 1974. Order II. *Cytophagales* nomen novum, p. 99. *In* R. E. Buchanan and N. E. Gibbons (ed.), Bergey's manual of determinative bacteriology, 8th ed. The Williams & Wilkins Co., Baltimore.

13. **Powers, R. D., W. M. Dotson, Jr., and F. G. Hayden.** 1982. Program Abstr. 22nd Intersci. Conf. Antimicrob. Agents Chemother., abstr. 448.

14. **Sacks, L. E.** 1972. Influence of intra- and extracellular cations on the germination of bacterial spores, p. 437–442. *In* H. O. Halvorson, R. Hanson, and L. L. Campbell (ed.), Spores V. American Society for Microbiology, Washington, D. C.

15. **Sigma Chemical Co.** 1989. Sigma manual. Sigma Chemical Co., St. Louis, Mo.

16. **Smith, J. C.** April 1970. U. S. patent 484,363,770.

17. **Smyth, D. R.** 1972. Ph.D. thesis. University of California, Los Angeles.

18. **Winshell, E. B., C. Cherubin, J. Winter, and H. C. Neu.** 1970. Antibiotic resistance of *Salmonella* in the eastern United States, p. 86–89. Antimicrob. Agents Chemother. 1969.

19. **Yagupsky, P., and M. A. Menegus.** 1989. Intraluminal colonization as a source of catheter-related infection. Antimicrob. Agents Chemother. 33:2025. (Letter.)

Illustrations and Tables: Write figure number and authors' names either in margin or on back (marked lightly with a soft pencil). For micrographs especially, indicate top. Do not use paper clips. Insert small figures in an envelope Do not submit illustrations larger than 8.5 × 11 in.

Continuous-Tone and Composite Photographs—Journal page width is 3.31 in. for a single column and 6.88 in. for a double column. Include only significant portion of an illustration. Photos must be of sufficient contrast to withstand loss of contrast and detail in printing. Submit one photograph of each continuous-tone figure for each copy of manuscript; no photocopies. If possible, submit figures full publication size so no reduction is needed. If reduction is needed, make sure all elements, including labeling, will remain legible.

If a figure is a composite of a continuous-tone photograph and a drawing or labeling, provide original composite for printer.

Electron and light micrographs must be direct copies of original negative. Indicate magnification with a scale marker on each micrograph.

Computer-generated Images—Produce computer-generated images with Adobe Photoshop or Aldus Freehand. For Aldus, one- and two-column art cannot exceed 20 picas or 41.5 picas. Use Helvetica (medium or bold) or Times Roman fonts. Adobe users, if online image density is below 1.25, enter as 1.4. If density is between 1.25 and 1.6, enter as 1.65. Enter all actual density readings above 1.65. Supply files with accepted manuscript. Include description of hardware/software used in figure legend.

Color Photographs—Color photographs are discouraged. If necessary, include an extra copy for a cost estimate for printing. Author(s) will bear cost of printing.

Drawings—Submit graphs, charts, complicated chemical or mathematical formulas, diagrams, and other drawings as glossy photographs made from finished drawings not requiring additional artwork or typesetting. Computer-generated graphics produced on high-quality laser printers are usually acceptable. Do not handwrite. Label both axes. Most graphs will be reduced to one-column width (3 5/16 in.); all elements must be large enough to withstand reduction. Avoid heavy letters and unusual symbols.

In figure ordinate and abscissa scales (table column headings), avoid ambiguous use of numbers with exponents. Use International System of Units (μ for 10^{-6}, m for 10^{-3}, k for 10^3, M for 10^6, etc.). See International Union of Pure and Applied Chemistry (IUPAC) "Manual of Symbols and Terminology for Physicochemical Quantities and Units" (*Pure Appl. Chem.* 21:3–44, 1970).

When powers of 10 must be used, show exponent power associated with number; e.g., 20,000 cells per ml, the numeral on the ordinate would be "2" and the label would be "10^4 cells per ml" (not "cells per ml $\times 10^{-4}$") and show enzyme activity of 0.06 U/ml as 6, with the label "10^{-2} U/ml." Preferred designation is "60 mU/ml" (milliunits per milliliter).

Figure Legends: Provide enough information so figure is understandable without reference to text. Describe detailed experimental methods in Materials and Methods section, not in figure legend. Report methods unique to one of several experiments in legend in one or two sentences. Define all symbols and abbreviations in figure not defined elsewhere.

Tables: Type each on a separate page. Arrange data so columns of like material read down, not across. Make headings sufficiently clear so data will be understood without reference to text. See Abbreviations section for those to be used in tables. Explanatory footnotes are acceptable, but table "legends" are not. Footnotes should not include detailed descriptions of experiment.

Avoid tables (or figures) of raw data on drug susceptibility, therapeutic activity, or toxicity. Analyze such data by an approved procedure, and present results in tabular form.

OTHER FORMATS

LETTERS TO THE EDITOR—500 words typed double-spaced. Include data to support argument. Comment only on articles published previously in journal.

MINIREVIEWS—6 printed pages of developments in fast-moving areas of chemotherapy. Base on published articles; not unpublished data. Submit three double-spaced copies.

NOTES—1000 words, 50 word abstract; keep tables and figures to a minimum. Describe materials and methods in text, not in figure legends or table footnotes. Do not use section headings; report methods, results, and discussion in one section. Use paragraph lead-ins. Do not use heading acknowledgments but present as in full papers. Use Reference format as stated above. Present brief observations that do not warrant full-length papers.

OTHER

Abbreviations: Use as an aid to reader, not as a convenience. Follow abbreviations in IUPAC-IUB (*Biochemical Nomenclature and Related Documents*, 1978). Use others only when a case can be made for necessity, such as in tables and figures.

Use pronouns or paraphrase a long word after its first use (e.g., "the drug," "the substrate"). Use standard chemical symbols and trivial names or their symbols (folate, Ala, Leu, etc.) for terms that appear in full in neighboring text.

Introduce all abbreviations except those listed below in first paragraph in Materials and Methods or define each abbreviation and introduce it in parentheses at first use. Eliminate abbreviations not used at least five times, including tables and figure legends.

In addition to abbreviations for Système International d'Unitès (SI) units of measurement, other common units (e.g., bp, kb, and Da), and chemical symbols for elements, use these without definition in title, abstract, text, figure legends, and tables: DNA; cDNA; RNA; cRNA; RNase; DNase; rRNA; mRNA; tRNA; AMP, ADP, ATP, dAMP, ddATP, GTP, etc. (add 2´-, 3´-, or 5´- when needed for contrast); ATPase, dGTPase; NAD; NAD⁺; NADH; NADP; NADPH; NADP⁺; poly(A), poly(dT), etc.; oligo(dT), etc.; P_i; PP_i; UV; PFU; CFU; MIC; MBC; Tris; DEAE; A_{260}; EDTA; PCR and AIDS. Abbreviations for cell lines (e.g., HeLa) need not be defined.

Abbreviations used without definition in tables:

amt (amount)	SD (standard deviation)
approx (approximately	SE (standard error)
avg (average)	SEM (standard error of the mean
concn (concentration)	sp act (specific activity)
diam (diameter)	sp gr (specific gravity)
expt (experiment)	temp (temperature)
exptl (experimental)	tr (trace)
ht (height)	vol (volume)
mo (month	vs (versus)
mol wt (molecular weight)	wk (week)
no. (number)	wt (weight)
prepn (preparation)	yr (year)

Pharmacokinetic Parameters—Introduce abbreviations and symbols at first occurrence in text.; most commonly used: α (or α phase), distribution phase; β (or β phase), elimination phase; A, zero-time intercept for α phase: B, zero-time intercept for β phase; AUC, area under the concentration-time curve; AUMC, area under the first moment of the concentration-time curve: AUC_{0-24}, $AUC0_{-\infty}$, etc., area under the concentration-time curve from 0 to 24 h, 0 h to ∞, etc.; CL, clearance; CL_R, renal clearance; CL_{NR}, nonrenal clearance; CL_{CR}, creatinine clearance; C_{max}, maximum concentration of drug in serum; T_{max}, time to maximum concentration of drug in serum; V_{max}, maximum rate of metabolism; X_u^{1-2}, drug concentration in urine between t_1 and t_2; V, volume of distribution; V_{ss}, volume of distribution at steady state; V_1, volume of distribution of the central compartment; k_{el}, elimination rate constant; k_{ss}, residence rate constant at steady state; $t_{1/2}$, half-life; $t_{1/2a}$, half life at α phase; $t_{1/2b}$, half-life at β phase. See also M. Rowland and G. Tucker (*J. Pharmokinet. Biopharm.* **8**:497–507,1980).

β-Lactamase Assays: Follow Bush and Sykes (*Antimicrob. Agents Chemother.* 30:6–10, 1986). Assays that measure hydrolysis of β-lactam antibiotics must be appropriate for substrate examined. Show reproducibility of results. When referring to β-lactamases, use Bush group designations (*Antimicrob. Agents Chemother.* 33:259–276, 1989).

Clinical Trials: State methods used to find and enroll patients and criteria for enrollment. Indicate, if appropriate, that written informed consent was obtained and that trial was approved by pertinent committee on human subjects.

Randomized, double-blind studies are preferred. Comparisons using historical controls are questionable unless differences in outcome between groups are dramatic and almost certainly the result of new intervention. Explain rationale for choice of control group. Justify sample size. State method of randomization.

State minimum criteria for evaluability. Criteria for evaluability are different from those for enrollment.

State number of patients in each group who were excluded from evaluation and reason(s) for each exclusion.

Define each outcome for each category of assessment (e.g., clinical: cure, improvement, and failure; microbiological: eradication, persistence, and relapse). State frequency and timing of such assessments in relation to treatment. Specify changes made in study regimen(s) during trial; state results for regimens with and without such modification separately. State criteria (questionnaires, results of specific laboratory tests) for evaluation of adverse effects and period encompassed in assessment and time of assessment in relation to time of treatment. State number of superinfections with each regimen and differentiate between superinfections and colonization. Mention duration of follow-up.

State type of statistical test and reason for choice of test. Give references for statistical procedures other than t test, chi-square test, and Wilcoxon rank sum test. Statistically evaluate comparability of treatment groups at baseline.

For trials which show no statistically significant difference between regimens, calculate probability (β) of a type II error and the power of the study ($1 - \beta$) to detect a specified clinically meaningful difference in efficacy between the regimens. See J. A. Freiman, T. C. Chalmers, H. Smith, Jr., and R. Kuebler (*N. Engl. J. Med.* 299:690–694, 1978). Indicate magnitude of difference between regimens that could have been detected at a statistically significant level with number of evaluable patients studied.

See editorial on guidelines for clinical trials (*Antimicrob. Agents Chemother.* 33:1829–1830, 1989).

Drugs and Pharmaceutical Agents: Do not use "nonstandard" abbreviations to designate names of antibiotics and other pharmaceutical agents. If used: define in abbreviation paragraph in Materials and Methods or at first use in text; be clear and unambiguous in meaning; and contribute to ease of assimilation by readers.

Use chemical or generic names of drugs, not trade names. When code names or corporate proprietary numbers must be used, provide either chemical structure of compound or a published literature reference illustrating structure. For compounds not identified by generic nomenclature, list all previous or concurrent identification numbers or appellations.

Experimentation, Human: Patient Identification—Do not identify isolates derived from patients in clinical studies using patients' initials, even as part of a strain designation. Change initials to numerals or use randomly chosen letters. Do not give hospital unit numbers; if a designation is needed, use only last two digits of unit. (Note: established designations of some viruses and cell lines, although they consist of initials, are acceptable [e.g., JC virus, BK virus, HeLa cells].)

In Vitro **Susceptibility Tests:** Tabulate results of determinations of minimal inhibitory and bactericidal concentrations according to range of concentrations of each antimicrobial agent required to inhibit or kill members of a species or of each group of microorganisms tested, as well as corresponding concentrations required to inhibit or kill 50 and 90% of strains (MIC_{50} and MIC_{90}, respectively). MIC_{50} and MIC_{90} values should be actual concentrations tested that inhibited 50% and 90%, respectively, of strains. They should not be values calculated from actual data obtained. When only six to nine isolates of a species are tested, tabulate only MIC range of each antimicrobial agent tested.

If more than a single drug is studied, insert a "Test agent" column between present columns and record data for each agent in same isolate order. Cumulative displays of MICs or MBCs in tables or figures not acceptable.

Perform bactericidal tests with a sufficient inoculum ($> 5 \times 10^5$ CFU/ml) and subculture volume (0.01 ml) to ensure accurate determination of the 99.9% killing endpoint (R. D. Pearson, R. T. Steigbigel, H. T. Davis, and S. W. Chapman, *Antimicrob. Agents Chemother.* 18:699–708, 1980 and P. C. Taylor, F. D. Schoenknecht, J. C. Sherris, and E. C. Linner, *Antimicrob. Agents Chemother.* 23:142–150, 1983). Inoculum size and subculture volume are critical to studies of combinations of antimicrobial agents. Synergy is defined in two-dimensional or checkerboard tests when fractional inhibitory concentration or fractional bactericidal concentration index is ≤ 0.5. In killing curves, synergy is defined as a ≥ 2-log_{10} decrease in CFU/ml between

the combination and its most active constituent after 24 h, and number of surviving organisms in the presence of the combination must be ≥ 2-log_{10} CFU/ml below starting inoculum. At least one drug must be present in a concentration which does not affect growth curve of test organism when used alone. Antagonism is defined by a ΣFIC or $\Sigma FBC > 4.0$.

In killing curve tests, state minimal, accurately countable number of CFU per milliliter and describe method used for determining number. In absence of any drug and with a sample size of 1 ml, this number is 30 (1.5 in log_{10}) CFU. If procedures for drug inactivation or removal have not been performed, state how drug carryover effects were eliminated or quantified. For drugs showing an inoculum effect, dilution below MIC obtained in standard tests is not sufficient.

Isotopically Labeled Compound: For simple molecules, indicate labeling in chemical formula (e.g., $^{14}CO_2$, ^3H_2O, $H_2^{35}SO_4$). Brackets are not used when isotopic symbol is attached to the name of a compound that in its natural state does not contain the element or to a word that is not a specific chemical name.

For specific chemicals, place isotope symbol in square brackets directly preceding the part of name that describes labeled entity. Configuration symbols and modifiers precede isotopic symbol.

Materials Sharing: Deposit strains in publicly accessible culture collections and refer to collections and strain numbers in text. Indicate laboratory strain designations and donor source as well as original culture collection identification numbers. Make available plasmids, viruses, and living materials newly described in article from a national collection or other way in a timely fashion and at reasonable cost to members of the scientific community for noncommercial purposes.

Nomenclature, Chemical and Biochemical: Use *Chemical Abstracts* (Chemical Abstracts Service, Ohio State University, Columbus) and its indexes and or *Merck Index* (11th ed., 1989; Merck & Co., Inc., Rahway, N.J.). For biochemical terminology, use *Biochemical Nomenclature and Related Documents* (1978; The Biochemical Society, London) and instructions to authors of *Journal of Biological Chemistry* and *Archives of Biochemistry and Biophysics* (first issues of each year).

Molecular weight is a unitless ratio. Express molecular mass in daltons.

For enzymes, use recommended (trivial) name assigned by Nomenclature Committee of the International Union of Biochemistry (*Enzyme Nomenclature,* Academic Press, Inc., 1992). If a nonrecommended name is used, place proper (trivial) name in parentheses at first use in abstract and text. Use EC number and express enzyme activity either in katals (preferred) or in micromoles per minute.

Nomenclature, Genetic:

Bacteria—Genetic properties of bacteria are described in terms of phenotypes and genotypes. Follow Demerec et al. (*Genetics* 4:61–76, 1966) for use of terms.

Use phenotype designations when mutant loci have not been identified or mapped. Also use to identify protein products of genes. Phenotype designations consist of three-letter symbols; do not italicize and capitalize first letter. Use Roman or Arabic numerals to identify a series of related phenotypes. Designate wild-type characteristics with a superscript plus (Pol$^+$) and, when necessary for clarity, use negative superscripts (Pol$^-$) to designate mutant characteristics. Use lowercase superscript letters to further delineate phenotypes, Define phenotype designations.

Indicate genotype designations by three-letter locus symbols. Use lowercase italic. If several loci govern related functions, distinguish by italicized capital letters following locus symbol. Indicate promoter, terminator, and operator sites as described by Bachmann and Low (*Microbiol. Rev.* 44:1–56, 1980).

Indicate wild-type alleles with a superscript plus (*ara*$^+$ *his*$^+$). Do not use superscript minus to indicate a mutant locus; thus, it is an *ara* mutant, not an *ara*$^-$ strain.

Designate mutation sites by placing serial isolation numbers (allele numbers) after locus symbol. If only a single such locus exists or if it is not known in which of several related loci mutation has occurred, use a hyphen instead of capital letter. Give allele numbers to mutations. For *Escherichia coli*, there is a registry of such numbers: *E. coli* Genetic Stock Center, Department of Biology, Yale University, New Haven, CT 06511-5188. For *Salmonella*: *Salmonella* Genetic Stock Center, Department of Biology, University of Calgary, Calgary, Alberta, T2N 1N4 Canada. For *Bacillus*: *Bacillus* Genetic Stock Center, Ohio State University, Columbus. A registry of allele numbers and insertion elements (omega [Ω] numbers) for chromosomal mutations and chromosomal insertions of transposons and other insertion elements has been established in conjunction with the ISP collection of *Staphylococcus aureus* at Iowa State University. Blocks of allele numbers and Ω numbers are assigned to laboratories on request. Obtain requests for blocks of numbers and additional information from Peter A. Pattee, Department of Microbiology, Iowa State University, Ames, IA 50011. A registry of plasmid designations is maintained by the Plasmid Reference Center, Department of Microbiology and Immunology, 5402, Stanford University School of Medicine, Stanford, CA 94305–2499.

Avoid use of superscripts with genotypes (other than + to indicate wild-type alleles). Designations indicating amber mutations (am), temperature-sensitive mutations (Ts), constitutive mutations (Con), cold-sensitive mutations (Cs), production of a hybrid protein (Hyb), and other important phenotypic properties should follow allele number. Define all other such designations of phenotype at first occurrence. If superscripts must be used, they must be approved by editor and defined at first occurrence.

Use subscripts to distinguish between genes (with same name) from different organism or strains. Use abbreviations with explanations. Use subscripts to distinguish between genetic elements with the same name. For example, promoters of the *gln* operon can be designated *glnAp*$_1$ and *glnAp*$_2$. This form departs slightly from that recommended by Bachmann and Low (e.g., *desclp*).

Indicate deletions by symbol Δ placed before deleted gene or region. Other symbols can be used with definitions.

Show a fusion of *ara* and *lac* operons as Φ(*ara-lac*)95 and Φ(*araB′-lacZ*$^+$)96 indicating that t fusion results in a truncated *araB* gene fused to an intact *lacZ* and Φ(*malE-lacZ*)97(Hyb) shows that a hybrid protein is synthesized. An inversion is shown as IN (*rrnD-rrnE*)1. An insertion of an *E. coli his* gene into plasmid pSC101 at zero kilobases (0 kb) is shown as pSC101 Ω(0kb::K-12*hisB*)4. An alternative designation of an insertion can be used in simple cases, e.g., *galT236*::Tn5. The number 236 refers to the locus of the insertion, and if the strain carries and additional *gal* mutation, it is listed separately. Additional examples, which utilize a slightly different format, can be found in the papers by Campbell et al. and Novick et al. cited below. It is important in reporting the construction of strains in which a mobile element was inserted and subsequently deleted that this latter fact be noted in the strain table. This can be done by listing the genotype of the strain used as an intermediate, in a table footnote, or by a direct or parenthetical remark in the genotype, e.g., (F-), ΔMu *cts*, *mal*::ΔMu *cts*::*lac*. In setting parenthetical remarks within genotype or dividing genotype into constituent elements, parentheses and square brackets are used without special meaning; square brackets are used outside parentheses. To indicate presence of an episome, parentheses (or brackets) are used (λ, F$^+$). Reference to an integrated episome is indicated as described above for inserted elements, and an exogenote is shown as, for example, W3110/F'8(*gal*$^+$).

Explain deviations from standard genetic nomenclature in Materials and Methods or in a table of strains.

For information about genetic maps of locus symbols in current use, consult Bachmann (in J. L. Ingraham, K. B. Low, B. Magasanik, M. Schaechter, and H. E. Umbarger, ed., *Escherichia coli* and *Salmonella typhimurium: Cellular and Molecular Biology*, 1987, American Society for Microbiology, Washington, D. C. p. 807-876) for *E. coli* K-12, Sanderson and Roth (*Microbiol. Rev.* 52:485-532, 1988) for *Salmonella typhimurium*, Holloway et al. (*Microbiol. Rev.* 43:73–102, 1979) for *Pseudomonas*, Piggot and Hoch (*Microbiol. Rev.* 49:158–179, 1985) for *Bacillus subtilis*, Perkins et al. (*Microbiol. Rev.* 46:426–570, 1982) for *Neurospora crassa*, and Mortimer and Schild (*Microbiol. Rev.* 49:181–213, 1985) for *Saccharomyces cerevisiae*. For yeasts, *Chlamydomonas*, and several fungal species, use symbols such as those given in the *Handbook of Microbiology* (A. I. Laskin and H. A. Lechevalier, ed., CRC Press, Inc., 1974).

Conventions for Naming Genes—Name new genes whose function is yet to be established by: (i) When applicable, give new gene the same name as a homologous gene already identified in another organism. (ii) Give gene a provisional name based on its map location in the style yaaA, analogous to the style used for recording transposon insertions (zef). (iii) A provisional name may be given in the style described by Demerec et al. (e.g., *usg*, for gene upstream of *folC*).

Strain Designations—Do not use a genotype as a name. If a strain designation has not been chosen, select an appropriate word combination (e.g., "another strain containing the leuC6 mutation").

Viruses—Use superscripts to indicate hybrid genomes. Genetic symbols may be one, two, or three letters. Delineate host DNA insertions into viruses by square brackets, and genetic symbols and designations for such inserted DNA should conform to those used for host genome. Genetic symbols for phage 1 can be found in Szyblaski and Szybalski (*Gene* 7:217–270, 1979) and in Echols and Murialdo (*Microbiol. Rev.* 42:577–591, 1978).

For transposable elements, plasmids, and restriction enzymes, follow Campbell et al. (*Gene* 5:197–206, 1979). The system of designating transposon insertions at sites where there are no known loci, e.g., *zef-123*::Tn5, has been described by Chumley et al. (*Genetics* 91:639–655, 1979). Use nomenclature recommendations of Novick et al. (*Bacteriol. Rev.* 40: 168–189, 1976) for plasmids and plasmid-specified activities, of Low (*Bacteriol. Rev.* 36:587–607, 1972) for F-prime factors, and of Roberts (*Nucleic Acids Res.* F-prime factors, and of Roberts (*Nucleic Acids Res.* 17:r347-r387, 1989) for restriction enzymes and their isoschizomers.

For recombinant DNA molecules, constructed in vitro, follow nomenclature for insertions in general. Describe DNA inserted into recombinant DNA molecules by using gene symbols and conventions for organism from which DNA was obtained. The Plasmid Reference Center (E. Lederberg, Plasmid Reference Center, Department of Microbiology and Immunology, 5402, Stanford University School of Medicine, Stanford, CA 94305-2499) assigns Tn and IS numbers to avoid conflicting and repetitive use and also clears nonconflicting plasmid prefix designations.

Nomenclature, Microorganisms: Use binary names (generic name and specific epithet). Names of higher categories may be used alone, specific and subspecific epithets may not. Precede specific epithets with a generic name at first use. Thereafter, abbreviate generic name to initial capital letter, if no confusion with other genera in paper. Underline or italicize names of all taxa (phyla, classes, orders, families, genera, species, subspecies).

For spelling of names follow *Approved Lists of Bacterial Names* (amended edition) (V. B. D. Skerman, V. McGowan, and P. H. A. Sneath, ed.) and *Index of the Bacterial and Yeast Nomenclatural Changes Published in the International Journal of Systematic Bacteriology since the 1980 Approved Lists of Bacterial Names* (*1 January 1980 to 1 January 1989*) (W. E. C. Moore and L. V. H. Moore, ed., American Society for Microbiology, 1989) and validation lists and articles published in *International Journal of Systematic Bacteriology* since 1 January 1989. If there is reason to use a name that does not have standing in nomenclature, enclose name in quotation marks and make appropriate statement about its nomenclatural status in text (for an example, see *Int. J. Syst. Bacteriol.* 30:547–556, 1980).

Since fungi classification is not complete, author(s) must determine accepted binomial for organism. Sources for names include: *The Yeasts: A Taxonomic Study*, 3rd ed. (N. J. W. Kreger-van Rij, ed., Elsevier Science Publishers B. V., Amsterdam, 1984) and *Ainsworth and Bisby's Dictionary of the Fungi, Including the Lichens*, 7th ed. (Commonwealth Mycological Institute, Kew, Surrey, England, 1983).

Use name for viruses approved by International Committee on Taxonomy of Viruses published in 4th Report of the ICTV, Classification and Nomenclature of Viruses (*Intervirology* 17:23–199, 1982), with modifications in 5th Report of the ICTV (*Arch. Virol.*, Suppl. 2, 1991). Add synonyms parenthetically when name is first mentioned. Approved generic (or group) and family names may also be used.

Give microorganisms, viruses and plasmids designations consisting of letters and serial numbers. Include a worker's initials or a descriptive symbol of locale, laboratory, etc., in designation. Give each new strain, mutant, isolate, or derivative a new (serial) designation distinct from those of the genotype and phenotype. Do not include genotypic and phenotypic symbols.

Nucleotide Sequences: Include GenBank/EMBL accession numbers for primary nucleotide and/or amino acid sequence data in original manuscript. Include accession number as a separate paragraph at end of Materials and Methods section for full-length papers or at end of text of Notes.

GenBank Submissions, National Center for Biotechnology Information, Bldg. 381, Rm 8N-803, 8600 Rockville Pike, Bethesda, MD 20894; e-mail (new submissions): gbsub@ncbi.nlm.nih.gov; e-mail (updates): update@ncbi.nlm.nih.gov.

EMBL Data Library Submissions, Postfach 10.2209, Meyerhofstrasse 1, 69012 Heidelberg, Germany; telephone: 011 49 (6221) 387258; fax: 011 49 (6221) 387306; electronic mail (data submissions): datasubs@embl.bitnet.

Nucleic Acid Sequences: Present limited length sequences freestyle in the most effective format. For longer nucleic acid sequences: Submit as camera-ready copy 8 1/2 × 1 in. (or slightly less) in portrait orientation. Print sequence in lines of 100 bases, each in a nonproportional (monospace) font, easily legible when published at 100 bases/6 in. Use uppercase and lowercase letters to designate exon/intron structure, transcribed regions, etc., if lowercase letters remain legible at 100 bases/6 in. Number sequence line by line; place numerals, representing first base of each line, to left of lines. Minimize spacing between adjacent lines of sequence, leaving room only for annotation of sequence. Annotation may include boldface, underlining, brackets, boxes, etc. Encoded amino acid sequences may be presented, if necessary, immediately above first nucleotide of each codon, using single-letter amino acid symbols. Comparisons of multiple nucleic acid sequences should conform to the same format.

Page Charges: $40 per printed page, if research was supported by special funds, grants (departmental, governmental, institutional, etc.), or contracts or if research was done as part of official duties. If research was not so supported, submit a request to waive charges to Journals Division, American Society for Microbiology, with manuscript. Request, separate from cover letter, must indicate how work was supported. Include a copy of Acknowledgment section.

Reporting Numerical Data: Use standard metric units for length, weight, and volume. For these units and molarity, use prefixes m, μ, n, and p for 10^{-3}, 10^{-6}, 10^{-9}, and 10^{-12}, respectively. Use prefix k for 10^3. Avoid compound prefixes such as mμ or μμ. Use μg/ml or μg/g in place of mg/liter or mg/kg or t ppm. Express temperature as: 37°C or 324 K.

When fractions are used to express units such as enzymatic activities, use whole units in denominator instead of fractional or multiple unit. Use unambiguous forms, such as exponential notation.

See *CBE Style Manual*, 5th ed., for detailed information about reporting numbers and appropriate SI units for reporting of illumination, energy, frequency, pressure, and other physical terms.

Reviewers: Suggest an appropriate editor. If another editor is assigned, corresponding author will be notified. Recommend two or three reviewers (not members of their institution(s) and never associated with them or their laboratory(ies). Provide names, addresses, phone and fax numbers, and areas of expertise.

Sensitivity and Susceptibility to Drugs: Use "sensitivity" in contexts that concern mechanisms of drug action or resistance. Use "susceptibility" in contexts that concern gross drug-organism interactions, such as death or inhibition of growth.

Style: Follow *CBE Style Manual* (5th ed., 1983; Council of Biology Editors, Inc., *ASM Style Manual for Journals and Books* (American Society for Microbiology, 1991), and Robert A. Day's *How to Write and Publish a Scientific Paper* (3rd ed., 1988; Oryx Press).

Use past tense to narrate particular events in the past, including procedures, observations, and study data. Use present tense for your conclusions, conclusions of previous researchers, and generally accepted facts. It may be necessary to vary tense in a single sentence.

SOURCE

January 1994; 38(1): i-xiii.

Applied and Environmental Microbiology
American Society for Microbiology

Journals Division, American Society for Microbiology
1325 Massachusetts Ave., N.W.
Washington, D.C. 20005-4171

AIM AND SCOPE

Applied and Environmental Microbiology (AEM) publishes descriptions of all aspects of applied research as well as research of a genetic and molecular nature that focuses on topics of practical value and basic research on microbial ecology. Topics include microbiology in relation to foods, agriculture, industry, biotechnology, and public health and basic biological properties of bacteria, fungi, protozoa, and other simple eucaryotic organisms as related to microbial ecology.

ASM publishes a number of journals covering various aspects of microbiology. AEM considers manuscripts describing properties of enzymes and proteins that are produced by either wild-type or genetically engineered microorganisms and that are significant or have potential significance in industrial or environmental settings. Studies dealing with basic biological phenomena of enzymes or proteins or in which enzymes have been used in investigations of basic biological functions are appropriate for *Journal of Bacteriology*. AEM considers papers describing the use of antimicrobial or anticancer agents as tools for elucidating aspects of applied and environmental microbiology. Papers on antimicrobial or anticancer agents, including manuscripts dealing with biosynthesis and metabolism of such agents, are appropriate for *Antimicrobial Agents and Chemotherapy*. Papers on biology of bacteriophages and other viruses are appropriate for *Journal of Virology* or *Journal of Bacteriology*. AEM considers manuscripts dealing with viruses in relation to environmental, public health, or industrial microbiology. Manuscripts dealing with the immune system or with topics of basic medical interest or oral microbiology are appropriate for *Infection and Immunity*. Submit reports of clinical investigations and environmental biology applied to hospitals to *Journal of Clinical Microbiology*. Submit papers that include mainly taxonomic material (e.g., descriptions of new taxa) to *International Journal of Systematic Bacteriology*. AEM will not consider reports that emphasize nucleotide sequence data alone (without experimental documentation of the functional and evolutionary significance of the sequence).

Manuscripts rejected by one ASM journal on scientific grounds or on the basis of its general suitability for publication are considered rejected by all other ASM journals.

MANUSCRIPT SUBMISSION

In cover letters state: journal to which manuscript is being submitted, most appropriate journal section, complete mailing address (including street), telephone and fax numbers of corresponding author, BITNET or other electronic mail address if available. Include written assurance of permission to cite personal communications and preprints.

Submit three complete copies, including figures and tables. Type every portion double-spaced (6 mm between lines), including figure legends, table footnotes, and References. Number all pages in sequence, including abstract, figure legends, and tables. Place last two items after References section. Keep 1 in. margins on all four sides. Use line numbers if possible. Make these characters easily distinguishable: numeral zero (0) and letter "oh" (O); numeral one (1), letter "el" (l), and letter "eye" (I); and multiplication sign (×) and letter "ex" (x). If cannot, mark items at first occurrence for cell lines, strain and genetic designations, viruses, etc., on modified manuscript to be identified by copy editor.

Enclose three copies of each "in press" and "submitted" manuscript.

Present only reports of original research. All authors must agree to submission and are responsible for content, including appropriate citations and acknowledgments, and must agree to the decisions of corresponding author. Submission is a guarantee that manuscript, or one substantially the same, was not published previously, is not being considered or published elsewhere, and was not rejected on scientific grounds by another ASM journal.

Follow definition of primary publication (*How to Write and Publish a Scientific Paper*, 3rd ed., Robert A. Day): "the first publication of original research results, in a form whereby peers of the author can repeat the experiments and test the conclusion, and in a journal or other source document readily available within the scientific community."

An author is one who made a substantial contribution to the "overall design and execution of the experiments". Acknowledge individuals who provided assistance, e.g., supplied strains or reagents or critiqued the paper, in the Acknowledgment section.

Obtain permissions from both original publisher and original author [i.e., copyright owner(s)] to reproduce figures, tables, or text (in whole or in part) from previous publications. Identify as to relevant item in manuscript (e.g., "permission for Fig. 1).

Do not submit work published in conference reports, symposium proceedings, technical bulletins, or any other retrievable source. Preliminary disclosures of research findings published in abstract form as an adjunct to a meeting is not considered prior publication.

Acknowledge prior publication of data in manuscript even though author(s) do not consider it in violation of ASM policy. Submit copies of relevant work.

Submit three copies of "in press" and "submitted" manuscripts that are important for judgment of the present manuscript.

MANUSCRIPT FORMAT

Title, Running Title, and Byline: Numbered series titles are not permitted. Avoid main title/subtitle arrangement, complete sentences, and unnecessary articles. On title page, include title, running title (54 characters and spaces), name of each author, address(es) of institution(s) at which work was performed, each author's affiliation, and footnote with present address of any author no longer where work was performed. Place an asterisk after name of author to whom inquiries should be directed, and give telephone and fax numbers. Complete mailing address, telephone and fax numbers and e-mail address of corresponding author is printed as a footnote if desired. Place in lower left of title page and label "Correspondent Footnote."

Abstract: 250 words. Concisely summarize basic content of paper without extensive experimental details. Avoid abbreviations. Do not include diagrams. When a reference is essential, use References citation but omit article title. Conclude with a summary statement. Make complete and understandable without reference to text.

Introduction: Supply sufficient background information to allow readers to understand and evaluate results without referring to previous publications. Provide rationale for study. Chose references carefully to provide the most salient background, not an exhaustive review.

Materials and Methods: Include sufficient technical information to allow experiments to be repeated. When centrifugation conditions are critical, give enough information to enable repetition of procedure: make of centrifuge, model of rotor, temperature, time at maximum speed, and centrifugal force (x g rather than revolutions per minute).

For commonly used materials and methods (e.g., media and protein determinations), a reference is sufficient. If several alternative methods are commonly used, identify method and cite reference. For example, state "cells were broken by ultrasonic treatment as previously described (9)" not "cells were broken as previously described (9)." Allow reader to assess method without constant reference to previous publications. Describe new methods completely. Give sources of unusual chemicals, equipment, or microbial strains. When large numbers of microbial strains or mutants are used, include tables identifying sources and properties of strains, mutants, bacteriophages, plasmids, etc.

Describe a method, strain, etc., used in only one of several experiments in Results section or very briefly (one or two sentences) in a table footnote or figure legend.

Results: Include rationale or design of experiments; reserve extensive interpretation of results for Discussion. Present results as concisely as possible as text, table(s), or figure(s). Avoid extensive use of graphs to present data more concisely or more quantitatively presented in text or tables. Limit photographs (particularly photomicrographs and electron micrographs) to those absolutely necessary to show experimental findings. Number figures and tables in order cited in text. Be sure all figures and tables are cited.

Discussion: Provide an interpretation of results in relation to previously published work and to experimental system at hand. Do not repeat Results or reiterate introduction. In short papers, combine Results and Discussion sections.

Acknowledgments: Acknowledge financial assistance and of personal assistance in separate paragraphs. For grant support state: "This work was supported in part by Public Health Service grant CA-01234 from the National Cancer Institute." Absence of such an acknowledgment assumes that no support was received.

Appendixes: Appendixes are permitted. Do not use titles, authors, and References sections that are distinct from those of primary article. Label equations, tables, and figures with letter "A" preceding numeral to distinguish from those cited in main body of text.

References: Include all relevant sources. All listed references must be cited in text. Arrange in alphabetical order, by first author. Number consecutively. Abbreviate journal names according to *Serial Sources for the BIOSIS Data Base* (BioSciences Information Service, 1992). Cite each listed reference by number in the text.

1. **Armstrong, J. E., and J. A. Calder.** 1978. Inhibition of light-induced pH increase and O_2 evolution of marine microalgae by water-soluble components of crude and refined oils. Appl. Environ. Microbiol. **35:**858–862.

2. **Berry, L. J., R. N. Moore, K. J. Goodrum, and R. E. Couch, Jr.** 1977. Cellular requirements for enzyme inhibition by endotoxin in mice, p. 321–325. *In* D. Schlessinger (ed.), Microbiology—1977. American Society for Microbiology, Washington, D.C.

3. **Cox, C. S., B. R. Brown, and J. C. Smith.** J. Gen. Genet., in press.*

4. **Dhople, A., I. Ortega, and C. Berauer.** 1989. Effect of oxygen on in vitro growth of *Mycobacterium leprae*, abstr. U-82, p. 168. Abstr. 89th Annu. Meet. Am. Soc. Microbiol. 1989.

5. **Finegold, S. M., W. E. Shepherd, and E. H. Spaulding.** 1977. Cumitech 5, Practical anaerobic bacteriology. Coordinating ed., W. E. Shepherd. American Society for Microbiology, Washington, D.C.

6. **Fitzgerald, G., and D. Shaw.** *In* A. E. Waters (ed.), Clinical microbiology, in press. EFH Publishing Co., Boston.

7. **Gill, T. J., III.** 1976. Principles of radioimmunoassay, p. 169–171. *In* N. R. Rose and H. Friedman (ed.), Manual of clinical immunology. American Society for Microbiology, Washington, D. C.

8. **Gustlethwaite, F. P.** 1985. Letter. Lancet **ii:**327.

9. **Jacoby, J., R. Grimm, J. Bostic, V. Dean, and G. Starke.** Submitted for publication.

10. **Jensen, C., and D. S. Schumacher.** Unpublished data.

11. **Jones, A.** (Yale University). 1990. Personal communication.

12. **Leadbetter, E. R.** 1974. Order II. *Cytophagales* nomen novum, p. 99. *In* R. E. Buchanan and N. E. Gibbons (ed.), Bergey's manual of determinative bacteriology, 8th ed. The Williams & Wilkins Co., Baltimore.

13. **Powers, R. D., W., M. Dotson, Jr., and F. G. Hayden.** 1982. Program Abstr. 22nd Intersci. Conf. Antimicrob. Agents Chemother., abstr. 448.

14. **Sacks, L. E.** 1972. Influence of intra- and extracellular cations on the germination of bacterial spores, p. 437–442. *In* H. O. Halvorson, R. Hanson, and L. L. Campbell (ed.), Spores V. American Society for Microbiology, Washington, D. C.

15. **Sigma Chemical Co.** 1989. Sigma manual. Sigma Chemical Co., St. Louis, Mo.

16. **Smith, J. C.** April 1970. U. S. patent 484,363,770.

17. **Smyth, D. R.** 1972. Ph.D. thesis. University of California, Los Angeles.

18. **Yagupsky, P., and M. A. Menegus.** 1989. Intraluminal colonization as a source of catheter-related infection. Antimicrob. Agents Chemother. **33:**2025. (Letter.)

Illustrations and Tables: Write figure number and authors' names either in margin or on back (marked lightly with a soft pencil). For micrographs especially, indicate top. Do not use paper clips. Insert small figures in an envelope. Do not submit illustrations larger than 8.5 × 11 in.

Continuous-Tone and Composite Photographs—Journal page width is 3.31 in. for a single column and 6.88 in. for a double column. Include only significant portion of an illustration. Photos must be of sufficient contrast to withstand loss of contrast and detail in printing. Submit one photograph of each continuous-tone figure for each copy of manuscript; no photocopies. If possible, submit figures full publication size so no reduction is needed. If reduction is needed, make sure all elements, including labeling, will remain legible.

If a figure is a composite of a continuous-tone photograph and a drawing or labeling, provide original composite.

Electron and light micrographs must be direct copies of the original negative. Indicate magnification with a scale marker on each micrograph.

Computer-Generated Images—Produce computer-generated images with Adobe Photoshop or Aldus Freehand. For Aldus, one- and two-column art cannot exceed 20 picas or 41.5 picas. Use Helvetica (medium or bold) or Times Roman fonts. Adobe users, if online image density is below 1.25, enter as 1.4. If density is between 1.25 and 1.6, enter as 1.65. Enter all actual density readings above 1.65. Supply files with accepted manuscript. Include description of hardware/software used in figure legend.

Color Photographs—Color photographs are discouraged. If necessary, include an extra copy for a cost estimate for printing. Author(s) will bear cost of printing.

Drawings—Submit graphs, charts, complicated chemical or mathematical formulas, diagrams, and other drawings as glossy photographs made from finished drawings not requiring additional artwork or typesetting. Computer-generated graphics produced on high-quality laser printers are usually acceptable. Do not handwrite. Label both axes. Most graphs will be reduced to one-column width (3 5/16 in.); all elements must be large enough to withstand reduction. Avoid heavy letters and unusual symbols.

In figure ordinate and abscissa scales (and table column headings), avoid ambiguous use of numbers with exponents. Use the International System of Units (μ for 10^{-6}, m for 10^{-3}, k for 10^3, M for 10^6, etc.). See International Union of Pure and Applied Chemistry (IUPAC) "Manual of Symbols and Terminology for Physicochemical Quantities and Units" (*Pure Appl. Chem.* 21:3–44, 1970).

When powers of 10 must be used, show exponent power associated with number; e.g., 20,000 cells per ml, the numeral on the ordinate would be "2" and the label would be "10^4 cells per ml" (not "cells per ml × 10^{-4}") and an enzyme activity of 0.06 U/ml would be shown as 6, with the label "10^{-2} U/ml." Preferred designation is "60 mU/ml" (milliunits per milliliter).

Figure Legends: Provide enough information so figure is understandable without reference to text. Describe detailed experimental methods in Materials and Methods section, not in a figure legend. Report methods unique to one of several experiments in legend in one or two sentences. Define all symbols and abbreviations in figure not defined elsewhere.

Tables: Type each on a separate page. Arrange data so columns of like material read down, not across. Make headings sufficiently clear so data will be understandable without reference to text. See "Abbreviations" sections for those to be used in tables. Explanatory footnotes are acceptable, but table "legends" are not. Footnotes should not include detailed descriptions of experiment.

OTHER FORMATS

LETTERS TO THE EDITOR—500 words typed double-spaced. Include data to support argument. Comment only on articles published previously in journal.

MINIREVIEWS—6 printed pages of developments in fast-moving areas of chemotherapy. Base on published articles; not unpublished data. Submit three double-spaced copies.

NOTES—1000 words, 50 word abstract; keep tables and figures to a minimum. Describe materials and methods in text, not in figure legends or table footnotes. Do not use section headings; report methods, results, and discussion in one section. Use paragraph lead-ins. Do not use heading acknowledgments but present as in full papers. Use Reference format as stated above. Present brief observations that do not warrant full-length papers.

OTHER

Materials Sharing: Make available plasmids, viruses, and living materials described in article from a national collection or other way in a timely fashion and at reasonable cost to members of the scientific community for noncommercial purposes. Deposit strain in a recognized culture collection when that strain is necessary for description of a new taxon (see *Bacteriological Code*, 1990 Revision, American Society for Microbiology, 1992).

Nomenclature, Chemical and Biochemical: Follow Chemical Abstracts (Chemical Abstracts Service, Ohio State University, Columbus) and its indexes or The Merck Index (11th ed., 1989; Merck & Co., Inc., Rahway, N.J.). For biochemical terminology, use Biochemical Nomenclature and Related Documents (1978; reprinted for The Biochemical Society, London) and instructions to authors of Journal of Biological Chemistry and the Archives of Biochemistry and Biophysics (first issues of each year).

Molecular weight is a unitless ratio. Express molecular mass in daltons.

For enzymes, use trivial name assigned by the Nomenclature Committee of the International Union of Biochemistry (*Enzyme Nomenclature*, Academic Press, Inc., 1992). If a nonrecommended name is used, place the proper (trivial) name in parentheses at first use in abstract and text. Use EC number and express enzyme activity either in katals (preferred) or in micromoles per minute.

Nomenclature, Microorganisms: Use binary names (generic name and specific epithet). Names of higher categories may be used alone, specific and subspecific epithets may not. Precede specific epithets with a generic name at first use. Thereafter, abbreviate generic name to initial capital letter, if no confusion with other genera in paper. Underline or italicize names of all taxa (phyla, classes, orders, families, genera, species, subspecies).

For spelling of names follow *Approved Lists of Bacterial Names* (amended edition) (V. B. D. Skerman, V. McGowan, and P.H.A. Sneath, ed.) and *Index of the Bacterial and Yeast Nomenclatural Changes Published in the International Journal of Systematic Bacteriology since the 1980 Approved Lists of Bacterial Names* (1 January 1980 to 1 January 1989) (W. E. C. Moore and L. V. H. Moore, ed., American Society for Microbiology, 1989) and validation lists and articles published in the *International Journal of Systematic Bacteriology* since 1 January 1989. If there is reason to use a name that does not have standing in nomenclature, enclose name in quotation marks and make

appropriate statement about its nomenclatural status in text (see *Int. J. Syst. Bacteriol.* 30:547–556, 1980).

Since fungi classification is not complete, author(s) must determine accepted binomial. Sources for names include: *The Yeasts: a Taxonomic Study*, 3rd ed. (N.J.W. Kregervan Rij, ed., Elsevier Science Publishers B. V., Amsterdam, 1984) and *Ainsworth and Bisby's Dictionary of the Fungi, Including the Lichens*, 7th ed. (Commonwealth Mycological Institute, Kew, Surrey, England, 1983).

Use names for viruses approved by International Committee on Taxonomy of Viruses (ICTV) and published in the 4th Report of the ICTV, Classification and Nomenclature of Viruses (*Intervirology* 17:23–199, 1982), with modifications contained in the 5th Report of the ICTV (*Arch. Virol.*, Suppl. 2, 1991). Add synonyms may be added parenthetically. Approved generic (or group) and family names may also be used.

Give microorganisms, viruses, and plasmids designations consisting of letters and serial numbers. Include a worker's initials or a descriptive symbol of locale, laboratory, etc., in designation. Give each new strain, mutant, isolate, or derivative a new (serial) designation distinct from those of the genotype and phenotype. Do not include genotypic and phenotypic symbols.

Nomenclature, Genetic:

Bacteria—Genetic properties of bacteria are described in terms of phenotypes and genotypes. Follow Demerec et al. (*Genetics* 54:61–76, 1966) for use of terms.

Use phenotype designations when mutant loci have not been identified or mapped. Also use to identify protein products of genes. Phenotype designations consist of three-letter symbols; do not italicized and capitalize first letter. Use Roman or Arabic numerals to identify a series of related phenotypes. Designate wild-type characteristics with a superscript plus (Pol^+) and, when necessary for clarity, use negative superscripts (Pol^-) to designate mutant characteristics. Use lowercase superscript letters to further delineate phenotypes, Define phenotype designations.

Indicate genotype designations by three-letter locus symbols. Use lowercase italic. If several loci govern related functions, distinguish by italicized capital letters following locus symbol. Indicate promoter, terminator, and operator sites as described by Bachmann and Low (*Microbiol. Rev.* 44:1–56, 1980).

Indicate wild-type alleles with a superscript plus (ara^+ his^+). Do not use superscript minus to indicate a mutant locus; thus, it is an *ara* mutant, not an *ara* strain.

Designate mutation sites by placing serial isolation numbers (allele numbers) after locus symbol. If only a single such locus exists or if it is not known in which of several related loci mutation has occurred, use a hyphen instead of capital letter. Give allele numbers to mutations. For *Escherichia coli*, there is a registry of such numbers: *E. coli* Genetic Stock Center, Department of Biology, Yale University, New Haven, CT 06511-5188. For *Salmonella*: *Salmonella* Genetic Stock Center, Department of Biology, University of Calgary, Calgary, Alberta, T2N 1N4 Canada. For *Bacillus*: *Bacillus* Genetic Stock Center, Ohio State University, Columbus. A registry of allele numbers and insertion elements (omega [Ω] numbers) for chromosomal mutations and chromosomal insertions of transposons and other insertion elements has been established in conjunction with the ISP collection of *Staphylococcus aureus* at Iowa State University. Blocks of allele numbers and Ω numbers are assigned to laboratories on request. Obtain requests for blocks of numbers and additional information from Peter A. Pattee, Department of Microbiology, Iowa State University, Ames, IA 50011. A registry of plasmid designations is maintained by the Plasmid Reference Center, Department of Microbiology and Immunology, 5402, Stanford University School of Medicine, Stanford, CA 94305–2499.

Avoid use of superscripts with genotypes (other than + to indicate wild-type alleles). Designations indicating amber mutations (am), temperature-sensitive mutations (Ts), constitutive mutations (Con), cold-sensitive mutations (Cs), production of a hybrid protein (Hyb), and other important phenotypic properties follow allele number. Define all other such designations of phenotype at first occurrence. If superscripts must be used, they must be approved by editor and defined at first occurrence.

Use subscripts to distinguish between genes (with same name) from different organism or strains. Use abbreviations with explanations. Use subscripts to distinguish between genetic elements with the same name. For example, promoters of the *gln* operon can be designated $glnAp_1$ and $glnAp_2$. This form departs slightly from that recommended by Bachmann and Low (e.g., *desclp*).

Indicate deletions by symbol Δ placed before deleted gene or region. Other symbols can be used with definitions. Show a fusior of *ara* and *lac* operons $\Phi(ara\text{-}lac)95$ and $\Phi(araB'\text{-}lacZ^+)96$ indicating that t fusion results in a truncated *araB* gene fused to an intact *lacZ* and $\Phi(malE\text{-}lacZ)97$(Hyb) shows that a hybrid protein is synthesized. An inversion is shown as IN (*rrnD-rrnE*)*1*. An insertion of an *E. coli his* gene into plasmid pSC101 at zero kilobases (0 kb) is shown as pSC101 Ω(0kb::K-12*hisB*)4. An alternative designation of an insertion can be used in simple cases, e.g., *galT236*::Tn5. The number 236 refers to the locus of the insertion, and if the strain carries and additional *gal* mutation, it is listed separately. Additional examples, which utilize a slightly different format, can be found in the papers by Campbell et al. and Novick et al. cited below. It is important in reporting the construction of strains in which a mobile element was inserted and subsequently deleted that this latter fact be noted in the strain table. This can be done by listing the genotype of the strain used as an intermediate, in a table footnote, or by a direct or parenthetical remark in the genotype, e.g., (F-), ΔMu *cts*, *mal*::ΔMu *cts*::*lac*. In setting parenthetical remarks within genotype or dividing genotype into constituent elements, parentheses and square brackets are used without special meaning; square brackets are used outside parentheses. To indicate presence of an episome, parentheses (or brackets) are used (λ, F$^+$). Reference to an integrated episome is indicated as described above for inserted elements, and an exogenote is shown as, for example, W3110/F'8(*gal*$^+$).

Explain deviations from standard genetic nomenclature in Materials and Methods or in a table of strains.

For information about genetic maps of locus symbols in current use, consult Bachmann (in J. L. Ingraham, K. B. Low, B. Magasanik, M. Schaechter, and H. E. Umbarger, ed., *Escherichia coli* and *Salmonella typhimurium*: *Cellular and Molecular Biology*, 1987, American Society for Microbiology, Washington, D. C. p. 807–876) for *E. coli* K-12, Sanderson and Roth (*Microbiol. Rev.* 52:485–532, 1988) for *Salmonella typhimurium*, Holloway et al. (*Microbiol. Rev.* 43:73–102, 1979) for *Pseudomonas*, Piggot and Hoch (*Microbiol. Rev.* 49:158–179, 1985) for *Bacillus subtilis*, Perkins et al. (*Microbiol. Rev.* 46:426–570, 1982) for *Neurospora crassa*, and Mortimer and Schild (*Microbiol. Rev.* 49:181–213, 1985) for *Saccharomyces cerevisiae*. For yeasts, *Chlamydomonas*, and several fungal species, use symbols such as those given in the *Handbook of Microbiology* (A. I. Laskin and H. A. Lechevalier, ed., CRC Press, Inc., 1974) should be used.

Conventions for Naming Genes—Name new genes whose function is yet to be established: (i) Give new gene the same name as a homologous gene already identified in another organism. (ii) Give gene a provisional name based on its map location in the style *yaaA*, analogous to the style used for recording transposon insertions (*zef*). (iii) A provisional name may be given in the style described by Demerec et al. (e.g., *usg*, for gene upstream of *folC*.).

Strain Designations—Do not use a genotype as a name. If a strain designation has not been chosen, select an appropriate word combination (e.g., "another strain containing the *leuC6* mutation").

Viruses—Use superscripts to indicate hybrid genomes. Genetic symbols may be one, two, or three letters. Delineate host DNA insertions into viruses by square brackets. Genetic symbols and designations for inserted DNA should conform to those used for host genome. Genetic symbols for phage λ can be found in Szybalski and Szybalski (*Gene* 7:217–270, 1979) and in Echols and Murialdo (*Microbiol. Rev.* 42:577–591, 1978).

For transposable elements, plasmids, and restriction enzymes, follow Campbell et al. (*Gene* 5:197–206, 1979). The system of designating transposon insertions at sites where there are no known loci, e.g., *zef*-123::Tn5, has been described by Chumley et al. (*Genetics* 91:639–655, 1979). Use nomenclature recommendations of Novick et al. (*Bacteriol. Rev.* 40: 168–189, 1976) for plasmids and plasmid-specified activities, of Low (*Bacteriol. Rev.* 36:587–607, 1972) for F-prime factors, and of Roberts (*Nucleic Acids Res.* 17:r347–r387, 1989) for restriction enzymes and their isoschizomers.

For recombinant DNA molecules constructed in vitro follow nomenclature for insertions in general. Describe DNA inserted into recombinant DNA molecules by using gene symbols and conventions for the organism from which the DNA was obtained. The Plasmid Reference Center (E. Lederberg, Plasmid Reference Center, Department of Microbiology and Immunology, 5402, Stanford University School of Medicine, Stanford, CA 94305–2499) assigns Tn and IS numbers to avoid conflicting and repetitive use and also clears nonconflicting plasmid prefix designations.

Nucleotide Sequences: Include GenBank/EMBL accession numbers for primary nucleotide and/or amino acid sequence data in original manuscript. Include accession number as a separate paragraph at end of Materials and Methods section for full-length papers or at end of Notes.

GenBank Submissions, National Center for Biotechnology Information, Bldg. 381 Rm 8N-803, 8600 Rockville Pike, Bethesda, MD 20894; e-mail (new submissions): gbsub@ncbi.nlm.nih.gov; e-mail (updates): update@ncbi.nlm.nih.gov.

EMBL Data Library Submissions, Postfach 10.2209, Meyerhofstrasse 1, 69012 Heidelberg, Germany; telephone: 011 49 (6221) 387258; fax: 011 49 (6221) 387306; electronic mail (data submissions): datasubs@embl.bitnet.

Nucleic Acid Sequences: Present limited length sequences freestyle in the most effective format. Present longer nucleic acid sequences: Submit as camera-ready copy 8 1/2 × 1 in. (or slightly less) in portrait orientation. Print sequence in lines of 100 bases, each in a nonproportional (monospace) font, easily legible when published at 100 bases/6 in. Use uppercase and lowercase letters to designate exon/intron structure, transcribed regions, etc., if lowercase letters remain legible at 100 bases/6 in. Number sequence line by line; place numerals, representing first base of each line, to left of lines. Minimize spacing between adjacent lines of sequence, leaving room only for annotation of sequence. Annotation may include boldface, underlining, brackets, boxes, etc. Encoded amino acid sequences may be presented, if necessary, immediately above first nucleotide of each codon, using single-letter amino acid symbols. Comparisons of multiple nucleic acid sequences should conform to the same format.

Page Charges: $40 per printed page (subject to change), if research was supported by special funds, grants (departmental, governmental, institutional, etc.), or contracts or research was done as part of their official duties. If research was not so supported, send a request to waive charges Journals Division, American Society for Microbiology, with manuscript. Request is separate from cover letter and must indicate how work was supported. Include a copy of Acknowledgment section.

Reviewers: Suggest an appropriate editor. If another editor is assigned, corresponding author will be notified. Recommend two or three reviewers (not members of their institution(s) and never associated with them or their laboratory(ies). Provide names, addresses, phone and fax numbers, and areas of expertise.

Style: Follow *CBE Style Manual* (5th ed., 1983; Council of Biology Editors, Inc., Bethesda, MD), *ASM Style Manual for Journals and Books* (American Society for Microbiology, 1991), and Robert A. Day's *How to Write and Publish a Scientific Paper* (3rd ed., 1988; Oryx Press).

Use the past tense to narrate particular events in the past, including the procedures, observations, and study data. Use the present tense for your own general conclusions, conclusions of previous researchers, and generally accepted facts. It may be necessary to vary the tense in a single sentence.

SOURCE

January 1994, 60(1): i-xi.

Applied Microbiology and Biotechnology
Springer International

Dr. Karl Esser
Fakultät für Biologie
Ruhr-Universitat Bochum
Allgemeine Botanik
Bochum D 44780
Federal Republic of Germany

AIM AND SCOPE

Papers deal with the following aspects of applied microbiology and biotechnology excluding pure biochemical articles: Biotechnology with reference to culturing of microorganisms, animal cells, plant cells or parts of these organisms (e.g., enzymes), corrosion of materials, leaching of metals and biodeterioration; Biochemical engineering including all biotechnological processes using microorganisms or enzymes, e. g., immobilized cells and enzymes, model building, computer applications, new bioreactors and their characterization and kinetics of microbial processes (Papers containing only purification of enzymes will be excluded.); Applied genetics end regulation including all aspects of research and development relating to biotechnology, further aspects of genetic control and development of microbial strains by classical and molecular techniques with a view to biotechnological demands, strain improvement, cell fusion and recombinant DNA technology; Applied microbial and cell physiology including all fields of applied microbiology with relation to biotechnology, excluding medical microbiology (Included are production of primary and secondary metabolites, transformations by microorganisms, microbial energy production, conversion of substrates, degradation of substances and control of microbial processes); Food biotechnology including all papers containing real biotechnological results on food and feed processing; Environmental biotechnology covering microorganisms and their physiology in processes such as sewage purification by activated sludge, biological filtration, aerobic sludge stabilization, anaerobic digestion of sludges and wastes, composting of refuse, sludge and wastes, biomass production and product formation from wastes of foods, crops and industry and covering the technology of processes in relation to the organisms involved, especially conditions of growth, kinetics of degradation, process control and other microbiological activities related to the environment, treatment of effluents from the food and fermentation industries, toxic wastes, bioremediation, pollution, biotechnological soil decontamination, degradation of xenobiotics and other problematic substances.

Theoretical treatments will be considered only if they have direct relevance to experimental studies in applied microbiology and biotechnology. Descriptions of technical equipment will be published only where they have a direct bearing on microbiological processes.

MANUSCRIPT SUBMISSION

Submission implies: that work described has not been published before (except in abstract form or as part of a published lecture, review, or thesis); that it is not under consideration for publication elsewhere; that its publication has been approved by all coauthors, if any, as well as by responsible authorities at institute where work has been carried out; that, if and when manuscript is accepted for publication, authors will transfer copyright to publisher; and that manuscript will not be published elsewhere in any language without consent of copyright holders. Submit in duplicate.

MANUSCRIPT FORMAT

Write in clear and simple English, without repetition. Type double-spaced throughout (including References, Acknowledgments, Footnotes, Tables, Figures, and Legends), with a 5-cm-wide margin on one side of paper only. Do not exceed 18 manuscript pages (6 printed pages), including figures and tables. Indicate in left margin where figures and tables are to be inserted.

Title Page: (Page one) title, complete name(s) of author(s), name(s) of institution(s), any footnotes referring to title (indicated by asterisks), and address to which proofs should be sent. Furnish a telex or fax number.

Abstract: 250 word summary of most important results.

Introduction: One page definition of scope of work in relation to other work in same field. Do not give an exhaustive literature review.

Materials and Methods: Provide enough information to permit repetition of experimental work.

Results: Present with clarity and precision.

Discussion: Interpret results. Do not recapitulate them. Omit conclusions.

Footnotes: Indicate title footnotes by asterisks. Number text footnotes consecutively. Indicate table footnotes with lower-case letters.

References: List only works cited in text and that have been accepted for publication in alphabetical order under first author's name. List works by two authors alphabetically according to second author's name. List works by three or more authors chronologically. For more than one work by same author or team of authors in same year, add a, b, c, etc. to year both in text and in reference lists.

JOURNAL

Dobbins WO, Bill J (1965) Diagnosis of Hirschsprung's disease excluded by rectal suction biopsy. N Engl J Med 272: 990-993

BOOK

Bohr H (1975) On calcium metabolism during immobilization. In: Kuhlencordt F, Kruse H-P (eds) Calcium metabolism, bone and metabolic bone diseases. Springer, Berlin Heidelberg New York, p 39

Wolf GH, Lehmann P-F (1976) Atlas der Anatomie, 4th edn, vol IV/3. Fischer, Berlin Gottingen

Authors are responsible for accuracy of references.

Cite in text by author and date in parentheses (Child 1941; Godwin and Cohen 1969; MacWilliams et al. 1970). When a paper has more than two authors, name only first in text (Komor et al. 1979).

Illustrations: Restrict to minimum needed to clarify text. Do not submit previously published illustrations. Color illustrations will be accepted.

Do not duplicate numerical data in graphs or tables. Mention all figures, whether photographs, graphs, or diagrams, in text. Mark top of figure, title of paper, author's name, and figure number lightly on back in soft pencil. Number all figures consecutively throughout and submit separately. Figures should either be column width (8.6 cm) or printing area size (178 × 240 cm).

When possible, group figures into a plate on one page.

Line Drawings: Submit high-quality glossy prints in final desired size. Make inscriptions clearly legible. Use letters 2 mm high.

Halftone Illustrations: Submit sharp, well-contrasted photographic prints trimmed at right angles in desired final size. Make inscriptions about 3 mm high. For plate, mount figures on regular bond paper, not cardboard. If reduction is necessary, state alternative scale desired.

Legends: Provide brief, self-contained descriptions of illustrations. Do not repeat information in text.

OTHER FORMATS

MINI-REVIEWS—5 printed pages. Contact an International Editor before submitting.

ORIGINAL PAPERS—6 printed pages including figures, tables and references (about 18 manuscript pages, typed double-spaced).

SHORT CONTRIBUTIONS—2-3 pages, including one table, one figure and references. Short reports of new results of special interest. Append a note explaining why text merits publication in this form.

OTHER

Color Charges: DM 1200. for first and DM 600. for each additional page.

Page Charges: DM 250. for each page exceeding six printed pages (18 manuscript pages, including figures and tables).

Style: Underline genus and species names once for italics.

SOURCE

April 1993, 39(1): inside front and back covers.

Archives Internationales de Physiologie, de Biochimie et de Biophysique

Société de Physiologie
Société de Biochimie
Societe de Biophysique
Vlaamse Vereniging voor Plantenfysiologie
Société de Physiologie Végétale de la Communauté Francophone de Belgique
Vaillant-Carmann s.a.

Mr. J. Lecomte, Editorial Secretary
Archives Internationales de Physiologie, de Biochimie et de Biophysique
BP 144,
B-4000 Liège

AIM AND SCOPE

The *Archives Internationales de Physiologie, de Biochimie et de Biophysique* publishes, in French and English, original research papers of an experimental nature, as well as reviews.

MANUSCRIPT SUBMISSION

Submit manuscripts in triplicate. Make papers as brief as possible; do not exceed 10 journal page. Send a text *ne varietur*. Changes to proof are charged to authors. Precede article with a brief and objective Summary, suitable for use as an "Abstract" or "Refract" by bibliographic organizations. Include an English translation with summaries in French. Add four keywords at end of summary.

MANUSCRIPT FORMAT

References: Cite within text by author and year. Include name of author (underlined twice to indicate small caps) and year of publication (in brackets). Examples: (HENQUIN, 1978a & b); (HENQUIN & MEISSNER, 1983); (HENQUIN *et al*, 1982)

Group full citations for references together at end of article under heading "References" ("Bibliographie" for articles in French). Arrange in alphabetic order of author's names.

Type author's names in small capitals (underlined twice); title of journal in italics (underlined once); and volume number in boldface (underlined with a wavy line). For books cited, indicate: names and initials of authors (underlined twice); date of publication; title of book (in italics, underlined once); name and town of publisher.

HENQUIN, J.-C. (1978, *a*) *Nature (London)* **271**, *271-273*.

HENQUIN, J.-C. (1978, *b*) *Endocrinology* **102**, *723-730*.

HENQUIN, J.-C. & MEISSNER, H. P. (1983) *Biochem. Biophys. Res. Commun.* **112**, 614-620.

HENQUIN, J.-C., MEISSNER, H. P. & SCHMEER, W. (1982) *Pflügers Arch. Eur. J. Physiol.* **393**, 322-327.

Figures: Keep number to minimum strictly necessary for understanding text. Supply text for figures (letters, words, numbers on abscissae and ordinates, etc.) separately. Do not include in figures; attach to figure on a sheet of transparent paper or plastic, or indicate in an accompanying photo or photocopy of figure. Make all figures of dimensions (maximum 11 cm × 18 cm) which can be incorporated directly into text. Organize figure legends on separate pages from text.

Tables: Keep number and dimensions to minimum necessary for comprehension of text. Do not include the same numerical data presented both as a table and as a figure.

SOURCE

January-February 1994, 102(1): back matter.

Archives of Biochemistry and Biophysics
Academic Press

Editorial Office
525 B Street, Suite 1900
San Diego, CA 92101-4495

AIM AND SCOPE

Archives of Biochemistry and Biophysics is an international journal dedicated to the dissemination of fundamental knowledge in all areas of biochemistry and biophysics. Research Reports and regular manuscripts that contain new and significant information of general interest to workers in these fields are welcome. Sufficient detail must be included to enable others to repeat the work.

MANUSCRIPT SUBMISSION

Submit one original and three photocopies, including four sets of original figures or good quality glossy prints.

Submit original papers only. The same work may not have been published or be under consideration for publication elsewhere. Submission must be approved by all authors and appropriate authorities at institution where work was done. Corresponding author must obtain agreement of all coauthors prior to submission. Abstracts of oral or poster presentations are not considered previous publication. Full paper must contain additional information that justifies publication and does not repeat presentation of same data. Include sufficient copies of all preliminary communications and all relevant manuscripts in press or under editorial consideration by another journal. For papers that include mathematical equations, submit derivations; derivations will not be published, but they facilitate review.

MANUSCRIPT FORMAT

Type double-spaced throughout on 8.5 × 11 in. paper, with 1 in. margins on all sides. Number all pages consecutively, including references, tables, and figure legends.

Title Page: Title; author(s); affiliation(s); short running title (an abbreviated form of title, 65 characters including letters and spaces); name, complete mailing address, and telephone and fax numbers of corresponding author; and appropriate Subject Area for listing in Table of Contents:

Biological Oxidations and P450 Reactions

Glycoconjugates and Oligosaccharides; Proteoglycans and Extracellular Matrices

Lipids, Lipid Mediators, and Glycolipids

Biophysical Chemistry and X-Ray Crystallography

Membrane Proteins and Transport: Intracellular Signals; Cytokines and Receptors

Enzyme Structure and Mechanisms; Cellular Regulation; Phosphorylation and Dephosphorylation

Plant Biochemistry and Molecular Biology

Protein Structure and Function; Proteases, Protein Turnover, and Post-translational Processing

Gene Expression, Transcription, and Translation

Indicate that paper is part of a larger series in a title footnote. Include reference to preceding paper(s).

Spell out in full either first or second given name(s) of author(s). Acknowledge financial support in numbered title footnotes. At title page bottom, give approximately six descriptive key words.

Abstract: Second page, 250 words, informative summary of contents and conclusions. Refer to all new information, mention incidental findings that may be valuable to others. Do not include nonessential details (e.g., methods) and unsupported information or claims. Make intelligible to nonspecialist; avoid specialized terms and abbreviations or symbols.

Text: Suggested organization: abstract; introductory statement; Materials and Methods or Experimental Procedures; Results; Discussion; Acknowledgments; References. Combining some sections may make presentation clearer and more effective.

In introductory statement, state investigation's purpose and relation to other work. Do not include an exhaustive literature review.

Give a brief description of experimental procedures (or materials and methods), adequate for repetition of work by a qualified investigator. Refer to previously published procedures employed by citation of both original description and pertinent published modifications. Do not include extensive details unless they present substantially new modifications.

In Materials and Methods, indicate any chemical or biological hazards involved. Describe relevant safety precautions; reference relevant practice standards. In experiments involving recombinant DNA molecules constructed *in vitro* and subsequently inserted into cells, describe methods used. Reference appropriate NIH guidelines and/or other pertinent regulations or comparable documents. Describe significant deviations from recommended practices. Note certification by the NIH Program Advisory Committee on Recombinant DNA of host system as disabled. Present results in tables or figures. Give simple findings in text. Interpret results in Discussion. Results and Discussion may be combined.

References: Cite in text by Arabic numerals in parentheses. List at end of paper in numerical order, typed double-spaced on a separate sheet, under heading References. Abbreviate journal titles as in *Chemical Abstracts Service Source Index*.

JOURNAL ARTICLE

1. Terry, M. J., Wahleithner, J. A., and Lagarias, J. C. (1993) *Arch. Biochem. Biophys.* **306**, 1-15.

BOOK

2. Bolander, F. F. (1989) Molecular Endocrinology, Academic Press, San Diego.

CHAPTER IN BOOK

3. Compton, T. (1990) in PCR Protocols: A Guide to Methods and Applications (Innis, M. A., Galfand, D. H., Sninsky, J. J., and White, T. J., Eds.), pp. 39-45, Academic Press, San Diego.

Give text citations to references written by more than two authors as, Smith *et al*. In references list, list all authors' names with initials.

For articles accepted for publication, but not yet in print, list reference:

4. Smith, J. (1994) *Arch. Biochem. Biophys.*, in press.

Do not use "in preparation," "private communication," and "submitted for publication" in reference list. Cite "personal communication" and "unpublished work" parenthetically following individual(s) name(s) in text. Verify wording of personal communication with suppliers and obtain and submit approval for use of names in connection with quoted information or for citation of unpublished work.

Do not cite abstracts of papers presented at scientific meetings as references unless they appear in publications included in *Biological Abstract List of Serials*. Indicate abstracts not citable as references in footnotes.

If the paper submitted is in a series, include immediately preceding paper in references.

Footnotes: Identify by superscript Arabic numerals. Type on separate sheet following references; number title footnotes, authors' names, and affiliations in sequence with text footnotes. Identify table footnotes superscript lowercase italic letters, *a, b*, etc., at bottom of table.

Tables: Include only essential data or data needed to illustrate or prove a point. Give each table a short explanatory title. Give experimental details in table footnotes; do not repeat details given in Materials and Methods or another table or figure.

Give each column an appropriate heading. For abbreviations, follow recommendations given Abbreviations and Symbols. State data units at the top of each column. Do not repeat on each line. Repeat words or numerals on successive lines; do not use ditto marks.

Indicate units of measure clearly. If an experimental condition is the same for all tabulated experiments, put information in a table footnote (not in a column of identical figures).

Avoid presenting large masses of essentially similar data. Replace lengthy tabulations by reporting mean values with some accepted measure of dispersion (standard deviation, range); indicate number of individual observations in a table footnote. Give probability values derived from appropriate statistical tests with significance of measures. Define all statistical measures. Do not include more significant digits than are justified by accuracy of determinations. Do not use vertical rules. Limit number of horizontal rules. Complex tables may be submitted as camera-ready copy typed single-spaced whenever possible, and submitted as 5×7 in. glossy prints.

Figures and Illustrations: Cite all figures consecutively in text. Type figure legends typed double-spaced consecutively on a separate sheet. Legends should contain sufficient experimental detail to permit figure to be interpreted without reference to text. Indicate quantities and units alongside ordinate and abscissa scales. Try to combine data into a single figure to keep number of illustrations to a minimum.

Identify all figures on back, in soft pencil, with authors' names and figure number. Indicate TOP.

Submit all figures and illustrations so as to permit photographic reproduction without retouching or redrawing, planned to fit 3.5 in. column width. Lettering should be professional quality or generated by high-resolution computer graphics and be large enough (10-12 points) to take a reduction of 50-60%. Make drawings with black India ink on tracing linen, smooth-surface white paper, or Bristol board. High-quality computer graphics may be acceptable. Use blue-ruled graph paper. Submit figures as originals no larger than 8.5×11 in. or, if originals are larger, as glossy prints no larger than 8.5×11 in. Avoid simple histograms. Tables or text is preferred.

Submit halftone photographs on glossy paper and rich in contrast. Submit only those parts of photograph necessary to illustrate the point. Cut off nonessential areas. Submit photographs as unmounted, single prints. Halftone artwork must be glossy prints or originals. Color illustrations are reproduced at author's expense.

OTHER FORMATS

MINIPRINT SUPPLEMENTS—Present supportive data. Include analyses relevant to studies of amino acid or polynucleotide sequences, details of chemical syntheses, preparation of enzymes, extended statistical or mathematical discussions, and X-ray diffraction data. Single space, except with subscripts and superscripts. Use easily legible type [not more than 10 characters per 25 mm (1 in.)] and a uniformly dark impression. Use figure lettering comparable to size of typewriter type.

Four typed pages, can be presented on one journal page after reduction to approximately 40%. These dimensions cannot be exceeded. Text pages will be assembled in two long columns. Have figures and tables of a size that can be inserted into text at appropriate points.

Papers that are a two-page condensation plus miniprint supplement should follow the same format as ordinary papers. They may include or omit usual divisions of Materials and Methods, Results, and Discussion. In a title footnote describe supplement material.

RESEARCH REPORTS—1000 words and six tables and/or figures. Provide rapid publication of complete manuscripts on outstanding and novel research findings. In first paragraph describe, without abbreviations, background, rationale, principal findings, and study conclusions. Report experimental methods in table or figure legends, not text. Include acknowledgments in a title footnote. Omit abstract and text subheadings, and keep literature citations to a minimum. Identify manuscript as a Research Report.

OTHER

Abbreviations and Symbols: Restrict use. Define all nonstandard abbreviations in a single footnote, inserted immediately after first abbreviation. Define abbreviations used only in a table or figure in legend. Abbreviations listed in Tables I, VIII, and XI of Instructions (*Arch. Biochem. Biophys.* 308(1): xvi, xixxx), and those indicated in other tables, may be used without definition. Special symbols are essential in studies on biopolymers.

When other abbreviations for chemical compounds are needed, use standard chemical symbols (C, H, O, N, P, etc.), numerical multiples (subscripts 2 and 3, not di or D or T, etc.), as in Me_2SO, Me_3Si-), and trivial names and

their symbols (e.g., folate, P, Me, Pr, Bu see Tables III-IX (*Arch. Biochem. Biophys.* 308(1): xvii- xix).

Combine symbols to represent more complex symbols, such as Tos-Arg-OMe, in which the basic structure (arginine) remains recognizable.

Do not abbreviate names of enzymes except substrates for which accepted abbreviations exist (e.g., ATPase and RNase). Follow *IUPAC-IUBMB Enzyme Nomenclature* (1992). Do not abbreviate class names, such as fatty acids, proteins, etc., or short terms (poly, furan, folate, etc.) or terms like "central nervous system," "red blood cells," or "extracellular fluid."

Designate monomeric units by three-letter symbols—capital followed by two lowercase letters. Do not use abbreviations for free monomers in test of papers. There is a standard treatment for the three groups of macromolecules built up from these units. When sequence of residues is known, write symbols in order and join by dashes or hyphens. When sequence is not known, enclose group of symbols, separated by commas, in parentheses.

For amino acid residues in polypeptides, place residue with the free α-amino group (if present) at left of sequence as written. For polysaccharides, join symbols for sugars by dashes or arrows to indicate links between units. The position and nature of the links are shown by numerals and anomeric symbols α and β.

Arrows point away from hemiacetal link. Use of dash assumes that the hemiacetal link is to the left. When indicating furanose or pyranose, use the letter f or p after the saccharide symbol.

Represent macromolecules composed of repeating sequences by prefix "poly" or subscript n, both indicating "polymer of." Enclose symbols for monomeric units of the sequence in parentheses. There is no space or hyphen between poly and parenthesis. The n may be replaced by a definite number, an average, or a range. "Oligo" may replace "poly" for short chains. The same rules apply to oligo- and polynucleotides.

Chemical and Mathematical Usage: Follow IUPAC rules on chemical nomenclature:

Sections A through F and H of the *Nomenclature of Organic Chemistry,* Pergamon Press, Oxford, 1979

Sections E (stereochemistry), F (natural products and related compounds), and H (isotopically modified compounds) in *Biochemical Nomenclature and Related Documents,* published by the Biochemical Society (London) for the IUBMB (see Table XII)

Nomenclature of Inorganic Chemistry, 3rd ed. (1990), and the *Compendium of Analytical Nomenclature,* 2nd ed. (1987), both from Blackwell Scientific Publications, Oxford.

Use formulas that can be printed in single horizontal lines of type to reference simple chemical compounds. Do not use two-dimensional formulas in running text. Center chemical equations, structural formulas, and mathematical formulas between successive lines of text. Prepare structural formulas for direct photographic reproduction and include on a duplicate sheet at end of paper. Long sequences of amino acids or nucleotides reproduce better if they are drawn in ink or typewritten Designate ionic charge as a superscript following chemical symbol, e.g., Mg^{2+}, S^{2-}.

Equilibrium and Velocity Constants: Write dissociation constants, association constants, and Michaelis constants in terms of concentration. Indicate units where equilibrium constant is defined and value is given. Specify values of rate constants, give first-order velocity as s^{-1} (other units of time may be used, always specify time unit). Give second-order rate constants in $M^{-1} s^{-1}$.

Experimentation, Animal: Follow standards for use established by the Institute of Laboratory Animal Resources, U.S. National Academy of Sciences. Justify experiments in which curariform agents are used. Provide details of steps taken to reduce or avoid distress to animals, particularly with regard to electrical stimulation.

Isotope Experiments and Isotopically Labeled Compounds: Isotope symbol is placed in square brackets directly attached to front of name, as in [^{14}C] urea. When more than one position in a substance is labeled by means of the same isotope and positions are not indicated, number of labeled atoms is added as a right subscript, [$^{14}C_2$]glycolic acid. Symbol U indicates uniform and G general labeling, e.g., [U-^{14}C]glucose (where each molecule has ^{14}C at all six positions), [G-^{14}C]glucose (where a molecule may have ^{14}C at any or all, but not necessarily all, of the six positions).

Isotopic prefix precedes part of name to which it refers, as in ethyl [^{14}C]formate, phenyl[$^{14}C_2$] acetic acid, and l-amino [^{14}C]methylcyclopentanol.

When isotopes of more than one element are used, arrange symbols in alphabetical order, including ^{2}H and ^{3}H.

When not sufficiently distinguished by these means, indicate positions of isotopic labeling by Arabic numerals, Greek letters, or prefixes, placed within square brackets and before element symbol attached by a hyphen.

The same rules apply when the labeled compound is designated by a standard abbreviation or symbol, other than the atomic symbol.

For simple molecules, indicate labeling by writing chemical formulas, e.g., $^{14}CO_2$, $H_2^{18}O$, ^2H_2O, and $H_2^{35}SO_4$ with prefix superscripts attached to proper atomic symbols in formulas.

Do not contract terms such as "^{131}I-labeled albumin" "[^{131}I]albumin" (since native albumin does not contain iodine); however, ^{131}I-albumin and [^{131}I]iodoalbumin are both acceptable.

Do not use square brackets when isotopic symbol is attached to a word that is not a chemical name, abbreviation, or symbol (e.g., ^{131}I-labeled proteins or ^{14}C-labeled amino acids).

Kinetic Constants: Represent velocity constants for forward and backward reactions in the nth step of an enzymatic reaction by k_{+n} and k_{-n} respectively, or by k_q and k_{q+1} respectively, where $q = 2n - 1$. The Michaelis constant is defined as $Km = [S]$ when $v = V/2$, where v is the initial rate of appearance of product or disappearance of substrate at a given substrate concentration $[S]$ and V (or V_{max}) is the initial rate when the enzyme is saturated with that substrate. When reactions with two substrates A and B are being considered K^A_m or $K_m(A) = [A]$ when $v = V/2$ and $[B]$ has been extrapolated to infinity; a value for $[A]$ when $v = V/2$ at a finite concentration (which must be specified) of B should be referred to as an apparent K_m for A. K_s (or

K_d) is the equilibrium constant of the dissociation of the substrate-enzyme complex.

Molecular Weight and Mass: Do not express molecular weight in daltons. Relative molecular mass (M_r) is defined as a mass ratio, a pure number. Dalton (Da) is a unit of mass equal to one-twelfth the mass of an atom of carbon-12; hence, it is correct to say "the relative molecular mass of X is 106," or "the molecular mass of X is 10^6 Da," or "the molecular mass is 10^6 g," or to use such expressions as "the 16,000-Da peptide." For entities that do not have a definable molecular mass do not say, for example, "the mass of a ribosome is 10^7 Da." Do not say "the molecular weight of X is 10^6 Da."

Naming Compounds: Chemical names are run together except for those of acids, esters, ethers, glycosides, ketones, and salts, which are printed as separate words. Use hyphens to separate numbers, Greek letters, or some configurational and italic prefixes from words. Refer to *Chemical Abstracts* for specific information.

Nomenclature, Biochemical: Follow *Biochemical Nomenclature and Related Documents* (Biochemical Society, London).

Nomenclature, Enzyme: Where an enzyme is the main subject of a paper, give its source, recommended name, reaction catalyzed by enzyme, and code number preceded by EC. Identify other enzymes at first mention by code numbers. Consult *IUPAC-IUBMB Recommendations on Enzyme Nomenclature* (Academic Press, San Diego, 1992) for code numbers of enzymes. Report new enzymes to IUBMB Commission on Biochemical Nomenclature, which will assign appropriate code number.

Nomenclature, Organisms:

Animals and Plants—Include full binomial Latin names as well as common names. State strain and source of plant material and laboratory animals.

Microorganisms—In title, abstract, and at first mention in text, give full binomial Latin names, underscored. Give collection number from recognized collection of microorganisms or quote strain number or name, not underscored. Do not underscore names of ranks higher than genus, generic names used adjectivally, and names of microorganisms used colloquially. Spell generic name with a capital letter. Single-letter abbreviations may be given for generic name; if two genera with same initial letter are studied, use abbreviations such as Strep. and Staph.

Identify microorganisms and tissue culture strains by a type culture collection number. If not possible, make strains used available to interested investigators on request.

Recommendations on nomenclature in bacterial genetics have been proposed by Demerec, Adelberg, Clark, and Hartman (1966) *Genetics* 54, 61-76.

Nucleotide Sequence Data: Submit original nucleotide or amino acid sequence data to a databank such as GenBank or EMBL. Include footnote with accession number on title page. Deposit sequence data either prior to submission or by acceptance. For further information contact editorial office or GenBank or EMBL directly.

Optically Active Isomers: Differentiate names of chiral compounds whose absolute configurations are known by prefixes R- and S- [see IUPAC (1970) *J. Org. Chem.* 35, 2849–2867]. When compounds can be correlated sterically with glyceraldehyde, serine, or another standard accepted for a specialized compound class, use small capital letters D-, L-, and DL- for chiral compounds and their racemates. Where direction of optical rotation is all that can be specified, (+)-, (–)-, and (±)-, or *dextro, levo,* and "optically inactive," are used, but specify conditions of measurement.

Powers in Tables and Figures: Use units that eliminate need for exponents in headings. Do not use compound prefixes. When impractical or undesirable, precede quantity expressed by power of 10 by which its value has been multiplied. The units in which the quantity is expressed may not be multiplied by a power of 10; however, the unit may be changed using an appropriate prefix, such as m or μ, or of another accepted scale. Multiplier must apply to quantity, not unit. The same principles apply to labeling coordinates. To distinguish quantities and units, state them both, with units in parentheses. If an exponential multiplier is used, it must be applied to the quantity and not the unit. Curie (Ci, not C) or becquerel (Bq) is used for radioactivity measurements; curie is equal to 3.7×10^{10} dis/s or 37 GBq (1 becquerel = 1 dis/s).

Prefixes: Use italics for certain prefixes, e.g., *cis, trans, o, m, p, dextro, levo, meso,* and also for *O, N,* etc., to indicate an element carrying a substituent. Do not use italics for allo, bis, cyclo, epi, iso, n (not *n*), neo, s (not *sec.*), t (not *tert.*), nor tris.

Follow alphabetical order for prefixes denoting substituents. Syllables indicating multiple substituents, e.g., bis, di, tri, do not count in deciding order.

Protein Structure: Follow "Recommendations for the Documentation of Results in the Determination of the Covalent Structure of Proteins" *J. Biol. Chem.* 251, 11-12 (1976).

Reviewers: Suggest competent reviewers in field and individuals whom they wish to exclude.

Solutions and Buffers: Specify composition of all solutions and buffers in sufficient detail to define concentration of each species. For ordinary buffers, such as 0.1 M sodium acetate, pH 5.0, it will be assumed that molarity refers to the total concentration of the species that buffers at indicated pH and concentration of the counterion is sufficient to neutralize the charge of the ionized buffer species. Indicate composition of mixtures by a colon or a slash (/). Do not use hyphens or dashes. Do not use abbreviations for buffers, specify these as Buffer A, B, etc., after defining under Materials and Methods.

Give complete unabbreviated name and source (or a reference that gives complete composition) for all culture media.

Spectrophotometric Data: Indicate relationships among symbols used. Follow IUPAC (1970, *Pure Appl. Chem.* 21, 1). State Beer's law as

$A = -\log T = \varepsilon lc$

where A is absorbance; T, transmittance $(=I/I_0)$; ε, molar absorption coefficient; c, concentration of absorbing substances in moles per liter; and l, length of optical path in centimeters. Under these conditions e has the dimensions liter mol^{-1} cm^{-1}, or M^{-1} cm^{-1} (not cm^2 mol^{-1}).

If Beer's law is not followed by a substance in solution, explicitly state so and characterize substance by absorbance at a specified concentration. When spectrophotometric measurements are made with a radiant energy source that is not confined strictly (as in a line spectrum) to wavelength or frequency specified, the exact value of ε will be ambiguous; report spectral characteristics of source.

SOURCE
January 1994, 308 (1): vii-xxii.

Archives of Dermatology
American Medical Association

Kenneth A. Arndt, MD
Beth Israel Hospital
330 Brookline Ave
Boston, MA 0221
Telephone: (617) 735-3200; Fax: (617) 735-4948

AIM AND SCOPE

The mission of the *Archives* is to publish clinical and laboratory studies that enhance the understanding of skin and its diseases. In addition to these studies, case reports that substantially add to our knowledge in a meaningful fashion will be published as Observations.

MANUSCRIPT SUBMISSION

Submit three complete copies of manuscript and all illustrative material. Neither manuscripts nor any essential part may have been previously published anywhere in any language and are not under simultaneous consideration by another publication.

In cover letter indicate corresponding author with address, telephone, and fax number. If first author was a dermatology resident or fellow at the time the work was done, include information in a footnote. Include one of the following statements on copyright or federal employment; statement on authorship responsibility; and statement on financial disclosure. All authors must sign each statement.

Copyright Transfer: "In consideration of the action of the American Medical Association (AMA) in reviewing and editing this submissions the author(s) undersigned hereby transfers, assigns, or otherwise conveys all copyright ownership to the AMA in the event that such work is published by the AMA."

Federal Employment: "I was an employee of the U.S. federal government when this work was investigated and prepared for publication; therefore, it is not protected by the Copyright Act and there is no Copyright of which the ownership can be transferred."

Authorship Responsibility: "I certify that I have participated sufficiently in the conception and design of this work and the analysis of the data (when applicable), as well as the writing of the manuscripts to take public responsibility for it. I believe the manuscript represents valid work. I have reviewed the final version of the manuscript and approve it for publication. Neither this manuscript nor one with substantially similar content under my authorship has been published or is being considered for publication elsewhere, except as described in an attachment. Furthermore, I attest that I shall produce the data on which the manuscript is based for examination by the editors or their assignees if requested."

Financial Disclosure: "I certify that I have no affiliation with, or financial involvement in, any organization or entity with a direct financial interest in the subject matter or materials discussed in the manuscript (e.g., employment, consultancies, stock ownership, honoraria) except as disclosed in an attachment."

MANUSCRIPT FORMAT

Type double-spaced, including title page, synopsis, abstract, references, figure legends, and tables, with nonjustified 1 in. right margins. Type on one side only, on heavy-duty white bond paper 8.5 × 11 in. Begin each section on separate pages: title page, synopsis, abstract, text, acknowledgments, references, legends, tables. Type page number in upper right corner of each page. Number consecutively beginning with title page.

Title Page: List title (short, clear, specific); authors' full names, academic degrees, institutional affiliations, and location; reprint request address; and, if the manuscript was presented at a meeting, name of organization and place and date of reading.

Abstract: Of Clinical and Laboratory Studies—250 words under headings: Background and design (overview of the topic, state study's main objective or hypothesis and describe its basic design, specifying key features (e.g., sample size, type of patients studied, prospective vs. retrospective, randomization, blinding, case control); Results (main study result, include confidence intervals and exact level of statistical significance); and Conclusions (only those supported by data and direct clinical application of findings; avoid speculation; give equal emphasis to positive and negative findings).

Of Observations—200 words under headings: Background (topic overview and main objective or reason for report); Observations (principal observations, findings, or results; include confidence intervals and levels of statistical significance for numerical results; Conclusions (conclusions supported by information, along with clinical applications, avoid overgeneralization).

References: Number in order of appearance (not alphabetically). Follow style of *Index Medicus*. Refer to recent issues for specific examples, preferred punctuation, and sequence style. Do not include unpublished data, personal communications, or manuscripts "submitted" in reference list. Incorporate such material, if essential, in body of article. List all authors and/or editors up to six; if more than six, list first three and "et al." Authors are responsible for bibliographic accuracy.

Figure Legends: 40 words each. Provide magnification and stain for photomicrographs.

Tables: Supplement, do not duplicate, information in text. Make self-explanatory with a brief title on a separate page. Do not use oversized paper.

Illustrations: Submit black-and-white illustrations as unmounted, high-contrast, glossy photographic prints. Preferred size is 5 × 7 in. Have line drawings, graphs, and charts professionally drawn, photographed, and sent as prints. Do not send original artwork.

Submit two sets of unmounted color glossy prints (accompanied by color slide) as well as three black-and-white glossy prints.

Number and cite each illustration in text. Affix label to back indicating figure number, name of senior author, and TOP. Submit photographic consent for recognizable photographs of patients. If photograph is of a minor, both parents (or guardian) must sign consent.

Acknowledgments: Acknowledge illustrative or textual material from other publications. Submit publishers' permission to reproduce.

OTHER FORMATS

CORRESPONDENCE—500 words, five references, and two figures. Double-space, submit in triplicate, clearly mark "for publication." Include a copyright transfer statement.

OTHER

Color Charges: Should editors feel there is no benefit to color printing, but authors wish to do so: $400 for up to six illustrations arranged on one page. State willingness to pay fee in cover letter.

Drugs: Use generic drug names; trade names may be included in parentheses if desired.

Experimentation, Human: Include a statement that informed consent was obtained after the nature of the procedure(s) had been fully explained.

Reviewers: Suggest two to four possible reviewers. Give full name and address.

Unit: Systéme International (SI) units.

SOURCE

January 1994, 130(1):131-133.

Archives of Disease in Childhood
British Paediatric Association
BMJ Publishing Group

Editors, *Archives of Disease in Childhood*
BMA House
Tavistock Square
London WC1H 9JR

MANUSCRIPT SUBMISSION

Submission implies that it contains original work not being offered elsewhere or published previously. Prepare manuscripts in accordance with Vancouver style (International Committee of Medical Journal Editors. Uniform requirements for manuscripts submitted to biomedical journals. *BMJ* 1991; 302:338–41).

Submit two copies of manuscript. Include copies of other papers on similar subject to assure the editors that there is no duplication.

MANUSCRIPT FORMAT

Type double-spaced throughout (including references and tables) with a 5 cm left margin. Do not justify right margin. Number pages in top right corner.

On title page give title (10 words; do not use 'child', 'children', or 'childhood'), name of author(s), place where work was carried out, and address of corresponding author. Keep number of authors to a minimum, include only those who have made a contribution to research; justify more than five authors. Acknowledge workers whose courtesy or assistance has extended beyond their paid work and supporting organizations. Give information about availability of reprints at end of references. Provide three key words for index.

Abstract: 150 words; for experimental or observational studies, state in sequence: main purpose of study; essential elements of study design; most important results illustrated by numerical data (not p values); implications and relevance of results. For case reports, summarize essential descriptive elements of case(s) and indicate relevance and importance. If structured abstract is more helpful, submit in that form.

Tables: Present separately. Type double-spaced without ruled lines.

Illustrations: Use only when data cannot be expressed clearly in any other way. Supply numerical data on which graphs are based. Trim to remove all redundant areas; mark top on back.

Conceal identity of patients shown in photographs or obtain written consent to publication.

Arrow ultrasound scans or other pictures on an overlay to indicate areas of interest or submit explanatory line drawings.

If any tables or illustrations submitted have been published elsewhere, submit written consent to republication from copyright holder and authors.

References: Number in order of appearance in text. Include all information (Vancouver style):

1. Donn, SM. Alternatives to ECMO. *Arch Dis Child* 1994; **70**:1626–8.

2. Hull D. Children's health. In: Smith R, ed. *The health of the nation: the BMJ view.* London:British Medical Journal, 1991: 64-70.

Cite abstracts, information from manuscripts not yet accepted, or personal communications only in text. Do not include in references. Verify references against original documents.

OTHER FORMATS

ANNOTATIONS—Commissioned by editors who welcome suggestions for topics or authors.

LETTERS—300 words, four references, signed by all authors, typed double-spaced. Send two copies.

MEDICAL AUDIT—Papers concerned with innovative medical audit, service evaluation, quality assurance, and outcome measures, especially models of good practice that include description of service before medical audit,

standards developed, professional training involved, and demonstration of improvement.

SHORT REPORTS—900 words, 7 in title, an abstract (50 words), one or two small tables or illustrations, and six references. If more illustrations are used, reduce text accordingly.

OTHER

Abbreviations: Use rarely. Precede by words in full at first appearance.

Color Charges: Cost of reproducing color figures will be charged to authors (contact editorial office for price).

Experimentation, Human: Refer to Editorial. Research involving children—ethics, the law, and the climate of opinion. *Arch Dis Child* 1978; **53**:441–2.

Statistical Analysis: Use 95% confidence intervals where appropriate.

Units: SI units; except blood pressure, which should be in mm Hg, and drugs in metric units.

SOURCE

July 1994: 71(1): 99 - Revised January 1993.

Archives of General Psychiatry
American Medical Association

Grayson S. Norquist, MD
UCLA Neuropsychiatric Institute
760 Westwood Plaza
Los Angeles, CA 90024
Telephone: (213) 206-0226

MANUSCRIPT SUBMISSION

Manuscripts may not be under simultaneous consideration by another publication. In cover letter, designate one author as correspondent and provide a complete address and telephone number. Identify special circumstances, vested interests, or sources of bias that might be deemed to affect integrity of reported information. Each author must have had enough substantial involvement in generating and formulating the published product to bear such accountability for integrity of scientific information. See Freedman DX. The meaning of full disclosure. *Arch Gen Psychiatry.* 1988; 45:689-691 and Freedman DX. Megamultiple authorship. *Arch Gen Psychiatry.* 1982; 39:351.

Following statement must be signed by all authors: "I have been sufficiently involved in the work to take public responsibility for its validity and final presentation as an original publication. I can provide documentation of my work upon reasonable request and I have fulfilled the obligations for full disclosure and authorship as described by *Archives of General Psychiatry...*" This signed statement must conclude with one of the two following sentences: "Accordingly, I hereby transfer, assign, or otherwise convey all copyright ownership to the AMA in the event such work is published by the AMA." [OR] "No copyright ownership can be transferred since this work was produced in the course of my official duties as a U.S. government officer or employee."

Submit copies of publications helpful for review or that bear on possible overlap of published data.

MANUSCRIPT FORMAT

Submit original and three copies. Type double-spaced throughout (including references) on one side of 8.5 × 11 in. white bond paper. If word processor is used, do not justify right margin and use a letter quality printer.

Title should not exceed 42 characters per line, including punctuation and spaces. Title page needs full names, academic affiliations, reprint request address, and, if manuscript has been presented at a meeting, name of organization, place, and date of presentation. In abstract (135 words) state problem considered, methods, results, and conclusions.

Do not use footnotes or appendixes. Incorporate materials into text or offer to interested readers on request.

Include name and affiliation of statistical reviewer, if appropriate.

Acknowledge material from other publications. Submit written permission from owner to reprint in any language or form without limitation.

References: Number in order mentioned in text; do not alphabetize. In text, tables, and legends, identify with superscript Arabic numbers. Follow AMA style, abbreviate names of journals according to *Index Medicus*. List all authors and/or editors.

JOURNAL

1. DuBois RN, Lazenby AJ, Yardley JH, Hendrix TR, Bayless TM, Giardiello FM. Lymphocytic enterocolitis in patients with 'refractory sprue.' *JAMA*. 1989;262:935-937.

CHAPTER IN A BOOK

2. Ritchie JM, Greene NM. Local anesthetics. In: Gilman AG, Goodman LS, Rall TW, Murad F, eds. *The Pharmacological Basis of Therapeutics*. 7th ed. New York, NY: Macmillan Publishing Co Inc; 1985: 309-310.

Tables: Double-space on separate sheets of 8.5 × 11 in. white bond paper. Give titles. Number in order of citation in text. If continued, repeat title on second sheet, with "(cont)." Print explanatory notes single-spaced beneath table.

Figures: Submit two sets of professionally prepared, high-contrast glossy prints (preferably 5 × 7 in.). Number according to order in text. Type figure number, name of senior author, and abbreviated title on a gummed label; affix to back of print.

If full-color figures are necessary, *Archives* may pay part of expense. Balance is borne by author.

Legends: Double-space (40 words each) on a separate sheet.

OTHER FORMATS

LETTERS—750 words; provide significant news or clarification of general interest. Follow manuscript preparation guidelines (e.g., double-spaced, copyright transfer). Illustrative material is not usually accepted; brief tables may be considered.

OTHER

Abbreviations: Do not use in title or abstract. Limit use in text.

Drugs: Use generic names, unless specific trade name is directly relevant to discussion.

Experimentation, Human: State that informed consent was obtained. Do not use real names or initials. Ensure that patient confidentiality was not breached.

Style: Iverson CI, Dan BB, Glitman P, et al. *American Medical Association Manual of Style*. 8th ed. Baltimore, Md: Williams & Wilkins; 1988.

International Committee of Medical Journal Editors. Uniform requirements for manuscripts submitted to biomedical journals. *Ann Intern Med*. 1988; 108:258–265.

Units: Systéme International (SI).

SOURCE

January 1994 51(1): 6-7.

Archives of Internal Medicine
American Medical Association

James E. Dalen, MD
2601 N Campbell Ave, Suite 202
Tucson, AZ 85719
Telephone: (602) 326-8334; Fax: (602) 326-6188

MANUSCRIPT SUBMISSION

Manuscripts must not be under simultaneous consideration by another publication. Submit completed and signed forms for transfer of copyright (page 16 of January 10, 1994 issue, volume 154).

Designate one author as correspondent and provide address and telephone number.

MANUSCRIPT FORMAT

Submit four copies: an original and three high-quality copies. Type all copy (including references, legends, and tables) double-spaced on 8.5 × 11 in., heavy-duty white bond paper. Keep ample margins (1 in.). If a word processor is used, do not justify lines.

Make titles short, specific, and clear. Do not exceed 42 characters per line, including punctuation and spaces. Limit to two lines. On title page include full names and academic affiliations of all authors, reprint request address, and, if the manuscript was presented at a meeting, name of organization, place, and date on which it was read.

Abstract: For Original Investigations (250 words): Background, Methods, Results, Conclusions. Abstracts are not required for Editorials and Commentaries. Begin Review Articles, Special Articles, and Clinical Observations with an appropriate summary (150 words).

Structured Abstract: Use for reports of original data from clinical investigations with human subjects. Include headings. Use phrases instead of full sentences. Headings: Objective, Design, Setting, Patients or Other Participants; Intervention(s); Main Outcome Measure(s); Results; and Conclusions. See *Arch Intern Med* 1994; 154(1): 18, 40, 111 for "Instructions for preparing structured abstracts." Instructions are based on Haynes RB et al. "More informative abstracts revisited," *Ann Intern Med*. 1990; 113:69-76.

References: List in consecutive numerical order (not alphabetically). Once cited, all subsequent citations should be to original number. Cite all references in text or tables. Do not list unpublished data and personal communications as references. References to journal articles include: author(s) (list all authors and/or editors up to six; if more than six, list first three and "et al"); title; journal name (as abbreviated in *Index Medicus*); year; volume number; and inclusive page numbers, in order. References to books include: author(s) (list all authors and/or editors up to six; if more than six, list first three and "et al"); chapter title (if any); editor (if any); title of book; city of publication; publisher; and year. Include volume and edition numbers, specific pages, and name of translator when appropriate. Author is responsible for accuracy and completeness of references and for their correct text citation.

Illustrations: Use only to clarify and augment text. Submit four copies, unmounted and untrimmed. Do not send original artwork. Send high-contrast glossy prints (not photocopies). Type figure number, name of senior author, and arrow indicating TOP on gummed label and affixed to back. Lettering must be legible after reduction to column size. Provide magnification and stain when pertinent. Make in a proportion of 5×7 in.

Employ an experienced medical illustrator for preparation of artwork. Template lettering or preset type is preferred to hand-lettered labels. Affix type and leaders to a clear acetate overlay registered to base drawing for halftone artwork with labels. Apply labels and leaders directly to drawing board surface if artwork is only line ink.

Submit positive color transparencies (35 mm preferred) for evaluation. Do not send color prints without original transparencies.

Legends: Type double-spaced, beginning on a separate sheet. Limit to 40 words each.

Acknowledgments: Acknowledge illustrations from other publications. Include following when applicable: author(s), title of article, title of journal or book, volume number, page(s), month, and year. Submit publisher's permission to reprint after manuscript has been accepted.

Tables: Type double-spaced, including all headings, on a separate sheet (8.5×11 in.). Do not use larger paper. If continued, use a second sheet and repeat all heads and stubs. Give each a title.

OTHER FORMATS

CLINICAL OBSERVATIONS—Concise reports that provide potential new insights into pathophysiology, diagnosis, or treatment are more likely to be published.

OTHER

Color Charges: $400 for up to six square-finished illustrations arranged on a one page layout. Additional illustrations or special effects are billed to author at cost.

Experimentation, Human: In "Methods" section state that informed consent was obtained after the nature of the procedure(s) had been fully explained.

Refer to patients by number (or, in anecdotal reports, by fictitious given names). Do not use real names or initials in text, tables, or illustrations.

Include letters of consent with photographs of patients in which a possibility of identification exists. It is not sufficient to cover eyes.

Statistical Review: Include name and affiliation of statistical reviewer.

Style: *AMA Manual of Style* (1988, Williams & Wilkins, Baltimore, MD).

Units: Systéme International (SI). Add milligrams per deciliter in parentheses for cholesterol levels.

SOURCE
January 10, 1994, 154(1): 14-15.

Archives of Microbiology
Springer International

Gerhart Drews
Lehrstuhl Mikrobiologie
Institut fur Biologie il der Universitat
Schanzlestr. 1
D-79104 Freiburg 1. Br. Germany

Hans-Gunter Schlegel
Universitat
Grisebachstr. 8
D-37077 Gottingen Germany

Donn J. Kushner
Department of Microbiology
University of Toronto
Toronto, Ontario M5S 1AB Canada

AIMS AND SCOPE

The *Archives of Microbiology* publishes basic results on molecular aspects of structure, function, cellular organization and eco-physiological behavior of prokaryotic and eukaryotic microorganisms.

MANUSCRIPT SUBMISSION

Submission implies that work has not been published before (except as an abstract or as part of a published lecture, review, or thesis); that it is not under consideration for publication elsewhere; that its publication has been approved by all coauthors, if any, as well as by the responsible authorities where work was done; that upon acceptance authors agree to automatic transfer of copyright to publisher; and that manuscript will not be published elsewhere in any language without the consent of copyright holders.

Do not exceed 18 pages (5500 words) double-spaced, including figures and tables. Restrict illustrations to minimum needed to illustrate points that cannot be described in text, to summarize, or to record important quantitative results. Do not present same data in both table and graph form.

Consult a copy of the journal and conform with its normal practice. Submit in duplicate, double-spaced with wide margins. Mark in margin where figures and tables may be inserted.

Submit copies of relevant papers that are not easily accessible or are "in press" elsewhere.

MANUSCRIPT FORMAT

Title Page: Title of paper; first name(s) and surname(s) of author(s); laboratory or institution; new affiliations (in footnote); title footnotes (indicated by asterisks); corresponding author's address, telex or fax number; and a list of non-standard abbreviations.

Abstract: 200 words. Write so relevance can be determined quickly. Present problem, scope of experiments, significant data, major findings, and conclusions. Use complete sentences, active verbs and third person. Write in past tense. Use standard nomenclature. Define unfamiliar terms, abbreviations and symbols at first mention.

Key Words: After abstract, key words indicating scope of paper.

Text: Make introduction (not so headed) concise. Define scope of work in relation to other work. Do not review literature exhaustively. In materials and methods give sufficient detail to allow experiments to be repeated. Include detailed descriptions only of genuine innovations.

Present results with clarity and precision. Write in past tense to describe experimental findings; use present tense to refer to established findings. Include interpretation and explanation of results. Confine discussion to interpretation of results without repeating them.

Number footnotes consecutively and type on a separate sheet. Indicate placement of tables and figures in margin. Use stoke (/) only once in a line.

When describing solutions, follow "...dissolved in 50 ml 13 mM phosphate buffer pH 6.8." Include water of crystallization when describing composition of culture media in g/l (e.g., 10 g $MgSO_4$ DOT $7H_2O$). Do not pluralize units. Write 0.3 g NaCl/ml, not 0.3 g/ml NaCl.

Acknowledgments: List sources of financial support. Make personal acknowledgements only with permission of named persons.

References: Cite literature in text by author and year; where two authors, name both, with three or more give first author's name plus "et al." Include only works mentioned in text in reference list. Arrange list alphabetically by author name. Do not mention citations regarding "unpublished results" or papers "in preparation."

JOURNAL

Names and initials of all authors, year in brackets, full title, journal as abbreviated in *Chemical Abstracts*, volume number, first and last page number.

Nicolary, K, Scheffers WA, Bruinenberg PM, Kaptein R (1982) Phosphorus-31 nuclear magnetic resonance studies of intracellular pH, phosphate compartmentation and phosphate transport in yeasts. Arch. Microbiol. 133:83-89

BOOKS

Names and initials of all authors, year, full title, edition publisher, place of publication.

Esser K, Kuenen R (1968) Genetics of fungi. Springer, Berlin Heidelberg New York

Priefer U (1984) Characterization of plasmid DNA by agarose gel electrophoresis. In: Puhler A, Timmis KN (eds) Advanced molecular genetics. Springer, Berlin Heidelberg New York, pp 26-37

Place on separate sheet. Make sure all references are cited and all citations are referenced. Verify accuracy of references.

When two or more papers by the same author are cited, list chronologically. Arrange papers by two authors alphabetically by first and second authors. Three or more, chronological order.

Figures: Use with discretion. Illustrations should clarify or reduce text. Do not repeat information from text in captions. Mention all figures, graphs and tables in text. Number with Arabic numerals. Make sure symbols in figures and legends correspond.

Submit figures in form suitable for reproduction separately from text. Indicate top of figure, author's name and figure number lightly in soft pencil on figure back. Either match column width (8.6 cm) or printing area (17.8×24.8 cm). Group several figures into a plate on one page (17.8×24.8 cm). If plates are submitted, mount on regular bond paper, not cardboard. Figures plus legends should not exceed printing area.

Submit line drawings as good-quality glossy prints in desired final size. Make inscriptions clearly legible. If reduction is necessary, state alternative scale.

Computer drawings must be of comparable quality to line drawings. Curves and lines must be smooth. Make letters (capitals) 2 mm high.

Half-tone illustrations should be well contrasted photographic prints (not photocopies) trimmed at right angles and in desired final size. Inscriptions should be about 3 mm high. If reduction is necessary, state alternative scale desired.

Legends: Provide a brief descriptive legend for each figure; legends are part of text. Append to text on a separate page. Do not type on illustrations.

OTHER FORMATS

SHORT COMMUNICATIONS—Six double-spaced pages, including tables; 100 word abstract. Present new results of special interest.

OTHER

Abbreviations: For standard abbreviations that do not need to be defined see *Arch. Microbiol.* 161(1): iii-v (1994) or *Eur. J. Biochem.* 1: 259-266 (1967).

Color Charges: Approximately DM 1200 for first and DM 600 for each additional page.

Taxonomic Names: Mark genus species and strains and any part of text requiring emphasis for italics by underlining. Refer to strain under investigation by: genus (initial capital), species (small initial letter) and variety name, designation of the strain, collection number or source. Give full scientific name in tile, abstract, in each section, figure and table the first time referred to. After, indicate generic name with first letter. Contact international authorities on nomenclature to propose new bacterial names. Inform *International Journal of Bacteriology*.

Units: International SI units. See *Arch. Microbiol.* 1994; 161: iii-vii (yellow pages) for "Units, symbols, abbreviations and conventions," including prefixes for SI units, special names and symbols for derived SI units, examples of poor abbreviations, examples of standard abbreviations for semi-systematic or trivial names; standard symbols for chemical groups, trivial names for buffers, names and epithets from personal names, and latin plurals.

SOURCE

January 1994, 161 (i): i-viii (yellow pages).

Archives of Neurology
American Medical Association

Robert J. Joynt, MD, PhD
University of Rochester School of Medicine and Dentistry
601 Elmwood Ave.
Rochester, NY 14642

MANUSCRIPT SUBMISSION

In a cover letter designate one author as correspondent and provide complete address, telephone and fax numbers. Manuscripts should have no more than six authors; more requires justification. Add a publishable footnote explaining order of authorship (see: The International Committee of Medical Journal Editors. Statements from the International Committee of Medical Journal Editors. *JAMA* 1991;265:2697-2698). Include statement on authorship responsibility, statement on financial disclosure, and one of the two following statements on copyright or federal employment Each must be signed by all authors.

Authorship Responsibility. "I certify that I have participated sufficiently in the conception and design of this work and the analysis of the data (where applicable), as well as the writing of the manuscript, to take public responsibility for it. I believe the manuscript represents valid work. I have reviewed the final version of the submitted manuscript and approve it for publication Neither this manuscript nor one with substantially similar content under my authorship has been published or is being considered for publication elsewhere, except as described in an attachment. If requested, I shall produce the data upon which the manuscript is based for examination by the editors or their assignees."

Financial Disclosure. "I certify that any affiliations with or involvement in any organization or entity with a direct financial interest in the subject matter or materials discussed in the manuscript (e.g., employment, consultancies, stock ownership, honoraria, expert testimony) are disclosed below." Research or project support should be listed in an acknowledgment.

Copyright Transfer. "In consideration of the action of the American Medical Association (AMA) in reviewing and editing this submission, the author(s) undersigned hereby transfers, assigns, or otherwise conveys all copyright ownership to the AMA in the event that such work is published by the AMA."

Federal Employment. "I was an employee of the U.S. federal government when this work was conducted and prepared for publication; therefore, it is not protected by the Copyright Act and there is no copyright of which the ownership can be transferred."

Manuscripts may not be under simultaneous consideration by another publication. All accepted manuscripts become permanent property of AMA and may not be published elsewhere without written permission from both author(s) and AMA.

Prepare manuscripts in accordance with *American Medical Association Manual of Style* (Iverson CL, Dan BB, Glitman P, et al., 8th ed. Baltimore, Md: Williams & Wilkins; 1988) and/or "Uniform Requirements for Manuscripts Submitted to Biomedical Journals" International Committee of Medical Journal Editors (*N Engl J Med* 1991;324:424-428).

Submit original manuscript and three copies, typed on one side of standard-sized white bond paper. Use ample margins. Double-space throughout, including title page, abstract, text, acknowledgments, references, legends for illustrations, and tables. Start each section on a new page, numbered consecutively in upper right corner, beginning with title page. Put authors names only on title page.

Provide copy that can be scanned by an optical character reader: no smudges or pencil or pen marks. Use only standard 10- or 12-pitch type and spacing. Do not use 10 pitch type with 12 pitch spacing. If prepared on a word processor, do not use proportional spacing; use unjustified (ragged) right margins and letter-quality printing.

On title page, type full names, highest academic degrees, and affiliations of all authors. If an author's affiliation has changed since the work was done, list new affiliation also.

Abstract: 250 word structured abstract with original contributions and observations. No abstracts for editorials, commentaries, and special features.

References: Number references in order of mention in text; do not alphabetize. In text, tables, and legends, identify references with superscript Arabic numerals. List references in AMA style, abbreviating names of journals according to *Index Medicus*. List all authors and/or editors up to six; if more than six, list first three and et al.

1. Lomas J, Enkin M, Anderson GM, Hannah WJ, Vayda E, Singer J. Opinion leaders vs audit and feedback to implement practice guidelines: delivery after previous cesarean section. *JAMA*. 1991;265:2202–2207.

2. Marcus R, Couston AM. Water-soluble vitamins: the vitamin B complex and ascorbic acid. In: Gilman AG, Rall TW, Nies AS, Taylor P. *Goodman and Gilman's The Pharmacological Basis of Therapeutics*. 8th ed. New York, NY: Pergamon Press: 1990:1530–1552.

Verify accuracy and completeness of references and for correct text citation.

Tables: Double-space on separate sheets of standard-sized white bond paper. Title all tables and number them in order of citation in text. If a table must be continued, repeat title on a second sheet, followed by (cont).

Illustrations: Submit, in triplicate, 5 × 7 in. glossy photographs for all graphs and black-and-white photographs; high-contrast prints for roentgenograms; and color transparencies (carefully mounted and packaged) for color illustrations. Computer-generated graphics produced by high-quality laser printers (300 dots per in.) also are acceptable. Number illustrations according to order in text. Affix a label with figure number, name of first author, short form of title, and an arrow indicating TOP to back of print. Never mark print or transparency.

Double-space legends (40 words) on separate pages. Indicate magnification and stain used for photomicrographs.

Acknowledge all illustrations and tables taken from other publications and submit written permission to reprint from original publishers.

OTHER

Abbreviations: Do not use in title or abstract; limit use in text.

Drugs: Use generic names, unless specific tradename of a drug used is directly relevant to discussion.

Experimentation, Animal and Human: State in methods section that an appropriate institutional review board approved project. For those investigators who do not have formal ethics review committees (institutional or regional), follow principles outlined in Declaration of Helsinki should be followed (41st World Medical Assembly. Declaration of Helsinki: recommendations guiding physicians in biomedical research involving human subjects. *Bull Pan Am Health Organ.* 1990;24:606-609). For investigations of human subjects, state in methods section manner in which informed consent was obtained from subjects.

Include a signed statement of consent to publish all case descriptions and photographs from all identifiable patients (parents or legal guardians for minors).

Units: Use Systeme International (SI) measurements (Lundberg GD. SI unit implementation the next step. *JAMA.* 1988;260:73-76).

SOURCE

January 1994, 51(1): 101-102.

Archives of Ophthalmology
American Medical Association

Editor, Morton F. Goldberg, MD
Maumenee 727, The Wilmer Institute
The Johns Hopkins Hospital, 600 N Wolfe St.
Baltimore, MD 21287-9278
Telephone: (410) 955-1358; Fax: (410) 550-5374

Manuscripts must not have been previously published anywhere in any language and cannot be under simultaneous consideration by another publication. Manuscripts following oral presentations that result in publication of substantive information elsewhere, including magazines, or "tabloids" may be ineligible. Manuscripts may be submitted following presentation or publication of preliminary finding (e.g., in an abstract) at either the Annual Meeting of the American Academy of Ophthalmology or the Association for Research in Vision and Ophthalmology, only if publication in other print media is not under consideration. Include copies of possibly duplicative material.

Designate corresponding author and provide address and telephone number. If not a North American resident, provide a fax number, if possible.

Do not list more than six authors. If more than six, describe each person's personal contributions. Contributions must be substantial. If authorship is attributed to a group, either solely or in addition to one or more individual authors, all group members must meet criteria and requirements for authorship and submit signed Authorship Responsibility, Financial Disclosure, and Copyright Transfer forms. One or more authors may take responsibility for a group (byline will read "Jane Doe for the Collaborative Study Group,") and only those need to sign Authorship Responsibility, Financial Disclosure, and Copyright Transfer forms. In both cases, participating group members will be listed in footnote at end of article.

All authors must sign following statement submitted in a cover letter.

Copyright Transfer: "In consideration of the American Medical Association's taking action in reviewing and editing my submission, the author(s) undersigned hereby transfers, assigns, or otherwise conveys all copyright ownership to the AMA in the event that such work is published by the AMA."

Federal Employment: In case the work was done by a federal employee, each author must include a signed statement that the work reported was done while he or she was employed by the federal government.

Author Responsibility: "I certify that I have participated sufficiently in the conception and design of this work and the analysis of the data (when applicable and within my area of expertise), or the writing of the manuscript, to take public responsibility for it. I believe the manuscript represents valid work. I have reviewed the final version of the manuscript and approve it for publication. Neither this manuscript nor one with substantially similar content under my authorship has been published or is being considered for publication elsewhere, except as described in an attachment. Furthermore, I attest that I shall produce data upon which the manuscript is based for examination by the editors or their assignees, if requested."

Financial Disclosure: If a device, equipment, an instrument, or a drug is discussed, state in a footnote whether they do or do not have any commercial or proprietary interest in the product or company. Reveal whether they have any financial interest or receive payment as a consultant, reviewer, or evaluator. Disclose any financial interest owned by a spouse, minor child, or relative same house-hold, or known to be held by author's employer, partner, or business associate. The following must be true and signed by all authors:

"All affiliations with or involvement in any organization or entity with a direct financial interest in the subject matter or materials discussed in the manuscript (e.g., employ-ment, consultancies, stock ownership, honoraria) have been disclosed in an attachment to the manuscript. I un-derstand that such affiliation(s) will not affect consider-ation of the manuscript but may be published as a footnote if the manuscript is accepted."

MANUSCRIPT FORMAT

Submit original typescript and two high-quality copies. Type all copy (including references, legends, and tables) double-spaced on 8.5 × 11 in. paper with 1 in. margins. If a word processor is used, do not justify right margin. Text should not exceed 14 pages. Begin each section on a new page in order: title page; abstract; text; acknowledgments; references; tables; and figure legends.

Title Page: Make tile short, specific, and clear. Do not ex-ceed 42 characters per line, including punctuation and spaces. Limit to two lines. Include first name, middle ini-tial, and last name of all authors with highest academic de-gree and professional affiliations with city location. Include reprint request address.

Abstract: Use structured abstract for: Expedited Publica-tions, Clinical Sciences, Laboratory Sciences, and Epide-miology and Biostatistics sections. Follow instructions in *Journal of the American Medical Association* (1992; 268:42-44). For Expedited Publications, Clinical Sciences and Epidemiology and Biostatistics, limit to 250 words and use headings in sequence: Objective; Design; Setting, Patients or Other Participants; Intervention(s); Main Out-come Measure(s); Results; and Conclusions. For Labora-tory Sciences, limit to 150 words and use headings: Objective; Methods; Results; and Conclusions.

Use traditional free-text abstracts (136 words) for Clinico-pathologic Reports, Ophthalmology in Other Countries, New Instruments, Surgical Techniques, and Special Arti-cles. Brief articles such as Correspondence, Case Reports, and Photo Essays do not require abstracts.

References: List in numerical order (not alphabetically). Once cited, all subsequent citations should be to original number. Cite all references in text or tables. Do not list un-published data and personal communications as referen-ces but parenthetically within text. References to journal articles include: author(s) (if more than six, write "et al" after third name); title; journal name (as abbreviated in *In-dex Medicus*); year; volume number; and inclusive page numbers, in that order. References to books include: au-thor(s); chapter title (if any); editors (if any); title of book; city of publication; publisher; year; and page, if indicated. Authors are responsible for accuracy of references

Illustrations: Use only to clarify and augment text. Justi-fy more than 10. Submit, preferably 5 × 7 in., in triplicate, unmounted and untrimmed, package each set separately. Do not send original artwork. Send high-contrast glossy prints or original laser prints. Type figure number, name of senior author, and arrow indicating TOP on a gummed la-bel and affixed to back of each. Provide magnification and stain. If a specific figure arrangement is desired, include a photocopied layout. Crop out all extraneous portions on clinical photographs, CT scans, and ultrasound images, or indicate crop marks on photo margins.

Illustrations in full color may be accepted if color will add significantly to manuscript. Submit one set of positive trans-parencies (35- mm slides) and two sets of color prints.

Legends: Type double-spaced, beginning on a separate sheet. Limit to 40 words that make illustration understand-able without recourse to text. Use arrows, letters, etc., for enhanced understanding.

Acknowledgments: List research or project support. If in-dividuals are acknowledged, submit following signed statement: "I have obtained written permission from all persons named in the acknowledgment." Do not list typists or clerical help. Acknowledge writing assistance from a professional writer, editor, or other person not participat-ing in original clinical or research work. Disclose their identity upon submission.

Tables: Type double-spaced, including all headings, on a separate sheet of 8.5 × 11 in. paper. Do not use larger pa-per. If continued, use a second sheet and repeat all heads and stubs. Give each a title.

OTHER FORMATS

CORRESPONDENCE AND CASE REPORTS—500 words (three references, and two figures or one figure plus one ta-ble). Comments on articles published in the *Archives*, in-formative case reports, or other matters of ophthalmic interest.. Submit in triplicate double-spaced with a ragged right margin. Include a signed copyright transmittal letter.

PHOTO ESSAY—Emphasize visual aspects of subject pre-sented. Picture(s) should be of high quality and self-ex-planatory and can be of clinical entities, laboratory studies and findings, diagnostic techniques, therapeutic proce-dures, or a combination. It should be a collection of photos that convey an important laboratory or clinical diagnosis and/or management story that becomes self-evident on in-spection of the illustrations. Submit a concise essay (250 words) describing the clinical or laboratory information, photograph(s), and a brief list of references (five) in tripli-cate. No limit on number of photos. Limit articles to one or two printed pages, and all material must fit into this format.

QUESTIONS AND ANSWERS—Submit questions to Stewart M. Wolff, Section Editor, Archives of Ophthalmology, Physician's Pavilion at GBMC, Suite 503, 6565 N. Charles St., Baltimore, MD. 21204. Do not exceed six double-spaced typed lines. Include signed copyright transmittal letter.

OTHER

Color Charges: $400 for up to six square-finished illus-trations that can be arranged on a one-page layout. Submit a statement indicating willingness to pay. Additional illus-trations or special effects are billed to author at cost.

Experimentation, Animal: Indicate in "Methods" sec-tion what animal-handling protocols were followed, e.g.,

"Institutional guidelines regarding animal experimentation were followed."

Experimentation, Human: State in "Methods" section that an appropriate institutional review board has approved project and/or informed consent has been obtained from all adult participating subjects and from parents or legal guardians of participating minors. Refer to patients by number, e.g., patient 1 (or, in anecdotal reports, by fictitious names).

Recognizable photographs require specific written consent from patient. Observe restrictions or limitations on consent. Include a specific statement that photograph and information related to a case may be published either separately or in connection with each other, in professional journals or medical books, provided that it is specifically understood that the patient shall not be identified by name.

For parental consent include signatures of both living parents or legally appointed guardian.

Reprinted Material: Illustrations from other publications are rarely published but if used must be acknowledged. Include: author(s), title of article, title of journal or book, volume number, page(s), month, and year. Include publisher's permission to reprint.

Statistical Consultation: Obtain statistical consultation for studies with statistical content. Include name and affiliation of statistical consultant.

Style: Follow: Uniform requirements for manuscripts submitted to biomedical journals. (*JAMA*, 1993; 269:2282-2286) and *AMA Manual of Style*, (1989 edition, Williams & Wilkins, Baltimore, MD).

Units: Use Systéme International (SI) measurements. Exceptions are visual acuity measurements and intraocular pressure recordings.

SOURCE

January 1994, 112(1): 12–14.

Archives of Oral Biology

Pergamon Press

Dr. D. B. Ferguson
University of Manchester, School of Biological Sciences
G38 Stopford Building, Oxford Road
Manchester M13 9PT, U.K.

Dr. E. J. Kollar
Department of Oral Biology
University of Connecticut Health Center
Farmington, CT 06032

AIM AND SCOPE

Archives of Oral Biology publishes papers concerned with advances in knowledge of every aspect of the oral and dental tissues and bone over a whole range of vertebrates, whether from the standpoint of anatomy, bacteriology, biophysics, chemistry, DNA biotechnology, epidemiology, genetics, immunology, molecular biology, paleontology, pathology, physiology, or otherwise.

MANUSCRIPT SUBMISSION

Submission implies that it has not previously been published, that it is not under consideration for publication elsewhere and that, if accepted will not be published elsewhere without consent of Editorial Board.

MANUSCRIPT FORMAT

Type double-spaced on one side of paper, with a left margin of not less than 40 mm. Submit three copies, complete with all illustrations. Draft quality print out is not acceptable. Divide into sections, i.e., Introduction; Materials and Methods; Results or Findings; and Discussion.

Title Page: Make titles as concise and informative as possible. Include animal species. Indicate type of method on which observations are based, e.g., chemical, bacteriological, electron-microscopic or histochemical, etc.

Supply a running title (40 letters and spaces).

Give addresses of authors in full. Repeat name, address, and fax number of corresponding author below running title.

Supply a key word index.

Summary: Give entire paper in miniature. Do not repeat what is in title. Avoid sentences such as "the findings are discussed."

Introduction: A succinct statement of problem investigated, within a brief review of relevant literature. Reserve literature directly relevant to inferences or arguments in Discussion for that section. Conclude with the reason for doing work. Do not state what was done nor findings.

Materials and Methods: Give enough detail so others can repeat procedures exactly. Where materials and methods were exactly as in a previous paper, do not repeat details but give sufficient information for readers to comprehend what was done without having to consult earlier work.

Results or Findings: Be clear and concise. Avoid drawing inferences that belong in Discussion. Data may be presented in various forms such as histograms or tables. Do not present same data in more than one form.

Analyze numerical results statistically. State number, mean value and appropriate measure of variability. Indicate method of analysis. With a statement that the difference between mean values of two groups of data is statistically significant, give probability level set as significant and indicate statistical test used. Do not quote use of a statistical package without naming tests used.

Discussion: Present inferences drawn from Results. Recapitulate sparingly, sufficient to make argument clear.

Acknowledgments: As appropriate.

Legends: Type on a separate page at end of paper. Limit to description of figure.

Tables: Type, with legend, on a separate sheet at end of paper. Avoid footnotes.

Illustrations: Do not be insert in text. All photographs, charts and diagrams are referred to as "Figures" (abbreviated to "Fig."), and are numbered consecutively in order referred to in text. Number tables separately.

Lightly write on back author's name, figure number and top of picture.

Photographs, including photomicrographs, should be glossy prints. Arrange illustrations so they can be examined without rotating page.

Publisher will insert numbers, pointers or lettering to appear on photographs. Draw exact position of pointers or lettering on an overleaf of tracing paper, firmly attached to photograph. Include registration marks to ensure that, if overleaf is displaced, it can easily be returned to correct register. Where positioning pointers is critical, place directly on one set of prints for guidance. Encircle specific areas of photograph of particular importance on an overleaf. Indicate magnifications by a scale bar on photomicrograph or electron-micrograph itself. Give magnification of photographic print in the legend.

Draw charts and line drawings boldly in India ink at least twice the size they are likely to be printed. If not convenient to submit originals, submit large size photograph. Leave lettering to be inserted by publisher.

Depict essential chromatographic and simple electrophoretic data as densitometric traces or diagrams unless photographs of sufficient quality can be submitted.

OTHER FORMATS

SHORT COMMUNICATIONS—1200 words without summary and references and excluding photographs and tables, to communicate brief but definitive investigations. Do not exceed two printed text pages (approximately). Begin with a brief summary and follow usual pattern, Introduction, Materials and Methods, Results, etc. Do not use actual headings.

OTHER

Abbreviations: Do not use in title. Spell out in full at first mention, except those universally understood. List as a footnote on title page.

ADP, AMP, ATP, DEAE-cellulose, DNA, RNA, EDTA, EMG, tris, mm, g, min, u.v., w/v may be used without definition.

Chemical symbols may be used for elements, groups and simple compounds, but avoid excessive use.

Experimentation, Animal: State that conditions of animal experiments were humane; specify mode of anaesthesia and killing.

Experimentation, Human: State briefly that subjects gave informed consent; and preferably that work was approved by an Ethical Committee.

Nomenclature, Microorganisms: Scientific names of bacteria are binomials, with generic name only with a capital, underlined once (for italics).

Give name in full at first mention; in subsequent mentions, abbreviate generic name unambiguously. Avoid single letter abbreviations (thus: *Staph. aureus, Strep. pyogens,* not *S. aureus, S. pyogens).* When generic name defines a group, it should have a capital and be italicized; trivial names, or generic names used as adjectives should not have capitals nor be italicized.

Proprietary Names: Use proper names instead of proprietary names. Where desirable to indicate a particular brand, give proprietary name and source in parentheses.

Spelling: Follow *Oxford English Dictionary*; thus, anaesthesia, dentine, haematoxylin, sulphate, technique.

Style: Use *Writing Scientific Papers in English* (1975, M. O'Connor and F. P. Woodford, Associated Scientific Publishers, Amsterdam).

Express findings in past tense. Use present tense to refer to existing knowledge, or to state what is known or concluded.

Units and Symbols: SI (Systéme International) units and symbols. Follow: *Symbols, Signs and Abbreviations* (1969) Royal Society, or "Metric and Decimal Systems" in *Council of Biology Editors Style Manual* (1978, 4th edn, Council of Biology Editors Inc). Write liter in full in place of SI dm^3 and ml in place of SI cm^3.

Define units of enzyme activity, preferably using SI units.

State centrifugal field in multiples of g, rather than rev/min.

SOURCE

January 1994, 39(1): iii-vi.

Archives of Otolaryngology—Head & Neck Surgery

American Academy of Facial, Plastic and Reconstructive Surgery

American Society for Head and Neck Surgery

American Society of Pediatric Otolaryngology

American Medical Association

Michael E. Johns, MD
The Johns Hopkins University School of Medicine
School of Medicine Administration, Room 100
720 Rutland Ave., Baltimore, MD 21205-2196
Telephone: (410) 550-5226; Fax: (410) 550-5228

MANUSCRIPT SUBMISSION

Manuscripts must not have been published previously in print or electronic format and may not be under consideration by another publication or electronic medium.

Photocopy and sign statements (copies or faxed signatures are not acceptable) on authorship responsibility, financial disclosure, and appropriate copyright or federal employee statement found in *Archives* January, 1994 issue, volume 120, page 16. (See also Lundberg GD, Flanagin A. New requirements for authors: signed statements of authorship responsibility and financial disclosure. *JAMA.* 1989; 262:2003–2004).

MANUSCRIPT FORMAT

Submit original manuscript and two photocopies, typed on standard-sized white bond paper, using 1 in. margins.

Double-space throughout, including title page, abstract, text, acknowledgments, references, legends, and tables. Number pages consecutively in upper right corner, beginning with top page. Use unjustified right margins and letter quality printing.

Titles should not exceed 75 characters, including punctuation and spacing. Avoid abbreviations in title or abstract and limit their use in text.

Title Page: Type full names, highest academic degrees, and affiliations of all authors. (If affiliation has changed, list new affiliation also.) Designate a corresponding author and include a complete mailing address, telephone (and fax number if available). Specify reprint request address. If manuscript was presented at a meeting, state meeting name, city where held, and exact date of reading.

Abstract (page 2): Include a structured abstract (250 words) for reports of original data from clinical or basic science investigations and reviews (including meta-analyses). Include headings. Use phrases rather than complete sentences.

For other manuscripts abstract is no more than 150 words.

See *Arch Otolaryngol Head Neck Surg*, 1994; 120(1): 12-165 for "Instructions for preparing structured abstracts." Instructions are included for Reports of Clinical Data, Reports of Basic Science, and Review Manuscripts (including Meta-analysis). A glossary of methodological terms is also included. Instructions are based on Haynes RB et al. "More informative abstracts revisited," *Ann Intern Med*. 1990; 113:69-76.

References: Number in order mentioned in text; do not alphabetize. Identify with superscript Arabic numerals. Follow AMA style, abbreviate names of journals according to *Index Medicus*. List all authors and/or editors up to six; if more than six, list first three and "et al."

Verify accuracy and completeness of references.

Illustrations: Submit three sets of all figures: unmounted, untrimmed (5 × 7 in.) high-contrast glossy prints for all graphs and black-and-white photographs; high-contrast glossy prints for roentgenograms; or color prints accompanied by original positive color (35-mm) transparencies. Do not send glass-mounted transparencies. Affix a typed gummed label with figure number, first author, and arrow indicating TOP on back of each figure or frame of slides. Acknowledge illustrations and tables reprinted from other publications. Submit written permission to reprint from original publishers.

Legends: 40 words each typed doubled-spaced, on a separate sheet. Indicate magnification and stains used for photomicrographs. Include specific postoperative intervals where applicable.

Tables: Double-space on separate sheets of standard-sized white bond paper. Title all tables and number in order of citation in text. If continued, repeat title, head, and stubs on a second sheet, followed by "(cont)."

Acknowledgments: List research or project support.

Obtain written permission from all persons named. In cover letter state: "I have obtained written permission from all persons named in the Acknowledgment."

OTHER FORMATS

CASE REPORTS—Six double-spaced pages, including references, with an unstructured abstract (150 words). Focus on a new disease state (with diagnostic documentation including pathologic findings), identification of a new complication from treatment or procedure, new diagnostic technique, or new technology or application from another field. Do not report an old disease at a new site.

RESIDENT'S PAGE: IMAGING—Five pages double-spaced, three figures. Quiz cases and letters commenting on cases presented from residents and fellows.

RESIDENT'S PAGE: PATHOLOGY—Quiz cases and letters commenting on cases presented from residents and fellows. Photomicrographs can include an outline drawing with important structures labeled. Submit illustrations as positive color transparencies (35 mm preferred).

State of the Art—Twelve double-spaced pages, including bibliography (20 references, first 4 are Suggested Readings), four tables and/or figures. Practical, up-to-date information with a core of most recent scientific knowledge on topic distilled through personal experience into a clinical perspective. Explain how to apply data to every day patient care. Do not review literature exhaustively.

OTHER

Drugs: Use generic names of drugs, unless specific trade name is directly relevant to discussion.

Experimentation, Animal: Specify in "Methods" section what animal-handling protocols were followed, e.g., "Institutional guidelines regarding animal experimentation was followed." For those who do not have formal ethics review committees (institutional or regional), follow principles of Declaration of Helsinki. See 41st World Medical Assembly. Declaration of Helsinki: recommendations guiding physicians in biomedical research involving human subjects. *Bull Pan Am Health Organ*. 1990;24:606–609. Include signed statements of consent to publish all case descriptions and photographs from all identifiable patients (parents or legal guardians for minors).

Experimentation, Human: State in "Methods" section that the appropriate institutional review board approved project. For those who do not have formal ethics review committees (institutional or regional), follow principles outlined in the Declaration of Helsinki. Specify in "Methods" section manner in which consent was obtained from all human subjects. Include signed statements of consent to publish all case descriptions and photographs from all identifiable patients (parents or legal guardians for minors).

Style: Iverson CL, Dan BB, Glitman P, et al. *American Medical Association Manual of Style*. 8th ed. Baltimore, MD: Williams & Wilkins; 1988. International Committee of Medical Journal Editors. Uniform requirements for manuscripts submitted to biomedical journals. *N Engl J Med*. 1991;324:424–428.

Units: Use Systéme International (SI) measurements. See Lundberg, GD. SI unit implementation: the next step. *JAMA* 1988;260:73-76.

SOURCE

January 1993, 119:9–12.

Archives of Pathology & Laboratory Medicine

College of American Pathologists
American Medical Association

William W. McLendon, MD
University of North Carolina School of Medicine
Department of Pathology, CB #7525
Chapel Hill, NC 27599-7525
Telephone: (919) 966-5902

MANUSCRIPT SUBMISSION

Manuscripts may not be under simultaneous consideration by another publication.

In cover letter designate one author as correspondent and provide complete address and telephone number. Limit authors to six; more require justification. Coauthors must have contributed to study and manuscript preparation, be familiar with final manuscript and be able to defend conclusions.

Include one of following statements on copyright and statement on the financial interests of authors, signed by all.

Copyright Transfer: "In consideration of the American Medical Association's taking action in reviewing and editing this submission, the author(s) undersigned hereby transfer(s), assign(s), or otherwise convey(s) all copyright ownership to the AMA in the event that this work is published by the AMA."

Federal Employment: "I was an employee of the United States Federal Government when this work was investigated and prepared for publication; therefore, it is not protected by the Copyright Act and there is no copyright of which the ownership can be transferred."

Financial Interest: List all affiliations with or financial involvement in any organization or entity with a direct financial interest in the subject matter or materials of the research discussed in the manuscript (e.g., employment, consultancies, stock ownership). All such information will be held in confidence during the review process. Should the manuscript be accepted, the editor will discuss with the author the extent of disclosure appropriate.

MANUSCRIPT FORMAT

Submit an original manuscript and two photocopies, typed on one side of standard-sized white bond paper. Use ample margins. No appendices; add data to text.

Double-space throughout, including title page, abstract, text, acknowledgments, references, legends, and tables. If a word processor is used, use a letter-quality printer and do not justify right margins.

Start each section on a new page, numbered consecutively in upper right corner, beginning with title page.

Title Page: Type full names, highest academic degree obtained, and affiliations of all authors. If affiliation has changed since work was done, list new affiliation also. Briefly acknowledge financial support. State name and affiliation of any statistical reviewer consulted.

Abstract: 135 words; state problem considered, methods, results, and conclusions. Do not use abbreviations in title or abstract, and limit use in text.

Tables: Double-space on separate sheets of standard-sized white bond paper. Title and number in order of citation in text. If continued, repeat title on a second sheet, followed by "(cont)." List abbreviations used in a key below table.

Illustrations: Submit in triplicate as professionally prepared glossy photographs. Submit high-contrast print roentgenographic photographs.

Number according to order in text. Affix figure number, name of senior author, short form of manuscript title, and an arrow indicating TOP to back of print. Never mark on print or transparency.

Color figures may be submitted (one transparency and two color prints).

Legends: Double-space (40 words) on separate sheets of standard paper. Indicate magnification and stain for photomicrographs.

Acknowledge previously published illustrations. Submit written permission to reprint from original publisher.

References: Number in order mentioned in text; do not alphabetize. Identify with superscript Arabic numerals. Submit preprints for items cited as in press. Follow AMA style, abbreviate journal names according to *Index Medicus*. List all authors and/or editors up to six, if more than six, list first three and "et al."

1. DuBois RN, Lazenby AJ, Yardley NJ, Hendrix TR, Bayless TM, Giardiello FM. Lymphocytic enterocolitis in patients with 'refractory sprue.' *JAMA*. 1989; 262:935–937.

2. Ritchie JM, Greene NM. Local anesthetics. In: Gilman AG, Goodman LS, Rall TW, Murad F, eds. *The Pharmacological Basis of Therapeutics*. 7th ed. New York, NY: Macmillan Publishing Co Inc.; 1985:309–310.

OTHER FORMATS

BRIEF REPORTS—1500 words, three figures and/or tables, and ten references. Double-spaced, without right justified margin. Submit in triplicate with assignment of copyright.

LETTERS TO THE EDITOR—500 words, five references. Double-spaced without justified right margin. Submit in duplicate with assignment of copyright.

OTHER

Color Charges: Author is charged $400 for up to six square-finished illustrations arranged on a one-page layout. Submit letter of intent to pay.

Drugs: Use generic names of drugs, unless specific trade name of drug used is directly relevant to discussion.

Experimentation, Human: State that consent was obtained from subjects after nature of procedure(s) was explained. Include signed statement of consent from patient (or, a minor, from both parents or legal guardian) with all identifiable photographs. In form state that photographs and information related to a case may be published separately or together, and that patient's name will not be disclosed.

Reagents: Give sources (name of company and location) for all special reagents.

Units: Use Systéme International (SI) measurements.

SOURCE
January 1994, 118(1): 6.

Archives of Pediatric & Adolescent Medicine
American Medical Association

Catherine D. DeAngelis, MD
The Johns Hopkins University School of Medicine
Office of the Senior Associate Dean
720 Rutland Ave., Suite 106
Baltimore, MD 21205-2196

MANUSCRIPT SUBMISSION

Manuscripts are considered with the understanding that they have not been published previously and are not under consideration by another publication. Give name, address, affiliation, and telephone and fax number of corresponding author in cover letter. Limit authors to six. Include all of the following statements signed by all authors (original signatures).

On Authorship: "I certify that I have participated sufficiently in the conception and design of this work and the analysis of the data (when applicable), as well as the writing of the manuscript, to take public responsibility for it. I believe the manuscript represents valid work. I have reviewed the final version of the manuscript and approve it for publication. Neither this manuscript nor one with substantially similar content under my authorship has been published or is being considered for publication elsewhere, except as described in an attachment. Furthermore, I attest that I shall produce the data upon which the manuscript is based for examination by the editors or their assignees if requested."

On Financial Disclosure: I certify that affiliations (if any) with or involvement in any organization or entity with a direct financial interest in the subject matter or materials discussed in the manuscript (e.g., employment, consultancies, stock ownership, honoraria, expert testimony) are disclosed below."

On Copyright Transfer: "In consideration of the action of the American Medical Association (AMA) in reviewing and editing this submission, the author(s) undersigned hereby transfer(s), assign(s), or otherwise convey(s) all copyright ownership to the AMA in the event that such work is published by the AMA."

Federal Employees only: "I was an employee of the U.S. federal government when this work was investigated and prepared for publication; therefore, it is not protected by the Copyright Act and there is no copyright of which the ownership can be transferred."

Submit four copies of manuscript typed double-spaced throughout, including title page, abstracts, text, acknowledgments, references, legends for illustrations, and tables. Number pages. Do not justify right margins.

MANUSCRIPT FORMAT

Title Page: Give full names, degrees, and academic affiliations of all authors, address for requests of reprints, and, if manuscript was presented at a meeting, organization, place, and exact date on which it was read. Title should be no more than 75 characters.

Abstract: Include a 250 word structured abstract. See *Arch. Pediatr. Adolesc. Med.* 148(1): 113 for instructions for preparing structured abstracts.

Tables: Type each table, with a title, on a separate sheet of paper, with each line, including headings, double-spaced. Put continuations on a second sheet and repeat all headings.

Illustrations: Submit high-contrast, glossy prints, in quadruplicate, in a proportion of 5×7 in. Type figure number, name of first author, and arrow indicating TOP on a gummed label on back of each illustration. Do not write directly on print.

Submit full color illustrations as 35-mm, positive color transparencies, mounted in cardboard, and carefully packaged.

All photographs in which there is a possibility of patient identification should be accompanied by a signed statement of consent from both parents (or guardians). Covering eyes to mask identity is not sufficient.

References: List in order of appearance in text, typed double-spaced in AMA format. List all authors and/or editors up to six; if more than six, list first three and "et al." Verify accuracy of references.

JOURNALS

Bier DM, Fulginiti VA, Garfunkel JM, et al. Duplicate publication and related problems. AJDC. 1990;144:1293-1294.

BOOKS

Naeye RL. How and when does antenatal hypoxia damage fetal brains? In: Kubli F, Patel N. Schmidt W, Linderkamp O, eds. *Perinatal Events and Brain Damage in Surviving Children.* New York, NY: Springer Verlag NY Inc; 1988:83-91.

Do not include unpublished data, personal communications, or manuscripts "in preparation" or "submitted" in reference list. Incorporate such material, if essential, in body of article.

Acknowledgments: Acknowledge illustrations and tables from other publications, with written permission from publisher and author.

OTHER FORMATS

EDUCATIONAL INTERVENTION—Educational efforts in the broad field of pediatrics.

PATHOLOGICAL CASE OF THE MONTH—Submit directly to Enid Gilbert-Barness, MD, Departments of Pathology, Tampa General Hospital, University of South Florida, Davis Island, FL 33606.

PEDIATRIC ADVOCACY—A forum for discussion of broad issues relating to children, policy, and the law in today's environment.

PEDIATRIC FORUM—A mixture of brief, peer-reviewed observations or investigations and letters to the editor and other short commentaries.

PEDIATRIC PERSPECTIVES—Current information on clinical and scientific pediatric medical advances and analyses of current controversies in pediatrics.

PEDIATRIC REVIEWS—Timely, comprehensive, well-referenced topics.

PICTURE OF THE MONTH—Submit directly to Walter W. Tunnessen, Jr., MD, Children's Hospital of Philadelphia, Room 2143, 34th and Civic Center Blvd., Philadelphia, PA 19104.

RADIOLOGICAL CASE OF THE MONTH—Submit directly to Beverly P. Wood, MD, Childrens Hospital of Los Angeles, Department of Radiology, 4650 Sunset Blvd., Los Angeles, CA 90054-0700.

REVIEW MANUSCRIPTS (INCLUDING META-ANALYSIS)—Include an abstract (no more than 250 words) under: Objective, Data Sources, Study Section, Data Extraction, Data Synthesis, and Conclusions.

OTHER

Color Charges: $400 for up to six square-finished illustrations that fit on one page. A letter of intent to pay fee must accompany submission.

Experimentation, Human: State in "Methods" section that an appropriate institutional review board approved project and/or that informed consent was obtained from both legal guardians and/or child, if appropriate. Submit signed consent statements from both parents (or guardians) for photographs in which there is a possibility of patient identification. Covering eyes to mask identity is not sufficient.

Style: Use proper English usage and syntax; consult the *American Medical Association Manual of Style* (Williams & Wilkins, Baltimore, MD).

Terms: See *Arch. Pediatr. Adolesc. Med.* 148(1): 114-115 for Glossary of Methodologic Terms.

Units: Systéme International (SI) measurements.

SOURCE

January 1994, 148(1): 111-112.

Archives of Psychiatric Nursing
Society for Education and Research in Psychiatric-Mental Health Nursing
W.B. Saunders

Judith B. Krauss, Editor
Archives of Psychiatric Nursing
Yale University School of Nursing
25 Park St., PO Box 9740
New Haven, CT 06536-0740

AIM AND SCOPE

The purpose of the *Archives of Psychiatric Nursing* is to disseminate knowledge to guide practitioners of psychiatric and mental health nursing. The *Archives of Psychiatric Nursing* considers psychiatric and mental health nursing in its broadest perspective, including theory, practice, and research applications related to all ages, special populations, settings, mental health disciplines, and both the public and private sectors. The *Archives of Psychiatric Nursing* is a medium for clinician-scholars to provide theoretical linkages among diverse areas of practice. Manuscripts are sought and published that will inform current practice and shape public policy for the delivery of psychiatric and mental health nursing services.

MANUSCRIPT SUBMISSION

Manuscripts are voluntary contributions submitted for exclusive use of *Archives of Psychiatric Nursing*.

Submit three copies of manuscript and three camera-ready copies of each illustration. Type manuscripts or print on a high-quality printer double-spaced, with 1.5 in. margins. Include two face sheets: one with manuscript title, name and address of author(s), institution affiliations, and reprint address; and the other with only title, and no identifying author information. Do not exceed 16 pages (4000 words), exclusive of references. Line drawings and illustrations must be camera-ready copy; photocopies are not acceptable. Photographs must be 5 × 7 in. glossy black-and-white prints.

MANUSCRIPT FORMAT

Make titles short. Include an abstract (100 words) on a separate page. Type or draw tables and figures on one page. Note relative placement in text.

Obtain and submit written permission for reproducing copyrighted materials.

OTHER

Abbreviations: Spell out first time used.

Style: Follow *American Psychological Association Publication Manual* (3rd ed,1983) for format for text and references.

Refer questions to production editor at W.B. Saunders Company, (215) 238-8321.

Symbols: Except standard statistical symbols, identify first time used.

SOURCE

February 1994, 8(1): front matter.

Archives of Surgery
American Medical Association

Editor, Claude H. Organ, Jr, MD
Professor, Dept of Surgery
University of California—Davis, East Bay
1411 E 31st St, Oakland, CA 94602

MANUSCRIPT SUBMISSION

Send manuscripts by first-class mail. Manuscripts must not be under simultaneous consideration by another publication. In cover letter designate a corresponding author, give address and telephone number. Specify reprint request author and address. Include statements on copyright and financial disclosure, dated and signed by all authors.

Copyright Transfer: "In consideration of the American Medical Association's taking action in reviewing and editing my submission, the author(s) undersigned hereby transfers, assigns, or otherwise conveys all copyright ownership to the AMA in the event that such work is published by the AMA."

Authorship Responsibility: "I certify that I (1) have participated sufficiently in this work to take public responsibility for the content of this manuscript; (2) believe the experimental design and method as well as the collection, analysis, and interpretation of the data are sound; and (3) have reviewed the final version of the manuscript and approve it for publication. This manuscript has not been published and is not being considered for publication elsewhere except as described in an attachment hereto."

Federal Employee: "I was an employee of the U.S. federal government when this work was investigated and prepared for publication."

Financial Disclosure: Sign the following statement: "I certify that all affiliations with or involvement in any organization or entity with a direct financial interest in the subject matter or materials discussed in the manuscript (e.g., employment, consultancies, stock ownership, honoraria) have been disclosed in an attachment. I understand that such affiliation(s) will not affect consideration of the manuscript but may be published as a footnote is accepted."

MANUSCRIPT FORMAT

Submit original typescript and two high-quality copies, including copies of all illustrations, legends, tables, and references. Type all copy, including references, legends, and tables, double-spaced on 8.5 × 11 in., heavy-duty white bond paper with 1 in. margins. Do not justify right margins.

Abstract: 250 words structured abstract, state problem, methods of study, results, and conclusions. See *Archives of Surgery*, 1993; 128: 377 for instructions on preparing a structured abstract.

References: List in consecutive numerical order as cited in text, not alphabetically. Once cited, all subsequent citations are to original number. Cite all references in text or tables. Do not list unpublished data and personal communication as references. References to journal articles include: authors (list all authors and/or editors up to six; if more, list first three and "et al"); title; journal name as abbreviated in *Index Medicus;* year; volume number; and inclusive page numbers in that order. References to books include: authors (list all authors and/or editors up to six; if more, list first three and "et al"); chapter title, if any; editor, if any; title of book; year; city; and publisher. Include volume and edition numbers, specific pages, and name of translator when appropriate. Verify accuracy and completeness of references and correct text citation.

Title: Short, specific, and clear. Limit to 42 characters per line, including punctuation and spaces; two lines only. On title page include full names and academic affiliations of all authors. If manuscript was presented at a meeting, indicate name of organization, place, and date it was read.

Illustrations: Clarify and augment text. Submit in triplicate, unmounted and untrimmed. Do not send original artwork.

Send high-contrast glossy prints (not photocopies). Computer-generated graphics produced by high-quality laser printers are acceptable. Type figure number, name of senior author, and arrow indicating TOP on a gummed label and affixed to back. Make lettering large enough to be legible after reduction to column size. Make illustrations in a proportion of 5 × 7 in. Color illustrations are accepted if color adds significantly to manuscript. Submit positive color transparencies (35 mm preferred) for evaluation. Send color prints with original transparencies.

Legends: Type double-spaced, 40 words each. Begin on a separate sheet.

Tables: Type double-spaced, including all headings, on a separate sheet of 8.5 × 11 in. paper. Do not use larger paper. If continued, use a second sheet and repeat all heads and stubs. Give each table a title.

Acknowledgments: Acknowledge illustrations from other publications; include when applicable: authors, title of article, title of journal or book, volume number, pages, month and year. Submit publisher's permission to reprint after acceptance.

OTHER FORMATS

BRIEF CLINICAL NOTES—400 words (synopsis-abstract 80 words), two references, and one illustration. Case reports that provide a new or unique contribution.

CORRESPONDENCE AND BRIEF COMMUNICATIONS—250 words, typewritten, double-spaced without an abstract, and clearly marked "for publication." No more than two references and only illustrations or tables essential to message. Letters pertain to material published in *Archives* and other matters of interest to our readers. An interesting case may be published as a brief communication.

OTHER

Color Charges: $400 for up to six square-finished illustrations arranged on a one-page layout. Additional illustrations or special effects are at cost.

Experimentation, Animal: Indicate that institution's or National Research Council's guide for or any national law on care and use of laboratory animals was followed.

Experimentation, Human: In "Methods" section state that institutional review board approved project and/or informed consent was obtained from all adult subjects and parents or guardians of minors.

Submit letters of consent with photographs of patients in which a possibility of identification exists. It is not sufficient to cover eyes to mask identity.

Use numbers or, in anecdotal reports, fictitious given names. Do not use real names or initials in text, tables, or illustrations.

Statistical Review: Include name and affiliation of statistical reviewer.

Style: *AMA Manual of Style* (1988, Williams & Wilkins, Baltimore, MD).

Units: Use Systéme International (SI) measurements.

SOURCE

June 1994, 129(6): 569-570.

Arteriosclerosis and Thrombosis
European Vascular Biology Association
American Heart Association

Editor, *Arteriosclerosis and Thrombosis*
Editorial Office, Division of Cardiology
UCLA School of Medicine, Room 47-123 CHS
Los Angeles, CA 90024-1679
European Editor, *Arteriosclerosis and Thrombosis*
Wallenberg Laboratory, Sahlgren's Hospital
S-413 45 Gothenburg, Sweden

MANUSCRIPT SUBMISSION

Manuscripts must be submitted solely to this journal. Do not submit articles in which a significant portion of data has been published elsewhere. If any form of prior publication, other than a short abstract, has occurred or is contemplated, submit reprint or copy. All authors must have seen and approved final manuscript. Cite all sources of support from both profit and nonprofit organizations.

Submit four copies of each manuscript and set of illustrations. Also submit diskette. Double-space throughout. Begin title page, summary, text, acknowledgments, references, legends, tables, and figures on separate sheets in that order.

A dated form letter, containing signatures of all authors, must state: "We, the authors, assign first and subsidiary rights to the American Heart Association in the event that our manuscript (title) is published by *Arteriosclerosis and Thrombosis.*"

MANUSCRIPT FORMAT

Title Page: Make title brief and specific. Include name, address, academic affiliation, and highest degree(s) for each author; acknowledge of grant support where appropriate; address for galley proofs and reprint requests; a short title (40 characters) for a running head; and three to 10 key words or phrases for indexing.

Abstract: 200 words on purpose, methods, results, and conclusions.

Text: Divide into "Introduction," "Methods," "Results," and "Discussion."

References: Cite consecutively by number (3) in text and list this way in Reference list. Cite all references. Verify accuracy of references. Style follows Uniform Requirements. Cite personal communications and unpublished data in parentheses in text. Submit letters with the direct quotation and author's signature. Give inclusive page numbers and all authors' names.

Tables: Cite in text. Give numbers and brief informative titles. Omit vertical or horizontal rules. Explain abbreviations in footnotes. For footnotes, use symbols in sequence: *, †, ‡, §, ‖, ¶, #, ††, ‡‡, §§, ‖‖.

Illustrations: Request world rights from publishers and authors for figures and tables reproduced or adapted from previously published material. Enclose in a separate envelope. Write sufficient identifying information lightly in pencil on backs.

Artists should prepare drawings with black India ink on a white background. High-quality laserjet illustrations are acceptable for line drawings. Letters, numbers, and symbols should be clear and even when reduced should be equal to 9 points in type size. Submit glossy photographs of original drawings.

Submit photomicrographs as unmounted glossy prints no larger than 2 columns wide. Submit composite illustrations as separate figures with a sketch to show desired layout. Place internal scale markers on photomicrographs. For color illustrations, send two sets of prints.

Legends: Brief, but sufficient description for figure interpretation on a separate sheet.

OTHER

Abbreviations: Consult *American Medical Association Manual of Style,* (Eighth Ed. American Medical Association, Chicago, IL). Define at first appearance.

Drugs: Use generic names of drugs.

Experimentation, Animal: State species, strain, number used, and other pertinent descriptive characteristics. Identify preanesthetic and anesthetic agents used and amount or concentration and route and frequency of administration for each. Use of paralytic agents, such as curare or succinylcholine, is not an acceptable substitute for anesthetics. For other invasive procedures, report analgesic or tranquilizing drugs used. If none, justify exclusion. When reporting studies on unanesthetized animals, indicate that procedures followed were in accordance with institutional guidelines.

Experimentation, Human: Describe subject characteristics. Indicate that procedures followed were in accordance with institutional guidelines.

Page Charges: $50 per printed page; expense for color reproduction of figures; $50 per printed page for excessive author alterations.

Reviewers: Submit names, addresses, telephone numbers, and fax numbers of four or five potential reviewers.

Style: Follow Uniform Requirements for Manuscripts Submitted to Biomedical Journals *(Ann Intern Med* 1982;96:766-771).

Units: Systeme International (SI) units. See "Transition to SI units," *Arteriosclerosis and Thrombosis* 14(1): A6.

SOURCE

January 1994, 14(1): A5.
Instructions to authors appear in January and July issues.

Arthritis and Rheumatism
American College of Rheumatology
J. B. Lippincott

Peter H. Schur, MD Editor
Arthritis and Rheumatism
Editorial Office Room 422
Richardson Fuller Bldg
221 Longwood Avenue
Boston, MA 02115
Telephone: (617) 732-7638; Fax: (617) 731-4964

AIM AND SCOPE

Articles should definitely pertain to the field of arthritis and rheumatism

MANUSCRIPT SUBMISSION

Send a cover letter signed by corresponding author stating type of manuscript (Full-Length Article, Brief Report, Concise Communication, or Letter to the Editor), and that manuscript is not being simultaneously submitted elsewhere. Include telephone and fax numbers of corresponding author.

Submit copies of related manuscripts that have been or will be published by or submitted to another journal or to *Arthritis and Rheumatism.*

Articles must not been published elsewhere (in part or in full, in other words or in the same words, in letter or article form, or otherwise), are not under consideration by another journal or publication, and will not be submitted elsewhere unless rejected.

MANUSCRIPT FORMAT

Submit 4 sets of manuscript, illustrations, and tables.

Type on 8.5 × 11 in. paper with 1 in. margins, double or triple spaced throughout, including references, tables, and figure legends. Do not exceed 15 typed pages (not including references, tables, and figure legends).

Number all sheets in succession. Title page is page 1; type title, name(s) of author(s) with major degrees, source of work or study, grant supporter(s), reprint request address, and corresponding author's telephone and fax numbers. Include 3 key words.

Second page, two to three sentence summary of how work is relevant to clinical rheumatologists. Third page, abstract, 75 words divided into: Objective, Methods, Results, and Conclusion. Fourth page begin introduction (no heading); continue with Materials and Methods (or Patients and Methods), Results, Discussion, References.

In Methods section include a short paragraph detailing proportion of patients who satisfy the ACR and classification criteria for disease described.

Illustrations: Send 4 sets of illustrations (camera-ready only), packaged as 4 sets, in labeled envelopes. Attach a Xerox copy of each figure at end of manuscript. Attach a label to each figure with number and first author's name. Do not use paperclips.

Submit radiographs unretouched, oriented with patient's left on observer's right, with same tonal relationship as original (i.e., bones white on a dark background). Crop prints to show only area of interest. Arrows or other identifying labels should be professional quality and placed on final print only, so removal is possible.

Radiographs, micrographs, other halftones, and linecuts should be no more than 3.25 in. square.

To print four-color photographs submit color slides. Author must pay all costs.

Tables: Refer to current issues for table style.

Provide each table with an explanatory title so that it is intelligible without reference to text. Give each column with an appropriate heading, which can be abbreviated. Indicate clearly any units of measure.

References: Compile numerically according to order of citation. Use title abbreviations from *Index Medicus.* Double space.

JOURNAL

For articles with multiple authors, list all contributors. Cite inclusive page numbers.

1. Barger BO. Acton RT. Koopman WJ. Alarcon GS: D'R antigens and gold toxicity in white rheumatoid arthritis patients. Arthritis Rheum 27:601–605, 1984

BOOKS

2. Gilman AG. Advances in Cyclic Nucleotide Research. Vol. II. New York, Raven Press, 1972

CONTRIBUTION TO BOOK

3. Castor CW Jr: Regulation of connective tissue metabolism, Arthritis and Allied Conditions. Ninth edition. Edited by DJ McCarty. Philadelphia, Lea & Febiger, 1979

OTHER FORMATS

BRIEF REPORTS—9 manuscript pages, including references, 3 tables and/or figures, and 15 references. Follow format above. Investigations into disease mechanisms, reports of clinical experience, therapeutic trials, or research and, for clinical contributions to diagnosis, treatment, etiopathology, and epidemiology of rheumatic diseases.

CASE REPORTS—9 manuscript pages including references, 3 tables and/or figures, and 15 references. Include an abstract (second page, 75 words). Page three is introduction (no heading) followed by description of case and general discussion. Single cases that contribute to the body of knowledge of rheumatic diseases by reporting in-depth observations and/or laboratory investigations that may reveal etiopathogenetic mechanisms of a broader significance, or by presenting a concise discussion of an infrequently encountered problem that emphasizes a particular clinical point not generally available in the literature.

LETTERS TO THE EDITOR/CONCISE COMMUNICATIONS—4 pages, including references, 1 table or figure. Double-space. Commentaries on previous articles in *Arthritis and Rheumatism* and issues affecting rheumatology and American College of Rheumatology. Concise Communications are very short reports of cases or research findings, with no abstract, subheads, or acknowledgments section.

OTHER

Abbreviations: Define first time used. A list of standard abbreviations appears in each January and July issue.

Experimentation, Human: Obtain releases from patients in pictures. Blinders on eyes are not a substitute for patient's permission.

Page Charge: $70 per page. Charge may be reduced or waived. Direct questions to Managing Editor, *Arthritis and Rheumatism,* 60 Executive Park South, Suite 150, Atlanta, GA 30329.

For estimates of cost of printing color photographs, write to *Arthritis and Rheumatism,* J.B. Lippincott Co., 227 East Washington Square, Philadelphia, PA 19106.

A $75 allowance per article is provided for tables, black and white figures, and authors' alterations; authors bear excess costs. Alterations to page proofs are charged against this allowance.

Style: Do not use new technical words, laboratory slang, words not defined in dictionaries, or abbreviations or terminology not consistent with internationally accepted guidelines.

SOURCE

January 1994, 37(1): front matter.

Behavioral Ecology and Sociobiology
Springer International

Tatiana Czeschlik
Editorial Office, *Behavioral Ecology and Sociobiology*
Springer-Verlag
Tiergartenstr. 17, D-69121 Heidelberg, Germany
Fax: (+49) 6221-409840; e-mail: bes@sprint.compuserve.com

AIM AND SCOPE

The journal publishes original contributions dealing with quantitative empirical and theoretical studies in the field of the analysis of animal behavior on the level of the individual, population and community. Special emphasis is placed on the proximate mechanisms, ultimate functions and evolution of ecological adaptations of behavior.

Aspects of particular interest: Intraspecific behavioral interactions, with special emphasis on social behavior; Interspecific behavioral mechanisms, e.g., of competition and resource partitioning, mutualism, predator-prey interactions, parasitism; Behavioral ecophysiology; Orientation in space and time; Relevant evolutionary and functional theory.

Purely descriptive material is not acceptable unless it is concerned with the analysis of behavioral mechanisms or with new theory.

MANUSCRIPT SUBMISSION

Submission implies: that work has not been published before (except as an abstract or as part of a published lecture, review, or thesis); that it is not under consideration for publication elsewhere; that its publication has been approved by all coauthors, if any, as well as by responsible authorities where work was carried out; that, if and when manuscript is accepted, authors agree to transfer of copyright to publisher; and that manuscript will not be published elsewhere in any language without consent of copyright holders.

Submit manuscripts in quadruplicate, typed double-spaced. Type original manuscript on one side of paper, three copies may be photocopied on both sides.

MANUSCRIPT FORMAT

Title Page: Include: Title; Authors' names and initials; Author's affiliation; Footnotes to title or author, marked by asterisks; author's fax number and e-mail address.

Summary: Of main results. Refer to important results expressed in figures by citing their number.

Key Words: 3-5.

Text: Divide into: Introduction; Methods; Results; Discussion; Acknowledgments; References.

Do not exceed 25 typescript pages. Outline main problem briefly in introduction. Avoid detailed historical introductions. Establish continuity with earlier work by reference to recent papers or reviews. Clearly set out experimental methods. Give detailed descriptions of methods only if substantially new information is conveyed. Do not divide papers into several parts to make them appear shorter. No more material should be included in one paper than can be treated in one coherent discussion. Briefly state each category of experimental result or each newly established fact. Summarize remaining material or give in tables. Avoid lengthy tables. State in a footnote where full supporting evidence can be obtained.

Footnotes: Keep to a minimum, place at foot of page to which they apply. Number consecutively throughout paper.

References: In text give author's surname with year in parentheses, e.g., Carlin (1992); Brooks and Carlin (1991); Carlin et al. (1992).

List references at end of paper in alphabetical order by first author's name; list all works referred to in text, and only those.

JOURNAL PAPERS

Names and initials of all authors; year; full title; journal abbreviated in accordance with international practice; volume number; first and last page numbers:

Seeley TD, Towne WF (1992) Tactics of dance choice in honey bees: do foragers compare dances? Behav Ecol Sociobiol 30:59–69

SINGLE CONTRIBUTION IN A BOOK

Names and initials of authors; year; title (of book or article); editor(s); (title of book); edition; volume number; publisher; place; page numbers:

Noirot C (1992) Sexual castes and reproductive strategies in termites. In: Engels W (ed) Social insects. Springer, Berlin Heidelberg New York, pp 5–35

BOOK

Holldobler B, Wilson EO (1990) The ants. Springer, Berlin Heidelberg New York

Tables: Number consecutively in Arabic numerals. Keep to a minimum. Type each on a separate sheet with a clear descriptive legend.

Illustrations: Mark approximate position of figures and tables in margin of manuscript. Color illustrations are accepted.

Keep to minimum needed to clarify text. Number all figures, photographs or diagrams, consecutively throughout in Arabic numerals. Refer to all illustrations in text. Submit on separate sheets, with figure number and author's name. Group figures within maximum display area for figures and legends (17.8 × 24.1 cm).

Give each illustration a concise, descriptive legend. Legends are part of text. Type consecutively on sheets separate from figures. Do not repeat data presented in tables in figures.

Submit photographs with good contrast. Insert arrows, letters, and numbers with template rub-on letters. State magnifications in legend. Refer to a bar in figure.

Line Drawings: Submit good-quality glossy prints or ink drawings in desired final size. Make inscription clearly legible. Make capital letters about 2 mm high in final version.

OTHER

Color Charges: DM 1,200 for first and DM 600 for each additional page.

Genus and Species Names: Underline (for italics). Do not capitalize common names of animals.

SOURCE

1994, 34(1): inside front and back cover.

Biochemical and Biophysical Research Communications

Academic Press

John N. Abelson
Division of Biology 147-75
California Institute of Technology
Pasadena, CA 91125
Fax: (818) 796-7066

Ernesto Carafoli
Laboratorium fur Biochemie
ETH-Zentrum
CH-8092 Zurich, Switzerland
Fax: 41-1-252-323

I.C. Gunsalus
Environmental Research Laboratory
1 Sabine Island Drive, Bldg. #37
Gulf Breeze, FL 32561-5299
Fax:(904) 934-9201

Akira Ichihara
2-7-55 Minamiyaso
Tokushima 770 Japan
Fax: 81-886-32-4104

James D. Jamieson
Department of Cell Biology
Yale School of Medicine
333 Cedar Street, P.O. Box 333
New Haven, CT 06510
Fax:(203) 785-7445

Yasuo Kagawa
Department of Biochemistry
Jichi Medical School
Minamikawachi-Machi Kawachi-gun
Tochigi-ken 329-04, Japan
Fax: 81-28-544-1827

M. Daniel Lane
Department of Biological Chemistry
The Johns Hopkins University
725 No. Wolfe Street
Baltimore, Maryland 21205
Fax:(410) 955-5759

William J. Lennarz
Department of Biochemistry and Cell Biology
SUNY at Stony Brook
Stony Brook, NY 11794-5215
Fax:(516) 632-8575

Masami Muramatsu
Department of Biochemistry
Saitama Medical School
Moroyama, Iruma
Saitama 350-04, Japan
Fax: 0492-94-9751

Claude Paoletti
Laboratoire de Biochimie
INSERM-CNRS
Institut Gustave-Roussy
94805 Villejuif, France
Fax: 33-1-46-78-4120

Bennett M. Shapiro
Merck Research Laboratories
RY80K, P.O. Box 2000
Rahway, New Jersey 07065-0900
Fax: (908) 594-6645

AIM AND SCOPE

Biochemical and Biophysical Research Communications is devoted to the rapid dissemination of timely and significant observations in the diverse fields of modern experimental biology. Manuscripts submitted should be so written as to emphasize clearly the novel aspects of the information reported.

MANUSCRIPT SUBMISSION

Send manuscripts in duplicate to any editor listed above. Submit a signed photocopy of copyright assignment form on page 4 of the August 30, 1994 issue. Manuscript must be approved by all authors and institution where work was carried out.

Manuscripts must be original. The same work may not have been submitted to another Editor of this journal and may have been published or submitted for publication elsewhere. Persons cited as a source of personal communications must approve such citation. Written authorization may be required. Submit copies of all preliminary communications and all relevant manuscripts in press or under editorial consideration by another journal.

MANUSCRIPT FORMAT

Submit as camera-ready copy. Use one-and-a-half spacing and 12-point (minimum) type on 8.5 × 11 in. high-quality paper. Typing area on page must not be larger than 6.5 × 9.5 in. Do not exceed 8 pages, including figures and tables. Single-space methods, references, footnotes, tables and figure legends. Type footnotes, references, tables, and figure legends on separate pages.

Manuscripts generated as word-processor or computer printouts must be "letter quality." Dot-matrix printouts will be returned. For typewritten manuscripts, use a black silk or carbon typewriter ribbon. Ensure a sharp clean impression of letters throughout. Use an electric typewriter. Avoid erasures and errors in spelling, word division, and punctuation. No changes may be made after acceptance.

Type page numbers at top of page. Type title in capital letters near top of first page, with name(s) and affiliation(s) of author(s) just below.

Summary: 100-150 words, single-spaced, summary of pertinent findings. At bottom of page include telephone and fax numbers of corresponding author.

Illustrations and Tables: Do not insert figures, figure legends, or tables into text. Label all drawings in black ink in a manner suitable for direct reproduction (Leroy lettering size 120 or 140). Do not use press-on lettering; typewritten lettering is not acceptable.

Figures will be reduced to fit type area of page (approximately 5×8 in.); original drawings should be no larger than 8.5×11 in.

Glossy photographs of drawings must be sharp, clear prints with an even black density. Authors must defray cost of color illustrations.

Ink carefully in black mathematical or chemical symbols that cannot be typed. Draw structural formulas in black ink. Submit illustrations and tables on separate sheets.

References: Cite by number in parentheses in text. List in numerical order in single-spaced list at end of manuscript. Use journal abbreviations from *Chemical Abstracts Service Source Index,* 1985.

JOURNAL

1. Graan, T., and Ort, D. R. (1986) Arch. Biochem. Biophys. 248, 445-451.

BOOK

2. Metzler, D. E. (1977) Biochemistry: The Chemical Reaction of Living Cells. p. 298. Academic Press, New York.

CHAPTER IN BOOK

3. Rosenfeld, R. G., and Hintz, R. L. (1986) In The Receptors (P. M. Conn, Ed.), Vol. III, pp. 281-329. Academic Press, Orlando, Fl.

Give inclusive pagination.

OTHER

Abbreviations and Symbols: Items in Tables I, VIII, and XI of Instructions to Authors of *Archives of Biochemistry and Biophysics* (1993, **300**, xiii-xxviii), and those indicated in other tables, may be used without definition. Avoid use of all other abbreviations and symbols. When essential, be consistent with form and style of above document. Define in a single footnote. Type on separate page. Do not use in Summary.

Data Submission: Use databases as repositories for detailed data on protein structure and DNA sequences and mapping assignments. Complete data entry and annotation forms and submit (in floppy disk, tape, or hardcopy form) to database prior to or within one week after submission of manuscript so that mapping or sequencing work can be accessed by data retrieval methods. Provide copies of data entry forms or other verification of submission and pertinent accession numbers. Article will include a footnote indicating database to which data have been provided and accession number.

Nomenclature: See Instructions to Authors of *Archives of Biochemistry and Biophysics* (1993) **300,** xiii-xxviii.

Nomenclature, Enzyme: Do not abbreviate except when substrate has an approved abbreviation. Use Enzyme Commission ("EC") code numbers.

Nomenclature, Isotope: Follow recommendations of IUBMB Commission of Editors of Biochemical Journals. Place symbol of isotope introduced in square brackets directly attached to front of name (e.g., $[^{32}P]CMP$). Use isotope number only as a (superior) prefix to atomic symbol, not to an abbreviation.

Safety: In Methods section, draw attention to chemical or biological hazards involved in experiments. Describe relevant safety precautions; reference accepted codes of practice followed.

SOURCE

August 30, 1994, 203(1): back matter.

Biochemical Journal
Portland Press

Managing Editor, *Biochemical Journal*
59 Portland Place
London W1N 3AJ United Kingdom
Telephone: +4471-637 5873; Fax: +4471-323 1136

AIM AND SCOPE

The *Biochemical Journal* publishes papers in all fields of biochemistry, provided that they make a sufficient contribution to biochemical knowledge. Papers may include new results obtained experimentally, descriptions of new experimental methods of biochemical importance, or new interpretations of existing results. Theoretical contributions will be considered equally with papers dealing with experimental work. All work presented should have as its aim the development of biochemical concepts rather than the mere recording of facts. Preliminary, confirmatory or inconclusive work will not be published.

MANUSCRIPT SUBMISSION

Do not submit material that has been wholly or largely published elsewhere, even as a preliminary communication or in unrefereed symposium proceedings; fragmentation of research into the "least publishable unit" is discouraged.

Submission implies approval by all named authors, that all persons entitled to authorship have been so named, that manuscript reports unpublished work that is not under consideration elsewhere, and that if the paper is accepted for publication authors will transfer copyright to the Biochemical Society, which will then not be published elsewhere in the same form, in any language, without the consent of the Society. Authors must sign an undertaking to these effects.

Submit copies of relevant published work and all related papers in press or under editorial consideration in this or other journals.

Submit three copies of typescript (four copies for Research Communications), with a brief covering letter and an additional copy of synopsis. On typescript indicate name, address, telephone, and fax numbers of corresponding author. Mark top copy of typescript as such and attach original artwork; for other copies, provide glossy prints (*not* photocopies) of half-tone figures. Photocopies of line drawings will suffice.

Fax Submissions: Criteria for Research Communications and BJ Letters, meet length requirement of Communications or Letter; not exceed 20 A4 pages, including cover letter, double-spaced text, tables, figures, legends and any supporting material; no half-tones. Retain original artwork until receipt acknowledged, then send to editorial office with manuscript number and additional copy of manuscript.

Do not number separate articles in a series; use subtitles. Indicate preferred section in table of contents for paper: Proteins; Enzymes; Carbohydrates and lipids; Gene structure and expression; Regulation of metabolism; Membranes and bioenergetics; Receptors and signal transduction; Cell biology and development.

MANUSCRIPT FORMAT

Consult a current issue to become familiar with general format, such as use of cross-headings, layout of tables and figures and citation of references. Double-space throughout (particularly references and table and figure legends) on numbered sheets of uniform size (preferably ISO A4) with wide margins. Low-quality dot matrix printouts are not acceptable. Research Papers are between six and eight printed pages. A page of text contains approximately 1200 words; when calculating printed length of papers, allow for space occupied by tables and figures; reduce number of text words accordingly.

Give paper a concise informative title with no nonstandard abbreviations. List authors' names with one forename in full, others as initials. Give name and address of establishments where work was done. If more than one, link author or establishment with symbols *, †, ‡, §, ||, ¶. Give a short (page-heading) title (75 characters).

Synopsis: 250 words; do not include inessential details or material not in text. References here are not part of numbered reference system; cite in full (name all authors, publication year, journal title, volume number and inclusive pages).

Main Body: Divide into: introduction; experimental, including materials and methods; results; discussion; acknowledgments, including financial support; and references. Results and discussion may be combined.

References: Use either Harvard or numerical system.

Harvard System—Cite references by one or two authors in text as: (Low, 1989) or Hooper and Turner (1989); for papers written by three or more authors as: Relton et al. (1983) or (Hooper et al., 1987). Cite more than one paper by same author in same year as Hooper et al. (1990a,b). Cite multiple references chronologically.

List references at end of paper in alphabetical order, except for papers by three or more authors, which are grouped in chronological order after any other papers by first author. Include authors' initials. Do not list titles of papers.

Hooper, N.M. and Turner, A.J. (1989) Biochem. J. **261**, 811-818
Hooper, N.M., Low, M.G. and Turner, A.J. (1987) Biochem. J. **244**, 465-469
Hooper, N.M., Keen, J.N. and Turner, A.J. (1990a) Biochem. J. **265**, 429-433
Hooper N.M., Hryszko, J. and Turner, A.J. (1990b) Biochem. J. **267**, 509-515

Numbering System—Cite references in text by sequential numbers in square brackets. List references in numerical order at end of paper. Use same style.

1 Hooper N.M., Hryszko, J. and Turner, A.J. (1990b) Biochem. J. **267**, 509-515
2 Hooper, N.M. and Turner, A.J. (1989) Biochem. J. **261**, 811-818
3 Hooper, N.M., Keen, J.N. and Turner, A.J. (1990a) Biochem. J. **265**, 429-433
4 Hooper, N.M., Keen, J.N. and Turner, A.J. (1990a) Biochem. J. **265**, 429-433

Both Systems—Abbreviate journal titles in accordance with *Chemical Abstracts Service Source Index* (1907-1989 Cumulative) and subsequent quarterly supplements.

For references to books and monographs, follow:

5 Turner, A.J. and Hooper, N.M. (1990) in Molecular and Cell Biology of Membrane Proteins: The Glycolipid Anchors of Cell-Surface Proteins (Turner, A.J., ed.), pp. 129-150, Ellis Horwood, Chichester

References to papers 'in the press' must be accompanied by evidence of acceptance. Include name of journal:

Smith, A. (1994) Biochem. J., in the press

For references to "personal communication" and "unpublished work" submit names of all persons concerned; obtain and submit written permission for quotation from person or persons concerned; both types of citation are permitted in text only, not in list of references.

Illustrations: Keep to a minimum. Put each on a separate sheet. Pack flat. Put author's name, paper title, and figure number on back. Indicate approximate position in manuscript margin.

Supply artwork in a form that can be reproduced directly by printer. Allow for a 40-50% reduction. Use black on white paper or card. Graph paper must have pale blue lines. Line thickness 0.4 mm is desired. Draw curves, lines, and symbols clearly. Do not extend axes beyond curves; give only relevant part of axis scale. Computer-generated curves must be smooth. Symbols must be large enough for reduction.

Preferred symbols for experimental points are: ○ ❑ △ ● ■ ▲. Do not use same symbols on two curves where points might be confused. Curves may also be distinguished by distinctive line forms, single-letter labels, or brief expository labels.

Do not exceed 17 cm width for single-column and 35 cm for double-column illustrations. Keep 3 cm margin.

Do not submit simple bar diagrams with only a few values. Present as table or incorporate in text.

Submit amino acid and nucleotide sequences in vertical alignment in high-quality camera-ready form.

Submit half-tone illustrations as glossy prints trimmed to intended reproduction size. Indicate magnification with a bar.

Legends: Give illustrations informative titles and explanatory legends. Make meaning comprehensible without reference to text. Group in a section at end of text. Start each on a new line. Type double-spaced.

Tables: Indicate position in text. Supply with informative headings and explanatory legends, starting on a new line. Type double-spaced on separate sheets. Heading and legend make meaning clear without reference to text. Use footnotes only to draw attention to a feature of a particular row, column, or value. State conditions specific to experiment. Give units of results at top of columns; do not repeat on each line. Do not repeat words, or numerals on successive lines; do not use 'ditto' or ditto marks.

OTHER FORMATS

BJ LETTERS—Two printed pages; brief items of scientific correspondence providing an opportunity to discuss or expand particular points made in published work, to comment on or criticize work previously published in the *Biochemical Journal*, to discuss features of novel nucleotide and/or amino acid sequence data of particular biological significance, or to present new hypothesis. Do not include extensive new data. Do not use to publish preliminary results. If a letter is polemical in nature, a reply may be solicited from other interested parties. No synopsis.

RESEARCH COMMUNICATIONS—4000 words, four printed pages, including references; short papers bringing particularly novel and significant findings to the attention of the research community. Include in letter of submission why Communication merits accelerated treatment. Follow usual style; 60 word synopsis; include enough experimental detail to permit replication of work. May be submitted by fax if no half-tone figures.

REVIEWS—Usually solicited, unsolicited reviews will be considered. Consult Reviews Editor, via editorial office, and enclose a short (one typed page) summary of proposal. No synopsis.

OTHER

Abbreviations: See Table 1. Abbreviations for systematic or semi-systematic names, or for method (*Biochem. J.* 1994; 297:5) for those that may be used without definition in title and in page-heading title. Keep use of others to a minimum. Define together in footnote on title page. Follow recommendations of Nomenclature Committee of IUBMB and IUPAC-IUBMB Joint Commission on Biochemical Nomenclature.

Amino Acid Nomenclature: Use full residue names or three-letter symbols. Symbolize a phenylalanine residue at position 231 as Phe-231 or Phe[231]. For polymers or sequences, join three-letter symbols by hyphens if sequence is known, with commas if not. For special consideration of spacing and punctuation of one-letter symbols, see *Biochem J* 1984, **219**, 366-368.

Centrifugation: Give sufficient information to repeat experiment. State quantitative composition of suspension medium. Identify centrifuge rotor and temperature of operation. State time of operation of rotor at sustained plateau speed. State centrifuge field in multiples of g, based on average radius of rotation of liquid.

For density-gradient centrifugation, plot results against distance from rotor center rather than against fraction numbers; it is then unnecessary to indicate top and bottom of gradient. If fraction numbers are used, indicate top and bottom of gradient.

For ultracentrifuge data: Sedimentation coefficient (not constant), s; sedimentation coefficient corrected at 20°C in water, $s_{20, w}$; sedimentation coefficient at zero concentration, $s^o, s^o_{20, w}$; Svedberg unit (10^{-13}s), S; partial specific volume, v; diffusion coefficient D, D^o, $D_{20,w}$, etc. as sedimentation coefficient. State temperature at which sedimentation and diffusion measurements made.

Chromatography: Express rate of movement of a substance relative to solvent front in paper or thin-layer chromatography as its R_F value, or if relative to a reference compound, by its $R_{compound}$ value. Describe solvents in form butan-l-ol/acetic acid (4:1,v/v).

Show elution diagrams for chromatographic columns with effluent volume increasing from left to right. Show units and volume of concentration. Quote column dimensions and give column void volumes (V_o). Characterize elution zone maxima by elution volumes (V_c) or by partition coefficients (α or $K\pi$). Indicate course of eluent gradients.

Color Charges: £1200 for first figure; £200 each additional. Appeal to Chair of Editorial Board if unable to pay.

Computer Programs: Document programs by reference to previously published original source, or deposition of program listing with a suitable depository.

Data Submission: Deposit computer programs, evidence for amino acid sequences, spectra and other information supplementing pagers with the British Library Document Supply Centre (DSC), Boston Spa, Wetherby, West Yorkshire, LS23 7BQ U.K.

Deposit structural data required to validate by X-ray crystallography proposed structure and its discussion. State in footnote that data has been deposited with Protein Data Bank, Brookhaven National Laboratory, Upton, NY 11973.

Electrophoresis: Quote electrophoretic mobilities (m), composition of electrophoretic medium, pH, temperature, and operative voltage. Use symbol pI for isoelectric point.

Enzymes: For nomenclature and EC numbers, follow latest edition of *Enzyme Nomenclature* (San Diego: Academic Press, 1992).

Define units of amount of enzyme. May be in terms of rate of reaction catalyzed under conditions specified. SI unit for rate is 1 mol of substrate transformed/s or 1 mol of measured product formed/s. This gives unit of amount of enzyme that has been given the name of katal (symbol: kat). Units of amount of enzyme may be expressed in terms of amount that can catalyze other rates.

Represent velocity constants for forward and backward reactions by k_{+n} and I_{-n}. Michealis constant is $K_m =$[S] when

$v=V/2$, where v is velocity of appearance of product or disappearance of substrate at a give substrate concentration [S] and V (or V_{max}) is velocity when enzyme is saturated with substrate. When reactions with two substrates A and B are being considered, $K_m=$[A] when $v=V/2$ and [B] has been extrapolated to infinity; a value for [A] when $v=V/2$ at a specified finite concentration of B should be referred to as apparent K_m for A. K_s is equilibrium constant of the dissociation of the substrate-enzyme complex. Catalytic-center activity (not turnover number) is defined as the number of molecules of substrate transformed/s per catalytic center.

Ethics, Publishing: See American Chemical Society's "Ethical Guidelines to Publication of Chemical Research," *Biochemistry* 25: 9A-10A, 1986.

Experimentation, Animal: Perform experiments in accordance with legal requirements of relevant local or national authority. Animals must not suffer unnecessarily. Detail experimental procedures and anesthetics used. See *Guidelines on the Use of Living Animals in Scientific Investigations* (London: Biological Council).

Experimentation, Human: Include a statement that research was carried out in accordance with Declaration of Helsinki of the World Medical Association, that the Ethical Committee of the institutional where work was performed approved it, and subjects gave informed consent.

"Homology": Preserve distinction between "homologous" (having a common evolutionary origin) and "similar." See Reeck, G. R. et al. *Cell* 40: 667, 1987 and R. Lewin, *Science* 237: 1570, 1987.

Isotopes: Express possible radioactivity in absolute terms; SI unit for radioactivity is Bq (becquerel) = 1 disintegration /s; Curie (Ci; 1 Ci=3.7×10^{10}); or disintegrations per unit of time.

Place symbol for isotope in square brackets attached directly to front of name. When more than one position in a substance is labeled by the same isotope and positions are not indicated, add number of labeled positions as a right-hand subscript. 'U' refers to uniform and 'G' to general labeling.

Isotopic prefixes precede that part of name to which it refers. Do not contract ^{131}I-labeled albumin to [^{131}I]-albumin or write ^{14}C-labeled amino acids as [^{14}C]amino acids.

Arrange isotope symbols for more than one element in alphabetical order, including 2H and 3H for deutrium and tritium.

If still not sufficiently distinguished, indicate positions of isotopic labeling by Arabic numerals, Greek letters or prefixes in square brackets before element symbol, attached by hyphen.

Use same rules when labeled compound is designated by a standard abbreviation or symbol.

For simple molecules, indicate labeling by writing chemical formulae with prefix superscripts attached to proper atomic symbol. Do not use square bracket.

See *Eur. J. Biochem.*86: 9-25, 1978 for a discussion of distinction between isotopically labeled and isotopically substituted compounds.

Materials Sharing: Make samples of unique biological material available to academic workers upon request. Deposit cell lines of more than local interest with appropriate collections at national centers, e.g., National Collection of Animal Cell Cultures, Centre for Applied Microbiology and Research, Porton Down, Salisbury, Wilts, SP4 0J6 or Amture Collections, 12301 Parklawn Dr., Rockville, MD 20852.

Molecular Mass: Relative molecular mass (M_r; not molecular weight) is ratio of mass of a molecule to 1/12 of mass of nuclide ^{12}C; it is dimensionless. Molecular mass is the mass of one molecule of a substance expressed in daltons (Da) or atomic mass units; dalton is 1/12 of the mass of one atom of ^{12}C. Do not express M_r in daltons or use K to represent M, 1000 or 1 kDa. Use M_r or molecular mass (kDa) consistently throughout.

Nomenclature, Animal: Use full binomial Latin name for all experimental animals except common laboratory animals. State strain and source. Specify source and composition of animal diet.

Nomenclature, Biochemical: Follow recommendations of IUBMB and IUPAC-IUBMB Joint Commission on Biochemical Nomenclature. See *Biochemical Nomenclature and Related Documents* (2nd ed, London: Portland Press, 1992).

Nomenclature, Chemical: Follow IUPAC recommendations in *Compendium of Chemical Terminology* (Oxford: Blackwell Scientific Publications, 1987).

Use chemical symbols for elements, groups and simple compounds. Use R, R', R" or R^1, R^2, R^3, if more than three, to denote variable substitutions in formulae. Use C_{20} acid to denote an acid containing 20 carbon atoms and C-3 or $C_{(3)}$ to denote carbon atom numbered 3. Use $C_{18:0}$, $C_{18:1}$ to denote number of double bonds in an unsaturated fatty acid.

Representations: Na^+, Zn^{2+}, Cl^-, PO_4^{3-}.

Run chemical names together except for acids, acetals, esters, ethers, glycosides, ketones, and salts; print as separate words. Use hyphens to separate numbers, Greek letters or some configurational and italic prefixes from words.

Use prefixes *R*- and *S*- to differentiate chiral compounds whose absolute configuration is known (see *Pure Appl. Chem.* 45: 11-30, 1976). Use small capital letters D-, L-, and DL- for chiral components and their racemates. When direction of optical rotation is all that can be determined, use (+)-, (–)- and (+/–)-, or *dextro, laevo*, and 'optically active.'

Use italics for *cis-, trans-,o-*, *m-, p-, dextro, laevo, meso* and *O-, N-*, etc to indicate an element carrying a substituent. Do not use italics for allo, bis, cyclo, epi, iso, n, neo, nor, s, t, tris. Follow alphabetical order for prefixes denoting substituents.

Nomenclature, Microbial: Give full binomial Latin names in title, synopsis, and first mention in text. Obtain from or deposit with a recognized collection. Give collection number. Or, quote strain name or number; do not underline. Do not italicize names of ranks higher than genus, generic names used adjectivally, and microbial names used colloquially. Spell generic name with a capital letter. Use single letter abbreviations for generic name in text; if two genera have the same letter, use abbreviations.

Verify identities. Deposit new organisms in pertinent culture collections.

Follow nomenclature recommendations in Demerec, M, et al. *Genetics* 54: 61-76, 1966. See also *Genetic Maps*, ed. S.J. O'Brien (Cold Spring Harbor, NY: Cold Spring Harbor Laboratory).

Nomenclature, Nucleosides, Nucleotides, and Polynucleotides: Symbols for ribonucleosides (use prefix r if possible ambiguity): A= adenosine; C= cytidine; G = guanosine; T = ribosylthymidine; I = inosine; U = uridine; X = xanthosine; and ψ = 5-ribosyluracil (pseudouridine). Designate 2′deoxyribonucleosides by same symbols, preceded by d.

Indicate 3′-phosphate by letter p (terminal phosphate) or hyphen (phosphodiester group) to left of nucleoside symbol. Indicate other points of attachment with numbers.

In sequences, oligonucleotides, or polynucleotides, indicate phosphate between nucleoside symbols by a hyphen, if sequence is known and a comma if not known.

See *Biochem. J.* 127: 753-756, 1972 for shorter forms for use in repetitive or obscure sequences.

Indicate associated chains or bases within chains by a center dot separating complete names or symbols; non-associated chains are separated by a plus sign, and unspecified or unknown association by a comma.

Use abbreviations kb (kilobases), nt (nucleotide) and bp (base pair) without definition.

Use single letters to designate a variety of possible nucleotides, applicable to both DNA and RNA. Use at all positions where U might appear in RNA. Sequences are assumed to have a deoxyribose backbone unless otherwise indicated. Use prefix lowercase d and r to distinguish between RNA and DNA.

G	guanine	S	G or C
A	adenine	W	A or T
T	thymine	H	A or C or T
C	cytosine	B	G or T or C
R	G or A	V	G or C or A
Y	T or C	D	G or A or T
M	A or C	N	G or A or T or C
K	G or T		

Nomenclature, Plants: Include full binomial Latin names. Specify variety and source.

Nucleotide Sequences: Fully determine nucleotide sequences in both senses of DNA. Give an explicit statement and a supporting diagram summarizing sequence data. Deposit primary nucleotide sequence data with European Molecular Biology Laboratory Data Library (EMBL) or an associated data library. EMBL database accession number will be included in paper.

Powers in Units and Figures: Avoid numbers with too many digits. Quantity expressed is preceded by power of 10 by which it has been multiplied. Units in which quantity is expressed may not be multiplied by a power of 10; unit may be changed using prefixes.

Reviewers: Suggest potential reviewers, Journal is not obligated to follow suggestions. Specify names of those wish excluded from review process.

Safety: Draw attention to chemical or biological hazards involved in carrying out experiments. Describe relevant safety precautions. Include accepted codes of practice followed. Reference relevant standards.

Solutions: Describe in terms of molarity (M), not normality (N). Express fractional concentrations in decimal system. Define % as w/w, w/v or v/v. For aqueous solutions of concentration less than 1, w/v need not be inserted if it is clear that concentration is in terms of weight of solute. Define incubation media by reference or composition. Symbol for ionic strength (mol/l) is I.

Specify buffers. Give complete composition of each solution at first mention or in Experimental section. See Table 4 for Abbreviations for common buffers (*Biochem. J.* 1994; 297:14).

Spectroscopic Data: Publish full spectra when novel or important. Deposit other spectra or spectral information with British Library Document Supply Centre. Give wavelength scale of u.v. and visible absorption, fluorescence, circular dichroism and optical rotation, whether or not a wavenumber scale is given. Use molar terms in absorption, circular dichroism and optical rotation.

Report circular dichroism as molar absorption coefficient $\Delta\varepsilon=\varepsilon_L-\varepsilon_R$ [or molar ellipticity]. For biopolymers, use molar concentrations in terms of mean residue M_r. Units of $\Delta\varepsilon$ are same as for ε. Specific ellipticity, molar ellipticity and mean residue ellipticity are directly analogous to terms used in optical rotation.

Give derivative spectra for electron spin resonance. Give a scale of magnetic-field strength (in mT) and/or g values. Describe peaks as "the $g=2$ peak."

For fluorescence excitation and emission spectra, state whether intensities, F, are relative, normalized or corrected (and nature of correction). Report fluorescence-polarization data and spectra as polarization ration, P, or preferably anisotropy ratio, A; both are dimensionless. Report infrared spectra as percentage transmittance, T, as a function of wavelength (μm) or frequency (cm^{-1}). When assigning bands, give units only for first value. Describe as: broad NH band.

Mass spectrometry spectra may be described as, e.g., m/z 300[M$^+$ (molecular ion)], 282 (M$^+$-H$_2$O), etc. If parenthetic values are quoted for percentage peak heights, state what these are relative to.

In Mossbauer spectroscopy, plot absorption (in %, arbitrary units or crude channel counts) against Doppler velocity, v (in mm/s). Quote chemical shift, δ, in units of mm/s, relative to a specified standard. Give temperature. Describe applied magnetic field.

Express nuclear magnetic resonance chemical-shift data, δ, as parts per million. Quote reference compound. Downfield shifts are positively signed. Express coupling constants in Hz. Report structural n.m.r. date as: δ (p.p.m.) (solvent) chemical-shift value [integration, peak type, coupling constant (in Hz), designation (relevant proton in italics)]. Singlet, doublet, etc. are abbreviated to s, d, etc.

without definition; other descriptions (broad and overlapping) in full.

Report optical rotation as specific rotation $[\alpha]_\lambda^t$ which is numerically equal to rotation in degrees of a 1 g/ml solution with a pathlength of 1 dm (10 cm) at wavelength λ and temperature t. Quote concentration (g/100 ml) and solvent. Define corresponding molar expression for molar rotation, $[M]=[\alpha] \times M_r$ and $[m]=[\alpha] \times M_r/100$. For biopolymers use mean residue M_r; $[m]_{m.r.w}$ is mean residue rotation. Report reduced mean residue rotation when a refractive-index correction is applied $[m']$. Dimensions of $[m]$ and $[m']$ are degrees·cm²·dmol⁻¹. Report optical rotary dispersion as variation of $[\alpha]$ or $[m]$ with wavelength or frequency.

In visible and ultraviolet-absorption spectroscopy, attenuance is the general name for quantity log(I0/I). This reduces to absorbance when there is negligible scattering or reflection. Use "attenuance" when scattering is considerable. Otherwise use "absorbance"; neither should be called extinction or optical density. Symbols used are: A, absorbance; D, attenuance; a, specific absorption coefficient (liter.g-1.cm-1); e, molar absorption coefficient (numerically equal to absorbance of a 1 mol/liter solution in a 1 cm light-path) (use units of liter.mol-1.cm-1 or M-1.cm-1 and not cm2.mol-1). Give wavelength in (nm) as subscripts without units. Do not use equal signs between e or A and its value.

Spelling: *Concise Oxford Dictionary of Current English* (Oxford, Clarendon Press)

Statistical Analyses: Report data from a sufficient number of independent experiments to permit evaluation of reproducibility and significance of results. To determine value of a quantity or statistical characteristics of a population, sufficient information is conveyed by: the number of individual experiments; the mean value; and the standard deviation, the coefficient of variation, or the standard error of the estimate of mean value. Indicate whether S.D. or S.E.M. is used.

Style: See Gowers, Ernest, *The Complete Plain Words* (London: H.M.S.O. and Penguin Press).

Sugar Symbols: Use to represent polymers or sequences and in tables and figures.

Ara	Arabinose
dTib	2-Deoxyribose
Fru	Fructose
Gal	Galactose
Glc	Glucose
Man	Manose
Rib	Ribose
Xyl	Xylose

Use f or p after saccharide to indicate furanose or pyranose.

Suffixes for derivatives:

A	uronic acid
N and NAc	2-amino-2-deoxysaccharides and their N-acetyl derivatives.

Use either extended or condensed system for representing oligosaccharide chains. In extended system, put configurational symbol (D or L) before symbol for monosaccharide, separated by a hyphen. Include anomeric symbol (α or β) before configurational symbol separated by a hyphen. Place locants that indicate respective position between the symbol (abbreviated name) of one monosaccharide group or residue and the next. Separate locants with arrow from locant corresponding to glycosyl carbon atom to locant corresponding to carbon atom carrying hydroxyl group enclosed in parentheses. Indicate position of a branch above or below main chain with numerals and arrow indicating glycosidic linkage. Hyphens may be omitted except those separating configurational and monosaccharide symbols. In condensed system symbol implies common configuration and ring. If ring size is different from common one or is emphasized, indicate with appropriate symbols from extended system. Anomeric descriptor indicates configuration of glycoside linkage and is placed before locant if bond direction is to right, or after locant if left. Locants are separated by hyphen. Use no hyphen between sugar symbol and parentheses; omit parentheses in branched oligosaccharides, when parentheses are used to indicate branches. Locants of anomeric carbon atoms may be omitted. Parentheses around specifications of linkage may be omitted. Hyphens may be omitted.

Ptd (phosphatidyl); Ins (id-*myo*-inositol); and P (phosphate) need not be defined.

Place multiple locants in parentheses.

Alternative "Chilton" forms may be used if defined.

Units: Use SI symbols. See *Quantities, Units and Symbols in Physical Chemistry* (Oxford: Blackwell Scientific Publications, 1988). Give SI equivalent with non-SI units. Give distance measurements at molecular level in angstroms (A). See Table 3, Physical quantities and their units for commonly used units (*Biochem. J.,* 1994; 297:12).

SOURCE

1994, 297: 1-15.

Biochemical Pharmacology
Pergamon Press

Prof. Alan C. Sartorelli
Yale University School of Medicine
Department of Pharmacology
Sterling Hall of Medicine, 333 Cedar Street
PO Box 208066, New Haven, CT 06520-8066
            From American Continent

Prof Jacques E. Gielen
Universite de Liege
Laboratoire de Chimie Medicale
Centre Hospitalier Universitaire
B-4000, Start-Tilman par Liege 1 Belgium
            From European Continent

From elsewhere: send manuscripts to either editor.

AIM AND SCOPE

Biochemical Pharmacology is an international journal which publishes research findings in pharmacology deriving from investigations that employ the disciplines of biochemistry, biophysics, molecular biology, genetics, structural biology, computer models and/or physiology. Reports of studies with intact animals, organs, cells, subcellular components, enzymes or other cellular molecules and model systems are acceptable if they define mechanisms of drug action. Descriptive mathematical models including those involving computer techniques are also welcome. Experiments involving the use of drugs to elucidate physiological and behavioral mechanisms in living organisms are also within the scope of the journal. In general, papers which record concentrations of drug and metabolites in body fluids will only be accepted if they contribute to an understanding of biochemical and biophysical mechanisms.

MANUSCRIPT SUBMISSION

Submit manuscript (including figure captions and references) in triplicate, typed on one side of paper, double-spaced with margins. Send copies of all relevant articles cited as in press or under consideration by another journal. Give assurance that material has not been published or submitted elsewhere.

MANUSCRIPT FORMAT

Include: Abstract, Introduction (with *no* heading), Methods, Results and Discussion. Provide a short running title (50 characters). Include first name of each author. Indicate on title page, name, address, telephone number and fax number of corresponding author.

Illustrations: Make about twice final size required. Keep separate from typescript. Do not paste on separate sheets. Present photographs as good quality glossy prints. Color photographs can be reproduced for a fee. Obtain estimate from publisher. Type captions together on separate sheet.

Tables: Keep tabular matter to a minimum. Do not reproduce same data in both tables and figures. Give each table and every column an explanatory heading. Indicate units of measure. Indicate table footnotes with same symbols used in text. Make tables and illustrations completely intelligible without reference to text.

References: Use consecutive numbers in square brackets in text. Cite full reference in numbered list at end of paper. List names and initials of all the authors with full title of paper, abbreviated journal title, volume number, first and last page numbers and year. Abbreviate journal titles in accordance with *Index Medicus.*

JOURNAL

1. Reiter R and Burk RF, Effect of oxygen tension on the generation of alkanes and malondialdehyde by peroxidizing rat liver microsomes. *Biochem Pharmacol 36: 925–929, 1987.*

BOOK

2. Winer BJ, *Statistical Principles in Experimental Design.* McGraw-Hill, New York, 1971.

CHAPTER IN A BOOK

3. Shatkin AJ, Colorimetric reactions for DNA, RNA, and protein determinations. In: *Fundamental Techniques in Virology* (Eds. Habel K and Salzman NP), pp. 231–237. Academic Press, New York, 1969.

SYMPOSIUM PROCEEDINGS

4. Wefers H and Sies H, Generation of photoemissive species during quinone redox cycling. In: *Bioreduction in the Activation of Drugs, Proceedings of the Second Biochemical Pharmacology Symposium, Oxford, UK, 25–26 July 1985* (eds. Alexander P, Gielen J and Sartorelli AC), pp. 22–24. Pergamon Press, Oxford, 1986.

Footnotes: Use symbols: *, †, ‡, §, ||, ¶, starting again on each page. Do not include in numbered reference list. Use for references to unpublished work (including work submitted for publication), personal communications (verify wording of quoted material with information supplier, and submit letter of permission to use), proprietary names of trademarked drugs and other material not appropriately referred to in text or reference list. Keep to a minimum.

OTHER FORMATS

COMMENTARIES—3000-5000 word commissioned review articles, designed as editorial statements on selected topics, intended to stimulate thought, may be controversial, draw attention to neglected areas, or present personal views on state and future of subject.

COMMENTS ON COMMENTARY—Invited authorities offer observations of a Commentary, published in same issue.

MEETING REPORTS—Short, factual two-page summaries of selected papers from small meetings or conferences where attendance is by invitation and published papers are refereed.

RAPID COMMUNICATIONS—4-6 pages. Emphasize novel aspects of research. Type "space and a half" on special laysheets available from editor. Send two copies. Type title in capital letters, centered 2 in. from top. Quadruple space. Type author's name, double-space, type author's address, capitalizing first letter of all main words. Sextuple space and begin abstract. Triple space and begin text. Include drawings of formulae, graphs and other figures in text. Paste illustrations (glossy prints) onto appropriate place or draw carefully onto typing area large enough for slight reduction. Make sure captions are not confused with text. Single space captions. Number each page at bottom in blue pencil.

SHORT COMMUNICATIONS—6 double-spaced typed pages including tables and figures. Include abstract. Sections may be combined.

OTHER

Abbreviations: Define in footnote on title page. Define abbreviations in abstract.

Nomenclature: Follow internationally agreed rules. Specify code number with systematic name (as distinct from trivial name) of enzyme or compound at first mention in text. Official names of drugs are preferred to trade names. Capitalize trade names and include trademark. Include name and location of company with source of compounds used.

Reviewers: Submit names, addresses, telephone and fax numbers of 5 or 6.

Statistical Analyses: Include probability values derived from appropriate statistical tests.

Style: Suggestions and Instructions to Authors, *Biochem J* **289**: 1–15, 1993. (Note instructions on isotopically labelled compounds.)

Terminology and Abbreviations, *J Biol Chem* **268**: 745–753, 1993.

International Union of Biochemistry, *Enzyme Nomenclature* (Academic Press, 1984).

CBE Style Manual Committee, *CBE Style Manual: A Guide for Authors, Editors and Publishers in the Biological Sciences* (5th Edn., Council of Biology Editors, Inc., Bethesda, MD).

Units: Specify mass as either relative molecular mass (M_r, not molecular weight) or molecular mass (expressed in kDa). Do not express in M_r daltons. Use either consistently. Temperatures with an unqualified degree symbol ° are assumed to be Centigrade. For solution strengths, express percentages by %, followed if ambiguous by w/w, w/v or v/v [e.g., 5% (w/v) means 5 g/100mL]. In spectrophotometry, use extinction (E), not optical density.

SOURCE

July 5, 1994, 48(1): i-iv.

Biochemistry
American Chemical Society

Gordon G. Hammes
Duke University Medical Center, P.O. Box 3673
081 Yellow Zone, Duke South
Durham, NC 27710
Telephone: (919) 681-3250; Fax: (919) 681-8344

AIM AND SCOPE

The purpose of Biochemistry is to publish the results of original research that contributes significantly to the understanding of the mechanism of biological phenomena in terms of molecular structure and/or function. Preference will be accorded to manuscripts that develop new concepts or experimental approaches particularly in the advancing areas of biochemical science. Manuscripts that are primarily descriptive, confirmative of previous work, or of a highly specialized nature, or those that in the judgment of the Editors are not of sufficient interest to the general readers of Biochemistry, will be returned to the author. Manuscripts dealing primarily with the isolation, sequencing, and expression of DNA segments usually are not appropriate for Biochemistry unless significant new information is obtained about the structure and/or function of the proteins.

MANUSCRIPT SUBMISSION

Do not submit preliminary or inconclusive results or manuscripts with overlapping content. Submission implies that work has not been published and is not under consideration for publication elsewhere.

Submit one original and three copies. Include glossy or clear photocopies of figures for each copy. Photocopies are acceptable if all details necessary for critical examination of data, especially gel patterns and halftone photographs, are shown.

Include a submittal letter, a signed copyright status form (see copy in first issue in January), and three copies of related papers submitted for publication or in press. Listing coauthors implies approval of manuscript and full agreement with content.

MANUSCRIPT FORMAT

Condense to utmost compatible with clarity. Double space throughout on one side of good grade 8.5 × 11 in. or A4 paper. Keep 2.5 cm margins on each side and 3 cm at top and bottom. Computer printouts must be clear and readable, preferably from a laser printer, with no smaller than 12-pitch print.

Assemble in order: title page (including full title, title footnote, byline, and running title), footnotes, abstract, introduction, experimental procedure (materials and methods), results, discussion, acknowledgments, references, tables, figure legends, and figures. Number all pages starting with title page.

Title Page: Brief and informative title. Do not use trade names of drugs or abbreviations. Use serial numbers only if consecutive papers appear in same issue.

Acknowledge financial aid in title footnotes (†).

In byline list full names and institutional affiliations of all authors. Consecutively identify footnote references to authors, including present addresses, with symbols:‡, §, ||, ⊥. Reserve symbol * for corresponding author. Provide a brief (50 letters and spaces), informative running title.

Key Words: Supply four words not in title.

Abbreviations and Textual Footnotes: On separate page, give list of abbreviations. Cite frequently used abbreviations in first footnote. Number textual footnotes consecutively. Be brief and keep to a minimum.

Abstract: 250 words; present problem and experimental approach. State major findings and conclusions. Make self-explanatory and suitable for abstracting. Do not use footnotes or undefined abbreviations. If a reference is cited, give complete publication data. See White, R. H. (1982) *Biochemistry* 21, 4271-4275.

Introduction (no heading): State purpose of investigation and relation to other work. Avoid extensive literature review.

Experimental Procedures (Materials and methods): Describe materials in sufficient detail to enable repetition of experiments. Include names of products and manufacturers only if alternate sources are unsatisfactory. Describe novel experimental procedures in detail. Refer to published procedures by literature citation of original work and modifications. Make materials otherwise not obtainable available to interested academic researchers.

Results: Present concisely. Use tables and figures only if essential for comprehension of data. Do not present same data in more than one figure or in both a figure and a table. Interpret results in discussion. Place supplementary data in a separate section for inclusion in microfilm edition of Journal.

Discussion: Interpret results and relate to existing knowledge. Be brief and clear. Avoid personal polemics. Do not repeat information given elsewhere. Avoid exhaustive literature reviews.

Acknowledgments: Include technical assistance, advice from colleagues, gifts, etc. Obtain permission from acknowledged persons.

Tables: Use only when data cannot be presented clearly otherwise. Number consecutively with Arabic numerals. Provide brief title for each table and a brief heading for each column. Indicate units of measure (preferably SI). Round data to nearest significant figure. Give standard deviations if appropriate. Put explanatory material in title footnotes (a). Identify table footnotes with superscript letters, lower case and italicized. Submit tables comparing amino acid or nucleotide sequences and tables that require special treatment as camera-ready copy. Cite all tables and figures in text. Mark approximate positions in margins of text.

Structures: Submit camera-ready copy with all component parts intact. Make economical use of space.

Figures: Keep to a minimum. Use only to document experimental results or methods that cannot be described in text. Do not present same data in different forms in more than one figure or in a figure and table(s). Avoid simple linear plots that can be described in text. When three-letter abbreviations for amino acid sequences are used, put symbols for 15 amino acids on each line (for reduction to single-column width). Separate symbols with a hyphen (or comma when applicable). Put 30 three-letter symbols for amino acid residues on each line for double-column presentation. For presentation of extended protein sequences, use one-letter abbreviations of amino acid residues.

Preferred illustration size is 20×28 cm; do not exceed. Paste smaller illustrations, particularly when part of same figure, on a 22×28 cm sheet. Print photographs on glossy, high-contrast paper. Submit stereo diagrams so that they can be reproduced full size, make interocular distance between similar points on original 5.5–6.5 cm. Draw graphs in India ink or with a high-quality laser printer on plain white drawing paper, blue drawing cloth, or coordinate paper printed in nonreproducing blue. Enclose graphs on all four sides by rules. Place numbers and lettering outside coordinates. Do not put headings or descriptive material in figures. Include in legends. Make symbols, letters, and numbers large enough to be readable after reduction to single-column width (8.4 cm). Use lettering set to label coordinates. Only one set of original figures is needed; reproductions suffice for other copies. For photographs of electron micrographs, gels, and similar primary data, provide glossy prints with all copies. Indicate if an illustration requires special treatment to show essential details. Indicate on figure back in soft pencil, figure number, TOP, and author's name. Number consecutively with Arabic numerals. Indicate approximate positions in margins of text.

Submit color illustrations only if essential for clarity of communication. Explain and justify need for color upon submission.

Figure Legends: Collect at end of manuscript. Avoid lengthy legends with detailed experimental protocols.

State pertinent experimental conditions. Avoid repetition of text. Explain all symbols and abbreviations. Use simple symbols such as O, ■, and Δ. Use error bars on data points.

OTHER FORMATS

ACCELERATED PUBLICATION—Six printed pages (25 typewritten pages, including references, tables, and illustrations, one illustration per page). Work must be of exceptional significance and of immediate interest to general readers. Must contain all elements of a regular manuscript and satisfy usual publication criteria. In transmittal letter state why work justifies accelerated publication. Include names of at least six scientists qualified to act as impartial reviewers. Mark ribbon copy, three additional copies, transmittal letter, and mailing envelope "For Accelerated Publication."

NEW CONCEPTS IN BIOCHEMISTRY—15 typed pages, double-spaced, including references, tables, and illustrations in usual format, without abstract and subheadings. Present personal views and new and opposing ideas.

OTHER

Atomic Coordinates: Deposit atomic coordinates for structures of biological macromolecules determined by X-ray, NMR, or other methods with Protein Data Bank, Chemistry Department, Brookhaven National Laboratories, Upton, Long Island, NY 11973. Manuscripts will be processed only after deposit. Obtain a File Name for macromolecule. Put in footnote or body of text.

Color Charges: $300 for first color illustration plus $160 for each additional.

Nucleotide Sequence Data: Submit sequence data, preferably in computer-readable form, plus information for annotation of data and a copy of paper to GenBank Submissions, National Center for Biotechnology Information (NCBI), Building 38A, Room 8N-805, 8600 Rockville Pike, Bethesda, MD 20894 (telephone: (301) 496-2475; fax: (301) 480-9241). Submission to GenBank ensures entry also into the EMBL Nucleotide Sequence Library.

Page Charges: $30 per printed page. Payment is not a condition of publication. Address requests for page charge waiver forms, to Journals Department, Publications Division, American Chemical Society, 1155 16th Street, N.W., Washington, D.C. 20036.

Protein Sequence Data: Deposit with Protein Identification Resource, National Biomedical Research Foundation, Georgetown University Medical Center, Washington, D.C. 20007. For documentation of sequence analysis of proteins, see *Biochemistry* (1983) 22, 2595.

Safety: State necessary precautions.

Supplementary Material: Submit three clear copies, (one high-contrast original). Computer printouts are acceptable only if individual characters are clearly legible. Optimum size of submitted material is 22×28 cm. Size of smallest character may not be less than 1.5 mm. Insert captions or other identifying information directly on material to be microfilmed. Indicate nature of supplementary material in a paragraph, preceding references, as follows:

"Supplementary Material Available" followed by a brief description of material, number of pages, and statement "Ordering information is given on any current masthead page." Clip supplementary material together and put at end of manuscript with heading: "Supplementary Material for the Microfilm Edition." Number figures and tables in supplement consecutively but independently of those in body of manuscript.

SOURCE

January 11, 1994, 33(1): 7A–10A.

Biochimica et Biophysica Acta
Elsevier

P.I. Dutton
c/o BBA Editorial Office
P.O. Box 9123
Brookline, MA 02146-9123
Telephone: (617) 277-7919; Fax: (617) 730-5698

G.K. Radda
c/o BBA Editorial Secretariat
P.O. Box 1345
1000 BH Amsterdam, The Netherlands
Telephone: +31 20 5803510; Fax: +31 20 5803506

Y. Kagawa
Department of Biochemistry
Jichi Medical School
Minamikawachi
Tochigi-ken 329-04 Japan
Telephone: +81 285 44 2111 ext. 3149;
Fax: +81 285 44 1827

AIM AND SCOPE

Biochimica et Biophysica Acta publishes papers reporting advances in our knowledge or understanding of any field of biochemistry or biophysics. Descriptions of new methods valuable for biochemists or biophysicists are also acceptable if their application is made obvious or a new principle is introduced. Negative results will be accepted only when they can be considered to advance our knowledge significantly. Papers providing only confirmatory evidence or extending observations firmly established in one species to another will not normally be accepted.

MANUSCRIPT SUBMISSION

Submission of a manuscript implies that work described has not been published before (except as an abstract or as part of a published lecture, review or academic thesis), that it is not under consideration for publication elsewhere, that its publication has been approved by all authors and tacitly or explicitly by responsible authorities in laboratories where work was carried out and that, if accepted, it will not be published elsewhere in the same form, in the same or another language, without consent of Editors and Publisher.

In submission letter, provide: full name, address, telephone and fax numbers of corresponding author; known address changes within 6 months of submission; type of paper (Regular Paper, Rapid Report, Short Sequence-Paper or Review) and preferred journal section; full title and suggested reviewers (optional).

Submit four copies of all material (including submission letter, if it contains information of relevance). Legible carbon copies or photocopies are adequate for second and subsequent copies, except for photographs of electrophoretic patterns, electron micrographs, etc; provide high-quality prints in quadruplicate. Submit copies of relevant unpublished results, papers submitted elsewhere or in press with each manuscript copy.

Journal sections are: Bioenergetics; Biomembranes; General subjects; Gene Structure and Expression; Lipids and lipid Metabolism; Molecular Basis of Disease; Molecular Cell Research; Protein Structure and Molecular Enzymology; and Reviews on Cancer.

Submit Short Sequence-Papers to Netherlands office only. Others may go anywhere.

MANUSCRIPT FORMAT

Consult a recent journal issue for conventions and layout of articles, especially title pages and reference lists. Begin each new paragraph by indentation.

Type entire text, including figure and table legends and reference list, double- or triple-spaced on pages of uniform size, on one side of paper, with a 5-cm margin. Number all pages consecutively starting with title page.

Title: Concise, informative title. Do not use abbreviations. Follow with names of authors (with first or middle names in full) and name(s) and address(es) of laborator(ies) where work was done. If work was done in more than one laboratory, indicate one in which each author worked during study with numeric footnotes. Provide name, full postal address, e-mail address, telephone, fax and telex numbers of corresponding author.

Key Words: 6 on title page for use in annual index.

Summary: 100–200 words for a Regular Paper and 50 words for a Rapid Report. If reference is cited, give complete publication data.

Introduction: State reasons for performing work. Make brief references to relevant previous work.

Materials and Methods: Make detailed enough to be reproduced. Refer to other work on same subject, indicating agreement or disagreement with results.

Results: Give conclusions drawn from experiments described in tables and figures.

Discussion: Overall conclusions based on reported work.

Acknowledgments: Make only with permission of person(s) named. Include funding organizations.

References and Citations: Cite references in text by consecutive number, thus [1]. Mention authors' names sparingly; for more than two authors 'et al.,' or 'and co-workers' is used. Refer to unpublished data or data in preparation in text ('unpublished data'), not in reference list. Provide unpublished results and obtain permission for use of personal communications. Papers submitted for publication may be included as a number in text and in reference section with name of journal. Submit copies of articles. Articles accepted for publication are cited in reference list as 'in press' and are included as numbered reference in text. Supply evidence of acceptance. Double or triple space reference list numbered consecutively in order of appearance. Give every reference a unique number and every number must refer to a single reference. In reference list, give names of all authors and editor(s) of multiple-author books. Abbreviate journal titles according to List of Serial Title Word Abbreviations (International Serials Data System, Paris). Give first and last page numbers.

JOURNALS

1 Spach, P.I., Herbert, J.S. and Cunningham, C.C. (1991) Biochim. Biophys. Acta 1056, 40-46.

BOOKS

2 Westerhoff, H.V. and Van Dam, K. (1987) Thermodynamics and Control of Biological Free Energy Transduction, pp. 209-212, Elsevier, Amsterdam.

3 Hider, R.C. and Hall, A.D. (1991) in Progress in Medicinal Chemistry (Ellis, E.P. and West, G.B., eds.) Vol. 28, pp. 41-173, Elsevier, Amsterdam.

Figures: Use to illustrate experimental results. Remember page and column dimensions (typesetting areas 17.7 and 8.4×23.5 cm, respectively). Make symbols and letters at least 1.5 mm high after reduction. One set of figures must be suitable for reproduction. Label continuous or half-tone figures on an overlay or supply in duplicate. Avoid already rastered half-tones. Avoid handwriting and typing (except for camera-ready sequences). Electron micrographs must carry a bar denoting unit length. Do not mount photographs. Letter photographs on an overlay (strip/press-on lettering). Discuss color figures with Managing Editor and Publisher. On reverse side list figure number, corresponding author, and short paper title. Do not fold or mount on heavy card. Supply good copies for reviews. Express units correctly on figure axes. Explain in legends all abbreviations not in text. Use decimal points, not commas. Use leading zeros, in figures and figure legends. See *Biochim Biophys Acta* 1075:275 for symbols to be used. Choose size of symbols in relation to lettering. Put wavelength scale in curves showing absorption in ultraviolet of visible range.

Photographs of gels, micrographs, etc. should be glossy and rich in contrast. Provide four copies of electrophoretic patterns and electron micrographs. Provide in size that will fit format without reduction. Indicate unit length by labeled bar in electron micrographs. Submit amino acid and nucleic acid sequences camera ready.

Legends: Type double- or triple-spaced, on separate sheets of paper. Give a preliminary sentence constituting a title, followed by a brief description of how experiment was carried out, and any other necessary material describing symbols or lines on figure. Collate on a separate sheet.

Tables: Use sparingly, only when data cannot be presented clearly in text. Type double-spaced on separate pages with legends. After title describe way in which experiment was carried out. Give details of items in table in footnotes. Avoid repetition of terms; list units at top of columns in parentheses under physical quantity measured. Express quantity or unit followed by a power of 10. Larger or smaller units may be used.

OTHER FORMATS

RAPID REPORTS—4 printed pages (12 double-spaced pages, including a total of 4 tables and figures). Do not divide into sections after Summary, except for reference list. Put acknowledgments in final paragraph before reference list.

REVIEWS—Contact responsible Managing Editor before contributing.

SHORT SEQUENCE—PAPERS—Submit to Amsterdam office marked 'Short Sequence-Paper and state in cover letter. Include data on structural RNAs and their genes, DNA sequences coding for a protein or regulatory regions (promoters, enhancers) and other DNA regions of interest. Present sequences camera-ready. Deposit in data bank before submission. Mention accession number in cover letter and in footnote on first page.

OTHER

Abbreviations: Define in footnote on first page.

Color Illustrations: At authors' expense.

Data Deposition: Deposit structural data in electronic database.

EMBL Data Library Submission
Postfach 10.2209
D-6900 Heidelberg
Germany
Tel: +49 (06221) 387 258
BITNET/EARN: datasub embl DDBJ Submissions

GenBank
Mailstop K710
Los Alamos National Laboratory
Los Alamos, NM 87545 U.S.A.
Tel: +1 (505) 665-3295
BITNET/EARN: genbank%life lanl.gov

PIR Data Submissions
National Biomedical Research Foundation
Georgetown University Medical Center
3900 Reservoir Road N.W.
Washington, D.C. 20007-2195
Tel: +1 (202) 687-2121
BITNET/EARN: pirsub gunbrf
Telex: 490 997 5103
Dialcom: 42:cdt0105

Laboratory of Genetic Information Analysis
Center for Genetic Information Research
National Institute of Genetics
Mishima, Shizuoka 411, Japan
Tel: +81 (559) 75 0771 ext. 647
Fax: +81 (559) 75 6040
BITNET/EARN: ddbjsub ddbj.nig.ac.jp
ddbj ddbj.nig.ac.jp (for inquiries)

International Protein Information Database in Japan (JIP-ID)
c/o Professor Akira Tsugita
Research Institute for Biosciences
Science University of Tokyo
Yamazaki, Noda 278, Japan

Tel: +81(471) 23 9777
BITNET/EARN: ex5292 jpnsut30 tsugita jpnsut31
Dialcom: cdt0079

Deposit sequences reported in Short Sequence-Papers before submission. Put accession number in footnote on title page. Deposit supplementary data of other kinds that are too lengthy or too specialized for inclusion in a data bank of choice. BBA has it own data bank also. Submit in duplicate in camera-ready form, or if not adequately copyable, provide five original prints. Number pages and figures. After acceptance, editor will deposit in requested data bank and add footnote to paper.

Experimentation, Animal: Follow institutional or national research council's guide for care and use of laboratory animals.

Experimentation, Human: Follow procedures in accordance with Helsinki Declaration of 1975 (revised 1983).

Footnotes: Keep to a minimum. Do not number. Collate on a separate page. Explain all abbreviations in footnote on title page.

Nomenclature: Follow IUPAC/IUB recommendations.

Nomenclature, Biochemical: Follow IUPAC-IUB Commission of Biological Nomenclature recommendations in *Biochemical Nomenclature and Related Documents* (London, Portland Press, 1991) and *Enzyme Nomenclature, Recommendations* (New York, Academic Press, 1984).

Nomenclature, Chemical: Follow IUPAC's *Nomenclature of Organic Chemistry* (Oxford, Pergamon Press, 4th ed., 1979), "the blue book"; IUPAC's *Nomenclature of Inorganic Chemistry* (1990), "the red book;" *Compendium of Analytical Nomenclature* (2nd ed, 1977), "the orange book;" *Compendium of Macromolecular Nomenclature* (1991), "the purple book;" and the *Compendium of Chemical Terminology* (1987), "the gold book;" all published by Blackwell Scientific, Oxford. Also IUB's *Biochemical Nomenclature and Related Documents* (London, Portland Press, 1991).

Nomenclature, Organisms: Include full binomial Latin names for all experimental organisms other than common laboratory species. State strain, variety, source and dietary or growth regimes. If name used does not conform to current usage, add accepted nomenclature in parentheses after first mention in Summary and text. Give different characteristics of organism. Identify microorganisms and tissue-culture strains by type culture collection number. Make strains available upon request.

Reviewers: Suggest suitable reviewers in cover letter.

Solutions and Buffers: Describe in terms of molarity (M) not normality (N). Express fractional concentrations as decimals. Restrict term % to w/v or v/v. For salt solutions expressed as percentages, indicate if anhydrous or hydrated compounds used.

Statistical Analyses: Representative experiments often suffice. Indicate number of experiments carried out and variability. When sufficient results are available, express as mean plus or minus standard error of the mean, followed by number of observations in parentheses. Distinguish between standard deviation and standard error. State type of statistical test and value of P. If insufficient data for statistical analysis, state mean followed by range (in parentheses) and number of experiments. Represent both values if only two determined.

Style: Indent paragraphs.

Name handwritten characters or Greek letters in margin at first mention.

Type Latin expressions in normal typeface. Do not underline or italicize.

Symbols: Avoid excessive use of chemical symbols and structural formulae.

Units: SI units. See IUPAC's *Quantities, Units and Symbols in Physical Chemistry* (Oxford, Blackwell Scientific, 1988), "the green book." Use M for concentration, and its subdivisions, Curie (Ci) . Angstrom is not recommended. Use Svedberg (S) to express sedimentation coefficient (s). Molecular weight is defined as a ratio. Do not express in daltons. Express molecular mass in daltons. Define units of amount of enzyme in terms of rate of reaction catalyzed under specified conditions. Represent multiplication of numbers by a central dot, multiplication of units by a space, and of unlike terms or dimensions by a multiplication sign.

SOURCE
1991, 1075: 267-280. (1993, 1149: refers to this issue).

Biological Psychiatry
Society of Biological Psychiatry

Wagner H. Bridger, M.D.
Medical College of Pennsylvania
at Eastern Pennsylvania Psychiatric Institute
3200 Henry Avenue, Philadelphia, PA 19129

AIM AND SCOPE
Biological Psychiatry welcomes original contributions from all sources and from all disciplines and research areas that may extend our knowledge of psychiatry.

MANUSCRIPT SUBMISSION
Submission implies that neither manuscript nor data have been previously published (except in abstract) or are currently under consideration for publication. Clearly indicate any possible conflict of interest, financial or otherwise. Upon acceptance, author(s) will be asked to transfer copyright to the Society of Biological Psychiatry. Submit five copies, including all illustrations and tables, typed double-spaced throughout on one side only of 8.5 x 11 in. paper. Number pages consecutively. Type corresponding author's name on each page.

MANUSCRIPT FORMAT
On title page, include full names of authors, academic or professional affiliations, and complete address with phone and fax numbers of corresponding author. List six key words. Indicate financial support as numbered footnote to

title or to individual author's name. If title exceeds 45 characters, supply an abbreviated running title.

Organize full-length manuscripts as: Abstract (150 words), Introduction, Methods and Materials, Results, Discussion, References, Tables, Figures, and Legends. Acknowledge personal and technical assistance after Discussion.

Type reference lists, tables, and figure legends on separate sheets. Number tables in order of mention with Arabic numerals and give brief descriptions. Spell out acronyms on first use in text and in table title or footnote.

Illustrations: For line artwork, submit black-ink drawings of professional quality or glossies of originals on separate sheets. Color illustrations are accepted; cost must be paid by author. On back, give first author's name, figure number, and TOP. Submit original photos of histological sections, tissues, chromatograms, etc., for each manuscript copy. Number figures with Arabic numerals in order of mention. Avoid 1a, 1b, 1c numbering; numerical sequence and separate captions are preferred. Include photocopies with original figures. Text mention of two or more figures to appear together should be cited together on same manuscript page.

References: List alphabetically at end of paper. Refer to in text by name and year in parentheses. Use periodical abbreviations from *Index Medicus*.

JOURNAL

Post RM, Weiss SRB (1992): Endogenous biochemical abnormalities in affective illness: Therapeutic versus pathogenic. *Biol. Psychiatry* 32:469-484.

BOOK

American Psychiatric Association (1987): *Diagnostic and Statistical Manual of Mental Disorders*, 3rd ed rev. Washington, DC: American Psychiatric Press.

EDITED BOOK

Martin JH (1985): Properties of cortical neurons, the EEG, and the mechanisms of epilepsy. In Kandel ER, Schwartz JH (eds), *Principles of Neural Science*, 2nd ed. New York: Elsevier, pp 461-471.

OTHER FORMATS

BRIEF REPORTS AND CASE STUDIES OF UNUSUAL SIGNIFICANCE—1000 words. No abstract; minimal figures, tables, and references.

CORRESPONDENCE—600 words with minimal figures, tables, and references,

SOURCE

July 1, 1994, 36(1): inside back cover.

Biology of Reproduction
Society for the Study of Reproduction
Elsevier

Dr. Fuller W. Bazer
442D Kleberg Center, Animal Science Department
Texas A&M University
College Station, TX 77843-2471
Telephone: (409) 862-2660; Fax: (409) 862-2662

AIM AND SCOPE

Biology of Reproduction publishes original papers on a broad range of topics in the field of reproductive biology. Authors are not required to be members of the Society.

MANUSCRIPT SUBMISSION

Submit signed copyright release and page charge agreement form published with Instructions to Authors in each journal issue. Each author must sign. In signing, each author attests that: manuscript represents original, unpublished material; manuscript is not under editorial consideration elsewhere; studies were conducted in accordance with Guiding Principles for the Care and Use of Research Animals promulgated by the Society for the Study of Reproduction; authors agree to page charges; and individuals included in acknowledgments are aware that they are named.

Print on one side of good quality standard size paper. Double- or triple-space throughout, including references, figure legends, and tables. Keep 1 in. margins on both sides and at top and bottom. Do not justify right margins. Number lines of text in margin (number every fifth line by hand). Manuscripts produced by dot-matrix printers should have large, dark type, with "descenders" of the letters p, g, y, and g below line.

MANUSCRIPT FORMAT

Arrange in order: title page, abstract, introduction, materials and methods, results, discussion, acknowledgments, references, figure legends, tables, illustrations (line drawings, graphs, and continuous-tone prints).

Title Page: Complete article title, author(s), complete name(s) and location(s) of laboratory(ies) or institution(s) of affiliation; a short title (53 characters), and three to six key words for indexing. Indicate corresponding and reprint request authors (include telephone and fax numbers, and electronic mail address). List financial support (grant or contract numbers) and name of funding institution or agency in a title footnote. Leave upper right corner of title page blank.

Abstract: 200 words, stating purpose of work, methods used, and summarizing results and conclusion. Make understandable in itself; do not cite references. Avoid using abbreviations.

Materials and Methods: Provide sufficient information to replicate work. Do not give details of published methodology if appropriately cited, but describe modifications. Include information regarding statistical analysis of data. Specify composition of all solutions, buffers, mixtures, and culture media. Provide source of reference for each. Use generic or chemical names of drugs and chemicals; if a trade name is mentioned or a particular instrument or tool is identified, put manufacturer's name and location in parentheses. Note individuals or companies who have donated supplies or reagents.

Results: Present findings in appropriate detail. Use past tense. Refer to tables and figures.

Discussion: Clear and concise interpretation of results.

Acknowledgments: Acknowledge assistance other than financial support, such as statistical review, technical support, editorial and/or clerical assistance.

Illustrative Materials: Send a complete set of figures for each manuscript copy. Mark one full set, consisting of glossy prints or originals of line art, graphs, and continuous-tone prints, "printer's copies." Paste adhesive label on back indicating top, author's name, and figure number.

Legends: Type double- or triple-spaced, on a sheet separate from figures; use numbers and letters consistent in style with those in illustration. Identify all symbols in legend or in figure itself. Indicate magnification in legends or by reference bars on figures.

Reduction and Enlargement: Line drawings and graphs are reduced or enlarged to appropriate dimensions. Continuous-tone prints are reproduced at 100% of original size. One-column material is 3.5 in. (21 picas) wide; two column material is 7.25 in. (43.5 picas) wide. Maximum length for a single figure or plate is 9 in. (54 picas). If figure dimensions are maximal (7.25×9 in. or 43.5 by 54 picas), legend is printed on opposite page. Do not submit larger continuous-tone or color figures.

Line Drawings and Graphs: Make lines dark and sharply drawn. Use solid black, white or bold designs for histograms. Light stippling or gray tones may be lost in printing. Make lettering and symbols large enough to remain legible if reduced. Indistinct hand lettering or drawing is not acceptable.

Continuous-tone Prints: Mount individual prints, use double-sided tape. Mount plates consisting of several prints on cardboard. Apply black or white tooling between mounted prints, or leave a thin, even margin. Align all mounted prints at right angles. Apply press-on numbers to lower right corner of individual prints mounted together as one plate; use letters to designate separate parts of a single figure.

Tables: Make as simple as possible. Type each on a separate sheet, titled, and numbered sequentially with Arabic numerals. Round numbers to nearest whole number or significant digit. No legends; put explanatory matter in double-spaced footnotes identified by superscript letters or symbols (not numbers) keyed to data in table. Do not use vertical lines and no more than three margin-to-margin horizontal lines per table. Tables are usually not printed lengthwise.

OTHER FORMATS

MINIREVIEW—5 printed pages. Concise information about a topic of current importance or an area of controversy.

OTHER

Color Prints: Authors are responsible for cost.

Experimentation, Animals (Adopted 1983): Follow highest possible standards for humane care and use of animals in research. In developing the research plan, consider use of in vitro models, appropriateness of animal species, and minimum number of animals needed to meet rigorous scientific and statistical standards. Use animals bred specifically for laboratory study whenever practical;

however, there are situations where wild, captive, random source, or pound animals are necessary.

Acquire, retain, and use all research animals in compliance with federal, state, and local laws and regulations. Properly house, feed, and keep animal surroundings in sanitary conditions in accordance with the National Institutes of Health Guide for the Care and Use of Laboratory Animals (DHEW Publication No. (NIH) 80-23, Revised 1985. Office of Science and Health Reports, DRR/NIH, Bethesda, MD 20205).

Give research animals appropriate anesthetics, analgesics, tranquilizers and care to minimize pain and discomfort during preoperative, operative, and postoperative procedures. Make choice and use of most appropriate drugs in strict accordance with NIH Guide. All procedures must be accepted veterinary medical practice. If study or animal's condition requires that animal be killed, use a humane method.

Use of animals shall be under the direct supervision of an experienced teacher or investigator.

See New York Academy of Sciences Interdisciplinary Principles and Guidelines for the Use of Animals in Research.

Publication Costs: Society of the Study of Reproduction members: $50 per page; nonmembers $75.

Reviewers: Provide four names, telephone and fax numbers, and addresses.

Style: Follow *CBE Style Manual*, 5th ed. (Council of Biology Editors, Bethesda, MD).

SOURCE

July 1994, 51(1): back matter.

Biometrics
The Biometric Society
Waverly Press

C. A. McGilchrisy
Department of Statistics
The University of New South Wales
P.O. Box 1, Kensington, NSW 2033, Australia
Biometric Society
1429 Duke St., Suite 401
Alexandria, VA 22314

AIM AND SCOPE

The Biometric Society is an international society devoted to the mathematical and statistical aspects of biology. The general objects of *Biometrics*, the journal of the Biometric Society, are to promote and to extend the use of mathematical and statistical methods in pure and applied biological sciences, by describing and exemplifying developments in these methods and their applications in a form readily assimilable by experimental scientists. It is also intended to provide a medium for exchange of ideas by experimenters and those concerned primarily with analysis and the development of statistical methodology.

MANUSCRIPT SUBMISSION

Submit four copies of each manuscript double-spaced throughout. Photocopies are acceptable if legible. It is understood that submitted manuscript is not identical or similar to a paper being handled by another journal. Do not exceed 20 manuscript pages.

MANUSCRIPT FORMAT

Summary: Completely summarize without repeating, verbatim, sentences from paper. Avoid mathematical symbols. Reference previous work in full. Type key words or phrases at bottom of first page. Include author's name followed by full postal address.

References: In text give author's surname and publication date. For publications by three authors, list all the first time; use *et al.* subsequently. For four or more, use *et al.* throughout. In list of references at end of paper, list authors alphabetically by last name. Give complete journal titles. Follow format in recent issues.

Tables and Figures: Present only essential material. Place on separate sheets identified by Arabic numerals and short descriptive titles. Execute original diagrams or graphs in India ink in black on white; rule coordinate lines in black. Good glossy prints are desirable. For printing, illustrations may be reduced to one half or one third original dimensions. Make lines sufficiently thick, and decimal points, periods and dots large enough to reproduce well. Insert all lettering and numbering on original large enough to be legible in reduced size. Note placement of all tables and figures in text. Use only horizontal lines in tables. Incorporate very small tables into text. Align all decimals and numbers in columns.

OTHER FORMATS

ABSTRACTS—Of papers presented at regional meetings should be sent to Editor, *Biometric Bulletin* (see address above) no later than six months after meeting. Do not exceed 150 words. Do not include display formula or multilevel symbols. Type in format of previous abstracts published in this journal.

CONSULTANT'S FORUM—Manuscripts should be sent to the Editor-Elect, as for Regular Papers. Four copies are needed. See March 1977 issue, page 230, for guidelines.

METHODOLOGY PAPERS—Discuss results and methods of use in biological sciences. Make accessible to biologists by including an introductory section on biological applications, use of real data with inclusion of intermediate steps in worked examples, and removal of mathematical derivations and proofs to an appendix.

PAPERS ON BIOLOGICAL SUBJECTS—Report conclusions of interest reached by mathematical or statistical analysis, or illustrate either use of less well-known analytical techniques or application of standard techniques in a new field; present analyses in sufficient detail for the reader to be able to reproduce calculations and adapt them for use in other fields.

SHORTER COMMUNICATIONS—Submit papers under 10 manuscript pages in length to the Editor, Byron J. T. Morgan, Institute of Mathematics, University of Kent, Canterbury, Kent CT2 7NF, England.

SOURCE

March 1994, 50(1): ii.

Biometrika
Bernoulli Society
Canadian Statistical Association
Institute of Mathematical Statistics

Ms. B. J. Sowan, Executive Editor, *Biometrika*
Department of Statistics, University of Oxford
1 South Parks Road, Oxford, OX1 3TG, U.K.
Telephone: 0865 272875

AIM AND SCOPE

Biometrika is primarily a journal of statistics in which emphasis is placed on papers containing original theoretical contributions of direct or potential value in applications.

MANUSCRIPT SUBMISSION

Submit three copies of manuscripts. Be as concise as possible without detracting from clarity. Papers must neither have been published nor be under consideration for publication elsewhere. Submit copies of unpublished references. Give an address for each author, including a department.

Type double-spaced, including references, on one side of paper. Keep 1.25 in. margins on all four sides. If typesetting or a word processor is used, make depth of white space between lines at least 5 mm. Typing or typesetting, including sub- and superscripts, should not be smaller than 12 characters per inch. Do not justify. Make all copies legible, including accents and sub- and superscripts; number pages consecutively.

MANUSCRIPT FORMAT

Examine a recent issue of *Biometrika*. Follow arrangements of formulae, references, tables, etc. as closely as possible.

Begin with a single paragraph summary; do not include formulae; follow with key words in alphabetical order.

Divide paper into numbered sections with suitable short verbal titles. Use subheadings, but not sub-subheadings. Give each subsection a number and title.

References: Follow current style in Biometrika. List of references should correspond to those in text,. Make book references to latest edition; include a page, section or chapter number. References to books of papers should include: title of book, editor(s), first and last page numbers of paper, publisher and where published. Give complete lists of authors and editors; do not use et al. Do not include unpublished reports in References unless accepted for publication.

Acknowledgments: At end of paper, be as brief as possible subject to politeness. Exclude information, such as contract numbers, of no interest to readers. Place acknowledgments to other institutions here.

Tables: Number in order of appearance. Refer to by number. Give self-explanatory titles. Type on separate sheets. Make effective use of *Biometrika* sized page. Avoid layouts that have to be printed sideways. Limit tables to 92 characters wide, including decimal points and brackets (1 character), and minus and other signs and spaces (at least 2 characters).

Figures: Submit professionally drawn diagrams with accepted manuscripts. Draw in India (black) ink about twice printed size; write labels for axes, etc. on a photocopy. Type titles on a separate sheet. Check for error-free original drawing and labels. Number figures, and refer to by number, consecutively. If computer-drawn, make lines at least 0·3 mm to allow a reduction of 0·5. Put on dimensionally stable white or transparent paper. Drawings should have good definition. Photocopies are not acceptable.

Appendix: Follow acknowledgments but precede reference. Give titles.

OTHER

Style: Do not use verbal phrases inside brackets or dashes, or in italic type.

Avoid footnotes, except for tables.

Do not use abbreviations like a.s., i.i.d., d.f., ANOVA and ML and special symbols like $\xrightarrow{d}$ and $\forall$.

Arrange brackets in order, $[\{()\}]$.

Follow usual conventions for e, exp, use of solidus, square root signs, etc. as in a recent issue. Do not use square root sign .

Avoid multiple overbars and symbols with underbars. Align subscripts and superscripts (and second-order sub- and superscripts) horizontally, or mark in colored ink. Avoid sub- and superscripts of third, and greater, order.

Do not use distinctive type for matrices and vectors. Mark script letters clearly. Distinguish between easily confused characters like w, ω; v,ν; k, κ, K; o, O, 0; l, 1; etc. Where ambiguous and if handwritten, indicate character by name in margin the first time it is used. Avoid capital Greek sigma, Σ, except as summation sign. Do not use symbols at start of a sentence.

Use: var (x) not var x or Var (x); cov not Cov; pr for probability not Pr or P; tr not trace; $E(X)$ for expectation not EX or $\mathcal{E}(X)$; log x not $\log_e x$ or $\ln x$. Avoid: '·' for product; a/bc, which should be written $a/(bc)$ or $a(bc)^{-1}$. Use: zeros preceding decimal points, 0·2 not ·2; the form $x_1, \ldots, x_n$ not $x_1, x_2, \ldots, x_n$; $\frac{1}{2}$, etc. to avoid bracketing, e.g., $\Gamma\left(\frac{1}{2}n + \frac{1}{2}\right)$ not $\Gamma\{(n+1)/2\}$ or $\Gamma\left(\frac{n+1}{2}\right)$.

Include equation numbers only when equations are referred to; place numbers on right. Display long or important equations on a separate line. Leave short formulae in text to save space where possible, but must not be more than one line high. Avoid equations involving lengthy expressions by introducing suitable notation.

SOURCE

March 1994, 81 (1): v-vi.

Biophysical Journal
Biophysical Society
American Physical Society-Division of Biological Physics
Victor Bloomfield
9650 Rockville Pike
Bethesda, MD 20814
Telephone: (301) 571-0663; Fax: (301) 571-0667

AIM AND SCOPE

The *Biophysical Journal* publishes original articles, letters, and reviews on molecular, cellular, and general biophysics.

MANUSCRIPT SUBMISSION

Complete and send Article Submission Form printed in each issue.

MANUSCRIPT FORMAT

Type manuscripts double-spaced, including references. Avoid inessential mathematical expressions. Submit four copies of papers. Upon acceptance of a manuscript, submit original drawings or clear photographs of figures, and a computer diskette copy of manuscript.

Follow conventions of *Council of Biology Editors Style Manual* (5th ed., 1983. Council of Biology Editors, Bethesda, MD).

Title: Titles (100 characters and spaces) should identify content of article; clarity and conciseness are essential for indexing, abstracting, and retrieval. Furnish a condensed title (40 units) on title page for running titleline.

Key Words: Six keywords or phrases not in title.

Abstracts: 200 words, at head of paper. Objectively describe work done in as general a language as practical.

Materials and Methods: Capitalize trade names and give manufacturers' full names and addresses (city and state).

References: In body of text, cite references by surname and year (Smith, 1992); (Smith and Jones, 1991); (Smith et al., 1990). Type double-spaced beginning on a sheet separate from text. List references alphabetically by surname of first author. Include all authors' names (do not use "et al."), year, complete article titles, and supply inclusive page numbers. Abbreviate names of journals as in *Serial Sources for the Biosis Data Base* (BioSciences Information Service of Biological Abstracts. Philadelphia, PA); spell out names of unlisted journals. Include citations such as "submitted for publication," "in preparation," and "personal communication" parenthetically in text ; do not include in Reference section. For personal communications include cited author's institutional affiliation and written permission to use material cited.

JOURNAL ARTICLES

Benditt, E. P., N. Ericksen, and R. H. Hanson. 1979. Amyloid protein SAA is an apoprotein of mouse plasma high density lipoprotein. *Proc. Natl. Acad. Sci. USA.* 76:4092–4096.

Brown, W. and A. Nelson. 1989. Phosphorus content of lipids. *J. Lipid Res.* In press.

Yalow, R. S. and S. A. Berson. 1960. Immunoassay of endogenous plasma insulin in man. *J. Clin. Invest.* 39:1157–1175.

COMPLETE BOOKS

Myant, N. B. 1981. The Biology of Cholesterol and Related Steroids. Heinemann Medical Books. London.

ARTICLES IN BOOKS

Innerarity, T. L., D. Y. Hui, and R. W. Mahley. 1982. Hepatic apoprotein E (remnant) receptor. *In* Lipoproteins and Coronary Atherosclerosis. G. Noseda, C. Fragiacomo, R. Fumagalli, and R. Paoletti, editors. Elsevier/North Holland, Amsterdam. 173–181.

Illustrations: Submit a complete set of figures with each copy of manuscript. Place original set in a separate envelope marked with primary author's name and number of figures enclosed. Label back of each figure with figure number, authors' names, and TOP. Do not write directly on back, mount on cardboard, bend, or damage with paper clips. Clearly label parts on figure itself, but give title in figure legend only, not on figure. Double-space figure legends in a separate section at end of references.

Line Drawings: Submit either glossy prints or plain paper prints from a laser printer. Drawings, including axis labels, should occupy one column (3.25 in. wide) when printed. Proportions of drawing should be approximately square, or slightly wider than high. If necessary for legibility, design figures to extend full width of a page (7 in.). Lettering should be 1/8 in. high (9 pt type size) after reduction. A useful recipe for a standard line graph on a laser printer is a plot area 4 in. wide and 3 in. high. Extend axis numbering and labeling, in a 14 pt sans serif plain (not bold) font such as Helvetica, to approximately 3/4 in. from left and bottom of plot borders. In metric units, plot area is 100×75 mm. and figure including lettering should be 120×95 mm. Lettering should be 5 mm high. Authors may be asked to prepare new figures if the figures submitted (including inserts) are not legible after reduction.

Halftone Figures and Color Figures: Submit four sets of sharp, glossy prints, 8.5×11 in. or less. If of sufficient quality, one original set of glossies will suffice, with three sets of photocopies. Figures may be resized by printer; include an internal scale marker if magnification is important. Size figures to fit either a one-column (3.25 in.) or two-column (7 in.) layout. Include layout with multi-part figures. If mounted, figures must not exceed 7×9 in. on 8.5×11 in. board. Illustrations may be submitted on diskette.

Cost of color separations is charged to author, if referees and Editor determine that color is necessary to convey desired information.

Provide color electronic images in Encapsulated PostScript (EPS) format. Provide black and white illustrations in Tagged Image File Format (TIFF) sized to appropriate column width. Submit original artwork with diskette.

Tables: Type double-spaced; give titles. Do not use vertical rules. A standard structure is a single box heading with all pertinent information located below in distinct columns. Place footnotes at bottom in sequence *, ‡, §, ‖, ¶.

OTHER FORMATS

BOOK REVIEWS, AND NEW & NOTABLE COMMENTARIES—Generally solicited.

LETTERS—Make specific scientific reference to papers published previously in *Biophysical Journal*. Letters will be reviewed.

PEDAGOGICAL ARTICLES—Articles on how to teach various topics in biophysics, particularly at the graduate level. Most articles are solicited; contributed ones are considered. Query Associate Editor for Education with a brief summary and outline.

REVIEWS—Reviews of topics of current interest in biophysics provide an overview of basic principles and contemporary research. Most are solicited; contributed ones are considered. Query Editor with a brief summary and outline.

OTHER

Data Submission:

X-ray Diffraction Studies—Submit to Protein Data Bank all of structural data required to validate discussion of new structure determinations, including both x-ray amplitudes and phases and derived atomic coordinates. If paper discusses a protein structure only at level of main chain alpha carbon atoms, then deposit only alpha carbon coordinates. If discussion involves higher resolution data, for example all atoms in active site of an enzyme, then deposit full set of x-ray data and coordinate list. Inform Editor, no later than completion of editorial process, that necessary information has been sent to Protein Data Bank.

If requested by authors, editor will ask Data Bank not to distribute information until a specific date. For coordinate lists, this may not be more than one year beyond acceptance date of manuscript. For full structure amplitude and phase data time interval before distribution may not exceed four years. Release date specified by author will appear in a footnote to paper with statement that information has been submitted to Protein Data Bank. In absence of a specified release date, it is assumed that information is available immediately on appearance of publication.

Nucleic Acid and Lipid Databases—Submit nucleic structures to Nucleic Acid Data Base (H. Berman, Rutgers University) and lipid phase transition and miscibility data to LIPIDAT (M. Caffrey, The Ohio State University)

Electronic Submission: Submission of manuscripts on diskette is expected after manuscript has been accepted in final form. Dsk file must be an exact representation of final accepted form of manuscript. Be sure there are no extraneous files on disk. Disk format is that of original word processing program. Do not send ASCII files. Label disk with name, title of article, type of computer used, and name and version number of word processing software. For further information contact Cadmus Journal Services at (410) 528-4096 or fax (410) 528-4204.

Equations: Type or carefully hand letter. Clearly indicate capital and lowercase letters. Label Greek and unusual symbols at first appearance. Use fractional exponents instead of root signs. Solidus (/) for fractions saves vertical space. Cite equation numbers in text without parentheses: e.g., Eq. 9.

Page Charges: $20 per page. Charges will be waived by Editor upon request. Payment is not a criterion for judging acceptability of a manuscript nor is non-payment cause for delay in publication.

Reviewers: Suggest four names to be added to list from which referees will be selected. Suggest an appropriate Editorial Board member from list on inside front cover.

SOURCE

July 1994, 67(1): 486-487.

Biosensors and Bioelectronics

Elsevier Advanced Technology

Mrs. Ann C. Barnett
Biosensors & Bioelectronics
Elsevier Advanced Technology
P.O. Box 150
Kidlington Oxford OX5 1AS, England
Telephone: +44-(0)865-512242; Fax: +44-(0)865-310981

Reinhard Renneberg
Institut fur Chemo- und Biosensorik (ICB)
Nottulner Landweg 90 D-48161 Munster, Germany
Telephone: (+49) 2534 80 01 94; Fax:(+49) 2534 80 01 96

Products & Innovations, Practical corner, Discussion department, Profile, Update, etc.

AIM AND SCOPE

The refereed papers in *Biosensors & Bioelectronics* cover research, technology and application of biosensors and the exploitation of biochemicals in electronic devices. The scope includes sensors that employ biological molecules or systems in the sensing elements and other devices that sense parameters in biological processes, biomimetics, etc. Many sorts of measuring device are covered, including: enzyme, whole organism, and immunoelectrodes (both amperometric and potentiometric); piezoelectric crystal detectors; novel chemical sensors; optoelectronic devices; ion-selective electrodes; specialist applications of mass spectrometry and nuclear magnetic resonance: types based on field effect transistors; and those which utilize the principles of biological fuel cells. Since the biosensor and bioelectronic field is multidisciplinary, spanning fundamental and applied aspects of biochemistry, electrochemistry and electronics, some articles are of an introductory nature, directed particularly at life scientists who are showing increasing interest in biosensors and bioelectronics but who might be new to the field. Others concern end-user requirements.

MANUSCRIPT SUBMISSION

Submission of a manuscript implies that it is not being considered contemporaneously for publication elsewhere. Submission of a multi-authored manuscript implies consent of all participating authors.

MANUSCRIPT FORMAT

In letter-headed covering letter include full address, telephone and fax numbers. Provide three copies, double-spaced on pages of uniform size, with a wide margin at left. Do not exceed 10,000 words or 25 printed pages. Include an Abstract (100-150 words), reporting concisely on purpose and results of paper. Add a list of key words.

Consult an issue of the journal for style and layout. Use Système International d'Unitès (SI) for all scientific and laboratory data; add other units in parentheses. Give temperatures in Kelvin scale. Do not use unit billion (10^9 in America, 10^{12} in Europe) as it is ambiguous.

Type tables, references and legends to illustrations on separate sheets. Place at end of paper. Avoid footnotes with information which could be included in text.

References: Indicate references to published work in appropriate places in text, according to Harvard system (i.e., using author(s) names(s) and date), with a reference list, in alphabetical order, at end of paper. In list give name(s) and initial(s) of author(s), year of publication and exact title of paper or book. For journals there should follow title, volume number, and initial and final page numbers of article. For books there should follow name(s) of editor(s) (if appropriate), name of the publisher, and town of publication. Where appropriate, quote initial and final page numbers. All references in this list should be indicated at some point in text and vice versa. Unpublished data or private communications should not appear in the reference list.

Illustrations: Provide original and two copies of each illustration. Copies may be reduced size, or photocopies. Submit line drawings suitable for direct reproduction, in Indian ink, with stencilled lettering (avoid dry transfer or typewritten lettering). High-quality computer-generated line diagrams, or glossy prints, are acceptable. The type area is 160 mm wide × 222 mm deep; make lettering large enough to be legible after reduction of illustration to fit. Submit photographs as clear black-and-white prints on glossy paper. Number each illustration with title and name(s) of author(s) on reverse side.

SOURCE

1994; 9(1): inside back cover.

Biotechnology & Bioengineering
John Wiley

Professor Eleftherios T. Papoutsakis
Dept. of Chemical Engineering
Northwestern University
2145 Sheridan Rd., Evanston, IL 60208-3120

AIM AND SCOPE

Biotechnology & Bioengineering publishes original articles, reviews, and mini reviews that deal with all aspects of applied biotechnology. Papers are considered based upon novelty, their immediate or future impact on biotechnological processes, and their contribution to the advancement of biotechnology and biochemical engineering science. Theoretical papers are judged based on novelty of approach and capability to predict and elucidate experimental observations. Papers delaing with routine aspects of bioprocessing, descriptions of established equipment, and routine applications of established methodologies are discouraged. Topics appropriate for the journal include: applied aspects of cellular physiology, metabolism and energetics of bacteria, fungi, animal and plant cells; enzyme systems and their applications, including enzyme reactors, purification, and applied aspects of protein engineering; animal-cell biotechnology, including media development, modeling, tissue engineering, and applied aspects of cell interactions with their environment and other cells; bioseparations and other downstream processes, including cell disruption, chromatography, affinity purifications, extractions, and membrane processing; environmental biotechnology, including aerobic and anaerobic processes, systems involving biofilms, algal systems, detoxification and bioremediation, and genetic aspects; applied genetics and metabolic engineering, including modeling molecular processes of applied interest; plant-cell biotechnology; biochemical engineering, inlcuding transport phenomena in bioreactors, bioreactor design, kinetics and modeling of biological systems, instrumentation and control, biological containment, and bioprocess design; biosensors; spectroscopic and other instrumental techniques for biotechnological applications, including NMR and flow cytometry; thermodynamic aspects of cellular systems and their applications; mineral biotechnology, including coal biotechnology; biological aspects of biomass and renewable resources engineering; and fundamental aspects of food biotechnology.

Manuscript Submission

Submit a copyright transfer agreement; a copy appears in most journal issues and is available from Editor. Agreement must be signed by author upon submission. If paper is a "work made for hire," agreement must be signed by employer. Manuscripts must represent original research or original review or mini review on subjects within scope of journal. All authors must agree to submission and to implications of submission. All authors are responsible for complete contents of manuscript. Submission implies that manuscript, in its present or a substantially similar form, has not been published or is not being considered for publication elsewhere. Publication in any reasonably retrievable source constitutes prior publication. Meeting abstracts or preprints do not constitute prior publication. Obtain written permission for reproduction of figures, tables, or text from published work from copyright owner, publisher of journal or book. Permission from authors is also encouraged as a professional courtesy.

Serial papers must be reviewed together, no more than two manuscripts at a time. Do not exceed 22 double-spaced pages of text (not including references) and no more than 15 figures and/or tables for full papers.

Submit four copies of manuscript (including all figures and tables), a separate set of original figures, and copyright agreement. In letter of transmittal include title of manuscript and all author(s) name(s), and a brief statement of important contributions of paper.

Manuscript Format

Type double-spaced (including References, Tables, Appendixes, and List of Figures) on one side only on paper as close as possible to U.S. standard 8.5 × 11 in. with 1 in. margins. Number all pages, including tables, lists of figures, and figures. Figure copies for review must clearly identify each figure by number. Submit photographs (of polyacrylamide gels, for example) with details important to review in triplicate in addition to originals.

Submit a separate (title) page with title of paper, names and affiliations of all authors, name, address, phone number (and fax number) of corresponding author, and a short running title (55 characters). Full papers contain: a Summary (300 words), Introduction, Materials and Methods (including computational methods), Modeling or Theoretical Aspects (optional), Results, Discussion, Conclusions (optional), Acknowledgments (optional), Nomenclature (optional), and References.

Provide 3-6 key words or phrases that best characterize manuscript following Summary.

For all types of manuscripts, list of References is followed by Tables, List of Figures with legends, and number copies of figures. Acknowledge personal or financial assistance in a separate paragraph following Discussion and Conclusions sections. Present supplementary material of theoretical or experimental nature in Appendixes following Discussion and Conclusions sections.

Make title specific, informative and as short as possible. Prepare Summary of main contributions in manuscript carefully. Divide main text of paper into subsections as necessary by inserting appropriate headings. Refer to any issue of *B&B* for examples. Do not use footnotes in text.

References: Include all relevant and necessary published work in reference list; cite all listed references in text. Arrange references in alphabetical order by first author, and number consecutively. Cite references in text by number. Abbreviate journals according to *Chemical Abstracts Service Source Index*. Listed references include only published journal papers, books, book chapters, and Ph.D. theses and dissertations. Do not list manuscripts in preparation or submitted, abstracts, meeting presentations, editorials, patents, M.S. theses, letters to the editor, published science briefings, and in general all other published material that has not been edited or refereed. Mention such material parenthetically in text. If an "in press" publication is included in list of References, submit two copies.

1. Cavallaro, J. F., Kemp, P. D., Kraus, K. H. 1994. Collagen fabrics as biomaterials. Biotechnol. Bioeng. 43: 781-791.

2. Chang, S. W. 1993. Studies of growth kinetics and estimation of oxygen uptake rate for *Bacillus thuringiensis*. Ph.D. thesis, University of Illinois, Urbana-Champaign, IL.

3. Tchobanoglous, G., Burton, F. L. 1991. Wastewater engineering: treatment disposal reuse. 3rd edition. McGraw-Hill, New York.

4. Trelstad, R. L., Kemp, P. D. 1993. Matrix glycoproteins and proteoglycans, pp. 35-57. In: W. N. Kelley, E. D. Hams, S. Ruddy. and C. B. Sledge (eds.), Textbook of rheumatology, 4th edition. W.B. Saunders, Philadelphia.

Tables: Number with capital Roman numerals in order of appearance in text. Give table columns explanatory headings. Avoid duplication of information in Tables and figures. Avoid long tables with large amounts of information; show only typical or most important results.

Figures and Illustrations: Supply legends for all figures; compile on a separate sheet following tables. Submit all figures in a form suitable for reproduction. Prepare graphs and photographs to be printed in a single column. Only if absolutely necessary figures may occupy both columns. Single column width is 3 5/16 in.; double column width is 6 7/8 (maximum) in.. Submit good glossy prints of all photographs. Continuous-tone photographs (e.g., of polyacrylamide gels) must have sufficient contrast. Submit four copies of continuous-tone photographs (one set of originals and three additional prints for review process). Photocopies are not acceptable. Submit continuous-tone photographs in final published size, so that no reduction is necessary.

Color photographs are not encouraged. If absolutely necessary, authors must bear cost of printing. Contact editor or publisher for printing cost.

Draw line drawings (graphs, etc.), clearly with India ink on heavy white paper.

Submit computer-generated graphs only if printed with a good quality laser or pen printer on good quality paper and have acceptable standard lettering. Good quality glossy photographs of original drawings are also acceptable. Lettering of graphs must be large, clear and "open" so that letters and numbers do not fill in when reduced. Submit drawings that are twice the size of final figures; test drawings for clarity and readability by reducing them 50% using a copy machine. Prepare graphs to be printed in one column. Two columns may be used for graphs or photographs upon request on an individual basis. Write figure number and last names of authors on figure originals, either in margin or on back with a soft pencil. Use an arrow to indicate top of photograph or graph. Submit figure originals in an envelope protected (especially glossy prints) from sticking together.

OTHER FORMATS

COMMUNICATIONS—9 double-spaced pages and 5 figures and/or tables. Short papers on a single issue or follow-up papers with comments on a previously published paper.

Include: a Summary (100 words), a brief Introduction, Materials and Methods, Results and Discussion, Acknowledgments (optional), and References. Subject to the same review process as full papers. Not for preliminary investigations.

REVIEWS AND MINI REVIEWS—Concise, critical assessments of recent literature on a fast-moving areas of contemporary interest. Reviews are same length as full papers. Mini reviews are 10 double-spaced pages of text, with 60 references and 5 figures and/or tables. Contact editor on suitability of a subject and qualifications of authors before submitting. Explain clearly, upon manuscript submission, reasons for longer manuscripts.

Include: a Summary (300 words for reviews and 100 words for mini reviews), a brief Introduction, appropriate subsections for various topics, Conclusions, Future Directions, Acknowledgments, and References.

Erratum—Allows authors to correct inadvertent (e.g., typographical) errors in published papers. Changes or additions to manuscript and its data are not permitted. Submit directly to publisher.

OTHER

Electronic Submission: Deliver final, accepted version of manuscripts on diskette. Submit a 5.25 or 3.5 in. diskette in IBM MS-DOS, Windows, or Macintosh format. WordPerfect 5.1 (or later) is preferred; manuscripts prepared on any microcomputer wordprocessor are acceptable. Do not use desktop publishing software (Aldus Pagemaker 1D, Quark Xpress, etc.); if used, export text to a wordprocessing format. Keep document as simple as possible. Refrain from complex formatting. Do not use footnote function.

Submission of electronic illustrations is encouraged, but not required. Submit on a separate diskette from text. Use TIFF and EPS files or native application files. For gray scale and color figure submissions contact editor for more detailed instructions. Submit files on SyQuest 44 or 88 megabyte cartridges.

Submit each article as a single file. Name each file with your last name (8 letters), followed by a period, plus three-letter extension BIT. If last name exceeds eight letters truncate to fit. If using a Macintosh, maintain MS-DOS filenaming convention of eight letters, a period, and three-letter extension. Label all diskettes with your name, file name, and wordprocessing program used.

Submit hard copy printout with disk. If disk and paper copy differ, paper copy is considered definitive.

For assistance contact: Gerry Grenier, Internet: GGrenier@JWiley.com; telephone: 212-850-8860; fax: 212-850-8888

Reviewers: Submit names and complete addresses of five possible independent reviewers outside institution. Request that certain reviewers be excluded from reviewing.

Symbols and Nomenclature: Type wherever possible all mathematical and chemical symbols, equations, and formulas. If handwritten, write clearly and leave ample space above and below for printer's marks; use only ink. Identify Greek or unusual symbols in margin at first use. Distinguish in margin between capital and small letters where confusion may arise (e.g., k, κ). Underline all vector quantities with a wavy line. Use fractional exponents to avoid root signs.

Use nomenclature sponsored by International Union of Pure and Applied Chemistry for chemical compounds. Place chemical bonds correctly, indicate double bonds clearly. Indicate valence by superscript plus and minus signs. Define all necessary quantities with units compatible with SI system. If many symbols are used in text and in equations, assemble an alphabetical list of definitions and units of all symbols in a Nomenclature section just before list of References. Assemble Greek and unusual symbols separately.

SOURCE

January 5, 1955, 45(1): back matter.

Blood

American Society of Hematology
W. B. Saunders

James D. Griffin, MD
1018 Beacon Street, Brookline, MA 02146
Telephone: (617) 738-9808; Fax: (617) 738-9868

AIM AND SCOPE

Blood provides an international forum for the publication of original articles describing laboratory and clinical investigations encompassed in the field of hematology.

MANUSCRIPT SUBMISSION

Submit original and two copies of all text, and three complete sets of figures and tables. Double-space on standard white paper with 1 in. margins.

Manuscripts must be original contributions and can not contain data that has been published elsewhere, including symposia volumes, etc. Meeting abstracts do not constitute prior publication.

MANUSCRIPT FORMAT

Organize as: Cover Letter, Title Page, Abstract, Introduction, Methods, Results, Discussion, Acknowledgments, References, Tables, Figure Legends, and Figures.

Cover Letter: Identify name, address, phone and fax number of Corresponding Author. Specify Table of Contents category: Clinical Intervention and Therapeutic Trials, Hematopoiesis, Hemostasis and Thrombosis, Immunobiology, Neoplasia, Phagocytes, Red Cells, Transplantation, and Transfusion Medicine. Specify category of manuscript (Rapid Communication, Regular Manuscript, etc.). Papers which Editors decide not to process as Rapid Communications will be handled as Regular Manuscripts unless authors request that manuscript be returned. Indicate where appropriate willingness to pay costs of publishing color photographs. Indicate if color photographs are being submitted for consideration as cover illustration, (no charge if selected). Disclose potential conflicts of interest.

Title Page: Title, authors' names, names of institutions in which work was done, research grant support (as a title footnote), and name, address, phone and fax number of corresponding author.

Abstract: 200 words stating rationale, objectives, findings, and conclusions of study. Do not include summary at end of Discussion.

References: Double-space in numerical order at end according to order of citation in text. Include all authors' names (do not use "et al."), complete article titles, and articles "in press." Abbreviate journal names according to latest edition of *Index Medicus*. Supply only first page number. Appropriately identify references to abstracts and letters. Reference personal communications, unpublished observations, and manuscripts submitted and "in preparation" parenthetically in text. Submit written permission to cite a personal communication.

JOURNALS–ONE OR MORE AUTHORS

1. Gitschier J, Kogan S, Levinson B, Tuddenham EGD: Mutations of factor VIII cleavage sites in hemophilia. A. Blood 72: 1022, 1988

JOURNALS–IN PRESS

2. Foroni L, Laffan M. Boehm T, Rabbitts TH, Catovsky D, Luzzato L: Rearrangement of the T-cell receptor delta genes in human T cell leukemias. (Blood, in press)

BOOK

3. Adams DO, Edelson PJ, Koren HS: Methods for Studying Mononuclear Phagocytes. San Diego, CA, Academic, 1981

CHAPTER IN BOOK

4. Sallan SE, Weinstein HJ: Childhood acute leukemia, in Nathan DG, Oski FA (eds): Hematology of Infancy and Childhood, vol 2. Philadelphia, PA, Saunders, 1987, p 1028

CHAPTER IN BOOK–PART OF PUBLISHED MEETING

5. Baron MH, Maniatis T: Stage-specific reprogramming of globin gene expression, in Stamatoyannopoulos G, Nienhuis AW (eds): Developmental Control of Globin Gene Expression, Proceedings of the Fifth Conference on Hemoglobin Switching. New York, NY, Liss, 1987

CHAPTER IN BOOK–PART OF UNPUBLISHED MEETING

6. Freireich EJ: Methotrexate to molecular genetics: Forty years of leukemia treatment. XXII Congress of the International Society of Hematology. Milan, Italy, 1988, p 27

ABSTRACT

7. Clift RA, Martin P, Fisher L, Buckner CD, Thomas ED: Allogenic marrow transplantation for CML in accelerated phase-risk factors for survival and relapse. Blood 70:291a, 1987 (abstr)

LETTER TO THE EDITOR

8. Arlin ZA: Complete remission in acute promyelocytic leukemia. Blood 72:1101, 1988 (letter)

Footnotes: Place at bottom of respective pages.

Tables: Print or type each on a separate page. Number appropriately. Cite in numerical order in text.

Figures: Cite in text in numerical order using Arabic numerals. Submit three complete sets of clear, glossy, black and white photographs. For line drawings, submit either a single set of photographs with two sets of copies, or three sets of high quality laser figures. Do not write on front or back of photographs. Do not mount, fasten with paper clips, or store in clasp envelopes. Type name of first author, figure number, and designation of TOP on a label and adhere to backs of figures.

Legends: Type table legends on same page. Give sufficient explanation to make data intelligible without reference to text. Double space figure legends on a separate sheet. Include a brief title and a concise explanation of each figure.

OTHER FORMATS

CLINICAL REVIEWS—Short, highly focused, reviews of topics related to therapy, diagnosis, or other clinical issues in hematology. Consult Editor-in-Chief regarding topic. No manuscript submission fee or page charges.

EDITORIALS—Concise, invited, commentaries on articles published in *Blood* or other topics in hematology.

LETTERS—3 double-spaced pages, including references. Comments on published articles and short commentaries on topics of general interest to hematologists and instructive case reports.

RAPID COMMUNICATION—Definitive papers of exceptional scientific importance within the broad discipline of hematology. Request consideration as a Rapid Communication upon submission. Criteria for selection include originality, importance, and potential value of accelerated publication to readers. No page limit. Do not submit case reports, methods papers, or preliminary studies.

REVIEW ARTICLES—Consult Editor-In-Chief regarding topic. Be thorough, detailed. Include appropriate references to literature. Use tables and figures to summarize critical points. No manuscript submission fee or page charges.

OTHER

Abbreviations: Define at first use and apply consistently throughout. Do not use non-standard abbreviations or abbreviate terms appearing fewer than three times. Give chemical name of compound after first use of trivial name. Trivial name may follow throughout. Abbreviate units of measure only when used with numbers.

DNA Sequence Data: Submit to GenBank. For instructions write: GenBank, Group T-10, Mail Stop K710, Los Alamos National Laboratory, Los Alamos, NM 87545 (505-665-2177). Give accession number in footnote.

Materials Sharing: All reported reagents, DNA clones, cell lines, and antibodies must be freely distributed to qualified scientists.

Page Charges: $50 per published page for Regular Manuscripts and Rapid Communications. Cost of publishing color photographs is borne by authors.

Submission Fee: There is a $50 processing fee for regular manuscripts and Rapid Communications.

SOURCE

July 1, 1994, 84(1): xxixi-xxxi.

BMJ

British Medical Association

The Editor, *BMJ*
BMA House
Tavistock Square
London WC1H 9JR
Telephone: 071 387 4499; Fax: 071 383 6418; e-mail: Editor@bmjedit.demon.co.uk

AIM AND SCOPE

The *BMJ* aims to help doctors everywhere practise better medicine and to influence the debate on health. To achieve these aims we publish original scientific studies, review and educational articles, and papers commenting on the clinical, scientific, social, political, and economic factors affecting health. We are delighted to receive articles for publication in all of these categories from doctors and others. We can publish only about 12% of the articles we receive, but we aim to give quick and authoritative decisions.

The usual reasons for rejection are insufficient originality, serious scientific flaws, or the absence of a message that is important to a general medical audience. Referees are asked for their opinion on the originality, scientific reliability (overall design, patients studied, methods, results, interpretation and conclusions, and references), clinical imporance, and overall suitability for publication in the journal.

MANUSCRIPT SUBMISSION

Submit material exclusively to *BMJ*. All authors must give signed consent to publication. Type double-spaced on numbered pages and conform to uniform requirements for manuscripts submitted to biomedical journals (International Committee of Biomedical Editors. *Uniform Requirements for Manuscripts Submitted to Biomedical Journals*. Philadelphia: ICMJE, 1993).

Give names and address and appointment of authors at time they did work, as well as a current address for correspondence (including telephone and fax numbers).

Include a paragraph (250 words) for This Week in the *BMJ*; up to five brief points for key messages box (messages may relate to clinical practice, research, public health, epidemiology, or health policy; original data if it will help reviewers; copies of related papers (important where details of study methods are published elsewhere); copies of any nonstandard questionnaires used in research; details of sources of funding for research; and copies of previous referees' reports on research (since other journals may have been tried before submitting to *BMJ*, include responses to previous referees' comments).

The uniform requirements for manuscripts submitted to medical journals state that "authorship credit should be based only on substantial contribution to (a) conception and design, or analysis and interpretation of data; and to (b) drafting the article or revising it critically for important intellectual content; and on (c) final approval of the version to be published. Conditions (a), (b), and (c) must all be met. Participation solely in the acquisition of funding or the collection of data does not justify authorship (International Committee of Biomedical Editors. *Uniform Requirements for Manuscripts Submitted to Biomedical Journals*. Philadelphia: ICMJE, 1993). Give assurance that all authors fulfil criteria of authorship and that there is no one else who fulfils criteria but has not been included.

Disclose conflict of interest capable of influencing judgments. Conflicts may be financial, personal, political, or academic. Editors may decide that readers should know and make up their own minds. Before publishing information authors will be consulted. Explicitly state all sources of funding for research; information will be included in published paper.

If submitted paper overlaps by more than 10% with previously published papers or papers submitted elsewhere, send copies of those papers.

Articles may be withdrawn from publication if given media coverage while under consideration or in press.

All authors (except some government employees) transfer copyright to *BMJ* just before publication.

MANUSCRIPT FORMAT

Send three copies. Original papers should be no longer than 2000 words, with six tables or illustrations reporting original research relevant to clinical medicine in a way that is accessible to readers of a general journal. Papers for general practice section cover research or any other matters relevant to primary care.

Include a structured abstract (250 words) with headings: objectives, design, setting, subjects, interventions, main outcome measures, results, and conclusions (Haynes RB, Mulrow CD, Huth EJ, Altman DG, Gardner MJ. More informative abstracts revisited. *Ann Intern Med* 1990;113:69-76).

Tables, Illustrations and Photographs: Submit separately from text of paper. Type legends to illustrations on a separate sheet. Make tables simple; do not duplicate information in text. Use illustrations only when data cannot be expressed clearly in any other way. Submit numerical data on which graphs, scattergrams, or histograms are based; convert data in histograms into tabular form. Present line drawings as photographic prints or good quality photocopies. Submit other black-and-white illustrations as prints not negatives or x ray films. Photographs should be of highest quality possible (reproduction reduces quality), no larger than 30 × 21 cm (A4), trimmed to remove redundant areas; and TOP marked on back. Color prints can be used.

State staining techniques and include an internal scale marker on photomicrographs. X ray films reproduce most clearly from a glossy black-and-white print; do not submit slides. Provide a sketch or tracing paper overlay highlighting features of interest; do not mark photographs. If features are small or subtle, indicate with arrows marked on overlay.

If tables or illustrations have been published elsewhere obtain written consent for republication from copyright holder (usually publisher) and authors.

References: Number in order of appearance in text. At end of article, full list of references gives names and initials of all authors (unless more than six, give first six followed by *et al .*). Follow authors names by title of article; title of journal abbreviated according to *Index Medicus,* year of publication; volume number; and first and last page numbers.

21 Soter NA, Wasserman SI, Auster KF Cold urticaria: release into the circulation of histamine and eosinophil chemotactic factor of anaphylaxis during cold challenge *N Engl J Med* 1976;**294**:687-90

References to books give names of any editors, place of publication, editor, and year:

22 Osler AG. *Complement: mechanisms and functions.* Englewood Cliffs: Prentice-Hall, 1976

Cite information from manuscripts not yet in press, papers reported at meetings, or personal communications only in text, not as a formal reference. Obtain permission from source to cite personal communications. Verify references against original documents.

OTHER FORMATS

DRUG POINTS—300 words with five references and one table or figure reporting new adverse drug reactions or drug interactions. Priority is given to drug points that report more than one case; those in which patient is rechallenged with drug; and those which exclude other possible causative factors (disease process, other drugs, environmental agents). Include relevant clinical photographs If reporting side effects of drugs, contact Committee on Safety of Medicines and drug manufacturer to inquire if they have had similar reports; submit replies. Include age and sex of subjects; suspected drug and all drugs currently being taken (with start, stop, and restart dates and outcome); prior experience with drug or adverse reactions to related drugs; other diseases, environmental factors, and timing; all published reports; and other relevant factors to adverse reactions.

EDITORIALS—800 words, usually commissioned, unsolicited editorials are considered.

EDUCATION AND DEBATE ARTICLES— 2000 words on all aspects of medicine and health including sociological aspects of medicine; polemical pieces; and educational articles. Include an unstructured summary (150 words). Mostly commissioned.

LESSONS OF THE WEEK—1200 words with a single sentence explaining lesson. Usually case reports or case series alerting readers to a potential clinical problem. Make lesson as specific as possible for a general audience.

LETTERS TO THE EDITOR— 400 words, five references and one illustration or table, typed double-spaced and signed by all authors. Priority is given to letters responding to articles published in journal within past six weeks. There is no deadline for letters from abroad, submit as soon as possible after publication of article, preferably by fax.

MEDICINE AND THE MEDIA CONTRIBUTIONS—Discuss with editors before submission.

OBITUARIES— 250 words; submit within three months of death; include cause of death. Self-written obituaries (written in first person) are welcome, as are good quality photographs.

PERSONAL VIEW ARTICLES—1100 words signed; anonymous pieces published by special arrangement.

SHORT REPORTS—600 words with one table or illustration and five references.

OTHER

Electronic Submission: Submit accepted articles on disk. Preferred disk format is WordPerfect 5.1 (MS-DOS) on a 3-5 in. disk but we can also handle versions of WordPerfect and Microsoft Word for DOS, Windows, or Apple Macintosh. Do not use automatic page numbering, referencing, or footnotes. Number pages of hard copy by hand.

Experimentation, Human: If patient may be identified from a case report, illustration, or paper, obtain written consent of patient for publication. See Smith J. "Keeping confidences in published papers", (*BMJ* 1991;302:1168). Do not use black bands across eyes to disguise patient; changing details to disguise patients is bad scientific practice.

Statistical Analyses: Define statistical methods; describe any not in common use in detail or give references. For guidelines on use of statistical methods and on interpretation and presentation of statistical material as well as specific recommendations on statistical estimation and significance see Gardner MJ, Altman DG, eds. *Statistics with Confidence* (London: BMJ, 1989).

Style: Abbreviations should not be used. Drugs should be referred to by their approved, not proprietary, names. Give source of any new or experimental preparations. Give scientific measurements in SI units, except for blood pressure (mm Hg).

SOURCE

January 7, 1995, 310(6971): 50-51.

Brain
Oxford University Press

Professor W. I. McDonald
Institute of Neurology
National Hospital for Neurology and Neurosurgery
Queen Square, London WC1N 3BG.
Telephone: 071 837 2144; Fax: 071 837 3032

AIM AND SCOPE

Brain publishes definitive papers on neurology and related clinical disciplines, and on basic neuroscience, including molecular and cellular biology and neuropsychology when they have a neurological orientation and are relevant to the understanding of human disease. Papers which are predominantly technical or methodological in nature are not suitable.

Brain does not publish preliminary reports of work in progress, or brief reports on single cases. More detailed studies on single cases will be considered only when they definitively resolve an important problem in the field, or when the data lead to a significant conceptual advance. Studies on single cases which can be readily performed on groups of patients will not be accepted.

MANUSCRIPT SUBMISSION

Send four copies of paper.

In cover letter, include following declaration. 'The work reported in the attached paper entitled . . . has not been and is not intended to be published anywhere except in *Brain*,' signed by all authors. Previous publication of results in abstract form will not preclude consideration.

MANUSCRIPT FORMAT

Type double-spaced, including text, tables, legends and references, on one side of paper. Supply a running title, (40 characters including spaces). Provide a short summary. List full address, telephone and fax number of main author on title page.

References: Limit to essential literature. List at end of paper in alphabetical order. Do not number. For multiple publications by same author, list those by author alone first, those with two authors listed after, and any with three or more up to a maximum of six; indicate any more by *et al*. If there is more than one paper for a given year, list by a, b, c, etc. Present in Vancouver style (see International Committee of Medical Journal Editors. Style matters: uniform requirements for manuscripts submitted to biomedical journals. *BMJ* 1991; 302: 338-41). See *List of Journals Indexed in Index Medicus* for journal title abbreviations.

Bostok H. Impulses in experimental neuropathy. In: Dyck PJ, Thomas PK, editors. Peripheral neuropathy. 3rd ed. Philadelphia: W. B. Saunders, 1993: 109-20.

Bushby KMD, Garnder-Medwin D. The clinical, genetic and dystrophin characteristics of Becker muscular dystrophy. I. Natural history. J Neurol 1993; 240: 98-104.

Handwerker HO, Kobal G. Psychophysiology of experimentally induced pain. [Review]. Physiol Rev 1993; 73: 639-71.

Scaravilli F, editor. The neuropathology of HIV infection. London: Springer-Verlag, 1993.

In text give author's name and year of publication in brackets. If three or more authors, name of first is followed by *et al*. Follow journal style (see current issue) for punctuation, both in Reference list and text. References to papers 'in preparation' or 'submitted' are not acceptable; if 'in press,' give name of journal or book. Do not include 'personal communications' or other inaccessible information. Refer to information derived from personal communications or from unpublished work in text.

Illustrations: Submit clearly lettered original line drawings or glossy prints, with number and author's name on back. Half-tone photographs, particularly for electron micrographs or CT or MR images, must be of good quality. All micrographs must carry a magnification bar. Color illustrations are accepted, but the authors will be required to contribute to cost; an estimate will be provided. Keep number of all illustrations to a minimum. Indicate desired position of figures and tables in typescript. List figure legends on a separate sheet. Title all tables. Footnotes may be used in tables, but not in text.

OTHER

Experimentation, Animal: Conform to accepted ethical standards.

Experimentation, Human: State that subjects' consent was obtained according to the Declaration of Helsinki

[*BMJ* (1991) 302: 1194]. Record consent when photographs of patients are shown or other details are given which could lead to identification of individuals.

Units: Use Systéme Internationale (SI units). See International Committee of Medical Journal Editors' standards in *Ann. Intern. Med.* 108, 266–273, 1988.

SOURCE

February 1994, 117 (pt1): 223-224.

Brain Research
Elsevier

Professor Dr. D.P. Purpura
Office of the Dean, Albert Einstein College of Medicine
Jack and Pearl Resnick Campus
1300 Morris Park Avenue
Bronx, NY 10461
Telephone: (718) 430-2367; Fax: (718) 597-4072

AIM AND SCOPE

Brain Research provides a medium for prompt publication of articles in the fields of neuroanatomy, neurochemistry, neurophysiology, neuroendocrinology, neuropharmacology, neurotoxicology, neurocommunications, behavioral sciences, molecular neurology, and biocybernetics. Clinical studies that are of fundamental importance and have a direct bearing on the knowledge of the structure and function of the brain, spinal cord, and peripheral nerves will also be published.

MANUSCRIPT SUBMISSION

Submit in quadruplicate, including four copies of all illustrations. Double-space with at least a 4 cm margin on pages of uniform size. Manuscript must be complete in all respects and deal with original material not previously published, or being considered for publication elsewhere. It is assumed that all listed authors concur with submission and final manuscript has been approved by all authors and tacitly or explicitly by responsible authorities in laboratories where work was carried out. If accepted, manuscript shall not be published elsewhere in the same form, in either the same or another language, without consent of Editors and Publisher.

Submit written permission of author and publisher of illustrations or other small parts of articles or books already published elsewhere used in paper with manuscript. Indicate original source in legend.

Clearly state section for which article should be considered. Review articles will only be published in *Brain Research Reviews*. Developmental studies, molecular studies and cognitive studies will be considered for *Developmental Brain Research, Molecular Brain Research* or *Cognitive Brain Research*, unless explicitly requested to be considered for *Brain Research*.

MANUSCRIPT FORMAT

Submit a list of 6–8 key words, 200 word summary for Research Reports, and 50–70 words for Short Communications, on a separate page. Divide Research Reports into sections headed by a caption (Summary, Introduction, Materials and Methods, Results, Discussion, Acknowledgments, References). Short Communications have no headings.

Include authors' full names, academic and professional affiliations, and complete addresses on a separate title page. Specify name, address, telephone and fax numbers of corresponding author.

References: Cite literature references in text at appropriate places by numbers in square brackets. List all references cited in text at end of paper on a separate page (also double-spaced) arranged in alphabetical order of first author and numbered consecutively (numbers in square brackets). Cite all items in Reference list in text; list all references cited in text. Literature references must be complete, including name and initials of authors cited, title of paper, abbreviated periodical title of periodical, volume, year, and first and last page numbers of article. Use journal title abbreviations from *List of Serial Title Word Abbreviations,* CIEPS/ISDS, Paris, 1985). Form of literature references to books is: author, initials, title of book, publisher and city, year, and page numbers. References to authors contributing to multi-author books or to proceedings printed in book-form should be similar to those for books. Cite this journal as *Brain Res.*

JOURNAL

1 Starr, A., Kristeva, R., Cheyne, D., Lindinger, G. and Deecke, L., Localization of brain activity during auditory verbal short-term memory derived from magnetic recordings, *Brain Res.*, 558 (1991) 181–190.

BOOK

2 Kowler, E (Ed.), *Reviews in Oculomotor Research, Vol. 4, Eye Movements and Their Role in Visual and Cognitive Processes*, Elsevier, Amsterdam, 1990, 486 pp.

CHAPTER IN BOOK

3 Kolb, B., Animal models for human PFC-related disorders. In H.B.M. Uylings, C.G. van Eden, J. P. C. de Bruin, M. A. Corner and M. G. P. Feenstra (Eds.), *The Prefrontal Cortex: its Structure, Function and Pathology, Progress in Brain Research, Vol. 85,* Elsevier, Amsterdam, 1990, pp. 501–519.

Illustrations: Submit in quadruplicate and in a form and condition suitable for reproduction either across a single column (= 8.3 cm) or a whole page (=17.6 cm). Put author's name and Arabic numeral according to sequence of appearance in text (refer to as Fig. 1, Fig. 2, etc.). Make line drawings with Black Ink on drawing or tracing paper or glossy sharp photographs. Make lettering clear and of adequate size to be legible after reduction. Professional labelling is preferable but, if not possible, letter in fine pencil. Supply photographs, including roentgenograms, electroencephalograms and electron micrographs as clear black-and-white prints on glossy paper, rather than copies, usually larger than final size but not more than 20×25 cm. Micrographs should have a scale bar in legend.

Color Reproduction: Must be approved by Editor. Submit as separate prints. Do not mount on cardboard. Professional lettering is preferred, but if not possible, provide an overlay displaying desired labelling. Avoid press-on lettering. Slides taken from labelled prints are acceptable.

Legends: Type double-spaced on a separate page. Begin with number of illustration referred to.

Tables: Type numerical data double-spaced on a separate page, numbered in sequence in Roman numerals (Table I, II, etc.). Provide with a heading, and refer to in text as Table I, Table II, etc.

OTHER FORMATS

SHORT COMMUNICATIONS—1500 words or equivalent space in tables and illustrations. Research that has progressed to the stage when it is considered that the results should be made quickly known to other workers.

OTHER

Color Charge: Approximately Dfl. 1800.00 for first page, and approximately Dfl. 1500.00 per page for all following pages.

Page Charge: U.S. $80.00 for each printed page in excess of 10 printed pages for Research Reports, and each page in excess of 5 printed pages for Short Communications. An approximate guide for judging length: 3 type-written pages = 1 printed page; 3 'average' figures + legends = 1 printed page; 3 'average' tables = 1 printed page; 35 references = 1 printed page.

SOURCE

September 12, 1994, 656(2): iii-iv.

Brain Research: Brain Research Reviews
Elsevier

Professor Dr. D.P. Purpura
Office of the Dean, Albert Einstein College of Medicine
Jack and Pearl Resnick Campus
1300 Morris Park Avenue, Bronx, NY 10461
Telephone: (718) 430–2387; Fax: (718) 597-4072
Professor Dr. P.J. Magistretti
Institut de Physiologie, Faculté de Médecine
Université de Lausanne
7, rue du Bugnon, 1005 Lausanne, Switzerland
Telephone: (41) (21) 3132831; Fax: (41) (21) 3132865

AIM AND SCOPE

Brain Research Reviews is a section of *Brain Research* which provides a medium for the prompt publication of full-length review articles, short topical (mini-) reviews and, occasionally, long research papers with extensive review components, which give analytical surveys that define heuristic hypotheses and provide new insights into brain mechanisms. Critical and consolidated reports of small, highly topical symposia will be considered for publication after prior review of the proposal by, and in agreement with, the Editors.

MANUSCRIPT SUBMISSION

Manuscripts must be complete in all respects; they should neither have been previously published, nor presently be under consideration for publication elsewhere. For manu-

scripts with multiple authorship it is assumed that all listed authors concur with submission and that a copy of final manuscript has been approved by all authors and tacitly or explicitly by responsible authorities in laboratories where work was carried out. If accepted, manuscript shall not be published elsewhere in the same form, in either the same or another language, without consent of Editors and Publisher. State section for which article should be considered. Review articles will only be published in *Brain Research Reviews*. Developmental studies, molecular studies and cognitive studies will be considered for publication in *Developmental Brain Research*, *Molecular Brain Research* or *Cognitive Brain Research*, unless explicitly requested to be considered for *Brain Research*.

MANUSCRIPT FORMAT

Include authors' full names, academic and professional affiliations, and complete addresses on a separate title page. Specify name, address, telephone and fax numbers of corresponding author.

List 6-8 key words. Include a summary of 200 words for Full-length Reviews, and 50-70 words for Short (or Topical) Reviews or for Reports, on a separate page. Submit four copies of each manuscript in double-spaced typing with at least a 4 cm margin on pages of uniform size. Submit four copies of all illustrations. Divide Reviews and Reports into sections headed by caption (Summary, Introduction, Materials and Methods. Results, Discussion, Acknowledgments, References).

References: Cite literature references in text at appropriate places by numbers as superscripts without parentheses. List all references cited in text at end of paper on a separate page (also double-spaced) arranged in alphabetical order of first author and numbered consecutively. Cite all items in Reference list in text. Literature references must be complete, including name and initials of authors, title of paper, abbreviated title of periodical, volume, year, and first and last page numbers of article. Abbreviate journal titles according to *List of Serial Title Word Abbreviations* (CIEPS/ISDS, Paris, 1985). The form of literature references to books should be: author, initials, title of book, publisher and city, year, and page numbers referred to). References to authors contributing to multi-author books or to proceedings printed in book-form should be similar to those for books.

1 Kowler, E. (Ed.), *Reviews in Oculomotor Research, Vol. 4, Eye Movements and Their Role in Visual and Cognitive Processes,* Elsevier, Amsterdam, 1990, 486 pp.

2 Starr, A., Kristeva, R., Cheyne, D., Lindinger, G. and Deecke, L., Localization of brain activity during auditory verbal short-term memory derived from magnetic recordings, *Brain Res.,* 558 (1991)181–190.

3 Kolb, B., Animal models for human PFC-related disorders. In H.B.M. Uylings, C.G. van Eden, J.P.C. de Bruin, M.A. Corner and M.G.P. Feenstra (Eds.), *The Prefrontal Cortex: its Structure, Function and Pathology, Progress in Brain Research, Vol. 85,* Elsevier, Amsterdam, 1990, pp. 501-519.

Illustrations: Submit in quadruplicate in a form suitable for reproduction either across a single column (= 8.3 cm) or a whole page (= 17.6 cm). Illustrations should bear author's name and be numbered in Arabic numerals according to sequence of appearance in text; refer to as Fig. 1, Fig. 2, etc. Line drawings should be in Black Ink on drawing or tracing paper or glossy sharp photographs. Make lettering clear and of adequate size to be legible after reduction. Professional labelling is preferable; if not possible, lettering should be in fine pencil. Supply photographs, including roentgenograms, electroencephalograms and electron micrographs as clear black-and-white prints on glossy paper, rather than copies, larger than final size of reproduction but not more than 20 × 25 cm. Micrographs should have a scale bar, rather than a magnification factor in legend. The same degree of reduction will be applied to all figures.

Color Reproduction: Must be approved by Editor. Submit color figures as separate prints; do not mount on cardboard. Professional lettering is preferred; if not possible, provide an overlay displaying desired labelling. Avoid press-on lettering. Slides taken from labelled prints are also acceptable. Each illustration must have a legend. Type with double-spacing on a separate page; begin with number of illustration they refer to. If illustrations or other small parts of articles or books already published elsewhere are used, include written permission of author and publisher. Indicate original source in legend.

Tables: Type tables of numerical data double-spaced on a separate page, number in sequence in Roman numerals (Table I, II, etc.), and provide with a heading. Refer to in text as Table I, Table II, etc.

OTHER FORMATS

FULL-LENGTH REVIEWS—No limit to length.

REPORTS—Reports of topical symposia may have two formats: a consolidated summary of papers presented. Highlight areas of controversy and future perspectives. Submit signed approval by all *active* participants: an edited collection of short, incisive summaries of the research written by each participant. 1500 words each, 15 references, and three figures. A brief introductory overview should accompany collection of summaries. Submit a proposal to Editors for review prior to symposium.

SHORT (OR TOPICAL) REVIEWS—5000 words including references. Critically address topical issues pertaining to current research. Illustrations should complement the text and provide a synthesis of data and mechanisms. Keep illustrations containing primary data to a minimum except those displaying morphological information.

OTHER

Color Charge: Approximately Dfl. 1800.00 for first page, and approximately Dfl. 1500.00 per page for other pages.

SOURCE

January 1994, 19(1): front matter.

Brain Research: Cognitive
Elsevier Science

Professor Dominick P. Purpura
Office of the Dean, Albert Einstein College of Medicine

Jack and Pearl Resnick Campus
1300 Morris Avenue
Bronx, NY 10461
Telephone: (718) 430-2387; Fax: (718) 597-4072

AIM AND SCOPE

Cognitive Brain Research will publish original experimental studies of neural processes underlying intelligent mental activity and its disturbances. Areas of higher nervous functions of particular interest are perception, learning, memory, judgement, reasoning, language and emotion. Original reports dealing with all aspects of computational neuroscience including neural networks with reference to brain mechanisms subserving cognition are encouraged. Heuristic application of chaos theory to the analysis of brain functions and aberrant neurobehavioural processes will also be considered for publication.

MANUSCRIPT SUBMISSION

Submit manuscript in quadruplicate including four copies of all illustrations that are complete in all respects. For manuscripts with multiple authorship it is assumed that all listed authors concur with submission and that final manuscript has been approved by all authors and tacitly or explicitly by responsible authorities in laboratories where work was done. If accepted, manuscript shall not be published elsewhere in the same form, in either the same or another language, without consent of Editors and Publisher. Submission of a paper implies that it has not previously been published (except in abstract form) and that it is not being considered for publication elsewhere.

MANUSCRIPT FORMAT

Include a list of 6–8 key words and a summary (200 words) on a separate page. Double-space with at least a 4 cm margin on pages of uniform size. Divide Research Reports into sections: Summary, Introduction, Materials and Methods, Results, Discussion, Acknowledgments, References.

Title Page: Give authors' full names, academic and professional affiliations, and complete addresses. Specify name and address, telephone and fax numbers of corresponding author.

Literature References: Cite literature references in text at appropriate places by numbers in square brackets. List all references cited in text at end of paper on a separate page, double-spaced, in alphabetical order of first author and numbered consecutively. Cite all items in Reference list in text and, conversely, cite all references in text in list. Literature references must be complete, including name and initials of authors, title of paper, abbreviated title of periodical, volume, year, and first and last page numbers of article. Abbreviate journal titles according to *List of Serial Title Word Abbreviations* (CIEPS/ISDS, Paris, 1985). Literature references to books: author, initials, title of book, publisher and city, year, and page numbers referred to. References to works in multi-author books or to proceedings in book-form are similar to those for books.

1 Kowler, E. (Ed.), *Reviews in Oculomotor Research, Vol. 4, Eye Movements and Their Role in Visual and Cognitive Processes,* Elsevier, Amsterdam, 1990, 486 pp.

2 Starr, A., Kristeva, R., Cheyne, D., Lindinger, G. and Deecke, L., Localization of brain activity during auditory verbal short-term memory derived from magnetic recordings, *Brain Res.*, 558 (1991)181–190.

3 Kolb, B., Animal models for human PFC-related disorders. In H.B.M. Uylings, C.G. van Eden, J.P.C. de Bruin, M.A. Corner and M.G.P. Feenstra (Eds.), *The Prefrontal Cortex: its Structure, Function and Pathology, Progress in Brain Research, Vol. 85*, Elsevier, Amsterdam, 1990, pp. 501–519.

Illustrations: Submit in quadruplicate in a form and condition suitable for reproduction either across a single column (8.3 cm) or a whole page (17.6 cm). Illustrations should bear author's name. Number in Arabic numerals according to sequence of appearance in text; refer to as Fig. 1, Fig. 2, etc. Submit line drawings in black ink on drawing or tracing paper or as glossy sharp photographs. Make lettering clear and of adequate size to be legible after reduction. Professionally label or letter in fine pencil.

Supply photographs, including roentgenograms, electroencephalograms and electron micrographs as clear black-and-white prints on glossy paper, larger than final size, but not more than 20 × 25 cm. Give micrographs scale bars, rather than magnification factor in legend.

Each illustration must have a legend, typed double-spaced on a separate page. Begin with number of illustration to which they refer.

If illustrations or parts of articles or books already published elsewhere are used in papers, submit written permission of author and publisher concerned. Indicate original source in legend of illustration.

Color Reproduction: Must be approved by Editor. Authors contribute towards extra costs. Submit as separate prints, not mounted on cardboard. Professionally letter or provide an overlay displaying desired labelling. Avoid press-on lettering. Slides from labelled prints are acceptable.

Tables: Type tables of numerical data double-spaced on separate pages with headings, numbered in sequence in Arabic numerals (Table 1, 2, etc.); refer to in text as Table 1, Table 2, etc.

OTHER FORMATS

SHORT COMMUNICATIONS— 1500 words or equivalent space in tables and illustrations; 50-70 word abstract, no section headings. Report on research which has progressed to the stage when results should be made known quickly to other workers in the field.

NEUROSCIENCE PROTOCOLS–Contact Dr. F.G. Wouterlood, Department of Anatomy, Amsterdam, The Netherlands. Fax: (31)(20)4448054; E-mail: fg.wouterlood.anat @med.vu.nl

OTHER

Color Charges: Dfl. 1800.00 for first page with color, and Dfl. 1500.00 per page for all following pages.

Electronic Submission: Preferred medium of final submission is on disk with accompanying reviewed and revised manuscript. Preferred storage medium is 5.25 or 3.5 in. disk in MS-DOS format; other systems are welcomed, e.g., NEC and Macintosh (save file in usual manner, do not use option "save in MS-DOS format"). After final acceptance, submit disk plus one final, printed and exactly matching hard copy. File on disk and printout must be identical. Specify type of computer and wordprocessing package used (do not convert text file to plain ASCII). Ensure that letter "l" and digit "1," letter "O" and digit "0" have been used properly. Format   article (tabs, indents, etc.) consistently. Do not leave characters not available on wordprocessor (Greek letters, mathematical symbols, etc.) open; indicate by a unique code (e.g., Greek letter alpha, @, #, etc. for Greek letter α); use codes consistently throughout text; list codes and provide a key. Do not allow wordprocessor to introduce word splits; do not use a 'justified' layout. Adhere strictly to general instructions on styles/arrangement, especially reference style of journal. Tables and illustrations are handled conventionally.

SOURCE

July 1994, 2(1): front matter.

Brain Research: Developmental Brain Research
Elsevier Science

Professor Dominick P. Purpura
Office of the Dean, Albert Einstein College of Medicine
Jack and Pearl Resnick Campus
1300 Morris Avenue
Bronx, NY 10461
Telephone: (718) 430-2387; Fax: (718) 597-4072

AIM AND SCOPE

Developmental Brain Research is a special section of *Brain Research* which provides a medium for prompt publication of in vitro and in vivo developmental studies concerned with the mechanism of neurogenesis, neuron migration, cell death, neuronal differentiation, synaptogenesis, myelination, the establishment of neuron-glia relations and the development of various brain barrier mechanisms.

MANUSCRIPT SUBMISSION

Submit manuscript in quadruplicate including four copies of all illustrations that are complete in all respects.With multiple authorship it is assumed that all listed authors concur with submission and that final manuscript has been approved by all authors and tacitly or explicitly by responsible authorities in laboratories where work was done. If accepted, manuscript shall not be published elsewhere in the same form, in either the same or another language, without consent of Editors and Publisher. Submission of a paper implies that it has not previously been published (except in abstract form) and that it is not being considered for publication elsewhere.

MANUSCRIPT FORMAT

Include a list of 6–8 key words and a summary (200 words) on a separate page. Double-space with at least a 4 cm margin on pages of uniform size. Divide Research Reports into sections: Summary, Introduction, Materials and Methods, Results, Discussion, Acknowledgments, References.

Title Page: Include authors' full names, academic and professional affiliations, and complete addresses. Specify name and address, telephone and fax numbers of corresponding author.

Literature References: Cite literature references in text at appropriate places by numbers in square brackets. List all references cited in text at end of paper on a separate page, double-spaced, in alphabetical order of first author and numbered consecutively. Cite all items in Reference list in text and, conversely, cite all references in text in list. Literature references must be complete, including name and initials of authors, title of paper, abbreviated title of periodical, volume, year, and first and last page numbers of article. Abbreviate journal titles according to *List of Serial Title Word Abbreviations* (CIEPS/ISDS, Paris, 1985). Literature references to books: author, initials, title of book, publisher and city, year, and page numbers referred to. References to works in multi-author books or to proceedings in book-form are similar to those for books.

1 Kowler, E. (Ed.), *Reviews in Oculomotor Research, Vol. 4, Eye Movements and Their Role in Visual and Cognitive Processes,* Elsevier, Amsterdam, 1990, 486 pp.

2 Starr, A., Kristeva, R., Cheyne, D., Lindinger, G. and Deecke, L., Localization of brain activity during auditory verbal short-term memory derived from magnetic recordings, *Brain Res.,* 558 (1991)181–190.

3 Kolb, B., Animal models for human PFC-related disorders. In H.B.M. Uylings, C.G. van Eden, J.P.C. de Bruin, M.A. Corner and M.G.P. Feenstra (Eds.), *The Prefrontal Cortex: its Structure, Function and Pathology, Progress in Brain Research, Vol. 85*, Elsevier, Amsterdam, 1990, pp. 501–519.

Illustrations: Submit in quadruplicate in a form and condition suitable for reproduction either across a single column (8.3 cm) or a whole page (17.6 cm). Illustrations should bear author's name. Number in Arabic numerals according to sequence of appearance in text; refer to as Fig. 1, Fig. 2, etc. Submit line drawings in black ink on drawing or tracing paper or as glossy sharp photographs. Make lettering clear and of adequate size to be legible after reduction. Professionally label or letter in fine pencil.

Supply photographs, including roentgenograms, electroencephalograms and electron micrographs as clear black-and-white prints on glossy paper, larger than final size, but not more than 20×25 cm. Give micrographs scale bars, rather than magnification factor in legend.

Each illustration must have a legend, typed double-spaced on a separate page. Begin with number of illustration to which they refer.

If illustrations or other small parts of articles or books already published elsewhere are used in papers, submit written permission of author and publisher concerned. Indicate original source in legend of illustration.

Particular figures may be published on a full page (i.e. bleeding). This must be specifically requested. Figure

should be: 29×21 cm (for 1:1 reproduction), or multiples of above dimensions if to be enlarged or reduced. Labelling, scale bars, etc. must be at least 1 cm from outer edges of 29×21 cm figure; approximately 0.5 cm will be trimmed off during printing and binding. Figures not able to fulfill these requirements will not be considered and will be printed in the normal way.

Color Reproduction: Must be approved by Editor. Authors contribute towards extra costs. Submit as separate prints, not mounted on cardboard. Professionally letter or provide an overlay displaying desired labelling. Avoid press-on lettering. Slides from labelled prints are acceptable.

Tables: Type tables of numerical data double-spaced on separate pages with headings, numbered in sequence in Arabic numerals (Table 1, 2, etc.); refer to in text as Table 1, Table 2, etc.

OTHER FORMATS

SHORT COMMUNICATIONS— 1500 words or equivalent space in tables and illustrations; 50-70 word abstract, no section headings. Report on research which has progressed to the stage when results should be made known quickly to other workers in the field.

NEUROSCIENCE PROTOCOLS–Contact Dr. F.G. Wouterlood, Department of Anatomy, Amsterdam, The Netherlands. Fax: (31) (20) 4448054; E-mail: fg.wouterlood.anat@med.vu.nl

OTHER

Color Charges: Dfl. 1800.00 for first page with color, and Dfl. 1200.00 per page for all following pages.

Electronic Submission: Preferred medium of final submission is on disk with accompanying reviewed and revised manuscript. Preferred storage medium is 5.25 or 3.5 in. disk in MS-DOS format; other systems are welcomed, e.g., NEC and Macintosh (save file in usual manner, do not use option "save in MS-DOS format"). After final acceptance, submit disk plus one final, printed and exactly matching version (as a printout). File on disk and printout must be identical. Specify type of computer and wordprocessing package used (do not convert text file to plain ASCII). Ensure that letter "l" and digit "1," letter "O" and digit "0" have been used properly. Format article (tabs, indents, etc.) consistently. Do not leave characters not available on wordprocessor (Greek letters, mathematical symbols, etc.) open; indicate by a unique code (e.g. Greek letter alpha, @, #, etc. for Greek letter α); use codes consistently throughout text; list codes and provide a key. Do not allow word processor to introduce word splits; do not use a 'justified' layout. Adhere strictly to general instructions on styles/arrangement, especially reference style of journal. Tables and illustrations are handled conventionally.

SOURCE

February 16, 1995, 84(2): front matter.

Brain Research: Molecular Brain Research
Elsevier Science

Professor Dominick P. Purpura
Office of the Dean, Albert Einstein College of Medicine
Jack and Pearl Resnick Campus
1300 Morris Avenue
Bronx, NY 10461
Telephone: (718) 430-2387; Fax: (718) 597-4072

AIM AND SCOPE

Molecular Brain Research is a special section of *Brain Research* which provides a medium for prompt publication of studies on molecular mechanisms of neuronal, synaptic and related processes that underlie the structure and function of the brain. Emphasis is placed on the molecular biology of fundamental neural operations relevant to the integrative actions of the nervous system.

MANUSCRIPT SUBMISSION

Submit manuscript in quadruplicate including four copies of all illustrations complete in all respects. With multiple authorship it is assumed that all listed authors concur with submission and that final manuscript has been approved by all authors and tacitly or explicitly by responsible authorities in laboratories where work was done. If accepted, manuscript shall not be published elsewhere in the same form, in either the same or another language, without consent of Editors and Publisher. Submission of a paper implies that it has not previously been published (except in abstract form) and that it is not being considered for publication elsewhere.

MANUSCRIPT FORMAT

Include a list of 6–8 key words and a summary (200 words) on a separate page. Double-space with at least a 4 cm margin on pages of uniform size. Divide Research Reports into sections: Summary, Introduction, Materials and Methods, Results, Discussion, Acknowledgments, References.

Title Page: Include authors' full names, academic and professional affiliations, and complete addresses. Specify name and address, telephone and fax numbers of corresponding author.

Literature References: Cite literature references in text at appropriate places by numbers in square brackets. List all references cited in text at end of paper on a separate page, double-spaced, in alphabetical order of first author and numbered consecutively. Cite all items in Reference list in text and, conversely, cite all references in text in list. Literature references must be complete, including name and initials of authors, title of paper, abbreviated title of periodical, volume, year, and first and last page numbers of article. Abbreviate journal titles according to *List of Serial Title Word Abbreviations* (CIEPS/ISDS, Paris, 1985). Literature references to books: author, initials, title of book, publisher and city, year, and page numbers referred to. References to works in multi-author books or to proceedings in book-form are similar to those for books.

1 Kowler, E. (Ed.), *Reviews in Oculomotor Research, Vol. 4, Eye Movements and Their Role in Visual and Cognitive Processes,* Elsevier, Amsterdam, 1990, 486 pp.

2 Starr, A., Kristeva, R., Cheyne, D., Lindinger, G. and Deecke, L., Localization of brain activity during auditory verbal short-term memory derived from magnetic recordings, *Brain Res.,* 558 (1991)181–190.

3 Kolb, B., Animal models for human PFC-related disorders. In H.B.M. Uylings, C.G. van Eden, J.P.C. de Bruin, M.A. Corner and M.G.P. Feenstra (Eds.), *The Prefrontal Cortex: its Structure, Function and Pathology, Progress in Brain Research, Vol. 85,* Elsevier, Amsterdam, 1990, pp. 501–519.

Illustrations: Submit in quadruplicate in a form and condition suitable for reproduction either across a single column (8.3 cm) or a whole page (17.6 cm). Illustrations should bear author's name. Number in Arabic numerals according to sequence of appearance in text; refer to as Fig. 1, Fig. 2, etc. Submit line drawings in black ink on drawing or tracing paper or as glossy sharp photographs. Make lettering clear and of adequate size to be legible after reduction. Professionally label or letter in fine pencil.

Supply photographs, including roentgenograms, electro-encephalograms and electron micrographs as clear black-and-white prints on glossy paper, larger than final size, but not more than 20×25 cm. Give micrographs scale bars, rather than magnification factor in legend.

Each illustration must have a legend, typed double-spaced on a separate page. Beginning with number of illustration to which they refer.

If illustrations or parts of articles or books already published elsewhere are used in papers, submit written permission of author and publisher concerned. Indicate original source in legend of illustration.

Color Reproduction: Must be approved by Editor. Authors contribute towards extra costs. Submit as separate prints, not mounted on cardboard. Professionally letter or provide an overlay displaying desired labelling. Avoid press-on lettering. Slides from labelled prints are acceptable.

Tables: Type tables of numerical data double-spaced on separate pages with headings, numbered in sequence in Arabic numerals (Table 1, 2, etc.); refer to in text as Table 1, Table 2, etc.

OTHER FORMATS

SHORT COMMUNICATIONS— 1500 words or equivalent space in tables and illustrations; 50-70 word abstract, no section headings. Report on research which has progressed to the stage when results should be made known quickly to other workers in the field.

NEUROSCIENCE PROTOCOLS–Contact Dr. F.G. Wouterlood, Department of Anatomy, Amsterdam, The Netherlands. Fax: (31) (20) 4448054; E-mail: fg.wouterlood.anat@med.vu.nl

OTHER

Color Charges: Dfl. 1800.00 for first page with color, and Dfl. 1500.00 per page for all following pages.

Electronic Submission: Preferred medium of final submission is on disk with accompanying reviewed and revised manuscript. Preferred storage medium is 5.25 or 3.5 in. disk in MS-DOS format; other systems are welcomed, e.g., NEC and Macintosh (save file in usual manner, do not use option "save in MS-DOS format"). After final acceptance, submit disk plus one final, printed and exactly matching version (as a printout). File on disk and printout must be identical. Specify type of computer and wordprocessing package used (do not convert text file to plain ASCII). Ensure that letter "l" and digit "1," letter "O" and digit "0" have been used properly. Format article (tabs, indents, etc.) consistently. Do not leave characters not available on wordprocessor (Greek letters, mathematical symbols, etc.) open; indicate by a unique code (e.g., Greek letter alpha, @, #, etc. for Greek letter α); use codes consistently throughout text; list codes and provide a key. Do not allow wordprocessor to introduce word splits; do not use a 'justified' layout. Adhere strictly to general instructions on styles/arrangement, especially reference style of journal. Tables and illustrations are handled conventionally.

SOURCE

January 1995, 28(1): front matter.

Brain Research Bulletin
International Behavioral Neuroscience Society
Pergamon Press

Matthew J. Wayner
Division of Life Sciences
The University of Texas at San Antonio
6900 North Loop 1064 West
San Antonio, TX 78249-0062
Telephone: (210) 691-4481; Fax: (210) 691-4510

AIM AND SCOPE

Brain Research Bulletin will publish original reports of new and significant information concerning all aspects of the nervous system: biochemistry, physiology, anatomy, ultrastructure, electrophysiology, neurology, pathology and behavior. Behavioral studies as such will not be published unless they are obviously pertinent and make a significant contribution to one of the other fields. Brief communications which describe a new method, technique, or apparatus and results of experiments which can be reported briefly with limited figures and tables will be included as Rapid Communications. A limited number of relevant reviews and theoretical articles, results of symposia, and more comprehensive studies as monograph supplements will also be published.

MANUSCRIPT SUBMISSION

Include complete address information and current phone and fax numbers.

MANUSCRIPT FORMAT

Submit manuscripts in triplicate, typed double-spaced with wide margins on good quality paper. If a word processor is used, use a letter quality printer. Computer generated illustrations must be of the same quality as professional line drawings or will not be accepted. Title page: title of paper; author(s); laboratory or institution of origin with city, state, zip code, and country; complete address for mailing proofs; a running head (40 characters including spaces). Type references, footnotes, and legends for illustrations on separate sheets, double-spaced. Identify illustrations (unmounted photographs) on reverse with figure number and author(s) name; when necessary mark TOP. Type each table on a separate sheet double-spaced. All dimensions and measurements must be specified in the metric system.

Title: 85 characters, including spaces.

Abstract: 170 words suitable for use by abstracting journals. Prepared as follows:
MYERS, R. D., C. MELCHIOR AND C. GISOLFI. *Feeding and body temperature; Changes produced by excess calcium ions* . . . BRAIN RES BULL. Marked differences in extent of diffusion have been . . .
Type 3–12 (or more) words or short phrases suitable for indexing at bottom of abstract page.

Footnotes: Number title page footnotes consecutively. Footnote the reprint request author if not senior author. Text footnotes should not be used; incorporate material into text.

References: Prepare literature cited according to Numbered/Alphabetized style of the Council of Biology Editors. Cite references by number, in parentheses, within text (one reference to a number). Listed in alphabetical order (double-spaced) on a separate sheet at end of manuscript. Do not recite names of authors within text. Journal citations in reference list contain: surnames and initials of all authors (surname precedes initials); title of article; journal title abbreviated as listed in *List of Journals Indexed in Index Medicus*; volume, inclusive pages and year.

JOURNAL

1. Banks, W. A.: Kastin, A. J. Peptides and the blood-brain barrier: Lipophilicity as a predictor of permeability. Brain Res. Bull. 15:287–292; 1985.

Book references in following order: author, title, city of publication, publisher, year and pages.

BOOK

1. Myers, R. D. Handbook of drug stimulation. New York: Van Nostrand Reinhold Company; 1974.

CHAPTER IN BOOK

2. Mello, N. K. Behavioral studies of alcoholism. In: Kissin, B.; Begleiter, H., eds. The biology of alcoholism, vol. 2, Physiology and behavior. New York: Plenum Press; 1972: 219–291.

Illustrations: Prepare for use in a single column width whenever possible. Draw and letter all drawings for reduction to a given size to the same scale. Refer to all illustrations as figures and number in Arabic numerals. Do lettering in India ink or other suitable material proportionate to size of illustrations after reduction. Size lettering so that its smallest elements (subscripts or superscripts) will be readable when reduced. Keep lettering within framework of illustration; put key to symbols on face of chart. Use standard symbols: ○ ● △ ▲ □ ■ +. Give actual magnification of all photomicrographs. Indicate dimension scale. Sharply contrasting unmounted photographs of figures on glossy paper are required. Submit illustrations in black and white unless color reproduction is requested. Submit color prints in actual size; authors are responsible for additional costs.

Tables: Give each a brief heading; place explanatory matter in footnotes, not in title. Indicate table footnotes in body of table in order of appearance with symbols: * †, ‡, §, ¶, # **, etc. Do not duplicate material in text or illustrations. Omit vertical rules. Use short or abbreviated column heads. Identify statistical measures of variation, SD, SE, etc. Do not submit analysis of variance tables, but significant F's where appropriate within text. The appropriate form for reporting F value is: $F(11, 20) = 3.05, p < 0.01$.

Formulas and Equations: Keep structural chemical formulas, process flow-diagrams, and complicated mathematical expressions to a minimum. Draw chemical formulas and flow-diagrams in India ink for reproduction as line cuts.

Identify all subscripts, superscripts, Greek letters, and unusual characters.

OTHER

Anesthesia: In describing surgical procedures on animals, specify type and dosage of anesthetic agent. Curarizing agents are not anesthetics; if used, provide evidence that anesthesia was suitable grade and duration was employed.

Drugs: Capitalize proprietary (trademarked) names. Chemical name should precede trade, popular name, or abbreviation of a drug the first time it occurs.

Style: Follow standard nomenclature; abbreviations and symbols, as specified by Royal Society Conference of Editors. Metrication in Scientific Journals, *Am. Sci.* 56:159–164; 1968. Do not use italics for emphasis.

SOURCE

1994, 35(10): 103.

British Heart Journal
British Cardiac Society
BMJ Publishing Group

Editor, *British Heart Journal*
9 Fitzroy Square
London W1P 5AH

MANUSCRIPT SUBMISSION

Send papers relating to the heart and circulation in triplicate with 3 sets of figures, prepared according to the Uniform requirements for manuscripts submitted to biomedical journals (Vancouver agreement) (*BMJ* 1988;296:401–5). Complete a copy of checklist provided (1993;69:97–8 January issue). Nominate three suitable reviewers. A covering letter (wherever possible giving a fax number) must be signed by all authors stating that they have seen and approved paper and that work has not been, and will not be, published elsewhere. Copyright of article will be transferred to journal before publication.

All authors must fulfil the criteria of the International Committee of Medical Journal Editors (Vancouver agreement) (see *BMJ* 1991; 302:338–41).

MANUSCRIPT FORMAT

Type double-spaced throughout on one side only on A4 opaque white bond paper with wide margins all around.

Arranged as follows: title page, abstract, text, references, legends, tables. Number pages consecutively, beginning with title page as page 1.

The last name of the first author is typed at the top right corner of each page.

Title Page: The title and authors' names are typed on the title page and in the journal style.

Case reports have only three authors.

The address(es) of the institutions from which the work originated with the author's names are listed underneath. (If there is more than one address, several authors' names can be grouped under each appropriate address; strict order of authors' names is not necessary here.)

The full name, exact postal address with postal code, telephone number, and fax number of the author to whom communications and proofs should be sent are typed at the bottom, and the editorial office must be told of any subsequent change of address.

Abstract: An abstract is typed double-spaced on a separate page.

The abstract (250 words) is structured according to the framework described on page 1 of the January 1991 issue. This may not always be necessary —for example, case reports.

Abbreviations other than standard SI units of measurement are not used.

Authors may be asked to produce the data upon which manuscript is based.

Text: Appropriate headings and subheadings are provided.

Every reference, figure, and table is cited in the text in numerical order. (Order of mention in text determines the number given to each.)

Acknowledgments and details of support in the form of grants, equipment, or drugs are typed at the end of the text, before references.

References: References are identified in the text by Arabic numerals; no more than three references are cited for any one statement. References are typed double-spaced on sheets separate from the text (numbered consecutively in the order in which they are mentioned in the text) in the Vancouver style. As a general rule, no more than three references should be cited for any one statement.

Journal references contain inclusive page numbers; book references contain specific page numbers. Citations of abstracts and letters should be indicated in parentheses. Personal communications, manuscripts in preparation, and other unpublished data are not cited in the reference list but are mentioned in the text in parentheses. Abbreviations of journals conform to those used in *Index Medicus*, U.S. National Library of Medicine. Authors are responsible for accuracy.

JOURNAL

List all authors if six or less; otherwise list first six and add et al; do not use full stops after authors' initials.

31 Balcon R, Brooks N. Layton C. Correlation of heart rate/ST slope and coronary angiographic findings. *Br Heart J* 1984;**52**:304–8.

CHAPTER IN BOOK:

28 Schiebler GL, Van Mierop LHS, Krovetz LJ. Diseases of the tricuspid valve. In: Moss AJ, Adams F, eds. *Heart disease in infants, children and adolescents*. Baltimore: Williams and Wilkins, 1968:134–9.

BOOK

Personal author or authors: (all book references should have specific page numbers)

36 Feigenbaum H. *Echocardiography*. 3rd ed. Philadelphia: Lea and Febiger, 1981:549–63.

Authors are responsible for the accuracy of any references cited: these should be checked at source or with the listing in Index Medicus.

Legends: Figure legends are typed double-spaced on sheets separate from the text, and figure numbers correspond with the order in which figures are presented in the text. All abbreviations appearing on the figures are identified at the end of each legend. Written permission from the publisher and author to reproduce any previously published figures is included.

Figures: Three sets of unmounted glossy prints (not originals) of each photograph and drawing are submitted in three separate envelopes. Figures, particularly half tones and electrocardiographic tracings, have been submitted with the following guidelines in mind: the detail of the figure is sufficiently clear to withstand reduction, and special features are designated by arrows. Black ink is used for all line drawings. Decimals, lines, etc. must be strong enough for reproduction. The first author's last name, figure number, and TOP are indicated on the back of each illustration in light black pencil, preferably on a gummed label. Figure title and caption material appear in the legend not on the figure. Figures are limited to the number necessary for clarity and do not duplicate data given in the tables or text. (Estimates for color work will be provided on acceptance of the manuscript for publication. Some of the cost of color printing will be charged to the author(s).)

Tables: Tables are typed double-spaced on separate sheets with the table number and title above the table and explanatory notes below. The table numbers are Arabic and correspond with the order in which the tables are presented in the text. A footnote to each table identifies all abbreviations used and gives them in alphabetical order. Tables are self-explanatory, and the data are not duplicated in the text or figures. Written permission from the publisher and author to reproduce any previously published tables is included.

OTHER

Abbreviations: Do not use, except for mathematical calculations and units of measurement.

Experimentation, Human: Comply with code of ethics in Declaration of Helsinki. Include a statement that research protocol has been approved by locally appointed ethics committee and that informed consent of subjects has been obtained.

Measurements and Abbreviations: Measurements are given in SI units. Blood pressure should be given in mm Hg. Abbreviations or acronyms are always written out in full (for example, ECG, electrocardiogram; LVH, left ventricular hypertrophy; CAD, coronary artery disease, MI, myocardial infarction). Only units of measurement and mathematical formulas and calculations are abbreviated and they follow the form recommended in *Uniform Requirements for Manuscripts Submitted to Biomedical Journals* (*B r Heart J* 1985;**51**:1–6).

Statistical Analyses: If the same variable is measured by two different methods, assess agreement between methods according to guidelines published in the *British Heart Journal* (1988; **60**:177–80). Specify statistical measures of variation, such as SD or SEM; give in parentheses.

Units: Give all haematological and clinical chemistry measurements as SI units. Give blood pressures in mm Hg.

SOURCE

July 1994, 72(1): inside front cover.

British Journal of Anaesthesia
Royal College of Anaesthetists
Professional and Scientific Publications

Professor G. Smith
University Department of Anaesthesia
Leicester Royal Infirmary
Leicester LE1 5WW
Telephone: 0533 470141; Fax: 0533 854487

AIM AND SCOPE

The purpose of *British Journal of Anaesthesia is* the publication of original work in all branches of anaesthesia, including the application of basic sciences. One issue each year deals mainly with material of postgraduate educational value.

MANUSCRIPT SUBMISSION

Papers must not have been published in whole or in part in any other journal, and are subject to editorial revision. Copyright becomes vested in the Journal and permission to republish must be obtained from the Editor.

In cover letter include: information on prior or duplicate publication or submission of any part of work; statement of financial or other interests that may lead to conflicts of interest; statement that manuscript has been read and approved by all authors; and name, address and telephone number of corresponding author. Include a formal statement of request for publication signed by all authors.

A table or illustration that has been published elsewhere should be accompanied by a statement that permission for reproduction has been obtained from author and publishers. Submit three copies of each manuscript. Indicate title of paper, name(s), qualifications and full address(es) of author(s). Submit in letter quality heavy type (not dot matrix), double-spaced on one side only of paper, with a wide margin (7 cm). Include a complete set of figures with each copy. One set of figures must be unmounted glossy prints; two other sets may be photocopies.

MANUSCRIPT FORMAT

Consult papers in recent issues for general and detailed presentation. Subdivide into: Title page; Summary, including Key Words; Introduction (not headed); Methods; Results; Discussion; Acknowledgments; List of references; Tables (including legends); Illustrations; and Legends to illustrations.

Title Page: Include name(s), degrees and address(es) of author(s). Make clear which address relates to which author. Give authors' present addresses differing from those at which work was carried out, or special instructions concerning address for correspondence, as a footnote on title page and reference at appropriate place in author list by superscript symbols (*, †, ‡, §, ||, ¶). If address to which proofs should be sent is not that of first author, give instructions in a cover note and not on title page. Number title page as page 1. Include a short running title (50 characters and spaces).

Summary: 50–150 words. On a separate sheet, in a single paragraph, give a succinct account of problem, broad objectives of enquiry, methods (outline only), results (positive and negative, numerical data and probability values sparingly), and conclusions (not implications). Avoid citing references.

List three to five key words or phrases (for indexing) below summary. See *Key Words in Anesthesiology*, (ed. Greene NM, 3rd Edn. Amsterdam:Elsevier, 1988).

Introduction: 200 words, giving a concise account of background of problem and object of investigation. Quote previous work only if it has a direct bearing on present problem.

Methods/Materials and Methods/Patients and Methods: Describe in sufficient detail to allow investigation to be interpreted and repeated by readers. Describe modifications of previously published methods and give reference. If methods are commonly used, reference to original source only. Indicate error of method. Give names of instrument manufacturers with catalogue numbers or instrument identification with address. State methods of preparation and concentration of laboratory solutions. Provide a table or figure to illustrate complex program of research. Present data on age, weight, sex, height, criteria for selection of patients. Indicate general state of health and type of operation. May be done in tabular form. Indicate if permission was sought from patients for study. Describe statistical methods sufficiently for a knowledgeable reader with access to data to verify.

Results: Describe results enough to permit repetition of investigation. Do not repeat data unnecessarily in text, tables and figures, and avoid unwarranted numbers of digits. Provide only essential values. Submit all data as tables. Make tables self-explanatory with accompanying legend. Use graphs or histograms only where figure presents data more clearly than table or when a trend or comparison is being made. Give significance as values of probability. Indicate desired positions of tables and figures by written instructions within lines above and below and brackets.

Discussion: Interpret results against the background of existing knowledge. Include a statement of any assumptions on which conclusions are based. Summarize major findings. Qualify remarks in relation to these findings. Make succinct comparison of present data and relevant previous data. Make deductions which may explain differences between present and previous data. Draw conclusions from present study and implications for anaesthetic practice and indications for further enquiry.

Acknowledgments: Be brief. Include reference to sources of support and sources of drugs not freely available commercially. Acknowledge all sources of financial assistance. Do not acknowledge routine assistance. Named individuals must be given opportunity to read paper and approve of their inclusion before submission.

References: Place a table of references at conclusion of paper, commencing on a new sheet. Long lists are inappropriate except for review articles.

Number consecutively in order of mention in text, with exception of review articles, when references should be arranged alphabetically and numbered accordingly in both list and text.

Identify references in text, tables and legends by Arabic numbers in text in square brackets.

Use style of references of the U.S. National Library of Medicine used in Index Medicus. Give journal titles in full.

If it essential to name authors of a study in text, in addition to identifying number, cite up to three names. If four or more, use "and colleagues," "and co-workers," or "and others," not et al. Informal reference to previous work is permitted only in a paragraph which contains the cited reference.

Do not include references to "unpublished observations" or "personal communications" in final list. Cite personal communication in text as: [Brown AB, personal communication]. Verify that wording of references to unpublished work is approved by persons concerned. Include papers submitted and accepted for publication in list; phrase "in press" replaces volume and page number. Cite information from manuscripts submitted but not yet accepted in text as unpublished observations.

JOURNALS (LIST ALL AUTHORS)

Brown BR jr, Gandolphi AJ. Adverse effects of volatile anaesthetics. *British Journal of Anaesthesia* 1987; **59:** 14–23.

CHAPTER IN A BOOK

Hull CJ. Opioid infusions for the management of postoperative pain. In: Smith G, Covino BG, eds. *Acute Pain*. London: Butterworths, 1985; 155–179.

MONOGRAPHS

Moore, DC. *Regional Block*, 4th Edn. Springfield, Illinois: Charles C Thomas, 1979.

Restrict references to those that have direct bearing on work described. Cite only references to books and articles published *Index Medicus* journals.

Verify content and detail of references against original articles.

Tables: Put each on separate sheets. Make capable, with captions, of interpretation without reference to text. Do not include information more appropriate to Methods. Number consecutively with Roman numerals. Give units in which results are expressed in brackets at top of each column, do not repeat on each line of table. Do not use ditto signs, footnotes, and vertical lines.

Illustrations: Photographs should be unmounted glossy prints; protect adequately for mailing. Do not mar surfaces with clips, pins or heavy writing on back. Drawings, charts and graphs should be in black India ink on white paper. If in sets, present at uniform magnification. Number illustrations on back, in soft pencil, using Arabic numerals. They should be accompanied on a separate sheet by a suitable legend. Lettering should be professional looking, uniform, preferably in a common typeface, large enough to read at a reduced size, and in proportion to illustrated material. It will reduce to 8-point type and single column width. Make lines in original thick enough to allow for reduction. Make all axes, graph lines, and lettering in a figure equally bold. Avoid typewriter line thickness of computer printouts to A4 size. Program for production of exact size diagram. Labeling of axes should not extend dimensions of figure. Avoid suppression of zero point. Indicate magnifications by a scale on photograph. Indicate length of bar in legend. Symbols which are to appear in the legend should be chosen from the following available types: O ● ■ □ ▼ ▲ △ × +.

Write name of author and title of paper in soft pencil on back of illustrations. A4-size figures generated by computer are usually unacceptable.

Headings in the Text: Six grades are available, and may be indicated by the following letters of identification:

A	PART 1	(capitals)
B	RESULTS	(small capitals)
C	Blood-Gas Analysis	(l.c. Roman)
D	*The Action of Drugs*	(italics, center)
E	*Lung function studies*	(italics, full out)
F	*Volume*. Large volumes...	(italics, indent)

OTHER FORMATS

SHORT COMMUNICATIONS—1200 words, 6 references, 1 table or 1 figure, one-and-a-half pages of printed text. Short manuscripts suitable for rapid publication. Conform to requirements outlined above, except:

Format. Summary; Introduction (not headed); Methods and Results; Comment.

OTHER

Drugs: At first mention give generic or official name, followed in parentheses by chemical formula only if the structure is not well known, and by capitalized proprietary name. Provide figures only for early reports of new drugs. Tell source of molecular configuration. Give drug dosages by name of drug, followed by dose.

Experimentation, Animal: Satisfy the Board that no unnecessary suffering has been inflicted. Studies from the U.K. should specify the Home Office Licence number; from elsewhere, a statement of approval from an appropriate Licensing Authority.

Experimentation, Human: Conform to ethical standards as set out in the Declaration of Helsinki. Include a statement of approval from an appropriate Ethics Committee. Avoid the use of names, initials and hospital numbers which might lead to recognition of a patient. A patient must not be recognizable in photographs unless written consent of the subject has been obtained.

Spelling: *Shorter Oxford English Dictionary.* British spelling should be used with "z" rather than "s" spelling.

Statistical Analyses: See Bailar, JC, Mosteller F. Guidelines for statistical reporting in articles for medical journals. *Annals of Internal Medicine* 1988; **108**: 266-273.

Style: Follow International Committee of Medical Journal Editors, Uniform requirements for manuscripts submitted to biomedical journals. *British Medical Journal* 1988; **296**: 4010405. See also Dudley, H. *The Presentation of Original Work in Medicine and Biology* (London: Churchill and Livingston, 1977).

Symbols and Abbreviations: Follow conventions described in *Units, Symbols* and *Abbreviations. A Guide for Biological and Medical Editors and Authors* (ed. D. N. Baron, 1988, The Royal Society of Medicine, London). Write out words for which abbreviations are not included in full at first mention in summary and again in text and followed by abbreviation in brackets. Use large capitals without separating points. For "Pappenheimer" system of abbreviations of respiratory terms, see Pappenheimer JR, et al. Standardization of definitions and symbols in respiratory physiology. *Federation Proceedings* 1950; **9**: 602-605.

Units: SI system, except: pH and intravascular and ventilatory pressure measurements; use units of calibration. See *Units, Symbols and abbreviations: A Guide for Biological and Medical Editors and Authors* (4th Edn. Baron DN, ed. London: Royal Society of Medicine, 1988).

SOURCE

July 1994, 73(1): ii, iv, vi.

British Journal of Biomedical Science
Institute of Biomedical Science
Royal Society of Medicine Press

Editor, *British Journal of Biomedical Science*
IMLS, 12 Queen Anne Street
London W1M OAU, England
           Send by first-class mail or air mail

AIM AND SCOPE

Contributions will be considered on any aspect of biomedical science especially including the following areas of subject interest: Cell biology; Cellular pathology (including histopathology and cytology); Clinical chemistry (including biochemistry); Clinical pathology (only articles commissioned by the Editor from clinicians or pathologists and relating laboratory findings to clinical condition); Communications studies; Data processing (including computer studies); Education and training (of medical scientists and technologists: mainly articles commissioned by the Editor); Hematology; History and philosophy of medical science and technology; Immunology (including immunopathology); Management studies (relating to medical laboratories: mainly articles commissioned by the Editor); Microscopy (including light and electron microscopy); Microbiology (including bacteriology, mycology, parasitology and virology); Pharmacology; Physiology (human); Safety (in the laboratory); Transfusion science (including blood group serology).

MANUSCRIPT SUBMISSION

Papers should conform to uniform Vancouver style of presentation (*BMJ* 1991; **302:**338–41) and should be typewritten, double-spaced throughout (including figure and table legends and literature references), on one side of A4 paper with a margin of 40 mm all round. If produced on a word processor, script must be typed on a letter-quality printer and on plain (unlined) paper.

Submit three copies, one must be original typescript. Include text of paper on a floppy disc (IBM compatible and either 5.25 or 3.5 in. size and of single, double, quad or high density). Text must have been produced using Wordstar™, Word Perfect™, or MS Word™, or a program compatible with one of these or preferably in ASCII format. Text on disc must match exactly that on accompanying hard copy. In cover letter state disc format and word processor used. Label disc with name of author and title of submission.

Submission implies that material has not been published, and is not under simultaneous consideration elsewhere. Include a statement to this effect, signed by all authors with cover letter. The *British Journal of Biomedical Science* conforms to the policy on multiple publication of the International Committee of Medical Journal Editors (*Med. Lab. Sci* .1984; 41:97).

In cover letter state clearly that material is being offered for publication in the British Journal of Biomedical Science. Indicate a name and full postal address for correspondence.

Provide details of any other work which has been submitted, published or presented elsewhere, or is in preparation, and which is based on the same data set as the accompanying submission—whether or not in the same language and/or directed at the same, or an essentially similar, audience.

All authors are responsible for entire contents of manuscript. All named authors of a multi-author paper must have participated directly in planning, execution or analysis of work reported or in writing the paper. Include a statement to this effect, signed by all the authors in cover letter.

Obtain prior permission in writing from those named in Acknowledgments for use of their names. Submission of a manuscript is taken to imply that this condition has been met. Obtain written consent for reproduction of any copyright material used from copyright holder; attach permissions to cover letter.

MANUSCRIPT FORMAT

On a separate title page indicate title of work, followed on a separate line by initials and surname(s) of author(s) in upper case type, and address of institution where work was carried out, in lowercase; all aligned to the left.

Number all pages consecutively. Indicate position of all illustrations and tables in text; type legends for tables and illustrations separately at end of manuscript. Following 'Abstract' provide up to four 'key words', taken from the Medical Subject Headings of *Index Medicus*.

Reports of original research or of major new developments. Articles should normally be of 2500 to 5000 words and may include a reasonable number of illustrations. Divide into sections with headings.

Abstract: 150 words outlining purpose of study and conclusions. Make complete and understandable without reference to text.

Introduction: State reason(s) for investigation. Give brief historical background.

Materials and Methods: Give sufficient details to enable others to repeat work.

Results: Selective rather than comprehensive, and in form of text, or tables or graphs, or histograms: do not present data in more than one form.

Discussion: A critical review of results, highlighting any discrepancies and assessing value of investigation.

References and Notes: Type separately from main body of manuscript. Reference in text by consecutive numbers typed as superscripts running through whole paper. Include both literature references and extraneous matter, such as the full names and addresses of equipment suppliers, etc. Include sources of 'personal communications'. Do not enclose reference citation numbers in brackets. Literature references (restricted to those strictly necessary to the point being made) should follow style of the U.S. National Library of Medicine in *Index Medicus*. Do not type authors' names in capital letters.

IN TEXT

It has been shown[1] that the work of Jones and Smith[2] with the 'Autosquirt'[3] was based upon faulty sample preparation.[4]

IN REFERENCES

1 Falstaff J. Measurement of serum enzymes. *J Med Prog* 1989; 5:124–9.

2 Jones AB, Smith CD. *Liquid measurement*. Vol. 1, 2nd edn. London: Interprint, 1991:259–60.

3 T. Shandy Ltd, 3 Station Rd, South Ham, Surrey SU1 2UK, England, UK.

4 Regina V. Sample collection and preparation. In: Disraeli B, ed. Laboratory procedures. London: Gladstone, 1989: 291–311.

Number references cited only in tables or in legends to figures in sequence according to the first mention of table or figure in text. Check accuracy of all references, and ensure that all references in text agree with those in list of references and notes.

Illustrations: Provide each with a suitable legend which identifies the subject matter but does not expand upon textual description. Type each illustration on a separate page. In the case of photomicrographs, indicate degree of magnification.

Make drawings two or three times larger than final appearance in journal. Include minimum lettering. If a curve is given merely for illustrative purposes no cross-lines are needed but the axes should be graduated and the diagram should be boxed in. Where possible, group related diagrams to form a single figure, uniform and on the same scale.

Photographs (refer to as 'Figures') should be sharp, glossy black-and-white prints, preferably 5×7 in. but no larger than 8×10 in. Light and electron photomicrographs should be first generation prints from original negative. Write names of authors and figure number (in Arabic script) in pencil, or typed on separate adhesive labels attached on back of each illustration, together with the indication TOP.

Tables: Type on separate pages. Number in Arabic script, in a sequence separate from that of figures. Type without vertical lines. Do not box in by a frame. Column headings should not contain undefined abbreviations or be typed in capital letters; make sufficiently clear to be understandable without reference to text. Columns of like material should read down, not across. Simple lists are not considered to be tables; incorporate in text. Tables should not exceed one per 500 words of text.

Do not present data both in tabular and in graphic form. Submit three sets of illustrations, one untouched and without lettering and the other two (which may be photocopies) marked. Color photographs can be considered with prior agreement of Editor and on the understanding that extra cost is met in full by author(s).

OTHER FORMATS

ANNOTATIONS—'State of the art' review may be descriptive rather than critical, and reflecting the personal views of the author. Like Review Articles, Annotations are generally commissioned by Editor. Format is similar, including 'Abstract' (150 words), but citation of references is less extensive.

BIOMEDICAL UPDATES—Commissioned by Editor. A combined review of a number of recent publications on related topics. Each *Biomedical Update* briefly reviews each of the cited papers in turn, the citation references of which are cited in full, and may include references from within the cited paper or related to it. The author comments on papers cited in context of one another.

BOOK REVIEWS—Reviews of new books, or new editions, are commissioned by the Editor.

LETTERS TO THE EDITOR—Briefly report new or important observations which either do not merit a fuller report or which will be followed by one later; or offer comment, criticism, or supplementary information relating to any item already published in the journal. Criticism must be objective and supported by experimental or other evidence. Personal criticism is not allowed.

Letters intended for publication are subject to process of peer review. Letters which comment upon any item which has been published in the journal will be shown to original author, who will be invited to offer a reply in the same issue.

Follow style of presentation of those in current journal issue. Include reference citations but no tables or illustrations. Do not exceed 350 words.

REVIEW ARTICLES—Generally commissioned by Editor. Consult Editor.

Up-to-date comprehensive—and essentially critical—exposition of existing knowledge in a defined subject area. It is preceded by an 'Abstract' (150 words), is divided into sections with subheadings appropriate to subject matter, and is supported by extensive references.

SHORT COMMUNICATIONS—2500 words, necessary references and a few tables and/or illustrations (usually not more than two of each). Do not divide 'abstract' (150 words) into sections with subheadings; should consist of continuous prose simply divided into paragraphs. Follow the logical order of an original article far as possible. General reports of new or modified techniques, reagents or apparatus, or of investigations or observations which would not be appropriate for a full-length paper.

THESIS ABSTRACTS—The Institute of Medical Laboratory Sciences retains the right to publish all or any part of any thesis submitted in support of an application for admission to its Fellowship. The Editor prepares edited or shortened versions of the Abstract of thesis for publication.

OTHER

Experimentation, Animal and Human: If any patients or live animals have been used in any experiments reported in a paper obtain a written statement confirming that relevant rules and procedures have been followed from the appropriate ethical committee or Home Office licence holder. Include with cover letter.

Nomenclature, Bacterial: Conform to Approved Lists of Bacterial Names in *J System Bacteriol* 1980; **30**:255–420. In title, abstract and first mention in the text, each microorganism should be given its full binomial Latin name, underlined. Subsequently, it should be abbreviated, e.g., *Salm.*, *Myco.*, *Staph.*, *Esch.* Type trivial names in lowercase throughout.

Nomenclature, Blood Group: Follow conventions and style given in: Issitt PD, Crookstone MC. Blood group terminology: current conventions. *Transfusion* 1984; **24**:2–7.

Nomenclature, Chemical and Physical Units: Adhere to the system formulated by the IUPAC-IUB Commission on Biochemical Nomenclature (see *Biochem J* 1972; 126:1–91).

Nomenclature, Histopathology: Note that 'Formalin' denotes 40% aqueous solution of the gas Formaldehyde. In saline solution reference should be made to 'formalin saline', and not using the short forms 'formal' or 'formol' saline.

Nomenclature, Human Leucocyte Differentiation Antigens: FOLLOW Committee on Human Leucocyte Differentiation Antigens of the IUIS-WHO Nomenclature Sub-

Committee. See Bernard A, *et al. Bull World Health Organ* 1984; **62**:809–11, or *Immunology Today* 1984; **5**:280.

Nomenclature, Isotopically Labelled Compounds: For simple molecules, indicate labelling in the chemical formula (e.g., ^{125}I). For specific chemicals the isotope symbol is placed in square brackets directly preceding the name of the labelled material. Brackets are not employed when the isotopic symbol is attached to a word that is not a specific chemical name.

Spelling: *The Shorter Oxford English Dictionary.*

Statistical Analyses: When dealing with statistical elements of their paper authors should follow the Guidelines for Statistical Reporting in Articles for Medical Journals (Blair JC and Mosteller F, *Ann . Intern . Med .* 1988: **108**:266–73 and preferably should have their calculations checked and confirmed by an independent statistician prior to submission.

Style: Be consistent in use of either first or third person. Use past tense of verbs when narrating procedures, observations, etc. and present tense for statements of conclusions, generally accepted facts, and discussion of results. Take care with plural words (e.g., 'the data are . . .').
Underline foreign language words and abbreviations (such as *e.g., i.e., in vivo* and *in vitro),* and Latin names of microorganisms, etc., to indicate italic script.

Units: Use SI (Systéme International) symbols for physical units (see *SI for the Health Professions*, Geneva: WHO, 1977). The unit ml is used rather than cm^3.

SOURCE

March 1993, 50(1): 75-78 and March 1994, 51(1): front matter.

British Journal of Cancer
Cancer Research Campaign
British Association for Cancer Research
Association of Cancer Physicians
British Oncological Association
Macmillan Press Ltd

Professor G. E. Adams
Editorial Office, *British Journal of Cancer*
MRC Radiobiology Unit
Chilton, Didcot
Oxfordshire OX11 0RD U.K.

AIM AND SCOPE

The *British Journal of Cancer* will accept papers which contain new information, constitute a distinct contribution to knowledge, and are relevant to the clinical, epidemiological, pathological or molecular aspects of oncology.

MANUSCRIPT SUBMISSION

Submit manuscripts plus three copies and all editorial correspondence. The criteria for acceptance of manuscripts to be published in the *British Journal of Cancer* are originality and high scientific quality. Manuscripts must report unpublished work that is not under consideration for publication elsewhere; all named authors must have

agreed to its submission; and that if accepted manuscript will not be published in the same form, in any language, without the consent of the Publishers.

Type manuscripts double-spaced with a wide margin on one side of paper only. Include a title page numbered as 1, bearing title, authors' full names, affiliation and complete addresses, including postal (zip) codes; and a summary (200 words). List 6 key words after summary. Indicate author and address to whom correspondence, proofs and reprint requests are to be sent. Editorials explaining the editorial policies and practices of the *British Journal of Cancer* were published in the Journal between November 1990 and March 1991. Composite reprints are available from the Editor-in-Chief on request.

MANUSCRIPT FORMAT

Divide full papers into sections: 1. Introduction; 2. Materials and methods; 3. Results; 4. Discussion; 5. Acknowledgments and References.

Titles: Compose with care, combine brevity with clarity; the shorter the better (100 letters). Include running title (50 letters).

References: Quote only papers closely related to author's work. Quote original papers not reviews. Avoid exhaustive lists. Avoid citing Conference Proceeding or Meeting Abstracts unless there is no other reference. References in text should be made by giving author's surname, with year of publication in brackets. When reference is to a specific part of a book, cite page number. When reference is to a work by 3 or more authors, use first name followed by *et al.* If several papers by the same first author and from the same year are cited add *a, b, c,* etc. after year of publication. Check accuracy of all references. References should be brought together at end of paper in alphabetical order; give titles of papers and all authors' names in full. Abbreviate names of journals as in *Index Medicus,* but with full stops after abbreviations (and after authors' initials), followed by volume number and initial and final page numbers.

McMANUS, M.J., DOMBROSKE, S.E., PIENKOWSKI, M.M. & WILSON, A.B. (1978). Successful transplantation of human benign breast tumors into the athymic nude mouse and demonstration of enhanced DNA synthesis by human placental lactogen. *Cancer Res.* **38**, 2243–2248.

MEANS, J.H. (1984). *The Thyroid and its Diseases.* Lippincott: Philadelphia.

STEVENSON, A.C. (1966). Sex chromatin and sex ratio in man. In Sex Chromatin, Moore, A. (ed) pp. 405–425. W.B. Saunders: Philadelphia.

Figures: In text, use Arabic numbers. All illustrations should be specifically referred to in text, e.g., (Figure 2). Submit all illustrations at about 1.5 times intended final size. Number as figures whether they are photographs, representational drawings or line diagrams and graphs. Copies of all Figures must accompany all 4 submitted copies of manuscript.

Photographs and Photomicrographs: Submit unmounted glossy prints; do not retouch; exclude technical artifacts. Indicate magnification by a line representing a defined length included within photograph. Indicate areas of key interest and/or critical reproduction on a flimsy overlay attached to photograph or on a photocopy. Indicate all annotation and lettering the same way; do not include on original print. Clearly contrasted and focused prints are essential for adequate reproduction. Submit four originals of all photographs.

Illustrations: Full color illustrations may be included within the text, at the discretion of the Editor-in-Chief. However, a charge will be made to the author to cover the extra costs incurred in originating and printing of color illustrations. Authors will be advised of any such charges, which depend on the size and quantity of color illustrations, prior to publication.

Line Diagrams/Graphs: Submit on separate sheets; draw with black ink on white paper or use computer-generated output printed with laser or ink jet printers with appropriate line-smoothing facilities. Lettering on figures should be minimal; do not duplicate legend. Use symbols consistently. Include explanation of symbols in caption, not on figure. Make symbols be sufficiently large that they will be clearly discernible following reduction.

Tables: Include as few as possible, presenting only essential data. Type on separate sheets. Give titles or captions, and Roman numbers.

OTHER FORMATS

SHORT COMMUNICATIONS—(~1500 words) Include section headings and a brief summary (~50 words).

OTHER

Abbreviations/Units: Should follow the standards laid down in *Units Symbols and Abbreviations* published by the Royal Society of Medicine (third edition, 1977).

Color Charges: Inclusion of full color illustrations is at the discretion of the Editor-in-Chief. Authors will be advised of charges for extra costs in originating and printing prior to publication.

Drugs: Use generic names.

Experimentation, Animal: Papers disregarding the welfare of experimental animals will be rejected. Follow U.K. CCCR "Guidelines for the Welfare of Animals in Experimental Neoplasia" (*Br J Cancer* 1988;58: 109-113).

Experimentation, Human: In clinical study reports include statement of approval by ethical committee. Single case studies are usually not acceptable.

SOURCE

July 1994, 70(1): back matter.

British Journal of Clinical Pharmacology
British Pharmacological Society
Blackwell Scientific Publications

The Editorial Secretary
British Journal of Clinical Pharmacology
25 John Street

London WC1N 2BL
Telephone: 071 404 4101; Fax: 071 831 6745

AIM AND SCOPE

Papers will be considered for publication if they are relevant to any aspect of drug action in man, including chemotherapy.

MANUSCRIPT SUBMISSION

Submit four copies. Papers must be submitted exclusively to the journal as they become copyright on acceptance. Include a letter from corresponding author stating that all co-authors have seen the paper and, satisfied with the content, have approved its submission to the Journal.

Submit a statement that paper has not been, and will not be, published in whole or in part in any other journal. Authors who have any doubt about possible overlap between a manuscript submitted and a published paper should draw this to the attention of the Editorial Secretaries in a letter and enclose a copy of the relevant published paper(s).

Type manuscripts double-spaced on one side of standard paper (A4–210 × 297 mm). On separate title page give names of authors; institution(s) of authors; and name and address of one author for correspondence. Supply key words and a short title (50 letters).

MANUSCRIPT FORMAT

Main papers should be divided into the following sections: title, page, numbered summary, introduction, methods, results, discussion, acknowledgments, references, legends to figures, tables and figures.

Title and Summary: In title give an informative and accurate indication of content of paper. In numbered summary describe content of paper accurately, and conclusions.

Figures: Make no larger than A4. Submit original drawings, recorded tracings or high quality photographic prints made from them. Use symbols: O ● ■ ▢ △ ▲ ▼.

Do not mount photographs and photomicrographs. Submit glossy bromide prints of good contrast, well matched and preferably with a transparent overlay for protection. Use overlay to indicate masking instructions, lettering or arrows.

Tables: 85 characters to a line (counting spaces between columns as 4 characters). Type figure legends and tables on separate sheets.

References: Cite references using the Vancouver style. References in manuscript text appear as numbers in square brackets, following authors' names in text if necessary. Reference list shows references in numerical (not alphabetical) order, and includes names of all authors (if more than six, give only first three followed by *et al*), full title of article, title of publication (abbreviated in accordance with *World List of Scientific Periodicals, 4th ed*), year, volume number and first and last page numbers. References to books also include names of editor, edition number, where appropriate, and town of origin and name of publisher. Accuracy of references is responsibility of author.

1 Davies RH, Sherdan DJ. The treatment of heart failure– what next? *Br J Clin Pharmacol* 1993; **35**: 557-563.

2 Hoffman BB, Lefkowitz RJ. Beta-adrenergic receptor antagonists. In *The Pharmacologic Basis of Therapeutics*, Eighth Edition, eds Gilman AG, Rall TW, Nies AS, New York: Pergamon Press, 1990; 229-243.

Acknowledgments: Acknowledge all support, financial or otherwise, for any work described with exception of support from employing institutions identifiable from title page.

OTHER FORMATS

GENERAL ARTICLES—Will be considered for publication under various headings, such as Review, Methods, Topic in Therapeutics, Personal View, or Meeting Report. Contact Editorial Secretaries in advance for advice on suitability of article proposed.

LETTERS TO THE EDITORS—800 words plus one figure or table. Comment on previously published papers, items of topical interest, and brief original communications. Do not divide into sections, and put names and addresses of authors at end.

SHORT REPORTS—1200 words, two figures or tables, or one of each. Set out in same format as main papers except do not number.

OTHER

Drugs: Avoid code names for drugs, e.g., those used by pharmaceutical companies. Where code names have to be used detail chemical name or structure of drug in introduction. Where a code name has been used and a proper name (e.g., BAN, INN) becomes available before publication, inform the Journal.

Experimentation, Human: The ethical aspects of all studies involving human subjects will be noted. Include a statement that subjects consented to study after full explanation of what was involved. Indicate whether or not consent was obtained in writing. Detail should be given of the approval of the study protocol by an ethics committee or similar body. Describe ethics committee which approved the protocol in sufficient detail to allow the committee to be identified.

Statistical Analyses: Describe clearly in the methods section with references when appropriate. Study must have sufficient statistical power for its purpose. Considerations of the power of the study should be laid out in the methods section. In results section, cite 95% confidence intervals whenever possible for all important endpoints of study.

Style: Consult recent issues of the Journal. Further information can be found in Instructions to authors *(Br. J. clin. Pharmac.,* 1982, 14, 3–6) and Symbols in Pharmacokinetics *(Br. J. clin. Pharmac.,* 1982, 14, 7–13).

Units: SI units. If other units are used include a conversion factor in Methods section. See Editorials, *Br. J. clin. Pharmac.* (1983), 16, 1, 3–7.

SOURCE

July 1994, 38(1): front matter.

British Journal of Dermatology
British Association of Dermatologists
Netherlands Association for Dermatology and Venereology
Blackwell Scientific Publications

Editor, Dr. D. A. Burns
Department of Dermatology
Leicester Royal Infirmary
Leicester LE1 5WW

AIM AND SCOPE

The *British Journal of Dermatology* publishes original articles on all aspects of the normal biology, and of the pathology of the skin. Originally the Journal, founded in 1888, was devoted almost exclusively to the interests of the dermatologist in clinical practice. However, the rapid development during the past twenty years of research on the physiology and experimental pathology of the skin has been reflected in the contents of the Journal, which now provides a vehicle for the publication of both experimental and clinical research and serves equally the laboratory worker and the clinician. To bridge the gap between laboratory and clinic the Journal publishes signed editorials reviewing, primarily for the benefit of the clinician, advances in laboratory research, and also reviews of recent advances in such aspects of dermatology as contact dermatitis, therapeutics and drug reactions. Other regular features include book reviews, correspondence and society proceedings.

MANUSCRIPT SUBMISSION

Submit three copies of all material for publication. Type manuscripts on one side of paper only, with a wide margin, double-spaced, and bear title of paper, name and address of each author, with name of hospital, laboratory or institution where work was carried out. Give name and full postal address of author responsible for reading proofs on first page. Papers are assumed to have been submitted exclusively to the *Journal*. All authors must give signed consent to publication in cover letter.

MANUSCRIPT FORMAT

Illustrations: Refer to in text as 'Figs'. Give Arabic numbers. Mark on back with name(s) of author(s) and title of paper. Mark TOP with an arrow. Each figure should bear a reference number corresponding to a similar number in the text. Color photographs are published free of charge. Photographs should be unmounted glossy prints. Provide three sets of illustrations. Identify specific features either directly on prints, using Letraset or similar dry transfers, or on an overlay or photocopy. Final width of figures will be either 82, 114 or 173 mm. Put diagrams on separate sheets; draw with black ink on white paper, or faint-ruled graph paper; and make approximately twice the size of final reproduction. Lines should be of sufficient thickness to stand reduction (no less than 4 mm high for a 50% reduction). Provide a legend for each illustration. For photomicrographs, state the original magnification. Group legends on a separate sheet.

Submit as few tables as possible, include only essential data; type on separate sheets; give Arabic numbers.

References: Refer only to papers closely related to the author's work; avoid exhaustive lists. Give references in Vancouver style, i.e., references appear as superscript numbers in text. '... our previous report[1] and that of Smith *e t a l.*[2] ...', and are brought together numerically in reference list at end of article in order of first mention in text. Include references to articles and papers name(s) followed by initials of author(s), up to four authors: if more than four, include first three authors followed by *et al.*; title of paper; title of journal abbreviated in standard manner; year of publication; volume; page numbers of article. Thus: Cunliffe WJ, Shuster S. The rate of sebum excretion in man. *Br J Dermatol* 1969: **81:**697–9. References to books and monographs include: author(s) or editor(s); paper (if necessary) and book titles; edition, volume, etc.; place; publisher; year; page(s) referred to. Thus: Beare JM, Wilson Jones E. Necrobiotic disorders. In: *Textbook of Dermatology* (Rock AJ. Wilkinson DS, Eblfing FJ, eds), 2nd edn., Vol. 2. Oxford: Blackwell Scientific Publications, 1972; 1066.

OTHER

Abbreviations/Units: Include full names with uncommon abbreviations with first mention; coin new abbreviations only for unwieldy names; do not use unless names occur frequently. Do not use abbreviations in title; use sparingly in introduction and discussion.

gram(s)	g
kilogram(s)	kg
milligram(s) (10^{-3}g)	mg
microgram(s) (10^{-6}g)	μg
nanogram(s) (10^{-9}g)	ng
picogram(s) (10^{-12}g)	pg
hours(s)	h
minute(s)	min
centimeter(s)	cm
second(s)	s
cubic millimeter(s)	mm^3
millimeter(s)	mm
millicurie(s)	mCi
milliequivalent	mmol
molar	mol/l
milliliter(s)	ml
gravitational acceleration	*g*
micrometer(s)	μm
per cent	%
isotopic mass number placed as	^{131}I

Follow the recommendations of: Units, Symbols and Abbreviations, A Guide for Biological and Medical Editors and Authors 1988, The Royal Society of Medicine, London.

SOURCE

July 1994, 131 (1): inside back cover.

British Journal of Haematology
British Society for Haematology
European Haematology Association
Blackwell Scientific Publications

Editor, Professor Ian Peake
Department of Medicine and Pharmacology
Royal Hallamshire Hospital, Glossop Road
Sheffield S10 2JF, U.K.

AIM AND SCOPE

The *British Journal of Haematology* invites papers on original research in clinical laboratory and experimental hematology. Such papers should include only new data which have not been published elsewhere. The majority of papers report original research into scientific and clinical hematology.

MANUSCRIPT SUBMISSION

Please provide telephone and fax numbers.

MANUSCRIPT FORMAT

Type on one side of paper only, with wide margins, double-spaced. Include title of paper and name and address of author(s), with name of hospital, laboratory, or institution where work was carried out. Restrict authorship to individuals who have made a significant contribution to study. Give name and full address of corresponding and reprint request author(s) on first page. This will appear as a footnote in the Journal.

Include a running short title (60 characters and spaces). Include an informative summary (200 words) at beginning of paper. Divide papers into summary, introduction, methods (and/or materials), results, discussion, acknowledgments and references.

Be brief, long papers with many table and figures may require shortening. Supply original typewritten manuscript and three complete copies. If photocopies of any photograph are not adequate for assessment, provide additional prints of the photographs.

Key Words: Supply 5, for indexing after summary.

Headings: Clearly indicate relative importance of headings. Main categories of headings are: side capitals, side italics and shoulder italics. If necessary, use small capitals for subsidiary main headings. For examples see articles in a recent issue.

Illustrations: Refer to in text as, e.g., Fig 2, Figs 2, 4–7, using Arabic numbers. Each figure should bear a reference number corresponding to a similar number in text. Mark on back with name(s) of author(s) and title of paper. Mark orientation of illustrations with arrows. Photographs and photomicrographs should be unmounted glossy prints; do not retouch. Color illustrations are acceptable when found necessary by the Editor; author will contribute towards the cost. Submit diagrams on separate sheets; draw with black India ink on white paper and about twice the size of final reproduction. Make lines of sufficient thickness to stand reduction. Each illustration should be accompanied by a legend clearly describing it; group legends on a separate sheet.

Tables: Submit as few as possible; include only essential data: type on separate sheets and give Roman numerals.

References: Cite only papers closely related to authors work. Give author's surname with year of publication in parentheses. Where reference contains more than two authors, give only first surname plus *et al* , e.g., Jones et al (1948). If several papers by same author(s) and from same year, or by same author but different subsequent authors in same year, are cited, place a, b, c, etc., after year of publication, e.g., Jones *et al* (1948a, b). Bring all references together at end of paper in alphabetical order, with all authors, titles of journals spelt out in full and with both first and last page numbers given.

OTHER FORMATS

ANNOTATIONS AND REVIEWS—Normally invited contributions, but suitable papers may be submitted for consideration. Consult previous issues for style and length.

ANNOUNCEMENTS—150 words. Information about scientific meetings of general interest to readers of the Journal. Send to Editor as early as possible prior to event.

LETTERS TO THE EDITOR—600 words of text, one figure or table, and six references. Correspondence which relates to papers which have recently appeared in the journal. Editor may invite response from original authors for publication alongside. Letters dealing with more general scientific matters of interest to hematologists will be considered.

RAPID PAPERS—12 pages including a summary (150 words), tables, figures and legends, acknowledgments and references. Short papers of special scientific importance. Must be definitive and original and not occupy more than four pages. Submit manuscript for 'rapid publication'. Include a title page followed by the text. Submit original manuscript and three copies.

SHORT REPORTS—1000 words of text, two figures or tables and up to 12 references. Short reports which offer significant insight into scientific and clinical haematological processes. A summary (100 words) should be followed by continuous text, subdivided if appropriate. Short reports will include case reports, important preliminary observations, short methods papers, therapeutic advances, and any significant scientific or clinical observations which are best published in this format. Publication of initial results which will lead to more substantial papers is discouraged.

OTHER

Style: Follow style in recent issues. Detailed 'Instructions to Authors' concerning abbreviations, symbols, conventions, etc. were published in *Br J Dermatol* 1978; 40:1-20.

Units: Use SI units throughout.

SOURCE

September 1994, 88(1): front matte.r

British Journal of Industrial Medicine
British Medical Journal
BMJ Publishing Group

The Editor, *British Journal of Industrial Medicine*
BMJ Publishing Group, BMA House
Tavistock Square, London WC1H 9JR

AIM AND SCOPE

The *British Journal of Industrial Medicine* is intended for the publication of original contributions relevant to occupational and environmental medicine, including fundamental toxicological studies of industrial and agricultural chemicals. Short papers relating to particular experience in occupational medicine may be submitted. Other short papers dealing with brief observations relevant to occupational medicine that do not warrant a full paper, such as individual case histories, preliminary reports, or modifications to analytical techniques, will also be considered. Review articles should not be sent in unless the Editor has first been approached as to their suitability for the Journal. Letters to the Editor are always welcome.

MANUSCRIPT SUBMISSION

Submit in triplicate. Each author must sign cover letter as evidence of consent to publication. Papers are accepted on the understanding that they are contributed solely to this journal and are subject to editorial revision. The editor cannot enter into correspondence about papers rejected as being unsuitable for publication, and his decision is final.

MANUSCRIPT FORMAT

Follow requirements of the International Steering Committee of Medical Editors (*Br Med J* 1979;i:532–5). Preface papers with abstract of the argument and findings (more comprehensive than a summary). Type on one side of paper only. Double space and use wide margin. Give both SI units and equivalents throughout (Baron *et al.*, *J Clin Pathol* 1974;27:590–7). Submit photographs and photomicrographs on glossy paper unmounted. Draw charts and graphs in black ink on tracing linen or Bristol board or stout white paper. Type legends to figures on a separate sheet.

References: Will not be checked by the editorial office; responsibility for the accuracy and completeness of references lies with the author. Number references consecutively in the order in which they are first mentioned in the text. Identify references in texts, tables, and legends by Arabic numerals above the line. Number references cited only in tables or in legends to figures by the first identification in text of a particular table or illustration. Keep number of references to the absolute minimum; include only those essential to the argument or to the discussion or if they describe methods too long for inclusion. Usually, one reference per typed page of manuscript should be sufficient.

Use form of references of *Index Medicus*; for a standard journal article: authors (list all authors when six or fewer, when seven or more, list only three and add *et al*), title, abbreviated title of journal, year of publication; volume number: first and last page numbers.

OTHER

Experimentation, Human: Papers reporting results of experiments on human subjects will not be considered unless the authors state explicitly that each subject gave his or her informed written consent to the procedure and that the protocol was approved by the appropriate ethical committee.

SOURCE

January 1993 50(1): inside front cover.

British Journal of Nutrition
Cambridge University Press

Prof. D. A. T. Southgate, *British Journal of Nutrition*
10 Cambridge Court, 210 Shepherds Bush Road
London W6 7NJ

AIM AND SCOPE

The *British Journal of Nutrition* publishes reports in English of original work in all branches of nutrition from any country. It does not print reviews of the literature or polemical articles, but original articles critically re-examining published information and the conclusions drawn from it will be considered. The aim of all work presented should be to develop nutritional concepts.

MANUSCRIPT SUBMISSION

Submit a signed statement to the effect that author accepts conditions laid down in Directions to Contributors. Typescripts that do not conform will be returned. Contributors of accepted articles will be asked to assign their copyright, on certain conditions, to The Nutrition Society.

Submission implies that it represents results of original research or an original interpretation of existing knowledge not previously published; that it is not under consideration for publication elsewhere; and that if accepted, it will not be published elsewhere in same form, in English or in any other language, without consent of the Editorial Board.

Give authors' names without titles or degrees and one forename in full. Give name and address of laboratory or institution where work was performed. Put necessary descriptive material about authors, e.g., Beit Memorial Fellow, in parentheses after author's name or at end of paper and not in a footnote.

Send name and address of person to whom proof is to be sent and a shortened version of title (45 letters and spaces).

MANUSCRIPT FORMAT

Consult a current issue for *Journal's* typographical and other conventions, use of cross-headings, layout of tables, etc. Define potentially ambiguous words such as 'available'. Present papers on specialized aspects of subject so as to be intelligible, without undue difficulty, to ordinary readers of the *Journal*. Give sufficient information to permit repetition of published work.

Submit three complete copies, including all plates and figures, packed flat. Type double-spaced on one side of paper with wide margins. Do not use thin paper. Use line-numbered paper. Do not hyphenate words at ends of lines, unless hyphens are to be printed. Leave a 50 mm space at top of first sheet. Be as concise as possible. Do not suppress useful results.

Key Words: Supply two or three key words or phrases (each containing up to three words) on title page.

Synopsis: 250 words, giving a picture in miniature of article. In one paragraph outline aims of work; experimental approach taken, mentioning specific techniques where relevant; principal results, emphasizing new information; conclusions from results and their relevance to nutrition science. Use past tense to refer to experimental work. Use present tense to refer to existing knowledge or stating what is shown or concluded. Change of tense differentiates author's contribution from what is already known. Keep numerical information to a minimum.

Introductory Paragraph: Indicate briefly nature of question asked and reasons for asking it.

Experimental Methods Adopted: After introduction.

Results: As concise as possible, with figures or tables.

Discussion: Results and Discussion may be combined. Additional or alternative sections such as 'objectives' or 'conclusions' may be used.

References: Give in text thus: Sebrell & Harris (1967), (Wallace & West, 1982); when paper has more than two authors, cite as (Peto *et al.* 1981). When more than one paper has appeared in one year for which first name in a group of three or more authors is the same, refer to as follows: Adams et al. (1962a, b, c); or (Adams *et al.* 1962a b, c; Ablett & McCance, 1971). In text, give references in chronological order. At end of paper, on separate page(s), list references in alphabetical order according to name of first author, enter names with prefixes under prefix. Include author's initials and title of paper. Give names and initials of authors of unpublished work in text. Do not include in References. List journal titles in full. In references to books and monographs, include town of publication and edition number.

Ablett, J. G. & McCance. R. A. (1971). Energy expenditure of children with kwashiorkor. *Lancet* ii, 517–519.

Adams, R. L., Andrews, F. N., Gardiner, E. E., Fontaine, W. E. & Carrick, C. W. (1962*a*). The effects of environmental temperature on the growth and nutritional requirements of the chick. *Poultry Science* 41, 588–594.

Adams. R. L., Andrews, F. N., Rogler, J. C. & Carrick, C. W. (I 962*b*). The protein requirement of 4-week-old chicks. *Journal of Nutrition* 77, 121–126.

Adams, R. L., Andrews, F. N., Rogler, J. C. & Carrick. C. W. (1962*c*). The sulfur amino acid requirement of the chick from 4 to 8 weeks of age as affected by temperature. *Poultry Science* 41, 1801–1806.

Agricultural Research Council (1981). *The Nutrient Requirements of Pigs*. Slough: Commonwealth Agriculture Bureaux.

Edmundson, W. (1980). Adaptation to undernutrition: how much good does man need? *Social Science and Medicine* 14 D, 19-126.

European Communities (1971), *Determination of Crude Oils and Fats, Process A*. Part 18, *Animal Feedingstuffs*, pp. 15-19. London: H.M. Stationery Office.

Hegsted, D. M. (1963). Variation in requirements of nutrients-amino acids. *Federation Proceedings* 22, 1424–1430.

Heneghan, J. B. (1979). Enterocyte kinetis, mucosal surface area and mucus in gnotobiotics. In *Clinical and Experimental Gnotobiotics. Proceedings of the VIth International Symposium on Gnotobiology*, pp. 19-27 [T. M. Fliedner, H. Heit, D. Niethammer and H. Pflieger, editors]. Stuttgart: Gustav Fischer Verlag.

Hill, D. (1977). Physiological and biochemical responses of rats given potassium cyanide or linamarin. In *Cassava as an Animal Feed, Proceedings of a Workshop held at University of Guelph, 1977. International Development Research Centre Monograph* 095e, pp. 33-42 [B. Nestel and M. Graham, editors]. Ottawa, Canada: International Development Research Centre.

Lau, E. M. C. (1988). Osteoporosis in elderly Chinese (letter). *British Medical Journal* **296**, 1263.

Louis-Sylvestre, J. (1987). Adaptation de l'ingestion alimentaire aux dépenses energétiques (Adaptation of food intake energy expenditure). *Reproduction Nutrition Dévelopment* **27**, 171–188.

Martens, H. & Rayssiguier, Y. (1980). Magnesium metabolism and hypomagnesaemia. In Digestive Physiology and Metabolism in Ruminants, pp. 447–466 [Y. Ruckebusch and P. Thivend, editors]. Lancaster: MTP Press Ltd.

Ministry of Agriculture, Fisheries and Food (1977). Energy Allowances and Feeding Systems for Ruminants. Technical Bulletin no. 33. London: H.M. Stationery Office.

Peto, R., Doll, R., Buckly, J. D. & Sporn, M.B. (1981). Can dietary beta-carotene materially reduce human cancer rates? *Nature.* **290**, 201–208.

Sebrell, W. H. Jr & Harris, R. S. (1967). *The Vitamins*, 2nd ed., vol. 1. London: Academic Press.

Statistical Analysis Systems (1985). *SAS User's Guide, Statistics.* Cary, NC: SAS Institute, Inc.

Statistical Package for Social Sciences (1988). *Base Manual +V2·0.* Chicago, Ill: SPSS Inc.

Technicon Instruments Co. Ltd. (1967). *Technicon Methodology Sheet* N-36. Basingstoke: Technicon Instrument Co. Ltd.

Van Dokkum, W., Wesstra, A. & Schippers, F. (1982). Physiological effects of fibre-rich types of bread. 1. The effect of dietary fibre from bread on the mineral balance of young men. *British Journal of Nutrition* **47**, 451–260.

Wallace, R. J. & West, A. A. (1982). Adenosine 5´triphosphate and adenylate energy charge in sheep digesta. *Journal of Agricultural Science, Cambridge* **98**, 523–528.

Wilson, J. (1965). Leber's disease. PhD Thesis, University of London .

World Health Organization (1965). *Physiology of Lactation.* Technical Report Series no. 305. Geneva: WHO.

Figures: These include graphs, histograms, complex formulas, metabolic pathways. Submit originals and photocopies, each on a separate sheet, the same size as text sheets, packed flat. Do not mount on heavy cardboard. Print photographs of line drawings on matte paper only. In curves presenting experimental results, show determined points clearly, use symbols in order: O ● Δ ▲ ▢ ■ × +. Draw curves and symbols with a mechanical aid and not free hand. Do not extend beyond experimental points. Put scale-marks on axes on inner side of each axis and should extend beyond last experimental point.

Write numbers and letters in correct position on a flyleaf of tracing paper firmly attached. Type or stencil: title of paper and names of authors; figure number. Type legends for all figures on separate sheet. Number corresponding to relevant figures. Each figure, with legend, should be comprehensible without reference to text. Indicate approximate position of each in margin of text thus: 'Fig. 1 near here'.

Plates: Submit glossy photographs accompanied by a legend. Size of photomicrographs may be altered in printing. Show magnification by a scale on photograph itself. Draw scale with appropriate unit on flyleaf with lettering to be inserted by Press. Do not write details on back of prints, bend, use paper-clips or mark in any way. Type plate number, title of paper and authors' names on a label and paste onto back of print.

Tables: Write headings describing content. Make comprehensible without reference to text. Give dimensions of values, e.g., mg/kg, at top of each column; do not repeat on each line. Do not include in body of text, type on separate sheets. Do not subdivide by ruled lines. Define abbreviations in footnotes. Use signs for footnotes in sequence: *, †, ‡, §, ||, ¶, then **, etc. (omit * or †, or both, used to indicate levels of significance). Indicate approximate position in margin of texts: 'Table 1 near here'.

Diagrams: Prepare diagrams to appear as tables (e.g., flow diagrams) as for Tables using Letraset or stencils. No flyleaf is required.

OTHER FORMATS

LETTERS TO THE EDITORS—500 words (no figures) that discuss, criticize or develop themes put forward in papers published in the *Journal* or deal with other matters relevant to the *Journal*.

OTHER

Biological Assays: Base biological assays in which, eg., potency of a nutrient in an ingredient is estimated by a biological response, on soundly conducted multi-point responses that allow validity of assay to be established and measures of variance to be associated with results.

Chemical Formulas: Write on a single horizontal line. With inorganic substances, use formulas from first mention. With salts, state whether anhydrous material is used, e.g., anhydrous $CuSO_4$, or which of the different crystalline forms is meant, e.g., $CuSO_4 \cdot 5H_2O$, $CuSO_4 \cdot H_2O$.

Experimentation, Animal: Papers reporting work carried out using inhumane procedures will be rejected. Follow criteria in *Guidelines on the Use of Living Animals in Scientific Investigations* (1987, the Biological Council, Institute of Biology, London).

Experimentation, Human: Follow guidelines in the Declaration of Helsinki (1964) (*British Medical Journal* (1964) ii, 177–178) and to the Report of ELSE as printed in *British Journal of Nutrition* (1973) **29**, 149, the Guidelines on the Practice of Ethics Committees Involved in Medical Research Involving Children (1972, British Paediatric Association, London). Include a statement that the Ethical Committee in the Institution in which the work was performed has approved it. Submit letter of approval. In final paragraph of Experimental section headed *Ethical considerations* discuss and justify experiments from an ethical standpoint.

Mathematical Modelling of Nutritional Processes: Papers in which mathematical modelling of nutritional processes forms the principal element must: be based on demonstrably sound biological and mathematical principles; advance nutritional concepts or identify new avenues likely to lead to advances; fully describe assumptions used in its construction and support by appropriate argument; be described in such a way that its nutritional purpose is clearly apparent; and clearly define model's contribution to the design of future experimentation.

Nomenclature, Enzymes: Follow Recommendations of the Nomenclature Committee of the International Union of Biochemistry, 1984 (*Enzyme Nomenclature*, London: Academic Press, 1992). Give relevant *EC* numbers.

Nomenclature, Fatty Acids and Lipids: To describe results for analysis of fatty acids by conventional gas-liquid chromatography, see Farquhar, J. W., Insull, W., Rosen, P., Stoffel, W. & Ahrens, E. H. (*Nutrition Reviews* (1959), 17, Suppl.) for shorthand designation for individual fatty acids used in text, tables and figures. Use 18:1 to represent a fatty acid with eighteen carbon atoms and one double bond; if position and configuration of double bond is unknown, do not refer to this fatty acid as oleic acid. Use shorthand designation in synopsis but construct sentences so it is clear to the non-specialist reader that 18:1 refers to a fatty acid. If positions and configurations of double bonds are known, and are important to discussion, then a fatty acid such as linoleic acid may be referred to as *cis*-9,*cis*-12-18:2 (positions of double bonds related to the carboxyl carbon atom 1). To illustrate metabolic relations between different unsaturated fatty acid families, number double bonds in relation to terminal methyl carbon atom, *n*. Preferred nomenclature is then: 18:3*n*-3 and 18:3*n*-6 for α-linolenic and γ-linolenic acids, respectively; 18:2*n*-6 and 20:4*n*-6 for linoleic and arachidonic acids, respectively and 18:1*n*-9 for oleic acid. Distinguish positional isomers such as α- and γ-linolenic acid. It is assumed that double bonds are methylene-interrupted and are of the *cis*-configuration (see Holman, R. T. in *Progress in the Chemistry of Fats and Other Lipids*, vol. 9, part I, p. 3. Oxford: Pergamon Press, 1966). Refer to groups of fatty acids that have a common chain length but vary in double bond content or double bond position, for example, as C_{20} fatty acids or C_{20} polyunsaturated fatty acids. Use modern nomenclature for glycerol esters, i.e., triacylglycerol, diacylglycerol, monoacylglycerol not triglyceride, diglyceride, monoglyceride. State form of fatty acids used in diet, i.e., whether ethyl esters, natural or refined fats or oils. State composition of fatty acids in dietary fat, expressed as mol/100 mol or g/100 g total fatty acids.

Nomenclature, Microorganisms: Use correct names of organisms, conforming with international rules of nomenclature: synonyms may be added in brackets when name is first mentioned. Names of bacteria must conform with the current Bacteriological Code and the opinions issued by the International Committee on Systematic Bacteriology. Names of algae and fungi must conform with the current International Code of Botanical Nomenclature. Names of protozoa must conform with the current International Code of Zoological Nomenclature. The following books may be found useful:

Bergey's Manual of Determinative Bacteriology (8th edn, 1974), eds. R. E. Buchanan and N. E. Gibbons, Baltimore: The Williams and Wilkins Co.)

The Yeasts, a Taxonomic Study (2nd ed. 1970, J. Lodder, ed. Amsterdam: North Holland Publishing Co.)

Ainsworth and Bisby's Dictionary of the Fungi (6th edn, 1971, Kew: Commonwealth Mycological Institute).

Nomenclature, Plants: Put Latin name in italics after first mention of common name. Give cultivar where appropriate.

Nomenclature, Vitamins: See *Nutrition Abstracts and Reviews* A (1978). **48**, 831–835. Details of the nomenclature for menaquinones and other naturally-occurring quinones should follow the Tentative Rules of the IUPAC-IUB Commission on Biochemical Nomenclature (see *European Journal of Biochemistry* (1975), **53**, 15–18).

Previous Name	Recommended Name
Vitamin A_1	Retinol
Retinene or retinal	Retinaldehyde
Vitamin A acid	Retinoic acid
Vitamin A_2 or 3-dehydro-retinol	Dehydroretinol
Retinene₂ or 3-dehydro-retinal	Dehydroretinaldehyde
Vitamin D_2 or calciferol	Ergocalciferol
Vitamin D_3	Cholecalciferol
Vitamins E	See Generic descriptors
Vitamin K_1 or phylloquinone	Phylloquinone
Vitamin K_2 series	Menaquinones
Vitamin K_3, menadione or menaphthone	Menadione
Vitamin B_1, aneurin(e) or thiamine	Thiamin
Vitamin B_2 or riboflavine	Riboflavin
Nicotinic acid or niacin	Nicotinic acid
Niacinamide or nicotinic acid amide	Nicotinamide
Folic acid or folacin(e)	Pteroylmonoglutamic acid
Vitamin B_6, adermin or pyridoxal	Pyridoxine
Pyridoxal	Pyridoxal
Pyridoxamine	Pyridoxamine

Previous Name	Recommended Name
Vitamin B_{12} or cobalamin	Cyanocobalamin
Vitamin B_{12a}, B_{12b} or hydamide	Hydroxocobalamin
Vitamin B_{12c}	Nitritocobalamin
Inositol or *meso*-inositol	ψ-Inositol
Pantothenic acid	Pantothenic acid
Biotin	Biotin
Choline	Choline
p-Aminobenzoic acid	*p*-Aminobenzoic acid
Vitamin C or L-ascorbic acid	Ascorbic acid
L-Dehydroascorbic acid	Dehydroascorbic acid

Use terms vitamin A, vitamin C and vitamin D where appropriate, for example phrases such as 'vitamin A deficiency', 'vitamin D activity'.

Vitamin E. Use as descriptor for all tocol and tocotrienol derivatives qualitatively exhibiting biological activity of α-tocopherol. Use term tocopherols as generic descriptor for all methyl tocols.

Vitamin K. Use as generic descriptor for 2-methyl-1,4-naphthoquinone (menaphthone) and all derivatives exhibiting qualitatively the biological activity of phylloquinone (phytylmenaquinone).

Niacin. Use as generic descriptor for pyridine 3-carboxylic acid and derivatives exhibiting qualitatively the biological activity of nicotinamide.

Folate. Due to the wide range of carbon-substituted, unsubstituted, oxidized, reduced and mono- or polyglutamyl side-chain derivatives of pteroylmonoglutamic acid which exist in nature, it is not possible to provide a complete list. Use either generic names or other correct specific name(s) of derivatives, as appropriate for each circumstance.

Vitamin B_{12}. Use as generic descriptor for all corrinoids exhibiting qualitatively the biological activity of cyanocobalamin. Use term corrinoids as generic descriptor for all compounds containing corrin nucleus and thus chemically related to cyanocobalamin. Corrinoid not synonymous with vitamin B_{12}.

Vitamin C. Terms ascorbic acid and dehydroascorbic acid normally refer to naturally occurring L-forms. If subject matter includes other optical isomers, include L- or 4 D-prefixes, as appropriate. The same is true for all those vitamins which can exist in both natural and alternative isomeric forms.

Amounts of vitamins and summation. Express all amounts of vitamins in terms of mass, or in SI units. See *Metric Units, Conversion Factors and Nomenclature in Nutritional and Food Sciences* (London: The Royal Society, 1972, paras. 8 and 14–20).

Other Nomenclature, Symbols and Abbreviations: Follow current issues *British Journal of Nutrition* and IUPAC rules on chemical nomenclature, and the Recommendations of the IUPAC-IUB Commission on Biochemical Nomenclature (see *Biochemical Journal* (1978) **169**, 11–14). Symbols and abbreviations, other than units, are listed in *British Standard* 5775 (1979–1982). *Specifications for Quantities, Units and Symbols*, parts 9–13. Abbreviate day

to d, for example 7 d; except for example, 'each day', '7th day' and 'day 1'.

Refer to elements and simple chemicals (e.g., Fe and CO_2) by their chemical symbol or formula from first mention in text, except in title. Use well-known abbreviations for chemical substances without explanation. Other frequently mentioned substances may also be abbreviated, place abbreviations in parentheses at first mention.

Spectrophotometric terms and symbols are those proposed in *IUPAC Manual of Symbols and Terminology for Physicochemical Quantities and Units* (1979) (London: Butterworths): m (= milli) = 10^{-3}, μ (= micro) = 10^{-6}, n (= nano) = 10^{-9} and p (= pico) = 10^{-12}. Note also that ml (milliliter) should be used instead of cc, μm (micrometer) instead of μ (micron) and μg (microgram) instead of γ.

Numbers. Use figures with units, for example 10 g, 7 d, 4 years (except when beginning a sentence); otherwise, words (except when 100 or more), thus: one man, ten ewes, ninety-nine flasks, three times (but with decimal 2·5, times), 100 patients, 120 cows, 136 samples.

Page Charges: Overlong papers are subject to page charges.

Solutions, Compositions and Concentrations: Define solutions of common acids, bases and salts in terms of molarity (M), e.g., 0·1 M-NaH_2PO4. Express values of compositions expressed as mass per unit mass (w/w) as ng, μg, mg, or g per kg; for concentrations expressed as mass per unit volume (w/v), denominator is the liter. Do not express concentrations or compositions on a percentage basis. Express common measurements used in nutritional studies, e.g., digestibility, biological value and net protein utilization, as decimals rather than percentages, so that amounts of available nutrients can be obtained from analytical results by direct multiplication. See *Metric Units, Conversion Factors and Nomenclature in Nutritional and Food Sciences* (London: The Royal Society, 1972, para 8).

Spelling: Use *Concise Oxford Dictionary* (8th ed., Oxford: Clarendon Press 1990).

Statistical Analyses: Discuss study design with a statistician. Consider number of subjects used, results from small experiments are unreliable. Give sufficient information about experimental design so repetition is possible.

Do not give data from individual replicates for large experiments, only for small studies. Describe methods of statistical analysis, discuss analysis of variance tables only if relevant. State number of replicates, their average value and some appropriate measure of variability.

Make comparisons between means using either confidence intervals or significance tests. The most appropriate measure is usually the standard error of a difference between means (SED), or the standard errors of the means (SE or SEM) when these vary between means. SD is more useful only when there is specific interest in the variability of individual values. State degrees of freedom associated with SEDS, SEMS or SDS. When presenting results, use forms such as 'mean 3·51 (SE 0·67) μmol, rather than notation '±'. Quote sufficient number of decimal places.

If confidence intervals (CI) are used for comparisons between means, present as, e.g., 'difference between means 0·73 g (95% CI 0·14, 1·36). If significance tests are used, state if difference between the means for two groups of values is (or is not) statistically significant. Include level of significance, as an explicit P value (e.g., $P = 0·016$ or $P = 0·32$) rather than as a range (e.g., $P < 0·05$ or $P > 0·05$). State whether significance levels are one-sided or two-sided. Where a multiple comparison procedure is used, give an unambiguous description or explicit reference. Where appropriate, use a superscript notation in tables to denote levels of significance; like superscripts denote lack of significant difference.

If analytical method is unusual, or if experimental design is complex, submit further details (e.g., experimental plan, raw data, confirmation of assumptions, analysis of variance tables, etc.).

Units: Present results in metric units according to International System of Units (see *Quantities Units and Symbols*. London: The Royal Society, 1971, and *Metric Units, Conversion Factors and Nomenclature in Nutritional and Food Sciences* (London: The Royal Society, 1972) and *Proceedings of the Nutrition Society* (1972) 31, 239–247).

Express energy measurements in joules.

For substances of known molecular weight, e.g., glucose, urea, Ca, Na, Fe, K, P, express values as mol/l; for substances of indeterminate molecular weights, e.g., phospholipids, proteins, and for trace elements, e.g., Cu, Zn, use g/l .

Time: Use 24 h clock.

SOURCE

July 1994, 72(1): i-vi.

British Journal of Obstetrics and Gynecology

Royal College of Obstetricians and Gynecologists Blackwell Scientific Publications

The Editor
British Journal of Obstetrics and Gynaecology
27 Sussex Place
Regent's Park
London NW1 4RG

AIM AND SCOPE

The *Journal* publishes reports of original work in obstetrics and gynecology and related subjects such as human reproductive physiology and contraception.

MANUSCRIPT SUBMISSION

Submit three copies of manuscript and illustrations. Manuscripts are considered for publication on the understanding that they have been submitted exclusively to the *Journal* unless otherwise stated, and that neither the report nor the data it contains have been previously published. All authors must sign consent to publish.

MANUSCRIPT FORMAT

Typed double-spaced on one side of paper, with 3 cm margins. Number pages consecutively. Make type size no smaller than 12 pitch. Use standard font, e.g., Times Roman, Elite or modern. Designate corresponding and reprint request author(s). Provide complete mailing address, telephone and fax numbers.

Title Page: Names of all authors, present positions and hospital or institute of affiliation. Do not use abbreviations in title.

Abstracts: Structured, 250 words. Include some or all of the headings: Objective, Design, Setting, Subjects, Interventions, Main Outcome Measures, Results and Conclusions, as appropriate. Do not use abbreviations.

Key Words: 3-5.

Text: 3000–5000 words preferred. Do not exceed 7000 words. Follow form: Subjects and Methods, Results, and Discussion as main headings.

Give manufacturer's name and city and country for all products cited.

Clearly describe design of investigations, methods of analysis and source of date. Specify all statistical methods. See: Bracken M. B., Reporting observational studies, *Br J Obstet Gynaecol* 1989, 96: 383–388; Wald N. & Cuckle H., Reporting the assessment of screening and diagnostic tests, *Br J Obstet Gynaecol* 1989, 96: 389–396; Grant A., Reporting controlled trials, *Br J Obstet Gynaecol* 1989, 96: 397–400. See Altman et al. Statistical guidelines for contributors to medical journals. *Br Med J Clin Res Ed* 1983, 286: 1489-1493.

References: Use Harvard style; alphabetical by surnames in reference list; in text give multiple citations in ascending order of year of publication and alphabetically within same years. List no more than six surnames and initials of authors before stating *et al.*

JOURNALS

Authors' surnames followed by initials, year of publication, title of article, abbreviation of journal as given in *Index Medicus*, volume and complete page range.

CHAPTERS OF BOOKS

Authors' surnames and initials, year of publication, chapter title, name of book, initials followed by surnames of editors, publisher, city of publisher and complete page range.

ENTIRE BOOKS

Names of editor(s)/author(s), year of publication, title and name and location of publisher.

PROCEEDINGS OF A CONFERENCE

Authors' names, year, title of article and complete title of Proceedings. Contributors must specify whether these reports are published (if so, by whom), unpublished (if not, how obtainable) or whether it is an abstract.

REPORTS OR PRESENTATIONS

Specification of author, title, year, what type of report and whether published (if so, by whom) or unpublished (and if not, how obtainable).

Tables/Figures: Do not duplicate information in text. For tables or figures from previous publications submit written permission to republish from publisher and author and include a complete reference to original work. Put on separate sheets from text and include correct corresponding number in text, brief descriptive heading and definition of all acronyms and numerical values.

Illustrations may be either line diagrams or black and white photographs (half tones). Do not submit photocopies of photographs. Color photographs are discouraged. Extra cost will be charged to author. Number illustrations. Mark TOP on back.

OTHER FORMATS

CASE REPORTS/SHORT COMMUNICATIONS—1800 words, one illustration or table and six references. No abstract. Describe new, unusual or innovative management techniques or significant results from smaller studies.

COMMENTARIES—1800 words, 12 references, invited, signed editorials on subjects proposed by Editor-in-chief.

LETTERS TO THE EDITOR—500 words, double-spaced, signed by all authors, relating to previously published matter in *Journal*. If more than three authors, only name of first will by given, followed by "on behalf of all co-authors."

REVIEWS—4000-5000 words, invited critical evaluations of topics of interest. Include a substantial review of current literature.

OTHER

Abbreviations/Units: Use only universally recognized abbreviations in text. Spell out in full common medical expressions or standard medical conditions at first appearance, with abbreviation in parentheses. Give measurements in SI units, except haemoglobin (g/dl) and blood pressure (mmHg). Do not use mM if mmol/l or molar mass is meant. See *Units, Symbols and Abbreviations* (Royal Society of Medicine, 1988). Abbreviations are used sparingly and only if a lengthy name or expression is repeated frequently.

Drugs: Use generic names unless trade name is directly relevant. Accompany trade names with ™ or ® as appropriate.

Experiments, Human: Include a statement that work was approved by institution's ethics committee and that subjects gave informed consent to work.

SOURCE

January 1994, 101(1): last page before cover.

British Journal of Ophthalmology
BMJ Publishing Group

Editor, *British Journal of Ophthalmology*
Department of Ophthalmology
University of Aberdeen Medical School
Foresterhill, Aberdeen AB9 2ZD, Scotland, U.K.
Telephone: 0224 663812; Fax: 0224 663832

AIM AND SCOPE

The *British Journal of Ophthalmology* is an international journal covering all aspects of clinical ophthalmology and the visual/ophthalmic sciences. Contributors should consider the widely varying readership and write clear, simple articles with the minimum of technical detail.

MANUSCRIPT SUBMISSION

Indicate in cover letter which category of paper article represents. Manuscripts are processed by section editors who deal with specific areas of ophthalmology including surgical retina, medical retina, neuro-ophthalmology, glaucoma, paediatric ophthalmology, ocular motility, orbital disease, anterior segment disease, oncology, lens, optics and visual sciences, laboratory sciences, pathology, and immunology.

Papers are accepted on the understanding that they have not been and will not be published elsewhere, and that there are no ethical problems with the work.

Submit manuscript in triplicate, typed double-spaced on one side of paper, with 1 in. margins. Each author must sign cover letter as evidence of consent to publication.

MANUSCRIPT FORMAT

Clinical sciences articles on clinical topics are research reports of a general or specialized nature. Approximately 3000 words and 4–6 display items (Figures and Tables).

Laboratory sciences articles on ophthalmic or visual sciences are research reports of experimental work generally the same size as clinical research reports.

Both should include: title; key words (up to five); name and address of corresponding author; abstract; introduction; materials and methods; results and discussion sections; references and acknowledgments; legends for display items (Figures and Tables).

Illustrations: Submit in triplicate. Send prints with transparencies. Include only salient details. Label with author's name; number, in order cited in text whether in color or black-and-white; and indicate top. Submit radiographs as prints. Clearly label diagrams. Make widths 6.7, 10.2, 13.2 or 17.4 cm (only in exceptional cases) to fit column layout. Insert lettering large enough for reproduction. Give stain used and a scale bar or magnification. Type legends on a separate sheet.

Tables: Put each on a separate sheet, with a heading. Do not use vertical rules.

References: Verify for accuracy and completeness. Follow Vancouver numerical system. Type double-spaced. Cite in numerical order of first appearance. Give references in list in numerical order in which first appear in text, not in alphabetical order of authors' names. For references with one to six authors, include all names; for references with more than six authors, give first six and then *et al*. Abbreviate journal titles according to *Index Medicus* or give in full. References to books include name of editor(s) if there is one, town where published, name of publisher, year, volume, page numbers.

1 Kaye SB, Shimeld C, Grinfield E, Maitland NJ, Hill TJ, Easty DL. Non-traumatic acquisition of herpes simplex virus infection. *Br J Ophthalmol* 1992; **76**:412–8.

2 Jakobiec FA, Font RL. Orbit. In: Spencer WB, ed. *Ophthalmic pathology: an atlas and textbook*. 3rd ed. Philadelphia: Saunders, 1986:2461–76.

OTHER FORMATS

CASE REPORTS—500–600 words written as a letter with a maximum of two display items (Figures and Tables). Include introduction section (without heading), case report (heading: Case report) and comment (heading: Comment), and maximum of six references

LETTERS TO THE EDITOR—200-300 words in form of scientific correspondence.

REVIEW ARTICLES—3000–5000 words including references, may contain display items (Figures and Tables). Address any aspect of clinical or laboratory ophthalmology. Most are commissioned but uninvited reviews are welcomed.

OTHER

Statistical Analyses: Discuss representativeness of sample. Justify handling of missing data. Check tables, etc., to ensure that missing data are accounted for, that percentages add up to 100 and that numbers in tables are not at variance with those in text. Follow statistical guidelines in Altman DG, Gore SM, Gardner MJ, Pocock SJ. Statistical guidelines for contributors to medical journals, *BMJ* 1983; **286**: 1489–93. Avoid blanket statements on use of statistical techniques; make clear context in which procedure is used. Simple analyses are often adequate to support arguments presented.

Units: Report in units used. If not SI units, give equivalent in SI units in parentheses.

SOURCE

January 1994, 78(1): inside front cover.

British Journal of Pharmacology
British Pharmacological Society
Macmillan Press Ltd.

Editorial Office, *British Journal of Pharmacology*
St. George's Hospital Medical School
Crammer Terrace
London SW17 0RE

AIM AND SCOPE

The *British Journal of Pharmacology* welcomes contributions in all fields of pharmacology for publication as full papers or as high priority Special Reports. Papers should normally be based on new results obtained experimentally and should constitute a significant contribution to pharmacological knowledge. Papers which reassess pharmacological concepts based on earlier results will also be considered as will purely theoretical papers. Papers dealing only with descriptions of methods are acceptable if new principles are involved.

MANUSCRIPT SUBMISSION

Contributions may not have been published, or accepted or be under consideration for publication, with essentially same content. This does not apply to results published as abstracts, letters to editors, or as contributions to symposia, provided that significant new information is added.

Keep manuscripts as short as possible by reducing discussion and number of figures. Avoid repetition of published information.

Copyright: Copy and submit following Statement, Declaration and Copyright Agreement.

STATEMENT

1. Authors have obtained permission to publish from their employers or institution, if they have a contractual or moral obligation to do so, and those whose unpublished work, including papers accepted for publication (i.e., in press), has been cited or those whom the Authors wish to acknowledge as having improved the content or presentation of the manuscript.

2. The Authors declare that the manuscript contents are original and that they have not already been published or accepted for publication, either in whole or in part or, in any form other than as an abstract or other preliminary publication in an unrefereed article. Furthermore, the Authors verify that no part of the manuscript is under consideration for publication elsewhere and it will not be submitted elsewhere if accepted by the *British Journal of Pharmacology* and not before a decision has been reached by the Editorial Board.

DECLARATION

I/We assign to Macmillan Press, on behalf of the British Pharmacological Society, the copyright of my/our manuscript, currently entitled '..', for publication in the *British Journal of Pharmacology*.

Furthermore I/We have read, understood and accepted the terms and conditions as set out in Statement and Copyright, Instructions to Authors *Br. J. Pharmacol.* 1994, **111**, 378-384.

Name..
Signed...
Name..
Signed...
Name..
Signed...
Date..

COPYRIGHT

1. The Authors agree that, when the above manuscript has been accepted for publication in this Journal, the worldwide copyright shall pass to the Macmillan Press Ltd. on behalf of the British Pharmacological Society, on the understanding that the assignment of copyright will not affect subsisting Patent Rights arrangements pertaining to it. The Authors also accept that, when accepted, the contents will not be published subsequently in the same or similar form in any language without the consent of the Publisher or Editorial Board of the Journal.

This Agreement shall not compromise the Authors' rights to reproduce their own work (see 3). For its part, the British Pharmacological Society will protect the interest of Authors in the matter of copyright.

2. The Authors declare that, where excerpts from copyrighted works have been included, the Authors have obtained written permission from the Copyright owners and have credited the sources in the manuscript. They must also warrant that the article contains no libellous or unlawful statements and does not infringe the rights of others.

3. The Authors will be entitled to publish any part of the paper in connection with any other work by them, provided adequate acknowledgment is given.

4. If it is appropriate, the Authors' employer may sign this Declaration. It is understood that proprietary rights, with the exceptions of Copyright and Patent Rights are reserved by the signee.

5. If an Author is a U.S. Government employee and the work was done in that capacity, the assignment applies only to the extent allowed by U.S. law. If an Author is an employee of the British Government, HMSO will grant a non-exclusive license to publish the paper in the Journal, provided British Crown Copyright and user rights (including Patent Rights) are reserved.

6. If for good reason a co-author is unable to sign this Declaration, the other Author or co-authors may sign on his or her behalf, provided that this is clearly stated and on the understanding that they will make every effort to inform the person concerned of the terms of the agreement.

If the manuscript is not accepted for publication, the assignment will be null and void.

MANUSCRIPT FORMAT

Type on one side of A4 paper. Do not hyphenate words at ends of lines. Spell out handwritten characters or symbols in full in margin. Consult recent issues for general layout. Use type not smaller than 12 pitch or 10 point. Type each section double-spaced with 2.5 cm margins all round. Number each page. Submit original and one copy.

Title Page: Limit title to 150 characters. Do not use a sentence (statement or conclusion) or a question. Give a short running title (50 characters and spaces). Include names of authors and appropriate addresses. Make clear which address relates to which author. Give authors' present addresses differing from those at which work was carried out as footnotes and references at appropriate place in author list by superscript numbers. Use footnote to indicate corresponding author. Do not use footnotes for any other reason. If first author is not corresponding author, indicate in a cover letter, not on title page. Number title page as page 1.

Summary: Do not exceed 5% length of paper. Brief account of problem, methods, results and conclusions. Arrange in numbered and concise paragraphs. Display up to ten keywords or phrases of two to three words (including names and terms used in title) at end. Refer to most recent Index of Journal. Avoid unhelpful or unqualified terms such as "rat," "drug," etc.

Introduction: A short, clear account of background of problem and rationale of investigation. Cite only work with a direct bearing on present problem.

Methods: Describe in sufficient detail to allow experiments to be interpreted and repeated. Avoid detailed repetition of previously described methods; give references. A brief outline is often helpful.

Results: Be succinct, but give sufficient detail to allow experiments to be repeated. Present typical single experiments with a clear statement that n number of similar experiments had similar results. Where appropriate, give mean results with confidence limits or with standard errors of means and number of observations. Perform statistical tests of significance where appropriate. State such results as numerical value of probability (P) that is calculated, with any necessary clarification (e.g., one-tail or two-tail test).

Avoid unnecessary repetition of data in text, tables and figures. Do not elaborate conclusions and theoretical considerations in this section.

Discussion: Brief and pertinent interpretation of results against background of existing knowledge. State assumptions on which conclusions are based. Do not just recapitulate results. Avoid a review-like treatment, which reduces impact of conclusions. Convey main conclusion in a final paragraph.

Acknowledgments: Be brief. Include sources of support. Acknowledge sources of drugs not available commercially.

References: In the text, refer to other work as: (Bolton & Kitamura, 1983) or, 'Bolton & Kitamura (1983) showed that . . .'. If more than two authors, give first author's name followed by *et al.* (Bülbring *et al.*, 1981).

Mention references to 'unpublished observations' or 'personal communications' in text only. Do not include in list of references. Include papers submitted and accepted for publication, in list of references with names of periodicals and 'in press'. Submit copies. If not possible, indicate whether work cited is abstract or full paper. Do not include papers in preparation or submitted but not yet accepted for publication in list of references.

Arrange reference list at end of manuscript alphabetically according to surname of first author. When author surnames are identical, alphabetical order of their initials takes precedence over year of publication. Follow authors' names by year of publication in brackets. If more than one paper by same authors in one year, place a, b, c, etc., after year of publication, both in text and reference list. Give title of article in full, abbreviated title of periodical (use *International List of Periodical Title Word Abbreviations*), volume number and first and last page numbers. References to articles in books consist of names of authors, year of publication, title of article followed by title of book, editors, volume numbers, if any, and page numbers, place of publication and publisher.

BOLTON, T.B. & KITAMURA, K. (1983). Evidence that ionic channels of smooth muscle may admit calcium. *Br. J. Pharmacol.*, 78, 405–416.

BRADING, A.F. (1981). Ionic distribution and mechanisms of transmembrane ion movements in smooth muscle. In *Smooth Muscle: An Assessment of Current Knowledge.* ed. Bülbring, E., Brading, A.F., Jones, A.W. & Tomita, T. pp. 65–92. London: Edward Arnold.

Tables: Put each on a separate page, paginated as part of paper. Number consecutively with Arabic numerals. Follow with a brief descriptive caption, not more than two lines, at head of table. Keep proportions of text area in mind when designing layout. Use no more than 120 character/line, count spaces between columns as four characters. Maximum is 180 characters/line. Give each column a heading with units of measurement in parentheses. Make tables self-explanatory; place necessary descriptions at bottom of table.

Figures: Avoid unnecessary Figures, particularly half-tones, illustrate only critical points. Discuss colored Figures with Secretaries before submission.

Indicate Authors' names and Figure number lightly in pencil on back of each: if necessary, use adhesive label to avoid damage to Figure.

With each manuscript copy submit one set of labelled Figures (i.e., complete with lettering, numbering, arrows, etc.), an original set and one high quality photocopy.

Submit another original set of Figures identical in size but without any letters or numbers. Arrows and event marks on experimental records may be retained, provided they are larger than 3 mm wide.

No Figure should exceed 210×297 mm (A4).

Give each Figure a legend typed on a separate sheet paginated as part of manuscript. Legends explain Figures in sufficient detail to be understood without reference to text.

Line Figures: Submit original drawings (black ink on heavy white paper or faint blue graph paper) prepared to conform with style and convention of Journal. Letter in pencil and make larger (up to twice as large) than intended size.

Printed symbols and lines should retain clarity. Make sharply defined in original drawings and even density and breadth. Computer-generated graph lines must not show noticeable stepping. Use heavier (broader) lines for curves than for axes of graphs.

Choose symbols: O ● ■ □ △ ▲ ▼ + ×.

Preferred order to shading of histogram columns is: open (clear), closed (solid), cross-hatched, heavily stippled and other (if required).

Explain symbols and column headings in Figure legend.

For composite Figures, minimize spaces between parts, without over-crowding Figure.

Photographs/Photomicrographs: Submit twice as large as published size, as good quality prints of high contrast especially where traces and records are illustrated. Originals must not contain arrows, lettering or numbering; place these accurately on a duplicate print (or photocopy). Mark critical areas of half-tone illustrations on second copy or an overlay. Mark maximum trim areas on a second copy or on a tracing overlay, i.e., show any parts of photographs that could be excluded from finished illustration. Provide calibration bar on photomicrograph.

OTHER FORMATS

SPECIAL REPORTS—Provides rapid publication for news and important results of special pharmacological significance. Have publication priority over all other material.

Two printed Journal pages, two illustrations (Figures or Tables, with legends). Maximum length is 1700 words. For each Figure or Table, deduct 200 words. Include Title page, Summary (a single short paragraph), key words (10), Introduction, Methods, Results, Discussion and References (10).

OTHER

Abbreviations, Chemical and Biological: Consult *Nomenclature Guidelines for Authors (Br. J. Pharmacol.* 1994; 111: 385-387, Listed abbreviations may be used without definition. Give chemical, drug and enzyme names in full at first mention in title, summary and again in text. At first mention, follow with abbreviation in brackets.

For lists of abbreviations for chemical quantities and chemical and biological abbreviations, see *Br. J. Pharmacol*. 1994; 111: 383-384.

Drugs: List drugs in separate paragraph. Use approved names in British Approved Names, 1990 (HMSO). If no 'approved name', use chemical name and follow rules in current *Handbook for Chemical Society Authors* (London, Chemical Society), or give structural formula. Abbreviate cumbersome chemical names for later reference.

Give drug doses as unit weight per body weight, e.g., mmol kg^{-1} or mg kg^{-1}; give concentrations in terms of molarity, e.g., nM or μM. Refer to statistical analyses performed on results.

Experimentation, Animals: Papers must describe experiments which follow current laws governing animal experimentation in United Kingdom. Make clear that procedures were as humane as possible; state doses (initial and subsequent) of anesthetics and analgesics; define method of assessing anaesthesia, particularly after administration of skeletal muscle relaxants (neuromuscular blocking drugs). Consult Society's Ethics Committee.

Experimentation, Human: Provide evidence of approval by Ethics Committee. Papers about clinical trials or investigations of effects of drugs on patients are not appropriate.

Solidus: Avoid use. Substitute negative index, e.g., mg kg^{-1} not mg/kg; pmol mm^{-2} min^{-1} not pmol/mm^2/min.

Symbols: Print symbols denoting physical quantities as italic capitals (single underline in typescript). A dash over a symbol indicates mean value; a dot indicates time derivative. Use suffixes, printed as inferiors on line, to indicate 'where' and 'what'. Avoid use of multiple suffixes if a simpler symbol adequately defined is unambiguous. If necessary, separate by commas e.g., PA, CO_2, denotes partial pressure of CO_2 alveolar air.

Units: For physico-chemical quantities use SI symbols for units. Use following prefixes for multiples of units:

Multiplier	Prefix	Symbol
10^{-1}	deci	d
10^{-2}	centi	c
10^{-3}	milli	m
10^{-6}	micro	μ
10^{-9}	nano	n
10^{-12}	pico	p

10^{-15}	femto	f
10^{-18}	atto	a
10^3	kilo	k
10^6	mega	M
10^9	giga	G
10^{12}	tera	T

Micron = μm; ångstrom = 0.1 nm. Do not use mixed prefixes, m μg = ng. Restrict symbols d (10^{-1}) and c (10^{-2}) to when there is a strong need for them (e.g., cm).

SOURCE

January 1994, 111:378-384, Revised January 1, 1994.

A checklist of essential requirements is summarized in each issue or as last page of issue.

British Journal of Psychiatry
Royal College of Psychiatrists

Editor, The *British Journal of Psychiatry*
17 Belgrave Square
London SW1X 8PG

AIM AND SCOPE

The *British Journal of Psychiatry* is published monthly by the Royal College of Psychiatrists. The *Journal* publishes original work in all fields of psychiatry.

MANUSCRIPT SUBMISSION

Contributions are accepted for publication on the condition that their substance has not been published or submitted for publication elsewhere. Published articles become Journal property and can be published elsewhere, in full or in part, only with Editor's written permission.

MANUSCRIPT FORMAT

Submit two high-quality copies. Limit paper to 3000-5000 words typed on one side of paper, double-spaced throughout (including tables and references) and with wide margins (at least 4 cm); number all pages, including title page.

Title and Authors: Give brief, relevant title. Subtitles may be used. Designate a corresponding author and indicate address. More than five authors may be credited to a paper only at Editor's discretion.

If anonymous peer review is desired, submit work without personal identification, with names and addresses in cover letter only. Otherwise, put names of authors on title page in form wished for publication. Give names, degrees, affiliations and full addresses at time work described in paper was carried out at end of paper.

Summaries: Place structured summary (150 words) at beginning. Use headings: Background, Method; Results; Conclusions. Outline questions investigated, design, essential findings, and main conclusion of study. Review articles require summaries; comments, annotations, lectures and points of view do not.

References: List alphabetically at end. Give journal titles in full. Print titles of books and journals in italic (or underline). Reference lists not in *Journal* style will be returned.

Personal communications need written authorization; do not include in reference list. No other citation of unpublished work, including unpublished conference presentations, is permissible.

Check that text references and list are in agreement as regards dates and spelling of names. Text reference take form '(Smith, 1971)' or 'Smith (1971) showed that . . .'. Note that *et. al.* is used after third author's name in a work by four or more.

ALDERSON, M. R. (1974) Self poisoning: what is the future? Lancet, i, 104–113.

AMERICAN PSYCHIATRIC ASSOCIATION (1980) Diagnostic and Statistical Manual of Mental Disorders (3rd edn) (DSM-III). Washington DC: APA.

AYLARD, P. R., GOODING, J. H., MCKENNA, P. S., *et al* (1987) A validation study of three anxiety and depression self assessment scales. Psychosomatic Research, 1, 261–268.

DE ROUGEMONT, D. (1950) *Passion and Society* (trans. M. Belgion). London: Faber and Faber.

FISHER, M. (1990) *Personal Love*. London: Duckworth.

FLYNN, C. H. (1987) Defoe's idea of conduct: ideological fictions and fictional reality. In *Ideology of Conduct* (eds N. Armstrong & L. Tennehouse), pp. 73–95. L ondon: Methuen.

FREUD, S. (1955) Some neurotic mechanisms in jealousy, paranoia and homosexuality. In *Standard Edition,* Vol. 18 (ed. & trans. J. Strachey) pp. 221–232. London: Hogarth Press.

JONES, E. (1937) Jealousy. In *Papers on Psychoanalysis*, pp. 469–485. London: Bailliere, Tindall.

MULLEN, P. E. (1990a) Morbid jealously and the delusion of infidelity. In *Principles and Practice of Forensic Psychiatry* (eds R. Bluglass & P. Bowden), pp. 823–834. London: Churchill Livingstone.

—(1990b) A phenomenology of jealousy. *Australian and New Zealand Journal of Psychiatry*, **24**, 17–28.

VAUKHONEN, K. (1968) On the pathogenesis of morbid jealousy. *Acta Psychiatrica Scandinavica* (suppl. 202).

Tables: Submit each on a separate sheet, numbered and with an appropriate heading. Mention in text but do not duplicate information. Make heading, footnotes and comments self-explanatory. Indicate desired position in manuscript. Do not tabulate lists, incorporate into text, to be displayed. Obtain permission for use of tables from other source; acknowledge in table footnote.

Figures: Submit individual glossy photographs, or other camera-ready prints, or good-quality output from a computer, not photocopies, clearly numbered and captioned below. Avoid cluttering figures with explanatory text better incorporated in caption. Make lettering parallel to axes. Indicate units clearly; present in form quantity: unit (note: spell "liter" out in full unless modified to ml, dl, etc.). Obtain permission to use figures from other sources; acknowledge in legend. Color figures are reproduced at authors' cost.

OTHER FORMATS

LETTERS TO THE EDITOR—350 words, edited for clarity and conformity with *Journal* style; no more than five references

SHORT REPORTS—2000 words; structured summary (100 words); either one table or one figure; 10 references.

OTHER

Abbreviations: Spell out on first usage.

Drugs: Use generic names. Indicate source of compounds not yet available on general prescription.

Experimentation, Human: Obtain consent of described patients. Have patient read report before submission. If patient is not able to give informed consent, obtain it from an authorized person. Keep personal details and dates to a minimum. Change personal details to disguise patient if this is consistent with accurate reporting of clinical data; add note to this effect.

Notes: Avoid use separate to text, whether footnotes or a separate section. A footnote to first page gives general information about paper.

Statistical Analyses: Not all papers, including case histories and studies with very small numbers, require statistical analysis. Describe statistical analyses in language comprehensible to numerate psychiatrists and medical statisticians. Clearly describe study designs and objectives. Give evidence that statistical procedures were appropriate for hypotheses tested and correctly interpreted. Plan statistical analyses before data are collected. Explain any *post-hoc* analyses. Give value of test statistics used (e.g., χ^2, t, F- ratio) and significance levels so that their derivation can be understood. Do not report standard deviations and errors as ±; specify and refer to in parentheses.

Do not report trends unless supported by appropriate statistical analyses for trends.

Do not use percentages to report results from small samples, except when it facilitates comparisons.

See Gardner & Altman (1990, *British Journal of Psychiatry,* **156,** 472–474) for an introduction to the place of confidence intervals.

Include estimates of statistical power where appropriate. Reporting a difference as statistically significant is insufficient; comment on the magnitude and direction of change.

Units: SI units; or give SI equivalent in parentheses. Do not use indices: i.e., report g/ml, not gml^{-1}.

SOURCE

July 1994, 165(1): front matter.

British Journal of Radiology
British Institute of Radiology

Honorary Editors, *The British Journal of Radiology*
36 Portland Place
London WIN 4AT

Dr. S. G. A. Chenery, Ph.D.
Physics Section, Radiation Oncology Center
5271 F Street, Sacramento, CA 95819
             Nondiagnostic papers
Professor B J Cremin, FRACR, FRCR
Department of Radiology
Red Cross Children's Hospital
Rondebosch, Cape Town 7700, South Africa
Dr. R Fox, D. Phil.
Department of Medical Physics
Royal Perth Hospital
Perth, Western Australia
             Nondiagnostic papers
Professor M Takahashi
Department of Radiology
Kumamoto University
School of Medicine
1 -Chome, Honjo, Kumamoto, Japan 860

AIM AND SCOPE

The *British Journal of Radiology* is published independently by the British Institute of Radiology, which is a multidisciplinary organization. The *BJR* aims to bring together all the professions in radiology and allied sciences to share knowledge thereby improving the prevention, detection and treatment of disease. *BJR* includes papers from all the radiological sciences, in keeping with this positive multidisciplinary philosophy. Material published in *BJR* includes full papers case reports, technical notes, short communications, correspondence and reviews.

MANUSCRIPT SUBMISSION

Material contained in paper must not have not been published and is not intended for publication elsewhere. If a paper has more than one author, corresponding author bears responsibility for obtaining agreement of all authors. All authors must sign cover letter.

MANUSCRIPT FORMAT

Type manuscript, including references, figure legends and tables, in English on one side only of paper double-spaced with 25 mm margins on both sides and at top. Number pages. Submit three clear copies. In title give address of department(s) or hospital(s) at which work was done. If any author has moved, give as a footnote. Give degrees and diplomas of authors, not more than two for each author. Put tables (numbered 1, Il, etc.) on separate sheets; give each a short descriptive title. Put captions for illustrations, including diagrams and graphs, together on a separate sheet, denoted Figure 1, 2, etc. Define all abbreviations at first appearance. Set out mathematical expressions clearly. Use up to three levels of headings.

Abstracts: 250 words. Supply an accurate and succinct precis of paper that answers questions: "Why did you do the study?", "How did you do it?", "What did you find?" and "What does it mean?" Include key statistics. No references.

References: All references must appear both in text and in reference list. Verify accuracy. Use Vancouver system. In text, cite references as numerals in square brackets; within brackets, separate by commas; give three or more consecutive references as ranges, e.g., [1, 2, 7, 1 0-1 2, 14]. Cite references in numerical order; references cited in tables or figure captions count as being cited where table or figure is first mentioned in text. In list at end of paper, give references in order cited in text; include: number, author names, paper title, abbreviated journal title, volume, first and last page numbers of paper, and year of publication.

For up to four authors, give all authors' names. For more than four authors, use first three names plus "Et Al." If reference is to a book, give number, authors' names, title of paper or chapter, name of book, names of editors, publisher, town of publication, first and last page of material cited and year of publication.

ARTICLES

1. COWEN, A R, HAYWOOD, J M, WORKMAN, A and CLARKE, O F, A set of X-ray test objects for image quality control in digital subtraction fluorography. 1: Design considerations, *Br. J. Radiol.*, 60, 1001-1009 (1987).

BOOKS

2. MICHAEL, B D and HARROP, H A, Time scale and mechanism of radiosensitization and radioprotection at the cellular level. In *Radiation Sensitizers*, ed by L W Brady (Masson, New York), pp. 164-170 (1980).

Illustrations: Supply three sets of artwork. Draw diagrams and graphs, including numerals and lettering, clearly in black ink for photographic reproduction to a width of 76 mm (exceptionally 156 mm). Supply good quality photographic prints or original artwork. Details such as numbering and experimental points must remain legible after photographic reduction. Put figure number, TOP, and corresponding author's name on back of each illustration, pencilled lightly or on a sticky label. For advice on drawing diagrams, particularly electrical circuits, see *Notes for Authors* (Institute of Physics and Physical Society, 1976, revised 1983).

Supply three sets of unmounted prints of radiographs not larger than whole plate, 216 mm × 152 mm, and not smaller than half plates. Tracings of sections of prints are acceptable when necessary for additional clarification.

OTHER FORMATS

CASE OF THE MONTH—Two pages (850 words per page, less 200 words for each illustration). No abstract. Short papers reporting a recognized but rare abnormality, or teaching cases with points of general interest that might be seen at an examination. Relevant history and initial photographs will appear on a right-hand page, presenting a specific problem to readers. Comments are on overleaf, with results of further investigations and a conclusion, followed by a brief and up-to-date review of subject with five references. Limit to three authors. Radiotherapy and oncology papers, as well as diagnostic, are welcomed for this feature.

CORRESPONDENCE—Letters for publication are on any matters of interest to readers.

SHORT COMMUNICATIONS—Two page reports of initial studies in new areas, and comments which are too speculative or provocative for a normal paper.

TECHNICAL NOTES—Two page reports of technical or procedural advances.

OTHER

Color Charges: £185 for first color illustration and £30 for each subsequent one.

Experimentation, Animal: Provide a statement that work was carried out in accordance with Biological Council Guidelines on the Use of Living Animals in Scientific Investigations (2nd edition, Institute of Biology, London, 1987).

Experimentation, Human: Where relevant, authors will be required to sign a statement that permission for clinical trials was obtained from appropriate Ethical Committee.

Units, Symbols and Abbreviations: Use International System of Units (SI) (National Physical Laboratory, *The International System of Units , SI* 5th edn, Ed. by D T Goldman and R J Bell, Her Majesty's Stationery Office, London, 1987).

Give units of radiation in SI with old units in brackets if necessary; thus I Sv (100 rem), 1 Gy (100 rad), 1 MBq (27 Ci). See BIR Working Party On SI Units, "Conversion to SI Units in radiology", *Br. J. Radiol.*, 54, 377-380 (1981) and Jennings, W A, "How much do you know about quantities and units?" *Br. J. Radiol.*, 54, 381-383 (1981).

For quantities and units for radiation dosimetry, consult recommendations from British Committee on Radiation Units and Measurements ("Quantities for reporting the calibration of therapy-level dosimeters in photon beams" *Br. J. Radiol.*, 53, 917-918 (1980); "Use of air kerma for photon dosimetry in air" *Br.J.Radiol.*, 55, 375-376 (1982); and "SI units for radiation protection-an interim measure" *Br. J. Radiol.*, 55, 376-377 (1982).

See also *Units, Symbols and Abbreviations: a Guide for Biological and Medical Editors and Authors* (Royal Society of Medicine, 1988).

SOURCE

January 1994, 67(793): 118-119.

British Journal of Surgery
Blackwell Scientific Publications, Ltd.

The Senior Editor, *The British Journal of Surgery*
Blackwell Scientific Publications, Ltd.
25 John Street
London WC1N 2BL
Telephone: 071 404 1831; Fax: 071 404 1927

AIM AND SCOPE

The British Journal of Surgery publishes original articles, reviews and leading articles. Short notes, case reports and surgical workshops are also accepted.

MANUSCRIPT SUBMISSION

Articles are reviewed for publication on the understanding that the work has not been submitted simultaneously to another journal, has not been accepted for publication elsewhere and has not already been published. Any detected attempt at dual publication will lead to automatic rejection and may prejudice acceptance of future submissions. Articles and their illustrations become property of the Journal unless rights are reserved before publication.

Send a cover letter signed by all authors, stating if an abstract of the work has been published and whether they wish manuscript and/or illustrations to be returned if is not accepted. First named author must ensure that all authors have seen and approved manuscript and are fully conversant with its contents.

MANUSCRIPT FORMAT

Follow uniform requirements for manuscripts, see *British Medical Journal* 1988; 296: 401–5 or *Annals of Internal Medicine* 1988; 108: 258–65, or to obtain 'Uniform Requirements for Manuscripts Submitted to Biomedical Journals' from The Editor, British Medical Journal, BMA House, Tavistock Square, London WC1H 9JR, U.K.

For original articles follow format of introduction, methods, results, discussion. Include a summary on a separate page. Lengthy manuscripts are likely to be returned for shortening. Make discussion clear and concise, limited to matters arising directly from results. Avoid discursive speculation.

Submit clearly reproduced manuscript with adequate space for editorial notes. Type or print on A4 paper (210×297 mm) on one side of paper, double-spaced with 4 cm margins. Manuscripts which do not conform will be returned. Begin each section on a new page: title page, summary, text, acknowledgments, references, tables, legends for illustrations. In text follow standard sequence of introduction, methods, results and discussion. Run sections sequentially through text, avoiding blank space between them.

Title Page: On title page give: article title; name and initials of each author; department and institution to which work is attributed; name and address of corresponding and reprint request authors; sources of financial support; and category submitted (original article, case report, etc.). Indicate if paper is based on a previous communication to a society or other meeting.

Summary: On second page, 150 word. embodying purposes of study or investigation, basic procedures (study material and observational and analytical methods), main findings (with specific data and statistical significance) and principal conclusions.

Tables and Illustrations: Submit three copies of all illustrations and tables. Type each table with an appropriate brief title on a separate page. Submit at least one set of line drawings, radiographs, photomicrographs, etc. as unmounted glossy prints, original transparencies, or negatives. Photocopies should be of sufficient quality to be judged for content and value. Paste a label on reverse side of each illustration giving its number (corresponding with its reference in text) and name(s) of author(s); indicate top. Draw and label illustrations appropriately for reduction to one or two column widths. Place explanations of symbols and shading in legend. Give actual numbers of patients involved in table with survival curves. Legends to illustrations and footnotes to tables should contain brief but comprehensive explanations of information contained. Refer to recent issues for examples of accepted layout. Authors must contribute to cost of color reproduction. Give original source of a reproduced table or an illustration in full. Obtain permission to use published work from original author and publisher before submission.

References: Type double spaced. Cite in Vancouver style (see above: Uniform Requirements for Manuscripts Submitted to Biomedical Journals). Do not reference abstracts from meetings or unpublished communications. Cite personal communications in text, in parentheses. Frst author must confirm acceptability of quoted material with originator.

In text, number references consecutively by superscript: [1, 2] or [1-3]. Number references cited only in tables or figures in sequence according to first mention of table or figure in text.

The sequence for a standard journal article is: author(s); title of paper; journal name abbreviated as in *Index Medicus* (in full if no abbreviation quoted) and year of publication; volume number; page numbers. For example: De Bolla AR, Obeid ML. Mortality in acute pancreatitis. *Ann R Coll Surg Engl* 1984; **66**: 184–6.

The sequence for chapters of a book is: author(s); chapter title; editors; book title; place of publication; publisher; year of publication; page numbers. For example: Calenoff L, Rogers L. Esophageal complications of surgery and lifesaving procedures. In: Meyers M, Ghahremani G, eds. Iatrogenic Gastrointestinal Complications. New York: Springer-Verlag, 1981: 123–63.

Design: Clearly set out objectives of study, identify primary and secondary hypotheses, and explain rationale for choice of sample size.

Presentation: Use graphical presentations to illustrate main findings. Clearly distinguish use of standard deviation and standard error. Avoid use of '±' symbol: present these statistics in parentheses after mean value.

OTHER FORMATS

CASE REPORTS—400 words (300 if accompanied by a figure), five references and one illustration.

LEADING ARTICLES—600–1000 words. Commissioned, signed leading articles addressing controversial topics of current interest and supported by key references. Subjected to external review and assessment by Editorial Board. Editors may alter style and shorten material for publication. An honorarium is payable on publication.

REVIEWS—Seek advice on suitability of review topics from: Mr. J. A. Murie, University Department of Surgery, The Royal Infirmary, Edinburgh EH3 9YW, U.K. Authors receive honorarium.

SHORT NOTES—500 words, five references and one table or one figure, of original observations presented in brief form, following standard format of introduction, methods, results and discussion, but no summary.

SURGICAL WORKSHOPS—500 words (two figures and five references) describing technical innovations or modifications useful in surgical practice.

OTHER

Abbreviations: Limit use. Frequently mentioned terms may be abbreviated only if it does not detract from comprehension. Use consistently throughout. Clearly define on first use.

Experimentation, Animal and Human: The design of all human investigation and animal experiments must have been approved by local ethical committees or conform to standards currently applied in country of origin. Publication may be denied on the grounds that appropriate ethical or experimental standards have not been reached. Obtain and submit written consent from patient, legal guardian or executor for publication of any details or photographs which might identify an individual.

Names: When quoting specific materials, equipment and proprietary drugs, state name and address of manufacturer in parentheses. Use generic names.

Numbers and Units: Use decimal point, not comma, e.g., 5·7. A space, not a comma follows thousands and multiples thereof, e.g., 10 000. Use SI units (International System of Units) except for the measurement of blood pressure (mmHg). If measurements were made in non-SI units, give actual values and units with SI equivalents in parentheses at appropriate points.

Statistical Analyses: See *The British Journal of Surgery* 1991; 78: 782–4. Describe which methods were used for which analyses. Support methods not in common usage with references. Report results of statistical test by stating value of test statistic, number of degrees of freedom and P value. For example, 't = 1·34, 16 d.f., $P = 0·20$' might be reported. Report actual P values, to two decimal places, especially when result is not significant, do not state 'not significant'.

Report results of primary analyses using confidence intervals instead of, or in addition to, P values.

Do not confuse statistical significance with clinical significance. Interpret 'negative' findings through use of confidence intervals. Do not place undue emphasis on secondary analyses, especially when suggested by inspection of data.

SOURCE

January 1994, 81(1): 159-160

British Journal of Urology
Blackwell Scientific Publications

The Editors, *British Journal of Urology*
25 John Street
London WC1N 2BL U.K.

AIM AND SCOPE

The *British Journal of Urology* welcomes original contributions on topic of interest and importance to urologists, whether written by urologists, nephrologists, radiologists, nurses or basic scientists.

MANUSCRIPT SUBMISSION

Submit two copies typed double-spaced on A4 paper (21 × 29.7 cm).

Include a cover letter, signed by all authors, stating that paper has not been submitted elsewhere. Indicate corresponding author. There should normally be a maximum of five authors.

MANUSCRIPT FORMAT

Title Page: Ttle of article; initials and names of each author; their main appointment; and name(s) of their institution(s).

Summary: Subdivide into: objective; subjects/patients (or materials) and methods; results and conclusion.

Key Words: Provide 3 to 6 key words.

Text: Subdivide into: introduction; subjects/patients (or materials) and methods; results; discussion and conclusions; acknowledgments; and references.

References: Follow Vancouver style. Number references in text consecutively in order of appearance, indicated by Arabic numeral in parentheses. List only first six authors. If more than six, list first three followed by *et al*. Except for review articles, limit to thirty.

1. Morris JA, Oates K, Staff, WG. Scanning electron microscopy of adenomatoid tumours. *Br J Urol* 1986; 58: 183–7

2. Wright FS, Howards SS. Obstructive injury. In Bremner BM, Rector FC eds, *The Kidney*, 2nd edn, Vol, II. Chapt 38. Philadelphia: Saunder, 1981: 2009-2044

Illustrations: Submit original and one copy. Line drawings should be in black indelible ink twice size required for publication. Do not mount photographs. Label on an overlay. Give magnification in caption or show in scale bar. If figures are to be mounted on one folio, provide sketch of layout. Letter on an overly or photocopy. Make no less than 4 mm high for 50% reduction. Label in pencil on reverse with number, author's name, and top indicated. Put keys in legends with details of staining techniques. Figures and tables reproduced from published works must have source quoted and permission from author and publisher.

Use of color illustrations and graphs is encouraged and are printed free of charge. Supply all histological illustrations in color.

OTHER FORMATS

CASE REPORTS—250 words, two illustrations and three references, divided into Case report and Discussion.

OPERATIVE TECHNIQUES—Divide into indications; methods; comparison with other methods; advantages and disadvantages; difficulties and complications.

LETTERS—Send questions or comments concerning published papers to Editor, who will refer them to authors. Readers' comments and authors' replies may be published together.

REVIEW ARTICLES—Usually commissioned.

OTHER

Abbreviations and Units: Use SI units except blood pressure (mmHg). Write numbers in full to nine; use numerals from 10 upwards. Limit use of abbreviations; use consistently.

SOURCE
July 1994, 74(1): inside back cover.

British Medical Journal
see BMJ

Bulletin of Environmental Contamination and Toxicology
Springer International

Yukata Iwata
Zeneca, Inc., Agricultural Products
Western Research Center
1200 South 47 Street, Box 4023
Richmond, CA 94804-0023
        Analytical Methodology
Anthony Calabrese
USDC, NOAA, National Marine Fisheries Service
Milford Laboratory, 212 Rogers Avenue
Milford, CT 06460
        Aquatic Toxicology: Metals
Dora R. Passino-Reader
National Biological Survey, Great Lakes Center
1451 Green Road
Ann Arbor, MI 48105
        Aquatic Toxicology: Organics
John Hylin
Norkons Consultants
P.O. Box 6323
Incline Village, NV 89450
        Environmental Distribution and Human Exposure
James Knaak
OxyChem, Occidental Chemical Center
Product Stewardship
360 Rainbow Blvd. South, P.O. Box 728
Niagara Falls, NY 14302
        General Toxicology
G. A. Ansari
Department of Human Biological Chemistry
and Genetics, Department of Pathology
University of Texas Medical Branch at Galveston
Route F-05, 301 University Blvd.
Galveston, TX 77555-0605
        Metabolism and Biochemistry
Barry W. Wilson
Department of Environmental Toxicology
University of California, Davis
Davis, CA 95616
        Pesticides and Wildlife Toxicology

AIM AND SCOPE

Bulletin of Environmental Contamination and Toxicology will provide rapid publication of significant advances and discoveries in the fields of air, soil, water, and food contamination and pollution, as well as articles on methodology and other disciplines concerned with the introduction, presence, and effects of toxicants in the total environment. Current research will be presented as clear, concise reports to provide information potentially useful to all those concerned with environmental contamination and toxicology. manuscripts that repeat old methods with different substrates will be considered on merit. Manuscripts presenting analytical techniques on compounds for which analytical techniques have been standardized and for which data are available with not be considered. New methods, procedures, or techniques must be sufficiently detailed to permit their use in other laboratories. Short research papers presenting a variation, extension, or confirmation of topics already studied are not considered to be of interest. A series of short papers on the same or closely related topics will not be accepted, neither will review articles, abstracts, or archival papers.

MANUSCRIPT SUBMISSION

Do not exceed 8 pages, including figures, tables, and references. Papers are reproduced by photographing typewritten manuscript. It is essential that manuscripts are neatly prepared following instructions.

Submission of a manuscript implies: that work has not been published before (except as an abstract or as part of a published lecture, review, or thesis); that it is not under consideration for publication elsewhere; that its publication has been approved by all coauthors, if any, as well as by responsible authorities at institute where work has been carried out; that, if and when manuscript is accepted for publication, authors agree to automatic transfer of copyright to publisher; that manuscript will not be published elsewhere in any language without consent of copyright holders; that written permission of copyright holder is obtained by for material used from other copyrighted sources; and that costs associated with obtaining permission are authors' responsibility.

Send original typescript, original or very clean copies of photographs or line drawing, and two complete copies of each to the Associate Editor responsible for particular subject category.

MANUSCRIPT FORMAT

Typed single-spaced on good quality DIN A4 (8.5 × 11 in.) white 20 lb bond paper. Use electric typewriter if available; set pressure to achieve dark, even type on original. Use black ribbon in good condition, preferably new. Type must be of sufficient size to withstand 13% page reduction. Allow opaque white correction fluid to dry before retyping to avoid smearing. Authors without access to specified bond paper or similar quality can obtain paper from Editor-in-Chief.

To facilitate preparation of camera-ready pages, Springer-Verlag has developed a set of styles using WordPerfect 5.1 on an IBM PC or compatible with a laser printer (PostScript, Hewlett-Packard Laserjet II or III, or similar laser printers). Free 3.5 or 5.25 in. disks are available: write to BECT, Springer-Verlag, 175 Fifth Ave., New York, NY 10010; telephone: (212)460-1707; Fax: (212)533-5977. These WordPerfect styles produce pages that will not need to be reduced.

The following sections constitute an article. Do not include any others. Do not include abstracts and/or conclusions. Begin all section headings and text at left margin.

Introduction: Do not use section heading. Give one or two paragraphs on relevance of study to environmental contamination and toxicology.

Materials and Methods: Do not use subheadings. Do not underline; use all capitals for section heading.

Results and Discussion: Do not use subheadings. Do not underline; use all capitals for section heading.

Acknowledgments: Set heading in caps and lower case. Make separate paragraph two spaces below last paragraph of text. Include funding sources and disclaimers. Limit to six lines.

References: Use all capitals for section heading. Single-space and place in alphabetical order. Cite papers in text in parentheses, e.g., (Child 1941; Godwin and Cohen 1969; MacWilliams et al. 1970) except when author is mentioned, e.g., "and the study of Hillman and Tasca (1977)." Limit references to one page.

Keep references to unpublished works and personal communications to a minimum; mention only italicized in text in parentheses. Give references to published works at end of text under first author's name, citing all authors (surnames followed by initials throughout: do not use "and").

PERIODICALS

Garcia-Bellido A, Merriam JR (1969) Cell lineage of the imaginal disc in *Drosophila* gynandromorphs. J Exp Zool 170:61-76

White JJ (1983) Woodlice exposed to pollutant gases. Bull Environ Contam Toxicol 30:245-251

BOOKS

Meltzer YL (1971) Hormonal and attractant pesticide technology. Noyes Data, Park Ridge, NJ

WORK IN AN EDITED COLLECTION

Letey J (1985) Relationship between soil physical properties and crop production. In: Stewart BA (ed). Advances in Soil Science, vol 1. Springer-Verlag, New York, p 277.

Arrange references by same author(s) chronologically. If more than one reference by same author(s) published in same year, use a, b, c, etc., after year of publication in both text and list.

Title Page: Article title, authors, and affiliations are typeset and inserted by printer unless authors use BECT WordPerfect styles. Supply: title (capitalize first letter of nouns, verbs, and adjectives); author(s); and affiliation(s) on a separate, unnumbered page, double-spacing between items.

First Page: Must be 5.5 × 6.5 in. including author correspondence note, with at least 2.5 in. space at top (no space need be left if using BECT). Provide information for correspondence only when more than one author. Type as an unnumbered footnote at bottom of first page. Format:

  Correspondence to: (name)

or, if mailing address is different from affiliation provided,

  *Present address: (address)

  Correspondence to: (name)

Number page "1" lightly in blue pencil in lower right corner.

Remaining Pages: Text area of each page, including tables and figures, must not exceed 5.5 × 9 in. Utilize full space available, do not type beyond text area. Number each page lightly in blue pencil in lower right corner.

Tables and Figures: Type to fit within text area (5.5 × 9 in.), and place prior to reference section. Place figures at top of text pages. Submit original line drawings or photographs; draw and/or label to withstand 13% page reduction, unless using BECT. Mount photographs from back; marks from tape placed on a photograph will appear in published photograph. Boldface figure and table numbers. Do caption texts in lightface.

Double-space between figure legends and table captions (single-spaced immediately below figure and above table) and text. Tables may appear lengthwise on a single page if necessary; figures may appear lengthwise if they are greater than 5.5 in. wide.

Footnote: Avoid, if possible. Number in text as superscripts. Type single-spaced at bottom of appropriate page. Separate from text by a short line immediately above.

OTHER

Abbreviations:

A	acre	rpm	revolutions per minute
bp	boiling point	sec	second(s)
cal	calorie	μg	microgram(s)
cm	centimeter(s)	μL	microliter(s)
cu	cubic (as in "cu m")	μm	micrometer(s)
d	day	mg	milligram(s)
ft	foot (feet)	mL	milliliter(s)
gal	gallon(s)	mm	millimeter(s)
g	gram(s)	mM	millimolar
ha	hectare	min	minute(s)
hr	hour(s)	M	molar
in.	inch(es)	mon	month(s)
id	inside diameter	ng	nanogram(s)
kg	kilogram(s)	N	normal
L	liter(s)	no.	number(s)
mp	melting point	od	outside diameter
m	meter(s)	oz	ounce(s)
ppb	parts per billion	sp gr	specific gravity
ppm	parts per million	sq	square (as in "sq m")
ppt	parts per trillion	vs	versus
pg	picogram	wk	week(s)
lb	pound(s)	wt	weight
psi	pounds per square inch	yr	year(s)

Data: Analytical data must include percent recovery from fortified samples; include quality assurance and quality control procedures and results from analytical field data.

Reviewers: Suggest two reviewers not affiliated with authors' institution or company.

Style: Do not indent paragraphs. Double space between paragraphs, section heads, and/or figure legends, and table captions. Refer to a recent issue of the *Bulletin* for general format. Use Webster's Third *New International Dictionary of the English Language*, Unabridged, for spelling and hyphenation. Publisher will not make any corrections or adjustments to typed pages. Any smears, smudges, or fading type on original will necessitate further revision.

Symbols: Insert symbols not found on typewriter by hand in black ink from a fine tip pen. Do not use felt tip pens or pencils. They will reproduce exactly as drawn.

SOURCE

July 1994, 53 (1): after page 170. Revised May 1994.

Bulletin of the World Health Organization

Editor, *Bulletin of the World Health Organization*
1211 Geneva 27, Switzerland

AIM AND SCOPE

The subject of the article should have a direct or indirect bearing on the health of populations and should be of international significance. Rare syndromes, the treatment of diseases of little or no public health importance, laboratory work without foreseeable practical application, and results applicable only to one country are examples of unsuitable subjects.

MANUSCRIPT SUBMISSION

Write papers in English or French. Limit number of authors to five. Submit a statement that paper has not already been published and that if accepted for publication, will not be submitted for publication elsewhere without agreement of the World Health Organization.

Papers are accepted on the understanding that they are subject to editorial revision, including condensation of text and omission of tabular and illustrative material.

Obtain permission from copyright holder(s) to reproduce any copyrighted material. Attach letter granting permission.

MANUSCRIPT FORMAT

Type on one side of paper, double-space, with a 3 cm left margin. Submit two clear copies, one an original top copy. Give an exact description of author(s') post(s) in a footnote.

Provide a 200 word abstract. Provide a full summary (400-500 words) for translation into French for papers in English and into English for papers in French.

Make paper as brief as possible. Do not exceed 6000 words for Research articles. Update articles, and Reviews, including tables or illustrations.

References: Keep number of bibliographical references to a minimum. Number consecutively as occur in text. Italicize or underline number and place in paraentheses. Thus: A recent study in India (3) showed...

List references in numerical order at end of text. Refer to works by more than three authors by first author only, followed by "et al." Verify all references.

Present references to books and articles without abbreviations. Leave titles in foreign languages or transliterate. Print titles of books and journals in italics or underlined.

1. **Paganini JM, Novaes HM**. *Hospital accreditation for Latin America and the Caribbean*. Washington, DC. Pan American Health Organization/WHO and Latin American Hospital Federation, 1992 (SILOS series 13).

2. *Cancer pain relief and palliative care. Report of a WHO Expert Committee*. Geneva, World Health Organization, 1990 (WHO Technical Report Series No. 804).

3. **Bittencourt PRM et al**. Parasitosis of the central nervous system. In: Pedley TA, Meldrum BS, eds. *Recent advances in epilepsy*, vol. 4. Edinburgh, Churchill Livingstone, 1988: 123-159.

4. **Ajjan A**. *La vaccination*. Lyons, Institut Merieux, 1990: 226-227.

5. **Chippaux JP**. La distribution de la dracunculose en Afrique. *Medicine d'Afrique noire*, 1989, 36: 320-322.

6. **Klyingi KS, Sime L**. Compliance profiles of paediatrics patients in an outpatient department. *Papua New Guinea medical journal*, 1992, 35: 95-100.

7. **Pertet AM et al**. Weaning food hygiene in , Kenya. In: Alnwick S et al., eds. *Improving young child feeding in eastern and southern Africa: Proceedings of a workshop held in Niarobi, Kenya, 12-16 October 1987*. Ottawa, International Development Centre, 1988: 234-238.

8. **Razum O**. *Low valid immunization coverage in a rural district of Zimbabwe: should the mothers be blamed?* Paper presented at the Second Inter-Country Meeting in MCHFP, Juliasdale, Zimbabwe, 3-7 February 1992.

9. **Chen R**. *A measles outbreak in a highly vaccinated population: Health Sector Muyinga, Burundi, 1988-1989*. Atlanta, GA, Centers for Disease Control, 1991 (unpublished document).

10. **Esrey S**. *Interventions for the control of diarrhoeal diseases among young children: fly control*. Unpublished document WHO/CDD/91.37, 1991.

Tables and Illustrations: Limit to those essential to an understanding of text. Do not present same data in both tables and graphs.

Refer to tables and figures in text by number (e.g., Table 1, Fig. 3), not "the table above" or "the figure below".

Tables: Type each on a separate sheet of paper. Number consecutively in Arabic numerals. Group in sequence at end of text, do not incorporate into text.

Clearly draw graphs and figures; identify all data; avoid or explain abbreviations. Submit photographs on glossy paper, preferably in black and white. Make figure titles short but adequately explanatory.

OTHER FORMATS

RESEARCH PAPERS—Report scientific findings; reports of studies sponsored by Organization are given priority.

REVIEW ARTICLES–SHORT REPORTS—News items, terminology notes, and a brief description of recent WHO technical publications.

UPDATE ARTICLES–MEMORANDA—From WHO scientific meetings.

OTHER

Experimentation, Human: Research must be conducted in full accordance with ethical principles, including provisions of the Declaration of Helsinki (as revised, 29th World Medical Assembly, Tokyo, Japan) and additional requirements, if any, of the country of origin. Submit a clear statement to this effect, specifying that free and informed consent of subjects was obtained.

SOURCE

1994, 72 (1): back matter.

Calcified Tissue International
Springer International

Louis V. Avioli, M.D.
Washington University Medical Center
The Jewish Hospital of St. Louis
216 South Kingshighway
St. Louis, Missouri 63110
Telephone: (314) 454-8410; Fax: (314) 454-5325
John D. Termlne, Ph.D.
Eli Lilly Research Laboratories Lilly Corporate Center
Indianapolis, Indiana 46285
        Molecular and Cellular Biology articles only

AIM AND SCOPE

Calcified Tissue International publishes original research emphasizing the structure and function of bone and other mineralized systems in living organisms. It includes reports on connective tissues and cells, ion transport, and metabolism of hormones, nutrition, mineralized tissue ultrastructure, molecular biology, and research on humans which reveal important facets of the skeleton or bear upon bone and mineral metabolism.

The Journal serves as a forum to explore biochemical, biophysical, and clinical aspects of bone structure, function, and metabolism. Original reports in either clinical or laboratory investigation, rapid communications, comments, and letters to the editor serve as a communication vehicle within the community. Editorials and original case reports are accepted to present controversial issues for review and analysis by the readers. The journal also publishes Technical Notes reporting the results of studies detailing new methods or procedures in either basic or clinical investigational studies as well as others detailing the use of different cell types or laboratory models.

MANUSCRIPT SUBMISSION

In cover letter state that it is authors' wish to have manuscript evaluated for publication in *Calcified Tissue International*; affirm that it has not been published and will not be submitted or published simultaneously elsewhere.

Submit original and 2 copies of manuscript and all figures and tables.

Submission of a manuscript implies: that work described has not been published before (except as an abstract or as part of a published lecture, review, or thesis); that it is not under consideration for publication elsewhere; that its publication has been approved by all coauthors as well as by responsible authorities at institute where work was done; that, if and when manuscript is accepted for publication, authors agree to automatic transfer of copyright to publisher; that manuscript will not be published elsewhere in any language without consent of copyright holders; that written permission of copyright holder is obtained by authors for material used from other copyrighted sources; and that any costs associated with obtaining this permission are authors' responsibility.

Write clearly and concisely in English. Ensure correctness of spelling, grammar, and typing. Do not use laboratory slang, clinical jargon, or colloquialisms. Type manuscripts double-spaced with wide margins, on 8.5 × 11 in. paper, and not more than 12 pages.

Put each on a separate sheet: title page with authors' names and institution(s) and abbreviated title; tables; references; footnotes; and legends.

MANUSCRIPT FORMAT

Original Studies are investigative reports initiated by author. Limit to 4 printed pages (12 pages of manuscript including illustrations, tables, and references.)

Title: Provide a concise statement of article's major contents with sufficient information to allow readers to judge relevance of paper to interests. Provide a short running title (76 characters, including spaces).

Summary: 250 words. Briefly describe in complete sentences purpose of investigations, methods utilized, results obtained, and authors' principal conclusions. Make summary easily understood without recourse to text or references. Avoid nonstandard abbreviations, unfamiliar terms, symbols or acronyms not easily understood by general scientific reader.

Key Words: List 5 key words characteristic of contents following summary. Separate with dashes.

Introduction: Orient reader to state of knowledge in specific area under investigation. It assumes that reader has a knowledge of mineral and calcified tissue metabolism. Cite recent important work by others. Do not repeat information in introduction and discussion.

Materials and Methods: Describe and reference in sufficient detail so that other workers can repeat study. State sources of unusual chemicals and reagents and special pieces of apparatus. For modified methods describe only the modification.

Results: Present experimental data briefly in text, tables, or figures; avoid redundant methods of presentation.

Discussion: Avoid repetition of material in introduction and detailed repetition of experimental findings. Focus on experimental findings and their interpretation. Do not include unsubstantiated speculations and plans for future study.

Tables: Submit as typed copy. Give each a concise heading; construct as simply as possible; make intelligible without reference to text. Plan 4 vertical rows for a small table (one column) and 8 to 10 rows for a large table (both columns).

Minimize redundant or repetitious entries in a data table. For instance, rather than stating "$P < 0.05$" for each of 20 comparisons within a table, use superscript lowercase letters to denote specified probability levels, defined in a footnote to table. For a table with 20 measurements, combine or "pool" the associated 20 standard deviations or standard errors, making it possible to report only a single estimate of "measurement error" rather than multiple individual entries. This facilitates presentation of a compact table and increases reliability of estimate of standard deviation provided that results pooled are homogeneous.

Illustrations: Provide brief legends on a separate page. Label illustrations lightly in soft pencil on back to show: TOP; figure number; author(s') name(s); desired linear reduction or magnification (optional). Illustrations in color are accepted at author's expense. Keep figures to a minimum.

Figures should either match size of column width (8 cm) or printing area (16.9 × 22.7 cm). Group several figures into a plate on one page (16.9 × 22.7 cm).

Photographs—Photographs should be sharp, well-contrasted glossy prints of original negatives, trimmed at right angles. Prepare to fit format of printed page (one column is 3 5/16 in., page width is 6 7/8 in.) so that 1:1 reproduction is possible. Publisher may reduce or enlarge illustrations. Make illustrations which are to appear together correspond in size and mount as plates. With X-rays, indicate significant parts on back of copy, or on an overlay. Make inscriptions on halftones about 3 mm high.

Line Drawings—Submit good quality glossy prints in desired final size (one column is 3 5/16 in., page width is 6 7/8 in.). Make inscriptions clearly legible with words in upper- and lowercase characters, not all capitals. Recommended sizes are: 2 mm for capital letters and numbers, and 1.6 mm for lowercase letters.

References: Cite in numerical order [in brackets] in text with only one reference to a number.

ARTICLES IN JOURNALS

Give in order: names of authors with initials, year, title of paper with only initial letter of initial word capitalized, abbreviated journal name using *Index Medicus* form, followed by volume number and first and last page of article. Cite journal titles as they existed at time of publication.

Peppler WW, Mazess RB (1981) Total body bone mineral and lean body mass by dual-photon absorptiometry. I. Theory and measurement procedure. Calcif Tissue Int 33:353–359

BOOKS

Author's last name with initial(s), year of publication, title of book, publisher, and city of publication.

Chayen J, Bitensky L, Butcher RG (1973) Practical histochemistry. John Wiley & Sons, London, New York

CHAPTERS IN BOOKS

Byers P (1977) The diagnostic value of bone biopsies. In: Avioli LV, Krane SM (eds) Metabolic bone disease. Academic Press, New York London, p 183

List a paper in references as "in press" only with name of accepting journal; if it has not yet reached this stage, do not include in reference list but put author(s') name(s) in text should be followed by: (in preparation). Treat "Personal Communications" the same way.

Footnotes: Number footnotes to text consecutively. Designate footnotes to tables by lowercase letters in alphabetical order. Designate by footnote on title page, name and address of person to whom offprint requests should be directed

OTHER FORMATS

COMMENTS—Brief manuscripts reporting data that are preliminary, negative, confirmatory, or minor. Follow usual form for manuscripts.

MOLECULAR AND CELLULAR BIOLOGY AND ORTHOPEDIC SURGICAL FORUM—Provide mechanisms for organization of papers in fields of molecular and cellular biology, and clinical orthopedic surgery.

ORTHOPEDIC SURGICAL FORUM—Specific problems in clinical orthopedics with respect to interesting case reports, responses to surgical interventions, and presentation of new and/or complicated clinical surgical procedures. Submit to Louis V. Avioli, M.D.,

RAPID COMMUNICATIONS—Two journal pages (4 special typewritten pages), containing important new observations of sufficient significance to warrant rapid publication.

OTHER

Abbreviations and Symbols: For a list of abbreviations and symbols, see Appendix A, *Calcif. Tissue Int.* 1995; 56(1): vii-viii. Define abbreviations and symbols for terms and units not on this list in a single footnote.

Chemical Compounds: To designate carbon-containing compounds, use C_x to refer to a compound containing x carbon atoms and C-x to refer to the xth carbon atom of a molecule. Designate unfamiliar chemical compounds other than drugs, when first used, by correct systematic names, either by giving trivial name followed by systematic name in parentheses, or, if chemical substances are numerous, by providing a glossary of trivial and corresponding systematic names in a footnote at beginning of manuscript. Systematic chemical names should conform to usage in indexes of *Chemical Abstracts*.

Drugs: Use generic names. Use trade names, spelled exactly as trademarked and with initial letter capitalized, after a drug has been identified once by its generic name or by its systematic chemical name.

Electronic Submission: Submit electronically prepared manuscripts preferably in WordPerfect 5.1 Microsoft Word (Macintosh), or Microsoft Word (PC). Follow instructions exactly; if, for example, reference section is incorrectly styled, value of diskette submission is reduced. Submit double-spaced hard copy of manuscript with software.

Experimentation, Human: For clinical studies, describe patient and control populations in detail. In many studies details of age, race, and sex are important. In experiments involving any significant risk to or discomfort of patients, state that informed consent was obtained from subjects and that investigations were approved by an institutional human research committee. State safeguards for protection of rights of minors and mentally defective subjects.

In text, tables, and figures, identify patients by number or serial letter rather than by initials or names. Include photographs of patients' faces only if scientifically relevant; conceal identity of patient by masking. Obtain written consent from patients for use of such photographs.

Histomorphometric Terminology: Consult *Calcif. Tissue Int.* 42: 284–286, 1988.

Isotopically Labeled Compounds: To designate isotopes: use ^yA to designate element A whose atomic mass is y. For a chemical formula, insert symbol for isotope into formula (e.g., $NaH^{14}CO_3$). For a chemical name or abbreviation, use following examples: for glucose labeled in C-1 position use $1-^{14}C$-glucose or $[1-^{14}C]$glucose; for diiodotyrosine labeled with ^{131}I use ^{131}I-diiodotyrosine or ^{131}I-DIT, or $[^{131}I]$diiodotyrosine or $[^{131}I]$DIT. Use brackets.

Nomenclature: For nomenclature used in this journal, see Appendix B, *Calcif. Tissue Int.* 1995; 56(1): viii. Conform to recommendations of International Union of Pure and

Applied Chemistry (IUPAC) and International Union of Biochemistry (IUB). For a complete listing, see *Handbook for Authors* (American Chemical Society Publications, Washington, D.C., 1967).

Statistical Analyses: For bioassay and radioimmunoassay potency estimates, submit an appropriate measure of precision of estimates. For bioassays these usually will be standard deviation, standard error of the mean, coefficient of variation, or 95% confidence limits. Include a measure of precision relating to within-assay variability and to variability of measurements obtained on subsequent independent assays. If all relevant comparisons are made within the same assay, latter may be omitted. Precision of measurement may depend upon position on dose-response curve.

When several *t* tests are employed, nominal probability levels no longer apply. Accordingly, employ multiple *t* test or multiple range test or similar techniques to permit simultaneous comparisons. In lieu of several *t* tests, utilize an analysis of variance (ANOVA), to permit pooling of data, increase number of degrees of freedom and improve reliability of results. Use appropriate nonparametric tests of significance when data depart substantially from a normal distribution.

In presenting results of linear regression analyses, show 95% confidence limits for line and/or 95% confidence limits for a single observation around the line. Obtain appropriate statistical consultation.

For statistical methods, refer to *Statistical Methods* (George W. Snedecor, Iowa State University Press, 6th ed., 1967).

Style: For general aid in preparing of manuscripts, consult: *CBE Style Manual* (American Institute of Biological Science, (5th ed., Washington, 1983); *Webster's New International Dictionary*, Unabridged, (Philip Babcock Gove and the Merriam Webster Editorial Staff, Springfield, Mass., 1971); *Handbook for Authors of Papers in Journals of the American Chemical Society*, (1st ed., American Chemical Society Publications, Washington, 1967).

Units: Use metric system for weights and measurement, Celsius degrees for temperature and 24 hour clock to express time.

SOURCE

January 1995, 56(1): iv-x.

Canadian Journal of Microbiology
National Research Council of Canada

Dr. Susan F. Koval
Department of Microbiology and Immunology
University of Western Ontario
London, Ontario N6A 5C1,Canada
Telephone: (519) 661-3439; Fax: (519) 661-34991
    Microbial structure and function

Dr. D. M. Benson
Department of Plant Pathology
North Carolina State University, Box 7629
Raliegh, NC 27695-7629 USA
Telephone: (919) 515-3966; Fax: (919) 515-5657

Dr. Marc-Andre Lachance
Department of Plant Sciences
University of Western Ontario
London, Ontario, N6A 5B7 Canada
Telephone: (519) 661-3752; Fax: (519) 661-3935
    Fungi and other eucaryotic protists

Dr. Jean-Gay Bilaillon
Institut Armand Frappier, Universite du Quebec
531, boulevard des Prairies, C.P. 100
Laval-des-Rapides, Quebec H7N 4Z3 Canada
Telephone: (514) 687-5010; Fax: (514) 686-5501

Dr. O. P. Ward
Microbial Biotechnology Laboratory
Department of Biology
University of Waterloo
Waterloo, Ontario N2L 3G1 Canada
Telephone: (519) 888-4523; Fax: (519) 746-4989
    Applied microbiology and biotechnology

Dr. B. B. Finlay
Biotechnology Laboratory
University of British Columbia
237-6174 University Boulevard
Vancouver, British Columbia V6T 1Z3 Canada
Telephone: (604) 822-2210; Fax: (604) 822-9830

Dr. D. J. Hoban
Health Sciences Center - MS 675G
Department of Clinical Microbiology
820 Sherbrook St.
Winnipeg, Manitoba, Canada R3A 1R9
Telephone: (204) 787-3191; Fax: (204) 787-4699
    Infection and immunity

Dr. P. M. Fedorak
Department of Microbiology
University of Alberta
Edmonton, Alberta T6G 2E9 Canada
Telephone: (403) 492-3670; Fax: (403) 492-2216

Dr. D. A. Zuberer
Department of Soil and Crop Science
Texas A & M University
College Station, TX 77843-2474
Telephone: (409) 845-5669; Fax: (409) 845-0456.
        Microbial ecology
Dr. Peter Loewen
Department of Microbiology
University of Manitoba
Winnipeg, Manitoba, Canada R3T 2N2
Telephone: (204) 474-9372; Fax: (204) 275-7615
Dr. Janet M. Wood
Department of Microbiology
University of Guelph
Guelph, Ontario, Canada N1G 2W
Telephone: (519) 824-4120; Fax: (519) 837-1802
        Physiology, metabolism, and enzymology
Dr. M. S. DuBow
Department of Microbiology and Immunology
McGill University
3775 University Street
Montreal, Quebec H3A 2B4 Canada
Telephone: (514) 398-3926; Fax: (514) 398-7052
Dr. R. G. Marusyk
Provincial Laboratory of Public Health
University of Alberta
Edmonton, Alberta S7N 0W0 Canada
Telephone: (403) 492-8903; Fax: (403) 492-8984
        Virology, genetics, molecular biology
If unsure as to appropriate editorial board member, submit
paper to:
Drs. J. J. Germida and L. M. Nelson
Department of Soil Science
University of Saskatchewan
Saskatoon, Saskatchewan,S7N 0W0 Canada
Telephone: (306) 966-6879; Fax: (306) 966-6949

AIM AND SCOPE

The *Canadian Journal of Microbiology* is a widely sub-
scribed-to international journal of general microbiology
and publishes articles, notes, minireview, reviews and letters in English or French. In co-operation with the Canadian Global Change Program, the journal welcomes
submissions of appropriate manuscripts dealing with microbial processes and global changes.

MANUSCRIPT SUBMISSION

Submit manuscripts directly to appropriate editorial board
member. Send original manuscript and three duplicate
copies, one set of original figures or glossy photographs
and three sets of clear copies, and in the case of photographs, four sets identical in quality. Check symbols, abbreviations, and technical terms for accuracy, consistency,
and readability. Examine *Journal* for details of layout, especially tables and reference lists.

In a cover letter formally submit paper for publication and
confirm compliance with requirements about Ethics, Conflict of Interest, and Use of Experimental Animals. Provide two preprints of relevant papers that have been
submitted, are *in press,* or have been recently published.
Each author should have made a substantial contribution

to the overall design and execution of experiments, all are
equally responsible for entire paper. Recognize individuals who provided assistance, but were not involved in the
intellectual process in Acknowledgments.

Affirm that identical or substantially similar reports have
not been published elsewhere nor are being considered or
published elsewhere. Identify any material, figures, or tables that have been previously published. Submit written
permission for use from copyright holder. Affirm that all
authors have read and approved manuscript.

Declare financial support from companies for research in
Acknowledgments. State any personal financial involvement with a company that has an interest in marketing or
using a product of the research reported in a title footnote.
Identify company affiliation by reporting address where
part or all of research was carried out.

MANUSCRIPT FORMAT

Type all parts of manuscript, including footnotes, tables,
captions for illustrations, and references, double-spaced,
on one side only of white paper 21.5×28 cm, with 3-4 cm
margins. Underline only to indicate italics. Use capital letters only when letters or words should appear in capitals
in printed paper. Indent first line of all paragraphs in text
and all captions and footnotes. Number each page and
each line if possible. Put tables and captions for illustrations on separate pages after text.

Title: On first page type title, authors' names, authors' affiliations, and footnotes. Addresses are institution(s)
where work was done. Group each address with respective
authors' names. Do not separate names and addresses.
Footnote present addresses, if different for correspondence and reprints. Give address and telephone and fax
numbers of corresponding author in a footnote. Title
should be clear, concise, and informative.

Abstract: 200 words on a separate page for articles, notes,
or minireviews. Reviews have a table of contents. Avoid
abbreviations. Do not include references unless absolutely
necessary; give complete bibliographic information. Provide three to five key words, directly below abstract.

Text: Introduction—Briefly explain context, significance,
and objective(s) of study. Methods, figures, and footnotes
to tables—Write so others can repeat or extend work. Results–Summarize principal findings and logic used in
reaching them. Discussion—Emphasize overall conclusions; make clear distinctions between those supported by
data and those suggested by them. Results and discussion
may be combined; retain essential features of both.

References: Denote in text by author and date in parentheses
in accordance with Harvard System. If two or more publications are listed for author(s) in the same year, differentiate
chronologically by *a, b. c,* etc., placed after year, without
space. One author–(Smith 1977), (Rogers 1969, 1979),
(Jones 1977; Smith 1965), (Brown 1965*a,* 1965*b*). Two authors–(Smith and Rogers 1976). Three or more authors–
(Smith *et al.* 1975). If author names form part of text, put date
only in parentheses: "Miller (1975) reported that ...."
Place reference list at end of text. List references in alphabetical order. If two or more publications are for same au-

thor(s), same year, differentiate using *a*, *b*, *c*, etc. References to papers in periodicals, serials books, or conference proceedings include titles and inclusive page numbers and name and location of publisher:

Sprott, G.D., Beveridge, T.J., Patel, G.B., and Ferrante, G. 1986. Sheath disassembly in *Methanospirillum hungatei* GP1. Can. J. Microbiol. 32: 847–854.

Krieg, N.R. (*Editor*). 1984. Bergey's manual of systematic bacteriology. Vol. 1. Williams & Wilkins Co., Baltlmore.

Simon, M. I., and Silverman, M. 1983. Recombinational regulation of gene expression in bacteria. *In* Gene function in prokaryotes. *Edited by* J. Beckwith, J. Davis, and J.A. Gallant. Cold Spring Harbor Laboratory, Cold Spring Harbor, N.Y. pp. 211–227.

McCready, R.G.L., and Gould, W.D. 1989. Bioleaching of uranium at Denison mines. *In* Biohydrometallurgy 1989. Proceedings of the International Symposium Biohydrometallurgy 89, Jackson Hole, Wyoming. *Edited by* J. Salley, R.G.L. McCready, and P.L. Wichlacz. Energy, Mines and Resources Canada, Ottawa. pp. 477–485.

Abbreviate journal names according to *CASSI* (*Chemical Abstracts Service Source Index,* Chemical Abstracts, Columbus, OH) or *Serial Sources for the BIOSIS Data Base* (BioSciences Information Service, Philadelphia, PA). If in doubt, write out name in full.

Waite, R.T., and Wood, D.O. 1991. Isolation of the *Rickettsia prowazekii gyrA* gene. Gen. Meet. Am. Soc. Microbiol. 91st, 1991. Abstr. D-160. p. 105.

Wang, I.K. 1991. 6-Methylsalicylic acid polyketide synthetase: enzyme purification and gene cloning. Ph.D. thesis, University of Calgary, Calgary, Alta.

Brown, M.C., and Johnson, L.W. 1991. A new method of cell breakage. J. Gen. Microbiol. In press.

Brown, M.C., and Johnson, L.W. 1992. A new method of cell breakage. Submitted for publication.

Romesburg, H.C., and Mohai, P. 1991. Letter. Can. J. For. Res. 21: 1297–1298.

Brown, M.C. 1990. (University of Alberta.) Personal communication.

Brown, M.C., and Johnson, L.W. 1991. (University of Alberta.) Unpublished data.

Footnotes: Do not use in text unless unavoidable. Designate by superscript Arabic numerals in serial order throughout except in Tables. Place at bottom of page where reference to it is made.

Tables: Number with Arabic numerals, give a brief title, and refer to in text. Type on separate pages. Make column headings and descriptive matter brief. Do not use vertical rules. Consult a *Journal* issue to see how tables are set up and where lines are placed. Designate footnotes by symbols (*, †, ‡, §, ||, ¶, #) or superscript lowercase italic letters. Place other descriptive material under a table as a NOTE.

Illustrations: Plan to fit into area of one or two columns of text. One column maximum finished size is 8.8 × 24 cm; two-columns is 18.3 × 24 cm. Number consecutively

(including half-tones) in Arabic numerals. Refer to each in text. Make self-explanatory. Terms, abbreviations, and symbols must correspond with those in text. Use only essential labeling; give detailed information in caption. Include descriptive statements in addition to legend. Identify figures by figure number, authors' names, and manuscript control number, written below illustration at left.

Line Drawings: Submit well focused prints. Line drawings should not exceed 21 × 28 cm; larger is unacceptable. Draw with India Ink on plain or blue-lined white paper or other suitable material. Rule in coordinate lines to appear. Make lines sufficiently thick to reproduce well. Make symbols, superscripts, subscripts, decimal points, and periods in proportion to drawing and large enough to allow for reduction. Make letters and numerals with a printing device (not typewriter) or with sheets of printed characters; smallest character may not be less than 1.5 mm high when reduced. Use same size and font lettering for all figures of similar size. Use a clear sans serif font. Do not use unusual symbols. Check legibility after reduction by examining a photocopier-prepared reduction of figures.

Photographs: Submit four sets; one set mounted on light, white cardboard, with no space between those arranged in groups. Submit high quality prints made on glossy paper, with strong contrast. Trim mounted copy for reproduction to show only essential features. Plan to fit photographs, or group of them, into area of one or two columns of text with no further reduction. Include a scale bar directly on electron micrographs or photomicrographs. Match contrast and density of all figures in a plate. For composite figures (halftone print and a drawing) mount original photograph with or on ink drawing; do not submit photograph of composite. Each section must be of sufficient contrast to withstand loss of detail and contrast in printing process.

Color illustrations may be accepted if Editor decides it is essential. Mount on heavy but flexible drawing paper.

OTHER FORMATS

CANADIAN GLOBAL CHANGE PROGRAM—Submissions deal with microbial processes and global change. Include and cite datasets on which work is based. To preserve and make datasets widely accessible, extensive datasets will be handled and maintained as Supplementary Material.

LETTERS/FORUM—500 words, no illustrations, and five references. Points of view or opinions on topical microbiological matters or on issues or details raised in papers published in the *Journal*. Letters that take issue with published papers will be faxed to author with an invitation to reply. Both letters will be either approved and published together, or rejected.

MINIREVIEWS—2–6 printed pages; include an abstract. Stimulating commentaries of limited scope which provide a novel synthesis or appraisal of research results or theories. Write for a general rather than a specialist readership.

NOTES—15 manuscript pages; include an Abstract; do not divide into Introduction, Materials and methods, Results, and Discussion sections. Report work that is largely confirmatory, advances in knowledge arising as by-products of broader studies, or descriptions of research techniques or developments in instrumentation.

REVIEWS—Focused, general, current interest articles providing a somewhat more comprehensive, although not exhaustive treatment. Include a table of contents (no abstract); appropriate section headings; and a carefully selected bibliography.

OTHER

Abbreviations, Nomenclature, and Symbols for Units of Measurements: Conform to international recommendations. Use metric units or give metric equivalents. Use SI units; see *Canadian Metric Practice Guide* (1989, Canadian Standards Association, Rexdale, Ontario).

As a general guide for biological terms, use *CBE Style Manual: A Guide for Authors, Editors, and Publishers in the Biological Sciences* (1983, 5th ed, Council of Biology Editors, Inc., Bethesda, MD).

For enzyme nomenclature, follow: *Enzyme Nomenclature (1992): Recommendations of the Nomenclature Committee of the International Union of Biochemistry and Molecular Biology* (1992, Academic Press, San Diego).

Define abbreviations and contractions of names of substances, procedures, *etc.* at first mention.

Identify symbols and unusual or Greek characters clearly; make superscripts and subscripts legible and carefully placed; explain with marginal notes when necessary.

Color Charges: $1000 per page or part.

Equations: Set up clearly in type, triple-spaced. Identify by numbers in square brackets placed flush with left margin. Do not distinguish between mathematical and chemical equations. Routine structural formula need not be submitted as figures

Experimentation, Animal: Employ the most humane methods on the smallest number of appropriate animals required to obtain valid results. In infectious disease investigations the appropriate guideline is that "in the face of distinct signs that such processes are causing irreversible pain or distress, alternative endpoints should be sought to satisfy both the requirements of the study and the needs of the animal." Include a statement in Materials and methods section which indicates: that animals were cared for in accordance with approved guidelines such as *Guide to the Care and Use of Experimental Animals* (Vol. 1, 2nd ed. 1993, and Vol 2,1984, Canadian Council for Animal Care, Ottawa, Ontario) or the *NIH Guide for the Care and Use of Laboratory Animals* (1985, National Institutes of Health, Bethesda, MD); and that use of animals was reviewed and approved by appropriate institutional animal care review committee.

Materials Sharing: Make any plasmids, viruses, and living materials, such as microbial strains and cell lines available from a national collection or in a timely fashion and at a reasonable cost to scientists for noncommercial purposes.

Nomenclature, Genetic:

BACTERIA—Describe genetic properties in terms of genotypes and phenotypes. Phenotype describes observable properties. Genotype refers to genetic constitution, usually in reference to a standard wild type. Use recommendations of Demerec *et al.* (*Genetics*, 54: 61–74,1966) as a guide in employing terms. Use phenotypic designations, consisting of three-letter symbols, not italicized, first letter capitalized, when mutant loci have not been identified or mapped. Superscript letters may be used. Genotypic designations are three-letter locus symbols in lowercase italics. Indicate wild-type alleles by positive superscripts. If several loci control related functions, distinguish by italicized capital letters following locus symbols. Indicate mutation sites by putting serial isolation numbers (allele numbers) after locus symbol. Where large numbers of strains are used, show a table of strains. Define deviations from normal use. For more information about symbols in current use, see Bachmann (*Microbiol. Rev.* 47: 180–230,1983) for *Escherichia coli* K-12; Sanderson and Roth (*Microbiol. Rev.* 52: 485–532,1988) for *Salmonella;* and Henner and Hoch (*Microbiol. Rev.* 44: 57–82,1980) for *Bacillus subtilis.*

VIRUSES—No distinctions are made between genotype and phenotype. Genetic symbols may be one, two, or three letters.

TRANSPOSABLE ELEMENTS AND PLASMIDS—Follow Campbell *et al.* (*Gene*, 5:197–206,1979), and Novick *et al.* (*Bacteriol. Rev.* 40:168–189,1976).

Nomenclature, Microorganisms: Follow revision of *International Code of Nomenclature of Bacteria* in 1990. New names are not valid until a note containing name is published in *International Journal of Systematic Bacteriology.* Give microorganisms and viruses strain designations, letters (usually two) followed by serial numbers. Use worker's initials or a descriptive symbol of locale or laboratory. Each new isolate is given a new (serial) designation (AB1, AB2, etc.). Do not include genotypic and phenotypic symbols.

Numbers: In long numbers separate digits into groups of three, counted from decimal marker to left and right. Use space as separator. Use point as decimal marker. Precede decimal point in all numbers between 1 and -1, except 0, by a 0. Use sign × to indicate multiplication.

Reviewers: In cover letter list potential referees and/or suggest those who should not referee.

Spelling: Use *Webster's Third New International Dictionary* or *Oxford English Dictionary.*

Style: For guidelines for publication of illustrations in scientific journals and books see *Illustrating Science; Standards for Publication* (1988, Council of Biology Editors, Inc., Bethesda, MD).

Supplementary Material: National Research Council of Canada maintains a depository in which supplementary material such as extensive tables or data, detailed calculations, and colored illustrations may be placed. Submit completed work for examination by editors and mark parts to be considered for deposition. Deposited material is indicated by a footnote to appropriate parts of paper. Obtain photocopies of depository material from: Depository of Unpublished Data, CISTI, National Research Council of Canada, Ottawa, Ont., Canada K1A 0S2.

SOURCE

January 1994, 30(1): ix -xiv.

Instructions to authors are published once a year, in first issue of each volume.

Canadian Journal of Physiology and Pharmacology
National Research Council Canada

Editors, *Canadian Journal of Physiology and Pharmacology*
Room 513, Botterell Hall
Queen's University
Kingston, ON K7L 3N6, Canada
Telephone: (613) 545-6335; Fax: (613) 545-6336
E-mail: cjppoff@qucdn.queensu.ca

AIM AND SCOPE

The *Canadian Journal of Physiology and Pharmacology* is an international journal and publishes, in English or in French, reports of original research on all aspects of physiology, pharmacology, nutrition, and toxicology.

MANUSCRIPT SUBMISSION

Submit original and two copies of paper. Do not submit double-sided copies. Give assurance that no part of manuscript reporting original work is being considered for publication, in whole or in part, by another journal. Affirm that all authors have read and approved manuscript. Submit written permission from copyright holder for reproduction of copyrighted material.

MANUSCRIPT FORMAT

Type double-spaced, with 4 cm margins. Underline only to indicate italics. Use capital letters only when letters or words should appear in capitals in print. Indent first line of all paragraphs in text and all captions and footnotes. On first page list title, authors' names, authors' affiliations, and any relevant footnotes. Give authors' addresses as institution(s) where work was done. If present addresses are different, give in footnote. Indicate name, telephone and fax number of corresponding author, as well as category in which paper is to be published: physiology/nutrition or pharmacology/toxicology. Number each page, beginning with title page followed by abstract on a separate page. Put tables and legends for illustrations on separate pages after list of references, which follows text.

Abstract: 200 words for a full paper, brief report, or review, much shorter for a rapid communication. Avoid abbreviations. Type all authors names and initials and complete title of paper at top of abstract page.

Key Words: Three to five directly below abstract.

References: Designate each reference in text by surnames of authors and year: (Green 1970) or Green and Brown (1970). Depending on sentence construction, names may or may not be in parentheses, but year always is. If three or more authors, cite name of first author followed by *et al.* When more than one reference in a given year with same first author, distinguish references chronologically by *a, b, c,* etc. after year and list in this order in reference list. In reference list put citations in alphabetical order according to surnames of first authors. When first author is the same, arrange second authors' names alphabetically. Put citations with three or more authors involving the same first author after dual-authored papers, arranged chronologically. Cite in order: year, title of paper, name of periodical abbreviated according to *CASSI (Chemical Abstracts Service Source Index),* or *Serial Sources for the BIOSIS Previews Database,* volume number, and inclusive page numbers. Materials in press, with name of journal, may be used as references. Give references to conference proceedings in full with complete title, editor's name, location and date of conference, and name and location of publisher. Reports not yet accepted for publication and private communications are not references; put in footnotes or in parentheses in text giving all authors' names and initials.

JOURNAL ARTICLE

Forster, C., Larosa, G., and Armstrong, P.W. 1992. Coronary artery responsiveness in pacing-induced heart failure. Can. J. Physiol. Pharmacol. **70:** 1417–1422.

BOOK

Werns, S.W., and Lucchessi, B.R. 1988. The role of the polymorphonuclear leukocytes in mediating myocardial reperfusion injury. *In* Oxygen radicals in the pathophysiology of heart disease. *Edited by* P.K. Singal. Kluwer Academic Publishers, Boston, Mass. pp. 122–144.

PAPER IN CONFERENCE PROCEEDINGS

Vihko, K.L., La Polt, P.S., Dargan, C., Nishimori, K., and Hsueh, A.J.W. 1991. Stimulatory effect of recombinant FSH. Proceedings of the 11th North American Testis Workshop. April 24-27, 1992, Montreal, Que. p. 24.

Check all parts of each reference listing against original documents.

Footnotes: Designate by superscript Arabic numerals in serial order throughout manuscript except in tables. Place at bottom of manuscript page where reference to it is made.

Tables: Number consecutively with Arabic numerals, give a brief title, and refer to in text. Make column headings and descriptive matter brief. Do not use vertical rules. Consult a Journal issue to see how tables are set up and where lines in them are placed. Designate table footnotes by symbols (*, †, ‡, §, ||, ¶, #) or superscript lowercase italic letters. Place descriptive material not designated by footnote under a table as a Note. Double-space all table parts (title, headings, stub, body, and footnotes).

Illustrations: Design each figure, or group of figures to fit into either one or two columns of text. Maximum finished size of a one-column illustration is 8.8×24 cm; two column illustration is 18.3×24 cm. Number (including halftones) consecutively in Arabic numerals. Refer to each one in text. Make each self-explanatory. Terms, abbreviations, and symbols must correspond with those in text. Use only essential labelling, give detailed information in caption. Identify each by figure number and authors' names, preferably written below illustration at left. Do not fold for mailing.

Line Drawings: Submit original drawings or one set of unmounted clear, well focused glossy photographs and two sets of clear copies. Originals should not be more than three times size of final reproduction. Make drawings with India ink on plain or blue-lined white paper or other suitable material. Drawings on colored drafting film are not acceptable. Rule in coordinate lines. Make lines sufficiently thick to reproduce well, and decimal points, periods, dots, etc., in good proportion to rest of drawing and large enough to allow for reduction. Make letters and numerals with a printing device, not a typewriter, or use sheets of printed characters of such size that smallest character will not be less than 1.5 mm high when reduced. Use same size and font of lettering for all figures of similar size. Avoid use of complex symbols; incorporate symbols or keys into legend on illustration itself. Present bar graphs in a simple, clear format. Do not make three dimensional. Avoid using faint or dense shading.

Computer-generated illustrations must be of professional artistic quality (lines and lettering must be smooth and continuous). They must be in black, on high quality paper. Make lines, lettering (in sans serif font), and symbols large enough for required reduction without loss of detail. Dot matrix lettering is not acceptable.

Photographs: Submit three sets of all photographs; mount one set on illustration board, covered, and ready for reproduction. Submit high quality prints, with strong contrast. Trim copies for reproduction square to show only essential features; mount on white cardboard, with no space between those arranged in groups. Plan to fit photograph, or group of them into area of either one or two columns with no further reduction. If a figure is a composite of a halftone print and a drawing, mount original photograph with or on ink drawing, i.e., do not submit a photograph of composite. Each section must be of sufficient contrast to withstand loss of some detail and contrast inherent in printing process.

For guidelines for publication of illustrations see *Illustrating Science: Standards for Publication* (1988, Council of Biology Editors. Inc., Bethesda, MD).

OTHER FORMATS

BRIEF REPORT—2000 words, 20 references, and four figures and/or tables describing well-documented results of smaller scope. May not be used for preliminary publication or as a progress report.

CRITICAL REVIEWS—Cover a subject that has not been recently reviewed and in which there are noteworthy new developments. Should not be a compendium or autobiographical (100 references maximum).

CURRENT COMMENTARY—Short reviews providing an outlook, insight, or new approach or direction to an area of research.

LETTERS TO THE EDITORS—Comments on scientific issues arising from papers published in Journal. Authors of original paper are invited to respond. Comments may be accompanied by response.

RAPID COMMUNICATION—Three printed pages (about 12 double-spaced typed pages), including tables and figures, but sufficiently descriptive to allow repetition of work. Intended for rapid preliminary publication of particularly novel and significant findings. Submit letter justifying publication as a rapid communication.

SURVEY REVIEWS—Comprehensive update of work in a field that has not been reviewed recently (300 references maximum).

OTHER

Abbreviations: Define abbreviations and contractions of names of substances, procedures, etc, at first occurrence. List extensively used abbreviations with full definitions as a footnote on first page of Introduction. Write symbols and Greek letters clearly and place superscripts and subscripts appropriately.

Color Charges: $1000 per page or part. Editor will decide if essential.

Drugs: Mention trade names of drugs in parentheses in first text reference, use generic names in text, tables, and figures. Capitalize trade names; do not capitalize generic or chemical names. Give chemical nature of new drugs when known. Indicate form of drug used in calculations of doses. When several drugs are used, include a separate paragraph in Methods or a separate table listing relevant information about all drugs employed.

Equations: Set up clearly in type, triple-spaced. Identify by numbers in square brackets placed flush with left margin. In numbering, do not distinguish between mathematical and chemical equations. Typeset routine structure formulas; do not submit as figures for direct reproduction.

Experimentation, Animal: Assure that animals were cared for in accordance with principles and guidelines of the Canadian Council on Animal Care (see *Guide to the Care and Use of Experimental Animals*. Vol. 1 (2nd ed., 1993) and Vol. 2 (1984) (Canadian Council on Animal Care, Ottawa, Ontario).

Experimentation, Human: Provide assurance that appropriate standards were followed, and that experiment has been reviewed and approved by institution's ethics review committee.

Nomenclature: Follow recommendations of the International Union of Biochemistry, such as those on enzyme nomenclature (*Enzyme Nomenclature 1992: Recommendations of the Nomenclature Committee of the International Union of Biochemistry and Molecular Biology* (Academic Press, San Diego, 1992). For a complete listing of recent IUPAC IUB bulletins, see *European Journal of Biochemistry* 151: A5–A11, 1985. For physiological and biological terms see, *CBE Style Manual: A Guide for Authors, Editors, and Publishers in the Biological Sciences* (5th ed, Council of Biology Editors, Inc., Bethesda, MD, 1983).

Reviewers: Submit names and addresses of several possible reviewers.

Spelling: Follow *Webster's Third New International Dictionary* or *Oxford English Dictionary*.

Supplementary Material: The National Research Council of Canada maintains a depository in which supplementary material, such as extensive tables, detailed calculations, and colored illustrations, is placed. Submit complete work and mark parts to be considered for deposition. Editors may require portions of papers to be deposited. Deposition is indicated by a footnote to an appropriate part of paper. Copies of deposited material may be purchased from: Depository of Unpublished Data, Document Delivery, CISTI, National Research Council of Canada, Ottawa, Ont., Canada K1A 0S2.

Units: SI units (Système international d'unités) or give SI equivalents. See *Canadian Metric Practice Guide* (1989, Canadian Standards Association, Rexdale, Ontario).

SOURCE

January 1994, 72(1): vii-ix.

Canadian Journal of Zoology
National Research Council of Canada

Dr. J.G. Eales
The University of Manitoba
Winnipeg, MB R3T 2N2 Canada

AIM AND SCOPE

The *Canadian Journal of Zoology* publishes, in English or French, articles, notes, reviews, and comments in the general fields of behavior, biochemistry, physiology, developmental biology, ecology, genetics, morphology, ultrastructure, parasitology, pathology, systematics, and evolution. Manuscripts must contain significant new findings of fundamental and general zoological interest and may not be considered if they do not meet these criteria. Surveys and descriptions of new species are published only where there is sufficient new biological information or taxonomic revision also involved to render the paper of general zoological interest. Low priority is given to confirmatory studies, investigations primarily of local or regional interest, techniques unless of broad application, and species range extensions.

MANUSCRIPT SUBMISSION

Submit original copy and two duplicates of papers.

Give assurance that no part of manuscript reporting original work is being considered for publication, in whole or in part, by another journal. Corresponding author must affirm that all authors have read and approved of manuscript. Limit co-authors of a paper to those who have made significant scientific contributions to work. All authors should be able to take public responsibility for content. Indicate additional contributions in a footnote or in Acknowledgments section.

Check symbols, abbreviations, and technical terms carefully for accuracy, consistency, and readability. Examine *Journal* for details of layout, especially tables and reference lists. Manuscripts that do not conform to requirements may be returned for modification.

Present an accurate account of research performed; give complete reports of observations made and data collected. Relate work to that of others and provide complete and accurate citations so that readers can objectively evaluate paper.

Avoid fragmentation of research reports or submission of trivial reports,

Obtain formal or informal approval or clearance of paper from institution or company before it is forwarded to Re-search Journals. Identify sources of all information quoted and material obtained privately. When a manuscript contains material (tables, figures, charts, etc.) protected by copyright, secure written permission from holder of copyright. Send letters of permission must be sent to Editorial Office before final acceptance of manuscript.

Disclose any information that may affect acceptance or rejection of paper. This includes indicating if work has been previously presented in any format (conference proceedings, abstract publication, etc.) and submitting a list of related manuscripts that author has in press or under consideration by another journal. Paper will be considered for publication only with understanding that it has not already been submitted to, accepted by, or published in another journal.

Once a paper is accepted for publication, author transfers copyright to National Research Council of Canada.

MANUSCRIPT FORMAT

Type all parts of manuscript, including title page, footnotes, references, tables, and captions for illustrations, double-spaced, on one side only of white paper 21.5×28 cm, with margins of 4 cm. Double-sided copies are not acceptable. Number every page in top right corner.

Start Abstract and Introduction on separate pages. After body of manuscript (usually Introduction, Materials and methods, Results, and Discussion), sequential numbering is continued on pages containing Acknowledgments, References, Tables, Figure captions, and Appendices in order. Type figure captions double-spaced on one or more sheets; captions for a group of figures should follow on same line. Indent first line of all paragraphs in text and all figure captions and footnotes.

Title Page: Title, names of authors, each followed immediately by affiliation, corresponding author's address, telephone and fax numbers, and E-mail address, and any necessary footnotes. Key words are not required; equivalent information should be readily retrievable from title and abstract. Title should be as short and simple as possible. Include common and correct taxonomic names in title if organism is not well known.

Abstract: 200 words for all contributions. Type all authors' names and initials and complete title of paper at top of page. Authors who can submit abstracts in both fluent English and fluent French are encouraged to do so. Avoid use of abbreviations and references; however, when it is essential to include a reference, use full literature citation but omit title of article.

Results: Include only enough explanation and interpretation to allow reader to understand why experiments or observations were carried out and what they mean. Ensure that number of significant digits used to describe data does not exceed accuracy with which measurement can be made. For numbers from -1 to 0 and from 0 to +1, precede decimal by a zero in text, tables, and figures.

Discussion: Include no new findings not already been mentioned under Results. Include conclusions; do not make a separate section.

Footnotes: Designate by superscript Arabic numerals in serial order throughout manuscript except in Tables. Place at bottom of manuscript page where reference to it is made.

Equations: Set up clearly in type, triple-spaced. Identify by numbers in square brackets placed flush with left margin. In numbering, no distinction is made between mathematical and chemical equations. Routine structural formulas can be typeset; do not submit as figures for direct reproduction, but they must be clearly depicted.

References: Check with original articles and refer to each in text by author and date, in parentheses. List at end of paper in alphabetical order in form used in current journal issues of the Journal.

JOURNAL ARTICLE

Sivak, J.G. 1991. Shape and focal properties of the cephalopod ocular lens. Can. J. Zool. **69**: 2501–2506.

BOOK

Sparks, A.K. 1985. Synopsis of invertebrate pathology exclusive of insects. Elsevier Science Publishers, Amsterdam.

ARTLCLE PUBLISHED IN A BOOK OR COLLECTION

Heisler, N. 1984. Acid-base regulation in fishes. In Fish physiology. Vol. 10A. *Edited* by W.S. Hoar and D.J. Randall. Academic Press, New York. pp. 315–400.

THESIS

Harvey, H.H. 1963. Pressures in the early history of the sockeye salmon. Ph.D. thesis, University of British Columbia, Vancouver.

CONFERENCE PROCEEDINGS

Spong, P., Bradford, J., and White, D. 1970. Field studies of the behavior of the killer whale (*Orcinus orca*). *In* Proceedings of the Seventh Annual Conference on Biological Sonar and Diving Mammals, Menlo Park, Calif., October 23 and 24, 1970. *Edited* by T.C. Poulter. Stanford Research Institute. Menlo Park, Calif. pp. 169–174.

Authors' initials follow surnames. Do not: underline names of journals or books; use ampersand in place of and; or use a comma in text citations between author(s) and date. In references to papers in periodicals and books, include titles and inclusive page numbers. Mention articles "submitted" and "in preparation" as footnotes; do not include in references. Confirm in cover letter that papers cited as "in press" have been accepted for publication. Abbreviate names of serials as in *CASSI (Chemical Abstracts Service Source Index,* Chemical Abstracts, Columbus, OH) or in *Serial Sources for the BIOSIS Data Base* (Bio-Sciences Information Service, Philadelphia, PA). In doubtful cases, write name in full. For citations of nonrefereed documents (e.g., environmental impact statements, contract reports), include address.

Tables: Number with Arabic numerals, give a brief title, and refer to in text. Type title, headings, stub, entries, and footnotes double-spaced. Begin each on a separate page; clearly indicate when it runs more than one page. Make column headings and descriptive matter in tables brief. Do not use vertical rules; do not use horizontal rules in body. Consult a copy of Journal for guidance on setting up tables and placing horizontal rules. Designate footnotes by symbols (*, †, ‡, §, ||, ¶, #) or superscript lowercase italic letters. Place descriptive material not designated for a footnote under a table as a NOTE.

Illustrations: Plan each figure, or group of figures to fit into area of either one or two columns of text. Maximum finished size of a one-column illustration is 8.8 × 24.0 cm and two-column illustration is 18.3 × 24.0 cm. Number figures (including halftones) consecutively with Arabic numerals; refer to each in text and make self-explanatory. All terms, abbreviations, and symbols must correspond to those in text. Use only essential labelling, with detailed information in caption. Label each illustration with figure number and authors' names. Send all illustrations with initial manuscript submission.

Line Drawings: Submit original drawings or laser prints (with diskette if possible) or one set of clear, well-focussed glossy photographs and two sets of clear copies. Do not substitute xerographic copies for original line drawings. Original drawings of actual publication size are preferred; if not possible, make originals not more than three times the size of final reproduction. Make drawings with India ink on plain or blue-lined white paper or other suitable material. Rule in coordinate lines to appear. Make lines sufficiently thick to reproduce well, and all symbols. superscripts, subscripts, decimal points, and periods in good proportion to rest of drawing and large enough to allow for reduction. Make letters and numerals neatly with a printing device (not a typewriter) or with sheets of printed characters and of such a size that smallest character will be not less than 1.5 mm high when reduced. Use same size and font of lettering for all figures of similar size. Use a clear sans serif font; heavy lettering tends to close up on reduction. Avoid unusual symbols, which printer may not be able to reproduce in figure caption. Use following symbols: (● ○ ■ □ ▲ △). Incorporate complex symbols or keys in a concise legend on illustration itself.

Computer-generated illustrations must be professional artistic quality (make lines and lettering smooth and continuous); follow guidelines above. Lettering made with dot matrix printers is not acceptable.

Photographs: Supply three sets of all photographs; mount one set on flexible white Bristol board, ready for reproduction. Two other sets can be photographic reproductions of mounted set suitable for review purposes. Prints must be of high quality, made on glossy paper, with strong contrasts. Trim copies for reproduction to show only essential features and mount on flexible white cardboard, with no space between those arranged in groups. Defining lines (routing) between photographs will be added by printer. Match contrast and density of figures in a plate. Plan photograph, or group to fit into area of either one or two columns of text with no further reduction. Indicate magnification wherever size is important. Include a scale bar in picture.

For guidelines for publication of illustrations in scientific journals and books see *Illustrating Science: Standards for*

Publication (1988, Council of Biology Editors, Inc., Bethesda, MD). Preparation of artwork, graphs, maps, computer graphics, halftones, and camera-ready copy, as well as color printing, are explained and illustrated.

Color Illustrations: Editor decides if color is essential. Authors are responsible for costs. Mount photographs on flexible white Bristol board. Obtain further details from Publishing Office.

OTHER FORMATS

ARTICLES—30 page reports of research (Abstract, Introduction, Methods, Results, Discussion).

COMMENTS—Short critiques of papers previously published in journal. Authors are invited to rebut critiques.

NOTES—Four printed pages (14 manuscript pages). Brief reports of original research. Organize like articles, with formal headings, or more simply.

REVIEWS—Deal with topics of general interest or current importance, synthetic rather than comprehensive in emphasis. They are considered only after invitation by or agreement with Editor.

OTHER

Abbreviations: Define abbreviations and contractions of names of substances, procedures, etc., individually at first use or together in a footnote on title page. Avoid abbreviations with more than one meaning.

Clearly identify symbols and unusual or Greek characters; make superscripts and subscripts legible and carefully placed; explain in marginal notes when necessary.

Color Charges: $1000 per page or part of a page.

Electronic Submission: Authors are encouraged to submit diskettes (containing approved version) with accepted manuscripts only. Diskettes (5.25 or 3.5 in.) may be prepared on either Macintosh or IBM-compatible personal computers. Identify word-processing software, version number, and type of computer (Macintosh or IBM) used.

Experimentation, Animal: Give assurance in Materials and methods that animals were cared for in accordance with principles and guidelines of Canadian Council on Animal Care (see *Guide to the Care and Use of Experimental Animals*, Vol. 1 (2nd ed., 1993) and Vol. 2 (1984); Canadian Council on Animal Care, Ottawa). Studies with unwarranted numbers of rare or endangered species may not be accepted.

Describe safeguards used to meet both formal and informal standards of ethical conduct of research (approval of a research protocol by an institutional committee, procurement of informed consent, proper treatment of animals, and maintenance.

Language: Research Journals program of National Research Council of Canada follows guidelines on publishing in both official languages as outlined by Treasury Board of Canada Secretariat (see Circular No. 1982-58, section 10). Articles are published in English or French according to author's choice, preceded by an informative summary or abstract in both English and French.

Make manuscripts free of any kind of prejudice, especially gender and racial stereotyping.

Nomenclature: Use *CBE Style Manual: A Guide for Authors, Editors, and Publishers in the Biological Sciences* (5th ed., 1983, Council of Biology Editors, Inc., Bethesda, MD). For enzyme nomenclature see *Enzyme Nomenclature (1992): Recommendations of the Nomenclature Committee of the International Union of Biochemistry and Molecular Biology* (Academic Press, San Diego, Calif.) should be followed.

Permissions: If manuscript contains material (tables, figures, charts, etc.) protected by copyright, secure written permission from holder of copyright. Send letters with manuscript. All material designated as "taken from..." must have a letter of permission. If material is not to be reproduced exactly as in original, designate as "modified from...." Include source of material in reference list.

Specimen Deposition: Consider depositing representative specimens in a recognized depository. Prior to commencing a study, make arrangements with a depository such as the Canadian Museum of Nature, P.O. Box 3443, Station D, Ottawa, ON K1A OM8, Canada, or a provincial museum. Include catalogue or accession numbers.

Spelling: Use Webster's *Third New International Dictionary* or *Oxford English Dictionary*. Ensure that spelling is consistent.

Style: Underline material to be set in italics. Use capital letters for letters and words that are to appear in capitals in printed paper. Material taken from theses must be thoroughly edited for brevity.

Supplementary Material: National Research Council of Canada maintains a depository in which supplementary material such as extensive tables of data, detailed calculations, and colored illustrations may be placed. Submit complete work for review; mark part to be considered for deposition. Editor may suggest that portions be placed in depository. Copies of material in depository may be purchased from Depository of Unpublished Data, National Research Council of Canada, Ottawa.

Units: Use metric units or give metric equivalents; SI units (Systeme international d'units) are required. System is explained in *Canadian Metric Practice Guide* (1989, Canadian Standards Association, Rexdale, Ontario).

SOURCE

January 1994, 72(1): vii-ix, xiii-xiv.

Canadian Medical Association Journal
see CMAJ

Cancer
American Cancer Society
J.B. Lippincott Co.

Robert V. P. Hutter, MD, Editor-in-Chief
Cancer Editorial Office
101 Old Short Hills Road, Suite 503
West Orange, New Jersey 07052-1023

AIM AND SCOPE

"The American Cancer Society is the nationwide voluntary health organization dedicated to eliminating cancer as a major health problem by preventing cancer, saving lives from cancer and diminishing suffering from cancer through research, education, and service."

Cancer is a peer-reviewed publication of the American Cancer Society integrating scientific information from worldwide sources for all oncologic specialties. The objective of *Cancer* is to provide an interdisciplinary forum for the exchange of information among oncologic disciplines concerned with the etiology and course of human cancer. *Cancer* accomplishes this objective by publishing original articles, as well as other scientific and educational documents, that support the mission of the American Cancer Society by facilitating the transfer of knowledge from the laboratory to the bedside; contributing to cancer prevention, early detection, diagnosis, cure, and rehabilitation; and diminishing suffering from cancer.

Cancer is pleased to receive original articles related to human cancer including, but not limited to: biologic response modifiers (such as growth factors, interferons, interleukins, lymphotoxins), clinical observations, chemotherapy, clinical trials, detection, epidemiology, ethical issues, etiology, genetics and cytogenetics, imaging, immunology and immunotherapy, oncogenes, pathology and clinicopathologic correlations, prevention, psychosocial studies, radiation therapy, screening, staging, and surgical therapy.

Publications fall within following categories: Tribute (highlights accomplishments of distinguished individuals for their contributions to oncology); Editorial (relates to a manuscript in the same issue or a topic of current interest); Commentary (presents a point of view of general interest); Review Article (a timely, in-depth treatment of an issue); Original Article by Anatomic Site or General Topic (not limited to an anatomic site, e.g., Kaposi sarcoma); and Correspondence (referring to a manuscript published in *Cancer* within six months. Authors of original publication will be given the opportunity to respond.) Special categories may be introduced for papers on selected topics (such as Communications from the Commission on Cancer of the American College of Surgeons, or American Joint Committee on Cancer). Case reports are considered only if justified by their unique significance.

MANUSCRIPT SUBMISSION

Do not list more than 10 authors; more than ten requires written justification and approval of Editor-in-Chief. Group authorship is permitted when: authorship can be attributed to an entire group when all group members meet criteria for authorship (group name is placed with manuscript title and names of all group members are listed in a footnote and their authorship acknowledged); names of up to 10 authors are listed with manuscript title, followed by name of group when individual authors, as well as all members of the group, meet criteria for authorship (group member names are listed in a footnote and their authorship acknowledged); specified authors (not more than 10) assume responsibility for an entire group, only specified authors must meet criteria for authorship (listed with manuscript title; group members are listed in a footnote but are not acknowledged as authors; corresponding author must state in cover letter that she/he has written permission from each group member to list her/his name as a member of group).

Follow "Uniform Requirements for Manuscripts Submitted to Biomedical Journals" (*JAMA* 1993; 269: 2282-6) to determine authorship. "Each author should have participated sufficiently in the work to take public responsibility for the content. Authorship credit should be based only on substantial contributions to (a) conception and design, or analysis and interpretation of data; and to (b) drafting the article or revising it critically for important intellectual content; and on (c) final approval of the version to be published. Conditions (a), (b), and (c) must all be met."

Manuscripts must not have been published nor submitted elsewhere in either identical or similar form or content, nor will be submitted elsewhere while under consideration for publication by *Cancer*. Include copies of related manuscripts under editorial consideration or in press. Submit journal's "Authorship Responsibility, Financial Disclosure, and Copyright Transfer Form" (in January 1 and July 1 issues) with original ink signatures from each author. In cover letter include complete title and category for manuscript (i.e., Tribute, Editorial, Commentary, Original Article, Correspondence). Suggest a specific anatomic site or general topic best suited for original articles. Provide address, telephone and fax numbers for corresponding author.

Properly cite information reproduced from other sources. Obtain and submit signed written permission from appropriate authors and/or copyright holders.

Obtain permission statements from at least one author when citing unpublished data, in-press articles, and/or personal communications.

MANUSCRIPT FORMAT

Provide four copies and four complete sets of tables and original illustrations. Print double-spaced on conventional 8.5 × 11 in. or ISO A4 paper, on one side only. Use standard 10- or 12-pitch type. Do not use proportional spacing or justified right margins. Beginning with title page, type page number and first author's last name in upper right corner. Begin each component on a separate page, identified with proper heading.

Title Page: Include: manuscript title; running title (short title — 40 characters); each author's name (first name, middle initials, and surname; or first initial, middle name, and surname) followed by highest academic degree; hospital/academic institution/or other site, and city where work was done; authors' affiliations; source of financial support; mailing address for correspondence and reprints; and acknowledgments as appropriate; identify scientific meeting where all or part of manuscript was presented, including dates and location; total number of each: text pages, including title page, references, legends, tables, and illustrations.

Précis for Table of Contents: 1 or 2 concise sentences; state significant conclusion or manuscript message.

Abstract: 250 words in four paragraphs, Background, Methods, Results, and Conclusions describing problem(s) addressed, strategy to solve problem(s), salient results, and conclusions. Not required for Tributes, Editorials, Commentaries, and/or Correspondence.

Key Words: Below abstract list four to ten key words; use terms from *Medical Subject Headings* list of *Index Medicus.* Document first report of any event with source and years of literature searched (eg., MEDLARS). Cite staging system with statements regarding cancer stage.

Footnotes: Use only for tables and figures. Do not use to elaborate on text.

References: List in a separate section following text. Verify all references. Follow style of "Uniform Requirements for Manuscripts Submitted to Biomedical Journals" (*JAMA* 1993; 269: 2282-6) for reference format, and *Index Medicus* for standard journal abbreviations. Number references sequentially in order cited in text; do not alphabetize. Number references cited only in tables or figures in sequence established by first mention in text of table or figure containing reference. Double-space references within and between entries.

Do not reference personal communications or manuscripts "in preparation" or "submitted for publication." If an essential written communication, cite source parenthetically in text with "unpublished data." List papers accepted but not yet published with references as "in press."

List names of all authors when six or fewer; if seven or more, list only first six, followed by "et al."

List journal references as: authors, article title and subtitle, journal abbreviation, year, volume number in Arabic numerals, and inclusive pages. List book references as: authors, title, edition (if other than first), volume (if more than one), city, publisher, year; chapter in a book: authors of chapter, title of chapter, "In:" editors/authors of book, title of book, edition (if other than first), volume, city, publisher, year, and inclusive pages of chapter.

1. Green-Gallo LA, Buivys DM, Fisher KL, Caporaso N, Slawson RG, Elias EG, et al. A protocol for the safe administration of debrisoquine in biochemical epidemiologic research protocols for hospitalized patients. *Cancer* 1991; 68:206–10.

2. Lever WF, Schaumburg-Lever G. Histopathology of the skin, 7th ed. Philadelphia: JB Lippincott, 1990:552-6.

3, Grover RF, Reeves JT, Rowell LB, Piantadosi CA, Saltzman HA. The influence of environmental factors on the cardiovascular system. In: Hurst JW, Logue RB, Rackley CE, Schlant RC, Sonnenblick EH, Wallace AG, et al. The heart. 6th ed. New York: McGraw-Hill, 1986: 1543–55.

Tables: A moderate number will be printed without charge. If tabular and illustrative materials exceed one page per four pages of printed text, charges are incurred. Group tables after reference section; number consecutively, using Arabic numerals, in order cited in text. Follow table number with a brief descriptive title. Double-space entire table, including title, headings, and footnotes. Use horizontal rules only to set off headings; do not use vertical rules. If tabular material

exceeds one page, identify succeeding pages with table number, *continued,* and repeat table subheadings. If tables contain special symbols, structures, or other art, submit as single-spaced, camera-ready copy.

Illustrations: Number all illustrations sequentially with Arabic numerals, in order cited in text. Affix a label to back of each figure, indicating TOP, figure number, and first author's last name. Double-space legends on a separate sheet, include figure number and a brief description. If illustrative and tabular materials exceed one page per four pages of printed material, authors may be charged. Color illustrations are printed at authors' expense.

Line Drawings: Submit in either India ink, original laser output, or in camera-ready form on a bright white background. Make labels, numbers, and symbols the same size for uniform reduction.

Photographs: Submit glossy black-and-white photographic prints with sharp contrast. Mask patients' identities or submit signed permission statements. For letters, arrows, and any other markings on photographs, use professional quality transfer type. Photographs may be cropped or reduced.

Figures: Use of previously published figures requires written permission from authors and copyright holders. Include credits for reproduced work in figure legend with authors, title, publisher or periodical name, volume, page, and year.

OTHER

Abbreviations: Use only standard abbreviations. Spell out in full preceding first use in text.

Drugs: State generic name, or trade name followed by generic name, and city (and country, if foreign) of manufacturer in parenthesis. Follow products cited by trade name by parenthetical designation of name and city of manufacturer.

Experimentation, Human: Submit a statement with cover letter confirming that informed consent was obtained from subject(s) and/or guardian(s). Also state in manuscript. Mask patients' identities in photographs whenever possible. If patients are identifiable, submit written permission to use photograph from patient or guardian.

Quotations: Enclose within quotation marks, be accurate, and give full credit to authors and source, including pages.

Randomized Control Trials: State how comparison groups were generated. Specify in title, precis and abstract that manuscript is a report of an RCT.

Reviewers: Provide names, addresses, and telephone numbers of potential reviewers.

Style: Use following sources: *CBE Style Manual* 5th ed. for spelling, capitalization, punctuation, hyphenation, and general style; "Uniform Requirements for Manuscripts Submitted to Biomedical Journals" (*JAMA* 1993; 269: 2282-6) for format of references; *Manual for Staging of Cancer* 4th ed. or *UICC* for citing stages of cancer; *International Histological Classification of Tumors* for histological typing of tumors; *Naming and Indexing of*

Chemical Substances for Chemical Abstracts for uses of chemical terms; *ICD-O: International Classification of Diseases for Oncology* 2nd ed., *Physicians' Current Procedural Terminology:* CPT 1994 and *SNOWMED International* for terms relating to diseases, operations, and procedures; *Cancer Treatment Reports* for presenting statistical material; and *Index Medicus* for abbreviating journals in references.

Units: Système International (SI) or metric system of measure.

SOURCE

July 1, 1994, 74(1): 21A–26A.

Cancer Genetics and Cytogenetics
Elsevier Science, Inc.

Dr. Avery A. Sandberg
The Cancer Center
Southwest Biomedical Research Institute
6401 East Thomas Road
Scottsdale, Arizona 85251
Telephone: (602) 945-4363, ext. 211; Fax: (602) 947-8220

AIM AND SCOPE

The aim of this journal is to publish papers of originality and high scientific quality in the various areas of genetics and cytogenetics (e.g., human, animal, molecular, population, biochemical), as they relate to the broad fields of cancer, in order to reach as multidisciplinary an audience as possible.

MANUSCRIPT SUBMISSION

Short Communications (3 printed pages) and Letters to the Editor (1 printed page) may be submitted. Review and Special articles are published from time to time, by invitation of Editor-in-Chief. Editorials may also be published. Book Reviews are published as space permits. Manuscripts are submitted with the understanding that they are original unpublished work and are not being submitted elsewhere. Upon acceptance of an article, author(s) must transfer copyright of article to publisher.

MANUSCRIPT FORMAT

Type double-spaced on 8.5 × 11 in. bond paper. Submit all materials in triplicate. On title page, include full names of authors (no degrees), academic or professional affiliations, complete address, phone and fax numbers for corresponding author, and an abbreviated title (45 characters and spaces). Except for Letters to the Editor, all articles include an Abstract (200 words). Type acknowledgments, reference lists, tabular material, and figure legends on separate sheets. Give tables, numbered with Arabic numerals, brief descriptive titles. Use horizontal rules only. Number footnotes to text consecutively with superior Arabic numerals. Follow nomenclature of ISCN 1985 and new ISCN 1991 nomenclature for cancer cytogenetics. Array karyotypes according to depictions and descriptions given in ISCN 1985 and ISCN 1991. Define abbreviations when first used; use consistently throughout. Follow generic with correct chemical names in parenthesis when first used. Avoid trade names. Provide normal laboratory values in parentheses when first used.

Illustrations: For line artwork, submit black-ink drawings of professional quality, or glossies of originals. Duplicates may be photocopies. On back of each, give first author's name, number of figure, and indication of TOP. Submit original chromosome karyotypes. Descriptions on legends must match information in illustrations. Type on a separate sheet of paper.

References: Include names of all authors for each reference. Verify completeness and accuracy of all references. Cite in text by number in brackets, corresponding to numbered reference list at end of article. Periodical abbreviations follow those in *Index Medicus*.

JOURNAL

Sandberg AA, Turc-Carel C (1987): The cytogenetics of solid tumors. Relation to diagnosis, classification and pathology. Cancer 59:387–395.

BOOK

Sandberg AA (1990): The Chromosomes in Human Cancer and Leukemia. 2nd Ed. Elsevier Science Publishing Co., New York, pp. 50–61.

EDITED BOOK

Ghez C (1991): The control of movement. In: Principles of Neural Science, 3rd Ed., ER Kandel. IH Schwartz, TM Jessel, eds. Elsevier Science Publishing, New York, pp. 533–547.

OTHER

Color Charges: $450 U.S. for one color figure on one page. Color is printed at author's expense. Enclose a letter agreeing to pay charge upon submission.

Karyotypes: Besides 3 glossy prints of karyotypes, submit original karyotype(s). Artwork larger than 7 × 9 in. is acceptable; excessively large artwork may not reproduce correctly. Make sure that all chromosomes are attached securely and karyotype appropriately protected. If indicated, original karyotypes will be returned. Label chromosomes accurately, both normal and abnormal; do not include annotations, put in legend. Do not use tape; remove all evidence of glue or similar material.

SOURCE

October 1, 1994, 77(1): inside back cover.

Cancer Research
American Association for Cancer Research
Waverly Press, Inc.

Dr. Carlo M. Croce, Editor-in-Chief, *Cancer Research*
American Association for Cancer Research, Inc.,
Public Ledger Bldg.,
620 Chestnut Street, Suite 816
Philadelphia, PA 19106-3483

AIM AND SCOPE

Cancer Research, the official journal of the American Association for Cancer Research, Inc., is devoted to the publication of significant, original research in all the areas of cancer research, including: biochemistry; biophysics; cell biology; chemical, physical, and viral carcinogenesis and mutagenesis; clinical investigations; endocrinology; epidemiology and prevention; experimental pathology; experimental therapeutics; immunology and immunotherapy; molecular biology and genetics; physiology; radiobiology and radiotherapy; and virology.

Only those papers reporting results of novel, timely, and significant studies and meeting high standards of scientific merit will be accepted.

MANUSCRIPT SUBMISSION

Submission implies understanding and acceptance of Journal policies. Neither submitted paper nor any similar paper, other than an abstract or preliminary communication, may have been published elsewhere or will be submitted for publication. All authors must agree to submission and content.

Reveal any relationships that could be construed as causing a conflict of interest.

Senior author should submit a cover letter stating: that paper should be considered for publication in *Cancer Research*; exact address of corresponding author with telephone and FAX numbers; that authorization has been given to use any information conveyed by either personal communication or release of unpublished experimental data; salient and novel findings of paper in a concise paragraph; five key words; and subject category:

    Biochemistry and Biophysics
    Carcinogenesis
    Clinical Investigations
    Endocrinology
    Epidemiology
    Experimental Therapeutics
    Immunology
    Molecular Biology and Genetics
    Tumor Biology
    Virology

Submit: cover letter in duplicate with above information; four copies of manuscript and four sets of original illustrations (indicate which set should be used by printer); and relevant papers in press or submitted for publication.

Copyright must be transferred to American Association for Cancer Research, Inc. It is understood that this material has not been published elsewhere, either whole or in part (except in abbreviated form as a preliminary communication). Have authorized agent of commercial firm or commissioning organization sign copyright transfer form if article was prepared as part of official duties as an employee. Papers from government laboratories do not require copyright transfer, provided that authors abide by the same provisions required of other authors and sign appropriate section of copyright transfer form. Forms are sent with acknowledgment of receipt of manuscript.

MANUSCRIPT FORMAT

Follow Journal style. Write in clear grammatical English. Do not use laboratory jargon or terminology and abbreviations not consistent with internationally accepted guidelines.

Type double-spaced on 8.5×11 in. paper, with ample margins. Number all pages consecutively; title page is page 1. Arrange in order: title, author(s) and complete name(s) and location(s) of institution(s) or laboratory(ies), running title, key words, footnotes, text and references, tables, legends for all illustrations, illustrations, and other material. Avoid numbered and lettered sections in text. Indicate appropriate location for each table and illustration by marginal notes. Present simple chemical formulas or mathematical equations so they can be reproduced in a single horizontal lines of type; submit more complicated mathematical formulas or chemical structures difficult to set in type as India ink drawings or glossy photographs for camera-ready reproductions.

Title: Brief but informative, 100 characters. List on title page five key words to identify subject matter, including, if applicable, species on which work is done. Avoid expressions such as "Studies on . . ." or "Observations of. . ." Do not use chemical formulas or abbreviations. Do not use Roman or Arabic numerals to designate paper as one in a series.

Authors: Include full names, with first and middle names or initials. Do not include academic degrees. Give full names of institutions and subsidiary laboratories, with address (including postal code). If several authors and institutions are listed, indicate with which department and institution each is affiliated.

Running Title: 50 characters. Do not use declarative or interrogative sentences.

Footnotes: Do not use lengthy footnotes. Present information in text. Designate title page and textual footnotes consecutively with superscript Arabic numerals. In title footnote give information on financial support, including source(s) and number(s) of grant(s). If paper is in a series, state in a footnote. Designate reprint request author and list all nonstandard abbreviation and definitions in footnotes.

Abstract: Concise summary indicative of paper's content. Recapitulate in abbreviated form study's purpose and experimental technique, results, and data interpretation. Include data such as number of test subjects and controls, strains of animals or viruses, drug dosages and routes of administration, tumor yields and latent periods, length of observation period, and magnitude of activity. Do not use vague, general statements. Incorporate important terms relevant to paper's content. Avoid abbreviations.

Introduction: Do not include all background literature. Acquaint reader with findings in field and with problem addressed with brief references to pertinent papers.

Materials and Methods: Explain experimental methods sufficiently for repetition by qualified investigators. Do not describe previously published procedures; cite appropriate references. Describe new and significant modifications of previously published procedures. Give sources of special chemicals or preparations with locations [city and state (country, if foreign)].

Results: Include a concise textual description of data presented in tables and illustrations. Avoid excessive elaboration of data in tables and illustrations. Combine Results and Discussion sections if space is saved or logical sequence of material is improved.

Discussion: Interpret data concisely without repeating material already presented in Results. Speculation is permissible, but must be well founded.

References: Number in order of first mention in text; cite by assigned number. Type double-spaced. Limit to citations essential to presentation. Comprehensive review articles are preferred to many separate references. Verify accuracy of references and check that all references have been cited in text. Supply all authors, complete titles of articles, and inclusive page numbers:

Saylors, R. L., III, Sidransky, D., Friedman, H.S., Bigner, S. H., Bigner, D.D., Vogelstein, B., and Brodeur, G.M. Infrequent *p53* gene mutations in medulloblastomas. Cancer Res., *51:*4721–4723, 1991.

Yuspa, S.H., Hennings, H., Roop, D., Strickland, J., and Greenhalgh, D. A. Genes and mechanisms involved in malignant conversion. *In:* C.C. Harris and L.A. Liotta (eds.), Genetic Mechanisms in Carcinogenesis and Tumor Progression, pp. 115–126. New York: Wiley-Liss, 1990.

JOURNAL ARTICLES AND SERIAL COMPENDIA
Give complete title, journal, volume number, inclusive pages, and year of publication. Cite numbered serial compendia, which appear annually in numbered sequence, as journals rather than books, omitting names of publishers and editors. Use *Serial Sources for the BIOSIS Previews Data Base* for abbreviations of journals and serials.

BOOKS AND CHAPTER CITATIONS
Cite a specific chapter or article in a book with author(s) of chapter, its title, editor(s) of book, book title, edition, volume, inclusive pages of chapter, location and name of publisher, and year of publication. For references to complete books, give all of above information that is pertinent.

PAPERS IN PRESS
List papers in press in references with journal name and tentative year of publication.

UNPUBLISHED MATERIAL
Cite papers in preparation or submitted for publication, unpublished data, and personal communications in footnotes, not in Reference section. Give names of all authors with manuscript titles if possible.

Addenda: Data acquired after acceptance of paper, by authors or others, cannot be added to text. A brief addendum may be added in proof with approval of Editor-in-Chief. Full expense will be charged to author.

Tables: Construct so that when typeset, they will fit within a single column (3.5 in.). Do not duplicate data. Do not include unnecessary columns of data that can easily be derived from results in tables. Avoid large groups of individual values: average and include appropriate designation of dispersion. Indicate significance of observations by appropriate statistical analysis.

Give each table a descriptive title and an explanatory paragraph with experimental details for understanding without reference to text. Give each column an appropriate heading. Include numerical measurements in column heading.

Number tables with Arabic numerals. Indicate table footnotes with superscript italic letters ([a, b, c], etc.). Clearly designate units of measurement and concentration. Avoid exponential terminology; mM is preferable to 10^{-3} M. If used in column headings, quantity expressed should be preceded, by power of 10 by which its value has been multiplied (10^{-3} × concentration (M)).

Illustrations: Line-cut (graphs and drawings) and halftone (photographs, photomicrographs, electrophoretic patterns, etc.) illustrations are figures. Use figures when salient points need illustration for better comprehension. Describe straight-line functions such as relationships between concentration and absorbance, or Lineweaver-Burk plots when these are linear, in a few lines in text.

Label each figure in pencil with first author's name and figure number on an adhesive label on reverse side. For halftones, indicate TOP.

Line-cut Illustrations (including flow diagrams and complex biochemical structures): Prepare with professional instruments, do not type. They may be on Bristol board, tracing paper or cloth, or coordinate paper printed in light blue. Do not mount on heavy cardboard. Clear, glossy prints are acceptable in lieu of original drawings, provided all parts are in focus. Do not send x-ray films or Polaroid photographs. Original drawings should not be larger than 8.5 × 11 in.

Computer-generated graphs are permissible provided that their quality a is acceptable, e.g., all labeling is clear and scaling is in proper proportion to reproduce legibly when reduced.

Except for especially complicated drawings showing large amounts of data, all line-cut illustrations are one-column (3.5 in.). Submit in this size. For larger ones, see that abscissas, ordinates, lines, and symbols are sufficiently large to permit reduction. When reduced to single column size, letters and numbers must be at least 1.5 mm high. Smallest part of illustration must be discernible. On original artwork, make minimum height for lowercase letters 5 mm; numerals and uppercase letters 6 mm; and symbols 5 mm. Thickness of ruled lines on graphs is also vital for clear presentation of data.

Define symbols in legend. Use only common symbols for which printer has type (× ○ ● ❑ ■ △ ▲). Do not extend lines connecting symbols beyond data points.

Rule graphs off close to area occupied by curve. Mark abscissas and ordinates with appropriate units. Do not extend explanations of coordinates beyond respective lines. Do not box-in graphs with top and right frame lines unless essential for reference. Titles outside drawing waste space; include information in legend. Put curves that may appropriately appear together in a single graph.

Avoid use of exponentials for labeling coordinates in graph. If exponentials must be used, precede quantity expressed by power of 10 by which its value has been multiplied, i.e., 10^3 × concentration (M). "Concentration (M × 10^3)" is not acceptable. If powers of 10 are used, designate in legend how quantity is to be calculated (whether multiplied or divided) to give correct value.

Halftone Illustrations: Submit unmounted and trimmed; exclude all but essential material. Halftone illustrations for printer are made from original negatives. Photographs from other prints are not acceptable. Present karyotypes as cardboard plates onto which chromosome sections from an original photomicrograph are pasted.

All halftones are published at either 1 (3.5 in.), 1.5 (5 in.), or 2 (7.25 in.) column width and placed as close as possible to first citation in text. Prepare halftones within these dimensions to be reproduced without reduction; they will be reduced to conform.

Do not put figure numbers on face of illustration. Group halftones to appear together for comparison under one figure number; letter each section "a," "b," "c," etc., in lower right corner on face of illustration. Composite figures may be mounted on a plate, with sections butted together and tooling (thin white lines) placed between parts. For optimal reproduction, make contrast among photographs on a plate consistent. Do not exceed overall dimensions of 7.25 × 9 in. Indicate minimum dimensions to which plate can be reduced on back.

Make symbols, arrows, or letters in photomicrographs contrast with background. India ink lettering is preferred. If using pressure-sensitive labeling such as Chartpak, Letraset, or Prestyped, place tissue overlays on halftones to protect surface. Indicate important areas of photographs that must be reproduced with greatest fidelity on overlays. Include internal scale markers on photographs themselves not in legend. Magnifications in legend reflects size before reduction.

Color Photographs: If illustration is a composite, mount parts together in a space-saving arrangement. Submit on flexible backing, including mounted composite figures. Submit prints that are of sufficient quality to permit accurate reproduction.

Legends: Required for all figures. Briefly describe data shown; do not repeat details in text. Include staining. Identify all symbols, abbreviations, mathematical expressions, abscissas, ordinates, units, and reference points used.

OTHER FORMATS

ADVANCES IN BRIEF— 3 printed pages (12 double-spaced typescript pages). Include an Abstract (100 words), a one-paragraph Introduction, an abbreviated Materials and Methods section, Results and Discussion sections (may be combined), a maximum of 20 references, and no more than 4 items for display of data (any combination of figures and tables). Short definitive reports of highly significant and timely findings in field.

BRIEF ANNOUNCEMENTS OF SCIENTIFIC MEETINGS AND COURSES IN CANCER-RELATED BIOMEDICAL SCIENCE— Submit at least 3 months prior to expected month of issue.

BRIEF LISTINGS OF RECENT DEATHS OF DISTINGUISHED CONTRIBUTORS TO FIELD OF CANCER RESEARCH.

BRIEF REPORTS OF MEETINGS, SYMPOSIA, AND CONFERENCES ON CANCER RESEARCH—3 printed pagers (12 double-spaced typescript pages). Include statement of purpose(s) of meeting, an integrated summary of findings presented, and recommendations for future research. Include names and affiliations of key speakers.

CONCISE REVIEWS—Submit an outline of proposed article for approval by Editorial Board.

LETTERS TO THE EDITOR—Correspondence about manuscripts published in journal. Do not write about articles not published in *Cancer Research*.

PERSPECTIVES IN CANCER RESEARCH—Invited articles analyzing either very active or undeveloped areas of research and presenting fresh insights and personal viewpoints on where research may or should be heading.

PROCEEDINGS OF SYMPOSIA—Published as external supplements to journal (*Supplements to Cancer Research*). Expenses are assumed by sponsoring agency. Proceedings are accepted for publication based on importance of topic covered, scope of presentations, and participants.

PUBLIC ISSUES—Brief reports on topics of interest to cancer researchers and general public; might include funding for cancer research, training in field, public education, etc.

OTHER

Abbreviations: Limit use to an absolute minimum. Do not abbreviate single words. Do not use in titles. Running titles may carry abbreviations for brevity. Abstracts may contain identified abbreviations for terms mentioned 3 or more times.

Follow recommendations of IUPAC-IUB Commission on Biochemical Nomenclature. Identify all nonstandard abbreviations in an inclusive abbreviation footnote to first abbreviation after Abstract.

Do not use abbreviations that form recognizable words. See *Cancer Res.* 54: 315-316, 1994 for abbreviations that may be used without definition.

For approved terms and abbreviations for chemical substances see: *Biochemical Nomenclature and Related Documents* (International Union of Biochemistry, Portland Press Ltd., 2nd Ed, 1992).

Color Charges: Expense is charged to author. Obtain estimates from AACR Publications Department. Price is dependent upon such factors as size and complexity of illustration.

Data Submission: Present concisely. Large masses of data of peripheral significance to the main thesis will not be published. Deposit with National Auxiliary Publications Service, c/o Microfiche Publications, P.O. Box 3513, Grand Central Station, New York, NY 10163-3513; (516) 481-2300. In a footnote indicate how ancillary material can be obtained. Submit data for review with manuscript.

Drugs: Use generic names; proprietary names may be used after first mention of generic name. Avoid use in titles unless both names can be listed easily. If a foreign proprietary name is used, give comparable U.S. product. If no generic name, give chemical name or formula or describe active ingredients.

Refer to formally adopted generic names listed in *USAN and the USP Dictionary of Drug Names* (1994).

Enzymes: Use Recommended Name in *Enzyme Nomenclature 1992: Recommendations of the Nomenclature Committee of the International Union of Biochemistry on the*

Nomenclature and Classification of Enzymes (Academic Press, Inc., Orlando, FL, 1992). Include Systematic Name or reaction catalyzed. State Enzyme Commission number at first mention.

For isozyme nomenclature, consult *Biochemical Nomenclature and Related Documents*.

Experimentation, Animal: Observe *Interdisciplinary Principles and Guidelines for the Use of Animals in Research, Testing, and Education* (New York Academy of Sciences).

Experimentation, Human: Perform investigations in accordance with principles of Declaration of Helsinki. Include a statement that investigations were performed after approval by a local Human Investigations Committee and in accordance with an assurance filed with and approved by Department of Health and Human Services, where appropriate. Include a statement that informed consent was obtained from each subject or subject's guardian.

General: Specify composition of all solutions and buffers in sufficient detail so that concentration of each component can be determined. Do not use "saline;" use "NaCl solution," with exact concentration. Do not use inexact terms such as "physiological saline" or "phosphate-buffered saline"; give exact contents and concentrations.

Use decimals not fractions; use 0.01, not .01 in text, tables, and illustrations.

Designate ionic charge by a superscript immediately following chemical symbol.

Histones: Label fractions H1, H1°, H2A, H2B, H3, and H4, not F1, F1°, F2a2, F2b, F3, and F2a1, respectively.

Inbred Strains: For designations of inbred mouse strains see "Standardized Nomenclature for Inbred Strains of Mice: Eighth Listing." *Cancer Res.*, *45:* 945–977, 1985; for designations of inbred strains of rats, see "Standardized Nomenclature for Inbred Strains of Rats: Fourth Listing," *Transplantation*, *16* (3): 221–245, 1973.

Interferon Assays: State name, identifying number, and assigned potency of international standard used to calibrate assay, along with observed geometric mean titer of standard, standard deviation of that value, number of titrations performed to obtain that value, and technical details of assay.

Isotopically Labeled Compounds: Indicate radioactive nuclide by mass number as a superscript to left of symbol (^{32}P); written form corresponds to spoken word (phosphorus-32).

For an isotopically labeled compound, place isotopic prefix in square brackets immediately preceding name (word) to which it refers. When more than one position in a substance is labeled by same isotope and positions are not indicated, add number of labeled atoms as a subscript to right of element. Symbol *U* indicates uniform labeling and *G*, general labeling.

Isotopic prefix precedes that part of name to which it refers. Do not contract terms such as "^{131}I-labeled albumin" "[^{131}I]albumin." Do not write ^{14}C-labeled amino acids" as "[^{14}C]amino acids."

With isotopes of more than one element, arrange symbols in alphabetical order. Designate deuterium and tritium as ^{2}H and ^{3}H or as D and T.

When not sufficiently distinguished by foregoing means, indicate positions of isotopic labeling by Arabic numerals, Greek letters, or prefixes in italics, as appropriate; placed within square brackets to appear before symbol of element and attached to it by a hyphen. Symbol indicating configuration precedes bracketed isotope; hyphen is used to separate it from brackets.

The same rules apply when labeled compound is designated by a standard abbreviation or symbol other than atomic symbol. Square brackets are not used with atomic symbols, or when isotopic symbol is attached to a word that is not a specific chemical name, abbreviation, or symbol. Proper usage is: ^{14}CO$_2$, ^{2}H$_2$O, H$_2$^{35}SO$_4$, ^{32}P$_i$, ^{131}I-labeled, ^{3}H-ligands, ^{14}CX-steroids.

Materials Sharing: Make freely available to other academic researchers any cells, clones of cells or DNA or antibodies, etc. used in research and not available from commercial suppliers.

Outbred Animal Stocks: Follow recommendations of Committee on Nomenclature, Institute of Laboratory Animal Resources:" A Nomenclature System for Outbred Animals," Lab. Animal Care, *20:* 903-906, 1970.

Page Charges: $75 nonrefundable submission fee. Page charge is $65 per printed page. Waiver may be applied for upon submission if no source of funds.

Reviewers: Suggest appropriate Associate Editors (from names listed in front of each issue) and reviewers.

Style: Refer to: *Stedman's Medical Dictionary* (25th Ed, 1990, Williams & Wilkins Co., Baltimore, MD); *CBE Style Manual* (5th Ed, 1983, Council of Biology Editors, Inc., Chicago, IL); and The *ACS Style Guide* (1st Ed, 1986, American Chemical Society, Washington, D.C.).

Tumors: Clearly describe and identify in acceptable terminology tumors used in experiments. If well known and identified in previous publications, extended photomicrographs are unnecessary.

In clinical papers use TNM staging system approved by International Union Against Cancer and American Joint Committee on Cancer.

SOURCE
January 1, 1994, 54: 312-318.

Carcinogenesis
Oxford University Press

Dr. A. Dipple and Dr. C.C. Harris,
Carcinogenesis Editorial Office,
NCI-FCRDC, PO Box B,
Frederick, MD 21702
        For The Americas
Dr. R.C. Garner,
Carcinogenesis Editorial Office
University Road,
Heslington, York YO1 5DU, U.K.
        For Europe, Japan and the rest of the world

AIM AND SCOPE

Carcinogenesis is a multidisciplinary journal designed to bring together all the varied aspects of research which will ultimately lead to the prevention of cancer in man. The journal will publish full papers and short communications which warrant prompt publication in the areas of viral, physical and chemical carcinogenesis and mutagenesis; factors modifying these processes such as DNA repair, genetics and nutrition; metabolism of carcinogens; the mechanism of action of carcinogens and promoting agents; epidemiological studies; and the formation, detection, identification and quantification of environmental carcinogens. The editors may, from time to time, invite commentaries or short reviews.

MANUSCRIPT SUBMISSION

Submit original and two copies of manuscript. Submission implies that paper reports unpublished work and is not under consideration for publication elsewhere. If previously published tables, illustrations or more than 200 words of text are included, obtain and submit copyright holder's written permission.

MANUSCRIPT FORMAT

Type on A4 or American quarto paper. Double-space all sections (6 mm between lines of type). Leave 1 in. margins on all sides of each page. Number each page top right (Title Page is l). Avoid footnotes; use parenthesis within brackets. Underline words or letters to appear in italics. Identify unusual or handwritten symbols and Greek letters. Differentiate between letter O and zero, and letters I and l and number 1. Mark position of each figure and table in margin.

Subdivide regular full-length papers into sections: Title page, Abstract, Introduction, Materials and methods, Results, Discussion, Acknowledgments, References, Tables, Legends to Figures.

Abstract: On second page, a 300 word single paragraph. Avoid abbreviations and reference citations.

Acknowledgments: Include at end of text, not in footnotes. Personal acknowledgments should precede those of institutions or agencies.

References: Verify accuracy. Include published articles and those in press (state journal which has accepted them). Arrange in numerical order with numbers in text in brackets and on line (not superscripts). Mention all authors' names. Initials follow name; date precedes title; give title in full; abbreviate name of journal according to *World List of Scientific Periodicals* and underline to indicate italics.

1. Saffhill, R., Margison, G., and O'Connor (1985) Mechanisms of carcinogenesis induced by alkylating agents. Biochim Biophys Acta, 823, 111–145.

2. Hatch, F.T., Felton, J.S., Stuemmer, D.H. and Bjeldanes, L.F. (1984) Mutagens from food. In de Serres, F.J. (ed.) Chemical Mutagens: Principles for their Detection, Plenum Press, New York, vol . 9, pp.111–164.

3. Bennett, M. V. L. and Spray, D. C. (1985) Gap Junctions. Cold Spring Harbor Laboratory Press, Cold Spring Harbor, NY.

Submit authorization for use of personal communications (J. Smith, personal communication). Cite unpublished data as (unpublished data). Use both sparingly and only when unpublished data is peripheral rather than central to discussion. Cite references to manuscripts in preparation, or submitted but not yet accepted, in text as (B. Smith and N. Jones, in preparation). Do not include in reference list.

Tables: Type on separate sheets; number consecutively with Roman numerals. Make self-explanatory and include a brief descriptive title. Indicate footnotes by lower case letters; do not include extensive experimental detail. Place arrow in text margin to indicate insertion in text.

Illustrations: Refer to line drawings and photographs in text as Figure 1, etc. Abbreviate 'Fig. 1.' in figure legend. Write title, name of first author and figure number lightly in blue pencil on back. Indicate TOP. Place arrow in text margin to indicate insertion in text.

Photographs: Submit in desired final size to avoid reduction. Page area is 248 (height) × 185 mm (width); photographs, including legends, must not exceed. Single column is 88 mm wide, double column 185 mm. Make fit either single or double column. Photographs should be of sufficiently high quality in detail, contrast and fineness of grain to withstand loss of contrast and detail inherent in printing. Indicate magnification by a rule on photographs. Color plates are subject to a special charge.

Line Drawings: Provide as clear, sharp prints, suitable for reproduction as submitted. No additional artwork, redrawing or typesetting will be done. Make labelling with a lettering set. Ensure lettering size is in proportion with overall dimensions of drawing. Submit line drawings in desired final size to avoid reduction (248 × 185 mm maximum including legends) to fit either single (88 mm) or double column width (185 mm). If reduction is required check that lettering will still be legible; not smaller than 1.5 mm in height.

Figure Legends: Use a separate, numbered manuscript sheet. Define all symbols and abbreviations used. Do not redefine common abbreviations and others in text.

OTHER FORMATS

SHORT COMMUNICATIONS—2000 words, excluding references and figures. Do not divide into headed sections. Present methods, results and discussion in a single section. Include a short abstract; present acknowledgments and references as in full-length papers. Complete pieces of work which do not justify full-length papers.

OTHER

Abbreviations: Restrict use to SI symbols and those recommended by IUPAC. Define in brackets after first mention in text. Use standard units of measurements and chemical symbols of elements without definition.

Conventions: Follow conventions of *CBE Style Manual* (5th ed, Council of Biology Editors, Bethesda, MD, 1983).

Follow *Chemical Abstracts* and its indexes for chemical names. For biochemical terminology follow recommendations of IUPAC-IUB Commission on Biochemical Nomenclature (*Biochemical Nomenclature and Related Documents,* Biochemical Society, U.K.). For enzymes use recommended name assigned by IUPAC-IUB Commission on Biochemical Nomenclature, 1978 (*Enzyme Nomenclature,* Academic Press, New York, 1980).

Italicize genotypes (underline in typed copy); do not italicize phenotypes. For bacterial genetics nomenclature follow Demerec *et al.* (1966) *Genetics,* **54,** 61–76.

Reviewers: Submit names and addresses of 3 or 4 potential referees in cover letter.

Units: Use SI (Système Internationale) units.

SOURCE

January 1994, 15(1): inside back cover.

Cardiovascular Research
British Cardiac Society
BNW Publishing Group

The Editor, *Cardiovascular Research*
Box 1, The Rayne Institute, St. Thomas's Hospital,
London SE1 7EH, U.K.

AIM AND SCOPE

Cardiovascular Research is a journal concerned with the link between the basic sciences, clinical physiology and clinical cardiology. Thus its purpose is to provide a forum for those engaged in the application of methods of the basic sciences to the understanding of clinical disease.
The Editor of *Cardiovascular Research,* an international journal, invites submission of manuscripts in the form of Original Articles, Letters to the Editor, and Review Articles. Emphasis is given to all aspects of the heart and circulatory system and all scientific disciplines—the journal strives to promote links between fundamental research and clinical cardiology.

MANUSCRIPT SUBMISSION

Submission of a paper implies that it contains original work which has not been published before and is not being submitted for publication elsewhere. Submit four sets of complete manuscript including all original figures.

In cover letter include: manuscript's title and names of all authors; declaration that "the manuscript has neither been published (except in the form of abstract or thesis) nor is it currently under consideration for publication either in whole or in part, by any other journal," and agreement to pay for the printing costs of any color figures submitted.

Either have each submitting co-author sign or above signature of first author state "the manuscript has been read by (each of) my co-author(s), who (have) has approved its submission to *Cardiovascular Research* for publication."

MANUSCRIPT FORMAT

Type double-spaced throughout.

Title Page: Include title (120 characters), a short title for running head, names of authors, place where work was undertaken, name, address and telephone and fax numbers of corresponding author, and (if appropriate) current address of co-authors.

Abstract: 300 words, subdivided into subsections: Objectives, Methods, Results, Conclusions.

Introduction: Outline background and rationale for study.

Methods: Make sufficiently detailed to permit replication of study. Describe published methods briefly with appropriate citation. Give full details of experimental material.

Results: Be concise. Do not repeat methods. Do not replicate data in tables or figures, or vice versa.

Discussion: Clearly distinguish between deduction and speculation.

Key Terms: Up to 10

References: Vancouver style; see journal issue. Except for review articles, total number should not exceed 40.

Figures and/or Tables: Follow journal style exactly; do not use more than one column width. Type figure legends on separate sheets. Keep tables to a minimum; do not include material not commented on in text. Color illustrations can be reproduced at authors' cost.

OTHER

Experimentation, Animal: Include a statement of ethical approval by relevant domestic regulatory authority. In the absence of such an authority, state either that "The investigation conforms with the *Guide for the Care and Use of Laboratory Animals* published by the U.S. National Institutes of Health (NIH publication No 85-23, revised 1985)" or "The investigation was performed in accordance with the Home Office *Guidance on the Operation of the Animals (Scientific Procedures) Act 1986,* published by Her Majesty's Stationery Office, London."

Experimentation, Human: Include statement: "The investigation conforms with the principles outlined in the Declaration of Helsinki" (*BMJ* 1964;ii:177).

Units: Use SI units.

SOURCE

January 1994, 28(1): inside back cover.

Cell
Cell Press, Inc.

Cell Press
50 Church Street
Cambridge, Massachusetts 02138
Fax: 617-661-7061
Laboratory of Eukaryotic Molecular Genetics
National Institute for Medical Research
The Ridgeway, Mill Hill
London NW7 1AA, England
Telephone: 81-906-3897; Fax: 81-913-8527
Manuscripts from Europe

AIM AND SCOPE

Cell publishes reports of novel results in any area of experimental biology. The work should be not only of unusual significance within its field but also of interest to researchers outside the immediate area.

MANUSCRIPT SUBMISSION

Provide four copies of each manuscript. If manuscript is closely related to papers in press or submitted elsewhere, provide copies of these papers. Papers will not be considered that contain any data that have been or will be submitted for publication elsewhere (including symposium volumes).

MANUSCRIPT FORMAT

Be as concise as possible. Text must contain fewer than 55,000 characters; submit no more than 7 figures (which together with legends should fit into 2 pages). Organization and general style of research articles:

Summary: A single paragraph, 120 words.

Running Title: 50 characters.

Introduction: Succinct, no subheadings.

Results and Discussion: May each be divided by subheadings or may be combined. Do not use footnotes; information will be transferred to text.

Italicize genetic loci; do not italicize protein products of the loci or journal names, foreign phrases, or species names. Define nonstandard abbreviations when first used in text. Text should be letter quality and double-spaced.

Experimental Procedures: Include sufficient detail so all procedures can be repeated, in conjunction with cited references.

References: Include only articles that are published or in press. Cite unpublished data, submitted manuscripts, or personal communications within text. Document personal communication by a letter of permission. Do not cite abstracts of work presented at meetings.

Miller, C. (1989). Genetic manipulation of ion channels: a new approach to structure and mechanism. Neuron 2, 1195–1205.

Fallon, J. H., and Loughlin, S. E. (1993). Anatomical localization of neurotrophic factors. In Neurotrophic Factors, S. E. Loughlin and J. H. Fallon, eds. (San Diego, California: Academic Press, Inc.), pp. 1–24.

Figures: Include with each copy of manuscript, a set of figures of sufficient quality for reviewers to judge data. Indicate magnification by a bar scale. Lettering on halftone figures generated by computer should be outside area of halftone. Color figures require reviewers' opinions that they are essential.

Figure Legends and Tables: Include in submitted manuscript as separate sections, following journal style.

OTHER FORMATS

LETTERS TO THE EDITOR—Present short, decisive observations of interest based on published data. They should not be preliminary observations that need a later paper for validation. Submit to Assistant Editor. For sequence analyses, provide database accession numbers and other relevant information.

MINIREVIEWS—Briefly discuss a sharply focused topic of recent experimental research and make it accessible to researchers in other areas. Provide a critical but balanced view of the field. Submit proposals to Reviews Editor.

OTHER

Color Charges: $1000 for first page, $500 for second, and $250 for each additional page.

Materials Sharing: Distribute freely to academic researchers for their own use any materials (e.g., cells, DNA, antibodies) used in published experiments. In cases of dispute, authors may be required to make primary data available to Editor. Nucleic acid and protein sequences as well as X-ray crystallographic coordinates should be deposited in appropriate databases.

SOURCE

October 7, 1994, 79(1): back matter.

Cell Calcium
Churchill Livingstone

Dr. A. Scarpa
Department of Physiology and Biophysics
Case Western Reserve University
School of Medicine
Cleveland, OH 44106
        North America
Prof. R. M. Case
Department of Physiological Sciences
University of Manchester
Manchester M13 9PT, U.K.
        All other countries

AIM AND SCOPE

The chief aim of *Cell Calcium* is to publish articles of high quality quickly and efficiently and so make it the principal forum for research on calcium. *Cell Calcium* publishes work from all branches of science and medicine on the regulation of cell calcium or the regulation by calcium of cell function. This includes the following areas of research: inorganic chemistry of calcium related to biology, calcium transport across membranes, calcium ionophores, Ca-ATPases, calcium accumulation and release by intracellular membranes and organelles, calcium binding proteins, the roles of calcium in contraction, secretion, metabolism, membrane function, cell division, cell communication and cell adhesion, interaction between calcium and cyclic nucleotides and prostaglandins, methodology. Articles on extracellular calcium which relate specifically to these areas will also be considered. Original, research papers will predominate. They may be of any length but must be completed pieces of work—preliminary communications will not be accepted. Hypotheses and Reviews are welcomed. Authors intending to submit a review are advised to communicate their intentions to the Editor so as to avoid possible duplication.

MANUSCRIPT SUBMISSION

Submit three copies of typescript and illustrations. Copyright must be explicitly transferred from author to publisher. Use copyright transfer agreement in second issue each year.

MANUSCRIPT FORMAT

Follow journal style. Type double-spaced throughout on one side of good, white A4 paper with 3 cm margins.

Set out paper as follows: title page, summary, text, acknowledgments, references, tables, captions to illustrations.

Title Page: Give: title of article; initials and name of each author; name and address of department or institution to which work should be attributed; name, address, telephone and fax number of corresponding and reprint request author(s); and sources of grant support.

Summary: 200 words summarizing contents of article.

Text: Use headings appropriate to nature of paper. For experimental papers, follow usual conventions. Other papers can be subdivided as desired; headings enhance readability. Use two categories of headings: major ones typed in capital letters in center of page and underlined; minor ones typed in lower case (with an initial capital letter) at left hand margin and underlined.

References: Verify accuracy. Enter consecutively with Arabic numerals in parentheses in text. List references in numerical order on a separate sheet double- or triple-spaced. For references to journals include author's name and initials (list all authors when six or fewer, when seven or more list only first three and add et al.), full title of paper, journal title abbreviated, using *Index Medicus* abbreviations, year of publication, volume number, first and last page numbers.

1. Mohr, F. C. and Fewtrell C. The relative contributions of extracellular and intracellular calcium to secretion from tumour mast cells. J. Biol. Chem., 1987; 262: 3440-3450.

2. Thastrup, O., Dawson, A. P., Scharff O. et al. Thapsigargin, a novel molecular probe for studying intracellular calcium release and storage. Agents Actions, 1989; 27: 17-23.

Tables: Type double-spaced on separate sheets. Use only horizontal rules. Do not submit as photographs. Place a short descriptive title above and any footnotes, suitably identified, below. Include all units. Ensure that each is cited in text.

Illustrations:

Mark all illustrations by a label pasted on back or by a soft crayon with figure number and author's name; indicate TOP by an arrow. Never use ink of any kind. Do not use paper clips. Type captions double-spaced on separate sheet.

Submit written permission to reproduce borrowed material (illustrations and tables) from original publishers and authors. Acknowledge borrowed material in captions: 'Reproduced by the kind permission of . . . (publishers) from . . . (reference)'.

Line Illustrations: Present a crisp black image on an even white background 5 × 7 in. or no larger than 8 × 10 in.

Photographic Illustrations and Radiographs: Submit as clear, highly contrasted black and white prints (unmounted), sizes as above. Show magnification and details of staining techniques on photomicrographs. Submit x-ray films as photographic prints, made to bring out detail to be illustrated, with an overlay indicating area of importance. Appropriately label figures in capitals. Keep size of letters appropriate to illustration; take into account necessary size reduction.

OTHER

Abbreviations: Avoid in title and abstract. Explain unusual abbreviations at first occurrence in text.

Color Charges: £625 for first page and £425 for subsequent pages.

Experimentation, Animal and Human: A high standard of ethics must be applied in carrying out investigations. In the case of invasive studies in humans, include a statement that research protocol was approved by a local ethical committee.

Proprietary Names: Indicate proprietary names of drugs, instruments, etc.; use initial capital letters.

Spelling: British or American.

Style: Follow International Committee of Medical Journal Editors "Uniform requirements for manuscripts submitted to biomedical journals," (*BMJ* 296: 401–405, 1991).

Units: Express in SI units. Use liter (l) as unit of volume and curie (ci) as unit of radioactivity. Imperial units are acceptable from U.S.A. contributors. See: *Units, Symbols and Abbreviations: A Guide for Biological and Medical Editors and Authors* (Royal Society of Medicine)

SOURCE

July 1994, 16(1): back matter.

Cell & Tissue Research
Springer International

Prof. A. Oksche
Institut fur Anatomie und Zytobiologie
Justus-Liebig-Universitat, Aulweg 123
D-35392 Giessen, Germany
Fax: (0) 641-702-3977

Prof. H. Altner
Institut fur Zoologie, Universitat Regensburg
Universitat 31,
D-93053 Regensburg, Germany
Fax: (0) 941-943-2305

Prof. M. J. Carey
Department of Biological Sciences
The University of Calgary
2500 University Drive NW
Calgary, Canada T2N 1N4
Fax: (403) 220-5055; E-mail: mcavey@acs.ucalgary.ca

Prof. D. E. Kelly
Association of American Medical Colleges
2450 N Street NW
Washington, D.C. 20037-1126
Fax: (202) 825-1125; E-mail: dekelly@aamc.org

Prof. B..Lofts
School of Biological Sciences
University of East Anglia
Norwich NR4 7TJ United Kingdom

Dr. J. F. Morris
Department of Human Anatomy
University of Oxford, South Parks Road
Oxford OX1 3QX United Kingdom
Fax: (865) 27 24 20; E-mail: jfmorris@vax.ox.ac.uk

Prof. Brenda Russell
Department of Physiology and Biophysics
University of Illinois at Chicago
901 S. Wolcott (M/C 901)
Chicago, IL 60612-7342
Fax: (312) 996-6312; E-mail: u24889@uicvm.uic.edu

Dr. Berta Scharret
Department of Anatomy
Albert Einstein College of Medicine
Bronx, NY 10461
Fax: (718) 518-7236

AIM AND SCOPE

Vertebrate and invertebrate structural biology and functional microanatomy, emphasizing neurocytology, neuroendocrinology, endocrinology, reproductive biology, morphogenesis, immune cells and systems, immunocytology, and molecular cell structure.

MANUSCRIPT SUBMISSION

Follow recommendations of International Committee of Medical Journal Editors [*Ann Intern Med* (1988) *108*:258-265] regarding criteria for authorship. Each author must meet all three criteria: conceived, planned, and performed work leading to report, or interpreted evidence presented, or both; written report or reviewed successive versions and shared in their revisions; and approved final version. Each author should have sufficient knowledge of and participation in work that s/he can accept public responsibility for report.

In cover letter, corresponding author must certify that all listed authors quality for authorship and that: no part of work has been published before [except in abstract form or as part of a published lecture, review, thesis, or dissertation (appropriately cited)]; that work is not under consideration for publication elsewhere; that, when accepted for publication, author(s) agree to automatic transfer of copyright to publisher; and that manuscript, or its parts, will not be published elsewhere in any language without consent of copyright holders.

Do not to separate fragments of a study into individual reports; combine for full development of a topic. Make reports efficient and concise. Normal length of accepted reports is less than 20 printed pages, including figures and references. Avoid reports published in series.

MANUSCRIPT FORMAT

Submit manuscripts and accompanying illustrations in triplicate (original typed on one side, photocopies double-sided). Double-space text with wide margins throughout. Daisy-wheel or laser printer output is preferred; do not use low-definition dot-matrix printers. Follow precise format of recent journal issues. Submit three accurate sets of figures; mounted layouts of originals and two sets of good-quality photographs (not photocopies).

Arrange text in sections: Summary; brief Introduction; Materials and methods; Results; Discussion; Acknowledgments; References; Figure legends: Tables. Make text concise and consistent as to spelling, abbreviations, etc. Number all pages consecutively. Indicate desired position of figures and tables in left margin.

First page: title; name(s) of author(s); departmental and institutional affiliation(s). If several institutes and addresses, indicate corresponding author. Furnish telex or fax number. Title should be brief and indicate species studied. Do not use subtitles. Numbered pairs of articles are acceptable only if submitted and accepted together.

Acknowledgments: Disclose all sources of financial support, including salaries, or other considerations emanating from outside and inside institution.

Abstract: 200 words, one paragraph; no references or abbreviations.

Key Words: Immediately after Summary, up to seven topical key words from Thesaurus, published in General Index (1988). List names of species used. If more than three, list only three most important.

Footnotes: If not referring to title or authors, number consecutively and keep to a minimum.

Introduction (not headed): Give pertinent background to study and questions study seeks to answer.

Materials and Methods: Give names and numbers of animals used, conditions under which animals were kept, and anesthetic or other methods used to prevent animal suffering during experimental procedures. For particular chemicals or equipment, give name and location of supplier of particular chemical or equipment in parentheses.

Results: Present research findings supported by statistical or illustrative validation of assertions. No discussion.

Discussion: Set new findings presented in Results in context; interpret with a minimum of speculation.

References: Follow examples: "Land (1975, 1976) has shown . . ." or "(Land 1975, 1976; Sandeman and Markl 1980; Hengstenberg et al. 1982)." References to "unpublished observations" are discouraged; use only for uncontroversial items. Fully attribute observations and "personal communications," give name and initials of source. Submit copies of "In press" or "Submitted" references.

List all references at end of paper:

Single author – list alphabetically and then chronologically; Author and one coauthor– list first alphabetically by coauthor, and then chronologically; first author and more than one coauthor — list chronologically (not alphabetically by second author) only first author's name and "et al." followed by publication year in text. For more than one paper by same author or team of authors published in the same year, put letters a, b, c, etc., after year in both text and reference list.

References to articles in periodicals include: names and initials of all authors, year of publication, complete title of paper, name of journal (abbreviated in accordance with *Index Medicus*), number of volume, and first and last page numbers.

References to books include: names of authors, year, full title, edition, publisher, and place of publication.

Hansen GN, Hansen BL, Jorgensen PN (1987) Insulin-, glucagon- and somatostatin-like immunoreactivity in the endocrine pancrease of the lungfish, *Neoceratodus forsteri*. Cell Tissue Res 248:181–185

Marder E (1987) Neutrotransmitters and neuromodulators. In: Selverston AI, Moulins M (eds) The crustacean stomatogastric nervous system. Springer, Berlin Heidelberg New York, pp 263–300.

Sternberger LA (1986) Immunocytochemistry, 3rd edn. Wiley, New York

Cite articles "in press" in reference list, but not articles "submitted" or "in preparation." Underline species and subspecies names in reference to indicate italicization.

Illustrations: Make number and size consistent with minimum requirement for clarification of text. Previously published figures cannot be accepted. Do not repeat numerical data given in graphs or tables in text.

Number all figures, whether layouts of micrographs, graphs, or diagrams, consecutively. Submit on separate sheets. Mark on reverse with figure number, author's name, and brief title.

Halftone Illustrations: Submit micrographs, shaded drawings, and wash drawings marked for reproduction as sharp, glossy originals, maximally trimmed at right angles to eliminate unnecessary areas in desired final size; for color illustrations, send slides.

Inscriptions: Make 3 mm high and clearly visible. Use Letraset (or equivalent instant lettering) for lettering micrographs, including figure numbers.

Express magnification numerically in legend.

Add calibration bars to microscopic figures even when illustrations are reduced or enlarged.

When possible, mount individual micrographs in economically arranged layouts, without necessarily taking numerical order into consideration. Separate individual figures by no more than 1 mm. Do not place illustrations and legends sideways.

Mount all illustrations on flexible white drawing paper, do not use thick cardboard.

Match single figure layouts to column width (8.6 cm), page width (17.8 cm) or make 12.8 cm wide with legend on side. Maximum layout length is 24 cm, including legend printed at foot. Use maximum possible size only to increase information content; avoid large peripheral areas with little information.

Indicate parts of figures that require particular care in reproduction on a transparent overlay with a soft pencil.

Stereo Illustrations: Refer to detailed instructions in previous issues (e.g., Vol. 266. No. 3, 1991) or consult editors.

Line Drawings: Submit as good-quality prints. Make letters 2 mm high in printed version. Make inscriptions clearly legible. If reduction is necessary, state scale desired.

Legends: Provide brief, self-sufficient explanations of illustrations. Avoid remarks like: "For explanation, see text." Type on a new page after References. Put magnifications in sentences in parentheses, e.g., ($\times$ 500). In legend state, e.g., "...cytoskeleton $\times$ 41000." Give length of calibration bars, e.g., *Bar:* 1.0 µm. Give explanation of labels as "... with Reissner's fiber (*RF*)" *or "RF Reissner's fiber."* Underline "arrows," "*arrowheads,*" and "asterisk."

Tables: Number consecutively in Arabic numerals. Type on separate sheets. At head provide brief, self-sufficient explanation of contents.

OTHER FORMATS

SHORT COMMUNICATIONS—Four printed pages, including tables and figures (no more than two-thirds of one printed page). Brief accounts containing new interesting results.

OTHER

Color Charges: DM 980.00 for first page and DM 500.00 for each additional page.

Experimentation, Animal: State that principles of Laboratory Animal Care (NIH Pub. No. 86-23, revised 1985) were followed, as well as specific national laws. In Materials and methods include description of analgesic and anesthetic procedures followed, as well as humane methods of maintenance and sacrifice.

Experimentation, Human: Include a statement that studies were reviewed by appropriate ethics committees and were performed in accordance with ethical standards of the 1964 Declaration of Helsinki. State that all persons gave informed consent prior to inclusion in study. Omit details that might reveal identity of subjects.

Nomenclature, Animals: Except for common domesticated forms, designate both accepted common name, if there is one, and scientific name. Follow International Rules of Zoological Nomenclature. Consult an expert to ensure it is properly identified and correct name is used. If no generally accepted specific common name, group name may be used with scientific name. For domesticated species use common names alone.

Page Charges: If too many halftone illustrations, authors must contribute to cost.

SOURCE

October 1994, 2 78(1): A7-A8.

Cell Motility and the Cytoskeleton
Wiley-Liss

Editor-in-Chief: B.R. Brinkley, CMC Journal Office
Department of Cell Biology, Baylor College of Medicine,
1 Baylor Plaza, Houston, TX 77030
Telephone: (713) 798-7599; Fax: (713) 790-0545
Jeremy S. Hyams, Department of Biology
Darwin Building, University College London
Gower Street, London WC1E 6BT, England
European editor

AIM AND SCOPE

Cell Motility and the Cytoskeleton is an international journal specializing in the rapid publication of articles concerning all phenomena related to cell motility, including structural, biochemical, biophysical, and theoretical approaches. Topics of interest will include molecular architecture; supramolecular structure; cell shape; interactions of motile systems (e.g., tubulin- and actin-based systems); interactions of cytoskeletal proteins; non-motile roles of motile or cytoskeletal proteins; biochemical, biophysical, and molecular aspects of cytoskeletal proteins and related control- or binding-proteins; genetics and molecular biological approaches to the study of motility and/or the cytoskeleton; the pathology of motile behavior and the cytoskeleton; nuclear-cytoplasmic transport and interchange; membrane structure and receptor transport; action of drugs on motility and the cytoskeleton; prokaryotic flagellar motility; prokaryotic gliding motility; changes in organelle shape (axostyles, costae); microtubule gliding and particle transport; axonal transport; reticulopodial movement in foraminifera, radiolaria; targeted intra-cellular particle transport; actin-based particle transport; bulk cytoplasmic streaming (in protists, plant, animal, and fungal cells); amoeboid motility; movement of tissue cells in vitro or in vivo; endo- and exocytosis; spreading of platelets, tissue cells; morphogenetic movements; the role of motile and cytoskeletal proteins in development; the nuclear matrix; intranuclear movements; mitotic movements (e.g., kinetochore, centrosomes, and particles); cytokinesis; muscular contraction; cytoplasmic contractility; spasmoneme and myoneme contraction; eukaryotic flagellar and ciliary movement; the centriole, centrosome, and flagellar rootlet derivatives.

Mini-reviews should not contain an exhaustive review of an area, but rather a focused brief treatment of a contemporary development or issue in a single area.

MANUSCRIPT SUBMISSION

In a cover letter signed by senior or responsible author state explicitly that manuscript contains original results of author's own work, and that results have been submitted solely to the journal. Results may not have been published in any part or form in another publication of any type, professional or lay. Manuscripts become property of publisher. Acknowledge material reproduced or adapted from any other published or unpublished source. Obtain and submit permission to reproduce copyrighted material. Upon acceptance for publication, author must sign an agreement transferring copyright to publisher, who reserves copyright.

Submit one original typescript and one reproduction set of figures and three photocopies of manuscript and three reviewer's sets of figures. A reviewer's set of figures consists of glossy prints of halftone illustrations and photocopies of line drawings. A reproduction set of figures consists of originals or glossy prints of halftone illustrations, trimmed and labeled, and originals (in India ink) or glossy prints of line drawings.

MANUSCRIPT FORMAT

Type double-spaced on one side of 8.5 × 11 in. good quality white paper with at least 1 in. margins on all sides. Begin each section on a separate page. Number all pages in sequence, beginning with title page.

Title Page: Give informative title, names and affiliations of all authors, institution at which work was performed, name, address, telephone, telex, and fax numbers for all contacts, and a short running title.

Abstract: 250 word factual condensation of entire work; include statements of problem, method of study, results, and conclusions.

Key Words: List 5 or 6 key words or phrases not used in title that will index subject material of article.

Text: Format: Introduction, Materials and Methods, Results, Discussion, Conclusions, Acknowledgments, and References. Define all unusual abbreviations at first mention.

References: Cite references in text by name and date system (when more than two authors, provide only first name followed by "et al."). Include citations such as "in preparation" and "personal communication" parenthetically in text; for personal communications, include institutional affiliation and date. Double-space reference list, put in alphabetical order, and chronological for more than one reference with same authorship. Use a letter suffix if more than one author reference is for same year. Begin each reference with names of all authors and year of publication. For references to journals, give titles of articles in full, abbreviate journal names according to *Index Medicus,* and provide inclusive pagination. For references to books, include all authors' names, year of publication, chapter title (if any), editor (if any), book title, city of publication, and publisher's name.

Blum, J. J., and Hayes, A. (1979): A high-affinity ATP-binding site on 30S dynein. J. Supramol. Struct. 11:117–122.

Ranney, D. E., and Pincus, J. H. (1977): Suppression of stimulating cell activity by microtubule-disrupting alkaloids. In Revel, J. P., Henning, U., and Fox, C.F. (eds.): "Cell Shape and Surface Architecture." New York: Alan R. Liss, pp. 287–294.

Tables: Type double-spaced on separate sheets. Give self-explanatory titles, number in order of appearance with Roman numerals, and key to text.

Illustrations: Make illustrations, with legends, self-explanatory. Cite sequentially with Arabic numerals within text. Type legends double-spaced and consecutively on a separate sheet. Use lowercase Roman letters to designate multiple parts within a figure. Affix a gummed label to back of each figure, indicating author's name, figure number, an arrow showing top, and desired final size. Mount illustrations composed of multiple pieces.

Photographs: Submit photographs, including gels and micrographs, as high-contrast black-and-white glossy prints. Plan micrographs for no reduction. Single-column figures must not exceed 3 5/16 × 9 in.; double-column 6 13/16 × 9 in. Micrographs should not exceed one-half page. Design figure with legend to fit on a single page.

Limit field of micrographs to structures specifically discussed in report. Place a tissue overlay over each micrograph, circle lightly those areas for which continuous tone should be most faithfully reproduced. Be sure that symbols and areas of special interest are not too close to edges.

Square corners. Do not put magnifications in legends; on all prints, indicate scale used (μm scale). Use sans serif style labeling, 3–3.5 mm high, in India ink or securely affixed transfer letters. Explain symbols used on micrographs in figure legends. Figures may be prepared to "bleed," or extend beyond edges of page. Prepare partial bleeds, extending beyond three edges, at 4 3/16 in. × 11 1/4 in. for single-column placement, or 8.5 in. width and between 3.5 and 9 in. length. Prepare full-page bleeds at 8.5 × 11.25 in. No critical detail or labeling should be less than 5/8 in. from any edge of photograph. Specify requests for bleeds.

When submitting gels at reproduction size (single-column width or less preferred), indicate "100%" on back; otherwise, indicate desired percentage of reduction. In absence of information, gels will be reduced at discretion of publisher.

Line Drawings: Provide original black India ink drawings or sharp glossy prints. They will be reduced to one column (6 13/16 in.) or less. Make lettering sans serif and of such size that capitals, numbers, and symbols will reduce to a height of 1.5–1.75 mm. Keep size of lettering in scale with figure.

OTHER FORMATS

VIDEO REPORTS—Manuscripts may be accompanied by video recordings illustrating both general and specific aspects of a motility phenomenon or a series of experimental results. Videos demonstrate dynamic processes only. Submit a videocassette contribution accompanied by a brief report of 1–4 printed pages summarizing important features shown on videocassette and referencing more extended published accounts of work. Do not exceed ten minutes. Contact Editor-in-Chief for details.

VIEWS AND REVIEWS—2–3 printed pages with 10–20 references and one display item (figure, diagram or table). Timely, crisp overviews, intended to highlight topics of current interest in the field. Contact Views and Reviews editor: Conly L. Rieder, Wadsworth Center for Labs and Research, Empire State Plaza, P.O. Box 509, Albany, NY 12201-0509 for further details.

OTHER

Abbreviations: Follow *CBE Style Manual* (5th Ed, Council of Biology Editors, Inc., Chicago).

Color Charges: $950; one to three additional pages are $500 each. Price cycle repeats with every four pages of color. Estimates will be sent after article is accepted.

Nomenclature: Follow Subject Index of *Chemical Abstracts*. Capitalize trade names and give manufacturers' names and addresses.

Units: All measurements are in metric units.

SOURCE

1994, 29(1): 98-100.

Cellular Immunology
Academic Press

Dr. H. Sherwood Lawrence
Infectious Disease and Immunology Division
Department of Medicine
New York University Medical Center
550 First Avenue
New York, New York 10016

AIM AND SCOPE

Cellular Immunology will provide an international medium devoted exclusively to publication of original investigations concerned with the immunological activities of cells in experimental or clinical situations. The scope of the journal will encompass the broad area of in vitro and in vivo studies of cellular immune responses. These include, for example, delayed type hypersensitivity or cellular immunity; immunologic surveillance and tumor immunity; transplantation immunology; autoimmunity, resistance to intracellular microbial and viral infection; immunologic deficiency states and their reconstitution; antigen receptor sites; thymus and lymphocyte immunobiology, etc.

MANUSCRIPT SUBMISSION

Manuscripts may not have been and will not be submitted elsewhere. Submission must be approved by all authors and institution where work was carried out. Obtain and submit approval of persons cited as sources of personal communications.

If accepted for publication, copyright of article shall be assigned exclusively to Publisher.

MANUSCRIPT FORMAT

Type double-spaced throughout (including tables, footnotes, references, and figure captions) on one side of good grade 8.5 × 11 in. white paper, with 1 in. margins on all sides. Mimeographed or duplicated manuscripts must be indistinguishable from good typed copies. Submit three complete copies. Each copy must include all figures and tables.

Number pages consecutively. On Page 1 list title, author(s), and affiliation(s); at bottom of page type short running head title (45 characters, including spaces), and name and complete mailing address (including zip code) of Corresponding author. Page 2: short abstract (100 to 150 words).

Headings: Indicate organization of paper by appropriate headings and subheadings.

Footnotes: Use only when absolutely necessary. Type consecutively, double-spaced, on a separate sheet in order of appearance in text. Use superscript Arabic numbers [1,2], etc. in text citations and list.

Tables: Number consecutively with Arabic numerals in order of appearance in text. Type double-spaced throughout on separate sheets; avoid vertical rules. Supply a short descriptive title below table number. Type footnotes, lettered [a,b], etc. at end of table.

Figures: Number consecutively in text with Arabic numerals. Type legends double-spaced together on a separate sheet. Make original illustrations no larger than 8.5 × 11 in. Submit in finished form suitable for reproduction, planned to fit proportion of printed page (5 × 8 in.). Make lettering professional quality or generated by high-resolution computer graphics and large enough (10–12 points) to be reduced 50–60%. Make drawings with black India ink on tracing linen, smooth surface white paper or Bristol board. High-quality computer graphics may be acceptable. Plot graphs with black India ink on light blue or white coordinate paper; label coordinates properly; ink grid lines to show in black. Submit photographs as glossy prints (5 × 7 or 8 × 10 in.). Do not use staples, paper clips, or pencil marks. Indicate groupings of photographs; if mounting is necessary, use 10- to 12-point flexible board. Authors must defray cost.

References: Cite in text by number in parentheses. List at end of paper in numerical order.

1. Leopardi, E., and Rosenau, W., *Cell Immunol.* 83, 73, 1984.

2. Buckley, R. H., *In* "Monoclonal Antibodies: Probes for the Study of Autoimmunity and Immunodeficiency" (B. F. Haynes and G. S. Eisenbarth, Eds.), pp. 83–95. Academic Press, Orlando, FL, 1983.

3. Katz, D. H., "Lymphocyte Differentiation, Recognition, and Regulation." Academic Press, New York, 1977.

Abbreviate journal names using *Chemical Abstracts Service Source Index*, 1985. Cite unpublished observations and personal communications in text as such; do not include in reference list.

OTHER FORMATS

LETTERS TO THE EDITOR—Must relate to articles in *Cellular Immunology*. Type as in other materials, limit to two pages.

OTHER

Abbreviations: Do not use final periods after abbreviations of measure (cm, sec, kg, etc.) in text or in tables, except for "in.". Use American Chemical Society *Style Guide*, 1986, as a reference.

Equations: Number consecutively; place number in brackets to extreme right of equation. Refer to equations as Eq. [3] or simply [3], except at beginning of a sentence; spell out word Equation. Punctuate equations to conform to their place in sentence syntax.

Spelling and Style: Either American or English style is acceptable; for former consult *Merriam-Webster*, for latter consult *Oxford Shorter Dictionary*.

Symbols: Underline letters that represent mathematical symbols to be set in *italic* type.

SOURCE

August 1994, 157(1): back matter.

Chemical & Pharmaceutical Bulletin
Pharmaceutical Society of Japan

Editorial Department
Pharmaceutical Society of Japan
2-12 15-201 Shibuya

Shibuya-ku, Tokyo 150, Japan
Telephone: 03 3406-3325; Fax: 03 3498-1835

AIM AND SCOPE

The subject matter of *Chemical & Pharmaceutical Bulletin* embraces the chemical and pharmaceutical sciences, including physical and inorganic chemistry, organic chemistry, medicinal chemistry, analytical chemistry, pharmacognosy (chemical), physical pharmacy and so on.

MANUSCRIPT SUBMISSION

At least one author must be a member of the Pharmaceutical Society of Japan and also a regular subscriber to *Chem. Pharm. Bull.* or *Biol. Pharm. Bull.* Qualifications do not apply to foreign contributors from outside Japan. Copyright belongs to Pharmaceutical Society of Japan.

Three types of manuscripts may be submitted: Regular Articles, Notes (3 printed pages), and Communications to the Editor (3 offset-printed pages).

Attach a "Submission Card," available free of charge from Editorial Department of Society on request. Send 2 copies of manuscript (3 copies of Communications).

MANUSCRIPT FORMAT

Submit type on A4-size (or international-size) paper, double-spaced (24 lines per page, 70 strokes per line). Number each sheet.

Title Page: Type title, author's name, research institute and its location (with postal code) on separate lines. Give full author's name. When authors are from different organizations, place superscripts *a, b*... at end of author's name to indicate corresponding research institute. Indicate corresponding author with an asterisk, and mailing address, telephone and fax numbers at bottom left corner of title page in Japanese.

[2nd Page] A 250 word summary and 3–6 key words in order of importance. First 3 key words need to be independent.

[3rd and Subsequent Pages] Present text, acknowledgments and references in order. Provide figures, tables and structural formulae separately on attached sheets.

Figures, Tables, Structural Formula: Figures, tables and structural formula provided by author are used for printing. Number figures and tables serially. Use bold Arabic numbers to identify compounds. List figure titles and legends together on a separate sheet(s).

References and Notes: Place in same section; number in order of appearance. Position each number after last relevant word as a superscript with a right-half parenthesis mark. Collect all References and Notes together at end of text. Journal title abbreviations conform with *Chem. Abstr.*

(1) Present address: *Pharmaceutical Society of* Japan, 2-12-15 *Shibuya, Shibuya-ku, Tokyo 150, Japan.*

(2) Blank R. J., Tojo K., Higuchi T., *J. Pharm. Sci.*, **58**, 1098–1112 (1969).

(3) Cornwell P., Dell C. P., Knight D. W., *J. Chem. Soc., Perkin Trans. 1*, **1993**, 2395–2405.

(4) Cook J. W. (ed.), "Progress in Chemistry," Vol. 4, Butterworths Scientific Publications, London, 1958.

(5) Robert A. B., Sporn M. B., "The Retinoids," Vol. 1. 2, ed. by Sporn M. B., Robert A. B., Goodman D. S., Academic Press, Orlando, 1984, pp. 209–286.

(6) Suzuki A., Yamada H., Abstracts of Papers, The 110th Annual Meeting of the Pharmaceutical Society of Japan, Sapporo, August 1990, Part 2, p. 45.

Indicate special typefaces: italic (underline), bold (wavy underline, small capital (double underline).

OTHER FORMATS

COMMUNICATIONS—3 pages, including figures and tables. Papers are offset-printed in actual size. Use A4-size paper (W 21.0 × H 29.5 cm) with type size 100-115 letters and spaces per line and 45-50 lines per page. Text area is approximately 17 cm wide and 24 cm high. Submit written statement of reason for selection of Communication style on A4-size paper, in English or Japanese.

Type title and headings in all capital letters. Type each author's name with given name first, followed by family name. Capitalize first letter of given name, second and subsequent letters in lowercase letters. Capitalize all letters in family name. Mark name of corresponding author with an asterisk as a superscript to right of name. See *Chem. Pharm. Bull.* 1995, 43(1): vi for style of typed manuscript.

Write or print figures, tables and structural formula directly in actual size in desired locations in text. They must have an appropriate size and line thickness.

For style of references and notes follow regular format, but type volume numbers either in bold or underline.

Write in pencil name, mailing address, telephone number and fax number at bottom left corner of title page.

OTHER

Abbreviations: These may be used without explanation: AIDS, ara-C, ATP, ATPase, cAMP, cDNA, CoA, Da, dansyl, DNA, ED_{50}, EDTA, EGTA, ELISA, ESCA, ESR, FAB-MS, FAD, FMN, GC-MS, G-protein, GTP, HEPES, HOMO, HPLC, IC_{50}, Ig, IR, LD_{50}, LUMO, MO, mRNA, MS, NAD, NADH, NADP, NADPH, NMR, P450, PCR, poly(A), RIA, RNA, rRNA, S.D., S.E., STO, TLC, Tris, tRNA, UV.

Spell out other abbreviations in full at initial appearance, followed by abbreviation in parentheses. For common abbreviations, see *Farumashia* [1992; 28(9): 1055].

Page Charges: ¥720 per printed page for Regular Articles and Notes, ¥5250 for communications to the Editor. Language Correction charge: ¥720. Photographic plates, color art and blocks: actual cost.

Spectral Data: ^{1}H-NMR ($CDCl_3$) δ: 1.25 (d, 3H, J =7.0 Hz, CH_3CH), 3.55 (q, 1H, J =7.0 Hz, $CHCH_3$), 6.70 (1H, m, C6-H). IR (KBr) cm^{-1}: 1720 (C=O), 1050, 910. UV λ_{max} (EtOH) nm (ε): 241 (10860), 288 (9380). UV λ_{max} (H_2O) nm (log ε): 280 (3.25). FAB-MS m/z: 332.1258 (Calcd for $C_{18}H_{20}O_6$: 332.1259). MS m/z (rel. int. %): 332 (M^+, 86),180 (100),168 (77). *Anal.* Calcd for $C_{19}H_{21}NO_3$: C, 73.29; H, 6.80; N, 4.50. Found: C, 73.30; H, 6.88; N, 4.65.

Units: Length (m, cm, mm, μm, nm, Å); mass (kg, g, mg, μg, ng, pg); volume (l, ml, μl); time (s, min, h, d); temperature (°C, K); radiation (Bq, Ci, dpm, Gy, rad); concentration, *etc.* (M, mol/l, N, μg/ml, %, %(v/v), % (w/v), ppm).

SOURCE

January 1995, 43(1): vi-vii.

Chemical Research in Toxicology
American Chemical Society

Professor Lawrence J. Marnett
Department of Biochemistry
Vanderbilt University School of Medicine
23rd Ave. at Pierce, Room 854 MRB
Nashville, TN 37232-0146

AIM AND SCOPE

Chemical Research in Toxicology publishes Articles, Communications, Invited Reviews, and Perspectives on structural, mechanistic, and technological advances in research related to the toxicological effects of chemical agents. In addition, a feature entitled Forum is published once a year. The Journal is intended to provide a forum for presentation of research relevant to all aspects of the chemical basis of toxicological responses. It emphasizes rigorous chemical standards and encourages application of modern techniques of chemical analysis to toxicological problems. It publishes papers devoted to (1) identification of novel toxic agents and reactive intermediates, (2) development of specific and sensitive new methods for detection of modification of biological macromolecules by toxic agents, (3) characterization of the alteration of macromolecular structure and function by interaction with chemical agents, (4) experimental and theoretical studies of chemical factors that control reactivity with specific macromolecules, and (5) metabolism of toxic agents as it contributes to their biological effects. Toxic effects are broadly defined to include toxicity, mutagenicity, carcinogenicity, teratogenicity, neurotoxicity, and immunotoxicity.

MANUSCRIPT SUBMISSION

Submit four copies of manuscript with a transmittal letter. Provide e-mail address with postal and express mail addresses and phone and fax numbers. Provide sufficient numbers of original photographs for reviewer use.

Complete and submit Copyright Status Form (back of first issue of each volume). If not submitted with manuscript, Editorial Office will send one to corresponding author.

Abstracts submitted to ACS Journals as part of an accepted manuscript are published in Advance ACS Abstracts up to 12 weeks before journal is published. Take this into account when planning intellectual and patent activities related to paper.

MANUSCRIPT FORMAT

Type manuscript (or reproduce on a high-quality printer) double-spaced on 22 × 28 cm (or A4) paper on one side only. Original manuscript includes original inked drawings or photographs of structural formulas for direct use. High-quality output from laser printers is acceptable.

Assemble Articles and Communications in order: title page (including full title, byline, and running title), abstract, footnotes, introduction, experimental procedures (materials and methods), results, discussion, acknowledgment, references, tables, figure legends, and figures. Begin each section on a separate page. Number all pages consecutively starting with title page. Invited Reviews, Perspectives, and Forum articles discuss published research and may employ a narrative style. Describe experimental procedures appropriately throughout text .

Title Page: Give a brief and informative title. Do not use trade names of drugs or abbreviations. List full names and institutional affiliations of all authors, and if differentiation is necessary, indicate affiliations of authors by superscript symbols †, ‡, §, etc. Use symbols to indicate author affiliations different from those stated on title page and present address information. Indicate corresponding author by an asterisk. Listing a person as an author implies that s/he has agreed to appear as an author. Provide a brief and informative running title (50 letters and spaces).

Abstract: 300 words in one paragraph, present problem and experimental approach; state major findings and conclusions. Make self-explanatory and suitable for reproduction without rewriting. Do not use footnotes or undefined abbreviations. If a reference must be cited, give complete publication data. Include with all Articles and Communications.

Footnotes: Type double-spaced; number in one consecutive series using superscript Arabic numerals. Cite nonstandard abbreviations as a footnote. Cite unpublished results and personal communications as footnotes. Do not acknowledge financial support in a footnote. Do not mix footnotes and reference citations.

Introduction: State purpose of investigation and its relation to other work. Briefly describe relevant background material. Avoid detailed or lengthy literature reviews.

Experimental Procedures: Describe in sufficient detail to enable others to repeat experiments. Include names of products and manufacturers (with city, state address) only if alternate sources are unsatisfactory. Describe novel experimental procedures in detail; refer to published procedures by literature citation of both original and published modifications. Include purity of key compounds and description(s) of method(s) used to determine purity. For buffers, use terminology such as 20 mM potassium phosphate buffer (pH7.7) containing.... State w/v or v/v when appropriate.

Results: Present concisely. Design tables and figures to maximize presentation and comprehension of experimental data. Do not present the same data in more than one figure or in both a figure and a table. Reserve interpretation of results for discussion section; under some circumstances results and discussion may be combined. It is sometimes desirable to place supplementary data (also subject to review) in a separate section for inclusion in microfilm edition of Journal.

Discussion: Interpret results and relate them to existing knowledge. Do not repeat information elsewhere in manuscript. Avoid extensive literature reviews.

Acknowledgment: Acknowledge financial support, technical assistance colleagues, gifts, etc. Obtain permission from persons whose contribution is acknowledged.

References: Number references to literature in one consecutive series in text. Assign each literature reference one number; place in text as an italicized Arabic numeral in parentheses. Type complete list double-spaced beginning on a separate page after Acknowledgment section.

Keller, G. M., Turner, C. R., and Jefcoate, C. R. (1982) Kinetic determinants of benzo[a]pyrene metabolism. *Mol. Pharmacol.* **22,** 451-458.

Abbreviate journal titles of journals according to *Chemical Abstracts Service Source Index.* List serial publications such as *Methods in Enzymology* and *CRC Critical Reviews in Toxicology* as journals.

References to chapters and monographs:

Koop, D. R., Morgan, E. T., and Coon, M. J. (1982) Purification of multiple forms of rabbit hepatic cytochrome P-450. In *Microsomes, Drug Oxidations, and Drug Toxicity* (Sato, R., and Kato, R., Eds.) pp 85–86, Wiley-Interscience, New York.

Designate submitted manuscripts as "in press" only if formally accepted for publication; place "unpublished results" after names of authors as a footnote in text.

Tables: Tabulate experimental results when this leads to more effective presentation or to more economical use of space. Number tables consecutively with Arabic numerals. Provide a brief title with each table and a brief heading for each column. Indicate units of measure (preferably SI). Round data to nearest significant figure. Include explanatory material referring to the whole table as a footnote to title. Cite footnotes in tables as italicized lowercase letter superscripts. Submit tables that require special treatment, such as insertion of arrows or other special symbols under or over alphanumeric characters, or that contain many structures as camera-ready copy. Cite all tables in text in consecutive order.

Illustrations: Journal pages are produced electronically. Artwork and photographs are scanned using a digital scanner. The scanner is very sensitive, faithfully copying all flaws such as smudges, uneven lines, incomplete erasures, etc.

Submit two sets of original figures (four sets for black-and-white photographs, such as gels, autoradiograms, and electron micrographs) on single sheets of paper no larger than 22 × 28 cm. Submit stereo diagrams in single units with left and right images correctly positioned. Designate blocks of structural formulas as schemes or charts rather than figures. Make figure legends sufficiently descriptive that figure can be understood without reference to text. Write number of figure, scheme, etc., and name of first author on back of each piece of artwork at top.

Line Art: Black and white contrast is important. Use dark black ink on high-quality, smooth, opaque white paper. Ordinary white bond works well. Avoid tracing paper or textured "artist" papers. Submit original artwork or photographic print of original, not photocopies. Artwork will be reduced to fit either a one- or a two-column format; one-column format is preferable Maximum width for single column is 3.25 in., for double column 7 in. Maximum depth is 9. 5 in. Submit artwork which does not have to be reduced. Make lettering no smaller than 9-point (3 mm) type. If artwork must be reduced, keep design simple and use thicker lines and larger lettering. Make symbols clearly discernible; use brackets to indicate magnitude of statistical variation. Do not use complex textures and shading to achieve a three-dimensional effect. If a pattern is needed, use a simple crosshatch design. Artwork produced by a high-quality graphics plotter is better than artwork produced on a dot-matrix or laser printer. If material must be prepared on a dot-matrix or laser printer, use high-quality paper and choose highest resolution available; select 600 dpi rather than 300 dpi.

Photographs: Submit high-contrast continuous-tone prints with a smooth or glossy finish. Submit four original black-and-white photographs, single- or double-column width that will not have to be reduced. Include scale bars on micrographs. Avoid negatives, slides, and viewgraphs. Do not submit screened artwork, halftones, and images having various shades of gray produced on a laser printer. Do not write on front or back of image area; marks may show through.

Color photographs and artwork are acceptable and may be used either in black-and-white drawing or in color. There is an extra charge for printing in color; an estimate will be given upon request. Include a letter acknowledging willingness to defray cost with revised manuscript. Original color figures, negatives, and proofs will not be returned.

Structures: Clearly draw structural formulas; construct for maximum clarity in minimum space. Group original drawings or high-quality reproductions at end of manuscript with other illustrations. Combine structures with the same skeleton but different functional groups into a single formula with an appropriate legend to specify different chain lengths, substituents, etc. Use line formulas whenever possible.

If using ChemDraw program, use preference items: fixed length, 14.4 pt 0.2 in.); bold width, 2.0 pt (0.0278 in.); line width, 0.6 pt (0.0083 in.); tolerance, 3 pt (0.0417 in.); margin width, 1.6 pt (0.0222 in.); hash spacing, 2.5 pt (0.0345 in.); bond spacing, 18% of width. Use single-width bold and dashed lines for stereochemical notation; use 10 pt Helvetica font both for atom labels and for text material. Prepare drawings with page setup at 100% and print with a laser printer on a good quality white paper. Mark each drawing to be reduced to 75% for publication."

OTHER FORMATS

COMMUNICATIONS—Four journal pages (4000 words or equivalent). Preliminary reports of sufficient importance and general interest to justify accelerated publication.

INVITED REVIEWS AND PERSPECTIVES—Invited short accounts of subjects of active current interest

FORUM—Articles highlight important issues in molecular toxicology where controversy exists. Investigators with differing viewpoints are invited to contribute.

OTHER

Abbreviations: Use without periods. Use standard abbreviations. Table 1 lists abbreviations considered standard (*Chem. Res. Toxicol.* 1995;8(1):9A-10A). Keep nonstandard abbreviations to a minimum; define in text following first use. Define nonstandard abbreviations in an "Abbreviations" footnote. Cite footnote once after first use of a nonstandard abbreviation. Some abbreviations considered standard in certain fields are nonstandard for *Chemical Research in Toxicology.* See Table 2 (*Chem. Res. Toxicol.* 1995;8(1):10A). Table 3 (*Chem. Res. Toxicol.* 1995;8(1):10A) lists abbreviations that should not be used.

Analyses: Provide adequate evidence to establish identity and purity for new compounds. Include elemental analysis. State purity of compounds used for biological testing with a description of method used to evaluate it.

Data, Biological: Reference biological test methods or describe in sufficient detail to permit experiments to be repeated. Place detailed descriptions of biological methods in experimental procedures section. Present data as numerical expressions or in graphical form. Give statistical limits (statistical significance) for biological data. If statistical limits cannot be provided, give number of determinations and some indication of variability and reliability of results. Include references to statistical methods of calculation. Express doses and concentrations as molar quantities (e.g., μmol/kg, mM, etc.) when comparisons of potencies are made on compounds having large differences in molecular weights. Indicate routes of administration of test compounds and vehicles used.

Data, Spectral: Include such data for representative compounds in a series, for novel classes of compounds, and in structural determinations. Do not include routine spectral data for every compound. Papers where interpretations of spectra are critical to structural elucidation and those in which band shape or fine structure needs to be illustrated may be published with spectra included. When such presentations are deemed essential, only pertinent sections are reproduced.

Electronic Submission: Manuscripts prepared with either WordPerfect or Microsoft Word can be used. Documents prepared with other word-processing packages will be handled on an experimental basis. Submit disk with final accepted version of manuscript. Version on disk must exactly match final version accepted in hard copy.

Use document mode or its equivalent in word-processing program, i.e., do not save files in a Text Only" (ASCII) mode. Do not include any page-layout instructions such as placement information for graphics in file. Left justify text; turn off automatic end-of-line hyphenation. Use carriage returns only to end headings and paragraphs, not to break lines of text. Do not insert spaces before punctuation. Make references conform to Journal format (*vide infra*). Ensure that all characters are correctly represented: for example, 1 (ones) and l (ells), 0 (zeros) and O (ohs), x (exs) and $\times$ (times sign). Check final copy carefully for consistent notation and correct spelling. Check disk with a virus detection program. Disks containing viruses will not be processed. Label disk with manuscript number and corresponding author name. Provide details of disk on Diskette Description form.

Include all of text (including title page, abstract, all sections of body of paper, figure captions, scheme or chart titles and footnotes, and references) and tabular material in one file, with complete text first followed by tabular material. Use fonts Times and Symbol. Other fonts may not translate properly. For tabular material, obtain column alignment using either single tabs or spaces but not both. If using proportional spacing, tabs are better. Use table format. Do not integrate graphic material into file.

Put graphics such as figures, schemes, etc., in a separate file. Save file in TIFF (tagged image file format), PostScript, or Encapsulated PostScript. Make filename for graphic descriptive of graphic, e.g., Figure 1, Figure 2.

Users with a World Wide Web browser such as Mosaic or Gopher can obtain up-to-date information on submission of softcopy manuscripts: Mosaic–use Universal Resource Locator (URL) address gopher://acsinfo.acs.orgr. Gopher –use address acsinfo.acs.org. Information on submission of softcopy manuscripts is in document "Manuscript Submission on Disk" located in directory "ACS Publications."

Nomenclature: Provide correct nomenclature. Be consistent and unambiguous and conform with current American usage. Use systematic names similar to those recommended by International Union of Pure and Applied Chemistry, the International Union of Biochemistry, Chemical Abstracts Service, and other appropriate bodies. Among the recommendations are *IUPAC Nomenclature of Organic Chemistry* (1978 ed., Pergamon Press, Elmsford, NY); *Enzyme Nomenclature* (1984, Academic Press, New York; *Compendium of Biochemical Nomenclature and Related Documents* (1992, Portland Press, Ltd., London, England. Other important references include "Chemical Substance Index Names" *Chemical Abstracts Index Guide* (1989, American Chemical Society, Columbus, OH); *Ring System Handbook* (1988, American Chemical Society, Columbus, OH). For CA nomenclature advice, consult Manager of Nomenclature Services, Department 64, P.O. Box 3012, Chemical Abstracts Service, Columbus, OH 43210.

In line with Nebert et al. [(1991) *DNA Cell Biol.* 10,1-14] and a desire to standardize P450 nomenclature, use Arabic instead of Roman numerals (i.e., P450 3A rather than P450IIIA, etc.). Cite this document (complete with article title) as a footnote for nomenclature.

Safety: Identify all hazardous chemicals as such. Explicitly state and reference precautions for handling dangerous materials or performing hazardous procedures at beginning of Experimental Procedures section. Include Caution: *The following chemicals are hazardous and should be handled carefully: (list of chemicals and handling procedures or references).*

Style: For general information on preparation of manuscripts for ACS journals see *The ACS Style Guide* (1986, Washington, D.C.).

Supplementary Material: Manuscripts submitted with extensive tabular, graphical, or spectral data should relegate such supplementary material to microfilm edition of journal. Use this resource in the interests of shorter articles, to save journal space and make clearer and more readable presentations. Do not "dump " material to this section; Editors may make downward adjustments in amount of material presented.

Separately identify materials to be placed in microfilm edition with authors' names and manuscript title; submit in a form easily handled for photoreproduction. Put figure captions, titles to tables, and other identifying captions on the same page as figure or table, not on a separate sheet. Number supplementary figures, tables, etc., in order S1, S2, etc. Preferable page size is 22 × 28 cm with readable material aligned parallel with 22 dimension. Figures and illustrative material should be original drawings or black-and-white matte prints (not glossies) of originals. Typed sheets should be originals; good, clear copies with clean characters and good contrast are acceptable. Make type or letter size large enough for easy reading. Refer to *The ACS Style* Guide for more specific information. Place a statement of availability of supplementary data after Acknowledgment section, using format: "Supplementary Material Available: Give description of material (x pages). Ordering information is given on any current masthead page."

SOURCE

January/February 1995, 8(1): 5A-13A

Chemico-Biological Interactions
Elsevier

Dr. T.A. Baillie
Dept. of Medicinal Chemistry
BG-20, School of Pharmacy, University of Washington
Seattle, WA 98195
Fax: (206) 685-3252
         from the Americas and Japan

Prof. P. Moldus
Division of Toxicology
Institute of Environmental Medicine
Karolinska Institutet
Box 210. S-171 77 Stockholm. Sweden
         from other parts of the world

Dr. R.P. Mason
Laboratory of Molecular Biophysics
NIEHS/NIH, P.O. Box 12233
Research Triangle Park, NC 27709
         Free Radicals

Dr. Maged Younes
WHO European Centre for Env. & Health
P.O. Box 1
3720 BA Bilthoven, The Netherlands
Fax: (31) 30 294252
         Review Editor

AIM AND SCOPE

Chemico-Biological Interactions publishes research reports, rapid communications, review articles and commentaries that examine: (a) molecular aspects of cytotoxicity, carcinogenesis, mutagenesis and teratogenesis; (b) molecular mechanisms by which drugs exert their therapeutic or toxic effects. Manuscripts may deal with synthetic or naturally oc-curring compounds and the molecular basis of their interactions with animal systems or components thereof. Manuscripts that are limited to descriptions of such interactions without experimental considerations of possible mechanisms will be considered for publication only if similar interactions have not been previously delineated.

MANUSCRIPT SUBMISSION

In Letter of Submission state that paper has not been published in part and is not being considered for publication elsewhere. State that all authors agree to submission. Include names and addresses of 3-5 potential referees. Submission of a paper implies exclusive authorization of publisher to deal with all matters concerning copyright.

MANUSCRIPT FORMAT

Submit manuscripts in quadruplicate double-spaced on pages of uniform size (preferably 20 × 28 cm). Divide papers sections, headed by a caption (e.g., Introduction, Materials and Methods, Results, Discussion, etc.). Include authors' full names, academic or professional affiliations and addresses on first page, also name, complete address and telephone number of corresponding author. Preferred medium of submission is on disk with accompanying manuscript.

Summary and Key Words: 200-300 words. List 3-6 key words for indexing.

Tables: Type double-spaced on separate pages. Provide with headings on the same page.

Figures: Submit in a form suitable for reproduction, drawn in India ink on drawing paper. Submit photographs, including autoradiographs and electron micrographs as clear black-and-white prints on glossy paper (no smaller than 10 × 12 cm and no larger than 20 x 25 cm). Degree of reduction is determined by publisher; assume that the same degree of reduction will be applied to all figures in the same paper. If illustrations should not be reduced, so state and do not exceed 14 × 20 cm. Type legends for figures double-spaced on a separate page. If illustrations from previous articles or books are used, obtain and submit written permission of author and publisher.

References: Give all references at end of paper. Number references in order of citation in text, not in alphabetical order. References must be complete, including initials of authors(s), full title of paper, volume number, year of publication and first and last page numbers. Abbreviate journal titles *International Series Catalogue*, 1978. To cite books, include: Author(s), chapter title, in: editor(s) (if applicable), complete book title, publisher, place of publication, year of publication, page numbers. All references cited in text must be represented in list of references and, conversely, cite all items in list of references in text.

OTHER FORMATS

ANNOUNCEMENTS—Organizers of relevant meetings may submit announcements for publication free of charge as space permits.

RAPID COMMUNICATIONS—State why paper deserved rapid handling.

REVIEW ARTICLES and COMMENTARIES—Submit outlines of papers for these sections to the Editors.

OTHER

Electronic Submission: Preferred storage medium is a 5.25 or 3.5 in. disk in MS-DOS format; other systems are welcome, e.g., Macintosh (save file in usual manner, do not use option 'save in MS-DOS format'). Submit disk and exactly matching printed version. In case of revision, follow same procedure such that, on acceptance of article, file on disk and printout are identical. Specify type of computer and wordprocessing package used (do not convert textfile to plain ASCII). Ensure that letter 'l' and digit '1' (also letter 'O' and digit '0') are used properly, and format article (tabs, indents, etc.) consistently. Do not leave characters not available on word processor (Greek letters, mathematical symbols, etc.) open, indicate by a unique code (e.g., Greek alpha, @, #, etc., for Greek letter alpha). Use codes consistently throughout text. List codes and provide a key. Do not allow word processor to introduce word split; do not use 'justified' layout. Adhere strictly to general instructions on style/arrangement and, in particular, journal reference style. Obtain further information from Publisher.

SOURCE

October 1994, 93(1): 87-89.

Chest
American College of Chest Physicians

Editor-in-Chief, A. Jay Block
American College of Chest Physicians
3300 Dundee Road
Northbrook, IL 60062-2348.

MANUSCRIPT SUBMISSION

Submit original manuscript on 8.5 × 11 in. heavy-duty white bond paper, Submit three duplicate copies, and four sets of figures. Submit diskette (properly labelled) in Macintosh, DOS, or Windows format in Microsoft Word or WordPerfect (3.5 and 5.25 in.).

Manuscripts may not have been submitted to other journals or news media.

Submit a signed release of copyright to the American College of Chest Physicians. Include title of article being submitted and date. Include signatures of coauthors. On Letters to the Editor include words "For Publication" to serve as assignment of copyright.

List authors' affiliations with or financial involvement in any organization with a direct financial interest in subject. Information will be confidential during review process. If manuscript is accepted, disclosures will be discussed by Editor-in-Chief.

Each author should have participated in work, analysis of data, and writing of manuscript and should attest to this responsibility.

MANUSCRIPT FORMAT

Double-space all copy, including references, legends, and footnotes. Begin each segment on a new page: title page;

synopsis; references; legends; tables. On title page include institution at which work was performed and academic titles of authors. Restrict number of authors to not more than five. Use subheads; format is flexible, but subheads ordinarily include such sections as: Methods and Materials, Case Reports, Results, Discussion.

Abstract: 200 words for articles less than 10 typed pages; 250 words for longer articles. Longer, structured abstract preferred. Start each component on a new line: Study objective; Design; Setting; Patients or Participants; Interventions; Measurements and results; Conclusions.

Key Words: Three to ten; consult *Medical Subject Headings-Annotated Alphabetic List* (National Library of Medicine).

References: Cite consecutively in text as superscript numerals. List on a separate sheet in numerical order. Each should contain, in order: author (last name, initials), title of article (lower case, no quotation marks), source, year of publication, volume, and inclusive page numbers. Abbreviate journal names using *Index Medicus* style.

JOURNAL ARTICLE

1 Codier JF, Chailleux E, Lauque D, Reynaud-Gaubert M, Dietermann-Molard A, Dalphin JC et al. Primary lymphomas: a clinical study of 70 cases of nonimmunocompromised patients. Chest 1993; 103:201-08.

BOOK

2 Cane RD, Shapiro BA, Davison R. Case studies in critical care medicine, 2nd ed. Chicago: Year Book Medical Publishers, 1990: 193

BOOK CHAPTER

3 Tuchschmidt J, Akil B. The lung and AIDS in developing countries. In: Sharma OP, ed. Lund disease in the tropics. New York: Marcel Dekker, 1991; 305-18

Illustrations: Use black India ink for original drawings or graphs. Do not use typewritten or freehand lettering. Have lettering done professionally. Do not send original art work, x-ray films, or ECG tracings. Glossy print photographs are preferred; good contrast is essential.

Number all illustrations and cite in text. Provide a legend for each, list on a separate sheet. Pencil lightly on back or type on a gummed label affixed to back: figure number, title of manuscript, name of senior author, and arrow indicating top. Do not use paper clips. Place in a separate envelope mailed with text.

Color illustrations are encouraged. Journal will pay part reproduction and printing cost. Author's share may be obtained from editorial offices.

Tables: Make self-explanatory; supplement, not duplicate, text. Number consecutively and give titles. Type double-spaced on separate sheets. Do not use more than 12 columns.

OTHER FORMATS

CASE REPORTS—500 words, one or two illustrations, and six references, 50 word unstructured abstract.

OTHER

Abbreviations: List on a separate sheet all abbreviations used with full definitions. Do not repeat in text, place, when appropriate, in footnotes to tables.

Experimentation, Human: Specify whether ethical standards were used in research. Include notation of approval by institutional committee on human research and that appropriate informed consent was obtained from subjects.

Submit signed statements of consent from patients or guardians with photographs if there is a possibility of identification.

Reviewers: Provide names of two or three particularly qualified reviewers in United States or Canada who have had experience in subject.

Style: Follow uniform requirements document published in *New England Journal of Medicine* 1991; 324: 424-28.

SOURCE

July 1994, 106(1): A4.

Chromosoma
Springer International

Peter B. Moens
Department of Biology, York University
4700 Keele Street
North York, Ontario M3J 1P3, Canada
Fax: (416) 736-5731

AIM AND SCOPE

Chromosoma publishes original contributions in the field of nuclear and chromosome research including biochemical, molecular and genetic approaches, and ultrastructural studies. Articles on cytotaxonomy and karyotype descriptions will be accepted only if they are of general interest or present unusual facts. Methodological papers will only be accepted if they are of fundamental importance to chromosome research. In the *Chromosoma Focus* invited mini reviews, comments and meeting reports will be published. Preliminary papers and short notes, as well as papers that merely repeat or confirm already published information will *not* be accepted.

MANUSCRIPT SUBMISSION

Submit one original manuscript typed on one side of sheet only and 2 double-sided photocopies. Send copy of "Manuscript Submission Form" (inside back cover). Submission implies: that work has not been published before (except in abstract form or as part of a published lecture, review, or thesis); that it is not under consideration for publication elsewhere, that its publication has been approved by all coauthors, if any, as well as by responsible authorities at institute where work was carried out; that, if and when manuscript is accepted, authors will transfer copyright to publisher; and that manuscript will not be published elsewhere in any language without consent of copyright holders.

MANUSCRIPT FORMAT

Put author's telex or fax number on title page. In abstract summarize results in a single paragraph. Do not include references or technical details.

References: Restrict number to essential minimum. Do not refer to meeting abstracts. Do not include references cited in text as (in preparation), (submitted), (personal communication), etc. in reference list.

Obtain permission to cite unpublished results of other authors; add a corresponding statement in Acknowledgments.

JOURNAL

Chen TL, Manuelidis L (1989) SINEs and LINEs cluster in distinct DNA fragments. Chromosoma 98:309–316

BOOK

van Holde KE (1989) Chromatin. Springer, New York Heidelberg Berlin

ARTICLE QUOTED FROM A BOOK

Scheer U (1987) Contributions of electron microscopic spreading preparations (Miller spreads) to the analysis of chromosome structures. In: Hennig W (ed) Structure and function of eukaryotic chromosomes. Results and problems in cell differentiation 14. Springer, Berlin Heidelberg New York, pp 147–165

Illustrations: Restrict number and size. Either match size of column width (8.6 cm) or printing width (17.8 cm). Group several figures into a plate which, including legends, should not exceed printing area (17.8 × 24.8 cm). Fold-out pages are accepted, but cost extra as will color illustrations.

Add to all microscopic figures (drawings and photos) a 10 μm bar to indicate magnification; state corresponding microscopic length in legend, not on photo or drawing.

Photomicrographs: Magnification of light microscopic photos must not exceed × 3000.

Present metaphases as karotypes or plates, not both. Supply photographs and other figures as three sets of faultless glossy prints or groups of prints suitable for reproduction without reduction. Photos must be either unmounted or mounted on flexible material. Trim to eliminate unnecessary background. Inscribe original set with detachable letters (Letraset, etc., no ink).

Make lettering about 2 mm high. Indicate on transparent overlays any parts of photos that need special care.

Graphs and Drawings: Submit original drawings on glossy or good-quality prints (no photostatic copies) in desired final size. Letter drawings to give a final height of 2–3 mm after reduction. In coordinate and other inscriptions of graphs capitalize only first letter of first word.

OTHER

Color Charges: DM 1200.00 for first page and DM 600.00 for each additional page.

DNA Sequences: Supply EMBL accession numbers. Manuscript will not be processed without.

SOURCE

March 1994, 103(1): A10.

Circulation
American Heart Association

Circulation Editorial Office
St. Luke's Episcopal Hospital/Texas Heart Institute
Rm B524 (MCI-267), 6720 Bertner Ave.
Houston, TX 77030-6585
Telephone: (713) 794-6585; Fax: (713) 794-6810

AIM AND SCOPE

Circulation publishes original communications of scientific excellence concerned with clinical research on cardiovascular diseases, including observational studies, clinical trials, and advances in applied and basic laboratory cardiovascular research; the journal accepts for publication meritorious basic research that contributes to an understanding of clinical cardiovascular medicine.

MANUSCRIPT SUBMISSION

Submit four compete sets of title page, four copies of manuscript, four sets of original illustrations, five address labels with name and address of corresponding author, and copyright transfer agreement signed by all authors (*Circulation* 90(1): A19, January 1994).

Indicate that all authors have read and approved manuscript and that material has not been published and is not being considered for publication elsewhere. Submit four copies of potentially overlapping prior work, in preparation, submitted, published or in press.

MANUSCRIPT FORMAT

Type double-spaced, including references, figure legends and tables, on one side of page only with 1 in. margins. Do not use proportional spacing or justified margins. Do not exceed seven typeset pages (6000 words; no more than one figure or table for every 750 words). Cite each figure and table in text in numerical order. Cite each reference in text in numerical order and list in Reference Section.

Assemble manuscript in order: title page, abstract page, text, acknowledgments, references, figure legends, tables, figures.

Title Page (page 1, not numbered): Full title of article, and further down page, first author's surname and a short title (50 characters, including spaces). Title, authors' names, authors' affiliations, complete address for correspondence and reprints. Fax and telephone numbers and e-mail addresses.

Abstract (Page 2): 250 words with headings: Background (rationale for study), Methods and Results (describe methods and presentation of significant results), and Conclusions (data interpretation). Do not cite references. Limit use of acronyms and abbreviations. Provide a condensed abstract (100 words) for annotated table of contents.

Text: Introduction, Methods, Results, and Discussion are typical main headings (no Introduction or Summary). Define abbreviations at first mention in text, tables and figures.

Methods section—Give complete name of manufacturers of apparatus used. State species, strain, number and other pertinent descriptive characteristics of animals used in experiments. For surgical procedures on animals, identify preanesthetic and anesthetic agents used with amount or concentration and route and frequency of administration. Paralytic agents, such as curare or succinylcholine, are not acceptable substitutes for anesthetics. For other invasive procedures on animals, report analgesic or tranquilizing drugs used. If none, justify exclusion. Give generic names of drugs.

Acknowledgments: List all sources of support, plus substantive contributions of individuals. Submit written permission from scientists acknowledged. State potential author conflicts of interest.

References: Verify accuracy of references against original sources. List all authors and inclusive page numbers. Cite in numerical order according to first mention in text. Cite personal communication, unpublished observations and submitted manuscripts in text as "(unpublished data, 19XX)." Cite abstracts only if sole source; identify in reference as abstract. "In press" citations must have been accepted for publication; include name of journal of book publisher.

Figures and Legends: Figures are either black and white drawings, graphs or halftones (photographs). Submit photographs as camera ready, unmounted, glossy prints. Good quality computer-generated prints (not photocopies) are acceptable for line art. Make letters and locants uniform in size and style. Indicate figure number, first author, short manuscript title, and TOP of figure lightly in pencil on back. Do not use a separate label. Enclose each set in a separate envelope. Submit four photocopies of each figure and attach one set of copies to each manuscript. Put figure numbers on photocopies. Put scale bar on photomicrographs. Type figure legends double-spaced on a separate page. Define abbreviations or symbols used in figures in figure or figure legend.

Tables: Give brief, informative titles. Number with Arabic number followed by period. Type double-spaced in same size type as text on separate sheets. Do not use vertical lines. Use horizontal lines above and below column headings and at bottom only. Use extra space to delineate sections. Indicate footnotes in order: *, †, ‡, §, ||, ¶, #, **, ††.

OTHER FORMATS

BASIC SCIENCE REPORTS—Studies in experimental animals and in vitro experiments.

BRIEF COMMUNICATIONS—10 double-spaced typed pages, including references, legends, and tables and 2 figures published within 2 months of acceptance. Present important laboratory or clinical findings. Articles on novel applications of established or previously published new method are considered. Do not submit incomplete or preliminary data. Indicate that observation will require validation of further elaboration of associated findings or that a larger experimental series or clinical trial may be desirable.

BRIEF RAPID COMMUNICATIONS—Short articles on clinical research.

EDITORIALS—Generally invited opinions of recognized leaders in cardiovascular medicine and research.

CLINICAL CARDIOLOGY FRONTIERS AND BASIC RESEARCH ADVANCES—Invited internationally recognized authorities prepare concise, state-of-the-art presentations on various topics.

CLINICAL INVESTIGATION AND REPORTS—Studies in human subjects. Case reports will not be accepted.

CORRESPONDENCE—500 words, pertinent, timely letters. Do not cite feature articles or articles appearing in *Circulation* supplement. Authors original *Circulation* articles and cited in letters will be given an opportunity to reply. Replies must be signed by all authors. Figures and tables cannot be included; maximum of 10 references.

CURRENT PERSPECTIVES—Generally invited articles by recognized authorities; spontaneously submitted position papers, reviews and special topics of special interest are considered.

OTHER

Abbreviations: Keep to a minimum. Write out abbreviated phrases at first mention, followed by abbreviation in parentheses.

Drugs: Use generic names of drugs.

Experimentation, Animal and Human: Indicate that study was approved by institutional review committee and that subjects gave informed consent. Indicate that procedures followed were in accordance with institutional guidelines.

Page Charges: $50 per printed page; expense for color reproduction of figures; reprints; $50 per printed page for author alterations.

Reviewers: List in cover letter with address and telephone number.

Style: See *American Medical Association Manual of Style*, 8th ed, Baltimore, MD, Williams & Wilkins, 1989. Follow "Uniform Requirements for Manuscripts Submitted to Biomedical Journals (*N Engl J Med*. 1991;324:424-428).

Units: Use SI units of measure. Change molar (M) to mol/L; mg/dL to mmol/L; and cm to mm. Express units of measure previously reported as percentages as decimal fractions. Measurements currently not converted to SI units in biomedical applications are blood and oxygen pressures, enzyme activity, H+ concentration, temperature and volume. Use SI unit in text followed by conventionally used measurements in parentheses.

SOURCE

January 1994, 90(1): A17–A18.

Circulation Research
American Heart Association

Stephen F. Vatner, MD
Harvard Medical School
New England Regional Primate Research Center
One Pine Hill Drive, PO Box 9102
Southborough, MA 01772-9102
Telephone: (506) 624-0014; Fax: (508) 624-0960

AIM AND SCOPE

Circulation Research provides a medium for bringing together basic research on the cardiovascular system from various disciplines including biology, biochemistry, biophysics, cellular biology, molecular biology, morphology, pathology, physiology, and pharmacology. The journal also will accept for publication manuscripts on clinical research that contribute to an understanding of fundamental problems.

MANUSCRIPT SUBMISSION

Submit four complete sets of title page; four copies of manuscripts and four sets of original figures. If prior work, submitted, published or in press, is potentially overlapping or provides information essential to understanding of a submitted manuscript, include four copies of that work. Submit copyright transfer agreement signed by all authors. Photocopy *Circ Res* 75(1): A9, but original signatures are required. Submit five address labels with name and address of corresponding author and stamped, self-addressed large envelope if manuscript and figures are to be returned.

Provide written documentation that all persons acknowledged have seen and approved mention of their names. Cite all sources of support. State potential conflicts of interest. Indicate that all authors have read and approved submission of manuscript. Indicate that material has not been published and is not being considered for publication elsewhere.

MANUSCRIPT FORMAT

Type double-spaced, including references, figure legends, and tables, on one side of page with 1 in. margins. Do not use proportional spacing or justified paragraphs. Submit no more than 1 figure or table for every 70 words. Do not exceed 6000 words.

Cite each figure and table in text in numerical order. Cite each reference in text in numerical order and list in References Section.

Assemble in order: title page, abstract, text, acknowledgments, references, figure legends, tables, and figures.

Title Page: Include full title and further down, first author's surname and a short title (50 letters). List title, authors' names, authors' affiliations, and complete address for corresponding and reprint request authors. Include telephone and fax numbers, and e-mail addresses.

Abstract: 250 words, state study's rationale, describe methods, present significant results, and a statement of data interpretation. Do not cite references. Limit use of acronyms and abbreviations. Provide three to five key words.

Text: Typical main headings are: Material and Methods, Results, and Discussion (no Introduction or Summary). Define abbreviations at first mention in text, tables, and figures.

Materials and Methods: Supply complete name of manufacturers for apparatus used. State species, strain, number and other pertinent descriptive characteristics of animals used. Describe characteristics of human subjects.

Acknowledgment: Recognize all sources of support for research plus substantive contributions of individuals. Obtain and submit written permission to express appreciation for assistance with research or manuscript. All persons acknowledged must have seen and approved mention of names in article.

References: Verify against original sources, especially journal titles, inclusive page numbers, publication dates, accents, diacritical marks, and spelling in languages other than English. List all authors. Cite in numerical order according to first mention in text. Cite personal communications, unpublished observations, and submitted manuscripts in text as "(unpublished data, 19XX)." Cite abstracts only if sole source; identify in reference as "Abstract." "In press" citations must have been accepted for publication; include name of journal or book publisher.

Figures and Legends: Figures are either black and white drawings, graphs or halftones (photographs). Prepare artwork using professional standards and photographed as camera-ready, unmounted, glossy prints. Good quality computer-generated prints (not photocopies) are acceptable. Make letters and locants uniform in size and style. Flaws will not be corrected.

Submit four sets of prints. Indicate figure number, first author, short title, and TOP of figure lightly in pencil on back, not on a separate label. Enclose each set of prints in four separate envelopes. Prepare four photocopies with figure numbers of each figure and attach one set of copies to each manuscript. Supply a scale bar with photomicrographs. Provide copy for figure legends on a separate sheet double-spaced. Define abbreviations or symbols in figure or figure legend.

Tables: Begin each on a separate sheet, double-spaced. Use Arabic number followed by a period and a brief informative title. Use same size type as in text. Indicate footnotes with symbols in sequence *, †, ‡, §, ||, ¶, #, and then double as necessary. Do not use vertical lines. Use horizontal lines above and below column headings and at bottom of table. Use extra space to delineate sections.

OTHER FORMATS

BRIEF DEFINITIVE COMMUNICATIONS—Six journal pages including illustrations (18 typed pages, double-spaced). Well documented experimental results of unusual interest but circumscribed scope. Must be definitive, not incomplete or preliminary; longer publications using the same data will not be accepted later. Format should resemble that of Rapid Communications.

EXPEDITED PUBLICATIONS—Selected full length articles are published within six months of receipt of manuscript. Do not submit as Expedited Publications. Qualification is based on peer review process.

MINI REVIEWS IN CELLULAR AND MOLECULAR BIOLOGY—3-4 printed pages and 20 references. Figures must be diagrammatic. Highlight timely topics in the area of cell and molecular biology of the cardiovascular system. Primarily invited succinct discussions of particular questions of current interest.

RAPID COMMUNICATIONS—15 typed pages, double-spaced, including references, legends, tables, and figures. Important new laboratory findings reported concisely.

Submit definitive, not incomplete or preliminary reports; longer publications using the same data will not be accepted later.

REVIEWS—Present, for some broad aspect of the cardiovascular or related systems, a timely analysis of its present state, existing problems, and future directions of research. Usually invited and written for general readership; avoid use of highly specialized terminology.

OTHER

Color Charges: Authors are responsible for cost.

Experimentation, Animal: State species, strain, number used, and other pertinent descriptive characteristics. Identify preanesthetic and anesthetic agents used; state amount or concentration and route and frequency of administration for each. Use of paralytic agents, such as curare or succinylcholine, is not an acceptable alternative to anesthesia. Report analgesic or tranquilizing drugs used. If none, justify. Indicate that procedures followed were in accordance with institutional guidelines.

Experimentation, Human: Describe characteristics. Indicate approval by institutional review committee and informed consent of subjects. Indicate that procedures followed were in accordance with institutional guidelines.

Page Charges: $50 per printed page.

Reviewers: Suggest potential reviewers in cover letter. Include address and telephone and fax numbers.

Style: Conform to "Uniform Requirements for Manuscripts Submitted to Biomedical Journals: (*N Engl J Med*, 1991;324:242-428) and use *American Medical Association Manual of Style* (8th ed, Baltimore: Williams & Wilkins)

Units: Use Système International. Express molar (M) as mol/L, mg/dL as mmol/L, cm as mm, and percentages as decimal fractions. Express units of measure previously reported as percentages as decimal fractions. Measurements not converted to SI units are blood and oxygen pressures, enzyme activity, H^+ concentration, temperature, and volume. Use SI unit in text, followed by conventionally used measurement in parentheses. Make conversions before submission.

SOURCE

July 1994, 75(1): A7-A8.
Instructions to Authors appear in January and July issues.

Circulatory Shock
Wiley-Liss

Dr. C. Victor Jongeneel
Ludwig Institute for Cancer Research,
Lausanne Branch, CH-1006
Epalinges, Switzerland
Telephone: +41-21-692-5851; Fax: +41-21-652-7705;
E-mail: vjongene@isrec-sunl.unil,ch

AIM AND SCOPE

Circulatory Shock will accept original contributions concerned with significant new developments in research on inflammation, shock and related processes. The research may deal with the physiology, cell biology, molecular biology and immunology of inflammation, shock and their mediators. Short papers on a unique finding, new phenomenon, or novel technique will also be considered. Concise review articles and position papers are invited.

MANUSCRIPT SUBMISSION

Manuscripts must be submitted solely to this journal, may not have been published in any part or form in another publication of any type, professional or lay, and become property of publisher. Upon acceptance, author will be requested to sign an agreement transferring copyright to publisher, who reserves copyright.

MANUSCRIPT FORMAT

Submit original and two copies of manuscript (including tables and illustrations), plus a diskette containing text and tables. Type text on one side of good quality 8.5 × 11 in. paper with at least 1 in. margins. Double-space everything. Start a new page for each major division. Number all pages in sequence beginning with title page. Arrange copy in order.

Title Page: Include complete title, names and affiliations of authors, institution at which work was performed, current name, and address for correspondence, and a running head (45 characters).

Abstract: 100–150 words summarizing major findings and conclusions.

Key Words: Submit five to ten key words for indexing. Do not repeat words or terms used in title.

Text: Follow format: introduction, materials and methods, results, discussion, and conclusions. Use subheadings and paragraph titles whenever possible. Mention trade names, with manufacturer and location in Methods section. Place acknowledgments as last element of text, before references.

References: Cite references in text consecutively by numbers in brackets. In final list, put in numerical order, include complete title of article cited, and names of all authors. Journal abbreviations should follow *Index Medicus* style. Do not use all capitals, do not underline.

JOURNAL ARTICLES

1. Zhang H, Vincent J-L: Oxygen extraction is altered by endotoxin during tamponade-induced stagnant hypoxia in the dog. Circ Shock 40:168-176, 1993.

BOOKS

2. Clark DA: Coronary Angioplasty. New York: Wiley-Liss, 1991.

ARTICLES IN BOOKS

3. Kern MJ, Aguirre FV, "Donohue T: Pressure wave artifacts. In Kern MJ (ed): "Hemodynamic Rounds: Interpretation of Cardiac Pathophysiology From Pressure Waveform Analysis." New York: Wiley-Liss, 1993, pp 199-206.

Tables: Number in order of appearance with Roman numerals. Each must have a title and be keyed into text.

Legends: Include with each illustration. Define all abbreviations used therein.

Illustrations: Glossy black-and-white photographs 3 & 5/16 × 8.75 in., single column, or 6 & 3/16 × 8.75 in. double-column in size are preferred. Color will be printed at author's expense. Do not submit original recordings, graphs, radiographic plates, or art work. They will be requested at a later date if needed. Lettering must meet professional standards and be legible after reduction; typewritten or hand lettering is unacceptable. Number in order of appearance with Arabic numerals. Identify each on back by affixing a gummed label which lists: number of illustration, name of illustration, name of first author, title of manuscript, and an arrow indicating TOP.

OTHER

Abbreviations: Follow Guidelines in *Council of Biology Editors Style Manual* (5th ed, Council of Biology Editors, Inc., Chicago, IL). Use generic names for all drugs and pharmaceutical preparations.

Color Charges: First page is $950. Second and subsequent pages, up to four, $500.

Experimentation, Animal: State adherence to NIH guidelines for use of experimental animals.

Experimentation, Human: State approval by institutional human experimentation review committees.

SOURCE

May 1994, 43(1): inside back cover; revised July 1994.

Cladistics
Willi Hennig Society

R. T. Schuh
Department of Entomology
American Museum of Natural History
Central Park West at 79th Street
New York NY 10024

AIM AND SCOPE

The Society publishes papers on theoretical or practical cladistics. It has wide scope and publishes papers in zoology, botany, morphology, molecular biology, ontogeny, biogeography, ecology and systematic philosophy.

MANUSCRIPT SUBMISSION

Manuscripts contents must not have appeared, nor will not appear, elsewhere in substantially the same or abbreviated form.

The Society does not specify length of manuscripts, but do not exceed length of an individual journal issue. If there is any doubts as to suitability, authors of long manuscripts should contact editor before submission.

MANUSCRIPT FORMAT

Type manuscripts on one side only, double-spaced, on A4 (208 × 298 mm) or equivalent paper. Make all margins at least 25 mm. Submit two copies, plus original, in final fully corrected form. With multiple authors, identify corresponding author.

Title Page: Include title, authors, institutions, and short running title. Make title concise but informative. Where appropriate, mention families or higher taxa. A subtitle may be used. Do not submit papers in numbered series. Include names of new taxa in titles.

Abstract: 100-200 words on a separate page in a form intelligible in conjunction with title. Do not include references.

Table of Contents: Provide list of first, second and third order headings on a separate sheet. A complete hierarchy of headings is desirable.

Subject Matter: Divide paper into sections under short headings. Except in systematic hierarchies, hierarchy of headings should not exceed three. Follow Botanical and Zoological codes, where appropriate. Underline names of genera and species to indicate italic; do not underline suprageneric taxon names. Use SI units and appropriate symbols (e.g., mm, not millimeter; μm, not micron; Myr for million years). Original published cladograms should be accompanied with a data matrix and a list of characters where appropriate. Avoid elaborate data matrices, species lists and character lists within text; include in appendices. Avoid footnotes; keep cross references to a minimum.

References: Citations in text should read: (Croizat et al., 1974; Mickevich, 1981) . . ., Farris (1983) . . ., Hennig (1983) said . . ., Nelson and Platnick (1981) . . ., Nobel (1976: 925) . . ., (Rosen, 1976, 1978). Names of joint authors are connected by "and" in both text and literature cited. Reference need not be cited when author and date are given as authority for a taxonic name. List references alphabetically; abbreviate journal titles according to *Serials Sources for the BIOSIS Data Base* (Biological Abstracts). For books, give title, name of publisher, place of publication and indication of edition if not first. Follow conventions used in recent journal issues for reference style.

Give foreign language references in ordinary English alphabetic form but copy accents in French, German, Spanish, etc. For Cyrillic alphabet use British Standard 2979 (1958). If only a published translation has been consulted, cite translation, not original.

Tables: Keep as simple as possible. Avoid vertical rules. Type legends in numerical order on a separate sheet.

Illustrations: Include black and white text figures reproduced from ink drawings or half-tone illustrations from black and white photographs. Use one consecutive set of Arabic numbers for all figures (do not separate "Plates" from "Text figures," treat all as "Figures"). Drawings, including cladograms, should be in India ink on smooth white card such as bristol board, or on draughtsman's tracing paper. Make lines clean and heavy enough to stand reduction; drawings should be no more than twice page size. Place explanations in legend; drawing itself should contain a minimum of lettering. Submit photographs in final size. Include scale line to indicate magnification. If lettering is necessary and cannot be undertaken competently, indicate on a duplicate print in exact positions required. Type legends for figures in numerical order on a separate sheet.

Obtain permission to reproduce figures from other published works before paper is submitted.

SOURCE

March 1993, 9(1): back matter.

Clinica Chimica Acta
Elsevier Science Publishers

Professor M. Werner
2819 McGill Terrace NW
Washington D.C. 20008
Professor D. W. Moss
Department of Chemical Pathology
Royal Postgraduate Medical School
Hammersmith Hospital
London W12 0NN U.K.

AIM AND SCOPE

Clinica Chimica Acta publishes original Research Communications in the field of clinical chemistry and medical biochemistry, defined as 'the application of chemistry, biochemistry, immunochemistry and molecular biology to the study of human disease in cells, tissues or body fluids'. The objective of the journal is to publish novel information leading to a better understanding of biological mechanisms of human diseases, their diagnosis and treatment. Reports of an applied clinical character are also welcome. Papers concerned with normal metabolic processes or with constituents of normal cells or body fluids, such as reports of experimental or clinical studies in animals, are considered if they are clearly and directly relevant to human disease.

MANUSCRIPT SUBMISSION

Submit one original and two copies of all typescripts. Place photographs and transparencies in a separate heavy paper envelope. Include a cover letter from corresponding author stating that typescript has been seen and approved by all authors. Give any additional information that may be helpful to editor and whether author(s) will meet cost of reproducing color illustrations. Include copies of permissions needed to reproduce published material or to use illustrations of identifiable subjects.

Submitted typescripts should not have been published previously and should not be under consideration for publication elsewhere. Consult any member of Editorial Board, if in doubt about any aspect of scope, format or content of a proposed paper.

MANUSCRIPT FORMAT

Use white bond paper, 8 × 10.5 or 8.5 × 11 in. or ISO A4 on one side, with wide margins of at least 2 in. Use double-spacing throughout. Begin each section on a separate page: title page and key words, summary, text, acknowledgments, references, individual tables and legends. Number pages consecutively, beginning with title page. Type page number in upper right corner of each page.

Title Page: Include title of article, concise but informative; first name, middle initial and last name of each author; name of department(s) and institution(s) to which work should be attributed; key (indexing) words, 3 to 6, use terms from *Medical Subject Headings* list from *Index Medicus;* disclaimers, if any; name and address of corresponding and reprint request authors, if other than first author, or statement that reprints will not be available.

Summary: 150 words on second page stating aims of investigation, basic procedures (study subjects or experimental animals and observational and analytic methods), main findings (give specific data and statistical significance, if possible) and principal conclusions. Emphasize new and important aspects of study or observations. Avoid use of abbreviations.

Text: Usually, but not necessarily, divided into the following sections Introduction, Materials and Methods, Results and Discussion. Use subheadings to clarify content of sections.

Introduction: State aims of investigation. Summarize rationale for study and hypothesis tested. Give only strictly pertinent references; do not review subject extensively.

Materials and Methods: Describe selection of experimental subjects, including controls. Identify methods, apparatus (manufacturer's name and address in parentheses) and procedures in sufficient detail to allow other workers to reproduce results. Reference established methods, including statistical methods; provide references and brief descriptions of methods that have been published but are not well known; describe new or substantially modified methods, give reasons for using them and evaluate their limitations.

Results: Present results in logical sequence in text, tables, and illustrations. Do not repeat in text data in tables or illustrations, or both; emphasize only important observations.

Discussion: Emphasize new and important aspects of study and conclusions that follow. Do not repeat in detail data in Results. Include implications of findings and their limitations. Relate observations to other studies. Link conclusions with aims of study. Avoid unqualified statements and conclusions not completely supported by data. Avoid claiming priority and alluding to incomplete work. State new hypotheses when warranted, but clearly label as such. Include recommendations, when appropriate.

Acknowledgments: Keep brief. Acknowledge only persons who have made substantial contributions to work. Obtain written permission from everyone acknowledged by name. Acknowledge source(s) of support in the form of grants, equipment, drugs.

References: Number consecutively in order first quoted in text. Identify references in text, tables and legends by Arabic numerals [square brackets]. Number references cited only in tables or legends in accordance with sequence established by first identification in text of table or illustrations.

Type list double-spaced on a separate sheet. Follow system used in "Uniform requirements for manuscripts submitted in biomedical journals" *(Br Med J* 1982;284:1766–1770).

List all authors when six or less; when seven or more, list only first three and add 'et al'. Do not include references to personal communications, unpublished data or manuscripts either 'in preparation' or 'submitted for publication'. When a reference is to data 'in press' submit a copy of accepted manuscript.

For abbreviations of journal titles follow *Bibliographic Guide for Editors and Authors* (American Chemical Society, 1974). Avoid using abstracts as references. Do not include 'unpublished observations' and 'personal communications' in reference list; insert references to written, not verbal, communications (in parentheses) in text. Include in references manuscripts accepted but not yet published; designate journal followed by 'in press'. Cite information from manuscripts submitted but not yet accepted in text as 'unpublished observations' (in parentheses). Submit copies of accepted or submitted manuscripts. Verify references against original documents.

JOURNALS

Standard Journal Article

1 Virii MA. Circulating prostatic acid phosphatase-immunoglobulin complexes in Sjogrens syndrome. Clin Chim Acta 1985;151:217–222.

Corporate Author

2 The Committee on Enzymes of the Scandinavian Society for Clinical Chemistry and Clinical Physiology. Recommended method for the determination of gamma-glutamyltransferase in blood. Scand J Clin Lab Invest 1976;36:119–125.

3 Anonymous. Epidemiology for primary health care. Int J Epidemiol 1976;5:224–225.

BOOKS AND OTHER MONOGRAPHS

Personal Author(s)

4 Akatsu T. Artificial heart: total replacement and partial support. Amsterdam: Elsevier/North Holland, 1980;217–240.

Corporate Author

5 American Medical Association Department of Drugs. AMA drug evaluations. 3rd edn. Littleton: Publishing Sciences Group, 1977.

Editor, Compiler, Chairman as Author:

6 Rhodes AJ, Van Rooyen CE, comps. Textbook of virology; for students and practitioners of medicine and the other health sciences. 5th edn. Baltimore; Williams & Wilkins, 1986.

Chapter in Book:

7 Winder AF. Factors influencing the variable expression of xanthelasmata and corneal arcus in familial hypercholesterolaemia In: Bergsma D, Berman ER, eds. Genetic diseases of the eye. New York: A.R. Liss, 1982;449–462.

Agency Publication:

8 National Center for Health Statistics. Acute conditions: incidence and associated disability. United States July 1968–June 1969. Rockville, Md.: National Center for Health Statistics, 1972. (Vital and health statistics. Series 10; Data from the National Health Survey, No. 69) (DHEW publication No. (HSM) 72–1036).

OTHER ARTICLES

Newspaper Article:

9 Shaffer RA. Advances in chemistry are starting to unlock mysteries of the brain: discoveries could help cure alcoholism and insomnia, explain mental illness. How the messengers work. Wall Street Journal 1977 Aug 12: 1(col 1),10(col 1).

Magazine Article:

10 Roueche B. Annals of medicine: the Santa Claus culture. The New Yorker 1971 Sept 4;66–81.

Cite key references only. Exhaustive lists are not accepted. Include about 20 references.

Tables: Type each double-spaced on a separate sheet. Do not submit as photographs. Number consecutively with Roman numerals. Supply a brief title for each. Give each column a short or abbreviated heading. Place explanatory matter in footnotes, not in heading. Explain in footnotes all non-standard abbreviations used. For footnotes, use symbols in sequence: *, **, ***, or ^{a}, ^{b}, ^{c}, ^{d}, etc. Identify statistical measures of variation such as S.D. and S.E.M. Omit vertical rules.

Cite each table in text in consecutive order.

If data from another published or unpublished source is used, obtain permission and acknowledge fully.

Examine issues of *Clinica Chimica Acta to* estimate how many tables to use per 1,000 words of text.

Editor may recommend that tables containing important backup data too extensive to be published be deposited with the National Auxiliary Publications Service or be made available by author(s). A statement will be added to text. Submit tables for consideration with manuscript.

Illustrations: Supply one original and two duplicate sets of professionally drawn figures presented as sharp, glossy black-and-white photographic prints, with high contrast, usually 5×7 in. but no larger than 8×10 in. Make letters, numbers and symbols clear and even throughout and of sufficient size to ensure legibility after reduction. Put titles and detailed explanations in legends.

Indicate magnification by a line representing actual scale of reproduction (0.1 μm, 1 μm or 10 μm).

Illustrations will not be redrawn. Provide line drawings suitable for direct reproduction. Prepare with black ink on white paper. They should be completely and consistently lettered. Make lettering appropriate to drawing, taking into account necessary reduction in size (preferably not more than one third).

Mark each illustration clearly on reverse side with name of author(s), number of illustration and TOP; use a soft pencil or felt-tipped pen.

Photomicrographs must have internal scale markers. Symbols, arrows, or letters should contrast with background. If photographs or persons are used, either subjects must not be identifiable or submit written permission to use photograph.

Cite each figure in text in consecutive order. If a figure has been published, acknowledge original source and submit written permission from copyright holder to reproduce. Permission is required regardless of authorship or publisher, except for documents in public domain.

For color illustrations in color, supply color negatives or positive transparencies and accompanying drawings marked to indicate region to be reproduced. Send two positive color prints to assist editors in making recommendations. Author must pay for extra cost.

Legends for Illustrations: An illustration with its legend should be understood without reference to text. Legends must contain sufficient information about methods and experimental design to make them self-explanatory.

Type double-spaced. Begin on a separate page. Use Arabic numerals corresponding to illustrations. Identify and explain symbols, arrows, numbers, or letters. Explain internal scale and identify method of staining in photomicrographs.

OTHER FORMATS

SHORT COMMUNICATIONS—Complete reports of limited scope offering conclusive results (1500 words, 2 figures and 2 tables and a compact current list of 15 references) and brief reports (1000 words, no summary) of new or improved methods.

Reports of new or improved methods. Identify element of novelty claimed and advantages over existing technology. Document performance characteristics, including effects of interfering substances, comparisons with results of accepted methods and reference values based on appropriate population samples by adequate data. Cite earlier publications; do not repeat details for reagents, procedures, etc., already in print. Provide sufficient information to allow readers to duplicate work or to compare technique with current practice. Instrument and kit evaluations usually will not be accepted unless a new principle is involved. State in cover letter why paper should receive priority handling.

OTHER

Abbreviations: Avoid in title and Summary. The full term for which an abbreviation stands precedes its first use in text unless it is a standard unit of measurement.

Drugs: Identify precisely all drugs and chemicals used, including generic name(s), dosage(s) and route(s) of administration.

Electronic Submission: Typescripts may be submitted on diskette. Submit typescript, plus one copy, floppy and a list of nonstandard characters. File format must be IBM compatible or Macintosh. Do not leave non-reproducible characters as a blank space; replace by character(s) not used elsewhere; use consistently. See: *Chicago Guide to Preparing Electronic Manuscripts* (The University of Chicago Press, Chicago, 1987).

Experimentation, Animal: Indicate whether institution's or *National Research Council's Guide for the Care and Use of Laboratory Animals* was followed.

Experimentation, Human: Indicate whether procedures followed were in accord with ethical standards of committee on human experimentation of institution in which experiments were done or in accord with the Helsinki Declaration of 1975. Do not use patients' names, initials, or hospital numbers.

Nomenclature: Follow *Information for Contributors to Biochimica et Biophysica Acta* (BA Editorial Secretariat, P.O. Box 1345, 1000 BH Amsterdam, The Netherlands).

Statistical Analyses: Detailed statistical analyses, mathematical derivations, etc. may sometimes be suitably presented as an appendix. Include numbers of observations and statistical significance of findings when appropriate.

Units: Follow international practice relating to use of SI units. State concentrations of solutes of known molecular mass in mol/l or recognized submultiples thereof (nmol/l, etc.). Express other solutes in g/l, mg/l, etc. Specify reagent composition either in molar terms or in mass or volume of each solute per liter of final solution. Do not use % or w%. Report enzyme activities in katals or U/l whenever possible, accompanied by a reference to, or a description of, procedure used for measurements.

SOURCE

October 14, 1994 230(1): inside back cover refers to 223(1993): 197-203.

Clinical and Experimental Immunology
British Society of Immunology
Blackwell Scientific Publications

Dr. R A Thompson
East Birmingham Hospital
P.O. Box 1696
Bordesley Green
Birmingham B9 5PZ

AIM AND SCOPE

Papers describing original research on the role of immunology in disease are welcome, including case reports of unusual patients with immunological disorders, the characterization of new, or the clarification of existing disease entities, the use of immunological investigations in clinical management, new forms of clinical therapy involving immune manipulation, important new clinical diagnostic methods, and studies which throw light on the immunopathogenesis of disease. These may include experimental studies of animal disease and relevant experimental forms of immunotherapy. Since a better understanding of normal immunity is a necessary prelude to understanding immunopathology, research which throws light on the normal human immune system will also be welcome.

MANUSCRIPT SUBMISSION

Send three complete copies of text, tables, and figures. Retain one copy or original. No substantial part of the paper may have been, or will be, published elsewhere. Once accepted, papers become copyright of journal.

MANUSCRIPT FORMAT

Articles may not exceed six printed pages (approximately 3000 words, three tables and three figures or equivalent).

Type double-spaced on one side of standard (A4—30 × 21 cm) paper with 30 mm margins all round. On a separate sheet, give title of paper, a short title, and name of institution where work was done; all authors with initials. List 5 keywords for indexing.

Divide into parts in order: Summary (250 words); Introduction (reasons for doing work); Materials and Methods (present sufficient information to permit repetition of experimental work); Results (give concisely; do not use tables and figures to illustrate the same results); Discussion (presentation of results is separate from a discussion of their significance; do not repeat results); Acknowledgments (including financial support); References.

References: References in text appear as numbers in square brackets,. Reference list is in numerical, not alphabetical order. Give full title of paper with first and last page numbers. Abbreviate journal names according to *Index Medicus.*

1 Shingu M, Hurd ER. Sera from patients with systemic lupus erythematosus reactive with human endothelial cells. J Rheumatol 1981; **8:**581–6.

2 Zavazava N, Halene M, Westphal E, *et al.* Expression of MHC class I and II molecules by cadaver retinal; pigment cells: optimization of post-mortem HLA typing. Clin Exp Immunol 1991; **84:**163–6.

3 Kearse KP, Kaplan AM, Cohen DA. Role of cell surface glycoproteins in the formation of T-Cell: APC conjugates. In: Schook LB, Tew JG, eds. Antigen presenting cells: diversity, differentiation, and regulation determined in mouse surrogates. New York: Alan R. Liss, 1988:221–34.

4 Virella G, Goust JM, Fudenberg HH. Introduction to medical immunology, 2nd Edn. New York: Marcel Dekker, 1990.

Do not put work not yet accepted for publication or personal communications in reference list. Include work accepted for publication with journal in which it is to appear.

Illustrations: Label with figure number and author's name in soft pencil on back. Identify TOP edge. Photographs should be glossy bromide prints of good contrast, suitable for reproduction. Indicate masking instructions or lettering, arrows, etc. on an overlay.

Submit three copies of all figures with lettering and shading. Number figures (Arabic numerals) *seriatim.* Type legends on a separate sheet. Authors will bear cost of color prints.

Type tables on separate sheets; give Arabic numerals. Put units in parentheses in column headings but not in body of table. Repeat words or numerals on successive lines; do not use 'ditto'.

OTHER

Experimentation, Human: Papers involving human subjects should receive local Ethical Committee approval.

Nomenclature: Use CD nomenclature for all cell differentiation antigens, with or without alternative names which give information about molecule's function. Name complement receptors CR 1, etc., followed by correct CD terminology in parentheses: CR1 (CD35); CR2 (CD21); CR3 (CD1 1b, CD18); CR4 (CD11c, CD18).

Publication Charge: $119 billed to corresponding author who will receive a year's personal subscription to the journal. Fee is waived for subscribers.

Units: Use Système International d'Unités (SI Units) proposed in *Units, Symbols and Abbreviations* (1972, Royal Society of Medicine, London). Use other abbreviations only for unwieldy names, and only when they occur frequently. Where such non-standard abbreviations are used, provide a glossary.

SOURCE
October 1994, 98(1): front matter.

Clinical Chemistry
American Association for Clinical Chemistry

Dr. David E. Bruns, Editor,
P.O. Box 3757,
University Station,
Charlottesville, VA 22903-0757
For regular mail
Towers Office Building,
Suite 609, 1224 West Main Street,
Charlottesville, VA 22903
For express delivery

AIM AND SCOPE
Clinical Chemistry welcomes contributions that contain original information, experimental or theoretical, that advances the science of clinical chemistry. Such information may concern basic materials or principles, analytical and molecular diagnostic techniques, instrumentation, data processing, statistical analyses of data, clinical investigations in which chemistry has played a major role, and laboratory animal studies of chemically oriented problems of human disease.

MANUSCRIPT SUBMISSION
Principal desiderata are: subject matter that is original and significantly advances the state of knowledge of clinical chemistry; conclusions that are justified from design of experiments and data presented; information that is sufficiently detailed to permit replication of work by a competent clinical chemist; writing that is clear, concise, and grammatical.

Each author must have participated significantly in work and assume responsibility for manuscript's content. Copyright is transferred to publisher under signatures of all authors. No paper presenting the same information—other than an abstract or preliminary report (cited in manuscript)—may have been or will be published elsewhere. Copy, complete, and submit Authors' Assurances and Assignment of Copyright form (*Clin Chem* 40(1): 6, January 1994). Even if an authorized agent for an employer or institution transfers copyright to AACC, all authors must sign form to verify authorship.

Acknowledge all funding sources supporting work and all institutional or corporate affiliations. Disclose other kinds of associations, such as consultancies, stock ownership or other equity interests, or patent-licensing arrangements, in a cover letter. Such information is held in confidence and does not influence editorial decision. Upon acceptance Editor will discuss how best to disclose relevant information.

Submit letters of permission for use of information from papers cited in press, personal communications, or other unpublished work confirming work and giving permission for citation.

In cover letter state title of paper; paper's main point; and name, address, telephone and fax numbers of corresponding author. Enclose three copies each of unpublished manuscripts that may be critical to evaluation of submitted paper. Submit four copies of manuscript. Include all figures and tables.

MANUSCRIPT FORMAT

Use wide (2 cm) margins and double- (with typewriter) or triple- (with word processor) spacing throughout abstract, text, references, tables, figure legends, and footnotes. Type on one side of paper. Use separate sheet for title page. Use another page for abstract and Indexing Terms (not part of title). Place references, tables, figure legends, and footnotes on separate pages, in that order. Number every page, starting with title page, in top right corner.

Most manuscripts include sections: title, abstract, introduction, materials and methods, results, discussion, and references.

Do not duplicate text in figures and tables. Do not use figures that show only linear relation or few data points. Combine several curves in a single figure. Replace lengthy tabulations with statistical statements.

Subject Heading: Select appropriate subject heading. Place on left corner of title page: Drug Monitoring and Toxicology; Endocrinology; Enzymes and Protein Markers; General Clinical Chemistry; Immunology; Laboratory Management and Utilization; Automation and Analytical Techniques; Liquids and Lipoproteins; Molecular Pathology; Other Areas of Clinical Chemistry.

Title Page: Brief but informative title. Give authors' first and last names in full. Use footnotes to indicate authors' laboratory affiliations, authors' current addresses if different from where work was done, and corresponding author with fax number and e-mail address.

Abstract: 150 word summary of study plan, basic procedures, main results, and principal conclusions.

Indexing Terms: Below abstract provide three to eight words or phrases selected from *Medical Subject Headings* list of *Index Medicus.*

Introduction: Describe purpose of study, its importance, and its relation to previous work, and state very briefly your main finding. If reporting a method, indicate why your method is preferable to older methods. Provide strictly pertinent references, not an extensive subject review.

Materials and Methods: Material dictates way in which this section is written.

Apparatus—Identify apparatus used, provide name and location (city and two-letter state abbreviation if U.S.; city and country if outside U.S.) of manufacturer. Provide model number if multiple versions. Add further details only if equipment is not standard. If standard apparatus has been modified or new apparatus built, describe in sufficient detail to allow a knowledgeable reader to understand operation or to duplicate construction.

Reagents—List all reagents used; describe in detail and indicate source of special or unusual ones. If purity or stability of reagents is critical, indicate degree of purity and method of preservation. In addition to molar units, provide mass units for expressing amounts or concentrations of uncommon chemical compounds. State pH and composition of buffers, citing standard references. Indicate temperature at which critical pH measurements are made. Indicate all components in a reaction, whether "active" or not. Make clear whether concentrations in a reaction mixture represent final or original concentrations of solutions mixed.

Procedures—Provide citations rather than descriptions of published procedures except where a brief summary of principle of a method or modifications is appropriate. Include sufficient details of critical steps, standardization, calibration, and calculation of results to enable others to understand work and repeat it. Mention potential hazards as appropriate.

Results: Present concisely. Use tables or graphs for clarification Do not repeat in text data in tables and illustrations.

Discussion: Discuss significance of results and show how they lead to a valid conclusion. Compare with those of previous studies. Distinguish between speculation and fact.

Acknowledgments: Before references, acknowledge financial support, gifts, technical help, or other assistance, including that rendered by a company in evaluation of a product.

References: Cite in text in numerical order, using underlined Arabic numerals in parentheses. Cite only essential references; refer to review articles when appropriate. Include in text (in parentheses) references to unpublished work, manuscripts in preparation, verbal reports given at meetings, and personal communications. Include in reference list citations of papers in press.

List names of all authors except, if more than seven, list first six and "et al." Indicate if references are Editorial, Abstract, Letter, Technical Brief, or Review.

Abbreviate journal names according to *Chemical Abstracts* and *Biological Abstracts.*

JOURNALS

1. Fiechtner M, Ramp J, England B, Knudson MA, Little RR, England JD, et al. Affinity binding assay of glycohemoglobin by two-dimensional centrifugation referenced to hemoglobin A_{1c}. Clin Chem 1992;38:2372–9.

2. Demers LM. New biochemical marker for bone disease: is it a breakthrough? [Editorial]. Clin Chem 1992;38:2169–70.

3. Davey L, Naido L.. Urinary screen for acetaminophen (paracrysmolA) in the presence of N-acetylcysteine [Letter]. Clin Chem 1993;39:2348-9.

BOOKS AND MONOGRAPHS (EXCEPT SERIAL VOLUMES, WHICH ARE TREATED AS JOURNALS)

4. Siminovitch KA. Molecular characterization of human anti-DNA antibodies. In: Farid NR, Bona CA, eds. The molecular aspects of autoimmunity. San Diego: Academic Press, 1991:59–72.

5. Bailar JC III, Mosteller F, eds. Medical uses of statistics, 2nd ed. Boston: NEJM Books, 1992:449pp.

6. Harley JB, Gaither KK. Autoantibodies. In: Klippel JH, ed. Systemic lupus erythematosus (Zvaifler NJ, ed. Rheumatic disease clinics of North America, Vol. 14). Philadelphia: WB Saunders, 1988:43–56.

7. Haughton MA. Immunonephelometric measurement of vitamin D binding protein [MAppSci thesis]. Sydney, Australia: University of Technology, 1989:87pp.

Text Footnotes: Use for auxiliary or explanatory material that cannot be incorporated into text. Indicate position of footnote in text by superscript Arabic numerals. Type on a separate page, indicating page number on which each appears.

Figures: Submit four copies of each. For halftones, submit glossy prints. For line drawings, submit glossy prints or laser prints on coated (nonabsorbent) laser printer paper or drawings in black India ink on white drawing paper. Verify that symbols and lettering, particularly subscripts and superscripts will be legible after reduction to one column (85 mm) or two column (176 mm) width. Recommended type size after reduction is 8 points (references are set in 8-point type). Use lowercase lettering with an initial capital letter. On back indicate figure number, first author, and TOP. Cite every figure in text, in numerical order, using Arabic numerals.

Figure Captions and Subcaptions: On a separate sheet provide short captions and subcaptions for each figure. Double- or triple-space between lines. For previously published figures, acknowledge original source and submit written permission from copyright holder.

Tables and Table Captions: Type each with its brief caption, on a separate page. Indicate explanatory footnotes by superscript lowercase italic letters in alphabetical order, reading across table. Number tables consecutively with Arabic numerals in order of mention in text. Give every column a heading with clearly defined units as appropriate.

OTHER FORMATS

Editorials and Reviews are ordinarily invited. Unsolicited manuscripts are also welcome. Submit, for review by the Editor, a brief outline proposed editorial or review before preparing in final form. Book Reviews also are ordinarily invited. Confer with Book Reviews Editor.

LETTERS TO THE EDITOR—750-words, no abstract (four double-spaced pages, including references) and one figure or one table concerning any subject relevant to clinical chemists.

REVIEWS—Include headings appropriate to text.

TECHNICAL BRIEFS—500-words, no abstract, three manuscript pages (typed double-spaced), with one figure or one table. Summaries of findings of interest to a limited audience. Readers requiring fuller details than can be provided should obtain them from author.

Other appropriate contributions include Articles, Case Reports, Case Conferences, Scientific Notes, Opinions, and History. Consult recent journal issues to determine appropriate category for their manuscript and for examples of acceptable style.

OTHER

Abbreviations: Avoid use of unnecessary new abbreviations. Define all nonstandard abbreviations in text at first use and in a single footnote on page 2. Following standard abbreviations need not be explained: CV, coefficient of variation (relative standard deviation); SD, standard deviation; EDTA, ethylenediaminetetraacetate; HEPES, 4-(2-hydroxyethyl)-1-piperazineethane sulfonic acid; Tris, tris(hydroxymethyl)methylamine; ELISA, enzyme-linked immunosorbent assay; HPLC, high-performance liquid chromatography; IgA, IgG, and IgM, immunoglobulins A, G, and M, respectively; RIA, radioimmunoassay; HPLC, high-performance liquid chromatography; IRMA, immunoradiometric assay.

Analytical Methods and Results: In describing development and evaluation of performance of methods and instruments, discuss chemical sensitivity and specificity, detection limits, precision, accuracy, recovery, interference, comparability with other analytical methods, and normal range; when appropriate, include clinical data. Document analytical advantages of new or modified method over existing methods.

Accuracy—Analytical recovery studies involve analyses after known amounts of analyte are added to biological fluid on which determination will be performed. Calculate recovery of added analyte.

Perform interference studies to assess effects of common interferents, e.g., lipids, hemoglobin, bilirubin, and components of uremic plasma. Test exogenous materials such as commonly used or commonly coadministered drugs that might interfere with determination for interferences.

Comparison-of-methods studies should compare results by new method with those by a reference-quality method or other generally accepted analytical method for which assay performance is documented. See: Carey RN, Garber CC. Evaluation of methods. In: Kaplan LA, Pesce AJ, eds. *Clinical Chemistry. Theory, Practice and Correlation* (2nd ed. St. Louis: CV Mosby, 1989:290-310) and Koch DO, Peters T Jr. Selection and evaluation of methods. In: Burtis CA, Ashwood ER, eds. *Tietz Textbook of Clinical Chemistry* (2nd ed. Philadelphia: WB Saunders, 1994:508-25). Test 100 to 200 different samples from patients selected to include a wide variety of pathological conditions and to present range of values for the analyte that includes those likely to be encountered in routine application. Appropriate statistical evaluation of data typically requires regression analysis with slopes and intercepts (and their standard deviations) and standard errors of estimates. Correlation coefficient has limited utility.

Analytical Sensitivity and Detection Limit—Terms are commonly confused. The International Union of Pure and

Applied Chemistry defines analytical sensitivity as the ability of an analytical procedure to produce a change in signal for a desired change of the quantity (i.e., the slope of the calibration curve). Detection limit (or limit of detection) is defined as the lowest concentration or quantity of an analyte that can be detected with reasonable certainty for a given analytical procedure. Supply operational definition of this limit; e.g., concentration at a specified signal-to-noise ratio or concentration corresponding to a 3 SD above mean for a calibrator that is free of analyte.

Calibration Curves and Linearity—Subject data to linear-regression analysis (if a linear response is obtained) and include slope, intercept, standard error of estimate (standard deviation about regression line), and standard deviations of slope and intercept. Include standard deviations of repeated points. For radioimmunoassay calibration curves, use any objective, statistically valid method, Specify method used (See: Rodbard D. Data processing for radioimmunoassays: an overview. In: Natelson S, Pesce AJ, Dietz AA, eds. Clinical chemistry and immunochemistry; chemical and cellular bases and applications in disease. Washington, D.C.: *Am Assoc Clin Chem*, 1978:477-94).

Chromatograms—Present gas-liquid and liquid chromatograms so reveal efficiency of separation and resolution from interfering substances in matrix.

Enzyme Activities—Express in international units (U). Describe temperature and other key assay features in text or by reference to a published method. When first mentioned in text, number enzymes (measured by activity or mass assays) in accordance with recommendations of the Nomenclature Committee of the International Union of Biochemistry and Molecular Biology on the Nomenclature and Classification of Enzymes (EC no.). See: International Union of Biochemistry and Molecular Biology Nomenclature Committee. *Enzyme Nomenclature 1992* (San Diego: Academic Press, 1992).

Precision—Include estimates of "within-run" and "total" standard deviations. Determine at low, normal, and above-normal concentrations with use of specimens that are in an appropriate biological matrix. One method of estimating both within-run and total standard deviation is analysis of variance experiment described in NCCLS EP5 (See: National Committee for Clinical Laboratory Standards. NCCLS Tentative Guideline EP5-T. *User evaluation of precision performance of clinical chemistry devices.* Villanova, PA: National Committee for Clinical Laboratory Standards, June 1984), which calls for two replicates per specimen per run and two runs per day for 20 days. This permits separate estimation of between-day and between-run, within-day standard deviations, as well as within-run and total standard deviations. Acceptable alternatives that include one run per day are discussed in cited document.

Reference Interval (normal range)—Depending on conclusions of accuracy studies, existing reference range may need modification. In description of reference interval, include details about sampling; selection of subjects, including their number, age, and sex distribution; statistical method for summarizing results (Solberg HE. Establishment and use of reference values. In: Burtis CA, Ashwood ER, eds. *Teitz Textbook of Clinical Chemistry,* 2nd ed.

Philadelphia: WB Saunders, 1994:454-84); and other factors that would influence values obtained.

Data Sharing: Retain raw data from laboratory or clinical studies for a minimum of 1 year. Present upon request of Editor. Make unique materials necessary to reproduce reported results, such as computer software, ref agents, cell lines, cloned DNA, and monoclonal antibodies, available for use by qualified investigators.

Experimentation, Animal: State whether care was in accordance with *Guide for the Care and Use of Laboratory Animals* (Committee on Care and Use of Laboratory Animals, Washington, DC: Institute of Laboratory Animal Resources, National Research Council, 1985) or with your institution's guidelines.

Experimentation, Human: State whether procedures followed were in accordance with ethical standards of your institution's responsible committee or with Helsinki Declaration of 1975 (revised 1983).

Statistical Analyses: Describe methods with enough detail to enable a knowledgeable reader with access to original data to verify reported results. When possible, quantify findings and present with appropriate indicators of measurement error or uncertainty (such as confidence intervals). Avoid sole reliance on statistical hypothesis testing, such as of P values. When appropriate, present confidence intervals. See: Harris EK. On P values and confidence intervals (why can't we P with more confidence) [Editorial]. *Clin Chem* 1993;39:927-8 and Henderson AR. Chemistry with confidence: should Clinical Chemistry require confidence intervals for analytical and other data? [Opinion]. *Clin Chem* 1993;39:929-35.)

Evaluation of Diagnostic Accuracy—In clinical studies, simple testing of significance of differences between mean values of patients' groups (e.g., by Student's t-test) is insufficient to assess diagnostic accuracy. Scatter plots of data, calculations of diagnostic sensitivities and specificities, and use of approaches such as receiver-operating characteristic (ROC) curves (see: Zweig MH, Campbell G. Receiver-operating characteristic (ROC) plots: a fundamental evaluation tool in clinical medicine [Review]. *Clin Chem* 1993;39:561-77, note error in Fig 4.12), cumulative distribution analyses (see: Krouwer JS. Cumulative distribution analysis graph, an alternative to ROC curves [Tech Brief]. *Clin Chem* 1987;33: 2305-6.), likelihood ratios (Albert A. On the use and computation of likelihood ratios in clinical chemistry. *Clin Chem* 1982;28:111~9), and discriminant analysis (Solberg HE. Discriminant analysis. *Crit Rev Clin Lab Sci* 1978;9:209-42) provide information appropriate to specific situations. Discussions of predictive values in illustrative settings may be useful to assess potential clinical utility of tests with known accuracy.

Style: Follow: International Committee of Medical Journal Editors. Uniform requirements for manuscripts submitted to biomedical journals. *N Engl J Med* 1991;324:424-8.

See: *The ACS Style Guide: A Manual for Authors and Editors* (Dodd JS, ed. Washington, D.C.: American Chemical Society Publications, 1986) or the *CBE Style Manual* (5th ed. Chicago: Council of Biology Editors, Inc., 1983).

Units: Use International System of Units (National Institute of Standards and Technology. The *International System of Units* (SI). Taylor BN, U.S. ed. NIST Special Publication 330, 1991.

Liter is L; mole is mol; international unit of enzyme activity is U; IU is units of peptide hormones and vitamins.

For mass: kg, g, mg, μg, ng, pg; mass concentration: kg/L, g/L, mg/L, μg/L; substance concentration: mol/L, mmol/L, μmol/L, nmol/L; temperature: °C; time: s, min, h (but days, months, years); density: kg/L (the term relative density replaces "specific gravity"); absorbance: *A*; radioactivity: Ci, mCi, μCi.

SOURCE

January 1994, 40(1): 1–6.

Clinical Endocrinology

Society for Endocrinology
Blackwell Science Ltd.

Editors, 25 John Street
London WC1N 2BL
Fax: 071 831 6745

AIM AND SCOPE

Clinical Endocrinology will publish papers which contribute to the understanding of human endocrine disease. It will also publish from time to time, commissioned or preferred review articles as rapid communications. Correspondence will also be considered.

MANUSCRIPT SUBMISSION

Submit four complete copies of text, tables and figures. If possible, submit disc also. Papers are accepted on the understanding that no substantial part has been, or will be, published elsewhere. Appearance of an author's name on a paper indicates that s/he has read and approved paper.

MANUSCRIPT FORMAT

Type double-spaced on one side of paper. Precede text with a structured summary including headings: objective (include background sentence setting study in context), design, patients, measurements, results, conclusions. Only abbreviations listed below may be used in summary. On a single 'front page' give: title; short title; name(s) of author or authors (qualifications not required); department(s) in which work was done; and name and full postal address of reprint request and corresponding author headed 'Correspondence'.

References: Check accuracy. Include title and both first and last page numbers, journal title in full and all authors. Cite references in text chronologically; give authors' surnames, if one or two authors, followed by date of publication; if three or more authors, quote first author's surname, followed by *et al*. List references alphabetically at end of text.

Edge, J.A., Matthews, D.R. & Dunger, D.B. (1990) The dawn phenomenon is related to overnight growth hormone release in adolescent diabetics. *Clinical Endocrinology*, **33**, 729–737.

Russell, W.E. & Van Wyk, J.J. (1989) Peptide growth factor. In *Endocrinology* (ed. L.J. De Groot). pp. 2504–2524. W.B. Saunders, Philadelphia.

Figures: Submit half-tone illustrations as well-contrasted prints on glossy paper; label on back with names(s) of author(s), figure number and TOP edge. Include a separate sheet containing figure legends.

Tables: Type on separate sheets, number with Arabic numbers. Give titles. Indicate approximate position of figures and tables in text in margin.

OTHER FORMATS

CASE OF THE MONTH—Single cases which convey a simple educational message. Preface with a brief abstract, highlighting salient message. Confine discussion to unusual aspects, not a major literature review. Limit references to six.

CASE REPORTS—Present a significant advance in therapy or clinical understanding or highlight substantial scientific advances in understanding mechanism(s) of disease process.

CORRESPONDENCE—Must contain constructive criticism on published articles, authors of which will be given the right of reply.

RAPID COMMUNICATIONS—New data of sufficient importance to warrant immediate publication, a self-contained paper and not a tentative preliminary communication.

OTHER

Abbreviations: Follow recommendations in *Units, Symbols and Abbreviations* (1988, Royal Society of Medicine, London). Use without definition: ACTH corticotrophin; ADH antidiuretic hormone, vasopressin; AVP arginine vasopressin; FSH follicle stimulating hormone; GH growth hormone; GnRH gonadotrophin releasing hormone; hCG human chorionic gonadotrophin; hMG human menopausal gonadotrophin; hPL human placental lactogen; IGF insulin-like growth factor; LH luteinizing hormone; OT oxytocin; PRL prolactin; PTH parathyroid hormone; T4 thyroxine; T3 triiodothyronine; rT3 reverse T3; TRH thyrotrophin releasing hormone; TSH thyrotrophin; VIP vasoactive intestinal peptide. Define all other abbreviations where introduced. No chemical symbols and formulae in text.

Biochemical Journal (1975) **145**, 10–11, enumerates accepted abbreviations for certain compounds (such as ATP, DNA, DEAE-cellulose, etc.) and describes correct use of chemical names and formula.

IRP International Reference Preparations

Steroids—See: *Journal of Endocrinology* (1980) **84**, 3–4.

Amino acids—Use abbreviations only in tables or for representing polymers or sequences in text [see *Biochemical Journal* (1975) **145**, II].

Isotopically Labeled Compounds—See *Biochemical Journal* (1975) 145, 13–14 and *Radiochemical Catalogue* (Radiochemical Centre, Amersham, Bucks).

Describe solutions in terms of molarity (M) not normality (N). Express fractional concentration in decimal system; for values less than 0·1 M use mM (50 mM not 0·05 M). Give buffers: composition, pH and method of adjustment, e.g., 0·1 M potassium dihydrogen phosphate adjusted to pH 7·4 with 2 M sodium hydroxide. Volume ratios—e.g., methanol:water, 8·2 v/v (use 'by vol' instead of v/v if more than two substances are involved).

Experimentation, Human: Indicate how experiments complied with recommendations of Declaration of Helsinki (*British Medical Journal*, 1964, **ii,** 177). Studies must have been approved by a local Ethical Committee. Include a statement to this effect.

Spelling: Follow *The Concise Oxford Dictionary of Current English* (Clarendon Press, Oxford).

Statistical Analyses: Use appropriate statistical methods in design and analysis of study. See: *British Medical Journal* (1983) **286,** 1489-1493.

Time: Use 24-hour clock.

Units: Use SI units not mass units. Pituitary and other hormones expressed in units per liter should quote appropriate standard reference preparation used.

SOURCE

July 1994, 41(1): inside back cover.

Clinical Immunology and Immunopathology

Academic Press

Clinical Immunology and Immunopathology
Editorial Office
525 B Street, Suite 1900
San Diego, CA 92101-6467
Telephone: (619) 699-6467; Fax: (619) 699-6859

AIM AND SCOPE

Clinical Immunology and Immunopathology is an international medium devoted to publication of original investigations of the molecular and cellular basis of immunological diseases. The scope of the journal will stress the following broad areas: Congenital and acquired immunodeficiencies; Disorders of immune regulation; Mediators of inflammation; Immunoglobulin disorders; Complement disorders; Lymphoreticular malignancies; Clinical immunogenetics; Autoimmune diseases; Immune complex diseases; Allergies and hypersensitivities; Cancer immunology; Animal models of human diseases; Immunohematology; Immunodermatology; Neuroimmunology; Immunology of infectious and parasitic diseases; and Immunotherapy.

MANUSCRIPT SUBMISSION

Same work may not have been published, nor may be under consideration for publication elsewhere. Submission for publication must have been approved by all authors and by institution where work was carried out. Persons cited as a source of personal communications must approve citation. Written authorization may be required.

If accepted for publication, copyright in article, including the right to reproduce the article in all forms and media, shall be assigned exclusively to the Publisher.

MANUSCRIPT FORMAT

Type double-spaced throughout (including tables, footnotes, references, and figure captions) on one side of good grade 8.5 × 11 in. white paper, with 1 in. margins on all sides. Mimeographed or duplicated manuscripts must be indistinguishable from good typed copies.

Submit four complete copies. Each copy must include all figures and tables. Number pages consecutively.

Be as concise as possible. Present data clearly without repetition. Refer to previously published material rather than including lengthy descriptions in Materials and Methods section, combine tables or illustrations, and avoid repetition in Introduction and Discussion sections.

Page 1: Article title, author(s), and affiliation(s); at bottom of page type a short running head (abbreviated title, 55 characters, including spaces), and name, complete mailing address (including zip code), telephone and fax or telex numbers of corresponding author. List 3–5 key words for indexing. Page 2: Short abstract (100–150 words).

Indicate organization by appropriate headings and subheadings. Do not use final periods after abbreviations of measure (cm, sec, kg, etc.) in text or in tables, except for "in." (inch). Use American Chemical Society *Style Guide,* 1986, as a reference. Underline in pencil letters that represent mathematical symbols; these will be set in italic type. Use footnotes only when absolutely necessary; type consecutively double-spaced, on a separate sheet in order of appearance in text. Use superscript Arabic numbers to refer to footnotes both in text and in typed list.

Equations: Number consecutively, with number in parentheses to extreme right of equation. Refer to equations as Eq. (3) or simply (3), except at beginning of a sentence; spell out word Equation. Punctuate to conform to their place in sentence's syntax.

Tables: Number consecutively with Arabic numerals in order of appearance in text. Type each, double-spaced throughout on a separate sheet; avoid vertical lines. Supply a short descriptive title below table number. Type table footnotes cited by superscript lowercase italic letters at end of table.

Figures: Number with Arabic numerals. Mention consecutively in text. Put author(s) name(s) on top left side on back of each figure in pencil. Number on top right side. Type legends together on a separate sheet. Place figures in an envelope with name of senior author. For halftone photographs, submit three complete sets. Put each set in a separate envelope, marked with name of corresponding author.

Submit figures and illustrations so they can be reproduced without retouching or redrawing. Plan to fit proportions of printed page (7 1/8 × 9 in.; column width 3.5 in.); no larger than 8.5 × 11 in. Lettering should be of professional quality or generated by high-resolution computer graphics. Make large enough (10–12 points) to be reduced 50–60%. Make drawings with black India ink on tracing linen, smooth-surface white paper, or Bristol board. Or submit high-quality computer graphics. Plot graphs with black India ink on light blue coordinate paper, no larger than 8.5 × 11 in.; label coordinates properly; ink grid lines that are to show in black. Submit photographs as glossy prints (5 × 7 or 8 × 10 in.; do not mar with staples, paper clips, or pencil marks. Make original drawings no larger than 8.5 × 11 in. Color Illustrations are accepted only if authors defray cost.

References: Cite in text by number in parentheses. List at end of paper in numerical order. Abbreviate journal names in style of *Chemical Abstracts Service Source Index*, 1985. Cite unpublished observations and personal communications in text as such; do not include in reference list.

1. Hooks, J. J., Detrick, B., and Evans, C. H., Leukoregulin, a novel cytokine enhances the anti-herpesvirus actions of acyclovir. *Clin. Immunol. Immunopathol.* **60**, 244–253, 1991.

2. Rose, N. R., and Mackay, I. R., Genetic predisposition to autoimmune diseases. *In* "The Autoimmune Diseases" (N. R. Rose and I. R. Mackay, Eds.), pp. 1–27, Academic Press, San Diego, CA, 1985.

3. Rose, N. R., and Bigazzi, P. E. (Eds.), "Methods in Immunodiagnosis," pp. 1–6, Wiley, New York, 1980.

OTHER FORMATS

BRIEF COMMUNICATIONS—3000 words and two tables and/or figures, considered on an accelerated schedule.

EDITORIALS AND MEETING REPORTS—Must be directly relevant to the themes of the Journal.

LETTERS TO THE EDITOR—Must relate to articles published in the Journal.

SHORT ANALYTICAL REVIEWS—15 manuscript pages, exclusive of references are published from time to time. They deal analytically with recent developments in basic immunity of immediate or potential interest to clinical immunologists and immunopathologists. Contact Editor-in-Chief or Associate Editor-in-Chief before undertaking a review in order to determine suitability.

SOURCE

November 1994, 73(2): back matter.

Clinical Infectious Diseases
Infectious Diseases Society of America
University of Chicago Press

The Editor, *Clinical Infectious Diseases*
VAMC, West Los Angeles, Building 114, Room 228
Wilshire and Sawtelle Boulevards
Los Angeles, California 90073

MANUSCRIPT SUBMISSION

Obtain and submit written permission from all investigators cited in a personal communication who are not coauthors and from copyright holders for previously published tables or figures to be reproduced.

MANUSCRIPT FORMAT

Type manuscripts double-spaced, including tables and references, on standard-sized paper. Number all pages, including title page. Submit original typescript and two copies, each with a complete set of original illustrations.

Order of appearance of material is: title page, abstract, text, references, tables, legends for figures, figures.

Title Page: Title (two printed lines, 160 letters and spaces); names and affiliations of authors; a running head (45 letters and spaces); and footnotes with sources of financial support, changes of address, and name and address of reprint request author. Include acknowledgments of persons who assisted authors on page preceding references.

Abstract: 150 words on second page. Abstracts for a review or historical article summarize data, ideas, and conclusions presented in text. Abstracts for a research report indicate purpose of research, methods used, results (with specific data given, if possible), and conclusions. Do not cite references. For lengthy review articles, supply a table of contents.

References: Ensure that information is accurate. Type double-spaced. List only works that have been published or accepted for publication as references. Include unpublished author observations and personal communications as parenthetical expressions in text. Number references in order of appearance; number those cited in tables or figures according to order in which table or figure is cited in text. Insert reference numbers in text in brackets (not parentheses).

Follow format of *Index Medicus* and "Uniform Requirements for Manuscripts Submitted to Biomedical Journals" (*Ann Intern Med* **1982**;96:766–70). Provide all authors' or editors' names except when seven or more; list first three names and add et al. Spell out in full titles of journals not listed in *Index Medicus*. References to doctoral dissertations include author, title, institution, location, year, and publication information, if published.

1. Kryger P, Pedersen NS, Mathiesen L, Nielsen JO. Increased risk of infection with hepatitis A and B viruses in men with a history of syphilis: relation to sexual contacts. J Infect Dis **1982**;145:23–6.

2. Reynolds DW, Stagno S, Alford CA. Laboratory diagnosis of cytomegalovirus infections. In: Lennette EH, Schmidt NJ, eds. Viral, rickettsial and chlamydial infections. 5th ed. Washington, DC: American Public Health Association **1979**;339–439.

3. Anderson LJ, Winkler WG, Baer GM. The Centers for Disease Control's experience with a human diploid rabies vaccine [abstract no. 475]. In: Program and abstracts of the 19th Interscience Conference on Antimicrobial Agents and Chemotherapy. Washington, DC: American Society for Microbiology, **1979**:109.

Footnotes: Do not use substantive footnotes (containing a comment, explanation, or other than textual matter).

Tables: Do not show the same data in both a table and a figure. Number in order of appearance in text. Do not use vertical rules. Place footnotes below tables designated by symbols (listed in order of location when table is read horizontally): *, †, ‡, §, ||, #, **, ††, etc. Give each column an appropriate heading; indicate units of measure clearly.

Figures: Submit figures, including line drawings and graphs, as glossy prints, preferably no larger than 5 × 7 in. Label with first author's name and figure number. Type legends double-spaced on a separate sheet. Color figures are published by special arrangement with editors. In photomicrographs show only most pertinent areas or mark for cropping to avoid reduction in size. Place a micron bar or appropriate scale marking on figure.

Use computer enhancement and other alterations of photographic material for purposes of clarification with discretion; manipulation may not alter scientific content, which constitutes scientific fraud. When a photograph has been altered, also submit original photograph. In cover letter, state that photographic material was altered, which illustrations were adjusted and why and what specific changes were made.

OTHER FORMATS

PHOTO QUIZ—Submit 3 glossy prints and 1 or 2 slides of each photograph. Photographs can be black-and-white or color. Provide a double-spaced, one-paragraph description of case. Do not give away diagnosis (i.e., do not give treatment details, but state whether treatment was successful).

PHOTO QUIZ ANSWER PAGE—Photographs must be black-and-white. Provide 3 glossy prints and a double-spaced figure legend for each. Send a separate set of black-and-white glossies with arrows indicating key elements. In legend, describe elements indicated by arrow(s). Spell out abbreviations on photograph in legend. On a separate page, give diagnosis, a brief overview of diagnosed illness, and treatment. Provide authors' names and affiliations and a complete address for correspondence.

For further guidance, see photo quizzes published in October 1993 and January 1994 regular issues. Send all photo quiz submissions directly to Dr. Philip A. Mackowiak, Medical Services (111), VA Medical Center, 10 North Greene Street, Room 5D143, Baltimore, Maryland 21201.

OTHER

Abbreviations: Use minimally. Spell out terms that appear only a few times. When terms are repeated many times, introduce abbreviations when terms are first used and use abbreviations thereafter. Use conventional or SI units of measure without definition.

Experimentation, Human: State in a footnote to be included on title page that informed consent was obtained from patients or their parents or guardians, and that guidelines for human experimentation of U.S. Department of Health and Human Services and/or those of authors' institution(s) were followed in the conduct of the clinical research. Do not use initials to refer to patients.

Nomenclature: Use latest widely accepted systems of nomenclature. For bacterial nomenclature used use *Approved Lists of Bacterial Names (International Journal of Systematic Bacteriology* 1980;30:225–420) or *Bergey's Manual of Determinative Bacteriology* (Baltimore: Williams & Wilkins). For enzyme nomenclature, follow *Enzyme Nomenclature Recommendations (1984) of the Nomenclature Committee of the International Union of Biochemistry* (Orlando, Florida: Academic Press, 1984).

For names and abbreviations of drugs and chemical compounds, use *Merck Index* (Rahway, NJ: Merck).

Statistical Analyses: Identify tests used in statistical analyses both in text and tables and figures where results of statistical comparisons are shown.

Style: Use *Council of Biology Editors Style Manual* (5th ed. Washington, D.C.: American Institute of Biological Sciences, 1983) and *The Chicago Manual of Style* (13th ed. Chicago: University of Chicago Press, 1982). Current journal issues are also useful. For commercially obtained products used in experiments (particularly those referred to by trademark), give full names and locations of suppliers; exceptions are media and compounds that are commonly available. Describe drugs by generic names.

SOURCE

July 1994, 19(1): back matter.

Clinical Orthopaedics and Related Research

J.B. Lippincott Co.

Carl T. Brighton, M.D., Ph.D.
1226 Penn Tower
Hospital of the University of Pennsylvania
3400 Spruce St.
Philadelphia, PA 19104

AIM AND SCOPE

Clinical Orthopaedics and Related Research publishes original articles offering significant contributions to the advancement of orthopedic knowledge.

MANUSCRIPT SUBMISSION

Have cover letter signed by responsible author. State that all coauthors have seen and agree with contents of manuscript. Indicate that work has not been submitted or published elsewhere.

Submit original and three high quality copies of manuscript. Type all sections double-spaced with wide margins. Number pages consecutively in order: Title page; Abstract; Introduction; Materials and Methods; Results; Discussion; Acknowledgments, References; Tables; and Legends. Type author's name and page number in upper right corner of each page. Use letter quality print in a readable typeface.

Accuracy of information and of citations of literature is sole responsibility of author(s).

MANUSCRIPT FORMAT

Title Page: Title (80 spaces); running title (40 spaces); author(s)'s full name, highest academic degree, institutional affiliation; and address; and names, address telephone and fax numbers of corresponding and reprint request author(s).

Abstract: 200 words (15-18 lines) one paragraph statement of main conclusions and clinical relevance of manuscript.

References: Type double-spaced in alphabetical order. Do not submit computerized literature search. References should have been read by authors and have pertinence to manuscript. Do not use boldface, underlining, or complete upper case. Do not include manuscripts in preparation, unpublished data, including articles submitted but not in press, and personal communications, in list. Cite in text, in parentheses, only if necessary to content and meaning of article

JOURNAL ARTICLES

Author(s) (up to 5, if more, cite 3, then et al); title; journal name (*Index Medicus* abbreviations); volume; inclusive page numbers; year.

Kaplan FS, August CS, Dalinka MK: Bone densitometry observations of osteoporosis in response to bone marrow transplantation. Clin Orthop 294:79-84, 1993.

CHAPTERS IN A BOOK

Author(s); chapter title; editors; book title; edition, if not first; city of publication; publisher; specific pages; year.

Young W: Neurophysiology of Spinal cord Injury. In Errico TJ, Bauer RD, Waugh T (eds). Spinal Trauma. Philadelphia, JB Lippincott Company 377-414, 1991.

BOOK

Hoppenfeld S, Zeide MA: Orthopaedic Dictionary. Philadelphia, JB Lippincott Company 39-41, 1994.

Tables: Type double-spaced on pages separate from text. Number in order of citation in text. Give titles. Give columns headings. Limit to 6 columns. Spell out abbreviations used in footnote at bottom of table. Do not use photocopy reduction.

Figures: Send four sets of unmounted black and white glossy prints. Remove all markings, such as patient identification, from radiographs before photographing. Apply arrows or lettering with professional products such as Letraset or Paratype. Make large enough to be seen after reduction in size. Submit signed consent from patient or guardian with photographs of recognizable persons. Submit written permission from author or copyright holder with previously published figures or tables. Attach a self-stick label to back of each indicating TOP, figure number, and first author's name. Have line or original drawings done by a professional medical illustrator. Indicate magnification, internal scale markers, and stains. Type legends double-spaced on separate page from text.

OTHER

Reviewers: Provide names of potential reviewers and/or requests that specific persons not serve as reviewers.

Submission Fee: $50.00 (U.S.), $75.00 (other countries), for each unsolicited manuscript.

SOURCE

November 1994, 308: iv.

Clinical Pharmacology & Therapeutics
American Society for Clinical Pharmacology & Therapeutics

American Society for Pharmacology & Experimental Therapeutics
Mosby

Dr. Marcus M. Reidenberg
Department of Pharmacology
Cornell University Medical College
1300 York Ave.
New York, NY 10021

AIM AND SCOPE

Clinical Pharmacology & Therapeutics is a monthly journal devoted to the publication of articles of high quality dealing with effects of drugs in human beings. These include studies of pharmacokinetics, pharmacodynamics, pharmacoepidemiology, and new technologies of molecular biology that directly affect drug action in human beings. *Clinical Pharmacology & Therapeutics* invites original articles in these fields and identifies several broad general categories of interest to our readers: Pharmacokinetics and Drug Disposition, Pharmacodynamics and Drug Action, Clinical Trials and Therapeutics, Pharmacoepidemiology, Commentaries, Special Case Reports, and Letters to the Editor.

MANUSCRIPT SUBMISSION

Articles must be contributed solely to *Clinical Pharmacology & Therapeutics*. All observations reported in a manuscript must be original, may not have been previously published, or be under consideration by another journal. Corresponding author must indicate any data previously published in any form or in press. For multicenter study reports include references to all papers that have presented any data in report. Note data that may be included in a multicenter study report in the future in a footnote. Commentaries and addresses submitted for publication may contain, in part, material that has been previously published.

Submit statement signed by one author: "The undersigned author transfers all copyright ownership of the manuscript (title of article) to Mosby Year Book, Inc. in the event that the work is published. The undersigned author warrants that the article is original, is not under consideration by another journal, and has not been published previously. I sign for and accept responsibility for releasing this material on behalf of any and all coauthors." Authors will be consulted, when possible, regarding republication of their material.

MANUSCRIPT FORMAT

Follow "Uniform Requirements for Manuscripts Submitted to Biomedical Journals" (Vancouver style) (*N Engl J Med* 1991;324:424–8). Type all material double-spaced on 8.5 × 11 in. paper (one side only), with liberal margins. Submit original typescript with photocopies of figures; one complete photocopy of manuscript and figures; and one glossy set of illustrations. Retain one copy. Designate a corresponding author to receive galleys and reprint requests; give complete mailing address and business and home telephone numbers.

Title Page: Make title specific and short (80 characters). Include first name and highest earned academic degree for each author. Insert name and address of institution from which work originated and sources of financial support. List name, address, and telephone numbers of corresponding author. List number of pages, figures, and of tables.

Lead Summary and Text: Supply an abstract (150 words) to appear under title. A structured abstract (see *Ann Intern Med* 1990;113:69–76 for details) of 250 words may be used instead. Either will replace a summary at end of article. Number all pages, including references, figure legends, and tables, in sequence. Use only standard abbreviations for· measurements and omit periods; explain in tabular form all other abbreviations and acronyms. Limit other abbreviations and acronyms to those in common use. Do not invent abbreviations. Use kinetic parameter abbreviations suggested by the American College of Clinical Pharmacology. Place acknowledgments at end of text before reference list. Include only names of those who have made a scientific contribution.

References: Place list immediately after acknowledgments. Type double-spaced in Vancouver style (*N Engl J Med* 1991;324:424–8). Abbreviate journal names as in *Index Medicus*. Use "in press" references with name of accepting journal. Do not include personal communications and other unpublished and nonarchival references in reference list; make these footnotes to text, stating source, and date.

If six or fewer authors, list all; if seven or more, list first three and add et al.):

PERIODICALS

You CH, Lee KY, Chey RY, Menguy R,. Electrogastrographic study of patients with unexplained nausea, bloating and vomiting. Gastroenterology 1980;79:311–4.

BOOKS

Eisen HN. Immunology: an introduction to molecular and cellular principles of the immune response. 5th ed. New York: Harper & Row Publishers, Inc., 1974:406.

CHAPTERS IN BOOKS

Weinstein L, Swartz MN. Pathogenic properties of invading microorganisms. In: Sodeman WA Jr, Sodeman WA, eds. Pathologic physiology: mechanism of disease. Philadelphia: WB Saunders Co., 1974:457–72.

Appendixes: Include extensive or very detailed descriptions of analytic methods or mathematic calculations appendixes, following references. A short legend may be necessary.

Figures: Include a photocopy of each figure with each manuscript copy. Submit one glossy photograph of each figure. Prepare drawings or graphs in black India ink or typographic (press-apply) lettering; typewritten or freehand lettering is unacceptable.

Make sure decimal points, dotted lines, etc. are dark enough to reproduce plainly. Have all lettering done professionally in proportion to drawing, graph, or photograph. Do not send original artwork, x-ray films, or electrocardiogram (ECG) strips. Submit glossy print photographs, 3×4 in. (minimum) to 5×7 in. (maximum), with good black-and white contrast or color balance. Note special instructions regarding sizing. Number figures in

order discussed in text. Make sure all figures are mentioned. Mark first author's last name, figure number, and an arrow indicating TOP edge on a label pasted on back of each figure. A reasonable number of black-and-white illustrations will be reproduced free of charge, special arrangements must be made for extra illustrations, color plates, or elaborate tables.

Figure Legends: Type double-spaced on a separate page marked "Legends for figures." If a figure has been previously published, give full credit to original source.

Tables: Number in Roman numerals in order of mention in text; mention all tables. Type double-spaced, including title, column heads, body, and footnotes. Title concisely describes content of table; column heads do the same for data in columns. If a table or any data have been previously published, give full credit to original source in a table footnote.

OTHER FORMATS

CLINICAL PHARMACOLOGY ROUNDS—Reports of cases in which new observations are made that can be used to illustrate a principle of clinical pharmacology that is applied in therapy. Limit discussion; focus on the specific teaching point of the case.

Papers on the following topics are encouraged: issues of quality assurance of drug therapy; new technologies of molecular biology that directly affect drug action in humans; studies of new classes of drugs such as recombinant proteins and antibodies; studies of more traditional classes of drugs; in vitro studies of human-derived materials and animal experimental data in addition to human data (primary focus must be on drugs in human beings). *Clinical Pharmacology & Therapeutics* welcomes papers that do not fit into a specific category but that contribute to our knowledge of the effects of drugs in human beings.

CLINICAL TRIALS AND THERAPEUTICS—Controlled clinical trials and studies that examine therapeutic effects of drugs in disease and adverse drug reactions, including case reports of confirmed novel adverse drug reactions.

COMMENTARY—Brief, thoughtful essays on topics relevant to clinical pharmacology.

LETTERS TO THE EDITOR—400 words, including references, comment on articles previously published in the Journal. Use the same form as original articles. Include copyright transferral statement quoted above.

PHARMACODYNAMIC STUDIES—Include descriptions of the actions of drugs in humans and studies of the mechanism of action of drugs.

PHARMACOEPIDEMIOLOGY AND DRUG UTILIZATION STUDIES—Including studies of drugs that are widely used or are of particular public health importance.

PHARMACOGENETIC STUDIES—In any of the above categories.

PHARMACOKINETICS AND DRUG DISPOSITION—Present the state of the art in this field as models with wide generalizability or advances in technical or conceptual methodology and understanding of drug disposition.

OTHER

Drugs: Refer to by accepted generic or chemical names only. Do not abbreviate; spell in full each time. Do not use code numbers. USAN, BAN, and INN are acceptable names. Give proprietary names in parentheses. Insert IUPAC name (if necessary).

When a compound can exist as stereoisomers, state exact nature of drug being studied in methods sections. Identify specific stereoisomers, if used, or a racemic mixture. If stereochemistry of material is not known, so state.

Experimentation, Human: Follow principles embodied in the Declaration of Helsinki. Papers must have followed all ethical and legal standards for pursuing original experiments on human beings as set forth by their home institutions and states.

Page Charges: $35 per page for first eight printed pages; $50 per page for pages 9-16; and $100 per each additional page. Authors may request approval for waiver from Dr. Reidenberg after bill is received.

Permissions: Submit written permission for use of direct quotations, tables, or illustrations that have appeared in copyrighted material from copyright owner and original author with complete source information. Submit signed releases that show informed consent for photographs of identifiable persons.

SOURCE

July 1994, 56(1): 8A–10A.

Clinical Research
see Journal of Investigative Medicine

Clinical Rheumatology

Prof. Dr. J. Dequeker
Division of Rheumatology
University Hospitals, K.U.Leuven
B-3212 Pellenberg
Belgium
Telephone: 32-16-338720; Fax: 32-16-338724

AIM AND SCOPE

Clinical Rheumatology is an international journal devoted to publishing, in the English language, original clinical investigation and research in the general field of rheumatology with accent on clinical aspects at post-graduate level. Studies carried out anywhere in the world will be considered, the basic criterion for acceptance being the medical and scientific standard of the work described.

Rheumatology in Europe is not restricted to inflammatory rheumatic diseases but also includes nonsurgical bone diseases, in general, and rehabilitation. The aim of the editorial board is for *Clinical Rheumatology* to reflect the field of rheumatology in its wider aspects. Hence, *Clinical Rheumatology* aims to cover all modern trends in clinical and experimental research as well as the management and evaluation of diagnostic and treatment procedures connected with the inflammatory, immunologic, metabolic, genetic and degenerative soft and hard connective tissue diseases.

Original publications, informative case reports, short communications, editorials, and reviews in relation to these topics will be accepted. Letters to the editor are welcome as enhancement to discussions.

MANUSCRIPT SUBMISSION

Submitted manuscripts must not have been published nor will not be simultaneously submitted or published elsewhere. By submitting a manuscript, authors agree that copyright for article is transferred to publisher, upon acceptance for publication.

MANUSCRIPT FORMAT

Submit manuscripts, complete with figures and tables in triplicate, typed on one side of paper, double-spaced with wide margins.

Title Page: Include title of manuscript, authors' names and affiliations, 100–200 word summary, 3–6 key words, running title and full address of corresponding author.

Divide manuscript into sections in order: Introduction, Materials and Methods, Discussion, followed by Acknowledgments (if any) and References.

Tables and Figures: Cite in text using Roman numerals for tables and Arabic numbers for figures. Type tables double-spaced, each on a separate sheet using horizontal lines only with title above and explanation below. Submit figures as original drawings in India ink suitable for immediate reproduction or as high contrast sharp photographs on glossy paper. Do not exceed 5×7 in. There is a fee for color photographs.

References: Compile numerically in order of appearance in text. For abbreviations of journal names, refer to *List of Journals Indexed in Index Medicus*.

De Lange, W.F., Gispen, W.H. Effect of lung diffusion capacity in sclerosis. J Rheumatol 1979, 5, 216–219.

Kopera H. Vasculitis. In: Arthritis and Allied Conditions, 8th ed., Eds.: Hollander, J.L., McCarthy, D.J.Jr., Philadelphia, Lea and Febiger, 1972, 230–240.

SOURCE

March 1994, 13(1): inside back cover.

Clinical Science
Medical Research Society
Biochemical Society

Managing Editor, *Clinical Science*
59 Portland Place, London W1N 3AJ, U.K.
Telephone: 071-637-5873; Fax: 071-323-1136

AIM AND SCOPE

Clinical Science publishes papers in the field of clinical investigation, provided they are of a suitable standard and contribute to the advancement of knowledge in this field. The term 'clinical investigation' is used in its broadest sense to include studies in animals and the whole range of biochemical, physiological, immunological and other approaches that may have relevance to disease in man. Studies which are confined to normal subjects, or animals, or are purely methodological in nature may be acceptable.

The material presented should permit conclusions to be drawn and should not be only of a preliminary nature. The journal publishes four types of manuscript, namely invited Editorial Reviews, Full Papers, Rapid Communications and Correspondence. Other features are the Glaxo/MRS Young Investigator Prizewinners, the Techniques in Clinical Research series and the Regional Focus series.

MANUSCRIPT SUBMISSION

Submission implies that manuscript has been approved by all named authors, that all persons entitled to authorship have been so named, that it reports unpublished work that is not under consideration for publication elsewhere, and that if the paper is accepted for publication copyright will be transferred to the Biochemical Society. Authors must sign an undertaking to these effects. The restriction on previous publication does not usually apply to previous publication of oral communications in brief abstract form; enclose three copies of the abstracts of previous publications.

In cover letter include author's telephone and fax numbers.

Submit four copies (three may be photocopies, except for half-tone figures) of typescript, Tables, Figures, etc. Retain one copy. Do not sent typescripts produced on dot-matrix printers that are not 'letter quality'.

Papers must be intelligible to non-specialist readers. Include a very brief résumé of the current state of knowledge. Give certain types of material, e.g., mathematical formulations requiring more than trivial derivations, in a separate Appendix.

If author's previous works with important details relevant to present work are named, send three copies, especially in relation to methodology.

Length may be justified by content. Do not exceed minimum length required for precision in describing experiments and clarity in interpreting them. Most papers are between six and eight printed pages. Deposit extensive tables of data with the Royal Society of Medicine.

Papers concerned primarily with computer modelling techniques must demonstrate that use of such techniques lead to a clear choice between two or more alternative hypotheses, or to formulation of a new hypothesis amenable to experimental challenge or verification, or provides some new insight into behavior of a particular physiological system. Do not give extensive technical details of hardware and software.

MANUSCRIPT FORMAT

Refer to a current issue of *Clinical Science* for general layout. Arrange typescripts as follows:

Title Page: Informative title; indicate species in which observations were made. Do not number paper as part of a series. List of authors' names (degrees and appointments are not required). Laboratory or Institute of origin.

Key Words: For indexing, select from current issues of *Medical Subject Headings'* of *Index Medicus*.

Short Title: 45 characters and spaces for use as a running heading; do not use abbreviations.

Corresponding and reprint request authors: give name and address of author.

Summary: 250 words arranged in numbered paragraphs of what was done, what was found and what was concluded. Avoid use of abbreviations; define those used. Avoid statistical and methodological details including exact doses unless essential to understanding Summary.

Introduction: Make comprehensible to general readers. Include reason for work, but not findings or conclusions.

Methods: Give sufficient information in text or by reference to permit work to be repeated without communicating with author.

In describing certain techniques, namely centrifugation (when the conditions are critical), chromatography and electrophoresis, follow recommendations of the Biochemical Society (*Biochem J* 1994; **297:** 1–15).

Results: Do not include material appropriate to Discussion.

Discussion: Do not include results. Make pertinent to data presented.

Acknowledgments: Be as brief as possible.

Figures and Tables: Keep number to a minimum. Indicate appropriate position in margin of text. Refer to figures and tables in Arabic numerals; number in order of appearance. Do not present the same data in both a figure and a table.

Supply figures, with captions attached, as original drawings or photographs together with photocopies. Put number and authors' names in pencil on back; indicate TOP with a pencilled arrow. Symbols for experimental points are ● ▲ ■ ○ △ ❏. Do not use × or +. Do not use same symbols for two curves where points might be confused. For scatter diagrams, use solid symbols. When a variable appears in more than one figure, use same symbol for it.

Do not draw curves beyond experimental points, nor axes beyond data. Write only essential information that cannot readily be included in legend within figure.

Computer-generated line diagram curves must be smooth (not 'stepped') and symbols must be large enough to be clearly distinguishable after reduction.

Submit figures for half-tone reproduction as glossy prints. Provide four copies (not photocopies) of each print. Indicate magnification with a bar representing a stated length. Type tables separately from text. Underline title, follow with a legend.

Avoid numbers with an inconvenient number of digits in table headings and figures.

Make captions for figures, and titles and legends for tables, readily understandable without reference to text. Include adequate statistical information, including that on regression lines, in figure captions where appropriate.

Footnotes: Avoid use. Identify in tables by symbols *, †, ‡, §, ‖, ¶, in order.

References: Use Vancouver system: number references in text consecutively in order of mention, give numerals in brackets. Number references cited in only figure legends or tables only in sequence determined by position of first mention in text of figure or table. List references in numerical order. Give names of all authors (except where seven or more, list only first three and add et al.), with full title of paper and source details in full including first and last page numbers.

2. Knox AJ, Britton JR, Tatterfield E. Effect of vasopressin on bronchial reactivity to histamine. Clin Sci 1989; **77**: 467–71.

When quoting a book, use following format, giving relevant pages or chapter number:

20. Armitage P. Statistical methods in medical research. Oxford: Blackwell Scientific Publications, 1983.

21. Ghatei MA, Bloom SR. Enteroglucagon in man. In: Bloom SR, Polak JM, eds. Gut hormones. Edinburgh: Churchill Livingstone, 1981: 332–8.

Give references to 'personal communications' and 'unpublished work' in text only and not in reference list. Give name and initials of source of information. For quotations from personal communications, submit written permission for use. Indicate references to material accepted for publication but not yet published, in reference list by 'In press' with name of journal and, expected date of publication. If citation is of major relevance to submitted manuscript, include a copy.

OTHER FORMATS

CORRESPONDENCE—750 words, one figure or table, six references, or 1000 words without a figure or table, of original observations or critical assessments or material published in *Clinical Science,* including Editorial Reviews. Submit letters relating to material previously published in *Clinical Science* within 6 months of publication. They will be sent to authors for comment; letter and reply will be published together. Consideration will also be given to publication of letters on ethical matters.

EDITORIAL REVIEWS—These are commissioned. Unsolicited reviews are considered. Submit a synopsis of proposed review.

RAPID COMMUNICATIONS—Four printed pages (3000 words, with appropriate deductions, at the rate of 1000 words/page for figures and tables). Submit in similar format to full papers. Editorial process is expedited for data that are novel and exciting, when rapid publication is of importance and when material can be presented concisely. Describe completed work, not a preliminary communication.

OTHER

Abbreviations: Avoid use; if used, define at first mention; coin new abbreviations only for unwieldy names which occur frequently. Do not use abbreviations, except those indicated by an asterisk in list (*Clin Sci* 86(1): vii-viii) in title, short title or Summary. List is of abbreviations, conventions, definitions, symbols and special comments and shows standard symbols and abbreviations that can be used without definition. Use numbers, not initials, for patients and subjects.

Buffers and Salts: Give acidic and basic components with pH. Alternatively, reference composition of buffer. See *Biochem J* 1994; **297**: 1–15.

When describing solutions containing organic anions and their parent acids, use salt designator instead of acid name unless it is certain that virtually all of the acid is in the undissociated form.

Describe composition of incubation or reference composition.

Data Submission: Very large sets of individual values or very large numbers of diagrams cannot be published. Include a summary of information only. Submit information from which summary was derived with manuscript. Upon acceptance Editors may ask for a copy of the full information and diagrams to be deposited with the Librarian, the Royal Society or Medicine, London, who will issue copies on request. Requests are frequently received.

Doses: Express doses of drugs in mass terms, e.g., milligrams (mg) or grams (g), and also in (parentheses) in molar terms, e.g., mmol, mol, where relevant. See *The Merck Index*, (11th ed. Rahway, NJ.: Merck and Co. Inc., 1989) for molecular masses of many drugs.

Enzymes: For nomenclature follow *Enzyme Nomenclature* (San Diego: Academic Press, 1992). Quote Enzyme Commission (EC) number at first mention. If enzyme has a commonly used informal name, use after first formal identification. Express unit of enzyme activity as that amount of material which will catalyse transformation of 1 μmol of the substrate/s under defined conditions, including temperature and pH. The unit of the amount of enzyme given is the katal (symbol kat). Alternatively, or when natural substrate has not been fully defined, express activity in terms of units or activity relative to that or a recognized reference preparation, assayed under identical conditions. Express activities of enzymes as units/ml or units/mg of protein.

Experimentation, Animal: Take care to ensure that experimental animals do not suffer unnecessarily. State anesthetic procedures used, and all precautions taken to ensure that animals did not suffer unduly during and after experimental procedure.

Experimentation, Human: State that research was carried out in accordance with the Declaration of Helsinki (1989) of the World Medical Association, and has been approved by Ethical Committee of institution in which work was performed. Obtain and record in paper consent from each patient or subject after full explanation of purpose, nature and risk of all procedures used. Papers where ethical aspects are open to doubt will not be accepted.

Homology: Homologous means having a common evolutionary origin'. It is often used to mean simply 'similar'. Use the more precise definition; see Reeck GR, et al. *Cell* 1987; **40**: 667 and Lewin R. *Science* 1987; **237**: 1570).

Isotope Measurements: Express radioactivity in absolute terms; the SI unit is the becquerel (Bq), defined as 1 disintegration/s; the curie (C_1; $1\ C_1$; $= 3.7 \times 10^{10}$ Bq) may also be used. Alternatively, express radioactivity as disintegrations (or counts) per unit of time, e.g., disintegrations/s (d.p.s.) or counts/min (c.p.m.).

Nomenclature, Anatomical: Follow recommendations of the International Anatomical Nomenclature Committee (*Nomina Anatomica*. 3rd ed. Amsterdam: Excerpta Medica Foundation, 1966).

Nomenclature, Animals, Plants and Microorganisms: Give full binomial specific names at first mention for all experimental animals other than common laboratory animals. State strain and source of animals. Thereafter in text, use single-letter abbreviations for genus; if two genera with same initial letter, abbreviate, e.g., *Staph.* and *Strep.*

Nomenclature, Biochemical: Follow recommendations of the Nomenclature Committee of IUBMB and the IUPAC-IUBMB Joint Commission on Biochemical Nomenclature (see *Biochemical Nomenclature and Related Documents*, 2nd ed., London: Portland Press, 1992).

Nomenclature, Disease: Follow *International Classification of Disease* (9th revision. Geneva: World Health Organization, 1979.

Precision of Measurement Procedures: When a new measuring procedure has been used, or when an established procedure has been applied in a novel fashion, estimate precision of procedure. Indicate what sources of variation have been included in estimate, e.g., variation of immediate replication, variation within different times of day, or from day to day, etc.

If precision of measurement varies in proportion to the magnitude of the values obtained, express as coefficient of variation; otherwise express by an estimate of the (constant) standard error of a single observation, or by estimates of several points within range of observed values.

When recovery experiments are described, state approximate ratio of amount added to amount already present and stage of procedure at which the addition was made.

For methods or assays crucial to understanding the paper, provide information on validity, accuracy and precision of methods.

Radionuclide Applications in Man: For new or modified radionuclide applications in man, estimate maximal possible radiation dose to the body and critical organs. Express in rems, with corresponding figure in sieverts (Sv) in parentheses after it.

Reviewers: Suggest potential referees in submission letter.

Solutions: Describe concentration of solutions in molar terms (mol/l and subunits thereof). State molecular particle weight if necessary. Do not express values as normality (N) or equivalents. Express mass concentration as g/l or subunits thereof. For solutions of salts, molar concentration, avoid ambiguity as to use of anhydrous or hydrated compounds. Give concentrations of aqueous solutions as mol/l or mol/kg (g/l or g/kg if not expressed in molar terms) rather than % (w/v) or % (w/w). Make clear whether concentrations of compounds in a reaction mixture are final concentrations or concentrations in solutions added.

Spectrophotometric Data: Quantity log (I_0/I) is called attenuance. It reduces to absorbance when there is negligible scattering or reflection. Use the more general term 'attenuance' when scattering is considerable. Otherwise use absorbance; neither should be called extinction or optical density. Symbols used are: A, absorbance: D, attenuance; a, specific absorption coefficient (liter g^{-1} cm^{-1}); ε molar absorption coefficient (absorbance of a molar solution in a 1 cm light-path) (liter mol^{-1} cm^{-1}, not cm^2 mol^{-1}).

Spelling: Use *Concise* or *Shorter Oxford Dictionary of Current English* (Oxford: Clarendon Press) and *Butterworth's Medical Dictionary* (London: Butterworths).

Statistical Analyses: Two common errors are the use of means, standard deviations and standard errors in describing and interpreting grossly nonnormally distributed data and the application of t-tests for significance of difference between means in similar circumstances, or when variances of the two groups are non-homogeneous. It may be more appropriate to provide a 'scattergram' than a statistical summary. Consult statistical guidelines of Altman et al. in 'Statistical guidelines for contributors to medical journals' (*Br Med J* 1983; **286:** 1489–93.

State type of statistical test used in Methods section. Give a reference for the less commonly used statistical tests. The format for expressing mean values and standard deviations or standard errors of the mean is, for example: mean cardiac output 10.4 liters/min (SD 1.2; n=11). Indicate degrees of freedom where appropriate. Express levels of significance as $P<0.01$.

Trade Names: Give name and address of suppliers of special apparatus and biochemicals. For drugs, give approved names with trade names and manufacturers in parentheses.

Units: Use recommended Systeme International (SI) units (see *Quantities, Units and Symbols in Physical Chemistry*, Oxford: Blackwell Scientific Publications Ltd, 1988). Use these units except for blood pressure values (mmH) and gas partial pressures (mmHg-with kPa in parentheses or kPa-with mmHg in parentheses). Express airways pressure in kPa. Where molecular mass is known, express amount of a chemical or drug in mol or in an appropriate subunit, e.g. mmol. Do not express energy in joules (J). For basic SI units and symbols and derived SI units see *Clin Sci* 86(1): vi. Liter is a special name for cubic decimeter (1 liter = 1 dm^3).

Basic and derived SI units, including symbols of derived units that have special names, may be preceded by prefixes to indicate multiples and submultiples. For prefixes see *Clin Sci* 86(1): vii. Use compound prefixes.

Full stops are not used after symbols.

Use minutes (min), hours (h), days and years in addition to SI unit of time [second(s)]. Use solidus in a unit as long as it does not have to be employed more than once; replace by ml min^{-1} kg^{-1}.

SOURCE

January 1994 86(1): i–ix.
Guidance for Authors is published in the January issue, and revised periodically.

CMAJ

Canadian Medical Association

Editor-in-Chief, *CMAJ*
1867 Alta Vista Dr.
Ottawa, ON KIG 3Y6 Canada

AIM AND SCOPE

The *Canadian Medical Association Journal* will consider original manuscripts in English or French on the basic and clinical aspects of medicine, medical education, medical economics or politics, medicolegal affairs, health care policy or delivery and the history of medicine.

MANUSCRIPT SUBMISSION

In a cover letter signed by all authors state that manuscript has not been published previously and is not under consideration by any other journal. Explain purpose of paper and intended readership.

Submit signed letters of permission from people identified as sources of personal communications and those mentioned in acknowledgments or identifiable in illustrative material, as well as from copyright holders of previously published material (e.g., tables, illustrations and lengthy quotations) reproduced, with or without modification, in submitted article.

Disclose sources of financial or material support, commercial interest they may have in subject of study and affiliation or involvement with an organization that has a financial interest in research materials or topic.

Indicate number of words in main text (excluding title page, abstract, reference list, tables and figure legends).

If reprint requests also can be received by fax or electronic mail, include pertinent fax number or e-mail address after mailing address on title page.

Before publication all authors must sign a document transferring copyright to *CMAJ*. Accepted manuscripts become permanent property of *CMAJ* and may not be published elsewhere, in whole or in part, without written permission from publisher.

MANUSCRIPT FORMAT

Consult "Uniform requirements for manuscripts submitted to biomedical journals" *(Can Med Assoc J* 1992; 146: 861–868) for instructions about preparing manuscripts. Manuscripts that do not adhere to these requirements and others will be returned. Submit an original printout and four high-quality copies. Authors wishing "blind" peer review must exclude all identifying information from three copies.

Double-space all pages (title page, abstract and key words [MeSH terms], text, references, tables and figure legends); use 10-cpi, letter-quality type (preferably Courier), without right justification or proportional spacing. Make top margin on title page 5 cm deep; all other margins are 2.5 cm. Text should not exceed 2500 words (13 double-spaced pages).

Place reference numbers between less-than and greater-than signs and closed up to preceding word or punctuation mark (e.g., medicine<4> and surgery,<5>). Use a hyphen to link first and last numbers in a series of three or more consecutive reference numbers but use commas, without spaces, to link nonconsecutive numbers (e.g., <1-4,7>).

Tables and Figures: Prepare according to criteria in "Uniform requirements" and "Illustrative material: What editors and readers expect from authors" *(Can Med Assoc J* 1990; 142: 447–449). Number references cited in tables and figure legends according to where table or figure is first cited in text. Authors are required to pay additional cost of color figures. Submit a positive transparency and four color prints of each color figure.

Abstracts: 250 word structured abstracts for original research articles. See format proposed by Haynes and colleagues *(Ann Intern Med* 1990;113: 69–76). For articles containing original data concerning course (prognosis), cause (etiology), diagnosis, treatment, prevention or economic analysis of a clinical disorder or an intervention to improve quality of health care, use these headings and information.

Objective. State main questions or objectives of study and main hypothesis tested.

Design. Describe basic design of study; indicate use of randomization, blinding, criterion standards for diagnostic tests, temporal direction (retrospective or prospective), etc.

Setting. Indicate study setting, including level of clinical care (e.g., primary or tertiary, private practice or institutional).

Patients (or participants). State selection procedures, entry criteria and numbers of participants entering and finishing study.

Interventions. Describe essential features of interventions, including method and duration of administration.

Main outcome measure(s). Indicate primary study outcome measure(s) as planned before data collection began. If hypothesis being reported was formulated during or after data collection, clearly say so.

Results. State main results of study using actual numbers. Describe measurements that are not evident from nature of main results; indicate any blinding. Include confidence intervals (most often the 95% interval) and exact level of statistical significance. For comparative studies relate confidence intervals to differences between groups. Indicate absolute values when risk changes or effect sizes are given. Do not report data in abstract that do not appear elsewhere.

Conclusions. State only those conclusions of study that are directly supported by evidence reported, along with their clinical application (avoiding speculation and overgeneralization) and an indication of whether additional study is required before information should be used in clinical settings. Give equal emphasis to positive and negative findings of equal scientific merit.

For additional statements on *CMAJ's* requirements for articles on original research see (1989;141:17–19), case series (1990;142:1205–1206), case-control studies (1990; 143: 17–18), cohort studies *(ibid:* 179–180) and randomized controlled trials *(ibid:* 381–382).

OTHER FORMATS

BIOMEDICAL REVIEW ARTICLES (AND META-ANALYSIS)— 3500 words (18 double-spaced pages). Consult "Biomedical review articles: What editors want from authors and peer reviewers" *(Can Med Assoc J* 1989; 141: 195–197).

Abstracts: 250 word structured format prepared according to format proposed by Haynes and colleagues *(Ann Intern Med* 1990; 113: 69–76). Use headings.

Objective: State primary objective of review.

Data sources. Summarize data sources used, including time restrictions, indexing terms and constraints.

Study selection: Identify number of studies reviewed and criteria for selection.

Data extraction: Describe guidelines used for extracting data and how they were applied.

Data synthesis: State main results of review and methods used to obtain results.

Conclusions: State primary conclusions and their clinical applications; avoid over-generalization. Suggest areas for additional research if needed.

CLINICAL PRACTICE GUIDELINES OR CONSENSUS STATE-MENTS–3500 words (18 double-spaced pages). Statements from associations, societies or consensus groups. Consult "Initiating, conducting and maintaining guidelines development programs" *(Can Med Assoc J* 1993; 148: 507–512).

Abstracts. 250 words structured according to format proposed by Hayward and associates *(Ann Intern Med* 1993; 118: 731-737). Use following headings.

Objective. State primary objective of guidelines, including health problem and targeted patients, providers and settings.

Options. Indicate clinical practice options considered in formulating guidelines.

Outcomes. Describe significant health and economic outcomes considered in comparing alternative practices.

Evidence. Describe how and when evidence was gathered, selected and synthesized.

Values. Disclose how values were assigned to potential outcomes of practice options and who participated in process.

Benefits, harms and costs. Indicate type and magnitude of expected benefits and harms to patients and expected costs of guidelines implementation.

Recommendation(s). Summarize key recommendations.

Validation. Report any external review, comparison with other guidelines or clinical testing of guidelines use.

Sponsor(s). Identify person(s) who developed, funded and endorsed guidelines.

CASE REPORTS—800 words (four double-spaced pages), excluding title page and references. Include title page, 100-word summary, introduction, case report, comments and references. See "Case reports: What editors want from authors and peer reviewers" *(Can Med Assoc J* 1989;141:379–380).

EDITORIALS AND PLATFORM ARTICLES—1500 words (eight double-spaced pages); include a 100-word summary, a statement of problem plus a tentative answer, a middle section that develops author's argument and a conclusion that proposes a clear solution. For more information see "Editorials and Platform articles: What editors want from authors and peer reviewers" *(Can Med Assoc J* 1989;141:666–667).

PROGRAM DESCRIPTIONS—1500 words (eight double-spaced pages) about new programs or services; include title page, 150-word nonstructured abstract, introduction, description of program or service, discussion and refer-

ences. For additional information see "Descriptive studies: What editors want from authors and peer reviewers" *(Can Med Assoc J* 1989; 141:879–880).

OTHER

Electronic Submission: If manuscript was prepared on a computer-based word-processing system, specify in cover letter the software program and version used. Indicate if willing to send a diskette with revised manuscript.

SOURCE

July 1, 1994, 151(1): 55-57.

Cognitive Brain Research
see Brain Research: Cognitive

Community Dentistry and Oral Epidemiology
Munksgaard International Publishers

Dr. O. Fejerskov
Royal Dental College
Vennelyst Boulevard
DK-8000 Aarhus, Denmark

AIM AND SCOPE

The aim and scope of *Community Dentistry and Oral Epidemiology* are to promote the exchange of scientific data and knowledge within the entire spectrum of community dentistry ranging from descriptive epidemiology and clinical sciences to behavioral sciences and sociology. The journal will deal with such data obtained from all age groups, and particular interest will be devoted to the growing problems of dental health care for the elderly. The journal encourages reports of analytical epidemiology, methodologically detailed, which link diseases with biological, social, or behavioral variable in ways that assist identification of causes, risk factors, determinants, and probabilities. Surveys will be published in short form. Studies are encouraged which address pertinent questions relating to improving the development of dental services in economically developing countries. It is also their desire to make the journal a forum for scientific debate and so are anxious to receive letters to the editor. Book reviews and short reports from important international meetings are also welcome.

MANUSCRIPT SUBMISSION

Submit two copies double-spaced, on ISO A4 size paper. Submit manuscript, whenever possible, in machine-readable form and accompanied by a diskette. Note software used, program version number, and format of diskette.

Paper must represent unpublished original research that is not being considered for publication elsewhere. Paper must have been read and approved by all authors. Submission means that authors automatically agree to assign exclusive copyright to Munksgaard International Publishers if and when manuscript is accepted for publication.

Follow Vancouver system for biomedical manuscripts, see International Committee of Medical Journal Editors: Uniform requirements for manuscripts submitted to biomedical journals. *Br Med J* 1982; *284:* 1766–70. Also see *Nordic Biomedical Manuscripts. Instructions and Guidelines* (Svartz-Malmberg, G. & Goldman, R., eds., Universitetsforlaget: Oslo, Norway).

MANUSCRIPT FORMAT

Title Page: Give following information in order: Full title of manuscript; Authors' full names; Authors' institutional affiliations including city and country; A running title (60 letters and spaces); and name and address of corresponding.

Abstract Page: On a separate page: authors' surnames and initials; title of manuscript; title of Journal, abbreviated as in reference list; the word Abstract followed by a summary of the complete manuscript; Key words according to *Index Medicus*; name and address of reprint request author.

References: Number consecutively in order of first mention in text. Identify references in texts, tables, and legends by Arabic numerals (in parentheses). Style is based on format used by U.S. National Library of Medicine in *Index Medicus*.

Avoid using abstracts of articles as references. Do not use "unpublished observations," "personal communications," and unaccepted papers as references; insert references to written, not verbal, communications (in parentheses) in text.

JOURNALS

Standard journal article (List all authors when six or less; with seven or more, list first three and add *et al.*).

MALTZ M, ZICKERT I. Effect of penicillin on Streptococcus mutans. *Scand J Dent Res* 1982; *90:*193–9.

CORPORATE AUTHOR

WHO COLLABORATING CENTER FOR ORAL PRECANCEROUS LESIONS. Definition of leukoplakia: an aid to studies on oral precancer. *Oral Surg* 1978; *46:* 518–39.

BOOKS AND OTHER MONOGRAPHS

Personal author(s)

PRADER F. *Diagnose and Therapie des infizierten Wurzelkanales.* Basel: Benno Schwabe, 1949; 123.

CHAPTER IN BOOK

BRANDTZAEG P. Immunoglobulin systems of oral mucosa saliva. In: Dolby AD, ed. *Oral mucosa in health and disease.* London: Blackwell, 1975; 137–214.

Illustrations: All graphs, drawings and photographs are considered figures; number in sequence with Arabic numerals and abbreviate Fig(s). Give each figure a legend. Type all legends together on a separate sheet numbered correspondingly. Type text on figures in capitals. Plan figures to fit proportions of the printed page. Submit original color transparencies and two sets of color prints. Cost must be paid by author. Current rates will be quoted by publisher.

Tables: Use only to clarify important points. Double documentation is not accepted. Number tables consecutively with Arabic numerals. Type each table on a separate sheet, with due regard to proportions of the printed page.

OTHER FORMATS

SHORT COMMUNICATIONS—1 page for quick publication. Usual division into Material and methods, etc. need not be followed. Include an abstract.

OTHER

Abbreviations, Symbols and Nomenclature: Consult CBE Style Manual Committee. *Council of Biology Editors Style Manual: a Guide for Authors, Editors and Publishers in the Biological Sciences* (4th ed. Arlington, Virginia: Council of Biology Editors, 1978) and O'Connor M, Woodford FP. *Writing Scientific Papers in English: an ELSE-Ciba Foundation Guide for Authors.* (Amsterdam: Elsevier-Excerpta Medica, 1975).

Experimentation, Human: It is assumed that ethical aspects of his (her) research have been considered and attention paid to the guidelines of the Helsinki declaration as revised in 1975.

Page Charges: D.kr. 750 for additional pages over 3 printed pages.

SOURCE

February 1994, 22(1): inside back cover.

Comparative Biochemistry and Physiology

Pergamon Press

Professor G. A. Kerkut
Department of Physiology and Biochemistry
University of Southampton
Southampton SO9 3TU, U.K.

AIM AND SCOPE

The journal will publish papers that contain the results of original research on the biochemistry, physiology and pharmacology of animals. Particular attention will be paid to those papers in which the subject is approached from a comparative point of view.

Part A (Comparative Physiology) deals with cellular physiology, respiration, circulation, neurophysiology, sensory physiology, muscle, digestion, endocrinology, nutrition, osmoregulation, excretion, etc.

Part B (Comparative Biochemistry) deals with metabolism, enzymology, amino acids, proteins, carbohydrates, lipids, nucleic acids, respiratory pigments, second messengers, hormones, receptors, venoms, etc.

Part C (Comparative Pharmacology and Toxicology) deals with the study of the action of drugs and chemicals (including pollutants) on cells, tissues and whole animals. In particular, it is concerned with the differences between species and organs in their responses to drugs. This section will carry papers of interest to workers on toxicology, insecticides, molluscicides, anthelmintics and antiprotozoan drugs. In addition, basic research will be published on nerve muscle transmitters, nerve-nerve transmitters, and other pharmacological studies on simpler animals.

Mini Reviews are welcome on any of the above subjects, for sections A, B or C.

MANUSCRIPT SUBMISSION

Submit two copies of each paper. Submission is held to imply that it has not previously been published; that it is not under consideration for publication elsewhere; and that if accepted it will not be published elsewhere in the same form, in English or in any other language, without the written consent of the Editor.

MANUSCRIPT FORMAT

Follow conventional form—introduction and literature, materials and methods, results, discussion and conclusions, and references. Include an abstract (50-100 words), immediately after title and author's name, consisting of numbered sentences that summarize paper's main facts and conclusions.

Type double-spaced on one side of A4 paper (approx. 12 in. × 8 in.) with a 1.5 in. left margin. Two typescripts are required. Retain a copy.

Keep communications as concise as possible. Indicate those parts of paper that might be printed in small type. Avoid footnotes. Do not use italics for emphasis. Supply a running head (shortened title, 45 letters and spaces).

It is not necessary to publish all the individual results of replicated experiments. State number, mean value, and some appropriate measure of variability. Indicate methods of analysis followed, but give statistical details such as tables of analysis of variance only if relevant to discussion. In a statement that the difference between mean values of two groups is statistically significant, indicate level of significance attained.

Tables: Tables are photo-offset directly from typed manuscript. Prepare tables for direct camera copy. Refer to current tables in journal, for required spatial layout. Type should be clear and even. If possible, use a laser printer with Times Roman font. Type tables, headings and legends on a separate sheet. Insert heavy rules at head and foot of each table, and fine rules below column headings. Leave minimum space required to avoid confusion between columns. Type genus and species names, and other words normally italicized, in italics or underlined.

For guidance on how to prepare tables for photographic reproduction, contact Photoreprographic Section of your institution.

Illustrations: Do not insert illustrations in text. All photographs, charts and diagrams are referred to as "Figs"; number consecutively in order referred to in text.

Photographs, including photomicrographs, should be glossy prints; restrict to minimum necessary. Lightly write on back, author's name, figure number and indicate TOP. Lines or lettering to appear on photographs should be in good quality stencil. Make illustrations no larger than foolscap size. All illustrations, diagrams and graphs should have the final labeling ready for press.

Give full Latin name of all animal species used.

Do not publish same data in tables and figures.

Figure Legends: Type captions consecutively on a separate page(s) at end of paper. Make caption or legend sufficient so that tables or figures are intelligible without reference to text.

References: Give full references at end of paper in alphabetical order. Include names of authors, date of publication, full title of paper, title of journal, volume number and first and last page numbers. For references to books, include edition number, volume, relevant pagination, publisher and town of publication. In text quote a reference by author's name and date in parentheses, as, for example: Smith (1960).

Ashby W. (1943) Carbonic anhydrase in mammalian tissue. *J. biol. Chem.* **151,** 521–527.

Evans P. D. (1985) Octopamine. In *Comprehensive Insect Physiology, Biochemistry and Pharmacology* (Edited by Kerkut G. A. and Gilbert L. I.), Vol. 11, pp. 499–530. Pergamon Press, Oxford.

OTHER

Abbreviations: Use only standard abbreviations. Where specialized terms are given, indicate a specific abbreviation as a footnote to paper.

SOURCE

1994, 108A(1): v-vi.

Critical Care Medicine
Society of Critical Care Medicine
Williams & Wilkins

Bart Chernow, MD, FCCM
8101 East Kaiser Boulevard
Anaheim, CA 92808-2259
Telephone: (714) 282-6009

MANUSCRIPT SUBMISSION

Submit original and four copies of manuscript, including copies of figures, double-spaced on 8.5 × 11 in. plain paper. Submit three sets of glossy figures or computer-generated original paper copies. Copyright ownership is transferred by the following written statement, which must be submitted with manuscripts and signed by all authors.

"The undersigned authors transfer all copyright ownership of the manuscript (title of article) to Williams & Wilkins in the event the work is published. The undersigned authors warrant that the article is original, is not under consideration by another journal, and has not been published previously."

Include a cover letter that designates a corresponding author with an address (including "zip plus four" zip code), phone and fax number. Restrict number of authors to only those persons who have truly participated in conception, design, execution, and writing of manuscript. Disclose potential financial or ethical conflicts of interest regarding contents of submission.

MANUSCRIPT FORMAT

Title Page: Include title, a shortened running title, full names of authors, including degrees and one fellowship designation, institutional affiliations, name of institution where work was performed, address for reprints, ten key words, and financial support for study.

Abstract: Not required for case reports or brief reports. Use structured abstract format. For details on preparing structured abstracts, see *JAMA* 1992; 268:42–44. Use headings: Objective, Design, Setting, Patients (for Clinical Investigations) or Subjects (for Laboratory Investigations), Interventions, Measurements and Main Results, and Conclusions.

For review papers and special articles use headings: Objective, Data Sources, Study Selection, Data Extraction, Data Synthesis, and Conclusion.

Text: Organize into sections: Introduction, Materials and Methods, Results, and Discussion. Acknowledgments, References, Figure Legends, and Tables. Secretarial and editorial assistance are not acknowledged. Present results in text, figures, or tables. Do not reiterate tabular data in text or use superfluous illustrations. Discussion section interprets results without unnecessary repetition. Include references to related studies in this section.

Key Words: Select ten key words from *Medical Subject Headings* of *Index Medicus*.

References: Cite references in sequential order in text. Type double-spaced on a separate sheet. Identify references in text, tables, and legends by full-size Arabic numerals on the line and in parentheses. Set journal titles in italics; abbreviate according to *Index Medicus*; spell out journal titles not listed. Note unpublished data or personal communications parenthetically within text. Follow format of "Uniform Requirements for Manuscripts Submitted to Biomedical Journals" (see February 1988 issue *Annals of Internal Medicine* or February 6, 1988 issue of *British Medical Journal*). Use inclusive page numbers for all references.

Tables: Type single-spaced on a separate sheet. Give each an Arabic number and a descriptive title. Do not repeat material in text. Do not use all capital letters in headings and text. Use only tabs and line returns to format tables. Do not use center, decimal tab, and justification commands. Separate columns by a single tab. Do not use spaces to separate columns. Use a single tab, not a space, on either side of ± symbol. Do not underline or draw lines within tables (do not use table creation functions): Reference footnoted information using italicized, superscript, lowercase letters. Identify each abbreviation in a paragraph in the footnote to the table (e.g., MAP, mean arterial pressure; HR, heart rate). Reference tables in text in sequential order.

Figures: Have illustrative material professionally drawn. Submit three sets of glossy black-and-white prints. Computer-generated original paper copies are also acceptable. Stack figures to save space. Details must be clear when reduced for a 3 5/16 in. single-column width or a 7 in. double-column width (or provide camera-ready). Radiographic prints may need arrows for clarity. Use overdots, overbars, and units according to journal style. Color illustrations will occasionally be published; authors will be charged costs . Label back of each illustration with senior author's last name, figure number, and an arrow indicating TOP. Type legends double-spaced on a separate sheet. Figures must be referenced sequentially in text.

OTHER FORMATS

LETTERS TO THE EDITOR—Type double-spaced with references; submit on a disk with two copies.

OTHER

Abbreviations: Except for clinically used abbreviations, such as adult respiratory distress syndrome (ARDS), mean arterial pressure (MAP), blood pressure (BP), and heart rate (HR), avoid use.

Electronic Submission: Submit manuscript on a diskette, 3.5 or 5.25 in., with four hard copies. If submission on a diskette is not possible, submit cleanly reproduced hard copies to enable accurate electronic scanning. Type double-spaced and leave 1 in. margins on all sides. Indent five spaces to begin new paragraphs. Number all pages consecutively beginning with title page. Label diskette with: date, name of author, word processor and version number, and filename(s). Do not write on a floppy diskette, except with a felt-tip pen. Do not paperclip diskette to its hard copy.

Keep tables in separate files;, FILE.T1, FILE.T2, etc. Type tables beginning flush left; separate data with tab spaces. Refer to specific instructions for table preparation.

Word Processing software accepted (WordPerfect preferred):AmiPro (all versions); CEOwrite through 2.0; DCA/RFT or DCA/FFT; DEC WPS PLUS (DX) through 4.1; DisplayWrite 2-5 through 1.0; First Choice through 3.0; IBM Writing Assistant through 1.01; MacWrite II 1.1; Microsoft Word through 6.0 DOS/WIN (Mac through 4.0); Microsoft RTF through 1.0; MultiMate through 4.0 NAVY DIF (from DIF only); Nota Bene through 3.0; PFS Write A through C; Professional Write 1.0, 2.0, 2.1, 2.2; Samna Word through Samna Word IV Plus; SmartWare II 1.5; Wang PC (IWP) through 2.6; WordMARC through Composer Plus; WordPerfect through 6.0 DOS/WIN (Mac through 2.2); WordStar through 6.0; WordStar 2000 3.5; XyWrite through III Plus. If word processing software is not listed, submit in ASCII format.

Experimentation, Animal: Include a statement within Materials and Methods section that study was approved by Institutional Review Board for the care of animal subjects, and that care and handling of animals were in accord with National Institutes of Health guidelines, or some other internationally recognized guidelines for ethical animal research.

Experimentation, Human: Include a statement within Materials and Methods section indicating approval of study by Institutional Review Board, that subjects have signed written informed consent, or that Institutional Review Board waived need for informed consent.

Manufacturers: Provide in parentheses model number, name of manufacturer, city, and state or country, for all equipment described. Use only generic drug names . Note exceptions in accompanying correspondence).

Permissions: Submit written letters of permission both from author and copyright holder (Publisher) for submitted materials that are to be reproduced (or adapted) from copyrighted publications.

Units: Use in standard American measurement units with SI units in parentheses. See *JAMA* 1986; 255:2329, for details regarding SI units for laboratory data.

For standard American units, do not use values that are more significant than analysis is capable of accurately measuring; e.g., Pao_2 84 torr (11.2 kPa), not 83.7 torr.

For hemodynamic measurements for pressure (e.g., MAP) use mm Hg and for gas measurements (e.g., Po_2) use torr with SI units in parentheses. Units for vascular resistance are dyne·sec/cm^5. Use mL instead of cc. Provide SI equivalents for blood chemistry.

SOURCE

November 1994, 22(ii): I–II.

Critical Reviews in Biochemistry and Molecular Biology
CRC Press

Dr. Gerald D. Fasman
Graduate Department of Biochemistry
Brandeis University
Waltham MA

AIMS AND SCOPE

Critical Reviews in Biochemistry and Molecular Biology publishes reviews that organize, evaluate, and present the current status of a particular topic in biochemistry and molecular biology. Critical surveys of specific topics of current interest are included. The reviews aim to give a better understanding of the trends of modern biochemistry and molecular biology. They permit the synthesis of fruitful hypotheses and new concepts and typically integrate and give interpretive evaluations of conflicting literature.

MANUSCRIPT SUBMISSION

Contact the editor to determine the suitability of a topic for review. Submit two copies of reviews, typed double spaced, with pages numbered and a running head on each sheet. On a separate cover page, give author names, affiliations, address, and phone with title of article. Obtain permission for use of copyrighted figures, tables, or text from previous publications.

MANUSCRIPT FORMAT

Use three heading weights. First section following abstract is an Introduction; final section is References.

Abstract: 200-500 words.

References: Within text, cite by author and year. List in References section is alphabetical.

JOURNAL ARTICLES

Chou, K. C. and Zhang, C. T., A correlation-coefficient method for evaluating protein structural classes. *Eur. J. Biochem.*, **207**, 429–433 (1992).

BOOKS

Cohen, B. B., *Biochemistry*, C. V. Mosby, St. Louis (1988).

ARTICLES IN BOOKS

Cohen, F. E., and Abarbanel I. D., Pattern based prediction of protein structure. Langone, J. J. , Ed., *Methods in Enzymology*, Vol. 202, pp. 252-264, Academic Press, San Diego (1991).

SOURCE

January 1995, 30(1): front matter.

Critical Reviews in Microbiology
CRC Press

Dr. Ronald M. Atlas
Department of Biology
University of Louisville
Louisville KY 40292

AIMS AND SCOPE

Critical Reviews in Microbiology publishes reviews in all areas of microbiology, including bacteriology, virology, phycology, mycology, and protozoology. Reviews cover all disciplines of microbiology, including molecular biology, microbial genetics, microbial physiology, microbial biochemistry, microbial structure, microbial taxonomy, food microbiology, industrial microbiology, medical microbiology, epidemiology, public health, diagnostic microbiology, microbial ecology, and environmental microbiology. The reviews describe the current status of a field, placing the various contributions to that field in perspective. Authors of reviews are encouraged to critique the articles cited, indicating the significance of each article, its strengths, and weaknesses. The reviews permit the authors latitude to express their own views of the field, the historical development of that field, its current status, and the future directions of that field.

MANUSCRIPT SUBMISSION

Submit two copies of reviews, typed double spaced, with pages numbered and a running head on each sheet. On a separate cover page, give author names, affiliations, address, and phone with title of article. Obtain permission for use of copyrighted figures, tables, or text from previous publications.

MANUSCRIPT FORMAT

Use three heading weights. First section following abstract is an Introduction; final section is References.

Abstract: 200-500 words.

References: Within text, cite by a number, set as a superscript. Number in order of appearance, not alphabetically. List in References section is numerical by order of citation in text.

JOURNAL ARTICLES

Miller, J. A. and Miller, E. C., The metabolic activation of carcinogenic aromatic amines and amides, *Prog. Exp. Tumor Res.* 11, 273, 1969.

BOOKS

Barrett, J. T., *Textbook of Immunology,* 3rd ed., C. V. Mosby, St. Louis, 1978, 261.

ARTICLES IN BOOKS

Walzer, P. D., Kim, C. K., and Cushn, M. T., *Pneumocystis carinii,* in *Parasitic Infections in the Compromised Host.* Walzer, P. D. and Genta, R. M., Eds., Marcel Dekker, New York, 1989, 83.

CITATION IN REFERENCED WORK

Snell, J. J. S., The distribution of nonfermenting bacteria, *cited in:* Alexander, M., Ed., *Adv. in Microbial Ecology,* Vol. 2, Plenum Press, New York, 1978, 272.

SOURCE

January 1995, 21(1): front and back matter.

Critical Reviews in Toxicology
CRC Press

Dr. Roger O. McClellan
Chemical Industry Institute of Toxicology
P.O. Box 12137
6 David Drive
Research Triangle Park, NC 27709

AIMS AND SCOPE

Critical Reviews in Toxicology publishes reviews in the field of toxicology. The Journal is aimed at critical assessments of subjects that are part of the advancing frontiers of toxicology and related basic scientific disciplines; or issues that form the core of some of the most complex and intractable problems with which the toxicologist has to grapple.

MANUSCRIPT SUBMISSION

Contact the editor to determine the suitability of a topic for review. Submit two copies of reviews, typed double spaced, with pages numbered and a running head on each sheet. On a separate cover page, give author names, affiliations, address, and phone with title of article. Obtain permission for use of copyrighted figures, tables, or text from previous publications.

MANUSCRIPT FORMAT

Use three heading weights. First section following abstract is an Introduction; final section is References.

Abstract: 200-500 words.

References: Within text, cite by a number, set as a superscript. Number in order of appearance, not alphabetically. List in References section is numerical by order of citation in text.

JOURNAL ARTICLES

Miller, J. A. and Miller, E. C., The metabolic activation of carcinogenic aromatic amines and amides, *Prog. Exp. Tumor Res.,* 11, 273, 1969.

BOOKS

Ward, J. T., *Textbook of Toxicology,* 3rd ed., C. V. Mosby, St. Louis, 1978, 261.

ARTICLES IN BOOKS

Walzer, P. D., and Kim, C. K., The toxic reactions due to toxins produced by the red tide algae, in *Toxins Produced by Algae,* Walzer, P. D. and Genta, R. M., Eds., Marcel Dekker, New York, 1989, 83.

SOURCE

January 1995, 25(1): front matter.

Current Genetics
Springer International

F. Kaudewitz, Managing Editor
Institut für Genetik und Mikrobiologie
Universität Munchen
Maria-Ward-Strasse, Munchen, Germany
Fax: 49 89 17 91 98 31

C.W. Birky, Jr. Regional Editor
Department of Molecular Genetics
484 West 12th Avenue
The Ohio State University
Columbus, OH 43210
Fax: (615) 292 4466
                    For The Americas

AIM AND SCOPE

Current Genetics is an international journal devoted to the rapid publication of original articles on genetics of eukaryotic organisms, with the emphasis on yeasts, other fungi, protists and cell organelles.

MANUSCRIPT SUBMISSION

Papers on formal genetics will only be accepted if they contribute to the understanding of specific genetic systems or lead into molecular approaches. All papers must be of sufficient immediate importance to justify urgent publication. They should present new scientific results. Omit any data which only confirm previously published findings.

Submit manuscripts and illustrations in triplicate to an editorial board member. Send just original typed on one side of sheet only. Photocopy additional copies on both sides.

Furnish telex or fax numbers on title page. Double-space papers, keep wide margin. Indicate desired positions of figures and tables in left margin. Restrict manuscript to 8 printed pages, including tables and figures (24 typewritten A4 pages with 28 lines per page and 57 strokes per line).

MANUSCRIPT FORMAT

Structure papers as follows: title page, Abstract, Introduction, Materials and methods, Results, Discussion, Acknowledgments, References, tables, legends for all figures listed together on a separate page, and figures. Underline genus or species names, gene symbols, mathematical formula and any other words to be given special emphasis for italics. Clearly distinguish similar signs, e.g., 1 (one) and l (letter) or 0 (zero) and O (letter).

Title Page: Include, in order: title of paper; names of authors and institutes; and mailing address of corresponding author. If more than one institute is involved, make affiliations of all authors clear.

Abstract: 100 words giving the main experimental results. List 4 key words corresponding to the main terms in article/title.

Introduction: Avoid a detailed historical Introduction; outline main problem briefly. Establish continuity with earlier work on subject by reference to recent papers and reviews.

Materials and Methods: Avoid repeating description of techniques already published elsewhere. Describe experimental methods clearly.

Discussion: Restrict to main results of experimental data.

References: List only works cited in text. List only works accepted for publication at end of paper in alphabetical order under first author's name. List works by two authors alphabetically according to second author's name. List works by three or more authors chronologically. If more than one work by the same author or team of authors in the same year, add a, b, c, etc. to year both in text and reference list. Use internationally accepted abbreviations of journal titles.

BOOKS

Attardi G, Gaines G, Montoya J (1983) Regulation of ribosomal RNAa. In: Schweyen RJ, Wolf K, Kaudewitz F (eds) Mitochondria 1983: nucleo-mitochondrial interactions. de Gruyter, Berlin New York

JOURNALS

Kanno A, Hirai A (1993) A transcription map of the chloroplast genome from rice, Curr Genet 23:166–174

Citations in text are by author and date. When two authors, name both; when more than two authors, give first author's name plus "et al."

Figures: Restricted to minimum needed to clarify text. Give all figures short legends and submit on separate sheets. Information in legends should not be repeated in text. Do not present the same data in both graph and table form. Previously published illustrations are not acceptable. Number all figures, whether photographs, graphs or diagrams, consecutively throughout. Put author's name and figure number on all illustrations. Size figures to match either column width (8.6 cm) or printing area (17.8 × 24.1 cm). Where possible, group figures into a plate on one page. For plate, mount figures on regular bond paper, not on cardboard.

Provide labeled line drawings ready for reproduction. Computer-generated drawings made with a laser printer are also accepted. Do not submit photocopies. Submit high-quality glossy prints in desired final size. Make inscriptions clearly legible.

Half-tone illustrations should be well contrasted and trimmed at right angles to desired final size with inscriptions about 3 mm high.

Tables: Submit on separate sheets. Give short titles.

OTHER FORMATS

SHORT COMMUNICATIONS—2 printed pages, including 1 table or 1 figure. Brief accounts of particularly interesting results.

OTHER

Abbreviations: Follow rules and recommendations of the IUPAC-IUB Commission on Biochemical Nomenclature (CBN) and the IUB Commission of Editors of Biochemical Journals (CEBJ).

Color Charges: DM 1.200.- for first and DM 600.- for each additional page.

Data Submission: Place extensive tables and data on file at author's institution. Give directions for obtaining them in a footnote.

Footnotes: Indicate footnotes to title by asterisks, to text by consecutive superscript numbers; to tables by superscript lowercase letters.

Gene Sequences: Obtain accession number from a database such as EMBL or Genbank. Give accession number in legend of figure showing corresponding sequence. Submit sequences with more than 100 nucleotides camera-ready copy.

SOURCE

1994, 26(1): inside back cover.

Cytogenetics and Cell Genetics
S. Karger Publishers

T. B. Shows
Department of Human Genetics
Roswell Park Memorial Institute
666 Elm Street
Buffalo, NY 14263
Telephone: (716) 845-3108; Fax: (716) 845-8449
        Gene mapping, cloning, molecular genetics

P. I. Pearson
Department of Medicine and the Welsh Medical Library
The Johns Hopkins University
1830 East Monument Street
Baltimore, MD 21205
Telephone: (301) 955-9705; Fax: (301) 955-0054
        Molecular cytogenetics, gene mapping, somatic
                cell genetics, informatics

O.J. Miller
Molecular Biology and Genetics
Wayne State University School of Medicine
540 East Canfield Avenue
Detroit, MI 48201
Telephone: (313)577-5323; Fax: (313)577-5218
        Comparative cytogenetics

M.M. LeBeau
Department of Medicine
University of Chicago
Joint Section Hematology/Oncology
5814 S. Maryland Ave., Box 420
Chicago, IL 60637
Telephone: (312)702-0795; Fax: (312)702-3163
        Cytogenetics and molecular genetics of cancer

T. W. Glover
Department of Human Genetics
4708 Med. Sci. II, Box 0618
University of Michigan
1137 E. Catherine
Ann Arbor, MI 48109-0618
Telephone: (313)763-5222; Fax: (313)763-3784
        Gene mapping and cloning

G. Klein
Microbiology and Tumor Biology Center
Karolinska Instutet, Box 280
S-171 77 Stockholm, Sweden
Telephone: (08)728 64 00; Fax: (08)33 04 98
    Genetic regulation of cell malignancy

J.R. Korenberg
Medical Genetics Birth Defects Center
Cedars-Sinai Medical Center
110 George Burns Road
Davis Bldg., Suite 2069
Los Angeles, CA 90048
Telephone: (310) 885-7627; Fax: (310) 652-8010
    Molecular cytogenetics and gene mapping

L.A. Cannizzaro
Department of Pathology
Albert Einstein College of Medicine
1300 Morris Park Avenue
Bronx, NY 10461
Telephone: (718) 430-2898; Fax: (718) 892-1720
    Molecular cytogenetics and molecular pathology

U. Francke
Department of Genetics
Howard Hughes Medical Institute
Beckman Center for Molecular and Genetic Medicine
Stanford University Medical Center
Stanford, CA 94305-5428
Telephone: (415) 725-8089; Fax: (415) 725-8112
    Somatic cell genetics

AIM AND SCOPE

Cytogenetics and *Cell Genetics* publishes original research reports in mammalian cytogenetics, molecular genetics including gene cloning and sequencing, gene mapping, cancer genetics, comparative genetics, gene linkage, and related areas.

MANUSCRIPT SUBMISSION

Send manuscript in duplicate, including illustrations, to Editor under whose investigative section the paper falls.

MANUSCRIPT FORMAT

Prepare manuscripts strictly in accordance with journal's style. Use 10- or 12-point type. Double-space throughout (including reference list, legends, etc.), with 1.25 in. margins on all sides. On first page give full names of authors and their affiliations, an abbreviated running title, full postal address, telephone and fax numbers. Provide a brief abstract for full reports and special section reports as noted.

References: Quote references in text as: single author: Jones (1990); two authors: Jones and Smith (1991); more than two authors: Jones et al. (1992). Double-space reference list throughout (even within references). Arrange alphabetically according to first author's surname. Give titles in full. For book references include book title, editor(s), publisher, and city of publication. Consult recent journal issues for reference list style. Check agreement between text citations and reference list. Verify accuracy of references.

Tables and Illustrations: Prepare on separate sheets. Give, respectively, headings and legends.. Original drawings and photographs are preferable; negatives cannot be used. Illustrations can be grouped as a block for reproduction. Submit illustrations at reproduction size. Do not exceed 180×236 mm. Do not send original figures larger than 240×300 mm without consulting Editorial Office. Draw figure numbers or letters on illustrations if lettering is adequately large and neatly executed. Otherwise, indicate lettering on a transparent overlay. Provide original figures and karyotypes, not photographic copies. Prepare karyotypes with a minimum of space between chromosome pairs and lines; mount on pure white cardboard. Number chromosomes only if essential. Obtain permission to reproduce illustrations, tables, etc. from other publications.

OTHER FORMATS

ANIMAL CYTOGENETICS AND COMPARATIVE MAPPING—Includes full or short reports of cytotaxonomic, comparative cytogenetic, and comparative molecular genetic studies; and reports of the chromosome complements of rare or endangered species (see Editorial, Vol. 53:177,1990).

BRIEF REPORTS—Three printed pages (1000 words to a printed page) with adequate allowance for title, tables, figures, and reference list (10 citations), and a 60 word abstract. Essentially complete but short presentations of significant new and original findings will be accepted for accelerated publication.

COMMENTARIES—2000 words, no abstact. A forum for observations, opinions, and comments outside the realm of conventional scientific papers. Present submissions completely original and unpublished. Include original data, tables, and illustrations, provided these meet established criteria (see Vol. 29:125-126, 1981). Preliminary reports or information to be published elsewhere will not be accepted.

GENE MAPPING, CLONING, AND SEQUENCING—2000 words, prepared as a brief report, including a 60 word abstract. Provides a space-efficient and rapid means of reporting new findings in these areas. Prepare papers so as to allow readers and database curators to quickly evaluate and extract essential information.

GENETIC LINKAGE DATA—Short, primarily tabular presentations of positive and negative gene linkages in man and other mammals (see Vol. 39:81, 1985). No abstract required.

HIGH PRIORITY REPORTS—Papers containing information of particular benefit to the scientific community will be published 8 to 12 weeks from acceptance. Authors who believe their papers qualify, and who can provide an electronic manuscript, should request consideration at submission. For full details see Editorial in Vol. 62, 1993.

OTHER

Color Charges: $400 per page for up to six per page.

Page Charges: No charge for first five printed pages (5000 words minus tables or figures). Additional pages cost $200 per page. Authors pay extra costs of half-tone engravings exceeding two printed pages at $400 per page. Submit application for dispensation from costs (except color plates) after notification of acceptance.

Processing Fee: $75 must accompany full reports.

SOURCE
1995, 69(1/2): front matter.

Cytometry
International Society for Analytical Cytology

Brian H. Mayall, MD
Department of Laboratory Medicine
University of California, San Francisco
MCB 230, Box 0808
San Francisco, CA 94143-0808
Telephone: (415)476-6683

AIM AND SCOPE

The scope of *Cytometry* embraces all aspects of analytical cytology, which is defined broadly as characterization and measurement of cells and cellular constituents for biological, diagnostic and therapeutic purposes. It includes components of cytochemistry, cytophysics, cell biology, molecular biology, physiology, pathology, image analysis, statistics, instrumentation, clinical laboratory practice, and other relevant subjects.

MANUSCRIPT SUBMISSION

Manuscripts may include new experimental results, descriptions of new methods or modifications of existing ones, or new interpretations of previously published data. Authors are not required to be members of the International Society for Analytical Cytology.

Submission implies that manuscript has been approved by all authors, that it reports unpublished work, and that it is not under consideration for publication elsewhere. If accepted for publication, it will become the sole property of the publisher and will be copyrighted in the name of the publisher. The manuscript, in whole or part, must not be published elsewhere in either English or another language. Submit three complete copies of manuscript. In cover letter state title of article, name(s) of author(s), and a brief statement of the paper's major point.

MANUSCRIPT FORMAT

Double-space manuscript throughout. Number each page consecutively. Begin each section on a new page. Page 1 contains complete title, name and affiliations of each author, a specific mailing address and telephone, telex, and/or fax numbers, and, as title footnotes, credits for research support. If manuscript is based on a conference presentation, acknowledge as a footnote. Except for Book Reviews and Letters to the Editor, page 2 must contain a short title (48 characters including spaces) for use as a running headline. Page 3 must contain Abstract and Key Terms.

Divide Articles, Reviews, Clinical Communications, Brief Reports, and Technical Note into sections:

Title: Be succinct and informative; avoid subtitles.

Abstract: 200 words; state subject, approach, and main conclusions of article. Must be understandable when read alone.

Key Words: Below Abstract provide three to ten key words or short phrases for indexing. Do not repeat terms in title. Include subject under investigation and technologies or methods employed. Use *Medical Subject Headings* from *Index Medicus* whenever possible.

Introduction: State aim of work and problem that stimulated it. Summarize relevant published investigations.

Materials and Methods: Present in sufficient detail to permit work to be repeated by others.

Results: Give concisely; avoid redundant use of both tables and figures to illustrate same data.

Discussion: Limit to interpretation of results with a minimum of recapitulation of findings.

Tables, Figures and Legends: Design to be self-explanatory without reference to text.

Literature Cited: For citation of references use Number Alphabetic System in which reference number is entered in parentheses in text. If an author's name is used in text, reference number follows name at an appropriate point in same sentence. Make references only to published works and papers in press. Do not include work in progress, unpublished experiments, and personal communications in reference list; acknowledge in parentheses in text. Number references serially and arrange alphabetically by last name of first author, by co-authors, and chronologically for same author. Abbreviate journal names according to *Index Medicus*. Verify accuracy of references.

JOURNAL ARTICLE

1. Laborde K, Bussieres L, De Smet A, Dechaux M, Sachs C: Quantification of Renal Na-k-ATPase activity by image analysing system. Cytometry 11:859–868,1990.

ARTICLE IN A BOOK OR COMPARABLE PUBLICATION

2. Gray JW, Cram LS: Flow karyotyping and chromosome sorting. In: Flow Cytometry and Sorting, Second Edition, Melamed MR, Lindmo T, and Mendelsohn ML (eds). J. Wiley & Sons, Inc., New York, 1990, pp. 503–529.

BOOKS, THESES, OR COMPARABLE PUBLICATIONS

3. Shapiro HM: Practical Flow Cytometry. Alan R. Liss, Inc., New York, 1985.

Figures: Number photographs, drawings, and graphs consecutively. Submit three sets of all figures. Label back of each figure with author's name, figure number, and TOP.

Figures must be high contrast glossy prints of black on white background; submit photographs of computer screens as negative images (black on white). Do not submit computer printouts as figures. Crop and edit photomicrographs and photographs to show only essential information. Indicate whether figure should be printed as one or two column width. Include color figures when appropriate.

OTHER FORMATS

ARTICLES—Full-length reports of original research in all fields of analytical cytology that contributes to the development, evaluation, or application of cytometric methods and their use in biology and medicine.

BRIEF REPORTS—Concise communications of preliminary findings. Submit only when rapid communication of results is of immediate importance. Justify special circumstances in cover letter.

CLINICAL COMMUNICATIONS—Address specific clinical applications of Cytometry, including case reports or clinical studies not appropriate as regular articles.

LETTERS TO THE EDITOR—Provide a forum for communication of opinion, interpretation, and new information on scientific/political matters. Be concise and pointed. State purpose clearly in text.

REVIEWS—Comprehensive appraisals of current cytometric issues and advances. Consult Editor about the suitability of a proposed Review.

TECHNICAL NOTES—Briefly describe improvements or helpful modifications in procedures for cytometric and related techniques. Include a brief statement of purpose; an adequately detailed description of procedure; statement of expected results, and references to pertinent literature. Include observations based on application of method.

OTHER

Color Charges: Upon editorial recommendation one or more pages of color figures will be published at no cost. Or authors pay $950 for one page; $500 for subsequent pages. Enclose color prints, and transparencies if available.

Page Charges: There is no upper or lower limit to manuscript length; length must be commensurate with scientific content. Articles longer than eight pages will be charged. Direct requests for exemption from page charges to the Editor.

Products and Companies: When referring to specific products of companies, identify company by citing, in parentheses, its name, city, and state.

Reviewers: Include names and addresses of several persons outside their own institutions who might be qualified to serve as reviewers. Authors can request particular reviewers not be used. Authors may request that their manuscript be send to specific Editorial Board members.

Spelling: Use *Webster's New Collegiate Dictionary.*

Style: For information on style, abbreviations, nomenclature, and preparation of copy, see *CBE Style Manual* (5th Ed, 1983). Spell out abbreviations and acronyms, when first used in text, and then given in parentheses.

Units: Measurement units must conform to International System of Units (SI). If measurement is made in other units, then give SI equivalent in parentheses.

SOURCE

March 15, 1994 18(1): front matter.

Development
The Society of Biologists, Ltd.

Development Editorial Office
The Company of Biologists Limited
Bidder Building, 140 Cowley Road
Cambridge CB4 4DL, UK
Telephone: (0223) 420007; Fax: (0223) 423353

AIM AND SCOPE

The aim of *Development* is to act as a forum for all research that offers a genuine insight into mechanisms of development. Manuscripts will be considered primarily with respect to this aim.

Studies on both plant and animal development are welcome and can be focused upon any aspect of the developmental process, at all levels of biological organization from the molecular and cellular to the tissue levels.

MANUSCRIPT SUBMISSION

Photocopy and include completed copy of *Development* check list from back of journal. Supply address, telephone and fax numbers for correspondence.

MANUSCRIPT FORMAT

Provide three copies of typescript and figures. Manuscripts must be fully ready for press, revision in proof is not possible. Do not exceed 8000 word total count including references and legends. Figures must not exceed the equivalent of four printed pages (page area is 235 mm × 174 mm). Type double-spaced throughout, in order: Title Page, Summary (500 words), Introduction, Materials and Methods, Results, Discussion, Acknowledgments, References, Figure legends (starting a new page), Tables with titles, etc. Number pages.

Supply a short title (40 characters). Retain original illustrations initially; send with revised manuscript.

Illustrations: Label to a professional standard. Create line drawings to fit within column widths of 84 mm, 127 mm or 174 mm; 8 pt Times should be used for text labeling; ensure that symbols are of suitable size for clarity. Supply on disk if possible; refer to disk sheet for details.

Supply original photographs for reproduction; submit at final journal size (maximum length 210 mm, widths as above), if possible. Use press-on symbols and letters (use 14 pt lower case letters for labeling).

Number or letter each figure in a plate. Use capital letters for figure parts.

Indicate magnification with a bar scale.

Write author(s) name(s) and figure number on backs of figures, plus orientation.

Summary: If references must be cited, include journal name, volume number, pages and year, all in parentheses. Do not use abbreviations.

Text: Note sequences submitted to EMBL Database Library (or other data library). Use s.e.m. and s.d. for standard errors, etc.

Cite each figure and table in text in numerical order.

List each reference in References and vice versa; check carefully. Cite literature in text as:

ONE AUTHOR

Jones (1990) or (Jones, 1990; Smith, 1991).

TWO AUTHORS

Jones and Jackson (1990) or (Jones and Jackson, 1990; Smith, 1991).

MORE THAN TWO AUTHORS

Jones et al. (1990) or (Jones et al., 1990a, b; Smith et al., 1986, 1989).

MANUSCRIPTS ACCEPTED FOR PUBLICATION BUT NOT YET PUBLISHED

List in References as (in press).

CITATION OF UNPUBLISHED WORK

Cite unpublished observations, including results submitted for publication, as such in text only, and not in reference list.

Submit written permission for personal communications, i.e., the unpublished observations of other scientists.

References: Cite in alphabetical order by first author's surname. List works by same author(s) chronologically, beginning with earliest publication. "In press" citations must have been accepted for publication; include name of journal or publisher. Include initials after all surnames in list of authors; insert full stop and space after each initial and brackets around date followed by a full stop. Use **bold** for authors' names.

Use USA National Standard abbreviations for Journals.

Jones, P., Beck, C. B. and Smith, J. (1991). Cell division in the sea urchin. Cell **99**, 197–223.

Rossant, J. (1977). Cell commitment in early rodent development. In Development in Mammals vol. 2. (ed. M. H. Johnson), pp. 119–150. Amsterdam: North Holland.

Text Figures/Photographs: Number in a single series (Fig. 1, Figs 2, 3, etc.), in order referred to in text. Type illustration legends together, separate from main text. Submit three copies. Copies of photographic plates must be suitable to judge quality of work; photocopies are not acceptable. Keep original figures; send with revised manuscript.

OTHER

Abbreviations: Define first time used in text—type uppercase without stops (USA, UK); lowercase with full stops (u.v.).

Color Charges: Free of charge at editor's discretion. Charge for excessive use of color is at editor's discretion.

Ions: Use Ca^{2+}, etc.

Isotopes: If isotope is of an element in compound, place symbol for isotope in square brackets; if compound does not normally contain isotopically labeled element, use either ^{131}I-labelled albumin or ^{131}I-albumin.

Reagents: Make specialized reagents available to other scientists.

Sequences: All nucleotide or amino acid sequences should have been accepted by a database library before publication of article. Forms for submission of sequences to EMBL Data Library are available from Editorial Office. See *Development* **108**, 727–734 for information on submitting sequences.

Style: Do not italicize Latin words. Use "and" not "&." Italicize or underline gene names; keep protein products of a gene in Roman type. If you do not have specified fonts, use <u>underline</u> for *italics* and <u>double underline</u> for **bold**. Use Fig. 6A, B or Figs 7, 8.

Units: Use SI units; write out units for time in full—days, hours, minutes, seconds. Type a space after a digit before a unit, e.g., 1 mM (except 1%, 4°C). Use relative molecular mass (M_r) and not MW. M_r is dimensionless; express as $\times 10^3$.

SOURCE

January 1994, 120(1): back matter; inside back cover.

Developmental Biology
Academic Press

Peter J. Bryant, Editor-in-Chief
Developmental Biology
Developmental Biology Center
University of California
Irvine, California 92717-2280
Telephone: (714) 856-4715; Fax: (714) 856-7399;
E-mail: PJBRYANT@UCI

AIM AND SCOPE

Developmental Biology will publish papers containing new and significant information bearing on developmental mechanisms of general interest. Analysis of development, growth, regeneration, and tissue repair in plants and animals at the molecular, cellular, and genetic levels are topics suitable for the journal. Descriptive work, at either the anatomical or molecular level, will not be published unless it leads to fundamentally new conclusions.

MANUSCRIPT SUBMISSION

Send three complete copies including illustrations. In cover letter or on title page, indicate name, address, and telephone number of corresponding author. Do not suggest reviewers or associate editors. Original papers only will be considered. Submission implies assurance that the same work has not been, will not be, nor is presently submitted elsewhere, and that its submission has been approved by all authors and by the institution where work was carried out; further, that any person cited as a source of personal communications has approved such citation. Explicit proof of these implied assurances may be required at the Editor's discretion.

Copyright in article, including the right to reproduce in all forms and media, shall be assigned exclusively to the Publisher.

MANUSCRIPT FORMAT

Do not exceed 15,000 words. Provide a running title (35 letters and spaces). Include a 200 word abstract that should be fully intelligible by itself. Be as concise as possible; combine full documentation of results with maximum brevity and clarity. Standard order of sections: abstract, introduction, materials and methods, results, and discussion. Present quantitative data so readers can judge statistical significance of differences. In discussion relate present work to previous knowledge; do not expand into a review of field. Results and discussion may be combined. Acknowledgments, references, footnotes, tables, and figure legends follow discussion in order.

Number all pages consecutively. Type double-spaced throughout. Use footnotes sparingly; cite with superscript numbers; type on a separate sheet. Do not italicize for emphasis.

Figures: Number consecutively in order of appearance. Give each a descriptive legend. Mark all figures with orientation and authors' names. All figure sizes are "preferred" sizes. Line drawings are reduced in accordance with figure complexity. Good quality prints are acceptable. Maximum size is 215 × 280 mm. Submit photographs as actual size for reproduction, fitting column (90 mm) or page (185 mm) width. Sets of micrographs may be mounted on 10- to 12-point flexible illustration board; combinations up to 203 × 267 mm are acceptable. Provide clear, glossy prints with adequate contrast. Indicate magnification by a bar on photograph. Figures in a plate should have comparable contrast and density; mount on 10- to 12-point flexible illustration board or paper as close together as possible. Do not combine line drawings and halftones into composite plates; if you wish them to appear together, submit unmounted original art and a layout sheet. The printer will place thin white lines between the figures on each plate. Provide an overlay that indicates the most critical areas of the micrograph.

Color Figures—Two color plates will be published *free* of charge in each article, provided color is deemed scientifically necessary by the reviewers and the Editorial Board. Additional color figures will be charged to the author. See the instructions above if mounting is necessary; color figures *must* be submitted on flexible board due to the nature of the reproduction process.

Tables: Type on separate pages. Number consecutively with Arabic numerals; give each a concise heading.

References: To the literature should be cited as: Doe (1987) has observed that . . .; (Doe, 1988). Use Doe *et al.* (1988) for articles with more than two authors. Use suffixes a, b, etc., following date to distinguish two or more works by same authors in same year (Doe, 1987a,b). Arrange literature citations in bibliography according to author surnames.

Colin, A. M. (1986). Rapid repetitive microinjection. *In* "Methods in Cell Biology" (T. E. Schroeder, Ed.), Vol. 27, pp. 395–406. Academic Press, San Diego, CA.

Holley, J. A., and Silver, J. (1987). Growth pattern of pioneering chick spinal cord axons. *Dev. Biol.* **123**, 375–388.

Abbreviate journal titles according to *Chemical Abstracts Service Source Index, 1985.* Do not refer to abstracts or papers not accepted for publication.

OTHER FORMATS

RAPID COMMUNICATIONS—Two pages, 1200 words, two tables or figures, and eight references. Include a 150 word abstract. Findings that will have an immediate and significant impact and should therefore be made known as quickly as possible. May be submitted by fax if no photographs.
REVIEW AND MEMORIAL ARTICLES—Usually solicited; inquiries and suggestions are invited.

OTHER

Abbreviations: Use consistently. Consult lists published in *Proceedings of the National Academy of Sciences (USA), Journal of Biological Chemistry,* and *Biochemistry* for those which need not be defined. Collect all abbreviations into a footnote and refer to it at first use of an abbreviation.

Data Submission: Deposit all genetic sequence information with GenBank. Indicate GenBank accession number with manuscript submission. Articles will include a footnote indicating database and accession number.

Experimentation, Human: Submit copy of document authorizing proposed research, issued and signed by appropriate official(s) of institution where work was conducted.

Materials Sharing: Authors will distribute freely to interested academic researchers for their own use any clone (of cells or DNA or antibodies) used in experiments reported.

Units: Follow Système International. Use prefixes, e.g., μM, not $10^{-6}M$. M stands for molar, not mole. Use hr, min, sec, rpm, and cpm without definition, without a period, and without an "s" in the plural (i.e., 20 min).

SOURCE

November 1994, 166(1): back matter.

Developmental Brain Research
see Brain Research: Developmental

Diabetes
American Diabetes Association

Philip E. Cryer, M.D.
Washington University School of Medicine
Box 8218, 660 South Euclid Avenue
St. Louis, MO 63110

AIM AND SCOPE

Diabetes publishes original research about the physiology and pathophysiology of diabetes mellitus. Submitted manuscripts can report any aspect of laboratory, animal, or human research. Emphasis is on investigative reports focusing on areas such as the pathogenesis of diabetes and its complications, normal and pathological pancreatic islet function and intermediary metabolism, pharmacological mechanisms of drug and hormone action, and biochemical and molecular aspects of normal and abnormal biological processes. Studies in the areas of diabetes education or the application of accepted therapeutic and diagnostic approaches to patients with diabetes mellitus are not published.

MANUSCRIPT SUBMISSION

Submit original manuscript and 2 originals with 4 photocopies of figures and photomicrographs.

Diabetes does not publish material that has been reported elsewhere, including symposia, proceedings, preliminary communications, books, and invited articles, unless presented in conjunction with American Diabetes Association annual meeting. Explain conflicts of interest or support of private interests. Acknowledge in cover letter that manuscript is not under consideration for another publication. Also provide complete address and telephone number of corresponding author.

Submit this copyright transfer statement signed by all authors: In consideration of ADA reviewing my (our) submission, the undersigned author(s) transfers, assigns, or otherwise conveys all copyright ownership to ADA in the event the work is published.

MANUSCRIPT FORMAT

Type double-spaced throughout on 8.5 × 11 in. nonerasable white bond paper, including (in order) title page, summary, text, acknowledgments, references, tables, and figure legends. Number pages consecutively, beginning with title page.

Title Page: Include title; short running title (40 characters, including spaces); first name, middle initial, and last name of each author; name of departments and institutions to which work is affiliated (in English); name, address, telephone number, and facsimile number of corresponding author; and 3–6 keywords (not diabetes).

Summary: 250 words, self-contained summary understandable without reference to text.

Main Text: State species, strain, and other pertinent information of experimental animals. When describing surgical procedures, identify preanesthetic and anesthetic used; state amount or concentration and route and frequency of administration. Do not substitute paralytic agents, e.g., curare or succinylcholine, for anesthesia. For invasive procedure, report analgesic or tranquilizing drugs used; if none, justify.

Use designations insulin-dependent diabetes mellitus (IDDM or type I) and non-insulin-dependent diabetes mellitus (NIDDM or type II) to refer to the two major forms of diabetes mellitus. Do not use diabetic as a noun.

Identify statistical methods.

Acknowledgments: Have acknowledgments of aid or criticism approved by person being recognized. Briefly state acknowledgments of assistance and financial support.

Tables: Type double-spaced on separate sheets. Title all tables and number in order of citation in text. For footnotes, use symbols in sequence: *, †, ‡, §, ||, ¶, #, **, ††, etc., in order from left to right and from top to bottom in body of table.

Figures: Have professionally drawn and photographed or produced on a laser printer. For laser-printed figures, use paper specially made for camera -ready copy (paper opacity and whiteness ≥ 90). Mark with soft pencil on back orientation (arrow pointing up), first author's name, and manuscript number. Do not mount figures. Do not make larger than 5 × 7 in. Crop photographs and gels to one or two column width.

Figures will be reduced to one column width (3.25 in.); produce accordingly. Reduce on photocopier to ensure that data points can be distinguished and labeling is readable. Make information on axes succinct; use abbreviations; label y-axis vertically. Place key information in available white space within figure; if no space, place in legend. Mark figures with multiple parts A, B, C, etc; place descriptions of panels in legend not on figure.

Make lines and symbols in graphs bold enough to be read after reduction. Mark data points with symbols: ○ ● □ ■ Δ ▲. Make bars black and white unless more than two data sets; draw extra bars with clear, bold hatch marks or stripes.

References: Number in order of appearance in text. Identify reference numbers in text by enclosing in parentheses. Do not include works submitted for publication in reference section; cite as unpublished observations in text with initials and last names of all authors. Type references double-spaced. Include all authors (do not use et al. except in text) and complete article titles. Abbreviate names of journals as in *Serial Sources for the BIOSIS Data Base*; spell out names of unlisted journals. Indicate abstracts and supplements. Supply inclusive page numbers. Verify accuracy of references.

1. Primhak RA, Whincup G, Tsankas JN, Milner RDQ: Reduced vital capacity in insulin-dependent diabetes. *Diabetes* 36:324–26, 1987
2. Nerup J, Christy M, Patz P, Ryder P, Svejgaard A: Aspects of the genetics of insulin-dependent diabetes mellitus. In *Immunology in Diabetes*. Andreani D, Dimario U, Federlin KF, Heding LG, Eds. London, Kimpton, 1984, p. 63–70
3. Seine S, Bell GI: Comparison of the 5′-flanking sequences of chimpanzee, African green monkey, and human insulin genes (Abstract). *Diabetes* 34 (Suppl. 1):20A, 1985
4. Permutt MA, Andreone TA, Chirgwin J, Elbein S, Rotwein P: Insulin gene polymorphism and type II or non-insulin-dependent diabetes mellitus (NIDDM). In Proc *Int Congr Endocrinology, 7th*. Labrie F, Proulx L, Eds. Amsterdam, Excerpta Med., 1985, p. 245–48

OTHER FORMATS

RAPID PUBLICATIONS—10 double-spaced typewritten pages, including figures, tables, and references. Observations considered of unusual importance that would lose scientific impact if not published promptly. Describe a completed, concise, and properly controlled investigation.

Submit a diskette and 4 accurate double-spaced paper copies of manuscript. Label diskette: author's name, article title, software and hardware used. Use IBM, IBM-compatible, or Apple computers (do not use "Fast Save" option).

Do not make output approximate or match typeset page. Format manuscript as usual. Mark special characters (including Greek and mathematical characters) on hard copies. Use extended character sets. Never type letter "l" for numeral "1." Never interchange letter "O" for numeral "0." Do not manually hyphenate words; let text wrap. If word processor has automatic hyphenation turn it off. Do not place figure captions and tables within text. Put figure legends after text. Put tables after figure legends. Prepare references in *Diabetes* style.

OTHER

Drugs: Use generic names of drugs. If a special item is obtained, include supplier, city, and state, or city and country if foreign.

Experimentation, Human: Describe characteristics of human subjects or patients. State formally that consent was obtained from subjects after nature of procedure was explained. When anesthetized humans are studied, indicate that procedure was in accord with institutional guidelines. Conduct all human investigations according to the principles of the Declaration of Helsinki.

Page Charge: $50/page for original articles and rapid publications. Identify author responsible for payment.

Style: Follow *CBE Style Manual* (Council of Biology Editors, Inc., Bethesda, MD) and *The Chicago Manual of Style* (The University of Chicago Press, Chicago).

Units: Use metric units. Use Système International (SI) units. See Young DS: *Ann Intern Med* 106:114-29, 1987. For insulin see: Volund A, Brange J, Drejer K, Jensen I, Markussen J, Ribel V, Sorenson AR, Schlichtkrull J: In vitro and in vivo potency of insulin analogues designed for clinical use. *Diabetic Med* 8:839-47, 1991.

For units of measurement see *CBE Style Manual*. Define other abbreviations at first use.

SOURCE

January 1994, 43(1): back matter.

Diabetes Care
American Diabetes Association

Allan L. Drash, MD
Children's Hospital, Rangos Research Center
3705 Fifth Avenue
Pittsburgh, PA 15213
Telephone: (412) 692-5851; Fax: (412) 692-5960

AIM AND SCOPE

Diabetes Care is a journal for the health-care practitioner that is intended to increase knowledge, stimulate research, and promote better management of people with diabetes mellitus. To achieve these goals, the journal publishes original articles on human studies in the areas of epidemiology, clinical trials, behavioral medicine, nutrition education, health-care delivery, medical economics, and clinical care. The journal also publishes clinically relevant review articles, clinical observations, letters to the editor, and health/medical news of interest to clinically oriented physicians, researchers, epidemiologists, psychologists, diabetes educators, and other health-care professionals.

MANUSCRIPT SUBMISSION

Submit five copies of entire manuscript, including tables and figure legends (original and 4 photocopies). If black and white graphs or charts are used, submit 3 original sets of prints, others may be photocopies. If photographs are used, submit 5 glossy sets. Do not submit material that has been printed previously or submitted elsewhere, except abstracts (400 words). The American Diabetes Association holds copyright on all material appearing in *Diabetes Care*. Sign a letter acknowledging no prior publication and copyright transfer to the ADA:

We approve the submission of this paper to *Diabetes Care* for publication and have taken due care to ensure the integrity of this work. We confirm that neither the manuscript nor any part of it has been published or is under consideration for publication elsewhere (abstracts excluded).

In consideration of ADA reviewing my (our) submission, the undersigned author(s) transfers, assigns, or otherwise conveys all copyright ownership to ADA in the event the work is published.

Follow definition and requirements of authorship in Uniform Requirements for Manuscripts Submitted to Biomedical Journals (*N Engl J Med* 324:424–28, 1991); authorship implies substantial contributions to conception and design or analysis and interpretation of data and drafting of article or critical revision for important intellectual content.

State conflict of interest or support of private interests.

Include in cover letter address and telephone and fax numbers of corresponding author.

MANUSCRIPT FORMAT

Original articles report clinical investigation in areas relevant to diabetes. Do not exceed 5000 words (~20 typewritten double-spaced pages), including tables, figure legends, and references necessary to support data and interpretation. Include: hypothesis testing, suitable controls, appropriate statistical methods, clear reporting of results, and conclusions supported by results.

Type double-spaced (including references, tables, and figure legends) on one side of 8.5 × 11 in. noneraseable white bond paper. Keep 1 in. margins at top, bottom, and both sides. Arrange manuscript in order: title page, abstract, introduction (no heading), research design and methods, results, conclusions, acknowledgments, references, tables, and figure legends. Number pages consecutively beginning with title page.

Title Page: Titles should be brief. Include a short running title (<40 characters); first name, middle initial, last name, and highest academic degree of each author; affiliation in English of each author during study; name and address of corresponding and reprint request author; and 3–6 key words for subject indexing.

Abstract: 250 words, self-contained and clear without reference to text. Write for a general journal readership. Use structured format: *Objective*, purpose or hypothesis of study; *Research Design and Methods*, basic design, setting, number of participants and selection criteria, treatment or intervention, and methods of assessment; *Results*, significant data found; *Conclusions*, validity and clinical applicability.

Text: Use designations insulin-dependent diabetes mellitus (IDDM or type I) and non-insulin-dependent diabetes mellitus (NIDDM or type II) when referring to the two major forms of diabetes mellitus. Do not use term diabetic as a noun.

Materials: Provide name and location of source for specified chemicals and other materials only if alternate sources are considered unsatisfactory.

Acknowledgments: Brief statements of assistance, financial support, and prior publication of study in abstract form, if needed.

References: Cite all authors and inclusive page numbers. Abbreviate journal titles according to *Serial Sources for* *the BIOSIS Data Base;* for unlisted journals, give complete journal titles. Verify accuracy of references.

JOURNAL ARTICLES

Banting FG, Best C: The internal secretion of the pancreas. *J Lab Clin Med* 7:251–66, 1922

ABSTRACTS

Seaborn J: Gastrointestinal side-effects of high-fiber diets in diabetic rats (Abstract). *Gut* 33:A4394, 1992

BOOKS

Allen FM: *Studies Concerning Glycosuria and Diabetes.* Cambridge, MA, Harvard Univ. Press, 1913

CHAPTERS IN BOOKS

Stauffacher W, Renold AE: Pathophysiology of diabetes mellitus. In *Joslin's Diabetes Mellitus.* 11th ed. Marble A, White P, Bradley RF, Krall LP, Eds. Philadelphia, PA, Lea & Febiger, 1971, p. 35–98

GOVERNMENT PUBLICATIONS

Fajans SS (Ed.): *Diabetes Mellitus.* Washington, DC, U.S. Govt. Printing Office, 1976 (DHEW publ. no. NIH 76–854)

Figures: Have professionally drawn and photographed or produced on a laser printer. For laser-printed copies, use paper specially made for camera-ready copy, such as Hammermill Laser-Print Plus (opacity ≥ 90 and whiteness ≥ 90). Mark each in soft pencil on back showing orientation (up arrow), first author's name, and manuscript number. Do not mount, staple, or make larger than 5 × 7 in. Crop photographs to one or two columns in width.
Figures will be reduced to one column width (2.25 in.); produce accordingly. Reduce on photocopier to ensure that relevant data points can be distinguished and that labeling is readable. Make information on axes succinct, use abbreviations where possible, label y-axis vertically, not horizontally. Place key information in white space within figure, if no space, place in legend. Mark figures with multiple parts A, B, C, etc. and place a description of each panel in legend, not figure.
Make lines and symbols in graphs bold enough to be read after reduction. Use ○ ● ◻ ■ △ ▲ for data points large enough to be reduced. Make bars black and white if two data sets; if more use clear, bold hatch marks or stripes, not shades of gray.

Tables: Double-space on separate pages with table number and title. Submit tables with internal divisions as individual tables. Limit symbols for units to column headings. Keep abbreviations to a minimum and define in table legend. For footnotes, use following symbols consecutively, left to right, top to bottom of table: *, †, ‡, §, ||, ¶.

OTHER FORMATS

CLINICAL PRACTICE OBSERVATIONS—1500 words based on original clinical findings that tested, refined, validated, or questioned aspects of clinical practice.
COMMENTARIES—1500 word short critical articles on topics in diabetes care and on other articles in issue. Do not include an exhaustive literature review, analyze a few carefully selected findings.

LETTERS TO THE EDITOR—500 word opinions on topics published in journal or relating to diabetes in general.
SHORT REPORTS—1500 word case reports, observations relating to practice of diabetology, and other brief communications.
TECHNICAL ARTICLES—5000 word descriptions and assessments of material and devices used for the care of patients with diabetes.

OTHER

Abbreviations: Use only when necessary, e.g., for long chemical or procedure names or terms used throughout article. At first use, precede by word for which it stands. Abbreviate units of measure only with numbers. Use in table and figures for space considerations. Define in accompanying legends. See *CBE Style Manual* for lists of standard scientific abbreviations.

Experimentation, Animal and Human: Follow principles of the Declaration of Helsinki. All studies involving animals must state that guidelines for use and care of laboratory animals of authors' institution or National Research Council or any national law were followed.

Page Charges: $50 per printed page for original articles and short reports.

Style: Write in clear, concise English following recommendations in the *CBE Style Manual* (5th ed., 1983, Council of Biology Editors). Editing will follow *CBE Style Manual* and *The Chicago Manual of Style* (12th ed., 1982, The University of Chicago Press).

Units: Follow Système International form (see SI table in each issue). Express glycosylated hemoglobin as percentage of total and as standard deviation from mean control levels. See Young DS: *Ann Intern Med* 106:114-29, 1987.

SOURCE

January 1994, 17(1): 102-106.

Diabetologia

European Association for the Study of Diabetes
Springer International

E. Ferrannini, MD
Editor-in-Chief, *Diabetologia*
P. O. Box 296, Via S. Maria, 26
I-56100 Pisa, Italy

AIM AND SCOPE

Diabetologia publishes reports of clinical and experimental work on all aspects of diabetes research and related subjects, provided they have scientific merit and represent an important advance in knowledge. The Editor-in-Chief will consider papers from any country whether or not the author(s) is a member of the EASD.

MANUSCRIPT SUBMISSION

Diabetologia does not publish material that has been printed previously or is under consideration for publication elsewhere. State in cover letter that neither the material, nor any part of it, is under consideration for publication nor has already been published elsewhere in any language or any form (except abstracts of <300 words). Criteria apply particularly to tables and figures and include publication of symposia, proceedings, preliminary communications, books and invited articles. Complete and submit manuscript submission form and signature statement (*Diabetologia* 37(1): A9-A10, January 1994).

Submit five copies of manuscript (original and four double-sided photocopies. Submit five copies of "in press" references, along with acceptance letter. Consult a current issue to ascertain general format, such as use of sub-headings, layout of tables and citation of references. Type double-spaced, on one side of paper only with 3 cm margins on left side of page. Number pages. Give fax number on all correspondence.

MANUSCRIPT FORMAT

Title Page: Include title of paper (including animal species if appropriate); shortened version for page heading (70 letters and spaces, including first author's name, initials, et al.); author's names (initials only—no qualifications); institution(s) of origin.

Summary: 250 words on a separate page describing background, methods, results and the conclusions reached. Use no abbreviations.

Key Words: Five to ten (suitable for *Index Medicus* listing) at end of Summary.

Introduction: A clear statement of aim and novelty of study. Do not include results or conclusions.

Subjects, Materials and Methods: Give sufficient information to permit repetition of experimental work.

Description of Patients: Use terms IDDM (insulin-dependent diabetes mellitus) and NIDDM (non-insulin-dependent diabetes mellitus). Provide detailed descriptions of patients' clinical characteristics upon which classification was made. Do not use diabetic as a noun. Express body weight in terms of body mass index, i.e., (weight in kg) ÷ (height in meters)^2 (not % ideal body weight).

Footnotes: Do not use.

Results: State concisely without comment. Do not present same data in both figure and table. Indicate desired position of figures and tables in margins.

Discussion: Interpret results, do not recapitulate them. Do not refer directly to tables and figures. Relate new information to corpus of knowledge in field. In Rapid Communications, Results and Discussion may be combined.

Acknowledgments: Be as brief as possible. Mention grant support. Write out names of funding organizations in full.

References: Follow ELSE-Ciba style. Keep references in numerical order in text; put number in parentheses on line,. Type and number in same order at end of manuscript on a separate sheet, double-spaced.

ARTICLES IN JOURNALS

Names of up to six authors with initials (abbreviate seven authors or more to et al. after third author's name); (year);

title of paper in full; abbreviated name of journal (according to *Index Medicus*); volume number; first and last page numbers:

Zaninetti D, Crettaz M, Jeanrenaud B (1983) Dysregulation of glucose transport in hearts of genetically obese (*fa/ fa*) rats. Diabetologia 25: 525–529

WHOLE BOOK

Names of *all* authors with initials; (year); title of book; edition; name and city of publisher:

Cobelli C, Bergman RN (1981) Carbohydrate metabolism. 2nd edn. Wiley, Chichester New York Brisbane Toronto.

CHAPTER FROM A BOOK

Names of *all* authors with initials; (year); chapter title; In: editor's names with initials (eds); title of book; volume number; name and city of publisher; pp first and last page numbers:

Mancini M, Rivellese A, Riccardi G (1984) Dietary and oral therapy. In: Belfiore F, Galton DJ, Reaven GM (eds) Diabetes mellitus: etiopathogenesis and metabolic aspects. Frontiers in diabetes, Vol 4. Karger, Basle Munich Paris London New York Tokyo Sydney, pp 160–167

LETTERS TO THE EDITOR

As for articles in journals with (Letter) after page numbers:

Wiseman M, Viberti GC (1983) Kidney size and glomerular filtration rate in Type 1(insulin-dependent) diabetes mellitus revisited. Diabetologia 25:530 (Letter)

ABSTRACTS

As for articles in journals with (Abstract) after page numbers:

Eckel J, Reinauer H (1982) Involvement of microtubules in insulin-receptor turnover in isolated cardiocytes. Diabetologia 23:165 (Abstract)

PAPERS QUOTED (IN PRESS)

Provide copy and written evidence of manuscript's acceptance.

Do not include references to personal communications, unpublished data and manuscripts either "in preparation" or "submitted for publication." Incorporate in parentheses in appropriate place in text and provide written consent to publish.

Verify references. Keep references to abstracts, letters to the editor, congress proceedings, and other non-peer-reviewed publications to a minimum.

Name and Address of Author for Correspondence: Give at end of references, not on title page.

Tables: Type each on a separate page. Number with Arabic numerals. Do not submit glossy prints. Give each a short informative heading which is comprehensible without reference to text. Keep footnotes to a minimum. Do not use footnote to describe method from Methods section. Conditions specific only to particular experiment can be stated. Give units of results in parentheses at head of each column. Do not repeat on each line. Avoid abbreviations. Use superscripts "a," "b," "c" for footnote. See recent issue for examples.

Do not submit extensive tabular material. Editorial Office may accept deposition of extensive tables of important data, which will be made available upon request as photocopies, for a modest fee.

Figures: Do not present same data in both figure and table. On back of each list figure number, author(s)' name and TOP.

Graphs and Diagrams—See a current issue for style and layout–omit headings, include keys to symbols and statistics in legend. Submit one set of original figures and four copies as laser quality computer prints or sharp photographs on white glossy paper. Line drawings and graphs should be of professional standard. Freehand or typewritten lettering and numbering are unacceptable. Make ruled lines, curves and symbols sufficiently bold to withstand reduction to one column width (8.6 cm). Do not draw curves beyond experimental points, or experimental points extend beyond scales of axes. Use solid symbols for scatter diagrams.

Photographs for Half-tone Reproduction—Submit five unmounted copies on white glossy paper printed with best possible contrast. Crop prints to one column width (8.6 cm), or be capable of being reduced to such size. If grouping into plates is desired, indicate layout as diagram. For plates, put figure numbers on lower left corner of photographs; make stand out against background. Make numbers, letters and symbols large enough to be read after reduction (final size 2 mm or larger for lettering and numbering). Indicate magnification in legend and where appropriate by an internal reference marker on photograph itself (as horizontal line 1 mm thick after reduction). If photographs of patients are used, either subject should not be identifiable, or submit written permission for use.

Legends for Figures: Must be informative and comprehensible without reference to text. Type double-spaced on a separate sheet(s). Include keys to symbols and statistical analysis in legend, not in figure.

OTHER FORMATS

FOR DEBATE PAPERS—Usually solicited, unsolicited articles will be considered.

FULL-LENGTH PAPERS—Twelve printed pages including references, tables, figures and legends.

LETTERS TO THE EDITOR—1 printed page, 1000 words, including references and one table or one figure containing critical assessment of papers recently published papers. Type double-spaced with a heading. Do not use abbreviations. Submit two copies.

RAPID COMMUNICATIONS—2 printed pages, 1700 words, including up to 10 references and two illustrations (tables and/or figures) reporting clinical and experimental work. Format is similar to full papers. Submit three copies.

REVIEWS—Usually solicited, unsolicited Reviews may be considered. Consult Editor-in-Chief.

WORKSHOP REPORTS—Workshop or meeting reports may be published. Do not exceed two printed pages of text; one page is preferred. Costs are paid by authors; request fee from publisher. Submit two copies.

OTHER

Abbreviations: A list of accepted abbreviations to be used without definition is shown at end of Instructions to Authors in January and July issues. Use other abbreviations only when necessary. Define in a separate list after key words. No abbreviations should appear in the title, short running title or Summary. Chemical symbols and formula should be used only in the Methods and Results sections. Endocrine pancreatic cells should be designated as beta cells (not B cells or β cells), alpha cells, delta cells or PP cells.

Bioassays and Radioimmunoassays: Give appropriate measures of precision in terms of SD, SEM, coefficient of variation or 95% confidence limits. Reference original technique.

Buffers and Incubation Media: Describe compositions of incubation media, or supply reference, with pH. Describe concentrations of solutions in molar terms (mol/l and subunits thereof), equivalents or percentage weight/ volume or weight/weight. Express mass concentration as g/l (or subunits thereof—mg/l or μg/l). Make clear whether concentrations in a mixture are final concentrations or those of solutions added. In the case of insulin, mU/l is acceptable.

Chemical Substances: Properly identify. Except for standard laboratory chemicals, give source of supply. Identify drugs by generic or official names. Avoid proprietary names.

Color Charges: For color illustrations authors are charged approx. DM 980.00 for first and DM 500.00 for each additional page.

Equipment: Give manufacturer, city and country.

Experimentation, Animals: Give age, sex, source and, where appropriate, genetic background of animals used. Describe animals as killed rather than sacrificed.

Experimentation, Human: Indicate that informed consent has been obtained from patients. Include a statement that responsible Ethical Committee has given approval, and/or that reported investigations have been performed in accordance with principles of the Declaration of Helsinki [*Diabetologia* (1978)15: 431–132].

Permissions: For verbatim material or illustrations taken from other published sources, submit written statement from author, and publisher if holding copyright, giving permission.

Radioisotopes: Identify isotopically labelled chemicals by atomic number and symbol of isotope and its location in molecule. Give specific activity of starting material in terms of curies (Ci) or becquerels (Bq: disintegrations/ second) per molar weight.

Solidus: Use in a unit as long as it does not have to be employed more than once (e.g., mmol/l is acceptable), but ml/min/kg is *not* acceptable; replace with ml·min^{-1}·kg^{-1}.

Spelling: Use *Concise Oxford Dictionary*.

Statistical Analyses: Identify statistical methods used. Identify computer software packages used for anything other than known standard statistical procedures by name or acronym, and by author or organization of origin. When variability is expressed in a standard error of a mean (SEM) or standard deviation (SD), give number of observations (n). Express levels of significance as more or less than or equal to a given probability (e.g., $p < 0.01$).

Avoid common errors: using one-tailed instead of two-tailed tests; inappropriate use of parametric analyses; and lack of correction (for multiple comparisons). Do not use terms 'significant' and 'not significant' in a statistical sense without providing level of significance in terms of p.

Units: Conform to SI convention, except for blood pressure (express in mmHg) and hemoglobin (g/l). Give gas or pressure values as mmHg with kPa in parentheses or vice versa. Where molecular weight is known, express amount of substance in mol or appropriate subunit (mmol). Express energy in kcal or joules (J).

SOURCE

January 1994, 37(1): A5–A11.
Full instructions to authors appear in January and July issues.

Digestive Diseases and Sciences
Plenum Publishing

Richard L. Wechsler, MD
GI Lab, Presbyterian-University Hospital
DeSoto at O'Hara Streets
Pittsburgh, Pennsylvania 15213-2582
Telephone: (412) 647-8444; Fax: (412) 647-8446

MANUSCRIPT SUBMISSION

Submission implies that the manuscript has not been published previously in any form, including symposia and proceedings of meeting, except in brief abstract form, and is not currently under consideration for publication elsewhere. A statement transferring copyright from the authors (or their employers, if they hold the copyright) to Plenum Publishing Corporation is required.

Detach or copy Instructions for Authors page and mail with manuscript. Include and sign following statement:

I certify that none of the material in this manuscript has been published previously in any form and that none of this material is currently under consideration for publication elsewhere. This includes symposia and proceedings of meetings, except in brief abstract form.

MANUSCRIPT FORMAT

Submit four complete sets of typescript (original and three copies), double-spaced throughout on 8.5 × 11 in. paper, with 1.5 in. margins all around. Arrange as: abstract, key words, introduction, methods, results, discussion, references, tables, figure legends, figures.

Title Page: Include title of article, name(s) of author(s) and highest academic degree(s), department and institution from which work originated, acknowledgment of support or sponsorship, postal address for proofs and reprint requests, and suggested running head (40 characters, including spaces). Put personal acknowledgments in a separate section at end of article.

Abstract: 150 words. State concisely purpose, results, and conclusions of study. Do not use abbreviations, footnotes, and references.

Key Words: 4–6 below abstract, selected from *Index Medicus Medical Subject Headings* (MeSH) list.

References: Cite in text by numerals enclosed in parentheses. Type references section double-spaced on separate sheets, numbered consecutively in order cited in text. Include last names and initials of all authors, title of article, name of publication, volume, inclusive pages, and year published. Use abbreviations from *Index Medicus*.

JOURNAL ARTICLE

1. Rinderknecht H, Renner IG, Koyama HH: Lysosomal enzymes in pure pancreatic juice from normal healthy volunteers tested in a Kentucky study. Dig Dis Sci 24:180–186, 1979

BOOK

2. Banks PA: Pancreatitis. New York, Plenum Medical, 1979

CONTRIBUTION TO A BOOK

3. Creutzfeldt W: Endocrine tumors of the pancreas of malnourished rats. *In* The Diabetic Pancreas, BW Volk, KF Wellmann (eds). New York, Plenum Medical, 1977, pp 551–590

Tables: Typed double-spaced on separate sheets. Center table number and title above table and explanatory notes below table.

Figure Legends: Place figure title and caption material in legend, not on figure. Type double-spaced on sheets separate from text. Number figures in one consecutive series in order cited in text. Include sufficient information in legends to interpret figures without reference to text.

Illustrations: Submit three complete sets of unmounted illustrations. Reproductions are clear, with sharp contrast, especially where fine lines such as those of grids or traces are to be retained. Make no larger than 8.5 × 11 in., preferably of a size not needing enlargement or reduction. Submit three sets of glossy print color illustrations (prints not slides). In cover letter state that author will bear cost of color reproduction. Indicate author's last name, figure number, and TOP, if otherwise not clear, on back of each illustration. Submit best quality photomicrographs. Photocopies of original photographs of routine histology and electron microscopy cannot be assessed.

OTHER

Experimentation, Animal: Document that animal experimentation was performed under appropriate circumstances.

Experimentation, Human: Obtain prior approval of appropriate institutional committee on human experimentation. Submit a signed statement as to date and details of appropriate review. In countries where such mechanisms do not exist, submit a signed declaration that research was carried out in accordance with Helsinki Declaration.

SOURCE

January 1994, 39(1): front matter.

Diseases of the Colon & Rectum

American Society of Colon and Rectal Surgeons

**National Society of University Colon
and Rectal Surgeons**

Williams & Wilkins

Robert W. Beart, Jr., MD

Editor in Chief, *Diseases of the Colon & Rectum*

201 First Avenue Building, Suite 240

Rochester, Minnesota 55902

Telephone: (507) 284-0712; Fax: (507) 284-0713

AIM AND SCOPE

Diseases of the Colon & Rectum is designed for the publication of original papers that constitute significant contributions to the advancement of knowledge within the special field designated by the name of this journal.

MANUSCRIPT SUBMISSION

Submit five copies, original and four duplicates with a cover letter (signed by all authors) including: "I (we) certify that I (we) have participated sufficiently in the work to take public responsibility for the appropriateness of the experimental design and method, and the collection, analysis, and interpretation of the data.

I (we) have reviewed the final version of the manuscript and approve it for publication. To the best of my (our) knowledge and belief, this manuscript has not been published nor is it being considered for publication elsewhere."

Name one author as correspondent, with address. Rejected manuscripts will not be returned. Mail required number of manuscript copies in a heavy paper envelope, enclose manuscript copies and figures in cardboard, if necessary, to prevent bending of photographs. Place photographs and transparencies in a separate heavy-paper envelope.

In cover letter state that manuscript has been seen and approved by all authors. Give any additional information that may be helpful, such as type of article, information on publication of any part of manuscript, and willingness to pay cost of reproducing color illustrations. Include copies of permissions needed to reproduce published material or to use illustrations of identifiable subjects. Include permissions to reproduce previously published materials or to use illustrations that may identify subjects.

Follow journal's instructions for transfer of copyright.

Submit copies of any possibly duplicative published material.

MANUSCRIPT FORMAT

Follow "Uniform Requirements for Manuscripts Submitted to Biomedical Journals." Type double-spaced, including title page, abstract, text, acknowledgments, references, tables, and legends. Begin each component on a new page: Title page; Abstract and key words; Text; Acknowledgments; References; Tables (each complete with title and footnotes, on a separate page); Legends for illustrations.

Illustrations must be good quality, unmounted glossy prints usually 5 by 7 in. but no larger than 8 by 10 in.

Type on white bond paper 8 by 10.5 in. or 8.5 by 11 in. or ISO A4 with 1 in. margins. Number pages consecutively, beginning with title page. Type page number in upper right corner.

Title Page: Include title of article (concise but informative); identify a short running head or footline (40 characters, count letters and spaces) placed at foot of title page; first name, middle initial, and last name of each author, with highest academic degree(s); name of department(s) and institutions(s) to which work should be attributed; disclaimers, if any; name and address of corresponding author; name and address of reprint request author or state that reprints will not be available; source(s) of support in form of grants, equipment, drugs, or all.

Abstract and Key Words: On second page place 250 word structured abstract stating purposes of study, basic procedures (study subjects or experimental animals and observational and analytic methods), main findings (give specific data and statistical significance), and principle conclusions. Emphasize new and important aspects of study or observation. Use only approved abbreviations.

Below abstract, provide and identify three to 10 key words or short phrases. Use terms from *Medical Subject Heading* list from *Index Medicus*.

Text: Observational and experimental articles may be divided into sections, Introduction, Methods, Results, and Discussion. Use subheadings in long articles to clarify content, especially in Results and Discussion sections.

Introduction: Clearly state purpose of article. Summarize rationale for study or observation. Give only strictly pertinent references; do not review subject extensively.

Methods: Describe selection of observational or experimental subjects (patients or experimental animals including controls). Identify methods, apparatus (manufacturer's name and address in parenthesis), and procedures in sufficient detail to allow reproduction of results. Reference established methods, including statistical methods; provide references and brief descriptions of published methods that are not well known; describe new or substantially modified methods, give reasons for using and evaluate limitations.

Results: Present in logical sequence in text, tables, and illustrations. Do not repeat data in tables and/or illustrations in text; emphasize or summarize only important observations.

Discussion: Emphasize new and important aspects of study and conclusions that follow. Do not repeat detailed data in Results. Include implications and limitations of findings. Relate observations to other relevant studies. Link conclusions with goals of study; avoid unqualified statements and conclusions not completely supported by data. Avoid claiming priority and alluding to incomplete work. State new hypotheses when warranted, labelling them as such. Appropriate recommendations may be included.

Acknowledgments: Acknowledge only persons who have made substantive contributions. Obtain written permission from everyone acknowledged by name.

References: Number consecutively in order mentioned in text. Identify in text, tables, and legends by superior Arabic numerals (without parenthesis). Number references cited only in tables or legends in accordance with sequence established by first mention in text of table or illustration.

Use form of U.S. National Library of Medicine in *Index Medicus*. Abbreviate journal titles according to "List of Journals Indexed," in January issue of *Index Medicus*.

Avoid using abstracts as references; do not use "unpublished observations" and "personal communications" as references; insert references to written, not verbal, communications in parenthesis in text. Include among references manuscripts accepted but not yet published; designate journal followed by "in press" (in parenthesis). Cite information in text from manuscripts submitted but not yet accepted as "unpublished observations" (in parenthesis). Verify references against original documents.

JOURNAL

Standard Journal Article: (List all authors when six or less; when seven or more list only first three and add *et al.*).

Pilepich MV, Munzenrider JE, Tak WK, Miller HH. Preoperative irradiation of primarily unresectable colorectal carcinoma. Cancer 1978;42:1077–81.

Corporate Author:

The Committee on Enzymes of the Scandinavian Society for Clinical Chemistry and Clinical Physiology. Recommended method for the determination of gammaglutamyltransferase in blood. Scand J Clin Lab Invest 1976;36:119–25.

Anonymous. Epidemiology for primary health care. Int J Epidemiol 1976;5:224–5.

BOOKS AND OTHER MONOGRAPHS

Personal Author(s):

Corman ML. Colon and rectal surgery. Philadelphia: JB Lippincott, 1984.

Corporate Author:

American Medical Association Department of Drugs. AMA drug evaluations. 3rd ed. Littleton: Publishing Sciences Group, 1977.

Editor, Compiler, Chairman as Author:

Rhodes AJ, Van Rooyen CE, eds. Textbook of virology: for students and practitioners of medicine and the other health sciences. 5th ed. Baltimore: Williams & Wilkins, 1968.

Chapter in Book:

Sugarbaker PH, Gunderson LL, Wittes RE. Colorectal cancer. In: DeVita VT Jr, Hellman S, Rosenberg SA, eds. Cancer: Principles & practice of oncology. 2nd ed. Philadelphia: JB Lippincott, 1985:795–866.

AGENCY PUBLICATION:

National Center for Health Statistics. Acute conditions: incidence and associated disability, United States July 1968–June 1969. Rockville Md.: National Center for Health Statistics, 1972. (Vital and health statistics. Series 10: Data from the National Health Survey no. 69) (DHEW publication no. (HSM)72–1036).

OTHER ARTICLE

Newpaper Article:

Shaffer RA. Advances in chemistry are starting to unlock mysteries of the brain: discoveries could help cure alcoholism and insomnia, explain mental illness. How the messengers work. Wall Street Journal 1977 Aug 12:1(col. 1) 10 (col. 1).

Magazine Article:

Roueche B. Annals of medicine: the Santa Claus culture. The New Yorker 1971 Sep 4:66–81.

Tables: Type each double-spaced on a separate sheet. Do not submit as photographs. Number consecutively and supply a brief title for each. Give each column a short or abbreviated heading. Place explanatory matter in footnotes not in heading. Explain in footnotes all nonstandard abbreviations. For footnotes use symbols in sequence: *, †, ‡, §, ¶, ‖, **, †† .... Identify statistical measures of variations such as SD and SEM.

Omit internal horizontal and vertical rules.

Cite each table in text in consecutive order.

If data from another published or unpublished source is used, obtain permission and acknowledge fully.

Too many tables in relation to length of text produces difficulties in layouts. Examine journal to estimate how many tables to use per 1000 words of text. Editor may recommend that additional tables containing important backup data too extensive to be published be deposited with National Auxiliary Publications Service or made available by author(s). An appropriate statement will be added to text. Submit such tables with manuscript.

Illustrations: Submit in duplicate. Submit required number of complete sets of figures. Have figures professionally drawn and photographed, freehand or typewritten lettering is unacceptable. Instead of original drawings, roentgenograms, and other material, send sharp, glossy black-and-white photographic prints, usually 5 × 7 in., no larger than 8 × 10 in. Make letters, numbers, and symbols clear and even throughout, and of sufficient size that when reduced will still be legible. Place titles and detailed explanations in legends, not on illustrations.

Paste a label on back indicating figure number, names of authors, and TOP of figure. Do not write on back of figures. Mount them on cardboard, or scratch or mar them using paper clips. Do not bend figures.

Place internal scale markers on photomicrographs. Make symbols, arrows, or letters in photomicrographs contrast with background.

If photographs of persons are used, either subjects must not be identifiable or submit written permission to use photograph.

Cite each figure in text in consecutive order. If a figure has been published, acknowledge original source and submit written permission from copyright holder. Permission is required, regardless of authorship or publisher.

For color illustrations, supply color negatives or positive transparencies and, accompanying drawings marked to indicate region to be reproduced; in addition, send two positive color prints. Author must pay extra cost.

Legends for Illustrations: Type double-spaced, starting on a separate page with Arabic numerals corresponding to illustrations. When symbols, arrows, numbers, or letters are used, identify and explain each. Explain internal scale and identify method of staining in photomicrographs.

OTHER FORMATS

THE TECHNICAL NOTES SECTION—Provide a means by which surgeons may describe a diverse array of variations and innovations in surgical technique without accompanying case reports, literature reviews, or extensive bibliographies. In cover letter request that manuscript be considered for the "Technical Notes" section.

Limit abstract to two or three sentences. Begin text on third page. Repeat title and authors at top of page. Type author's name and abbreviated running title in right upper corner of all pages. Make introductory paragraph(s) concise and specific. State problems simply. Avoid philosophic dissertation, unnecessary literature referencing, and other "padding." If possible, summarize solution in final sentence of introductory paragraph(s). Entitle second part of text "Technique," "Procedure," or other appropriate term. Describe technique, instrument, method, etc. in a precise manner. Knowledgeable readers should be able to duplicate procedure without further information. Where unique instruments, drugs, or materials are required, use generic name and provide manufacturer's name and address in a footnote. Separate "results," "discussions," "conclusions," or "summary" sections are usually not employed. If necessary and appropriate, add a brief paragraph on results at end of Techniques. Include only most pertinent references. Type references and legends on separate pages. Include at least one, not more than four, black and white or line and wash drawings to demonstrate procedure.

OTHER

Abbreviations: Use only standard abbreviations. Consult: *Council of Biology Editors Style Manual: A Guide for Authors, Editors, and Publishers in the Biological Sciences* (4th ed. Arlington: Council of Biology Editors., 1978) and O'Connor M, Woodford FP. *Writing Scientific Papers in English: an ELSE-Ciba Foundation Guide for Authors* (Amsterdam, Oxford, New York: Elsevier-Excerpta Medica, 1975.) Avoid use in title. Full term for which abbreviation stands precedes abbreviation with its full term at first use unless a standard unit of measurement.

Drugs: Identify drugs and chemicals used including generic name(s) dosage(s) and route(s) of administration.

Experimentation, Animal: Indicate whether institution's or National Research Council's guide for care and use of laboratory animals was followed.

Experimentation, Human: Indicate whether procedures followed were in accord with ethical standards of the institutional Committee on Human Experimentation or in accord with the Helsinki Declaration of 1975. Do not use patients' names, initials, or hospital numbers.

Statistical Analyses: Include numbers of observations and statistical significance of findings when appropriate.

Present detailed statistical analyses, mathematical derivations, etc., as appendixes.

Units: Report measurements in units they were made.

SOURCE

January 1994, 37(1): 110–113.
Complete "Notice to Contributors" is published annually in January issue of Journal.

DNA and Cell Biology
Mary Ann Liebert, Inc.

Editor-in-Chief: Dr. Mark I. Greene
Department of Pathology and Laboratory Medicine
University of Pennsylvania School of Medicine
Philadelphia, PA 19104–6082
Telephone: (215) 898-2947

Senior Editor: Dr. David B. Weiner
Department of Medicine, 574 Maloney Building
University of Pennsylvania
Philadelphia, PA 19104–4283
Telephone: (215) 662- 2352

West Coast Editor: Michael Karin
Department of Pharmacology
UCSD School of Medicine
9500 Gilman Drive
LaJolla, CA 92093-0636
Telephone:(619) 534-1361; Fax: (619) 534-8158

Walter L. Miller
Department of Pediatrics
Building MR-IV, Rm. 209
University of California at San Francisco
San Francisco, CA 94143-0978
Telephone: (415) 476-2598
            Molecular Endocrinology

George Weinstock
Dept. of Biochemistry and Molecular Biology
University of Texas Medical School
P.O. Box 20708
6431 Fannin Street
Houston, TX 77225
Telephone: (713) 792-5266
            Prokaryotic Systems

European Editor: Peter H. Seeburg
Zentrum fur Molekulare Biologie
Im Neuenheimer Feld 282
6900 Heidelberg, Germany
Telephone: (49) 6221-566-891; Fax: (49) 6221-565-894

Hilary Koprowski
Thomas Jefferson University
1020 Locust Street
462 JAH
Philadelphia, PA 19107
Telephone: (215) 955-4761
            Virology

AIM AND SCOPE

DNA and Cell Biology publishes papers, short communications, reviews, laboratory methods, and editorials on any subject dealing with eukaryotic or prokaryotic gene structure, organization, expression, or evolution. Papers studying genes or their expression at the level of RNA or proteins and by cell biology approaches are as appropriate as papers dealing directly with DNA. Manuscripts may be of any length.

MANUSCRIPT SUBMISSION

Manuscripts submitted to *DNA and Cell Biology* may not be under consideration for publication elsewhere. Submit original and three copies, including illustrations and tables.

MANUSCRIPT FORMAT

Type entire manuscript, including figure legends, tables, and references, double-spaced on regular typing paper (not erasable bond). Leave 1.5 in. margins on both sides, top and bottom.

On first page, give article title, name(s) of author(s), institutional affiliation(s), and name (with complete address and phone number) of corresponding author; supply a running title (45 characters). On second page, supply a 200-word abstract, stating aims, results and conclusions drawn from study. Follow with introduction, materials and methods, results, discussion, acknowledgments, references, tables and figure legends. Keep results and discussion as separate sections. Begin each sections on separate page. Number pages consecutively; put first author's last name on each page.

Tables and Illustrations: Type each with its caption on a separate sheet. Number with Arabic numerals. Do not repeat information given in text. Do not make a table for data that can be given in text in one or two sentences. Supply original glossy photographs for illustrations. Submit DNA sequence data as glossy photos or as computer printout provided a letter-quality laser printer (not dot matrix printer) is used. Submit all nucleic acid sequences in camera-ready format. Submit all figures in size wish it to appear. Number all illustrations and write first author's name and article title on back. Indicate TOP. Supply a legend for each illustration; number legends consecutively; type double-spaced. Number figures in order cited in text. Submit a complete, separately collated set with each manuscript copy.

References: Cite by authors and date within text: The ribosomal RNA (rrn) operons in *E. coli* contain single copies of each of the 16S, 23S, and 5S r(RNA) genes (Pace, 1973). If more than two authors are involved, use et al. after first author's name: (Sogin et al., 1971). If several papers by same authors are cited in one sentence, use lowercase letter designation to indicate individual papers: (Dunn and Studier, 1973a, 1973b). Use same designation in reference list.

Double-space references in Bibliography. List in alphabetical order at end of article. Include complete titles of cited articles.

JOURNAL CITATION

PACE, N. (1973). Structure and synthesis of the ribosomal ribonucleic acid of prokaryotes. Bacteriol. Rev. 37, 562–603.

BOOK CITATION

BROSIUS, J. (1987). Expression vectors employing λ, *trp*, *lac*, and *lpp*-derived promoters. In *Vectors, a Survey of Molecular Cloning Vectors and Their Uses*. R. L. Rodriguez and D.T. Denhardt, eds. (Butterworth Publishers, Stoneham, MA) pp. 205–225.

When dates from an unpublished source are given, supply researcher's name. If work is in press, give journal in which it is to be published or name of publisher. Abbreviate journal names following *Index Medicus*.

OTHER

Abbreviations: Rules for abbreviations and nomenclature follow those detailed annually in first January issue of *J. Biol. Chem.*

Color Charges: $200 per piece of color art plus $1000 per page of color. The publisher will provide a more precise cost estimate when figures are received.

Page Charges: $40 per printed page from authors who have funds available from research grants or from their institutions.

Permissions: Obtain written permission to reproduce material such as figures and tables from copyrighted material from publisher of journal or book concerned. List publication from which material is taken in references. Footnote a reprinted table, or write legend of a reprinted figure, to read, "reprinted by permission from Jones et al." and list appropriate reference. Show permissions listings in manuscript; they cannot be entered on proofs.

SOURCE

January 1994, 13(1): inside back cover.

Drug Metabolism and Disposition: The Biologic Fate of Chemicals

American Society for Pharmacology and Experimental Therapeutics
Williams & Wilkins

Dr. Vincent G. Zannoni, Editor
Drug Metabolism and Disposition
Department of Pharmacology
University of Michigan Medical School, MSI
Ann Arbor, MI 48109-0626
Fax: (313) 763-4450

AIM AND SCOPE

Drug Metabolism and Disposition will review *in vitro* and *in vivo* experimental results that contribute significant and original information on xenobiotic metabolism and disposition. The term xenobiotic includes pharmacologic agents as well as environmental chemicals. Pharmacokinetic and pharmacodynamic manuscripts and those involving mechanisms are invited. Manuscripts concerned with factors which affect the biological fate of chemicals such as genetic, nutritional or hormonal are of interest. Papers addressing toxicological consequences of xenobiotic metabolism are appropriate.

MANUSCRIPT SUBMISSION

Submit three copies (original and two copies) of each manuscript. Submission implies that material has not previously been published except as an abstract for a scientific meeting, and that it is not being submitted elsewhere.

MANUSCRIPT FORMAT

Type double-spaced with ample margins, on one side of 8.5 × 11 in. pages. Number all pages in sequence, starting with title page.

Arrange full-length papers as follows:

Title Page: Title of paper, names of all authors, and institution(s) where work was done. Do not use footnote numbers with title. Title should briefly yet explicitly indicate contents of paper; include names of chemicals or chemical classes studied, species used, etc.

Running Title: 50 total characters and spaces. Put name and address of person to whom editorial correspondence and galley proofs should be sent at bottom of page.

Abstract: 250 words.

Introduction: Briefly summarize pertinent literature and state aims of work.

Materials and Methods: Give species, strains, sexes, and ages or sizes of animals, with Latin names where required for distinction. Indicate sources and purities of chemicals other than common reagents. Specify equipment used and conditions of use. Bibliographic reference of published methods is sufficient; describe minor modifications. For extensive modifications, describe entire new procedure. Describe work so peers will be able to repeat. Indicate variations in conditions for similar experiments in legends to figures and tables. Give properties and proof of structure for reference compounds used for metabolite identification.

Results: Present as much as possible in graphic and tabular form. If a table includes only two or three values, present data in sentence form in text. Avoid using several tables describing very similar experiments; combine wherever possible, unless overcomplicated, unwieldy tables result. Do not repeat same data in tables and figures. Use text to describe and summarize data and to draw primary conclusions; do not repeat numerical data. Do not include extended discussion of results.

Discussion: Assemble major conclusions to be drawn from work. Discuss with respect to existing body of knowledge in immediate area. Use graphic schema to clarify conclusions. Identify speculation as such. Discuss questions raised by work, or those inherent in experiments. Results and Discussion sections may have to be combined occasionally.

Acknowledgments: Of technical assistance, gifts of materials, and other aid. Do not include financial support here; put in an unnumbered title footnote.

References: Number in order of citation in text.

1. S. S. Lau, G. D. Abrams, and V. G. Zannoni: Activation and detoxification of bromobenzene. *J. Pharmacol. Exp. Ther.* **214,** 703–708 (1980).

2. T. C. Butler: The distribution of drugs. In "Fundamentals of Drug Metabolism and Drug Disposition" (B. N. La Du, H. G. Mandel, and E. L. Way, eds.), pp. 44–62, Williams & Wilkins, Baltimore, 1971.

Cite papers that have been accepted for publication among references; give authors, journal name, and "in press." Cite work not published or accepted, or personal communications by footnote. Where knowledge of unpublished work is crucial to evaluation of a paper, send duplicate copies of pertinent data.

Ensure accuracy of references. Do not include references that have not been examined personally If primary reference could not be examined, also indicate secondary source:

4. K. Maemoto, N. Seike, and M. Hirata, *Kobunshi Kagaku* **15,** 660 (1958); *Chem. Abstr.* **54,** 14775 (1960).

Cite references in text by numbers within parentheses. Abbreviate journal names as in *Index Medicus.*

Footnotes: Present in order: Unnumbered footnote giving source of financial support, thesis information, citation of abstracts of meetings where work was presented, etc., and, in a separate paragraph, name and full address (with street address or P.O. box where applicable, and zip code numbers of reprint request author. Numbered footnotes, starting with those to authors' names. Numbers in sequence throughout text, using superscript numbers.

Tables: Each on a separate page. Number with Arabic numbers. Type title in italics (or underlined) with only first word and proper names capitalized. Follow with general statements about table in paragraph form. Indicate footnotes by italicized lower case superscript letters, starting with *a* for each table. Type footnotes immediately below each table.

Figure Legends: Number figures with Arabic numbers, followed by title in italics (or underlined), with only first word and proper nouns capitalized. Place remainder of explanatory material in paragraph form below title.

Index Terms: Type list of index terms on last typed page.

Figures: Submit as unmounted glossy photographic prints no larger than 8 × 10 in. Submit two xerox or other good reproductions of line prints; send three photographic prints of halftone material. Make letters, numbers, and symbols large enough to be legible after reduction to single-column size; at least 1.5 mm after reduction. Make style of figures uniform throughout. Make figures simple and uncluttered. In line graphs with multiple lines, use different symbols for experimental points for different lines and explain symbols in legend; do not label individual lines. Label abscissa and ordinate with scale, name, and dimensions of quantities expressed.

OTHER FORMATS

LETTERS—Discuss relevant work described in this or other journals, raise interesting points as suggestions for further work, or make new interpretations of existing data. Where appropriate, authors of papers to which reference is made are invited to reply. Letters presenting new data will not normally be accepted.

NATIONAL AUXILIARY PUBLICATIONS SERVICE (N.A.P.S.)—For deposit with National Auxiliary Publications Service, submit data in triplicate with manuscript. The original copy for the documentary material will be deposited with N. A. P. S. and the accession number will be published.

REVIEWS—Occasional brief reviews covering limited aspects of a subject in area of interest of this journal, typically critical reviews that present author's view of the current state of the subject, or prospective articles intended to stimulate discussion and research.

SHORT COMMUNICATIONS—1000 words, Do not divide into sections appropriate for full papers. No abstract. In first paragraph introduce work. Describe experiments and results in narrative fashion. Present a moderate amount of tabular and graphic material, but space will not exceed three printed pages. Give structural formula of parent compounds and metabolites as general formula with variable groups identified in legend.

OTHER

Abbreviations: Do not punctuate units of measurement. Do not distinguish between singular and plural forms. Indicate multiples of units as: kg, g, mg, μg (not ug, mcg, or γ), ng, pg. Abbreviations for units of measurement are:

Mass—g (gram); mol (mole); eq (equivalent). Do not use M for mole, only for concentration term molar.

Concentration—M (molar); N (normal); % (percent). For latter, indicate basis of formulation as % (w/w),% (w/v), or % (v/v) to signify g/100 g, g/100 ml, or ml/100 ml. Do not use mg%. Show mixtures as, e.g., acetone 0.5 M KCl/glacial acetic acid, 1:5:2 (v/v).

Length—m (meter); μm (micrometer). Å (Ångstrom) may be used.

Volume—Spell out liter to avoid confusion with numeral 1. Abbreviate compound words containing root liter as ml, μl, etc.

Time—hr (hour); min (minute); sec (second). Spell out days and longer units of time.

Radioactivity—Ci (curie); r (roentgen); cpm or dpm (counts or disintegrations per min).

Electricity—V(volt); amp (ampere); Hz (cycles/sec).

Spectrometry—A_{000} [absorbance (not OD or E) at 000 nm (not mμ) wavelength]; $\mathcal{E}$ (molar absorption coefficient, with units M^{-1} cm^{-1}); UV (ultraviolet); IR (infrared); ESR (electron-spin resonance); NMR (nuclear magnetic resonance); δ [chemical shift, with units ppm (parts per million)]; s (singlet); d (doublet); t (triplet); m (multiplet); amu (atomic mass units); m/z (mass/charge ratio).

Chromatography—TLC (thin-layer chromatography); R_F (retardation factor); GLC (gas-liquid chromatography); R_T (retention time); GC/MS (coupled gas chromatography-mass spectrometry); HPLC (high-pressure liquid chromatography).

Equilibrium and Kinetic Constants—K_d (dissociation constant); K_s or K_i (dissociation constant of enzyme-substrate or enzyme-inhibitor complex); K_M (Michaelis constant); V_{max} (maximum initial velocity); k (rate constant); pK_a (negative logarithm of acidic dissociation constant); $t_{1/2}$, half-life; AUC,

area under curve of plasma concentrations vs. time.

Statistics—p (probability of chance observation); N (number of experiments); SD (standard deviation of the series); SE (standard error of the mean).

Other Abbreviations—°C (degrees of temperature); g (acceleration due to gravity, as in 9000g); rpm (revolutions per min); LD$_{50}$ and ED$_{50}$ (median lethal and effective doses); iv (intravenous); ip (intraperitoneal); im (intramuscular); sc (subcutaneous); po (peroral); m.p. (melting point); sp.g. (specific gravity).

Drugs and Chemicals—Use without definition: NAD+, NADH, NADP$^+$, NADPH, FMN, FAD, ATP et al., RNA, DNA, UDPglucuronic acid, Tris, EDTA; SKF 525-A; explain all other abbreviations in a single footnote at point of use of first one. Name cytochromes as: cytochrome b_5, cytochrome c, cytochrome P450. Use generic names of drugs; give trade-name, capitalized, in parentheses after first use of generic name. If no generic name, use a code number with full chemical name in parentheses following first use of number. Do not use trade-names and code numbers in title.

Bring composite character of drugs which are mixtures of stereoisomers to reader's attention. Use prefix rac-, e.g., rac-propranolol (for racemates) and (Z/E)- or cis/trans- in the case of that type. Make implications of composite nature of drugs studied for interpretation of data measured and conclusions drawn explicit.

Page Charges: $30 per printed page unless involve personal financial hardships to authors. Payment is not a condition for publication.

Submission Fee: $40.00 (in U.S. funds payable to ASPET).

SOURCE

January/February 1994, 22(1): front matter.
Instructions to Authors appears in every issue.

Ecological Monographs
Ecological Society of America

Managing Editor, Lee N. Miller
Ecological Society of America
328 E. State Street
Ithaca, New York 14850-4318

AIM AND SCOPE

We invite contributions from scientists working on the full spectrum of ecological problems. Included within this spectrum are studies of the physiological responses of individual organisms to their abiotic and biotic environments, ecological genetics and evolution, the structure and dynamics of populations, interactions among individuals of the same or differing species, the behavior of individuals and groups of organisms, the organization of biological communities, landscape ecology, the processing of energy and matter in ecosystems, historical ecology and paleoecology, and the application of ecological theory to resource management and the solution of environmental problems. Reports of ecological research on all kinds of organisms in all environments are welcome.

The pages of *Ecology* and *Ecological Monographs* are open to research and discussion papers that develop new concepts in ecology or test ecological theory with data from field and laboratory experiments, observations, or simulations. New methodologies with a potential for broad use in ecology are also of interest. Papers that are well grounded in ecological theory and that have broad implications for environmental policy or resource management are welcome in *Ecological Applications* as well as in *Ecology* or *Ecological Monographs*.

MANUSCRIPT SUBMISSION

Submit original research or reanalysis of published research that results in new insights. Treatment should lead to generalizations that are potentially applicable to other species, populations, communities, or ecosystems.

Papers are no longer than 20 printed pages. Assignment of an accepted paper to *Ecology* or *Ecological Monographs* depends primarily on length. Three pages of typescript equal approximately one printed page. Each figure counts as one page. Illustrations and tabular material together may constitute up to 20% of total printed length without extra cost.

Submit four photocopies of manuscript; retain original. Do not submit originals or photographic prints of illustrations; photocopies are preferred. Include address and telephone and fax numbers in cover letter. Describe extent to which data or text have been used in other papers or books that are published, in press, submitted, or soon to be submitted elsewhere. Adhere to ESA Code of Ethics (*Bulletin of the Ecological Society of America, 74(4), December 1993*) on authorship, plagiarism, fraud, authorized use of data, copyrights, errors, confidentiality, intellectual property, attribution, willful delay of publication, and conflicts of interest.

MANUSCRIPT FORMAT

Write with precision, clarity, and economy; use active voice and first person. Double-space all material (text, quotations, figure legends, tables, literature cited, etc.) at three lines per in. (12 lines/10 cm). Number lines of text if option is available. Type on one side of standard-sized paper, approximately 22 × 28 cm. Leave 2.5 cm margins on all sides. Do not use very thin paper. Print all parts at 10 characters per in. (4 characters/cm), or use a 12-point font.

Early in manuscript, identify type(s) of organism or ecosystem studied. Avoid descriptive terms familiar only to specialists. Provide scientific names of all organisms. Common names may be used when convenient after stating scientific names.

Title Page: Provide a running head (40 letters and spaces) at top of title page.

Title should tell what article is about. Make it informative and short (13 words or 100 characters; longer titles will be shortened). Do not include authority for taxonomic names in title or abstract. Do not include numerical series designations.

For each author, state relevant address (usually institutional affiliation during research period). If present address is different, give as a footnote at bottom of title page.

Abstract: Brief summary of research, including purpose, methods, results, and major conclusions. Avoid speculation. Do not include any literature citations.

Key Phrases and Key Words: Following abstract, list key words and 8–10 key phrases for annual index. Extract from topic sentences that tell main conclusions or "take-home lessons." Each should describe a relationship or fact.

Make key words supplement, not duplicate key phrases. Title words may be used. Make each key word or phrase useful as an entry point for a literature search.

Body: Organize in sections: Introduction, Methods, Results, and Discussion. In long articles Conclusions may be added. Brief articles do not require an Introduction label. If nature of research requires a different organization, specify level of each section heading (1st order head, 2nd order head, etc.) in margin. Give motivation or purpose of research in introduction, state questions you sought to answer. In Methods section provide sufficient information to allow repetition of work. A clear description of experimental design, sampling procedures, and statistical procedures is especially important in papers describing field studies, simulations, or experiments. State results concisely and without interpretation. In Discussion differentiate results from data obtained from other sources. Distinguish factual results from speculation and interpretation.

Acknowledgments: Place acknowledgments or dedications just before Literature Cited. Do not include on title page.

Footnotes: Avoid use; incorporate in text.

Literature Cited: Double-space. Check each citation in text against Literature Cited to see that they match exactly. Make list conform in sequencing and punctuation to recent journal issues. Spell out journal titles completely. In article titles, capitalize common bird names and spell all words exactly as in original publication. Provide publisher's name and location for symposia or conference proceedings; distinguish between conference date and publication date if both are given. Do not list abstracts or unpublished material. Include in text as personal observations (by an author of present paper), personal communications (from others), or unpublished x, where x = data, manuscript, or report; provide author names and initials for all unpublished work and abstracts.

Tables: Double-space all parts. Start each on a separate page. Number in order discussed in text. Provide a short descriptive title at top; do not repeat labels on columns and rows, reveal the point of grouping certain data in the table in title. Provide statistical and other details as footnotes, not in title. Do not rule tables unless essential to avoid ambiguity. Consult recent journal issues for style. Never repeat material in figures and tables; figures are preferable. Do not include any class of information in tables that is not discussed in text.

Illustrations: Include photocopies of illustrations with each copy of manuscript. Number in order discussed in text. Group figure legends in numerical order on one or more pages, separate from figures. Do not submit photographic prints until requested. If important details cannot be distinguished on a photocopy, submit a glossy print for review purposes. Most illustrations will be reduced to single-column width; symbols and lettering must be legible.

Examine for legibility after photo-reduced to 76 mm. After reduction, lettering must be at least 6 points. Uppercase letters are preferred except where SI requires lowercase letters for unit abbreviations. Avoid very large letters and lettering styles in which portions of letters are very thin. Use italic lettering for variables, constants, and scientific names in illustrations, consistent with text. Typewritten lettering is not acceptable. Solid black bars in bar graphs overwhelm adjacent text; use white, shaded, or hatched bars.

Do not submit prints larger than 22×28 cm unless asked to do so. Protect drawing surfaces that include rub-on lettering or other attachments; include a reference photocopy of each original illustration. Identify each black-and-white figure by number with a label at top of reverse side. Identify color figures with a label on top of a photocopy; do not attach anything to back.

Assembly of Manuscript Copies: Assemble parts of each copy of manuscript in order: title page, abstract, key words and phrases, text, acknowledgments, literature cited, appendices, tables, figure legends, figures. Number all pages consecutively.

OTHER

Equations: Equations set separately from text will be broken into two or more lines if exceed width of one column; mark equations for appropriate breaks. Clarify subscripts and superscripts by marginal notes.

Manufacturer Names: Supply name and location of product manufacturers. Give model number for equipment used. Supply complete citations, including author (or editor), title, year, publisher, and version number, for computer software mentioned.

Spelling: Use American spellings.

Statistical Analyses: State probability values without leading zeroes (e.g., $P < .01$).

Style: See *CBE Style Manual* (5th edition, Chicago: Council of Biology Editors).

Underlining indicates italics. Underline scientific names and symbols for all variables and constants except Greek letters. Italicize symbols in illustrations to match text. Do not use italics for emphasis.

Supplements: Extensive data sets, mathematical proofs, and other information of interest to a small subset of readers are too costly to publish, but can be made available to readers as a supplementary publication on paper or computer diskette. A footnote giving instructions for obtaining a copy from author or journal is printed in article.

Units: Use International System of Units (SI) for measurements. Consult *Standard Practice for Use of the International System of Units* (ASTM Standard E-380-92, Philadelphia: ASTM) for guidance on unit conversions, style, and usage.

Use terms mass or force rather than weight; when one unit appears in a denominator, use the solidus (e.g., g/m^2); for two or more units in a denominator, use negative exponents, and use international spelling of metre and litre, and a capital L as symbol for litre.

SOURCE
February 1994, 64(1):110–112.

Ecology
Ecological Association of America

Lee N. Miller
Ecological Society of America
328 E. State Street
Ithaca, New York 14850-4318

AIM AND SCOPE

The pages of *Ecology* and *Ecological Monographs* are open to research and discussion papers that develop new concepts in ecology or test ecological theory with data from field and laboratory experiments, observations, or simulations. New methodologies with a potential for broad use in ecology are also of interest. Papers that are well grounded in ecological theory and that have broad implications for environmental policy or resource management are welcome in *Ecological Applications* as well as in *Ecology* or *Ecological Monographs*. Each paper should report original research or a reanalysis of published research that results in new insights. Treatment of the research topic should lead to generalizations that are potentially applicable to other species, populations, communities, or ecosystems.

We invite contributions from scientists working on the full spectrum of ecological problems. Included within this spectrum are studies of the physiological responses of individual organisms to their abiotic and biotic environments, ecological genetics and evolution, the structure and dynamics of populations interactions among individuals of the same or differing species, the behavior of individuals and groups of organisms, the organization of biological communities, landscape ecology, the processing of energy and matter in ecosystems, historical ecology and paleoecology, and the application of ecological theory to resource management and the solution of environmental problems. Reports of ecological research on all kinds of organisms in all environments are welcome.

MANUSCRIPT SUBMISSION

Papers occupy up to 20 printed pages, including illustrations. Notes, Comments, and Replies to Comments do not exceed 16 pages of typescript, including cover page, text, literature cited, tables, and figures. Assignment of an accepted paper to *Ecology* or *Ecological Monographs* depends primarily on length of paper after it is set in type. Three pages of typescript equal approximately one printed page. *Ecological Applications* considers papers of any length. Illustrations and tabular material may constitute up to 20% of total printed length of article without extra cost to author.

Assemble parts of each copy of manuscript in order: title page, abstract, key words and phrases, text, acknowledgments, literature cited, appendices, tables, figure legends, figures. Number all pages consecutively.

After manuscript is accepted for publication, submit to Managing Editor: two papers copies of manuscript, one copy on diskette, name and version of word processing program used to prepare diskette, and two photocopies of each illustration.

MANUSCRIPT FORMAT

Submit four photocopies of manuscript; retain original. Do not submit originals or photographic prints of illustrations; photocopies are preferred until manuscript has been accepted. Include address and telephone and fax numbers in cover letter. At time of submission provide information describing extent to which data or text have been used in other papers or books that are published, in press, submitted, or soon to be submitted elewhere. Adhere to ESA Code of Ethics (*Bulletin of the Ecological Society of America,* 71(4), December 1990), which deals with authorship, plagiarism, fraud, authorized use of data, copyrights, errors, confidentiality, intellectual property, attribution, willful delay of publication, and conflicts of interest.

Write with precision, clarity, and economy; use active voice and first person whenever appropriate. Use American spellings (e.g., behavior, not behaviour). Use *CBE Style Manual* (5th ed., Council of Biology Editors, Chicago, IL) for details of style.

Double-space all material (text, quotations, figure legends, tables, literature cited, etc.) at three lines per inch (12 lines/l0 cm). Number lines of text if a line-numbering option is available in word processing program. Type on one side of standard sized paper, approximately 22 × 28 cm. Leave 2.5 cm margins on all sides of each page. Do not use very thin paper. Use a 12-point font (proportionally spaced type) or 10 characters/inch (4 characters/cm) if letter spacing is uniform.

Underlining indicates italicization. Underline scientific names and symbols for all variables and constants except Greek letters. Italicize symbols in illustrations to match text. Do not use italics for emphasis.

Identify type(s) of organism or ecosystem studied. Avoid descriptive terms familiar only to specialists. Provide scientific names of all organisms. Common names may be used when convenient after stating scientific names.

Organize article in sections: Introduction, Methods, Results, and Discussion. In long articles, add Conclusions. Brief articles do not require a labelled Introduction. If nature of research requires a different organization, specify level of section headings (1st order head, 2nd order head, etc.) in margin. Include motivation or purpose of research in introduction, state questions you sought to answer. In Methods provide sufficient information to allow repetition of work. Clearly describe experimental design. sampling procedures, and statistical procedures in papers describing field studies, simulations, or experiments. Supply name and location of manufacturers of products (e.g., animal food, analytical device) mentioned. Give model number for equipment used. Supply complete citations, including author (or editor), title, year, publisher, and version number, for computer software mentioned. State Results concisely and without interpretation. In Discussion section, carefully differentiate results from data obtained from other sources. Distinguish factual results from speculation and interpretation.

Title Page: Provide a running head (40 letters and spaces) at top of title page. Title should tell what article is about; be informative and short (13 words or 100 characters); longer titles will be shortened by editor. Do not include authority for taxonomic names in title or abstract. Titles may not include numerical series designations. For each author, state relevant address, usually institutional affiliation of author when research was done. Author's present address, if different, should appear as a footnote at bottom of title page.

Abstract, Key Phrases, and Key Words: Provide an abstract, key phrases, and key words for all articles, including Notes, Comments, and Special Features. Abstract should provide a brief summary of research, including purpose, methods, results. and major conclusions. Avoid speculation; if included, speculation about possible interpretations or applications of results should play a minor role. Do not include literature citations. Following abstract, list key words and 8-10 key phrases for annual index. Extract key phrases from topic sentences that tell main conclusions or "take-home lessons" of article. Key phrases describe a relationship or fact. Example: TOPIC SENTENCE: Pollination in a sexually dimorphic species was affected by petal length, nectar, sexual type (hermaphrodite or female), and pollinator visitation rates. KEY PHRASES: Floral characteristics of hermaphrodites vs. females; Pollinator visitation rates vs. flower type; Sexual dimorphism of flowers: effects on pollination. KEY WORDS: nectar; petal length. Key words supplement but do not duplicate key phrases. Include title words in key words. Make key words or phrases as entry points for literature searches. Use topic sentences only for creating key phrases.

Acknowledgments: Place heading just before Literature Cited; include dedication. Do not include on title page.

Literature Cited: Double-space list of Literature Cited. Check each citation in text against Literature Cited to see that they match exactly. Delete citations from list if not cited in text. List should conform in sequencing and punctuation to that in recent journal issues. Spell out all journal titles. Capitalize common names of birds in titles; spelling of all words should agree exactly with that in original publication. Provide publisher's name and location for symposia or conference proceedings; distinguish between conference date and publication date if both are given. Do not list abstracts or unpublished material in Literature Cited; list in text as personal observations (by an author), personal communications (from others), or unpublished x, where x = data, manuscript, or report; provide author names and initials for all unpublished work and abstracts.

Tables: Double space all parts. Start each on a separate page. Number in order discussed in text. Provide a short descriptive title at top of each. Do not repeat labels on columns and rows; title should reveal the point of grouping data in table. Provide statistical and other details as footnotes, not in title. Do not rule tables unless essential to avoid ambiguity. Consult recent journal issues for style. Never repeat material in figures and tables; figures are preferable. Do not include any class of information in tables that is not discussed in text.

Illustrations: Each manuscipt copy should include photocopies of illustrations. Number in order discussed in text. Group figure legends in numerical order on one or more pages, separate from figures. Do not submit photographic prints until requested. If important details cannot be distinguished on a photocopy, submit glossy prints.

Illustrations will be reduced to single-column width; make symbols and lettering clearly legible at that size. Examine illustrations for legibility after photoreduction to 76 mm. After reduction, all lettering should be at least 6 points. Use uppercase letters except for Sl unit abbreviations. Avoid very large letters and lettering styles in which portions of letters are very thin. Use italic lettering for variables, constants, and scientific names in illustrations consistent with text. Typewntten lettering is not acceptable. Solid black bars in bar graphs overwhelm adjacent text; use white, shaded, or hatched bars.

Editor may ask for modifications of illustrations. After modifications have been made, provide either original drawings or sharply focused photographic prints. Do not submit prints larger than 22×28 cm unless asked. Protect drawing surfaces with rub-on letterring or other attachments; include a reference photocopy of each original illustration. Identify each black-and-white figure by number with a label at top of reverse side. Identify color figures with a label on top of a photocopy; do not attach anything to backs of color illustrations.

Footnotes: Avoid use; incorporate material into text.

OTHER

Equations: Equations to be set separately from the text are broken into two or more lines if they exceed width of one column; mark equations for appropriate breaks. Clarify subscripts and superscripts by marginal notes. State probability values without leading zeroes (e.g., $P < .01$).

Supplementary Material: Extensive data sets, mathematical proofs, and other information of interest to a small subset of readers can be made available as a supplementary publication on paper or computer diskette. A footnote giving instructions for obtaining a copy of supplement from author or journal office will be printed in artiicle.

Units: Use International System of Units (Sl) for measurements. Consult *Standard Practice for the Use of the International System of Units* (ASTM Standard E-38092) for guidance on unit conversions, style, and usage. (ASTM, Philadelphia, PA). Sl requires use of terms mass or force rather than weight, when one unit appears in a denominator, use the solidus (e.g., g/m^2); for two or more units in a denominator, use negative exponents (e.g., $g \cdot m^{-2} \, d^{-1}$). Use tinternational spelling of metre and litre and a capital L as symbol for litre.

SOURCE

May 1993, 74(1): 1614-1615 (referred from January 1994, 75(1): 270.

Electroencephalography and Clinical Neurophysiology

International Federation of Clinical Neurophysiology
Elsevier Scientific Publishers Ireland, Ltd.

Dr. G.G. Celesia
Loyola University Stritch School of Medicine
Department of Neurology, 21600 South First Ave,
Maywood, IL 60153
    From North and South America, Australasia and Japan
Dr. F. Mauguiere
Hopital Neurologique, Service EEG
59 Boulevard Pinel, 69003, Lyon, France
    From Europe, Asia (except Japan) and Africa

AIM AND SCOPE

Electroencephalography and Clinical Neurophysiology publishes articles on all aspects of clinical neurophysiology, both normal and abnormal. Research articles on animals must have clear relevance and applicability to humans.

MANUSCRIPT SUBMISSION

Material may not have been published, except in abstract form, nor be simultaneously under consideration by another journal. Editors may require original data to compare with illustrations or results. All listed authors must concur with submission and approve final manuscript. Manuscript must have approval of responsible authorities in laboratories where work was done.

Submit original and three copies of manuscripts.

Indicate if wish publication in 'Evoked Potentials,' 'EMG and Motor Control,' or in 'EEG Journal' proper.

MANUSCRIPT FORMAT

Double-space all typing with 3 cm margins; do not justify right margin, This includes reference lists, figure legends and tables. Place all footnotes on a separate page. Place a superscript [1] in text and before footnote. Use 8.5×11 in. or European A4 bond paper. Submit only originals or Xerox copies. Do not use italics or script fonts throughout. Type each table on a separate page with Roman numeral heading. Type figure legends on a separate page(s); number separately using Arabic numerals (Fig. 1, Fig. 2, etc.). Cite Tables and Figures in numerical order of their mention in text.

Organize manuscripts: Abstract, Introduction, Methods and Materials, Results, Discussion, Reference, Tables, Figures and Legends, Acknowledgments for personal and technical assistance follow Discussion. Indicate financial support on title page as numbered footnotes to title or to individual author's name. Write a 200 word Abstract that should eliminate the need for a summary in main text.

Title page: (in capital letters) manuscript's title, initials and names of authors, and place where work was done. Do not use phrase 'with the technical assistance of ...'. List assistants as co-authors, or mention their help under Acknowledgments. Include a running title (50 letters) and name, address, telephone and fax numbers of corresponding author, and up to 6 key words for indexing.

References: Ensure accuracy of reference citations and exact agreement between text and Reference List. Reference listing should conform with original spelling (including accents), punctuation, hyphenation, etc. Cite references in text by author(s) name(s) followed by year in chronological order (Møller and Jannetta 1982; Moore 1985). With more than two authors, name only first followed by 'et al.' (Gardi et al. 1979). When two or more papers by the same author(s) appear in one year, distinguish with a, b, etc., after date (Babb et al. 1984a, b). Include papers 'in press' (accepted in final form by a journal) in Reference List. Send one copy of manuscript in press with submitted manuscript. Do not list manuscripts in preparation or submitted but not accepted in Reference List. Cite these data in text by author(s) name(s) followed by 'in preparation' or 'personal communication.' Put Reference List in alphabetical order of first author's names. Place two or more references to same author(s) in chronological order, according to Harvard system. Abbreviate journal titles according to *Index Medicus*. For examples of forms of citation of papers, books and chapters in books, see recent issues of Journal. Cite Journal as *Electroenceph. clin. Neurophysiol.*

Illustrations: Send 4 copies of each figure as high-contrast black-and-white prints on glossy paper. Originals are preferred, but photocopies are acceptable for 3 manuscript copies if of equivalent quality to originals. Do not mount illustrations; keep between 12.5 × 17.5 cm and 20 × 25 cm.

Color figures are acceptable, but authors will bear part of costs. Submit either as 24 × 36 mm transparencies or as color prints the same size as black and white figures.

Label each print on back with figure number and author(s) name(s).

Legends (typed on a separate page) should be clear, brief and self-explanatory. All artwork (graphs, diagrams, line drawings, etc.) should be done professionally.

Illustrations may be reduced in size. Make symbols, lettering and numbering sufficiently large to remain legible after reduction to fit a single column.

OTHER FORMATS

SHORT COMMUNICATIONS—1500 words, 100 word abstract, 15 references, 3 illustrations or tables.

SOCIETY PROCEEDINGS—Abstracts from proceedings of Member Societies of the Federation must be received within 3 months of meeting. Abstracts must be in correct English form, typed double-space on a separate numbered page, and no longer than 200 words. Submit two complete sets of abstracts. On a separate page give title of member Society, place and date of meeting, and name and address of Secretary. Precede each abstract by number, title, author(s) name(s), name of one institution where work was done (not more than 5 words), city and country.

TECHNICAL NOTES—1500 words, 100 word abstract, 15 references, 3 illustrations or tables.

OTHER

Abbreviations: Consult recent Journal issues conventional use of abbreviations.

Experimentation, Animal: Indicate steps taken to eliminate pain and suffering. Show recognition of moral issues involved in experiments with live subjects in writing.

Experimentation, Human: Specify that research received prior approval by appropriate institutional review body and that informed consent was obtained from each subject or patient. With photographs of recognizable persons submit a signed release from patient or legal guardian authorizing publication. Masking eyes to hide identity, while desirable, is not sufficient.

Permissions: Submit written permission from both author and publisher to use non-original material (quotations not exceeding 100 words, any graph, table or figure). Credit source in text or table footnote, or at end of a figure legend, as appropriate.

Units: Use metric system. Give all laboratory numerical data in Système International units. When other units are unavoidable, give S.I. equivalents in brackets.

SOURCE

July 1994, 91(1): inside back cover.

EMBO Journal
European Molecular Biology Organization
IRL Press

Dr. J. Tooze
The EMBO Journal, ICRF, 44 Lincoln's Inn Fields
London WC2A 3PX U.K.

Dr. I Mattaj
The EMBO Journal, EMBL, Mererhofstrasse 1
Heidelberg, D-69117 Germany

AIM AND SCOPE

The EMBO Journal provides for rapid publication of full length papers describing original research of general rather than specialist interest in molecular biology and related areas. The journal selects those manuscripts which merit urgent publication because they report novel findings of wide biological significance. Reports that are only preliminary and describe, for example, partial gene or cDNA sequences, new monoclonal antibodies or new methods are reviewed in the context of this policy. Papers devoted solely to nucleic acid sequences are published only if they are judged to be of wide interest and biological significance; reports of nucleic acid sequences that serve primarily to confirm known protein sequences are not considered. Papers should be intelligible to as wide an audience as possible; particular attention should be paid to the Introduction and Discussion sections, which should clearly draw attention to the novelty and significance of the data reported.

MANUSCRIPT SUBMISSION

Keep within length limit, ideally six printed pages in total (approximately 22 typed pages, including references, figure legends and tables). Manuscripts are not extensively copy-edited, check texts carefully before submitting. Submit original manuscript plus three copies. Submission implies that paper reports unpublished work that it is not under consideration for publication elsewhere. If previously published tables, illustrations or more than 200 words of text are included, obtain and submit copyright holder's written permission.

Manuscripts may be submitted on 3, 3.5 and 5.25 in. disks. Mark format and word processor on disk. Include three hard copies, as above.

Type legibly on A4 or American quarto paper. Type all sections double-spaced (space between lines of type not less than 6 mm). Keep 1 in. margins at sides, top and bottom. Number each page at top right (Title page is 1). Avoid footnotes; use parenthesis within brackets. Underline only words or letters to appear in italics. Clearly identify unusual or handwritten symbols and Greek letters. Differentiate between letter O and zero, and letters I and l and number 1. Mark position of figures and tables in margin.

MANUSCRIPT FORMAT

Subdivide manuscripts into sections: Title page, Abstract, Introduction, Results, Discussion, Materials and methods, Acknowledgments, References, Legends to figures, Tables, Running heads.

Title Page: Make title short, specific and informative. Serial titles are not accepted. Follow surname and initials of each author by his or her department, institution, city with postal code and country. Give changes of address in numbered footnotes. Indicate corresponding author(s). Provide a running title (50 characters).

Key Words: Give five key words (words may or may not appear in title), in alphabetical order, below title, separated by a slash (/).

Abstract: On second page include only Abstract, a single paragraph not exceeding 200 words. Make Abstract comprehensible to readers before reading paper. Avoid abbreviations and reference citations.

Acknowledgments: Include at end of text, not in footnotes. Precede institutional or agency acknowledgments with personal acknowledgments.

References: Verify accuracy of References. Include published articles and those in press (state journal which has accepted them and enclose a copy of manuscript). In text cite references by author and date; do not cite more than two authors per reference; if more than two, use *et al*. At end of manuscript type citations in alphabetical order, with authors' surnames and initials inverted. Include, in order: authors' names, year, article or journal title, editors (books only), journal or book title, name and address of publisher (books only), volume number and inclusive page numbers. Abbreviate journal name according to *World List of Scientific Periodicals*. Underline to indicate italics.

Holdsworth, M. J., Bird, C.R., Ray, J., Schuch, W. and Grierson, D. (1987) *Nucleic Acids Res.,* 15, 731–739.

Bernard, A., Boumsell, L., Dauset, J., Milstein, C. and Schlossman, S.F. (1984) *Leukocyte Typing*. Springer Verlag, Berlin.

Huynh, T.V., Young, R.A. and Davis, R.W. (1985) In Glover, D.M. (ed.), *DNA Cloning* A *Practical Approach*. IRL Press, Oxford, Vol. I, - pp. 49–78.

Personal communications (J. Smith, personal communication) should be authorized in writing by those involved. Cite unpublished data as (unpublished data). Use both as sparingly as possible and only when unpublished data referred to is peripheral rather than central to discussion. Cite references to manuscripts in preparation or submitted, but not yet accepted, in text as (B. Smith and N. Jones, in preparation). Do not include in list of references.

Tables: Type on separate sheets and number consecutively with Roman numerals. Make self-explanatory; include a brief descriptive title. Indicate footnotes by lower case letters. Do not include extensive experimental detail. Use arrows in text margin to indicate insertion in text.

Illustration: If possible submit in desired final size, to fit single column width (82 mm). Avoid double-column figures (173 mm maximum width). Make lettering approximately 2 mm high and in proportion with overall drawing dimensions.

Photographs—Must be of sufficiently high quality with respect to detail, contrast and fineness of grain to withstand loss of contrast and detail inherent in printing. Indicate magnification by a bar on photograph. When submitting several prints of the same figure, indicate which should be used for reproduction.

Line Drawings—Provide as clear sharp prints. No additional artwork, redrawing or typesetting will be done. Print all labelling. Faint shading or stippling will be lost upon reproduction and should be avoided. Heavy shading or stippling may appear black on reproduction.

Figure Legends—Place on a separate, numbered manuscript sheet. Define all symbols and abbreviations used. Do not define common abbreviations or redefine those used in text.

OTHER

Abbreviations: Use only SI symbols and those recommended by IUPAC. Define abbreviations defined in brackets after first mention in text. Standard units of measurements and chemical symbols of elements may be used without definition in text.

Color Charges: $400 per figure.

Data Submission: Publication implies an obligation to deposit novel nucleic acid sequence data referred to in text with EMBL Data Library. Make deposition immediately after paper has been accepted for publication; add accession number at page proof stage. Find submission details on data submission in *Nucleic Acids Research* (issue 1 of each volume).

Deposit structural data, including derived atomic coordinates, with an appropriate Protein Data Bank; include Data Bank reference number in manuscript. Specify agreed date for release of data (maximally two years after publication). Primary DNA sequence data is published only if it is essential to the understanding of paper. Otherwise provide only data bank accession number.

Materials Sharing: Authors agree to make freely available to colleagues in academic research any cells, nucleic acids, antibodies, etc. used in research reported and that are not available from commercial suppliers.

Nomenclature: Follow *CBE Style Manual* (5th ed, Council of Biology Editors, Bethesda, MD, 1983).

Follow *Chemical Abstracts* and its indexes for chemical names.

For biochemical terminology follow IUPAC-IUB Commission on Biochemical Nomenclature's *Biochemical Nomenclature and Related Documents* (Biochemical Society).

For enzymes, use recommended names assigned by IUPAC-IUB Commission on Biochemical Nomenclature, 1978, *Enzyme Nomenclature* (Academic Press, New York, 1980).

Nomenclature of plant genes should follow International Society for Plant Molecular Biology Commission on Plant Gene Nomenclature, which are posted regularly on the public databases and published annually in Plant Molecular Biology Reporter, starting with December 1993 issue. Italicize genotypes (underline in typed copy); do not italicize phenotypes. For bacterial genetics nomenclature follow Demerec *et al. (1966) Genetics,* **54,** 61–76.

Page Charges: $160 per additional page over eight. Essential changes of an extensive nature may be made only by insertion of a Note added in proof. A charge (minimum $80) for amendments to text at page-proof stage.

Reviewers: List four possible referees.

SOURCE
January 3, 1995, 14(1): back matter.

Emerging Infectious Diseases
Centers for Disease Control and Prevention

Editor, *Emerging Infectious Diseases*
National Center for Infectious Diseases
Centers for Disease Control and Prevention
1600 Clifton Road, Mailstop C-12
Atlanta, GA 30333
Telephone: (404) 639-3967; Fax: (404) 639-3039; E-mail: eideditor@cidod1.em.cdc.gov.

AIM AND SCOPE
The goals of *Emerging Infectious Diseases* (EID) are to promote the recognition of new and reemerging infectious diseases and to improve the understanding of factors involved in disease emergence, prevention, and elimination. EID has an international scope and is intended for professionals in infectious diseases and related sciences. We welcome contributions from infectious disease specialists in academia, industry, clinical practice, and public health as well as from specialists in economics, demography, sociology, and other disciplines whose study elucidates the factors influencing the emergence of infectious diseases.

MANUSCRIPT SUBMISSION
Send documents in hardcopy (Courier 10-point font), on diskette, or by e-mail. Acceptable electronic formats for text are ASCII, WordPerfect, AmiPro, DisplayWrite, MS Word, MultiMate, Office Writer, WordStar, or Xywrite. Send graphics documents in Corel Draw, Harvard Graphics, Freelance, TIF (TIFF), GIF (CompuServe), WMF (Windows Metafile), EPS (Encapsulated Postscript), or CGM (Computer Graphics Metafile). Preferred font for graphics files is Helvetica. Convert Macintosh files into a suggested format. Submit photographs in camera-ready hardcopy.

MANUSCRIPT FORMAT
Prepare manuscripts according to "Uniform Requirements for Manuscripts Submitted to Biomedical Journals" (Vancouver Style) *JAMA* 1993;269(17):2282-6.

Begin each section on a new page in order: title page, abstract, text, acknowledgments, references, each table, figure legends, and figures. On title page, give complete information about each author (full names and highest degree). Give current mailing address for correspondence (include fax number and e-mail address). Follow Uniform Requirements style for references. Consult *List of Journals Indexed in Index Medicus* for accepted journal abbreviations. Number tables and figures separately (each beginning with 1) in order mentioned in text. Double-space everything, including title page, abstract, references, tables, and figure legends. Italicize scientific names of organisms from species name all the way up, except for vernacular names (viruses that have not really been speciated, such as coxsackievirus and hepatitis B; bacterial organisms, such as pseudomonads, salmonellae, and brucellae).

OTHER FORMATS
PERSPECTIVES—3500 words, 40 references. Address factors known to contribute to the emergence of infectious diseases, including microbial adaption and change; human demographics and behavior; technology and industry; economic development and land use; international travel and commerce; and breakdown of public health measures. Begin with an introduction outlining relationship of issues discussed in paper to emergence of infectious diseases. Use additional subheadings in main body of text. If detailed methods are included, follow body of text with a separate section on experimental procedures. Photographs and illustrations are optional. Provide a short abstract (150 words) and a brief biographical sketch.

SYNOPSES—3500 words and 40 references. Submit concise reviews of infectious diseases or closely related topics. Preference is given to reviews of new and emerging diseases; timely updates of other diseases or topics are welcome. Begin with an introduction outlining relationship of issues discussed in paper to emergence of infectious diseases. Use additional subheadings in main body of text. If detailed methods are included, follow body of text with a separate section on experimental procedures. Selective use of illustrations is encouraged. Provide a short abstract (150 words) and a brief biographical sketch.

DISPATCHES—1000 to 1500 words providing brief updates on trends in infectious diseases or infectious disease research in a letter to the editor format; do not divide into sections. Begin with a brief introductory statement about relationship of topic to emergence of infectious diseases. Include methods development; references (5); and figures or illustrations (2).

SOURCE

Internet: http://www.cdc.gov.

Endocrine Reviews
The Endocrine Society

Andres Negro-Vilar, M.D., Ph.D.
121 South Estes Drive, Suite 203-D, P.O. Box 3560
Chapel Hill, NC 27514
Telephone: (919) 929-6273; Fax: (919) 929-6287

AIM AND SCOPE

The Endocrine Society publishes papers describing the results of original research in the field of endocrinology and metabolism in the *Journal of Clinical Endocrinology and Metabolism, Endocrinology, Molecular Endocrinology* and *Endocrine Reviews. Endocrine Reviews* features in-depth review articles on both experimental and clinical endocrinology. Consult with editor concerning the most appropriate journal to which to submit manuscript.

Endocrinology is defined in its broadest sense as the study of hormones and local regulatory factors and their influence(s) on biologic functions in normal or diseased states, at the organismal, cellular or biomedical level in all species. Articles dealing with the clinical practice of endocrinology may be suitable when warranted by significant change or controversy in a field. Manuscripts should present a timely, authoritative review of a topic in endocrinology that is of general interest. Authors should provide sufficient background information to allow both endocrinologists and other scientists to gain both knowledge and perspective of the subject. While a limited amount of new data may be included to buttress the author's point of view, manuscripts that deal primarily with new findings should be submitted to *Endocrinology* or *The Journal of Clinical Endocrinology and Metabolism*, or *Molecular Endocrinology*.

MANUSCRIPT SUBMISSION

Proposals for manuscripts may be submitted. Include a full outline of proposed review, listing all topics to be covered; a *curriculum vitae* for each author; a bibliography of author's published work; and a completion date.

Provide assurance, on form in journal, or a copy thereof, that no substantial part of work has been published or is being submitted or considered for publication elsewhere. If some results are to appear in another journal, or elsewhere, provide details to Editor, and a copies of other paper(s) submitted, with expected date of publication.

Affidavit affirming originality and assigning copyright to The Endocrine Society signed by all authors must be submitted with all manuscripts (see copy in journal issues).

MANUSCRIPT FORMAT

Type copy double-spaced with wide margins. Submit original manuscript and three copies, including tabular material. Only one set of original illustrations is required.

Title Page: Give full and abbreviated manuscript titles; full names, degrees, and institutions for all authors; name and address of corresponding author; and supporting grants or fellowships.

Title: Prepare a concise statement of article's major contents; give sufficient information to allow judgement of reader interest. Include key words or phrases useful for indexing and information retrieval. Give a running title (40 letters and spaces) for page headings. Include an outline of manuscript's contents.

Introduction: Indicate essence of review; provide historical perspective; describe scope of review; provide insight into perspective and conclusions. Cite previous reviews. Avoid non-standard abbreviations, unfamiliar terms, symbols and acronyms.

Body: Organize to suit, but make logical and readily comprehensible. Indicate organization in table of contents.

References: Give in numerical order (in parentheses) in text. List in same numerical order at end of manuscript on a separate sheet(s). Only one reference to a number.

ARTICLES IN JOURNALS

Cite all author names with initials, year, title of paper (capitalize only initial letter of first word), abbreviated journal name, volume, and fully annotated, inclusive pages.

Fraser IS, Challis JRG, Thorburn GD 1976 Metabolic clearance rate and production of oestradiol in conscious rabbits. J Endocrinol 68:313–320

CHAPTERS IN BOOKS

Heard RDH 1948 Chemistry and metabolism of the adrenal cortical hormones. In: Pincus G, Thimann KV (eds) The Hormones. Academic Press, New York, vol 3:549–555

Greenwald GS 1974 Role of follicle-stimulating hormone and luteinizing hormone in follicular development and ovulation. In: Knobil E, Sawyer WH (eds) Handbook of Physiology, sect 7, pt 2. American Physiological Society, Washington DC, vol 4:293–298

BOOKS

Turner CD 1960 General Endocrinology, ed 3. Saunders, Philadelphia, p 426–429

References should be comprehensive for current or recent citations, but need not include all previous references. Cite appropriate recent reviews whenever possible. Do not cite unpublished observations, personal communications and manuscripts in preparation or submitted for publication in bibliography. Insert at appropriate places in text, in parentheses and without serial number, or present in footnotes. Cite manuscripts in press in bibliography; supply name of journal. For references to personal communications, obtain written proof of exchange. If citing an abstract is necessary, so designate at end of reference. Verify accuracy of references.

Illustrations: Give each figure a title and a legend, typed together on a separate sheet. Identify all figures on back, in soft pencil, with authors' names and figure number; indicate TOP.

Submit graphs and diagrams as sharp photographs on white glossy paper. Submit one set of unmounted illustrations with original manuscript. Additional sets may be prepared by whatever means available; prints of photographs for halftone reproduction are required.

Figures will be reduced to one column width. Make letters or numbers at least 1.5 mm high after reduction, symbols at least 1.0 mm. Submit photographs for halftone reproduction on white glossy paper, trimmed or able to be reduced to one column width.

For graphs with very large or very small numbers on an axis, examples of conventional notation: For an ordinate for counts per minute, with values between 1,000 and 20,000, the scale may run from 1 to 20 and label ordinate "cpm $(\times 10^{-3})$," *i.e.,* the true value has been multiplied by 10^{-3} to give number on ordinate. In a Scatchard plot with values of 0.1 to 2 femtomolar $M(\times 10^{15})$, the scale may run from 0.1 to 2 and label abscissa with the < symbol aligned above the > symbols followed by $(\times 10^{15})$.

Color photographs must be approved by Editor-in-Chief, and require 4–6 weeks additional lead time prior to publication. Authors pay cost of color separations and printing and will receive an estimate when manuscript is submitted to printer, at which time they must provide written acceptance of charges.

Tables: Submit as typed copy, not photographs. Give each a concise heading. Construct as simply as possible; make intelligible without reference to text. Description of experimental conditions may appear at foot of Table with footnotes. Do not duplicate material in text or figures. Relate table width to size of journal page. Do not put more than 4 vertical rows in a small one column table and not more than 8 to 10 rows in a large double-column table.

OTHER FORMATS

COUNTERPOINT ARTICLES—Companion pieces to standard reviews in order to give a wider perspective on a given topic. Provide a clinical perspective to basic science reviews, or vice versa. Follow instructions for standard reviews.

OTHER

Abbreviations, Symbols and Nomenclature: Use International Système of Units (SI) (Page, C. H. and P. Vigoureux (eds.), NBS Special Publication 330, U.S. Government Printing Office.) Use metric system for weights and measurement, degrees Celsius for temperature and 24 hour clock for time.

Do not express molecular weights in daltons. Molecular weight is relative molecular mass of a substance; molecular weight is dimensionless. The dalton is a unit of mass equivalent to l/12 of the mass of one atom of carbon 12.

For systematic chemical names follow usages in indexes of *Chemical Abstracts*.

For nomenclature and symbols follow recommendations of the International Union of Pure and Applied Chemistry (IUPAC) and the International Union of Biochemistry (IUB). See *J Biol Chem* 252:10,1977. For a complete listing see *Handbook for Authors* (American Chemical Society Publications: Washington, D.C., 1967).

For peptide nomenclature follow IUPAC rules (*J Biol Chem* 242:555,1967; 247:323,1972).

For isotopes use ^yA to designate element A whose atomic mass is y.

Use C_x to refer to a compound containing x carbon atoms. Use C-x is to refer to the xth carbon atom of a molecule. To designate isotopically labeled chemical compounds, for a chemical formula, insert symbol for isotope into formula; for a chemical name or abbreviation, use: for glucose labeled in C-1 position use $[1^{-14}C]$glucose; for diiodotyrosine labeled with ^{131}I use $[^{131}I]$diiodotyrosine. Designate proteins labeled with ^{125}I or ^{131}I as iodoproteins. As a generalization, recognize any group used to label a molecule which produces a derivative of that molecule in naming the molecule.

Drugs: Designate by generic names; use trade names (spelled exactly as trademarked) after a drug has been identified by its generic or systematic chemical name. Identify this way if trade name appears in title or summary.

Materials Sharing: Provide propagative materials such as DNA clones, cell lines, and hybridomas developed to qualified investigators who wish to duplicate published experiments.

Permission: Obtain permission from copyright holder for use of figures or tables that are not original. Copyright holder is either publisher or author(s). Publisher may require permission of first author as well. If author, obtain permission from first author (check with publisher to make sure author actually holds copyright). If figure is from your articles, still obtain permission from publisher if they hold copyright.

Adapting a figure or table involves using a significant portion of another a original figure to convey your own information. Derived figures are based on data contained in another author's article and do not require permission; mention reference in legend.

When requesting permission, include: citation for article, chapter, etc. in which figure appeared; original figure number; name of your article and journal in which it will appear and acknowledgment as it will appear in *Endocrine Reviews.*:

Article: First author *et al.*: *Journal* vol:page, year (reference #).

Book: First author *et al.*: *Title* (editors), publisher, city, place, year, page (reference #).

After obtaining permission, send originals to *Endocrine Reviews* Editorial Office, preferably with submitted manuscript.

SOURCE

February 1994, 15(1): front matter.
Complete instructions to authors appear in first (February) and fourth (August) issues of each volume.

Endocrinology
The Endocrine Society

Dr. Shlomo Melmed
Cedars-Sinai Medical Center
8700 Beverly Boulevard B-138
Los Angeles, CA 90048-1865
Telephone: (310) 855-3371; Fax: (310) 657-9656

AIM AND SCOPE

The Endocrine Society publishes papers describing the results of original research in the field of endocrinology and metabolism in the *Journal of Clinical Endocrinology and Metabolism, Endocrinology, Molecular Endocrinology*, and *Endocrine Reviews*.

Endocrinology publishes nonprimate biochemical and physiological studies, although work on material of primate origin is not excluded. Consult with Editors concerning most appropriate journal.

MANUSCRIPT SUBMISSION

Provide assurance, on form in each issue, (or copy) that no substantial part of work has been published or is being submitted or considered for publication elsewhere. If some results are to appear in another journal, or elsewhere, provide details to Editor, and submit copies of other submitted paper(s), with expected date of publication.

Submit affidavit (see copy in journal issues) affirming originality and assigning copyright to The Endocrine Society signed by all authors with manuscript.

Two types of manuscripts are considered: Original Studies and Rapid Communications. Original Studies are original investigative reports of any length; their main emphasis cannot be methodological. Rapid Communications must contain new observations of unusual interest and importance to the field of endocrinology. They may be either preliminary reports or completed studies. Rapid Communications describe methodological or technical innovations of immediate benefit.

Present a clear, honest, accurate, and complete account of research performed. Describe a complete study or a completed phase of an extended study. Avoid fragmentation of reports. Supply details about other results appearing in other journals, in publications of congresses, symposia, workshops, etc.; supply copies of other submitted paper(s)

Describe work in sufficient detail to allow repetition. Include all relevant data, including those which may not support hypothesis being tested. Cite those publications which have a direct bearing on novelty and interpretation of results.

List only individuals who made significant contributions to intellectual and procedural aspects of study as authors. An author should have participated in conception and planning of work, interpretation of results, and writing of paper. An acknowledgment is appropriate recognition for other contributions. Signatures on the Affirmation of Originality and Copyright Release form indicate that authors have approved final version of manuscript and take public responsibility for work. Failure to notify editor of previous or simultaneous publication of results will result in placement of a notice in the journal that the authors have violated the Ethical Guidelines for Publication of Research in Endocrine Society Journals.

MANUSCRIPT FORMAT

Submit five copies typed double-spaced, including references and all tabular material, with wide margins on 8.5 × 11 in. paper. All five copies must be complete with all tabular and illustrative material.

Title Page: Include authors' names and institutions, abbreviated title, tables, references, footnotes and legends on separately.

Title: Carefully prepared, concise statement of article's major contents. Provide sufficient information to allow reader to judge interest. Include key words or phrases for indexing and information retrieval. Do not exceed three lines (150 letters and spaces).

Provide a short running title (40 letters and spaces) for page headings.

Abstract: 200 words. Briefly describe in complete sentences purpose of investigation, methods utilized, results obtained and principal conclusions. Make easily understood without recourse to text or references. Avoid nonstandard abbreviations, unfamiliar terms, symbols or acronyms not easily understood by general scientific reader, references, and uninformative sentences.

Introduction: Acquaint reader with state of knowledge of topic. Assume knowledge of basic endocrinology, cite recent important work of others, use references to reviews for convenience, and make a succinct statement of problem.

Materials and Methods: Describe and reference in sufficient detail so that others can repeat. Describe source of hormones, unusual chemicals and reagents, and special pieces of apparatus. For modified methods, describe only modification. For a new method that is fully described only in a paper in press or accepted for publication, describe briefly so that readers understand its principle and application. If data were subjected to statistical analysis, present method(s) used at end of Materials and Methods.

Results: Present experimental data without redundancy in tables and figures. Explain succinctly but completely in text.

Discussion: Avoid repetition of material in introduction and experimental findings. Be brief. Focus on experimental findings and interpretation.

References: Cite in numerical order (in parentheses) in text. List in same numerical order at end of manuscript on a separate sheet(s). Only one reference to a number.

Give references to articles in journals in order: names of all authors with initials, year, title of paper (capitalize only initial letter of first word), abbreviated name of journal, volume, and pages. Follow style for journal abbreviations in National Library of Medicine's *Index Medicus*.

Lewin N, Blum M, Roberts JL 1989 Modulation of basal and corticotropin-releasing factor-stimulated proopiomelanocortin gene expression by vasopressin in rat anterior pituitary. Endocrinology 125:2957–2966

CHAPTERS IN BOOKS

Baulieu E-E, Mester J 1989 Steroid hormone receptors. In: DeGroot LJ (ed) Endocrinology. WB Saunders, Philadelphia, vol 1:16–39

ABSTRACTS FROM MEETINGS

Kaptein EM, Grieb D, Wheeler W, Peripheral thyroid hormone kinetics in critical illness. Program of the 62nd Annual Meeting of The Endocrine Society, Washington DC, 1980, p 189 (Abstract).

Keep number of references cited to less than 50; cite appropriate recent reviews whenever possible. Do not cite unpublished observations, personal communications, or manuscripts in preparation or submitted for publication in bibliography. Insert at appropriate places in text, in parentheses and without serial number, or present in footnotes. Cite manuscripts in press in bibliography; supply name of journal. Submit three copies of preprints for pertinent quoted papers in press. For references to personal communications, submit written permission from investigator concerned. If citing an abstract, designate at end of reference. Verify accuracy of references. Ensure correctness of spelling, capitalization and diacritical marks in foreign-language titles.

Illustrations: Give each figure a title and a legend; type legends together on a separate sheet. Identify all figures on back, in soft pencil, with authors' names and figure number; indicate TOP. Figure legends are not part of figure proper.

Submit graphs and diagrams as sharp photographs on white glossy paper. Submit one set of unmounted illustrations with original manuscript. Additional sets may be prepared by whatever means available; glossy prints of photographs for halftone reproduction (*e.g.*, photomicrographs and pictures of gels) are required.

Line drawings and graphs will be reduced to one column width. Make letters or numbers to be at least 1.5 mm high after reduction, symbols at least 1.0 mm. Keep wording on figures to a minimum. Include explanations in figure legends. Do not give figures titles. Make insets large enough to be reduced clearly. Limit labeling to one vertical and one horizontal side of figure. Place groupings of figures in such a way that they can be printed within one column width and be uniform.

Have line drawings and graphs professionally drawn and lettered; freehand or typewritten lettering is unacceptable. Make ruled lines and curves thick enough to withstand reduction to one column width. Size recommendations for graphs prepared with lettering sets: #1 Leroy (or equivalent) for graph grids bonds and arrows; #2 Leroy for graph borders and reference lines; #5 Leroy for graph curves and emphasis lines.

For graphs with very large or very small numbers on axis, examples of conventional notations: For an ordinate for counts per minute, with values between 1,000 and 20,000, the scale may run from 1 to 20, and label ordinate "cpm $(\times 10^{-3})$," *i.e.*, the true value has been multiplied by 10^{-3} to give number on ordinate. In a Scatchard plot with values of 0.1 to 2 femtomolar (10^{-15} M), the scale may run from 0.1 to 2 and label abscissa "M ($\times 10^{15}$)." Three-dimensional bar graphs will not be published unless information is given in the third dimension.

Submit photographs for halftone reproduction on white glossy paper printed with good contrast. Crop prints to be one column wide, or capable of being reduced to such

size. Do not mount prints. Indicate layout for grouping into plates in a diagram. Apply figure numbers to lower right corner of photographs so it stands out against background. Make numbers, letters and symbols large enough to be easily read after reduction (final size 5 mm or more for letters and numbers). Indicate magnification in legends and by internal reference markers in photographs, as horizontal lines at least 1 mm thick after reduction, placed in lower left corner. Make length represent fraction or multiple of a micrometer appropriate to magnification.

Tables: Submit as typed copy, not photographs. Give each a concise heading. Construct as simply as possible; make intelligible without reference to text. Describe experimental conditions at foot of table with footnotes. Do not duplicate material in text or figures. Relate table width to size of journal page. Do not put more than 4 vertical rows in a small one column table and not more than 8 to 10 rows in a large double-column table.

OTHER FORMATS

RAPID COMMUNICATIONS—Follow same format requirement. Type on special forms obtainable from Editorial Office. Type camera ready; pay attention to legibility of graphs. Be sure figures fit exactly in spaces provided. Upper limit of length is four special form pages, including text, references, and illustrative material. Leave room for Date of Receipt at bottom of first column.

Submit to Editor-in-Chief in triplicate (original and two duplicates, preferably photocopies reduced to page size; provide half-tone illustrations as glossy prints in triplicate). Pack carefully to avoid damage in transit.

Criteria for acceptance are: novelty, exceptional significance, and interest to more than a highly specialized small sector of the endocrine community.

OTHER

Abbreviations, Symbols and Nomenclature: See list in Appendices A and B (*Endocrinology* 135(1): 14A-15A, 1994). Define all abbreviations and symbols for terms and units that do not appear on this list. Use metric system for weights and measurement, degrees Celsius for temperature and the 24-hour clock for time.

Do not express molecular weight in daltons. Molecular weight is relative molecular mass of a substance; molecular weight is dimensionless. The dalton is a unit of mass equivalent to 1/12 of the mass of one atom of carbon 12. For systematic chemical names follow usages in indexes of *Chemical Abstracts*. For nomenclature and symbols follow recommendations of the International Union of Pure and Applied Chemistry (IUPAC) and the International Union of Biochemistry (IUB). See *J Biol Chem* 252:10, 1977. For a complete listing, see *Handbook for Authors* (American Chemical Society Publications: Washington, D.C., 1967). For peptide nomenclature follow IUPAC rules (*J Biol Chem* **242**: 555,1967; **247**: 323,1972).

For steroid nomenclature see Appendix C (*Endocrinology* 135(1): 15A, July 1994).

For nomenclature of vitamin D metabolites see Appendix D (*Endocrinology* 135(1): 16A, July 1994).

For isotopes use ^yA to designate element A whose atomic mass is y.

Use C_x to refer to a compound containing x carbon atoms. Use C-x to refer to the xth carbon atom of a molecule.

To designate isotopically labeled chemical compounds, for a chemical formula insert symbol for isotope into formula; for a chemical name or abbreviation, use: for glucose labeled in the C-1 position use [1-^{14}C]glucose; for diiodotyrosine labeled with ^{131}I use [^{131}I]diiodotyrosine. Designate proteins labeled with radioactive iodine as [^{125}I]iodoprotein or ^{125}I-protein. As a generalization, recognize any group used to label a molecule which produces a derivative of that molecule in naming that molecule.

Bioassays and Immunoassays—Express potency estimates for protein and polypeptide hormones, whether made by bioassays or immunoassays, in terms of reference preparations such as International Standards, NIDDK Preparations, etc. Describe nature of reference preparation, its species and tissue of origin and some evidence for its similarity to hormonal activity being measured under Materials and Methods. If a suitable recognized standard does not exist, give a full description for preparation and standardization of reference material used. For bioassays, stipulate species and strain of animal used.

For specimens of human origin, express results of bioassays and of radioimmunoassays of FSH and LH in urine and urinary extracts in International Units, in terms of the First International Standard of Urinary FSH and LH (IC-SH), human, for bioassay. The former Second International Reference Preparation of Human Menopausal Gonadotropins, urinary, for bioassay, now replaced, is available as a working standard and not as the 2nd IRP. For human materials other than urine, express results of radioimmunoassay of FSH or LH may be in terms of weight of preparation LER 907 (National Pituitary Agency, NIADDK).

State method of assay and standard employed in Materials and Methods and other places where appropriate. See Albert et al., J Clin Endocrinol Metab 28: 1214, 1968.

For an up-to-date list of International Standards and Reference Preparations of Endocrinological and Related Substances, see Biological Substances International Standards, Reference Preparations and Reference Reagents (World Health Organization, 1977). Obtain reference standards for hormones from U.S. Pharmacopeial Convention, Inc., Rockville, MD.

Color Charges: Approximate cost for one page is $1800. Authors will receive an estimate of expense upon submission to printer for composition. Provide written or verbal acceptance of charges before production.

Drugs: Designate by generic names; use trade names (spelled exactly as trademarked) after a drug has been identified by its generic or systematic chemical name. Identify this way if trade name appears in title or abstract.

Experimental, Animals: Include a statement that the following principles and procedures were followed. All studies must be conducted in accordance with the highest standards of humane care. Consider appropriateness of experimental procedures, species and number of animals used in study design. Acquire and use research animals in compliance with federal, state, and local laws and institutional regulations. Maintain animals in accordance with the NIH Guide for the Care and Use of Laboratory Animals.

Research animals must receive appropriate tranquilizers, analgesics, anesthetics, and care to minimize pain and discomfort during preoperative, operative, and postoperative procedures. Make choice and use of drugs in accordance with NIH Guide. Where use of anesthetics would negate experimental results, justify protocol and obtain approval of local Committee on Animal Care and Use. Monitor animal health. If either study or condition of animals requires that they be killed, it shall be done in a humane manner.

Materials Sharing: Make any clones, whether of cells or genes, available to other researchers on request. Provide propagative materials such as DNA clones, cell lines, and hybridomas developed by the authors and used in the studies to qualified investigators who may wish to duplicate the published experiments.

Page Charges: $45 per printed page in the journal. Additional charges may be levied for excessive or unusual illustrative material. Publications Committee may waive charges upon appeal.

Statistical Analyses: Document that experiments presented are reproducible, and that any differences noted between experimental groups are not due to random variation. In general, quantitative data should be from no fewer than three replicative experiments. Illustration of "representative" experiments is discouraged. If this is done, documented how experiment chosen for illustration represents others with similar results, even if pooling of results is not feasible. Use appropriate statistical methods to establish significance of quantitative differences between findings compared. Do not use term "significant" unless statistical analysis supporting it was performed. Specify probability value used as cutoff between "significant" and "not significant."

When data points are fitted with lines, specify method used for fitting (graphical, least squares, computer program). If differences in slopes and/or axis intercepts are claimed for plotted lines, support by statistical analysis.

For bioassay and radioimmunoassay potency estimates, include an appropriate measure of precision of estimates. For bioassays, use standard deviation, standard error of the mean, coefficient of variation, or 95% confidence limits. For both bioassays and radioimmunoassays, include a measure of precision relating to within-assay variability and also to variability of measurements obtained on subsequent independent assays. If all relevant comparisons are made within same assay, latter may be omitted. The precision of a measurement depends upon its position on the dose-response curve. Descriptions of new assays must include: within-assay variability; between-assay variability; slope of dose-response curve; mid-range of assay; least detectable concentration (concentration resulting in a response two standard deviations from the zero dose response); data on specificity; data on parallelism of standard and unknown; comparison with an independent method for assay of compound; and data on recovery. When several *t* tests are employed, nominal probability levels no longer apply. Use the multiple *t* test or multiple range test or similar techniques to permit simultaneous comparisons. In lieu of using several *t* tests, utilize analysis of variance to permit pooling of data, increase number of degrees of freedom and improve reliability of results. Use appropriate nonparametric tests of significance when data depart substantially from a normal distribution.

In presenting results of linear regression analyses, show confidence limits.

Useful references for statistical methods are: McArthur, J. W. and T. Colten (eds.), *Statistics in Endocrinology*, MIT Press, Cambridge, 1970, and Finney, D. J., *Statistical Method in Biologic Assays*, ed. 2, Griffin, London, 1967.

Submission Fee: $70 for each typed manuscript accompanied by a computer diskette containing complete manuscript text (most word processing languages are acceptable). For manuscripts without a computer diskette, processing fee is $170. Diskettes may be submitted with final, revised version of manuscript (i.e., after it has been reviewed and returned). Requirements for computer disk do not apply to authors of Rapid Communications; submission fee remains $70.

In cover letter state wish to be evaluated for publication in *Endocrinology*. Suggest names of appropriate reviewers; also request that particular individuals not serve in that capacity.

SOURCE

July 1994, 135(1): 9A-16A.

Environmental Health Perspectives (EHP)

U.S. Department of Health and Human Services, Public Health Service, National Institutes of Health

Editor-in-Chief, *Environmental Health Perspectives*
National Institute of Environmental Health Sciences
111 T. W. Alexander Dr., P.O. Box 12233
Research Triangle Park, NC 27709
Telephone: (919) 541-3406; Fax: (919) 541-0273

AIM AND SCOPE

Environmental Health Perspectives (EHP) is a forum for the discussion of issues in environmental health. The primary qualifications for publication are environmental significance and scientific quality. Environmental science is made up of many fields and therefore scientific progress in all of them will be considered. Articles ranging from the most basic molecular biology to environmental engineering will be considered. Articles on mechanisms of toxic action and new approaches for detecting or remedying environmental damage are encouraged. Factual articles about issues that affect the environment and human health are considered. Legislative and regulatory developments, grant information from NIEHS and other granting agencies, new research areas, environmental problems, technological advances, and information about the National Toxicology Program and other important programs are summarized.

MANUSCRIPT SUBMISSION

Manuscripts must be typed, double-spaced on 8.5 × 11 in. paper with 1 in. margins. Type on one side of paper only. Number pages consecutively beginning with title page. Submit four copies.

In cover letter provide assurances that manuscript is not being considered for publication elsewhere. Obtain permission to reprint figures or tables from other publications prior to submission. State that all authors have read manuscript and agree that work is read for submission and that they accept responsibility for contents.

MANUSCRIPT FORMAT

Title Page: List title, authors (first or second names spelled out in full), full address of institution at which work was done, and affiliation of each author. Indicate corresponding author and give complete address and telephone and fax numbers.

Second Page: Provide a short title (50 characters and spaces) for a running head. List 5-10 key words for indexing. List and define all abbreviations.

Abstract: Double-space on third page. 250 words stating purpose of study, basic procedures, main findings, and principal conclusions. Emphasize new and important aspects of study of observations.

Introduction: On a new page, state purpose of research, give brief overview of background information. Do not include data or conclusions.

Methods: On a new page, describe materials used and their sources. Include enough detail to allow repetition of work or cite appropriate references.

Results: On a new page, present in a logical sequence. Do not repeat materials and methods or data in tables or figures. Summarize only important observations; may be combined with Discussion.

Discussion: On a new page, emphasize new and important aspects of study and conclusions. Relate results to other relevant studies. Do not recapitulate data from Results.

References: On a separate page. Number in order of citation in text. Cite numbers in text in parentheses.

JOURNAL ARTICLE

1. Canfield RE, O'Connor JF, Birken S, Kirchevsky A, Wilcox AJ. Development of an assay for a biomarker of pregnancy in early fetal loss. Environ Health Perspect 74:57-66 (1987).

BOOK CHAPTER

2. Lohman AHM, and Lammer AC. On the structure and fiber connections of the olfactory centres in mammals. In: Progress in brain research: sensory mechanisms, vol 23 (Zotterman Y, ed), New York: Elsevier, 1967; 65–82.

BOOK

3. Harper R, Smith ECB, and Land DB. Odour description and classification. New York: Elsevier, 1968.

EDITOR AS AUTHOR

4. Doty, RL, ed. Mammalian olfaction, reproductive processes, and behavior. New York: Academic Press, 1976.

REPORT

5. NCTR. Guidelines for statistical tests for carcinogenicity. Biometry technical report 81-001. Jefferson, AR: National Center for Toxicological Research, 1981.

Abbreviate journal names according to *Index Medicus* or *Serial Sources for the BIOSIS Previews Database*. List all authors; do not use et al. Include title of journal article or book chapter and inclusive pagination. Cite reference to papers accepted for publication but not yet published in the same manner as other references, with name of journal followed by "in press." Do not list personal communication, unpublished observations, or manuscripts in preparation. Insert at appropriate places in text, in parentheses, without a reference number.

Figures: Submit three sets of publication quality figures. Submit graphs and figures as original drawings in black India ink, laser-printed computer drawings or glossy photograph. Electronic versions of figures are encouraged; submit in addition to hardcopies of figures. Do not send dot matrix computer drawings. Make style of figures uniform throughout. Draw letters, numbers and symbols to be at least 1.5 mm high after reduction. Choose a scale so that each figure may be reduced to two-column width or one-column width. Identify figures on back with authors' names, figure number and indicate TOP. Color figures will be considered if color facilitates data recognition and comprehension.

Legends: Type on separate sheet following references. Number with Arabic numerals.

Tables: Put each on a separate page. Number with Arabic numerals. Indicate footnotes by lowercase superscript letters beginning with a for each table. Indicate footnotes with statistical significance by *, **, #, ##. Type footnotes directly on table. Submit complex tables as glossy photographs.

OTHER FORMATS

COMMENTARIES—Short articles offering ideas, insight, or perspectives.

INNOVATIONS—Short articles describing novel approaches to the study of environmental issues. Some speculation on the potential usefulness of new techniques or novel processes may be included. Do not include references; include a suggested reading list.

MEETING REPORTS—2400 words; begin with title of meeting and authorship of report. Start text on next page. Detail when and where meeting was held, number of participants, sponsors, and special arrangements. List sponsors and principal participants on a separate page. Summarize contribution of meeting to scientific knowledge, insight, and perspective. Emphasize novel ideas, perspectives and insights. Send an electronic copy and four hard copies.

RESEARCH ADVANCES—Concise articles addressing only the most recent developments in a scientific field. Lengthy historical perspectives are not appropriate. Submit photograph and brief biography.

REVIEWS AND COMMENTARIES—Brief, up-to-date narrowly focused review articles with commentaries offering perspective and insight.

SUPPLEMENTS—Results from conferences, symposia, or workshops. Include Perspective Reviews, in-depth, comprehensive reviews of a specific area.

OTHER

Experimentation, Animal: In cover letter state that animals were treated humanely according to institutional guidelines, with due consideration to the alleviation of distress and discomfort.

Experimentation, Human: In cover letter, state that participation followed informed consent.

SOURCE

January 1995, 103(1): 116-118.

Epilepsia
International League Against Epilepsy
Raven Press, Ltd.

Dr. Timothy Pedley
Neurological Institute, 710 W 168 St., Rm 1401
New York, NY 10032

AIM AND SCOPE

Original manuscripts, i.e., those that have not been published elsewhere except in abstract form, on any aspect of epilepsy (clinical, experimental, biochemical, etc.) will be considered. Laboratory, clinical, social and historical notes (no more than 1,000 words and two figures), announcements of meetings and awards, and book reviews are also published.

MANUSCRIPT SUBMISSION

Submit three copies of each manuscript in typewritten form, double-spaced with a 2 in. left margin. Two of three copies may be photocopies.

MANUSCRIPT FORMAT

Title Page: Page 1: title; authors' names; institution where work was done; footnotes to title or authors; five key words (refer to *Medical Subject Headings,* from the National Library of Medicine); running title (40 letters and spaces); name, address, and telephone number of corresponding author.

Summary: On a separate sheet, 200 words suitable for abstracting journals. Concisely and specifically describe purpose, methods, results, and conclusions of study. If possible, supply translations in French, German, and Spanish.

Introduction: Make object of research clear with reference to previous work.

Methods: Describe in sufficient detail so work can be duplicated, or reference previous descriptions if readily available.

Results: Describe clearly, concisely, and in logical order. When possible give range, standard deviation or mean error, and significance of differences between numerical values.

Discussion: Interpret results and relate to previous work.

Acknowledgments: Type minimum compatible with courtesy on a separate sheet.

References: Cite all references in reference list in text. Make text citations correspond exactly with List of References, including spelling and year of publication. Ensure accuracy of reference data.

In text supply names of authors and year of publication; if more than two authors, supply first followed by et al. and year.

List of References: Double-space on separate sheets. Arrange alphabetically by author(s). Arrange more than one paper by same author(s) by year; if same year, distinguish by a, b, etc., after year. For abbreviations of journal names, refer to *List of Journals Indexed in Index Medicus*.

JOURNAL ARTICLE

Bossi L, Munari C, Stoffels C, et al. Somatomotor manifestations in temporal lobe seizures. *Epilepsia* 1983;25:70–6.

BOOK

Delgado-Escueta AV, Wasterlain CG, Treiman DM, Porter RJ, eds. *Status epilepticus: mechanisms of brain damage and treatment*. New York: Raven Press, 1982. (Advances in neurology; vol 34).

CHAPTER IN A BOOK

Dana-Haeri J, Trimble MR, Oxley J. Serum prolactin levels after generalized and partial seizures in epileptic patients. In: Parsonage M, Grant RHE, Craig AG, Ward AA Jr, eds. *Advances in epileptology: the XIVth Epilepsy International symposium*. New York: Raven Press, 1983:127–31.

Tables: Type each on a separate sheet; keep as few and as simple as possible. Put title above and notes or description below. Explain all abbreviations. Do not give same information in tables and figures.

Figure Legends: Number sequentially; type double-spaced on a separate sheet. List meaning of all symbols and abbreviations.

Figures: Submit as clear, glossy prints (two duplicate copies may be photocopies), professionally drawn and lettered, with lettering large enough to be legible when reduced. Maximum final size is 17 by 22.5 cm. Maximum size of artwork is 8.5 × 11 in. Label ordinate and abscissa; give calibration; explain symbols and abbreviations in legends. Submit photomicrographs in final size. Excessive half-tone plates will be charged. On back indicate figure number, authors' names, and TOP (use arrow).

Photographs: Obtain and submit signed release from patient or family.

OTHER FORMATS

PUBLICATION OF PROCEEDINGS OF MEETINGS—Abstracts of annual meetings of national constituents of International League Against Epilepsy may be published; costs to be borne by national Society. 200 words per abstract.

OTHER

Abbreviations: These do not need to be written out at first mention: CNS, CSF, EEG. See *CBE Style Manual* (Council of Biology Editors, Bethesda, MD) or other standard sources.

Drugs: Use generic names; give trade names in parentheses after first mention. See *Epilepsia* 27(3): 311.

Experimentation, Animal and Human: Indicate that clinical experiments conform to Declaration of Helsinki and animal experiments to policy of American Physiological Society.

SOURCE

January/February 1994 35(1): front matter.

European Heart Journal
European Society of Cardiology
W.B. Saunders

ECOR, European Heart House
2035 rue des Colles
Les Templiers B.P. 179
06903 Sophia Antipolis Cedex, France
Att: Tina Oldenburg
       Papers for regular issues
Dr. D.L. Brutsaert
University of Antwerp, Physiology Laboratory
Groenenbrogerlaan 171
2020 Antwerp Belgium
       Basic cardiology papers

European Heart Journal, W.B. Saunders Co., Ltd., 24-28 Oval Road, London NW1 7DX, U.K.
       Papers for supplements

AIM AND SCOPE

The *European Heart Journal* is the official Journal of the European Society of Cardiology. It is particularly concerned with publishing work from Europe, but will accept manuscripts from all parts of the world by members and non-members of the Society. Papers are accepted on the understanding that they have not been or will not be published elsewhere in the same form, in any language.

MANUSCRIPT SUBMISSION

If accepted for publication, copyright is assigned to the European Society of Cardiology. The Society will not put any limitations on the personal freedom of authors to use material in other works.

Individual case reports (two pages of text) must shed new light on some aspect of cardiology.

Submit four copies (one original and three photocopies). Provide four sets of artwork.

MANUSCRIPT FORMAT

Type on one side of paper only; double-space; use 2-5 cm margins.

Text: Include in sequence: Key Words; Abstract (200 words);. Introduction;. Methods; Results; Discussion; Acknowledgments; References; Tables; Figure legends. Format may be altered for review articles.

Tables: Type on separate sheets. Number with Arabic numerals and give titles.

Illustrations: Identify on back with figure number in Arabic numerals, title of paper and name of author.

Submit figures unmounted on glossy paper; mount if several are grouped together in a composite figure. Submit black and white ECGs; protect originals with transparent paper; do not exceed 12 cm.

Legends: Type on a separate sheet.

References: Vancouver style, (see *Br Med J* 1979; 1: 532–5). Cite in text by number. Number in order cited. Double-space reference section at end of text. Abbreviate journal titles according to *Index Medicus* (see list printed annually in January issue). Give complete information for each reference including title of article, abbreviated journal title and inclusive pagination. List all authors up to 6; list first 3 authors for 7 or more and add *et al.*

OTHER FORMATS

CASE REPORTS—1000 words and one figure. On front page: Title, Name(s) of authors, Institution where work was done; footnote title and all addresses of authors, and corresponding author with complete address and telephone number.

OTHER

Abbreviations: Do not use except with explanation. SI units or traditional units may be used; quote alternative in brackets where appropriate.

Statistical Analyses: Seek expert statistical advice before submission.

SOURCE

January 1994, 15(1): 148.

European Journal of Biochemistry
Federation of European Biochemical Societies
Springer International

Prof. Philipp Christen
Editorial Office, *European Journal of Biochemistry*
Apollostrasse 2, Postfach, CH-8032 Zurich, Switzerland
Telephone: + 41 1 383 00 02; Fax: + 41 1 383 07 16

AIM AND SCOPE

The *European Journal of Biochemistry* was first published in 1967 and continues the tradition of *Biochemische Zeitschrift,* founded in 1906. The Journal publishes original papers in all fields of molecular biosciences, including Bioenergetics, Carbohydrates, Developmental biochemistry, Enzymology, Glycoproteins, Immunology, Inorganic biochemistry, Lipids, Membranes, Metabolism, Metabolic regulation, Molecular cell biology, Molecular evolution, Molecular neurobiology, Molecular genetics, Nucleic acids, Physical biochemistry, Protein synthesis, Protein chemistry, Protein structure, and Signal transduction.
The Journal also publishes reviews.

MANUSCRIPT SUBMISSION

Authors do not have to belong to a member society of the *Federation of European Biochemical Societies* (FEBS). Submit four copies of manuscript. Submission implies that work described has not been published before (except as an abstract or as part of a lecture, review or thesis), that it is not under consideration for publication elsewhere, that its publication has been approved by all authors and responsible authorities in laboratories where work was carried out, and that if accepted, it will not be published elsewhere in the same form, in any language, without the consent of FEBS, the copyright holder. Make reference to previously published abstracts, etc., in introduction. Submission also implies willingness to make available to *bona fide* academic researchers, for their own use, cell lines, DNA clones, antibodies or similar materials, used in experiments reported.

MANUSCRIPT FORMAT

Consult a recent issue of the Journal. Type in normal-size print throughout. Do not exceed 21 cm × 30 cm (A4 format), with a 3 cm margin. Number pages starting with title page. Type top copy triple-spaced throughout (1-cm blank space between lines), including title page, reference list, figure legends, tables and supplements. Three additional copies may be less widely spaced and typed on both sides. Provide two copies of any unpublished references.

Title: Head page 1 with a concise but informative title, with a subtitle if necessary. Do not use abbreviations.

Authors' Names: Appear below title, with first or middle name of each author spelled out in full.

Addresses: Give laboratory (or laboratories) where work was done below authors' names. If work was done at more than one laboratory, then follow names of authors with superscript numbers, which should precede name(s) of corresponding laboratory (or laboratories).

Subdivision: Give subdivision of Table of Contents, under which paper is to appear after address(es): Molecular cell biology and metabolism; Nucleic acids, protein synthesis and molecular genetics; Protein chemistry and structure; Enzymology; Carbohydrates, lipids and other natural products; and Membranes and bioenergetics.

Corresponding Author: At end of page 1, give full name and address of corresponding author. Include telephone and fax numbers. Indicate if wish fax number published.

Summary: Begin on page 3. Summary should constitute 3–4% of text. It should be intelligible without reference to other parts of paper. Write out references in full, and define abbreviations.

Introduction: Briefly state purpose of work, in relation to other work in field, without an extensive literature review.

Materials and Methods or Experimental Procedures: Give enough detail to permit reproduction of experiments.

Results: In short papers, combine with Discussion.

Discussion: Interpret results; do not recapitulate. Avoid excessive speculation.

Acknowledgments: Write out names of funding organizations in full.

References: Cite in text, starting in introduction, in either style: By numbers in square brackets, e.g. [1], in numerical order of citation in text (list references after acknowledgments in numerical order); or by author's name and year of publication [one author (Pettersson, 1993); two authors (Goold and Baines, 1993); three or more authors (Lawrence et al., 1993)]. In this style, list references with the same first authors chronologically. Where citations in text would be identical, differentiate by lowercase letters after year, e.g., 1993a,b.

In either style, serial publications in reference list should include names and initials of all authors, year of publication in parenthesis, title of article, abbreviated title of periodical, as given in *Chemical Abstracts* or *Biological Abstracts,* volume number and first and last page numbers:

Martin, P. T., Tsu-Ping Chung, B. & Koshland, D. E., Jr (1993) Regulation of neurosecretory habituation by cAMP. Role of adaptation of cAMP signals, *Eur. J. Biochem. 217,* 259–265.

ARTICLES FROM NON-SERIAL PUBLICATIONS

Sambrook, J., Fritsch, E. F. & Maniatis, T. (1989) *Molecular cloning: a laboratory manual,* 2nd edn, Cold Spring Harbor Laboratory, Cold Spring Harbor, NY.

ARTICLES FROM NON-SERIAL PUBLICATIONS WITH EDITORS

Stone, K. L. & Williams, K. R. (1993) Enzymatic digestion of proteins and HPLC peptide isolation , in *A practical guide to proteins and peptide purificaion for microsequencing* (Matsudaira, P., ed.) 2nd edn, pp. 45-69, Academic Press Inc., London.

Ensure that references cited in text correspond exactly to those in reference list. Titles must be provided for all serial publications, or will be deleted. Include title and name of journal for references to articles cited as 'in the press.' Provide written evidence that paper has been accepted. Keep references to unpublished work, including papers in preparation, to a minimum. Mention in parenthesis in text as unpublished work, not in reference list. Mention personal communications only in text and provide evidence provided of agreement of person(s) quoted.

Tables and Figures: Present data that cannot easily be described in text. Number with Arabic numerals and indicate positions in margin of top copy.

Give each table a title. Type each on a separate page. Include experimental conditions and general remarks in a paragraph between title and table. Use footnotes only if information cannot be included in legend; indicate by superscript lowercase letters. Give columns headings; put units under column heading(s), identical units for successive columns need not be repeated.

Label figures on back to indicate TOP of figure, authors' names and figure number. Do drawings and graphs in India ink; computer print-outs should be black (not grey) on white. Photocopies of most types of figures are sufficient, but glossy prints of gels, electron micrographs, etc., are required. Letter abscissa or ordinate(s) of each graph with numerical scale and measured quantity with units according to Journal style, using Letraset, type or stencils. A charge may be made for color prints.

Separate figure legends from figures, give each a title, and make comprehensible without reference to text.

OTHER FORMATS

PRIORITY PAPERS—Four printed pages (20 pages with 1 cm between lines, four figures plus tables). Articles of sufficient importance to merit expedited handling and publication; complete concise manuscripts, written in the style of the Journal. Indicate submission as a Priority Paper in cover letter and on first page of each copy of manuscript. Provide names of four possible referees with telephone and fax numbers. Send four copies of manuscript, triple-space top copy and single-space others.

OTHER

Abbreviations: Define on page 2, unless included in Tables 3– 5 (*Eur. J. Biochem.* 1994; 219(1): A21-22). Use only when essential, i.e., due to repetition or excessive length of full name. See *Biological Nomenclature and Related Documents, A Compendium* (1992, 2nd ed., International Union of biochemistry and Molecular Biology, London: Portland Press) for naming chemical compounds.

Data Bank Submission:

Sequence Data—Determine nucleotide sequences from both DNA strands. Describe sequencing strategy employed and justify why any regions of the sequence have been determined from only one strand. Submit nucleotide and amino acid sequence data to *European Molecular Biology Laboratory* (EMBL) Sequence Data Library in computer-readable format. Upon acceptance of a manuscript with sequence data, Editorial Office will send necessary information and forms to corresponding authors; send completed forms directly to data bank. Insert accession number into page proofs.

X-ray Data—Submit structural data, including both X-ray amplitudes and phases and derived atomic coordinates to an appropriate data bank. Send data bank reference number and expected release date to Editorial Office with corrected proofs. Maximum delay before release of data is two years; if no release date, immediate availability of data is assumed. Preliminary crystallographic data will only be accepted if they contain features of special interest.

Electronic Submission: Submission of manuscript on diskette is encouraged but not compulsory. Send diskette with manuscript with a word-processor file with entire text, followed by tables, figure legends, which is identical to printed manuscript. Prepare file using any word-processing system on any size of diskette. Diskettes will not be returned.

Enzymes: List enzymes mentioned in text on page 2, followed by current Enzyme Commission (EC) number in parenthesis.

Experimentation, Animal: Perform in accordance with ethical standards stipulated in 1964 Declaration of Helsinki. State measures taken to protect animals from pain or discomfort.

Experimentation, Human: State steps taken to avoid unnecessary risk or discomfort, including peer review of investigation and informed consent of subjects.

Nomenclature, Symbols, Units and Abbreviations: Follow rules defined in International Union of Pure and Applied Chemistry's *Nomenclature of Organic Chemistry,* (3rd edn, 1979), IUPAC's *Nomenclature of Inorganic Chemistry* (2nd edn 1988), and IUPAC's *Quantities, Units and Symbols in Physical Chemistry,* (Blackwell Scientific Publishers, Oxford, England). More detailed information can be found in *Information for Contributors* (1982, BBA Editorial Office, Postbus 1345, NL-1000 BH Amsterdam, The Netherlands).

For biochemical nomenclature, including abbreviations, see *Biochemical Nomenclature and Related Documents,* (2nd edn, 1991, Portland Press Ltd., Colchester, Essex, England) and Table 1 (*Eur. J. Biochem.* 219(1): A20, 1994). For commonly used symbols for quantities and units see Table 2 (*Eur. J. Biochem*. 219(1): A21, 1994).

Reviewers: In cover letter include names and addresses of four possible referees, as well as persons whose participation is not desired.

Supplementary Material: Supplementary material that will not be published in the Journal may be submitted with manuscript. Send four copies of material including one set of originals of all figures; type text and tables single-spaced throughout on one side of A4 paper. Use of unlisted (see *Eur. J .Biochem*. 219(1): A21-22, 1994) abbreviations and symbols is discouraged. If used, define at first appearance in text. Number figures Fig. S1, Fig. S2, etc. and Table S1, Table S2, etc., respectively. Do not cite in main manuscript. Indicate availability of supplementary material in main manuscript by a paragraph before References, stating, 'Supplementary material' and providing titles of figures and tables. Supplementary material is available on request from Editorial Office.

January 1994, 219(1): A9-A22.

European Journal of Cancer (EJC)

Pergamon Press, Ltd.
European Organization for Research and Treatment of Cancer
European School of Oncology
European Association for Cancer Research
Federation of European Cancer Societies

The Editor in Chief
European Journal of Cancer
British Postgraduate Medical Federation

33 Millman Street
London WC1N3EJ, U.K.
Telephone: +44 71 430 0100; Fax: +44 71 430 0072

The *European Journal of Cancer* is an international journal that publishes original research, editorial comment, review articles, book reviews, news, and letters on cell and molecular oncology, on clinical oncology, and on cancer epidemiology and prevention. Research data may be submitted as a paper or letter. The journal will consider negative results from well-designed studies, which will be published as letters. Letters may be published as short communications. Letters that comment on an article previously published in the *EJC* or that express a viewpoint about oncology are also invited.

MANUSCRIPT SUBMISSION

Submit four copies of manuscript with four sets of figures (all copies of photomicrographs and gels should be of good quality).

In cover letter, signed by all authors, identify person responsible for editorial correspondence (address, and telephone and fax numbers). Include details of previous or concurrent submission. Give any information to support submission (e.g., original or confirmatory data, relevance, topicality) or whether any text, figures, or tables can be omitted.

By submitting a manuscript, authors agree that copyright is transferred to publisher if and when accepted for publication. Assignment of copyright is not required from authors who work for organizations which do not permit such assignment.

Manuscripts must not have been published and may not be simultaneously submitted or published elsewhere. No substantial part of manuscript may have been or may be published elsewhere. This restriction does not apply to abstracts or press reports published in connection with scientific meetings.

MANUSCRIPT FORMAT

Write in English double-spaced on one side of single A4 sheets with high-quality print. Limits on lengths: papers—3000 words (150 word abstract, 30 references); Short Communications—1000 words (100 word abstract, 2 tables/figures, 20 references); Fax Communications—750-1000 words (10–20 references); Letters—500 words (10 references, 1 table/figure). There are about 250 words on an A4 page of double-spaced typing.

Prepare manuscripts according to the Vancouver guidelines ("Uniform requirements for manuscripts submitted to biomedical journals,: *N. Engl. J .Med.* 1991, 324, 424–428). Read these guidelines, especially when deciding on who qualifies as an author. See also: "Guidelines for writing papers," *BMJ* 1991, 302, 40–42.

For manuscripts containing research data, follow order: Introduction, Patients (or Subjects or Materials) and Methods, Results, and Discussion. Start each section at top of a new page; nmber all pages. On front page include: concise but informative title; authors' names; department/institution and an address for each author, with a symbol to link authors with their address; name and address of corresponding author; a short title (40 characters/spaces); details of sources of support in form of grants, equipment, and drugs.

On next page type: Abstract (include methods in brief and main findings and conclusions); up to 10 key words from *Medical Subject Headings* from *Index Medicus*.

Next pages: Introduction, Methods, Results, and Discussion.

References: List on a new sheet and number consecutively in parentheses in text. "Unpublished data" and "Personal communications" are not allowed; rather, state in text "(data not shown)" or "(Dr F. G. Tomlin, Karolinska Institute)." Reference accepted but unpublished papers (but not submitted manuscripts) as "in press."

Use format for references from Vancouver guidelines. Include: names of all authors when six or fewer, followed by initials. Otherwise list only first three and add *et al*; title of article or chapter; journal name abbreviated as in *Index Medicus*, year and volume, and first and last pages; for a book, names of any editors (as for authors), city and name of publisher, and year and pages.

1. Jiang FN, Liu DJ, Neydorff H, Chester M, Jiang S-Y, Luy JG. Photodynamic killing of human squamous cell carcinoma cells using a monoclonal antibody-photosensitizer conjugate. *J Natl Cancer Inst* 1991, **83**, 1218–1225.

2. Gullick WJ, Venter DJ. The c-*erb*B2 and its expression in human tumors. In Waxman J, Sikora K, eds. *The Molecular Biology of Cancer*. Oxford, Blackwell Scientific Publications, 1989, 38–53.

3. Lumley JSP, Green CJ, Lear P, Angell-James JE. *Essentials of Experimental Surgery*. London, Butterworths, 1990.

Figures: Submit original glossy photographs or professional drawings (high-quality computer-generated artwork is acceptable). Mark back of each figure with its number and first author's name; indicate TOP. Figures are generally reduced to 84 mm wide or less when published. Type legends in order on a separate page; limit to 40 words or less. Cite figures consecutively in text.

Color illustrations can be published, although there may be a charge to the author.

Tables: Type on separate pages with titles. Avoid white spaces by using footnotes. Explain all symbols or abbreviations. Distinguish between "zero" and "not done" as a table entry. Cite tables consecutively in text.

OTHER

Abbreviations and Units: Avoid abbreviations in Title and abbreviating single words. Explain all abbreviations at first mention in Abstract and text, except for: DNA, RNA, AIDS, and HIV. Standard SI abbreviations for units do not need to be spelled out.

Drugs: Use generic names for drugs. Commercial names may be included in parentheses at first mention in text. Abbreviate complicated drug names or regimens, with abbreviation in parentheses after first mention.

Experimentation, Human: For clinical trials, state details of ethical committee approval and type of informed consent. Do not use patient's and volunteers' names, initials, and hospital numbers.

SOURCE

January 1994, 30A(1): front matter.

European Journal of Cell Biology
European Cell Biology Organization
Deutsche Gessellschaft für Zellbiologie
Deutsche Gessellschaft für Elektronenmikroskopie
Wissenschaftliche Verlagsgesellschaft mbH

Karin Klingsberg
Wissenschaftliche Verlagsgesellschaft mbH
Birkenwaldstr. 44, D-70191 Stuttgart Germany, or
P.O. Box 101061, D-70009 Stuttgart Germany
Telephone: 0711/2582-312; Fax: 0711/2582-390
Dr. Reinhard Jahn
Howard Hughes Medical Institute
Depts. of Pharmacology and Cell Biology
Yale University School of Medicine
P.O. Box 9812
New Haven, CT 06536, or
295 Congress Ave.
New Haven, CT 06510
            U.S. and Canada only

AIM AND SCOPE

The *European Journal of Cell Biology*, a Journal of experimental cell research, publishes papers on the structure, function and macromolecular organization of cells and cell components. Aspects of cellular dynamics, differentiation, biochemistry, immunology, and molecular biology in relation to structural data are preferred fields for contributions. Methodologically oriented manuscripts concerning significant technical advances are also welcomed. In addition, biomedical manuscripts which are of general cell biological interest will be published.

MANUSCRIPT SUBMISSION

Write in concise and grammatically correct English; keep length minimum required for precision in describing experiments and clarity in interpreting them.

With submission of a manuscript, authors declare that results or parts of them have not been published elsewhere. Submit manuscripts in triplicate, with three sets of figures (one unmounted and unlabeled and two mounted and labeled sets), typed double-space with generous margins. In Introduction and Discussion sections state novelty and significance of observations. Suggest expert scientists in their fields for possible referees. Color plates are published at no extra charge if particularly important for understanding article.

MANUSCRIPT FORMAT

Arrange text in sections: Introduction Materials and methods, Results, Discussion, Acknowledgments, References. In Materials and methods section give full address of suppliers of chemicals and equipments.

Title Page: Title of paper, first name(s) written in full. Surnames of authors, laboratory or institution. name of town and country. Give name, address, fax and telephone numbers of corresponding and reprint request author(s) in footnote. Include a short running title (40-50 characters). List abbreviations in alphabetical order.

Abstract: 20-30 lines, preceded by 5 key words. Avoid quoting scientific papers; if mentioned, give full citation of source.

References: References to literature appear in text as numbers either between angular brackets [3] or slanting strokes /3/. Numbers in text correspond to alphabetical order of authors in list of references at end of paper. Make Reference sources, such as "unpublished data, personal communication, manuscript in preparation, and manuscript submitted," parenthetically in text; do not include in list of references. Include accepted and "in press" papers in list of references only by quoting relevant journal and stating month of publication.

JOURNALS

Koch, P. J., M. J. Walsh, M. Schmelz, M. D. Goldschmidt, R. Zimbelmann, W. W. Franke: Identification of desmoglein, a constitutive desmosomal glycoprotein, as a member of the cadherin family of cell adhesion molecules. Eur. J. Cell Biol. **53**, 1–12 (1990).

BOOKS

Tartakoff, A. M.: The Secretory and Endocytic Paths. Mechanism and specificity of vesicular traffic in the cell cytoplasm. pp. 7–39. John Wiley & Sons. New York 1987.

Lucocq, J. M., J. Roth: Colloidal gold and colloidal silver-metallic markers for light microscopical histochemistry. In: G. R. Bullock, P. Petrusz (eds.): Techniques in Immunocytochemistry. Vol 3. pp. 203–236. Academic Press. London 1985.

Figures, Graphs and Diagrams: Limit to essential minimum. Supply three copies, two sets mounted (heavy cardboard undesirable), labeled and numbered. including a bar defining a dimension (a μm or fractions of multiples of a μm). Submit original set to be used for printer unmounted and without lettering. Mark each piece in original set on reverse side with author's name, figure number and orientation (top or bottom).

Mention all figures and tables in text. Number figures with Arabic numerals, tables with Roman numerals. Type legends for figures and tables on separate pages at end of text. Make headings for tables clear; short exploratory footnotes are also acceptable. Define abbreviations and symbols in figures in legend. Use symbols: → ⇒ ▷ ► ○ ● ▢ ■ △ ▲ × + #.

Draw line drawings, including graphs and diagrams in black ink. They will be published in smallest appropriate size. Supply electron micrographs or halftone illustrations as white glossy prints (preferably mounted as plates). Mounted plates are 17 cm wide and between 6 and 22.5 cm high. Single figures are 8.1 or 17 cm wide with a maximum height of 22.5 cm.

OTHER FORMATS

SHORT COMMUNICATIONS—Five pages of double-spaced typescript. including figures and tables.

REVIEW ARTICLES and MINIREVIEWS—Consult editorial office in advance.

OTHER

Nomenclature: Use sandard nomenclature, for example, recommendations of IUPAC-IUB Combined Commission on Biochemical Nomenclature.

Spelling: Follow usage in Merriam-Webster dictionaries.

SOURCE

October 1994, 65(1): inside back cover.

European Journal of Clinical Pharmacology

Springer International

Prof. Rune Dahlqvist
Division of Clinical Pharmacology
University Hospital
S-90185 Umea, Sweden

AIM AND SCOPE

The *European Journal of Clinical Pharmacology* publishes original papers, short communications, and letters to the editors on all aspects of clinical pharmacology and drug therapy in humans. Data from animal experiments are accepted only in the context of parallel experiments in man reported in the same paper.

The Journal also accepts review articles on special problems related to these areas, and encourages debate on controversial issues. Manuscripts are welcomed on the following topics: therapeutic trials; pharmacokinetics; drug metabolism; adverse drug reactions; drug interactions; all aspects of drug development; prescribing policies; pharmacoepidemiology, and matters relating to the safe use of drugs.

Methodological contributions relevant to these topics are also welcomed.

MANUSCRIPT SUBMISSION

Submit manuscripts in triplicate. Submit one set of original figures; each copy should contain copies of all figures. Type original on one side of paper only; copies may be photocopied on both sides.

Submission of a manuscript implies: that work has not been published before (except as an abstract or as part of a published lecture, review, or thesis); that it is not under consideration for publication elsewhere; that its publication has been approved by all coauthors, if any, as well as by responsible authorities at institute where work was carried out; that, if and when manuscript is accepted for publication, authors agree to automatic transfer of copyright to the publisher; and that manuscript will not be published elsewhere in any language without consent of copyright holders.

MANUSCRIPT FORMAT

Type double-spaced with 2-cm margins. Number pages, including title page, figures, and tables, consecutively.

Title Page: Include title, authors' names, institution where work was carried out (include city and country), and name of corresponding author with fax number.

Summary: Make understandable without reference to text; set out in separate subsections. Provide six key words.

Introduction (without separate heading): Be brief; state relevant background and main purposes of study.

Materials, Methods, and Subjects Studied: Use relevant subheadings where possible.

Results

Discussion and Conclusions

Acknowledgments

Reference: Give as either: cited by author and year in text (e.g., Hammer 1969; Hammer and Sjoqvist 1967; Hammer et al. 1969) and listed in alphabetical order in reference list; or numbered consecutively in text in order of appearance and listed in numerical order in reference list. Abbreviate journal titles according to *Index Medicus*.

JOURNALS

Lange H, Eggers R, Bircher J (1988) Increased systemic availability of albendazole when taken with fatty meal. Eur J Clin Pharmacol 34: 319–321

BOOKS

Tallarida RJ, Raffa RB, McGonigle P (1988) Principles in general pharmacology. Springer, Berlin Heidelberg New York (Springer series in pharmacological science)

CHAPTORS IN BOOKS

MacIntyre I, Zaidi M, Milet C, Bevis PJR (1988) Control of calcium. In: Baker PF (ed) Calcium ub drug actions. Springer, Berlin Heidelberg New York, pp 411-439

Tables: Type double-spaced, each on a separate sheet. Make self-explanatory without reference to text. Type table number and title at top. Give additional information as footnotes on same page. Use only horizontal lines to divide sections. Show data as mean (with SD in parentheses) rather than mean plus or minus SD. Other acceptable ways of displaying data include median (range) or mean (95%Cl).

Figure Legends: Give on each copy of figures and also collected on a separate page.

Figures: Number. Submit in a form ready for reproduction. Make lettering and symbols clear and large enough to be easily readable after size reduction. Use following symbols in order: ○ ● □ ■ △ ▲ ▼. Size figures to fit column (8.6 cm) or page (17.8 X 24 cm) size. Submit good quality prints for line drawings. Make lettering about 3 mm high. Computer drawings must be of comparable quality to line drawings. Make lines and curves smooth. Lines and numbers should be of professional quality and proper dimensions. Trim halftone illustrations photographic prints of good contrast, at right angles; submit in desired final size (recommended height of letters, 3 mm).

OTHER FORMATS

LETTERS TO THE EDITOR—600 words.

SHORT COMMUNICATIONS—200 words plus up to two figures or two tables, or one figure and one table. Submit pharmacokinetic studies involving standard methods as short communications.

OTHER

Color Charges: Approx. DM 1,200.00 for first and DM 600.00 for each additional page.

Style: Follow a recent copy of the Journal for style and form of presentation.

SOURCE

1994, 47(1): inside back cover.

European Journal of Endocrinology

European Federation of Endocrine Societies
Scandinavian University Press

Prof. A. G. Burger
Thyroid Unit, Hopital Cantonal Universitaire, CP 30
CH-1211 Geneva 14, Switzerland

AIM AND SCOPE

The *European Journal of Endocrinology* publishes original research papers, short communications, and reviews within clinical and experimental endocrinology. In addition, correspondence related to published work is published at the discretion of the board.

MATERIAL SUBMISSION

Submit one original and three copies of the complete manuscript. Include a cover letter with a statement signed by all authors: "The author(s) hereby confirms that neither the manuscript nor any part of it, except for abstracts less than 400 words, has been published or is being considered for publication elsewhere. By signing this letter, each of us acknowledges that he or she participated sufficiently in the work to take public responsibility for its content". If an author is added or removed, submit written acceptance by all authors to the editorial office. Also include: information on prior or duplicate publication or submission elsewhere of any part of work; statement of financial or other potential conflict of interest relationships; written permission from persons mentioned by name; permisssions to reproduce previously published materials or to use illustrations with identifiable human subjects; and name, address, telephone and fax numbers of corresponding author.

Material will be copyrighted by the Journal. A copyright transfer form will be sent to corresponding author with proofs.

Manuscripts not conforming to instructions will be returned.

MANUSCRIPT FORMAT

Follow "Uniform Requirements for Manuscripts Submitted to Biomedical Journals" (*Br. Med. J.* 1991; 302: 338-41). Write in proper scientific English. Begin each manuscript component on a new page in sequence: Title page, Abstract, Text, References, Tables (each with title and footnotes on a separate page), Legends for illustrations. Double space; keep 5 cm left margin.

Title Page: Concise, informative article title; subtitle; first name, middle initial and last name of each author with institutional affiliation; department(s) and institution(s) to which work is attributed; name, address, and fax numbers of corresponding author; and 4-9 keywords for indexing.

Abstract: 200 words stating purpose, objective, hypotheses of work; procedures, main findings and principal conclusions, emphasizing new or modified methods. State experimental design and conditions. Detail main findings that mention important incidental findings. State major conclusion. Use structured abstract (250 words) for original research articles on cause, course, diagnosis, prevention and therapy of endocrine diseases.

Introduction: State purpose of article. Summarize rationale for study or observation. Give only pertinent references. Do not review subject extensively.

Subject and material: Describe selection of observational and expermental subjects, including controls.

Methods: Identify methods, apparatus (manufacturer's name and address in parentheses) and procedures in enough detail to allow repetition. Reference established methods, including statistical methods. Describe published methods that are not well known. Describe new or modified methods; give reasons for using and evaluate limitations. Identify drugs and chemicals; include generic name(s), dose(s), and route(s) off administration.

Results: Present in logical sequence in text, tables, and illustrations. Do not repeat data in tables or illustrations. Emphasize or summarize only important observations. Avoid discussion and reference to other work.

Discussion: Emphasize new and important aspects of study and conclusions that follow. Do not repeat data or other previously given material. Include implications of findings and their limitations. Give implications for future work. Relate observations to other studies. Link conclusions to study goals; avoid unqualified statements and conclusions not supported by your data.

Acknowledgements: Acknowledge only those who have made substantial contributions to study. Specify contributions that need acknowledging but do not justify authorship; technical help; financial and material help; and financial relationships that may pose a contflict of interest. Written permission from acknowledged persons may be requested.

References: Verify against original documents. Number consecutively in order mentioned in text, tables, and legends by Arabic numerals within parentheses, e.g., 'Smith (1) has reported...', 'Smith and Green (2) and Smith et al. (3) have found...' or give appropriate number(s) in parentheses, e.g., 'It has been reported (11-3) that ...'. In reference list include all authors referred to in text, tables, and legends. Number references cited only in tables and legends in order of identification in text of table or legend.

Abbreviate journal titles according to *Index Medicus*. Include papers accepted but not yet published; give journal and add (in press) in reference list, not in text. Provide copies of these papers and acceptance notes. For unpublished work of others, include letter of permission; cite information from submitted manuscripts in text as '(unpublished observations)'; do not include 'unpublished observations' and 'personal communications' in reference list; insert references to written, not oral communications in parentheses in text.

JOURNAL

Yano S, Tanigawa K, Suzuki S, Kobayashi Y, Shimada, Morioka S eet al. Effect of diabetes mellitus on levels of atrial natriuretic hormone in plasma and the right atrium in the non-obese diabetic mouse. Acta Endocrinol (Copenh) 1991;124:595-601

Garibaldi LR, Acrto Jr. T, Weber C. The pattern of gonadotropin and estradiol secretion in exaggerated thelarche. Acta Endocrinol 1993 (in press)

The Royal Marsden Hospital Bone-Marrow Transplantation Team. Failure of syngeneic bone-marrow graft. Lancet 1977;2:242-4

Anonymous. Coffee drinking and cancer of the pancreas (Editorial). Br Med J 1981;283:628.

Heickendorff L. The basement membrane and arterial smooth muscle cells. Acta Pathol Microbiol Immunol 1989;97(Suppl 9):1-32

Storz P, Peters F. Beer-induced prolactin secretion is mediated by malt (Abstract). Acta Endrinol (Copenh) 1990;122(Suppl 1):46

Seaman WB. The case of the pancreatic pseudocyst. Hosp Pract 11981;16(Sep):24-5

Adams EF, Winslow CL, Mashiter K. Pancreatic growth hormone releasing factor stimulates growth hormone secretion by pituitary cells. Letter. Lancet 1983;1:1100-1

BOOKS AND OTHER MONOGRAPHS

Pedersen J. The pregnant diabetic and her newborn. 2nd ed. Copenhagen: Munksgaard, 1977

Diener HC, Wilkinson M, eds. Drug-induced headache. New York: Springer-Verlag, 1988

Jackson IMD, Cobb WE. Disorders of the thyroid. In: Kohler PO, ed. Clinical endocrinology. New York: John Wiley & Sons, 1986;73-166

Ellahou HE, Cohen D, Herzog D. Obesity and hypertension: what correlation at long-term follow-up? In: Berry EM, Blondheim SH, Ellahou HE, Shafrir E, eds. Proc 5th Int Cong Obes: 1986 Sept 14-19; Jerusalem, Israel. London: John Libbey & Co, 1987:42-5

Toth BE, Kacsoh B, Grosvenor CE. Natural prolactin-antagonist in rat milk. 72nd Ann Meet Endocr Soc, Atlanta, GA, 1990. Abstract 234

Dinan TG, O'Keane V. The premenstrual syndrome: a psychoneuroendocrine perspective. In: Grossman A, Psychoneuroendocrinology. London: Ballièrre Tindall 1991:143-65 (Ballière's Clinical Endocrinology and Metabolism; vol 5)

Akutsu T. Total heart replacement device. Bethesda (MD): National Institutes of Health, National Heart and Lung Institute; 1974 Apr. Report No.: NIH-NHLI-69-2185-4

WHO Expert Committee on Diabetes Mellitus. Second Report, 1978: WHO Tech Rep Ser No. 546

Schaper NC. Growth hormone in type I diabetic and healthy man. (Thesis), Groningen, Holland: University of Groningen, 1990. 48 pp.

Rensberger B, Specter B. CFCs may be destroyed by natural process. The Washington Post 1989 Aug 7; Sect A:2 (col 5)

Figures: Number in Arabic numerals; refer to as Fig. in text. Have professionally designed. Computer-generated figures must be of professional standard. Submit one set of clear, well-focused unmounted, glossy prints and four sets of clear photocopies. Single-column figures are 80-180 mm. Rule coordinate lines thick enough to reproduce. Center axis labels along outside of each axis; use initial capitals. Typeset letters and numbers (do not typewrite) or use preprinted transfer sheets or handletter with a stencil or mechanical lettering device. Use the same lettering on all figures. Use medium weight type style, not bold or light, e.g., Helvetica or Trade Gothic. Keep lettering in proportion to drawing for good reduction. Smallest character is not less than 1.5 mm when reduced. Include titles, keys to symbols and statistics in legend, not on figure.

Photographs must be glossy prints with sharp focus, good contrast and a full range of tonal values. Submit four unmounted copies. Crop prints to 82 mm or they will be reduced. Do not request enlargements. Include a scale on prints of photomicrographs and election micrographs. State staining technique. Color illustrations are subject to Editor's decision. Author is responsible for costs.

Legends: Make informative and comprehensive. State whether variation is standard deviation (SD), standard error of the mean (SEM), fractile or range.

Tables: Make as simple as possible and intelligible without reference to text. Type double-spaced on separate sheets. Number in Arabic numerals; give concise headings. Do not use vertical or horizontal rules in body. Place explanatory material in footnotes; use small superscript letters for footnotes. Identify statistical measures of variation. Arrange data so tables read in same order as text.

OTHER FORMATS

LETTERS—One printed page, including references, one table or figure.

SHORT COMMUNICATIONS—Two pages, one figure, one table, and 15 references.

OTHER

Abbreviations: Use standard abbreviations only. Abbreviation may be used for repetitive use of complicated substances. Use only three. Abbreviation must give proper allusion to original. Avoid use in title and abstract. Full term must precede first use of abbreviation, except for standard units of measurement. Redefine abbreviations used in legends. For abbreviations of medical terms, use *Dorland's Medical Dictionary*, 27th ed.

Biological and Chemical Substances: Abbreviate biological compounds according to *Journal of Biological Chemistry*. Give chemical name of compound after first use of trivial name in accordance with Index of *Chemical Abstracts*.

Experimentation, Animal: Indicate whether institution's or National Research Council's guide, or any national law for care and use of laboratory animals was followed.

Experimentation, Human: Indicate whether procedures used were in accordance with ethical methods of the responsible institutional or regional committee on human experimentation or with Helsinki Declaration. Do no use patients names, initials, or hospital numbers.

Immunoassays and Bioassays: Give appropriate measures of precision in terms of SD or coefficient of variation and reference preparation.

Nomenclature, Enzymes: See "Enzyme Nomenclature. Recommendations 1984" (*Eur. J. Biochem.* 1986; 157: 11-26 and 1989; 179: 489-533).

Nomenclature, HLA: See "Nomenclature for Factors of the HLA System, 1989" (*Hum. Immunol.* 1990; 28: 326-42).

Nomenclature, Steroids: See *Pure Appl. Chem.* 1989; 61: 1783-822.

Page Charges: NOK 900 for each additional page over six for full papers and eight for reviews.

Radioisotopes: Indicate as carbon-14, hydrogen-3; abbreviate as ^{14}C, ^{3}H. For isotopically labelled compounds, indicate radionuclide and position within square bracket on compounds such as [4-^{14}C]progesterone, [^{3}H]dihydroalprenolol. Use ^{131}I-labelled albumin or ^{131}I-albumin since native albumin does not contain iodine. Do not use brackets for simple compounds. State purity.

Reviewers: Suggest three non-national scientists. Identify persons not to be consulted; state reasons.

Statistical Analyses: Describe methods sufficiently to permit verification of results. Use standard works (state pages) to reference study design and statistical methods. State general-use computer programs used.

Units: Use International System of Units (S.I. units). Use litre as unit of volume. Express amount of substance in moles (mol. not M) when molecular weight is known.

SOURCE

January 1995, 132(1): 3-5.

European Journal of Immunology
European Federation of Immunological Societies

H. von Boehmer
Basel Institute for Immunology
487 Grenzacher Straβe
CH-4005 Basel 5 Switzerland
Fax: 41 61 6 05 12 22

M. Kazatchkine
Unité d'Immunopathologie
Hopital Broussais
96, Rue Didot
F-75674 Paris Cedex 14 France
Fax: 331 45 45 90 59

D. Mathis
Institut de Chimie Biologique
Faculté de Médicine
11, Rue Humann
F-67805 Strasbourg Cedex France
Fax: 33 88 37 01 48

N. A. Mitchison
Deutsches Rheuma-Forschungszentrum Berlin
Forschungslaboratorium
Robert Koch Institut, Haus 11
Nordufer 20
D-1000 Berlin 65 FRG
Fax: 49 30 805 59 36

L. Morreta
Immunopathology Laboratories
Instituto Scientifico Tumori
Viale Benedetto XV n-10
I-16132 Genova Italy
Fax: 39 10 354123

M. S. Neuberger
Laboratory of Molecular Biology
Hills Road
Cambridge
CB2 2QH Great Britain
Fax 44 223 41 21 78

R. M. Perlmutter
Department of Immunology
University of Washington
1264 Health Sciences Building, SL-05
Seattle,WA 98195
Fax: 1 (206) 543-1013

H. L. Ploegh
Center for Cancer Research
Massachusetts Institute of Technology
77 Massachusetts Ave, E17-322
Cambridge, MA 02139-4307
Fax: 1 (617) 253-9891

M. Sela
Department of Chemical Immunology
The Weizmann Institute of Science
Rehovot 76100 Israel
Fax: 97 28 46 69 66

P. Vassalli
Département de Pathologie
Centre Médical Universitaire
1, Rue Michel Servet
CH-1211 Genéve Switzerland
Fax: 4 12 24 73 334

R. Zinkernagel
Institut fur Pathologie der Universitat Zurich
Universitatsspital
Schmelzbergstr. 12
CH - 8091 Zurich Switzerland
Fax: 41 12554420

Submit papers to member of Executive Committee whose area of expertise is closest to topic of paper.

AIM AND SCOPE

The *European Journal of Immunology* is intended to collect papers on the various aspects of immunological research. It will accept original articles and short communications from the following fields of experimental and human immunology: molecular immunology, immunobiology, immunopathology, immunogenetics and clinical immunology. Manuscripts concerning B and T cell repertoire can only be accepted if they make a new biological point.

MANUSCRIPT SUBMISSION

Follow instructions carefully. Editors reserve the right to return manuscripts that are not in accordance with these instructions. Submit original manuscript (with original figures and tables) plus three complete copies of good quality. Include copies of manuscripts to be published or submitted elsewhere if content impinges on results in present manuscript.

Work must not have been published elsewhere, either completely, in part, or in another form. Manuscript may not have been submitted to another journal nor will be published elsewhere within one year after its publication in this journal.

MANUSCRIPT FORMAT

Two types of manuscripts are accepted: Papers of normal length (25 manuscript pages including references, tables and figure legends) and Short Papers (maximum length 12–14 manuscript pages including tables and figure legends, marked as "Short Paper" immediately above title on first page). Include sufficient information to permit repetition of experimental work. Use letter-quality printers (not dot matrix) using double-spacing (including footnotes, references, tables, legends, etc.) with a 5-cm left margin. Paper should not exceed 30 cm in height (standard A4 paper). Maximum of 30 lines, with an average of 270 words, per page of text.

First Page: Title of paper (include most important key words pertaining to subject matter; do not use abbreviations); Full names (including forenames) of authors and name of institute(s) (If publication originates from several institutes, clearly state affiliation of all authors); A shortened title (70 letters); Key words (maximum 5); Name and full postal address of corresponding author (include fax and telephone numbers); A list of abbreviations used (if used more than five times in text; exclude standard abbreviations); and footnotes regarding funding sources, additional addresses, etc.

Divide manuscripts into sections. Indicate subdivisions of sections by numbered subheadings.

Summary: Self-explanatory and intelligible without reference to text. Type on second and (if necessary) third page of manuscript. Write out citations in full.

Introduction: Describe problem under investigation. Give a brief survey of existing literature.

Materials and Methods: For special materials and equipment, provide manufacturer's name and location.

Results

Discussion: Enrich, do not repeat Introduction or Results. Results and Discussion may be amalgamated and followed by a short section entitled "Concluding remarks."

References: Number sequentially, including those in tables and figure legends, in order of appearance in text. Set numbers in brackets. List in numerical order at end of manuscript under heading "References." Type double-spaced throughout. Abbreviate journal titles of journals according to CASSI (Chemical Abstracts Service Source Index). Underline abbreviated title and volume number.

JOURNALS

Lidsky, M. D., Sharp, J. T. and Rudel, M. L., *Arch. Biochem. Biophys.* 1967. *121:* 491.

Cite other serial publications such as "Advances in Immunology" in the same manner as journals.

BOOKS

a) Elves, M.W., *The Lymphocytes,* 2nd Edn., Lloyd-Luke Ltd., London 1972, p. 274.

b) Moller, E. and Greaves, M. F., in Mäkelä, O., Cross, A. and Kosunen, T. U. (Eds.), *Cell Interactions and Receptor Antibodies in Immune Responses,* Academic Press, London and New York 1971, p. 101.

Include allusions to "unpublished observations," papers "to be published" or "submitted for publication" and the like in text, either in parentheses or as footnotes. Enter material "in press" under references. Verify accuracy of bibliographic references.

Footnotes: Indicate by an asterisk in parentheses (*). Write at bottom of page on which asterisk appears in text.

Tables: Collect tables with suitable captions at top and number with Arabic numerals at end of text on separate sheets (one page per table). If possible, provide camera-ready. Make understandable without frequent reference to text. Keep column headings as brief as possible. Indicate units if appropriate. Mark footnotes a) b) c) etc. and type on page with table. Indicate position at which table is to appear in the text.

Figures: Submit in a format which can be reduced to a width of 5–8.5 cm or 11–17.5 cm, and symbols and captions to a height of 1.5–2.0 mm. As far as possible, make lettering the same size. Make numbers, letters and symbols large enough to be legible (i.e., 1.5–2 mm after reduction).

Submit diagrams and photographs on separate pages at end of article (new page for each figure). Number in order of appearance with Arabic numerals. Submit photographs in quadruplicate as glossy prints of high contrast. Polaroid photographs are not acceptable. Electron micrographs should contain a scale.

Cost of colored figures is charged to authors.

Each figure has a legend, which should be understandable without frequent reference to text. Collect legends and type double-spaced on a separate page. Indicate position at which figure is to appear in text.

OTHER

Abbreviations: Standard abbreviations may be used without definition. See *Eur. J. Immunol.* 24(1): Instructions to Authors, 1994

Nomenclature: For immunoglobulins, follow recommendations of World Health Organization (*Eur. J. Immunol.* 1973, *3:* 62). For allotypic markers of immunoglobulins follow *Eur. J. Immunol.* 1976, *6:* 599. For HLA system factors follow recommendations of WHO Nomenclature Committee (*Immunol. Today* 1990, *11:* 3). Italicize genetic symbols only when necessary to avoid ambiguity. For complement and interferon nomenclature follow *Immunochemistry* 1970, *7:* 137 and *Eur. J. Immunol.* 1980, *10:* 660, respectively. For biochemical compounds follow recommendations of the nomenclature committees of IUB, of the IUB Committee of Editors of Biochemical Journals and of other international committees as detailed in *Eur. J. Biochem.* 1989, *180:* A9–A11.

For chemical compounds, follow IUPAC nomenclature in *Chemical Abstracts*.

Page Charges: DM 250 for each page over 6.

Spelling: Use American spelling.

SOURCE

January 1994, 24(1): front matter.

European Journal of Pharmacology
Elsevier

Editorial Office, *European Journal of Pharmacology*
Universiteitsweg 100
3584 CG Utrecht, The Netherlands
Telephone: +31.30.538833; Fax: +31.30.539033

MANUSCRIPT SUBMISSION

Send original and 2 copies complete with 3 sets of figures and tables. Submissions sent by facsimile transmission will not be considered.

Type or print with double-spacing (at least 6 mm between lines) and with wide margins (26 lines per page). Use a laser-type, a similar quality printer or a high-quality typewriter with a black carbon ribbon to ensure proper contrast. Use a standard (at least 12-point) type face. Manuscripts printed with low-quality printers will be returned. Manuscripts can also be submitted in electronic form accompanied by a print-out.

MANUSCRIPT FORMAT

Arrange manuscript: title; surname(s) of author(s), preceded by one name spelled out in full; name and address where work was done; name, full postal address, telephone, telex and fax numbers of corresponding author (all on 1 page); abstract and key words (indexing terms: normally 3–6 items); sections: 1. Introduction; 2. Materials and methods; 3. Results; 4. Discussion; Acknowledgments; References; figure legends and figures; tables. Number subdivisions of a section 2.1, 2.2, 2.3. etc. Number all pages consecutively, with title page as p. 1.

Abstracts and Key Words: 150 words typed on a separate sheet. Present problems, suggest scope and plan of experiments, indicate significant data, and point out major findings and conclusions. Make completely self explanatory. Do not use footnotes. Cite references in full. Use standard terms and scientific nomenclature. Do not use abbreviations and contractions, except for weights and measures and those explained. Below abstract, type 3–6 key words or short phrases suitable for indexing. Select from *Index Medicus* or *Excerpta Medica Index*. First category key words (e.g., Hyperphagia; G-Strophantidin; Fiber shortening) will be listed and cross-indexed. Second category key words (e.g., Rat, Cold, Metabolites) will be listed under index entry for first category key words.

Introduction, Materials and methods, Results, Discussion: Do not give an extensive review of literature in introduction. Refer to previous work that has direct bearing on topic. Write Materials and methods clearly and in such detail that work can be repeated. Refer to previously published procedural detail by citation. When a modified procedure is used, give only author's modification in detail. Describe results concisely. Make text, tables, and figures internally consistent. Include only significant findings in discussion.

References: Verify accuracy. List alphabetically. Two or more references to same first author with same publication year should have a,b, c, etc. suffixed to the year indicating alphabetical order of second or third author, etc. References to journals contain names and initials of author(s), year, full title, abbreviation of periodical title (follow *List of Serial Title Word Abbreviations*, Paris: International Serials Data System), volume and page number. For books include title and name and city of publisher. Cite references in text by author(s) name and year of publication. For 3 or more authors, first author's name is followed by et al. Downie and Larsson (1990) or (Stoof and Kebabian, 1984; Hicks et al., 1988, 1989; Seeman et al., 1990, 1991a,b,c).

Harrison, J.K., W.R. Pearson and K.R. Lynch, 1991, Molecular characterization of α_1- and α_2-adrenoceptors, Trends Pharmacol. Sci. 12, 62.

De Graan, P.N.E., P. Schotman and D.H.G. Versteeg, 1990, Neural mechanisms of action of neuropeptides: macromolecules and neurotransmitters, in: Neuropeptides: Basics and Perspectives, ed. D. De Wied (Elsevier, Amsterdam) p. 139.

Appenzeller, O., 1990, The Autonomic Nervous System (Elsevier, Amsterdam) p. 419.

ORDER OF REFERENCES

De Groat, W., 1990,
Maggi, C.A., 1988,
Maggi, C.A. and A. Lecci, 1987,
Maggi, C.A. and A. Meli, 1986,
Maggi, C.A., P. Santicoli, R. Patacchini, P. Rovero, A. Giachetti and A. Meli, 1989a,
Maggi, C.A., R. Patacchini, P. Rovero, A. Giachetti and A. Meli, 1989b,
Maggi, C.A., Giuliani, R. Patacchni, P. Santicioli, A. Giachetti and A. Meli, 1990,
Monsma, Jr., F.J., 1989,
Van der Giessen, 1990,

Keep number of references to a minimum; cite appropriate recent reviews when possible. Refer to unpublished observations, personal communications, and manuscripts in preparations or submitted for publications in text; do not include in list of references. Manuscripts in press may be included; give name of journal.

Illustrations and Graphs: Submit either original drawings plus two good photographs or three good photographs of drawings (unmounted on glossy paper). Photocopies may not be used instead of originals. Photographs should have adequate definition and contrast. Limit number of illustrations.

Allow for reduction to fit a single (8.4 cm) or double (17.6 cm) column width. Explain lettering and key symbols in legend. See *Eur J Pharm* 251(1): ix for symbols that can be typeset and should be used in preference to non-standard forms. Use Letraset for letters and numerals; alternatively use India ink. Make all letters and numerals in a particular illustration the same size. Comparable illustrations should have same size letters, figures and numerals when reduced to 8.4 cm width. If abscissa and ordinate of a graph require labelling, place abscissa label below related numerals and place ordinate label horizontally and above ordinate. If not, all labels will appear in legend only.

Have graphs, electrocardiograms and oscillograms prepared by a skilled photographer so that the dark, cross-hatched background is eliminated, the faint portions of the graphs intensified, and a sharp print obtained. Use blue-ruled instead of black-ruled recording paper for originals. Kymograph records on sooted paper should be reversed photographically to ensure graphs in black on a white background.

Prepare drawings of complicated chemical structures in the same way as graphs.

Draw calibration bars on micrographs instead of giving magnification rate in figure legend.

Refer to all illustrations as figures. Number in Arabic numerals.

Legends to figures should make these comprehensible without reference to text.

Color reproduction is available at cost.

Tables: Prepare for use in single (8.4 cm) column or page (17.6 cm) width. Do not duplicate material in text or illustrations. Give brief explanatory headings and sufficient experimental detail after title to be intelligible without reference to text. Use short or abbreviated column headings. Explain if necessary in footnotes, indicated as [a], [b], [c], etc. Identify statistical measures of variation. Number separately in Arabic numerals (table 1, 2 etc.).

Formulas and Equations: Clearly present structural chemical formulas, process flow diagrams and complicated mathematical expressions. Identify all subscripts, superscripts, Greek letters, and unusual characters.

OTHER FORMATS

RAPID COMMUNICATIONS—Submit original manuscript and 2 copies. 700 word report (8 references; one simple table or figure of not more than 8.4 cm = 1 column width) on exciting new results within the scope of the journal. Type double-spaced, with wide margins, without subheadings. Arrange manuscript: title (85 characters including spaces); surname(s) of author(s), preceded by one name spelled out in full; name and address of establishment where work was done (all on 1 page); abstract (75 words) and key words (3); text; acknowledgment(s); references; figure legend and figure; table. Give name, full postal address, telephone, fax and telex numbers of corresponding author on title page.

SHORT COMMUNICATIONS—2000-2300 words, 100 word abstract, 15 references, 2 figures and/or tables, arrange like a full length paper.

OTHER

Abbreviations: Use as few as possible. Write out names of compounds, receptors, etc., in full throughout text. Do not use unnecessary and nonsense abbreviations. Do not abbreviate generic names. Abbreviations that have come to replace full terms may be used, provided term is spelled out in abstract and text at first use. Unwieldy chemical names may be abbreviated. Give full chemical name once in text and if needed in abstract followed in both cases by abbreviation. Code names may be used, but give full chemical name in text. See *Eur J Pharm* 251(1): vi, 1994 for abbreviations that can be used without definition, units of measurements and other terms.

Place isotope mass number before atomic symbol ($[^3H]$norradrenaline, $[^{14}C]$choline). Write ions: Fe^{3+}, Ca^{2+}, Mg^{2+}. Use term absorbance (A) not extinction or optical density.

See *Units, Symbols and Abbreviations, A Guide for Biological and Medical Editors and Authors* (1988, London: Royal Society of Medicine) or *CBE Style Manual: A Guide for Authors, Editors and Publishers in the Biological Sciences* (1983, Bethesda, MD: Council of Biology Editors).

Drugs: Use generic and chemical names. Proprietary equivalents may be indicated once in parentheses. See *Pharmacological and Chemical Synonyms* (1990, Amsterdam: Elsevier).

Nomenclature, Chemical: Make consistent, clear and unambiguous. Follow usage of American Chemical Society and convention of International Union of Pure and Applied Chemistry. Consult indexes of *Chemical Abstracts*, reports and pamphlets of American Chemical Society Committee on Nomenclature, Spelling and Pronunciation, and *Biochemical Nomenclature and Related Documents* (London: Biochemical Society).

Nomenclature, Enzymes: Trivial names may be used in text. Quote systematic name and classification number according to *Enzyme Nomenclature* (1984, New York: Academic Press) at first use.

Nomenclature, Peptides: See *Neuropeptides* 1: 231, 1981.

Nomenclature, Receptors: Follow TIPS Receptor Nomenclature Supplement (*Trends Pharmacol. Sci.* Supplement, January 1991).

Nomenclature, Stereoisomers: When drugs are mixtures of stereoisomers, make their composite nature clear and consider implications for interpretation and conclusions. Use appropriate prefixes. Generic name alone without prefix means an agent with no stereoisomers. See *IUPAC, Nomenclature of Organic Chemistry* (J. Rigaudy and S.P. Klesney, eds., 1979, London: Pergamon Press) and "Signs of the Times: the need for a stereochemically informative generic name system" (M. Simonyi, J. Gal and B. Testa, 1989, *Trends Pharmacol. Sci.* 10: 349).

Spelling: Consult *Webster's New International Dictionary* or the *Oxford English Dictionary*. Do not use Latin plurals if English equivalent is accepted form. Make use of hyphens, capital letters, numbers (written or spelled out) consistent. Do not divide words at ends of lines.

Style: See *CBE Style Manual: A Guide for Authors, Editors and Publishers in the Biological Sciences* (1983, Bethesda, MD: Council of Biology Editors) and *How to Publish a Scientific Paper* (R. A. Day, 1979, Philadelphia: ISI Press).

SOURCE

January 4, 1994, 251(1): v-x.

European Journal of Surgery

Society for the Publication of Acta Chirurgia Scandinavica

Swedish Surgical Society
Association of Surgery of the Netherlands
Scandinavian University Press

European Journal of Surgery
Furugatan 1, S-582-45 Linkoping, Sweden
Telephone: +46-(0)13-12 54 87; Fax: +46-(0)13-12 19 93

AIM AND SCOPE

The *European Journal of Surgery* is published by the Society for the Publication of Acta Chirurgica Scandinavica, founded by Professor Axel Key in 1869 and sponsored by the Key Foundation, Sweden. It is also a forum for the Association of Surgery of the Netherlands. We consider for publication Original Scientific Reports (clinical and experimental), descriptions of surgical techniques and Case Reports from all parts of the world within the field of general surgery: Gastrointestinal surgery; Hepatopancreatobiliary surgery; Colorectal surgery; Endocrine surgery; Pediatric surgery; Surgical oncology; Trauma; Vascular surgery.

MANUSCRIPT SUBMISSION

Manuscripts will be reviewed on the understanding that they have not been published, simultaneously submitted or accepted for publication elsewhere. Only papers judged to offer significant contributions to the advancement of surgical knowledge will be accepted for publication.

Submit 3 good quality photocopies of all written material and illustrations; retain the originals. Include a cover letter signed by all authors. Note if a professional linguistic reviewer has revised the manuscript.

If manuscript is word processed, submit floppy disc and describe program used. Include an ASCII file on disc. Also supply two print out copies.

MANUSCRIPT FORMAT

Write in British English. Type double-spaced including references, tables and figure legends, on one side of white paper, size ISO A4 with a margin of 5 cm. Number pages consecutively, beginning with abstract, then text, acknowledgments, references and legends to figures.

Submit three consecutive title pages. On first page print a short title (45 characters) for use as a running head. Indicate whether manuscript is intended as an "original paper" or a "case report". Put complete title on second page; on third page list name of authors (first name, middle initial and last name of each) and addresses of authors (department, hospital, country). Give full address of corresponding and reprint request authors in a footnote.

Text: Recommended length is 5–6 printed pages (10–12 typed pages, including tables and illustrations).

On first page of text include a structured abstract (150 words). Write in note-form under subheadings, start each one on a separate line: Objective, Design, Setting, Subjects, Interventions, Main outcome measures, Results, Conclusion.

Make text brief, succinct, well prepared and expressed in clear, concise English. Do not use abbreviations.

Divide text under headings: Introduction, Material(s) (or Patients or Subjects) and Method(s), Results and Discussion. Long articles may need subheadings to clarify content, especially in Results and Discussion sections.

References: Arrange list in alphabetical order, number consecutively by first letter of first author's surname. Identify references in text by these numerals (in parentheses). List cited journals as abbreviated in *Index Medicus*. Sequence for a standard journal article is: author(s), title, journal, year, volume, first and last page numbers. Sequence for a book is: author(s), title of reference (if distinct from that of book), editor(s) or compiler(s), title, edition number, place of publication, publisher's name, year of publication, first and last pages of reference (if relevant).

Tables and Illustrations: State number of patients–or animals–studied and the mean (SD), mean (SEM), and median (range or interquartile) tables consecutively (Roman numerals). Supply a brief title for each. Explain all nonstandard abbreviations used.

Have figures professionally drawn and photographed. Instead of original drawings, roentgenograms, and other material, send sharp glossy black-and-white photographic prints. Paste a label on back of each figure indicating figure number, names of authors, and TOP.

OTHER FORMATS

CASE REPORTS—2–3 printed pages (4–6 pages of typescript, including illustrations). Divide text under headings: Introduction, Case Report(s) and Conclusion(s).

SUPPLEMENTS—Doctorate theses, monographs or proceedings of symposia may be accepted as supplements; apply to Editor. Printing and distribution costs for 3000 copies are borne by authors. Use same format and front-cover as journal.

OTHER

Color Charges: Must be borne by authors; application can be made for contribution from the Society for Publication of Acta Chirurgica (via the Editor).

Statistical Analyses: Specify the mean (SD), mean (SEM), and median (range or interquartile). The $\pm$ sign is not acceptable. State statistical test used, result of that test, and p value; include 95% confidence intervals where possible.

Units: Use international system of units (SI).

SOURCE

January 1994, 160(1): inside back cover.

Experientia
Birkhauser

Experientia, Birkhauser Verlag
P.O. Box 133
CH-4010 Basel, Switzerland
Telephone: Switzerland-61-271 7400
Fax: Switzerland-61-271 7666

AIM AND SCOPE

Experientia is a monthly journal for life sciences devoted to publishing articles which are interdisciplinary in character and which are of general scientific interest. Hitherto unpublished manuscripts that fall within the following categories will be considered:

Research Articles; Mini-reviews; Review Articles; Multi-author Reviews (collections of reviews on one topic assembled by an appointed Coordinator); and Scientific correspondence.

The subjects covered in the journal fall into the following broad groups, but the journal also publishes papers on other topics in life sciences, including ethics of scientific experimentation. Indicate in cover letter to which subject or subjects the paper belongs: Anatomy, Physiology, neurobiology; Endocrinology, neuroendocrinology; immunology; pharmacology and toxicology; cell and molecular biology, biochemistry, biophysics; Microbiology, virology; Natural product chemistry; Botany, plant physiology; Zoology (including invertebrate physiology), parasitology; genetic, developmental biology, evolution; Ecology, ethology, epidemiology.

MANUSCRIPT SUBMISSION

Do not submit reports of preliminary results. In order to emphasize interdisciplinary character of papers submitted as *Research Articles,* make significance of research clear in both Abstract and Introduction.

Submit in triplicate, typed double-spaced. Manuscripts may also be submitted on diskette together with three hard copies. Diskette and hard copy must contain exactly the same text.

MANUSCRIPT FORMAT

For research articles follow standard scientific format (Materials and methods; Results; Discussion). On title page: author's name and address, telephone and fax numbers, a short Abstract, and 8 Key Words. Avoid use of footnotes.

Tables and Figures: Submit on separate pages. List explanatory legends (captions) together on separate pages.

Illustrations: Submit drawings on good quality paper marked clearly in black. Computer graphs must be of highest quality and high contrast.

Supply photographs as glossy positive prints. Color photographs are not usually published, but can be accepted at the author's expense.

References: For Research Articles and Mini-reviews, use an abbreviated bibliography. List and number references in order of appearance; titles and last pages are omitted. For Reviews, list references alphabetically and number. Cite in text by number. In references give full article titles and both first and last pages of cited article.

OTHER FORMATS

PRIORITY PUBLICATION—Superior manuscripts whose subject matter is judged to be of immediate urgency will be given accelerated handling.

SCIENTIFIC CORRESPONDENCE—Letters to the editors concerning articles published in this journal. They will be printed together with any reply or rebuttal from authors of articles in question.

OTHER

Experimentation, Animal: Conform to the *Ethical principles and guidelines for scientific experiments on animals* of the Swiss Academy of Medical Sciences (see January issues of *Experientia*).

Units: Express data in units conforming to Système International (SI).

SOURCE

January 15, 1994, 50(1): inside back cover.

Experimental Brain Research
Springer International

P. Andersen
Institute of Neurophysiology
University of Oslo, P.B. 1104 Blindern
N-0317 Oslo, Norway
Fax: 22-85 12 49; E-mail: andersen@basal med.uio.no

V.J. Wilson
The Rockefeller University
1230 York Avenue
New York, NY 10021
Fax: (212) 327-8530;
E-mail: vwilson@rockvax.rockefeller.edu
Neurophysiology

R. Shapley
New York University
Faculty of Arts and Science
Center for Neural Science
4 Washington Place, R. 912
New York, NY 10003
Fax: (212) 995-4011; E-mail: shapley@cns.nyu.edu

R.F. Schmidt
Institute of Physiology
University of Würzburg
D-97070 Würzburg, Germany
Fax: 931-54553
Sensory Physiology

G.L. Collingridge
University of Birmingham
Dept. of Pharmacology
The Medical School
Edgbaston, Birmingham B15 2TT, U.K.
Fax: 21-414-4509
Neuropharmacology

L.J. Garey
Dept. of Anatomy
Charing Cross & Westminster Medical School
Fulham Palace Road, London W6 8RF, U.K.

Fax: 81-846-7025

C. Sotelo
INSERM – U 106 –
Hopital de la Salpetriere
47 Bd. de l'Hopital
F-75651 Paris Cedex 13, France

Fax: 1-45709990
Neuroanatomy and Developmental Neurobiology

A. Bjorklund
Dept. of Histology, University of Lund
Biskopsgatan 5, S-22362 Lund, Sweden
Fax: 46-107927
Histochemistry and Neuroplasticity

C. Marzi
Dept. of Neurological Science and Vision
Physiology Section
8 Strada Le Grazie, I-37134 Verona, Italy
Fax: 45-580881; E-mail: marzic@borgoroma.univr.it
Behavioural Sciences and Neuropsychology

AIM AND SCOPE

Experimental Brain Research accepts original contributions on many aspects of experimental research of the central and peripheral nervous system in the fields of morphology, physiology, behavior, neurochemistry, developmental neurobiology, and experimental pathology relevant to general problems of cerebral function

MANUSCRIPT SUBMISSION

Furnish a telex or fax number on title page. Submit three copies to a Section Editor.

The Editors subscribe to recommendations formulated by the International Committee of Medical Journal Editors [*Ann Int Med* (1988) *108:*258–265] regarding criteria for authorship. Each person listed as an author or coauthor must meet all of three criteria: conceived, planned, and performed work leading to the report, or interpreted evidence presented, or both; written or reviewed successive versions of report and shared in their revisions; and approved final version. Each author should have sufficient knowledge of and participation in work that he or she can accept public responsibility for report. In cover letter, senior or corresponding author must certify that all listed authors qualify for authorship. Also, certify that no part of work described has been published before [except as an abstract or as part of a published lecture, review, thesis, or dissertation (appropriately cited)]; that work is not under consideration for publication elsewhere; that, if and when manuscript is accepted for publication, author(s) agree to automatic transfer of copyright to publisher; and that manuscript, or its parts, will not be published elsewhere subsequently in any language without consent of copyright holders.

MANUSCRIPT FORMAT

Type manuscript double-spaced throughout with wide margins. Be clear and concise. Follow typographical conventions of recent journal issues.

Each article must have an abstract, followed by five key words, characterizing scope of paper. Place each main heading (Introduction, Materials and methods, etc.) and subheading on separate lines. Mark Materials and methods and sections of lesser importance for small print. Underline genus and species names once for italics.

Number footnotes in text consecutively; place at bottom of page to which they refer.

References: Include in list only publications cited in text. Cite in alphabetical order under first author's name; list all authors (surnames followed by initials-do not use and); give complete title (capitalize only initial letter of title).

JOURNAL ARTICLES

Name(s) and initials of all authors, year in parentheses, full title, journal name as abbreviated in *Index Medicus*, volume; colon; first and last page numbers.

Jacobson SG, Eames RA, McDonald WI (1979) Optic nerve fibre lesions in adult cats: patterns of recovery of spatial vision. Exp Brain Res 36: 491-508

BOOKS

Name(s) and initials of all author(s), year in parentheses, title, edition (ed), publisher, place of publishing.

Chan-Palay V (1977) Cerebellar dentate nucleus: organization, cytology, and transmitters. Springer, Berlin Heidelberg New York

MULTIAUTHOR BOOKS

Name(s) and initials of all author(s), year in parentheses, title of paper, name(s) and initials of all editors, title of book, publisher, place of publishing, first and last pages.

Oscarsson O (1973) Functional organization of spinocerebellar paths. In: Iggo A (ed) Somatosensory system. Handbook of sensory physiology, vol II. Springer, Berlin Heidelberg New York, pp 339-380

Tables: Number consecutively with Arabic numbers. Indicate table footnotes by lowercase superscript letters, beginning with ^{a} in each table.

Illustrations: Limit to materials necessary for text. Do not submit previously published illustrations. Submit good quality prints for line drawings. Computer drawings must be of comparable quality. Lines and curves must be smooth. Number figures consecutively. Submit separately from text. Mark figure number and author(s) name(s) on back. Mount photo- or micrographs together to save space; do not take numerical order into consideration. For plates, mount figures on regular bond paper, not cardboard.

Supply line drawings as clear back-and-white drawings suitable for reproduction. Make lines of uniform thickness. Make letters and numbers of professional quality and proper dimensions. Make drawings twice final size in indelible black ink. Note thickness of lines, size of inscriptions, size of measuring points, adequate spacing in shaded and dotted areas. Words should be in upper and lowercase characters, not capitals. Optimal size of letters as appear in journal: Capital letters and numbers 2 mm, lowercase letters 1.6 mm.

Submit all illustrations as sharp, glossy, high-quality photographs of a size permitting direct printing (no reduction). If possible, arrange for one column (8.6 cm) or full type area (17.5 cm) and 23.6 cm high, maximum. Avoid intermediate sizes.

Submit photographs and electron micrographs as sharp, high-contrast prints on glossy paper, trimmed at right angles and in a size permitting direct printing. Insert arrows, letters and numbers with template rub-on letters or make inscriptions on a transparent overlay. Give micrographs internal magnification markers and state magnification in caption. Do not combine photographs and line drawings in one figure.

Make captions brief (4-5 lines), self-sufficient explanations of illustrations; avoid phrases like "For explanation, see text." Captions are part of text; append to text on separate page. Indicate magnification in micrographs by scale bars; do not quote a factor.

OTHER FORMATS

RESEARCH NOTES—4 printed pages, including references and 1 or 2 figures; 250 word abstract. Quick publication of interesting findings of ongoing research. Do not submit material for publication elsewhere as part of a full paper.

OTHER

Color Charges: Approximately DM 1200,- for first and DM 600,- for additional pages.

Electronic Submission: Submit text on diskettes formatted for DOS or Macintosh systems. If possible, store text in two versions: in standard data file format offered by your word processing system; in data interchange formats (listed in declining order of preference): RTF (Microsoft Rich Text Format); DCA/RFT (Document Containment Architecture/Revisable Form Text); DCA/FFT (Document Containment Architecture/Final Form Text); ASCII or "text only".

Keep a copy of diskette. Make sure diskette is adequately packed. Enclose a printout of final text. Text file and printout must correspond exactly.

Do not incorporate any special page layout in text. Input text continuously; only insert hard returns (⏎) at ends of paragraphs or headings, subheadings, lists, etc. Do not use space bar to make indents; use a tabulator or an indent command.

Use automatic pagination function in word processing system; do not insert page numbers manually.

Indicate special emphasis on words or phrases in text by italic script or underlining. Use boldface type only in running text for certain mathematical symbols, e.g., vectors. In table titles, boldface "Table" and table number. In figure legends, boldface abbreviation "Fig.", figure number, and letters referring to figure parts (a, b, etc.). Boldface headings for visual emphasis.

Place all tables at end of file. Separate individual columns using tabulators, not space bar.

Delete annotations or comments from final text file. Send only final updated version.

Experimentation, Animal: State that "Principles of laboratory animal care" (NIH publications no. 86-23, revised 1985) were followed, as well as specific national laws. Confirm that principles approved by Council of the American Physiological Society were followed: that only lawfully acquired animals were used and their retention and use was in compliance with state and local laws and regulations; that animals received every consideration for their bodily comfort and that they were kindly treated, properly fed, and in sanitary surroundings; that appropriate anesthetics were used during operative procedures and recovery (if N_2O is used, supplementary anesthetics must be available and precautions taken to monitor level of anesthesia described–Papers based on use of curarizing agents will be accepted only if adequate evidence is offered that no distress was caused to animals); that when study does not require recovery from anesthesia, that animals are killed in a humane manner at end of observations; and that postoperative care minimized discomfort and pain and was equivalent to accepted practices in schools of veterinary medicine.

Experimentation, Human: Include a statement that studies were approved by appropriate ethics committees and were therefore performed in accordance with ethical standards of the 1964 Declaration of Helsinki. State that all persons gave informed consent prior to inclusion in study. Omit details that might disclose subjects' identities.

Terminology: See *J Neurochem* and *Eur J Biochem* for biochemical terminology.

Units: Express temperatures in degrees Celsius; times apart from seconds (s), in minutes (min), hours (h), etc. Use International System of Units (SI) when possible. See *Metric Practice Guide: A Guide to the Use of SI - the international system of units* (Publ. E380-70, Philadelphia: American Society for Testing and Materials). Give other units in parentheses when first appearing in text.

SOURCE

1994, 102(1): A5-A6.

Experimental Cell Research
Academic Press

Editorial Office, *Experimental Cell Research*
Karolinska Institute, CMB, Doktorsringen 2 D
17177 Stockholm Sweden
Dr. Renato Beserga, Associate Editor
Experimental Cell Research
525 B Street, Suite 1900
San Diego, CA 92101-4495

AIM AND SCOPE

The chief purpose of *Experimental Cell Research* is to promote the understanding of cell biology by publishing experimental studies on the general organization and activity of cells. The scope of the journal includes all aspects of cell biology, from the molecular level to the level of cell interaction and differentiation. While it is recognized that a variety of cell types, tissues, and organisms provide the experimental materials for cell research, the journal does not include within its scope those investigations whose results are more specifically of morphological, histological, pathological, pharmacological, and microbiological interest, nor which are concerned with biochemical mechanisms per se. The relevance of the aims and results of the contributions to the universal attributes of cells will be the foremost consideration in the evaluation of papers.

Examples of areas of research that fall within the scope of ECR are: physical and chemical aspects of cellular and intercellular structure, biosynthesis with reference to cell growth, reproduction and differentiation, mechanism of meiosis and mitosis, cell cycles, membrane function, motility, interactions between cells in tissues or in culture, modulations in cultured cells, environmental relations and adaptations, interactions between genome and cytoplasmic factors, functional role of subcellular particles, energetic aspects of cell structure and function, control and regulation of cellular processes.

Reports of original methodological or theoretical work having a direct bearing on the experimental approach are welcome.

MANUSCRIPT SUBMISSION

Original papers only. Manuscripts may not contain work that has been published, may not be under consideration for publication elsewhere, must be approved by all authors and by appropriate authority at institution where work was done. Verify wording and obtain and submit permission to cite personal communications or unpublished work.

Submit three copies. Include all figures, original or glossy prints, and tables.

MANUSCRIPT FORMAT

Type double-spaced on one side of 8.5 × 11 in. white paper with 1 in. margins on all sides. Number pages consecutively. Page 1: Article title, author(s), affiliation(s), short running title (abbreviated title, 55 letters and spaces), name, address, telephone and fax numbers of corresponding author. Page 2: Short abstract (50-250 words).

In Methods, draw attention to any particular chemical or biological hazards involved in carrying out experiments. Describe relevant safety precautions. State codes of practice followed. Reference relevant standards.

Footnotes: Avoid use.

Figures: Number consecutively with Arabic numerals in order of mention in text; give each a descriptive legend. Type legends together on a separate sheet, double-spaced. Submit illustrations in finished form suitable for photo reproduction planned to fit proportion of printed page (7 1/8 × 9 in.; column width, 3.5 in.). Mark on back in soft pencil with author's last name, title of paper, and figure number; indicate TOP.

A convenient size for drawings is 8.5 × 11 in. Lettering should be professional quality or generated by high-resolution computer graphics; make large enough (10–12 points) to be reduced 50–60%. Make with black India ink on tracing linen, smooth-surface white paper, or Bristol board. Use blue ruled graph paper. Ink grid lines in black. Properly prepared glossy prints are acceptable.

Submit photographs actual size for reproduction; label with letters 2 to 4 mm high. Provide clear, sharp, glossy prints with adequate contrast. Providing an overlay indicating the most critical areas of a micrograph ensures high-quality reproduction. Match contrast and density of all figures on a single plate. Mount figures to appear together as one plate on paper. Give magnifications in legends or indicate with a scale marker on photograph.

Tables: Type double-spaced on separate pages; number consecutively with Arabic numerals in order of mention in text. Give each table a short explanatory title.

References: Cite in text by Arabic numerals in square brackets; list at end of paper in consecutive order. Abbreviate journal titles as in *Chemical Abstracts Service Source Index,* 1985.

1. Stoddart, J. H., Lane, M. A., and Niles, R. M. (1989) *Exp. Cell Res.* **184,** 16–27.

2. Bovey, F. A., and Jelinski, L. W. (1983) Chain Structure and Conformation of Macromolecules, p. 150, Academic Press, New York.

3. Britton, G. (1985) *in* Methods in Enzymology (Law, J. H., and Rilling, H. C., Eds.), Vol. 111, pp. 113–149, Academic Press, San Diego.

OTHER FORMATS

SHORT NOTES—2000 words inclusive of tables, illustrations with captions, and references. One printed page of text contains about 500 words.

OTHER

Color Costs: $800 for first figure and $400 for each subsequent figure.

Data Submission: Submit to a gene data bank. Make novel sequences available to scientific community.

Reviewers: Suggest reviewers and those to be excluded.

SOURCE

November 1994, 215(1): back matter.

Experimental Eye Research
Academic Press

Dr. Berndt Ehinger
Dept. of Ophthalmology, University of Lund
S-221 85 Lund, Sweden
Telephone: (46) 46-171690; Fax: (46) 46-11574

Dr. Ramesh Tripathi
South Carolina Eye Institute
University of South Carolina School of Mediciine
Four Richland Medical Park, Suite 100
Columbia, South Carolina 29203
Telephone: (803) 254-4398; Fax: (803) 252-5452
        General section papers

Dr. Joe G. Hollyfield
Cullen Eye Institute, Baylor College of Medicine
1 Baylor Plaza, Houston, Texas 77030
Telephone: (713) 798-5964; Fax: (713) 798-5866

Dr. Ann H. Milan
Dept. of Ophthalmology, RJ-10
University of Washington
Seattle, Washington 98195
Telephone: (206) 543-9188; Fax: (206) 543-4414
        Retina section papers

Dr. Frank J. Giblin
Eye Research Institute, 422 Diodge Hall
Oakland University
Rochester, Minnesota 48309-4401
Telephone: (313) 370-2390; Fax: (313) 370-2006
        Lens section papers

AIM AND SCOPE

Experimental Eye Research is published monthly and contains the results of original research on all eye tissues. The Journal is subdivided into three Sections: General Section, Retina Section and Lens Section.

The *General Section* will embrace studies on eye tissues (except retina and lens) and all research material for the entire eye as an organ.

The *Retina Section* will include studies dealing with the structure and function of the retina and pigment epithelium from the tissue through the molecular level of organization.

The *Lens Section* will include all studies dealing with the normal and cataractous lens.

MANUSCRIPT SUBMISSION

Submit original manuscript and two copies.

MANUSCRIPT FORMAT

Divide into sections: Summary, Introduction, Materials, Methods, Results, Discussion, Acknowledgments and References.

Summary: 500 words intelligible to the general reader without reference to main text. Do not use abbreviations. Do not cite references.

Key Words: 5–15 key words.

Layout: Type double spaced throughout (including summary, footnotes, tables and legends) on one side of quarto paper, with wide margins. Type footnotes, tables and legends for illustrations separately at end of manuscript.

Running Headline: 40 characters (including spaces), suitable for page headings, if full title is longer.

Tables: Type each on a separate page. Number with Roman numerals, in order of mention in text. Include a brief title; give all columns headings. Give explanatory information essential to understanding the table as footnotes indicated by superscript symbols (*, †, ‡, §, ||, ¶); type at bottom of page.

Scale illustrations, graphics and halftone (photographic) illustrations to fit page format of journal; publisher may make adjustments. Non-photographic figures may be, but do not have to be glossy prints. Print photographic illustrations on glossy paper and make either 80 mm (single column) or 168 mm (double column). Length may not exceed 248 mm, optimally no longer than 204 mm. Do not place lettering, shading or hatching on original drawings. Indicate on accompanying photocopies. Write lettering that is to appear on a halftone on a duplicate photograph or semi-transparent paper marked or fixed to register with original photograph. Affix machine printed transparent paper to original.

Authors must pay cost of color illustrations. An estimate will be sent; price may be reduced if color separated negatives are supplied.

Number each figure and plate, using Arabic numerals, in order mentioned in text. Identify figures by number, name(s) of author(s), running title, and, if necessary, indicate top. Submit original figures with manuscript and clear duplicate sets.

Provide a legend for each figure; include a key to symbols used. Type legends in numerical order as a separate list; append to manuscript after Footnotes.

References: List alphabetically at end of paper; abbreviate according to *Index Medicus*.

When more than one paper by the same author(s) is cited, list chronologically. With two or more authors, arrange references alphabetically and then chronologically within each alphabetical group. Give titles of papers in full.

Smith, J., Brown, F. and Robinson, S. (1957). Alphacrystallin in the human senile cataract. *Exp. Eye Res.* **6**, 203–12.

Verify accuracy of references. Citations in text should read: Smith, Brown and Robinson (1957), or (Smith, Brown and Robinson, 1957), and subsequently as Smith et al. (1957), or (Smith et al., 1957). When a citation has more than three authors cite in text: (Smith et al., 1957),

Use (Brown, 1957a), (Brown 1957b) where more than one paper by the same author(s) has appeared in one year. Cite first and last page number of papers.

Footnotes: Avoid use. Indicate essential footnotes by superscript figures in text; collect on a single page at end of manuscript.

OTHER FORMATS

LETTERS TO THE EDITOR—Short (four printed pages) communications on current research or remarks on recently published papers.

OTHER

Abbreviations: Use customary abbreviations. Refer to *Chemical Abstracts, Biochemical Journal* **273**, 1 (1991), *Handbook for Chemical Society Authors, Style Manual for Biological Journals,* British Standards Institution recommendations and *Notes on the Preparation of Papers to be Communicated to the Royal Society* (1951). Certain non-standard abbreviations of chemical substances are convenient, and those listed in the *Journal of Biological Chemistry* may be used without further definition (e.g., DNA, ATP). Use any other such abbreviations sparingly; define where first use occurs.

Conventions: See current journal issues for style of headings, references, etc., to follow.

Experimentation, Animal and Human: Abusive treatment of human subjects or experiments on animals are grounds for refusal of manuscripts.

SOURCE

July 1994, 59(1): inside back cover.

Experimental Hematology
International Society for Experimental Hematology

Peter J. Quesenberry, MD
University of Massachusetts Cancer Center
Two Biotech, Suite 202, 373 Plantation Street
Worcester, MA 01605
Telephone: (508) 856-3931; Fax: (508) 856-1310

MANUSCRIPT SUBMISSION

Submit original manuscript and three copies. Type manuscript entirely double-spaced in black ink on white paper, with wide margins (2 cm all sides). Use standard-size paper (between 20×26 cm and 22×31 cm).

Include cover letter signed by all authors. If original work is described include: "This manuscript has not been submitted previously for publication. All data contained herein are original data of the authors and have not been published previously except as part of published abstracts of this work or as clearly indicated and referenced in the text." Identify any data previously published in any form, except abstracts, in text.

MANUSCRIPT FORMAT

Title Page: Title; authors' full names, including academic degrees; authors' institutional affiliations (key each affiliation to its author); a running head (short title); 2 to 5 key words (mention each in abstract); and name, address, telephone and fax numbers of corresponding author.

Abstract: Be as concise as possible.

Text: Introduction; Materials and Methods; Results; Discussion; Acknowledgments; Figure Legends.

Tables: Number consecutively; include a title for each.

Figures: Include a title for each figure. Number in order mentioned in text. Submit half-tones as unmounted, glossy, high-contrast prints. Include a copy with each manuscript. Submit one negative if possible. Submit line art as unmounted, high-quality printouts. For digitally generated figures, submit figures on disk with name of graphics program with high-quality print.
Use similar size type within each figure. Match illustrations grouped together for height. Make uniform in size and style of lettering. Color photographs will be printed in black and white unless authors pay charges. Write first author's name, figure number, and arrow indicating top on back of each slide or glossy print.

References: Cite consecutively in text with Arabic numerals in square brackets.

1. Tobias JM, Lipton MA, Le Pinat A (1946) Effect of anesthetics and convulsants on brain acetylcholine. Proc Soc Biol Med 61:51

2. Umezawa H (1971) Enyzme Inhibitors of Microbial Origin. Tokyo, Japan: University of Tokyo Press

3. De Lorento C (1949) Cerebral cortex; architectures, intracortical connections, motor projections. In: HF Folton (ed) Physiology of the Nervous System, 3rd edition. New York: New York University Press, 288

Do not include unpublished observations, personal communications and papers in press in References; cite in text and explain in parentheses. Papers that rely on such citations for details that are essential for critical review may be rejected. Submit preprints of material in press. For personal communications, submit permission from individual who communicated data.

OTHER FORMATS

ANNOUNCEMENTS—Must be of interest to readership.

CONCISE REPORTS—3 journal pages, 6 typed columns, including space for figures or tables. Camera ready manuscripts reporting important new observations of sufficient significance and urgency to warrant both rapid review and publication.
Use legal-size paper. Type article title, author(s) name(s), affiliations, and an abbreviated version of title (running head) on separate sheet. Type text including tables single-spaced in single columns 4 1/8 × 11 5/8 in. Do not type outside this area. Number pages consecutively in light pencil at bottom. Material will be reduced 20%. Leave 3.3 in. space at top of first two pages for title and author name. Key textual references to reference list at end of paper by numbers in square brackets. Typing must be sharp and regular. Use laser printer with new cartridge. Have figures ready for reproduction. Do not fold sheets. Place sheets

between stiff cardboard. If possible, have pages professionally typeset at 1200 dpi (8.5 point Stone Serif with 10 point leading; column size 3.31 × 7.25 in.; leave extra 3.3 in. at top of first two columns).

LETTERS TO THE EDITOR—May relate to material published in journal, to ISEH business, editorial policy, or anything else pertinent to experimental hematology. Letters about specific papers will be sent to authors for possible reply.

MINIREVIEWS

RAPID COMMUNICATIONS—Report data of unusual scientific importance to hematology. Fully document results; give experimental design and methods, all pertinent data, and ample reference to literature.
In cover letter explain nature and significance of observations and request review as rapid communication. All authors must sign letter, indicating approval of data, that manuscript is an original report not presented elsewhere and that manuscript is not under consideration elsewhere.

OTHER

Abbreviations: Define first time used in text.

Drugs: Designate by generic names. Use trade names, spelled exactly as trademarked and with initial letter capitalized, after a drug has been identified once by its generic name or by its systematic chemical name. Designate other unfamiliar compounds, when first issued, by correct systematic names. Conform systematic chemical names to usage in indexes of *Chemical Abstracts*.

Experimentation, Human: Include details of age, race and sex of subject(s). State that informed consent was obtained from subjects and that investigations had been approved by an institutional Human Research Committee. State safeguards for protection of rights of minors and mentally defective subjects. Identify patients by number or serial letter, not by initials or names. If photographs of patients are required, obliterate faces and obtain written consent from patients for use of photographs.

Manufacturers: Include name of pharmaceutical or equipment company, with city, state and country, in parentheses after first mention of every material used.

Measurements: Use metric system and Celsius degrees. Use M for mole, L for liter. Do not type "m" for mu (μ); spell out "mu" or use "u."

Nomenclature and Symbols: Follow recommendations of International Union of Pure and Applied Chemistry and International Union of Biochemistry. In respect to molecular regulators of hemopoiesis that are either essential for proliferation and differentiation of a given cell line or potentiate the action of a specific regulator, give sufficient information in respect to its molecular weight, chemical characteristics, stability, resistance to enzymatic digestion, method of assay, etc., so that others can appreciate the basic properties. When a regulatory molecule has been purified to homogeneity, sequenced, cloned and produced as a pure entity, use the name then given to the humoral regulator. When there are numerous synonyms, list once. Until there is a consensus in respect to nomenclature, use names you prefer.

Page Charges: $30 per page for first five published pages and $180 for each additional page over five. $65 non-refundable fee for rapid full-length communications. Check must accompany submission. In unusual circumstances and upon request, the Editor-in-Chief may waive some or all of charges.

Recombinant DNA: Include statement regarding containment category of genetically engineered organisms. Specify procedures and safeguards for construction and handling of recombinant DNA molecules and organisms and viruses containing recombinant DNA.

Statistical Analyses: Describe novel methods, models, and approaches to statistical analysis concisely, but in sufficient detail to allow evaluation of reported results.

SOURCE

January 1994, 22(1): 105-107.

Experimental Neurology
Academic Press

John R. Sladek, Jr., Ph.D.
Editor-in-Chief, *Experimental Neurology*
The Neuroscience Institute
University of Health Sciences
The Chicago Medical School
3333 Green Bay Road
North Chicago, Illinois 60064-3095
Telephone: (708) 578-3000

AIM AND SCOPE

Experimental Neurology publishes the results and conclusions of original research in neuroscience with a particular emphasis on novel findings in neural development, regeneration, plasticity, and transplantation. Emphasis is also placed on basic mechanisms underlying or related to neurological disorders. Although the journal does not consider case reports for publication, information that bridges basic and clinical questions is of a high priority. Brief communications of important new data and scholarly reviews of important questions will be considered for publication. Manuscripts in other areas of neuroscience will be considered if they show relevance to the primary mission of the journal.

MANUSCRIPT SUBMISSION

Submit original manuscript and three copies. Send one original glossy print of each illustration and three photocopies or laser copies. Provide four original glossy prints if machine copies are not clear and sharp. Manuscript content must not have been published previously except as a preliminary report and is not submitted elsewhere for publication. Manuscripts must have received appropriate clearance and approval of all authors.

If accepted for publication, copyright in article, including right to reproduce in all forms and media, shall be assigned exclusively to the Publisher.

MANUSCRIPT FORMAT

Prepare typewritten or word-processed manuscripts double-spaced throughout, including abstract, acknowledgments, and references, on one side of 8.5 × 11 in. white paper and in *Experimental Neurology* format. Number pages consecutively. Limit length to about 20 pages, including tables and figures (about 6 printed pages).

Page 1: Article title, authors' names, and complete affiliations. At bottom place footnotes to title, indicated by superscript *, †, ‡. Include 6-8 key words at bottom of page.

Page 2: Proposed running head (abbreviated form of title, 50 characters), name, mailing address, and telephone and/or fax number of corresponding author, and a short abstract (250 word summary of problem, methods, and results and their significance in one paragraph).

Illustrations: Number Arabic numerals in order of appearance in text. Type legends consecutively on a separate sheet. Plan all figures to fit proportions of printed page (7 1/8 × 9 in.). Submit illustrations in finished form suitable for reproduction and no larger than 8.5 × 11 in. Identify all figures on back lightly in soft pencil with author(s) name(s) and figure number; indicate TOP. Letter each component of a composite figure consistently on lower or upper left face and refer to in text and caption.

Make lettering on drawings professional quality or generate by high-resolution computer graphics. Make large enough (10–12 points) to take 50-60% reduction. Freehand, penciled, or typewritten lettering is not acceptable. Make drawings in black India ink on tracing linen, smooth-surface white paper, or Bristol board or use high-quality computer graphics. Make symbols used to identify graph points large enough to be easily distinguishable after reduction. Plot graphs on blue coordinate or white paper. Grid lines must be black. Drawings may be submitted as glossy prints. Photocopies submitted as original artwork are not acceptable.

Submit clear, sharp halftone photographs on glossy paper as rich in contrast as possible. Submit only those parts of photograph that are necessary to illustrate the point. Cut off nonessential areas. Match contrast and density of all figures on a single plate. If not possible, provide unmounted original art. Provide an overlay that indicates critical areas of micrographs.

Color illustrations are accepted if authors defray cost. If mounting is necessary, mount color figures on 10- to 12-point flexible illustration board or paper.

Tables: Organize horizontally if possible. Number consecutively with Arabic numerals in order of appearance in text; collect at end of manuscript. Type double-spaced on a separate page with short descriptive captions directly above and footnotes (indicated by superscript lowercase italic letters) typed directly below table.

References: Use numerals in parentheses in text. List references alphabetically and number consecutively; do not incorporate "personal communications" or "unpublished observations" into list. Abbreviate journal titles as in *Index Medicus*.

1. BÉDARD, P. J., R. BOUCHER, and T. DI PAOLO. 1990. Effect of dopamine D$_1$ receptor stimulation in MPT monkeys. In *Disorders of Movement* (N. Quinn and P. Jenner, Eds.), pp. 187–192. Academic Press, London.

2. BRODAL, A. 1981. *Neurological Anatomy in Relation to Clinical Medicine,* 3rd ed. Oxford Univ. Press, New York.

3. VACA, K., AND E. WENDT. 1992. Divergent effects of astroglial and microglial secretions on neuron growth and survival. *Exp. Neurol.* **118:** 62–72.

OTHER FORMATS

BRIEF COMMUNICATIONS—6 or 7 double-spaced typed pages and 1 figure and/or table (2 printed pages). Include an abstract of a few sentences, but no section headings.

CRITICAL REVIEWS—Communicate with Editor-in-Chief; describe topic and intent of potential review articles prior to submission. Emphasize emerging fields or critically analyze existing data and questions. These reviews may be accompanied by scientific dialogue from other contributors.

OTHER

Experimentation, Animal: Experiments on unanesthetized animals must conform with standards for use of laboratory animals established by Institute of Laboratory Animal Resources, U.S. National Academy of Sciences. Justify experiments in which curariform agents are used. Give details as to steps taken to reduce or avoid distress to animal, particularly with regard to electrical stimulation.

Style: Follow recommendations of the Council of Biology Editors *CBE Style Manual,* 5th ed. (Council of Biology Editors, Bethesda, MD). Follow *Webster's New International Dictionary,* Third Edition, Unabridged (1986) for spelling and division of words. Nonstandard abbreviations should be minimal. Use numerals with standard units of measurement and for numbers above nine. Use metric system. Abbreviate according to International System *(Standard Metric Practice Guide,* American Society for Testing Materials, Philadelphia, 1970).

SOURCE

September 1994, 129(1): back matter.

Experimental Parasitology
Academic Press

Experimental Parasitology
Editorial Office, Third Floor
1250 Sixth Avenue
San Diego, CA 92101

AIM AND SCOPE

Experimental Parasitology publishes research papers, research briefs, and capsule reviews on topics that are at the experimental forefront of parasitology. Research papers and briefs are not solicited but should be submitted to the editors. Minireviews are invited and topics should be submitted to the Editors for consideration.

MANUSCRIPT SUBMISSION

Submit only original papers. Manuscripts must not have been and will not be submitted elsewhere. Submission must have been approved by all authors and by institution where work was carried out. Persons cited as a source of personal communications must approve citation; written authorization may be required. Copyright of accepted articles will be assigned exclusively to Publisher, who will not refuse any reasonable request by author for permission to reproduce any of his or her contributions.

MANUSCRIPT FORMAT

Write in clear, concise, and grammatical English. Type double-spaced with wide margins on standard size white paper (about 21.5×28 cm). Submit one original typescript and one reproduction set of figures along with three photocopies of manuscript and three reviewers' sets of figures. A reproduction set of figures consists of originals or glossy prints of halftones, cropped, labeled, and mounted on 10–12 point flexible board, and originals (in black India ink) or glossy prints of line drawings. Allow 1–2 mm space between mounted halftone parts. A reviewer's set of figures consists of mounted glossy prints of halftones and photocopies of line drawings.

Title: If possible, precede rest of title with specific parasite, followed by a colon. Put common or scientific name of host at end of title (e.g., *Neoaplectana glaseri*: Infectivity of Clones Reared in Species Isolation for Larvae of the Insect Weevil, *Hylobius pales*). For review manuscripts, precede title, as described above, by heading, in capital letters: A REVIEW. Make titles short; detailed qualifiers should appear in list of Index Descriptors. Suggest a short version of title as a running head.

Authors and Their Addresses: Include full professional mailing address of all authors with their names. When at different institutions, put different addresses after each author's name sequentially, or sectionally with identifying symbols. Use a footnote only if a change of address has occurred to indicate present address. Manuscript title, author's name and address, and running head are only material on page 1.

Research Paper Sections and Headings: Abstract, Index Descriptors and Abbreviations, Introduction, Materials and Methods, Results, Discussion, Acknowledgments (if any), References. Number pages consecutively from title-author page through references. Within each section, avoid subtitles. Justify departure from format in a separate letter.

Abstract: Make factual, not indicative. Do not use abbreviations.

Index Descriptors and Abbreviations: Key words and phrases follow abstract. Reflect contents of paper accurately and help to describe it to scientific nonexpert readers. Include taxonomic designations of all organisms discussed. Mention larger taxonomic categories (e.g., in a paper on *Watsonius watsoni*, the term "trematode" should appear as an index descriptor). Define all abbreviations used in text; use only abbreviation thereafter. Avoid lab jargon abbreviations; use only those that have general usage and genuinely save space.

References: Cite literature in text by author's surname and date (e.g., Jones and Smith 1987). Arrange reference list first alphabetically by name and secondarily by date. Each reference should consist of names of all authors, date of publication, exact title, full name (do not abbreviate) of journal, volume, and inclusive pages.

VINAGHI, R. A., VENTURIELLO, S. M., AND BRUSCHI, F. 1984. Resistance to cytotoxicity of *Trichinella* larva. *In* "Trichinellosis" (C.W. Kim, Ed.), pp. 42–46. The State Univ. of New York Press, Albany, NY.

CAULFIELD, J. P., AND CIANCI, C. M. L. 1985. Human erythrocytes adhering to schistosomula of *Schistosoma mansoni* lyse and fail to transfer membrane components to parasite. *Journal of Cell Biology* 101, 158–166.

Illustrations and Tables: Number all figures consecutively with Arabic numerals and give each one a descriptive legend. Use lowercase letters a, b, c, etc. for multiple parts of a single figure. Type legends double-spaced in a list at end of manuscript. Illustrations must be able to withstand photographic reproduction without retouching or redrawing. Have line drawings professionally drafted. Plan figures to fit the proportions of the printed page (5.5 × 8 in.; column width, 2 and 5/8 in.). Make lettering on drawings and gels large enough (10–12 points) to take a reduction of 50 to 60%.

Photographs should not be marred by staples, paper clips, or pencil marks. Crop figures to contain only relevant information. Areas of major interest should not be too close to edges of micrograph. Give magnifications in legends or place a magnification scale bar within micrograph and indicate dimension in legend. Labeling should be sans serif style, 3–3.5 mm high, in India ink or transfer letters. Explain symbols in micrographs in figure legends. Provide sharp photographs with good contrast since contrast is lost during reproduction. Match contrast and density of all figures on a single plate. An overlay indicating most critical areas of micrograph will help ensure high quality reproduction. Illustrations in color are accepted only if authors defray cost.

Type tables on separate pages, number consecutively with Roman numerals, and collect at end of manuscript. Give tables descriptive headings and make understandable without reference to text. Type table footnotes directly beneath tables.

When illustrations or tables refer to parasite or host, spell out name of organism completely at least once in each table heading or figure legend.

Acknowledgments: Include disclaimers, funding support, and other circumstantial information in Acknowledgments.

OTHER FORMATS

RESEARCH BRIEFS—Do not subdivide into sections; do not exceed one printed page.

REVIEWS—Format is identical with that for experimental papers up to and including Introduction. Divide body of review in a manner appropriate to subject matter. Acknowledgments (if any) and references follow body of review; pagination is continuous.

OTHER

Abbreviations: Keep to an absolute minimum; define in Index Descriptors and Abbreviations section.

Data Submission: Submit accession numbers from EMBL/GenBank or NBRF for nucleic acid and/or protein sequence information. GenBank Submissions, Mail STOP K710, Los Alamos National Laboratory, Los Alamos, NM 97545, telephone (505) 665-2177. Include following footnote to article title on manuscript page 1: The sequence data reported herein has been submitted to GenBank and assigned the accession number ___.

Nomenclature, Abbreviations, etc.: Consult *CBE Style Manual* (5th ed., 1983, Bethesda, MD: Council of Biology Editors) for principles of style and such details as appropriate abbreviations.

Whenever electron microscopy is the subject of reporting, the accepted unit is the micrometer, μm. The magnification of all electron and light micrographs should be indicated by a bar on the print.

Nomenclature, Enzymes: Use Enzyme Commission (EC) number for accurate identification and retrieval purposes in Index Descriptors and Abbreviations section. EC numbers may be found in: Recommendations (1984) of the Nomenclature Committee of the International Union of Biochemistry, *Enzyme Nomenclature* (1984, San Diego, CA: Academic Press).

SOURCE

August 1994, 79(1): back matter.

FASEB Journal
Federation of American Societies for Experimental Biology

Editor-in-Chief, Dr. W. J. Whelan, *The Faseb Journal* P.O. Box 016129, Miami, FL 33101-6129
University of Miami School of Medicine
Gautier Building, Room 317, 1011 N.W. 15th Street
Miami, FL 33136-1019
          Private courier address

AIM AND SCOPE

The aim of *FJ* is to illustrate the unity of biology and the interdependence of its constituent disciplines. Therefore, in keeping with this policy, and to qualify for acceptance, an original communication must not only be of outstanding scientific quality but must also be of fundamental interest.

FJ publishes two types of articles: brief, definitive, and essentially final research communications of fundamental interest and significance that are considered to warrant prompt publication; and state-of-the-art reviews, drawn, as far as possible, from the topics of the *Faseb* symposia.

The subject coverage of *FJ* is illustrated by the following disciplinary areas: biochemistry, biophysics, cell biology, developmental biology, genetics, immunology, neurobiology, nutrition, pathology, pharmacology, and physiology.

MANUSCRIPT SUBMISSION

Submit original and four copies, with figures and tables.

Authors understand that if accepted for publication, copyright of article, including right to reproduce in all forms and media, shall be assigned exclusively to publisher, who will not refuse any reasonable request by authors for permission to reproduce any of their contributions.

Certify submission has not been published other than as an abstract and is not being considered for publication elsewhere, and that it will not be submitted for publication elsewhere until its acceptability for *FJ* has been decided. If submission depends significantly on articles in press, send copies of those articles.

MANUSCRIPT FORMAT

Type double-spaced with 1 in. margins, on 8.5 × 11 in. bond paper. Computer printouts must be readable; dot-matrix printer is generally unacceptable. Do not exceed 5000 words, inclusive of illustrations and diagrams. Arrange pages and number consecutively in order: title page, footnotes, abstract (200 words) and indexing key words (five), text, references (double-spaced), figure legends, tables, and illustrations.

Begin with an abstract. Continue with an introduction followed by results and discussion; conclude with a succinct bibliography. Methods may be within figure legends and tables or as separate section.

Title Page: Title of article; author(s); laboratory or institution of origin with city and state or country; complete address for mailing proofs and telephone and fax numbers for corresponding author; and shortened title (50 characters and spaces) for running foot.

Title: Brief (90 characters, including letters, spaces, and punctuation) and informative. Do not use phrases in which more than three words modify another word. Do not use serial titles, except as a footnote.

Abstract: 200 word paragraph written for the general readership and free from jargon, self-explanatory and suitable for use by abstracting services. State purpose, major findings and conclusions of study. Avoid citing references; if used, include bibliographic information.

Footnotes: Double-space, assemble on one or more separate sheets; number consecutively throughout.

Text: Readable, clear and concise. Make corrections neat and legible. Use standard nomenclature; define unfamiliar or new items at first mention. Underline foreign words not in general use in English for italic type; do not use italics for emphasis. Do not use Latin plurals if English equivalent has been accepted.

Use active voice rather than passive voice. Use present tense for references to existing knowledge or accepted concepts, and proven conclusions from the present work; use past tense when describing experimental work on which paper is based.

References: Cite in text in numerical order; place numeral in parentheses. Type separately with inclusive pages and titles, double-spaced, with one reference per number. Verify accuracy and completeness of references.

Do not enter citations to unpublished work in reference list unless paper has been accepted for publication. Include in text as "(unpublished observations)" or "(personal communications)," with authors' initials and surnames.

For titles of journals, use abbreviations in *Serial Sources for the BIOSIS Data Base*. Use article titles and inclusive pages.

6. Fraker, P. J., Gershwin, M. E., Good, R. A., and Prasad, A. (1986) Interrelationships between zinc and immune function. *Federation Proc.* 45, 1474–1479.

For book references include in order: author(s), year of publication, title, pages, publisher, and city of publication. Underline book title for italic type. To cite a chapter give its title and page numbers and name book's authors or editors.

1. Dixon, M., and Webb, E. C. (1964) *Enzymes*, 2nd Ed, pp. 565–567, Longman Green, London

2. Hohmann, C., Antuono, P., and Coyle, J. T. (1988) Basal forebrain cholinergic neurons and Alzheimer's disease. In *Psychopharmacology of the Aging Nervous System* (Iversen, L. L., Iversen, S. D., and Snyder, S. H., eds) pp. 69–106, Plenum, New York

Illustrations: Keep to a minimum. Identify lightly with pencil on reverse with figure number and author name(s); mark TOP clearly. Refer to as figures in text, numbered with Arabic numerals; give each a legend.

When preparing figures, particularly graphs, see H. G. Hers (*Nature* 307: 205, 1984). Suggestions include:

Illustrations should be sharp, contrasty, unmounted photographs on glossy paper. Photographs should be one column (3.5 in.) or two columns (7.13 in.) width. Draw and letter all drawings for reduction to a given size to the same scale.

Do lettering in black ink and legible after reduction (i.e., at least 1.5 mm high). Make smallest elements (subscripts or superscripts) readable when reduced. Typewritten or computer-generated lettering is not preferred.

Prepare graphs such as electrocardiograms, kymograms, and oscillograms so that dark cross-hatched background is eliminated, faint portions of graphs are intensified, and sharp prints are obtained. Use blue-ruled recording paper for original records.

Denote figures with several panels with same axes, a, b, etc.: indicate on each panel its experimental specificity and label axes as precisely as possible. Express results in mol rather than cpm or absorbance units. If results are in percent, define 100% in standard units in legend.

If possible, keep all lettering within framework of illustration; place key to symbols on face of chart. Use one symbol for the same experimental conditions in all comparable figure. When figure is so filled that it is necessary to explain symbols in legend, use these standard characters: ❏ ■ ○ ● △ ▽ ▲ ▼ ▼ ×.

Give actual magnification of all photomicrographs. Editorial Office will correct for reduction. An appropriate scale on photomicrograph is preferable and more accurate.

Arrange with Editorial Office for reproduction of color illustrations. Authors pay full cost of color plates and their printing.

FJ reproduces figures and charts in the smallest size consistent with readability and purpose of illustration. Make recommendations for reduction or enlargement.

If illustrations have been published elsewhere, obtain and submit written permission from publisher and author.

Type figure legends double-spaced, consecutively on one or more sheets with sufficient information to provide adequate description without reference to text.

Tables: Type each double-spaced, on a separate sheet with a brief title and numbered with Arabic numerals. Place explanatory matter in footnotes. List table footnotes in order of appearance with consecutive superior letters.

Do not duplicate material in text or illustrations. Prepare for printing either 3.5 or 7.13 in. wide. Omit nonsignificant figures. Use short or abbreviated column heads. Identify statistical measures of variation as such.

Indicate approximate positions in margin of text.

Formulas and Equations: Arrange structural chemical formulas, process flow diagrams, and complicated mathematical expressions precisely and carefully and keep to a minimum. Glossy prints of complicated formulas and expressions suitable as line drawings are preferred. Identify all subscripts, superscripts, Greek letters, and unusual characters.

OTHER FORMATS

HYPOTHESES—These articles cater to hypotheses on topics of significant current interest that lie within the scope of the journal. They should conform to the style of the journal; include a 100-200 word abstract; 3000 words or equivalent in terms of illustrations and tabular matter and including bibliography.

STATE-OF-THE-ART REVIEWS—These research reviews originate in three ways. The Editorial Board selects from symposia presented at the annual Experimental Biology or other *Faseb* Society meetings cutting edge topics. The organizers of the symposia are invited to review the topic in a succinct form, emphasizing the growing points and directions of the research. Plenary lecturers receive an automatic invitation to record their lectures in the *FJ*. The Editorial Board also invites reviews on topics not currently covered by Society activities. Volunteered reviews are also welcomed that embody the principles of timeliness, topicality, and broad interest. Submit proposals for reviews, not a completed review, should be sent to the Editor-in-Chief, who will advise on acceptability.

OTHER

Abbreviations: Include, as a first page footnote, a list of any new or special abbreviations used, with spelled-out form and definition if necessary for clarity.

Drugs and Trade Names: Chemical or generic names precede drug name abbreviation at first use. Capitalize proprietary (trademarked) names and check spelling. Capitalize trade names of chemicals or equipment. Supply an

acceptable scientific name in every case as an alternative to the trade name. Do not use trade names in titles.

Experimentation, Animal: Follow "Guiding Principles in the Care and Use of Animals."

Experimentation, Human: Follow principles embodied in the Declaration of Helsinki.

Nomenclature: Keep chemical and biochemical terms and abbreviations in accordance with recommendations for usage by the International Union of Pure and Applied Chemistry (IUPAC), the International Union of Biochemistry (IUB), and their Committee on Nomenclature. See *Bio-*

chemical Nomenclature and Related Documents, a compendium of IUPAC-IUB documents, (The Biochemical Society, Essex, U.K.). For isotope specifications follow IUPAC system, with mass number placed as a superscript preceding the chemical symbol as ^{14}C. Italicize genotypes, not phenotypes. Identify enzymes with their EC number and recommended IUB name: see *Enzyme Nomenclature: Recommendations (1984) of the Nomenclature Committee of the International Union of Biochemistry* (Orlando, FL: Academic; 1984). For specialized fields, see "Glossary on respiration and gas exchange" (*J. Appl. Physiol.* 34: 549–558; 1973); "Glossary of terms for thermal physiology" (*J. Appl. Physiol.* 35: 549–558; 1973); "Glossary of terms for thermal physiology" (*J. Appl. Physiol.* 35: 941–961; 1973); *The ACS Study Guide: a Manual for Authors and Editors*, edited by J. S. Dodd and M. C. Brogan (Washington, D.C.: American Chemical Society; 1986); *A Manual for Authors of Mathematical Papers* (Providence, RI: American Mathematical Society; 1980); *Style Manual for Guidance in the Preparation of Papers for Journals Published by the American Institute of Physics and its Member Societies*, 3rd ed. (New York: American Institute of Physics; 1978).

Spelling: Use *Webster's New Collegiate Dictionary* (1977) for spelling, compounding, and word separation.

Style: Use *CBE Style Manual*, 5th ed. (1983, CBE Style Manual Committee, Bethesda, MD).

Units: Use metric units.

SOURCE

January 1994, 8(1): 139-142.

FEBS Letters

Federation of European Biochemical Societies
Elsevier Science B. V.

G. Semenza, Managing Editor
Swiss Federal Institute of Technology
Department of Biochemistry
ETH-Zentrum, Universitatstrasse 16
CH-8092 Zurich, Switzerland
Fax: 41 1 632 1089

M. Saraste, Reviews Editor
EMBL, Meyerhofstrasse 1,
Postfach 10.2209, D-69012
Heidelberg, Germany
Fax: 49 6221 387306

J.E. Celis
Department of Medical Biochemistry
Ole Worms Allé, Building 170
University Park, Aarhus University
DK-8000 Aarhus, Denmark
Fax: 45 86 131 160 (Editor of book reviews also)

M. Baggiolini
Th. Kocher-Institut
Universitat Bern, Freiestrasse 1
CH-3012 Berne, Switzerland
Fax: 41 31 631 3799

U.-I. Flugge
Lehrstuhl Botanik II, Gyrofstr. 15,

50932 Kolin, Germany
Fax: 49 221 470 5039

J.M. Gancedo
Instituto de Investigaciones Biomedicas, CSIC
Calle Arturo Duperier 4, 28029 Madrid, Spain
Fax: 34 1 585 4587

B. Halliwell, Pharmacology Group
King's College, Chelsea Campus
Manresa Rd., London SW3 6LX, England
Fax: 44 71 333 4949

J. Hanoune, Unite de Recherches
INSERM U-99, Hopital Henri Mondor
94010 Creteil, France
Fax: 33 1 48 98 09 08

P.M. Harrison
Department of Molecular Biology and Biotechnology
University of Sheffield
P.O. Box 594, Firth Court
Western Bank, Sheffield, S10 2UH, England
Fax: 44 742 728 697

M. Hatanaka
Institute for Virus Research
Kyoto University, Sakyo-ku
Kyoto 606-01, Japan
Fax: 81 75 751 3998

H. Holzer
Biochemisches Institut der Universitat
Hermann-Herder-Str. 7
D-79104 Freiburg i. Br., Germany
Fax: 49 761 203 5253

Th. L. James
UCSF Magnetic Resonance Laboratory
Department of Pharmaceutical Chemistry
926 Medical Science
San Francisco, CA 94143-0446
Fax: 1 415 476 0688

P. Jolles
Laboratory of Proteins/Enzymes/Glycoconjugates
CNRS/Universite de Paris V
45, rue des Saint-Peres, F 75270
Paris Cedex 06, France
Fax: 33 1 4015 9296

C. Klee
NIH, National Cancer Institute
Laboratory of Biochemistry, Bldg. 37
Room 4E-28, Bethesda, MD 20892
Fax: 1 301 402 3095

N. Mantel, Neurobiologie
ETH-Honggerberg, CH 8093
Zurich, Switzerland
Fax: 41 1 633 1046

P. Mathis
Dept. de Biologie Moleculaire et Cellulaire
Section de Bioenergetique, Bat. 532
C.E. Saclay, F91191 Gif-sur-Yvette, France
Fax: 33 1 69 08 87 17

A. D. Mirzabekov
Engelhardt Institute of Molecular Biology
Russian Academy of Sciences
Vavilov str. 32, 117984

Moscow B-334, Russia
Fax: 7 095 135 1405; e.mail: amir (@imb.msk.su.internet)

Sh. Mizushima
School of Life Sciences
Tokyo College of Pharmacy
1532-1 Horinouchi, Hachioji
Tokyo 192-03, Japan
Fax: 81 426 77 7497

M. Montal
Department of Biology, Science Building
University of California San Diego
9500 Gilman Drive, La Jolla, CA 92093,
Fax: 1 619 534 0931

M.J. Owen
Imperial Cancer Research Fund
P.O. Box 123, 44 Lincoln's Inn Fields
London WC2A 3PX, England
Fax: 44 71 269 3479

P.J. Randle
Nuffield Department of Clinical Biochemistry
Radcliffe Infirmary, Woodstock Rd.
Oxford OX2 6HE, England
Fax: 44 865 224 000

T.A. Rapoport
Max-Delbruck-Zentrum fur molekulare Medizin
Robert Rossle-Strasse 10,
D-13122 Berlin-Buch, Germany
Fax: 49 30 949 4161

S. Shaltiel
The Weizmann Institute of Science
Rehovot 76100, Israel
Fax: 972 8 465 488

V.P. Skulachev
A.N. Belozersky Laboratory of Molecular
Biology and Bioorganic Chemistry
Moscow State University, 119899 Moscow, Russia
Fax: 7 095 939 0338

A.V. Somlyo, Department of Molecular Physiology
and Biological Physics, University of Virginia
Box 449, 1300 Jefferson Park Ave.
Charlottesville, VA 22908
Fax: 1 804 982 1616

G. Tettamanti
Universita degli Studi de Milano
Facolta di Medicina e Chirurgia
Dipartimento di Chimica e Biochimica Medica
Via Saldini, 50, I 20133 Milano, Italy
Fax: 39 2 236 3584

M. Van Montagu
Laboratorium voor Genetika
Universiteit Gent, Ledeganckstraat 35
B9000 Gent, Belgium
Fax: 32 9 264 5349

G. Von Heijne
Dept. of Biochemistry, Arrhenius Laboratory
Stockholm University, S-106 91 Stockholm, Sweden
Fax: 46 8 15 36 79

F. Wieland
Institut fur Biochemie I

Im Neuenheimer Feld 328
D-69120 Heidelberg, Germany
Fax: 49 6221 564 366

Sh. Yamamoto
Dept. of Biochemistry
Tokushim University School of Medicine
Kuramoto-cho, Tokushima 770, Japan
Fax: 81 886 33 6409

AIM AND SCOPE

FEBS Letters is intended to be a journal for the fast dissemination of significant and novel work in an essentially complete form in the field of Biochemistry, Biophysics and Molecular Cell Biology. It is not the vehicle for preliminary or fragmentary observations, or for 'leftovers' from larger papers, or for material which, although sound, does not have to appear quickly, or is addressed to a small audience only. As a rule, for example, we do not publish incomplete sequences, or sequences which are known in other species; or incomplete NMR or other spectroscopic assignments; or the conventionally achieved expression of a gene in bacteria or yeast; or just negative observations. Also, methodological papers are not usually published by us, unless they are truly novel and significant. The overriding criterion is that a paper must be of sufficient immediate importance to justify urgent publication.

Subject coverage: protein chemistry, enzymology, biophysical chemistry, nucleic acids, protein synthesis, biochemical and molecular genetics, cell biology, metabolism, immunochemistry and natural products.

MANUSCRIPT SUBMISSION

Submit original manuscript and figures and two copies of both to Editor closest to field of interest (rather than geographically). Submit, if possible, a floppy disk of manuscript. Papers may be submitted in English, French, or German, with an abstract in English.

Include 3 self-addressed sticky labels and three extra copies of title page. Provide authors' fax, telex, and/or telephone numbers on title page. Indicate corresponding author with full address, telephone and fax numbers in a footnote on title page.

If paper is a dedication, part of a numbered series, or has been reported in part previously, indicate in a footnote on title page.

Submission implies that this has not been published in part nor is being considered elsewhere and that all authors agree to submission.

MANUSCRIPT FORMAT

On first page include title, authors' names and affiliations (with full address, telephone and fax numbers), abstract (100 words), 6 key words and appropriate footnotes. On subsequent pages, follow Introduction 1. with numbered sections: 2. Materials and Methods, or 2. Experimental; 3. Results; 4. Discussion. These may be combined. Acknowledgments precede References; do not include as footnote. Put Figure legends and tables on separate sheets after References.

For key words consult annual cumulative subject index of *Biochimica et Biophysica Acta* or subject headings list from *Index Medicus* for preferred synonyms and abbreviations. Use American spellings. Avoid plural terms. Do not use general terms.

Type double-spaced with wide margins on one side of paper. Printed version should not exceed 5 printed pages, including figure legends, tables, and references (3 double-spaced typed pages=1 printed page; 3 average figures plus legends=1 printed page; 30 references=.5 printed page, one full-length column).

References: Number in square brackets in order of citation in text. List names and initials of all authors, publication year in parentheses, journal title abbreviation, volume number and inclusive page numbers. Submit copies of acceptance letters for articles cited as 'in press.' For books include names and initials of authors, year of publication, title of book, name and initials of editors in parentheses, volume, edition, inclusive page numbers, and name and location of publisher. Do not underline. Do not use &. Keep author's names in lowercase preceding initials. Place citations as 'unpublished' and 'personal communication' in main text.

[1] Diller, T., Stockberger, M., Oesterhelt, D. and Tittor, J. (1987) FEBS Lett. 217, 297-304.

[2] Gomperts, B.D. and Fewtrell, C.M.S. (1985) in: Mechanisms of Transmembrane Signalling (Cohen, P. and Houslay, M.D. eds.) Mol. Aspects Cell Regul, vol. 4, pp. 377-409, Elsevier, Amsterdam, New York.

[3] Robeson, B. and Garnier, J. (1986) Introduction to Proteins and Protein Engineering, Elsevier, Amsterdam, New York.

Figures: Draw line figures, including graphs, in black ink on white paper. Machine letter to allow for reduction. Use graph paper with pale blue rule. Show grid lines in black ink. Design figures to economize space. Do not include half-tones in line drawings.

Submit nucleic and amino acid sequences as sharp clean prints legibly reduced to fit single- (8.4 cm) or double-column (17.6 cm) width. Do not use dot-matrix printer to produce. In reporting DNA sequences, do not present sequencing strategy. Confirm by statement in Materials and Methods that sequence was determined on both strands.

Submit half-tones as sharp, high contrast photoprints. Display relevant information only; avoid disproportionately large labeling of lines. For labeling, supply initial photographs mounted on white paper with lettering.

Use symbols: $\triangle \oplus \square \; \stackrel{\wedge}{\rightsquigarrow} \blacktriangle \blacktriangledown \blacklozenge \bullet \blacksquare \star \times +$. Other symbols may also be used. See FEBS Lett. 1994; 350:348 for whole list.

For preparing figures and graphs, see: H.G. Hers (1984) *Nature* 307,205. Label lines in graphs if more than one. On figures with several panels with same axes, indicate experimental specificity on each panel. Use same symbol for same experimental condition in all figures. When using several symbols, use open circle for control. Label axes precisely. If results are in percent, define 100% in standard units in legend. Keep legends short; restrict to title if possible. Figure should be intelligible without reference to legend.

OTHER FORMATS

HYPOTHESES—Submit 3 paper copies and diskette copy to Prof. M. Saraste. Present novel ideas or new interpretations of established observations based on sound data. Avoid excessive speculation.

MINIREVIEWS—Submit 3 paper copies and diskette copy to Prof. M. Saraste. 2000 words, 40 references, and 1-3 figures or tables, 100 word abstract and 4-6 key words. Make topical and of interest to specialists and general readers. Include sufficient information in Introduction to explain background to non-specialists. Use short descriptive subtitles.

OTHER

Color Charges: Dfl.1500.00 single printed full color page; Dfl 1200.00 two, three or four full color pages in combination.

Electronic Submission: Submit 3 copies of manuscript, floppy disc and list of nonstandard characters. Preferred storage medium is 5.25 or 3.5 in. disc in MS-DOS format. Do not split article into separate files. Format disc correctly; ensure that no other files are on disc. Replace non-reproducible characters with characters not used elsewhere; use consistently; state use of these non-standard symbols. See: *Chicago Guide to Preparing Electronic Manuscripts* (1987, Chicago: University of Chicago Press).

Experimentation, Animal: State steps taken to protect animals from pain.

Experimentation, Human: State steps taken to avoid unjustified risk or discomfort, including peer review of proposed investigation and informed consent by subjects.

Nomenclature and Abbreviations: Follow rules in 'Information to Contributors' to *Biochimica et Biophysica Acta* (1982, Amsterdam: Elsevier Science Publishers) and in *Biochemical Nomenclature and Related Documents* (1978, Colchester, England: Biochemical Society Book Depot). Expand abbreviations in footnote on title page.

SOURCE

August 22, 1994, 350(2,3): 347-348.

Details on preparation of manuscripts appear at end of each tenth volume (at end of Vol. 360, 370, 380, etc.).

FEMS Microbiology Letters

Federation of European Microbiological Sciences
Elsevier Science Publishers

Chief Editor, Professor C.A. Fewson
Department of Biochemistry
University of Glasgow
Glasgow G12 8QQ, U.K.

MANUSCRIPT SUBMISSION

Submit on disk with manuscript if possible. Send Research Letters to any Receiving Editor (addresses, telephone numbers, and fields of interest are listed in each issue) preferably to Editor who has closest field of interest. If necessary, papers may be sent to the Chief Editor.

For both Research Letters and Mini Reviews send: original typescript plus two copies; one set of original line drawings suitable for direct reproduction plus two copies; three sets of half-tone Figures; two copies of any papers cited as 'in press'; three sets of self-addressed sticky labels; telephone and fax numbers of corresponding author.

MANUSCRIPT FORMAT

Papers should preferably be in English (either British or American spelling), but may also be submitted in French or German. Type double-spaced (including references, Tables and legends for Figures) with wide margins, on one side only of good-quality paper. Consult a recent issue of the journal to see conventions and layout of articles, especially title pages and reference lists. Indicate beginning of each new paragraph by indentation. Number all pages consecutively.

Priority will be given to short papers. Most papers are three or four journal pages. Avoid repetition of information in the text and illustrations and excessively long reference lists. A journal page of only text has about 700 words; make allowance for headings, illustrations, etc. The normal maximum length is eight full pages of double-spaced text including references (but excluding title and authors' affiliations and legends to Figures) plus a combined total of six Figures and Tables. If text exceeds ten pages and/or combined total of Figures and Tables exceeds six, paper will be returned unless compelling reasons for unusual length are provided.

Title, Authors and Key Words: Do not submit part of a numbered series. Give a concise informative title followed by names of authors (with first or middle names in full and including all initials) and by name(s) and address(es) of laboratory(ies) where work was done. If work was done in more than one laboratory, indicate the one in which each author worked during the study. Provide name and full postal address, telephone and fax number of corresponding author in a footnote.

List three to six key words on first page. Consult subject indices of *FEMS Microbiology Letters* or list of subject headings from *Index Medicus* for preferred synonyms and standard abbreviations. Avoid plural terms. Important words and phrases may be in both title and key words. Do not use general terms unless qualified.

Abstract: 100 words that must be able to stand independently. Do not cite references. Avoid abbreviations, if used, define.

Introduction: State objectives; do not summarize results. Give sufficient information to explain background to non-specialists; do not do a comprehensive literature survey.

Materials and Methods: Provide sufficient detail to allow work to be repeated. Mention suppliers of materials only if this may affect results.

Results: State how many times experiments were repeated and whether mean or representative results are shown. Indicate variability statistically. If results are expressed as percentages, state absolute value corresponding to 100%. Avoid unjustified significant figures.

Discussion: Do not simply recapitulate Results. Combine Results and Discussion.

Acknowledgments

References: Number sequentially in order of citation in text and insert between square brackets. Reference list follows order of citation. Type double-spaced. References to journals contain names and initials of all authors, year of publication in parentheses, title of paper, and abbreviation of journal title according to *Index Medicus*. These are followed by volume number and first and last page numbers. Do not reference work 'in press' unless it has been accepted for publication; submit two copies. References to books include book title, publishing company and year of publications.

1 O'Donnell, C.M. and Edwards. C. (1992) Nitrosating activity in Escherichia coli. FEMS Microbiol. Lett. 95, 87–94.

2 Dinter, Z. and Morein, B. (1990) Virus Infections in Ruminants. 592pp. Elsevier, Amsterdam.

3 McCarthy, A.J. (1989) Thermomonospora. Bergey's Manual of Systematic Bacteriology (Williams, S.T., Sharpe, M.E. and Holt, J.G., Eds.), Vol. 4, pp. 2552–2572. Williams and Wilkins, Baltimore, MD.

Unpublished results and personal communications may be mentioned in text provided that names and initials of all persons involved are listed, and they have all granted permission for citation. For unpublished accession numbers for nucleotide sequences and similar information give sufficient details to allow retrieval of relevant information.

Tables and Figures: Select to illustrate specific points. Do not tabulate or illustrate points that can be adequately and concisely described in text. Type tables and their legends double-spaced.

Make line-drawn Figures black ink on white paper. Letter ready for direct reproduction. Figures can also be produced on a good-quality laser printer.

Design Figures to economize on space and fit dimension of page and column. Symbols and letters should be at least 1.5 mm high after reduction. See H.G. Hers (1984) *Nature* 307, p. 205, for suggestions about preparing Figures.

Submit half-tone Figures as very sharp, glossy prints with maximum contrast. Indicate magnification with a bar marker. Cut off non-essential areas of photographs. Replace photographs of electrophoretograms, etc. in which there is poor contrast by line drawings, but also submit photographs. If photographs have been digitally processed for quality, so state. Color photographs are not published unless essential for understanding of work and have been discussed with Chief Editor; authors bear additional costs of color printing and process may lengthen publication time.

Type legends for both line-drawings and half-tones double-spaced, on a separate sheet. Legends consist of a preliminary sentence constituting a title, followed by a brief description of the way experiment was carried out, and any other information describing symbols or lines.

OTHER FORMATS

MINI REVIEWS—These are concise but authoritative articles covering topics of current interest or controversial aspects of microbiology with emphasis on the same areas as Research Letters. The style is the same as for Research Letters, except maximum length is 3000 words (9-12 manuscript pages), a combined total of 6 Figures and Tables and 25 references. No rigid format, include a brief Introduction in which background is presented. Arrange rest of text under one or two levels of subheading, finishing with a Conclusions or Outlook section. Mini Reviews are normally invited but prospective authors are encouraged to discuss possible submissions with Dr. F.G. Priest, Department of Biological Sciences, Heriot-Watt University, Edinburgh EH14 4AS, U.K.

RESEARCH LETTERS—The Editors give priority to concise papers that merit urgent publication by virtue of their originality and general interest and their contribution to new developments in microbiology. All aspects of microbiology are covered. Areas of special interest include: molecular biology and genetics; microbial biochemistry and physiology; structure and development; pathogenicity medical and veterinary microbiology; plant-microbial interactions; applied microbiology and microbial biotechnology; systematics. Papers can deal with any sort of microorganisms: bacteria, filamentous fungi and yeasts, protozoa or viruses. Papers should be complete in themselves and adequately supported by experimental detail; they should not be preliminary versions of communications to be published elsewhere. Descriptions of new methods are acceptable and the Editors are also prepared to consider papers that put forward new hypotheses. However, papers that provide confirmatory evidence or merely extend observations firmly established in one species to another will not be accepted unless there are strong reasons for doing so.

OTHER

Electronic Submission: Preferred storage medium is a 5.25 or 3.5 in. disk in MS-DOS format although other systems are welcome, e.g., Macintosh (save file in usual manner, do not use 'save in MS-DOS format'). Submit disk and exactly matching printed version. Do not split article into separate files. Format disk correctly and have only article on disk. Specify type of computer and word-processing package used. Label disk with your name and name of file. Ensure that letter 'l' and digit '1' (also letter 'O' and digit '0') have been used properly. Format article (tabs, indents, etc.) consistently. Do not leave characters not available on your word processor (Greek letters, mathematical symbols, etc) open; indicate by a unique code. Use codes consistently throughout. Make a list of codes and provide a key. Do not allow word processor to introduce word splits. Do not use a 'justified' layout.

Experimentation, Animal: Perform in accordance with legal requirements of relevant local or national authority. Procedures should be such that animals do not suffer unnecessarily.

Experimentation, Human: Include a statement that the Ethical Committee of institution in which work was done has approved it, and that subjects gave informed consent to work.

Materials Sharing: Make new and variant organisms, viruses and vectors described in FEMS Journals available, upon written request and for their own use, to all qualified members of the scientific community. Indicate if delays in strain or vector distribution are anticipated or if they are available from other sources. Deposit important strains in publicly accessible culture collections. Refer to collections and strain numbers in text. Indicate laboratory strain designations and name and address of donor as well as original culture collection identification number, if any.

Nomenclature, Abbreviations and Units: Follow internationally accepted rules and conventions. See Nomenclature sections of current 'Instructions to Authors' of the *Biochemical Journal* and *Journal of Bacteriology*. Take care with genetic nomenclature. Use international system of units (SI); ml^{-1} is acceptable in place of cm^3 for liquid measures. Preferred form for units is μg ml^{-1} and not μg/ml, do not use brackets to improve clarity. Indicate multiplication by a multiplication sign with spaces on either side and of units by a space.

Nucleotide and Amino Acid Sequences: Fully determine nucleotide sequences in both senses of the DNA. Sequence information will be accepted for publication only if it is relevant to a question of more general interest, there is additional, complementary information, or there is some particular, explicit reason for urgent publication. Deposit all nucleotide and amino acid sequences in an appropriate data bank. Obtain an accession number before submission. Mention in cover letter. Use EMBL Data Library or other archives, such as GenBank. For complete information on how to submit data see: *Nucleic Acids Res.* 19, i-viii (1991) or contact: Data Submissions, EMBL Data Library, Postfach 10.2209, 6900 Heidelberg, Federal Republic of Germany (telephone: (06221) 387258; fax: (06221) 387306; telex: 461613 (embl d); electronic mail: DATALIB@EMBL-Heidelberg.de). Include accession number in appropriate Figure legends.

Nucleic acid and amino acid sequence figures must be as sharp and clean as possible and retain legibility after reproduction. Print with a letter-quality computer-printer, make character size of capital letters 2 mm after reduction and add additional markings clearly in black ink or with a computer drawing package.

Safety: Draw attention to chemical or biological hazards that may be involved in materials and methods used in experiments.

SOURCE

January 1, 1994, 115(1): v-xii.

Fertility and Sterility

American Fertility Society
Society of Reproductive Endocrinologists
Society of Reproductive Surgeons

Society of Alternative Reproductive Technology
Pacific Coast Fertility Society
Canadian Fertility and Andrology Society

Roger D. Kempers, MD
200 First Street SW, 505 Norwest
Rochester, Minnesota 55905
Telephone: (507) 284-3850; Fax: (507) 284-0780

AIM AND SCOPE

Original papers, contributed solely to *Fertility* and *Sterility* and limited to the fields of fertility, sterility, or the physiology of reproduction, will be considered for publication.

MANUSCRIPT SUBMISSION

Provide original and two copies. In cover letter identify corresponding author with address, telephone, and fax numbers. State that material has not been published, has not been submitted, or is not being submitted elsewhere. Published manuscripts become property of the journal, and copyright will be taken out in the name of The American Fertility Society.

All authors must qualify for authorship, having participated sufficiently in work to take public responsibility for content. No more than 6 authors is considered appropriate.

MANUSCRIPT FORMAT

Follow "Uniform Requirements for Manuscripts Submitted to Biomedical Journals." (See January issue for complete instructions). Manuscripts must adhere to format. A Checklist is available from Editorial Office.

Type double-spaced on 8.5 × 11 in. white bond paper with 1 in. margins. Begin each component on a separate page: Page 1, Running Title (40 spaces and letters); Page 2, Title page (full title, full name(s) of author(s), highest awarded academic degree, institutional affiliation(s), address(es), and acknowledgment of financial support-grant number, institution, and location); Page 3, Capsule (30 word summary of abstract and description of final conclusions; Page 4, Structured Abstract (200 words presenting study objective, design, setting, patients, interventions, main outcome measure(s), results, and conclusions emphasizing new and important aspects of study or observations and Key Words (3 to 10 key words or short phrases to assist indexers). Continue with Introduction, Materials and Methods, Results, Discussion, Acknowledgments, References, Tables and Legends.

Acknowledgments: Obtain written permission and name those who have contributed intellectually to paper, but do not justify authorship. Describe their function or contribution.

References: Number consecutively in order of first mentioned in text. If appear for the first time in tables and figures, number in sequence with those cited in text when table or figure is mentioned. References are identified by Arabic numerals in parentheses.

Do not use abstracts, unpublished observations, or personal communications as references; reference to written, not oral communications may be inserted (in parentheses) in text. Include among references papers accepted but not yet published; designate journal and add "in press." Cite information from manuscripts submitted but not yet accepted in text as "unpublished observations" (in parentheses). If a foreign manuscript is included, provide two copies of a translated English abstract. If author's(s') own papers are quoted as "In Press," provide two copies.

Verify references against original documents. Abbreviate titles of journals according to style in *Index Medicus*. Limit references to a maximum of 25 (not applicable for manuscripts submitted as Communications-in-Brief or Modern Trends.) List all authors, but if exceed six, give six followed by et al.

JOURNAL

1. Takihara H, Sakatoku J, Cockett ATK. The pathophysiology of varicocele in male infertility. Fertil Steril 1991; 55:861-8.

BOOK

2. Colson JH, Armour WJ. Sports injuries and their treatment. 2nd rev. ed. London: S. Paul, 1986.

3. Diener HC, Wilkinson M, editors. Drug-induced headache. New York: Springer-Verlag, 1988.

4. Weinstein L, Swartz MN. Pathologic properties of invading microorganisms. In: Sodeman WA Jr, Sodeman WA, editors. Pathologic physiology: mechanisms of disease. Philadelphia: Saunders, 1974:457-72.

LETTER

5. Kremer J. Yardsticks for successful donor insemination [letter]. Fertil Steril 1991;55:1023-4.

IN PRESS

6. Lillywhite HB, Donald JA. Pulmonary blood flow regulation in an aquatic snake. Science. In press.

Tables: Title, number in Arabic numerals in order of first citation in text, and type double-spaced on separate pages. Do not submit as photographs. Give each column a short heading. Place explanatory matter in footnotes, not in heading. For footnotes use symbols in sequence: *, †, ‡, §, ||, ¶. Do not use internal horizontal and vertical rules.

Illustrations: Have professionally drawn and photographed; freehand or typewritten lettering is unacceptable. Submit unmounted, clear glossy prints not to exceed 5×7 in. Make lettering and identifying marks clear and type size consistent. Use capital letters for specific areas of identification in a figure. Make symbols, lettering, and numbering distinctly recognizable so that when reduced each item will still be legible. Put titles and detailed explanations legends, not on illustrations. Place an adhesive label on back designating top, first author's name, and figure number. Submit at least two sets of illustrations. Do not send slides. There is a page charge for color photographs. One illustration and/or table is accepted for every 600 words (approximately 3 typewritten pages).

Legends: Type double-spaced in consecutive order on a separate page.

Permissions. Submit a written statement from copyright holder giving permission to *Fertility and Sterility* for reproduction of material from other sources.

OTHER FORMATS

BOOK REVIEWS—Send to: Steven J. Ory, M.D., 200 First Street SW, 505 NW, Rochester, MN 55905.

COMMUNICATIONS-IN-BRIEF, CASE REPORTS, AND TECHNIQUES AND INSTRUMENTATIONS—Three double-spaced typewritten pages, four references, and one illustration. Abstract is omitted in lieu of a brief Summary section. Limit Techniques and Instrumentations four double-spaced pages, four references, and three illustrations.

EDITOR'S CORNER—Brief manuscripts in editorial fashion on current social, political, and medical issues pertinent to reproductive medicine. Type double-spaced 1000-1500 words, four references.

LETTERS-TO-THE-EDITOR—400 word (5 citations) critical comments directed to a specific article recently published in the journal. Inclusion of illustrations is at the discretion of the editor. Send double-spaced letter in journal format to Paul G. McDonough, M.D., Department of Obstetrics and Gynecology, Medical College of Georgia, Augusta, Georgia 30912.

MODERN TRENDS—15 to 20 journal pages; do not overreference. Send to: Edward E. Wallach, M. D., Dept. of Gynecology and Obstetrics, The Johns Hopkins Medical Institutions, 600 North Wolfe Street, 264 Houck Bldg., Baltimore, Maryland 21205 (Telephone (301) 955-7800).

OTHER

Statistical Analyses: Quantify findings and present with appropriate indicators of measurement error or uncertainty. Define statistical terms, abbreviations, and symbols.

Units: Use metric units or decimal multiples for length, height, weight, and volume. Give temperature in degrees celsius. Report hematological and clinical-chemistry measurements in metric system in conventional units.

SOURCE

July 1994, 62(1): back matter.

Free Radical Biology & Medicine
Oxygen Society
Pergamon Press

Dr. Kelvin J.A. Davies
Department of Biochemistry & Molecular Biology
The Albany Medical College
New Scotland Avenue
Albany, NY 12208
Telephone: (518) 262-5315

AIM AND SCOPE

Free Radical Biology & Medicine is an international, interdisciplinary publication encompassing chemical, biochemical, physiological, pathological, pharmacological, toxicological, and medical approaches to free radical research. The Journal welcomes original contributions dealing with all aspects of free radical production, metabolism, damage, protection, and measurement in biological systems.

MANUSCRIPT SUBMISSION

Submission implies that work described has not been published before (except in abstract form or as part of a published lecture, review or academic thesis), that it is not under consideration for publication elsewhere, that corresponding author has ensured that publication has been approved by all authors and tacitly or explicitly by responsible authorities in laboratories where work was carried out, that all persons entitled to authorship have been so named, and that, if accepted, it will not be published elsewhere in the same form in either the same or another language, without consent of Editors and Publisher. Refer to previously published abstracts, etc. at end of Introduction.

Submit full-length papers in quadruplicate, typed double-spaced, with an abstract summarizing essential contents. Submit one set of camera-ready figures or photographs (original artwork or glossy prints). In cover letter include name, address, and phone number of corresponding author. Send copies of any published reports that may duplicate material in manuscript, and written permission of author(s) and publisher(s) to use any previously published material (figures, tables, or quotations of more than 100 words). Upon acceptance for publication, a copyright transfer will be sent. It must be signed and dated by all authors.

MANUSCRIPT FORMAT

Type on standard A4 or 8.5 × 11 in. white bond paper with broad margins. Use double-spacing throughout, including references and figure legends. Organize in order indicated below, with each component beginning on a separate page. Type a running title and page number in upper right corner of each page.

Title Page: Page 1: title of article (80 spaces); authors' full names (first name, middle initial(s), and surname); affiliations (name of department, if any, institution, city, and state or country where work was done), indicating which authors are associated with which affiliations; (acknowledgments of grant support and of individuals who were of direct help in preparing of study; name, address, and telephone number of corresponding author and, if different, reprint request author; and running title (30 spaces).

Abstract and Key Words: Page 2: title of article, followed by 200 word abstract, stating purpose of study, basic procedures, most important findings, and principal conclusions; emphasize new aspects of study. List up to 8 keywords or phrases for indexing.

Text: Type on one side of paper, double-spaced and with wide margins. Organize original contributions and brief communications: Introduction, Materials and Methods (or Experimental Procedures), Results, Discussion, Acknowledgments, Abbreviation List, References. A short Summary. Conclusions section may be inserted following Discussion section. Results and Discussion sections may be combined. Use other descriptive headings and subheadings if appropriate. Avoid jargon; spell out all nonstandard abbreviations on first mention; present contents of study as clearly and as concisely as possible.

References: Type double-spaced and number consecutively in order first mentioned in text, not alphabetically. Identify in text, tables, and legends by Arabic numerals typed as superscripts. Number those cited only in table or figure legends in accordance with sequence established by first mention in text of table or figure. Abbreviate journal titles according to *List of Journals Indexed in Index Medicus*.

JOURNAL

1. Davies, K. J. A.; Doroshow, J. H. Redox cycling of anthracyclines by cardiac mitochondria: I. anthracycline radical formation by NADH dehydrogenase. *J. Biol. Chem.* 261:3060–3067; 1986.

BOOK

2. Anderson, D. C.; Winter, J. S., eds. Adrenal cortex. London: Butterworth; 1985.

CHAPTER IN EDITED BOOK

3. Tappel, A. L. Lipid damage to membranes. In: Trump, B.; Arstila, A., eds. Pathobiology of cell membranes. Vol. 1. New York: Academic Press; 1975:145–172.

ABSTRACT

4. Davies, K. J. A.; Lin, S. W. *E. coli* proteases selectively degrade oxidized proteins. *Fed. Proc.* 45:1597 abstr. (1986).

Cite manuscripts accepted for publication may be cited in reference list using estimated year of publication. Submit copy of such manuscripts:

5. Winterbourn, C. C. The ability of scavengers to distinguish ·OH production in the iron-catalyzed Haber-Weiss reaction: comparison of four assays for ·OH. *Free Radic. Biol. Med.* in press; 1987.

List manuscripts that have been submitted (but not yet accepted) or that are "in preparation," in footnotes.

Footnotes: Indicate by *, †, etc. Type at end of reference list, keyed to appropriate page. Use for references to unpublished work (including work submitted for publication), personal communications, proprietary names of trademarked drugs, and other material not appropriately referred to in text or in numbered reference list. Keep to a minimum.

Tables: Keep each to a reasonable size; very large tables, packed with data confuse readers. Give each table and every column an explanatory heading, indicating units of measure. Do not include same data in both table and figures. Indicate footnotes by superscript, lowercase letters. Make tables and illustrations, with their footnotes or captions, intelligible without reference to text.

Figures: Submit illustrations about twice final size required. Have professionally drawn and photographed. Submit as glossy, high-contrast, black-and-white photographs. Include photographs of tissues, cells, or subcellular components only when essential. Label back of each figure to indicate article's title and TOP of figure. Do not write directly on back of photographs. Do not trim, mount, clip, or staple illustrations. Securely package all artwork in a protective envelope.

Figure Legends: Type double-spaced and number with Arabic numerals corresponding to illustrations and submit on a separate page. Explain symbols, arrows, numbers, or letters used to identify parts of illustrations in legend. For photomicrographs, define internal scale markers and give method of staining. Legends permit understanding of figures without reference to text. If figure has been previously published, include a credit line and supply a permission letter.

OTHER FORMATS

BRIEF COMMUNICATIONS—Peer-reviewed, brief studies that are reviewed rapidly (4 printed pages or less).

FORUM—Opinions on scientific questions of societal importance by authorities in the field. Often personal statements of views and may not be peer-reviewed.

HYPOTHESIS PAPERS—Current state-of-the-art reviews of significant frontier areas of free radicals in biology and medicine. Include an Introduction and a Summary or Conclusions section.

LETTERS TO THE EDITOR—Short (400 words) letters dealing with published articles or matters of interest to free radical researchers. Type double-spaced; include references where appropriate. Where a published article is involved, original author(s) are invited to respond.

REVIEW ARTICLES—Organize: Outline (use main and second-order section headings), Introduction, Text, Conclusions or Summary, Acknowledgments (optional), Appendices (optional), List of Abbreviations, References. Include a brief biography of each author (150 words total for all), focusing on educational background, research areas, honors, and personal interests.

THE RADICAL VIEW—Editorials and news of general interest.

OTHER

Nomenclature and Abbreviations: Keep stylistic details constant. For example, P450 is not typed P_{450}; electron spin resonance is abbreviated either ESR or EPR (for electron paramagnetic resonance). Either can be used; give both and state as equivalent at first mention. Formulae for radicals follow IUPAC recommendations and contain a superscripted (not centered) dot that precedes a charge, if any. Thus, superoxide is represented by $O_2^{-\cdot}$, not $O_2^{\cdot -}$, O_2^{-} $\cdot$ or some other permutation. Other examples are: $HO^{\cdot}$ or $^{\cdot}OH$ (not $OH^{\cdot}$), $RO^{\cdot}$, $ROO^{\cdot}$, $^{\cdot}NO_2$, $^{\cdot}CH_2OH$, etc. In text, use names of radicals, not formulas. For names of radicals, use alkoxyl, peroxyl, and hydroxyl; correct nomenclature requires the "L" on the end of radicals. Use , not t-, etc., for abbreviations.

Follow internationally agreed rules for nomenclature and abbreviations. When an enzyme or compound is first mentioned, specify its code number and its systematic name (as distinct from its trivial name). Official names of drugs are preferred to trade names.

Use standard three-letter codes for common amino acids without definition. Use one-letter codes only for comparisons of long protein sequences. Similar considerations apply to nucleosides and nucleotides. Use standard three-letter codes for carbohydrates and for purine and pyrimidine bases. Define all other abbreviations when first appear in text. If an extensive list of abbreviations is used

furnish an alphabetical list with definitions following references at end of article.

Temperatures denoted by an unqualified degree symbol are assumed to be Celsius. For solution strengths, express percentages by % sign, followed in cases of ambiguity by w/w, w/v, or v/v. Do not use periods following common abbreviations.

Reviewers: Submit names and addresses of five or six individuals who could expertly review manuscript.

SOURCE

July 1994, 62(1): back matter.

Gastroenterology
American Gastroenterological Association
W. B. Saunders

Nicholas F. LaRusso, MD, Editor, *Gastroenterology*
Mayo Foundation
200 First Street SW
Rochester, MN 55905
Telephone: (507) 284-9019; Fax: (507) 284-9184

AIM AND SCOPE

Gastroenterology publishes clinical and basic studies of all aspects of the digestive system, including the liver and pancreas, as well as nutrition.

Original articles are full-length reports of original research and will be considered for either the Alimentary Tract section or the Liver, Pancreas and Biliary section of the journal. Articles cover topics relevant to clinical and basic studies of the digestive tract, liver, and the pancreas. They may discuss nutrition, immunology, cell biology, molecular biology, morphology, physiology, pathophysiology, epidemiology, or therapy. Both adult and pediatric problems are included. To be published, the work presented in the manuscript must be original; on occasion, confirmatory studies of timely and important observations will also be acceptable.

MANUSCRIPT SUBMISSION

Submit four sets of manuscripts and figures. Do not submit more than one article dealing with related aspects of same study. In cover letter state reasons for deviations, if any, from standard format and provide clarification of any potential conflicts related to the exclusive nature of the publication. Outline (3-4 lines) novel or important aspects of manuscript.

Complete *Gastroenterology's* Copyright Assignment, Authorship Responsibility, Financial Disclosure, and Institutional Review Board/Animal Care Committee Approval form in every issue thereby affirming: that none of the material in the manuscript is included in another manuscript, has been published previously, or is currently under consideration for publication elsewhere (this includes symposia proceedings, transactions, books, articles published by invitation, and preliminary publications of any kind except an abstract of less than 400 words. If there is any potential overlap with a manuscript previously published by the authors, include related manuscripts.); that any commercial associations that might pose a conflict of interest in connection with the submitted article have been disclosed; that only people who contributed to intellectual content, analysis of data, and writing of manuscript are listed as authors and that all authors take public responsibility for the research results being reported; and that ethical guidelines were followed by in studies on humans or animals. Cite approval of institutional review board or animal care committee in Methods section. Copyright portion of agreement is valid only upon acceptance of article for publication.

MANUSCRIPT FORMAT

Follow "Uniform Requirements for Manuscripts Submitted to Biomedical Journals" of the International Committee of Medical Journal Editors (*N Eng J Med* 1991; 324: 424-428). Type manuscripts double-spaced throughout, including references, tables, figure legends, and footnotes, on 8.5 × 11 in. paper with 1 in. margins all around. Place page number and first author's name at top of each page. Arrange manuscripts: title page, abstract, introduction, materials and methods, results, discussion, references, figure legends, and tables. Begin each on a separate page. Attach full-page photocopies of all figures to back of each manuscript. Cite references, tables, and figures consecutively as they appear in text.

Title Page: Title: 120 characters; include animal species; do not use abbreviations. Short title: 45 characters. Authors: include first names of all authors and name and full location of department and institution where work was performed. Limit authorship to those who contributed substantially to study design, data analysis, or writing of the paper. Grant support: list all assistance. Correspondence: provide name, complete address, telephone and fax numbers for corresponding author.

Abstract: 200 words organized into: Background/Aims, Methods, Results, and Conclusions. Do not include abbreviations, footnotes, or references.

Introduction: Be concise, to the point, and state reason for and specific purpose of study. Carefully choose references to provide most important background information.

Methods: Describe in sufficient detail to permit evaluation and replication of work by others. Outline controls and steps taken to avoid observer bias. Outline statistical methods. Describe ethical guidelines followed in studies of humans or animals. Cite approval of institutional human research or animal welfare committee in text. Describe procedural or chemical hazards in detail, including safety precautions. Give manufacturer's name and city for any equipment or brand name drugs mentioned.

Results: State clearly and concisely. Include minimum number of tables and figures necessary for proper presentation. Do not include data appearing in tables in text.

Discussion: An exhaustive literature review is not necessary. Discuss major findings in relation to other published work. Explain differences between results of present study and those of others. Identify hypotheses and speculative statements. Do not restate results. Do not introduce new results.

References: Identify in text by Arabic numerals in parentheses. Number in order first cited in text. Do not put personal communications and unpublished data in reference list; mention in text. Abbreviate journal according to *Index Medicus*. Check for accuracy, completeness and nonduplication.

ARTICLE (LIST ALL AUTHORS):

13. Meltzer SJ, Ahnen DJ, Battifour H, Yokokota J, Cline MJ. Protooncogene abnormalities in colon cancers and polyps. Gastroenterology 1987;92:1174–1180.

BOOK:

18. Day RA. How to write and publish a scientific paper. 3rd ed. Phoenix: Oryx, 1988.

ARTICLE IN A BOOK:

22. Costa M, Furness JB, Llewellyn-Smith IF. Histochemistry of the enteric nervous system. In: Johnson LR, ed. Physiology of the gastrointestinal tract. Volume 1. 2nd ed. New York: Raven, 1 987: 1–40.

Figure Legends: Type double-spaced on pages separate from text. Number in order of appearance in text. Include enough information to permit interpretation without reference to text.

Tables: Type double-spaced on pages separate from text. Put table number and title above and explanatory notes below table.

Figures: Submit four complete sets of high-quality glossy or laser generated prints in four separate envelopes. Attach photocopies of figures to each manuscript copy. Use least number of figures required to convey message of paper. Do not duplicate data presented in tables or text. On back of each figure list first author's last name, figure number, and an arrow indicating TOP of figure typed on a self-adhesive label.

Make illustrations 1 column (3.3 in.) wide unless there is good reason to be larger, either 1.5 columns (5.25 in.) or 2 columns (6.8 in.) wide. Make labeling 1.5 mm high and no closer than 3 mm from margin. Designate special features by arrows. Make symbols, arrows, or letters in photomicrographs contrast with background. Do not mount multiple part figures; include suggested layout.

Authors must pay for printing color illustrations. Submit patients written permission to publish with photographs of identifiable patients.

OTHER FORMATS

BOOK REVIEWS—Informative analyses of recently published books relevant to clinical practice and research. Submit unsolicited reviews to Book Review Editor.

CASE REPORTS—Include case studies of four patients that describe a novel situation or add important insights into mechanisms or diagnosis or treatment of a disease.

CLINICAL CHALLENGES—Articles are case studies of clinical problems covering difficulties in diagnosis or management of patients with digestive diseases.

CORRESPONDENCE—Three pages, double-spaced and 10 references. Opinions on papers published in *Gastroenterology*. Include a copy of article.

EDITORIALS—Usually solicited by Editor.

RAPID COMMUNICATIONS—15 double-spaced, typewritten pages, including tables, figures, and references (two figures equal one page). Papers representing concise, original, and definitive studies of exceptional scientific importance will be considered as Rapid Communications upon request. This applies to basic scientific reports as well as clinical observations that point out new and important pathogenetic insights, diagnostic developments, or therapeutic advances.

SELECTED SUMMARIES—Concise overviews of articles recently published in other journals that are of potential interest to readers. Submit unsolicited summaries to Selected Summaries Editor.

SPECIAL REPORTS AND REVIEWS—In-depth, comprehensive, state-of-the-art reviews of important clinical topics solicited by the Editor. Reviews may be invited by the Editor or they may be unsolicited.

THIS MONTH IN GASTROENTEROLOGY—Written by special sections editor. Previews of specialized articles.

OTHER

Abbreviations: Use is discouraged; alphabetically list those not mentioned in Style Guide (found in every issue) on title page.

Drugs: Use generic names wherever possible. With trademarks, give manufacturers name and city.

Greek Letters: Use Greek letters over English equivalents.

Manufacturers: When specific scientific equipment or other products is cited, give manufacturer's full name, city, and state (or country) in parentheses. Capitalize and cite manufacturer's name and location.

Molecular Weight: Molecular weight is a pure number, defined as molecular weight ratio; it is not expressed in daltons. Dalton is a unit of mass equal to 1/12 the mass of one atom of carbon 12.

Names: Spell out chemical names and style according to *Merck Index*, 10th edition.

Reviewers: In cover letter state three to five referees and Associate Editor best qualified to review paper.

Radioisotopes: Follow recommendations adopted by the IUB Committee of Editors of Biochemical Journals. Place symbol for isotope in square brackets directly attached to front of name or formula labeled. Attach isotopic prefix to the part of the name to which it refers. Exceptions are: When native chemical or substance does not contain any isotope of the radiolabel, use hyphenated form, e.g., ^{131}I-albumin. When radiolabeled chemical or substance is not a specific chemical name, use hyphenated form, e.g., 3H-ligands, ^{14}C-steroids.

Use or omit square brackets in short chemical formulas or when isotope stands alone.

When known, indicate positions of isotopic labeling by Arabic numerals, Greek letters, or italicized prefixes (as appropriate) placed within square brackets and before isotope symbol.

See Instructions to Authors of *Biochimica et Biophysica Acta* and the *Journal of Biological Chemistry*, or in the IUPAC-CNOC Recommendations on Isotopically Modified Compounds (*Eur J Biochem* 1978;86:9–25).

Style: American Medical Association *Manual of Style*, 8th ed.

SOURCE

July 1994 107(1): before p. 1.

Gene
Elsevier Science Publishers B.V.

Prof. W. Szybalski
McArdle Laboratory for Cancer Research
University of Wisconsin
1400 University Ave
Madison, WI 53706
Telephone: (608) 262-1259; Fax: (608) 262-2824;
Email: Szybalsk @ macc. wisc. edu

AIM AND SCOPE

Gene is an international journal focusing on gene cloning and gene structure and function. *Gene* covers analysis of gene structure, function and regulation, including reconstruction of simple and complex genomes. Mechanisms of interaction between proteins and nucleic acids. Purification of genes by genetic, biophysical and biochemical methods, including recombinant DNA techniques.

Molecular mechanisms of genetic recombination, including site-specific, general and other recombinational events. Analysis and applications of restriction and modification systems. Basic and practical applications of genetic engineering and recombinant nucleic acid technology, including development of vectors and gene products, and cloning, mapping, sequencing and study of plant, animal, human and microbial genomes.

MANUSCRIPT SUBMISSION

Submission implies that manuscript is not being published or submitted elsewhere (including symposia volumes or other types of publication). If paper was already handled by another journal, include all pertinent correspondence, including referees' comments and authors' replies. Submit manuscript to an Editor with appropriate area of expertise. List of Editors with addresses, telephone and fax numbers, and areas of expertise are listed in the front-matter of every issue.

Corresponding author (preferably senior author with a permanent address) is responsible for all authors having seen and approved original paper and all of its following versions. Submit papers and figures in triplicate (one set of original figures ready for printer + two sets of photocopies, but three sets of half-tone photos of gels). Include a letter of transmittal, specifying names of five or more potential referees (including complete addresses and telephone numbers), plus any other special requests.

Send Editor-in-Chief: a copy of title page and summary; a copy of transmittal letter, total number of pages in paper (two figures/tables=one page), plus number of figures and tables, and six typed adhesive labels with address of corresponding author.

MANUSCRIPT FORMAT

Editorial style is extremely important. Papers should be in perfect English, conform to *Gene's* style, typed double-spaced and proofread by all authors. In text or legends, refer to specific section of paper, as 'see section a or b or . . .'. Number all pages. Explain all abbreviations first time used.

Paper should not be over 20-double-spaced, typewritten (6–10 printed) pages, including references, tables, figures and legends divided into Summary, Introduction, and Results and Discussion. Do not prepare a separate Materials and Methods section, include all pertinent experimental details in corresponding figure legends and/or table footnotes.

Title Page: Page No. 1. Double-space in order: brief descriptive title (lowercase); 6–10 key words (not already in title; no obscure abbreviations); author(s)' name(s) (lowercase; circle complicated or multiple family names); (short) address(es) and telephone number(s) of laboratory where work was performed (underline or italicize; specify country); corresponding author (Correspondence to: Dr. X. Y. Doe, complete mailing address, telephone and Fax numbers); single combined footnote with present address(es) and telephone number(s) of all author(s) who have moved; abbreviations (alphabetically; consult *Gene's* List of Abbreviations (in each issue) and pertinent literature, for accepted abbreviations, especially for genes and proteins; list all used, even if already in *Gene's* list; create new abbreviations if necessary).

Summary: (on separate page) Concise, description of purpose, results and specific conclusions of study. Explain abbreviations when first used. Use *Gene's* style for reference in Summary (give complete source of references).

Introduction: Short (≤ 1 typed page), specify background and aims of work (not another Summary).

Materials and Methods: Not a separate section; list experimental details together with specific experiments, in figure legends and/or tables footnotes, as either: a separate paragraph in legend and/or footnote entitled Methods; simply incorporate all pertinent experimental details into legends and/or footnotes. In rare cases, describe a specific novel method (especially in methodological papers) in a special detailed subsection in Results (or Experimental) and Discussion.

Experimental and Discussion, or Results and Discussion: Point to data in figures and tables, and briefly discuss significance. Divide into subsections (a), (b), (c), etc., with short subtitles (lowercase, Roman, not underlined). Do not prepare a separate Discussion section; subtitle last subsection, (x) Conclusions. Make it short and divide into points, (1), (2), (3). . .[see, e.g., *Gene* 100 (1991) 48–49, or 110 (1992) 5–6].

References: Assemble alphabetically. Type double-spaced. Refer to by name and year (Harvard System). Identify more than one citation by same author(s) and year by letters, a, b,

c, etc., after year of publication. In text to refer to a paper by more than two authors, with name of first author, followed by non-italicized 'et al.' When referring to a personal communication or unpublished paper, cite initials and name(s) of all author(s) andsubmit written permission from authors to your Editor (copy to Editor-in-Chief).

Westmoreland, B. C., Szyblaski, W. and Ris, H.: Mapping of deletions and substitutions in heteroduplex DNA molecules of bacteriophage lambda by electron microscopy and confirmation by sequencing reactions. Science 163 (1969) 1343–1348.

Rodriguez, R. L. and Tait, R. C.: Recombinant DNA Techniques, Addison-Wesley, Reading, MA. 1983.

Daniels, D.I., and Blattner, F. R.: A molecular map of coliphage lambda based upon PCR sequencing. In: Hendrix, R. W., and Weisberg, R. A. (Eds.). Lambda II. Cold Spring Harbor Laboratory, Cold Spring Harbor, NY, 1983, pp. 469–676.

Verify references. Abbreviate journal titles according to *List of Serial Title Word Abbreviations* [1985; International Serials Data System, CIEPS, Paris]. For reviews with a great number of references, see system used in *Gene* 100 (1991) 13–26.

Tables: Number I, II, III, etc. Place a short descriptive title at top. Type detailed footnote(s) identified by superscript, [a,b,c,] usually one for each major column, double-spaced under table.

Figures: (line drawings, including graphs) Prepare as laser-quality computer printouts. Send original printouts, not photos. If laser-quality printer is not available (dot matrix is unsatisfactory), prepare professionally drawn figures in black ink on white paper. Use large and bold lettering, and heavy smooth lines, to permit photographic reduction. Make all symbols large. Match explanations in legend. Make sequence figures either 60 nucleotides (or amino acids) in width (to fit single column), or 120–150 nt (or aa) wide (to fit across entire page). Make figures compact, but readable. Submit half-tone illustrations (e.g., gel photos) as black-and-white glossy prints with good contrast. Add bp, nt, kb or kDa symbols above numerals in marker lanes. Minimum size is 10×12 cm, maximum is 20×25 cm.

Type legends double-spaced on separate pages. Specify pertinent technical details in figure legends and/or table footnotes, with enough detail to permit repetition of experiments; when appropriate, add a bold subtitle, Methods. Add all details [temp., time(s), voltage, concentrations, type and % of gel, markers, sizes (bp, kb, kDa), etc.] to legends (for figures) or to footnotes (for tables).

OTHER FORMATS

BRIEF NOTES—6 double-spaced, typewritten (2 printed) pages short communication presenting a restriction map, new clones or vectors, or a sequence. Include a very short Summary, a brief text (not divided into sections), map or sequence with a short legend, including a short statement about standard methods used, and pertinent references.

Provide three copies of detailed experimental evidence. Include following footnote: 'On request, the author(s) will supply detailed experimental evidence for the conclusions reached in this brief note.'

LETTERS TO THE EDITOR/COMMENTARY—Send letters concerning papers already published in *Gene* to original Editor, Editor-in-Chief, and original author to give them the opportunity to reply.

MINI-REVIEWS—Generally solicited by Editors.

SHORT COMMUNICATIONS—12 double-spaced, typewritten (6 printed) pages divided into Summary, Introduction, and Experimental and Discussion. Submit papers representing new vectors or routine cloning, sequencing, and characterization of genes and neighboring regions or as Brief Notes.

OTHER

Data Submission: Deposit new nucleotide and amino acid sequences with GenBank before publication [GenBank, National Center for Biotechnology Information, Bldg. 38A, Room 8N-803, 8600 Rockville Pike, Bethesda, MD 20894; telephone: (301) 496-2475; fax: (301) 480-9241; email inquiries: info@ncbi.nim.nih.gov; email submission: gb-sub@ncbi.nim.nih.gov]. Include accession number with all sequences in figure legend.

Page Charge: Color illustrations are reproduced at author's expense (publisher provides a cost estimate upon submission).

Symbols: For proteins, use Roman letters with at least first letter capitalized; for genes (or DNA) use italicized letters. Consult nomenclature (or usage) for your particular organism to use (or create) proper gene/protein abbreviations. DNA, RNA, cDNA, clones and genes are expressed; proteins are produced or synthesized [e.g., use *cat* expression or CAT production (not CAT expression). Do not italicize mutant (allele) symbols or numbers. Arrange λ genes in same order as on map. Avoid common mistakes: use μ (not u); *Hin*dll (italicize only first three letters); k (not K) for 'kilo' (kDA, kb); Pollk (not Klenow); Ap^R (superscript, small capital ^{R}, not Ap^r).

SOURCE

December 30, 1994, 151 (1/2): after p. 345.

General and Comparative Endocrinology

American Society of Zoologists-Division of Comparative Endocrinology
European Society for Comparative Endocrinology
Asia and Oceana Society for Comparative Endocrinology
Japan Society of Comparative Endocrinology
Academic Press

Fran C. L. Moore
Department of Zoology
Oregon State University
Corvallis, Oregon 97331-2914
Telephone: (503) 737-5346; Fax: (503) 737-4720

Ian W. Henderson
Institute of Endocrinology
Department of Animal and Plant Sciences, Western Bank
University of Sheffield
Sheffield S10 2TN, United Kingdom
Telephone: 742-768-555 ext. 4672; Fax: 742-760-159

AIM AND SCOPE

General and Comparative Endocrinology will publish articles based on studies on cellular mechanisms of hormone action, and on functional, developmental, and evolutionary aspects of vertebrate and invertebrate endocrine systems. The journal will serve as a place for publication of fundamental research in endocrinology, but occasional brief reviews, dealing with particular fields or problems are invited.

Comparative endocrinology is the area in the field of regulatory biology in which one compares the structure, function, and mechanisms of action of endocrine glands, and their hormones either among various groups of vertebrate and invertebrate animals or as elements in adaptive mechanisms. *General and Comparative Endocrinology* will consider all manuscripts of high quality but first priority will be given to those which deal with comparative aspects of endocrinology. Manuscripts will also be considered favorably which do not include such comparisons if they deal with species other than laboratory mammals. Such manuscripts are of interest and contextual importance to comparative endocrinologists; few appropriate journals are available for their publication.

MANUSCRIPT SUBMISSION

Do not exceed 20 double-spaced typewritten pages; longer ones are considered if special circumstances justify. Original papers only are considered. The same work may not have been, will not be, nor is presently being submitted elsewhere. Submission must be approved by all authors and institution where work was done; and persons cited as sources of personal communications must approve such citation. Written authorization may be required.

If accepted for publication, copyright in article, including right to reproduce it in all forms, and media, shall be assigned exclusively to Publisher, who will not refuse any reasonable request by author for permission to reproduce any of his or her contributions to journal.

MANUSCRIPT FORMAT

Be concise and consistent in style, spelling, and use of abbreviations. Submit two copies, complete with all figures and tables. Type original copy double-spaced, on one side of white bond paper, about 8.5 × 11 in. in size, with 1 in. margins on all sides. Include a short title (35 letters and spaces). Indicate address to which proof should be mailed. Number all manuscript pages.

Abstract: A brief (3% of total length of paper), concise statement of results precedes text. Use descriptive subheadings in order: Methods, Results, Discussion, Acknowledgments, and References.

Footnotes: Avoid if possible; if essential, cite in consecutive order in text and indicate by superscript numbers.

Figures: Number consecutively with Arabic numerals in order of mention in text; give each a descriptive legend. Type legends together on a separate sheet, double-spaced. Submit in finished form ready for reproduction. Maximum size is 5 × 7.5 in. for double column and 2.5 × 7.5 in. for single column. Illustration in color are accepted if authors defray cost.

Photographs: Keep to a minimum; submit as glossy prints with strong contrast; indicate magnification by a scale where possible. Group photographs as plates.

Line Drawings: Make with India ink on tracing linen, smooth white paper, or Bristol board.

Tables: Type double-spaced on separate pages; number consecutively, with Arabic numerals in order of mention in text; and collect at end of manuscript. Give each a descriptive heading, typed above it.

References: Cite references to literature in text, e.g., as: Doe (1988) has observed that . . .; (Doe, 1986); (Doe *et al.,* 1987); (Doe, 1988, p. 250). Use suffices *a, b,* etc. following date to distinguish two or more works by same author(s) in same year; e.g., Doe, 1988a, b. Arrange literature citations in bibliography alphabetically according to surname of author. Order of items: author(s), date of publication, title of article or book, journal name, volume number, inclusive pages, publisher and city (for books only). Abbreviate journal titles in accord with *Chemical Abstracts Service Source Index, 1985.*

OTHER

Electronic Submission: Submit manuscripts on personal computer disks after manuscript has been accepted and revisions have been incorporated onto disk. Label disk with type of computer used, type of software and version number, and disk format. Tabular and mathematical material is typeset conventionally. Prepare art as camera-ready copy. Supply hard copy printout of manuscript that exactly matches disk file. File names must clearly indicate content of each file.

For further information on preparing disk, contact: Editorial Supervisor, Journal Division, Academic Press, 525 B Street, Suite 1900, San Diego, California 92101-4495; telephone: (619) 699-6415; fax: (619) 699-6800).

Experimentation, Animals: Avoid cruelty to animals by use of proper anesthetics. Certain experiments require use of unanesthetized animals, perform such experiments in a manner which assures a minimum of discomfort to animals.

Units: Abbreviate units of weight, measure, etc., when used in conjunction with numerals. Do not punctuate.

SOURCE

January 1995, 97(1): back matter.

Genes & Development
The Genetical Society of Great Britain
Cold Spring Harbor Laboratory Press

Cold Spring Harbor Laboratory Press
Box 100, 1 Bungtown Road
Cold Spring Harbor, NY 11724-2203
Telephone: (516) 367-8492; Fax: (516) 367-8532

AIM AND SCOPE

Genes & Development welcomes high-quality research papers of general interest and biological significance in molecular biology, molecular genetics, and related areas.

MANUSCRIPT SUBMISSION

Submit five copies; with at least four copies of original art. In a cover letter include: name, address, and telephone and fax numbers of corresponding author; statement that manuscript has been seen and approved by all authors; specific requirements for reproduction of art; and status of any permissions needed. Papers must present original research that has not previously been published. Submission implies that a paper is not currently being considered for another journal or book. Submit closely related papers in press elsewhere or that have been submitted elsewhere. Researchers must be prepared to make available to qualified academic researchers materials needed to duplicate research results (DNA, cell lines, antibodies, microbial strains, and the like). Submit nucleic acid and protein sequences to appropriate data bank.

MANUSCRIPT FORMAT

Be as concise as possible. Papers will occupy 5 to 10–12 journal pages. Short papers (5-8 pages) are encouraged. A manuscript of 28–32 typed, double-spaced pages total (including methods, references, tables, and figure legends), with 27 lines of text per page (63,000 characters), and with six single-column figures and one single-column table is 10–12 journal pages. Type entire paper (including tables, figure legends, references, footnotes) double-spaced on standard-sized European or American bond paper with at least 1 in. margins on all sides. Computer printouts should be letter quality with 11 point typeface. Label each page with first author's name and page number.

Preferred order: Title page, Abstract, Introduction, Results, Discussion, Methods, Acknowledgments, References, Tables, Figure legends.

On Title page include: title; all authors' full names; all affiliations; a shortened title for a running head (45 characters); and key words (up to 6) for indexing. A 200 word Abstract should summarize aim of report, methodological approach, and significance of results. Methods appear at end of paper. Make detailed enough to allow qualified researchers to duplicate results.

Figures and Legends: Supply five sets of figures as high-quality glossy prints. Half-tones should be high-contrast, particularly gels. Have line drawings, graphs, charts and chemical formula professionally prepared and labeled. Submit multiple-part figures as mounted, camera-ready composites. Authors pay publication costs of four-color artwork. Price estimates are supplied upon acceptance.

Label back of each figure with first author's name, figure number, and TOP. Number figures consecutively in order referred in text. Size is adjusted to fit Journal format;

keep labels, symbols, etc. in proportion to figure size and detail.

Make figure legends brief; do not include methods. Identify symbols in figure in legend text. If figures are reprinted from another source, submit permission to reprint.

Tables: Present tabular data concisely and logically. Number consecutively in order cited in text. Each should have a title. Use only horizontal rules. Make sure column headings are unambiguous in indicating columns to which they refer. Include table legends and footnotes where needed. If tables are reprinted from another source or include data from another source, submit permission to reprint.

References: Use name/date citations in text; do not cite by number. For undated citations (unpublished, in preparation, personal communication) include first initials and last names of authors. Supply bibliographic information in order. For journal articles: Author(s), year, article title, journal title, volume, inclusive page numbers. For books: Author(s), year, chapter title, book title, editors' names, volume, inclusive page numbers, publisher, city of publication.

OTHER

Page Charges: $25 per page. If unable to pay, submit a letter of explanation upon acceptance for publication.

SOURCE

January 1994, 8(1): after p. 132.

Genetics
Genetics Society of America

Genetics Editorial Office
P.O. Box 2427
Chapel Hill, North Carolina 27514
Telephone: (919) 942-3466

AIM AND SCOPE

Genetics accepts contributions that present the results of original research in genetics and related scientific disciplines. Although *Genetics* is an official publication of the Genetics Society of America, membership in the Society is not required of contributors: publication in the journal is open to members and nonmembers.

MANUSCRIPT FORMAT

Use American spelling; pay attention to correct grammar and punctuation. Submit three copies typed, printed or photocopied without blemish. Use 1 in. margins on all four edges. Use 8 × 11 in. paper of good quality and full weight. Double-space throughout, including footnotes, all parts of tables (especially boxheads), literature cited and appendices, as well as abstract and body. Number pages consecutively at bottom, beginning with cover page. If unpublished companion papers (whether submitted or in press) are referenced, include two copies of each.

Follow general usage in recent issues with particular regard to genetic symbols, references to literature and arrangement of Literature Cited. Check accuracy and clarity of symbols and abbreviations. Specify ambiguous letters

at first use, or always if typewriter does not provide a unique character. (Write "zero" above 0, "oh" above O, "one" above 1 and "el" above 1.) Identify Greek letters with a marginal note.

Provide title, names of authors and institutional affiliations on cover page. Make title compact and informative; indicate organism used. If different authors have different affiliations, key institutional addresses using *, †, ‡, and §; symbol follows comma after author name, but precedes address. Give compact institutional addresses; omit street addresses and districts.

Page 2: Provide a running head, a shortened version of title (35 characters including spaces). Provide five key words for indexing. Identify corresponding author with name, full mailing address (including street address and district), telephone and fax numbers, and e-mail address. Display mailing address line by line as on an envelope.

Abstract: 200 words in a single paragraph summarizing subject. Do not cite references.

Text: Follow abstract without repagination. Do not begin subdivisions (including acknowledgments) on new pages. Use four levels of heads in text. First level: centered, light-face full caps as in MATERIALS AND METHODS; use for blocks of text that constitute a major fraction of article. Second level is "freestanding flush-left bold face;" use infrequently to group two or more closely related third-level heads in long papers. Third level is "paragraph-initiating bold face" followed by a colon (most frequent subhead). Fourth level is "paragraph-initiating italic" followed by a colon. Do not number heads except for fourth level, and then only when numbers must be cited frequently in text.

Indicate desired locations of tables and figures in margin. In text citations with two authors include both names. Limit citations with three or more authors to first author and "*et al.*" Cite only articles that are published or in press. When citing personal communications or unpublished results, list all names, including initials; do not use "*et al.*"

Acknowledgments: Place immediately following Discussion; include all references to grant support and institutional publication codes. Write out names of granting agencies.

Literature Cited: Begin on new page. Cite only articles that are published or in press. Give citations of unpublished work (including authors initials) in text but not in Literature Cited: for instance (A. COLLEAGUE, personal communication) or (I. AUTHOR, unpublished results).

Citations follow format of a recent issue of *Genetics*. Names of individuals are printed in "full and small caps." (MENDEL, G.). Authors may type them thusly or in "full caps" (MENDEL, G.). Space between author's initials.

Put citations in alphabetical order by first author. For multiple citations with same first author, first list single-author entries by year using 1776a, 1776b (etc.) as needed. Then list two-author entries alphabetically by second author. Finally, list entries by three or more authors (cited in text as "FIRST *et al.* 1776") only by year and without regard to number of authors or alphabetical rank of authors beyond first; for papers with more than five authors, list five and then write "*et al.*"

JOURNAL ARTICLE

BRIDGES, C. B., and E. G. ANDERSON, 1925 Crossing over in the *X* chromosomes of triploid females of *Drosophila melanogaster*. Genetics **10**: 418–441. (Note spaces between authors' initials and after the boldfaced colon.)

BOOK

STURTEVANT, A. H. and G. W. BEADLE, 1939 *An Introduction to Genetics*. W. B. Saunders, Philadelphia.

CHAPTER IN BOOK

BEADLE, G. W., 1957 The role of the nucleus in heredity, pp. 3–22 in *The Chemical Basis of Heredity*, edited by W. D. MCELROY and B. GLASS. John Hopkins Press, Baltimore.

Appendix: Reserve for large bodies of data or arcane mathematical derivations that would disrupt main text. Do not use for items less than two typed pages.

Tables: Do not present tabular data as part of text; prepare numbered tables. Present very large arrays of tabular data as appropriately formatted figures. Type each on a separate page and give each a concise title. Double-space, especially boxheads. Conform to printed column or page size (3.25 or 7 in. wide); type very wide tables sideways or on oversize paper (which may consist of pasted sheets).

Number consecutively with Arabic numerals. Do not number consecutive tables 1A, 1B, etc., interior parts of a table can be labeled A, B, etc. for easy reference in text.

Define boxhead and bottom of table with horizontal rules. Use shorter horizontal rules within boxhead to indicate unambiguously which subheads are subordinate to a higher-level heading. Vertical and diagonal rules are not set in type. Give each column a title in boxhead; make each boxhead entry refer to underlying material and not to material to its right.

Indicate table footnotes with lowercase, superscript italic letters; type directly below table, not on footnote page. Use *, **, and *** to indicate conventional levels of statistical significance; explain below table. Use table legends in lieu of or in addition to, but preceding, footnotes.

Do not include line drawings or other illustrations within tables unless they clearly can be set in type. Prepare a separate, numbered figure to accompany table.

Figure Legends: Type double-spaced together on a new page. Begin legends with a brief title leading into unparagraphed text. Use conventional symbols to indicate figure data points.

Footnote Page: Place all footnotes on a final footnote page. Exclude symbols linking authors to institutions; these are not footnotes. Place footnotes to tables directly below each table. Order of footnotes: Dedicatory footnotes, unnumbered and printed in bold face; Leave several lines of space for a corresponding author's address (added by Editorial Office from information on page two; Numbered footnotes, beginning with present-address footnotes; Key numbered footnotes with superscript Arabic numbers. Footnotes in text are distracting to reader and should be avoided.

Figures: Number consecutively using Arabic numerals. Do not number consecutive figures 1A, 1B, etc.; label parts A, B, etc., if necessary for reference in text. If a figure is submitted in unattached parts, include a sketch of arrangement of printed version. Lightly write name of author(s) and figure number on back of each figure. Indicate UP with an arrow on back of each figure.

Submit halftones as high-contrast, glossy prints with sharp detail; some loss of detail and contrast is inevitable in printing. Avoid color photographs. Good contrast is particularly important for chromatographs, such as gel separations. Indicate important aspects of photographs with arrows or numbers; remove unnecessary parts. Trim photographs to be grouped; printer will cut in thin spacing. Make size and proportion of each group suitable for reproduction at 3.25 or 7 in. wide. Include three prints of each photograph; do not submit photostatic copies.

Treat drawings, graphs, mating-type charts, complex chemical formulas and other sketches as numbered figures. Use italics for symbols that are normally italicized such as genotypes, chromosomes, and first three letters of names of restriction nuclease sites; do not underline Roman characters. Italicize all mathematical variables. Distinguish between similar characters, such as letter l and number 1 or letter O and number 0. Be sure that axes of graphs are exactly perpendicular, that all lines are of regular density and that sizes of all numbers and letters in a graph are appropriate. Typed or dot-matrix graphs are generally inadequate, laser-printed graphs are encouraged. Make drawings with pen and undiluted India ink or equivalent; make 50% larger than anticipated printed size. Submit original drawings, not photographic reproductions.

OTHER FORMATS

LETTERS TO THE EDITOR—Letters deal with research and theory in basic genetics or with social issues of particular interest to geneticists. Submit constructive comments on subjects of papers from recent journal issues. Avoid figures, complex tables and complex mathematical formulas.

PERSPECTIVES—Consult editors James F. Crow Genetics Department) and William F. Dove (McArdle Laboratory), University of Wisconsin, Madison, Wisconsin 53706.

OTHER

Abbreviations: Abbreviate customary units of measurement only when preceded by a number. Write percent as one word, use % with a number. Indicate temperature in centigrade (37°): include a following letter only when some other scale is intended (45°K).

Binomial Names: Italicize names of organisms only when species is indicated. Italicize first three letters of names of restriction sites (*Hin*dIII). Write names of strains in Roman except when incorporating specific genotypic designations.

Electronic Submission: Disk should contain only text of paper, omit files of tables. Do not send disk with original submission. Provide both a disk and three printed copies after revision; describe hardware and software (plus version number). Do not include numerous complex mathematical expressions. Illustrations on diskette may be submitted. Provide color electronic images in Encapsulated PostScript (EPS) format, black and white illustrations in Tagged Image File Format (TIFF) sized to appropriate column width. Submit original artwork with disk.

Genotypes: Italicize genotype names and symbols (including chromosomes and all components of alleles, but not "+" indicating wild type, and not when name of a gene is same as name of an enzyme). Distinguish between genotype and phenotype in writing and symbolism. If italics cannot be printed, indicate by underlining except in figures. Distinguish between phenotype and genotype for readers not well versed in subject. Italicize or underline genotypes in typed sections; italicize genotypes in figures. Genotype consists of chromosome names or numbers, gene names, allele designations in short, all elements of their proper names. Present strain names and linkage groups in Roman unless strain designation is merely genotype.

Materials Sharing: Authors describing unique research materials agree to provide them at reasonable cost to colleagues. Examples of unique research materials are strains, gene clones, antibodies (including cell lines producing monoclonal antibodies) and computer programs. A colleague is any active investigator, Donor may require recipient to agree neither to use such materials for commercial purposes nor to transfer them to a third party without consent of donor.

Numbers: In text, write out numbers nine or less except as part of date, fraction decimal, percent or unit of measurement. Use Arabic numbers for those larger than nine, except as first word of a sentence; avoid starting a sentence with such a number.

Page Charge: $30 per printed page if a computer disk is supplied or $35 if it is not. No paper will be refused because of an author's inability to pay page charges. Authors citing grant support are assumed able to pay charges. Authors who lack funds should explain their situation when submitting papers and provide a computer disk. No charge for color plates or special processing required for data presentation, reprint charges may be slightly higher.

Sequences: Enter DNA, RNA or protein sequences corresponding to ≥50 nt into an appropriate data bank; provide accession numbered before publication.

Long sequences, more than about two pages, will not be published unless reviewers and editor agree it is necessary. Indicate that an unpublished sequence will be made available on disk or as hard copy for anyone requesting it. Authors wishing to include a long sequence not recommended for publication must pay a page charge (in excess of $100 per page).

SOURCE

January 1995, 139(1): front matter.

Genomics
Academic Press

Genomics, Editorial Office
525 B Street, Suite 1900
San Diego, California 92101-4495
Telephone: (619) 699-6469; Fax: (619) 699-6859

AIM AND SCOPE

Genomics publishes original articles on the comparative analysis of the hierarchical organization and evolution of the human and other complex genomes. The editorial philosophy emphasizes the integration of basic and applied research in human and comparative gene mapping, molecular cloning and large-scale restriction mapping, and DNA sequencing and computational analysis. *Genomics* welcomes full-length articles, short communications, and brief reports and invites timely reviews and editorial commentary. Selected relevant book reviews are also included.

MANUSCRIPT SUBMISSION

Send original and three copies of manuscript, including four sets of figures. State which Associate Editor authors think appropriate. Suggest possible reviewers (up to five).

Submit only original papers. The same work must not have been published and must not be under consideration for publication elsewhere. Submission must have been approved by all authors and by institution where work was done. Persons cited as sources of personal communication must approve citation. Written authorization may be required.

If accepted for publication, copyright for article, including right to reproduce article in all forms and media, shall be assigned exclusively to Publisher, who will not refuse any reasonable request by author for permission to reproduce any of his or her contributions to journal.

MANUSCRIPT FORMAT

Type manuscripts (point size, 12), double-spaced throughout, on one side of good-quality 8.5 × 11 in. white paper, with 1 in. margins on all sides. Number all pages consecutively, including references, tables, and figure legends.

Specify on page 1 Subject Category of article: Reviews and mini-reviews; New technology and resources; Chromosome structure; Gene structure and gene organization; Gene structure/function relationships; Markers and polymorphisms; Linkage and genetic maps; Physical maps; Expression maps (cDNA); Gene discovery and disease genes; Disease mutations; Comparative genomics; or Special features.

Page 1 also contains article title, author(s), complete affiliation(s), a short running title (50 character abbreviated form of title), and name, complete mailing address, and telephone and fax numbers of corresponding author.

Page 2 contains an abstract (150 words), a brief but informative summary of paper's contents and conclusions. Make intelligible to nonspecialists; avoid specialized terms and abbreviations or symbols that require definition. Organization of paper: Abstract, Introduction, Materials and Methods, Results, Discussion, Acknowledgments, and References.

In Methods section, describe experimental procedures in sufficient detail to enable work to be repeated. Give references to other papers describing techniques. Give correct chemical names; specify strains of organisms. Mention names and addresses for suppliers of uncommon reagents or instruments. Draw attention to chemical or biological hazards that may be involved in carrying out experiments. Dscribe relevant safety precautions; if an accepted code of practice has been followed, reference relevant standards.

References: Arrange reference list in alphabetical sequence by author's last name. Cite references in text by author and date, for example, (Doe *et al.,* 1989). Include only articles that have been published or are in press in references. Cite unpublished results or personal communications as such within text. Abbreviate journal names according to *Chemical Abstracts Service Source Index,* 1985.

Messing, J. (1983). New M13 vectors for cloning. *In* "Methods in Enzymology" (R. Wu, L. Grossman, and K. Moldave, Eds.), Vol. 101, pp. 10–89, Academic Press, New York.

Ray, A, Cheah, K-C., and Skurray, R. (1986). An F-derived conjugative cosmid: Analysis of *tra* polypeptides in cosmid-infected cells. *Plasmid* **16:** 90–100.

Rodriguez, R. L., and Tait, R. C. (1983). "Recombinant DNA Techniques: An Introduction," pp. 1–3, Addison-Wesley, Reading, MA.

Tables: Number consecutively with Arabic numerals in order of appearance in text and collect at end of manuscript. Type double-spaced on separate pages with short descriptive titles typed directly above table and with any necessary footnotes (indicated by superscript lowercase italic letters) typed directly below. Complex tables may be submitted as 7 1/8 × 9 in. glossy prints for camera-ready reproduction; type single-spaced. Submit extensive lod score tables with manuscript for review purposes; archive in LODSOURCE. They will not be published as part of article.

Figures and Illustrations: Submit figures so as to permit photographic reproduction without retouching or redrawing; cite consecutively in text with legends typed double-spaced on a separate page. Identify all figures on back lightly in soft pencil with author(s) name(s) and figure number; indicate TOP. Plan figures to fit proportions of printed page (7 1/8 × 9 in.). Have artwork professionally drafted or generate by high-resolution computer graphics. Make lettering large enough (10-12 points) to be reduced 50-60%. Make symbols used to identify points within a graph large enough to be distinguishable after reduction. Figures should be no larger than 8.5 × 11 in. or provide as glossy prints. Ink grid lines to be shown in black.

Submit halftone photographs on glossy paper with rich contrast. Do not submit photocopies for review. Submit only those parts of photograph that are necessary to illustrate matter under discussion. Submit photographs as single prints; do not mount on cardboard.

Color illustrations may be accepted if authors defray cost.

Footnotes: Identify footnotes in text by superscript Arabic numerals; type on a separate sheet; number footnotes to

author(s) name(s) and affiliation(s) in sequence with text footnotes.

OTHER FORMATS

BRIEF REPORTS—One printed page (four double-spaced typed pages, including references). Ideal for single gene mapping by study of somatic cell hybrids, *in situ* hybridization, and genetic linkage and descriptions of molecular nature of allelic variants, normal and pathologic. Submit figures and tables for review; submit only one figure or table. No abstract or section headings except Acknowledgments and References. Arrange references in alphabetical order by author's last name and number in that order. Cite references in text by Arabic numerals, on line, in parentheses. Incorporate main conclusion of research into title. For example, use titles "Assignment of the human insulin gene (INS) to 11p15.5 by *in situ* hybridization" and "Detection of a missense mutation (G-to-C) in exon 10 of the cystic fibrosis gene," not "Chromosomal assignment of the human insulin gene" or "Detection of a point mutation in the cystic fibrosis gene."

REVIEWS—Concise, highly focused review articles summarizing recent progress in very active areas of research involving analysis of genomes. Submit a short proposal, with expected date of submission, to Editor-in-Chief

SHORT COMMUNICATIONS—Three printed pages (six double-spaced typed pages, including references). Do not use section headings except Acknowledgments and References. Include an abstract. Two tables or figures. Use same reference style as Brief Reports. Succinct reports of important findings. Section is designed to offer opportunity to present in brief form noteworthy results of work in progress, confirming reports, and essential descriptions of original data submitted for peer review and archiving by genetic databases.

SPECIAL FEATURES—Minireviews, special reviews of research programs for mapping and sequencing, historical sketches, genomlcs news, etc., are published periodically.

OTHER

Abbreviations: Spell out at first occurrence in text.

Data Submission: Complete data entry and annotation forms for appropriate database(s). General information about and addresses are available from the Editorial Office for: GenBank/EMBL, Genome Data Base (GDB), GBASE (genetic database of the mouse), and LODSOURCE (lod score tables for pairs of genetic markers, including disease loci and DNA polymorphisms). Provide copies of data entry forms or other verification of submission and, when available, pertinent accession numbers. Published articles include a footnote indicating database to which data have been supplied and accession number. Deposit of data in appropriate database(s) is a condition of publication. Submit full sequence data, extensive lod score data, and other supporting data to assist review process and deposit in appropriate database.

Electronic Submission: Submit manuscripts on personal computer disks after manuscript has been accepted and after all revisions have been incorporated. Label disk with type of computer used, type of software and version number, and disk format. Tabular and mathematical material is typeset conventionally. Prepare art as camera-ready copy. Supply an exactly matiching hard copy printout. File names must indicate content of each file. For further information on preparing disk for typesetting conversion and a list of generic codes, contact publisher (Editorial Supervisor, Journal Division, Academic Press, 525 B Street, Suite 1900, San Diego, California 92101-4495; telephone: (619) 699-6415; fax: (619) 699-6800).

Nomenclature: Use HGMW-approved nomenclature for human genes. Obtain approved gene symbols prior to submission from P. J. McAlpine (Nomenclature Editor, HGMW Nomenclature Committee), Department of Human Genetics, University of Manitoba, T250-770 Bannatyne Avenue, Winnipeg, Manitoba, Canada R3E 0W3; telephone: (204) 789-3393; fax: (204) 786-8712; e-mail: andersow@lgenmap.hgen.umanitoba.ca.

Obtain approved mouse nomenclature prior to submission in United States and Canada from M. T. Davisson, The Jackson Laboratory, Bar Harbor, Maine 04609-0800; fax (207) 288-8982; e-mail: mtd@morgan.jax.org; and in all other countries from J. Peters, MRC Radiobiology Unit, Genetics Division, Chilton, Didcot, Oxon, U.K. OX1 ORD; fax: 44-235-834-918; e-mail: mlfl@uk.ac.rl ib.

Obtain D-numbers for cloned segments from Genome DataBase (M. Chipperfield, telephone: (410) 9550884; fax: (410) 955-0054; e-mail: chipper@library.welch.jhu.edu).

Statistical Analyses: In giving the frequency of polymorphisms, e.g., RFLPs, state number of chromosomes assayed in unrelated persons and ethnic background of subjects.

Symbols: Use SI units and follow guidelines for abbreviations and symbols of the IUPAC-IUBMB Joint Commission on Biochemical Nomenclature.

SOURCE

January 1, 1995, 25(1): back matter.

Gut
British Society of Gastroenterology
BMJ Publishing Group

Editor, *Gut*
BMA House, Tavistock Square
London WC1H 9JR, U.K.

AIM AND SCOPE

Gut publishes original papers, short rapid communications, leading articles, and reviews concerned with all aspects of the scientific basis of diseases of the alimentary tract, liver, and pancreas. Case reports will only be accepted if of exceptional merit. Letters related to articles published in *Gut* or with topics of general professional interest are welcomed.

MANUSCRIPT SUBMISSION

Submit two copies of manuscript and figures. Follow Vancouver conventions (see *BMJ* 1979, i: 532–5. *Gut* 1979; 20: 651–2). Use of abbreviation is discouraged. In cover letter signed by all authors state that data have not been published elsewhere in whole or in part and that all authors agree to publication in *Gut*. Disclose previous publication in abstract form in a footnote. Papers must not be published elsewhere without prior permission.

MANUSCRIPT FORMAT

Type double-spaced on one side of paper only. On title page include name of author with initials or distinguishing first name only, and name and address of hospital or laboratory where work was performed. Include a precise summary (200 words).

Illustrations: Provide unmounted photographs on glossy paper. Do not insert in text. Mark on back with figure numbers, title of paper and name of author. Refer to all photographs, graphs, diagrams as figures and number consecutively in text in Arabic numerals. Type legends for illustrations on a separate sheet.

References: Follow Vancouver system. Number references consecutively in text and list numerically with journal titles abbreviated in style of *Index Medicus*. List up to six authors, then add *et al*.

OTHER FORMATS

SHORT RAPID COMMUNICATIONS—10 double-spaced A4 pages including references, tables, and figures. In cover letter request paper be considered in this category and give valid reasons.

OTHER

Experimentation, Human: See Medical Research Council's publications on ethics of human experimentation, and World Medical Association's code of ethics, known as the Declaration of Helsinki (See *BMJ* 1964; ii: 177).

Reviewers: Submit names and addresses of four experts whom authors consider suitable to peer review work.

Units: Express all measurements except blood pressure SI units. Give values in tables and illustrations in SI units. For guidance and useful conversion factors, see *The SI for Health Professions* (WHO, 1977).

SOURCE

January 1994, 35(1): inside back cover.

Gynecologic Oncology
Society of Gynecologic Oncologists
Society of Gynecologic Oncologists of Canada
Felix Rutledge Society
Western Association of Gynecologic Oncologists
The Society of Memorial Gynecologic Oncologists
New England Association of Gynecologic Oncologists
Mid Atlantic Gynecologic Oncologic Association
Academic Press

Gynecologic Oncology, Editorial Office
525 B Street, Suite 1900
San Diego, California 92101-4495

AIM AND SCOPE

Gynecologic Oncology, An International Journal, will provide an international archive devoted to the publication of clinical and investigative articles that concern tumors of the female reproductive tract. Reports of investigations relating to the diagnosis and treatment of these diseases, their etiology and histogenesis will be considered. Such contributions may come from any of the disciplines related to this field of interest, including gynecologic surgery, radiotherapy, pathology, cytology, endocrinology, chemotherapy, immunology, genetics, cell kinetics, cell biology, and epidemiology. All aspects of scholarship related to tumors of this region are welcome, with clarity, quality, and originality the chief criteria for acceptance.

MANUSCRIPT SUBMISSION

Submit only original papers. The same work may not have been published or be under consideration for publication elsewhere. Submission must have been approved by all authors and by institution where work was done; persons cited as sourcse of personal communications must approve citation. Written authorization may be required.

If it is accepted for publication, copyright in article, including right to reproduce article in all forms and media, shall be assigned exclusively to Publisher, who will not refuse any reasonable request by author for permission to reproduce any of his or her contributions to journal.

MANUSCRIPT FORMAT

Submit manuscripts in quadruplicate (one original and three photocopies), including four sets of original figures or good quality glossy prints. All copies of halftone artwork must be glossy prints or originals. Type manuscipts double-spaced on one side of 8.5×11 in. white paper with 1 in. margins on all sides.

Page 1: Article title, author(s), affiliation(s), a short running head (abbreviated title, 50 characters including letters and spaces), and complete mailing address and telephone number of corresponding author. Page 2: Short abstract (100-200 words).

Article Précis: On a separate page following references, include a brief (25 words) preview of article to be included beneath article title in table of contents.

Footnotes: Indicate footnotes to title by superscript Arabic numerals; type double-spaced at bottom of title page. Do not use footnotes in text; incorporate information in body of article.

Tables: Number with Arabic numerals in order of appearance in text. Type double-spaced on separate pages. Give short descriptive captions. Type table footnotes (indicated by superscript lowercase italic letters) at bottom of table.

Figures: Number in order of appearance (with Arabic numerals); type short descriptive captions double-spaced on a separate sheet. On back of each indicate (in soft pencil) author's last name, figure number, and TOP of illustration.

Line drawings must be originals, in black India ink; make lettering large enough to be legible after 50-60% reduction. Photographs must be glossy prints. Provide an overlay that indicates critical areas of halftones. Figures should not exceed 8.5×11 in. in size. Illustrations in color can be accepted only if authors defray cost.

References: Cite in numerical order in text by Arabic numerals in square brackets. List numerically and type double-spaced on a separate sheet at end of manuscript. Abbreviate journal names according to *Chemical Abstracts Service Source Index,* 1985. Cite all authors' names; do not use *"et al."* in numerical list.

1. Ross, W. M., Carmichael, J. A., and Shelley, W. E. Advanced carcinoma of the ovary with central nervous system relapse, *Gynecol. Oncol.* **30,** 398–406 (1988).

2. McFalls, J., and McFalls, M. H. *Disease and fertility,* Academic Press, New York (1984).

3. Griffiths, C. T. Surgical treatment of advanced ovarian cancer, in *Ovarian cancer* (C. N. Hudson, Ed.), Oxford Univ. Press, Oxford, pp. 213–238 (1985).

OTHER

Electronic Submission: Submit manuscripts on personal computer disks after manuscript has been accepted and after all revisions have been incorporated. Supply file as a straight ASCII file with generic codes specified by publisher. Label disk with type of computer used, type of software and version number, and disk format. Tabular and mathematical material are typeset conventionally. Prepare art as camera-ready copy. Supply an exactly matching hard copy printout of manuscript. For further information on preparing disk for typesetting conversion and a list of generic codes, contact publisher (Editorial Supervisor, Journal Division, Academic Press, 525 B Street, Suite 1900, San Diego, California 92101-4495; telephone: (619) 699-6415; fax: (619) 699-6800).

Style: Conform to style recommended in the *CBE Style Manual* (5th ed., 1983, Council of Biology Editors).

SOURCE

January 1995, 56(1): back matter.

Hearing Research
Elsevier Science Publishers

Professor Aage R. Møller
P. O. Box 99187
Pittsburgh, PA 15233-4187
Telephone: (412) 647-6783; Fax: (412) 647-5559
Department of Neurological Surgery
F948 Presbyterian-University Hospital
Lothrop and DeSoto Streets,
Pittsburgh, PA 15213
       Courier Mail

AIM AND SCOPE

The aim of the journal is to provide a forum for papers concerned with basic auditory mechanisms. Emphasis is on experimental studies, but theoretical papers will also be considered.

Papers submitted should deal with auditory neurophysiology, ultrastructure, psychoacoustics and behavioral studies of hearing in animals, and models of auditory functions. Papers on comparative aspects of hearing in animals and man, and on effects of drugs and environmental contaminants on hearing function will also be considered. Clinical papers will not be accepted unless they contribute to the understanding of normal hearing functions.

MANUSCRIPT SUBMISSION

Submit original research not previously published or being considered for publication elsewhere. Submission carries with it the right to publish that paper and implies transfer of Copyright from author to publisher.

Submit original plus three copies of all items including figures and tables: Submit four original sets of all micrographs. Submit 3 copies of articles by authors cited as "in press."

MANUSCRIPT FORMAT

Type double-spaced with wide margins on one side of paper only. Divide full-length papers into sections (e.g., Introduction, Materials and Methods, Results, Discussion, etc.). Give authors' full names and academic addresses on title page, address, telephone and fax (or telex) numbers for corresponding author.

Title: 85 characters including spaces. Make informative. Avoid extraneous words such as 'study', 'investigation', etc.

Summary: 200 words at beginning of paper followed by 3–6 indexing terms (key words).

Tables: Type double-spaced, each on a separate sheet, numbered consecutively with Roman numerals (TABLE 1, etc.). Use only horizontal lines. Give a short descriptive heading above each and footnotes and explanations underneath.

Figures: Submit original line drawings in India ink or very sharp well-contrasting prints on glossy paper suitable for immediate reproduction or laser prints of very high quality. Half-tone figures should be black-and white, very sharp, well-contrasting, and on glossy paper. Lettering should be large enough to stand reduction. Prepare figures for either one column (84 mm) or entire page width (175 mm). Maximum height is 240 mm. Publisher will determine reduction or enlargement; line drawings are usually reduced to one column width. Authors may request larger reproduction. Type request on relevant figure legend page. Micrographs will usually not be reduced. Type figure legends double-spaced on a separate sheet.

References: Assemble in alphabetical order on a separate sheet. In text refer to by name and year (Harvard System). Identify more than one paper from the same author in the same year by letters a, b, c, etc., after year of publication In text, refer to a work by more than two authors, by name of first author, followed by *et al.* Literature references consist of names and initials of all authors, year, title of paper, abbreviated title of periodical, volume number, and first and last page numbers of paper.

Sellick, P. M., Patuzzi, R. and Johnstone, B. M. (1982) Measurement of basilar membrane motion in guinea pig using the Mössbauer technique. J. Acoust. Soc. Am. 72, 131–141.

Békésy, G. von (1960) Experiments in Hearing. McGraw-Hill, New York.

Wilson, J. P. and Evans, E. F. (1983) Observations on mechanics of cat basilar membrane. In: W. R. Webster and L. M. Aitkin (Eds.), Mechanisms of Hearing, Monash University Press, Clayton, Australia, pp. 30–35.

Abbreviate journal titles according to *List of Serial Title Word Abbreviations* (International Serials Data System).

OTHER FORMATS

LETTERS TO EDITORS—4 printed pages of comments and clarifications on articles published in *Hearing Research*.

REVIEW ARTICLES— Survey, evaluate and critically interpret recent research data and concepts in the fields covered by the journal.

SHORT COMMUNICATIONS—4 printed pages, 1 figure, 50 word summary. Brief but complete account of a particular piece of work.

OTHER

Experimentation, Animal: At end of Methods section state that: The care and use of the animals reported on in this study were approved by . . . (either) a grant application agency or a specific university's Animal Care and Use Committee provided that the committee adheres to the guidelines of the Declaration of Helsinki. Provide grant title name/number.

If such approval cannot be provided, state that experiments were performed in accordance with guidelines of Declaration of Helsinki (copy available from Editor's office). Manuscripts not complying will be returned.

Experimentation, Human: Indicate that investigations have been performed in accordance with principles of Declaration of Helsinki.

Reviewers: One or two referees from Editorial Board may be suggested.

SOURCE

November 1994, 80(2): inside back cover.

Hepatology
American Association for the Study of Liver Diseases

Paul D. Berk, M.D., *Hepatology* Journal Office
Box 1039, Mt. Sinai Medical Center
1 Gustave L. Levy Place
New York, NY 10029-6574

AIM AND SCOPE

Hepatology publishes original papers on clinical and research topics of general interest to those specializing in the liver and biliary tract. *Hepatology* publishes original scientific and/or clinical papers in all subject areas relating to liver structure, function and disease.

MANUSCRIPT SUBMISSION

Submit four copies of manuscript, including illustrations. In a cover letter provide explicit written assurance that paper is an original and previously unpublished work and that no other submission or publication of the same material or portions thereof has been or will be made before completion of the review process or before its publication. Abstracts of presentations are not considered previous publications. Preliminary communications or prior publication of minor portions may not preclude publication. Cite preliminary or prior report in submitted manuscript, provide four copies and specify what important additional data are in the new paper to justify publication.

For multiple authored manuscripts, all listed authors must have participated meaningfully in study, concur with submission and have seen and approved final manuscript.

Copyright is vested in the American Association for the Study of Liver Diseases. Include following statement, signed by principal author: "The undersigned author transfers all copyright ownership of the manuscript (insert the title of article) to the American Association for the Study of Liver Diseases in the event the work is published. The undersigned author warrants that the article is original, is not under consideration by another journal and has not been previously published. I sign for and accept responsibility for this material on behalf of any and all coauthors."

Photocopy and complete Author's Checklist from Instructions to Authors in every issue.

MANUSCRIPT FORMAT

Write concisely. Include only those figures and tables deemed essential to illustrate or supplement textual material. Typed double-spaced throughout, including references, tables, footnotes and figure legends.

Include in order: title page, running-title page, footnote page, abstract, introductory statement (without heading), description of experimental procedures or methods, description of results, discussion of experimental findings and, on separate pages, lists of references, figure legends and tables. Presentation may be more effective if sections are combined, especially for shorter papers. Avoid using subheadings; when required for complicated studies, limit subheadings to one level and avoid redundancy.

Title Page: Create an informative title. Make title as short as is consistent with clarity (two printed lines). Do not include formulas or arbitrary abbreviations.

Give full name(s) of author(s) and title of institution(s) to which authors belong. Indicate each author's affiliation with Arabic numeral superscripts. If an author belongs to more than one department, number each department separately. If addresses have changed, indicate author's current addresses in a footnote.

Footnote Page: Type footnotes to title, author names or text in sequence on a separate page. Include a listing of grant support and address for reprint requests.

State number of pages, figures and tables.

Give name, address, telephone and fax or telex numbers for corresponding author. Unless specified, all communication will be with first author.

Abstract: 250 words succinctly stating objective of paper, procedures used, findings and significance. Do not cite references, tables and figures or use abbreviations.

Introduction: (No separate heading) Indicate objectives of study and hypotheses tested. Give enough background information to make clear why study was undertaken; do not give lengthy reviews of past literature.

Materials and Methods: Be brief but make sufficient to allow repetition of work. Make reference to published procedures by citation. Do not include extensive descriptions of common experimental procedures unless they represent substantially new modifications.

Results: Present concisely. When possible, refer to appropriate tables summarizing experimental findings or to figures illustrating important points.

Discussion: Include concise statements of principal findings, validity of observations, findings in light of other published work and possible significance of work. Avoid extensive discussion of literature.

Acknowledgments: Place personal acknowledgments at end of text. List grant support on footnote page.

References: Cite unpublished references within the text as follows: "The procedures were those described previously (1)." Or: "Our observations are in agreement with those of Ronstadt and Brown (2) and of White et al. (3)." Abstracts, identified as such may be included. Letters to the editor and material submitted for publication (not yet in press) are not acceptable. Refer to in text by first author and reference (without title) and identify type of material: (Forker EL, et al., *Hepatology* 1986;6:543-544, Correspondence) and (Doe JP, et al., Unpublished observations, May 1990).

Reference section follows text and acknowledgments. Include only literature cited in paper. List and number references in sequence used in text: name(s) of author(s), full title of paper, source of reference, year, volume number and first and last pages. List names of first seven authors, followed by "et al." when more than seven. Use abbreviations of journal titles from *Cumulated Index Medicus*. Verify accuracy of references.

1. Cheng S, Levy D. Characterization of the anion transport system in hepatocyte plasma membranes. J Biol Chem 1980;255:2637-2640.

2. Zuckerman AJ. Hepatitis B: nature of the virus and prospects for vaccine development. In: Popper H, Schaffner F, eds. Progress in liver disease. Vol V. New York: Grune & Stratton, 1976:326-337.

3. Takahashi K, Imai M, Tsuda F, Takahashi T, Miyakawa Y, Mayumi M. Association of Dane particles with e antigen in the serum of asymptomatic carriers of hepatitis B virus surface antigen. J Immunol 1976;117:102-105.

4. Such J, Guarner C, Soriano G, Teixido M, Tena F, Mendez C, Sainz S, et al. Oral antibiotics increase the ascitic fluid total proteins and C3 levels in cirrhotic patients [Abstract]. *Hepatology* 1988;8:1351.

Permissions: Submit written permission for use of direct quotations, tables or illustrations that appeared in copyrighted material from copyright owner and original author, along with complete information as to source. Submit signed releases showing informed consent with photographs of identifiable persons.

Tables: Present on separate pages and refer to in text. Number consecutively (Arabic numerals) in order mentioned in text. Include only data needed to illustrate important points. Give each an explanatory title and sufficient experimental detail in legend to make intelligible without reference to text. Do not duplicate data in text and table. Give each column an appropriate heading. Set out material clearly enough to be readily understood by compositor. If table has appeared in or has been adapted from copyrighted material, include full credit to original source in legend and provide written permission for use.

Illustrations and Legends: Submit as high-quality glossy photographic prints, appropriately lettered and labeled. Make lettering visible against background and legible if reduced. Typewritten or hand lettering is unacceptable. Photographs should have good black-and-white contrast or color balance. Size may range from 3 × 4 in. to 5 × 7 in. Consistency in size is preferred.

Provide four complete sets of figures. One set of publication quality figures is for publisher. Others may be prints or photocopies. Photocopies of photomicrographs or electron micrographs are not acceptable for review purposes.

Submit each set of figures in a separate envelope. Do not use paper clips,. Identify each figure on reverse with first author's name, figure number and orientation (top, bottom). Note special instructions regarding sizing or placement. For multiple part figures, do not mount publisher's set.

Cite all figures in text. Give each a title and explanatory legend. Number (Arabic numerals) in order mentioned in text. Make title part of legend; do not letter on figure itself. Type legends double-spaced on a separate page; put legends on same page. Include sufficient experimental detail to make intelligible without reference to text. Do not to duplicate data in text and figure legend. If figure has previously appeared in a copyrighted source, include full credit to original source in legend and provide written permission for use.

There is no limit on illustrations, but use sparingly. No charge is made for illustrations unless special handling (e.g., color plates) is required.

OTHER FORMATS

CASE REPORTS—Submission of routine case reports is not encouraged. Case reports that illustrate an especially important point will be considered as brief original articles.

LETTERS TO THE EDITOR—Brief scientific commentary about articles published in *Hepatology*; subject to peer review.

RAPID COMMUNICATIONS—Four printed pages, including text, references, figures and tables (one printed page equals 3.5 pages double-spaced typescript. Clearly identify complete, self contained reports of particular scientific importance to Editor upon submission. Revised articles may be resubmitted as ordinary original articles.

OTHER

Abbreviations: Express magnitudes of variables in numerals. Use metric system for all measurements. Express temperature in degrees Celsius (centigrade). Express metric abbreviations in lowercase without periods and with no distinction between singular and plural.

Nonstandard abbreviations may be used if used at least three times in text (not abstract). When there are many nonstandard abbreviations, include an alphabetical list as a footnote inserted immediately after first abbreviation. Define abbreviations used only in table or figure in legend.

Drugs: Use generic names. Give trade names in parentheses if appropriate, with generic drug name.

Experimentation, Animal: Include a statement in "Materials and Methods" giving assurance that all animals received humane care in compliance with institution's guidelines or National Research Council's criteria for humane care, as outlined in "Guide for the Care and Use of Laboratory Animals" prepared by the National Academy of Sciences and published by the National Institutes of Health (NIH publication no. 86-23, revised 1985).

Experimentation, Human: Include assurance that informed consent was obtained from each patient and that study protocol conformed to ethical guidelines of 1975 Declaration of Helsinki as reflected in a priori approval by institution's human research review committee.

Reviewers: Suggest names of reviewers whose expertise qualifies them to review paper.

SOURCE

July 1994, 20(1): front matter.

Histochemistry
see Histochemistry and Cell Biology

Histochemistry and Cell Biology
Society for Histochemistry
Springer

Professor D. Drenckhahn
Institute of Anatomy
Julius Maximilian University
Koellikerstrasse 6
D-97070 Wurzburg, Germany
Telephone: (4 99 31)3 17 03; Fax: (4 99 31) 1 59 88
Professor J. Roth
Division of Cell and Molecular Pathology
Department of Pathology
University of Zürich
Schmelzbergstrasse 12
CH-8091 Zurich, Switzerland
Telephone: (4 11) 2 55 50 90; Fax: (4 11)2 55 44 07

AIM AND SCOPE

Histochemistry is a journal devoted specifically to the field of molecular histology and cell biology. Only original articles dealing with the localization and identification of molecular components, metabolic activities and cell biological aspects of cells and tissues will be published. Papers are also welcome that make a substantial contribution to the development, application, and/or evaluation of methods and probes that can be used in the entire area of histochemistry and cell biology. Preference will be given to studies involving experimental approaches and/or a variety of different methods that help to characterize and quantify molecular constituents, dynamic properties and functional aspects of cellular and extracellular structures.

MANUSCRIPT SUBMISSION

Submission implies that work described has not been published before (except as an abstract or as part of a published lecture, review, or thesis); that it is not under consideration for publication elsewhere; and that its publication has been approved by all authors and by responsible authorities at institution where work done. If paper is accepted for publication, authors agree to automatic transfer of Copyright to publisher, and that manuscript will not be published elsewhere in any language without consent of copyright holder.

MANUSCRIPT FORMAT

Submit manusciprts in triplicate. Divide into sections: title page; Abstract; Materials and methods; Results; Discussion; Acknowledgments; Reference lists, figure legends and tables. Type original manuscript double-spaced with wide margins on one side of paper only; photocopy duplicate copies on both sides of paper.

Title Page: Give first name(s) and surname(s) of author(s); title of paper; initial(s) of given name(s) and last name(s) of author(s) with full address(es) of institute(s); footnotes to title; address to which proofs should be sent and telephone and fax numbers of corresponding author.

Abstract: 200 word summary precedes article.

References: List of references includes only works cited in text. Refer to "unpublished work" and "personal communications" in text; do not include in reference list. List references at end of text as follows:

Single author- list alphabetically and then chronologically. Author and one coauthor - list alphabetically by first author and then by coauthor and then chronologically. First author and more than one coauthor- list alphabetically by first author and then chronologically; only first author's name and "et al." followed by year of publication is used in text. If more than one paper by same author or team of authors published in same year is cited, insert letters a, b, c, etc., following year, e.g., Schwartz (1971a), in both text and reference list.

JOURNALS

Name(s) of author(s) followed by initials (do not use "and"); year of publication; complete title; journal as abbreviated in *Index Medicus;* volume number, first and last page numbers.

Roth J, Taatjes DJ, Tokuyasu KT (1990) Contrasting of Lowicryl K4M thin sections. Histochemistry 95:132–136

BOOKS

Name(s) of author(s) followed by initials; year of publication; complete title (in English and French titles do not capitalize nouns, adjectives, etc.); edition and volume if appropriate; publisher; place of publication.

Pearse AGE (1985) Histochemistry. Theoretical and applied, 4th edn, vol 2. Churchill Livingstone, Edinburgh

BOOK CHAPTERS

Raap AK, Hopman AHN, Ploeg M van der (1989) Hapten labeling of probes for DNA in situ hybridization. In: Bullock GR, Petrusz P (eds) Techniques in immunocytochemistry, vol 4. Academic Press, London, pp 167-197

Figures: Submit in triplicate. Restrict figures to minimum number needed to clarify text. Do not repeat information in legends in text; do not present same data in both graph and table form. Previously published illustrations are not usually accepted. Number all figures, whether photographs, graphs or diagrams, in a single continuous sequence in order of citation in text. Mount each figure on a separate sheet. Indicate author's name, number of figure and TOP on back or on sheet it is mounted on.

Group figures into plates with 1 mm spaces between individual illustrations. Mount all figures on regular bond paper, not cardboard. Layouts or single figures should match width of a single column (8.6 cm) or printing area (17.6 × 23.6 cm).

Line Drawings— Submit good-quality prints. Inscriptions must be clearly legible. Use letters 2 mm high. Computer drawings must be of comparable quality to line drawings.

Halftone Illustrations:—Submit well-contrasted photographic prints, trimmed at right angles and in desired final size. Inscriptions should be about 3 mm high.

Legends: Give each figure a short title; follow with a concise description. Give magnification of micrographs either in legend or by a scale bar on micrograph. Avoid remarks like: "For explanation, see text." Legends are part of text; append to text in a list starting on a separate page.

Tables: Number consecutively with Arabic numerals. Type each on a separate sheet. Title should provide a brief, self-sufficient explanation of content.

OTHER FORMATS

SHORT COMMUNICATIONS—Two or three printed pages. Brief accounts of particularly interesting results.

OTHER

Color Chargess: No charge for one page. DM 1,200 for second page and DM 600 for each additional page.

Experimentation, Animal: State that "Principles of laboratory animal care" (NIH publication No. 85-23, revised 1983) were followed as well as specific national laws.

Experimentation, Human: State that all human studies were approved by appropriate ethics committee and were performed in accordance with ethical standards of the 1964 Declaration of Helsinki. State that all persons gave informed consent prior to inclusion in study. Omit details that might disclose identity of subjects.

Genus and Species Names: Mark with a single straight underline for italics. Words may be emphasized in body of text and set in italics; mark in same way. Mark approximate desired positions of figures and tables in margin of manuscript. For literature citations in text, use name/year system.

SOURCE

January 1995, 103(1): inside back cover.

Histopathology

British Division of the International Academy of Pathology

Blackwell Scientific Publications

Professor R. N. M. MacSween
Department of Pathology, University of Glasgow
Western Infirmary, Glasgow G11 6NT, Scotland
Telephone: 041 339 8822 ext. 4732; Fax: 041 337 2494

AIM AND SCOPE

Histopathology is an international journal whose primary purpose is to publish original histopathological material having clinical application in the study of human disease. It is designed to be of practical importance to the diagnostic histopathologist and to investigative workers in clinicopathological fields of study contributing to the better understanding of disease in man. It publishes critical and authoritative articles on current research relating directly to clinical disease including electronmicroscopic, histochemical, immunological and experimental studies.

MANUSCRIPT SUBMISSION

Papers may not have been nor will not be published elsewhere, as they become copyright of the journal. In cover letter from corresponding author, state that all co-authors have seen the paper and, satisfied with the content, approve its submission to *Histopathology*. Submit two copies typed with a wide margin, double-spaced throughout, on one side of standard paper.

MANUSCRIPT FORMAT

On title page include author's name(s), place of work, address for correspondence (with telephone and fax numbers where available), full title and short running title. Give a small number of 'key words'.

Precede text with a short summary (200 words) and proceed to sections of Introduction, Materials and methods, Results and Discussion. Number pages consecutively in Arabic numerals. Submit tables, footnotes, figure legends and acknowledgments on separate sheets. Indicate approximate position of tables and figures in text margin.

References: Cite using Vancouver style. In text number in order of appearance as follows: "The method used was as previously described[1]"; or "The method used was that of Brown and Smith [2]"; or "The method used was that of Jones *et al.*[3]"Keep reference list in sequence used in text. Include: names and initials of all authors (if more than six, give only first three, followed by et al.); full title of article; source of reference using abbreviations for journal titles in *Index Medicus*; year; volume number; and first and last pages. Identify abstracts as such; do not use 'unpublished observations' and 'personal communications' in reference list. For references cited from books, follow title of chapter by names and initials of editors, title of book, edition, place of publication, publisher, year of publication and first and last pages. Verify accuracy of references.

4. Bartolo DCC, Warren BF. Refractory soliary rectal ulcer syndrone. In Dorbrilla G, Bardhad KD, Steele A eds. Nonresponders in Gastroenterology. Milan: Raven Press, 1991; 169-180.

5. Morente M, Piris MA, Orradre JL, Rivas, C, Villuendas R. Human tonsil intraepithelial B cells: a marginal zone-related subpopulation. *J. Clin. Pathol.* 1992; 45;668-672.

Illustrations: Label with figure number and author's name in soft pencil on back; identify top edge. Submit photographs as glossy bromide prints of good contrast and well matched, preferably mounted on card, with a transparent overlay for protection. Use overlay to indicate masking instructions, lettering or arrows. Crop photographs as close as possible to area of interest. Make either 8.5 or 17 cm wide (half or full page width) not exceeding 22.5 cm in depth. Authors must contribute to cost of printing if space required is considered excessive. Use full page width for electronmicrographs; identify specific features either directly on prints or overlay. Indicate magnifications, optional for photomicrographs but mandatory for electronmicrographs, in legends; do not insert scales on prints. Color photographs are allowed but author must contribute to cost. Draw line diagrams with black ink on tracing paper or white card or supply as glossy prints.

OTHER FORMATS

BRIEF REPORTS—500 words, with 3 illustrations and 6 references.

Review articles, significant collections of case material and analyses of technical innovations are published occasionally. Book reviews, correspondence relating to articles in previous issues and notices of forthcoming conferences are regular features.

SINGLE CASE REPORTS—Must be of sufficient interest in a wider field.

OTHER

Spelling: Follow *The Concise Oxford Dictionary of Current English*.

Units: For units of measurement, symbols and abbreviations use *Units, Symbols and Abbreviations* (1972, Royal Society of Medicine, London). This specifies use of Systéme International Units.

SOURCE

July 1994, 25(1): inside back cover.

Human Gene Therapy
Mary Ann Liebert, Inc., Publishers

Journal Office
University of Southern California School of Medicine
1975 Zonal Avenue, KAM 304
Los Angeles, CA 90033
Telephone: (213) 342-GENE (4363); Fax (213) 342-3618
Claudio Bordignon, M.D.
H.S. Raffaele, Gene Therapy Program
Service of Hematology, via Olgettina
60, 201 32 Milano, Italy
Telephone: 39-2-26443235 1; Fax 39-2-26432285
      From Europe

AIM AND SCOPE

Human Gene Therapy publishes scientific papers on original investigations into the transfer and expression of genes in mammals, including man. Improvements in vector development, delivery systems, and animal models are covered. Ethical/legal/regulatory papers relating directly to the area of gene transfer into humans are welcome. In addition, the Journal will publish Abstracts and Commentaries, Point/Counterpoint, and News and Comments, including ongoing Clinical Protocols, Regulations, Meetings, Industrial News, Patents, Publication Announcements, and Book Reviews.

MANUSCRIPT SUBMISSION

Obtain written permission for reproduction of material from copyrighted material from publisher. Include publication from which figure or table is taken in reference list. A footnote of a reprinted table or legend of a reprinted figure, should read, "reprinted by permission from Jones et al." and list appropriate reference. Show all permissions listings in manuscript.

MANUSCRIPT FORMAT

Type entire manuscript including figure legends, tables, and references double-spaced on regular typewriter paper (not erasable bond). Leave ample margins on both sides, top and bottom (1.5 in.). Submit original and two copies, each with a set of original illustrations. On first page, give title of article, name(s) of author(s), institutional affiliation(s), and name with complete address and phone number of corresponding authors; supply a running title (45 characters). On second page, supply an abstract (200 words), stating aims, results, and conclusions drawn from study. Follow with introduction, materials and methods, results, discussion, acknowledgments, references, tables, and figure legends. Make results and discussion separate sections. Begin each section on a separate page. Number pages consecutively; put first author's last name on each page. Suggest potential reviewers.

Tables: Type each with its caption on a separate sheet. Number with Arabic numerals. Do not repeat information given in text, and do not make a table for data that can be given in text in one or two sentences.

Illustrations: Supply original glossy photographs. Photocopies cannot be used for reproduction. Submit DNA sequence data as glossy photos or computer printout provided a letter-quality laser printer (not dot matrix) is used. Submit nucleic acid sequences in camera-ready format. Number illustrations amd write first author's name on back. Indicate TOP of illustration.

Supply legend for each illustration. Number consecutively and type double-spaced. Number figures in order cited in text. Submit a complete, separately collated set with each copy of manuscript. Authors subsidize cost of color printing. For further details, contact Editor or Publisher.

Abstracts: 250 words and an overview summary following (150 words).

References: Cite references by authors and dates within text. If more than two authors, use et al. after first author's name, i.e., (Sogin et al., 1971).

If several papers by the same authors are cited in one sentence, use a lowercase letter designation to indicate individual papers, i.e., (Dunn and Studier, 1973a, 1973b). Use same designation in reference list.

Double-space references in Bibliography; list in alphabetical order at end of article. Include complete titles of cited articles.

JOURNAL CITATION

PACE, N. (1973). Structure and synthesis of the ribosomal ribonucleic acid. Bacteriol. Rev. 37, 562–603.

BOOK CITATION

BROSIUS, J. (1987). Expression vectors employing λ, *trp, lac,* and *lpp*-derived promoters. In *Vectors, a Survey of Molecular Cloning Vectors and Their Uses.* R.L. Rodriguez and D.T. Denhardt, eds. (Butterworth Publishers, Stoneham, MA) pp. 205–225.

When dates from an unpublished source are given, supply researcher's name. If work is in press, give journal in which it is to be published or name of publisher. Abbreviations of journal names as in *Index Medicus.*

SOURCE

December 1994, 5(12): inside back cover.

Human Genetics
Springer International

Professor A. G. Motulsky, M.D.
Division of Medical Genetics
University of Washington, School of Medicine
Seattle, WA 98195
Fax: (206) 543-3050
Professor C. J. Epstein, M.D.
Division of Medical Genetics, Rm. U585L
Department of Pediatrics, Box 0748
University of California, 505 Parnassus Avenue
San Francisco, CA 94143
Fax: (415) 476-9976

AIM AND SCOPE

Human Genetics publishes original investigations in the field of human and medical genetics.

MANUSCRIPT SUBMISSION

Submission implies: that work has not been published before (except as an abstract or as part of a published lecture, review, or thesis); that it is not under consideration for publication elsewhere; that its publication has been approved by all coauthors, if any, as well as by responsible authorities at institute where work was carried out; that, if and when manuscript is accepted for publication, authors agree to automatic transfer of copyright to publisher; and that manuscript will not be published elsewhere in any language without consent of copyright holders.

Submit manuscripts in quadruplicate: one original typed on one side of paper plus three photocopies copied on both sides. Illustration copy should include original drawings, etc., plus three copies.

MANUSCRIPT FORMAT

Type double-spaced with wide margins on one side of paper only. Consult a copy of journal and conform to its normal requirements. Mark Materials and methods section and Case reports for printing in small type to distinguish from rest of text. Mark in left margin for insertion of figures and tables.

Limit papers dealing with frequency of genetic polymorphisms to one printed page.

Single cytogenetic case reports will not normally be published. Publish series of cases or submit interesting cases under "cases observed" category.

Title Page: Title of paper, first name(s) and surname of author(s), institute, any footnotes referring to title (indicated by asterisks), and address to which proofs should be sent. Furnish a telex or fax number.

Abstract: Keep short.

Footnotes: Number consecutively.

References: Include only works referred to in text. Use name-date system: (Child 1941; Godwin and Cohen 1969: MacWilliams et al. 1970) or "and the study of Hillman and Tasca (1977)."

List at end of paper in alphabetical order by surname of first author. If several works with same (first) author, enter in order within list: all by single author in chronological order; same author and one coauthor alphabetically by coauthor then chronologically; same author plus two or more coauthors strictly chronologically. Distinguish works by same author/s or by several authors with same one name first in same year by addition of lowercase a, b, c, etc. to year of publication in both text citations and list. Abbreviate journal titles in accordance with *Index Medicus*.

Epstein CJ (1987) Aneuploidy in mouse and man. In: Vogel F, Sperling K (eds) Human genetics. Proceedings of the 7th International Congress, Berlin 1986. Springer, Berlin Heidelberg New York, pp 260-268

Kalla AK, Khanna S, Singh IP, Sharma S, Schnobel R, Vogel F (1989) A genetic and anthropological study of atlantooccipital fusion. Hum Genet 81:1051 12

Vogel F, Motulsky AG (1986) Human genetics, 2nd edn. Springer, Berlin Heidelberg New York

Figures: Use with discretion, only to clarify or reduce text. Do not present same data in both graph and table form. Do not repeat information in captions in text. Do not use previously published illustrations. Guarantee that reproduction of illustrations in which a patient is recognizable has been approved either by patient or by his/her legal representative.

Mention figures, graphs, and tables in text. Number with Arabic numerals. Provide brief descriptive legend for each; append to text on separate page. Submit figures suitable for reproduction separately from text. Lightly write top of figure, author, and figure number in soft pencil on back of each.

Do not exceed 17.8 cm in width and 24 cm in height. Group smaller figures with narrow spaces between them in blocks that fill page width as nearly as possible. Single, small figures should not exceed column width (8.6 cm).

Photographs: Supply in quadruplicate (originals plus three photocopies), good, original glossy prints suitable for reproduction with minimal reduction. Trim at right angles to eliminate unnecessary background.

Inscribe one set with rub-on template lettering (no ink) about 2 mm in height after reduction. Indicate portions of photographs that need special care during reproduction on a transparent overlay. If plates are submitted, mount figures on regular bond paper, not cardboard.

For line drawings, submit good quality prints. Letter drawings in a size that will yield a final height of 2 mm after reduction. For coordinate inscriptions, capitalize only first letter of first word. Computer drawings must be of comparable quality to line drawings; curves and lines must be smooth.

OTHER FORMATS

CASES OBSERVED—Rare chromosome aberrations and clinical genetic observations. Submit findings in one sentence with a query as to whether similar cases have been observed elsewhere.

CLINICAL CASE REPORTS—3-5 manuscript pages, including one illustration and pedigree; a short clinical discussion and interpretation of case. Cite only literature essential to understanding case.

CYTOGENETIC CASE REPORTS—Submit to Prof. Dr. A. Schinzel, Institut fur Medixinische Genetik der Universitat, Ramstrasse 74, CH-8001 Zurich, Switzerland (Fax: 0-12620470). Contact editor prior to submission. It is the aim of the journal to collect similar cases for collaboratory publications and to publish single cases only if they contribute particularly interesting new information.

DNA VARIANTS—One printed page, including figures and references. Give source, restriction enzymes used, chromosomal location, frequency in a defined population, and evidence for Mendelian inheritance.

RARE GENETIC VARIANT REGISTER—New, rare variants of blood groups, serum proteins, enzymes, etc. Such variants usually do not justify extensive publication. Announce essential properties in two or three sentences.

REVIEW ARTICLES—Consult editors before submitting.

SHORT COMMUNICATIONS—6 manuscript pages (2 printed pages), including bibliography, figures, and tables. Significant new and original findings; can be accepted for accelerated publication in a separate section.

OTHER

Color Charges: Approx. DM 1,200.00 for first and DM 600.00 for each additional page.

Experimentation, Animal: State that "Principles of laboratory animal care" (NIH Publication No. 86-23, revised 1985) were followed, as well as specific national laws.

Experimentation, Human: State that all human studies were reviewed by appropriate ethics committees and were performed in accordance with ethical standards laid down in an appropriate version of 1964 Declaration of Helsinki. State in text that all persons gave informed consent prior to inclusion in study. Omit details that might disclose identities of subjects.

SOURCE

July 1994, 94(1): inside front and back covers.

Human Pathology
W.B. Saunders Co.

Fred Gorstein, MD
Department of Pathology
Vanderbilt University Medical Center, C-3322 MCN
Nashville, TN 37232
Telephone: (615) 322-2123
Frederick T. Kraus, MD
Department of Pathology
St. John's Mercy Medical Center
615 S New Ballas Rd.
St. Louis, MO 63141-8221
         Case Studies

AIM AND SCOPE

Human Pathology is designed to bring to the laboratory and clinical physician, and to the student, authoritative information of clinico-pathologic significance to human disease. Its primary goal is the presentation of information drawn from morphologic and clinical laboratory studies having direct relevance to the understanding of diseases of humans. Accordingly, it is intended that the journal embrace articles concerned with morphologic and clinico-pathologic observations, reviews of a disease or group of diseases, analyses of problems in pathology, significant collections of case material, and new advances in concepts or techniques of value in the analysis and diagnosis of disease. Theoretical and experimental pathology and molecular biology pertinent of the diseases of humans will be included. The journal articles will be of interest to all physicians engaged in the clinical and laboratory practice of medicine as well as to teachers of pathology and clinical medicine. They will be critical and authoritative. Above all, they will embody a point of view.

MANUSCRIPT SUBMISSION

Submit only original articles that have never been published and are not under simultaneous review by another publisher. Provide original manuscript and illustrations suitable for publication with two copies and two extra sets of illustrations. Unless otherwise indicated, follow requirements of the International Committee of Medical Journal Editors' Uniform Requirements for Manuscripts Submitted to Biomedical Journals (*Ann Intern Med* 108:258–265, 1988).

MANUSCRIPT FORMAT

Do not exceed 30 double-spaced typewritten pages, or equivalent, including tables, illustrations, and references. Use standard manuscript form. Type manuscript, including references and legends typed double-spaced with ample margins throughout. Computer printouts must be of

letter quality. On title page list authors' names, highest earned degrees, and institutional affiliations, four to five key words or phrases, and mailing address and telephone number of corresponding author. Disclose commercial associations that might pose a conflict of interest. Acknowledge all funding sources in a footnote, with all institutional or corporate affiliations of authors. Use separate pages for title, abstract, text, acknowledgments, references, tables, and legends. Begin with a brief abstract (150–200 words) with factual information, not a description of manuscript's contents.

Fully identify borrowed material, illustrations, tables, and lengthy verbatim quotations, as to original author and source. Obtain and submit written permission from both author and publisher.

References: Arrange in order of citation and number consecutively. For journal articles include in order: authors' names with initials, article title and subtitle, journal name as abbreviated in *Index Medicus*, volume number, inclusive page numbers, and year. Indicate "Personal communications" and "unpublished observations" within text; exclude from reference list. Cite information from manuscripts submitted but not yet accepted in text as "unpublished observations."

In book citations, include, when appropriate, chapter authors' names and chapter title along with book editors' names, title and edition, place of publication, publisher, year, and inclusive page numbers.

Illustrations: Make single column width (3.31 in.), with an additional white border. Do not mount. Larger illustrations will be printed at Editor's discretion. Reduction or cropping may be necessary.

Best results are obtained from original artwork and original photographs. Number each illustration on back with a very soft pencil; mark to show author's name and TOP. Insert reference to illustration in consecutive numerical order, in text.

Cost of color illustrations, including Kodachromes, must be subsidized by author. Apply code letters, symbols, arrows, and labels professionally. Make sure spellings and abbreviations correspond to those in text. Make style and size of labels consistent. Make labels on drawings large enough to be legible when reduced.

Legends: Type double-spaced on a separate sheet and attach to manuscript. Specify stains and magnifications for photomicrographs.

OTHER FORMATS

CASE STUDIES—Provide new information. Clinically significant observations based on new or developing technology will receive special consideration. Imaginative applications of established methods are encouraged. Limit reference list and illustrations to two.

LETTERS TO THE EDITOR—500 words and no more than five pertinent references published at Editor's discretion. Type double-spaced and submit in triplicate.

SOURCE

January 1994, 25(1): front matter.

Hybridoma
Mary Ann Liebert, Inc.

Zenon Steplewski
The Wistar Institute of Anatomy and Biology
36th and Spruce Streets
Philadelphia, PA 19104
Telephone: (215) 898-3924

MANUSCRIPT SUBMISSION

Obtain pemmission to reproduce figures, tables, and text from previously published material from original copyright holder (generally publisher, not author or editor) of journal or book concerned. Include an appropriate credit line in figure legend or table footnote; include full publication information in reference list. Obtain and submit written permission from author of unpublished material cited from other laboratories. List all permissions in manuscript, they cannot be entered on proofs.

MANUSCRIPT FORMAT

Send an original plus two copies including artwork. Type manuscripts double-spaced on regular 8.5 × 11 in. typewriter paper (not erasable bond). Leave ample margins on sides, top, and bottom. Submit original and two copies, including illustrations and tables.

Title page includes authors' names and affiliations, source of a work or study (if any), and a running title (45 characters). Indicate name and address of author to whom correspondence should be addressed.

On second page include an abstract (250 words); make self-explanatory without reference to text. Follow format: abstract, introduction, materials and methods, results, discussion, acknowledgments, references. Number pages consecutively. At end of article, give name and address of reprint request author.

Tables: Type each with its title on a separate sheet. Use Arabic numerals to number. Each table must stand alone, i.e., include all necessary information in caption. Make table understandable independent of text. Include details of experimental conditions in table footnotes. Do not repeat text information within tables; tables should not contain data that can be given in text in one or two sentences.

Illustrations: Supply original pen drawings and charts (drawn and lettered in India ink), glossy photostats of drawings or graphs, and original glossy photographs for halftone illustrations with original manuscript. Identify all illustrationson back; give author's name. Indicate TOP of illustration. Supply alist of figure legends at end of manuscript, double-spaced. Include magnifications when appropriate.

Acknowledgments: List collaborations, sources of research funds, and other acknowledgments in a separate section at end of text ahead of References section.

References: Cite all references in text by an Arabic number within parentheses or brackets. Use brackets when parentheses are used for numbers for equations. List references by numbers in order of appearance in text. Type double-spaced; separate from each other by one line of space. Use *Chemical Abstracts* format for references. For journal citations: surname of author(s), initials, article title, journal, volume number, first and last page of citation, and year in parentheses. For book citations: surname of author(s), initials, title of book, editor of book (if applicable), edition of book (if applicable), publisher, city of publication, year of publication (not in parentheses), and first and last page reference (if applicable). For patents: country of original registration, (patent number, month, day, and year of issuance.

For data from an unpublished source, supply researcher's name and location. If work is in press, give journal in which it is to be published or publisher.

OTHER FORMATS

MONOCLONAL ANTIBODIES—Prompt publication of current information on new hybrid cell lines, methodology, and other information arising from use of monoclonal antibodies. Gatherings of brief presentations of data which allow clear identification of new hybrid cell lines and their products and characteristics. Concise descriptions of antigens defined by monoclonal antibodies are published.

Give full information about methods of immunization, selection, and screening for specificity. Describe specific innovations in methodology; note pertinent information such as toxicities of sera, chemicals and PEG is noted. Include table or chart describing specificity.

Format when submitting data: 1. Antigen used for immunization; 2. Method of immunization; 3. Parental cell lines of the hybrid; 4. Selection and cloning procedures; 5. Class of immunoglobulin; 6. Specificity in terms of binding to pertinent targets; 7. Specific antigen identified; 8. Availability; 9. Name, address and phone number of source of material; 10. References (limit to 10).

SOURCE

February 1, 1994, 13(1): back matter.

Hypertension
American Heart Association

Edward D. Frohlich, MD
Alton Ochsner Medical Foundation, BH514
1516 Jefferson Highway
New Orleans, LA 70121
Telephone: (504) 842-4103; Fax: (504) 842-4128

AIM AND SCOPE

Hypertension is a forum for the presentation of scientific investigation of the highest quality in the broad field of blood pressure regulation and the pathophysiological mechanisms underlying the hypertensive diseases. The editors are interested in receiving original articles that deal with fundamental and clinical research in the fields of biochemistry, cellular biology, molecular biology, genetics, immunology, physiology, pharmacology, epidemiology, health care analysis, outcomes research, prevention research, or innovative scientific hypotheses.

MANUSCRIPT SUBMISSION

Photocopy and complete checklist and sign copyright form (July 1994; 24(1): A7-A9). All authors must sign copyright transfer agreement. Original signatures are required. Send five address labels with name and address of corresponding author. Send stamped, self-addressed enveloped if wish manuscript and figures returned.

State if multiple manuscripts from a single study are planned; provide a copy of other manuscripts with their cross-references. Provide written documentation that all persons acknowledged have seen and approved mention of their names. Cite all sources of support. State potential conflicts of interest. Indicate that all authors have read and approve submission of manuscript. Indicate that material has not been published and is not being considered for publication elsewhere in whole or in part except as an abstract.

For manuscripts with more than four authors, state each author's contribution. Each must have contributed substantively to manuscript.

Send four copies of manuscripts in press or submitted to other journals that are relevant to review of submitted manuscript or potentially overlapping.

MANUSCRIPT FORMAT

Follow "Uniform Requirements for Manuscripts Submitted to Biomedical Journals" (*N Engl J Med.* 1991;324:424-428). Submit four copies of manuscript and four sets of original illustrations. Send four complete sets of title page.

Type one side of page only, double-space with 1 in. margins. Do not use proportional spacing or justified margins. Do not exceed 7 typeset pages (6000 words) unless justified in cover letter and approved by editors. Include no more than 1 figure or table for every 750 words. Assemble manuscript in order: title page; abstract page; text; references; figure legends; tables; figures. Do not staple.

Title Page: (Page 1, do not number) Include: full title, first author's surname and short title (50 characters including spaces). Title, author(s)' names, authors' affiliation(s), name and complete address for corresponding and reprint request authors. Fax and telephone numbers and e-mail addresses.

Abstract: (Page 2) 250 words maximum. Include rationale for study, briefly describe methods, present significant results and state data interpretation. Do not cite references. Limit use of acronyms and abbreviations. List 5-7 key words from *Medical Subject Headings* list in *Index Medicus.*

Text: Typical main headings: Methods, Results, and Discussion, no Introduction or Summary. Define abbreviations at first mention in text, tables, and figures. Give complete name of manufacturers of apparatus used in Methods. If methods have been previously published, refer to paper and submit copies as reference material.

Acknowledgments: List all sources of support for research, plus substantive contributions of individuals. Enclose written permission from persons acknowledged.

References: Verify against original sources, especially journal titles, inclusive page numbers, publication dates, diacritical marks, and foreign spellings. List all authors. Cite in numerical order of mention in text. Cite personal communications, unpublished observations, and submitted manuscripts in text as "(unpublished data, 19XX)."

Cite abstracts only if sole source; identify in reference as "Abstract," 'In press" citations must have been accepted for publication; include name of journal of book publisher.

Figures and Legends: Figures are either black and white drawings, graphs, or halftones. Prepare artwork using professional standards and photographed as camera-ready, unmounted, glossy prints. Good-quality computer-generated prints (not photocopies) are acceptable for line art. Make letters and locants uniform in size and style. Submit 4 sets of prints. Indicate figure number, first author, short manuscript title, and top of figure lightly in pencil on back, not on a separate label. Enclose each set in a separate envelope. Prepare 4 photocopies of each figure and attach one set to each manuscript. Put figure numbers on all copies. Supply a scale bar with photomicrographs. Define abbreviations or symbols in figure or figure legend. Provide copy for figure legends on a separate sheet double-spaced.

Tables: Type double-spaced on separate sheets. Use Arabic numbers followed by a period and a brief informative title. Use same size type as text. Indicate footnote in order: *, †, ‡, §, ||, ¶, #, * *. Use horizontal rules above and below column headings and at bottom of table; elsewhere use extra space to delineate sections. Do not use vertical rules.

OTHER FORMATS

BRIEF REVIEWS—24 double-spaced typewritten pages. Evenhanded, thorough reviews that summarize present state of knowledge concerning a particular aspect of a special field. Cite work of all relevant contributors.

EDITORIALS AND EDITORIAL COMMENTARIES—Generally invited pieces concerned with opinions of recognized leaders in hypertension and cardiovascular medicine and research.

LETTERS TO THE EDITOR—1000 pertinent and timely words with 10 references. Do not comment on feature articles or articles in *Hypertension* supplements. Do not use figures or tables.

RAPID COMMUNICATIONS—10 double-spaced typewritten manuscript pages of unusual scientific value and importance, reporting original, complete, and definitive research of particular significance to the field. In cover letter request consideration as a Rapid Communication. Submit names of five possible reviewers.

OTHER

Experimentation, Animal: State species, strain, number and other pertinent descriptive characteristics of animals used. When describing surgical procedures on animals, identify preanesthetic and anesthetic agents used and state amount or concentration and route and frequency of administration for each. Use of paralytic agents, such as curare or succinylcholine, is not an acceptable substitute for anesthetics. For other invasive procedures on animals, report analgesic or tranquilizing drugs used. If none, justify exclusion.

Experimentation, Human: Describe characteristics of human subjects. Manuscripts that describe studies on humans must indicate that study was approved by an institutional review committee and that subjects gave informed consent. Reports of studies on both animals and humans must indicate that procedures followed were in accordance with institutional guidelines.

Page Charges: $50 per printed page; expense for color reproduction of figures; $50 per printed page for excessive author alterations.

Reviewers: Suggest potential reviewers in cover letter. Include addresses and telephone and fax numbers.

Style: *AMA Manual of Style,* (8th ed., Baltimore, Maryland, Williams & Wilkins, 1989). Define abbreviations on first appearance in text, tables, and figures. Use generic names of drugs.

Units: Use Systéme International units of measurement. Use mol/L for molar (M); use mmol/L for mg/dL; use mm for cm; and change percentages to decimal fractions. Measurements not converted to SI units are blood and oxygen pressures, enzyme activity, H^+ concentration, temperature and volume. Use SI unit in text followed by conventionally used measurement in parentheses.

SOURCE

July 1994, 24 (1): A7-A8.

Immunogenetics
Springer International

Prof. Dr. Jan Klein
Max-Planck-Institut für Biologie
Abteilung Immungenetik
Corrensstrasse 42
W-7400 Tübingen 1, Germany
Telephone: 0 70 71/60 12 90; Fax: 0 70 71/60 04 37

AIM AND SCOPE

Immunogenetics publishes original full-length articles, brief communications, and reviews on research in the following areas: immunogenetics of cell interaction, immuogenetics of tissue differentiation and development, phylogeny of alloantigens and of immune response, genetic control of immune response and disease susceptibility, and genetics and biochemistry of alloantigens.

MANUSCRIPT SUBMISSION

Submission of a manuscript implies: that work described has not been published before (except as an abstract or as part of a published lecture, review, or thesis); that it is not under consideration for publication elsewhere; that its publication has been approved by all coauthors, if any, as well as by responsible authorities at institute where work was done; that, if and when manuscript is accepted for publication, authors agree to automatic transfer of copyright to publisher; and that manuscript will not be published elsewhere in any language without consent of copyright holders.

Submit 3 complete copies, original and 2 copies of manuscript including Figures/illustrations. Type original on one side, photocopies may be double-sided. Type on standard bond paper, double-spaced with wide margins, including Abstract, Footnotes, References, Tables and Figure legends. Arrange as: title page (including complete and short titles), Abstract, Introduction, Materials and methods, Results, Discussion, Acknowledgments (including grant and/or other sources of support), References, Tables, Figure legends, and Figures or illustrations. Number all pages consecutively, beginning with title page.

MANUSCRIPT FORMAT

Title Page: On separate page, type complete title, first name(s), middle initial(s), and surname(s) of author(s), laboratory or institution of origin, including city, state, and country and complete address and telex or fax number of corresponding author.

Title: 85 characters, including spaces. Short title is 55 characters including spaces.

Abstract: (page 2) 250 words. Do not include references. Do not divide into paragraphs. Describe precisely what was done, results obtained and conclusions drawn.

Introduction: Do not summarize results. State research question asked. Put answer in Results and summary in Abstract.

Discussion: Do not start with summary of data just described. Tie up loose ends and put data in proper context in terms of previously published work. Acknowledge all work reported by others.

References: List only those cited in text and accepted for publication. List at end of paper in alphabetical order under first author's name. List works by two authors according to second author's name after initial mention (e.g., Brown and Clark 1989, Brown and Hart 1985, Brown and Lipski 1973). List works by 3 or more authors chronologically. If more than one work is by the same author or team in the same year, add letters a, b, c, etc. to year both in text and reference list. Use internationally accepted journal title abbreviations.

Gorer, P. A. The antigenic basis of tumour transplantation. *J Pathol Bacteriol 47:* 231-252, 1938

Gorer, P. A. Sessile and humoral antibodies in homograft reactions, *In* G. E. W. Wolstenholme and M. O'Connor (eds.): *Symposium on Cellular Aspects of Immunity,* pp. 303-347, Little, Brown and Co, Boston, 1960

Gorer, P. A., Mikulska, A. B., and O'Gorman, P. The time of appearance and isoantibodies during the homograph response to mouse tumors. *Immunology 2:* 211-219, 1959

Cite in text by author and date. If two authors, name both. If more than two, mention only first plus "et al." or "and co-workers."

Do not include data not yet in press in reference list. Refer to parenthetically in text as "(unpublished data)" or "(personal communications)."

Tables: Type double-spaced on separate sheets. Number with Arabic numbers in order of appearance in text (e.g., **Table 1.** The effector of...). In text refer to as: "The results are summarized in Table 1;" or The results (Table 1) indicate...." Make headings concise and descriptive of content.

Figures: Supply as unmounted, glossy photographs. Number with Arabic numerals in order of appearance in text. Make lettering large and clear enough to be legible after reduction. Do not write on illustrations. On back indicate figure number.

Figure Legends: Type together on separate sheet. Read, for example, "**Fig. 1.** Cytotoxicity of lymphocytes...."

OTHER FORMATS

BRIEF COMMUNICATIONS—10 typed pages, including Tables and Figures. Do not divide into sections. Make last paragraph a conclusive paragraph. Do not submit preliminary results.

GENE-ANTIGEN REGISTER—8 typed pages, including Tables and Figures.

MHC PEPTIDE MOTIF REGISTERS—6-8 typed pages (see Editorial *Immunogenetics* 38: 81, 1993).

REVIEW ARTICLES—Solicited by Editorial Board.

SEQUENCE REGISTER—3 typed pages, including Tables and Figures.

OTHER

Data Sharing: If hybridoma, cell line or gene clone is described, state willingness to share material. Manuscript will not be published without statement.

If original nucleotide or animo acid sequence data are described, obtain an accession number from any of the database groups before submission.

DDBJ: DNA Data Bank of Japan, DNA Research Center, National Institute of Genetics, Mishima, Shizuoka 422 Japan, Fax: +81-559-75-6040

EMBL: EMBL Data Laboratory, Postfach 10.2209, D69012, Heidelberg, Germany, Fax: +49-6221-387-519

GenBank: GenBank Submissions, National Center for Biotechnology Information (NCBI), 8600 Rockville Pike, Bldg. 38A, Rm 8N-805, Bethesda, MD; Fax: +1-301-480-9241

Experimentation, Animal: State that "Principles of laboratory animal care" (NIH Publication No, 85-13, revised 1985) were followed as well as specific national laws.

Experimentation, Human: State that studies have been reviewed by appropriate ethics committee and have been performed in accordance with the ethical standards of an appropriate version of the 1964 Declaration of Helsinki. State in text that all persons gave informed consent prior to inclusion in study. Omit details that might disclose identities of subjects.

Nomenclature, Genetic: Use official or standardized genetic nomenclature whenever available, e.g., revised *H-2* nomenclature, immunoglobulin and T-cell receptor nomenclature, and CD nomenclature for human cell surface alloantigens. When introducing a new symbol, make sure it complies with established or officially sanctioned rules and is not a duplication. For mouse genetic nomenclature, complete version is available online from The Jackson Laboratory through GBASE, Gopher and World Wide Web (see also *Mouse Genome* 92(2), June 1994 and *Genetic Variants and Strains of the Laboratory Mouse* (3rd ed, 1994).

Italicize all genetic (not protein or other phenotypic) symbols. Do not use Greek letters.

Page Charges: DM 100 for full-length papers exceeding 7 printed pages (1 printed page is approximately 3 typed pages). If Table and Figures are excessive, less may be assessed.

SOURCE

1994, 40(1): A3-A4.
Instructions to authors appears in approximately every second issue.

Immunology
British Society for Immunology
Blackwell Scientific Publication

Professor M. W. Steward
P. O. Box 830
Epping, Essex CM16 4 BB, U.K.

AIM AND SCOPE

Immunology publishes papers describing original work in all areas of immunology including cellular immunology, immunochemistry, immunogenetics, allergy, transplantation immunology, cancer immunology and clinical immunology. The Journal also occasionally publishes solicited and unsolicited review articles on subjects of topical interest to immunologists. It does not in general publish papers of a primarily methodological nature.

MANUSCRIPT SUBMISSION

Papers will be accepted for publication provided they have not been and will not be published elsewhere, in whole or in part. Papers accepted become copyright of Journal. Send three complete copies of manuscript, figures and tables. Type on one side of A4 paper only, with a 4-cm margin, double-space. Include title of article, name and postal address of author, and name of Institute where work was done.

Limit manuscripts to 3000 words of text, up to four figures and four tables (or equivalent), and approximately 30 essential references. Accepted articles which would exceed 8 printed pages will be returned for reduction in length.

MANUSCRIPT FORMAT

Divide paper into parts in order: Summary (250 words); Introduction (give reasons for doing work, not an extensive review of previous work in area; Materials and Methods (concise but sufficient detail to allow work to be repeated; where appropriate, reference previously published methodology in place of experimental detail); Results (concise as possible; do not use tables and figures for same results); Discussion (discuss significance of results, do not repeat results; Acknowledgments; and References.

Illustrations: Consult *Suggestions to Authors* and *Notes on Preparation of Illustrations*, from Editorial Office, *Biochemical Journal*, 59 Portland Place, London W1, U.K.

Make figures and tables comprehensible without reference to text. Include a list of figures and tables. Refer to illustrations and diagrams as 'Figs' and give Arabic numbers; make about twice size of finished block, each on a separate sheet, bearing author's name, paper's short title and figure number on back. Draw diagrams in India ink on plain white paper, with letters, numbers etc., written lightly in pencil.

Photographs should be well contrasted on glossy paper and be about the same size as the finished block. Mark TOP with an arrow. Color illustrations will be accepted, when found necessary by Editors; authors will pay for all aspects of the color reproduction.

Legends: Type on one sheet, separate from figures. Include statement of magnification for drawings and photographs.

Tables: Type on separate sheets. Give Arabic numbers. Indicate approximate position in text.

References: Refer only to papers related to work: avoid exhaustive lists. Number references sequentially as appear in text, number those cited in figure legends and tables according to position of citation of figure or table in text. In numerical list of references to articles and papers at end of paper indicate: number of citation; name(s) followed by initials of author(s); year of publication in parentheses; title of paper; abbreviated title of Journal according to current edition of *Index Medicus*; volume; first page number of article.

1. PEMBERTON R. M., WRAITHE D.C. & ASKONAS B. A. (1990) Influenza peptide-induced self-lysis and down-regulation of cloned cytotoxic T cells. *Immunology*, 70, 223.

2. BOLTON A. E. & HUNTER W. M. (1986) Radioimmunoassay and related methods. In: *Handbook of Experimental Immunology* (ed. D. M. Weir), Vol. 1, 4th edn, p. 26.1. Blackwell Scientific Publications, Oxford.

References to books and monographs include: number; author(s) or editor(s); year of publication; title; edition; page referred to; publisher; place.

3. GERMUTH F. G. & RODRIGUEZ E. (1973) *Immunopathology of the Renal Glomerulus*, p. 209. Little, Brown & Company, Boston.

List only papers that have been published or accepted for publication (in press) in references. Cite unpublished observations and personal communications as such in text.

OTHER FORMATS

RAPID COMMUNICATION—6 typed pages, 4 illustrations (figures and/or tables); cite only essential references. Short papers of high quality describing results of research in novel and topical areas. Do not structure as for conventional papers, include: Summary followed by text consisting of brief accounts of background to work and reasons for study; experimental design; results obtained; and discussion of significance. Do not use headings. Include concise methodological information in figure or table legends.

OTHER

Abbreviations: Use abbreviations in Instructions to Authors *Biochemical Journal*. Abbreviations not listed may be used for unwieldy terms, when occur frequently. List abbreviations and meanings on a separate page; list will appear as a footnote on first page.

SOURCE
September 1994, 83(1): inside back cover.

Infection and Immunity
American Society for Microbiology

Journals Division
American Society for Microbiology
1325 Massachusetts Ave., N.W.
Washington, D.C. 20005-4171

AIM AND SCOPE

IAI is devoted to the advancement and dissemination of fundamental knowledge concerning: infections caused by pathogenic bacteria, fungi, and parasites; ecology, epidemiology, and evolution of pathogenic microbes; mechanisms of pathogenicity and virulence factors such as toxins and microbial surface structures; nonspecific factors in host resistance and susceptibility to infection; immunology of microbial infection; and development and evaluation of vaccines against pathogens.

IAI will consider papers concerned with the ecology of pathogenic microbes. Clinical descriptions and papers concerning the microbiology of hospital environments should be submitted to the *Journal of Clinical Microbiology*. Papers concerned with environmental ecology should be submitted to *Applied and Environmental Microbiology*. Papers dealing with aspects of laboratory diagnosis or animal models relevant to the diagnosis of human disease should be sent to *Clinical and Diagnostic Laboratory Immunology*.

IAI will consider papers concerned with specific and nonspecific immunity to microorganisms, including the function of phagocytes, lymphocytes, immunoglobulins, and other factors. Studies of basic immunology and tumor immunology are more appropriate for non ASM journals. *IAI* will consider papers describing experimental models of infection and the pathological consequences of infection. In addition, the journal will consider papers describing microbial products that are or may be related to pathogenesis. Submit papers describing microbial products or activities that are related to diagnosis to the *Journal of Clinical Microbiology*. Submit papers containing extensive taxonomic material to the *International Journal of Systematic Bacteriology*. Submit papers concerned with antimicrobial therapy to *Antimicrobial Agents and Chemotherapy*. Submit papers concerned with viral infections to the *Journal of Virology*. In most cases, *IAI* will not consider reports that emphasize nucleotide sequence data alone (without experimental documentation of the functional and evolutionary significance of the sequence). Papers describing methodology are not encouraged.

MANUSCRIPT SUBMISSION

Send a covering letter stating: journal to which manuscript is being submitted; most appropriate journal section; complete mailing address (including street), telephone and fax numbers of corresponding author; electronic mail address if available; and former ASM manuscript number and year if resubmission. Include written assurance that permission to cite personal communications and preprints has been granted. Submit three complete copies of each manuscript, including figures and tables. Enclose three copies of "in press" and "submitted" manuscripts important for judgment of present manuscript. Manuscripts must represent reports of original research. All authors must have agreed to its submission and are responsible for its content, including appropriate citations and acknowledgments, and must also have agreed that the corresponding author has authority to act on their behalf. Submission implies that manuscript, or one substantially the same, was not published previously, is not being considered or published elsewhere, and was not rejected on scientific grounds by another ASM journal.

Primary publication is defined in *How to Write and Publish a Scientific Paper,* third edition, by Robert A. Day, to wit: ". . . the first publication of original research results, in a form whereby peers of the author can repeat the experiments and test the conclusions, and in a journal or other source document readily available within the scientific community." Papers published in a conference report, symposium proceeding, technical bulletin, or any other retrievable source are unacceptable on grounds of prior publication. Preliminary disclosure of research findings published in abstract form is not considered "prior publication."

Acknowledge prior publication of data in a manuscript. Submit copy of relevant work.

Obtain and submit permissions from both original publisher and original author to reproduce figures, tables, or text (in whole or in part) from previous publications. Identify each as to relevant item manuscript.

An author is one who made a substantial contribution to the "overall design and execution of the experiments"; all authors are responsible for the entire paper. Recognize individuals who provided assistance in Acknowledgment section. Authors must agree to name order in byline. Footnotes regarding attribution of work are not permitted. If necessary, include statements in Acknowledgment section.

Corresponding author must sign copyright transfer agreement on behalf of all the authors upon acceptance. If authors were U.S. government employees when work was performed, corresponding author should instead, attach to agreement a statement attesting that manuscript was prepared as a part of their official duties and, as such, is a work of the U.S. government not subject to copyright. If some authors were government employees but others were not, corresponding author should sign copyright transfer agreement as it applies to that portion performed by non-government employee authors.

MANUSCRIPT FORMAT

Type every portion double-spaced (6 mm between lines), including figure legends, table footnotes, and References. Number all pages in sequence, including abstract, figure legends, and tables. Place last two items after References section. Keep 1 in. margins on all four sides. Number lines if possible. Ensure character sets are easily distinguishable: numeral zero (0) and letter "oh" (O); numeral one (1), letter "el" (l), and letter "eye" (I); and a multiplication sign ($\times$) and letter "ex" (x). If distinctions

cannot be made, mark items at first occurrence for cell lines, strain and genetic designations, viruses, etc., on modified manuscript.

Title, Running Title, and Byline: Present results of an independent, cohesive study; numbered series titles are not permitted. Avoid main title/subtitle arrangement, complete sentences, and unnecessary articles. On title page, include title, running title (54 characters and spaces), name of each author, address(es) of institution(s) at which work was performed, each author's affiliation, and a footnote indicating changes in present address(es) of any author(s). Place an asterisk after name of corresponding author and give telephone and fax numbers and electronic mail address.

Correspondent Footnote: If desired, complete mailing address, telephone and fax numbers, and e-mail address of corresponding author will be published as a footnote. Include information on lower left corner or title page labeled "Correspondent Footnote."

Abstract: 250 words; concisely summarize basic content of paper without extensive experimental details. Avoid abbreviations; do not include diagrams. When essential to include a reference, use References citation but omit article title. Make complete and understandable without reference to text.

Introduction: Supply sufficient background information to allow understanding and evaluation of results of present study without reference to previous publications. Provide rationale for study. Choose references to provide salient background, not an exhaustive topical review.

Materials and Methods: Include sufficient technical information to allow experiments to be repeated. When centrifugation conditions are critical, give enough information to enable repetition of procedure: make of centrifuge, model of rotor, temperature, time at maximum speed, and centrifugal force ($\times g$ rather than revolutions per minute). For commonly used materials and methods, a reference is sufficient. If several alternative methods are commonly used, identify method briefly and cite reference. Allow reader to assess method without constant reference to previous publications. Describe new methods completely; give sources of unusual chemicals, equipment, or microbial strains. When large numbers of microbial strains or mutants are used, include tables identifying sources and properties of strains, mutants, bacteriophages, plasmids, etc.

Describe a method, strain, etc., used in only one of several experiments reported in Results section or very briefly (one or two sentences) in a table footnote or figure legend. Statements disclaiming governmental or any other type of endorsement or approval will be deleted.

Results: Include rationale or design of experiments as well as results; reserve interpretation for Discussion section. Present concisely as either text, table(s), or figure(s). Avoid use of graphs to present data more concisely presented in text or tables. For tabular data include either standard deviation values or standard errors of the means. Include number of replicate determinations (or animals) used for making calculations. Give statements concerning significance of differences observed with probability values in parentheses. State statistical procedure used in Materials and Methods. Limit photographs (particularly photomicrographs and electron micrographs) to those absolutely necessary to show experimental findings. Number figures and tables in order cited in text; cite all figures and tables.

Discussion: Provide an interpretation of results in relation to previously published work and to experimental system; do not repeat Results or reiterate introduction. In short papers, Results and Discussion may be combined.

Acknowledgments: Acknowledge financial assistance and personal assistance in separate paragraphs. Format for acknowledgment of grant support: "This work was supported by Public Health Service grant CA-01234 from the National Cancer Institute."

Appendixes: Appendixes contain supplementary material to aid readers. Titles, authors, and References sections distinct from those of primary article are not permitted. If not feasible to list author(s) of appendix in byline or Acknowledgment section of primary article, rewrite appendix so it can be considered for publication as an independent article. Label equations, tables, and figures with letter "A" preceding numeral to distinguish from those cited in main body of text.

References: Include all relevant sources; cite all listed references in text. Arrange citations in alphabetical order by first author and number consecutively. Abbreviate journal names according to *Serial Sources for the BIOSIS Data Base* (BioSciences Information Service, 1992). Cite each listed reference by number in text.

1. **Alderete, J. F., and D. C. Robertson.** 1978. Purification and chemical characterization of the heat-stable enterotoxin produced by porcine strains of enterotoxigenic *Escherichia coli.* Infect. Immun. **19:**1021–1030.

2. **Berry, L. J., R. N. Moore, K. J. Goodrum, and R. E. Couch, Jr.** 1977. Cellular requirements for enzyme inhibition by endotoxin in mice, p. 321–325. *In* D. Schlessinger (ed.), Microbiology—1977. American Society for Microbiology, Washington, D.C.

3. **Cox, C. S., B. R. Brown, and J. C. Smith.** J. Gen. Genet., in press.*

4. **Dhople, A., I. Ortega, and C. Berauer.** 1989. Effect of oxygen on in vitro growth of *Mycobacterium leprae*, abstr. U-82, p. 168. Abstr. 89th Annu. Meet. Am. Soc. Microbiol. 1989.

5. **Finegold, S. M., W. E. Shepherd, and E. H. Spaulding.** 1977. Cumitech 5, Practical anaerobic bacteriology. Coordinating ed., W. E. Shepherd. American Society for Microbiology, Washington, D.C.

6. **Fitzgerald, G., and D. Shaw.** *In* A. E. Waters (ed.), Clinical microbiology, in press. EFH Publishing Co., Boston.

7. **Gill, T. J., III.** 1976. Principles of radioimmunoassay, p. 169–171. *In* N. R. Rose and H. Friedman (ed.), Manual of clinical immunology. American Society for Microbiology, Washington, D.C.

8. **Gustlethwaite, F. P.** 1985. Letter. Lancet **ii:**327.

9. **Jacoby, J., R. Grimm, J. Bostic, V. Dean, and G. Starke.** Submitted for publication.

10. **Jensen, C., and D. S. Schumacher.** Unpublished data.

11. **Jones, A. (Yale University). 1990.** Personal communication.

12. **Leadbetter, E. R.** 1974. Order II. *Cytophagales* , p. 99. *In* R. E. Buchanan and N. E. Gibbons (ed.), Bergey's manual of determinative bacteriology, 8th ed. The Williams & Wilkins Co., Baltimore.

13. **Miller, J. H.** 1972. Experiments in molecular genetics, p. 352–355. Cold Spring Harbor Laboratory, Cold Spring Harbor, N.Y.

14. **Powers, R. D., W. M. Dotson, Jr., and F. G. Hayden.** 1982. Program Abstr. 22nd Intersci. Conf. Antimicrob. Agents Chemother., abstr. 448.

15. **Sacks, L. E.** 1972. Influence of intro- and extracellular cations on the germination of bacterial spores, p. 437–422. *In* H. O. Halvorson, R. Hanson, and L. L. Campbell (ed.), Spores V. American Society for Microbiology, Washington, D. C.

16. **Sigma Chemical Co.** 1989. Sigma manual. Sigma Chemical Co., St. Louis, Mo.

17. **Smith, J. C.** April 1970. U.S. patent 484,363,770.

18. **Smyth, D. R.** 1972. Ph.D. thesis. University of California, Los Angeles.

19. **Yagupsky, P., and M. A. Menegus.** 1989. Intraluminal colonization as a source of catheter-related infection. Antimicrob. Agents Chemother. **33:**2025. (Letter.)

Note: For an "in press" reference to an ASM publication state control number or name of publication if a book.

Illustrations and Tables: Write figure number and authors' names on either margin or back (lightly with a soft pencil). For micrographs especially, indicate TOP. Do not use paper clips. Insert small figures in an envelope. Do not submit illustrations larger than 8.5 by 11 in.

Continuous-Tone and Composite Photographs: Keep journal page width, 3 5/16 in. (single column) and 6 7/8 in. (double-column, maximum), in mind. Include only significant portions of illustrations. Photos must be of sufficient contrast to withstand loss of contrast and detail inherent in printing. Submit one photograph (no photo copies) of each continuous-tone figure for each copy of manuscript. Size figures as they will appear so no reduction is necessary. If they must be reduced, make sure that all elements, including labeling, can withstand reduction and remain legible. If a figure is a composite of a continuous-tone photograph and a drawing or labeling, provide original composite for printer. Electron and light micrographs must be direct copies of original negative. Indicate magnification with a scale marker on each micrograph.

Computer-Generated Images: Produce with Abode Photoshop or Aldus Freehand software. Images produced with other types of software may not be acceptable.

For Aldus, one- and two-column art cannot exceed 20 picas (3 5/16 in.) and 41.5 picas (6 7/8 in.), respectively. Use either Helvetica (medium or bold) or Times Roman text font.

Adobe users should check densities of images on-line. If image's shadow density reads below 1.25, enter density as 1.40. If between 1.25 and 1.60, enter as 1.65. Enter density reading above 1.65 as actual density reading.

Provide computer files (along with prints for sizing) on a floppy disk (Macintosh) with accepted manuscript. For large images, use 40- or 80-megabyte Syquest cartridges or magneto-optical cartridges. For transfer from UNIX systems, submit either 9-track or 8-mm "tar" archive. Incorporate all final lettering, labeling, tooling, etc., into final supplied material. It cannot be added at a later date. Do not include figure numbers.

'Electronic image" telephone hotline: The William Byrd Press, Richmond, VA, 800-888-2973, ext. 3361; outside U.S. 804-264-2828, ext. 3361) to assist authors in producing gels digitally for publication.

Include a description of software/hardware used in figure legend(s).

Drawings: Submit graphs, charts, complicated chemical or mathematical formulas, diagrams, and other drawings as glossy photographs made from finished drawings not requiring additional artwork or typesetting. Computer-generated graphics produced on high-quality laser printers are acceptable. Do not handwrite any parts. Label both axes of graphs. Most graphs will be reduced to one-column width (3 5/16 in.); make all elements large enough to withstand reduction. Avoid heavy letters and unusual symbols. Avoid ambiguous use of numbers with exponents.

Figure Legends: Provide enough information so figure is understandable without reference to text. Place detailed experimental methods in Materials and Methods section, not in figure legend. Report a method unique to one of several experiments in a legend only if discussion is very brief (one or two sentences). Define all symbols and abbreviations not defined elsewhere.

Tables: Type each on a separate page. Arrange data so columns of like material read down, not across. Make headings clear so meaning of data will be understandable without reference to text. Explanatory footnotes are acceptable, extensive table "legends" are not. Do not include experimental details in footnotes. Include enough information to warrant table format; incorporate those with fewer than six pieces of data into text.

Camera ready tables can be photographically reproduced for publication without further typesetting or artwork. Do not hand letter.

OTHER FORMATS

AUTHOR'S CORRECTIONS—Provides a means of correcting or adding citations that were overlooked in a published article. Author who failed to cite a reference and uncited author must agree.

ERRATA—Provides a means of correcting errors that occurred during writing, typing, editing, or printing. Send directly to Journals Division.

LETTERS TO THE EDITOR—500 word comments on articles published previously in journal. Must include data to support argument. Send to Journals Division. A reply from corresponding author of article may be solicited. Type double-spaced.

MINIREVIEWS—4 printed pages. Brief summaries of developments in fast-moving areas, based on published articles, addressing any subject within scope of *IAI*. Submit three double-spaced copies.

NOTES—1000 words. 50 word abstract. Present brief observations that do not warrant full-length papers. No section headings; report methods, results, and discussion in a single section. Paragraph lead-ins are permissible. Keep figures and tables to a minimum. Describe materials and methods in text, not in figure legends or table footnotes. Present acknowledgments (no heading) and References as in full-length papers.

OTHER

Abbreviations: Limit use. Use those recommended by IUPAC-IUB *(Biochemical Nomenclature and Related Documents,* 1978). Use others only when necessary, such as in tables and figures. Use pronouns or paraphrase a long word after first use. Use standard chemical symbols and trivial names or their symbols for terms that appear in full in neighboring text. Introduce all abbreviations except those listed below in first paragraph in Materials and Methods or define each abbreviation and introduce it in parenthesis at first use. Eliminate abbreviations not used at least five times in text (including tables and figure legends).

For abbreviations that do not require abbreviation in title, abstract, text, figure legends and tables and abbreviations that can be used without definition in table, see *Infect. Immun*; 1994;62(1): ix-x.

Amino Acid Sequences: Use single-letter, rather than three-letter, designations.

Color Photographs: Use is discouraged. If necessary, include an extra copy at submission to get cost estimate.

Drugs: Use chemical or generic names; do not use code numbers or trade names.

Isotopes: For simple molecules, indicate labeling in chemical formula. Do not use brackets when isotopic symbol is attached to name of compound that in its natural state does not contain the element or to a word which is not a specific chemical name.

For specific chemicals, place isotope symbol in square brackets directly preceding part of name that describes labeled entity. Note that configuration symbols and modifiers precede isotopic symbol.

Follow conventions for isotopic labeling of *Journal of Biological Chemistry;* see instructions to authors of that journal (first issue of each year).

Materials Sharing: Plasmids, viruses, and living materials such as microbial strains and cell lines newly described must be made available from a national collection or will be made available in a timely fashion and at reasonable cost to the scientific community for noncommercial purposes.

Nomenclature, Chemical and Biochemical: Use *Chemical Abstracts* and its indexes and *The Merck Index* (11th ed., 1989; Merck & Co., Inc., Rahway, N.J.) as authority for names of chemical compounds. For biochemical terminology, including abbreviations and symbols, consult *Biochemical Nomenclature and Related Documents* (1978; The Biochemical Society, London) and instructions to authors of *Journal of Biological Chemistry* and *Archives of Biochemistry and Biophysics* (first issues of each year).

Do not express molecular weight in daltons; express molecular mass in daltons.

For enzymes, use recommended (trivial) name assigned by Nomenclature Committee of the International Union of Biochemistry as described in *Enzyme Nomenclature* (Academic Press, Inc., 1992). If a nonrecommended name is used, place proper (trivial) name in parentheses at first use in abstract and text. Use EC number if assigned; express enzyme activity either in katals (preferred) or in micromoles per minute.

Nomenclature, Genetic–Bacteria: Use recommendations of Demerec et al. (*Genetics* 54:61–76,1966) as a guide to use of terms phenotype and genotype. See Instructions to Authors in January issue of *Journal of Bacteriology*.

Conventions for naming genes. Name new genes whose function is not established by either: giving it the same name as a homologous gene already identified in another organism or a provisional name based on its map location in the style *yaaA*; or give a provisional name in the style described by Demerec et al. (e.g., *usg,* for gene upstream of *folC*).

"Mutant" vs. "mutation." Distinguish between a *mutation* (an alteration of the primary sequence of the genetic material) and a *mutant* (a strain carrying one or more mutations). A mutant has no genetic locus, only a phenotype.

Transposable elements, plasmids, and restriction enzymes. Follow recommendations of Campbell et al. (*Gene* 5:197–206,1979). For system of designating transposon insertions at sites where there are no known loci see Chumley et al. (*Genetics* 91:639–655, 1979). Use nomenclature recommendations of Novick et al. (*Bacteriol. Rev.* 40:168–189, 1976) for plasmids and plasmid-specified activities, of Low (*Bacteriol. Rev.* 36:587–607,1972) for F-prime factors, and of Roberts (*Nucleic Acids Res.* 17:r347–r387, 1989) for restriction enzymes and their isoschizomers. Recombinant DNA molecules, constructed in vitro, follow nomenclature for insertions in general. Describe DNA inserted into recombinant DNA molecules by using gene symbols and conventions for organism from which obtained. The Plasmid Reference Center (E. Lederberg, Plasmid Reference Center, Department of Microbiology and Immunology, 5402, Stanford University School of Medicine, Stanford, CA 94305–2499) assigns Tn and IS numbers to avoid conflicting and repetitive use and also clears nonconflicting plasmid prefix designations.

Nomenclature, Microorganisms: Use binary names, consisting of a generic name and a specific epithet. Names of higher categories may be used alone, but specific and subspecific epithets may not. Precede specific epithet by a generic name the first time. Thereafter, abbreviate generic name to initial capital letter, provided there is no confusion with other genera. Italicize names of all taxa (phyla, classes, orders, families, genera, species, and subspecies); do not italicize strain designations and numbers. For spelling of names follow *Approved Lists of Bacterial Names* (amended edition) (V. B. D. Skerman, V. McGowan, and P. H. A. Sneath, ed.) and *Index of the Bacterial and Yeast Nomenclatural Changes Published in the International Journal of Systematic Bacteriology since the 1980 Approved Lists of Bacterial Names (1 January 1980 to 1 January 1989)* (W. E. C. Moore and L. V. H. Moore, ed., both American Society for Microbiology, 1989), and validation lists and articles published in *International Journal of Systematic Bacteriology* since 1 January 1989. If there is reason to use a name that does not have standing in nomenclature, enclose name in quotation marks and make an appropriate statement concerning nomenclatural status in text (see *Int. J. Syst. Bacteriol.* 30:547-556,1980). Deposit strain in a recognized culture collection when necessary for description of a new taxon (see *Bacteriological Code,* 1990 Revision, American Society for Microbiology, 1992).

Determine accepted binomial for fungi. Sources for names include *The Yeasts: a Taxonomic Study,* 3rd ed. (N. J. W. Kreger-van Rij, ed., Elsevier Science Publishers B.V., Amsterdam, 1984) and *Ainsworth and Bisby's Dictionary of the Fungi, Including the Lichens,* 7th ed. (Commonwealth Mycological Institute, Kew, Surrey, England, 1983). Give microorganisms, viruses, and plasmids designations consisting of letters and serial numbers. Include a worker's initials or a descriptive symbol of locale, laboratory, etc., in designation. Give each new strain, mutant, isolate, or derivative a new (serial) designation distinct from those of genotype and phenotype; do not include genotypic and phenotypic symbols. A registry of plasmid designations is maintained by the Plasmid Reference Center, Department of Medical Microbiology, Stanford University, Stanford, CA 94304.

Nucleic Acid Sequences: Present nucleic acid sequences of limited length which are the primary subject of a study freestyle in the most effective format. Present longer nucleic acid sequences in the following format to conserve space. Submit sequence as camera-ready copy of dimensions 8.5 by 11 in. (or slightly less) in standard (portrait) orientation. Print sequence in lines of 100 bases, each in a nonproportional (monospace) font which is easily legible when published at 100 bases/6 in. Use uppercase and lowercase letters to designate exon/ intron structure, transcribed regions, etc., if lowercase letters remain legible at 100 bases/6 in. Number sequence line by line; place numerals, representing the first base of each line, to the left of the lines. Minimize spacing between adjacent lines of sequence, leaving room only for annotation of sequence. Annotation may include boldface, underlining, brackets, boxes, etc. Present encoded amino acid sequences immediately above first nucleotide of each codon, using single-letter amino acid symbols. Use same format for comparisons of multiple nucleic acid sequences.

Nucleotide Sequences: Include GenBank/EMBL accession numbers for primary nucleotide and/or amino acid sequence data. Include accession number as a separate paragraph at end of Materials and Methods section for full-length papers or at end of text of Notes.

GenBank Submissions, National Center for Biotechnology Information, Bldg. 38A, Rm 8N-803, 8600 Rockville Pike, Bethesda, MD 20894; e-mail (new submissions): gbsub@ncbi.nlm.nih.gov; e-mail (updates): update@ncbi.nlm.nih.gov. The EMBL Data Library may be contacted at: EMBL Data Library Submissions, Postfach 10.2209, Meyerhofstrasse 1, 6900 Heidelberg, Germany; telephone: 011 49 (6221) 387258; fax: 011 49 (6221) 387306; electronic mail (data submissions): datasubs@embl.bitnet.

Page Charges: $40 per printed page for authors whose research was supported by special funds, grants (departmental, governmental, institutional, etc.), or contracts or whose research was done as part of their official duties. If research was not so supported, request a waive of charges from Journals Division with manuscript (not in cover letter). Indicate how work was supported and include copy of Acknowledgment section. Minireviews and Letters to the Editor have no page charges.

Patient Identification: Do not use patient initials for identification, even as part of a strain designation. Change initials to numerals or use randomly chosen letters. Do not give hospital unit numbers; if a designation is needed, use only last two digits of unit. Established designations of some viruses and cell lines, although they consist of initials, are acceptable.

Reviewers: Suggest an appropriate editor. Recommend two or three reviewers who are not members of your institution(s) and have never been associated with them or their laboratory(ies). Provide name, address, phone and fax numbers, and area of expertise for each.

Style: Use *CBE Style Manual* (5th ed., 1983; Council of Biology Editors, Inc., Bethesda, MD.), *ASM Style Manual for Journals and Books* (American Society for Microbiology, 1991), and Robert A. Day's *How to Write and Publish a Scientific Paper* (3rd ed., 1988; Oryx Press).

Units: Use International System of Units. See International Union of Pure and Applied Chemistry (IUPAC) "Manual of Symbols and Terminology for Physicochemical Quantities and Units" (*Pure Appl. Chem.* 21:3–4, 1970). When using powers of 10, associate exponent power with number shown. In representing 20,000 cells per ml, the ordinate is "2" and the label is "10^4 cells per ml." Show enzyme activity of 0.06 U/ml as 6, with label 10^{-2} U/ml. Preferred designation is 60 mU/ml labeled as mU (or milliunits) per ml.

Use standard metric units to report length, weight, and volume. For these units and for molarity, use prefixes m, μ, n, and p. Use prefix k for 10^3. Avoid compound prefixes. Use μg/ml or μg/g in place of ppm. Present units of temperature as: 37°C or 324 K. When fractions are used to express units such as enzymatic activities, use whole units in denominator. Use unambiguous forms such as exponential notation.

See *CBE Style Manual,* 5th ed., for information about reporting numbers and information on SI units for reporting of illumination, energy, frequency, pressure, and other physical terms. Report numerical data in applicable SI units.

Verb Tense: Use past tense to narrate past events, including procedures, observations, and data. Use present tense for conclusions, conclusions of previous researchers, and generally accepted facts. For a discussion of tense in scientific writing, see p. 158–160 in *How to Write and Publish a Scientific Paper,* 3rd ed.

SOURCE

January 1994, 62(1): i-x.

International Archives of Allergy and Immunology

Karger

Prof. Dr. G. Wick, Editor-in-Chief
p.a. S. Karger AG, Editorial Department
International Archives of Allergy and Immunology
P. O. Box CH-4009 BASEL (Switzerland)

AIM AND SCOPE

International Archives of Allergy and Immunology appears monthly and provides a forum for publication of work from all aspects of modern immunology. It aims at bridging the gap between basic and clinical aspects of immunology. Papers considered of special interest will be treated as 'hot topics' and every attempt will be made to publish these especially fast. *International Archives* will publish original work in the fields of clinical immunology, allergy, immunopathology and transplantation, cellular immunology immuno-genetics, molecular biology, immunopharmacology and immunoendocrinology, mucosal immunity, phylogeny, ontogeny and aging, immunology of infectious diseases and immunology of connective tissue diseases. In addition, reviews, minireviews, commentaries and opinions on controversial subjects will be published regularly.

MANUSCRIPT SUBMISSION

Manuscripts must not be under simultaneous consideration by any other publication. In cover letter, give name, address, and telephone and fax numbers of corresponding author. Include a statement that all authors concur with submission. Submission implies transfer of copyright from author to publisher upon acceptance. Obtain permission to reproduce illustrations, tables, etc. from other publications.

Submit manuscripts in English in triplicate (with three sets of illustrations of which one is an original), typewritten double-spaced on one side of paper, with a wide margin. Submit fully documented reports of original research, describing significant and original observations to be critically evaluated and, if necessary, repeated.

MANUSCRIPT FORMAT

Title Page: On first page indicate title (main title underlined), authors' names, institution where work was conducted, and a short title for use as running head.

Key Words: 3–10 key words for indexing purposes.

Abstract: Up to 10 lines.

Text: Arrange in sections: Introduction, Material and Methods, Results, Discussion, Acknowledgment, References.

Small Type: Indicate paragraphs which can or must be set in smaller type (case histories, test methods, etc.) with a 'p' in left margin.

Footnotes: Avoid footnotes. When essential, number consecutively and type at foot of appropriate page.

Tables and Illustrations: Prepare on separate sheets and number in Arabic numerals. Tables require a heading and figures a legend, also on separate sheet. Only good drawings and original photographs can be accepted; negatives or photocopies cannot be used. When possible, group several illustrations on one block for reproduction (max. size 181 × 223 mm) or provide crop marks. On back of each illustration, indicate number, author's name, and TOP. Color illustrations are reproduced at author's expense.

References: Identify in text by Arabic numerals [in square brackets]. Note material submitted for publication but not yet accepted as 'unpublished data'; do not include in reference list. In list of references include only those publications cited in text. Do not alphabetize; number references in order mentioned in text. Give surnames of authors followed by initials. Use no punctuation other than a comma to separate authors. Cite all authors. Abbreviate journal names according to *Index Medicus*. See International Committee of Medical Journal Editors: Uniform requirements for manuscripts submitted to biomedical journals. *Br Med J* 1988;296:401–405.

PAPERS PUBLISHED IN PERIODICALS

Kauffman HF, van der Heide S, Beaumont F, Blok H, de Vries K: Class-specific antibody determination against *Aspergillus fumigatus* by means of the enzyme-linked immunosorbent assay. III. Comparative study: IgG, IgA, IgM ELISA titers, precipitating antibodies and IgE binding after fractionation of the antigen. Int Arch Allergy Appl Immunol 1986;80:300–306.

MONOGRAPHS

Matthews DE, Farewell VT: Using and Understanding Medical Statistics. Basel, Karger, 1985.

EDITED BOOKS

Hardy WD Jr, Essex M: FeLV-induced feline acquired immune deficiency syndrome: A model for human AIDS: in Klein E (ed): Acquired Immunodeficiency Syndrome. Prog Allergy. Basel, Karger, 1986. vol 37, p 353–376.

OTHER FORMATS

LETTERS TO THE EDITOR—2 manuscript pages including one table or figure concerning work published in the journal.

REVIEW ARTICLES, MINIREVIEWS AND COMMENTARIES—Contact Editor-in-Chief. Minireviews are focused, brief reports on topics of current interest. Commentaries offer a more personalized perspective on a topic of interest to the general readership.

SHORT COMMUNICATIONS—2 printed pages (5 manuscript pages), including an abstract, essential references and 3 tables or figures. Present complete, original studies, written in a continuous style without subdivisions and documented by experimental procedures and references.

CASE REPORTS—Must be of general immunological interest.

OTHER

Abbreviations: Avoid overuse. Introduce only when same term occurs three or more times.

Color Charges: SFr 600 (U.S. $40) per page for up to 6 per page.

Experimentation, Animal: Carefully and thoroughly justify use of painful or otherwise noxious stimuli.

Page Charges: No charge for papers of 5 or fewer printed pages. Each additional page is SFr.225 (U.S. $150).

Style: Consult our leaflet 'Rules for the Preparation of Manuscripts', available from Publisher, S. Karger AG, P. O. Box. CH-4009 Basel (Switzerland).

SOURCE

September 1994, 105(1): inside back cover.

International Journal of Cancer
International Union Against Cancer
Wiley-Liss, Inc.

Prof. N. Odartchenko
Swiss Institute for Experimental Cancer Research
CH-1066 EPALINGESs/LAUSANNE
Switzerland
Fax: 41-21-652 7705

AIM AND SCOPE

The *International Journal of Cancer* publishes original articles concerning all aspects of experimental and clinical cancer research.

MANUSCRIPT SUBMISSION

Submit original and two copies (text and illustrations). Consult a recent issue for article titles, average length, and general layout.

MANUSCRIPT FORMAT

Type text and references double-spaced with wide margins, 25–30 lines per page on A4 size (21 × 29 cm) paper. Divide text into: Title, Running Title (40 characters or spaces), Summary (250 words), Introduction, Material and Methods, Results, Discussion, Acknowledgments and References (limit to minimum). Do not use footnotes.

Figures: Submit three original sets of figures, in particular of photomicrographs, gels, and other documents. Incorporate a scale bar in all photomicrographs. Lightly indicate name of author, Figure number, and TOP of Figure in pencil on back of each photograph. Do not write on front. Figures will be reduced: 84, 100, 126, or 172 mm; take this into account when adding titles, letters, and other symbols. Use Helvetica medium type characters of suitable size to allow for reduction.

Figure Legends: Type on separate sheets. For karyotypes with individually glued chromosomes, ensure that original disposition remains intact (adequate rubber cement, enough paper space around chromosomes, protection, glossy prints of originals, etc.).

References: Do not number. In text, give name(s) of author(s) and year of publication (Harvard System). If more than two authors, give only first, followed by *et al.* List references in strict alphabetical order on a separate sheet, double-spaced.

ARVIS, G. and TOBELEM, G., Prostatic cancer; histological grade and indication for chemotherapy. *In*: G. P. Murphy, S. Khoury, R. Küss, C. Chatelain and L. Denis (eds.), *Prostate cancer, Part A: Research, endocrine treatment, and histopathology*, pp. 511–512, A. R. Liss, New York (1987).

HEIM, S. and MITELMAN, F., *Cancer cytogenetics*, p. 41, A. R. Liss, New York (1987).

NABI, I. R. and RAZ, A., Cell shape modulation alters glycosylation of a metastatic melanoma cell-surface antigen. *Int. J. Cancer*, **40**, 396–402 (1987).

OTHER FORMATS

LETTERS TO THE EDITOR—2500 words (2 printed pages) of pertinent observations on articles published in the Journal, or on cancer research in general.

REPORTS ON SCIENTIFIC MEETINGS—4 printed pages. Short reports on scientific meetings of general and current interest. See example: Fidler, I. J. and Burger, M. M., UICC Study Group in basic and clinical cancer research: cancer metastasis. *Int. J. Cancer*, **33**, 1–2 (1984). Reports may not be submitted more than one month following meeting.

OTHER

Color Charges: At authors' expense. Obtain estimate of cost from: T. Gilmartin, Wiley-Liss, Inc., 605 Third Avenue, New York, NY 10158-0012.

Page Charges: US$150.00 per page over 6 printed pages. No page charges for 6 pages or less.

Style: Consult *CBE style manual: a guide for authors, editors, and publishers in the biological sciences* (5th ed., Chicago, IL: 1983).
Italicize genotypes (indicate by underlining); do not italicize phenotypes.

SOURCE

October 1, 1994, 59(1): back matter.

International Journal of Mass Spectrometry and Ion Processes
Elsevier Science Publishers

M. T. Bowers
Department of Chemistry
University of California
Santa Barbara, CA 93106
Fax: (805) 893-8703

AIM AND SCOPE

The journal contains papers which consider fundamental aspects of mass spectrometry and ion processes, and the application of mass spectrometric techniques to specific problems in chemistry and physics. The following topics, amongst others, can be found in the journal: theoretical and experimental studies of ion formation (i.e., by electrons, laser or other forms of radiation, heavy ions, high-energy particles, etc.), ion separation and ion detection processes; the design and performance of instruments (or their parts) and accessories; measurements of natural isotopic abundances, precise isotopic masses, ionization, appearance and excitation energies, ionization cross-sections; development of techniques related to determining molecular structures, geological age determination, studies of thermodynamic properties, chemical kinetics, surface phenomena, radiation chemistry, and chemical analyses to mass spectral data; chemistry and physics of cluster ions; spectroscopy of gaseous ions including studies related to interstellar chemistry; mechanistic and studies of unimolecular processes and ion/molecule reactions in the gas phase including computational aspects (ion trajectory calculations, quantum mechanical studies of potential energy surfaces); physical organic chemistry of isolated ions; biological applications of mass spectrometry.

The journal is of interest to all mass spectrometrists and other scientists interested in the chemistry and physics of charged particles.

MANUSCRIPT SUBMISSION

Submission of an article implies that the article is original and unpublished and is not being considered for publication elsewhere. Upon acceptance of an article, author(s) will transfer copyright to the publisher.

MANUSCRIPT FORMAT

Submit three copies, double-spaced with adequate margins on pages of uniform size. Put acknowledgments and references at end of paper.

Abstract: 200 words using standard systematic terminology. Include essential numerical data unless too extensive.

References: In text number (on line and in square brackets) in order of appearance. Keep reference list at end of article in numerical order of appearance in text. Abbreviate journal titles according to Chemical Abstract Service's *Bibliographic Guide for Editors and Authors* (The American Chemical Society, Washington, D.C., 1974). For multi-author references, name all authors and give initials in reference list. Use Smith et al., in text.

1 J.J. Lingane and A.M. Hartley, Anal. Chim. Acta, 11 (1954) 475.

2 F. Feigl, Spot Tests in Organic Analysis, 7th edn., Elsevier, Amsterdam, 1966, p. 516.

Tables: Type on separate pages; number with Arabic numerals in order mentioned in text. Give descriptive titles. Use chemical formula and conventional abbreviations in tables and figures; do not use chemical formula in text unless necessary for clarity. Abbreviate units of weight, volume, etc., used with numerals; do not punctuate.

Figures: Draw in India ink on drawing or tracing paper. Use standard symbols in line drawings, including:

○ ⊕ ● × △ ▲ ▢ ■ ☆ ★ –

Photographs: Submit as clear black-and-white glossy prints the same size as typed pages. Type legends on a separate page. Number figures with Arabic numerals in order mentioned in text.

OTHER FORMATS

RAPID COMMUNICATIONS—4–5 printed pages, published within 2–3 months after receipt. State why rapid publication is desirable.

REVIEWS—30-40 printed pages. Focus on recent developments with minimal historical documentation. Often solicited, contact editors regarding appropriateness of subject matter.

SHORT COMMUNICATIONS—4-5 printed pages, no abstract. Complete but concise articles, fully documented both by reference to literature and description of experimental procedures.

OTHER

Nomenclature and Symbols: See IUPAC Manual of Physicochemical Symbols and Terminology, *J. Am. Chem. Soc.*, 82 (1960) 5517, or IUPAC document U. I. P. 11 (S. U. N. 65-3) 1965, *Symbols, Units, and Nomenclature in Physics*. See also IUPAC Recommendations for Symbolism and Nomenclature for Mass Spectrometry, *Pure Appl. Chem.*, 50 (1978) 65.

SOURCE

Oct. 2, 1992, 120 (1/2): inside front and back covers.

International Journal of Pharmaceutics
Elsevier Science Publishing

P. F. D'Arcy
School of Pharmacy, Medical Biology Center
The Queen's University of Belfast
97 Lisburn Road, Belfast BT9 7BL, U.K.
Fax: +44 (232) 247-794
            For Europe (except Scandinavia and Finland), Africa, Near East

J. M. Newton
Dept of Pharmaceutics, The School of Pharmacy
University of London, 29–39 Brunswick Square
London WC1N 1AX, U. K.
Fax: +44 (1) 753-5920
            For Scandinavia and Finland

J. H. Rytting
Pharmaceutical Chemistry Dept.
School of Pharmacy
University of Kansas
Lawrence, KS 66045
Fax: (913) 749-7393
            For The Americas and Australia

T. Nagai
Department of Pharmaceutics
Faculty of Pharmaceutical Sciences
Hoshi University, Ebara 2-4-41, Shinagawa-ku
Tokyo 142, Japan
Fax: +81 (3) 3782-7849
            For Japan

AIM AND SCOPE

The *International Journal of Pharmaceutics* provides a medium for the publication of research results dealing with all aspects of Pharmaceutics including physical, chemical, analytical, biological and engineering studies related to drug delivery in its broadest sense. Such areas include: Drug delivery systems; Physical pharmacy; Biopharmaceutics; Pharmacodynamics and pharmacokinetics; Membrane function and transport; Biotechnology; Drug and prodrug design; Radiopharmaceuticals; Sterility and sterilization; Quality control of pharmaceuticals; Pharmaceutical analysis; Drug stability; Pharmaceutical technology; Descriptions/reviews of computer programs relating to the above areas; and Application of cell and molecular biology in drug delivery.

MANUSCRIPT SUBMISSION

For manuscripts with multiple authors, it is assumed that all listed authors concur with submission and that final manuscript copy has been approved by all authors and tacitly or explicitly by responsible authorities in laboratories where work was done. If accepted, manuscript shall not be published elsewhere in the same form, in either the same or another language, without consent of Editors and Publisher.

Submit original (plus computer disk version if available) and 3 copies, complete with 3 sets of figures (including originals or duplicates of sufficient quality for clarity of reproduction) and tables. Supply all data that would help referees evaluate papers.

MANUSCRIPT FORMAT

Type double-spaced with adequate margins on one side of sheet (not more than 26 lines per page). Number all pages sequentially.

Title: 85 characters including spaces.

Authors and Affiliations: List initial(s) (one given name may be used) followed by surname of author(s) with affiliations. If work was carried out at more than one address, indicate affiliation of each author using superscript, lowercase letters. Indicate corresponding author with asterisk. Include telephone, telex and fax numbers.

Abstract: 200 words, one paragraph on a separate sheet. Key Words: 6 key words or phrases suitable for indexing. If possible, select from *Index Medicus* or *Excerpta Medica*. Refer to Master Indexes in *International Journal of Pharmaceuticals*.

Text: Divide into main sections, such as: 1. Introduction, 2. Materials and methods, 3. Results, 4. Discussion, Acknowledgments, References, figure legends, tables and figures. Number sections consecutively as indicated. Number subdivisions of sections, e.g., 2.1. Materials, 2.2. Relative humidity measurement, 2.3. Sample preparation.

Figure Legends, Table Legends, Footnotes: Type double-spaced on separate sheets. Number footnotes consecutively in superscript throughout text; limit use.

References: Use Harvard System of citation. Cite references in text in parentheses. Arrange multiple citations within parentheses in ascending order of year of publication. When more than one reference with the same publication year is cited, arrange in alphabetical order of first author's names. If more than two authors, give name of first author and et al. Examples: (Gesztes et al., 1988; Chesnut et al., 1989; Legros et al., 1990; Mhando and Li Wan Po, 1990; Korsten et al., 1991; Langerman et al., 1991, 1992a,b; Masters et al., 1991; Bonhomme, 1992; Kolli et al., 1992). (Sahw et al., 1978; Nakano and Arita, 1990b; Nakano et al., 1990a,b; Bone et al., 1992).

List all references cited in text alphabetically at end of paper typed double-spaced. Identify more than one paper from same author in same year; must be by letters a, b, c, etc., after year of publication. References consist of names and initials of all authors, year, title of paper, abbreviated title of periodicals and volume, and first and last page numbers of paper. Cite 'personal communications' and 'unpublished data' in text only. Include name of journal for papers referred to as 'submitted for publication.' Abbreviate journal titles according to *List of Serial Title Word Abbreviations* (International Serials Data System, Paris).

ARRANGEMENT OF REFERENCE LIST

Crowe, J.H., Crowe, L.M. and Chapman, D., Infrared spectroscopic studies of water and carbohydrates with a biological membrane. *Arch. Biochem. Biophys.*, 232 (1984a) 400-407.

Crowe, J.H., Crowe, L.M., and Hoekstra, F. A., Phase transitions and permeability changes in dry membranes during rehydration. *J. Bioenerg. Biomembr.*, 21 (1989) 77-92.

Crowe, J.H., Crowe, L.M., Carpenter, J.F. and Aurell Wistrom, C., Stabilization of dry phospholipid bilayers and proteins by sugars. *Biochem. J.*, 242 (1987) 1-10.

Crowe, J.H., Crowe, L.M., Carpenter, J.F., Rudolph, A.S., Wistrom, C.A., Spargo, B.J. and Anchordoguy, T.J. Interactions of sugars with membranes. *Biochim. Biophys. Acta,* 861 (1986) 131-140.

Crowe, L.M., Crowe, J.H., Womersley, C., Reid, D., Appel, L., and Rudolph, A., Prevention of fusion and leakage in freeze-dried liposomes by carbohydrates. *Biochim. Biophys. Acta,* 861 (1986) 131-140.

Crowe, L.M., Mouradian, R., Crowe, J.H., Jackson, S.A. and Womersley, C., Effects of carbohydrates on membrane stability at low water activities. *Biochim. Biophys. Acta,* 769 (1984b) 141-150.

TYPES OF PUBLICATIONS

Langerman, L., Chaimsky, G., Golomb, E., Tverskoy, M., Kook, A.I. and Benita, S., A rabbit model for evaluation of

spinal anesthesia: chronic cannulation of the subarachnoid space. *Anesth. Analg.*, 71 (1990) 529-535.

Timsia, M.P., Martin, G.P., Marriott, C., Ganderton, D. and Yianneskis, M., Drug delivery to the respiratory tract using dry powder inhalers, *Int. J. Pharm.*, 101 (1994) 1-13.

Gibaldi, M. and Perrier, D., *Pharmacokinetics*, 2nd Edn, Dekker, New York, 1982.

Deppeler, H.P., Hydrochlorothiazide. In Florey, K. (Ed.), *Analytical Profiles of Drug Substances*, Vol 10, Academic Press, New York, 1981, pp. 405-441.

US Pharmacopeia XXII, US Pharmacopeial Convention, Rockville, MD, 1990, pp. 1434-1435.

Muelller, L.G., Novel anti-inflammatory esters, pharmaceutical compositions and methods for reducing inflammation. *UK Patent GB 2 204 869 A*, 23 Nov., 1988.

DuPlessis, J., Topical liposomal delivery of biologically active peptides. Ph.D. Thesis, Potchefstroom University for CHE, South Africa (1992).

Tables: Number consecutively with Arabic numerals and cite in text. Compile on separate sheets with a short descriptive titles, and legends and/or footnotes identified by superscripts [a, b, c,] etc. Do not use vertical lines. Keep horizontal rules to a minimum.

Figures: Draw line drawings (including graphs) in black ink on white paper or on tracing paper with blue or faint grey rulings; graduation will not be reproduced. Make lettering large enough to permit photographic reduction. If figures are not to be reduced, do not exceed 16×20 cm. Submit photographs (or half-tone illustrations) as sharp, high-contrast black and white prints on glossy paper. Indicate magnification of micrographs by a scale bar in figure. Mark on reverse side with number, orientation (TOP) and author's name; use a soft pencil or felt-tipped pen. Number with Arabic numerals. Type legends double-spaced on separate sheets.

OTHER FORMATS

DESCRIPTIONS/REVIEWS OF COMPUTER PROGRAMS—Include information on hardware compatibility and software availability (including costs, ordering address, etc.). Send manual describing software validation; operating instructions (in greater detail than in manuscript) and illustrations of input and output data sets. Submit three copies of manual and software (diskettes, tapes, etc.). Software and manuals will be returned only upon request by the Authors.

NOTES—1500 words or equivalent, including figures/tables prepared as full length manuscripts; do not divide into sections; include an Abstract and reference list.

RAPID COMMUNICATIONS—1500 words or equivalent space. Do not divide text into sections. Include an Abstract and a full reference list. Do not include figures.

REVIEWS—Suggestions will be considered by the Editors in Chief.

OTHER

Nomenclature: Use standard nomenclature. Define unfamiliar or new terms and arbitrary abbreviations when first used. Avoid unnecessary or ambiguous abbreviations and symbols. Express data in SI units.

SOURCE

May 30, 1994, 106 (2): 175-178.

International Journal of Radiation Biology

Taylor & Francis

Editor, *International Journal of Radiation Biology*
Paterson Institute for Cancer Research
Christie Hospital NHS Trust
Manchester M20 9BX, U.K.
Fax: +44 (0) 61 448 1770

AIM AND SCOPE

The *IJRB* publishes original research and review papers on the effects of ionizing, ultraviolet and visible radiation, accelerated particles, microwaves, ultrasound, heat and related modalities. The focus is on the biological effects of such radiations; from radiation chemistry to the whole spectrum of responses of living organisms and the underlying mechanisms, including genetic abnormalities, repair phenomena, cell death, dose-modifying agents, and tissue responses. Application of basic studies to medical uses of radiation extend the scope of the journal to a variety of practical problems such as physical modifications and the chemical adjuvants which improve the effectiveness of radiation in cancer therapy. Assessment of the hazards of low doses of radiation is also considered.

MANUSCRIPT SUBMISSION

Papers may not be considered for publication elsewhere. Corresponding author must have prior agreement from authors to submit manuscript.

Send five copies of contributions, typed 10 or 12 characters per in. on one side of A4 sized paper, double-spaced with wide margins, each page numbered.

MANUSCRIPT FORMAT

Provide a running title and up to five indexing phrases. Include telephone and fax numbers on title page. Include a 200 word abstract and Introduction, Materials and methods, Results, Discussion, and References sections. Provide original data of papers involving animals or humans when feasible as reduced-size tables either in text or in an Appendix, for archival purposes.

References: Give in text by author(s) and date [Smith and Brown (1992), or Smith *et al.* (1992) if more than two authors]; list alphabetically at end of paper, giving titles of all articles and journals in full:

JONES, G. D. D. AND O'NEILL, P. 1991, Kinetics of radiation-induced strand break formation in single-stranded pyrimidine polynucleotides in the presence and absence of oxygen; a time-resolved light-scattering study. International Journal of Radiation Biology, 59, 1127–1146.

TUBIANA, M., DUTREIX, J. and WAMBERSIE, A., 1991, *Introduction to Radiobiology* (Taylor & Francis, London).

BRYANT, P. E., 1991, DNA strand breaks and chromosomal aberrations. In: *New Developments in Fundamental and Applied Radiobiology*. Edited by: C. Seymour and C. Mothersill. (Taylor & Francis, London), pp. 84–94.

Figures: Provide either as glossy photographs of high contrast, laser reproductions, or as line drawings in Indian ink on white board or tracing paper, lettered ready for reproduction after reduction. Design to be legible when reduced to 8.5 cm (single column) or 17.5 cm (double-column) width. For diagrams, submit one set of prints and four sets of photocopies; for half-tone photographic items, submit 5 sets of original prints. Where possible, include symbol and curve identification, and cell types, etc. in diagram; do not repeat in legend. Show magnification by scale bars where possible, check stated magnifications on proofs. Put author's name and figure number on all figures. Type legends on a separate sheet. Color photographs will be charged to authors.

OTHER FORMATS

CURRENT TOPICS—Two or four printed pages, providing overviews on a wide range of subject areas including aspects of radiation in the environment, its effect on chemical and biological systems, and its use for example in therapy. Authors are invited; suggestions are welcome.

LETTERS TO THE EDITOR—Comment on topical subjects, matters arising from published articles, but not presenting primary data; may be sent for comment to the authors of articles being discussed.

MEETING OR SYMPOSIUM REPORTS—Contact Editor in advance for details.

RAPID COMMUNICATIONS—Two printed pages, preliminary accounts of important observations where early publication is highly desirable. Submit a statement justifying urgent publication. Supplementary data, not to be published, may be submitted for referees' consideration. Separate headings may not be needed; provide a one-sentence summary.

REVIEWS—10 to 15 printed pages accompanied by a 200 word abstract. Authors are invited; suggestions are welcome.

TECHNICAL NOTES (up to two printed pages) or TECHNICAL REPORTS (up to four printed pages)—Describe new or modified techniques, assays, or methods of analysis.

OTHER

Experimentation, Animal: Include a statement concerning national or institutional regulations under which studies were performed. Papers may be rejected if procedures disregard welfare of experimental animals.

Experimentation, Human: Include a statement concerning permission obtained for studies from an appropriate Ethical Committee.

Units and Abbreviations: Comply with *Units, Symbols and Abbreviations* (Royal Society of Medicine, London). Use SI units (see *IJRB* 46:99 for tables of SI units).

SOURCE

July 1994, 66(1): back matter.

International Journal of Radiation Oncology/Biology/Physics

American Society for Therapeutic Radiology and Oncology

Pergamon Press

Philip Rubin, M. D.
Division of Radiation Oncology
Strong Memorial Hospital
601 Elmwood Avenue
Rochester, N.Y. 14642

AIM AND SCOPE

The *International Journal of Radiation Oncology/Biology/Physics* publishes original scientific research and clinical investigations related to radiation oncology, radiation biology and medical physics. The clinical studies submitted for publication include experimental studies of combined modality treatment, especially chemoradiotherapy approaches and relevant innovations in hyperthermia, brachytherapy, high LET irradiation, nuclear medicine radiosensitizers, and radioprotectors. Technical advances related to tumor imaging, dosimetry, and 2-D/3-D conformal radiation treatment planning are encouraged.

MANUSCRIPT SUBMISSION

Submit original and 4 copies of each manuscript, double-spaced and typewritten on 8.5×11 in. white bond paper and 5 copies of illustrations. Do not submit more than 10 tables and/or illustrations per article. Include senior author's telephone number in cover letter. Original papers must be contributed solely to the Journal. Accepted manuscripts become property of the Journal and may not be published elsewhere without written permission from both editor and publisher.

MANUSCRIPT FORMAT

Title Page: Make title brief and specific. State author and coauthors, academic degree and department for each author, hospital or academic institution and city where work was done, source of financial support, mailing address for reprints, and if manuscript was presented at a meeting, name of organization, place and date of presentation. Supply corresponding author's telephone and fax numbers and e-mail address.

Acknowledgments: Place on a separate sheet of 8.5×11 in. white bond paper.

Abstract: 200–300 words, typed double-spaced on a separate sheet. Concisely define significant aspects of article. Include clearly labeled sections: Purpose, Methods and Materials, Results, and Conclusion. For review articles and special features, a non-structured abstract can be submitted.

Key Words: Place same page as Abstract.

Text Style: Include a running head (shortened form of title and name of senior author) on each page of manuscript (including legends, tables, and references). Number each page consecutively, beginning with title page. If appropriate, organize text to include: Introduction, Methods and Materials, Results, and Discussion. Quotations must be accurate; give full credit to source. Obtain permission to quote and reproduce text previously published. Include references to personal communications in text, not in reference list; use following form after indicating person(s) with whom one has communicated in text: (written or oral communication, month, year). For unpublished data, give (name, unpublished data, month, year).

Tables: Type double-spaced on separate pages of 8.5×11 in. white bond paper. Follow table number (in Arabic) with a brief specific title. Number consecutively with respect to citation in text. Make data self-explanatory; do not duplicate text. Include acknowledgments for previously published material; include sources in reference list with reference number cited in acknowledgment. Obtain permission for to use previously published material. Define abbreviations in a key.

Illustrations: Use only those illustrations (photographs, line drawings and graphs) that clarify and augment text. Full-color illustrations will be accepted if editors believe use of color is appropriate; cost will be quoted and invoiced directly to author. Submit 5 sets of illustrations, unmounted and untrimmed. Send high-contrast 5×7 in. glossy prints (not photocopies) or laser prints. All lettering must be legible after reduction to column size. Template lettering or preset type is preferred. In photographs, mask identities of patients; submit consent forms for photographs in which there is any possibility of identification of patient.

Legends: Type double-spaced; number consecutively on a separate sheet; be brief and specific. Indicate magnification and staining materials in photomicrograph legends. List acknowledgments for previously published material in legends; include source in reference list; cite reference numbers in acknowledgments. Define abbreviations in figures in legends.

References: Type triple-spaced, arrange alphabetically and number consecutively on a separate sheet. Give journal references in order: author(s), initial(s); article title (with subtitle, if any), journal abbreviation, volume number in Arabic numerals, inclusive pages, and year. If paper is an abstract, add (Abstr.), followed by name, volume, pages and year of abstract journal.

Order for book references: author(s), initial(s); title, volume (if more than one), edition number (if other than first), city, publisher, year. Chapter in a book: author(s) of chapter, initial(s); title of chapter. In: editor(s) names, ed(s), book title, city; publisher; year; inclusive pages of chapter. Ensure bibliographic accuracy; cite every reference in manuscript and proofread again in page proof.

CBE NUMBERED/ALPHABETIZED STYLE

1. Kuban, D. A.; El-Mahdi, A. M.; Schellhammer, P. F. The effect of TURP on prognosis in prostatic carcinoma. Int. J. Radiat. Oncol. Biol. Phys. 13:1653–1659;1987.

2. Singh, K.; Reinhold, N. The genesis life. Elmsford, NY; Pergamon Press; 1984.

3. Wasserman, T. H.; Kligerman, M. Chemical modifiers of radiation effects. In: Perez, J., Brady, J. B., eds. Principles and Practice of Radiation Oncology. Philadelphia, PA: Lippincott Co; 1987:360–376.

OTHER

Abbreviations: Define first time used in text.

Experimentation, Human: Include statement to the effect that informed consent was obtained after the nature of the procedure(s) was fully explained.

Units: Express radiation doses in Gray units rather than rad (1 rad = 1 cGy). List all radiation treatment factors. Do not use NSSD, TDF or CRE.

SOURCE

August 30, 1994, 30(1): back matter.

International Journal of Systematic Bacteriology

International Committee on Systematic Bacteriology of the International Union of Microbiological Societies
American Society for Microbiology

Journals Division
American Society for Microbiology
1325 Massachusetts Ave., N.W.
Washington, D.C. 20005-4171

AIM AND SCOPE

The *International Journal of Systematic Bacteriology* publishes papers dealing with all phases of the systematics of bacteria, yeasts, and yeastlike organisms, including taxonomy, nomenclature, identification, characterization, and culture preservation. Papers dealing only with characteristics that are differential below the subspecies level, primarily of epidemiological interest, will not be considered. As required by the Bacteriological Code, the *IJSB* publishes Validation Lists containing new names validly published in other journals.

MANUSCRIPT SUBMISSION

Include a cover letter stating journal to which manuscript is being submitted, complete mailing address (including street) and telephone and fax numbers of corresponding author, and former manuscript number (if resubmission). Include written assurance that permission to cite personal communications and preprints has been granted.

All authors must agree to submission and are responsible for content, including appropriate citations and acknowledgments, and agree that corresponding author has authority to act on their behalf. Authors guarantee that manuscript, or one substantially the same, was not published previously and is not being considered or published elsewhere. A manuscript accepted and published by the *IJSB* must not be published again in any form without consent of the International Union of Microbiological Societies.

An author is one who made a substantial contribution to the "overall design and execution of the experiments"; all authors are responsible for entire paper. Recognize individuals who provided assistance in Acknowledgment section. Authors must agree to order of names in byline. Footnotes regarding attribution of work are not permitted. If necessary, include such statements in Acknowledgment section.

Follow definition of primary publication in *How to Write and Publish a Scientific Paper,* 3rd ed, by Robert A. Day, ". . . the first publication of original research results, in a form whereby peers of the author can repeat the experiments and test the conclusions, and in a journal or other source document readily available within the scientific community." A scientific paper published in a conference report, symposium proceeding, technical bulletin, or any other retrievable source is unacceptable on grounds of prior publication. A preliminary disclosure of research findings published in abstract form as an adjunct to a meeting, is not considered "prior publication."

Acknowledge prior publication of data contained in a manuscript with a copy of relevant work.

Obtain and submit permissions from both original publisher and author to reproduce figures, tables, or text (in whole or in part) from previous publications. Identify each as to relevant item in manuscript.

Corresponding author must sign copyright transfer agreement on behalf of all authors upon acceptance of manuscript.

Submit three complete copies, including figures and tables.

MANUSCRIPT FORMAT

Type every portion double-spaced (6 mm between lines), including figure legends, table footnotes, and References. Keep 1 in. margins on all sides; use line numbers if possible. Include following sections: Abstract, Introduction, Materials and Methods, Results, Discussion, Acknowledgments, and References. Number all pages in sequence, including abstract, figure legends, and tables. Place last two items after References section. Make sure these sets of characters are easily distinguishable: numeral zero (0) and letter "oh" (O); numeral one (1), letter "el" (l), and letter "eye" (I); and a multiplication sign (×) and letter "ex" (x). If distinctions cannot be made, mark items at first occurrence for cell lines, strain and genetic designations, viruses, etc.

Correspondent Footnote: If desired, complete mailing address, telephone and fax numbers, and e-mail address of corresponding author will be published as a footnote. Include information on lower left corner or title page labeled as "Correspondent Footnote."

Title: Concise statement of contents of manuscript. If an opinion of Judicial Commission is requested, use "Request for an Opinion" as subtitle. Include any proposed new names or combinations. Exercise care in using scientific names of bacteria. Scientific names used by themselves or unless otherwise qualified refer to all the constituent elements of the taxa to which the names refer. When only selected elements of a taxon were studied, have title reflect this. Use strain designations when applicable. Never use name of a bacterial taxon in a general sense in a title if the nomenclatural type of the taxon was not included in study.

Abstract: 250 words at beginning. Cite all new names and combinations proposed and all type and neotype strains designated if practical.

References: Include all relevant sources; cite all listed references in text. Arrange citations in alphabetical order by first author, then second author, etc., and number consecutively. Abbreviate journal names according to *Serial Sources for the BIOSIS Data Base* (BioSciences Information Service, 1992). Cite each listed reference by number in text.

1. **Berry, L. J., R. N. Moore, K. J. Goodrum, and R. E. Couch, Jr.** 1977. Cellular requirements for enzyme inhibition by endotoxin in mice. p. 321-325. *In* D. Schlessinger (ed.), Microbiology—1977. American Society for Microbiology, Washington, D.C.

2. **Cox, C. S., B. R. Brown, and J. C. Smith.** J. Gen. Genet., in press.*

3. **Dhople, A., I. Ortega, and C. Berauer.** 1989. Effect of oxygen on in vitro growth of *Mycobacterium leprae*, abstr. U-82, p. 168. Abstr. 89th Annu. Meet. Am. Soc. Microbiol. 1989.

4. **Finegold, S. M., W. E. Shepherd, and E. H. Spaulding.** 1977. Cumitech 5, Practical anaerobic bacteriology. Coordinating ed., W. E. Shepherd. American Society for Microbiology, Washington, D.C.

5. **Fitzgerald, G., and D. Shaw.** *In* A. E. Waters (ed.), Clinical microbiology, in press. EFH Publishing Co., Boston.

6. **Gustlethwaite, F. P.** 1985. Letter. Lancet **ii:**327.

7. **Jacoby, J., R. Grimm, J. Bostic, V. Dean, and G. Starke.** Submitted for publication. [Note: must not involve data essential to nomenclatural proposals.]

8. **Jensen, C., and D. S. Schumacher.** Unpublished data.

9. **Johnson, J. L.** 1973. Use of nucleic-acid homologies in the taxonomy of anaerobic bacteria. Int. J. Syst. Bacteriol. **23:**308–315.

10. **Jones A. (Yale University).** 1990. Personal communication.

11. **Leadbetter, E. R.** 1974. Order II. *Cytophagales* , p. 99. *In* R. E. Buchanan and N. E. Gibbons (ed.), Bergey's manual of determinative bacteriology, 8th ed. The Williams & Wilkins Co., Baltimore.

12. **Miller, J. H.** 1972. Experiments in molecular genetics, p. 352–355. Cold Spring Harbor Laboratory, Cold Spring Harbor, N.Y.

13. **Powers, R. D., W. M. Dotson, Jr., and F. G. Hayden.** 1982. Program Abstr. 22nd Intersci. Conf. Antimicrob. Agents Chemother., abstr. 448.

14. **Sigma Chemical Co.** 1989. Sigma manual. Sigma Chemical Co., St. Louis, Mo.

15. **Smith, J. C.** April 1970. U.S. patent 484,363,770.

16. **Smyth, D. R.** 1972. Ph.D. thesis. University of California, Los Angeles.

17. **Yagupsky, P., and M. A. Menegus.** 1989. Intraluminal colonization as a source of catheter-related infection. Antimicrob. Agents Chemother. **33:**2025. (Letter.)

Note: An "in press" reference to an ASM publication should state control number or name of publication if a book.

Tables: Type each on a separate page. Arrange data so that columns of like material read down, not across. Make headings sufficiently clear so that meaning of data is understandable without reference to text. Explanatory footnotes are permitted, but detailed descriptions of experiments are not. Place materials and methods used to gain data in that section. Simple lists are not considered tables; incorporate into text. Include enough information to warrant table format; incorporate those with fewer than six pieces of data into text. If possible, submit tables that can be photographically reproduced for publication without further typesetting or artwork, that are "camera ready." Do not hand letter; carefully prepare to conform with journal style.

Continuous-Tone and Composite Photographs: Write figure number and authors' names on all figures, either in margin or on back (marked lightly with a soft pencil). For micrographs especially, indicate TOP. Do not use paper clips. Insert small figures in an envelope. Do not submit photographs larger than 8.5 by 11 in. Keep in mind journal page width: 3 5/16 in. for a single column and 6 7/8 in. for a double-column (maximum). Include only significant portions of illustrations. Photos must be of sufficient contrast to withstand loss of contrast and detail in printing. Submit one photograph of each continuous-tone figure for each copy of manuscript; not photocopies. If possible, submit figures in final size. If must be reduced, make sure all elements, including labeling, can withstand reduction and remain legible. If a figure is a composite of a continuous-tone photograph and a drawing or labeling, provide original composite for printer. Provide direct copies of original negatives for electron and light micrographs. Indicate magnification with a scale marker on each micrograph.

Computer-Generated Images: Produce with Adobe Photoshop or Aldus Freehand software.

For Aldus, one- and two-column art cannot exceed 20 picas (3 5/16-in) and 41.5 picas (6 7/8-in), respectively. Make text font either Helvetica (medium or bold) or Times Roman.

Adobe users should check densities of images on-line. If image's shadow density reads below 1.25, enter density as 1.40. If 1.25–1.60, enter density as 1.65. Enter density readings >1.65 as the actual density reading.

Provide computer files, along with prints to be used for sizing, on a floppy disk (Macintosh) with accepted manuscript. For large images, 40- or 80-megabyte Syquest cartridges or magneto-optical cartridges may be used. For transfer from UNIX systems, either 9-track or 8-mm "tar" archives may be submitted. All final lettering, labeling, tooling, etc., must be incorporated; it cannot be added later. Do not include figure numbers.

Designate software/hardware used in figure legend(s).

For aid, call "electronic image" telephone hotline: The William Byrd Press, Richmond, VA, 800-888-2973, ext. 3361, and outside U.S. 804-264-2828, ext 3361.

Color Photographs: These are discouraged. However, if necessary, include an extra copy upon submission to obtain a cost estimate. Cost is borne by author.

Drawings: Submit graphs, charts, diagrams, and other drawings as glossy photographs made from finished drawings not requiring additional artwork or typesetting. Computer-generated graphics produced on high-quality laser printers are also usually acceptable. Do not handwrite any part of graph or drawing. Label both axes of graphs. Most graphs will be reduced to one-column width (3 5/16 in.); all elements must be large enough to withstand reduction. Avoid heavy letters and unusual symbols. Do not submit drawings larger than 8.5 by 11 in.

OTHER FORMATS

AUTHOR'S CORRECTIONS—Provides a means of correcting errors of omission and errors of a scientific nature that do not alter the overall basic results or conclusions of a published article. Send corrections of a scientific nature to the editor who handled the article. Addition of new data is not permitted.

ERRATA—Provides a means of correcting errors that occurred during t writing, typing, editing, or printing of a published article. Send directly to Journals Division.

LETTERS TO THE EDITOR—500 words, must include data to support writer's argument and are intended only for comments on articles published previously in the journal. A reply from the corresponding author of the article may be solicited. Type double-spaced.

MINIREVIEW—4 printed pages, brief summaries of developments in fast-moving areas, based on published articles, addressing any subject within the scope of the *IJSB*. Provide three double-spaced copies.

MINUTES OF MEETINGS—Secretaries of subcommittees should submit four copies of meeting minutes to ASM Journals Division as soon as possible after meeting. Minutes should be concise and to the point. Include brief statements of taxonomic opinions, but exclude details of any taxonomic discussions; these should be published in the *IJSB* as a report of the subcommittee.

Title:
International Committee on Systematic Bacteriology
Subcommittee on the Taxonomy of . . .
Minutes of the Meeting, day–month–year–City, Country: Number minutes consecutively. Head by as few words as possible to indicate content. Head first minute "call to order"; mention time, place, and person who called meeting to order. Head second minute "record of attendance;" differentiate between subcommittee members and invited guests. Other minutes deal with significant business conducted during meeting. Include in last minutes a complete listing of current membership, time and place of next meeting, if known, and a statement of adjournment.
See *Int. J. Syst. Bacteriol.* **39**:500-501, 1989, and **40**:213-215, 1990.

NOTES—Submit as full-length papers. Brief observations that do not warrant full-length papers. Include a 50 word abstract. Do not use section headings; report methods, results, and discussion in a single section. Paragraph lead-ins are permissible. Keep text to a minimum, 1000 words, if possible; keep figures and tables to a minimum. Present acknowledgments as in full-length papers, do not use heading. References section is identical to full-length papers.

REQUESTS FOR OPINIONS—Exceptions to rules of bacteriological nomenclature may be requested of the Judicial Commission of the ICSB. Submit fully documented statement of the relevant facts. State basis of challenge and support it by a documented statement of relevant facts. Submit requests for opinions or challenges of requests for opinions (or challenges of proposed or issued opinions) to the ASM Journals Division.

RETRACTIONS—Reserved for major errors or breaches of ethics that, for example, may call into question the source of data or the validity of results and conclusions of an article. Send a retraction and an accompanying explanatory letter signed by all authors to editor in chief.

TAXONOMIC NOTES—Presents material or proposals in advance of formal discussion at a meeting of the International Committee for Systematic Bacteriology so that there may be international awareness of the item. Notes may also transmit items of importance to procaryotic systematics arising from publications other than *IJSB*, from the ICSB, or from individual scientists. These may be communicated as for papers or directly to the editor-in-chief.

OTHER

Abbreviations: Limit use. Use abbreviations other than those recommended by the IUPAC-IUB *(Biochemical Nomenclature and Related Documents,* 1978) only when necessary, such as in tables and figures. Use pronouns or paraphrase a long word after its first use. Use standard chemical symbols and trivial names or their symbols (folate, Ala, Leu, etc.) for terms that appear in full in neighboring text. Introduce all abbreviations except those listed below in first paragraph in Materials and Methods or define each abbreviation and introduce it in parentheses the first time it is used. Eliminate abbreviations not used at least five times in text (including tables and figure legends).

In addition to abbreviations for Systeme International d'Unités (SI) units of measurement, other common units (e.g., bp, kb, and Da), and chemical symbols of elements, use following without definition: DNA (deoxyribonucleic acid); cDNA (complementary DNA); RNA (ribonucleic acid); cRNA (complementary RNA); RNase (ribonuclease); DNase (deoxyribonuclease); rRNA (ribosomal RNA); mRNA (messenger RNA); tRNA (transfer RNA); AMP, ADP, ATP, dAMP, ddATP, GTP, etc. (for the respective 5′ phosphates of adenosine and other nucleosides) (add 2′-, 3′-, or 5′- when needed for contrast); ATPase, dGTPase, etc. (adenosine triphosphatase, deoxyguanosine triphosphatase, etc.); NAD (nicotinamide adenine dinucleotide); NAD+ (nicotinamide adenine dinucleotide, oxidized); NADH (nicotinamide adenine dinucleotide, reduced); NADP (nicotinamide adenine dinucleotide phosphate); NADPH (nicotinamide adenine dinucleotide phosphate, reduced); NADP+ (nicotinamide adenine dinucleotide phosphate, oxidized); poly(A), poly(dT), etc. (polyadenylic acid, polydeoxythymidylic acid, etc.); oligo(dT), etc. (oligodeoxythymidylic acid, etc.); P_i (orthophosphate); PP_i (pyrophosphate); UV (ultraviolet); PFU (plaque-forming units); CFU (colony-forming units); MIC (minimal inhibitory concentration); MBC (minimal bactericidal concentration); Tris [tris(hydroxymethyl)aminomethane]; DEAE (diethylaminoethyl); A_{260} (absorbance at 260 nm); EDTA (ethylenediaminetetraacetic acid); and AIDS (acquired immunodeficiency [or immune deficiency] syndrome).

Do not define abbreviations for cell lines.

Use following abbreviations without definition in tables:

amt (amount)	SE (standard error)
approx (approximately)	SEM (standard error of the mean)
avg (average)	sp act (specific activity)
concn (concentration)	
diam (diameter)	sp gr (specific gravity)
expt (experiment)	temp (temperature)
exptl (experimental)	tr (trace)
ht (height)	vol (volume)
mo (month	vs (versus)
mol wt (molecular weight)	wk (week)
no. (number)	wt (weight)
prepn (preparation)	yr (year)
SD (standard deviation)	SE (standard error)

DNA Base Composition: Report moles percent guanine plus cytosine to nearest whole number (accuracy equal to ±1%) in different runs for all methods.

Names, Scientific: Species and subspecies may not be referred to by the specific or subspecific epithet alone; use epithet in combination with the generic name. Spell out a generic name followed by a specific epithet at first use in text; subsequently, it may be abbreviated to its capitalized initial letter if context makes meaning clear. If there are several generic names in text with the same initial letter, spell out names each time.

Names, Vernacular: Generic names are singular Latin nouns; do not use plural verbs. Avoid use of generic names alone when reference is to members of the genus. Many microorganisms are known by both their vernacular (provincial, common) and scientific names. There are no rules for the use of vernacular names. It is often convenient to use vernacular names coined from generic names. Do not use initial capital letters or italics. For plural forms of vernacular names, Latin or other plural endings are used.

Neotype Strains: A neotype strain is a strain that replaces, by international agreement, a type strain which is no longer in existence. Neotype should possess characteristics given in original description; explain any deviations. A proposal of a neotype strain for consideration by the Judicial Commission must be published in the *IJSB* with a reference (or references) to first description and name for microorganism, a description (or reference to a description) of proposed neotype strain, and a record of the author's designation for neotype strain and of at least one culture collection from which cultures are available. The neotype strain becomes established 2 years after publication provided there are no objections. Propose a neotype strain only after a careful search has shown that none of the strains on which the original description was based is extant. If an original strain is subsequently discovered, refer matter immediately to Judicial Commission.

New Taxon: For a description of a new taxon, include: a list of strains included in taxon; a statement or tabulation of characteristics of each strain (see Strain Data); a list of characteristics considered essential for membership in taxon; a list of characteristics which qualify taxon for membership in the next higher taxon; a list of diagnostic characteristics which distinguish it from closely related taxa; and give reactions of type strain for all characteristics that vary among strains within species. Use photomicrographs and electron micrographs as part of description to show morphological or anatomical characters pertinent to classification.

Nomenclature, Bacteria: Use only names included in *Approved Lists of Bacterial Names* (V. B. D. Skerman, V. McGowan, and P. H. A. Sneath [ed.]) and the *Index of the Bacterial and Yeast Nomenclatural Changes Published in the International Journal of Systematic Bacteriology since the 1980 Approved Lists of Bacterial Names (I January 1980 to I January 1989)* (W. E. C. Moore and L. V. H. Moore [ed.]), American Society for Microbiology, and those published in *IJSB* since 1 January 1989. If there is reason to use another name, enclose name in quotation marks and make an appropriate statement concerning the nomenclatural status of name in text (see *Int. J. Syst. Bacteriol.* **30**:547-556, 1980).

Nomenclature, Yeasts and Yeastlike Organisms: Use only correct names of taxa. Although an organism may have a number of correct names, use one particular name consistently; if there are objections to its use, cite this name as a synonym. Follow *The Yeasts: a Taxonomic Study* (3rd ed. N. J. W. Kreger-van Rij, ed., Elsevier Science Publishers B.V., Amsterdam, 1984) and *Yeasts: Characteristics and Identification,* (2nd ed., New York: Cambridge University Press, 1990). If an author disagrees with nomenclature, follow first use of scientific name in text and in Abstract by name, in parentheses, as given in *The Yeasts*.

Nucleic Acid Sequences: Present nucleic acid sequences of limited length freestyle in the most effective format. Submit longer nucleic acid sequences as camera-ready copy 8.5 by 11 in. in standard (portrait) orientation. Print sequence in lines of 100 bases, in a nonproportional (monospace) font which is easily legible when published at 100 bases/6 in. Use uppercase and lowercase letters to designate exon/ intron structure, transcribed regions, etc., if lowercase letters remain legible. Number sequence line by line; place numerals, representing the first base of each line, to left of lines. Minimize spacing between adjacent lines of sequence, leave room only for annotation of sequence. Annotation may include boldface, underlining, brackets, boxes, etc. Encoded amino acid sequences may be presented, if necessary, immediately above first nucleotide of each codon, using single-letter amino acid symbols. Comparisons of multiple nucleic acid sequences should conform to the same format.

Nucleotide Sequences: Include a GenBank/EMBL accession number for primary nucleotide and/or amino acid sequences when used for taxonomic purposes. Quote accession numbers when using existing sequence from a data bank for comparison or to calculate relationships. Place in a separate paragraph in Materials and Methods section of papers or at end of text of Notes. The availability of taxonomically significant sequences from defined bacterial strains is considered crucial to modern bacterial systematics; acceptance of a manuscript requires acceptance of this principle.

GenBank Submissions, National Center for Biotechnology Information, Bldg. 38, Rm 8N-803, 8600 Rockville Pike, Bethesda, MD 20894; e-mail (new subscriptions): gb-sub@ncbi.nlm.nih.gov; e-mail (updates): update@ncbi.nlm.nih.gov.

EMBL Data Library Submissions, Postfach 10.2209, Meyerhofstrasse 1, 6900 Heidelberg, Germany; telephone: 011 49 (6221) 387258; fax: 011 49 (6221) 387306; electronic mail (data submissions): datasubs@embl.bitnet. telephone (202) 737-3600.

Strain Data: Present characteristics of each strain in text if practical or in a strain table if list is complex. List: characters for which all strains were positive; characters for which all strains were negative; and characters for which there were strain differences, with a listing of strains that gave the less common result (see Holmes et al., *Int. J. Syst. Bacteriol.* 27: 330–336, 1977, Tables 2 through 4). With very large numbers of strains, it is not practical to provide individual strain data; instead, cite percentage of strains that gave a positive or negative result for each character determined.

Style: Use *CBE Style Manual* (5th ed., 1983; Council of Biology Editors, Inc., Bethesda, Md.), *ASM Style Manual for Journals and Books* (American Society for Microbiology, 1991), and Robert A. Day's *How to Write and Publish a Scientific Paper* (3rd ed., 1988, Oryx Press), as interpreted and modified by the editor and the ASM Journals Division. Follow recommendations of Demerec et al. (*Genetics* 54:61–74,1966) for genetic symbols. Express enzyme activity in terms of international units (*Enzyme Nomencla*ture, Academic Press, Inc., 1992) and give EC number parenthetically at first use in text. To express lengths, weights, and volumes, use prefixes nano (n) and pico (p), not milli-micro (mμ) and micro-micro ($\mu\mu$). Express lengths in nanometers (nm; 10^{-9} m) or micrometers (μm; 10^{-6} m) instead of millimicrons (mμ; 10^{-9} m), microns (μ; 10^{-6} m), or Angstroms (Å; 10^{-10} m). Express parts per million (ppm) as micrograms per milliliter (μg/ml), micrograms per gram (μg/g), or microliters per liter (μl/liter), as appropriate.

Type Strains: Indicate all type strains at each occurrence in text, tables, and figures by a capital superscript T.

Units: Use standard metric units for reporting length, weight, and volume. For these units and for molarity, use prefixes m, μ, n, and p for 10^{-3}, 10^{-6}, 10^{-9}, and 10^{-12}, respectively. Use prefix k for 10^3. Avoid compound prefixes . Use μg/ml or μg/g in place of ppm. Present units of temperature as 37°C or 324 K. When fractions are used to express units such as enzymatic activities, use whole units, such as "g" or "min," in denominator instead of fractional or multiple units, such as μg or 10 min. Use unambiguous forms such as exponential notation instead of multiple slashes. See *CBE Style Manual*, 5th ed., for more detailed information about reporting numbers and information on SI units for reporting illumination, energy, frequency, pressure, and other physical terms. Report numerical data in applicable SI units.

Valid Publication of Names of Bacterial Taxa: Follow principles and rules of nomenclature published in the *International Code of Nomenclature of Bacteria (1990 Revision)*, American Society for Microbiology, 1992.

SOURCE

January 1994, 44(1): i-x.

Investigative Ophthalmology & Visual Science
Association for Research in Vision and Ophthalmology

Investigative Ophthalmology & Visual Science
550 North Broadway, Suite 412
Baltimore, MD 91205
Telephone: (410) 614-0392; Fax: (410) 614-0389

AIM AND SCOPE

Investigative Ophthalmology & Visual Science (IOVS) welcomes the submission of manuscripts describing laboratory and clinical investigations of the eye and the visual processes.

MANUSCRIPT SUBMISSION

Manuscripts may not been published previously anywhere in any language and nor be under simultaneous consideration by another publication. Authors are responsible for all statements in work, including changes made by copy editor.

In cover letter include following written statement, signed by each author:

The undersigned author(s) transfers all copyright ownership of the manuscript (title of article) to the Association for Research in Vision and Ophthalmology, Inc., in the event the work is published. The undersigned author(s) warrants that the article is original, is not under consideration by another journal, and has not been published previously other than as an abstract.

U. S. government employees use following statement:

This is to certify that the above-named article is a work of authorship prepared as part of the undersigned author's (authors') official duties as an officer or employee of the U. S. government and is therefore in the public domain. Should, however, the article ever be determined to be copyrightable, I (we) hereby transfer, assign, or otherwise convey all copyright ownership in the above-named article to the Association for Research in Vision and Ophthalmology, Inc.

Provide appropriate disclosure of financial interests according to the ARVO Proprietary Interest Policy.

MANUSCRIPT FORMAT

Type on one side only of 8.5 × 11 in. paper. Use double-spacing throughout text, tables, legends, and references. Leave 1 in. margins. Do not justify right margin. Single space after periods. Put first author's name on each page, and number all pages consecutively, beginning with title page. Submit five copies of typescript and four sets of illustrations. Do not send original art. Attach each set of illustrations to a copy of manuscript.

Arrange manuscript in order: title page; structured abstract; text; key words; acknowledgments, references, footnotes, figure legends and tables.

Submit manuscripts on disk. Contact *IOVS* editorial office in Baltimore for disk preparation guidelines.

Titles/Subheadings: Use descriptive clauses, not complete sentences or questions. Completely capitalize major headings; subheads have initial caps only.

Title Page: Title is 86 characters, including spaces, explicit for indexing. Provide a short title (55 characters). Give full names, degrees, affiliations, and addresses of author(s). In a footnote, provide name and mailing address of corresponding and reprint request author(s). Acknowledge financial support and sponsoring organizations. Disclose Proprietary Interest as: "Proprietary interest category: Cc7." For codings see *Invest Ophthalmol Vis Sci* 1994;35(8): iv. Select section code into which manuscript fits best. See *Invest Ophthalmol Vis Sci* 1994; 35(8): iv for codes.

Abstracts: 250 word structured abstract for articles and reports arranged under headings: Purpose, Methods, Results, Conclusions. Use a minimum of acronyms and no references.

Key Words: List five selected from Reviewer Topic List (*Invest Ophthalmol Vis Sci* 1994; 35(8): viii).

Text: 14 double-spaced typewritten pages. Direct style toward wide spectrum of vision researchers in general. In a brief introduction (without a subheading) describe research rationale and objectives without extensive review of literature. In "Methods" include experimental design, subjects used, and procedures (identify previously published procedures by reference only).

Use standard chemical or nonproprietary pharmaceutical nomenclature. Identify specific sources by brand name, company, city, and state, or country in parentheses. Present "Results" with a minimum of discussion. Cite all tables and figures in numerical order. Limit "Discussion" to significance and limitations of data.

Obtain permission to reprint illustrations or tables from other publications in writing.

Acknowledgments: Write in third person.

References: List in numerical order. Cite all references in text or tables and show as superscript numbers in text. List unpublished data (including material in preparation or submitted) or personal communications parenthetically in text. References to journal articles include: author(s) (if more than six, write "et al." after third name); title; journal name (as abbreviated in *Index Medicus*); year; volume number; and inclusive page numbers. References to books include: author(s); chapter title (if any); editors (if any); title of book; city of publication; publisher; year; and page, if indicated. Verify accuracy of references.

JOURNALS

1. Bill A. Aqueous humor dynamics in monkeys (*Macaca irus and Cercopithecus ethiops*). *Exp Eye Res.* 1971;11:195–200.

2. Bito L Z, DeRousseau CJ, Kaufman PL, Bito LZ. Age-dependent loss of accommodative amplitude in rhesus monkeys: an animal model for presbyopia. *Invest Ophthalmol Vis Sci.* 1982;23:23–31.

3. Sommer A, Tielsch JM, Katz J, et al. Racial differences in the cause-specific prevalence of blindness in east Baltimore. *N Engl J Med* 1991;325:1412-11417.

4. Engerman RL, Kern TS. Effect of sorbinil on retinopathy in diabetic dogs and galactosemic dogs. ARVO Abstracts. *Invest Ophthalmol Vis Sci.* 1990;31(suppl):124.

BOOKS

1. Stryer L. *Biochemistry.* 2nd ed. San Francisco: WH Freeman Co; 1981:559-596.

2. Gouras P, Lopez R, Brittis M, Kyeldbye HM, Fasano MK. Transplantation of retinal epithelium. In: Agardh E, Ehinger B, eds. *Retinal Signal Systems, Degenerations and Transplants.* Amsterdam: Elsevier; 1986:271-286.

Tables: Type each double-spaced on a separate page. Number consecutively in Arabic numerals, and supply brief titles. Give each column a short heading. For footnotes in tables, use symbols in sequence: *, †, ‡, §, ||, ¶ and #. Do not present data that can be given in text in one or two sentences in a table. Limit legends to 40 words. Make fully understandable without recourse to text.

Illustrations: Use minimum number and size to clarify text. Submit four sets. For one unmounted set, type figure number, name of first author, and arrow indicating TOP on a gummed label and affix to back of each illustration. Do not write on illustrations with ballpoint pen. Mount other sets as they will appear in journal or submit photographic composite of entire figure with labels preferably on front, including author, manuscript number, and figure number.

Make letters, numbers, bars, and markers large and solid for maximum legibility and proportionality. Illustrations will be reduced if not 3 1/8 in. (19 picas), 5 1/16 in. (30.5 picas), or 6 5/8 in. (40 picas). If necessary, submit an instructional sketch detailing set-up desired. Illustrations may be relettered for uniformity of style. Provide magnification and stain when pertinent. Use magnification bar on original photographs. If no bar, recalculate magnification in legend.

OTHER FORMATS

ANNOUNCEMENTS—Submit general information such as obituaries and notices of research funding opportunities to Editorial Office.

LETTERS TO THE EDITOR—Either relevant to material published in *IOVS* or to problems of general interest to vision scientists. Correct errors, provide support or agreement, or offer different points of view, clarifications, or additional information. Provide a brief title of less than 45 spaces.

RECENT DEVELOPMENTS IN VISION RESEARCH—Solicited short reviews of new research findings or new general methodologies that are of broad interest.

REPORTS—10 double-spaced typed pages; 10 references and four figures and tables. Rapid communication of important new information submitted in same format as regular articles, except each major section does not start on a new page.

OTHER

Abrreviations, Symbols, and Acronyms: Define first time used.

Color Charges: Between $1,100 and $2,000. Estimate will be sent before decision is made to print in color.

Experimentation, Animal: Include confirmation of adherence to ARVO Statement for the Use of Animals in Ophthalmic and Vision Research (see *Invest Ophthalmol Vis Sci* 1994; 35(8): v).

Experimentation, Human: State in "Methods" section that research followed tenets of the Declaration of Helsinki (*Invest Ophthalmol Vis Sci* 1994; 35(8): vi-vii); informed consent was obtained after nature and possible consequences of study were explained; and where applicable, research was approved by institutional human experimentation committee.

Style: Follow guidelines of style, terminology, measurement, and quantitation in *American Medical Association Manual of Style.*

Units: Use Systeme International (SI) measurements throughout the manuscript. Exceptions are visual acuity measurements and intraocular pressure recordings.

SOURCE

July 1994, 35(8): i-x.
"Information for Contributors" appears in January and July.

In Vitro Cellular & Developmental Biology
Society for In Vitro Biology

Wallace L. McKeehan
Albert B. Allkek Institute of Biosciences and Technology
Texas A&M University
2121 W. Holcombe Blvd.
Houston, TX 77030-3303
Telephone: (713) 677-7522

AIM AND SCOPE

Original manuscripts describing results of cellular, molecular and developmental biology research using in vitro grown or maintained organs, tissues or cells derived from multicellular organisms, will be considered for publication. The complete scope of the journal is embodied in the following matrix: Biotechnology—Industrial and applied applications, large scale production, technology transfer, in vitro fertilization, bioprocessing, genetic engineering, in vitro alternatives, transgenes; Cellular and Molecular Toxicology—In vitro alternatives, industrial and environmental concerns, genotoxicity, chemical carcinogenesis in vitro; Cellular Models—Developmental biology, cardio-

vascular studies, anatomy and cell biology, in vitro alternatives, cell culture of tissues and organs, development of cell culture systems, cellular aging; Growth, Differentiation and Senescence—Embryology, teratology, signal transduction, oncogenes, oncology, developmental biology, nutrition and metabolism, in vitro alternatives cellular aging, growth, growth factors, apoptosis; Infectious Diseases/Cellular Pathology—Protozoology, virology/AIDS, mycoplasmology, in vitro alternatives, host parasite interaction; Immunology—Immunogenetics, hematopoesis, in vitro alternatives, hybridomas, recombinant antibodies; and Genetics—Gene therapy, cytogenetics, immunogenetics, genetic engineering, stem cell isolation and cultivation, in vitro fertilization, transgenes, molecular genetics, apoptosis, in vitro alternatives.

MANUSCRIPT SUBMISSION

Select one of the 7 matrix headings, which best covers the manuscript's scope. Submission implies that all authors have agreed to its submission and that it is neither under consideration for publication elsewhere nor has appeared previously in part or in whole. Send a Copyright Transfer Form assigning copyright to the society for In Vitro Biology. Photocopy from January issue or obtained from editor. Submit manuscript on 8.5 × 11 in. paper, typed, double-spaced (including tables, references, and legends for figures), with 1 in. margins. Also submit 5.25 or 3.5 in. HD disk, along with three hard copies complete with all tables and illustrative material. Use either: WordPerfect 5.1 or 5.2 for DOS or Windows, Microsoft Word for Windows, or Microsoft Word for Macintosh.

MANUSCRIPT FORMAT

Start each section on a new page. Number all pages. First page (page 1): Title, Author's name, Institution, condensed title for running head (35 letters and spaces), and full address of corresponding and reprint request author(s). Second page: summary (250 words) and six key words for indexing.

Use descriptive subheadings, e.g., SUMMARY, INTRODUCTION, MATERIALS AND METHODS, RESULTS, DISCUSSION, and REFERENCES. Results and Discussion may be combined. Use footnotes only in tables as superscript lower case letters placed at bottom of page. Place acknowledgments on a separate sheet preceding references.

References: Follow style of *Council of Biology Editors Style Manual,* 5th ed. which is an alphabetical listing of references numbered sequentially. Text citation is to number assigned. Alternatively, references may be cited in text in either form recommended by the IUB Committee of Editors of Biochemical Journals (1973), *J. Biol. Chem.* 248:7279–7280; e.g., "Rat lung organ cultures were prepared as previously described (4)" or "Rat lung organ cultures were prepared as previously described (Weinhold et al., 1979)". For latter method, list is compiled in alphabetical order without numbers. For each manuscript, choose one citation system. Use *Index Medicus* journal title abbreviations; provide inclusive pages of work cited; provide title in original language (when reproducible in English alphabet) and state if translated. Verify accuracy of references with original publications. Cite as references only papers that have been accepted for publication. Cite "manuscripts in preparation", "unpublished results", "personal communications", etc., in text. Verify personal communications with information supplier.

1. Council of biology editors style manual. CBE style manual committee, 5th ed. Bethesda, MD: Council of Biology Editors; 1983.

2. Mokul'skaya, T. D.; Smetznina, E. P.; Mychko, G. E.; Mokul-skii, M. A. Secondary structure of DNA from phages T_4 and T_6. Mol. Biol (Moscow) 9:445–449; 1976. Translation of Mol. Biol. (Moscow) 9:552–555; 1975.

3. Trowell, O. A. Tissue culture in radiobiology. In: Willmer, E. N. ed. Cells and tissues in culture. Methods, biology and physiology. Vol. 3. London: Academic Press; 1966:63–149.

4. Weinhold, P. A.; Burkel, W. E.; Fischer, T. V.; Kahn, R. H. Adult rat lung in organ culture: maintenance of histopathic structure and ability to synthesize phospholipid. In Vitro Cell. Dev. Biol. 15:1023–1031; 1979.

Illustrations, Figures and Tables: Number photographs, text figures, and tables in agreement with text, identify on reverse side with title of manuscript and author's name. Number and lace after last page of text.

Print photographs on glossy paper. Multiple photographs forming a plate should not exceed 7×8 in. Single photographs should measure 3.37 in. wide (single column) or 7 in. wide (double column). State magnification of photomicrographs.

Give tables short titles and make intelligible without reference to text.

OTHER

Abbreviations: Use minimally. Where necessary, spell out abbreviated term followed by abbreviation in parentheses where first cited. Do not use in title or summary.

Cell Line and Reagent Data: Indicate source of cells utilized, species, sex, strain, race, age of donor, whether primary or established. State name, city, and state abbreviation of source of reagents within parentheses when first cited. Identify specific tests used for verification of cell lines and novel reagents. Specific tests for the presence of mycoplasmal contamination of cell lines are recommended. If not performed, clearly state. Include other data relating to unique biological, biochemical and/or immunological markers if available. Make unique reagents available to qualified investigators.

Nomenclature: Follow recommendations of Tissue Culture Association Committee on Terminology: Schaeffer, W. I. Terminology Associated with Cell, Tissue and Organ Culture, Molecular Biology and Molecular Genetics. *In Vitro Cell. Dev. Biol.* 26:97–101; 1990.

Page Charges: $45.00 per page. Each article is permitted the equivalent of one full page of tables, photographs, and figures at no additional cost. Excess material is $55 per page. Cost of color illustrations will be paid by author.

Reviewers: Suggest desired reviewers in field.

SOURCE
August 1994, 30A(8): inside back cover.

Japanese Journal of Cancer Research
Japanese Cancer Association
Elsevier Science Publishers

Editorial Office
Japanese Cancer Association, c/o Cancer Institute
Kami-lkebukuro 1-37-1
Toshima-ku, Tokyo 170, Japan

AIM AND SCOPE

The purpose of the Journal is to publish the results of original research relevant to the field of cancer research. The Journal accepts manuscripts describing original observations in laboratory as well as clinical research on all aspects of cancer from all over the world. Each manuscript should be complete, and the submission of preliminary or inconclusive results is discouraged. Manuscripts should consist of work which has not been published in any language, and which is not under consideration for publication elsewhere. There are three types of publications: Rapid communications that are brief and contain significant new results and ideas; Regular papers, and Review articles on subjects of importance to cancer researchers. Such review articles are generally prepared at the request of the Editor. In principle, authors of regular papers and rapid communications must be members of the Japanese Cancer Association. However, membership in the Association is not a prerequisite for publication in the journal. Copyright of articles printed in the Japanese Journal of Cancer Research lies with the Japanese Cancer Association.

MANUSCRIPT SUBMISSION

Submit manuscripts by registered mail. Submit one original and two clear copies. For photographs, submit two additional glossy photographs for review copies. Include copies of related papers submitted for publication elsewhere, or in press, and, published manuscripts relevant to work described.

MANUSCRIPT FORMAT

Double-space manuscripts with pica typescript (10 letters/2.5 cm) on a good quality paper of A4 or international size. For computer assisted printers, use type font such as Courier. Do not use non-standard characters (e.g., italic or boldface, etc.) or dot matrix printers. Keep a 2.5 cm margin on each side and 3 cm at top and bottom of each sheet. Make papers intelligible to the general Journal reader. Condense to the greatest extent compatible with clarity. Conform to Journal style. Look at a recent issue for assistance in proper arrangement.

Restrict length of a regular paper to 8 printed pages (3 typed pages correspond to 1 printed page) including figures and tables; rapid communication-4 printed pages including figures and tables. Number pages consecutively starting with title page. Assemble regular papers in order: Title page (including full title, author(s), affiliation(s), key words, running title and abbreviations), Summary, Introduction, Materials and Methods, Results, Discussion, Acknowledgments, References, Tables, Figure legends and Figures.

Title: 120 letters. Omit unnecessary words such as "Study of," "Results on," etc. Include key words which would serve as index words. Do not use abbreviations.

Authors and Affiliations: Follow title by full name(s) of author(s) and full names of institutions and subsidiary laboratories with complete address including postal code number. If several authors and institutions are given, indicate with which department and institution each author is affiliated with numerical superscripts. If author(s) have moved since, indicate present address, using a symbol and footnote.

Key Words: List five key words from title, summary, or text below affiliation as an aid to indexing. Avoid nonspecific terms that may be broadly interpreted.

Running Title: 40 letters, including space.

Summary: 250 words on a separate sheet of paper. Briefly state problem and experimental approach. Summarize important new results and findings, and give final conclusion. Indicate species, strain and sex of animals used. Write in one paragraph suitable for reproduction by an abstract service.

Introduction: Mention purpose of investigation, its relation on other work in the field, and reasons for undertaking research. Do not include all literature related to research. Reference the most relevant papers.

Materials and Methods: Briefly describe methods. Give sufficient detail to enable others to reproduce experiments. Cite appropriate references for published procedures. Completely describe only new and significant modifications of previously published procedures. Give names of products and manufacturers with their locations only if products are very specialized or may not be obtainable from other sources,. Always indicate sources of animals and tumors.

Results: Describe concisely. Use tables and figures only if indispensable for comprehension of data. Indicate appropriate location for each table and figure by marginal notes. Avoid excessive explanation of data in tables and figures. Combine Results and Discussion sections only if space is saved or presentation of results is improved.

Discussion: Interpret results and relate to existing knowledge in field. Do not repeat information already given in Introduction or Results.

Acknowledgments: Acknowledge financial support, aid in technical fields, animal care, photography, stenography, advice from colleagues, personal gifts, etc.

References: Verify accuracy of references; check against original reports. Number consecutively in order cited in text with numerical superscripts, such as [1], [2], [3]...... arrange list so that entries are in numerical order. List serial publications, such as *Advances in Cancer Research* and *Gann Monograph,* as journals.

1) Ogawa, Y., Sagata, N., Tsuzuku-Kawamura. J., Onuma, M., Izawa, H. and Ikawa, Y. Methylation pattern of the bovine leukemia provirus genome in bovine leukemic cells. Jpn. J. Cancer Res., 76, 5–8 (1985).

2) Harnden, D., Morten, J. and Featherstone, T. Dominant susceptibility to cancer in man. Adv. Cancer Res., 41, 185–256 (1984).

3) Henson, R. A. and Urich, H. "Cancer and the Nervous System," pp. 171–183 (1982). Blackwell Scientific Publications, Oxford.

4) Sugimura, T. A view of a cancer researcher on environmental mutagens. In "Environmental Mutagens and Carcinogens," ed. T. Sugimura, S. Kondo and H. Takebe, pp. 3–20 (1982). University of Tokyo Press. Tokyo.

If paper is in English, French or German, write in original language. Translate title of an essential citation in another language into English, indicate original language in parentheses. For journal abbreviations use international system given in ISO 833 and *Chemical Abstracts*. Cite papers accepted for publication as "in press." Cite papers in preparation or have been submitted but not yet accepted in text as "unpublished results" or "personal communications." Obtain permission to cite unpublished work of others.

Tables and Figures: Keep to a minimum.Write titles and descriptions of tables and figures in the same font as text.

Tables: Number consecutively with Roman numerals. Give each one a title; capitalize all words except articles, prepositions and conjunctions. Make column headings brief. For footnotes, use alphabetic superscripts, such as [a], [b] ...Give numerical values with Units in appropriate columns.

Figures: Use to document experimental results or methods that cannot be described adequately in text. Number consecutively in Arabic numerals (Fig. 1, Fig. 2). Size graphs and photographs to fit one column (8.2 cm) or two columns (16.4 cm). Materials of inappropriate size will be reduced. Type titles and legends to figures on a separate sheet(s).

Draw graphs in black ink on tracing paper or graph paper with light blue lines; do not hand-write. Make symbols, letters and numbers large enough to be legible.

Print photographs on glossy paper, with good contrast. Do not paste on board; place in an envelope, with author's name, figure number and orientation indicated on back. Arrange a group of photographs together in a single block (composite figure), if desired. Make letters, symbols and arrows applied to surface of photographs sufficiently large to be recognized easily. Use dry transfer lettering. Use scale markers, in image, in electron micrographs. For color illustrations, submit original color transparencies along with three sets of color prints. Their necessity will be decided by Editor; cost will be charged to authors.

OTHER FORMATS

RAPID COMMUNICATION—4 printed pages, including figures and tables,120 word summary, introduction, materials and methods, results and discussion, and acknowledgments in a single section without section headings. References are the same as in regular papers.

OTHER

Abbreviations: If there are many frequently used non-standard abbreviations, list in a footnote. Be consistent with recommended international nomenclature. If only a few abbreviations, give in parentheses in text.

Experimentation, Animal: Follow guidelines for care and use of laboratory animals.

Experimentation, Human: Carry out experiments involving either human subjects or the use of materials of human origin in accordance with the principles of the Declaration of Helsinki.

Measurement, Signs, and Symbols: Use units of measurements in international usage. Explain new units or symbols in text. Give units of measurements in SI units and temperatures in degrees Centigrade. Omit periods after units of measurement. Do not use plurals (e.g., cm, mm, ml, h, min, s, 37°C, etc.). Use % rather than per cent or percent.

Nomenclature: For names of chemical compounds, use International Union for Pure and Applied Chemistry (IUPAC) nomenclature; for biochemical terms use the International Union of Biochemistry (IUB) nomenclature. Avoid use of trade or proprietary names; if necessary add chemical name in parentheses. Begin trade names, with a capital letter; place name of producer in parentheses immediately following first mention of name. For names of enzymes use nomenclature recommended by IUB; follow name by enzyme number (EC) in parentheses. Give chemical structures only for uncommon compounds or where structural details are discussed. Draw structure clearly. Use chemical formulae for common inorganic compounds in experimental section. Write labeled organic compounds according to IUPAC rules.

Page Charges: $30 contribution to the cost of processing the manuscript. If length exceeds 8 printed pages including figures and tables, ¥5,000 per page for extra pages.

Safety: Perform experiments in recombinant DNA research in accordance with guidelines set up by authorized agency in country where research was done. When hazardous materials or dangerous procedures are used in experiments, describe precautions in "Materials and Methods."

SOURCE

January 1994, 85(1): iii-v.

JGM - Journal of General Microbiology
see Microbiology

Journal of Allergy and Clinical Immunology

American Academy of Allergy and Immunology
Mosby

Philip S. Norman, MD
5501 Hopkins Bayview Circle
Baltimore, MD 21224–6821
Telephone: (410) 550-2152; Fax: (410) 550-2684

MANUSCRIPT SUBMISSION

Items must be contributed solely to the *Journal of Allergy and Clinical Immunology* and have not been or will not be published elsewhere except in abstract form. Authors will be consulted, when possible, regarding republication of their material.

Acknowledge on title page, all funding sources and institutional or corporate affiliations of authors. In cover letter disclose any commercial associations that might pose a conflict of interest, including consultant arrangements, stock or other equity ownership, patent licensing arrangements, or payments for conducting or publicizing study. Disclosure will be held in confidence; if paper is accepted, Editor will determine how conflict of interest will be disclosed.

In cover letter include the following written statement, signed by one author: "The undersigned author transfers all copyright ownership of the manuscript (title of article) to Mosby-Year Book in the event the work is published. The undersigned author warrants that the article is original, is not under consideration by another journal, and has not been previously published. I sign for and accept responsibility for releasing this material on behalf of any and all coauthors."

MANUSCRIPT FORMAT

In original manuscripts describe fully, but concisely, the results of original clinical and/or laboratory research. Print in an easily readable font. Double-space all sections, including references, with 1 in. margins on all sides. Submit an original and two copies (three sets) ensuring that each copy is good quality. On each page place last name of first author and page number in upper right corner. Title page is page 1. Organize manuscript in order.

Title Page: Full title should be concise. Follow with list of authors, including full name, highest academic degree, and affiliation. Restrict number of authors to those who have made material contributions to the research, writing and review of manuscript. In bottom half of title page, list name, address, telephone and fax numbers of corresponding and reprint request authors.

Abstract Page: 200 words; summarize results and conclusions concisely. Do not include tabular data or acronyms. List 10 key words following abstract.

Organize abstracts of clinical studies papers as: Background (problem that prompted study); Objective (purpose of study); Methods (how study was done); Results (most important findings); Conclusions (most important conclusion).

Text: Write clearly and concisely in English. Organize text in sections: Introduction, Materials and Methods, Results, and Discussion. Begin each section on a new page. Use generic terms for all drugs and chemicals.

Acknowledgments: List general acknowledgments for consultations, statistical analyses, and the like at end of text, including names of individuals involved. Acknowledge financial support from external agencies, including commercial companies.

References: Journal names should follow *Cumulated Index Medicus* style. Do not cite manuscripts in preparation, personal communications, and other unpublished information in reference list; mention in text in parentheses. Identify references in text by superscript Arabic numerals. Number in consecutive order as mentioned in text. Type reference list double-spaced at end of article in numeric sequence. Use format in "Uniform Requirements for Manuscripts Submitted to Biomedical Journals" (Vancouver style) (*N Engl J Med* 1991;324:424-8).

List all authors when six or fewer; when seven or more, list first three and add *et al.*

JOURNAL ARTICLES

Salvatierra O, Vincenti F, Armend W, et al. Four-year experience with donor-specific blood transfusions. Transplant Proc 1983: 15:924–6.

BOOKS

Eisen HN. Immunology: an introduction. 5th ed. New York: Harper & Row, 1974:406.

CHAPTERS

Fromer PL. Workshop on congestive heart failure. In: Braunwald E, Mock MB, eds. Congestive heart failure. New York: Grune & Stratton, 1982:1-2.

Tables: Data in tables should supplement, not duplicate, text. Type on pages separate from the text, one table per page. Make tables self-explanatory. Number in Roman numerals in order of mention in text. Provide a brief title for each. Define any abbreviations in footnote form at bottom of table. Do not submit glossy prints or reduced versions of typewritten tables.

Figures: A reasonable number of black-and-white figures will be published free of cost. Make special arrangements with publisher for color plates, elaborate tables, or extra illustrations. Cite all figures in text. Number in order of mention in text. Number lightly on back with author's name and an arrow indicating TOP.

Make black-and-white graphs legible. Print clearly in jet-black ink on heavy paper. Design to fit a single 3 in. column. Unless printed by a high resolution printer (1200 dpi or better), furnish in larger size, 6 in. or more wide. Make value labels and legends in easily legible type font such as Times Roman or Helvetica Regular. Use bold or italic type faces only for special emphasis. For labels capitalize first letter only. Place labels on graph rather than in legend. Use 14 point or more type size and line width 1 point or more for 6 in. graphics and proportionately larger for larger graphics. Fill bar graphs and pie charts with shades of gray at least 25% different in density for adequate contrast. Use three-dimensional graphics only if three coordinates. Submit one original of each laser or plotter printed figure consisting of a graph or line drawing. If computer-generated originals of figures are not available, submit one set of professionally prepared, glossy or matte finish, prints. Attach a photocopy to each copy of manuscript. Submit three originals of each figure consisting of a photograph, microphotograph, or immunoblot photograph. Prints should have good black-and white-contrast and may be matte or glossy finish. Mark originals clearly on back of each figure, sort into sets, and place in envelopes with name of first author. Do not mount original prints on paper. Note special instructions regarding sizing or grouping. Type suitable legends double-spaced on a separate sheet before tables.

OTHER FORMATS

ALLERGY GRAND ROUNDS—Discussions of important clinical subjects in a Grand Rounds format. Include: introductory comments; brief illustrative case presentations; multidisciplinary discussion by members of relevant specialties, with helpful graphic illustrations; and summary. Follow style used by Goldstein MF, Atkins PC, Cogen FC, Kornstein MJ, Levine RS, Zweiman B: Allergic *Aspergillus* sinusitis. *J Allergy Clin Immunol* 1985;76:515–24.

BRIEF REPORTS—750 words, five references and one table or figure and a short title, no abstract. Instructive case reports or short scientific reports. Submit three copies with a copyright statement and cover letter.

LETTERS TO THE EDITOR—500 words, including tabular data and three references; start with "To the Editor,"; and give a short title. Brief letters concerning recent publications in the *Journal* and other subjects of particular interest. A response from original authors of article will be requested when appropriate. Submit three copies with a cover letter and a copyright statement.

REVIEW ARTICLES—Reviews are requested in collaboration with the programs of the Postgraduate Education and Continuing Medical Education Committees of the American Academy of Allergy and Immunology. Unsolicited review manuscripts are not accepted.

ROSTRUM ARTICLES—Opinion articles about subjects of particular interest and/or debate are published at the request of the Editor.

OTHER

Acronyms/Abbreviations: Use only standard abbreviations. See *Council of Biology Editors Style Manual* or *AMA's Style Manual.* Do not use in title; avoid in abstract. Spell out laboratory and chemical terms or disease processes at first mention, with abbreviation/acronym following in parentheses. List all such abbreviations on a separate page.

Experimentation, Animal: Include a statement in Methods section indicating which guidelines for care and use of animals were followed (e.g., "Principles of Laboratory Animal Care," formulated by the National Society for Medical Research, or the "Guide for the Care and Use of Laboratory Animals" (NIH Publication No. 86–23, as revised).

Experimentation, Human: Include a statement describing approval by Institutional Review Board.

Nucleotide Sequence Data: Submit data to GenBank (Mail Stop K710, Los Alamos National Laboratory, Los Alamos, NM 87545 (505-665-2177). Include accession number in footnote.

Protein Sequence Data: Submit data to: Protein Identification Resource (PIR), National Biomedical Research Foundation, Georgetown University Medical Center, 3900 Reservoir Rd., N.W., Washington, D.C. 200007 (202-687-1672). Include accession number in footnote.

SOURCE

July 1994, 94(1): 21A–23A.

Journal of Analytical Toxicology
Preston Publications

Editorial Offices, *Journal of Analytical Toxicology*
P.O. Box 48312
Niles, Illinois 60714

AIM AND SCOPE

The *Journal* accepts full-length original manuscripts, review articles, and short communications relating to the isolation, identification, and quantitation of potentially toxic substances and their biotransformation products in specimens of human, animal, or environmental origin. The articles should pertain especially to the monitoring of therapeutic drugs and environmental and industrial contaminants, clinical reports of poisonings with analytical data), the development of analytical techniques, and the results of forensic toxicological investigations.

MANUSCRIPT SUBMISSION

Correspondence will be with person who submits manuscript, unless instructed otherwise. Include complete address and telephone number.

Submission of a paper implies that it has not been published in, or submitted to, any other journal. Mention previous oral presentation in a footnote.

Submit four copies of manuscript written in third person, past tense and typed double-spaced with wide margins. Style should conform to currently acceptable grammar and syntax. For further information see *ACS Style Guide* (American Chemical Society, Washington, D.C., 1986).

MANUSCRIPT FORMAT

Abstracts: 150 words stating objectives of study, techniques used, and what was accomplished.

Text: Use some of the following headings: Introduction, Experimental, Apparatus, Methods, Results, Discussion, Conclusions, Acknowledgment, and References. Describe equipment and method(s) in sufficient detail to permit other toxicologists to duplicate results.

Figures: Provide graphs, diagrams, chromatograms, photos, etc., as clear, original positives, suitable for reproduction. Use chart paper without coordinates, or coordinates should be nonreproducing light blue or green. Make lettering and/or numbering clear and in proportion to figure. Use a lettering guide or press-type. Computer printout (dot matrix) is not reproducible. Optimum size for figures is 5×7 in. Captions are typeset; clearly indicate on a separate sheet. One original set of artwork is necessary; photocopies are acceptable for informational purposes.

Tables and Equations: Type clearly on separate pieces of paper. Use for tabulation of numerical data; include other information in text. Type equations clearly showing superscripts and subscripts, with Greek symbols written out in margin. Mark equations Eq. 1, Eq. 2, etc.

Bibliography and References: Cite all references at end of paper listed consecutively as they appear in text. List references in text by number in parentheses.

JOURNALS

First author's initials followed by last name; list additional authors in order of appearance in original work; title of article (no subtitles) in lowercase; *Chemical Abstracts* Journal abbreviation, underscored; volume number, colon; inclusive page numbers; year of publication in parentheses.

1. R.H. Cravey and N.C. Jain. Current status of blood alcohol methods. J. *Chromatogr. Sci.* 12: 209-13 (1974).

BOOKS

First author's initials followed by last name; additional authors (as above); if author is editor, Ed. should follow name; title of book, underscored and upper case; editor, if not listed with authors; publisher; city and state or country of publication; year of publication; specific page numbers or chapters referred to.

1. 1. Sunshine. *Manual of Analytical Toxicology.* CRC Press, Cleveland, OH, 1971, p. 85.

2. A.S. Curry. In *Progress in Chemical Toxicology, Vol. 1.* A Stolman, Ed., Academic Press, New York and London, 1963, pp. 135-55.

UNPUBLISHED WORKS

If an article has been submitted but has not been published, include as much information as possible, such as authors, title, journal, and year. Volume and page numbers may be added shortly before publication on galley proofs.

PATENTS

List initials followed by last name of person who applied for patent; country where patent application was filed; patent number; year in parentheses.

1. S.T. Preston, U.S. Patent 1234 (1976).

OTHER

Abbreviations and Units: Use abbreviations that are accepted and recognized as common scientific terminology without definition. Define all nonstandard abbreviations at first appearance. Use international system of units for expressing measurements. Examples of abbreviations for units are: Area: m^2, cm^2, mm^2 Length: m, cm, mm; Mass: kg, g, mg, ug, ng, pg; Mass concentration: g/L, mg/L, μg/L Volume: L, mL, μL.

SOURCE

January/February 1995, 19(1): 8A.

Journal of Anatomy
Anatomical Society of Great Britain and Ireland
Cambridge University Press

Professor P. K. Thomas
Department of Neurological Science
Royal Free Hospital School of Medicine
Rowland Hill Street, London NW3 2PF

AIM AND SCOPE

The *Journal of Anatomy* publishes original work that advances the study of the structure, mechanical functions and development of man and animals. The scope of the *Journal* embraces papers on human and comparative anatomy, including functional aspects, neuroanatomy, histology and cytology, histochemistry and electron microscopy. Papers on experimental pathology will be accepted if they throw light on normal structure and function. Occasional review papers will be included if they are topical and authoritative. Short communications dealing, for example, with aspects of descriptive anatomy, anomalies, important preliminary observations or new technical procedures will be considered. Book reviews are also included and correspondence arising out of papers published in the *Journal* or drawing attention to matters of general anatomical interest will be considered.

MANUSCRIPT SUBMISSION

Submission implies that paper is unpublished and is not being considered for publication elsewhere. In cover letter include: 'The work reported in the accompanying paper entitled..... has not been, and is not intended to be, published anywhere except in the *Journal of Anatomy*' signed by all authors. Previous publications of results in abstract form will not preclude consideration.

MANUSCRIPT FORMAT

Type manuscripts, including references, double-spaced, in English and fully corrected.

On a preliminary page state title of paper, names and addresses of authors, a short title for running headlines, and full address of reprint request author. List five key words.

Begin with a brief summary (headed 'Abstract') of chief descriptions or observations and conclusions drawn therefrom. Submit three sets of typescript and illustrations.

References: For references having three or more authors, give name of first author only, followed by et al.

Study reference lists in a journal issue to ensure that list is in correct format. Write out names of journals in full. Give first and last page numbers. Make sure all references given in text (and no others) appear in reference list and that spelling of authors' names and dates are correct in both text and reference list.

Illustrations: Submit in finished form ready for reproduction. Number consecutively, in bottom left corner. Letter subdivided figures in lowercase without brackets. Where possible, make half-tone illustrations and line diagrams into blocks; contrast should be similar in all photographs in a plate. Maximum figure size is 165 mm (width) by 230 mm (height). Refer to figures in text as (Fig. 10) or Figure 10. Apply lettering, pointers or scale bars directly to illustrations with Letraset or similar adhesive label; protect plate by an overlay. Include a scale bar in electron micrographs. Make letters 2-3 mm in height, capitals, with lower case letters added if necessary. Letters may be white or outlined in white on a dark background. There is no charge for black-and-white illustrations if necessary to illustrate text. If illustrations are excessive in number, authors will be asked to delete some or pay additional costs.

Editor will decide if color illustrations are essential. There is no charge if essential, if not cost must be paid by authors. Ask for an estimate when paper is submitted. Try to group figures. If reduction in size is necessary, make reduction factor the same for all figures.

Type legends on a separate page. Do not place on or appended to an illustration.

Lightly pencil author's name and figure number on back of each illustration.

OTHER

Experimentation, Animal: Experiments must conform with U.K. legal requirements. Provide a full description of anesthetic and surgical procedures and, when appropriate, state how animals were killed.

Experimentation, Human: Include a statement that experiments were performed with the understanding and consent of each subject. Refer to the Code of Ethics of the World Medical Association (Declaration of Helsinki) and to the Medical Research Council pamphlet *Responsibility in Investigations on Human Subjects* (reprinted in *British Medical Journal* 18 July 1964).

Nomenclature: Conform to *Nomina Anatomica* (5th edn., 1983), and *Nomina Anatomica Veterinari* (3rd edn., 1983) and *Nomina Anatomica Avium* (1979). Avoid use of initial letters for abbreviation.

SOURCE

August 1994, 185(1): inside back cover.

Journal of Antibiotics
Japan Antibiotics Research Association

Editorial Office, *Journal of Antibiotics*
Japan Antibiotics Research Association
2-20-8 Kamiosaki
Shinagawa-ku, Tokyo 141, Japan

AIM AND SCOPE

The *Journal of Antibiotics* is a monthly international journal devoted to the publication of research on antibiotics and other types of microbial products. The *Journal* will publish microbiological, biochemical, chemical and pharmacological studies but not those involving human therapy. Reports on recently discovered antibiotics are especially encouraged. Some of the specific subjects that are appropriate for the *Journal* are; discovery of new antibiotics and related microbial products, production, isolation, characterization, structural elucidation, chemical synthesis and derivatization, biological activities, mechanisms of action and structure-activity relationships of antibiotics and related microbial products, biosynthesis, bioconversion, taxonomy, and genetic studies on producing microorganisms, as well as improvement of production of antibiotics and related microbial products, novel physical, chemical, biochemical, microbiological or pharmacological methods for detection, assay, determination, structural elucidation and evaluation of antibiotics and related microbial products, and newly found properties, mechanisms of action and resistance-development of antibiotics and related microbial products.

Reports on known antibiotics and other microbial products, describing standard susceptibility tests with laboratory and clinically isolated microorganisms, pharmacological, biochemical and analytical studies by established methods, production detected in microorganisms belonging to the same genus as known producers, biochemical studies on action and resistance involving previously published mechanisms, and clinical or pharmacological evaluation in humans will not normally be accepted for publication.

MANUSCRIPT SUBMISSION

Work described may not be submitted for publication elsewhere. Consult recent journal issues; prepare manuscripts in an identical form and style.

Type or make with word processor in triplicate double-spaced throughout, including references, legends to tables, figures and plates, on good quality white bond paper, 21 cm by 30 cm (A4 size), with wide margins.

Indicate on manuscript type of publication desired; i.e., Original Article, Note or communication to the Editor.

MANUSCRIPT FORMAT

Present data as concisely as possible consistent with clarity. Extensive literature reviews are not accepted. Introduction: include only sufficient background material to show why work was done. Present experimental results clearly and logically. All reports on newly discovered substances should show the fundamental and definitive data which support the novelty of the described substances.

Original articles, six printed pages or less, include an abstract or summary, and appropriate experimental details to support conclusions. Biological studies include Title, Abstract Summary, Introduction, Materials and Methods, Results, and Discussion sections. Chemical studies include Title, Abstract/Summary, Description (Introduction+Results+Discussion), and Experimental sections. Reports on newly discovered substances include Title, Abstract/Summary, Introduction, Fermentation/Production, Taxonomy of Producers, Isolation/Purification, Physico-chemical and Biological Properties, and Discussion sections. A separate Experimental section is preferred.

Title: Reflect purpose and findings of work. Delete all nonfunctional words. Avoid abbreviations or trade names of drugs. Serial numbers may be used for consecutive papers. Follow with author's full first name, middle initials, and last name and addresses of all contributing laboratories. Mark name of corresponding author by an asterisk. Give present addresses of authors as a footnote indicated by a dagger (†).

Abstract/Summary: 100-150 words briefly indicating experimental approach and major findings and conclusions. Do not include diagrams. Define abbreviations. When a reference is essential, use full literature citation without title. Make abstract complete and understandable without reference to text.

Introduction (no heading): Supply sufficient background information to understand and evaluate results without referring to previous publications. Chose references to provide the most important background; avoid an extensive literature review.

Materials and Methods/Experimental: Describe materials in sufficient detail to enable others to repeat experiments. Include names of products and manufacturers only if alternative sources are unsatisfactory. Give source of experimental animals and microbial strains. List large numbers of microbial strains and mutants in a table. Indicate accession number of deposition in publicly accessible culture collections (e.g., American Type Culture Collection, Fermentation Research Institute of The Agency of Industrial Science & Technology, Japan, etc.). describe composition of culture media as:

1) the seed medium consisted of glucose 2%, glycerol 0.5%, Casamino Acids (Difco) 0.5%, K_2HPO_4, 0.1% and $MgSO_4 \cdot 7H_2O$ 0.1%, pH 7.2.

2) the medium contained glucose 20 g, soluble starch 20 g, Pharmamedia (Traders Protein) 10 g, K_2HPO_4, 2 g and $CaCO_3$, 3 g in 1 liter tap water.

Give instruments used and standard techniques and procedures at beginning of section. Describe novel experimental procedures in detail; refer to published procedures by literature citation of original articles and published modifications. Include in experiment title: chemical name and compound number (indicated by bold face numeral **1, 1a**) of any product prepared; subsequently identify compound by number. Use Standard Abbreviations, chemical formulas for simple compounds and compound numbers for more complex substances. Include molar equivalents of all reactants and percentage yields of products.

Results: Include rationale and design of experiments; present concisely. Reserve interpretation of results for Discussion. Use tables and figures only if essential for comprehension of data. Do not present same data in more than one form. Avoid extensive use of graphs, schemes, and illustrations to present data that can be more concisely or quantitatively presented in text or tables. Limit photographs (particularly photomicrographs and electron micrographs) to those absolutely necessary to demonstrate experimental findings.

Discussion: Interpret results and relate them to existing knowledge in field. Observe the utmost brevity consistent with clarity. Avoid polemics. Do not repeat information elsewhere in manuscript. Avoid exhaustive literature reviews.

Results and Discussion sections may be combined.

Acknowledgments: Acknowledge financial support, personal assistance, advice from colleagues, gifts, etc., in this section, not in footnotes.

References: Number consecutively. Indicate by superscript. Number each individually. List at end of manuscript.

1) PERLMAN, D.; A. J. VLIETINCK, H. W. MATTHEWS & F. F. LO: Microbial production of vitamin B_{12} antimetabolites. I. N^5-Hydroxy-L-arginine from *Bacillus cereus* 439. J. Antibiotics 27: 826– 832, 1974

2) WAKSMAN, S. A.: Microbial Antagonisms and Antibiotic Substances. pp. 200–214, Commonwealth Fund, New York, 1947

3) UMEZAWA, H.: Trends in antibiotic research and its expanded area. Antibiotics and low moleculur weight immuno-modifiers. *In* Trends in Antibiotic Research. Genetics, Biosyntheses, Actions & New Substances. *Ed.*, H. UMEZAWA *et al.*, pp. 1–15, Japan Antibiotics Res. Assoc., Tokyo, 1982

4) SLUSARCHYK, W. A. & F. L. WEISENBORN: The structure of the lipid portion of the antibiotic prasinomycin. Tetrahedron Lett. 1969: 659–662,1969

5) HERR, R. R. & M. E. BERGY: Lincomycin, a new antibiotic. II. Isolation and characterization. Antimicrob. Agents Chemother. -1962: 560–564,1963

6) KATAGIRI, K. & K. SUGIURA: Antitumor action of the quinoxaline antibiotics. Abstracts of Papers of 1st Intersci. Conf. on Antimicrob. Agents Chemother., No. 30, New York, 1961

7) FOSTER, J. W. & H. B. WOODRUFF (Merck & Co., Inc.): Production of streptothricin. U.S. 2,422,230, June 17, 1947

8) MEZES, P. S. F.: Inhibition of β-lactamases. Ph. D. Thesis, Univ. Waterloo, 1981

Abbreviate journal names of journals according to "International List of Periodical Title Word Abbreviations (Chemical Abstracts Service, Columbus, OH, 1970)."

Cite "unpublished data" and "S. A. WAKSMAN, personal communication" in text parenthetically. Supply written proof for "personal communication" and preprints for "in press" items. Footnote references other than bibliographic ones.

Tables and Illustrations: Type on separate sheets. Give headings that describe contents. Make comprehensible without reference to text.

Prepare figures on white drawing paper, tracing paper, or coordinate paper printed in blue for NMR, UV and IR charts. Use black India ink. Avoid hand-writing. Photographs should be high-contrast glossy prints. Make figures and photographs large enough to allow for reduction to half-size. Number tables, figures, plates, charts, schemes and other illustrations. Indicate desired positions in text.

Legends to tables and figures provide necessary descriptive information; do not repeat experimental methods section. Define all symbols and abbreviations in tables and figures except standard abbreviations. Do not define common abbreviations used in text.

Submit fine drawings, charts, chemical formulas, etc. that can be photographically reproduced for publication without further typesetting or art work. Such materials may be prepared for complicated diagrams, flow schemes, stereoscopic drawing of molecules, formulas and equations and portions of genetic maps and figures in which the location of materials, spacing and angles are important.

Use abbreviations identical to those in text. Mark graduations by short grids outside of axes. Make curves follow experimental points and not extend beyond. Draw one symbol upon another when more than one point has the same value. Leave axis and curve blank upon symbol. Do not minimize length of axes. In captions capitalize first letter, center, and place parallel to axis. Use minimum portion of scale covered by a curve. Place label below and centered on its grid mark on horizontal axis and left and centered on vertical axis. Use smaller type for numbering atoms in a structure. Draw axial and equatorial linkages and stereoisomers distinctly. Choose letter size balanced with drawing size. See *J. Antibiotics, 1994*;47(1): 7 for examples.

OTHER FORMATS

COMMUNICATIONS TO THE EDITOR—Two printed pages; preliminary reports of unusual urgency, significance and interest, whose subjects may be republished in expanded form. Experimental details are not required; include a general outline of experimental methods. Sectional format and Abstract/Summary are not required.

NOTES—Two printed pages; short experimental reports, scope more limited scope than Original Articles. Abstract/Summary is not required.

REVIEW ARTICLES—Critical reviews on subjects related to antibiotics and other microbial products are invited. Length depends on subject. Contact Managing Editor concerning suitability of topic.

OTHER

Abbreviations: Define at first appearance. Do not use abbreviations that are not employed at least five times in text, tables and figure legends. Use as an aid to readers, not a convenience to authors; limit use. Use standard chemical symbols and symbols for trivial names. Avoid abbreviations of antibiotics and pharmaceutical drugs; different usage of abbreviations for a single agent causes confusion.

Non-standardized abbreviations will be accepted if: it is defined at first appearance, it is clear and unambiguous in meaning, and it contributes to ease of assimilation by readers.

For abbreviations that may be used without definition see *J. Antibiotics*, 1994;47(1): 9-10.

Isotopically Labeled Compounds: For simple molecules, indicate isotopic labeling in chemical formula. For specific chemicals, place symbol for isotope introduced in square brackets directly preceding the part of name that describes labeled entity.

Report radioactivity by dpm. Show specific activity of labeled compound; 1 µCi/mmol, 0.5 mCi/mM, 10 µCi/mg.

Nomenclature, Actinomycetes: Adhere to standardized methods for characterization of new actinomycetes. Cite a depository number for culture described. For generic determination: describe morphology, physiology, chemical properties according to Lechevalier, H. A.: The actinomycetes III. A practical guide to generic identification of actinomycetes. *In* BERGEY'S Manual Systematic Bacteriology. Volume 4. *Ed.*, S. T. WILLIAMS *et al.*, pp. 2344–2347, Williams & Wilkins, 1989.

Species Determination— Follow guidelines in Lechevalier, 1981 (above). New methods may be introduced.

Media—Grow and characterize aerobic actinomycetes on ISP media (Shirling, E. B. & D. Gottlieb: Methods for characterization of *Streptomyces* species. *Int. J. Syst. Bacteriol.* 16: 313–340,1966). May be supplemented with specific media cited in generic descriptions.

Availability—Make all cultures available from a recognized culture collection.

Nomenclature, Chemical and Biochemical: Refer to new substances by name rather than code-number or symbol. Refer to all synonymous names and code-numbers previously used for subject substance.

Conform to conventions recommended by "International Union of Pure and Applied Chemistry (IUPAC), Nomenclature of Organic Chemistry, Sections A, B, C, D, E, F and H, 1979" and usage of *Chemical Abstracts*. Use generic names for compounds listed in *International Nonproprietary Names* (World Health Organization), *United States Adopted Names, Japan Accepted Names*, or *British Approved Names*. Avoid proprietary and trade names.

For guidelines to biochemical terminology, consult *International Union of Biochemistry: Biochemical Nomenclature and Related Documents* (The Biochemical Society, London, 1978), and most recent edition of *Handbook of Biochemistry and Molecular Biology* (CRC Press Inc.). For enzymes, use recommended names of the Nomenclature Committee of the International Union of Biochemistry as described in *Enzyme Nomenclature 1978* (Academic Press Inc., 1979). Use EC numbers.

If reported antibiotic is similar to, or identical with, an antibiotic previously reported in a published patent or in the published proceedings or abstracts of a scientific meeting, comment accordingly. Editorial Board may comment at end of manuscript under heading Editorial Comment.

Nomenclature, Genetic: Use phenotypic designations when mutant loci have not been identified. Do not italicize three-letter symbols; capitalize first letter. Define phenotypic designations of symbols. Use Arabic or Roman numerals to identify a series of related phenotypes. use superscript letters for further description of phenotype.

Indicate genotypic designations by three-letter locus symbols; use lowercase italics with italicized capital letter and Arabic numeral to indicate a series of loci. Recommended nomenclature for genetic loci of *Escherichia coli* (*Microbiol. Rev.* 44: 1– 56,1980), *Streptomyces coelicolor* (*Bacteriol. Rev.* 37: 371– 405,1973) and *Neurospora crassa* (*Microbiol. Rev.* 46: 426–570,1982) may be applicable to other microorganisms.

Recommended nomenclature of plasmids (*Bacteriol. Rev.* 40: 168–189, 1976) may be applicable for plasmids, including recombinant plasmids constructed *in vitro*.

Nomenclature, Microorganisms: Spell out binary name, generic name and specific epithet in italics at first time use. Names of genera and higher categories may be used alone, but specific and subspecific epithets may not. Subsequently, abbreviate generic name to initial capital letter if meaning is clear. In tables and figure-legends, abbreviate generic names of microorganisms belonging to same genus as the spelled out strain:

Strain No.	MIC (μg/ml)
Staphylococcus aureus ATCC 6538 P	0.01
S. epidermidis ATCC 12228	0.39
Streptococcus pyrogenes S-23	0.10
S. pneumoniae Type I	0.05

Strain No.	MIC (μg/ml)
S. pneumoniae Type II	12.5
Streptococcus sp. MK 1001	0.78
Enterococcus faecalis ATCC 10541	100
○ *Proteus mirabilis* MB-838	
Δ *P. vulgaris* 345-57G	
● *Proteus* sp. A-1	

Designate microorganisms by names in 8th Edition of *Bergey's Manual of Determinative Bacteriology, Ainsworth & Bisby's, Dictionary of the Fungi* and *The Yeasts, a Taxonomic Study.* 2nd Ed. (J. Lodder *Ed.,* North-Holland Publishing Co., Amsterdam, 1970). New scientific names proposed for microorganisms follow "International Code of Nomenclature of Bacteria (*Int. J. Syst. Bacteriol.* 23: 83–108, 1973)" and *International Code of Botanical Nomenclature* (A. Oosthoek's Uitgeversmaatschappij N. V., Utrecht, Netherlands, 1972).

Page Charges: $115 per printed page excess page charge.

Units: Use metric system units for all quantities of length, weight, volume, area and pressure. Use International System of Units (M for 10^6, k for 10^3, m for 10^{-3}, μ for 10^{-6}, n for 10^{-9} and p for 10^{-12}). Abbreviate each unit as: MHz; km, m, cm, mm, μm, nm; kg, g, mg, μg, ng, pg; liter(s), ml, μl; m^2, cm^2, mm^2; m^3, cm^3, mm^3; *etc.*

Use percent, %; avoid mg%. Use μg/ml or μg/g in place of ppm (use ppm only for NMR studies). Indicate temperature degree centigrade, °C, and pressure by torr, mmHg or kg/cm^2.

Convert numerals from other measuring systems into metric system units.

Express concentrations of compounds as: 1 mmol/ml, 3.5 mol/kg, 1% (w/v) $NaHCO_3$, 50% MeOH, 20 mg/ml ampicillin trihydrate, etc.

Express ratio and dilution as: 0.2 N HCl - MeOH (5:1); EtOH - $CHCl_3$, 1: 2 (v/v); BuOH - AcOH - H_2O, 4: 1: 1 (in volume); 1/10 vol of EtOH, concd to 1/20 vol, etc.

X-Ray Crystallographic Data: Submit tables of atomic coordinates for X-ray crystallography with "Crystal Data Form" (*J. Antibiotics, 1994;*47(1): 13). Data will be transmitted to Crystallographic Data Center, Cambridge, England, when paper is published.

SOURCE

January 1994, 47(1): 3–14 (tan pages).

Journal of Antimicrobial Chemotherapy
British Society for Antimicrobial Chemotherapy
Academic Press

Professor Roger C. Finch
Dept. of Microbial Diseases
Nottingham City Hospital
Nottingham NG5 1PB, U.K.

AIM AND SCOPE

The *Journal* will consider for publication Original Papers in English on all aspects of antimicrobial chemotherapy as well as Leading Articles and longer Reviews of subjects of current interest.

MANUSCRIPT SUBMISSION

Papers must be accompanied by a letter signed by corresponding author indicating that current Advice to Contributors has been read (published in first issue of each volume), instructions complied with and conditions accepted. Indicate that all named authors have seen and agreed to submitted version of paper; that all included in Acknowledgments section, or as providers of personal communications, have agreed to inclusion; and that material is original and unpublished and has not been simultaneously submitted elsewhere. Declare any previous or other publication of material (including publication in conference proceedings letters to journals and brief communications).

Submit two copies of paper, double-spaced throughout (including reference list) on one side only of A4 paper with a 4 cm left margin. Number pages consecutively. Use Roman typeface and bold, italic or other types where appropriate. Indistinct printing will not be accepted. Manuscripts by fax require prior agreement of Editor-in-chief. Include telephone and fax number for corresponding author. If accepted for publication exclusive copyright in paper shall be assigned to the Society.

MANUSCRIPT FORMAT

Arrange text: Title, Authors, Address, Running heading (50 characters, including spaces), Summary (separate unheaded paragraph), Introduction, Methods, Results, Discussion, Acknowledgments, References. Put Title, Authors, Address, and Running heading on a separate sheet. In summary explain briefly what was done, what was observed and what was concluded. Introduction is a brief statement outlining purpose and context of paper. Leave discussion for Discussion section. Restrict Methods, results and discussion to sections so named; preliminary results may be included in Methods section. Information in Results section may be conveyed in text or in figures or tables but not both. Heading and subheadings may be used, avoid using footnotes. Number hard copy pages of manuscript consecutively in order: title page, text, references, figure legends, tables, figures. State statistical methods used.

Acknowledge sources of financial support at end of text.

References: Verify accuracy of all references; check against original material. Restrict reference citations to those essential for introducing the paper's purpose and context, describing methods not given in detail, and for discussing results and issues raised by them. Quote references accurately; do not take out of context.

Include in first citations in text surnames of all authors up to three, followed by year of publication, e.g., Smith, Jones & Brown (1990). For subsequent citations of three, and all citations of four or more authors, refer only to first name followed by *et al.* Distinguish different papers by the same author in the same year as Smith (1990*a*), Smith (1990*b*), etc. Where two or more publications are referred to consecutively, use chronological order, and alphabetical order of first author's name for references dating from the same year.

Reference anonymous leading articles: Leading article (1990). Cite references to abstracts in conference proceedings in text by authors' names and dates; give fully in reference list. Cite accepted papers, not yet published, in usual way in text; list as 'in press', with name of journal and probable date of publication. Quote submitted, but not yet accepted, work in text as 'unpublished'. Personal communications or unpublished work, referred to in text, must include names of all responsible workers with initials; workers must agree to be quoted.

List all references alphabetically at end of text.

PERIODICALS

Author's surnames with initials of first names (Give all authors names up to six, if more give six authors with *et al.*); Year of publication in brackets; Full title of paper; Journal name in full and in italics; Volume (in Arabic numerals, bold and followed by a comma); First and last page numbers.

JOURNAL

Hawkey, P. M., Birkenhead, Kerr, K. G., Newton, K. E. & Hyde, W. A. (1993). Effect of divalent cations in bacteriological media on the susceptibility of *Xanthomonas maltophilia* to imipenem, with special reference to zinc ions. Journal of Antimicrobial Chemotherapy 31, 47–55.

JOURNAL SUPPLEMENT

Williams, I., Gabriel, G., Cohen, H., Williams, P., Tedder, R. S., Machin, S. *et al.* (1989).

Zidovudine—the first year of experience. *Journal of Infection* 18, *Suppl. 1*, 23–31.

LEADING ARTICLE/EDITORIAL WITH NO NAMED AUTHOR(S)

Leading article. (1987). Antibiotics by design. *Lancet ii*, 312–3.

CORPORATE AUTHOR

Royal College of Physicians Working Party. (1986). Research on healthy volunteers. Journal of the Royal College of Physicians **20**, 243–57.

BOOKS

Author's surnames, with initials of first names; Year of publication in brackets; Article or chapter title; Full book title in italics; Editor(s) in brackets (if applicable); Pages quoted; Name and city of publishers.

BOOK

Odds, F. C. (1988). Candida and Candidosis, 2nd edn. Bailliére Tindall, London.

CHAPTER IN A BOOK

Peutherer, J. F. (1992). Retroviruses. In *Medical Microbiology*, 14th edn (Greenwood, D., Slack, R. C. B. & Peutherer, J. F., Eds), pp. 627–38. Churchill Livingstone, Edinburgh.

MONOGRAPH

Committee for Clinical Laboratory Standards. (1990). *Dilution Antimicrobial Susceptibility —Second Edition: Approved Standard M7-A2.* NCCLS.

ABSTRACTS

Knapp, C. C., Ludwig, M. D., Barthel, J. S. & Washington. J. A. (1990). Variables affecting the susceptibility of *Helicobacter pylori* to metronidazole. In Program and Abstracts of the Thirtieth Interscience Conference on Antimicrobial Agents and Chemotherapy, Atlanta, GA, 1990. Abstract 877, p. 232. American Society for Microbiology, Washington, DC.

Tables: Type each on a separate sheet. Number consecutively in Roman numerals in order of mention in text. Table I, etc.... Give each a brief descriptive heading. Column headings explain meaning of each entry in table. Avoid footnotes. Use – to contrast with +, and zero to contrast with numerals.

Figures: Submit in finished form suitable for reproduction, as large as or larger than final size on page. Photographs should be glossy prints with strong contrast, trimmed to exclude unnecessary background. Plan figures to fit proportions of Journal pages; details should be easily discriminated at final size. Enquire at Editorial Office concerning color illustrations.

When tracings and photographs are re-drawn to provide figures, send originals for editorial review.

Number all illustrations with Arabic numerals as Figure 1, etc.... without abbreviation, in order of mention in text. Indicate author's name, number of illustration, Fig. 1, etc...., on reverse, with orientation.

Provide a short explicit legend for each figure. Collect all legends together in one section.

OTHER FORMATS

ANTIMICROBIAL PRACTICE AND FOR DEBATE—Topical or controversial issues.

BRIEF REPORTS—1500 words, 50 word summary abstract, two figures or tables and ten references. Follow format of a standard paper.

CORRESPONDENCE—800 words, one figure or table and ten references. Letters discussing topics of interest and concern in antimicrobial chemotherapy, particularly arising from papers or letters published in *Journal*.

OTHER

Abbreviations: Make unambiguous. Use an authorative scientific or medical dictionary as a guide to approved abbreviations.These may be used without explanation: Use capitals for MIC, MBC, LD_{50}, ID_{50}, MLD, WBC, RBC, DNA, RNA, group A, B, etc. for antigenic or other groups, CDC, PHLS, CDSC, NIH, MRC, WHO, CSF, MSU, EMU, CSU, NCCLS, AIDS.

Use: cfu, pfu, om, bd or bid, qds, tds, tid, od, rpm, iu, im, iv, po, PR, PV, ppm, ft, in, m, sq.m, kg, L, mmol, mmol, mequiv., M, mM, μm, g, mol. wt., Da, E, sp.gr., h, min, 37 C, 98 F, P(probability), *c.* (circa), cf., e.g., i.e. Use in-vitro (adjective) but *in vivo* (adverb), post-mortem (adjective) but *post mortem* (adverb). Use sp. (species, singular) and spp. (species, plural). Use Gram's stain and Gram-negative bacillus.

Electronic Submission: The *Journal* will typeset manuscripts using a file prepared on disc. All known word-processing packages are acceptable; versions of Microsoft Word, Wordperfect or Wordstar are preferred. Save text as a word-processing file, use ASCII and/or RTF file.

Order on disc: Running title (50 characters, including spaces), Title, Authors, Addresses, Summary, Introduction, Methods, Results, Discussion, Acknowledgments, References, Figure legends, Tables, Footnotes: abbreviations.

Use two carriage returns to end headings and paragraphs but not to rearrange lines or to introduce Space between headings and paragraphs. Enter references in *Journal* style, with space between names and initials or individual authors, and journal titles italicized and volume numbers bold. Type text without hyphenation, except for compound words. Do not use lower case 'l' (ell) for '1' (one) or 'o' for '0' (zero).

Label disc with JAC reference number (when assigned), name of first author, machine and program used to generate disc file. Adequately protect discs for mailing to avoid damage by bending or X-ray inspection. Submit two hard copies, including all tables and figures, with all text (including reference list) double-spaced. The edited hard copy of the manuscript is the definitive version.

Experimentation, Animal: Conform to U.K. standards of animal experimentation and experimental care in the Animal (Scientific Procedures) Act 1986 and the Code of Practice for the Housing and Care of Animals used in Scientific Procedures 1989, respectively, and embracing standard veterinary animal care.

Experimentation, Human: Indicate attention to ethical considerations. Note approval of an ethical committee. Do not identify patients by their initials or hospital numbers.

Nomenclature, Bacterial: When genus and species are together, use a capital for genus, a lowercase letter for species and italicize both. When genus alone is used as noun or adjective use lowercase Roman, unless genus is specifically referred to, e.g., 'staphylococci and streptococci' but 'organisms of the genera *Staphylococcus* and *Streptococcus*'.

Order names have an initial capital; do not italicize. For genera in the plural, use lowercase Roman, e.g., salmonellae. When species is used alone, use lowercase. For trivial names, use lowercase Roman.

After initial use in text of full name of a bacterium, abbreviate generic name to initial letter.

Nomenclature, Chemical: Conform names to current practice. Make sure they are correct. Proprietary names have a capital; use only when official name is inappropriate; or, in Methods section, to indicate a particular preparation; or once, in brackets, when proprietary name is more familiar. Give contents of proprietary mixtures.

Spelling: Use British spellings: haemophilus, haematology, paediatrics, leucocyte, leukaemia, bacteraemia, sulphonamides, aetiology; but *note:* neutropenia, fetal.

Units: Express measurements and quantities in SI or non-metric units.

SOURCE

July 1994, 34(1): back matter.

Journal of Applied Bacteriology
The Society for Applied Bacteriology
Blackwell Scientific Publications

Professor D.E.S. Stewart-Tull
Department of Microbiology
University of Glasgow
Glasgow G12 8QQ U.K.

MANUSCRIPT SUBMISSION

Submit two copies of manuscript with, if appropriate, copies of paper(s) cited as 'in press' and evidence in support of 'personal communications' or 'unpublished observations'. Type manuscript double-spaced on one side of A4 (21 × 29.5 cm) paper.

Do not justify right margin in word-processed text. Avoid artificial word breaks at end of line. Leave a 2.5 cm margin at top, bottom and sides of each page. Number pages consecutively.

On first page give only title, name(s) of author(s) and place(s) where work was done; abbreviated title (running headline–35 letters and spaces); name and address of corresponding author; include on second page three self-adhesive labels with name and address of corresponding author. Accepted papers become copyright of *Journal*.

MANUSCRIPT FORMAT

Papers should develop concepts as well as record facts. Present results of a substantial program of research. Sequential publication of numbered papers is not permitted.

Full-length paper sections: Summary: (150 words) brief self-contained, give major findings of investigation; Introduction: balance 'pure' and 'applied' aspects of subject; Materials and Methods: be sure work can be repeated according to details provided; Results: include well-prepared tables and figures that do not repeat text, but focus on importance of principal findings; Discussion: do not recapitulate results, do not combine with results; Acknowledgments; and References: cite references with three or more authors in text as Jones *et al.* (1992) unless this causes confusion, then, quote in full. Give a series of references in date order. Distinguish different publications with the same author(s) and year by, for example, 1992a, 1992b. This also applies to Bibliography. Cite papers or other publications with no obvious author as Anon. with year in text and bibliography. Do not list papers not freely available without charge to the public.

Materials and Methods: Hierarchy of headings: **MATERIALS AND METHODS** (first order) **Sample preparation** (second order); *The media* (third order-subheading begins paragraph). Do not indent first paragraph; indent second and subsequent paragraphs.

Tables: Type on separate sheets without ruled vertical lines. Indicate approximate position in pencil in left margin of text. Use explanatory footnotes; mark by *, †, ‡, §, ||, ¶.

Figures: Prepare line diagrams in black on a white background or on tracing paper. Insert numerals, points and essential lettering on graphs clearly. Type legends on a separate sheet. Use symbols: ○ △ ▢ ● ▲ ■ × +.

Photographs: Provide black-and-white unmounted glossy photographs of good quality and contrast (two prints of each). Indicate extent of reduction without loss of definition. Indicate magnification with a bar, representing a stated length. Composite photographs avoid the need for two. Cost of printing color photographs is borne by author.

Caption photographs as for figures. On back of all photographs note figure number, name of authors, title of paper, J.A.B., and orientation (viz. '↑top').

Footnotes: Use only first page to show change of address of an author and address for correspondence.

References: Use Harvard system. Place names with prefixes de, do, van, von, etc. in alphabetical order of first letter of prefix. Indicate (in pencil) in left margin of bibliography page number(s) in which reference is cited. For italics, either type in Roman and underline or print in italics.

Crook, B., Higgins, S. and Lacey, J. (1987) Gram-negative bacteria associated with handling of waste. In *Advances in Aerobiology* ed. Boehm, G. and Leuschner, R.M. pp. 371-373. Basel: Birkhauser-Verlag.

Garner, J.S. and Favero, M.S. (1985) *Guidelines for Handwashing and Hospital Environment Control*. US Public Health Service, Centers for Disease Control HHS No. 99-117. Washington DC: Government Printing Office.

Rutala, W.A. and Wever, D.J. (1991) Infectious waste—mismatch between science and policy. *New England Journal of Medicine* **325**, 578-582.

OTHER FORMATS

NOTES—Four printed pages, not a vehicle for publishing interim research reports (submit to *Letters in Applied Microbiology*). Present results of research projects having narrow, readily defined limits—the quality of the research equal to that of a full paper; observations which contribute to knowledge but not development of concepts or new methods, techniques or apparatus.

OBSERVATIONS—Six printed pages; short articles discussing current concepts and developments in applied microbiology. Discuss topic with Editors before submitting.

REVIEW ARTICLES—10-12 journal pages. Present a substantial survey with an adequate historical perspective of the literature on some facet of applied microbiology. Consult Editors before preparing a manuscript. Design own headings; list under heading 'Contents' on page 2 of manuscript.

OTHER

Experimentation, Animal: Show evidence of the ethical use of animals or harmful substances.

Nomenclature: Give binomial Latin name of microorganisms, plants and animals (other than farm animals) at first mention in text; thereafter abbreviate generic name so that confusion is avoided when dealing with several genera all beginning with the same letter. Italicize subspecies; print groups and types in Roman designated by capital letters or Arabic figures. Common names do not have an initial capital letter; do not underline in manuscript. Give specific name in full in table and figure captions. Write major ranks in Roman with an initial capital.

For plant pathogenic bacteria, refer to list of pathovars compiled by the International Society for Plant Pathology Committee on Taxonomy (Dye, D.W., Bradbury, J.F., Goto, M., Hayward, A.C. & Lelliott, R.A. 1980 International standards for naming pathovars of phytopathogenic bacteria and a list of pathovar names and pathotype strains. *Review of Plant Pathology* **59**, 153-168). In this, many species names not included in the Approved Lists *(International Journal of Systematic Bacteriology* 1980, **30**, 225–420) are reduced to the rank of pathovar so that the original names are retained in a trinomial form. Where the pathovar name is cited it may subsequently be abbreviated: *Pseudomonas syringae* pv. *phaseolicola* becomes *Ps. syr. phaseolicola*. Reference to these lists avoids the need for citing past authors who named or renamed pathogens but, for completeness or clarity, consider synonyms suggested by more recent work.

For a list of abbreviations of common generic names, see *J. Appl. Bacteriol.* 1994; 76: 2.

Nomenclature, Gnotobiotic Animals: Germ-free implies freedom from any detectable microorganisms or viruses; it is limited by tests used for detection of contaminants. Conventional animals have a full complement of associated microbes. Open conventional animals are housed in a standard animal house. Isolator conventional animals are maintained in isolators and associated with full flora. Ex-germ-free animals are those associated with a flora which have become conventional.

Nucleotide Sequences: Deposit in EMBL/GenBank/DDBJ Nucleotide Sequence Data Libraries. Reference accession number in manuscript. Include sequence data only if new (unpublished), complete (no unidentified nucleotides) and if sequence information itself provides important new biological insights of direct relevance to the question addressed in manuscript. Do not submit sequences if gene has been reported in another species, unless a comparison with related sequences contributes important new information. Include clear indications of nucleotide numbers on points of interest in presentation of nucleotide sequences. In comparisons, make difference in sequences readily visible. Use a font size that will facilitate appropriate reduction of figure.

Safety: Explain chemical or microbiological hazards involved in experiments. Describe relevant safety precautions adopted or cite an accepted code of practice.

Statistics: Present tests clearly to allow readers with access to data to repeat them. Do not describe every statistical test fully; make clear from context what was done. Clearly state null hypotheses. Consider assumptions underlying statistical tests used. Assure readers that assumptions are at least plausible. Use non-parametric tests if assumptions do not seem to hold.

Style: Numbers in text: write out nine and below in full, above nine as numerals. Use 'z' spelling, except analyse, dialyse, hydrolyse, etc.

Units: Use SI units. Use gl^{-1} not g/l; d, h, min, s (time units) but week and year in full; mol^{-1} (not M or N); probability is P. Refer to *Units, Symbols and Abbreviations* (1988, Royal Society of Medicine, London) and *Biochemical Journal* 281: 1-19, 1992.

SOURCE

January 1994, 76(1): 1-4.

Journal of Applied Physiology
American Physiological Society

Editor, *Journal of Applied Physiology*
American Physiological Society
9650 Rockville Pike
Bethesda, MD 20814-3991

AIM AND SCOPE

Journal of Applied Physiology publishes original papers on normal or abnormal physiological function, and integrative and adaptive mechanisms, in these often-related areas: 1) respiratory physiology; 2) nonrespiratory functions of the lungs; 3) environmental physiology; 4) temperature regulation; 5) exercise physiology, including cardiovascular, respiratory, endocrine, renal, and muscular responses; 6) space physiology; and 7) interdependence between the heart and respiratory system. The wide scientific span of the Journal rests on physiology as its keystone. However, the boundaries of physiology have enlarged as demarcations between concepts and techniques in the physiological, pharmacological, and biochemical sciences become increasingly blurred. Moreover, theoretical articles dealing with research at any level of biological organization ranging from molecules to humans fall within the broad scope of the Journal. Papers dealing with topics in other basic sciences that impinge on physiology are also welcome.

MANUSCRIPT SUBMISSION

In cover letter state that research reported is original and will not be submitted for publication elsewhere until a publication decision has been made by APS. Except in reviews and invited editorials, APS will not accept manuscripts in which, other than in abstracts, a significant portion of data, as figures and tables has been published elsewhere. If letter is signed only by corresponding author, he/she is acting as agent for all authors. Each author acknowledges: that he/she has made an important scientific contribution to study and is thoroughly familiar with the primary data; that manuscript is a truthful, original piece of work. Use of other investigators' data or ideas is acceptable if carefully documented, and, when appropriate, permission of other investigator(s) is given; provide appropriate reprints or preprints; that s/he has read complete manuscript and takes responsibility for its content and completeness; and understands that if a paper or part of a paper is found to be faulty or fraudulent, all authors share responsibility.

APS journals seek definitive papers that present entire contents of a research project. Present all data from a group of subjects, animals, or samples together in a single paper. If not, then cross-reference manuscript. Use identical subject, animal, and sample numbers in different manuscripts to identify their commonality. If a paper depends critically on another unpublished paper, include three copies for reviewers.

A copyright transfer form will be sent to submitting author. Transfer form must be completed and returned before work is typeset.

MANUSCRIPT FORMAT

Type manuscript double-spaced with wide margins on 8.5 × 11 in. paper. Submit four copies, including figures. Send glossies for photomicrographs, gels, and other halftones. Photocopies of line drawings are acceptable at submission as long as two sets of glossies are provided when manuscript is accepted; for computer-generated laser prints, use paper recommended for camera-ready copy.

Number pages in upper right corner (beginning with first text page). Arrange in order: title page, abstract and index terms, text, text footnotes, acknowledgments, references, figure legends, tables, illustrations.

Title Page: Title of article; author(s); department and institution where work was done, city, state or country, and zip code; abbreviated title for running head (55 characters including spaces); name and address for mailing proofs; and contact telephone and fax numbers. Put abstract and index terms on a separate sheet, double-spaced.

Text, footnotes, acknowledgments, references, and figure legends begin on separate sheets, all lines double-spaced. Type each table on a separate sheet, double-spaced.

Make text clear and concise; conform to accepted standards of English style and usage. Define unfamiliar or new terms when first used. Do not use jargon, cliches, and laboratory slang.

Identify illustrations on reverse (lightly with a soft pencil) with figure number and name of first author; when necessary, mark TOP.

Title: 110 characters including spaces. Make informative. Use no unnecessary words like "Studies in . . . ,"

Abstract: 170 word informative one-paragraph abstract. State concisely what was done and why (including species and state of anesthesia), what was found (in terms of data).

Key Words: Append 3-5 words or short phrases not included in title to abstract.

Promissory Notes: Do not include either implicit or explicit promises that future work will be published.

Footnotes: Number text footnotes consecutively throughout. Assemble, double-spaced, on one sheet.

References: Limit to 30 directly pertinent published works or papers that have been accepted for publication. Cite abstracts, properly identified (Abstract), only if sole source. Verify accuracy. Type separately, double-spaced (do not single-space any line), alphabetically by author; numbered serially, with only one reference per number. Include number appropriate to each reference in parentheses at proper point in text.

JOURNAL ARTICLES

Last name of first author, followed by initials, initials and last names of each coauthor; title of article (first word only capitalized); name of journal abbreviated as in *Serial Sources for the BIOSIS Data Base* (BioSciences Information Service), volume, inclusive pages, and year.

1. Chisholm, D. J., J. D. Young, and L. Lazarus. The gastrointestinal stimulus to insulin release. *J. Clin. Invest.* 48:1453–1460,1969.

BOOK REFERENCES

Author(s) as above; title of book (main words capitalized); city of publication; publisher; year and pages.

Include references to government technical documents only when availability is assured. For style of citation of these documents, congress proceedings, chapters in books, etc., consult recent Journal issues.

Do not include citations such as "unpublished observations" or "personal communication" in reference list; add in parentheses in text. Secure permission of person cited for "personal communications."

Illustrations: Submit sharp, unmounted glossy photographic prints (or computer-generated laser prints on camera-ready paper) not larger than 8.5 × 11 in. Number illustrations consecutively with Arabic numerals; refer to as figures. Prepare figures for single-column width (3.5 in); otherwise, for double-column width (7 in.). Draw and letter all drawings for reduction to a given size to same scale.

Prepare graphs such as electrocardiograms, kymograms, and oscillograms so that crosshatched background is eliminated. Use blue-ruled instead of black-ruled recording paper for original records.

Give actual magnification of photomicrographs. Editorial Office will make corrections for reduction. A length scale on print is preferable.

Designate special features on photomicrographs by letters, numerals, arrows, and other symbols that contrast with background.

Have lettering done in India ink by a draftsman, with graphic arts transfer lettering, or with a graphic arts lettering system. If transfer lettering is used, photograph figures for final print. Freehand, typewritten, or computer-generated dot-matrix lettering is not acceptable. Lettering and symbols must be proportionate to size of illustration to be legible after reduction. Size lettering so that smallest elements will be not less than 2 mm high after reduction.

When possible, place all lettering within framework of illustration; insert key to symbols on face of figure. When figure is so filled that symbols must be explained in legend, use only standard characters: ❑ ■ ○ ● ◐ △ ▲ ×.

Use photographs of equipment sparingly; good line drawings are more informative. Photographs of animals are not acceptable; good line drawings are more effective.

Figures in color are accepted if author assumes all printing costs. Supply three positive glossy color prints.

Indicate approximate position of each figure in margin of manuscript.

Give each figure a legend. Group legends in numerical order and type double-spaced on one or more sheets.

Tables: Submit illustrations rather than tables. Do not duplicate material in text or illustrations. Indicate approximate position in margin of manuscript.

Submit statistical summary tables when possible rather than tables with many lines of individual values. On Editor's recommendation and with author's approval, Editorial Office will deposit lengthy tables of data.

Number tables consecutively with Arabic numerals; prepare with size of Journal page in mind: 3.5 in. wide, single column; 7 in. wide, double column. Type double-spaced on a separate sheet. Give each a brief title; place explanatory matter footnotes, not in title. Omit horizontal and vertical rules and nonsignificant decimal places in tabular data. Use short or abbreviated column heads; explain if necessary in footnotes. Identify statistical measures of variations, SD, SE, etc. List table footnotes in order of appearance; identify by standard symbols *, †, ‡, § for four or fewer; for five or more, use consecutive superior letters throughout.

OTHER FORMATS

HISTORY ARTICLES—Length should be comparable to a research report in Journals.

LETTERS TO THE EDITOR—500 words. Type letters, including an informative title, double-spaced. Submit three copies. If a letter is acceptable, a copy is sent to original author, if applicable; author will have an opportunity to provide a rebuttal with new material that will be considered for publication with letter.

MODELING IN PHYSIOLOGY—Submitted via APS Publications Office directly to Editor of Modeling in Physiology (MIP). Application of models in physiology or related areas are emphasized. Submit original research contributions, critiques, reviews, survey papers, or tutorials.

Submit letters to the editor, highlight controversies, ambiguities, or misapplication of theory or method. Mathematics and technical jargon are welcome, but must be relevant and clearly explained and presented. Make articles reasonably self-contained, not dependent on a series of highly technical previous publications, unless well known.

Modeling developments should conform to standard modeling practice. Specify appropriate measures of variability for quantitative results based all or in part on a model and experimental data. For example, if a model is used to estimate model parameter values from data report variability measures for these estimates (e.g., confidence limits, standard deviations, coefficients of variation) as well as (point) estimates themselves. Indicate how such values were estimated (what algorithms, programs, etc.), including method for calculating variability estimates. Discuss meaning (or lack thereof) of reported parameter values, in context of purpose of modeling effort (e.g., physiological significance). If a model includes parameter (or variable) values estimated from data, and model is used to predict or explain something (e.g., a physiological implication), include an analysis/discussion of how variability in estimated values affects predictions, explanations, or conclusion. If a computer simulation, or simulation model is used, and numerical values are used to generate simulated solutions pertinent to reported results, evaluate sensitivity of such solutions/results to these parameters for numerical values used in simulations. Simulations are usually used to explain or predict real system behavior, which normally depends on numerical values of parameters used in simulation. Therefore, some form of parameter sensitivity analysis is needed to support results based on model.

RAPID COMMUNICATION—Four journal pages, including figures, tables, and references (one printed page equals four double-spaced typewritten pages or three figures or tables). Submit short manuscripts containing results of unusual interest identified as such.

SPECIAL COMMUNICATIONS—Manuscripts that describe new methods, new apparatus, techniques with physiological applicability, and critiques of methods and techniques.

SUPPLEMENTARY MATERIAL—Extensive tables of data, appendixes, mathematical derivations, extra figures, computer printouts, and other supplementary material too costly to include may be submitted for deposition (without charge) with National Auxiliary Publications Service (NAPS), c/o Microfiche Publications, P. O. Box 3513, Grand Central Station, New York, NY 10017. Submit material with manuscript for review. On acceptance, it will be deposited by Editorial office with NAPS. A footnote will added noting availability of material on microfiche and giving NAPS Document Number.

OTHER

Abbreviations, Symbols, and Terminology: Include a list of new or special abbreviations used, with spelled-out form or definition. Internationally accepted biochemical abbreviations such as ADP, NADH, and P_i, do not need to be defined; define other frequently used abbreviations only at first mention. A list of accepted abbreviations appears following contents of January and July issues of APS Journals. For commonly accepted abbreviations, word usage, symbols, etc., see *CBE Style Manual* (5th ed., 1983). Use chemical and biochemical terms and abbreviations in accordance with recommendations of IUPAC-IUB Combined Commission on Biochemical Nomenclature. Isotope specification should conform to IUPAC system. For style in specialized fields see: "Glossary on respiration and gas exchange" (*J. Appl. Physiol.* 34: 549-558, 1973); "Glossary of terms for thermal physiology" (*J. Appl. Physiol.* 35: 941-961, 1973).

Drugs: Capitalize proprietary (trademarked) names; check spelling carefully. Chemical or generic name precedes trade name or abbreviation of a drug at first use. Capitalize and check spelling of trade names of chemicals or equipment.

Electronic Submission: $100 charge for accepted manuscripts for which a disk is not provided for final revised, accepted version of the paper. Any popular word-processing software can be used on 3.5 or 5.25 in. low- or high-density diskettes.

Experimentation, Animal and Human: The Society endorses the principles embodied in Declaration of Helsinki (last page of instructions to authors). Conduct all investigations involving humans in conformity with these principles. Conduct all animal experimentation must be conducted in conformity with "Guiding Principles for Research Involving Animals and Human Beings" (last page of instructions to authors). In describing surgical procedures on animals, specify type and dosage of anesthetic agent. Curarizing agents are not anesthetics; if used, provide evidence that anesthesia was of suitable grade and duration. Papers in which evidence of adherence to these principles is not apparent will be refused.

Materials Sharing: Work must necessarily be independently verifiable. Authors describing results derived from the use of antibodies, recombinant plasmids and cloned DNAs, mutant cell lines or viruses, and other similarly unique materials are expected to make such materials available to qualified investigators on request.

Submit published nucleic acid/amino acid sequences to a widely accessible data bank. Sequence data submission forms for the National Biomedical Research Foundation-Protein Identification Resource Database (NBRF-PIR) are available from APS Publications Office.

Mathematical Formulas and Equations: Address mathematical aspects to readers who are not mathematicians. Present mathematical strategy, assumptions on which mathematics are based, and a summary of meaning of final mathematical statement and limitations. Lengthy or complex mathematical developments central to an article are often put in an appendix. Submit or note for referees filling in details of mathematics not explicitly stated and not needed in article proper or in appendix.

Simplify structural chemical formulas and complicated mathematical equations; carefully check. Clearly identify all subscripts, superscripts, Greek letters, and other unusual characters in penciled notes in margin where they first appear. Distinguish between 1 (one) and letter l (el), 0 (zero) and letter O, × (multiplication sign) and letter x. Use slant line (/) for simple fractions (a+b)/(x+y) in text rather than built-up fraction $\frac{a+b}{x+y}$ which should be used if equation is offset from text.

Use subscripts or superscripts wherever feasible and appropriate, because they simplify equations by eliminating extraneous operations [$R_A R_D$ instead of $RA\cdot RD$ or (RA)(RD)]. Use circles for pools in compartmental or flow-type models and whole arrows $\longrightarrow$ for interconnections or flows (not arrows with half-heads $\longrightarrow$, as in reversible chemical equations). Do not use nonstandard mathematical notations; e.g., do not use computer symbols in equations (* for multiplication or ** for exponentiation). Use lowercase letters for time-varying symbols in compartmental model equations, preferably q(t) for masses, c(t) for concentrations, with subscripts as needed. Convention for numerical subscripts for rate constants (k_{21}) is as in most life sciences but opposite to that in pharmacokinetics; i.e., k_{ij} is fractional rate of transfer from compartment j to compartment i (or to compartment i from compartment j). Notation is consistent with standard nomenclature in applied mathematics for matrices and matrix manipulation algorithms in commercial software packages for scientific/mathematical computations involving matrices. In addition to defining symbols as they appear in text, include a table of nomenclature in articles that utilize several or more different symbols, specifying units (dimensions) as well as each definition.

Page Charges: $50 per printed page. Editorial consideration is not related to acceptance of page charge; it is expected that charge will be paid by author's research funds or institution that supported research.

Style: Follow *Webster's Third New International Dictionary* for spelling, compounding, and word division.

SOURCE

December 1994, 77(6): back matter.

Journal of Bacteriology
American Society for Microbiology

Journals Division
American Society for Microbiology
1325 Massachusetts Ave., N.W.
Washington, D.C. 20005-4171

AIM AND SCOPE

The *Journal of Bacteriology* publishes descriptions of basic research on bacteria and other microorganisms, including fungi and other unicellular eukaryotic organisms. Topics that are considered include structure and function, biochemistry, enzymology, metabolism and its regulation, molecular biology, genetics, plasmids and transposons, general microbiology, plant microbiology, chemical or physical characterization of microbial structures or products, and basic biological properties of organisms.

ASM publishes a number of different journals covering various aspects of microbiology. Each journal has a prescribed scope that must be considered in determining the most appropriate journal for each manuscript. The *Journal of Bacteriology* will consider papers that describe the use of antibiotics and antimicrobial agents as tools for elucidating the basic biological processes of microorganisms.

The *Journal of Bacteriology* will consider manuscripts that emphasize the interrelationship of the bacteriophage and the host cell, manuscripts about work in which viruses were used as tools for elucidating the structure or biological processes of microorganisms, and manuscripts that concern phages that are related to transposable elements or plasmids.

Manuscripts describing new or novel methods or improvements in media and culture conditions will not be considered unless they are applied to the study of basic problems in microbiology.

In most cases, reports that emphasize methods and nucleotide sequence data alone (without experimental documentation of the functional and evolutionary significance of the sequence) will not be considered.

Genome maps of microorganisms will not be considered unless at least two restriction enzymes that yield identical results are used, the number of DNA fragments in each ethidium bromide-staining band is addressed, and the appropriate location(s) and number of *rrn* operons are positioned.

Manuscripts describing work, with a new organisms, that largely repeats published research done with a different organisms will be considered if they significantly increase the understanding of the original property, if they provide and extensive basis for evolutionary comparison, or if the work is of unusual importance because of its relationship to other properties of the new organism.

MANUSCRIPT SUBMISSION

In a cover letter state: journal to which manuscript is being submitted, most appropriate journal section, complete mailing address (including street), telephone and fax numbers of corresponding author, electronic mail address, and former ASM manuscript number and year if a resubmission. Include written assurance that permission to cite personal communications and preprints has been granted. Send three copies of "in press" and "submitted" manuscripts.

Submit three complete copies of each manuscript, including figures and tables. Type every portion double-spaced (6 mm between lines), including figure legends, table footnotes, and References. Use 1 in. margins on all sides. Number all pages in sequence, including abstract, figure legends, and tables. Place last two items after References section. Make characters easily distinguishable: numeral zero (0) and letter "oh" (O); numeral one (1), letter "el" (1), and letter "eye" (I); and multiplication sign (×) and letter "ex" (x). If distinctions cannot be made, mark these items at first occurrence for cell lines, strain and genetic designations, viruses, etc., on modified manuscript.

Manuscripts must represent reports of original research. All authors must have agreed to its submission and are responsible for its content, including appropriate citations and acknowledgments, and must also have agreed that corresponding author has authority to act on their behalf on all matters pertaining to publication of the manuscript. Submission guarantees that manuscript, or one substantially the same, was not published previously, is not being considered or published elsewhere, and was not rejected on scientific grounds by another ASM journal.

The American Society for Microbiology accepts the definition of primary publication in *How to Write and Publish a Scientific Paper* (third ed., Robert A. Day), to wit: ". . . (i) the first publication of original research results, (ii) in a form whereby peers of the author can repeat the experiments and test the conclusions, and (iii) in a journal or other source document readily available within the scientific community."

A scientific paper published in a conference report, symposium proceeding, technical bulletin, or other retrievable source is unacceptable for submission on grounds of prior publication. Preliminary disclosure of research findings published in abstract form as an adjunct to a meeting, is not considered "prior publication."

Acknowledge prior publication of data and send copies of relevant work.

Obtain and submit permissions from both original publisher and original author, copyright owner(s), to reproduce figures, tables, or text from previous publications. Identify each item as to relevant item in manuscript (e.g., "permissions for Fig. 1 in JB 123-93").

An author is one who made a substantial contribution to the "overall design and execution of the experiments"; all authors are responsible for entire paper. Do not list individuals who provided assistance as authors; recognize them in Acknowledgment section.

Authors must agree to order in which names are listed in byline. Footnotes regarding attribution of work are not permitted. If necessary, include statement in Acknowledgment section.

Corresponding author must sign copyright transfer agreement on behalf of all authors upon acceptance of manuscript. ASM will not publish manuscript without agreement.

If all authors were employed by U.S. Government when work was performed, corresponding author should attach a statement attesting that manuscript was prepared as a part of their official duties and, as such, is a work of the U.S. government not subject to copyright. If some authors were employed by the U.S. government, but others were not, corresponding author should sign copyright transfer agreement as it applies to that portion performed by the non-government employee authors.

MANUSCRIPT FORMAT

Title, Running Title, and Byline: Present results of an independent, cohesive study; numbered series titles are not allowed. Avoid main title/subtitle arrangement, complete sentences, and unnecessary articles. On title page, include title, running title (54 characters and spaces), name of each author, address(es) of institution(s) at which work was performed, each author's affiliation, and a footnote indicating present address of any author no longer there. Place asterisk after name of corresponding author with telephone and fax numbers. If electronic mail address is supplied on title page, it will be included as a footnote.

If desired, complete mailing address, telephone and fax numbers, and e-mail address of corresponding author will be published as a footnote. Include information in lower left corner of title page labeled as "Correspondent Footnote."

Abstract: 250 words concisely summarizing basic content of paper without presenting extensive experimental details. Avoid abbreviations; do not include diagrams. When essential to include a reference, use References citation but omit article title. It must be complete and understandable without reference to text.

Introduction: Supply sufficient background information to allow reader to understand and evaluate results of present study without referring to previous publications. Provide rationale for present study. Use only those references required to provide the most salient background rather than an exhaustive review of topic.

Materials and Methods: Include sufficient technical information to allow experiments to be repeated. When centrifugation conditions are critical, give enough information to enable repetition of procedure: make of centrifuge, model of rotor, temperature, time at maximum speed, and centrifugal force (x g rather than revolutions per minute). For commonly used materials and methods, a simple reference is sufficient. If several alternative methods are commonly used, identify method briefly and cite reference. Allow reader to assess method without reference to previous publications. Describe new methods completely and give sources of unusual chemicals, equipment, or microbial strains. When large numbers of microbial strains or mutants are used, include tables identifying sources and properties of strains, mutants, bacteriophages, plasmids, etc.

Describe enzyme purifications, describe results of procedures in Results section.

Describe a method, strain, etc., used in only one of several experiments in Results section or very briefly (one or two sentences) in a table footnote or figure legend.

Results: Include results of experiments. Reserve extensive interpretation of results for Discussion. Present results as concisely as possible in either text, table(s), or figure(s). Avoid extensive use of graphs to present data more concisely presented in text or tables. Limit photographs (particularly photomicrographs and electron micrographs) to those absolutely necessary to show experimental findings. Number figures and tables in order cited in text; cite all figures and tables.

Discussion: Provide an interpretation of results in relation to previously published work and to experimental system at hand. Do not extensively repeat Results section or reiterate introduction. In short papers, Results and Discussion may be combined.

Acknowledgments: Acknowledge financial assistance and personal assistance in separate paragraphs. Format for acknowledgment of grant support is: "This work was supported by Public Health Service grant CA01234 from the National Cancer Institute."

Appendixes: Appendixes contain supplementary material to aid reader. Do not use titles, authors, and References sections distinct from those of primary article. If author(s) of appendix cannot be listed in byline or Acknowledgment section of primary article, rewrite appendix so that it can be considered for publication as an independent article, either full-length or Note style. Label equations, tables, and figures with letter "A" preceding numeral to distinguish them from those cited in main body of text.

References: Include all relevant sources; cite all listed references in text. Arrange citations in alphabetical order, by first author, and number consecutively. Abbreviate journal names according to *Serial Sources for the BIOSIS Data Base* (BioSciences Information Service, 1992). Cite each listed reference in text by number.

1. **Anagnostopoulos, C., and J. Spizizen.** 1961. Requirements for transformation in *Bacillus subtilis.* J. Bacteriol. **81:**741–746. 2. **Berry, L. J., R. N. Moore, K. J.** 2. **Goodrum, and R. E. Couch, Jr.** 1977. Cellular requirements for enzyme inhibition by endotoxin in mice, p. 321–325. *In* D. Schlessinger (ed.) Microbiology—1977. American Society for Microbiology, Washington, D.C.

3. **Cox, C. S., B. R. Brown, and J. C. Smith.** J. Gen. Genet., in press.*

4. **Dhople, A., I. Ortega. and C. Berauer.** 1989. Effect of oxygen on in vitro growth of *Mycobacterium leprae*, abstr. U-82, p. 168. Abstr. 89th Annu. Meet. Am. Soc. Microbiol. 1989.

5. **Finegold, S. M., W. E. Shepherd, and E. H. Spaulding.** 1977. Cumitech 5, Practical anaerobic bacteriology. Coordinating ed., W. E. Shepherd. American Society for Microbiology, Washington, D.C.

6. **Fitzgerald, G., and D. Shaw.** *In* A. E. Waters (ed.) Clinical microbiology, in press. EFH Publishing Co., Boston .

7. **Gustlethwaite, F. P.** 1985. Letter. Lancet **ii:**327.

8. **Jacoby, J., R. Grimm, J. Bostic, V. Dean, and G. Starke.** Submitted for publication.

9. **Jensen, C., and D. S. Schumacher.** Unpublished data.

10. **Jones, A. (Yale University).** 1990. Personal communication .

11. **Leadbetter, E. R.** 1974. Order II. *Cytophagales* , p. 99. *In* R. E. Buchanan and N. E. Gibbons (ed.), Bergey's manual of determinative bacteriology, 8th ed. The Williams & Wilkins Co., Baltimore.

12. **Miller, J. H.** 1972. Experiments in molecular genetics, p. 352–355. Cold Spring Harbor Laboratory, Cold Spring Harbor, N.Y.

13. **Powers, R. D., W. M. Dotson, Jr., and F. G. Hayden.** 1982. Program Abstr. 22nd Intersci. Conf. Antimicrob. Agents Chemother., abstr. 448.

14. **Sigma Chemical Co.** 1989. Sigma manual. Sigma Chemical Co., St. Louis, Mo.

15. **Smith, J. C.** April 1970. U.S. patent 484,363,770.

16. **Smyth, D. R.** 1972. Ph.D. thesis. University of California, Los Angeles.

17. **Yagupsky, P., and M. A. Menegus.** 1989. Intraluminal colonization as a source of catheter-related infection. Antimicrob. Agents Chemother. **33:**2025. (Letter.)

An "in press" reference to an ASM publication should state control number or name of publication if it is a book.

Illustrations and Tables: Put figure number and authors' names on all figures, either in margin or on back (marked lightly with a soft pencil). For micrographs especially, indicate TOP.

Do not use paper clips. Insert small figures in an envelope. Do not submit illustrations larger than 8.5 × 11 in.

Continuous-Tone and Composite Photographs—When submitting continuous-tone photographs (e.g., polyacrylamide gels), keep in mind journal page width: 3 5/16 in.-single column and 6 7/8 in. - double-column (maximum). Include only significant portion of an illustration. Photos must be of sufficient contrast to withstand loss of contrast and detail inherent in printing. Submit one photograph of each continuous-tone figure for each manuscript copy; photocopies are not acceptable. Submit figures the size they will appear when published. If must be reduced, make sure that all elements, including labeling, can withstand reduction and remain legible.

If a figure is a composite of a continuous-tone photograph and a drawing or labeling, provide original composite for printer. Send original, labeled "printer's copy," with modified manuscript to editor.

Electron and light micrographs must be direct copies of original negative. Indicate magnification with a scale marker on micrograph.

Computer-Generated Images:—Use Adobe Photoshop or Aldus Freehand. For Aldus, one and two column art cannot exceed 20 picas (3 5/16 in.) and 41.5 picas (6 7/8 in.) respectively. Use Helvetica (medium or bold) or Times Roman font. Adobe users: check image densities on-line. If image shadow density reads below 1.25, enter density as 1.40; if 1.25–1.6, enter as 1.65; enter readings >1.65 as actual reading. Provide computer file with prints on floppy disk (Macintosh) with accepted manuscript.

For large images, use 40- or 80-megabyte Syquest cartridges or magneto-optical cartridges. For transfer from UNIX systems, submit either 9-track or 8-mm "tar" archives. Incorporate all final lettering, labeling, tooling, etc., in final supplied material. Do not include figure numbers on image.

Electronic image telephone hotline: William Byrd Press, (800) 888-2973, ext 3361 (nonU.S.: (804) 264-2828, ext 3361).

Include a description of software/hardware used in figure legend(s).

Color Photographs—Use is discouraged. If necessary, include an extra copy at manuscript submission for a cost estimate. Cost must be borne by author.

Drawings—Submit graphs, charts, sequences, complicated clinical or mathematical formulas, diagrams, and other drawings as glossy photographs made from finished drawings not requiring additional artwork or typesetting. Computer-generated graphics produced on high-quality laser printers are also acceptable. No part of graph or drawing should be handwritten. Label both axes of graphs. Graphs will be reduced to one-column width (3 5/16 in.); make elements large enough to withstand reduction. Avoid heavy letters and unusual symbols.

In figure ordinate and abscissa scales and table column headings, avoid ambiguous use of numbers with exponents. Use appropriate SI symbols (μ for 10^{-6}, m for 10^{-3}, k for 10^3, M for 10^6, etc.). For a complete listing of SI symbols see IUPAC "Manual of Symbols and Terminology for Physicochemical Quantities and Units" (*Pure Appl. Chem.* **21**:3–44, 1970).

When powers of 10 is used, associate the exponent power with the number shown. In representing 20,000 cells per ml, numeral on ordinate is "2" and label is "10^4 cells per ml" (not "cells per ml $\times 10^{-4}$").

Figure Legends: Provide enough information so that figure is understandable without reference to text. Describe detailed experimental methods in Materials and Methods section, not in a figure legend. Report a method unique to an experiment if it is one or two sentences. Define all symbols and abbreviations that have not been defined elsewhere.

Tables: Type each on a separate page. Arrange data so columns of like material read down, not across. Make headings clear so meaning of data will be understandable without reference to text. See Abbreviations for those that can be used in tables. Explanatory footnotes are acceptable, extensive table "legends" are not. Footnotes should not include detailed descriptions of experiment. Tables must include enough information to warrant table format; incorporate fewer than six pieces of data into text. Tables that can be photographically reproduced for publication without further typesetting or artwork are "camera ready." Do not hand letter.

OTHER FORMATS

AUTHOR'S CORRECTION—Provides a means of adding citations that were overlooked in a published article. The author who failed to cite a reference and the author whose paper was not cited must agree to publication.

ERRATUM—Provides a means of correcting errors that occurred during writing, typing, editing or printing of a published articles. Changes in data and addition of new material are not permitted. Send errata directly to the Journals Division.

MINIREVIEWS—4 printed pages. Brief summaries of developments in fast-moving areas; base on published articles; address any subject within scope. Provide three double-spaced copies.

NOTES—1000 words. Brief observations that do not warrant full-length papers. Submit the same way as full-length papers. Include a 50 word abstract. Do not use section headings in body; report methods, results, and discussion in a single section. Paragraph lead-ins are permissible. Keep figures and tables to a minimum. Describe materials and methods in text, not in figure legends or table footnotes. Present acknowledgments as in full-length papers; do not use a heading. References section is identical.

OTHER

Abbreviations: Use as an aid to reader, not a convenience to author; their use should be limited. Use abbreviations other than those recommended by the IUPAC-IUB (*Biochemical Nomenclature and Related Documents*, 1978) only as a necessity, such as in tables and figures.

Use pronouns or paraphrase a long word after first use. Use standard chemical symbols and trivial names or their symbols for terms that appear in full in neighboring text.

Introduce all abbreviations except those listed below in first paragraph of Materials and Methods. Or, define each abbreviation and introduce it in parentheses at first use. Eliminate abbreviations not used at least five times in text (including tables and figure legends).

In addition to abbreviations for Système International d'Unités (SI) units of measurement, other common units (e.g., bp, kb, and Da), and chemical symbols for elements, use the following without definition in title, abstract, text, figure legends, and tables: DNA (deoxyribonucleic acid); cDNA (complementary DNA); RNA (ribonucleic acid); cRNA (complementary RNA); RNase (ribonuclease); DNase (deoxyribonuclease); rRNA (ribosomal RNA); mRNA (messenger RNA); tRNA (transfer RNA); AMP,

ADP, ATP, dAMP, ddATP, GTP, etc. (for the respective 5′ phosphates of adenosine and other nucleosidcs) (add 2′-, 3′-, or 5′- when needed for contrast); ATPase, dGTPase, etc. (adenosine triphosphatase, deoxyguanosine triphosphatase, etc.); NAD (nicotinamide adenine dinucleotide); NAD (nicotinamide adenine dinucleotide, oxidized); NADH (nicotinamide adenine dinucleotide, reduced); NADP (nicotinamide adenine dinucleotide phosphate); NADPH (nicotinamide adenine dinucleotide phosphate, reduced); NADP⁺ (nicotinamide adenine dinucleotide phosphate, oxidized); poly(A), poly(dT), etc. (polyadenylic acid, polydeoxythymidylic acid, etc.); oligo(dT), etc. (oligodeoxythymidylic acid, etc.); P_i (orthophosphate); PP_i (pyrophosphate); UV (ultraviolet); PFU (plaque-forming units); CFU (colony-forming units); MIC (minimal inhibitory concentration); MBC (minimal bactericidal concentration); Tris [tris(hydroxymethyl)aminomethane]; DEAE (diethylaminoethyl); A_{260} (absorbance at 260 nm); EDTA (ethylenediaminetetraacetic acid); and AIDS (acquired immunodeficiency [or immune deficiency] syndrome). Abbreviations for cell lines (e.g., HeLa) also need not be defined.

The following may be used without definition in tables:

amt	amount	SD	standard deviation
approx	approximately	SE	standard error
avg	average	SEM	standard error of the mean
concn	concentration	sp act	specific activity
diam	diameter	sp gr	specific gravity
expt	experiment	temp	temperature
exptl	experimental	tr	trace
ht	height	vol	volume
mo	month	vs	versus
mol wt	molecular weight	wk	week
no.	number	wt	weight
prepn	preparation	yr	year

Data Submission: Deposit important strains in publicly accessible culture collections and refer to collections and strain numbers in text. Indicate laboratory strain designations and donor source as well as original culture collection identification numbers.

Isotopically Labeled Compounds: For simple molecules, indicate isotopic labeling in chemical formula. Do not use brackets when isotopic symbol is attached to the name of a compound that in its natural state does not contain the element or to a word that is not a specific chemical name. For specific chemicals, place symbol for the isotope introduced in square brackets directly preceding the part of the name that describes labeled entity. Configuration symbols and modifiers precede isotopic symbol.

Follow conventions for isotopic labeling of the *Journal of Biological Chemistry*; for more detailed information see instructions to authors of that journal (first issue of each year).

Materials Sharing: Make plasmids, viruses, and living materials such as microbial strains and cell lines newly described in the article available from a national collection or in a timely fashion and at reasonable cost to members of the scientific community for noncommercial purposes.

Nomenclature, Chemical and Biochemical: The recognized authority for names of chemical compounds is *Chemical Abstracts* (Chemical Abstracts Service, Ohio State University, Columbus) and its indexes. *The Merck Index* (11th ed., 1989; Merck & Co., Inc., Rahway, NJ) is also an excellent source. For guidelines to the use of biochemical terminology, consult *Biochemical Nomenclature and Related Documents* (1978; The Biochemical Society, London) and instructions to authors of *Journal of Biological Chemistry* and *Archives of Biochemistry* and *Biophysics* (first issues of each year).

Do not express molecular weight in daltons; molecular weight is a unitless ratio. Express molecular mass in daltons. For enzymes, use recommended (trivial) name assigned by the Nomenclature Committee of the International Union of Biochemistry (see *Enzyme Nomenclature*, Academic Press, Inc., 1992). If a nonrecommended name is used, place proper (trivial) name in parentheses at first use in abstract and text. Use EC number when one has been assigned; express enzyme activity either in katals (preferred) or in micromoles per minute.

Nomenclature, Genetic

Bacteria—Genetic properties of bacteria are described in terms of phenotypes and genotypes. Phenotype describes observable properties of an organism. Genotype refers to genetic constitution of an organism, usually in reference to some standard wild type. Use recommendations of Demerec et al. (*Genetics* **54**:61–76,1966) as a guide to terms.

Use phenotypic designations when mutant loci have not been identified or mapped. Use to identify protein products of genes. Phenotypic designations consist of three-letter symbols, not italicized, first letter capitalized. Use Roman or Arabic numerals to identify a series of related phenotypes. Designate wild-type characteristics with a superscript plus (Pol⁺). Use negative superscripts (Pol⁻) to designate mutant characteristics. Use lowercase superscript letters to further delineate phenotypes. Define phenotypic designations.

Indicate genotypic designations by three-letter lowercase italic locus symbols. If several loci govern related functions, distinguish by italicized capital letters following locus symbol. Indicate promoter, terminator, and operator sites as described by Bachmann and Low (*Microbiol. Rev.* 44: 1–56, 1980.

Indicate wild-type alleles with a superscript plus. Refer to an *ara* mutant rather than an *ara⁻* strain.

Designate mutation sites with serial isolation numbers (allele numbers) after locus symbol. If only a single such locus exists or if it is not known in which of several related loci the mutation has occurred, use a hyphen instead of a capital letter. In reporting isolation of new mutants, give allele numbers to mutations. For *Escherichia coli*, there is a registry of such numbers: *E. coli* Genetic Stock Center, Department of Biology, Yale University, P.O. Box 6666, New Haven, CT 06511-7444. For *Salmonella: Salmonella* Genetic Stock Center, Department of Biology, University of Calgary, Calgary, Alberta, T2N 1N4 Canada. For *Bacillus*: the *Bacillus* Genetic Stock Center, Ohio State University, Columbus. A registry of allele numbers and insertion elements (omega [Ω] numbers) for chromosomal mutations and chromosomal insertions of transposons and other insertion elements is at Iowa State University. Blocks of allele numbers and Ω numbers are assigned to laboratories on request. Obtain requests for blocks of numbers and additional information from Peter A. Pattee, Department of Microbiology, Iowa State University, Ames, IA 50011. A registry of plasmid designations is maintained by the Plasmid Reference Center, Department of Microbiology and Immunology, 5402, Stanford University School of Medicine, Stanford, CA 94305-2499.

Avoid use of superscripts with genotypes (other than + to indicate wild-type alleles). Designations indicating amber mutations (Am), temperature-sensitive mutations (Ts), constitutive mutations (Con), cold-sensitive mutations (Cs), production of a hybrid protein (Hyb), and other important phenotypic properties follow allele number. Define all other such designations of phenotype at first occurrence. Use of superscripts must be approved by editor; define at first occurrence.

Use subscripts to distinguish between genes (having the same name) from different organisms or strains. Use abbreviation if explained. Use a subscript to distinguish between genetic elements with the same name.

Indicate deletions by symbol Δ before deleted gene or region. Use other symbols with appropriate definition. Inversion is shown as IN(*rrnD-mlE*)*l*. Insertion of an *E. coli his* gene into plasmid pSC101 at zero kilobases (0 kb) is shown as pSC101 Φ(0kb::K12*hisB*)*4*. An alternative insertion designation is *galT236*::Tn*5*, where 236 refers to the locus of the insertion, and if the strain carries an additional *gal* mutation, it is listed separately. For additional examples, see Campbell et al. and Novick et al. cited below. In reporting construction of strains in which a mobile element was inserted and subsequently deleted, note latter fact in strain table by listing genotype of strain used as an intermediate, in a table footnote, or by a direct or parenthetical remark in the genotype. In setting parenthetical remarks within genotype or dividing genotype into constituent elements, parentheses and square brackets are used without special meaning; square brackets are used outside parentheses. To indicate presence of an episome, use parentheses (or brackets). Indicate reference to an integrated episome as described for inserted elements, and an exogenote is shown as, for example, W3110/F'8(*gal⁺*).

Explain deviations from standard genetic nomenclature in Materials and Methods or in a table of strains. For symbols in current use, consult Bachmann (B. J. Bachmann, p. 807–876, *in* J. L. Ingraham, K. B. Low, B. Magasanik, M. Schaechter, and H. E. Umbarger, ed., *Escherichia coli and Salmonella typhimurium: Cellular and Molecular Biology*, 1987, American Society for Microbiology, Washington, D.C.) for *E. coli* K-12, Sanderson and Roth (*Microbiol. Rev.* 52:485–532, 1988) for *Salmonella typhimurium*, Holloway et al. (*Microbiol. Rev.* 43:73–102, 1979) for *Pseudomonas*, Piggot and Hoch (*Microbiol. Rev.* 49:158–179, 1985) for *Bacillus subtilis*, Perkins et al. (*Microbiol. Rev.* 46:426–570, 1982) for *Neurospora crassa*, and Mortimer and Schild (*Microbiol. Rev.* 49:181–213, 1985) for *Saccharomyces cerevisiae*.

Bacteriophages—The genetic nomenclature tends to have separate conventions for each phage. Genetic symbols may be one, two, or three letters and are italicized. Do not italicize phenotypic symbols and designations of gene products; use superscript plus and minus symbols to indicate wild-type and mutant phenotypes, respectively. Delineate host DNA insertions into phages by square brackets, and make genetic symbols and designations for such inserted DNA conform to those used for host genome. For lists of gene symbols for phages see *Genetic Maps*, 5th ed. (S. J. O'Brien, ed., Cold Spring Harbor Laboratory, Cold Spring Harbor, N. Y., 1990). Relevant references include: for phage λ, Daniels et al. (D. L. Daniels, J. L. Schroeder, W. Szybalski, F. Sanger, and F. R. Blattner, p. 469–515, *in* R. W. Hendrix, J. W. Roberts, F. W. Stahl, and R. A. Weisberg, ed., *Lambda II*, Cold Spring Harbor Laboratory, Cold Spring Harbor, N. Y., 1983); for phage T4, Mosig (G. Mosig, p. 362–374, *in* C. K. Mathews, E. M. Kutter, G. Mosig, and P. B. Berget, ed., *Bacteriophage T4*, American Society for Microbiology, Washington, D. C., 1983); and for phage T7, Dunn and Studier (*J. Mol. Biol.* 166:477–535, 1983).

"Mutant" vs. "Mutation"—A mutation is an alteration of the primary sequence of genetic material and a mutant is a strain carrying one or more mutations.

Naming Genes—When applicable, give new gene same name as a homologous gene already identified in another organism. Gene may be given a provisional name based on its map location in the style *yaa*A, analogous to the style used for recording transposon insertions (*zef*). A provisional name may be given in the style described by Demerec et al. (e.g., *usg*, for gene upstream of *folC*).

Designations—Do not use genotype as a name. If a strain designation has not been chosen, select an appropriate word combination.

Transposable Elements, Plasmids, and Restriction Enzymes—For nomenclature of transposable elements follow Campbell et al. (*Gene* 5:197–206, 1979). The system of designating transposon insertions at sites where there are no known loci, e.g., *zef-123*::Tn*5*, is described by Chumley et al. (*Genetics* 91:639–655, 1979). Follow nomenclature recommendations of Novick et al. (*Bacteriol. Rev.* 40:168–189, 1976) for plasmids and plasmid-specified activities, of Low (*Bacteriol. Rev.* 36:587–607, 1972) for F-prime factors, and of Roberts (*Nucleic Acids Res.* 17:r347–r387, 1989) for restriction enzymes and their isoschizomers. Recombinant DNA molecules constructed in vitro follow the nomenclature for insertions in general. Describe DNA inserted into recombinant DNA molecules by using gene symbols and conventions for organism from which DNA was obtained. The Plasmid Reference Center (E. Lederberg, Plasmid Reference Center, Department of Microbiology and Immunology, 5402, Stanford University School of Medicine, Stanford, CA 94305–2499) assigns Tn and IS numbers to avoid conflicting and repetitive use and also clears nonconflicting plasmid prefix designations.

Nomenclature, Microorganisms: Use binary names, consisting of a generic name and a specific epithet. Names of higher categories may be used alone, specific and subspecific epithets may not. Precede specific epithet by a generic name at first use. Thereafter, abbreviate generic name to initial capital letter provided there is no confusion with other genera in paper. Italicize or underline names of all taxa (phyla, classes, orders, families, genera, species, subspecies). Spelling of names follows *Approved Lists of Bacterial Names* (amended edition) (V. B. D. Skerman, V. McGowan, and P. H. A. Sneath, ed.) and *Index of the Bacterial and Yeast Nomenclatural Changes Published in the International Journal of Systematic Bacteriology since the 1980 Approved Lists of Bacterial Names (1 January 1980 to 1 January 1989)* (W. E. C. Moore and L. V. H. Moore, ed., American Society for Microbiology, 1989), and validation lists and articles published in *International Journal of Systematic Bacteriology* since 1 January 1989. If there is reason to use a name that does not have standing in nomenclature, enclose name in quotation marks and make an appropriate statement concerning nomenclatural status of name in text (for an example, see *Int. J. Syst. Bacteriol.* 30:547–556, 1980). Since classification of fungi is not complete, determine accepted binomial for a given organism. Sources for names include *The Yeasts: a Taxonomic Study* (3rd ed., N. J. W. Kreger-van Rij, ed., Elsevier Science Publishers B.V., Amsterdam, 1984) and *Ainsworth and Bisby's Dictionary of the Fungi, Including the Lichens* (7th ed., (Commonwealth Mycological Institute, Kew, Surrey, England, 1983).

Give microorganisms, viruses, and plasmids designations consisting of letters and serial numbers. Include worker's initials or a descriptive symbol of locale, laboratory, etc., in designation. Give each new strain, mutant, isolate, or derivative a new (serial) designation. Make designation distinct from those of genotype and phenotype. Do not include genotypic and phenotypic symbols.

Nucleic Acid Sequences: Present limited length nucleic acid sequences which are the primary subject of a study freestyle in the most effective format. Present longer sequences in the following format to conserve space. Submit sequence as 8.5 × 11 in. camera-ready copy in standard (portrait) orientation. Print sequences in lines of 100 to 120 nucleotides in a nonproportional (monospace), legible font to be published with a line length of 6 inches. If possible, subdivide lines of nucleic acid sequence into blocks of 10 or 20 nucleotides by spaces within sequences or by marks above it. Use uppercase and lowercase letters to designate exon/intron structure, transcribed regions, etc., if lowercase letters remain legible at a 6-inch line length. Number sequence line by line; place numerals, representing the first base of each line, to the left of lines. Minimize spacing between adjacent lines, leave room only for sequence annotation. Annotation may include boldface, underlining, brackets, boxes, etc. Present encoded amino acid sequences, if necessary, immediately above first nucleotide of each codon, using single-letter amino acid symbols. Use same format for comparisons of multiple nucleic acid sequences.

Nucleotide Sequences: Include GenBank/EMBL accession numbers for primary nucleotide and/or amino acid sequence data in original manuscript or insert when manuscript is modified. Include accession number as a separate paragraph at end of Materials and Methods section for full-length papers or at end of text of Notes.

GenBank Submissions, National Center for Biotechnology Information; Bldg. 38A, Rm. 8N-803, 8600 Rockville Pike, Bethesda, MD 20894; e-mail (submissions): gbsub@ncbi.nlm.nih.gov, e-mail updates: update@ncbi.nlm.nih.gov.

EMBL Data Library Submissions, Postfach 10.2209, Meyerhofstrasse 1, 6900 Heidelberg, Germany; telephone: 011 49 (6221) 387258; fax: 011 49 (6221) 387306; electronic mail (data submissions): datasubs@embl.bitnet.

Numerical Data: Use standard metric units for reporting length, weight, and volume. For these units and molarity, use prefixes m, μ, n, and p for 10^{-3}, 10^{-6}, 10^{-9}, and 10^{-12}, respectively. Use prefix k for 10^3. Avoid compound prefixes such as mμ or μμ. Use μg/ml or μg/g in place of ppm. Units of temperature are: 37°C or 324 K.

When fractions are used to express units such as enzymatic activities, use whole units, such as "g" or "min," in denominator instead of fractional or multiple units, such as μg or 10 min. Use an unambiguous form such as the exponential notation.

See *CBE Style Manual,* 5th ed., for more detailed information about reporting numbers and information on SI units for reporting illumination, energy, frequency, pressure, and other physical terms. Report numerical data in the appropriate SI unit.

Page Charges: $50 per printed page if reported research was supported by special funds, grants (departmental, governmental, institutional, etc.). If not, request a waiver of charges to Journals Division, American Society for Microbiology with manuscript. Separate request from cover letter; indicate how work was supported and include a copy of Acknowledgment section.

Minireviews and Letters to the Editor are not subject to page charges.

Reviewers: Suggest an appropriate editor for new submissions. Recommend at least two or three reviewers who are not members of your institution(s) and have never been associated with you or your laboratory(ies). Provide name, address, phone and fax numbers, and area of expertise for each.

Style: Follow *CBE Style Manual* (5th ed., 1983; Council of Biology Editors, Inc., Bethesda, Md.), *ASM Style Manual for Journals and Books* (American Society for Microbiology, 1991), and Robert A. Day's *How to Write and Publish a Scientific Paper* (3rd ed., 1988; Oryx Press), as interpreted and modified by the editors and the ASM Journals Division

Use past tense to narrate events in the past, including procedures, observations, and data of study. Use present tense for your conclusions, conclusions of previous researchers, and generally accepted facts. Most of abstract, Materials and Methods, and Results sections will be in past tense, and most of introduction and some of Discussion will be in present tense. It may be necessary to vary tense in a single sentence.

For an in-depth discussion of tense in scientific writing, see p. 158–160 in *How to Write and Publish a Scientific Paper,* 3rd ed.

SOURCE

January 1994, 176(1): i-xi.

Journal of Biochemistry
The Japanese Biochemical Society

The Japanese Biochemical Society
Ishikawa Building-3f, 25–16
Hongo 5-chome, Bunkyo ku, Tokyo 113, Japan

AIM AND SCOPE

The Journal of Biochemistry publishes the results of original research in the fields of Biochemistry, Molecular Biology, and Cell Biology (not submitted for publication or previously published elsewhere) written in clear grammatical English, in the forms of Regular Papers and Communications. A Communication is not a Preliminary Note but is, although brief, a complete and final publication. The materials described in Communications should not be included in a later paper.

MANUSCRIPT SUBMISSION

Submit manuscripts in triplicate (one original and two clear copies). Select one of the following fields, under which the paper should be reviewed; indicate selection on title page: Carbohydrates, Lipids, Nucleic Acids, Proteins, Peptides, and Amino Acids, Physical Properties of Biomolecules, Biosynthesis of Nucleic Acids and Proteins, Gene Structure and Function, Enzymes, Metabolic Regulation, Biological Oxidation and Bioenergetics, Muscles, Cell Motility and Cytoskeleton, Membrane Structure and Function, Cellular Organelles, Receptors and Signal Transduction, Differentiation, Development and Aging, Biochemical Genetics, Neurochemistry, Molecular Immunology, Gene and Protein Engineering, Biotechnology, Physiological and Clinical Chemistry, and Nutrition, Microbial Physiology, Biochemical Pharmacology, Other Topics.

Make manuscript as concise as possible. A Communication should not exceed 3.5 printed pages including space for figures, tables, and references. One single printed page is approximately 3.5 pages typed double-spaced.

A manuscript describing primary structures of biological macromolecules (proteins and nucleic acids) without enough data for deductions within limited page space is not acceptable as a Communication.

A manuscript describing only partial sequences of biological macromolecules is unacceptable as a Communication unless it involves valuable biochemical information. In the case of a Communication, submit two copies of a free style letter describing urgency or necessity for rapid publication.

Manuscripts will be published only after agreement by author(s) to pay costs of publication including page charges. Provide "materials for the preparation of indices" upon request. Illustrations, photographs, electron micrographs, color plates, and other special illustrations will be reproduced at author's expense. Prices will be sent after final decision.

MANUSCRIPT FORMAT

Conform to style and usage of *Journal* as exemplified in current issues. Type on good grade white bond paper double-spaced throughout with 65 strokes × 25 to 28 lines including references, and figure legends. Use separate sheets for: title page(s); summary; text; footnote(s) to text; references; table(s); figure legend(s); and figures or other subsidiary matters. Arrange manuscripts in that order. Number all sheets in succession except figure(s); title page is page 1. Indicate appropriate location in text of tables, figures, and other subsidiary materials by marginal notes. Underline italicized print; show boldfaced print by a wavy underline; italicize Latin words.

Organize a Regular Paper as follows: Summary, Introduction with no heading, Experimental Procedures or Materials and Methods, Results, Discussion, References. Presentation may be clearer and more effective if some sections are combined. For a Communication, include a brief summary and omit headings and subheadings.

Title Page: State form of paper (Regular Paper or Communication) and field under which it is to be reviewed. List title; make informative and short. Do not include chemical formula or arbitrary abbreviations; use chemical symbols to indicate structures of isotopically labeled compounds. Do not number papers as parts in a series; titles and subtitles may be used if necessary. List full names of all authors in by-line. Footnote changes of author address. List institution(s) in which work was carried out (with Zip Code) in from-line. Leave 4-lines between From-line and Running title for date of reception of paper. Provide a short running title (60 strokes). List name, complete mailing address, telephone and fax numbers of corresponding author. (Japanese author(s): list in Japanese name and address of proofreader.) In title page footnotes acknowledge financial aid. Other Acknowledgment(s) appear in a separate section at end of Discussion. Obtain written authorization for acknowledgments; copies of manuscript must have been sent to persons mentioned. Define nonstandard abbreviations, even if familiar. List all nonstandard abbreviations used in alphabetical order in a footnote.

Summary: 200 words (Communications 100 words). Briefly present problem, scope of work and plan of experiments, mention significant data and state major findings and conclusions. Avoid statements such as "The significance of these results is discussed." Make intelligible to nonspecialists; avoid specialized terms and abbreviations. (Japanese author(s): submit 3 copies of summary in Japanese on A4 Japanese manuscript paper headed by Author(s) and Institution(s), give title both in Japanese and English.)

Introduction: Begin text of a Regular Paper with a short introduction with no heading. State reasons for performing work. Briefly reference previous work; avoid an extensive literature review.

Methods, Results, and Discussion: Arrangement after introduction is not fixed. Sections may be separated with italicized subheading; give careful thought to division of paragraphs. Give sufficient details to enable reader to repeat work exactly. Refer to previously published procedures; cite both original description and pertinent published modifications; do not include extensive description unless substantially new. Present Results in tables or figures; write simple findings in text without tables or figures. Make Discussion concise; interpret results. Do not repeat information already given under Results. Combining Results and Discussion may give a clearer and more compact presentation.

References: Number references cited in text in parentheses with italicized Arabic numerals in order of appearance. Refer to "unpublished experiments" and "personal communications" parenthetically in text following name(s) of source of information [(Yamada, T., personal communication), (Suzuki, M. & Yoshida, M., unpublished observations), etc.]. Verify wording of personal communication with person who supplied information and get approval for use of name with quoted information. List all references in numerical order typed double-spaced on a separate sheet under heading References.

JOURNAL

1. Suzuki, K., Hibino, K., & Imahori, K. (1976) *J. Biochem.* **79**, 1287–1295

10. Thomas, D.A. & Kuramitsu, H.K. (1969) *Arch. Biochem. Biophys.* **145**, 96–108

20. Edmond, J. (1974) *J. Biol. Chem.* **249**, 72–80

CHAPTER IN A BOOK

6. Hiromi, K. (1972) in *Kosogaku no Kiso* (in Japanese) (Ogura, Y., ed.) pp. 6–72, Asakura Shoten, Tokyo

31. Umbarger, H.E. (1969) in *Current Topics in Cellular Regulation* (Horecker, B.L. & Stadtman, E.R., eds.) Vol. 1, pp. 57–69, Academic Press, New York

BOOK BY ONE OR MORE AUTHORS

9. Loewy, A.G. & Siekevitz, P. (1969) *Cell Structure and Function* pp. 83–89, Holt, Rinehart and Winston, Inc., New York

18. Kavanau, J.L. (1965) *Structure and Function of Biological Membranes* Vol. 1, pp. 145–148, Holden-Day Inc., San Francisco

27. Yamada, T. (1971) *Byotai Seikagaku* (in Japanese) pp. 203–229, Iwanami Shoten, Tokyo

Cite references written by more than two authors in text as Smith *et al*. In reference list, give names of all authors (with initials). If an article has been accepted for publication, but has not yet appeared, list reference as:

23. Yamashita, H. (1986) *J. Biochem.* 100, in press

If available by galley proofreading, include publication year and volume number of journal. Do not use "in preparation" and "submitted for publication" in reference list.

Footnotes: Put footnote(s) to title, author's name(s), and affiliation(s) on title page. Put footnotes to text on a separate sheet. Number all in succession with superscript, Arabic numerals, starting from title page footnote(s).

Identify table footnotes with superscript lowercase letters at bottom of table.

Tables: Draw on separate sheets numbered consecutively in Roman numerals. For aid in designing tables, refer to current *Journal* issues. Give each an explanatory title and sufficient experimental detail, usually in a paragraph immediately following title, to be intelligible without reference to text. Give each column an appropriate heading. If typed on several pages be sure columns will align when put together. Indicate units of measure clearly. Extend horizontal lines above and below headings the full width of column or page. Do not draw vertical lines between columns or at outer edges of table. Keep table footnotes to a minimum; indicate by superscript lowercase letters at bottom. Do not exceed width of tables.

When a table (or figure) involves very large or small numbers, and powers of ten are used, make numbers on axis (or in column) products of original figure and factor on axis label (or column head); e.g., to represent 1,200 units as "1.2," label column head "unit $\times 10^{-3}$." If possible employ units that do not require exponents in heading; for instance, appropriate choice of mM, μM, or nM may reduce number of digits or exponential terms.

Submit large complex tables as "camera-ready" copies for offset printing. Type with a fresh ribbon to be published precisely as typed.

Illustrations: Place each figure (Scheme, Diagram) on a separate sheet numbered with an Arabic numeral (Fig. 1, Fig. 2, etc.). Figures will be reduced to fit 17.5×23.5 cm.

Identify in margin or on back, with author's name and figure number and indicate TOP.

Keep number of graphs to a minimum. Do not use a figure if information can be expressed by a mathematical formula or in a few words in text. Do not draw separate figures for curves that could appear together in one chart.

Make drawings on section paper or coordinate paper printed in light blue. Use black waterproof drawing ink. Mount figures drawn on tracing paper and photographs on cardboard or a sheet of paper used for text. Do not mount on heavy cardboard. Make figures preferably all the same size as text.

Include a title and an explanatory legend (Legends to Figures) with sufficient experimental detail to make figure intelligible without reference to text. Type Legends to Figures double-spaced, in numerical order, on a separate page. Put at least two blank lines between one legend and the next.

Photographs should be glossy and as high in contrast as possible. Duplicate copies should be same quality as original. Either trim or make reducible to one column width (8.5 cm). Submit highest quality possible to permit high quality reproductions. Indicate magnification of photomicrographs in legend or include a bar indicating scale in figure (or both).

Choose scale of figures and sizes of letters, points, symbols, etc., so figures may be reduced to one column width, or one-half column width with two figures side by side. Figures will be set in single column width unless there is a compelling reason to make larger. Draw figures to be reduced to less than one column width with proportionately larger lettering.

Enclose top and right side of figure to complete rectangle indicated by abscissa and ordinate. Indicate scales used in plotting data by short index lines; repeat index lines on opposite sides, unless more than one scale is used on ordinate.

Employ standard characters, for which printer has type (O ● ◻ ◼ ∆ ▲ ×). If other symbols are necessary, indicate meaning on face of chart; otherwise define all symbols on legend, not on figure.

Label scales of abscissa and ordinate with quantity measured, units and numerical scale.

Avoid "wasted white space"; extend abscissa and ordinate only as far as content of graph demands. Do not extend explanatory lettering beyond ends of lines.

Do not include title or figure legend.

Present flow diagrams and amino acid or nucleotide sequences as direct photographic reproduction.

OTHER FORMATS

MINIPRINT SUPPLEMENT—Use as an effective way of presenting supporting data and experimental details. Types of data that are particularly suited include analyses relevant to amino acid or nucleotide sequences, details of chemical syntheses, preparation of enzymes, extended statistical or mathematical discussions, and X-ray diffraction data. Miniprint supplement is published in *Journal* immediately following parent paper, and is included in reprints. It is reproduced directly from camera-ready copy supplied by authors. Number pages M-1, M-2, . . . to differentiate from those of Main Manuscript. Put Tables and Figures separate sheets numbered Table IS, Fig. 1S, *etc.*

OTHER

Abbreviations: Use abbreviations with specific meanings for convenience for complex chemical substances, particularly in equations, tables, or figures. Avoid in titles and summaries. Use abbreviations and symbols sparingly in text. In chemical equations, which traditionally depend upon symbols, use an abbreviation or symbol for a term that appears in full in neighboring text. Trivial names do not require abbreviations.

Use abbreviated names or symbols in a column heading in a table, figure, or photograph from "accepted" list in Section VII-9 (*J Biochem* 1993; 114(1): vi). Otherwise, introduce abbreviations only when absolutely necessary, as in tables, figures, and other illustrations where space is limited. Avoid use in text by use of pronouns or paraphrasing. Define all nonstandard abbreviations in text in alphabetical order in a single footnote on title page.

For spelling of chemical names consult current *Journal* issues. For chemical terms follow usages and rules recommended by Nomenclature Committee of IUBMB (IUBMB: International Union of Biochemistry and Molecular Biology) and IUPAC-IUBMB Joint Commission on Biochemical Nomenclature (JCBN, IUPAC: International Union of Pure and Applied Chemistry). Recommendations published before 1978 are found in *Biochemical Nomenclature and Related Document~*

(1978), (The Biochemical Society, London). See *J Biochem* 1993; 114(1): iv-v for list of references on the nomenclature of amino acids, peptides and proteins, nucleotides and nucleic acids, lipids and related compounds, carbohydrates and related compounds, phosphorus-containing compounds.

Chemical and Mathematical Formula: Refer in text to simple chemical compounds by their formula when these can be printed in simple horizontal lines of type. Do not use structural formula in running text. Center chemical equations, and mathematical formula between successive lines of text. Indicate formulas(1), (2) or Eq. 1, Reaction 2, *etc.* when cited in text. Prepare large structural formula and long mathematical equations in a form suitable for direct photographic reproduction. Include as a Diagram at end of paper.

Show ionic charge as a superscript following chemical symbol, e.g., Fe^{3+}, SO_4^{2-}.

Cytochromes: Designate by small italicized letters.

Enzymes: When one or more enzymes is important in manuscript, use recommended (trivial) name or systematic name assigned by Nomenclature Committee of IUBMB and IUPAC-IUBMB Commission on Biochemical Nomenclature. See *Enzyme Nomenclature, Recommendations (1984)* (Academic Press, New York), Nomenclature of multiple forms of enzymes *Eur J Biochem* 82: 1-3, 1978, Units of enzyme activity (1978). *Eur J Biochem* 97: 319-320 [erratum: *Eur J Biochem* 104: 1, 1980; and Symbolism and terminology in enzyme kinetics (1981) *Eur J Biochem* 128: 281-291, 1982. When an enzyme name is not the subject of a paper, include source, trivial name, systematic name (or reaction it catalyzes) and EC code number. Do not abbreviate names of enzymes except when substrate has an approved abbreviation.

Isotopically Labeled Compounds: Show symbol for an isotope in square brackets directly before name. When more than one position in a substance is labeled with the same isotope and positions are not indicated, indicate number of labeled atoms as a right-hand subscript ([$^{14}C_2$] - glycolic acid). Symbol U indicates uniform, e.g., [U-^{14}C] - glucose (where ^{14}C is uniformly distributed among all six positions). Isotopic prefix precedes that part of name to which it refers. Do not contract terms such as ^{131}I-labeled albumin to [^{131}I]albumin. When isotopes of more than one element are introduced, arrange symbols in alphabetical order. Use symbols ^{2}H and ^{3}H or D and T for deuterium and tritium respectively.

For simple molecules, indicate labeling by writing chemical formula with prefix superscripts attached to correct atomic symbols in formula: e.g., $^{14}CO_2$, $H_2{}^{18}O$, ^2H_2O. Do not use square brackets or when isotopic symbol is attached to a word that is not a specific chemical name, abbreviation or symbol: e.g., ^{131}I-labeled, ^{14}C-sugar, ^{14}C-steroids, $^{32}PO_4{}^{3-}$, but [^{32}P] phosphate.

Nomenclature, Animals, Plants, and Microorganisms: Give scientific names as Latin binomials in full in title and summary and on first mention in text. Subsequently use generic name contracted to first letter. State strain and source of laboratory animals.

Nucleotide Sequence: DNA Data Bank of Japan (DDBJ) data submission form will accompany notification of acceptance of manuscript. Deposit nucleotide sequence data in one of the data banks, DDBJ, GenBank™, or EMBL. Submission to one is sufficient because data are exchanged between them. If data are already deposited, indicate accession number in title page footnote.

Page Charges: ¥600 and ¥6000 per printed page of a Regular Paper and a Communication, respectively. If there is no source of grant or other support apply, at submission, for a grand-in-aid to the Editor-in-Chief.

Spectrophotometric Data: State Beer's law as $A = -\log T = \varepsilon l c$ where A is absorbance; T, transmittance (= I/I_o); ε, molar absorption coefficient; c, concentration of absorbing substances in moles per liter; and l, length of optical path in centimeters. Under these conditions ε has dimensions liter·mol^{-1}·cm^{-1} or more briefly M^{-1}·cm^{-1} (not cm^2·mol^{-1}). Do not use "O.D." and "E."

Units: For Abbreviations of Units of Measurement and Physical and Chemical Quantities that may be used without definitions see: *J Biochem* 1993; 114: v-vi.

SOURCE

July 1993, 114(1): i-vi.

Journal of Bioenergetics and Biomembranes
Plenum Publishing Corp.

Dr. Peter L. Pedersen
Department of Biological Chemistry
The Johns Hopkins University School of Medicine
725 North Wolfe Street
Baltimore, Maryland 21205

AIM AND SCOPE

Journal of Bioenergetics and Biomembranes is an international journal devoted to the publication of original research that contributes to fundamental knowledge in the areas of bioenergetics, membranes, and transport. The subspecialities represented include membrane transport, electron transport, ATP synthesis by oxidative or photophosphorylation, muscle contraction, and biomembranes.

MANUSCRIPT SUBMISSION

Submit original and two copies (including copies of all illustrations and tables). Submission is a representation that manuscript has not been published previously and is not currently under consideration for publication elsewhere. A statement transferring copyright from authors (or their employers, if they hold copyright) to Plenum Publishing Corporation is required before manuscript can be accepted for publication. Editor will supply forms for transfer.

MANUSCRIPT FORMAT

Type double-spaced on one side of 8.5 × 11 in. white paper using generous margins on all sides.

Title Page: Include title of article, author's name (no degrees), author's affiliation, and suggested running head. Affiliation is department, institution (usually university or

company), city, and state (or nation); type as a footnote to author's name. Suggested running head (80 characters including spaces); comprise article title or an abbreviated version thereof. Include complete mailing address and telephone number of corresponding author.

Abstract: 150 words.

Key Words: 10 key words (descriptive words or phrases used as indexing terms) directly below abstract.

Illustrations: Number illustrations (photographs, drawings, diagrams, and charts) in one consecutive series of Arabic numerals. Type captions on a separate sheet. Photographs should be large, glossy prints, with high contrast. Prepare drawings with India ink. Submit either original drawings or good-quality photographic prints. Identify figures on back with author's name and number of illustration. Color illustrations can be reproduced by special arrangement; total cost, which is significant, is borne author.

Tables: Number consecutively with Roman numerals; refer to by number in text. Type on separate sheets. Center title above table; type explanatory footnotes (indicated by superscript lowercase letters) below table.

References: List alphabetically at end of paper; refer to in text by name and year in parentheses. Where three or more authors, give only first author's name in text, followed by *et al.* References include (in order): last names and initials of *all* authors, year published, name of publication, volume number, and preferably inclusive pages.

JOURNAL ARTICLE

Inesi, G., Sumbilla, C., and Kirtley, M. E. (1990). *Physiol. Rev.* **70**, 749–760.

BOOK

DeFelice, L. J. (1981). *Introduction to Membrane Noise,* Plenum Press, New York.

CONTRIBUTION TO A BOOK

Bisson, R. (1990). In *Bioelectrochemistry III: Charge Separation Across Biomembranes* (Milazzo, G., and Blank, M., eds.), Plenum Press, New York, pp. 125–175.

Footnotes: Avoid use. When absolutely necessary, number consecutively using Arabic numerals and type on a separate sheet placed directly after References. Use appropriate superscript numeral for citation in text.

SOURCE

February 1994, 26(1): inside back cover.

Journal of Biological Chemistry
American Society for Biochemistry and Molecular Biology, Inc.

Editor, *The Journal of Biological Chemistry*
9650 Rockville Pike
Bethesda, MD 20814

AIM AND SCOPE

The *Journal of Biological Chemistry* publishes papers on a broad range of topics of general interest to biochemists. Manuscripts may come from any country but must be in

English. No subject is too specialized to be considered for publication, but because of the need to limit the number of pages published annually, it is often necessary to arbitrarily decline a manuscript that is scientifically sound, but is judged to be of limited interest to readers. The *Journal* also does not publish papers that are merely confirmatory or that report on well known processes in tissues or organisms not previously studied. Manuscripts that describe primarily a new method or the preparation of a reagent (such as monoclonal antibody) are usually not accepted.

MANUSCRIPT SUBMISSION

Submit manuscripts of full papers and Communications in triplicate. Include a letter indicating address, telephone and fax numbers of submitting author.

All submitted manuscripts should contain original work that has not been published previously, and is not under consideration for publication elsewhere. Oral or poster presentations are not considered previous publications. Studies reported in publications of symposia or in short communications will not be published as a full paper or Communication unless justified by additional unpublished information.

Include copies of manuscripts by authors on related subjects that are in press or currently under editorial review. Failure to provide these papers may delay review.

Authors must transfer copyright to the American Society for Biochemistry and Molecular Biology, Inc. All authors must sign or signing author must hold permission from any coauthors. Copyright for papers reporting research by U. S. Government employees as part of their official duties will only be transferred to the extent permitted by law.

MANUSCRIPT FORMAT

Type manuscript double- or triple-spaced throughout including references, tables, footnotes, and figure legends, except single-space "camera ready tables." Use 10 point font size. Submit three copies of manuscript, with parts arranged in order: title, author(s), and complete name(s) of institution(s); running title; summary; results; discussion; references; footnotes; figure legends (more than one legend can be on a page); tables; and figures. Number all pages; title page is page 1. Start a separate sheet for each section. Indicate by marginal notes suggested locations in text for tables and figures.

On first page give name of corresponding author with telephone and fax numbers. If institutional address is inadequate for postal delivery, include required address in a footnote. Also include in cover letter.

Organize manuscript as: Summary, Introduction (no heading), Experimental Procedures or Materials and Methods, Results, Discussion, and References. Results and Discussion may be combined in short papers and Communications.

Keep manuscript below six printed pages (24 double-spaced typed pages including figures and tables).

Title: Make as informative as possible, and as short as consistent with clarity. Do not exceed two printed lines. Do not include arbitrary abbreviations, formulas, or chemical symbols.

Running Title: A short title (60 characters and spaces) for a running head.

Names of Authors: Spell out first or second given name of each author

Summary: 200 words succinctly and clearly stating significant findings described. Do not use abbreviations or specialized terms. Make understandable in itself.

Introduction: Two typed pages giving purpose of studies reported and their relationship to earlier work. It should not be a literature review.

Experimental Procedures: Be as brief as possible but sufficiently descriptive to permit a qualified worker to repeat experiments. Refer to previously published procedures with any published modifications. Give new modifications in full only if they are substantial and necessary for repetition of work.

Results: Use figures or tables to present results. Describe simple results in text without figures or tables. Avoid extensive discussion of results in Results section. Explain why subsequent experiments were performed.

Discussion: Interpret results in less than four typed pages; do not summarize Results.

References and Footnotes: Cite references by number rather than name. Number list of references consecutively in order given in manuscript.

Hudson. T. H., and Grillo, F. G. (1991) *J. Biol. Chem.* **266**,18586–18592

Sambrook, J., Fritsch, E. F., and Maniatis, T. (1989) *Molecular Cloning: A Laboratory Manual*, 2nd Ed., Cold Spring Harbor Laboratory, Cold Spring Harbor, NY

Abbreviate journal names according to *Chemical Abstracts* or *Biological Abstracts*. Include first and last page numbers. Verify accuracy of references.

List papers accepted for publication in references, with authors' names, estimated year of publication, and abbreviated journal name, followed by "in press." Footnote citations to "manuscript in preparation," "unpublished observations," "personal communication," or similar information; do not include in references. Do not cite abstracts of papers presented at meetings in references unless they appear in *Biological Abstracts List of Serials*. Footnote abstracts not citable in references.

Send copies of manuscript to persons cited as "personal communications" and acknowledgments. Obtain their written approval. Proof of approval may be requested.

Use footnotes to give other information not contained in references, including a list of abbreviations and name of reprint request author.

Tables and Figures: Use to present essential data needed to illustrate or prove a point. Keep number to a minimum. Do not use if information can be expressed in a few words in text or in a mathematical formula or chemical equation. Do not put the same data in a table and a figure; justify need for double presentation in cover letter. Submit complex or large tables and figures in "camera ready" format typed or drawn in single space.

Give tables explanatory titles. Put sufficient experimental detail, in a paragraph following title, to make intelligible without reference to text (unless procedure is given in Experimental Procedures or under another table or figure).

Give each column an appropriate heading. Do not use abbreviations unless absolutely necessary. Indicate units of measure. If an experimental condition is the same for all tabulated experiments, give information in a statement accompanying table, not in a column of identical figures in table. Avoid presenting large masses of essentially similar data. To save space, replace extended tabulations by reporting mean values with an accepted measure of dispersion (standard deviation, range); indicate numbers of individual observations. Accompany statements of significance by probability values derived from appropriate statistical tests. Define all statistical measures clearly and unambiguously.

Do not include more significant figures in data than are justified by accuracy of determinations.

Submit a complete set of figures as photographic prints or line drawings on paper the same size as the manuscript with each manuscript copy. Identify figures with authors' names, figure number, and indicate TOP. One set of top quality is needed for printer; others may be prints or photocopy. Electron micrographs or halftone figures must be good quality prints. Do not mount on heavy cardboard. Do not submit fragile or oversized original drawings. Send upon acceptance, if necessary.

Give each figure a title and explanatory legend, do not letter on figure. Type legends together. Put sufficient experimental detail in legend to make figure intelligible (unless detail has appeared in a previous figure legend or "Experimental Procedures"). Combine related curves in a single figure. A composite of related figures often saves space and conveys more information. With electron or light micrographs, isolate essential part of figure and trim it; a second small figure at higher magnification may be more informative than a single large figure of either magnification.

Figures will be reproduced in single column width (8.4 cm), unless there is reason to make larger. Draw letters, numbers, and symbols to be 1.5 –3 mm high after reduction. Simple figures can be reduced to a smaller size than more complex figures or those conveying numerical information. Make lettering on simple figures proportionately larger. Do not label every index line; space numbers to avoid crowding. Draw lines and lettering evenly and heavily in black ink.

Use standard symbols, if possible, in order of preference: ● ○ × ■ ▲ □ △. Identify symbols and curves in legend or in figure itself, whichever is clearer.

Avoid "wasted white space." Extend abscissa and ordinate only as far as content of graph demands. Do not extend explanatory lettering beyond ends of lines.

Complete rectangle indicated by abscissa and ordinate; enclose top and right sides of figure, but do not draw a box around entire figure nor around a halftone figure. Indicate scales used in plotting data by short index lines; repeat index lines on opposite sides unless more than one scale is used. Choose intervals so that interpolation will permit reasonably accurate evaluation of experimental points.

OTHER FORMATS

COMMUNICATIONS—Four printed pages (1000 words per page), including figures, tables, and references. Provide rapid communication of novel ideas and results in new and developing areas of biochemistry that are of special interest or wide applicability; may be a preliminary report or a full account of research. Criteria for acceptance are more stringent than for full papers.

If figures or tables are included, reduce length of text correspondingly. One printed page equals four double-spaced typed pages or three figures or tables.

Submit three copies. Give title of manuscript, running title, and names and addresses of authors at beginning. Precede article with a brief Summary. Headings and subheadings used in full papers can be omitted.

OTHER

Abbreviations, Symbols, and Nomenclature: See Tables I-VI (*J. Biol. Chem.* 1994; 269(1): 782-784) for abbreviations of units of measurement and physical and chemical quantities, semisystematic or trivial names, amino acids, carbohydrates, nucleosides and nucleotides, and nucleic acids that may be used without definition. Define all other abbreviations used in text in a single footnote, inserted immediately after first abbreviation. Keep use to a minimum. If symbols or abbreviations are used in Summary, define in Summary as well as text. Well known abbreviations such as ATP, DNA, and RNA can be used in Title, but other should not be used.

Use short abbreviations for some important biochemical compounds, e.g., ATP, NAD$^+$, and RNA, do not create new abbreviations of this kind. Trivial names are sufficiently short. Do not use abbreviations for phrases such as "central nervous system," "red blood cells," or "extracellular fluid."

Do not abbreviate names of enzymes, except in terms of substrates for which accepted abbreviations exist e.g., DNase, and RNase, and ATPase. Use trivial names of enzymes recommended by the Nomenclature Committee of the IUB in *Enzyme Nomenclature, Recommendations, 1984* (1984, Academic Press). Give reaction catalyzed for an enzyme except for very well known enzymes.

Designate linear sequences (primary structures) of proteins and nucleic acids with one- or three-letter abbreviations for constituent amino acids or nucleotides given in Tables III and V (*J. Biol. Chem.* 1994; 269(1): 783, 784). Designate sequences of polysaccharides by three-letter abbreviations of monosaccharide (Table IV: *J. Biol. Chem.* 1994; 269(1): 784).

For protein sequences, write amino acid residues from the NH$_2$ terminus on the left to the COOH terminus on the right; separate each amino acid in three-letter code separated by a dash or hyphen. Put residues in unknown sequences in parentheses separated by a comma. Use same convention for one-letter abbreviations of amino acids except that no symbol separates each residue.

For nucleic acid sequences, write nucleosides from the 5´-end on the left to the 3´-end on the right with no symbol separating each nucleoside.

For oligosaccharide sequences, write three-letter symbols for constituent monosaccharides from the nonreducing end on the left to the reducing end on the right; indicate groups forming the glycosidic linkage by numerals separated by short dashes or an arrow and anomeric linkage by α and α.

Point arrow away from hemiacetal group. Designate furanose and pyranose rings by letters *f* and *p*, respectively, if necessary.

For bacterial genetics see Demerec, M., Adelberg, E. A., Clark, A. J., and Hartman, P. E. (1966) Bacterial genetics. In: *Genetic Maps* (Stephen J. O'Brien, Ed.), Cold Spring Harbor Laboratory, Cold Spring Harbor, NY. For bacterial nomenclature see *Bergey's Manual of Determinative Bacteriology* (8th Ed, Waverly Press, Baltimore, MD). For nomenclature for transposable elements in prokaryotes see Campbell *et al.* (1979) *Gene* **5**, 197–206 or Szybalski and Szybalski (1979) *Gene*, **7**, 217–270.

Consult references in Table VII (*J. Biol. Chem.* 1994; 269(1): 785) for further details about nomenclature of chemical compounds recommended by International Scientific Unions.

Amino Acid Sequences: Manuscripts dealing with amino acid sequences of proteins whose sequences are known from other tissues or species will be considered only if the new sequence contributes significant new biochemical insights. Partial sequence data will not be considered unless they contribute important information about the protein. This applies equally to amino acid sequences derived from nucleotide sequences.

Chemical and Mathematical Usage: In text refer to simple chemical compounds by formulas when these can be printed in single horizontal lines of type. Do not use two-dimensional formulas in running text. Center chemical equations, structural formulas, and mathematical formulas between successive lines of text. Prepare such structural formulas and mathematical equations in a form suitable for direct photographic reproduction and include on a duplicate sheet at end of paper.

Indicate ionic charge as a superscript following chemical symbol, e.g., Mg^{2+}, S^{2-}. Notation Mg(II) is also acceptable.

Crystallographic Studies: Submit manuscripts reporting new macromolecular structures by x-ray crystallographic methods. Manuscripts that only describe conditions for crystallization of a macromolecule or the diffraction pattern and space group of the crystals do not contain sufficient information to warrant publication. New structure determinations must be submitted to the Protein Data Bank at Brookhaven National Laboratory with all structural data required to validate discussion, including both x-ray amplitudes and phases and derived atomic coordinates. A footnote will state that necessary data have been deposited. If paper discusses a protein structure only at level of their main chain α-carbon atoms, then deposit only α-carbon coordinates. If discussion involves higher resolution data, then deposit the full set of x-ray data and coordinate list. Manuscript will not be sent to printer until deposition in the Protein Data Bank is confirmed.

Data in the Data Bank is usually made available upon publication of paper. Authors may ask the Protein Data Bank not to release data until one year after publication; provide an explanation to the Editor in writing.

Equilibrium and Velocity Constants: Write dissociation constants, association constants, and Michaelis constants in terms of concentration; indicate units when equilibrium constant is defined, and value given.

Specify values of rate constants, give first order velocity constants as s^{-1} (other units of time may be used, but, specify time unit). Give second order rate constants in $M^{-1}\ s^{-1}$.

Experimentation, Animal and Human: Research must have been carried out in accordance with the recommendations from the Declaration of Helsinki and the appropriate NIH guidelines. Research protocols must be approved where necessary by appropriate institutional committees.

Isomers: Differentiate names of chiral compounds whose absolute configuration is known by prefixes *R*- and *S*- (see IUPAC (1970) *J. Org. Chem.* 35, 2849–2867). When compounds can be correlated sterically with glyceraldehyde, serine, or another standard accepted for a specialized class of compound, use small capital letters D-, L-, and DL for chiral compounds and their racemates. Where direction of optical rotation is all that can be specified, use (+)-, (−)-, and (±)- or *dextro, laevo,* and "optically inactive;" specify conditions of measurement.

Isotopes: Follow guidelines adopted by the IUB Committee of Editors of Biochemical Journals (CEBJ). See IUPAC-CNOC Recommendations on Isotopically Modified Compounds (1978) *Eur. J. Biochem.* **86**, 9–25.

For most biochemical usage, indicate an isotopically labeled compound by placing symbol for isotope introduced in square brackets directly attached to front of name (word) or formula as in [^{14}C]urea, [α-^{14}C]leucine, L-[methyl-^{14}C]methionine, [^{2}H]CH$_4$. If specific position of labeling is known, indicate at first mention or in Materials and Methods section; thereafter, use less specific notation.

Isotopic prefix precedes that part of the name to which it refers. Do not contract terms such as ^{131}I-labeled albumin to [^{131}I]albumin since native albumin does not contain iodine; ^{131}I-albumin and [^{131}I]iodoalbumin are both acceptable.

Symbol indicating configuration should precede symbol for isotope.

The same rules apply when the labeled compound is designated by a standard abbreviation or symbol, other than atomic symbol.

When isotopes of more than one element are introduced, arrange symbols in alphabetical order. Use symbols ^{2}H and ^{3}H for deuterium and tritium, respectively. When more than one position in a substance is labeled by means of the same isotope and positions are not indicated, add number of labeled atoms is added as a right subscript. Symbol U indicates uniform and G general labeling.

When known, indicate positions of isotopic labeling by Arabic numerals, Greek letters, or prefixes (as appropriate) within the square brackets and before the element symbol, attached by a hyphen.

Forms $^{14}CO_2$, $^{32}PO_4$, $^{32}P_i$, rather than the more formally correct $[^{14}C]O_2$ or $[^{14}C]CO_2$, $[^{32}P]P_i$, etc. are acceptable. Do not use square brackets when isotopic symbol is attached to a word that is not a specific chemical name, abbreviation, or symbol. Abbreviation for Curie is Ci. The new SI unit is bequerel (Bq). 1 Bq = 1 disintegrations per sec. or 60 disintegrations per min. 1 Ci = 37×10^9 disintegrations per sec. = 37 GBq. 1000 dpm = 0.45 nCi = 16.7 Bq. Use either unit.

Materials Sharing: Make unique propagative materials, including, cell lines, hybridomas, and DNA clones available to any qualified investigator. Submission of a paper constitutes an agreement to abide by this principle.

Molecular Weight and Mass: "Molecular weight" (M_r, relative molecular mass) is the ratio of the mass of a molecule to 1/12 of the mass of carbon 12. It is dimensionless. Molecular mass (m) is not a ratio, and can be expressed in daltons (Da) or in atomic mass units (u). Molecular mass is the mass of one molecule of a substance, and thus is M (molar mass) divided by Avogadro's number. Dalton is 1/12 of the mass of carbon 12.

Both "the molecular mass of X is 10,000 daltons" (or "10 kDa") or "the relative molecular mass (molecular weight) (M_r = 10,000)" can be used; do not express M_r in daltons. Use expressions such as "the 10-kDa peptide" and "the mass of a ribosome is 2.6×10^7 daltons" (or "26 MDa"), even for an entity that is not a definable molecule. Avoid use of K as a shorthand for 1000 or for kDa (kilodalton).

When presenting estimates of relative molecular mass from gel electrophoresis data, include scale used to estimate molecular mass as one of the ordinates on the figure, not just location of various standards used.

Nucleotide Sequences: Deposit specific cloning and sequencing data in a nucleotide sequence data bank, such as GenBank or EMBL. Use lowercase letters for nucleotides in long sequences to improve legibility.

Data obtained by well established techniques that support assignment of sequences will not be published; do not submit with manuscript. Data may be requested by reviewers. For a manuscript describing a DNA sequence, state that final sequence was determined from both strands. Give other relevant information concerning quality of reliability of sequence data. For a new RNA sequence, submit data on which it is based with the paper in sufficient clarity for reviewers to evaluate the correctness of assignments.

Submit nucleic sequence data, preferably in computer-readable form or by electronic mail, and a copy of the paper to either GenBank™, GenBank Submissions, National Center for Biotechnology Information 8600 Rockville Pike, Bldg. 38A, Bethesda, MD 2894; telephone: 301-496-2475; e-mail (submissions): gb-sub@ncbi.nlm.nih.gov; e-mail (information): info@ncbi.nlm.nih.gov; or EMBL Nucleotide Sequence Library, Postfach 10.2209, Meyerhof-strasse 1, 6900 Heidelberg, Germany, 6221-387-258; telephone: +49(6221) 387258; e-mail: databasesub@embl-heidelberg.de. Obtain accession number before manuscript is sent to printer. A footnote will be included indicating deposit. Submission to either data bank ensures entry into both. When nucleotide probes are used, identify ends of probes by reference to published nucleotide number or restriction maps, or, if unpublished, include information in Experimental Procedures section.

Oligosaccharide Structures: *Journal* usually does not publish reports on oligosaccharide structure in a glycoprotein or glycolipid that does not differ substantially from one described in another glycoconjugate. Include additional studies which give new and novel information about the biochemistry of the oligosaccharides or the glycoconjugate. Manuscripts concerning the function of a carbohydrate in molecular and cellular processes are of interest. Purely descriptive articles that report only how removal of carbohydrate affects a glycoprotein are not. Manuscripts must give new insights into biochemical mechanisms involved in a particular carbohydrate function, or reveal some biological significance of the carbohydrate.

Solutions and Buffers: Specify composition of all solutions and buffers in sufficient detail to define concentration of each species. For ordinary buffers, it is assumed that molarity refers to total concentration of species that buffer at indicated pH and the concentration of counterion is sufficient to neutralize charge of ionized buffer species. Indicate composition of mixtures by a diagonal (/) or a colon; do not use hyphens or dashes. Give complete unabbreviated name and source (or a reference that gives complete composition) for all culture media.

Spectrophotometric Data: Indicate relation between symbols used. Adhere to symbols and terminology adopted by IUPAC ((1970) *Pure Appl. Chem.* **21**, 1). State Beer's law as: $A = -\log^{10} T = \varepsilon l c$, where A is absorbance; T, transmittance ($= I / I_o$), ε, molar absorption coefficient; c, concentration of absorbing substances in moles per liter; and l, length of optical path in centimeters. Under these conditions ε has dimensions liter mol^{-1} cm^{-1}, or M^{-1} cm^{-1} (not cm^2 mol^{-1}). Use "absorbance" not "optical density."

If Beer's law is not followed by a particular substance in solution, explicitly state; substance may be characterized by reporting absorbance at a specified concentration. When spectrophotometric measurements are made with a radiant energy source that is not confined strictly (as in a line spectrum) to wavelength or frequency specified, exact value of ε is ambiguous; report spectral characteristics of source.

Page Charges: There is a charge per page. If funds are not available, apply upon submission for a grant-in-aid for cost. Applications must be endorsed by a senior institutional official. Charges are made for publication of halftones, electron micrographs, color plates, and other special illustrations, currently $20 per plate. A special estimate will be made for color reproduction. Micrographs and other illustrations that require high quality reproduction are $100 per paper.

Reviewers: Authors may request that a specific individual with a possible conflict of interest not be involved in reviewing the manuscript. Provide names and addresses of potential reviewers.

Units and Numerical Values: See Table I (*J. Biol. Chem.* 1994; 269(1): 782) for abbreviations for units of measurement and physical and chemical quantities used by the *Journal* without definition. Prefixes that can be added to names of units and multipliers indicated by each prefix are included. Use of prefixes avoids numbers with excessively small or large numbers of digits. Also use multipliers of the power of 10, make clear what number is to be multiplied. The same applies to figure ordinates and abscissas.

SOURCE

January 7, 1994, 269(1):. 777–785.

Journal of Bone and Joint Surgery (American Edition)

American Orthopaedic Association
British Orthopaedic Association
American Academy of Orthopaedic Surgeons
Australian Orthopaedic Association
Canadian Orthopaedic Association
New Zealand Orthopaedic Association
South African Orthopaedic Association
Western Orthopaedic Association
Mid-America Orthopaedic Association
Eastern Orthopaedic Association
British Orthopaedic Research Society
Canadian Orthopaedic Research Society

Journal of Bone and Joint Surgery
10 Shattuck St.
Boston MA 02115-6095

AIM AND SCOPE

The *Journal of Bone and Joint Surgery* welcomes articles that contribute to orthopedic science from all sources in all countries.

MANUSCRIPT SUBMISSION

Articles are accepted only for exclusive publication in *The Journal of Bone and Joint Surgery*. Articles and their illustrations become property of *The Journal*. Submit a letter of transmittal containing the following, signed by all authors: "In consideration of *The Journal of Bone and Joint Surgery*, Inc. reviewing and editing my (our) submission, the author(s) undersigned hereby transfer(s), assign(s) or otherwise convey(s) all copyright ownership to *The Journal of Bone and Joint Surgery*, Inc., and represent(s) that he (they) own(s) all rights in the material submitted. The author(s) further represent(s) that the article is original, that it is not under consideration by another journal, and that this material has not been previously published. This assignment is to take effect only in the event that such work is published in *The Journal*."

If more than one author, include: "Each of the authors represents that he, or she, has read and approved the final manuscript."

Each author must have participated in the design of study, contributed to collection of data, participated in writing of manuscript, and assume full responsibility for content of manuscript. List no more than six authors, credit individuals who have contributed to only one segment of manuscript or who have contributed only cases in a footnote. Under extenuating circumstances, explain in cover letter why there are additional authors and state how each author contributed to manuscript.

A conflict-of-interest statement is required for accepted manuscripts. Statement has no bearing on editorial decision to publish a manuscript. Conflict-of-interest statements will be sent to corresponding author with edited manuscript when a decision has been made to publish. The statement selected will be indicated on first page of published article. Each author must sign. No article will be published until completed form is in *The Journal* office.

Submit original and three duplicate manuscripts complete with illustrations (four complete sets). Identify corresponding author with address and telephone number. Include two cover sheets. On first sheet list title, authors' names, and name and address of corresponding author; on second sheet list title only. Page headers include title but not authors' names. Do not mention institution where study was done in text. Insert after manuscript is accepted.

Include a bibliography, alphabetical and double-spaced, of references made in text. Follow style of bibliographies in *The Journal* exactly.

Include legends for all illustrations submitted, listed in order and typed double-spaced.

MANUSCRIPT FORMAT

Type manuscripts double-spaced with wide margins. Write out figures under 100 except percentages, degrees, or figures expressed in decimals. With direct quotations include exact page number on which it appeared in book or article. Give all measurements in SI metric units. In reporting results of surgery, only in rare instances can cases with less than two years' follow-up be accepted.

Abstract: 200 words, giving factual essence of article. No abstract for case reports. In basic science articles, in final paragraph, briefly and clearly describe clinical relevance of information in article.

Illustrations: Submit either black-and-white glossy prints of photographs (give magnification of photomicrographs) or original drawings or charts. Color illustrations cannot be used unless they convey information not available in a black-and-white print. Send both color and black-and-white prints.

Number all illustrations and indicate TOP. Write author's name on back of each. Send prints unmounted or mounted only with rubber cement. Do drawings, charts, and lettering on prints usually in black; use white on black backgrounds. Put dates or initials in legends, not on prints. Make lettering large enough to be read when drawings are reduced. When submitting a previously published illustration, give full information about original publication and credit to be given; state whether permission to reproduce has been obtained.

SOURCE

January 1994, 75-A(1): 2.

Journal of Bone and Joint Surgery (British Edition)

John Goodfellow, British Editor
The Journal of Bone and Joint Surgery
35-43 Lincoln's Inn Fields
London WC2A 3PN England

MANUSCRIPT SUBMISSION

Original articles are welcome from any part of the world and should be sent to the Editor. Submit three copies of article with Letter of Transmittal signed by all authors. Include: "In consideration of the Journal of Bone and Joint Surgery reviewing and editing my (our) submission, the author(s) undersigned hereby transfer(s), assign(s) or otherwise convey(s) all copyright ownership to the British Editorial Society of Bone and Joint Surgery, and represent(s) that he (they) own(s) all rights in the material submitted. The author(s) further represent(s) that the article is original, that it is not under consideration by another journal, and that it has not been previously published. This assignment is to take effect only if work is published in Journal."

Each copy should have a complete set of illustrations, captions, tables, references and an abstract. Type manuscripts or print with generous margins. Double-space throughout text, references and captions. Number pages consecutively. Separate tables from text.

A conflict-of-interest statement is required for accepted manuscripts. Statement has no bearing on editorial decision to publish a manuscript. Conflict-of-interest statements will be sent to corresponding author with edited manuscript when a decision has been made to publish. The statement selected will be indicated on first page of published article. Each author must sign. No article will be published until completed form is in *The Journal* office.

MANUSCRIPT FORMAT

Choose a title with care: a short one has more impact and may be supplemented by a subtitle. Use words that facilitate finding subject in an index. Put authors' names (normally no more than six) and initials in the same form as in their other publications. State name and address of hospital or institution where work was done. In a footnote include, for each author, major degrees, current appointment and full postal address.

Abstract: 150 words, summarize and highlight the most important points. Statistical minutiae are not appropriate. Divide main text by headings, in an order that best suits the article. For many papers use a brief Introduction, Material (or patients), Method, Results and then Discussion. Avoid repetition of material in introduction and discussion, or in results and discussion. Text should comment on, but not repeat, details in tables or captions.

Tables: Use short descriptive headings. Avoid small tables by writing material into text. Avoid very large ones or split into manageable size. Make numbers and descriptive details consistent with those elsewhere in article.

References: Include only those that are important and which have been read. Give name and date, in date order, not superscript numbers. List references at end of article in alphabetical order.

JOURNAL

Allen MJ, Stirling AJ, Crawshaw CV, Barnes MR. Pressure monitoring of injuries: an aid to management. *J Bone Joint Surg [Br]* 1985; 67–B: 53–7.

BOOK

Watson-Jones R. *Fractures and joint injuries*. Vol. 2, 4th ed. Edinburgh: E & S Livingstone Ltd, 1955: 744–5.

CONTRIBUTION TO EDITED WORK

Winquist RA, Frankel VH. Complications of implant use. In: Epps CH Jr. ed. *Complications in orthopaedic surgery*. Vol. 1. Philadelphia, etc: JB Lippincott Company, 1978: 99–129.

Check references at original sources for accuracy.

Illustrations: Do not mount. Mark each on back with soft pencil, with first author's name, figure number and word TOP. Do not use ink and paper clips. Submit either halftone photographs or line drawings, charts or graphs. Submit halftones as glossy prints (18cm by 12 cm). Similar illustrations should have similar magnification and show similar areas of interest. For histological slides inset a scale or state magnification of print; report stain. Make lettering on diagrams and graphs simple and accurate, and an appropriate size after reduction for printing. Make consistent in size and style throughout a series of pictures. Color will be accepted and used only when essential. Obtain permission to reproduce borrowed illustration from author and publisher; report permission in submitted article.

OTHER FORMATS

BRIEF REPORTS—600 words, no abstract, two or three figures and four or five references. A more detailed Guide is available from Editorial Office.

OTHER

Style: Spelling and hyphenation should follow *Butterworths Medical Dictionary* and *Chambers Twentieth Century Dictionary*. Methods and results are usually best expressed in past tense, technique in present tense. Discussion may need both, but not in the same sentence.

SOURCE

January 1994, 76-B(1): iii-iv.

Journal of Bone and Mineral Research
American Society for Bone and Mineral Research
Mary Ann Liebert, Inc.

Lawrence G. Raisz, M.D.
Head, Division of Endocrinology and Metabolism
University of Connecticut Health Center
Farmington, CT 06030

AIM AND SCOPE

Journal of Bone and Mineral Research proves a forum for papers dealing with all areas of metabolic bone diseases and reports on the increasingly large body of research in this area.

MANUSCRIPT SUBMISSION

Submit manuscripts in triplicate (original plus two copies, including illustrations and tables). The *Journal of Bone and Mineral Research* will accept original manuscripts which contain material that has not been reported elsewhere, except in abstract form (400 words). Describe prior abstract presentations in a title footnote. Include a letter, signed by all authors, requesting evaluation for publication in *JBMR* and indicating that work has not been published or submitted elsewhere.

MANUSCRIPT FORMAT

Type manuscripts double-spaced on regular 8.5 × 11 in. typewriter paper (not erasable bond). Leave ample margins on sides, top, and bottom.

Title Page: Include authors' name and affiliations, source of a work or study (if any), and a running title (45 characters). Indicate name and address of corresponding author. Second page: Abstract (250 words), self-explanatory without reference to text. Follow format: abstract, introduction, materials and methods, results, discussion, acknowledgments, references. Number pages consecutively. At end of article, give name and address of reprint request author.

Tables: Type each with its title on a separate sheet. Use Arabic numerals to number. Each must stand alone; contain all necessary information in caption, and be understood independently of text. Include details of experimental conditions in table footnotes. Do not repeat information in text. Tables should not contain data that can be given in text in one or two sentences.

Illustrations: Supply original pen drawings and charts (drawn and lettered in India ink), glossy photostats of drawings or graphs, and original glossy photographs for all halftone illustrations. Identify illustrations on back, indicate author's name and TOP. Supply a list of figure legends at end of manuscript, double-spaced. Include magnifications when appropriate.

Color illustrations can be printed by special arrangement with Publisher. Contact Editor or Publisher for details.

References:

JOURNAL ARTICLES

Horton MA, Rimmer EF, Chambers TJ 1986 Giant cell formation in rabbit long-term bone marrow cultures: Immunological and functional studies. J Bone Min Res **1**:5-14.

BOOKS

Boyde A 1972 Scanning electron microscope studies of bone. In: Bourne GH (ed) The Biochemistry and Physiology of Bone, 2nd ed., vol 1. Academic Press, New York, pp 259-310.

If necessary to cite an abstract, so designate. Verify accuracy of references.

OTHER

Abbreviations and Nomenclature: Follow recommendations of the International Union of Pure and Applied Chemistry and the International Union of Biochemistry [see *J. Biol. Chem.* **252**:10 (1977)]. International System of Units (SI units) is recommended, conventional units may also be used. Include appropriate conversion factors. List of abbreviations in *Endocrinology* **116**:12A (January 1985) is acceptable. Drug names should always be generic. For bone histomorphometry use nomenclature, symbols, and units described in the Report of the American Society of Bone and Mineral Research Committee in *JBMR*, 12: 595-610, 1987.

Experimentation, Animal: Ascertain that experimental procedures are in compliance with principles in the "Care and Use of Animals" published each month in Information for Authors of *American Journal of Physiology*.

Experimentation, Human: Describe experimental subjects in detail and document informed consent.

Page Charges: $50 nonrefundable manuscript submission charge. Page charges, $55 per printed page ($75 per page for rapid publications) from authors who have funds from research grants or from their institutions. Payments are not a prerequisite for publication.

Permissions: Obtain written permission to reproduce figures, tables, and text from previously published material from original copyright holder (generally publisher, not author or editor) of journal or book. Include an appropriate credit line in figure legend or table footnote, and full publication information in reference list. Obtain and submit written permission from author of any unpublished material cited from other laboratories.

Reviewers: Suggest names of appropriate reviewers and request that particular individuals not serve.

SOURCE

January 1994, 9(1): inside back cover.

Journal of Cardiovascular Pharmacology

Raven Press, Ltd.

Paul M. Vanhoutte, M.D., Ph.D.
Center for Experimental Therapeutics
One Baylor Plaza, Rm 826E
Houston, Texas 77030
Telephone: (713) 798-6452
    From North and South America, Australia, and New Zealand

Paul M. Vanhoutte, M.D., Ph.D.
I.R.I.S., 6 place de Pleiades
92415 Courbevoie, France
Telephone: (33) 1-46-41-60-00; Fax: (33) 1-46-41-60-11
    From Europe, Africa, and Asia

AIM AND SCOPE

Basic and clinical aspects of cardiovascular pharmacology; appropriate subjects include new drug development and evaluation, physiological and pharmacological bases of drug action, metabolism, drug interactions and side effects, and clinical results with new and established agents.

MANUSCRIPT SUBMISSION

Original manuscripts that have not been published elsewhere except in abstract form, will be considered for publication. Copyright will be held by Raven Press. Each co-author must sign a statement expressly transferring copyright in the event paper is published. A copyright transfer form will be sent to corresponding author by Editors-in-Chief when receipt of a manuscript is acknowledged.

MANUSCRIPT FORMAT

Submit three typewritten copies of each manuscript, double-spaced throughout with a 1 in. left margin. Put text in sequence: Introduction, Methods, Results, Discussion, Acknowledgments, References, tables, and figure legends. Number pages in succession, title page is page one.

Title Page: Include article title, author's names and affiliations, name, address, and telephone number of corresponding and reprint requests author, any footnotes to these items, and a running title (45 letters and spaces). Indicate affiliation of each author. Put information concerning sources of financial support in Acknowledgment section between Discussion and References.

Page following title page: brief abstract (200 words) describing purpose, methods, results, and conclusions of study, and up to six key words or terms.

Tables and Illustrations: Type tables neatly, each on a separate sheet, with title above and notes below. Explain all abbreviations. Do not give the same information in tables and figures. Submit illustrations as clear glossy prints (two duplicate sets may be photocopied), with lettering large enough to be legible when reduced. On back give author's name, figure number, and indicate TOP with an arrow. Give each figure a separate, fully explicit legend; define all sections of figure and all abbreviations and symbols. Excessive black-and-white illustrative material and all color illustrations will be charged to authors.

References: Use style in "Uniform Requirements for Manuscripts Submitted to Biomedical Journals" (see *Ann Intern Med* 1982;96:766–71 or *Br Med J* 1982; 284:1766–70).

Cite references in text by number; number in order cited. Type reference section typed double-spaced at end of text. Abbreviate journal names according to *List of Journals Indexed in Index Medicus.*

Provide all authors' names when fewer than seven; when seven or more, list first three and add *et al*. Provide article titles and inclusive pages. Verify accuracy of reference data.

JOURNAL ARTICLE

1. Horgan JH, O'Callaghan WG, Teo KK. Therapy of angina pectoris with low-dose perhexiline. *J Cardiovasc Pharmacol* 1981;3:566–72.

BOOK

2. Vanhoutte PM, Leusen I, eds. Vasodilatation. New York: Raven Press, 1981.

CHAPTER IN BOOK

3. Patrono C, Ciabattoni G, Pugliese F, et al. Effect of dietary variation in linoleic acid content of platelet aggrega-

tion and the major urinary metabolites of the E prostaglandins and (PGE-M) in infants. In: Hegyeli RJ, ed. *Prostaglandins and cardiovascular disease.* New York: Raven Press, 1981:111–22. (Atherosclerosis reviews; vol. 8).

OTHER FORMATS

RAPID COMMUNICATION—Four printed paged (12 double-spaced manuscript pages, including illustrations, tables, and references). Brief reports of new ideas or data of particular originality and timeliness. Submit by fax or air express service to both editors simultaneously with a letter stating why paper merits special treatment. Include traditional abstract, methods, results, discussion and reference sections. Include corresponding author's telex and fax numbers on title page.

SHORT COMMUNICATIONS—Three printed pages (9 double-spaced manuscript pages, including illustrations, tables, and references). Short papers capable of standing alone as completed work. Include usual abstract, methods, results, discussion and reference sections.

OTHER

Abbreviations: Follow list in "Uniform Requirements for Manuscripts Submitted to Biomedical Journals" (References section). For additional abbreviations, consult *CBE Style Manual* (Council of Biology Editors, Bethesda, MD) or other standard sources.

Drugs: Use generic names only, followed in parentheses after first mention by any commonly used variant generic.

Electronic Submission: Submit in conjunction with paper version in MS-DOS-compatible format on 5.25 in. or 3.5 in. disks. Macintosh-compatible format is also accepted. Submit files in: Microsoft Word, WordPerfect, WordStar, or XYWrite. Do not submit ASCII files. Label disk with author's name, item title, type of equipment used, word processing program (with version number), and filenames. File must be final corrected version of manuscript and must agree with final accepted version of submitted paper manuscript. Delete all other material from disk. Do not include extraneous formatting instructions. Use hard carriage returns only at end of paragraphs and display lines (titles, subheadings, etc.). Do not use extra hard return between paragraphs. Do not indent runover lines in references. Turn off line spacing. Turn off hyphenation and justification. Do not specify page breaks, page numbers, or headers. Do not specify typeface. Correctly enter one (1) and lower case l. and zero (0) and capital O. Use single hyphen with space between for a minus sign; use a double-hyphen with space before and after to indicate a long dash, and a triple hyphen (no extra space) to indicate a range of numbers. Include illustrations and tables at end of file. Code nonstandard characters consistently. Provide a list of such characters and codes.

SOURCE

January 1994, 24(1): A4, inside back cover.

Journal of Cell Biology
American Society for Cell Biology
Rockefeller University Press

Dr. Norton B. Gilula
Dept. of Cell Biology
The Scripps Research Institute
10666 North Torrey Pines Rd.
La Jolla, CA 92037
Telephone: (619) 554-9880; Fax: (619) 554-9960

AIM AND SCOPE

The Journal of Cell Biology publishes papers that report substantial and original findings on the structure and function of cells, organelles, and macromolecules. It is particularly interested in bringing to the attention of its readers new knowledge arising from the application of molecular, biophysical, genetic, immunological, or ultrastructural techniques as they apply to the correlation of structure with function. In addition to continuing to promote the disciplines traditional to the Journal, the Editors wish to encourage publication of papers describing cellular organization and regulation at the molecular and supramolecular levels. Areas of interest include, but are not restricted to, gene organization and expression, somatic cell genetics, cell motility, cellular architecture, membrane and organelle dynamics, and cell differentiation. Papers treating such topics in the contexts of developmental biology, immunology, neurobiology, or virology are welcome, as are papers dealing with procaryotic and plant cell systems. Manuscripts reporting new techniques will be published only when judged by the Editors to represent significant advances. Manuscripts reporting novel sequence information for proteins of fundamental interest to cell biology are encouraged; however, they will be prioritized on the basis of their novelty with respect to understanding the structure/function relationship of the molecule. In addition, manuscripts containing partial sequences should not be submitted unless there is some significant biological insight that results from the available sequence. The criteria for acceptance of manuscripts are scientific excellence, significance, and interest for the general readership.

MANUSCRIPT SUBMISSION

Alert Editors in a cover letter of a manuscript deserving of an unusually high priority and expeditious treatment based on the criteria for prioritization. Submission is taken to mean that data are original and that no similar paper has been or will be submitted for publication elsewhere.

Submitted manuscripts should be fully documented reports of original research. Make as concise as possible without compromising documentation of results. Lengthy literature surveys, detailed methods, or exhaustive bibliographies will not be published.

Submit one original typescript and one reproduction set of figures along with three xerographic copies of manuscript and three reviewers' sets of figures. A reviewer's set of figures consists of mounted glossy prints of halftone illustrations and xerographic copies of line drawings. A reproduction set of figures consists of originals or glossy prints of halftone illustrations, trimmed, labeled, and mounted on mounting board, and originals (in India ink) or glossy prints of line drawings. Mount color prints on flexible board. Reproduction set may be withheld until notification of acceptance; submit with revised version of paper.

Several papers may be published in tandem; transmit as one unit if possible. In cover letter explicitly request publication as a series.

MANUSCRIPT FORMAT

Type double-spaced on one side only of 8.5 × 11 in. white bond paper with 1 in. margins. Number each page in upper right corner.

Title Page: Provide complete names of institutions where work was done, and name, mailing address, telephone and fax numbers of corresponding author. If a change of address is imminent, indicate change and date effective. Furnish a condensed title (50 units) for running footline.

Abstract: 250 word synopsis of work reported—including materials, methods, and results—that is self-explanatory and suitable for use by abstracting services. May be paragraphed. Longer abstracts may be shortened in proof at author's expense. Cite references in full (not by number).

References: Type double-spaced beginning on a sheet separate from text. Alphabetize and either number and cite by number in text or cite by surname and year. Include all authors' names (do not use "et al."), year, complete article titles, and inclusive page numbers. Abbreviate journal names as in *Serial Sources for the Biosis Data Base* (BioSciences Information Service of Biological Abstracts, Philadelphia, PA); spell out names of unlisted journals. Include citations such as "submitted for publication," "in preparation," and "personal communication" parenthetically in text; do not include in Reference section. For personal communications include cited author's institutional affiliation and his/her written permission. do not cite unpublished work in Materials and Methods; incorporate citation parenthetically into text; try to substitute with a full-length paper; if not "deleted in proof will appear in place of deleted abstract citation."

JOURNAL ARTICLES

Two authors: 1. Yalow, R. S., and S. A. Berson. 1960. Immunoassay of endogenous plasma insulin in man. *J. Clin. Invest.* 39:1157–1175.

More than two authors: 2. Benditt, E. P., N. Ericksen, and R. H. Hanson. 1979. Amyloid protein SAA is an apoprotein of mouse plasma high density lipoprotein. *Proc. Natl. Acad. Sci. USA.* 76:4092-4096.

In press: 3. Brown, W., and A. Nelson. 1983. Phosphorus content of lipids. *J. Lipid Res.* In press.

COMPLETE BOOKS

4. Myant, N. B. 1981. The Biology of Cholesterol and Related Steroids. Heinemann Medical Books, London. 882 pp.

ARTICLES IN BOOKS

5. Innerarity, T. L., D. Y. Hui, and R. W. Mahley. 1982. Hepatic apoprotein E. *In* Coronary Atherosclerosis. G. Noseda, C. Fragiacomo, R. Fumagalli, and R. Paoletti, editors. Elsevier/North Holland, Amsterdam. 173–181.

Tables: Double-space on sheets separate from text; make self-contained and self-explanatory. Do not use vertical rules. Label at top with Roman numeral followed by title. Insert explanatory material and footnotes below table. Designate footnotes by symbols in order (horizontally across table): *, ‡, §, ||, ¶, **, ‡‡, etc. Supply units of measure at heads of columns.

Illustrations: Make self-explanatory; cite in numerical order in text. Indicate location in margin of typescript. Type legends double-spaced and consecutively on a separate sheet. Use Arabic numerals for figures and lowercase italic letters for multiple parts of a single figure. Do not exceed journal page dimensions (17.5 × 24 cm) unless arrangement is made with editor. Mount illustrations composed of multiple pieces on mounting board. Plan micrographs so they can be reproduced without reduction. Single-column figures must not exceed 8.25 × 24 cm; double-column figures 17.5 × 24 cm. Submit gels at reproduction size (single-column width or less preferred); indicate on back or indicate desired percent reduction. Plan labeling so that upon reduction height of numbers and capital letters will be 1.75 mm, lane width 5 mm. Micrographs should not exceed one-half page. Fit figures and legends on the same page.

Micrographs—Limit field of micrographs to structures specifically discussed. Place a tissue overlay over each micrograph; on overlay, circle lightly those areas for which continuous tone should be most faithfully reproduced. Be sure that symbols and areas of special interest are not too close to edges and corners are squared. Do not place magnifications in legends; indicate scale on prints. Use sans serif style, 3–3.5 mm high, in India ink or securely affixed transfer letters for labeling. Explain symbols used on micrographs in figure legends. On back of each figure in reproduction set, indicate TOP. Give figure number and first author's name. Use a soft pencil, not a pen; press lightly. Mail flat, using cardboard stiffeners. Never fold, roll, or use paper clips. Halftones will be reduced to one-column width unless use of additional space is justified. Obtain approval of Editor who accepted manuscript for printing figures at greater than one-column width.

Line Drawings—Provide original black India ink drawings or sharp glossy prints to be reduced to one column (8.25 cm) or less. Make lettering sans serif and of such size that capitals, numbers, and symbols will reduce to a height of 1.5–1.75 mm. Make lettering in scale with figure. If authors are unable to provide satisfactory drawings or lettering, Press will prepare them at cost to author.

Color Illustrations—Use bendable, reflective art. Do not submit color figures mounted on stiff boards. Authors are charged for separation and printing costs of color figures.

OTHER FORMATS

COMMENTARIES—Offer a personalized perspective or synthesis of information on a topic of interest to the general readership.

MINI-REVIEWS—3–4 printed pages. Contact a member of Editorial Board. Do not include an exhaustive review of an area, provide a focused brief treatment of a contemporary development or issue in a single area. Limit references to 75.

OTHER

Abbreviations: See January issue for list of *Journal's* abbreviations. Terms not on this list must be used three times or more to qualify as an abbreviation. Spell out term on first mention and follow with abbreviated form in parentheses. Thereafter use abbreviated form. Supply a footnote of nonstandard abbreviations used, in alphabetical order; give each abbreviation followed by its spelled-out version.

Color Charges: $1,000 for first page; $500 each additional page. If unable to pay, appeal in writing to Associate Editor, who will recommend either a reduction of charges or a change to a black-and-white image.

Data Submission: Submit manuscripts containing sequence information, provided that they offer novel insights concerning the structure/function relationships of cellular molecules. Nucleotide sequences must have an EMBL/GenBank/DDBJ database accession number. Accepted manuscripts without a number by page proof stage will be held until number is provided. Obtain blank sequence submission form while writing manuscript; send completed form to EMBL when manuscript is submitted. Last line of sequence's legend should read "These sequence data are available from EMBL/ GenBank/DDBJ under accession number___." EMBL can provide an accession number in one week or less, with either an electronic mail address or fax or telex number.

Submit sequence information to EMBL, not GenBank or DDBJ. Data submitted to EMBL will be shared with GenBank and DDBJ. To obtain EMBL submission forms, contact Data Submissions, EMBL Data Library, Postfach 10.2209, 6900 Heidelberg, Germany; telephone: 49-6221-387-258; fax: 49-6221-387-306; telex: 461613 (embl d). Form is also available with all releases of EMBL database. *Nucleic Acids Research* prints the sequence data submission form in the first issue of each year; copy this form.

Materials Sharing: The free exchange of all clones, cell lines, and biological reagents published in the *Journal* is encouraged in order to promote the transmission of scientific information and facilitate the progress of research in cell biology.

Page Charges: $40 per page; surcharge for each page of halftone illustrations is: 1–4 pages, no charge; S–7 pages, $55; over 7 pages, $80.

Reviewers: Provide names, addresses, and phone and fax numbers of four or five reviewers.

Style: Follow abbreviations and other conventions of *Council of Biology Editors Style Manual* (5th edition, 1983, Council of Biology Editors, Inc., Bethesda, MD). For chemical nomenclature, follow Subject Index of *Chemical Abstracts*. Capitalize trade names and give manufacturers' names and addresses. Mark actual gene names for typesetting in italics. Italicize only actual gene names.

SOURCE

October 1994, 127(1): back matter.

Journal of Cell Science
The Company of Biologists Ltd.

Dr. Fiona M. Watt
Keratinocyte Laboratory, Room 602
Imperial Cancer Research Fund
Lincoln's Inn, Fields, London WC2A 3PX, U.K.

Dr. Gary Borisy
Laboratory of Molecular Biology
University of Wisconsin
1525 Linden Drive
Madison, WI 53706
Bitnet no.: BORISY @ WISCMACC

Dr. David M. Glover
Department of Anatomy and Physiology
University of Dundee
Dundee, DD1 4HN, Scotland

Dr. Daniel Louvard
Biologie des Membranes
Institut Pasteur, 25 rue du Dr Roux
Paris F-75724, Cedex 15, France

Dr. David I. Meyer
Department of Biological Chemistry
UCLA School of Medicine
Los Angeles CA 90024-1737

AIM AND SCOPE

Journal of Cell Science publishes critical work over the full range of cell biology. Papers are called for that deal with all aspects of both prokaryotic and eukaryotic cells, the single most important criterion for acceptance being scientific excellence. Short surveys of topical subjects appear in each issue under the general title Commentary.

MANUSCRIPT SUBMISSION

Copyright to articles published in the *Journal* is assigned to the Company of Biologists Limited.

MANUSCRIPT FORMAT

Submit manuscripts in English, typed double-spaced, on one side of paper; number pages in order: Title Page, Summary, Introduction, Materials and Methods, Results, Discussion, References, Figure Legends (new page), Tables with titles, etc. Papers should be ready for press; revision in proof is not possible. Provide three copies of typescript and figures. Supply a short title (40 characters) for use as page headings, and at least 3 key words. Provide a summary (500 words) at beginning of text. If references are cited, include journal name, volume number, pages and year in parentheses. Do not use '&'; do not italicize Latin words. Place acknowledgments before list of references.

Illustrations: Submit originals only with revised manuscript. Label to a professional standard. Design to fit column widths of 84, 124, or 174 mm. Use 8 pt Times for text labelling. Ensure that symbols are sized for clarity. Submit original photographs in final journal size (210 mm max. length, width as above). Use press on symbols and letters (lowercase, 14 pt size) on photographs. Number or letter each figure in a plate. Use capital letters for figure parts. Indicate magnification with a scale bar. Write author(s) names(s) on back of figure with orientation. Send files of figures on disk with final version, if possible. Use Fig. 6A,B or Figs 7,8.

References: List each reference cited in text in References and vice vera; check carefully.
One author—Jones (1990) or (Jones, 1990; Smith, 1991)
Two authors—Jones and Jackson (1990) or (Jones and Jackson, 1990; Smith, 1991)
More than two authors—Jones et al. (1990) or (Jones et al., 1990a,b; Smith et al., 1986, 1989). List manuscripts accepted for publication in References as (in press). Cite unpublished observations, including results submitted for publication in text only. Submit written permission for use of personal communications

Cite references in alphabetical order by first author's surname. List works by same author(s) chronologically, beginning with earliest publication. "In press" citations must have been accepted; include journal or publisher. Initials follow surnames in author list; insert a full stop and space after each initial and brackets around date followed by a full stop. Use bold for authors names. Use U.S. National Standard abbreviations for Journals.

Jones, P., Beck, C. B. and Smith, J. (1991). Cell division in the sea urchin. *Cell* **99**, 197-223.

Rossant, J. (1977). Cell commitment in early rodent development. In *Development in Mammals* vol. 2 (ed. M. H. Johnson), pp. 119-150. Amsterdam: North Holland.

Tables: Submit on separate sheets with titles and legends.

Color Plates: Reproduced free of charge at discretion of the Editor. Charges for excessive use of color is also at Editor's discretion.

OTHER FORMATS

COMMENTARY—3000-5000 words, short review articles. Check with an Editor to make sure topic is suitable. A fee of £50 is paid. Follow the same instructions for text, layout and style as for regular articles, except that an abstract is not required and only two copies of typescript and figures should be submitted.

RAPID PUBLICATION—Revised manuscript must be accompanied by a word-processor disk. No color plates. Author cannot see proofs. Manuscript must conform to journal style, no subediting necessary. Figures must be labelled in journal style.

OTHER

Abbreviations: Define at first use; type uppercase without stops, lowercase with full stops.

Data Submission: Submit sequences to EMBL Database Library or other data library.

Ions: Use Ca^{2+}, etc.

Isotopes: If isotope is of an element in the compound, place symbol for isotope in square brackets; if compound does not normally contain the isotopically labelled element, use either ^{131}I-labelled albumin or ^{131}I-albumin.

Materials Sharing: Make specialized reagents (antibodies, DNA probes) available to scientists who require them for bona fide research.

Style: Italicize or underline gene names. Put protein products of genes in Roman type. If you do not have specified fonts, underline for italics and double underline for bold.

Units: SI units only; write out units for time in full. Use relative molecular mass (M_r) and not MW. M_r is dimensionless; kDa is acceptable for molecular mass.

SOURCE

January 1994, 107(1): inside back cover.

Journal of Cellular Biochemistry
Wiley-Liss, Inc.

C. Fred Fox
Department of Microbiology and Molecular Genetics
University of California, Los Angeles
Los Angeles, CA 90024-1489
Telephone: (310) 825-9329; Fax: (310) 206-1703

Gary S. Stein
Department of Cell Biology
University of Massachusetts Medical Center
55 Lake Avenue North
Worcester, MA 01655
Telephone: (508) 856-5625; Fax: (508) 856-6800

Max M. Burger
Friedrich-Miescher Institut
Posffach 2543
CH-4002 Basel, Switzerland
Telephone: (061) 697-68-50; Fax: (061) 697-67-83

Garth L. Nicolson
Department of Tumor Biology (108)
MD Anderson Cancer Center
1515 Holcombe Boulevard
Houston, TX 77030
Telephone: (713) 792-7481; Fax: (713) 794-0209

W. Jackson Pledger
Department of Cell Biology
Vanderbilt University School of Medicine
Nashville, TN 37232
Telephone: (615) 322-2104; Fax: (615) 343-4539

James Watson
Department of Pathology
University of Auckland School of Medicine
Auckland, New Zealand
Telephone: 0064 09 379-717; Fax: 0064 09 735-215

AIM AND SCOPE

The *Journal of Cellular Biochemistry* publishes descriptions of original research in which complex cellular, pathologic, clinical, or animal model systems are studied by molecular biological, biochemical, quantitative ultrastructural or immunological approaches. The areas covered include conditions, agents, regulatory processes, or differentiation states that influence structure, structure-function relationships, or assembly mechanisms in cells, viruses, or supramolecular constructs. This scope extends to cell structure and function; organelle assembly; regulation of cell organization, reproduction or differentiation; and to the development, organization or remodeling of tissues.

MANUSCRIPT SUBMISSION

Articles are full-length papers presenting complete descriptions of original research which have not been published and are not being considered for publication elsewhere.

Submit three copies of all materials. Type manuscript double-spaced, on one side of white bond paper.

MANUSCRIPT FORMAT

Organize in sequence: Abstract, Introductory Statement (without heading), Methods, Results, Discussion, Acknowledgments, References, Tables, Legends, and Figures.

Illustrations: Submit 3 high quality prints of all halftone figures, e.g., micrographs, acrylamide gels, etc. For electron micrographs, indicate on overlay sheets details of particular importance. Page space for illustrations is up to 3 × 9 in. for 1 column illustrations and up to 6 5/16 × 9 in. for 2 column illustrations. Figures and tables must fit these proportions; lettering size must allow for this reduction. Submit all figures and tables on separate sheets. Prepare drawings in black ink; glossy photographs are acceptable. Color is printed in at author's expense; inquire prior to submitting.

References: In text, cite by name and date. For more than two authors, use first surname and et al. If more than one work by an author in the same year, include a, b, c, for successive entries. In references section, list all entries alphabetically, include complete title of article, and names of all authors. Journal abbreviations follow *Index Medicus* style.

JOURNAL ARTICLES

Pockwinse SM, Wilming LG, Conlon DM, Stein GS, Lian JB (1992): Expression of cell growth and bone specific genes at single cell resolution during development of bone tissue-like organization in primary osteoblast cultures. J Cell Biochem 49:310–323.

BOOKS

Lamb CJ, Beachy RN (1990): "Plant Gene Transfer." New York: Wiley-Liss.

ARTICLES IN BOOKS

Evans GA. Evans KC (1990): Strategies for physical mapping of complex genomes. In Lamb CJ, Beachy RN (eds): "Plant Gene Transfer." New York: Wiley-Liss, pp 39–46.

Indexing: Supply 5–10 key words or phrases (not in title) and a shortened title (40 characters) for a running head.

OTHER FORMATS

PROSPECTS—Topical overviews on emerging areas of research, summarizing key problems, concepts, experimental approaches, and research opportunities which characterize a subject area. Do not include previously unpublished research results. Generally invited by Editors; consult with Editors.

VIEWPOINTS—Include news items, meeting summaries and announcements, book reviews and letters to the Editor.

OTHER

Symbols and Abbreviations: Unless common, define at first use. Do not include in abstract. For nomenclature, symbols, and abbreviations follow recommendations in *J*

Biol Chem 262:7–11 (1987).

SOURCE

September 1994, 56(1): inside back cover.

Journal of Cerebral Blood Flow and Metabolism

International Society of Cerebral Blood Flow and Metabolism

Raven Press, Ltd.

Dr. Myron D. Ginsberg
Department of Neurology (D4-5)
University of Miami School of Medicine
P.O. Box 016960, Miami, FL 33101
Telephone: (305) 547-6449; Fax: (305) 547-5830

AIM AND SCOPE

The *Journal* is devoted to the publication of original manuscripts, i.e., those that have not been published previously, except in abstract form, on experimental, theoretical, and clinical aspects of brain circulation and metabolism.

MANUSCRIPT SUBMISSION

Submit manuscript in triplicate, with a disk, typewritten with double-spacing throughout and 2,5 cm margins. Copyright on articles published in the *Journal of Cerebral Blood Flow and Metabolism* will be held by The International Society of Cerebral Blood Flow and Metabolism. Include a statement signed by author(s) expressly transferring copyright to The Society in the event paper is published. A copyright transfer form will be sent to the senior author when receipt of manuscript is acknowledged by the Editor-in-Chief.

MANUSCRIPT FORMAT

Title Page: Include title, authors' names, laboratory of origin, name, address, and telephone number of corresponding and reprint request author(s), and any necessary footnotes. Indicate specific affiliations of each author. Put information on sources of financial support in Acknowledgment section, between Discussion and References.

On page after title page include a brief abstract (200 words), up to six key words or terms, and a running title (45 letters and spaces).

Organize in standard fashion, i.e., introduction (give reasons for investigation), methods, results, and discussion.

In mathematical copy, be precise about alignment of numerals, letters, and symbols; hand-letter any that cannot be typewritten. Make a circled notation in nearest margin to identify possibly ambiguous characters such as 1 (one or el), 0 (zero or oh), X (multiplication sign or ex), Greek letters, script letters, diacritical marks, and prime, and certain letters (e.g., C, K, P) in which capitals and lowercase are nearly indistinguishable when handwritten. Use simplest form of equation that can be made by ordinary mathematical calculation; for instance:

$$\frac{((x+y)/c)^2}{(x+y)^2/d} \quad \text{and not} \quad \frac{\left(\dfrac{x+y}{c}\right)}{\dfrac{(x+y)^2}{d}}$$

Name simple chemical compounds in text of Experimental and Results sections by formulae when these can be printed in single horizontal lines of type.

Isotopic specifications should conform to IUPAC system (*Biochem J* (1978), 169:1–27). Give full details of specific activities of isotopes used. Express results in absolute terms, e.g., Ci, dps, dpm.

For tracer kinetic terminology, see guidelines in Brownell et al. (1968), "Nomenclature for tracer kinetics." *Int J Appl Radiat Isot* 19:249–262.

Tables and Figures: Construct so that they, together with their captions and legends, will be intelligible with minimal reference to text. Put each table on a separate page at end of manuscript. Give each figure an explanatory legend typed double-spaced. Do not use a separate sheet for each legend. Explain all symbols used in figure.

Indicate units of measure in column headings of a table or coordinates of a figure. If unit of measure is the same for all data in a table, give in table title or in a footnote below table.

Present data as clearly defined mean values. Use standard deviation as the measure of dispersion throughout manuscript, tables, and figure error bars. Do not use standard error of the mean. Give appropriate indications of statistical significance of differences from control values. Indicate number of individual values represented by a mean.

Submit line drawings as at least two sets of clear glossy prints, with lettering of high standard and large enough to be legible when reduced. Submit halftone photographs in triplicate as original photographic prints. Provide an overlay showing areas of photograph that require greatest definition. Provide scale bars on photograph.

Maximum final size of any figure is 7×9.5 in. On back of each figure give figure number, authors' names, and indicate TOP with an arrow. Color illustrations will be charged to author; cost is given upon request.

Insert symbols on graphs, etc., using following standard characters: ○ ● △ ▲ □ ■ × +.

Plan figures to avoid wasted "white" space. Do not extend coordinates appreciably beyond curves. It is not necessary for scale on a coordinate to start at 0.

References: Cite references in text by names of authors and year of publication; for papers written by three or more authors use "et al." At end of paper list references (typed double-spaced) in alphabetical order, group papers by three or more authors, in chronological order after any other papers by first author. Give opening and closing page numbers, full titles of papers, and names of all co-authors. Abbreviate journal titles according to *Index Medicus*.

Tenne M. Finberg JPM, Youdim MBH, Ulitzur S (1985) A new rapid and sensitive bioluminescence assay for monoamine oxidase activity. *J Neurochem* 44:1, 378–1,384

Siesjo BK, Wieloch T (1983) Fatty acid metabolism and the mechanisms of ischemic brain damage. In: *Princeton Conferences on Cerebrovascular Diseases, Vol. 13: Cerebrovascular Diseases* (Reivich M, Hurtig HI, eds), New York, Raven Press, pp 251–268

Mention unpublished experiments, papers in preparation, etc., only in text; do not include in list of references.

Cite papers accepted for publication but which have not appeared in references with abbreviated name of journal followed by "in press." Indicate date of acceptance of each paper upon submission. For references to unpublished papers by authors (e.g., in press or submitted) submit two copies of paper for use of referees.

Submit written authorization from communicator for use of personal communications with original manuscript; mention only in text, giving name of communicator.

OTHER FORMATS

RAPID COMMUNICATIONS—Five printed pages (14 double-spaced manuscript pages, 6500 characters/page, including references, tables, and illustrations). Concise papers that present innovative ideas and data of particular originality and timeliness. Criteria for acceptance are more stringent than for full papers.

Submit in quadruplicate. In cover letter, explain why paper warrants particular consideration for rapid publication.

Adhere to conventional form; include abstract, introduction, methods, results, and discussion. Results and discussion may be combined

SHORT COMMUNICATIONS—1500 word reports of concise pieces of work that lead to positive conclusions. Include 100 word abstract and 2 figures. Section headings may be omitted.

OTHER

Abbreviations: See list of abbreviations that may be used without definition in January issue.

Drugs: Use official or approved names; give trade names or common names in parentheses where drug is first mentioned.

Electronic Submission: Submit in conjunction with paper version in MS-DOS-compatible format on 5.25 in. or 3.5 in. disks. MacIntosh-compatible format is also accepted. Submit files in: Microsoft Word, WordPerfect, WordStar, or XYWrite. Do not submit ASCII files. Label disk with author's name, item title, type of equipment used, word processing program (with version number), and filenames. File submitted must be final corrected version of manuscript and must agree with final accepted version of submitted paper manuscript. Delete all other material from disk. Do not include extraneous formatting instructions. Use hard carriage returns only at end of paragraphs and display lines (titles, subheadings, etc.). Do not use an extra hard return between paragraphs. Do not use tables or extra space at start of paragraphs or for list entries. Do not indent runover lines in references. Turn off line spacing. Turn off hyphenation and justification. Do not specify page breaks, page numbers, or headers. Do not specify typeface. Correctly enter one (1) and lower case l. and zero

(0) and capital O. Use single hyphen with space between for a minus sign; use a double-hyphen with space before and after to indicate a long dash, and a triple hyphen (no extra space) to indicate a range of numbers. Include illustrations and tables at end of file. Code nonstandard characters consistently. Provide a list of such characters and codes.

SOURCE

January 1995, 15(1): front matter.

Journal of Chromatography
Elsevier Science Publishers

Editorial Office, *Journal of Chromatography* (A or B)
P.O. Box 681
1000 AR Amsterdam, Netherlands
Fax: (+31-20) 5862 304

AIM AND SCOPE

The *Journal of Chromatography* publishes papers on all aspects of chromatography, electrophoresis and related methods. Contributions consist mainly of research papers dealing with chromatographic and electrophoretic theory, instrumental developments and their applications.

Journal of Chromatography B: Biomedical Applications, under separate editorship, deals with the following aspects: developments in and applications of chromatographic and electrophoretic techniques related to clinical diagnosis or alterations during medical treatment; screening and profiling of body fluids or tissues related to the analysis of active substances and to metabolic disorders; drug level monitoring and pharmacokinetic studies; clinical toxicology; forensic medicine; veterinary medicine; occupational medicine; results from basic medical research with direct consequences in clinical practice.

MANUSCRIPT SUBMISSION

Submission of an article implies that article is original and unpublished and is not being considered for publication elsewhere.

Upon acceptance, author(s) will be asked to transfer copyright of article to publisher.

The preferred medium of submission is on disk with accompanying manuscript. Submit four copies of complete manuscript with illustrations and tables attached to each copy.

In a letter from the senior author, state that paper is being submitted for publication in the *Journal Of Chromatography*.

MANUSCRIPT FORMAT

Type manuscripts double-spaced on one side of consecutively numbered sheets of uniform size paper with 2 cm margins on each side, in easily readable 12 pt font and (in case of computer-processed manuscripts) using a letter-quality printer or equivalent. Precede manuscript with a sheet of paper carrying paper's title and name, full postal address and fax number of corresponding author. Divide papers into sections, headed by captions (e.g., Abstract, Introduction, Experimental, Results, Discussion). Put illustrations, photographs, tables, etc. on separate sheets.

Title: Make concise and informative; include key words. Follow with authors' full names, academic or professional affiliations, and address of laboratory where work was done. If present address of an author is different, give in a footnote. Acknowledge financial support in Acknowledgments at end of paper, not in a footnote to title or author names.

Abstract: 50–100 word clear and brief indication of what is new, different and significant. No references.

Introduction: Mention what has been done before on topic, with appropriate references. Sate what is new in this paper.

Experimental: Include sufficient information for others to repeat Experiments. Give general conditions here with specific details in figure captions.

References: Number in order cited in text. List in numerical sequence on a separate sheet at end of article. Put numbers in text at appropriate places in square brackets.

1 D. P. Ndiomu and C. F. Simpson, *Anal. Chim. Acta*, 213 (1988) 237.

2 T. Paryjczak, *Gas Chromatography in Adsorption and Catalysis*, Wiley, Chichester, 1986.

3 M. Saito, T. Hondo and Y. Yamauchi, in R. M. Smith (Editor), *Supercritical Fluid Chromatography*, Royal Society of Chemistry, London, 1988, Ch. 8, p. 203.

4 F. I. Onushka and K. A. Terry, in P. Sandra, G. Redant and F. David (Editors), *Proceedings of the 10th International Symposium on Capillary Chromatography, Riva del Garda, May 1989*, Hüthig, Heidelberg, 1989, p. 415.

Use journal title abbreviations from *Chemical Abstracts*. Give articles not yet published as "in press" (specify journal), "submitted for publication" (specify journal), "in preparation" or "personal communication."

Tables and Illustrations: Do not repeat data already given in text.

Type tables double-spaced on separate pages; number in Roman numerals according to sequence in text. Give a brief descriptive heading above each table. Describe experimental conditions below heading.

Submit figures in a form suitable for reproduction, either drawn in Indian ink on drawing or tracing paper, or as sharp prints [either photographic (glossy) prints or prints from a high-resolution laser printer]. Label axes of graphs and chromatograms; provide full quantitative data or equivalent information in legend. Lettering should be in a form suitable for reproduction, should be kept to a minimum, and should be such that numbers, etc., remain legible after reduction in size. Submit one reproducible copy and three photocopies. Make figures so that the same degree of reduction can be applied to all of them. Do not exceed size of text pages. Simple straight-line graphs can readily be described in text by means of an equation or a sentence. Support claims of linearity by regression data that include slope, intercept, standard deviations of the slope and intercept, standard error and the number of data points; correlation coefficients are optional. Use standard symbols in line drawings: ○ ● ❑ ■ △ ▲ + ×.

Submit glossy photographs with good contrast and intensity. Include references to illustrations at appropriate places in text by Arabic numerals; indicate approximate position of illustration in margin of manuscript. Give each illustration a caption; type all captions double-spaced together on a separate sheet.

If structures are given in text, provide original drawing.

Colored illustrations are reproduced at author's expense; cost is determined by number of pages and colors needed.

Obtain written permission of author and publisher for use of any figure already published; indicate source in legend.

OTHER FORMATS

DISCUSSIONS—One or two pages. Explain, amplify, correct or otherwise comment substantively upon articles recently published in journal.

REVIEW ARTICLES—Invited or proposed in writing to Editor. Submit an outline for preliminary discussion prior to preparation.

SHORT COMMUNICATIONS—Five printed pages. Descriptions of short investigations, or reports of minor technical improvements of previously published procedures.

OTHER

Abbreviations: Use widely accepted symbols, abbreviations and units (SI). If there is any doubt about a particular symbol or abbreviation, give full expression followed by abbreviation at first use in text. Explain abbreviations used in tables and figures in captions.

For a list of abbreviations and symbols that may be used without definition see *J. Chromatogr.* 1993A:, 657:468-469. Do not use abbreviations and symbols in titles. Most abbreviations should only be used in combination with a value, or in structural formula.

Chemicals: State supplier (with city/town, state, country) and degree of purity of all less common chemicals. Give EC number of enzymes and optical purity of enantiomers.

Chromatography, Column Liquid: State column dimensions (length × internal diameter), manufacturer and location, packing material (for non-commercial columns or columns that are not widely used, specify chemical composition), particle diameter, pore diameter, column temperature.

Give complete and unambiguous description of mobile phase composition or procedure for its preparation; pH; flow-rate; gradient program.

When reporting k' values, describe method for determining hold-up time (t_0).

Chromatography, Gas, and Supercritical Fluid Chromatography: In addition to parameters mentioned for column liquid chromatography, specify type of column (packed, capillary, etc.) support material, film thickness of the stationary phase, and surface modification, if applicable.

For carrier gas, state type, purity, flow-rate or inlet pressure (bar or MPa).

Detail all relevant temperatures (or temperature programs).

Chromatography, Planar: Describe internal dimensions, manufacturer and location, saturation, temperature, humidity of chamber.

State manufacturer and location, material, dimensions, type (laboratory-prepared or commercially precoated) and thickness of layer, additives (fluorescent indicator, binder), position of starting line, development mode, method of activation of thin layer or paper.

Give composition of solvent, monophasic or upper or lower phase of two-phase mixture, total volume of solvent.

For each sample, describe application method, size of spot or streak, solvent and amount of solute and volume of solution applied. Detail spray reagent, wavelength, details of colors, R_F values of detection.

Electronic Submission: Preferred storage is a 5.25 or 3.5 in. disk in MS-DOS format (for Macintosh, save file in the usual manner, do not use option "save in MS-DOS format"). Submit disk and (exactly matching) printed version. Specify type of computer and word-processing package used (do not convert textfile to ASCII). Ensure that letter "l and digit "1" (also letter "O" and digit "0") are used properly; format article (tabs, indents, etc.) consistently. Do not leave characters not available on word processor open; indicate by a unique code used consistently throughout text. List codes and provide a key. Do not allow word splits; do not use a "justified" layout.

Electrophoresis: For matrix, tell if cellulose acetate, agarose, polyacrylamide; gel concentration; percentage cross-linker; dimensions and material of tube, sheet, etc., surface modification, length between column inlet and detector, temperature.

Describe buffers used, pH and how pH was set or adjusted. Also describe injection method, voltage, current. In electropherograms, indicate anode and cathode.

Equipment: State model and manufacturer (with city/town, state, country) of commercial instruments. For instruments that are not commercially available, give sufficient detail (or a reference) to allow others to construct their own. Include detection parameters (e.g., type, wavelength, attenuation, linearity range, limit of detection at a specified signal-to-noise ratio).

Nomenclature: Follow recommendations of the International Union of Pure and Applied Chemistry (IUPAC) and the journal *Pure and Applied Chemistry*. See also *IUPAC Compendium of Analytical Nomenclature, Definitive Rules*, 1987).

Sample Preparation: In application papers include full details or a reference of method of sample preparation. For centrifugation steps, give details of g value and time. Specify injection device and volume and concentration of injected sample.

Units: The use of some non-SI units has been accepted, including units for time (min, h), volume (l), pressure (1 bar = 10^5 Pa), temperature (°C), energy (1 eV ≈ 160 219· 10^{-21}J), mass (1 u ≈ 1.66053·10^{-27}kg) and activity 1 Ci = 3.7·10^{10}Bq). *Journal* also accepts Å (- 0.1 nm). Express concentration in mol dm^{-3} or mol l^{-1}; symbol M is accepted; do not use normality (N). "Daltons" are not compatible with SI system; give relative molecular mass (M_r) as a value only (dimensionless). Express gravitational force in g; rpm is not allowed for centrifugation (but is for Vortex mixing). See *J. Chromatogr.* 1993A:, 657:467 for some conversion factors. For ppm, ppb, and ppt, the American billion (10^9) and trillion (10^{12}) are meant.

Use ppm, ppb and ppt only when referring to weight/weight or volume/volume ratios; do not use for weight/volume ratios. At first use indicate whether it refers to weight/weight or to volume/volume.

Indicate decimal points by full stops. Include a leading zero for all decimal numbers smaller than unity.

SOURCE

1993 A:, 657: 463–469.
Referred to in January 6, 1995 issue.

Journal of Clinical Endocrinology & Metabolism
The Endocrine Society

Lewis E. Braverman, M.D.
Massachusetts Biotechnology Research Park
373 Plantation Street
Worcester, MA 01605
Telephone: (508) 753-6105; Fax: (508) 752-9939

AIM AND SCOPE

The Endocrine Society publishes papers describing the results of original research in the field of endocrinology and metabolism in the *Journal of Clinical Endocrinology and Metabolism*, *Endocrinology*, and *Molecular Endocrinology*. The *Journal of Clinical Endocrinology and Metabolism* publishes endocrine and metabolic studies related to human and primate physiological, biochemistry and disease. Consult with Editors concerning most appropriate journal to which to submit manuscript.

MANUSCRIPT SUBMISSION

Present a clear, honest, accurate, and complete account of research performed. Describe a complete study or a completed phase of an extended study. Avoid fragmenting reports. If some results appear in another journal or publication, supply copies of other paper(s) to editor.

Describe work in sufficient detail to allow others to repeat. Include all relevant data, including those which may not support hypothesis. Cite publications which have a direct bearing on novelty and interpretation of results. Make any clones of cells and genes available to other researchers on proper request. Provide propagative materials such as DNA clones, cell lines, and hybridomas used in studies.

List only individuals who made significant contributions to intellectual and procedural aspects of study as authors. Authors should participate in conception and planning of work, interpretation of results, and writing paper. Recognize those who contributed to a lesser extent in an acknowledgment. Submit signed Affirmation of Originality and Copyright Release form (in each issue) to indicate that authors approved final version of manuscript and is prepared to take public responsibility for work. Failure to notify editor that some results are being or have been previously published will result in placement of a notice in the journal that the authors have violated the ethical guidelines for publication of research in Endocrine Society journals.

MANUSCRIPT FORMAT

Prepare in accordance with "Uniform Requirements for Manuscripts Submitted to Biomedical Journals" of the International Committee of Medical Journal Editors (*Ann Intern Med.* 1988;108:258-65, and *Br Med J.* 1988; 296:401-5).

Write clearly and concisely. Ensure accuracy of spelling, grammar, and typing. Avoid laboratory slang, clinical jargon, and colloquialisms.

Submit manuscripts as double-spaced typed copy with wide margins on 8.5 × 11 in. paper in quadruplicate, including tables and photocopies of figures.

Put title page, tables, references, footnotes, legends, and bibliography on separate pages.

Title: Concise statement of article's major contents; provide sufficient information to allow reader to judge relevance of paper to interests.

Abstract: 200 words. Describe briefly in complete sentences purpose of study, methods utilized, results obtained, and principal conclusions. Make easily understood without recourse to text or references. Avoid nonstandard abbreviations, unfamiliar terms, symbols, and acronyms.

Introduction: Orient reader to state of knowledge in area under investigation; assume a basic knowledge of clinical endocrinology and metabolism. Cite recent important work by others.

Materials and Methods: Describe and reference in sufficient detail so others can repeat work. State source of hormones, unusual chemicals and reagents, and special pieces of apparatus. For modified methods, describe only modifications.

Results: Present experimental data briefly in text, tables, or figures; avoid redundant methods of presentation. Prepare figures to minimize space required. Combine several figures into one composite figure whenever possible.

Make lettering on figures as large as possible to permit maximal reduction in size. In photographs show only a close-up of region of interest. Insert tabular material into "blank" areas of a figure. Five × 7 in. prints are preferred.

Graphs must be of professional quality; submit as sharp photographs on glossy paper. Do not mount photographs; trim to exclude all but essential areas. Number each figure on back; indicate TOP. Submit one set of unmounted glossy figures and a clear set of photocopies with figures numbered—one figure to a page with each manuscript copy. Submit glossy prints of photomicrographs, not photocopies.

At author's expense, halftones such as photo- or electron micrographs may be custom printed on special paper from engravings approved by author; prices are quoted on request. Authors pay cost of color separations and printing for color photographs or illustrations. A cost estimate is sent upon submission to printer for composition; authors must provide written or verbal acceptance of charges before production can begin.

Submit tables as typewritten copy. Give each table a concise heading; construct as simply as possible; make intelligible without reference to text. Use no more than four vertical rows for a small table (one column width); no more than eight to ten rows for a large table (both columns of a page).

Minimize redundant or repetitious entries in a table. For instance, rather than stating "P < 0.05" for each of 20 entries, use an asterisk or other special symbol to denote specified probability levels defined in a table footnote.

Discussion: Focus on interpretation and significance of findings. Make brief, objective comments describing their relation to other work. Avoid repetition of material in Introduction and experimental findings. Do not include unsubstantiated speculations and plans for future study.

References: Cite in numerical order (in parentheses) in text; one reference to a number.

Keep number to a reasonable minimum; cite appropriate recent reviews whenever possible. Do not cite unpublished observations, personal communications, or manuscripts in preparation or submitted for publication in bibliography. Insert at appropriate places in text, in parentheses and without serial number, or present in footnotes.

Abbreviate journal titles according to style in *Index Medicus*.

JOURNAL ARTICLES AND ABSTRACTS

List all authors when six or fewer; when seven or more, list only the first three and add et al.

1. Binoux M, Hossenlopp P. 1986 Insulin-like growth factor (IGF) and IGF-binding proteins: comparison of human serum and lymph. J Clin Endocrinol Metab. 67:509–14.

2. MacLaughlin DT, Cigarros F, Donahoe PK. 1988 Mechanism of action of Mullerian inhibiting substance. Prog 70th Meeting of the Endocrine Soc. 19.

BOOKS

List all authors or editors when six or fewer; when seven or more list only the first three and add et al.

3. Bonneville F, Cattin F, Dietemann J-L. 1986 Computed tomography of the pituitary gland. Heidelberg: Springer-Verlag; 15–6.

4. Burrow GN. 1987 The thyroid: nodules and neoplasia In: Felig P, Baxter JD, Broadus AE, Frohman LA, eds. Endocrinology and metabolism. 2nd ed. New York: McGraw-Hill; 473–507.

OTHER FORMATS

COMMENTS—Brief manuscripts reporting data that are preliminary, negative or confirmatory. Follow usual form for manuscripts.

RAPID COMMUNICATIONS—Three journal pages with important new observations of exceptional significance to endocrinologists in general to warrant rapid publication. In cover letter justify need for rapid publication and why publication as a comment or as an original study would be inappropriate. Prepare on special forms for photo-offset reproduction from Editorial Office. Instructions are sent with forms. Submit original and three copies.

OTHER

Abbreviations and Symbols: For a list of abbreviations and symbols see Appendices C (Abbreviations and Symbols) and D (Abbreviations for Polypeptide Hormones) *J Clin Endocrinol Metab* 1994; 79(1): 28A-29A. Clearly define those not on lists in abstract and text; if a large number, place in a footnote to page 1 labeled Abbreviations used in this paper.

Drugs: Designate drugs by generic names.

Experimentation, Animal: Conduct in accord with highest standards of humane animal care, as outlined in Appendix B (Guidelines for the Care and Use of Experimental Animals) *J Clin Endocrinol Metab* 1994; 79(1): 28A. Include a statement that studies were conducted in accord with principles and procedures outlined in Appendix.

Experimentation, Human: Conduct in accordance with guidelines proposed in The Declaration of Helsinki (see Appendix A: Declaration of Helsinki: Recommendations for conduct of Clinical Research, *J Clin Endocrinol Metab* 1994; 79(1): 27A-28A). In many studies details of race, age, and sex are important. In experiments involving any significant risk or discomfort to subjects, document that informed consent was obtained from subjects and that investigations was approved by an institutional human research committee.

In text, tables, and figures identify subjects by number or letter rather than by initials or names. Include photographs of patients' faces only if scientifically relevant. Obtain written consent from patient for use.

Nomenclature, Chemical: Designate unfamiliar chemical compounds at first use by correct systematic names either by giving trivial name followed by systematic name in parentheses, or, if chemical substances are numerous, by providing a glossary of trivial and corresponding systematic names in a footnote at beginning of manuscript. Conform to recommendations of the International Union of Pure and Applied Chemistry and the International Union of Biochemistry. See *Handbook for Authors* (American Chemical Society Publications, Washington, D.C., 1978).

For nomenclature for thyroid hormones in serum follow recommendation of American Thyroid Association (*J Clin Endocrinol Metab* 1987; 64:1089–94).

For steroid nomenclature see Appendices E (Nomenclature of Steroids) and F (Nomenclature of Vitamin D Metabolites: Analogous and Structurally Related Compounds) *J Clin Endocrinol Metab* 1994; 79(1): 29A.

For designating isotopes use ^{y}A to designate element A whose atomic mass is y.

To designate compounds containing carbon: use C_x to refer to a compound containing x carbon atoms; use C-x is to refer to the xth carbon atom of a molecule.

To designate isotopically labeled chemical compounds: for a chemical formula, insert symbol for isotope into formula (e.g., $NaH^{14}CO_3$); for a chemical name or abbreviation, follow form example: for glucose labeled in C-1 position use $[I-^{14}C]$-glucose; for thyroxine (T) labeled with ^{131}I use $[^{131}I]T$.

Express potency estimates for protein and polypeptide hormones in terms of reference preparations such as International Standards, NIDDKD preparations, etc. Include nature of reference preparation, its species and tissue of origin, and evidence for its similarity to hormonal activity being measured. If a suitable recognized standard does not exist, fully describe preparation and standardization of reference material used. Express results of bioassays for FSH and LH, and of radioimmunoassays for these hormones in human urine and urinary extracts in International Units of the Second International Reference Preparation of Human Menopausal Gonadotropins. For human materials other than urine, express results of radioimmunoassays for FSH or LH in $\mu g/L$ NPA LER 907 reference preparation or in IU/L. State method of assay and standard employed under Materials and Methods and in other places where appropriate.

Page and Other Charges: $35 per printed page and additional charges for color illustrations, as well as excessive or other unusual illustrative material. On appeal by author, Publications Committee may waive page or other charges.

Submit a manuscript processing fee of $70 with each typewritten manuscript accompanied by a computer diskette containing the complete text of the manuscript (most word processing languages are acceptable). For manuscripts without a computer diskette, the processing fee is $170. Diskettes may be submitted with the final, revised version of manuscript. Manuscripts will not be considered for publication until fee is received. The fee also is required if a previously rejected manuscript is resubmitted. Requirements for computer disk do not apply to Rapid Communications; submission fee is $60.

Statistical Analyses: Document that results are reproducible and that differences are not due to random variation. Use appropriate statistical methods to test significance of differences in results. Do not use term "significant" unless statistical analysis was performed; specify probability value used to identify significance(e.g., $P < 0.05$).

When several t tests are employed, nominal probability levels no longer apply. Use multiple t test, multiple range test, or similar techniques to permit simultaneous comparisons. Instead of several t tests, use an analysis of variance (ANOVA) to permit pooling of data, increase number of degrees of freedom, and improve reliability of results. Use appropriate nonparametric tests when data depart substantially from a normal distribution.

Do not include analysis of variance tables. F values with degrees of freedom as subscripts together with P values are sufficient.

In presenting results of linear regression analyses, show 95% confidence limits.

Style: Consult: *CBE Style Manual: A Guide for Authors, Editors and Publishers* (5th ed. Bethesda., MD: Council of Biology Editors; 1983).

Validation of Data, Assays: For bioassay and radioimmunoassay potency estimates include an appropriate measure of precision of estimates; for bioassays, usually standard deviation, standard error of the mean, coefficient of variation, or 95% confidence limits. In both bioassays and radioimmunoassays, include data relating to within-assay and between-assay variability. If all relevant comparisons are made within the same assay, omit latter.

In presenting results for new assays, include data on: within-assay variability; between-assay variability; slope of the dose-response curve; mid-range of the assay; least-detectable concentration (concentration resulting in a response two standard deviations away from the zero dose response); data on specificity; data on parallelism of standard and unknown and on recovery; and comparison with an independent method for assay of compound. When radioimmunoassay kits are used or hormone measurements conducted in other laboratories and assay is central to study, include data regarding performance characteristics.

Validation of Data, Pulse Analysis: Analyze data from studies of pulsatile hormone secretion using a validated, objective pulse detection algorithm. Algorithm should require that false-positive rates of pulse detection be defined in relation to the measurement error of data set being analyzed; describe methods used to determine measurement error. Describe methods used: to deal with missing or undetectable values; to determine peak frequency, interpeak interval, and pulse amplitude; and for statistical comparisons of peak parameters.

Units: Express results in Système International (SI) units throughout. Metric units may be added in parentheses. For a complete listing of SI units, see Lundberg GD. Iverson C, Radelescu. *JAMA*. 1986;255:2239–39 or Young DS. *Ann Intern Med*. 1987;106:114–29. Express temperature in degrees Celsius and time of day using 24 hour clock (e.g., 0800 h, 1500 h).

SOURCE

July 1994, 79(1): 23A-29A.

Journal of Clinical Investigation
American Society for Clinical Investigation, Inc.
The Rockefeller University Press, New York

Ajit P. Varki
UCSD School of Medicine
VA Medical Center, Room 5142
3350 La Jolla Village Drive
San Diego, CA 92161
        Courier service
P. O. Box 85182
San Diego, CA 92186-5182
        Regular mail

AIM AND SCOPE

The Journal of Clinical Investigation publishes original papers on research pertinent to human biology/physiology and mechanisms of disease. Work with animals or in vitro techniques is acceptable if it is relevant to normal or abnormal human biology.

MANUSCRIPT SUBMISSION

No substantial part of the paper may have been or may be published elsewhere. Single, complete papers are preferred to a series of interdependent ones. Submit Manuscript Submission Form (see July 1994 issue) with one original manuscript (photocopies of figures appended), plus one reproduction set of figures. Send three review copies of manuscript with photocopies of figures attached. By signing Manuscript Submission Form, corresponding author verifies that all authors approve of material submitted for publication, that material has not been previously reported, and that it is not under consideration for publication elsewhere.

Submit copies of any related preliminary report other than an abstract of 400 words or less has been published or submitted. Submit copies of closely related papers that might be considered as duplicate publication; mark as such.

Disclose in a cover letter any commercial affiliations as well as consultancies, stock or equity interests, and patent-licensing arrangements that could be considered to pose a conflict of interest. Specifics of disclosures remain confidential. Statements in acknowledgments regarding such disclosures may be recommended by editors. Credit all funding sources, institutional and corporate, in Acknowledgments section.

MANUSCRIPT FORMAT

Manuscripts should be as concise as possible and should ordinarily not exceed seven to eight printed pages including references, tables, and figures.

Determine length of manuscript by total number of characters. To determine approximate printed length: [(average number of characters/line) × (average number of lines/page) × (number of pages)] ÷ (7,320 characters per printed page) = (number of printed pages). Each line should contain no more than 72 characters, including spaces. Each double-spaced page should contain no more than 25 typed lines. Count each line drawing or table as 37 lines or 2,500 characters. To shorten manuscript, incorporate material into text rather than present in figure or tabular form. Electron micrographs and color figures will not be reduced. Count according to page space actually occupied.

Regular submissions should be no longer than 30–35 pages, Rapid Publications, 20–22 pages.

Type on 8.5 × 11 in. paper (no heavier than 20 lb) with 1 in. margins except title page (2 in. top margin). Double-space entire manuscript (including footnotes, references, tables, and figure legends). Do not staple original and one copy; staple two review copies.

Organize each section beginning on a new page in order: Title page, Abstract (key words listed beneath). Introduction, Methods, Results, Discussion, Acknowledgments, Footnotes, References, Tables, figure legends, photocopies of figures, and original figures. Number all pages in bottom right corner beginning with title page.

Title Page: List title, first and last name(s) of author(s), and department(s), institution(s), city, state, and zip code where work was performed. Include institutional affiliation of each author, if different from where work was performed. Furnish a condensed title (72 characters and spaces, no abbreviations) as a running title (if main title fits size requirement, list again as running title). Refer to a preliminary report or abstract in title footnote. State complete name, address, telephone, and fax number of corresponding author.

Title: Mention if observations have been made on animals: all other titles refer to studies on humans or in tissues derived from humans. If study was conducted in human tissue, add an appropriate qualifying phrase, such as "in vitro," "in cultured fibroblasts," "in tissues slices," etc. Do not use nonstandard abbreviations or chemical formula.

Abstract: 200 words or 1,500 characters. State rationale, objectives, newly observed findings, and conclusions drawn from observations. Write for the readership of a general journal. Make self-explanatory without recourse to text. Do not use footnotes, nonstandard abbreviations, or references.

Key Words: At bottom of abstract page list five key words not included in title. Use standard MeSH-Medline major subject headings. See *Medical Subject Headings: Supplement to Index Medicus* that contains these headings.

Acknowledgments: State source(s) of support in the form of grants, equipment, or drugs on separate Acknowledgments page. Include other appropriate acknowledgments, e.g., to other scientists for their help or advice, here.

Footnotes: Number footnotes to text consecutively. Assign footnote symbols to table and to author/affiliation sections by superscript symbols in order: *, ‡, §, ||, ¶, **, ‡‡.

References: Type double-spaced, beginning on a separate page. Number in order of appearance in text. Include all authors' names (do not use "et al."), complete article titles and indicate articles in press. Abbreviate journal names according to *Serial Sources for the Biosis Data Base* (BIOSIS, Philadelphia, PA). Spell out names of unlisted journals. Supply inclusive page numbers. Refer to manuscripts "submitted," "in preparation," "unpublished observations," and "personal communications" parenthetically in text; list as footnotes. Obtain and submit written permission when citing a personal communication or unpublished observation. Submit copies of articles cited as in press identified by reference number.

JOURNAL ARTICLES

1. Yalow, R. S., and S. A. Berson. 1960. Immunoassay of endogenous plasma insulin in man. *J. Clin. Invest.* 39:1157–1175.

MORE THAN TWO AUTHORS

2. Hobbs, H. H., E. Leitersdorf, C. C. Leffert, D. R. Cryer, M. S. Brown, and J. L. Gofldstein. 1989. Evidence for a dominant gene that suppresses hypercholesterolemia in a family with defective low density lipoprotein disorders. *J. Clin. Invest.* 84:656–664.

(Initials of first author follow surname, all other initials precede last name.)

IN PRESS

3. Gardner, W., and H. D. Schultz. 1990. Prostaglandins regulate the synthesis and secretion of the atrial natriuretic peptide. *J. Clin. Invest.* In press.

COMPLETE BOOKS

4. Myant. N. B. 1981. The Biology of Cholesterol and Related Steroids. Heinemann Medical Books. London. 882 pp.

ARTICLES IN BOOKS

5. Innerarity, T. L., D. Y. Hui, and R. W, Mahley. 1982. Hepatic apoprotein E (remnant) receptor. *In* Lipoproteins and Coronary Atherosclerosis. G. Noseda, S. Fragiacomo, R. Fumagalli, and R. Paoletti, editors. Elsevier/North Holland, Amsterdam. 173–181.

ABSTRACT

6. Packman, C. H., S. I. Rosenfeld, and J. P. Leddy. 1981. Inhibition of the C8/C9 steps of compliment lysis by a high density lipoprotein (HDL) of human serum. *Fed. Proc.* 40:967a. (Abstr.)

Tables: Double-space on separate pages. Make self-contained and self-explanatory. Provide titles. Use superscript symbols for footnotes (i. e., *, ‡, §, ||, ¶, **, ‡‡). Give sufficient explanation and detail in legend to make experiment or procedure intelligible without reference to text. Do not use vertical lines; do not submit photographs of tables.

Submit tables with different material in box heads as individual tables, not as sections of one table.

Figure Legends: Figure 1. Provide a short title. Give sufficient detail to make the whole intelligible without reference to text (unless a similar explanation has been given in another figure). Provide a key to symbols.

Figures: Cite in numerical order in text using Arabic numerals. Give title in figure legend, not on figure. Do not exceed 8.5 × 11 in. Submit one complete reproduction-quality set of original figures, unmounted and in an envelope clearly labeled "For Reproduction." High quality laser prints are acceptable as original line drawings. Original artwork and glossy reflective photographs are acceptable for all other types of figures meant for reproduction as original figures. Photocopies of figures will be used for review; attach a set to each copy of manuscript including original. If some photocopies are inadequate for review, provide three glossy sets of those figures, each sorted set in a separate envelope marked "For Review," with first author's name and first three words of title. Affix a label indicating figure number, name of first author, and an arrow indicating TOP to back of all original laser prints and glossy figures. Do not write in ink, use paper clips, or do anything that might mar figures. Clearly label parts (e.g., A, B) on face of figure. Do not send oversized or valuable original artwork with initial submission.

Line Drawings—For reproduction set, submit glossy prints or high quality laser prints. Graphs and other line drawings are reduced to single column width or smaller. Symbols, lettering and numbers must allow for reduction to an 8.25 cm width. After reduction capital letters should be 1.5 to 1.75 mm high; lowercase and numbers should be 1-15 mm high; symbols should be 3-4 times wider than curve lines. Use only standard symbols (○ ● □ ■ △ ▲). Make original graphs 2-3 times larger than reduced size. Use units of measure and abbreviations consistent with text. Do not give figures titles. Put symbol keys in legends not on figures. Put zeros in front of decimal points and commas in numbers with 4 or more digits. Include "air space" between solid curve and symbols. Include brake points where curves intersect. Use upper and lowercase letters in axis labels. Evenly space dashes in dashed curve.

Photographs (electron micrographs, gels)—Gels are reduced to single-column width or smaller. Figures with one or several gel lanes are reduced to half-column with legend on the side. Multiple vertical lanes (up to 15) will be reduced to a single column, plan labeling so that upon reduction height of numbers and capital letters will be 1.75 mm, lane width 5 mm. Figures with more than 15 lanes will be reduced to 1.5-2 column width.

Halftone Illustrations—Unless use of additional space is justified, halftones will be reduced to one column width. Crop out blank background areas and repetitive or irrelevant areas. Place tissue overlay over each micrograph; circle areas that must be most faithfully reproduced.

Color Photographs—Reproduced at actual size; do not exceed 17.5 cm width. Prints or transparencies may be used, not slides. Do not mount on flexible board. Cost will be borne by author.

Alternatives to Glossy Photographs for Review—Use a color xerographic copier to reproduce color or monochrome photomicrographs/gels; use a scanner to reproduce these figures onto quality copy paper; a plain paper facsimile machine usually can copy photographs by setting print quality to "halftone" and using fax "copy" mode.

Other Formats

RAPID PUBLICATIONS—Five printed pages of unusual scientific importance that represent definitive and original studies. Request Rapid Publication processing on submission form.

Other

Abbreviations: Use standard abbreviations without definition. See abbreviation list in July 1994 issue. List all other abbreviations used in a footnote. Do not use undefined abbreviations. Do not abbreviate terms used less than three times in text. Abbreviate units of measure only when used with numbers. Use metric system and Système International d'Unités (SI) as noted in the JCI abbreviations list in July 1994 issue. For biochemical abbreviations, use those currently recommended by *The Journal of Biological Chemistry* in instructions to authors.

Color Charges: $1,600 per page of color. Several photographs may be printed per page. Each additional page, in the same signature, costs $60.00 Each additional color separation costs $175. Publisher will provide more exact estimate after figures are received. If unable or unwilling to bear cost, figures can usually be printed as half-tone black-and-white photographs. Production Office must receive written agreement to cover cost before production begins.

Data Submission: Submit original nucleotide or amino acid sequence data to GenBank. A statement to this effect and an accession number will appear as a footnote at end of relevant figure legend. Obtain instructions on how to submit data from GenBank, Group T-10, Mail Stop K710, Los Alamos National Laboratory, Los Alamos, NM 87545; telephone: (505) 665-2177.

Gene Names: Mark actual gene names for typesetting in italics. Italicize only actual gene names, to distinguish from gene products of same or similar name; ad hoc designations for genes; gene segments; and gene clusters, families, complexes, or groups. Italicize genotypes, not phenotypes.

Experimentation, Animal and Human: Have all human and animal studies formally approved by Institutional Review Board. Clinical investigations must be conducted according to principles in the Declaration of Helsinki (see *J. Clin. Invest.* 46:1140).

Materials Sharing: Cell lines, DNA clones, hybridomas, and biological reagents described in published articles are to be made readily available to qualified scientists interested in obtaining them.

Page Charges: $125 for each page over eight.

Processing Fee: Submit completed submission form with $50 check or purchase order sent under separate cover. Manuscripts will not be sent for review until this is received. Invited revisions do not require a fee.

Reagents: Format for reporting: RPMI 1640, insulin-transferrin-selenite media supplement, mevalonic acid lactone, N^6-[isopentenyl]adenine, and Farnesol were obtained from Sigma Chemical Co., St. Louis, MO. Trypsin-EDTA and collagenase were from Gibco Laboratories, Grand Island, NY. Fetal bovine serum was supplied by Bioproducts for Science, Inc., Indianapolis, IN. Lovastatin was graciously provided by Merck Sharp and Dohme, Inc., West Point, PA.

Reviewers: Provide names, addresses, phone and fax numbers of up to five potential reviewers. These should not be recent collaborators or coauthors (within last three years), nor should they have provided substantial advice or critiques to authors in conduct of work. List reviewers that authors would prefer to exclude with reason.

Statistical Analyses: Specify estimates of variance (e.g., SE, SD) and give references for statistical methods used.

Style: Follow conventions set forth in *Council of Biology Editors Style Manual* (5th edition, Council of Biology Editors, Inc., Bethesda, MD, 1983). Capitalize trade names and provide manufacturers' names and locations. Give formal, chemical name of a compound as established by international convention following first use of trivial name.

List chemicals and scientific instruments used and their respective manufacturers' locations.

Source

July 1994, 94(1): i-iv.

Journal of Clinical Microbiology
American Society for Microbiology

Journals Division
American Society for Microbiology

1325 Massachusetts Ave., N.W.
Washington, D.C. 20005-4171

AIM AND SCOPE

The *Journal of Clinical Microbiology* is devoted to the dissemination of new knowledge concerning the microbiological aspects of human and animal infections and infestations, particularly their etiological agents, diagnosis, and epidemiology. Case reports will be considered if they are novel, add to existing knowledge, and are oriented toward microbiology. Manuscripts which describe the "normal" microbiota of humans which in turn become involved in disease production or complication, or manuscripts dealing with the interactions of hospitalized patients and the microbial environment of the hospital, may also be submitted for consideration.

ASM publishes a number of different journals covering various aspects of the field of microbiology. Each journal has a prescribed scope which must be considered in determining the most appropriate journal for each manuscript. The *Journal of Clinical Microbiology* will consider manuscripts: that describe the use of antimicrobial, antiparasitic, or anticancer agents as tools in the isolation, identification, or epidemiology of microorganisms associated with disease; that are concerned with quality control procedures for diffusion, elution, or dilution tests for determining susceptibilities to antimicrobial agents in clinical laboratories; and that deal with applications of commercially prepared tests or kits to assays performed in clinical laboratories to measure the activities of established antimicrobial agents or their concentrations in body fluids.

The *Journal of Clinical Microbiology* will consider manuscripts dealing with the isolation or identification of viral agents from humans and animals, with viral pathogenesis and immunity, and with the etiology and diagnosis of viral diseases. In addition, epidemiological studies of viral diseases or those involving the use of bacteriophages as a typing system or to identify bacteria will be considered.

Reports of clinical microbiology investigations or studies of the hospital population and the environment as they relate to nosocomial infections should be submitted to the *Journal of Clinical Microbiology*.

Note that rejection by one ASM journal on scientific grounds or on the basis of its general suitability for publication is considered rejected by all other ASM journals.

MANUSCRIPT SUBMISSION

Submit a cover letter stating: journal to which manuscript is being submitted, most appropriate journal section, complete mailing address (including street), telephone and fax numbers of corresponding author, an electronic mail address if available, and former ASM manuscript number and year if a resubmission. Include written assurance that permission to cite personal communications and preprints has been granted.

Submit three complete copies of each manuscript, including figures and tables, Type every portion double-spaced (6 mm minimum between lines), including figure legends, table footnotes, and References. Number all pages in sequence, including abstract, figure legends, and tables.

Place last two after References section. Keep 1 in. margins on all four sides and line number if possible. Make following character sets easily distinguishable: numeral zero (0) and letter "oh" (O); numeral one (1), letter "el" (l), and letter "eye" (I); and multiplication sign ($\times$) and letter "ex" (x). If distinctions cannot be made, mark items at first occurrence for cell lines, strain and genetic designations, viruses, etc., on modified manuscript.

Enclose three copies of each "in press" and "submitted" of importance for judging present manuscript.

Manuscripts must represent reports of original research. All authors must have agreed to submission and are responsible for content, including appropriate citations and acknowledgments, and must also have agreed that corresponding author has the authority to act on their behalf on all matters pertaining to publication. By submission of a manuscript, authors guarantee that manuscript, or one substantially the same, was not published previously, is not being considered or published elsewhere, and was not rejected on scientific grounds by another ASM journal.

The American Society for Microbiology accepts the definition of primary publication in *How to Write and Publish a Scientific Paper* (third ed, Robert A. Day) to wit: "... (i) the first publication of original research results, (ii) in a form whereby peers can repeat the experiments and test the conclusions, and (iii) in a journal or other source document readily available within the scientific community."

A scientific paper published in a conference report, symposium proceeding, technical bulletin, or any other retrievable source is unacceptable on grounds of prior publication. Preliminary disclosure of research findings in abstract form as an adjunct to a meeting, e.g., part of a program, is not considered "prior publication." Acknowledge prior publication of data in manuscript even if you do not consider it in violation of ASM policy. Include copies of relevant work.

Obtain and submit signed permissions from both original publisher and original author [i.e., the copyright owner(s)] to reproduce figures, tables, or text (in whole or in part) from previous publications. Identify each as to relevant item in ASM manuscript (e.g., "permissions for Fig. 1 in JCM 123-93").

An author is one who made a substantial contribution to the "overall design and execution of the experiments"; all authors are responsible for the entire paper. Do not list individuals who provided assistance, e.g., supplied strains or reagents or critiqued paper, as authors; recognize them in Acknowledgment section.

Agree to order in which author names are listed in byline. Footnotes regarding attribution of work (e.g., X. Jones and Y. Smith contributed equally to. . . .) are not permitted; include in Acknowledgment section.

Corresponding author must sign copyright transfer agreement on behalf of all authors upon acceptance of article. Unless this agreement is executed (without changes and/or addenda), ASM will not publish manuscript.

If all authors were U.S. government employees when work was performed, corresponding author should not sign copyright transfer agreement but should, instead attach to agreement a statement attesting that the manuscript was prepared as a part of their official duties and, as such, is a work of the U.S. government not subject to copyright.

If some authors were U.S. government employees when work was performed but others were not, corresponding author should sign copyright transfer agreement as it applies to that portion performed by non-government employee authors.

MANUSCRIPT FORMAT

Title, Running Title, and Byline: Present results of independent, cohesive studies; numbered series titles are not permitted. Avoid main title/subtitle arrangement, complete sentences, and unnecessary articles. On title page include title, running title (54 characters and spaces), name of each author, address(es) of institution(s) at which work was performed, each author's affiliation, and a footnote indicating present address(es) of any author(s) no longer at institution where work was done. Place an asterisk after name of corresponding author with telephone and fax numbers and electronic mail address.

Correspondent Footnote: If desired, complete mailing address, telephone and fax numbers, and e-mail address of corresponding author will be published as a footnote. Include in lower left corner of manuscript title page labeled as "Correspondent Footnote."

Abstract: 250 words concisely summarizing the paper's basic content. Do not present extensive experimental details. Avoid abbreviations; do not include diagrams. When essential to include a reference, use Reference citation but omit article title. Conclude with a summary statement. Make complete and understandable without reference to text.

Introduction: Supply sufficient background information to allow reader to understand and evaluate results without referring to previous publications on topic. Provide rationale for present study. Choose references carefully to provide the most salient background rather than an exhaustive review of topic.

Materials and Methods: Include sufficient technical information to allow experiments to be repeated. Provide sources of all media (i.e., name and location of manufacturer) or components of a new formulation. When centrifugation conditions are critical, give enough information to enable another investigator to repeat procedure: make of centrifuge, model of rotor, temperature, time at maximum speed, and centrifugal force ($\times g$ rather than revolutions per minute). For commonly used materials and methods), a simple reference or specifically recommended product or procedure is sufficient. If several alternative methods are commonly used, identify method briefly and cite reference. Reader should be able to assess the method without constant reference to previous publications. Describe new methods completely. Give sources of unusual chemicals, reagents, equipment, or microbial strains. For large numbers of microbial strains or mutants, include tables identifying sources and properties of strains, mutants, bacteriophages, plasmids, etc.

Describe a method, strain, etc., used in only one of several experiments in Results section or very briefly (one or two sentences) in a table footnote or figure legend.

Results: Include rationale or design of experiments and results; reserve extensive interpretation of results for Discussion section. Present results as concisely as possible in either: text, table(s), or figure(s). Avoid extensive use of graphs to present data which might be more concisely presented in the text or tables. All tabular data must be accompanied by either standard deviation values or standard errors of the means. Include number of replicate determinations (or animals) used for making calculations. for all statements concerning significance of differences observed, include probability values in parentheses. State statistical procedure used in Materials and Methods. Limit illustrations (particularly photomicrographs and electron micrographs) to those that are absolutely necessary to show experimental findings. Number figures and tables in order cited in text; cite all figures and tables.

Discussion: Provide an interpretation of results in relation to previously published work and to experimental system at hand. Do not repeat Results section or reiterate Introduction. In short papers, Results and Discussion sections may be combined.

Acknowledgments: Give acknowledgments of financial assistance and of personal assistance in separate paragraphs. Format for acknowledgment of grant support: "This work was supported by Public Health Service grant CA01234 from the National Cancer Institute."

Appendixes: Appendixes contain supplementary material to aid readers. Do not submit titles, authors, and References sections different from primary article. If not feasible to list author(s) of appendix in byline or Acknowledgment section of primary article, rewrite appendix so it can be considered for publication as an independent article, either full-length or Note style. Label equations, tables, and figures with letter "A" preceding numeral to distinguish from those cited in main body of text.

References: Include all relevant sources. All listed references must be cited in text. Arrange citations in alphabetical order by first author and number consecutively. Abbreviate journal names according to *Serial Sources for the BIOSIS Data Base* (BioSciences Information Service, 1992). Cite each listed reference by number in text.

1. **Agger, W. A., and D. G. Maki.** 1978. Efficacy of direct Gram stain in differentiating staphylococci from streptococci in blood cultures positive for gram-positive cocci. J. Clin. Microbiol. **7:**111–113.

2. **Berry, L. J., R. N. Moore, K. J. Goodrum, and R. E. Couch, Jr.** 1977. Cellular requirements for enzyme inhibition by endotoxin in mice, p. 321–325. *In* D. Schlessinger (ed.), Microbiology—1977. American Society for Microbiology, Washington, D.C.

3. **Cox, C. S., B. R. Brown, and J. C. Smith.** J. Gen. Genet., in press.*

4. **Dhople, A., I. Ortega, and C. Berauer.** 1989. Effect of oxygen on in vitro growth of *Mycobacterium lep-*

rae, abstr. U-82 p. 168. Abstr. 89th Annu. Meet. Am. Soc. Microbiol. 1989.

5. **Finegold, S. M., W. E. Shepherd, and E. H. Spaulding.** 1977. Cumitech 5, Practical anaerobic bacteriology. Coordinating ed., W. E. Shepherd. American Society for Microbiology, Washington, D.C.

6. **Fitzgerald, G., and D. Shaw.** *In* A. E. Waters (ed.), Clinical microbiology, in press. EFH Publishing Co., Boston .

7. **Gill, T. J., III.** 1976. Principles of radioimmunoassay, p. 169–171. *In* N. R. Rose and H. Friedman (ed.), Manual of clinical immunology. American Society for Microbiology, Washington, D.C.

8. **Gustlethwaite, F. P.** 1985. Letter. Lancet **ii**:327.

9. **Jacoby, J., R. Grimm, J. Bostic, V. Dean, and G. Starke.** Submitted for publication.

10. **Jensen, C., and D. S. Schumacher.** Unpublished data.

11. **Jones, A. (Yale University).** 1990. Personal communication .

12. **Leadbetter, E. R.** 1974. Order II. *Cytophagales* nomen novum, p. 99. *In* R. E. Buchanan and N. E. Gibbons (ed.), Bergey's manual of determinative bacteriology, 8th ed. The Williams & Wilkins Co., Baltimore.

13. **Miller, J. H.** 1972. Experiments in molecular genetics, p. 352–355. Cold Spring Harbor Laboratory, Cold Spring Harbor, N.Y.

14. **Powers, R. D., W. M. Dotson, Jr., and F. G. Hayden.** 1982. Program Abstr. 22nd Intersci. Conf. Antimicrob. Agents Chemother., abstr. 448.

15. **Sacks, L. E.** 1973. Influence of cations on the germination of spores, p, 437–442. *In* H. O. Halvorson, R. Hanson, and L. L. Campbell (ed.), Spores V. American Society for Microbiology, Washington, D.C.

16. **Sigma Chemical Co.** 1989. Sigma manual. Sigma Chemical Co., St. Louis, Mo.

17. **Smith, J. C.** April 1970. U.S. patent 484,363,770.

18. **Smyth, D. R.** 1972. Ph.D. thesis. University of California, Los Angeles.

19. **Yagupsky, P., and M. A. Menegus.** 1989. Intraluminal colonization as a source of catheter-related infection. Antimicrob. Agents Chemother. **33**:2025. (Letter.)

For "in press" references to ASM publication, state control number or name of publication if it is a book.

Illustrations: Write figure number and authors' names on all figures, either in margin or on back (marked lightly with a soft pencil). For micrographs especially, indicate TOP.

Do not clasp figures to each other or to manuscript with paper clips. Insert small figures in an envelope. Do not submit illustrations larger than 8.5 × 11 in.

Published illustrations will not be returned.

Continuous-Tone and Composite Photographs—Keep journal page width in mind: 3 5/16 in. single column and 6 7/8 inches double-column (maximum). Include only significant portions of illustration. Photos must be of sufficient contrast to withstand inevitable loss of contrast and detail inherent in printing. Submit one photograph of each continuous-tone figure for each copy of manuscript; photocopies are not acceptable. Submit figures in final size so no reduction is necessary. If they must be reduced, make sure all elements, including labeling, can withstand reduction and remain legible.

If a figure is a composite of a continuous-tone photograph and a drawing or labeling, provide original composite for printer. Send with modified manuscript to editor.

Electron and light micrographs must be direct copies of original negative. Indicate magnification with a scale marker on each micrograph.

Computer-Generated Images—Produce computer-generated images with Adobe Photoshop or Aldus Freehand software. For Aldus, one and two-column art cannot exceed 20 picas (3 5/16 in.) and 41.5 picas (6 7/8 in.). Text should be either Helvetica (medium or bold) or Times Roman. Adobe users check densities of images on-line. If image's shadow density reads below 1.25, enter density as 1.40; if 1.25-1.60, enter as 1.65; if >1.65, enter as actual density reading. Submit computer files (Macintosh floppy disk) with prints. For large images, use 40 or 80-megabyte Syquest cartridges or magneto-optical cartridges. For transfer from UNIX systems, submit either 9-track or 8-mm "tar" archives. Incorporate all final lettering, labeling, tooling, etc., in final supplied material. It cannot be added later. Do not include figure numbers on images. Electronic image hotline: William Byrd Press, Richmond, VA (800) 888-2973, ext 3361, outside U.S. (804) 264-2828, ext 3361).

Describe software/hardware used in figure legend(s).

Color Photographs—Use is discouraged. If necessary, include an extra copy at submission for a cost estimate. Cost must be borne by author.

Drawings—Submit graphs, charts, complicated chemical or mathematical formulas, diagrams, and other drawings as glossy photographs made from finished drawings not requiring additional artwork or typesetting. Computer-generated graphics produced on high-quality laser printers are acceptable. No part of the graph or drawing should be handwritten. Label both axes of a graph. Graphs will be reduced to one-column width (3 5/16 in.); make all elements large enough to withstand reduction. Avoid heavy letters and unusual symbols.

In figure ordinate and abscissa scales (as well as table column headings), avoid ambiguous use of numbers with exponents. Use appropriate SI symbols (μ for 10^{-6}, m for 10^{-3}, k for 10^3, M for 10^6, etc.). For a complete listing of SI symbol, see International Union of Pure and Applied Chemistry's "Manual of Symbols and Terminology for Physicochemical Quantities and Units" (*Pure Appl. Chem.* **21**:3–44, 1970).

When powers of 10 must be used, show exponent power associated with number. In representing 20,000 cells per ml, numeral of ordinate is "2" and label is "10^4 cells per ml" (not "cells per ml $\times 10^{-4}$"). Show an enzyme activity of 0.06 U/ml as 6, with label 10^{-2} U/ml. Preferred designation is 60 mU/ml labeled as mU (or milliunits) per ml.

Figure Legends: Provide enough information so that figure is understandable without reference to text. Describe detailed experimental methods in Materials and Methods section. Report a method unique to one experiment in legend in one or two sentences. Define all symbols and abbreviations in figure that have not been defined elsewhere.

Tables: Type each on a separate page. Arrange data so columns of like material read down, not across. Make headings sufficiently clear so meaning of data will be understandable without reference to text. Explanatory footnotes are acceptable, extensive table "legends" are not; do not include detailed descriptions of experiment. Include enough information to warrant table format; incorporate those with fewer than six pieces of data into text.

Tables that can be photographically reproduced for publication without further typesetting or artwork are "camera ready." Do not hand letter; prepare carefully to conform to journal style.

OTHER FORMATS

AUTHOR'S CORRECTIONS—Provides a means of adding citations that were overlooked in a published article. The author who failed to cite a reference and the author whose paper was not cited must agree to publication. Submit letters from both authors with corrections to the Journals Division.

ERRATA—Provides a means of correcting errors (e.g., typographical) in published articles. Changes in data and the addition of new material are not permitted.

LETTERS TO THE EDITOR—500 word comments on articles published in the Journal; include data to support argument. If published, a reply will be solicited from the corresponding author of article. Final approval for publication rests with the editor-in-chief. Type double-spaced.

MINIREVIEWS—Four printed pages; brief summaries of developments in fast-moving areas based on published articles addressing any subject within the scope of JCM. May be solicited or proffered by authors. Provide three double-spaced copies.

NOTES—1000 words with minimum tables and figures. Submit as full-length papers. Format is intended for the presentation of brief observations that do not warrant full-length papers. Notes contain firm data; observations alone are not acceptable.

Include 50 word abstract. Do not use section headings; report methods, results, and discussion in a single section. Paragraph lead-ins are permissible. Describe materials and methods in text, not in figure legends or table footnotes. Present acknowledgments as in full-length papers, but with no separate heading. References section is identical.

OTHER

Abbreviations: Use as an aid to reader, not a convenience for author; limit use. Use abbreviations other than those recommended by the IUPAC-IUB (*Biochemical Nomenclature and Related Documents*, 1978) only when necessity, such as in tables and figures.

Use pronouns or paraphrase a long word after its first use (e.g., "the drug," "the substrate"). Use standard chemical symbols and trivial names or their symbols for terms that appear in full in neighboring text.

Introduce all abbreviations except those listed below in first paragraph in Materials and Methods. Or, define each abbreviation and introduce it in parentheses at first use. Eliminate abbreviations used at least five times in text, tables, and figure legends.

Système International d'Unités (SI) units of measurement, other common units, chemical symbols for elements, and the following may be used without definition in title, abstract, text, figure legends, and tables: DNA (deoxyribonucleic acid); cDNA (complementary DNA); RNA (ribonucleic acid); cRNA (complementary RNA); RNase (ribonuclease); DNase (deoxyribonuclease); rRNA (ribosomal RNA); mRNA (messenger RNA); tRNA (transfer RNA); AMP, ADP, ATP, dAMP, ddATP, GTP, etc. (for the respective 5′ phosphates of adenosine and other nucleosides) (add 2′-, 3′-, or 5′- when needed for contrast); ATPase, dGTPase, etc. (adenosine triphosphatase, deoxyguanosine triphosphatase, etc.); NAD (nicotinamide adenine dinucleotide), NAD+ (nicotinamide adenine dinucleotide, oxidized); NADH (nicotinamide adenine dinucleotide, reduced); NADP (nicotinamide adenine dinucleotide phosphate); NADPH (nicotinamide adenine dinucleotide phosphate, reduced); NADP+ (nicotinamide adenine dinucleotide phosphate, oxidized); poly(A), poly(dT), etc. (polyadenylic acid. polydeoxythymidylic acid, etc.); oligo(dT), etc. (oligodeoxythymidylic acid, etc.); P_i (orthophosphate); PP_i (pyrophosphate); UV (ultraviolet); PFU (plaque-forming units); CFU (colony-forming units); MIC (minimal inhibitory concentration); MBC (minimal bactericidal concentration); Tris [tris(hydroxymethyl)aminomethane]; DEAE (diethylaminoethyl); A_{260} (absorbance at 260 nm); EDTA (ethylenediaminetetraacetic acid); and AIDS (acquired immunodeficiency [or immune deficiency] syndrome). Abbreviations for cell lines (e.g., HeLa) also need not be defined.

The following abbreviations should be used without definition in tables:

amt (amount)	SD (standard deviation)
approx (approximately	SE (standard error)
avg (average	SEM (standard error of the mean)
concn (concentration)	sp act (specific activity)
diam (diameter)	sp gr (specific gravity)
expt (experiment)	temp (temperature)
exptl (experimental)	tr (trace)
ht (height)	vol (volume)
mo (month)	vs (versus)
mol wt (molecular weight)	wk (week)
no. (number)	wt (weight)
prepn (preparation)	yr (year)

Drugs: Use generic names drugs; trade names are not permitted.

Experimentation, Human: Do not identify individuals by initials, even as part of a strain designation. Change initials to numerals or use randomly chosen letters. Do not give hospital unit numbers; if a designation is needed, use only last two digits. Established designations of viruses and cell lines with initials, are acceptable.

Isotopically Labeled Compounds: For simple molecules, indicate labeling in chemical formula (e.g., $^{14}CO_2$, ^3H_2O, $H_2^{35}SO_4$). Do not use brackets when isotopic symbol is attached to name of a compound that in its natural state does not contain the element or to a word that is not a specific chemical name.

For specific chemicals, place symbol for isotope introduced in square brackets directly preceding part of name that describes labeled entity. Configuration symbols and modifiers precede isotopic symbol. Follow conventions for isotopic labeling in *Journal of Biological Chemistry*; more detailed information can be found in instructions to authors of that journal (first issue of each year).

Materials Sharing: Deposit important strains in publicly accessible culture collections; refer to collections and strain numbers in text. Indicate laboratory strain designations and donor source as well as original culture collection identification numbers.

Make plasmids, viruses, and living materials such as microbial strains and cell lines described in article available from a national collection or in a timely fashion and at reasonable cost to members of the scientific community for noncommercial purposes.

Nomenclature, Chemical and Biochemical: The authority for names of chemical compounds is *Chemical Abstracts* (Chemical Abstract Service, Ohio State University, Columbus) and its indexes. *The Merck Index* (11th ed., 1989; Merck & Co., Inc., Rahway, N.J.) is also an excellent source. For biochemical terminology, including abbreviations and symbols, consult *Biochemical Nomenclature and Related Documents* (1978; The Biochemical Society, London) and instructions to authors of *Journal of Biological Chemistry* and *Archives of Biochemistry and Biophysics* (first issues of each year).

Do not express molecular weight in daltons; molecular weight is a unitless ratio. Molecular mass is expressed in daltons.

For enzymes, use recommended (trivial) name assigned by Nomenclature Committee of International Union of Biochemistry as described in *Enzyme Nomenclature* (Academic Press, Inc., 1992). If a nonrecommended name is used, place proper (trivial) name in parentheses at first use in abstract and text. Use EC number, if assigned; express enzyme activity in katals (preferred) or micromoles per minute.

Nomenclature, Genetic:

Bacteria—Genetic properties of bacteria are described by phenotype and genotype. Phenotype describes observable properties of an organism. Genotype refers to genetic constitution of an organism, usually in reference to some standard wild type. Follow recommendations of Demerec et al. (*Genetics* **54**:61–64,1966) for use of these terms. If manuscript contains information including genetic nomenclature, refer to Instructions to Authors in January issue of *Journal of Bacteriology*.

"Mutant" vs. "Mutation"—A mutation is an alteration of the primary sequence of genetic material; a mutant is a strain carrying one or more mutations. A mutant has no genetic locus, only a phenotype.

Viruses—Distinctions between phenotype and genotype cannot be made. Use superscripts to indicate hybrid genomes. Genetic symbols may be one, two, or three letters.

Nomenclature, Microorganisms: Use binary names, consisting of a generic name and a specific epithet. Names of higher categories may be used alone, specific and subspecific epithets may not. Precede specific epithet by a generic name at first use. Thereafter, abbreviate generic name to initial capital letter, provided there can be no confusion with other genera. Print names of all taxa (phyla, classes, orders, families, genera, species, subspecies) in italics or underline. Do not italicize strain designations and numbers.

Spell names according to *Approved Lists of Bacterial Names* (amended edition) (V. B. D. Skerman, V. McGowan, and P. H. A. Sneath, ed.) and *Index of the Bacterial and Yeast Nomenclatural Changes Published in the International Journal of Systematic Bacteriology since the 1980 Approved Lists of Bacterial Names (1 January 1980 to 1 January 1989)* (W. E. C. Moore and L. V. H. Moore, ed., American Society for Microbiology, 1989), and validation lists and articles published in *International Journal of Systematic Bacteriology* since 1 January 1989. If there is reason to use a name that does not have standing in nomenclature, enclose name in quotation marks and make an appropriate statement concerning its nomenclatural status in text (for an example, see *Int. J. Syst. Bacteriol.* **30**:547-556, 1980).

Deposit strain in a recognized culture collection when necessary for describing a new taxon (see *Bacteriological Code*, 1990 Revision, American Society for Microbiology, 1992).

Determine accepted binomial for a given fungi; some sources include *The Yeasts: a Taxonomic Study* (3rd ed., N. J. W. Kreger-van Rij, ed., Elsevier Science Publishers B.V., Arnsterdam, 1984) and *Ainsworth and Bisby's Dictionary of the Fungi, Including the Lichens,* 7th ed. (Commonwealth Mycological Institute, Kew, Surrey, England, 1983).

Use names for viruses approved by the International Committee on Taxonomy of Viruses (*Intervirology* **17**:23-199,1982, with modifications in *Arch. Virol.*, Suppl. 2, 1991). If desired, add synonyms parenthetically at first mention. Approved generic (or group) and family names may also be used.

Give microorganisms, viruses, and plasmids designations consisting of letters and serial numbers. Include a worker's initials or a descriptive symbol of locale, laboratory, etc., in designation. Give each new strain, mutant, isolate, or derivative a new (serial) designation distinct from those of genotype and phenotype; do not include italicized genotypic and phenotypic symbols.

Nucleic Acid Sequences: Present limited length nucleic acid sequences freestyle in the most effective format. Present longer nucleic acid sequences to conserve space;

submit as camera-ready copy 8.5 × 11 in. in standard (portrait) orientation. Print in lines of 100 bases in an easily legible nonproportional (monospace) font when published at 100 bases/6 inches. Use uppercase and lowercase letters to designate exon/ intron structure, transcribed regions, etc., if lowercase letters remain legible at 100 bases/6 inches. Number sequence line by line; place numerals, representing first base of each line, to left of lines. Minimize spacing between adjacent lines; leave room only for sequence annotation. Annotation may include boldface, underlining, brackets, boxes, etc. Present encoded amino acid sequence, if necessary, immediately above first nucleotide of each codon, using single-letter amino acid symbols. Comparisons of multiple nucleic acid sequences should conform to same format.

Nucleotide Sequences: Include GenBank/EMBL accession numbers for primary nucleotide and/or amino acid sequence data. Include accession number as a separate paragraph at end of Materials and Methods section for full-length papers or at end of text of Notes.

GenBank Submissions, National Center for Biotechnology Information, Bldg. 38A, Rm. 8N-803, 8600 Rockville Pike, Bethesda, MD 20894; e-mail (submissions): gbsub@ncbi.nlm.gov; e-mail update: update@ncbi.nlm.gov. EMBL Data Library Submissions, Postfach 10.2209, Meyerhofstrasse 1, 6900 Heidelberg, Germany; telephone: 011 49 (6221) 387258; fax: 011 49 (6221) 387306; electronic mail (data submissions): data subs@embl.bitnet.

Page Charges: $30 per printed page to be paid by authors whose research was supported by special funds, grants (departmental, governmental, institutional, etc.), or contracts or whose research was done as part of their official duties. If the research was not supported, submit a request to waive charges to Journals Division with manuscript in a note separate from cover letter. Indicate how work was supported; include a copy of Acknowledgment section.

Minireviews and Letters to the Editor are not subject to page charges.

Style: Use past tense to narrate procedures, observations, and data. Use present tense for general conclusions, conclusions of previous researchers, and generally accepted facts. Abstract, Materials and Methods, and Results sections are in past tense, and introduction and Discussion are in present tense. It may be necessary to vary tense in a sentence. For an in-depth discussion of tense in scientific writing, see p. 158–160 in *How to Write and Publish a Scientific Paper*, 3rd ed.

Follow *CBE Style Manual* (5th ed., 1983; Council of Biology Editors, Inc., Bethesda, Md.), *ASM Style Manual for Journals and Books* (American Society for Microbiology, 1991), and Robert A. Day's *How to Write and Publish a Scientific Paper* (3rd ed., 1988; Oryx Press).

Units: Use standard metric units for reporting length, weight, and volume. For these units and for molarity use prefixes m, μ, n, and p for 10^{-3}, 10^{-6}, 10^{-9}, and 10^{-12}, respectively. Use prefix k for 10^3. Avoid compound prefixes such as mμ or μμ. Use μg/ml or μg/g in place of ppm. Present units of temperature as: 37°C or 324 K.

When fractions are used to express units such as enzymatic activities, use whole units, in denominator instead of fractional or multiple units. Use an unambiguous form such as exponential notation.

See *CBE Style Manual*, 5th ed., for more information about reporting number appropriate SI units for reporting illumination, energy, frequency, and other physical terms. Report numerical data in appropriate SI units.

SOURCE

January 1994; 32(1): i-x.

Journal of Clinical Oncology
American Society of Clinical Oncology
W. B. Saunders Co.

George P. Canellos, MD
454 Brookline Ave., Suite 28
Boston, MA 02115

MANUSCRIPT SUBMISSION

When manuscripts are accepted, authors must provide following signed statement: "The writer/author represents and warrants that his/her part of the work as submitted will in no way violate any copyright or any other right and will contain nothing libelous or otherwise unlawful."

Copyright will be assigned exclusively to the American Society of Clinical Oncology (ASCO). ASCO will not refuse any reasonable request by authors for permission to reproduce any of his or her contributions.

Fully disclose primary financial relationships between authors and commercial companies under investigation in manuscript. Include statements regarding a company's support for research and/or related financial involvement of authors on manuscript title page. State other relationships that might pose a conflict of interest such as holding equity or paid consultationship in cover letter.

MANUSCRIPT FORMAT

Submit original and two complete copies of manuscript typed or printed double- or triple-spaced on good quality 8.5 × 11 in. white paper with 1 in. margins. Do not use erasable bond.

Title Page: On first page include: title of report; authors' names; name of institution where work was done; acknowledgments for research support; name and address of corresponding author; a running head (45 characters).

Abstract: Second page: 250 word abstract organized and formatted according to: Purpose; Patients and Methods (or Materials and Methods, Methods, etc., following structure of manuscript),; Results; and Conclusion. Substitute "Design" for "Patients and Methods" in abstracts of review articles.

Text: Research manuscripts should have clearly identified Introduction, Patients and Methods (or Materials and Methods, Methods, etc., as appropriate), Results, and Discussion sections. Do not include a summary. Write body of paper as concisely as possible.

Figures: Cite in order in text using Arabic numerals. Submit all figures in triplicate sorted sets (one per manuscript). Identify on back with corresponding author's name. Submit electron micrographs, halftones, and photographs as glossy prints; submit line drawing, charts, and graphs as either glossy prints or laser printer-generated figures. Contributors will pay costs of color photographs. Figures are generally reduced for publication. Indicate crop marks. Make numbers, letters, and symbols large enough so that when reduced they will remain at least 1/12 inch (2 mm) high. Figures not properly prepared will be returned.

Provide a letter of permission from copyright holders for reprinted figures.

Figure Legends: Type or print double-spaced on a separate sheet. Include all relevant and explanatory information extraneous to actual figure. Do not exceed 45 words.

Tables: Cite chronologically in text using Arabic numbers. Type each appropriately numbered table on a separate sheet with table legend on the same page; include definitions of abbreviations used.

Provide letters of permission from copyright holders of reprinted tables.

References: Compile at end of text; list and number in order cited in text. Type or print double-spaced, under heading REFERENCES. Use journal abbreviations from *Index Medicus's* "List of Journals Indexed in Index Medicus." Give inclusive page numbers. Identify references to abstracts, editorials, and letters-to-the-editor as such.

JOURNAL REPORT, ONE AUTHOR

1. Greenbaum BH: Transfusion-associated graft-versus-host disease: Historical perspectives, incidence, and current use of irradiated blood products. J Clin Oncol 9:1889–1902, 1991

JOURNAL REPORT, THREE AUTHORS

2. Omura G, Blessing JA, Ehrlich CE: A randomized trial of cyclophosphamide and doxorubicin with or without cisplatin in advanced ovarian carcinoma. Cancer 57:1725–1730, 1986

JOURNAL REPORT, MORE THAN THREE AUTHORS

3. Gulino A, Barrera G, Vacca A, et al: Calmodulin antagonism and growth-inhibiting activity of triphenylethyl antiestrogens in MCF-7 human breast cancer cells. Cancer Res 46:6274–6278, 1986

JOURNAL ARTICLE IN PRESS

4. Smith JW II, Urba WJ, Clark JW, et al: Phase I evaluation of recombinant tumor necrosis factor given in combination with recombinant interferon gamma. J Immunother (in press)

COMPLETE BOOK

5. Jandl H: Blood Textbook of Hematology. Boston, MA, Little, Brown, 1987

CHAPTER OF BOOK

6. Young RC, Fuks Z, Hoskins WJ: Cancer of the ovary, in DeVita VT Jr, Hellman S, Rosenberg SA (eds): Cancer: Principles and Practice of Oncology (ed 3). Philadelphia, PA, Lippincott, 1989, pp l162–1196

CHAPTER OF BOOK THAT IS PART OF A MEETING

7. Karnofsky DA, Burchenal JH: The clinical evaluation of chemotherapeutic agents in cancer, in Macleod CM (ed):Evaluation of Chemotherapeutic Agents. Symposium, Microbiology Section, New York Academy of Medicine. New York, NY, Columbia University, 1949, pp 191–205

ABSTRACT

8. Abeloff MD, Gray R, Tormey DC, et al: A randomized comparison of CMFPT versus CMFPT/VATHT and maintenance versus no maintenance tamoxifen in premenopausal, node positive breast cancer: An ECOG study. Proc Am Soc Clin Oncol 10:43, 1991 (abstr 47)

SUPPLEMENT

9. Thigpen JT, Blessing JA, Vance RB, et al: Chemotherapy in ovarian carcinoma: Present role and future prospects. Semin Oncol 16:58–65,1989 (suppl 6)

EDITORIAL

10. Livingston RB: Stage IV non-small-cell lung cancer: The guides are perplexed. J Clin Oncol 7: 1591–1 593, 1989 (editorial)

LETTER TO THE EDITOR

11. Dahan R, Espie M, Mignot L. et al: Tamoxifen and arterial thrombosis. Lancet 1:638, 1985 (letter)

OTHER FORMATS

EDITORIALS, REVIEW ARTICLES—Prepare in a manner appropriate for other papers.

LETTERS-TO-THE-EDITOR—Type or print two double-spaced pages; submit in duplicate. Give titles.

RAPID PUBLICATIONS—Particularly meritorious manuscripts that are of unusually high priority and significance and that represent definitive and original studies.

OTHER

Experimentation, Human: Follow precepts of the Helsinki Declaration. Include a statement that investigations were performed after approval by a local Human Investigations Committee and were in accord with an assurance filed with and approved by the Department of Health and Human Services, where appropriate. Include a statement that informed consent was obtained from each subject or subject's guardian.

Recombinant DNA: Include a description of physical and biologic containment procedures practiced, in accord with the National Institutes of Health Guidelines for Research Involving Recombinant DNA Molecules.

Style: Follow American Medical Association style guidelines. Comply with item 2.8.2. in *AMA Manual of Style*, Eighth Edition; provide a statement regarding any proprietary interests related to article.

Include generic drug names with proprietary drug names and name and location of manufacturer.

For clinical trials see guidelines in Zelen M: Guidelines for published papers on cancer clinical trials: Responsibilities of editors and authors. *J Clin Oncol* 1:164–169, 1983 and Simon R, Wittes RE: Methodologic guidelines for reports of clinical trials. *Cancer Treat Rep* 69:1–3, 1985.

SOURCE

January 1994, 12(1): xxiii-xxiv.

Journal of Clinical Pathology
Association of Clinical Pathologists
BMJ Publishing Group

Editors, *Journal of Clinical Pathology*
BMJ Publishing Group
BMA House, Tavistock Square
London WC1H 9JR
Telephone: 071 383 6209 6214; Fax: 071 383 6668.

MANUSCRIPT SUBMISSION

Submit in duplicate typed double-spaced on one side of paper only. Follow names of authors (with initials or one forename) by name of institution where work was performed. Indicate position held by each author in cover letter; include signatures of all authors. Guidelines on authorship are detailed in *J Clin Pathol* 1986;**39**:110.

MANUSCRIPT FORMAT

Papers should be about 2000 words long and report original research of relevance to the understanding and practice of clinical pathology. Use standard format and a structured abstract. Abstract (300 words) should contain headings Aims, Methods, Results and Conclusions. Body of paper should have separate sections for introduction, methods and results, and discussion. If statistics are used, state methods and confidence intervals. Seek expert advice if in doubt.

Do not show results as both tables and graphs; do not use histograms where tabular information would be more appropriate.

Illustrations: Submit diagrams as photographs.Type legends for illustrations double-spaced on a separate sheet. Photographs and photomicrographs must be of high quality in full tonal scale on glossy paper, and unmounted. Include only salient details. Background should be as near white as possible. Width of illustrations should be 6·7 cm, 10·2 cm, 13·7 cm or, in exceptional circumstances, 17·4 cm, to fit column layout. Color reproduction is welcomed and is subsidized; for details of costs to author, contact Journal office.

References: Use Vancouver style.

OTHER FORMATS

LETTERS TO THE EDITOR—Refer to previously published papers or make some point about the practice of pathology. Not a vehicle for the presentation of new data unrelated to earlier *Journal* articles.

OCCASIONAL ARTICLES—1500-2000 words in a less rigid format usually invited by editors; unsolicited submissions are considered

SHORT REPORTS—Single case reports and brief papers, such as those describing negative findings. Format: unstructured 150 word summary, 1500 word text, two tables or figures (or one of each) and 10 references.

OTHER

Abbreviations: Spell out at first use or explain in text. Avoid use of nonstandard abbreviations and acronyms. Symbols and abbreviations should be in Vancouver style.

Ethics: The critical assessment of papers will include ethical considerations.

Units: Give all measurements in SI units.

SOURCE

January 1994; 47(1): inside front cover.

Journal of Clinical Periodontology
Belgian Society of Periodontology
British Society of Periodontology
Dutch Society of Periodontology
French Society of Periodontology
German Society of Periodontology
Hellenic Society of Periodontology
Irish Society of Periodontology
Italian Society of Periodontology
Scandinavian Society of Periodontology
Spanish Society of Periodontology
Swiss Society of Periodontology
Munksgaard International Publishers

Journal of Clinical Periodontology
Department of Periodontology
Faculty of Odontology
Fack, S-400 33 Gothenburg 33, Sweden

AIM AND SCOPE

The aim and scope of this *Journal* is to convey scientific progress in periodontology to whose concerned with application of this knowledge for the benefit of the dental health of the community. It will address itself primarily to clinicians, general practitioners, periodontists, as well as to teachers and administrators involved in the organization of prevention and treatment of periodontal disease.

The *Journal* will publish original contributions of high scientific merit in the field of physiology and pathology of the periodontium, diagnosis, epidemiology and prevention and therapy of periodontal disease, review articles by experts on new developments in basic and applied periodontal science, advances in periodontal technique and instrumentation, and case reports which illustrate important new information.

MANUSCRIPT SUBMISSION

Submit manuscripts in triplicate to one of the Associate Editors representing the various Societies or to the editorial office.

Submission is with the understanding that work has not been published before, is not being considered for publication elsewhere and has been read and approved by all authors. Submission means that authors automatically agree to assign exclusive copyright to Munksgaard International Publishers if and when manuscript is accepted. The work shall not be published elsewhere in any language without written consent of publisher. Articles published in this journal are protected by copyright.

Acceptable material includes original investigations, reviews, and case reports (must provide new fundamental knowledge and use language understandable to clinicians). Manuscripts must represent unpublished original research.

MANUSCRIPT FORMAT

Submit manuscripts in triplicate typed on one side of paper only (quarto - A4) with wide margins and double-spacing. On first page, on separate lines: title of article and authors' names and institutions. If title is longer than 40 letters and spaces, submit a short title, not exceeding that limit, as a running head. On last page include full postal address of corresponding author. Divide papers into sections: Abstracts (200 words) - Key words - Introduction - Material and Methods - Results - Discussion - Acknowledgments (if any) - Summaries in German and French - References.

References: In text quote last name(s) of author(s) and year of publication (Brown & Smith 1966). Refer to 3 or more authors as, for example, (Brown et al. 1966).

List references at end of paper. Follow recommendations in *Units, Symbols and Abbreviations: A Guide for Biological and Medical Editors and Authors*, (1975, p. 36, London: The Royal Society of Medicine). Arrange references alphabetically by author's surname. Order of items in each reference for journal references: name(s) of author(s), year, title of paper, title of journal, volume number, first and last page numbers; for book references: name(s) of author(s), year, title of book, edition, volume, chapter and/or page number, town of publication, publisher. Arrange author's names thus: Smith, A. B., Jones, D. E. & Robinson, F. C. Note use of ampersand and omission of comma before it. Spell out author's names when repeated in next reference in full. Surround year of publication by parentheses: (1967). Include title of paper, without quotation marks. Write out journal title in full, italicize or underline. Follow with volume number in bold type (or double underline) and page numbers.

Engfeldt, B. & Hjertquist, S.-O. (1967) Effect of various fixatives on the preservation of acid glycosaminoglycans in tissues. *Acta Pathologica et Microbiologica Scandinavica* **71**, 219–232.

Shafer, W. G., Hine, M. K. & Levy, B. M. (1963) *A textbook of oral pathology*, 2nd edition, p. 138. Philadelphia: Saunders.

Bodansky, O. (1960) Enzymes in tumor growth with special reference to serum enzymes in cancer. In *Enzymes in health and disease*, eds. Greenberg, D. & Harper, H. A., pp. 269–278. Springfield: Thomas.

Illustrations: Print graphs, drawings or photographs on transparent or glossy paper. Number in sequence with Arabic numerals. Identify each Figure with Figure number on back; indicate orientation. In text, indicate approximate location by a circled marginal note.

Figures should clarify text. Keep number to a minimum. Make details large enough to retain clarity after reduction in size. Illustrations should fill single column width (54 mm) after reduction, in some cases 113 mm (double-column) and 171 mm (full page) widths will be accepted. Design micrographs to be reproduced without reduction; incorporate a linear size scale.

Submit photographs individually, unmounted or group together in a single block, mounted in such a manner as to ensure that full use is made of space. Best results are obtained with photos the same size as final blocks; submit as glossy prints with good detail and moderate contrast. Color illustrations must be paid for by author. Submit original color transparencies and three sets of color prints. Type captions to Figures (legends) consecutively on a separate page(s) at end of paper.

Tables: Type on sheets separate from text; make titles self-explanatory. Number consecutively with Arabic numerals. Keep in mind proportions of printed page. Indicate approximate location in text by a circled marginal note. Do not double document in tables and figures.

OTHER FORMATS

ANNOUNCEMENTS—Send material for announcements of important meetings and other Society matters to Publisher: MUNKSGAARD, International Publishers Ltd., 35 Norre Sogade, Postbox 2148, DK-1016 Copenhagen K, Denmark.

SHORT COMMUNICATIONS—1–3 pages for quick publication. Need not follow usual division into material and methods, etc.; include an abstract.

OTHER

Abbreviations and Symbols: Use only generally accepted standardized terms. Use *Units, Symbols and Abbreviations* (The Royal Society of Medicine, London) as reference source. Define unfamiliar abbreviations at first use. Use "two digit-system" for tooth identification (see Fédération Dentaire Internationale (1971) *International Dental Journal 21*, 104–106).

Color Charges: U.S. $1100. Final quotation will be given by publisher.

Drugs: Use generic names instead of proprietary names. If a proprietary name is used, attach ® at first use.

Electronic Submission: If possible, supply a disk with hard copy; state format/system used.

Nomenclature: Scientific names of bacteria are binomials. Capitalize initial letter of generic name only; underline or italicize. Give name in full upon first mention; generic name may be abbreviated thereafter, abbreviation must be unambiguous.

Page Charges: DKK 1800 (appr. U.S. $288) per page over 8.

Style: See *Style Manual for Biological Journals* (3rd edition, Washington, D.C.: American Institute of Biological Science 1972). Use no Roman numerals in text. In decimal fractions, use a decimal point, not a comma. Show minus sign as −.

SOURCE

January 1994, 21(1): inside back cover.

Journal of Clinical Psychiatry
Physicians Postgraduate Press

Physicians Postgraduate Press, Inc.
P.O. Box 240008
Memphis, TN 38124

AIM AND SCOPE

Submissions to *The Journal of Clinical Psychiatry* should be relevant and interesting to practicing clinical psychiatrists. We strive to publish academically sophisticated, methodologically sound manuscripts geared more toward the practitioner than the researcher. Manuscripts should be concisely written, appropriately referenced, and coherently focused. Conclusions should flow logically from the data presented, and methodological flaws and limitations should be acknowledged.

Manuscripts eligible to be published as articles include controlled studies, clinical observations of wide importance, critical overviews, pilot studies, open trials, chart reviews, and case series with literature reviews. Experimental drug trials involving a compound not currently available in the United States may be considered if (1) the compound is expected to be released soon in the United States or (2) it offers some unique and interesting clinical features. Manuscripts should deal with the epidemiology, classification, and treatment of psychiatric disorders and should not exclusively emphasize laboratory techniques, biostatistical models, validity studies, or the development of measurement instruments.

MANUSCRIPT SUBMISSION

Submit original manuscript and two copies. Follow requirements of "Uniform Requirements for Manuscripts Submitted to Biomedical Journals" of the International Committee of Medical Journal Editors (*N Engl J Med* 1991;324:424–428) summarized below.

Manuscripts must represent original material, have never been published before, not be under consideration for publication elsewhere, and have been approved by each author. Prior publication is any form of publication other than an abstract and includes invited articles, proceedings, symposia, and book chapters. Inform Editor if manuscript contains data or clinical observations published or submitted for publication elsewhere; supply copies and explain differences.

Acknowledge all forms of support, including drug company support, in Title Page Acknowledgments. In cover letter disclose all affiliations and financial interests directly related to subject or materials discussed in manuscript (e.g., employment, consultancies, patent ownership, stock ownership). Information may be shared with reviewers at Editor's discretion. If accepted, Editor will discuss extent of disclosure appropriate for publication. Involvements are not grounds for automatic rejection.

Submit a cover letter signed by all coauthors that includes: information on prior or duplicate publication or submission elsewhere of any part of the work; statement of financial or other relationships that might lead to a conflict of interest; statement that manuscript has been read and approved by all authors; name, address, telephone and FAX numbers of corresponding author; and express transfer of copyright.

Submit letters of permission from publishers for use of previously published tables, figures, and extensive quotations.

The Journal of Clinical Psychiatry requires express transfer of copyright to Physicians Postgraduate Press, Inc. Include a statement signed by all authors and worded as follows:

In consideration of Physicians Postgraduate Press, Inc., reviewing and editing my (our) submission [Put Title Here], the author(s) undersigned hereby transfer(s), assign(s), or otherwise convey(s) all copyright ownership to Physicians Postgraduate Press, Inc., in the event that such work is published by Physicians Postgraduate Press, Inc.

Work done as part of duties as a federal employee is in the public domain. Word statement as follows:

The work described in [Put Title Here] was done as part of my (our) employment with the federal government and is therefore in the public domain.

MANUSCRIPT FORMAT

Type manuscript on one side of white, nonerasable bond paper (8.5 × 11 in.), with 1 in. margins. Double-space throughout, including title page, abstract, text, references, tables, and legends for figures. Number pages consecutively in upper right corner, beginning with title page. Begin each section on a separate page; arrange in order: title page and acknowledgments, abstract and key words, text, references, and tables and legends for figures.

Title Page: Title should be concise but informative. For each author, include first name, middle initial, and last name with highest academic degree(s) and institutional affiliation. Give full address, telephone and fax numbers of corresponding author.

Assign authorship only to persons who contributed to intellectual content of paper and can take public responsibility for that content. Authors must meet criteria: conceived and designed work or analyzed and interpreted data; drafted article or reviewed it for intellectual content; and approved final version to be published. Participation solely in acquisition of funding or collection of data does not justify authorship.

At bottom of title page, list contributions that need acknowledging but do not justify authorship, such as general support by a departmental chairperson, critical review of study proposal, or data collection; acknowledgments of technical help; and acknowledgments of financial and material support (specify nature of support). Indicate previous oral presentation. Authors are responsible for obtaining written permission from persons acknowledged for other than financial or technical support.

Abstract and Key Words: On second page: structured abstract (250 words), four paragraphs with headings: Background—question addressed in study; Method—how study was performed (selection of study subjects, observational and analytic methods, criteria for diagnosis); Results—key findings (give specific data and statistical significance); and Conclusion—what authors conclude from results.

Below abstract, provide up to five key words or short phrases for indexing.

Text: Divide text of observational and experimental articles into sections with headings: Introduction, Method, Results, and Discussion. Lengthy articles may need subheadings.

Introduction: State purpose. Summarize rationale for study or observation. Give only strictly pertinent references; do not review subject extensively. Do not include data or conclusions from work being reported.

Method: Describe selection of observational or experimental subjects (including controls) clearly. Identify methods, apparatus (manufacturer's name and address within parentheses), and procedures in sufficient detail to allow other workers to reproduce results. Give references to established methods, including statistical methods. Describe new or modified methods, give reasons for using, and evaluate limitations. Identify drugs and chemicals used, including generic name(s), dose(s), and route(s) of administration. Justify use of diagnostic criteria other than DSM-III.

Results: Present results in logical sequence. Do not repeat in text data in tables or figures; emphasize or summarize only important observations.

Discussion: Emphasize new and important aspects of study and conclusions that follow. Do not repeat in detail material in Introduction or Results. Include implications of findings and limitations, including implications for future research. Relate observations to other relevant studies. Link conclusions with goals of study; avoid unqualified statements and conclusions not completely supported by data.

References: Include only references to published material in reference list. Verify accuracy and completeness of references against original documents.

Number references consecutively in order cited in text. Identify by superscript Arabic numerals. Number references cited only in tables or figure legends following sequence at point of identification in text of table or figure.

Cite "in press" references if title and journal name or book publisher are given. Do not include "unpublished observations" and "personal communications" in reference list; cite parenthetically in text. Cite information from manuscripts submitted but not yet accepted in text as "unpublished observations" (within parentheses). Secure permission from cited persons to cite such unpublished works. Cite symposium papers in reference list only if published proceedings of meeting are available.

Double-space throughout reference list. Abbreviate journal names according to *Index Medicus*.

1. O'Rourke D, Wurtman JJ, Wurtman RJ, et al. Treatment of seasonal depression with *d*-fenfluramine. J Clin Psychiatry 1989;50:313–347

2. Davis JM. Antipsychotic drugs. In: Kaplan HI, Sadock BJ, eds. Comprehensive Textbook of Psychiatry, vol 2, 4th ed. Baltimore, Md: Williams & Wilkins; 1985:1481–1513

3. Black DW, Noyes R, Goldstein R, et al. Family study of OCD. In: New Research Program and Abstracts of the 144th annual meeting of the American Psychiatric Association; May 14,1991; New Orleans, La. Abstract NR359:134

4. Huth EJ. How to Write and Publish Papers in the Medical Sciences. 2nd ed. Philadelphia. Pa: ISI Press; 1986

5. Dubovsky SL. Generalized anxiety disorder: new concepts and psycho-pharmacologic therapies. J Clin Psychiatry 1986;47(4, suppl):46–66

6. Lieb J. Antidepressant tachyphylaxis [letter]. J Clin Psychiatry 1990;51:36

Tables: Type double-spaced on separate page(s). Number consecutively in order of first citation in text. Do not duplicate text or figures; make self-explanatory. Identify by a brief descriptive title. Give each column a short heading. Place explanatory matter in footnotes, not in heading. Explain abbreviations; specify units of measurement. Designate footnotes by superscript lowercase letters. Do not use internal horizontal and vertical rules.

Obtain written permission from copyright holder for use of previously published material. Give credit in a table footnote.

Figures: Submit three complete sets of figures. Present as camera-ready glossy prints that can withstand reduction to 3.25 in. Do not include figure title on figure; explain symbols in body of figure. Reserve figures for indications of important trends; do not duplicate data presented in tables or text. Number figures consecutively according to order first cited in text. Type legends double-spaced on a separate page; identify with Arabic numerals corresponding to figure. Paste a label pasted on back indicating number, author name, and TOP.

If figure has been published, acknowledge original source and submit written permission from copyright holder to reproduce. Permission is required, irrespective of authorship or publisher, except for documents in public domain.

OTHER FORMATS

LETTERS TO THE EDITOR—500 words, no tables or figures. Provide a forum for single case reports. Describe novel, well-documented findings that will be of help and interest to practitioners. Type double-spaced throughout, including references. Letters that pertain to recent articles in the *Journal* may be sent to author(s) for response.

OTHER

Abbreviations: Use only standard abbreviations. Avoid abbreviations in title and abstract. Full term for which an abbreviation stands should precede its first use in text unless it is a standard unit of measurement.

Experimentation, Human: Include a statement that subjects (or parents/ guardians) gave informed consent after procedure(s) and possible side effects were fully explained. Ethical and legal considerations dictate protection of patients' anonymity. Do not use patients' names, initials, or hospital numbers in text or illustrative material. Avoid dates and disguise characteristics and personal history that would identify a patient.

Statistical Analyses: Describe methods with enough detail to enable a knowledgeable reader with access to original data to verify results. Quantify findings and present with appropriate indicators of measurement error or uncertainty (such as confidence intervals).

Avoid sole reliance on statistical hypothesis testing, such as use of p values, which fails to convey important quantitative information. Discuss eligibility of experimental subjects. Give details about randomization. Describe methods for, and success of, any blinding of observations. Report treatment complications. Give numbers of observations. Report losses to observation (such as dropouts from a clinical trial). Reference statistical tests that are not well known. Specify any general-use computer programs used.

SOURCE

January 1994, 55(1): 3–4.

Journal of Comparative Neurology
Wiley-Liss, Inc.

Dr. Clifford B. Saper
Department of Neurology
Harvard Medical School
220 Longwood Avenue
Boston, MA 02115

John S. Edwards
Department of Zoology
University of Washington
Seattle, WA 98195

Thomas E. Finger
Department of Cell and Structural Biology
University of Colorado Medical Center
B-111, 4200 East Ninth Avenue
Denver, CO 80262

Jan H. Kaas
Department of Psychology
Vanderbilt University
301 Wilson Hall
Nashville, TN 37240

Jeff W. Lichtman
Department of Anatomy and Neurobiology
Washington University School of Medicine
660 Euclid Ave.
St. Louis, MO 63110

George F. Martin
Department of Cell Biology, Anatomy and Neurobiology
Ohio State University College of Medicine
333 West 10th Avenue
Columbus, OH 43210

Paul E. Sawchenko
Laboratory of Neuronal Structure and Function
The Salk Institute for Biological Sciences
P.O. Box 85800
San Diego, CA 92186

Oswald Steward
Department of Neuroscience
University of Virginia School of Medicine
Box 230 Medical Center
Charlottesville, VA 22908

MANUSCRIPT SUBMISSION

Submit a statement certifying that all authors have read paper and have agreed to being listed as authors. Append similar statements for colleagues acknowledged in footnotes as having contributed to or criticized paper.

Upon acceptance for publication, author will be asked to sign a copyright transfer agreement, transferring rights to publisher, who reserves copyright.

MANUSCRIPT FORMAT

Title Page: Title of paper; Author's name(s); Institutional affiliation(s) with city, state, and Zip Code; Number of text pages, figures, graphs, and charts; Abbreviated title (running head, 48 letters and spaces); Key words (5 from *Index Medicus* list of accepted key words that do not repeat title words; Name, address, and telephone, telex, and fax numbers of corresponding author.

Abstract: 250 words to precede introductory section of text. Write in complete sentences. Succinctly state nature of biological problem investigated, paper's objectives, experimental design, and principal observations and conclusions. Make intelligible without reference to rest of paper. Do not use abbreviations. Do not cite references to literature without complete citation.

Literature Cited: Each reference in text must appear in literature list. Each reference in literature list must be cited in text. In text, cite references to literature by author's name followed by year of publication:

... studies by Westbrook ('90) reveal...

... studies by Westbrook and Bollenbacher ('90) reveal...

... studies by Westbrook et al. ('90) reveal...

List references in chronological order, when more than one is cited:

... earlier reports (Bunt et al., '80; Briggs and Porter, '85; Laemle, '90)

When references are made to more than one paper by the same author, published in the same year, designate in text as (Kuroda and Price, '91 a,b) and in literature as:

Kuroda, M., and J.L. Price (1991a) Synaptic organization of projections from basal forebrain structures to the mediodorsal thalamic nucleus of the rat. J. Comp. Neurol. *303:*513–533.

Kuroda, M., and J.L. Price (1991 b) Ultrastructure and synaptic organization of axon terminals from brainstem structures to the mediodorsal thalamic nucleus of the rat. J. Comp. Neurol. *313:*539–552.

Arrange literature list alphabetically by author's surname: Author's name (or names), year of publication, complete title, volume, and inclusive pages:

Houser, C.R., P.E. Phelps, and J.E. Vaughn (1988) Cholinergic innervation of the rat thalamus as demonstrated by the localization of choline acetyltransferase. In M. Bentivoglio and R. Spreafico (eds): Cellular Thalamic Mechanisms. New York: Elsevier, pp. 387–398.

Hreib. K.K., D.L. Rosene. and M.B. Moss (1988) Basal forebrain efferents to the medial dorsal thalamic nucleus in the rhesus monkey. J. Comp. Neurol. *277:*365–390.

Isseroff, A., H.E. Rosvold, T.W. Galkin, and P.S. Goldman-Rakic (1982) Spatial memory impairments following damage to the mediodorsal nucleus of the thalamus in rhesus monkeys. Brain Res. *232*:97–113.

Jones, E.G. (1985) Thalamus. New York: Plenum Press.

Kievit, J., and H.G.J.M. Kuypers (1977) Organization of thalamo-cortical connections to the frontal lobe in the rhesus monkey. Exp. Br. Res. *29*:299–322.

Abbreviate journal titles as in *Index Medicus*.

For text references beginning with 1901, and thereafter, abbreviate date of publication after author's name ('04, not 1904). Do not abbreviate references in text to papers published before 1901. In Literature Cited section never abbreviate year.

Papers "In Preparation" are not legitimate references; do not include in Literature Cited.

Footnotes: Number footnotes to text consecutively. Indicate corresponding reference numbers in text. Number additional references to identical footnotes with next following consecutive number. Type table footnotes directly beneath table; number 1, 2, 3, etc. Do not number in sequence with text footnotes.

Tables: Number and cite consecutively in text. Make simple and uncomplicated, with as few horizontal rules as necessary and no vertical rules. Indicate where tables are to appear in text. Make titles complete but brief. Present information other than that defining data in footnotes.

Figures: Number and cite all figures, including charts and graphs, in text. Number figure legends consecutively as: Fig. 1 . . ., Fig. 2. . . When possible, integrate figures into text. Group figures to fit a single page with legends. Reference to relevant text passages reduces legend length and avoids redundancy.

Illustrations: Limit figures to number necessary to document findings. Printer works best directly from original drawings or high-quality photographic prints. Sharp-contrast photocopies are acceptable for review purposes. Submit all illustrations in complete and finished form, with adequate labeling. Prints must be of adequate contrast; when more than one print in a figure, they should be of uniform tone. Black-and-white prints should be on white, nonmatte paper.

Indicate on each illustration reduction desired, Keep in mind: illustrations cannot be reduced to less than 50% of submitted size; excessively large figures are difficult to handle, ship, and store (do not exceed 11 × 16 in.; lettering and labels must be readable after reduction (when reduced, minimum height of a capital letter is 2.5 mm for a photomicrograph and 1 mm for graph or chart; maximum 4 mm and 3 mm, respectively); individual printed figure or group of figures cannot exceed of 7 × 9 in. or 3 3/8 in. × 9 3/8 in. single-column placement.

Line Drawings—Draw figures with black ink on medium-weight white paper or lightweight art board. Submit photographic prints in lieu of original drawings. Make artwork sharp and black for maximum contrast.

Use "stippling" and "hatching" techniques to achieve tonal quality. Avoid "shading" (pencil, wash, or air-brush) for tonal effect unless drawing will be reproduced as a half-tone with gray tint background. If original graphs are submitted, draw on blue-ruled paper; other colors reproduce.

Color Illustrations—Make from good-quality transparencies or color prints. Do not use silk finish or matte surface papers. Affix color plates to flexible substrate. Supply color or slide and color prints or transparencies. Mark frame of transparency to indicate area that can be cropped to arrive at critical image area.

Mounting Figures: Trim figures straight on all sides and "squared." Mount on strong, bristol board 1.5 points (0.4 mm) thick with a 1in. margin around figure or grouping of figures. Attach figures to bristol board with appropriate "dry mounting" materials, or a cement or glue that is white or colorless when set. When two or more figures are assembled, mount close together, evenly separated by no more than 1mm. Illustrations grouped in a single figure should be of similar density. Do not use black borders.

Lettering and Labels—Letter and number illustrations with printed paste-on or transfer labels. Make labels large enough to allow for suitable reduction and sturdy enough to withstand mailing and handling. Spray labeling with clear adhesive to prevent it from becoming scratched or being torn off. Do labeling directly on drawing or photographic print, never on an overlay. Place labeling at least .25 in. from edges of illustration. Place white labels over dark backgrounds and black labels over light backgrounds, or shadow labels with an appropriately light or dark highlight. Provide orientation as needed. Provide each micrograph or line drawing with a scale bar, within picture space.

General Illustration Instructions—Submit original illustrations, and three sets of sharp-contrast photocopies for review purposes. Number figures, including charts and graphs consecutively. Do not extend figures over more than a single page. If originals are too large for shipment, submit photographic prints. Do not submit photocopies of illustrations made on office duplicating machines. On reverse of each illustration indicate: author's name; figure number; TOP; reduction requested; and "review copy" on copies intended for reviewers. Do not fasten with paper clips, staples, etc. Ship flat, protected by heavy cardboard.

OTHER

Abbreviations: Do not used excessively in text. Use only standard abbreviations; frequently used nonstandard abbreviations are acceptable. At first occurrence of an abbreviated term in text, write out completely and follow with abbreviation in parentheses. Use same abbreviations in both text and figures. Do not use in abstract. If a few abbreviations are used in text, list in a first-page footnote, "Abbreviations." If many neuroanatomical abbreviations are used, list separately at beginning of paper. List abbreviations used repeatedly in illustrations with first figure to facilitate reading.

Color Charges: $950 for one page. Second and subsequent pages, up to four, $500 each.

Experimentation, Animal: Include a clear description of method of anesthesia and killing of any animals.

Experimentation, Human: Provide adequate documentation to certify that appropriate ethical safeguards and protocols were followed.

Nomenclature, Anatomical: Adhere to *Nomina Anatomica* (5th ed., Williams and Wilkins, 1983).

Numbers: Spell out numbers when first word in a sentence; do not follow with abbreviations. Use Arabic numerals when numbers indicating time, weight, and measurements are followed by abbreviations. Treat numbers applicable to the same category alike throughout a paragraph (e.g., 2 male rats and 14 female rats; two male rats and fourteen female rats).

Page Charges: $18 per page for text pages and tables and $9 per page for illustrations, both line drawings and halftones.

Spelling: Use current *Webster's International Dictionary* for spelling of nontechnical terms. Either American or British spelling is acceptable; maintain one usage throughout paper. Use "parvicellular" instead of "parvocellular"; "brainstem" is one word instead of two.

Units: Use metric system for all measurements, weight, etc. Express temperatures in degrees Celsius (centigrade). Metric abbreviations, as listed below, should be lowercase without periods. See *J Comp Neurol* 1995;351(1): 163 for table of metric abbreviations. Make lowercase; do not use periods.

SOURCE

January 1995, 351(1): 161-167.
Guide to authors published in first issue of each year.

Journal of Computer Assisted Tomography

Raven Press

Giovanni Di Chiro, M.D.
Neuroimaging Branch
The Warren G. Magnuson Clinical Center
Building 10, Room 1C453
National Institutes of Health
Bethesda, Maryland 20892

AIM AND SCOPE

Original manuscripts, i.e., those that have not been published elsewhere except in abstract form, on basic and clinical aspects of reconstructive tomography will be considered for publication.

MANUSCRIPT SUBMISSION

Copyright on all articles will be held by the publisher, Raven Press. Each co-author must sign a statement expressly transferring copyright in the event of publication. A copyright transfer form will be sent when receipt of manuscript is acknowledged.

Submit three copies of each manuscript and a disk double-spaced, typed with 2 in. left margin. Number all manuscript pages, page 1 is title page.

MANUSCRIPT FORMAT

Page 1 (title page): title; name(s) of author(s) including first names, middle initials, and degrees of all authors; institution where work was done (indicate which author is in which department); short title (45 letters and spaces); corresponding and reprint request author(s) with complete address (including postal codes) and telephone number. Page 2: abstract (200 words) subdivided by headings "Objective" (introduction), "Materials and Methods," "Results," and "Conclusion," followed by five key words or terms for indexing. Place information concerning financial support in Acknowledgment section between Discussion and References.

Figures and Tables: Submit figures as two sets of clear glossy prints, with lettering large enough to be legible when reduced. Do not mount figures. Maximum final figure size is 7×9.5 in. On back of each give figure number and authors' names; indicate TOP with an arrow. Make legends fully explicit; include explanations for all figure sections and all abbreviations and arrows used. Excessive black-and-white illustrative material and all color illustrations will be charged to authors. Provide tables on separate sheets with title above and descriptions below.

References: Cite in text by number. Number in order cited. Type reference section double-spaced at end of text. Abbreviate journal titles according to *Index Medicus*. Spell out single-word journal titles. Give complete information for each reference, including titles of journal articles and names of all authors. Provide all authors' names when fewer than seven; when seven or more, list first three and add et al.

JOURNAL ARTICLE

1. Lawler GA, Pennock JM, Steiner RE, Jenkins WJ, Sherlock S, Young IR. Nuclear magnetic resonance (NMR) imaging in Wilson disease. *J Comput Assist Tomogr* 1983;7: 1–8.

CHAPTER IN BOOK

2. Deck MCF. Imaging techniques in the diagnosis of radiation damage to the nervous system. In: Gilbert HA, Kagan AR, eds, *Radiation damage to the nervous system: A delayed therapeutic hazard*. New York: Raven Press. 1980:107–27.

BOOK

3. Lee JKT, Sagel SS, Stanley RJ, eds. *Computed body tomography*. New York: Raven Press, 1983.

OTHER FORMATS

ANATOMICAL IMAGES AND VARIANTS—Six typed double-spaced pages, including references and figure legends, 8 references and 2-3 figures. Short descriptions of routinely seen but not noted anatomical structures.

CLINICAL IMAGES—Six typed double-spaced pages, including references and figure legends, 8 references and 2-3 figures. Short case reports with no abstract.

OTHER

Electronic Submission: Preferred storage medium is 5.25 or 3.5 in. disk in MS-DOS compatible format. Files in Macintosh-compatible format are also accepted. Submit in Microsoft Word, WordPerfect, WordStar, or XY-Write. Do not submit ASCII files. Label disk with author name, item title, journal title, type of equipment used, word processing program (with version number), and filenames. File on disk must be final corrected version of manuscript and must agree with accepted version of paper manuscript. Delete other files from disk.

Do not include extraneous formatting instructions. Use hard carriage returns only at ends of paragraphs and display lines (titles, subheads). Do not use an extra hard return between paragraphs. Do not use tabs for extra space at start of paragraphs or for list entries. Do not indent runover lines in references. Turn off line spacing. Turn off hyphenation and justification. Do not specify page breaks, page numbers or headers. Do not specify typeface. Correctly enter "one" (1) and lowercase "el" and 'zero' (0) and capital "oh" (O). Code nonstandard characters (Greek letters, mathematical symbols) consistently. List characters and codes used.

Use a single hyphen with space before it for a minus sign; use a double hyphen with space before and after to indicate a long dash in text; use a triple hyphen with no extra space to indicate a range of numbers.

Illustrations and tables are handled conventionally. Include figure and table legends at end of file.

Mathematical Material: In mathematical copy, be precise about alignment of numerals, letters, and symbols; hand-letter any that cannot be typewritten. Make a circled notation in nearest margin to identify possibly ambiguous characters, such as 1 (one or el), 0 (zero or oh), × (multiplication sign or ex), Greek letters, script letters, diacritical marks, and prime, and certain letters (e.g., C, K, P) in which capitals and lowercase are nearly indistinguishable when handwritten. Use simplest form of equation that can be made by ordinary mathematical calculation. Authors may be charged for substantial amounts of mathematical composition.

NMR-MR Terms: Use NMR-MR symbols and notations listed in "Glossary of NMR Terms," (American College of Radiology, 1983). Suggested orientations for NMR-MR image presentation are: Display axial and coronal sections with patient's right side on left side of image. Display head sagittal sections as if patient were erect and facing to viewer's left; no standards have been set yet for body sagittal views.

Radiation Units: Use either new International System of Units (SI) or old system. Report measurements in units in which they were made.

Scan Orientation: Orient axial CT images as if "seen from below" with patient's left on viewer's right. This is standard for both head and body scans. Present coronal views, whether direct scans or computer reconstructed, with patient's left on viewer's right.

SOURCE
January/February 1995, 19(1): front matter.

Journal of Dental Research
International Association for Dental Research
American Association for Dental Research

Dr. Mark C. Herzberg
School of Dentistry, University of Minnesota
17-164 Moos Tower, 515 Delaware ST., S.E.
Minneapolis, MN 55455
Fax: (612) 624-8958

AIM AND SCOPE

The *Journal* seeks to publish definitive research reports of wide interest to the research community. Topical, concise reviews of the state of the art will be considered. Reports of observations and the development of new methods of techniques may be considered if they are of broad and fundamental interest. Submission of case reports is discouraged.

MANUSCRIPT SUBMISSION

In cover letter certify that research is original; not under consideration for publication elsewhere; free of conflict of interest; and conducted by principles of human subject and animal welfare. All authors must sign, indicating approval of complete content.

Include complete mailing addresses, telephone and fax numbers, and electronic mail addresses. Include copies of "in press" and "submitted" manuscripts that provide important information.

All rights in manuscripts shall be transferred to the *Journal of Dental Research* upon submission. Submission constitutes each author's agreement that the *Journal* holds all proprietary rights in manuscript submitted, including all copyrights. On acceptance, authors will be asked to sign a formal transfer of copyright, before publication.

MANUSCRIPT FORMAT

Submit manuscripts in triplicate, including figures. Prepare manuscript, tables, legends and footnotes as double-spaced text on 8.5 × 11 in. bond paper. Top, bottom, and side margins should be 1 in., with no indented paragraphs. Use a laser printer. Figures and tables should not exceed 8.5 × 11 in. Provide additional copy on computer diskette. Use bold type for title on page 1 and headings: Abstract, Introduction., Materials and methods, Results., Discussion., Acknowledgments, References. Use upper and lowercase letters. Use bold type for section subheads. Italicize section headings of still lower rank. Type no more than 10 characters per inch.

Show a clear chronological progression and logic to development of ideas throughout manuscript and within paragraphs and sentences. Use a direct and straight forward voice. State purpose, provide background, date and conclusions. Illustrate with examples. Stick to the subject. Make each sentence a single thought. Write short simple sentences. Avoid jargon and needless words. Review manuscript with computer grammar and spelling check software and colleagues expert in grammar and syntax. Ensure that all listed references, figures and tables are cited in text and that all cited references, figures and tables are in appropriate sections.

Manuscript components are: title page, Abstract, Introduction, Materials and methods, Results, Discussion, Acknowledgment, References, Tables, and Figure legends. Arrange manuscript in that order. Number pages consecutively in top right corner, including title page. Label Figures on back with authors' abbreviated names and number corresponding to citation in text. Indicate TOP, if not obvious.

Title Page (page 1): Use bold type; capitalize only first letters of main words. Type authors' initials and last names in upper- and lowercase letters. Use a superscript number system to relate authors to different departments or institutions or to indicate a change in address.

Give full postal address (including ZIP or Postal Code), e-mail address and telephone and fax numbers of reprint request author. If not first author, indicate by a number superscript, with phrase "To whom correspondence and reprint requests should be addressed."

Include short title (running head, 45 characters) and three to five key word. If applicable, include source footnotes on page 1 to indicate prior preliminary publication.

Report all sources of funding in Acknowledgments.

Abstract (page 2): 250 words (one typed page) self-standing summary of text describing background, hypothesis or study objective, design and key methods, essential results, and conclusions. Avoid abbreviations.

Introduction (page 3): Give background and rationale for hypothesis or study objective. Give sufficient detail to allow interdisciplinary readers to evaluate results without review of other publications. Describe and cite only most relevant earlier studies. Avoid an exhaustive review of the field.

Materials and Methods: Provide sufficient technical information so experiments can be repeated. State experimental or study design, specific procedures, and type of statistical analyses performed. Use section subheadings in a logical order to title each category or method. Name and cite previously published methods. Describe new methods completely. Present data that validate new methods. Describe a method used only for part of one experiment in Results section, Table footnote, or figure legend. Present descriptive information about large numbers of experimental reagents, microbes, test materials, etc. in tabular form with a brief explanation in text. Give proprietary names and sources of supply of all commercial products in parentheses in text (name and model of product company, city, and state or country).

Results: Introduce data in text, Tables and Figures. Call attention to their significant parts. Include rationale and design of experiments as needed. Reserve subjective comments, interpretation, or reference to previous literature for Discussion. Report results concisely, use Tables and Figures to present important differences or similarities that cannot otherwise be presented or summarized in text. Number Tables and figures in order described and cited in text.

Tabular data should report either standard deviation values or standard errors of the means; number of replicate determinations or animals; and probability values and name of statistical test for all differences reported to be statistically different. Restrict photo- and electron-micrographs to those essential to results. If essential for results, color can be published; call 202-898-1050 for estimates.

Discussion: Explain and interpret results with a scientifically critical view of previously published work. Highlight advances made by new data. Indicate limitations of findings. State conclusions and why they are merited by data. This is the only proper section for subjective comments.

References: List all sources cited in paper. Arrange citations in alphabetical order by last name of first author without numbering. When citing a reference in text, provide attribution for subject under discussion. For example: "Cold fusion has been difficult to replicate Williams and Jones (1988), but recent modifications (Jones *et al.,* 1982) continue to . . ." Avoid "Jones et *al.* (1989) found . . ." or "In a recent study, Jones (1990) found. . .." Use "et *al.*" (in italics) when work is by three or more authors. When cited work is by two authors, use both surnames separated by "and." List multiple references by same author in same year, using "a," "b," etc. (*e.g.*, Jones, 1980b). List multiple references in chronological order, separated by semi-colons. Insert and cite "unpublished observations" and "personal communications" in parentheses in text; do not use as references. Obtain written permission from correspondent.

Abbreviate journal titles according to *Index Medicus.* Avoid using abstracts as references. Include among references papers accepted but not yet published; designate journal and add "(in press)." Cite information from manuscripts submitted but not yet accepted in text as "unpublished observations" (in parentheses). Verify against original documents and check for correspondence between references cited in text and in reference list.

ARTICLES IN JOURNALS

Standard journal article (List all authors, if more than six, give six followed by *et al.*)

West DJ, Snavely DB, Zajac BA, Brown GW, Babb CJ (1990). Development and persistence of antibody in a hgih-risk institutionalized population given plasma-derived hepatitis B vaccine. *Vaccine* 8:111-114.

Organization as author

The Royal Marsden Hospital Bone-Marrow Transplantation Team (1977). Failure of syngeneic bone-marrow graft without preconditioning in post hepatitis marrow aplasia. *Lancet* 2:742–744.

No author given

Coffee drinking and cancer of the pancreas (editorial) (1981). *Br Med J* 283:628.

Article in a foreign language

Massone L, Borghi S, Pestarino A. Piccini R, Cambini C (1987). Localizations palmaires purpuriques de la dermatite herpetiforme. *Ann Dermatol Venereol* 114:1545–1547.

Volume with supplement

Magni F, Rossoni G, Berti F (1988). BN-52021 protects

guinea pig from heart anaphylaxis. *Pharmacol Res Commun* 20(Suppl 5):75–78.

Issue with supplement

Gardos G, Cole JO, Haskell D, Marby D, Paine SS, Moore P (1988). The natural history of tardive dyskinesia. *J Clin Psychopharmacol* 8(4Suppl):31S–37S.

Volume with part

Hanly C (1988). Metaphysics and innateness: a psychoanalytic perspective. *Int J Psychoanal* 69(Pt 3):389–399.

Issue with part

Edwards L, Meyskens F, Levine N (1989). Effect of oral isotretinoin on dysplastic nevi. *J Am Acad Dermatol* 20(2 Pt 1):257–260.

Issue with no volume

Baumeister AA (1978). Origins and control of stereotyped movements. *Monogr Am Assoc Ment Defic* (3):353–384.

No issue or volume

Danoek K (1982). Skiing in and through the history of medicine. *Nord Medicinhist Arsb*:86–100.

Pagination in Roman numerals

Ronne Y (1989). Ansvarsfall. Blodtransfusion till fel patient. *Vardfacket* 13:XXVI-XXVII.

Type of article indicated as needed

Spargo PM, Manners JM (1989). DDAVP and open heart surgery (letter). *Anaesthesia* 44:363–364.

Fuhrman SA, Joiner KA (1987). Binding of the third component of complement C3 by *Toxoplasma gondii* (abstract). *Clin Res* 35:475A.

Article containing retraction

Shishido A (1980). Retraction notice: Effect of platinum compounds on murine lymphocyte mitogenesis (Retraction of Alsabti EA, Ghalib ON, Salem MH. In: *Jpn J Med Sci Biol* 1979;32:53–65). *Jpn J Med Sci Biol* 33:235–237.

Article retracted

Alsabti EA, Ghalib ON, Salem MH (1979). Effect of platinum compounds on murine lymphocyte mitogenesis (Retracted by Shishido A. In: *Jpn J Med Sci Biol* 33:235–237,1980). *Jpn J Med Sci Biol* 32:53–65.

Article containing comment

Piccoli A, Bossatti A (1989). Early steroid therapy in IgA neuropathy: still an open question (comment). *Nephron* 48:12–17. Comment on: *Nephron* 51:289–291, 1989.

Article commented on

Kobayashi Y, Fujii K, Hiki Y, Tateno S, Kurokawa A, Kamiyama M (1988). Steroid therapy in IgA nephropathy: a retrospective study in heavy proteinuric cases (see comments). *Nephron* 48:12–17. Comment in: *Nephron* 51:289–291, 1989.

Article with published erratum

Schofield A (1988). The CAGE questionnaire and psychological health (published erratum appears in *Br J Addict* 84:701, 1989). *Br J Addict* 83:761–764.

BOOKS AND OTHER MONOGRAPHS

Personal author(s)

Colson JH, Armour WJ (1986). Sports injuries and their treatment. 2nd rev. ed. London: Butterworth-Heiniemann.

Editor(s), compiler as author

Diener HC, Wilkinson M, editors (1988). Drug-induced headache. New York: Springer-Verlag.

Organization as author and publisher

Virginia Law Foundation (1987). The medical and legal implications of AIDS. Charlottesville: The Foundation.

Chapters in a book

Weinstein L, Swartz MN (1974). Pathologic properties of invading microorganisms. In: Pathologic physiology: mechanisms of disease. Sodeman WA Jr, Sodeman WA, editors. Philadelphia: Saunders, pp. 457–472.

Conference proceedings

Vivian VL, editor (1985). Child abuse and neglect: a medical community response. Proceedings of the First AMA National Conference on Child Abuse and Neglect: Mar 30–31, 1984, Chicago. Chicago: American Medical Association.

Conference paper

Harley NH (1985). Comparing radon daughter dosimetric and risk models. In: Indoor air and human health. Proceedings of the Seventh Life Sciences Symposium, Oct 29–31, 1984, Knoxville (TN). Gammage RB, Kaye SV, editors. Chelsea (MI): Lewis, 69–78.

Scientific and technical report

Akutsu T (1974). Total heart replacement device. Apr. Report No.: NIH-NHLI-69-2185-4. Bethesda (MD): National Institutes of Health, National Heart and Lung Institute.

Dissertation

Youssef NM (1988). School adjustment of children with congenital heart disease (dissertation). Pittsburgh (PA): Univ. of Pittsburgh.

Patent

Harred JF, Knight AR, McIntyre JS, inventors (1972). Dow Chemical Company, assignee. Epoxidation process. US patent 3,654,317. Apr 4.

OTHER PUBLISHED MATERIAL

Newspaper article

Rensberger B, Specter B (1989). CFCs may be destroyed by natural process. *The Washington Post* Aug 7:Sect. A:2 (col. 5).

Audiovisual

AIDS epidemic: the physician's role (videorecording) (1987). Cleveland (OH): Academy of Medicine of Cleveland.

Computer file

Renal system (computer program) (1988). MS-DOS version. Edwardsville, KS: Medi-Sim.

Legal material

Toxic Substances Control Act: Hearing on S. 776 Before the Subcomm. on the Environment of the Senate Comm. on Commerce. 94th Cong., 1st Sess. 343 (1975).

Map
Scotland (topographic map) (1981). Washington: National Geographic Society.

Book of the Bible
Ruth 3:1–18. The Holy Bible. Authorized King James version (1972 ed.). New York: Oxford Univ. Press, 1972.

Dictionary and similar references
Ectasia. Dorland's illustrated medical dictionary. 27th ed. (1988). Philadelphia: Saunders, p. 527.

Classical material
The Winter's Tale: act 5, scene 1, lines 13–16. The complete works of William Shakespeare (1973). London: Rex, 1973.

UNPUBLISHED MATERIAL

In press
Lillywhite HB, Donald JA. Pulmonary blood flow regulation in an aquatic snake. *Science* (in press).

Tables: Type one per page. Number consecutively with Arabic numerals in order of mention in text. In heading follow table number with a brief descriptive title. Avoid overlong titles and cumbersome Tables, use explanatory footnotes. In Table or title indicate order of footnotes with superscript a,b,c,d... If needed in footnotes, cite in parentheses short form of all references. In tabular columns and text, decimals less than unity must have decimal point preceded by a zero. Report only number of significant digits appropriate to sensitivity and discrimination of measure and differences to be illustrated. Make column headings simple and clear so Tables will be understandable without consulting text. Tables will be printed either 3.14 or 7 in. wide.

Figures: Figures are illustrative materials, including photomicrographs, radiographs, charts, and graphs. When possible submit electronic copy. Discuss figures thoroughly in text. Black-and-white photographic prints, laser-quality reproductions, and original drawings on opaque white paper are preferred. Color is published at discretion of editor; call 202-898-1050 for cost estimate.

Submit original and two copies of each Figure. Place each set in a protective folder or envelope, one labeled "for printer" and two "for review." On back of each, identify TOP, and label with authors' names and Figure number and letter in order of mention in text. Figures will be printed approximately 3.25 or 7 in. wide. Crop out extraneous material. Photomicrographs must include a scale, clearly labeled with a convenient unit of length. Label graphs at abscissa and ordinate, including units of measure. Make height of lettering and numbers on axes of graphs no smaller than 3 millimeters after reduction. Title and identification may appear in legend.

Legends: Type together on a separate page. Make understandable without reference to text. Include a key for symbols or abbreviations used in Figure.

OTHER FORMATS

CONFERENCE REPORTS—Topical and brief, highlight important new data or findings.

GUEST EDITORIALS—2000 words, describing a clear and substantiated position on issues of interest to the community.

LETTERS TO THE EDITOR—500 words. Include evidence to support a position about scientific or editorial content of *Journal*.

RAPID COMMUNICATIONS—Brief, definitive reports of unusual significance. Request review for rapid publication in cover letter. Suggest four potential referees who are not members of their institution or present or former collaborators with names, addresses, telephone and fax numbers, and are of expertise.

OTHER

Electronic Submission: Use either Macintosh (Framemaker, MacWrite, Word, WordPerfect, Works WP, or WriteNow) or IBM PC (DCA-RFT, Framemaker, Multimate, Office Writer, Text, Word for Windows, WordPerfect, WordStar, Works WP, or XYWrite). Manuscripts should be clean, free of tabs and codes. Use bold type and italic type where they will be used on printed page. Italicize genus and species of an organism, g (gravitational force), Latin words and abbreviations, and journal names in References section. Use tabs to separate columns in tables.

Experimentation, Animal: Indicate level of review and assurance that protocol ensures humane practices.

Experimentation, Human: Indicate succinctly that subject's rights were protected and informed consent obtained.

Nomenclature: Use S.I. units; see *The International System of Units (SI)*, D.T. Goldman and R.J. Bell, Eds., NBS Special Publication 330 (1981, U.S. Department of Commerce, National Institute of Standards and Technology, Washington, D.C.). Use correct symbol for μ, and L (not l) for liter (as in mL, μL, etc.). Express grams as g (not gm), hours as h, seconds as s, and centrifugal force as g. Use nm rather than Angstroms. Express concentrations as mol/L or mmol/L, etc. Insert leading zeros in all numbers less than 1.0 in text, Tables, and Figures.

Give complete names of individual teeth in full in text (e.g., "permanent upper right first premolar"). In Tables, abbreviate names by Viohl's Two-digit System. According to the Two-digit System (approved by the International Standards Organization), the first digit indicates the quadrant and the second digit the type of tooth within the quadrant. Starting at upper right side and rotating clockwise, quadrants are assigned digits 1 to 4 for permanent and 5 to 8 for deciduous teeth; within the same quadrant teeth from the midline backward are assigned digits 1 to 8 (deciduous teeth, 1 to 5)

Refer microorganisms in accordance with International Rules of Nomenclature. Nomenclature for bacteria follows *Bergey's Manual of Systematic Bacteriology* (current edition). Make first reference to an organism by genus and species in full (e.g., *Lactobacillus casei*); subsequent mention may abbreviate genus (*L. casei*). When a common name of a bacterium or group is mentioned, do not italicize ("some lactobacilli").

Write out numbers of ten and fewer, except when indicating inanimate quantities; numbers greater than ten should appear as digits. Use digits to express dates, dimensions, degrees, doses, periods of time, percentages, proportions, ratios, sums of money, statistical results, weights, and measures, or to enumerate animals (but not people), culture cells and organisms, organs, and teeth. Leave a space between numbers and units and around = and ± signs.

Nucleotide Sequences: Deposit information in appropriate database (e.g., GenBank/EMBL). Provide accession numbers. A footnote will indicate accession number and database.

Reviewers: In cover letter nominate appropriate independent scientific referees; include names, mailing addresses, telephone and fax numbers, and area of expertise.

Spelling: *The Random House Dictionary of the English Language* (Unabridged) is authority for nonmedical terms. Where two plural forms are provided, use American English form. For anatomical nomenclature, see *Nomina Anatomica* (5th ed.) or *Dorland's Illustrated Dictionary*.

Style: Consult *Council of Biology Editors Style Manual* (current edition) as well as current issues of *Journal* for examples of format not specified here.

SOURCE
January 1994, 73(1): 75-78.

Journal of Endocrinology
Society for Endocrinology

Editor, *Journal of Endocrinology*
17 North Court, The Courtyard
Woodlands, Almondsbury
Bristol BS12 4NQ, U.K.
Telephone: 0454 616046

AIM AND SCOPE
All aspects of the nature and functions of the endocrine systems.

MANUSCRIPT SUBMISSION
Submit four copies of manuscript and four sets of plates. One set of line drawings needs to be publishable quality; others may be photocopies.

Give written assurance that work has not and will not be submitted for publication elsewhere until the *Journal* has reached a decision on whether to publish paper. If accepted the assurance automatically extends indefinitely. It is assumed that submitted manuscripts are approved by all authors. Provide fax number.

MANUSCRIPT FORMAT
Print manuscripts double-spaced, at least 10-point and in a Times font. Provide duplicate copies of papers quoted as In Press, particularly those concerned with methodology.

Do not supply document on computer disk. A word processor file will be requested if manuscript is accepted. Provide files on PC (DOS)-formatted disks. Preferred format is WordPerfect; other packages can be translated. Supply an ASCII (text only) file also. Put text, references, figure legends, and table headings on disk. Supply diagrams, photographs and other tabular material separately.

Title Page: Complete title, names of all authors, full addresses of all institutions and a short title for page headings.

Abstract: 5% of length of paper. Summarize main findings and conclusions reached.

Text: Follow format: Introduction, Materials and Methods, Results, Discussion.

References: In text, cite by name of first author *et al.* where more than two authors. In Reference List give all authors; write out titles of journals in full.

Tables: Produce separate pages. Give each a full and informative heading.

Figures: Number all illustrations. Put on separate pages, with first author's name. Make line drawings good quality originals. Do not photograph originals of line drawings. Avoid three-dimensional graphs to show two-dimensional data. Photo- and electron-micrographs should contain scale bars.

Cost of color reproduction over and above that for black-and-white is charged to authors. Enable readers to understand information in figures and plates without reference to text.

OTHER FORMATS
COMMENTARIES—Four printed pages (2500 words) of commentary on topical subjects.

RAPID COMMUNICATIONS—Observations must be novel and of immediate relevance. Describe methods so as to permit repetition; results must be adequate to justify claims. Include a brief statement of reasons for requesting rapid publication.

These papers are printed from author's typescript. Submit paper in usual format, if accepted, provide manuscript either pasted on forms provided by Editorial Office or as laser printed original, printed in two columns (except for title and author) within 6.5 × 8.5 in. text area. Position figures and tables within text in either format. Do not exceed four pages.

REVIEWS—Short reviews will be considered and are subject to the usual refereeing process. There are no special requirements for organization of material.

OTHER
Experimentation, Animal: Conform with *Guidelines on the Handling and Training of Laboratory Animals* (Universities Federation for Animal Welfare, U.K.). Papers which do not conform with these guidelines or with U.K. legal requirements cannot be accepted.

Experimentation, Human: Include a statement that consent has been obtained from each patient or subject after full explanation of purpose and nature of procedures. State that investigation was approved by local Ethical Committee, functioning according to guidelines of the Royal College of Physicians of London in September 1984, and amended in November 1984 and August 1986. Obtain copies from Medical Research Council, London, and

Guidelines on the Practice of Ethical Committees in Medical Research (Royal College of Physicians, London).

Spelling: Follow *Shorter Oxford English Dictionary.* Define all abbreviations.

SOURCE

January 1995, 144(1): back of each issue.

Journal of Eukaryotic Microbiology
Society of Protozoologists

Edna S. Kaneshiro
Department of Biological Sciences
University of Cincinnati
Cincinnati, OH 45221-0006
Telephone: (513) 556-4351; Fax: (513) 556-5280

AIM AND SCOPE

The Journal of Eukaryotic Microbiology is a publication of the Society of Protozoologists. Six issues are published per year on original research on protists, including lower algae and fungi, and covering all aspects of such organisms, e.g., cell biology, ultrastructure, morphogenetics, biochemistry, physiology, behavior, molecular biology, genetics, parasitology, chemotherapy, systematics, ecology and evolution. Multicellular microorganisms such as nematodes and rotifers per se are not included unless they are relevant to the article's major topic—i.e., protists. Manuscripts are solicited from members and nonmembers of the Society of Protozoologists from around the world.

MANUSCRIPT SUBMISSION

Photocopy checklist in any current issue; do not tear out. Check items and submit with four sets of manuscript and four sets of illustrations (original and three copies). Photographic illustrations must be glossy prints (not Xerox or laser copies).

Send a cover letter including a statement regarding prior publication and other works in review. With the exception of invited reviews, *Journal* will not publish articles that contain data that have been published or submitted for publication elsewhere. State which authors are members of the Society of Protozoologists and whether page charges should be billed to member authors in any order other than as the authors are listed on title page. If none are Society members, specify whom to bill, if not corresponding author.

MANUSCRIPT FORMAT

Type manuscripts on 8.5 × 11 in. paper (do not use dot-matrix printer). Use an easy-to-read typeface, that distinguishes ones from els and ohs from zeroes. Leave a 1 in. margin on all sides. Double-space throughout, including Title Page, Tables, Footnotes, Figure Legends, and Literature Cited. Do not use proportional spacing or justified margin. Avoid hyphens at ends of lines. Number pages consecutively in upper right corner. Use an unnumbered footnote for extensive/unusual abbreviations in text. Underline words to be italicized, even if italic type is used. Do not italicize common Latin words or phrases. Cite each figure and table in text. Organize text so they are cited in numerical order. Note table and figure insertion in left margin at first mention.

Cite each reference listed in Literature Cited section in text and vice versa. Double check spelling and details of publication. Include strain and subspecies designations in Materials & Methods section. Use "Figure" to start a sentence; otherwise use "Fig." even if plural.

Assemble manuscript in order: Title page; Abstract page; Text; Acknowledgments; Literature Cited; Footnotes; Tables; Figure legends; and Figures.

Title Page (Page 1, do not number): Place running head near top of page in all capital letters: author's surname(s), a long dash, and a short title (total characters/spaces ≤ 60). Center title near middle of page. Capitalize first letter of all words except articles, prepositions and coordinate conjunctions. Below title, include authors' names (First name, middle initial, last name); capitalize all letters. Denote affiliations using superscript asterisks (* ** ***). Use superscript number(s) following authors' names to designate corresponding author and current addresses in footnotes. Below names, list addresses in order of affiliation (* ** ***). Include street addresses or P.O. boxes if necessary. Include full mailing address, telephone and fax numbers, for corresponding author. Place information near bottom of title page.

Abstract Page (Page 2): Abstract (≤ 200 words) must be one paragraph and self-explanatory. Begin with "ABSTRACT." Include brief statements as to intent, methods, results, and significance. Do not cite references or use abbreviations/acronyms. Following abstract, alphabetically list key words or phrases (< 10) not already in title under heading "Supplementary key words."

Text (begin on page 3): Begin with a brief introduction, do not use heading "INTRODUCTION." Use headings in order: MATERIALS AND METHODS, RESULTS, DISCUSSION. Use all capital letters and center on one line. Indent subheads and draw a wavy line underneath to indicate boldface type. Capitalize first letter of first word only. Lower level subheads are typeset italic (straight underline).

Reference published materials (cited in Literature Cited section) within text by number, in square brackets. Reference unpublished materials as: (ESK, pers. observ.), or (ESK, unpubl. data) for an author; (J. Jones, unpubl. data) or (J. Jones, pers. commun.) for other. Cite Ph.D. dissertations, Master's theses, and abstracts in parentheses within text.

Literature Cited: Do not include: abstracts, articles submitted or in preparation, master's theses, Ph.D. dissertations. Cite within text. List all authors. (Do not use et al. or a long dash. Arrange references alphabetically by surname of first author. References by a single author precede multi-authored works by same senior author, regardless of date. List works by same author(s) chronologically, beginning with earliest date. List multi-authored works as:

1. Bruns, P. J. 1984.
2. Bruns, P. J. & Brussard, T. B. 1974a.
3. Bruns, P. J. & Brussard, T. B. 1974b.
4. Bruns, P. J. & Brussard, T. B. 1981.
5. Bruns, P. J. & Sanford, Y. M. 1978.

6. Bruns, P. J., Brussard, T. B. & Kaveka, A. V. 1976.

7. Bruns, P. J., Brussard, T. B. & Merriam, E. V. 1983.

8. Bruns, P. J., Katzen, A. L., Martin, L. & Blackburn, E. H. 1983.

JOURNAL ARTICLES

Koroly, J. J. & Conner, R. L. 1976. Unsaturated fatty acid biosynthesis in *Tetrahymena*. Evidence for two pathways, *J. Biol. Chem.*, **251**:7588–7592.

ARTICLES IN BOOKS

Van Houton, J., Hauser, D. C. R. & Levandowsky, M. 1981. Chemosensory behavior in protozoa. *IN*: Levandowsky, M. & Hutner, S. H. (ed.), Biochemistry and Physiology of Protozoa, 2nd ed. Academic Press, New York, New York, **4**:67–124.

BOOKS

Fenchel, T. 1987. Ecology of Protozoa. The Biology of Free-living Phagotrophic Protists. Science Tech Publishers, Madison, Wisconsin.

Verify all entries against original sources, especially journal titles, accents, diacritical marks, and spelling in languages other than English. Capitalize all nouns in German. "In press" citations must have been accepted for publication; include name of journal or publisher. Abbreviate and underline journal names. Spell out if one word.

Tables: Start each on a separate sheet; double-spaced. Use legal-size paper if necessary. Do not reduce type size of tables; use same size type as in text. Indicate footnotes by lowercase superscript letters. Do not reduce type size. Do not use vertical lines. Do not combine a table and a figure.

Figure Legends: Make self-explanatory. Explain all symbols in figure in legend, and vice versa. (If there are a great number of letters/symbols in figures, include a section headed "Figure Abbreviations" and list alphabetically. Put information just below first figure legend separated by a quadruple space.)

Double-space legends and group according to figure arrangements. Quadruple space between groups. (Do not use a separate page for each group.) Group legends for plates with multiple figures together. Provide a title for the entire plate; follow with a description for each individual figure in the plate.

Type legends in paragraph form, starting with the inclusive numbers AND an inclusive title that covers all figures of plate:

Fig. **3-6.** Life cycle of *Gregarina coronata* n. sp. **3.** Trophozoites. **4.** Immature gamonts. **5.** Mature gamonts. **6.** Oocysts.

Figures: Illustrations may be either black-and-white photographs or half-tones or line drawings, or graphs. Do not combine line drawings and half-tones. If both are necessary, place line elements on photograph or rephotograph figure and send that photograph as original.

Figures should appear in order cited in text. Arrange consecutively and assign Arabic numbers.

Mount original (printer's set) half-tones and color illustrations on thin white poster board or heavy white paper. Do not use heavy boards such as matting board. Affix with dry mount adhesive or equivalent, making sure all corners and edges are glued down. Leave a 1 in. margin on all sides. Do not write on back of image area. Protect illustrations with nonabrasive dust covers. Prepare illustrations using professional standards. Flaws will be returned.

Do not mount graphs, line drawings, and reviewers' copies of photographs.

Submit original figures in final journal size or slightly larger; if oversize, they must fit in a 10 × 13 in. envelope. Indicate on back of illustrations such as plates of micrographs whether one column (89 mm) or two columns (181 mm) width is required. Figures may be resized by Editor.

Maximum published length is 231 mm (minus space for legend; maximum width is 89 mm (1 column) and 181 mm (2 column).

Add symbols with press-on symbols and letters (preferably Helvetica or Franklin Gothic); handwritten or typed symbols are not acceptable.

Group related drawings to form a plate of drawings. Group related photos into one or more plates, each with glossy photos butted together with no space between photos. Each figure grouped in a plate requires a number or capital letter (at least 3 mm after reduction); number single figures in legend only (NOT on figure itself). Make letters and numbers on figure itself large enough to be 2 mm in height after reduction. Do not place "legends" within a graph or figure.

Illustrations of highly magnified areas require a bar scale; include a numerical magnification in caption.

Reviewers copies of half-tone figures and plates must be photographic reproductions approaching quality of originals; machine-reproduced copies are unacceptable.

Write author(s) name(s) and figure number(s) on back of each figure or plate, on both originals and review copies.

Consult editor about color reproductions for which authors will be charged.

OTHER FORMATS

RAPID COMMUNICATIONS—Regular articles or short communications prepared camera-ready using desk-top publishing procedures and high-quality post-script laser printer with fonts and formatting that produce printing similar to a type-set page in *Journal*. They will be fully reviewed and are not preliminary reports. Contact *Journal* office for detailed instructions.

SHORT COMMUNICATIONS—1500 words, 100 word abstract; if they include one figure and one table limit is 1200 words. Style and format are identical to full-length papers except use a single RESULTS AND DISCUSSION section. Papers must be complete in themselves, present sufficient information to allow repetition of results.

OTHER

Abbreviations: Write out acronyms and abbreviations first time used in text, abbreviate thereafter unless at beginning of sentence. Do not pluralize. Use these abbreviations without spelling out; h, min, s, yr. mo, wk, d, diam, cm, mm; designate temperature as 30° C.

Classification: Become familiar with classification system for protists. See Levine et al. *The Journal of Protozoology* (Feb. 1980) and Lee et al. *The Illustrated Guide to the Protozoa*, (Society of Protozoologists, Allen Press, Lawrence, Kansas, 1985). Authors are not required to adopt names or scheme of classification used/proposed therein but, if using or proposing different suprafamilial ranks or names (including spellings), provide rationale for changes.

New Taxa: Use standard abbreviations after proposed name (generally in title, abstract and following at least its first usage in text and in figure explanations), as appropriate (n. sp., n. g., n. fam., n. ord., n. cl., etc.). Deposit a fixed specimen in a recognized repository and provide an accession number in manuscript. See Bandoni & Duszynski, 1988, *J. Parasitol.*, **74:**519–523. Deposit fixed specimens with:

U.S. National Parasite Collection, Dr. J. Ralph Lichtenfels, Curator, USDA, ARS, BARC-East No. 1180, Beltsville, MD 20705, Telephone: (301) 504-8444.

International Protozoan Type Slide Collection, Dr. Klaus Ruetzler, Curator, Department of Invertebrate Zoology, National Museum of Natural History, MRC 263, Smithsonian Institution, Washington, D.C. 20560, Telephone: (202) 786-2130.

University of Nebraska State Museum, Prof. Mary Hanson Pritchard, Curator, 16th and W Streets, Lincoln, Nebraska 68588, Telephone: (402) 472-3334.

For living cultures, deposit strain in a recognized reference collection such as:

American Type Culture Collection, 12301 Parklawn Drive, Rockville, MD 20852, teelephone: (301) 231-5516, Culture Center for Algae and Protozoa, The Ferry House, Ambleside, Cumbria, LA22 OLP, England, U.K.

Numbers: Write out one to nine unless a measurement. Exceptions are made for standard usage (e.g., 9 + 2). Use 1,000 not 1000; 0.13 not .13% not percent. Spell out numbers and abbreviations when they occur at beginnings of sentences.

Do not hyphenate a measurement (unless used as a compound adjective).

Page Charges: $50 per page. Regular subscribing, honorary, and emeritus members of the Society receive five free pages per year (one paper only). No page charges for symposia, past-President's addresses, invited reviews, book reviews, and in memoriam articles.

Reviewers: Provide a list of suggested competent reviewers and potential reviewers constituting conflicts of interest in cover letter.

Style: Consult 5th ed. *Council of Biology Editors Style Manual* for general information on style and abbreviations. For chemical and mathematical terms, formula, abbreviations, etc. follow style in *Journal of Biological Chemistry* (see "Instructions to Authors" in the first issue of each volume in that journal).

Subspecies and Strain Designations: Acknowledge sources (e.g., American Type Culture Collection, Culture Center for Algae and Protozoa, etc.) in Materials and Methods section or give in appropriate tables. If not obtained from a recognized reference collection, indicate from whom and where strain was obtained and provide a history (e.g., obtained from Ching Kung, University of Wisconsin, who isolated the mutant).

Units: Use metric system in all instances. Express speed of centrifugation in terms of gravitational force (g), not revolutions per minute (rpm).

SOURCE

January/February 1994, 41(1): 90-93,

Journal of Experimental Biology
The Company of Biologists, Ltd.

R. G. Boutilier
M. Burrows
Department of Zoology
Downing Street
Cambridge CB2 3EJ, U.K.
Telephone: (01223) 311788 or 336628; Fax: (01223) 353980 or 336676
H. Hoppeler
Department of Anatomy
University of Bern
Buhlstrasse 26
Bern 9 CH-3000, Switzerland
Telephone: 010-41-31-631-4637; Fax: 010-41-31-631-3737

AIM AND SCOPE

The *JEB* publish papers on the form and function of living organisms at all levels of biological organization, from the molecular and subcellular to the integrated whole animal. Our authors and readers reflect a broad interdisciplinary group of scientists who study molecular, cellular and organismal physiology in an evolutionary and environmentally based context.

MANUSCRIPT SUBMISSION

Supply address, telephone and fax numbers, and E-mail address for correspondence about manuscript. Enclose three copies of text and figures. Authors are required to assign to Company of Biologists Limited copyright of articles published in Journal.

For Customs purposes, mark all packages as 'No Commercial Value,' otherwise charges are levied.

MANUSCRIPT FORMAT

Type double-spaced, with numbered pages in order: Title page, Summary (150–200 words), Introduction, Materials and methods, Results, Discussion, References, Figure legends, Tables.

Tables: Type each on a separate page with heading and footnotes. Figures and Tables should be numbered in order of appearance in text.

References: List alphabetically. Abbreviate titles according to *World List of Scientific Periodicals*, 4th edn.

KNOWLES, R. G. AND MONCADA, S. (1994). Nitric oxide syntheses in mammals. *Biochem. J.* **298**, 249–258.

JONES R. (1994). Cell division. In *The Cell,* vol. 6 (ed. J. Brachet and A. E. Mirsky), pp. 77–90. New York, London: Academic Press.

Citations in text read: Jones and Smith (1994) or (Jones and Smith, 1994; Jones *et al.* 1994). When more than two authors, use Jones *et al.*

llustrations: Submit originals as unlabeled art work on heavy white paper or as well-defined photographic copies, not on tracing or graph paper. Computer printouts are accepted if quality is sufficient to show fine detail. Do not use fine stippling or shading.

Number in a single series (Fig. 1., Figs 2, 3, etc.), in order cited in text. Type legends for all illustrations together, separately from main text. Submit three copies. Copies of photographic plates must be suitable for reviewers. Send original figures with revised manuscript. Color illustrations are reproduced free of charge at discretion of editor. Charges for excessive use of color is also at editor's discretion

Observe measurements: Maximum depth: 210 mm. Single column width: 88 mm; 2/3 page width: 122 mm; full page width: 183 mm.

Computer-Generated Line Drawings—Create to fit within column widths. Save as separate files: Fig. 1, etc. If artwork is final size, use 9 pt Times for text labelling, except for asterisk (12 pt). Ensure that symbols are of suitable size for clarity.

Subscript and superscript characters: 7 pt, or if application allows it: Superscript: 9 pt at 75%, offset 3 pt; Subscript: as above, offset 1.5 pt; 2nd-order subscript: 6 pt (75%) offset 2.5 pt. Minus sign: select Symbol font and use hyphen key (Times and Symbol font numbers are same). Make numbers horizontal not vertical.

When space is tight, use pointers or arrows, so lettering can be moved into a space. Avoid cross hatching unless program has a 'Preferences for printing resolution' option that can be set at 300 dpi. If not available, use ▢ and ■.

Grayscale—Do not use less than 20% or they may not be visible when printed. Try to keep a 10% difference in the levels of gray, so they are easily distinguished.

Rules—Do not use hairline rule, as this may be almost invisible when printed. Line drawings may have to be reduced by 50% to fit column widths if not prepared to final size.

For Macintosh (Canvas, CricketDraw, CricketGraph, ClarisDraw, ClarisWorks, DeltaGraph, Freehand, Illustrator, Kaleidagraph, MacDraw, MS Excel, Pagemaker, Persausion, Powerpoint, Quark Xpress, Superspaint, submit labelled computer files and one printout. Save ClarisCAD, Dream, MacDraft in PICT format. If not listed, change file format to one of above or PICT or PICT2. For PCs, save as EPS (with header or Adobe Illustrator (Macintosh) 88, 1.1 or later or POSTSCRIPT. Data can also submitted on Bernoulli storage device (not 20mb), Syquest 44 or 88mb, DAT drive (90 or 120m), magneto optical disks (3.5 in., 5.24 in. 600mb) or CD-ROM disks (HFS, High Sierra Format, CD-ROM Xa). Supply artwork at final size to fit column/page dimensions. Wrap disk in

alluminum foil, surround with a stiff card, write "Magnetic Media" on package.

Micrographs—Original photographs are needed for reproduction; submit at final journal size. Plates (including mount) should not be larger than 11.6×16.5 in. Oversize mounts will be cut for scannings. Write author(s) name(s) and figure number on back of each figure, and orientation.

Photographs should be glossy bromide prints. Insert asterisks, arrowheads, pointers, etc. using press-on symbols. If photographs are final size, use 14pt Helvetica italic lowercase. To identify Fig. parts, use Helvetica, bold A. B, etc. If not, lettering will be inserted by printer; indicate on a tracing paper overlay. Indicate magnification with a bar scale. If possible, provide electronic copies of micrographs.

Do not use color unless Editor has sanctioned it. Use CMYK process colors, not RGB or Pantone.

If scanned Images are resized, altered (image resolution 300 dpi preferred) or cropped, use Photoshop, or similar scanning software not applications such as Pagemaker/ Freehand/Illustrator or Quark.

Pagemaker (Links option): do not 'store copy in publication' (pictures remain externally linked).

Freehand/lllustrator/Quark: it is important to send the externally linked images as well as the labelled file, which only contains a 'low resolution screen preview picture'.

OTHER FORMATS

REVIEWS–4-8 printed pages. Contact Editor before writing to make sure topic is suitable. Follow instructions for text, layout and style for regular articles.

PERSPECTIVES—1-3 pages. Contact Editor before writing to make sure topic is suitable. Follow instructions for text, layout and style for regular articles.

OTHER

Electronic Submission: Full instructions on formatting and saving manuscript on disk are sent with letter of provisional acceptance. Send a copy of disk with printed copy of final revised manuscript.

Experimentation, Animal: Papers describing experimental procedures which may be reasonably presumed to have inflicted unnecessary pain, discomfort or disturbance of normal health on living animals will not be accepted. Make clear that advances made in physiological knowledge justified procedures, that appropriate anesthetic and surgical procedures were followed, and that adequate steps were taken to ensure that animals did not suffer unnecessarily at any stage of experiment. Experiments on vertebrates must comply with legal requirements in the U.K. (see *Handbook for the Animal Licence Holder*, Institute of Biology; *Notes on the Law Relating to Experiments on Animals in Great Britain*, Research Defence Society; *UFAW Handbook on the Care and Management of Laboratory Animals*) or Federal legislation and the National Institutes of Health Guidelines in the U.S. (Dept. of Health Education and Welfare Pub. No. NIH 78-23, 1978).

Materials Sharing: Publication implies willingness to make available specialized reagents (antibodies, DNA probes) to other scientists for bona fide purposes.

Sequences: All nucleotide or amino acid sequences in submitted manuscripts should have been accepted by a database library before publication. Submission forms for EMBL Data Library are available from Editorial Office.

Style: Papers should be concise and written in English (spelling as in *Oxford English Dictionary* or *Webster's Dictionary*). Use SI units.

SOURCE

January 1995, 198(1): inside front and back covers.

Journal of Experimental Medicine
Rockefeller University Press

Editors, *The Journal of Experimental Medicine*
The Rockefeller University
Box 289, 1230 York Ave.
New York, New York 10021
Telephone: (212) 327-8575; Fax: (212) 327-8511

AIM AND SCOPE

The Journal of Experimental Medicine emphasizes experimental studies that provide new information on physiologic or pathogenetic mechanisms. *The Journal of Experimental Medicine* emphasizes studies on physiologic and pathogenetic mechanisms in such areas as microbial pathogenesis, immunology and inflammation, connective tissue and vascular biology, oncology, and hematopoiesis. Sufficient space is provided to document conclusions and to describe insights at the level of the whole animal and patient.

MANUSCRIPT SUBMISSION

Submission of a manuscript means that no similar paper has been or will be submitted elsewhere. In cover letter list name, address, and telephone and fax numbers of corresponding author. Include a statement from corresponding author that affirms that all authors concur with submission and that material has not been previously reported and is not under consideration for publication elsewhere. Include pertinent articles cited as "submitted" or "in press"; clearly mark as such.

Submit three copies of each manuscript: original, double-spaced including references and legends, and two fully legible copies. Paginate manuscripts and clearly label figures.

Limit Regular Contributions and Brief Definitive Reports to 12 and 6 Journal pages, respectively, including illustrations (approximately 36 and 18 typed pages, double-spaced).

MANUSCRIPT FORMAT

Double space throughout. Include: a Summary that provides a clear statement of findings and thrust of manuscript; an Introduction that explains origin and goal of research but is not an exhaustive literature review nor a detailed summary; a Materials and Methods section that provides sufficient guidance to repeat and extend experiments (this implies that all reagents are accessible to others); Results in which text is presented with accurate subheadings

and in which all illustrative material is new, fully labeled, and essential; and a Discussion that enriches but does not repeat material covered in Results. Results and Discussion may be fused, especially in Brief Definitive Reports. Paginate manuscripts. Clearly label figures.

Title Page: List title, author(s), department(s), institution(s), city, state, postal code, and country where work was performed. Furnish a condensed title (72 characters and space), without abbreviations, as a running foot. Refer to preliminary reports or abstracts in a title footnote. State complete name and address of reprint request author. State source(s) of support as grants, equipment, or drugs on a separate Acknowledgments page.

Footnotes: Designate footnotes to tables and to author/affiliation sections by superscript symbols in order: *, ‡, §, ||, ¶, **, ‡‡, etc.

References: Type double-spaced on a sheet separate; number in order of appearance in text. Include all authors' names (do not use "et al."), complete article titles, and articles in press with inclusive pagination. Abbreviate names of journals as in *Biosis List of Serials*. Spell out names of unlisted journals. Manuscripts submitted or in preparation, unpublished observations, and persona communications appear parenthetically in text; do not list with references. Submit written permission from authors cited as personal and private communications.

JOURNALS

TWO AUTHORS

1. Bevan, M., and P. Fink. 1978. Thymus H-2 antigens and maturing of helper cells. *Immunol. Rev.* 42:3.

MORE THAN TWO AUTHORS

2. Julius, M., E. Simpson, and L. Herzenberg. 1973. A rapid method for the isolation of functional thymus-derived lymphocytes. *Eur. J. Immunol.* 3:645.

IN PRESS

3. Myers, L. K., J. M. Stuart, and A. H. Kang. 1990. A CD4 cell is capable of transferring suppression of collagen-induced arthritis. *J. Clin. Immunol.* In press.

COMPLETE BOOKS

4. Myant, N. B. 1981. The Biology of Cholesterol and Related Steroids. Heinemann Medical Books, London. 882 pp.

ARTICLES IN BOOKS

5. Pink, J. R. L., O. Lassila, and O. Vainio. 1987. B-lymphocytes and their self-renewal. *In* Avian Immunology. A. Toivanen and P. Toivanen, editors. CRC Press, Inc., Boca Raton, FL. 65–78.

ABSTRACT

6. Packman, C. H., S. I. Rosenfeld, and J. P. Leddy. 1981. Inhibition of the C8/C9 steps of complement lysis by a high density lipoprotein (HDL) of human serum. *Fed. Proc.* 40:976a. (Abstr.)

Illustrations: Must be self-explanatory; cite in numerical order in text. Do not exceed *Journal* page dimension of 17.5 × 24 cm. Submit originals whenever possible. Mount illustrations composed of multiple pieces on mounting board. Plan micrographs so they may be reproduced without reduction. Single-column figures must not exceed 8.25 × 24 cm; double-column 17.5 × 24 cm.

Gels—Plan labeling so that upon reduction height of numbers and capital letters will be 1.75 mm and lane width will be 5 mm.

Micrographs—Limit field of micrograph to structures specifically discussed in report. Be sure that symbols and areas of special interest are not too close to edges and that corners are squared. Include magnifications in legend. Make labeling sans serif, 3–3.5 mm high, in India ink or securely affixed transfer letters. Explain symbols used on micrograph in legend. On reverse of each reproduction figure, indicate TOP with a soft pencil, never a pen, and press lightly. Mail flat, using cardboard stiffeners. Never fold, roll, or use paper clips.

Line Drawings—Provide original black India ink or computer-generated drawings or sharp glossy prints. They will be reduced to one column (8.25 cm) or less. Make lettering sans serif and of such size that capitals, numbers, and symbols will reduce to a height of 1.5–1.75 mm. Use standard JEM abbreviations (h, min) in axis labels.

Color Illustrations—Use bendable, reflective art. Authors pay for separation and printing of color figures. Production Office will provide an estimate and require written agreement to pay all costs.

FACS Tracings—Label both axes and provide some measure of quantitation, e.g., $\log_{10}$ fluorescence, to make ratio of fluorescence intensity between peaks clear.

Legends: Include a short title after figure number with a short explanation in sufficient detail to make data intelligible without reference to text.

Tables: Type double-spaced on sheets separate from text. Make self-contained and self-explanatory. Provide titles. Use superscript symbols for footnotes. Give sufficient explanation and detail in legend to make experiment or procedure intelligible without reference to text. Do not use vertical lines; do not submit photographs of tables. Submit tables with different material in boxheads as individual tables, not as sections of one table.

OTHER

Abbreviations: The *Journal* publishes a standard abbreviations list at the front of the January and July issues, and in other issues on a space-available basis. These do not need to be spelled out within papers. For nonstandard abbreviations, spell out full name on first appearance and use abbreviation consistently thereafter.

Nucleic Acid Sequences: The *Journal* will publish nucleic acid sequences only if data have been submitted to EMBL/GenBank/DDBJ and have a database accession number. Obtain blank sequence submission form while writing manuscript, and send completed form to EMBL when manuscript is submitted to the Editors' office for review. The last line of the sequence's legend should read: "These sequence data are available from EMBL/GenBank/DDBJ under accession number _____." If

number is not available in time to include on typescript, fill in at page proof stage. EMBL can provide number in a week or less, provided author includes an electronic mail address, fax or telex number.

Data submitted to EMBL will be shared with GenBank and the DDBJ in Japan. See *Nucleic Acids Research*, which prints sequence submission form in first issue of each year; this may be photocopied. Forms are also available from: Data Submissions, EMBL Data Library, Postfach 10.2209 6900 Heidelberg, Germany, PH: 49-6221-387-258; Fax 49-6221-387-306; Telex 461613 (embl d).

Page Charges: $40 for each page ($50 January 1995). An author's inability to honor page charges will not affect publication. Four pages of micrographs are included, but a surcharge of $100 will be made for each additional page.

Style: Capitalize trade names. Give formal, chemical name of a compound as established by international convention after first use of trivial name. Thereafter, use trivial name. For special materials and equipment, give manufacturer's name and location. Use metric system and Systeme International d'Unites (SI) as noted in Abbreviations List. For biochemical abbreviations, follow recommendations in *Journal of Biological Chemistry's* instructions to authors. Abbreviate units of measure only when used with numbers. Specify estimates of variance (e.g., SD, SE) and give references for statistical methods.

For mouse H-2 genetics, consult revised nomenclature in *Mouse Genome* 92(2) June 1994 and in *Genetic Variants and Strains of the Laboratory Mouse* (3rd ed., 1994). Current nomenclature is available online from The Jackson Laboratory through GBASE, Gopher and World Wide Web.

Submission Fee: $25 handling fee for each new manuscript submission. A check must accompany submission and be made payable to: The Rockefeller University Press-JEM.

TCR Nomenclature: Follow recommendations of WHO-IUIS Sub-Committee on TCR Designation (*Bull. WHO* 71(1):113; 1993).

SOURCE

July 1, 1994, 180(1): front matter.
Detailed Instructions for Authors appear in first issue of each volume (January and July issues).

Journal of Experimental Zoology
American Society of Zoologists Division of Comparative Physiology and Biochemistry

Frank H. Ruddle
Department of Biology
Yale University
New Haven, Connecticut 06511
Telephone: (203) 432-3526; Fax: (203) 432-5690

AIM AND SCOPE

The *Journal of Experimental Zoology* will publish the results of original research of an experimental or analytical nature in zoology, including investigations of all levels of biological organization from the molecular to the organismal.

MANUSCRIPT SUBMISSION

Submit an original and three copies of manuscript and figures. One copy of oversize figures is sufficient. Make certain that photographic copies, especially oversized figures, are high quality to ensure that reviewers can judge materials. Do not submit papers which have already been published; simultaneous publication elsewhere is not allowed. Indicate preference of division for review and publication of work.

Send manuscripts with illustrations, packed flat, by express, prepaid or registered mail to Editor. Upon acceptance, authors must sign a form transferring copyright to publisher, who reserves copyright.

MANUSCRIPT FORMAT

Type manuscript double-spaced throughout on one side of bond or heavy-bodied paper, 8.5 × 11 in. with a 1 in. margin on all sides. Submit manuscript exactly as to appear in print. Manuscripts and illustrations that are not in proper finished form may be returned for revision. Use subdivisions in sequence: Title Page; Abstract; Text; Literature Cited; Footnotes; Tables; Legends. Start each subdivision on a new page.

Do not exceed twenty-six pages. Number all pages consecutively. Type double-spaced. Center titles and column heads of tables; arrange material to be esthetically pleasing and to give proper emphasis to pertinent data. Footnotes, follow Literature Cited. Figure legends are last section. Illustrations appear on one page as single or multiple figures occupying a total maximum area of 6 3/16 × 9 in.

Title Page: On first page include: Complete title of paper; Author's name or names; Institution from which paper emanated with city, state, and zip code; Total number of text figures, graphs, and charts; Abbreviated title (running headline, 48 letters and spaces); Name, address, telephone, and fax numbers of person to whom proofs are to be sent.

Abstract: 250 words preceding introductory section. Write in complete sentences. Succinctly state objectives, experimental design, principal observations and conclusions. Make intelligible without reference to rest of paper.

Literature Cited: Reference literature in text by author's name followed year of publication; e.g., studies by Briggs ('75) reveal . . or studies by Briggs and Porter ('75) reveal . . or studies by Briggs et al. ('75) reveal...

When references are made to more than one paper by the same author, published in the same year, designate in text as (Kelley, '70a,b) and in literature list as

Kelley, R.O. (1970a) An electron microscopic study of mesenchyme during development of interdigital spaces in man. Anat. Rec., *168*:43–54.

Kelley, R.O. (1970b) Fine structure of apical, digital and interdigital cells during limb morphogenesis in man. In: Proceedings of the Fifth International Congress of Electron Microscopy, Vol. III: 831–832.

Abbreviate journal titles as in *Index Medicus*. Beginning with 1901, and thereafter, for literature references in text use abbreviated date of publication after author's name:

('01), ('04), ('75) not (1901), (1904), (1975). Do not abbreviate years before 1901: (1784), (1889), (1900). Do not abbreviate year in list of Literature Cited.

Arrange Literature Cited alphabetically by authors as: Author's name (or names), year of publication, complete title, volume and inclusive pages.

JOURNAL ARTICLES

Andersen, H., N. Ehlers, M.E. Matthiessen and M.M. Cläesson (1967) Histochemistry and development of the human eyelids. II. A cytochemical and electron microscopical study. Acta Opthalmol. Kbh., *45*:288–293.

Humphrey, T. (1970) Palatopharyngeal fusion in a human fetus and its relation to cleft palate formation. Ala. J. Med. Sci., 7:398–429.

ABSTRACTS

Waterman, R.E. (1972a) Fine structure of oral membrane formation and rupture in the hamster. Anat. Rec., *172*:421

CHAPTERS IN BOOKS

Zwilling, E. (1961) Limb morphogenesis. In: Advances in Morphogenesis. M. Abercrombie and J. Brachet, eds. Academic Press, New York, Vol. 1, pp. 301–330.

Footnotes: Number footnotes to text consecutively; indicate corresponding reference numbers in text. Number additional references to identical footnotes with next following consecutive number. Type footnotes to a table directly beneath table and letter a, b, c, etc.

Tables: Make simple and uncomplicated, with as few vertical and horizontal rules as possible. Indicate in text where tables appear. Number with Arabic, not Roman, numerals.

Legends: Number consecutively. Abbreviate Fig. 1. List abbreviations in figure labels once alphabetically and place before first figure containing these abbreviations.

Illustrations: Limit figures to those which will adequately present findings. Printer must work directly from original photoprints or drawings with as little reduction as possible. Illustrations will be reproduced either as halftones or as line drawings. Submit all illustrations in complete and finished form, with adequate labeling.

Black-and-white prints for halftone reproduction. Photo paper must be blue-white with smooth finish.

Indicate reduction desired on each illustration. Illustrations cannot be reduced to less than 20% submitted size. Submitted line drawings cannot exceed 30 × 40 in. Lettering and labels must be readable after reduction. When reduced, minimum height of a capital letter should not be less than 2.5 mm for a photomicrograph and 1 mm for a graph or chart. When printed, an individual figure or group of figures may not exceed 6 3/16 wide by 9 in. long or 3 5/16 wide by 9 in. long for single column placement.

Color Prints—Color illustrations will be printed only if paid for by author. Firm quotes are provided after acceptance; author may approve both costs and proofs. Submit color photoprints or transparencie. Mark frame of transparency to indicate the area that can be safely cropped to arrive at critical image area. Do not use silk finish or other dimensional-surface photoprint papers which cause distortion in reproduction from random light reflections.

Line Drawings—Draw figures with India ink on medium-weight white paper or art board of at least postcard weight. Photoprints may be submitted in lieu of original. Make artwork sharp and black to achieve maximum contrast. Use "stippling" and "hatching" techniques to achieve tonal quality. Avoid use of "shading" (pencil, wash, or airbrush) unless drawing is to be reproduced as a halftone. For graphs use blue-ruled paper.

Make corrections with opaque white ink, or white patches, to cover areas to be deleted provided illustration is to be reproduced as a line drawing. Do not use to correct drawings reproduced as halftones.

Mount photomicrographs and illustrations with artifact labels. Trim figures on all sides and, "square." Mount figures on hard, strong, nonflexing bristol board of approximately 15 points (0.4 mm) thickness with, at least, a 1 in. margin surrounding figure or grouping of figures. Mount color figures on a lighter weight flexible bristol paper approximately 7 to 8 points (0.2 mm) thick to conform to curved cylindrical surface of color separation scanner. Attach figures to bristol board using appropriate "dry mounting" materials, or a cement or glue, which when set, is white or colorless. When assembling two or more figures, mount close together, separated by approximately 1/8 in. open space. Figures grouped to form a plate should be of equal density and highlighting.

Letter and number illustrations either by hand, using India ink, or with transfer type. Make labels large enough to allow for suitable reduction and sturdy enough to withstand mailing and handling in production process. For protection, spray labeling with an appropriate clear adhesive. Label illustrations directly on drawing or photographic print. Do not do labeling on an overlay. Place labeling at least 1/8 in. in from edges of illustration. Crate contrast between label or letter and its background, placing white labeling over dark backgrounds and black labeling over light backgrounds.

Number figures, including charts and graphs, consecutively. Indicate approximate position of each figure in margins of text.

Submit original illustrations, and three sets of sharp contrast photocopies. For oversized figures, send original figure and three photographic copies, reduced to either letter (8.5 × 11 in.) or legal (8.5 × 14 in.) size.

If original drawings are too large for shipment, photocopies are acceptable if there is no loss of critical detail. Machine copies of photographic illustrations are not acceptable. On reverse side of illustration indicate: Author's name, Figure number, TOP, Reduction requested, and "Review copy" on copies for reviewers. Do not fasten with paper clips, staples, etc. Ship illustrations flat, protected by heavy cardboard.

OTHER FORMATS

RAPID COMMUNICATIONS—Eight to ten typed pages and one page of illustrations. For short reports of timely and unusual interest.

OTHER

Dates: Write as: October 11, 1975; 11th of October.

Greek Letters: Label those in manuscript which could be confused with English alphabet characters.

Hyphenation: Do not divide words at ends of lines. Type or print corrections to manuscript legibly in ink.

Numbers: Spell out when first word in a sentence; do not follow with an abbreviation. Use Arabic numerals to indicate time, weight, and measurements when followed by abbreviations (e.g., 2 mm; 1 sec; 3 ml). Write numbers one to ten in text. Give higher numbers in Arabic numerals.

Page Charges: $50 per page for tabular and illustration pages that exceed 50% of total number of printed text pages. Payment is not a prerequisite for publication. Because of the high cost of color work, work will be initiated only at author's request and expense.

Spelling: Do not use abbreviations to begin sentences. Do not abbreviate word "Figure" in text except in parentheses. Spell nontechnical terms according to current *Webster's International Dictionary*.

Units: Use metric system to express all measurements. Express temperatures in degrees Celsius (centigrade). Metric abbreviations are lowercase without periods. When preceded by a digit, use % for percent, ° for degree.

SOURCE

September 15, 1994, 270(1): 71-75.

Journal of General Microbiology
see Microbiology

Journal of General Physiology
Society of General Physiology
Rockefeller University Press

Dr. Paul F. Cranefield
The Rockefeller University
1230 York Ave.
New York, NY 10021

AIM AND SCOPE

The Journal of General Physiology has traditionally published articles that deal with basic biological, chemical, or physical mechanisms of broad physiological significance. *The Journal* publishes theoretical articles only if they are of unusual interest and deal with subjects about which the *Journal* has often published experimental studies. A theoretical article can also be considered if it is submitted as a companion to an experimental article that depends upon the theoretical article in some significant way. The *Journal* does not publish articles that deal only with methods, but can consider articles that deal largely with a new method if they also report significant new findings obtained by using that method.

MANUSCRIPT SUBMISSION

Submit four high-quality photocopies of manuscript and a complete set of photocopied figures with each copy. Send original prints only if their detail is lost in photocopies, as with an electron micrograph.

If an article has more than one author, each must either sign letter of submission or notify editor in writing that he wishes to have his name appear as an author. The *Journal* does not publish articles reporting material that has been reported at any length elsewhere. State in a cover letter that material has neither been published elsewhere nor is being submitted for publication elsewhere. If any form of publication other than an abstract that contains no figures and is less than 400 words in length has occurred or is contemplated, submit three copies of a reprint or typescript of such other publication(s). Material will be sent to reviewers, who may be asked whether there is an overlap between submitted articles and other material. Indicate in cover letter the relationship between submitted article and other material.

MANUSCRIPT FORMAT

Type on 8.5 × 11 in. bond paper of good quality. Double-space all material, including abstract, text, legends, tables, bibliography, and footnotes. Keep 1 in. margins on sides, top, and bottom. Number all pages in sequence. Do not staple. Poor-quality photocopies or draft-quality dot matrix computer print-outs are not acceptable. Insertions or corrections should be few; type or print legibly in ink.

Title Page: Include name(s) of authors, department and institution where work was done, city, state, and zip code of institution, and name, address, and telephone number of corresponding and reprint request author(s). Include a running title (75 letters and spaces) with author's name. Do not include notices of grant support, place between acknowledgments and bibliography.

Abstract: 300 words to serve as a summary. Make meaningful to a wide range of readers. Write out sources cited in full; enclose in parentheses.

Illustrations: Illustrations will be requested if article is accepted. They may be high-quality glossy prints in sharp focus. Original drawings may be preferable unless they are excessively large, mounted on cardboard, or in poor condition. Design figures to fit *Journal* pages, 5 in. wide by 7.5 in. deep. Make figures consistent in overall format, size, and style. Abbreviations and units of measure should be those used in text. Never use two different sizes or styles of lettering in one figure. Figures that have broken, crooked, or excessively narrow or angular lettering are not acceptable. Use Leroy lettering with capital and lowercase letters (not all capitals), drawn with India ink. Use standard symbols (○ ● □ ■ △ ▲). Reduction will be determined by size of lettering. Capital letters and numbers will be approximately 1.75 mm high when reduced. Choose size of symbols accordingly; if symbols are overly small in relation to lettering, they will be so small when figure is reduced that they will not easily be distinguished. Avoid figures with white lettering on a black background. Figures should not have titles; place keys to symbols in figure legend, not on figure itself. Separate numbers and units. Label figures divided into parts A and B with letters approximately the same size as the capital letters in figure. Relettering is at author's expense.

Citations: Make references by giving author's name and year of publication in parentheses, except when author's name is part of sentence [e.g., "Osterhout (1918) showed that . . ."]. When citing a paper by two authors, give both names. If more than two authors, give all names the first time cited, thereafter, unless ambiguous, give first name with "et al." added. If ambiguous, distinguish references with lowercase italic letters after year. Do not include unpublished material in reference list; cite by giving authors' names and initials followed by "unpublished observations," "personal communication," "manuscript in preparation," or "manuscript submitted for publication."

If personal communications are cited, a submit signed permission from person cited, stating that person has seen and approved actual wording of citation. Obtain permission to thank others or acknowledges their help.

Bibliography: List citations alphabetically by author. Alphabetization takes precedence over publication date and number of authors:

Armstrong, C. M. 1975.
Armstrong, C. M., F. Bezanilla, and E. Rojas. 1973.
Armstrong, C. M., and L. Binstock. 1965.

Give journal name in full. Double-space bibliography.

Knight, B. W., J. -I. Toyoda, and F. A. Dodge, Jr. 1970. A quantitative description of the dynamics of excitation and inhibition in the eye of *Limulus. Journal of General Physiology.* 56:421–437.

To cite an article or chapter in a book:

Hume, J. R. 1989. Properties of myocardial K+ channels and their pharmacological modulation. *In* molecular and Cellular Mechanisms of Antiarrhythmic Agents. L. Hondeghem, editor. Futura Publishing Co., Inc., Mount Kisco, NY. 113–131.

Tables: Double-space on separate, numbered pages, with a brief title at top and legend and footnotes at bottom. Put legend and footnotes on same page as table. Design tables to fit *Journal's* 5 in. page width. to determine order in which footnote symbols are used read horizontally across table. Order of symbols: *, ‡, §, ||, ¶, **, ‡‡, etc. Use same abbreviations in tables as in text and figure legends; define abbreviations used for the first time in a table in a table footnote. Do not use vertical rules and ditto marks.

OTHER FORMATS

LETTERS TO THE EDITOR—Comment upon, criticize or explain findings published in *Journal*. Provide a title. Place authors names and affiliations between end of text and bibliography, not on title page. Letters are subject to review; one reviewer will be person whose work is commented upon who also has the option of submitting a reply (also subject to review).

NOTES AND BRIEF COMMUNICATIONS—The *Journal* does not publish notes or brief communications, but does publish articles that are as short as eight printed pages.

OTHER

Equations, Abbreviations, and Units of Measure: Define all symbols and abbreviations. Identify mathematical symbols and Greek and script letters in margin of manuscript opposite line in which they first appear. Distinguish between numeral 1 (one) and letter l (el), and between zero and letter o (oh). With san-serif typefaces, distinguish between capital I and lowercase l (el). Define abbreviations when they first appear; use consistently thereafter. Use same abbreviations in figures and text. The *Journal* uses the following abbreviations for common units of measure:

ampere, milliampere	A, mA
centimeter, square centimeter	cm, cm^2
curie, millicurie, microcurie	Ci, mCi, μCi
gram, kilogram, milligram, microgram	g, kg, mg, μg
hour	h (not hr)
liter, milliliter, microliter	liter (not l or L), ml, μl
micrometer	μm
minute	min
mole, millimole, micromole	mol, mmol, μmol
molar, millimolar, micromolar	M, mM, μM
ohm	Ω
second, millisecond, microsecond	s, ms, μs (not sec, msec, or μsec)
siemens, millisiemens	S, mS

For other units and abbreviations see *Council of Biology Editors Style Manual: A Guide for Authors, Editors, and Publishers in the Biological Sciences* (5th edition, 1983, Council of Biology Editors, Inc., Bethesda, MD).

Experimentation, Animal: Experiments must be conducted in accordance with *Guide for the Care and Use of Laboratory Animals* (NIH Publication 85-23, 1985).

Note following statement, "Guiding Principles in the Care and Use of Animals," approved by the Council of the American Physiological Society:

"Animal experiments are to be undertaken only with the purpose of advancing knowledge. Consideration should be given to the appropriateness of experimental procedures, species of animals used, and number of animals required.

"Only animals that are lawfully acquired shall be used in this laboratory, and their retention and use shall be in every case in compliance with federal, state, and local laws and regulations, and in accordance with the NIH Guide (Guide for the Care and Use of Laboratory Animals, NIH Publication 85-23, 1985).

"Animals in the laboratory must receive every consideration for their comfort; they must be properly housed, fed, and their surroundings kept in a sanitary condition. Appropriate anesthetics must be used to eliminate sensibility to pain during all surgical procedures. Where recovery from anesthesia is necessary during the study, acceptable technique to minimize pain must be followed. Muscle relaxants or paralytics are not anesthetics and they should not be used alone for surgical restraint.

They may be used for surgery in conjunction with drugs known to produce adequate analgesia. Where use of anesthetics would negate the results of the experiment, such procedures should be carried out in strict accordance with the NIH Guide. If the study requires the death of the animal, the animal must be killed in a humane manner at the conclusion of the observations. The postoperative care of animals shall be such as to minimize discomfort and pain, and in any case shall be equivalent to accepted practices in schools of veterinary medicine. When animals are used by students for their education or the advancement of science, such work shall be under the direct supervision of an experienced teacher or investigator. The rules for the care of such animals must be the same as for animals used for research."

Page Charges: $25 for each printed page of type or illustrations. An author's inability to pay will not affect publication. Four full pages of halftones are included at this rate; a surcharge of $100 will be made for each halftone page over four. Color illustrations will not be published unless e reviewers advise editors that it is necessary; full cost must be borne by the author.

SOURCE
July 1994, 104(1): 191-195.

Journal of General Virology
Society for General Microbiology

Editorial Office, *Journal of General Virology*
Marlborough House, Basingstoke Road
Spencers Wood, Reading RG7 1AE, U.K.

AIM AND SCOPE

The *Journal of General Virology* aims to publish papers which describe original research in virology and contribute significantly to their field. It is concerned particularly with fundamental studies. Papers must be in English. Standard papers, short communications and review articles are published. Papers that describe new materials or methods without applying them to research are generally not acceptable. Preliminary or inconclusive data will not be published. Data that differ only in a minor way from previously published results are not acceptable, this applies for instance to comparisons of nucleotide sequences with closely similar published sequences. Papers dealing with clinical or epidemiological aspects of virology would be more appropriately published elsewhere.

MANUSCRIPT SUBMISSION

Submit three sets of typescript and figures and three copies of relevant papers cited as "in press." Type on one side of paper only, double-spaced throughout, including Methods and Reference sections, figure legends and tables, with wide margins. Do not submit draft quality dot matrix printing.

Underline words or letters which are to appear in italics when printed. Where ambiguity is possible, distinguish between capital letter O and numeral 0, and between letters I, l and numeral 1. Identify Greek letters and other symbols not typewritten at first occurrence.

Submission implies that it reports unpublished work which is not under consideration for publication elsewhere, that all authors have agreed to its submission and that, if accepted, it will not be published again in the same form in any language without the consent of the Society for General Microbiology. Copyright is assigned to the Society for General Microbiology.

Work reported must have been carried out in a manner meeting local and national regulations regarding safety, ethics and use of animals.

MANUSCRIPT FORMAT

Examine a current copy of *Journal* for layout and conventions used.

In Reference section, list papers alphabetically by first author; list papers with three or more authors in chronological order after any other papers by first author. Use form: surnames and initials of authors, year (in parentheses), title of paper, full title of journal, volume number, first and last page numbers. Citations in text must match those in list.

Tables: Employ sparingly. Make generally comprehensible without reference to text. Type double-spaced on a separate sheet. Number in order cited in text. Footnotes may be used, but not table legends. Make footnotes brief; do not include experimental detail that could be in text. Footnote symbols are *, †, ‡, §, ||, and ¶; use in that order.

Figures: Employ sparingly to demonstrate important specific points. Design for clarity and economy of space.

Type legends double-spaced; do not repeat Methods section. Number in a single series in order referred to in text. Design to fit into either one or two columns. Maximum printed sizes (width × height), including lettering and legends, are 85 × 225 mm or 175 × 225 mm.

Lettering on diagrams and photographs may be supplied by authors if executed to a good standard. Otherwise, supply original figure lettering; indicate lettering on a copy to be inserted by printers.

Submit three copies of drawings, two of which may be photocopies, at approximately twice publication size. Each copy should be on a separate sheet with authors' names, figure number and TOP on back. Original diagrams should suitable for direct reproduction. Box graphs in. Show scale marks on opposing axes. Preferred point symbols: ○ ● □ ■ ▲ △ ▼ ▽. Define symbols in legend, not axes. Where possible, use same point symbols for comparable variables in different figures. Chose symbol sizes and line thicknesses for clear visibility after reduction.

Prepare nucleotide and amino acid sequence diagrams "camera-ready" with a good quality printer. Low quality dot matrix printing is not acceptable. For long nucleotide sequences, preferred format is 120 residues/line with no space between characters, which allows presentation of approximately 4800 residues on each page. Use single letter amino acid code.

Supply three copies of photographs approximately final printed size. Photocopies are not acceptable. Photographs should be sharply focused and well contrasted, on glossy paper. Do not screen or mount; remove nonessential areas. Where photographs are grouped to form a composite, match tone as nearly as possible and provide a layout guide. If mounted, allow little or no space between photographs. Show magnification of photomicrographs by a bar marker.

Color photographs can be reproduced at author's expense. The most satisfactory reproduction is obtained from original transparencies (positive or negative).

OTHER FORMATS

REVIEW ARTICLES—Occasional reviews are published giving an overview of a field suitable for a wide audience. Most are invited. Unsolicited reviews are considered, consult Editor-in-Chief.

SHORT COMMUNICATIONS—Four printed pages. Report completed work, not preliminary findings; an alternative format for describing smaller pieces of work. Include Summary and Reference section; main text is not divided into sections. Describe methods briefly in text, not within figure legends.

To estimate length: Allow 0·2 pages for title, names and addresses; allow 5900 characters per page of main text, and 7500 characters per page of table footnotes, figure legends, acknowledgments and references; for tables, allow 70 lines per page; or figures, estimate height after reduction (page height is 225 mm); and for tables and figures, whether they will be one column in width, or wider.

OTHER

Abbreviations: Use nonstandard abbreviations only where multiple use is made; define on first appearance. Obtain a list of allowed standard abbreviations not requiring definition from Editorial Office.

Color Charges: £750 per page.

Data Submission: Papers reporting new protein or nucleic acid sequence data should have a footnote stating that data have been deposited with a (named) library; give accession number.

Spelling: Follow usage of *Concise Oxford Dictionary*.

Style: Use past tense to describe results of paper, and present tense to refer to previously established and generally accepted results.

Units: Where possible, use SI units.

SOURCE

January 1994, 75(1): front matter.
Detailed information for intending contributions is published in January and July issues of Journal.

Journal of Gerontology: Biological Sciences

Edward J. Masoro, PhD
Department of Physiology
University of Texas Health Science Center
7703 Floyd Curl Drive
San Antonio, TX 78284-7756

AIM AND SCOPE

The *Journal of Gerontology: Biological Sciences* publishes articles on the biological aspects of aging in areas such as biochemistry, genetics, molecular biology, morphology, neuroscience, nutrition, pathology, pharmacology, and physiology. All submissions are peer-reviewed.

MANUSCRIPT SUBMISSION

Submission implies that manuscript has not been published and is not under consideration elsewhere. If accepted, it is not to be published elsewhere without permission. Corresponding author is responsible for certifying that permission has been received to use copyrighted instruments or software employed in research and that human or animal subjects approval has been obtained. For co-authored manuscripts, corresponding author is responsible for submitting a letter, signed by all authors, indicating that they actively participated in the collaborative work leading to publication and agree to be listed as authors. Assurances will be requested when paper has been formally accepted for publication.

MANUSCRIPT FORMAT

Print double-spaced, including references and tables, on 8.5 × 11 in. white paper using 1 in. margins. Number pages consecutively beginning with title page and including all pages. Submit five copies. Be concise. Follow abbreviations and other conventions of the *CBE Style Manual* (5th Ed, 1983, Council of Biology Editors, Inc.). Use IBM-compatible software to facilitate electronic typesetting; a disk will be requested upon acceptance.

Title Page: Include title of manuscript; author's full name and affiliation; and a short running page headline (40 letters and spaces) at foot of page.

Abstract: 150 words typed, double-spaced, on a separate page. State purpose of study, basic procedures (study participants or experimental animals and observational or analytical methods), main findings, and conclusions.

Text: Divide text of observational and experimental articles into sections with headings: Introduction, Methods, Results, and Discussion. Articles may need subheadings within some sections to clarify content. Discussion should interpret results.

Text References: Show all references using name-and-year system in *CBE Style Manual* (5th ed., pp. 47–48). Make references in text with author's name followed by year of publication, e.g., " . . . Studies by Smith (1975) . . ." When more than two authors are cited, use first author and et al. (Smith et al., 1975). List several references in one place chronologically (Yu, 1981; Allen, 1985).

References: Type double-spaced. Follow style in *CBE Style Manual,* (5th Ed, 1983).

JOURNALS

Steele, R. D. Role of 3-ethylthiopropionate in ethionine metabolism and toxicity in rats. J. Nutr. 112:118–125;1982.

BOOKS

Osler, A. G. Complement: mechanisms and functions. Englewood Cliffs, NJ: Prentice-Hall, Inc.; 1976.

Arrange entries alphabetically by author and then chronologically when authors of two or more entries are the same.

Footnotes: Describe sources of research support and other acknowledgments in a footnote on a separate page before references. Indicate where correspondence should be addressed. Do not use reference footnotes.

Tables: Type on separate sheets double-spaced. Number consecutively and give brief titles. Type footnotes immediately below table. Reference marks are superscript small letters ([a,b,c] . . .). Arrange footnotes alphabetically by superscripts. Use asterisk and double asterisk for probability levels of tests of significance.

Illustrations: For a graph or drawing, submit one glossy print and four photocopies; for photomicrograph, five glossy prints. Color plates are at author's expense. Submit sharp, glossy prints. Have figures professionally lettered in a sans-serif type (e.g., Univers or Helvetica) or use a laser printer. Typewritten/dot matrix lettering is not acceptable. Make letters sufficiently large that, when reduced, they will be legible.

Captions for Illustrations: Type double-spaced on a separate page with numbers corresponding to illustrations. Explain symbols, arrows, numbers, or letters used. Explain internal scale and identify staining method in photomicrographs. Interpret content of table or figure without reference to text.

OTHER FORMATS

BRIEF REPORTS—Six double-spaced typed pages, one table, graph, or illustration. Submit five copies.

GUEST EDITORIALS—One printed page. May be invited; unsolicited materials may also be submitted. Follow guidelines in Manuscript Format.

LETTERS TO THE EDITOR—500 to 700 words typed double-spaced. Submit five copies. A copy will be sent to author of original article for rebuttal. Letters and rebuttals are reviewed and subject to editing. Usually both letter and rebuttal are published together.

RAPID COMMUNICATIONS—If findings are considered to be of immediate, significant importance, request consideration as a rapid communication.

REVIEW ARTICLES—Usually invited by editor. Prepare unsolicited review articles as shown in Manuscript Format.

OTHER

Style: Use *CBE Style Manual* (5th Ed., 1983, Council of Biology Editors, Inc.).

SOURCE

January 1995, 50A: B58.

Journal of Gerontology: Medical Sciences
Gerontological Society of America
Waverly Press

Kenneth L. Minaker, M.D., F.R.C.P. (C)
Geriatric Research, Education and Clinical Center
Brockton/West Roxbury VAMC (182)

1400 Veterans of Foreign Wars Parkway
West Roxbury, MA 02132

AIM AND SCOPE

The *Journal of Gerontology: Medical Sciences* publishes articles representing the full range of medical sciences pertaining to aging. Appropriate areas include, but are not limited to, basic medical sciences, and nursing. It publishes articles on research pertinent to human biology and disease. Studies involving animals or in vitro techniques are acceptable if they are relevant to normal or abnormal human biology.

MANUSCRIPT SUBMISSION

In a letter of transmittal, signed by all authors, indicate that each author has participated in reported research reported, and has agreed to be an author. Certify that this paper has not been published elsewhere, nor is under consideration elsewhere and that if accepted for publication is not to be published elsewhere (except as Society Proceeding) without permission. If paper is accepted, a copyright transfer agreement will be sent to first author. Corresponding author is responsible for certifying that permission has been received to use copyrighted instruments or software employed in research and that human or animal subjects approval has been obtained.

Disclose any financial arrangement with a company whose product figures prominently in manuscript or with a company making a competing product. Information will be held in confidence while paper is under review and will not influence editorial decision; if article is accepted, editors will formally arrange with authors the manner in which such information is to be communicated.

MANUSCRIPT FORMAT

Prepare manuscripts according to "Uniform Requirements for Manuscripts Submitted to Biomedical Journals" of the International Committee of Medical Journal Editors. See *Annals of Internal Medicine* 1982; 96:766–771; summarized here. For abbreviations and other aspects of style follow recommendations of *Council of Biology Editors Style Manual* (5th ed., 1983). Type manuscript double-spaced, including references and tables, on 8.5 × 11 in. white paper using 1 in. margins. Number pages consecutively beginning with abstract and including all pages. Put word count (3000 or less) on last page. Submit four copies of entire manuscript including figures and tables. Order of presentation: title page, abstract, text, acknowledgments, references, tables, legends to figures, figures.

Divide text into sections with headings: Introduction, Methods, Results and Discussion. Be concise. Discussion may include conclusions derived from study and supported by data; avoid excessive reviews of literature. Number references in text in order cited with Arabic numbers in parentheses.

Title Page: Include title manuscript, author's full name and degrees, and name and location of institution(s) where work was performed. Place a short running headline (40 letters and spaces) at foot of page.

Abstract: 250 words on a separate page, consisting of four paragraphs, labeled Background, Methods. Results. and Conclusions. Briefly describe, respectively, problem being addressed in study, how study was performed, salient results, and what authors conclude from results.

Acknowledgments: List sources of research support, acknowledgments, preliminary reports or abstract presentation, and current location of authors, if different from title page. Give name and address of corresponding author to whom correspondence should be addressed.

References: 25 or less. List by number in order first cited in text. Follow reference style in "Uniform Requirements for Manuscripts Submitted to Biomedical Journals." For periodicals, abbreviate title as in *Index Medicus*. List all authors when six or fewer, when seven or more, list only first three and add et al.

STANDARD JOURNAL

You CH, Lee KY, Chey RY, Menguy R. Electrogastrographic study of patients with unexplained nausea, bloating and vomiting. Gastroenterology 1980;79:311-4.

BOOKS AND OTHER MONOGRAPHS

Eisen HN. Immunology: an introduction to molecular and cellular principles of the immune response. 5th ed. New York: Harperand Row, 1974:406.

CHAPTER IN A BOOK

Weinstein L, Swartz MN. Pathogenic properties of invading microorganisms. In: Sodeman WA Jr, Sodeman WA, eds. Pathologic physiology: mechanisms of disease. Philadelphia: WB Saunders, 1974:457–72.

Submit documentation for notations of "unpublished work" or "personal communications."

Tables: Type double-spaced, on separate sheets. Number consecutively using Arabic numbers; supply a brief title at top of each. Type legends and footnotes immediately below table; follow sequence: *, †, ‡, §, ||, ¶,**, ††. . . .

Figure Legends: Start on a separate page; type double-spaced; and number consecutively to conform with number on back of each figure. Explain all symbols used; identify all abbreviations. Explain internal scales; provide magnification and staining methods for photomicrographs.

Figures: For graphs or drawings, submit one glossy print and three copies; for photomicrographs, four sharp glossy black and white photographic prints. Color photos are at author's expense. Affix a label to back indicating figure number, name of first author and TOP of figure. Make no larger than 8 × 10 in. Have letters and symbols professionally lettered or made on a laser printer and of sufficient size that, when reduced, will be legible.

OTHER FORMATS

ARTICLES OR BRIEF REPORTS RELATING METHODOLOGICAL DEVELOPMENTS—Concentrate more heavily on the methodologies involved, with relatively brief introduction and discussion.

BRIEF REPORTS—Six double-spaced typewritten pages (including references) with one table, graph, or illustration. Submit four copies in above style.

GUEST EDITORIALS—One printed page, may be invited. Unsolicited editorials may also be submitted.

OTHER

Reviewers: Provide names of three to five potential reviewers with addresses and telephone numbers.

SOURCE

January 1995, 50A(1): M59.

Journal of Gerontology: Psychological Sciences

Gerontological Society of America
Waverly Press

Richard Schulz, PhD
University Center for Social and Urban Research
   and Department of Psychiatry
University of Pittsburgh
Room 607, 121 University Place
Pittsburgh, PA 15260

AIM AND SCOPE

The *Journal of Gerontology: Psychological Sciences* publishes articles on applied, clinical and counseling, experimental, and social psychology of aging. Appropriate topics include, but are not limited to, animal behavior, attitudes, cognition, educational gerontology, health psychology, industrial gerontology, interpersonal relations, neuropsychology, perception, personality, physiological psychology, psychometric tests, and sensation. Manuscripts reporting work that relates behavioral aging to neighboring disciplines are also appropriate.

MANUSCRIPT SUBMISSION

Submission of a manuscript implies that it has not been published and is not under consideration elsewhere. If accepted, it is not to be published elsewhere without permission. Corresponding author is responsible for certifying that permission has been received to use copyrighted instruments or software employed in research and that human or animal subjects approval has been obtained. For co-authored manuscripts, corresponding author must submit a letter, signed by all authors, indicating that they actively participated in collaborative work leading to publication and agree to be listed as an author. Assurances will be requested upon acceptance.

Anonymous review is available on request. Prepare manuscripts, to conceal author's identity. Omit cover page and footnotes that identify author(s) from two copies of manuscript. Manuscripts not prepared in this manner receive open review.

MANUSCRIPT FORMAT

Print paper double-spaced, including references and tables, on 8.5 × 11 in. white paper using 1 in. margins. Number pages consecutively beginning with title page. Submit four complete copies including tables and figures. Use IBM-compatible software to facilitate electronic typesetting; a disk will be requested upon acceptance.

Title Page: Include title, author's full name, and author's institution. Put a short running head (40 letters and spaces) at foot of page.

Abstract: 150 words, one-paragraph typed, double-spaced, on a separate page. State purpose of study, basic procedures used (study participants or experimental animals and observational or analytical methods), principal findings, and conclusions.

Text: Divide observational and experimental articles into major sections with headings such as Introduction, Methods, Results, and Discussion. Articles may require subheadings within sections for clarity. Discussion may include conclusions derived from study and supported by data. While full exposition is desirable, conciseness is imperative. Avoid sexist or ageist language.

Text References: Show references in text by citing in parentheses author's surname and year of publication. Example: " . . . a recent study (Jones, 1987) has shown . . ." If two authors, include surnames of both each time. If more than two and fewer than six authors, cite all first time reference occurs; in subsequent citations, and for all citations having more than six authors, include only surname of first author followed by "et al." If more than one publication in the same year by a given author or multiple authors is cited, distinguish by adding, a, b, etc. to year of publication. Separate multiple references cited at same point in text by semicolons enclosed in one pair of parentheses. Order within a pair of parentheses is alphabetical by first author's surname or, if all by the same author, in sequence by year of publication.

Reference List: Type double-spaced; arrange alphabetically by author's surname; do not number. Include only references cited in text; do not exceed 25 entries. Do not include references to private communications. Consult *Publication Manual of the American Psychological Association* for correct form.

Footnotes: Place footnotes related to title, each author's affiliation and address, source of research support, and other acknowledgments on a separate page before references. Indicate where correspondence should be addressed. Do not use reference footnotes.

Tables: Type double-spaced on separate sheets. Number consecutively and give brief titles. Type footnotes immediately below table referenced by superscript letters ([a,b,c] . . .); arrange footnotes alphabetically by superscripts. Use asterisks to indicate probability levels of tests of significance.

Illustrations: Provide one glossy print and three copies for each graph and drawing. Color plates are at author's expense. Sharp, laser printed or glossy prints are required. Have figures professionally lettered in a sans-serif type (e.g., Univers or Helvetica) or from a laser printer. Typewritten/dot matrix lettering is not acceptable. Make letters of sufficient size that, when reduced, they will be legible.

Captions for Illustrations: Type double-spaced on a separate page with numbers corresponding to illustrations. Explain scale used in figure.

OTHER FORMATS

BRIEF REPORTS—Six double-spaced typed pages (including references), one table, graph, or illustration. Submit four copies of manuscript.

REVIEW ARTICLES AND POSITION PAPERS—Solicited by editor only.

THEORETICAL OR METHODOLOGICAL ARTICLES—Include an integration and critical analysis of existing views in a specific area as well as proposed resolution(s) of controversial positions. Support methodological contributions with examples based upon empirical data if possible.

OTHER

Style: Other than as specified above, follow *Publication Manual of the American Psychological Association* (3rd ed., APA, Washington, D.C.).

SOURCE

January 1995, 50B(1): front matter.

Journal of Gerontology: Social Sciences
Gerontological Society of America
Waverly Press

David J. Ekerdt, PhD
Center on Aging
Room 5021 Wescoe
University of Kansas Medical Center
3901 Rainbow Blvd.
Kansas City, KS 66106-7117

AIM AND SCOPE

The *Journal of Gerontology: Social Sciences* publishes articles dealing with aging issues from the fields of anthropology, demography, economics, epidemiology, geography, political science, public health, social history, social work, and sociology.

MANUSCRIPT SUBMISSION

Submission implies that manuscript has not been published and is not under consideration elsewhere. If accepted, it is not to be published elsewhere without permission. Corresponding author is responsible for certifying that permission has been received to use copyrighted instruments or software employed in research and that human or animal subjects approval has been obtained. For co-authored manuscripts, corresponding author is responsible for submitting a letter, signed by all authors, indicating that they actively participated in collaborative work leading to publication and agree to be listed as an author. Assurances will be requested upon acceptance.

Anonymous review is available upon request. Prepare manuscripts to conceal author's identity. Omit title page that identifies author(s) from four copies of manuscript. Manuscripts not prepared in this manner will not receive this type of review.

MANUSCRIPT FORMAT

Limit text to 5000 words. Print paper double-spaced, including references and tables, on 8.5 × 11 in. white paper using 1 in. margins. Number pages consecutively, beginning with title page, including all pages. Submit five copies. Use IBM-compatible software to facilitate electronic typesetting.

Title Page: Include title of manuscript and full name and (with footnotes) affiliation of each author. Place a short running headline (40 letters and spaces) at foot of page.

Abstract: 150 words, one-paragraph typed, double-spaced, on a separate page. State purpose of study, basic procedures, main findings, and conclusions.

Text: Divide into sections headed: Introduction, Methods, Results, and Discussion. Subheadings may be used to clarify section content.

Text References: Cite references in text with author's surname and year of publication in parentheses. If a reference has three authors, includes surnames of each author each time citation appears. If more than three, include only surname of senior author and "et al." If reference list includes more than one publication in the same year by a given author or multiple authors, distinguish citations by adding, a, b, etc. to year of publication. Cite multiple references at same point in text separated by semicolons and enclosed in one pair of parentheses. Order is alphabetical by author's surname or, if all are by the same author, in sequence by year of publication.

References: Type double-spaced alphabetically by author's surname; do not number. Include only references cited in text; do not include private communications or unpublished work. Use full names; do not abbreviate journal titles.

JOURNALS

Yeatts, Dale E., Jeanne C. Biggar, and Charles F. Longino, Jr. 1987. "Distance Versus Destination: Stream Selectivity of Elderly Interstate Migrants." *Journal of Gerontology* 42:288–294.

CHAPTER

Fry, Christine L. and Jennie Keith. 1982. "The Life Course as a Cultural Unit." In Matilda W. Riley (Ed.), *Aging from Birth to Death: Sociotemporal Perspectives*. Boulder, CO: Westview Press.

BOOKS

Binstock, Robert H. and Ethel Shanas (Eds.). 1986. Handbook of Aging and the Social Sciences (2nd ed.). New York: Van Nostrand Reinhold.

Footnotes and Acknowledgments: Do not use reference or content footnotes. Place acknowledgments, including source of research support, on a separate page before references. Indicate corresponding author.

Tables: Type double-spaced on a separate sheet. Number consecutively and supply a brief title at top for each. Type footnotes immediately below. Reference marks are superscript small letters (^{a,b,c} . . .); arrange footnotes alphabetically by superscripts. Use asterisks to indicate probability level of tests of significance. Indicate in text preferred placement for each table by noting [Table 1 about here].

Illustrations: Have figures professionally lettered in a sans-serif type (e.g., Univers or Helvetica) or from a laser printer. Typewritten/dot matrix lettering is not acceptable. Do not send original copy with a manuscript. Include 5 photocopies of original or roughly drawn, legible draft. Upon acceptance of article, submit two glossy prints. Type captions double-spaced on a separate page.

OTHER FORMATS

BRIEF REPORT—2000 words prepared as above with 1 table, graph, or illustration. Submit 5 copies.

LETTERS TO THE EDITOR—500 to 750 words, typed double-spaced. Submit five copies. If appropriate, copy will be sent to author of original article for rebuttal. Letters and rebuttals will be reviewed and are subject to editing. Letter and rebuttal will be published in same issue.

REVIEW ARTICLES—Solicited by *Journal*.

THEORETICAL OR METHODOLOGICAL ARTICLES—Integrate and critically analyze existing positions on a specific topic, as well as proposed resolution(s) of controversial positions. Support methodological papers with data.

SOURCE

January 1995, 50B(1): back matter.

Journal of Heart and Lung Transplantation

International Society for Heart and Lung Transplantation
Mosby-Year Book, Inc.

Maria Rosa Costanzo, MD
Rush Presbyterian /St. Luke's Medical Center
Section of Cardiology, Suite 439
1725 W. Harrison St.
Chicago, IL 60612
Telephone: (312) 563-2121; Fax: (312) 850-0913

AIM AND SCOPE

The *Journal* publishes proceedings of the annual scientific sessions of the International Society for Heart and Lung Transplantation and will consider for publication suitable articles on all topics relevant to intrathoracic transplantation, support, and/or replacement.

MANUSCRIPT SUBMISSION

Articles are accepted for publication on the condition that they are original, are not under consideration by another journal, or have not been previously published. Manuscripts must be accompanied by statement signed by principal author: "The undersigned author transfers all copyright ownership of the manuscript (title of article) to the International Society for Heart and Lung Transplantation in the event the work is published. The undersigned author warrants that the article is original, is not under consideration by another journal, and has not been previously published. I sign for and accept responsibility for this material on behalf of any and all co-authors."

All persons designated as authors should qualify for authorship, having participated sufficiently in work to take public responsibility for its content.

MANUSCRIPT FORMAT

Submit manuscripts in triplicate, one original and two photocopies (not carbon copies), typed double-spaced on good-quality white paper, with liberal margins on all sides.

Title Page: Include authors' full names, degrees, and hospital and academic affiliations; and address, business and home telephone numbers, and fax number of corresponding author. References, figure legends, tables, and appendixes follow, in order.

Abstract: 250 word structured abstract, on a separate page, divided into sections: Background, Methods, Results, and Conclusions. Begin each section with heading and include relevant information.

References: Do not cite manuscripts in preparation, personal communications, and other unpublished information in reference list; mention in text in parentheses; append note of approval from source to manuscript. Identify references in text by superscript Arabic numerals. Number in consecutive order as mentioned in text. Type reference list double-spaced at end of article in numeric sequence. Conform to format of "Uniform Requirements for Manuscripts Submitted to Biomedical Journals" (Vancouver style) (*JAMA* 1993;269:2282-6). Abbreviate journal names following style of *Index Medicus*. List all authors when six or fewer; when seven or more, list first three and add *et al.*

JOURNAL ARTICLES

Salvatierra O, Vincenti F, Armend W, et al. Four-year experience with donor-specific blood transfusions. Transplant Proc 1983;15:924–6.

BOOKS

Eisen HN. Immunology: an introduction to molecular and cellular principles of the immune response. 5th ed. New York: Harper & Row, 1974:406.

CHAPTERS

Fromer PL. Workshop on congestive heart failure. In: Braunwald E, Mock MB, eds. Congestive heart failure. New York: Grune & Stratton, 1982:1–2.

Illustrations: Number illustrations (three sets of glossy prints) in order of mention in text. Mark lightly on back with author's last name and an arrow to indicate TOP. Prepare drawings or graphs in black India ink or use typographic (press-apply) lettering. Typewritten or freehand lettering is unacceptable. Lettering must be done professionally and be in proportion to drawing, graph, or photograph. Letter-quality, computer-generated illustrations are acceptable. Do not send original art work, x-ray films, or ECG strips. Glossy print photographs, 3 × 4 in. (minimum) to 5 × 7 in. (maximum), with good black-and-white contrast or color balance are preferred. Make special arrangements with publisher at (800) 325-4177, ext. 4317, or (314) 453-4317, for color plates or numerous illustrations. Make illustrations consistent in size. Note special instructions regarding sizing.

Figure Legends: Type legends double-spaced on a separate sheet numbered to correspond with figure. If an illustration has been previously published, give full credit to original source.

Tables: Make self-explanatory. Supplement, do not duplicate, text. Type double-spaced throughout, including column heads, data, and footnotes. Put each table on a separate sheet. Supply concise titles describing table's content. Give columns concise headings describing data in column. Type all footnotes immediately below table; define abbreviations. If a table has been previously published, give full credit to original source in a table footnote.

Appendixes: Prepare in same manner as tables.

OTHER FORMATS

BRIEF COMMUNICATIONS AND COMMENTARIES—Five to seven pages presenting new ideas, techniques, or findings of special significance or original elements in a debated question. Type double-spaced, including tables, figures, and references with a four sentence abstract.

LETTERS TO THE EDITOR—In cover letter explain circumstances relevant to "Letter to the Editor." Give letter a title; double-space; and submit in triplicate.

Brief "Letters to the Editor" contain substantive information concerning experimental or clinical studies, accompanied by a brief bibliography (no more than five) and one or two tables and/or figures.

OTHER

Abbreviation: Define at first appearance in text, tables, and figure legends. Do not use in title or abstract. Use generic names of all drug.

Experimentation, Animal: Include a statement in Methods section giving assurance that all animals received humane care in compliance with "Principles of Laboratory Animal Care" formulated by the Institute of Laboratory Animal Resources and "Guide for the Care and Use of Laboratory Animals" prepared by the Institute of Laboratory Animal Resources (National Institutes of Health, NIH Publication No. 86-23, revised 1985).

Experimentation, Human: Include a statement in Methods section indicating approval by institutional review board and affirming that informed consent was obtained from each patient. Submit signed releases showing informed consent for photographs of identifiable persons.

Page Charges: $50 per published page.

Units: Report measurements of length, height, weight, and volume in metric units or decimal multiples. Give temperatures in degrees Celsius; blood pressure in millimeters of mercury (mm Hg); hematologic and clinical chemistry measurements in metric system in terms of International System of Units (SI). Rejection grading must conform to ISHLT guidelines (*Journal* November/December 1990).printed in the November/December 1990 issue of the *Journal*.

SOURCE

January/February 1995, 14(1): 10A-12A.

Journal of Histochemistry and Cytochemistry
The Histochemical Society, Inc.

Paul J. Anderson, MD
Mount Sinai School of Medicine
New York, NY 10029-6574
Telephone: (212) 362-1801; Fax: (212) 874-8313

AIM AND SCOPE

The primary purpose of the *Journal* is to publish articles reporting original research in all fields of histochemistry and cytochemistry, provided the reported investigations contribute to the development, evaluation, or application of histo- and cytochemical methods. Histo- and cytochemistry are interpreted in a broad sense, covering analysis of the chemical and structural organization of cell and tissue components. Chemical or biochemical analyses that provide data directly pertinent to cell and tissue structures clearly fall within these limits. Papers describing the application for histochemical methods are acceptable only if they illustrate a unique contribution of success to the solution of a problem, or are of general interest in biology or pathology.

Full length Articles report original research in all fields of histochemistry and cytochemistry. The investigation may contribute to the development, evaluation, or application of histochemical and cytochemical methods and their use in the study of problems of general interest in bioscience, medicine, and pathology.

MANUSCRIPT SUBMISSION

Manuscripts are accepted from investigators in any country, whether or not they are members of the Histochemical Society, Inc. Submission implies that manuscript reports unpublished work, and that, if accepted for publication, it will not be published elsewhere without consent of the *Journal*. Manuscripts must be submitted only to the *Journal* and must not be published, simultaneously submitted, or already accepted for publication elsewhere. Submit copies of any possibly duplicated material.

Prepare a letter of transmittal signed by each author. Include: title of article; name(s) of author(s); identify one author as correspondent and include a complete address, telephone and fax numbers; provide justification if more than six authors; type of manuscript (Original Article, Rapid Communication, Brief Report, Technical Note, Review, Letter to the Editor); a statement that manuscript is not under consideration for publication in another journal, that no parts of manuscript have been previously published (or, if published, that publishing credits are provided in manuscript), and that all co-authors agree to share equal responsibility for manuscript's content; a brief statement summarizing main points; if manuscript contains color figures, state willingness and ability to pay costs of reproducing color figures; and identify affiliations with or financial involvement in any organization with a direct financial interest in subject matter or materials of research discussed (such as employment, consultancy, stock ownership). This information is held in confidence during review. If accepted for publication, Editor will discuss appropriate extent of disclosure.

Study a current journal issue to become familiar with format. Manuscripts that do not conform to instructions will be returned. Submit four complete copies (including illustrations).

Check to ensure that every reference is given a citation in text and that every cited reference appears in Literature Cited. Verify accuracy of references against original documents. Cite figures and tables in sequence order; check for correct citation. Manuscripts with errors in reference or figure citations will be returned.

MANUSCRIPT FORMAT

Use hard-surfaced white bond 8.5 × 11 in. paper. Do not use eraseable bond. Justify left margins, leave right margins ragged; keep 1.5 in. on all sides. Double-space throughout. Use a straightforward, easily legible typeface. Do not use "fancy" ones such as Orator or Gothic. Do not use italic or boldface anywhere. Type quality must be sharp, clear, and legible. A carbon film ribbon is preferred for typewritten manuscripts. Produce computer-generated manuscripts using a letter-quality or near-letter quality printer set at 10 or 12 cpi (characters per inch).

Conform to following order. If manuscript is in an unusual format, give reasons in cover letter. Number each page.

Running Headline (Page 1): Short title (48 characters including spaces). Use capital letters.

Title Page (Page 2): Leave a 3 in. space at top. Give complete title, first name, middle initial, and last name of each author, and name(s) of department(s) and institution(s) to which work is attributed. If more than one institution or department is involved, put initials of author(s) associated with each one after in parentheses. Credit research grants in lower left corner. Give specific mailing address, telephone and fax numbers for corresponding and reprint request authors in lower right corner.

Abstract (Page 3): 200 words. Specify objective of study, briefly describe methods used, and state conclusions reached and possible significance.

Below provide three to ten key words or short phrases suitable for indexing. Include subject under investigation, technologies or methods employed, and animal species studied. Use *Medical Subject Headings List* for *Index Medicus*.

Begin each section on a new page.

Introduction: State purpose or aim of work, problem that stimulated it, and a brief summary of relevant published reports.

Materials and Methods: Present in sufficient detail to enable work to be repeated by others; identify methods, materials, and apparatus (with manufacturer's name and location in parentheses); provide generic identification for all compounds used (trade or brand names alone are not acceptable); reference established methods, including statistical methods; provide references and a brief description of methods that have been published but are not well

known; describe new or substantially modified methods; give reasons for using and evaluate limitations.

Results: Describe concisely. Do not use both tables and graphs to illustrate the same results.

Discussion: Discuss results, with a minimum of recapitulation of findings. Emphasize new and important aspects of study and conclusions that follow. Include implications of findings and their limitations. Relate observations to other relevant studies. Link conclusions with goals of study. Avoid unqualified statements and conclusions not supported by data. Avoid claiming priority. State new hypotheses identified as such.

Literature Cited: List only published works and papers in press. Exclude work in progress, manuscripts submitted but not yet accepted, unpublished experiments, and personal communications from reference list. Acknowledge (in parentheses) in text. Abbreviate journal names as in *Index Medicus*.

One of three systems for citing references may be used: Vancouver System, Number System, and Name-and-Year System (Harvard System). In both Number System and Name-and-Year System, authors' names are arranged alphabetically in reference list.

Vancouver System—Number Literature Cited consecutively corresponding to order references are first cited in text. Identify references in text by consecutive numbers in parentheses, flush with line.

JOURNAL ARTICLES

Reference number, period, last name of author followed by initials (no internal punctuation), period, complete title of article (with only initial letter of first word capitalized), period, Journal abbreviation (no internal punctuation), year of publication, semicolon, volume number, colon, first page of article (no final punctuation).

18. Menten NL, Junge J, Green MH. A coupling histochemical azo dye test for alkaline phosphatase in the kidney. J Biol Chem 1944;153:471

BOOKS, THESES, AND COMPARABLE PUBLICATIONS

Reference number, period, last name of author followed by initials (no internal punctuation), period, complete title of book (capitalize only initial letter of first word), period. Volume number (abbreviated), period, edition number (abbreviated), period, city of publication, colon, complete name of publisher, comma, year of publication, colon, first page of chapter if needed (no final punctuation).

23. Smith W, Jones H, Watson M. Recent advances in immunology. New York: Plenum Press, 1987

23. Smith W, Jones H, eds. Studies in immunology. 2nd ed. Orlando, FL: Academic Press, 1988

CHAPTERS WITHIN A BOOK

23. Smith W, Jones H. The immune system in nude mice. In Hunter J, Hill W, Watson M, eds. Experimental immunology. New York: Raven Press, 1989:275

Number System—Cite numbers in text. Arrange list alphabetically, including names of co-authors, and numbered serially by surname of first author and chronologically, beginning with most recent publication, for the same author.

Cite references in text with reference number in parentheses, flush with line. If author's name is mentioned in text, reference number follows name at an appropriate point in sentence with name.

JOURNAL ARTICLES

Reference number, period, last name of author followed by initials (no internal punctuation), colon, complete title of article (capitalize only initial letter of first word), period. Journal abbreviation (no internal punctuation), volume number, colon, first page of article, comma, year of publication (no final punctuation).

1. Gomori G: Microtechnical demonstration of phosphatase in tissue section. Proc Soc Exp Biol Med 42:23, 1939

2. Menten NL, Junge J, Green MH: A coupling histochemical azo dye test for alkaline phosphatase in the kidney. J Biol Chem 153:471, 1944

BOOKS, THESES, AND COMPARABLE PUBLICATIONS

Reference number, period, last name of author followed by initials (no internal punctuation), colon, complete title of book (capitalize only initial letter of first word), period. Volume number (abbreviated), period, edition number (abbreviated), period, city of publication, comma, complete name of publisher, comma, year of publication, comma., first page of chapter if needed (no final punctuation).

1. Smith W, Jones H, Watson M: Recent advances in immunology. New York, Plenum Press, 1987

1. Smith W, Jones H, eds: Studies in immunology. 2nd ed. Orlando, FL, Academic Press, 1988

CHAPTERS IN BOOKS

1. Smith W. Jones H: The immune system in nude mice. In Hunter J, Hill W, Watson M, eds. Experimental immunology. New York, Raven Press, 1989, 275

Name-and-Year System—Use only for Original Articles and Reviews, preferred for Reviews. Arrange Literature Cited alphabetically, including names of co-authors, and chronologically for same author beginning with most recent date. Do not number entries. Follow authors' names by year of publication (with subdivision into a, b, etc. for publications in the same year), full title, and standard bibliographic information on source.

Enclose references cited in text in parentheses with author's last name followed by year of publication. Name only two authors for a given reference; use "et al." (not italicized) for three or more authors in text citations but not in references (give names of all authors).

JOURNAL ARTICLES

Last name of author followed by initials (no internal punctuation), year of publication in parentheses, colon, complete title of article (capitalize only initial letter of first word), period. Journal abbreviation (no internal punctuation), volume number, colon, first page of article (no final punctuation).

Gomori G (1939): Microtechnical demonstration of phosphatase in tissue section. Proc Soc Exp Biol Med 42:23

BOOKS, THESES, AND COMPARABLE PUBLICATIONS

Last name of author followed by initials (no internal punctuation), year of publication in parentheses, colon, complete title of book (with only initial letter of first word capitalized), period. Volume number (abbreviated), edition number (abbreviated), period, city of publication, comma, complete name of publisher, comma, first page of chapter if needed (no final punctuation).

Smith W, Jones H, Watson M (1987): Recent advances in immunology. New York, Plenum Press

Smith W, Jones H, eds (1988): Studies in immunology. 2nd ed. Orlando, FL, Academic Press

CHAPTERS WITHIN BOOKS

Smith W, Jones H (1989): The immune system in nude mice. In Hunter J, Hill W, Watson M, eds. Experimental immunology New York, Raven Press, 275

Tables: Use to compare and relate appropriate data sets in a visually economical manner. Help readers recognize relationships and patterns of data. Supplement, do not duplicate, text. Refer to every table in text and identify significant points; textual discussion of each tabular element makes table unnecessary.

Type double-spaced on separate sheets. Do not submit photographs or use paper larger than 8.5 × 11 in. Keep a 1 in. margin on all four sides. Number consecutively (Arabic numerals), and include a brief but informative title above table. Construct concise titles for stub heading (left column) and field headings (data columns); make telegraphic in style and no longer than the number of characters required by widest entry beneath heading.

Explain non-standard abbreviations, technical terms, statistical measures, etc. in notes at bottom. Arrange notes in order: general notes providing information relating to table as a whole (abbreviations, symbols, etc.); specific notes pertaining to a particular column or entry (use superscript lowercase letters in column or entry); statistical or probability notes to indicate results of tests of significance (use asterisks to indicate level of significance, one asterisk for lowest level of significance, two for next level, etc.).

Do not use horizontal or vertical rule lines to separate column or stub headings. Obtain permission and acknowledge source of tables or data borrowed from another published or unpublished source.

Illustrations: High-resolution graphics (line drawings, graphs, photographs, and color figures) are essential. Consult *Illustrating Science: Standards for Publication* (Council of Biology Editors, Bethesda, MD, 1988) for details concerning preparation of scientific illustrations.

Submit four sets of figures as sharp, critically focused, glossy prints. Do not submit photocopies. Mount each set on white cardboard no larger than 8.5 × 11 in. for each figure or composite. Trim and mount composite figures for reproduction as a single plate with an even amount of space (2-5 mm) between adjacent figures.

Label back in soft lead pencil (never ball-point or ink pen), indicating TOP, figure number, and first author's name. Mark back of best set "use for reproduction." Cover with protective overlays of tissue paper or clear acetate; use to indicate areas of importance in figure by light pencil marks and comments.

Mail illustrations mailed flat with cardboard stiffeners. Do not fold, roll, or use paper clips.

Have line drawings and graphs professionally drawn and lettered; freehand or typewritten lettering is unacceptable. Line drawings and graphs will be reduced to one-column width (3.25 in.); minimal height of lettering must be 2 mm. Make height of letters on an original 8×10 in. illustration at least 5 mm for lowercase and 6 mm for uppercase letters and numerals. Make symbols at least 5 mm in diameter. Make ruled lines and curves sufficiently thick to withstand reduction. Size recommendations for graphs prepared with lettering sets are: #1 Leroy (or equivalent) for graph grids, bonds, and arrows; #2 Leroy for graph borders or reference lines; #5 Leroy for graph curves and emphasis lines. Mark abscissas and ordinates with appropriate units, preferably International System of Units (SI). Avoid exponentials in labeling; if must be used, precede quantity expressed by the power of 10 by which its value has been multiplied.

Submit four sets of sharp, uniformly focused glossy prints or "photostats" of line drawings and graphs no larger than 5×7 in.; print will be reduced to one-column width (3.25 in.). Submit original illustration if best reproduction. Mount each photograph on 8.5×11 in. paper with rubber cement, dry-mount, or similar adhesive. Label back in soft lead pencil, indicating figure number, TOP, and first author's name. Provide legends for all line drawings and graphs. Identify units, abbreviations, mathematical expressions, abscissas, ordinates, and symbols.

Drawings and graphs produced by computer programs must meet same standards as professionally drawn copy. Do not use dot-matrix printers. Do not submit illustrations with discontinuous lines, step defects, and broken lettering.

Halftone illustrations (photographs, photomicrographs, electron micrographs, etc.) are reproduced at one-column width (3.25 in.), one and one-half-column width (4.75 in.), or two column width (6 7/8 in.). Halftones are reduced by cropping and sizing. Submit illustrations pre-trimmed to appropriate widths. Trim composite illustrations (multiple photographs to be reproduced as one plate) so all photographs are of equal size. Maximum plate area published without reduction is $6 \ 7/8 \times 8.75$ in.

Apply figure numbers to lower left corner of each photograph. Make number contrast with background. Make figure numbers, letters, symbols, and arrows applied to photograph surface large enough to withstand reduction to one- or two-column width (5 to 7 mm in letter or numeral height and from 1 to 2 mm in line or arrow thickness). Dry transfer lettering (such as Letraset) may be used; avoid scratches or other damage to figure surface.

Give photomicrographs and electron micrographs internal scale markers, a straight line or bar that represents a linear measurement in micrometers (μm) corresponding to mag-

nification of photographic image. Apply bar to surface of photograph, preferably in lower right corner in a horizontal position, and standing out from background. Make bars approximately 1 mm thick (#2 Leroy lettering pen). Dry transfer lines may be used.

Color Figures: Trim to eliminate unnecessary marginal detail. Mount on firm but highly flexible smooth-surfaced drawing paper. Do not mount on stiff cardboard. Submit four sets.

Extra cost is by author. An invoice will be sent and payment must be received before article is scheduled.

Figure Legends: Make informative and concise without duplicating material presented in body of text. Describe procedures and techniques in Materials and Methods, not in Figure Legends. Identify subject matter, symbols, arrows, insets, and method of preparation (e.g., histochemical or staining method). State size of internal scale marker in photomicrographs and electron micrographs in micrometers (μm); give original magnification given. Formats:

Figure 3. Electron micrograph of tissue from (a) treated and (b) untreated animals. Original magnification × 3,000. Bar = 10 μm.

Figure 3. (a) Electron micrograph of tissue from treated animals. (b) Similar tissue from untreated animals. Original magnification × 30,000. Bar = 1 μm.

Footnotes: Do not use; incorporate footnoted material into body of text.

OTHER FORMATS

BRIEF REPORTS—Short reports on current research and reports of preliminary findings that may be of interest to other investigators. Follow described format. Details in Materials and Methods should be specific enough to permit replication of work by others. Arrange references in either Vancouver or Number system.

LETTERS TO THE EDITOR—Provide a forum for communication of opinion, interpretation, criticism, and new information on scientific or science-related political matters. Be concise and to the point. State purpose of Letter clearly sin text. Accepted Letters that challenge previously published results or interpretations are sent article's author with an invitation to respond. Letter and response are published together as quickly as possible.

RAPID COMMUNICATIONS— Accelerated publication of manuscripts that report important new information which will be immediately useful to our readers. Subject matter ranges over the entire field of histochemistry and cytochemistry, including reports of methods and application to currently important biological and biomedical problems. Conform to order of manuscript presentation detailed above, except do not use Name-and-Year system of references. Incorrect formatting, including failure to double-space all text, inaccurate or omitted literature citations, incorrect format for Literature Cited, unacceptable assembling and labeling of illustrations are grounds for exclusion.

REVIEWS—Discuss historical development or critical analysis of selected topics in histochemistry and cytochemistry. Consult with Editor while preparing manuscript.

TECHNICAL NOTES—Brief descriptions of new histochemical and histochemically related techniques, or descriptions of significant modification of existing techniques. Present as concisely as possible. Follow format above, with modification where necessary. Include a brief statement of purpose, an adequately detailed stepwise description of procedure, a statement of expected results, and reference to pertinent literature. Include observations based on application of methods.

OTHER

Abbreviations and Nomenclature: For chemical abbreviations of biological compounds, use those recommended by *Journal of Biological Chemistry*. *Chemical Abstracts* index is the authority for names of chemical compounds. Use nomenclature published by the Biological Stain Commission in *Biological Stains* (9th Edition, The Williams & Wilkins Company, Baltimore, 1977) for dyes. Scientific abbreviations should conform to *CBE Style Manual* or its recommended sources. For terminology and classification for enzymes follow *Commission on Biochemical Nomenclature on the Nomenclature and Classification of Enzymes.*

Color Charges: $850 to $1200, depending on size, complexity of color separations, and number of proofs.

Experimentation, Animal: Indicate whether institution's or National Research Council's guides for care and use of laboratory animals were followed.

Experimentation, Human: Indicate whether procedures followed were in accordance with ethical standards of Committee on Human Experimentation of institution at which experiments were done, or in accord with Helsinki Declaration of 1975.

Reviewers: Include names and addresses of several persons outside institution qualified to review paper. Identify unqualified reviewers; state grounds for disqualification.

Spelling: Use *Webster's Third New International Dictionary* a and *Dorland's Medical Dictionary*.

Style: Refer to *CBE Style Manual*, (5th Edition, 1983, Council of Biology Editors, Bethesda, MD).

SOURCE

January 1994, 42(1): 131-136.
Editorial Policy and Practices and Guidelines to Authors are published annually in January issue.

Journal of Hypertension
Current Science Limited

The Editor, *Journal of Hypertension*
34-42 Cleveland Street
London W1P 5FB, U.K.

AIM AND SCOPE

The *Journal of Hypertension* publishes papers reporting original clinical and scientific research which are of a high standard and which contribute to the advancement of knowledge in the field of hypertension. The *Journal* publishes full papers and reviews (normally by invitation), full papers and Rapid communications.

MANUSCRIPT SUBMISSION

Paper must not have previously submitted to another journal or have been published elsewhere. Submit four copies, three photocopies and three copies of all figures and tables. Type on white bond paper with 3 cm margins. Double-space throughout, including following sections, each on a separate sheet: title page, summary and keywords, text, acknowledgments, references, individual tables and captions. Number pages consecutively, beginning with title page. Place page numbers in top right corner of each page. Define abbreviations, if used, at first appearance in text; avoid those not accepted by international bodies.

MANUSCRIPT FORMAT

Instructions comply with those formulated by the International Committee of Medical Journal Editors. Consult: "Uniform requirement for manuscripts submitted to biomedical journals," (*N Engl J Med* 1991, **324**:424–428).

Title Page: Include full title of paper; a short title (40 characters including spaces), for a 'running head' so identified; first name, middle initial and last name of each author; and if work is attributed to a department or institution, include its full name.

Place disclaimers on title page. Give name and address of corresponding and reprint request author(s). If reprints are not available, so state.

List sources of support for work as grants, equipment, drugs, or any combination. Disclose any affiliations, including financial, consultant or institutional associations that might lead to bias or conflict of interest.

Structured Abstract: (Second page) 250 words. Articles containing original data concerning course, cause, diagnosis, treatment, prevention or economic analysis of a clinical disorder or an intervention to improve quality of health care should include a structured abstract with headings and information:

Objective—State main question or objective of study and major hypothesis tested, if any.

Design—Describe study design indicating, as appropriate, use of randomization, blinding, criterion standards for diagnostic tests, temporal direction (retrospective or prospective), etc.

Setting—Indicate study setting, including level of clinical care (for example, primary or tertiary; private practice or institutional).

Patients, participants—State selection procedures, entry criteria and numbers of participants entering and finishing study.

Interventions—Describe essential features of any interventions including method and duration of administration.

Main outcome measure(s)—Indicate primary study outcome measures as planned before data collection began. If hypothesis was formulated during or after data collection, clearly state.

Results—Describe measurements that are not evident from nature of main results; indicate blinding. Indicate absolute values when risk changes or effect sizes are given.

Conclusions—State only those that are directly supported by data, along with clinical application (avoiding overgeneralization). Give equal emphasis to positive and negative findings of equal scientific merit.

For articles concerning original scientific research include a structured abstract with headings and information:

Objective—State primary objective (if appropriate).

Design—State principal reasoning for procedures adopted.

Methods—State procedures used.

Results—State results of study. Include numerical data but keep to a minimum.

Conclusions—State conclusions that can be drawn from data given.

For review articles include an abstract which, if appropriate, may be structured with headings: purpose, data identification, study selection, data extraction, results of data analysis, conclusions.

Obtain more information on structured abstracts and a glossary of terms from publisher.

Key Words: Follow abstract with 3–10 keywords or short phrases for cross-indexing. Use headings from *Index Medicus*.

Text: Divide full papers of an experimental or observational nature into sections headed Introduction, Methods, Results, and Discussion; Editorial reviews may require a different format.

Acknowledgments: Should be made only to those who have made a substantial contribution to study. Obtain written permission from people acknowledged by name.

References: Number consecutively in order of first appearance in text. Assign Arabic numerals and give in brackets. References include names of all authors when six or less; when seven or more list only first six and add *et al.;* give full title and source information. Abbreviate journal names as in *Index Medicus*.

JOURNAL

OTSUKA Y, CARRETERO OA, ALBERTINI R, BINIA A: **Sodium balance role in chronic two-kidney Goldblatt hypertension.** *Hypertension* 1979 **1**:389–396.

BOOK

GROSS F: **Renin stores in the kidney and plasma renin activity.** In *Kidney Hormes.* Edited by Fisher JW. New York Academic Press; 1971:103–107

Do not include personal communications and unpublished work in reference list; include in parentheses in text. Include unpublished work accepted for publication but not yet released in reference list with words 'in press' in parentheses beside name of journal. Verify references against original documents.

Illustrations: Reference figures and tables in order of appearance in text and with Arabic numerals in parentheses. Paste a label on back of all illustrations with figure number, title of paper, author's name and a mark indicating

TOP. Do not mount illustrations. Present half-tone illustrations as glossy prints (82 mm wide) and line illustrations as original artwork or prints. For line illustrations, make both print and original artwork 82 mm wide or where illustration demands it, 173 mm. Photomicrographs must have internal scale markers. In photographs of people, obscure identities or submit permission to use photograph. If a figure has been published before, acknowledge original source and submit written permission from copyright holder. Permission is required regardless of authorship or publisher, except for documents in public domain. Figures may be reduced, cropped or deleted by Editor.

Tables: Type each double-spaced on a separate sheet. Do not submit as photographs. Assign each table an Arabic numeral and a brief title. Do not use vertical rules.

Captions for Illustrations: Type double-spaced, beginning on a separate sheet. Give each an Arabic numeral corresponding to illustration or table to which it refers. Explain internal scales and identify staining methods for photomicrographs.

OTHER

Nomenclature, Experimental Renovascular Hypertension: Follow Special Report of the Nomenclature Committee of the Council for High Blood Pressure Research of the American Heart Association (Page IH, Oparil S, Bohr DF, Tobian L: Nomenclature for experimental renovascular hypertension. *Hypertension* 1979, **1**:61).

GOLDBLATT HYPERTENSION

One-kidney, one clip hypertension: one kidney removed, the other clipped

Two-kidney, one clip hypertension: both kidneys intact, one clipped.

Two-kidney, two clip hypertension: both kidneys intact, both clipped

PAGE HYPERTENSION

One-kidney, one wrapped hypertension

Two-kidney, one wrapped hypertension

Two-kidney, two wrapped hypertension

GROLLMAN HYPERTENSION

One-kidney, one figure-8 hypertension

Two-kidney, one figure-8 hypertension

Two-kidney, two figure-8 hypertension

Nomenclature, Hypertensive Rat Strains: Follow nomenclature accepted at the Eleventh Meeting of the International Society of Hypertension in Heidelberg, West Germany, 1986 (*Hypertension* 1987, **9**:110).

Nomenclature, Renin-angiotensin System and Animal Peptides: Follow Report of the Joint Nomenclature and Standardization Committee of the International Society of Hypertension, The American Heart Association and the World Health Organization (*J Hypertens* 1987, **5**:507–511).

Summaries of Papers (with Large Amounts of Information): It is impractical to publish very large amounts of data (e.g., from technical reports, phase 2 and phase 3 clinical trials, new study protocols, etc.). Editor may ask that a summary be published in the *Journal* and the full paper be published in a supplement. Reference to supplement and page numbers of full text version will appear as a footnote to summary, with information on how to obtain offprints of full paper. Authors will be asked to contribute page charges on full text version. There is no charge for summary in *Journal*.

Authors who wish to use this option should indicate this in cover letter; they will be informed of page charges.

Units: Use SI Units (see *Quantities, Units,* and *Symbols,* 2nd edn. London: The Royal Society of Medicine; 1975). Depart only when long-established clinical usage demands (e.g., measurement of blood pressure in mmHg and drug dosage in mg). Include other units of measurement in parentheses. Express renin in terms of International Standard Renin Unit [Bangham et al.: *Clin Sci* 1975 **48** (suppl):135s–159s].

Derived SI units may be used; for basic and derived units, use prefixes to denote multiples and submultiples.

SOURCE

December 1994, 12(12): xli–xlii.

Journal of Immunological Methods
Elsevier Science Publishers

V. Nussenzweig
Department of Pathology
New York University Medical Center
School of Medicine, 550 First Avenue
New York, NY 10016 USA

M.W. Turner
Division of Cell and Molecular Biology
Molecular Immunology Unit
Institute of Child Health
University of London
30 Guilford Street
London WC1N 1EH, U.K.

J. C. Unkeless
Department of Biochemistry, Box 1020
Mount Sinai School of Medicine
1 Gustave Levy Place
New York, NY 10029
            Review articles (usually by invitation)

AIM AND SCOPE

The *Journal of Immunological Methods* is devoted to covering techniques for: quantitating and detecting antibodies and/or antigens and haptens based on antigen-antibody interactions; fractionating and purifying immunoglobulins, lymphokines and other molecules of the immune system; isolating antigens and other substances important in immunological processes; labelling antigens and antibodies with radioactive and other markers; localizing antigens and/or antibodies in tissues and cells, in vivo or in vitro; detecting, enumerating and fractionating immunocompetent cells; assaying for cellular immunity; detecting cell-surface antigens by cell-cell interactions; initiating immunity and unresponsiveness; transplanting tissues; and studying

items closely related to immunity such as complement, reticuloendothelial system and others. In addition the *Journal* will publish articles on novel methods for analyzing the organization, structure and expression of genes for immunologically important molecules such as immunoglobulins, T cell receptors and accessory molecules involved in antigen recognition, processing and presentation. Articles on the molecular biological analysis of immunologically relevant receptor binding sites are also invited. Submitted manuscripts should describe new methods of broad applicability to immunology and not simply the application of an established method to a particular substance.

MANUSCRIPT SUBMISSION

Submission implies that manuscript has not previously been published and that it is not being considered for publication elsewhere. It is assumed that submission has the approval of all authors. Preferred medium of final submission is on disk with accompanying reviewed and revised manuscript.

Provide original and two copies.

MANUSCRIPT FORMAT

Restrict length to eight printed pages, i.e., approximately 20 manuscript pages. Include a title page and an abstract.

Type double-spaced with wide margins. Underline words to be printed in italics. Use metric system throughout.

Title Page: On a separate sheet include: title, name(s) of author(s), affiliations, a footnote indicating corresponding author with her/his full address, telephone and fax numbers, and a footnote providing abbreviations used in paper.

Abstract: 5% of article, on a separate sheet to appear at beginning of paper.

Key Words: 3–6 items at foot of abstract.

References: In text references start with name of author(s), followed by publication date in brackets.

Put reference list in alphabetical order, typed double-spaced on sheets separate from text.

References to journals contain names and initials of author(s), year of publication (between brackets), article title, abbreviated journal name (use *List of Serial Title Word Abbreviations,* International Serials Data System, Paris), volume number and page number. References to books include title (of series and volume), initials and names of the editor(s) and, between brackets, publisher and place of publication.

Troncone, R., Vitale, M., Donatiello, A., Farris, E., Rossi, G. and Auricchio, S. (1986) A sandwich enzyme immunoassay for wheat gliadin. J. Immunol, Methods 92, 21.

Strauss, J.H. and Strauss, E.G. (1985) Antigenic structure of togaviruses. In: M.H.V. Van Regenmortel and A.R. Neurath (Eds.), Immunochemistry of Viruses. Elsevier, Amsterdam, p. 407.

Campbell, A.M. (1984) In: R.H. Burdon and P.H. Van Knippenberg (Eds.), Laboratory Techniques in Biochemistry and Molecular Biology, Vol. 13, Monoclonal Antibody Technology. Elsevier, Amsterdam, p. 75.

Tables: Compile on separate sheets; number in Roman numerals.

Figures: Submit three sets of figures. At least one set must be of publication quality; other two sets must be clear enough to be useful to referees. Completely letter figures; make size of lettering appropriate ; take into account reduction (about one third). Consider page format in designing figures. Number in Arabic numerals.

Photographs must be of good quality; print on glossy paper. Photocopies of photomicrographs are not acceptable. Color photographs are accepted; contributors pay extra costs incurred.

Mark tables, figures and photographs on reverse side with number, author's name and orientation (TOP); use a soft pencil or a felt-tipped pen. Supply legends on separate sheets.

OTHER FORMATS

SOFTWARE NOTICES—Software must be noncommercial, of use to immunologic laboratories, involve methods of data analysis and/or visualization, or utilize a novel approach to techniques within the scope of the *Journal*. By submitting software, authors agree to unlimited distribution of software and documentation, if accepted. Distribution is coordinated through the software notice editor. A directory of available software is listed periodically in *Journal*.

Submit to: Stephen J. Merrill, Department of Mathematics, Statistics and Computer Science, Marquette University, Milwaukee, WI 53233. Send: two copies of source code and/or binaries (compiled version); one printed and one machine-readable (disk) version of documentation; a 1-2 page description of software including title of software package, primary purpose, type of computer, hardware/memory requirements, a short example illustrating its use, and all necessary references. 'Notice' will be an edited version of this description and is as an item which may be referenced in other papers.

To access Software Archive see *Journal of Immunological Methods*, 153 (1992) 5–6.

OTHER

Electronic Submission: Preferred storage medium is 5.25 or 3.5 in. disk in MS-DOS format; other systems are welcome. Macintosh users: save file in usual manner, do not use option 'save in MS-DOS format'; NEC users: use double or high density 5.25 in. disk or double density 3.3 in. disk, not high density 3.5 in. disks. Do not split article into separate files. Ensure that letter 'l' and digit '1' (also letter 'O' and digit '0') have been used properly. Structure article (tabs, indents, etc.) consistently. Do not leave characters not available on word processor (Greek letters mathematical symbols, etc.) open; indicate by a unique code (e.g., gralpha, @, #, etc., for the Greek letter α). Use codes consistently throughout entire text. List codes and provide a key. Do not allow word processor to introduce word splits; do not use a 'justified' layout. Strictly follow general instructions on style/arrangement and reference style. Save file in word processor format; do not use save files in flat ASCII. Format disk correctly. Put only article on disk. Specify type of computer and word processing package used. Label disk with your name and name of file on disk. After final acceptance, submit disk plus one final printed and exactly matching version. File on disk and print out must be identical. Obtain further information from publisher.

Experimentation, Animal: State that their care was in accordance with institutional guidelines. For animals subjected to invasive procedures, state anesthetic, analgesic and tranquilizing agents used, as well as amounts and frequency of administration.

Experimentation, Human: Include a statement indicating that informed consent was obtained after nature and possible consequences of studies had been fully explained.

SOURCE

1994, 177(1/2): 277–278.

Journal of Immunology
American Association of Immunologists

Dr. Peter Lipsky
9650 Rockville Pike
Bethesda, MD 20814-3994
Telephone: (301)530-7197; Fax: (301)571-1813

MANUSCRIPT SUBMISSION

Each manuscript must be accompanied by a copy of the current *Journal of Immunology* submission form printed in each issue.

Manuscripts must be original contributions that do not contain data that have been or will be submitted for publication elsewhere, including but not limited to symposia volumes, proceedings, transactions, etc. Abstracts of oral or poster presentations do not constitute previous publication. Chief criteria for acceptance are quality, originality, clarity, and brevity. Results must be verifiable; authors are expected to make unique materials available to qualified investigators.

If manuscript is published, it will become sole property of The American Association of Immunologists, and all rights in copyright are reserved to The American Association of Immunologists. Corresponding author signs Copyright Assignment form for all authors upon acceptance. For previously copyrighted material reproduced in manuscript, including modified figures and tables., submit documents giving permission.

Make length of manuscript commensurate with scientific content.

For manuscripts accepted for publication, submit a computer diskette containing current version of manuscript, exclusive of tables and equations.

MANUSCRIPT FORMAT

Submit original and three copies of manuscript and four sets of original figures. Type double-spaced including abstract, references, figure legends, and tables, (3/16 in. spacing from bottom of one line to top of a capital of a succeeding line) on one side of 8.5 × 11 in. good quality paper. Do not use compressed type formats.

Running Title: 60 characters, typed double-spaced on a separate page.

Full Title and Authors' Full Names: First name, middle initial(s), surname and affiliations typed double-spaced. When appropriate, indicate author's family name. Identify affiliations of all authors and institutions, departments, or organizations with: *, †, ‡, §, ||, #, **, ††, ‡‡, §§, ||||, ##.

Abstract: 250 words, one page, double-spaced. Do not use references.

Acknowledgment: Place immediately after text, and before references. Cite grant support as a footnote to the title.

References: Number as appear in text. If cited in tables or in figure legends, number according to position of citation of table or figure in text. List all authors. Use *BIOSIS* for primary source of journal name abbreviations and *Index Medicus* secondary.

PERIODICALS

1. Krensky, A. M., F. Sanchez-Madrid, E. Robbins, J. A. Nagy, J. A. Springer, and S. J. Burakoff. 1983. The functional significance, distribution, and structure of LFA-1, LFA-2, and LFA-3: cell surface antigens associated with target interactions. *J. Immunol. 131:611.*

BOOKS

2. Hauptfeld, V., M. Hauptfeld, M. Nahm, J. Trial, J. A. Kapp, and D. C. Shreffler. 1983. Partial characterization of eight anti-I-J and three anti-Ia monoclonal reagents. In *Ir Genes, Past, Present, and Future, vol. 2.* C. W. Pierce, S. E. Cullen, J. A. Kapp, B. D. Schwartz, and D. C. Shreffler eds. The Humana Press, Clifton, NJ, p. 51.

List only papers accepted for publication in references. Cite manuscripts in preparation, unpublished observations, and personal communications parenthetically in text. Cite submitted papers as footnotes to text. Substantiate personal communications by letters of permission.

Footnotes: Number sources of support, abbreviations used, correspondence address, current address, references submitted for publication, etc. consecutively in order of appearance in text. Group and place on separate page(s) between References and figure legends.

Tables: Number with Roman numerals in order of appearance in text.

Figures: Submit unmounted photomicrographs on glossy paper in quadruplicate. Layout requirements may necessitate mounting multicomponent figures. Submit illustrations, photographs, and line drawings (photographs or computer-generated figures are acceptable) as original artwork plus three copies; graphs should be in black ink on unlined or blue-lined paper. For graphs larger than 9 × 12 in. include smaller glossy photographs. Number figures and figure legends as appear in text. Mark on reverse side with figure number, first author's name, and TOP.

Color Photomicrographs: Request inclusion in a cover letter with submitted manuscript. Author will be notified of cost and must confirm acceptance of charges in writing.

OTHER FORMATS

COMMENTARIES—Invited brief summaries of and commentary on controversial subjects of broad interest to immunologists.

OTHER

Abbreviations: Use consistently throughout article. Do not use nonstandard, undefined abbreviations or abbreviations appearing only once or twice in text. See first issue of each volume for a list of standard abbreviations to be used without definition.

Cell and Tissue Cultures: Adhere to established standard in work with cell cultures. Include, at minimum, reference to source and quality control measures used to authenticate cell lines (Wolf, K. 1979. *Methods Enzymol.* 58:116; Freshney, R. I. 1994. *Culture of Animal Cells*, p. 251, Alan R. Liss, Inc., New York; Hay, R. J. 1994. *Atlas of Human Tumor Cell Lines.* Hay, R. J., Park, J. G., and Gazdar, A. eds. Academic Press, San Diego, p. 1).

Chemical Names and Enzymes: Use *The Merck Index* and the *IUPAC-IUB Commission on Biochemical Nomenclature-Chemical Abstracts* as the primary references for proper spelling and style of chemical names. Use *Enzyme Nomenclature* as the source for style and spelling of enzyme names.

Data Submission: Submit original nucleotide or amino acid sequence data to GenBank or EMBL Data Library at time of manuscript submission; an accession number from GenBank or the EMBL Data Library is required for publication. Instructions on submission of data may be obtained directly from GenBank (Mail Stop K710, Los Alamos National Laboratory, Los Alamos, New Mexico 875450) or European Molecular Biology Library, Nucleotide Sequence Library (Postfach 10.2209, Meyerhofstrasse 1, 6900 Heidelberg, Germany.

Isotopes: When the radioactive element normally occurs in the compound, place symbol for isotope in square brackets directly attached to front of the name or formula: $[a\text{-}^{14}C]$leucine, $[^{2}H]CH_4$. When element is not normally found in the compound, hyphenate the word "labeled" to the isotope, e.g., ^{131}I-labeled albumin. Do not use square brackets. The non-naturally occurring radioactive isotope may be hyphenated directly to the compound if the construction is compact and particularly if it occurs frequently, e.g., ^{125}I-CD20.

Molecular Weight and Mass: Molecular weight (m.w.) and relative molecular mass (M_r) are equivalent expressions that are dimensionless. Express molecular mass (symbol m) in daltons (Da).

Nomenclature, CD: For murine molecules, use nomenclature in *J. Immunol.* 149:10, 3129–3134, 1992 and for human molecules, standard CD nomenclature will be used as updated (*J. Immunol.* 152:1, 1–2, 1994).

Nomenclature, Genetic (Mice): See *Mouse Genome* 92(2), June 1994.

Nomenclature, TCR: Follow recommendations of WHO-IUIS Subcommittee on TCR Designation ([i]Bull. WHO[r] 71 (1):113, 1993).

Scientific Misconduct: Cases of possible scientific misconduct (suspected misrepresentation of data, double publication, or plagiarism) will be dealt with by Editor-in-chief. If matter is not resolved, corresponding author's institution will be contacted, which should then make inquiry and report back to Editor. If misconduct is confirmed, appropriate action will be decided by Publications Committee in consultation with the Council of the American Association of Immunologists.

Style: Follow *The CBE Style Manual* (Fifth Edition, Council of Biology Editors).

SOURCE

July 1, 1994, 153(1): 464-468.

Journal of Infectious Diseases
Infectious Diseases Society of America
University of Chicago Press

Editor, *The Journal of Infectious Diseases*
1910 Fairview Avenue E., Suite 210
Seattle, WA 98102-3699

AIM AND SCOPE

Reports of research related to any aspect of the fields of microbiology and infection, whether laboratory, clinical, or epidemiologic, will be considered for publication in the *Journal*. Major articles and concise communications are, in general, peer reviewed; correspondence is reviewed by the Editors.

MANUSCRIPT SUBMISSION

Describe original investigations that are important advances in the field and that have been brought to an acceptable degree of completion. Submit descriptive or primarily methodologic manuscripts as concise communications.

Submission is a tacit declaration that the same material has not been submitted or accepted for publication elsewhere. Obtain and submit written permission from all investigators cited in a personal communication who are not coauthors and from copyright owner for reproduction of a previously published table or figure. Submit three copies of "in press" references and articles by same authors on same subject submitted elsewhere. Include complete address, telephone and fax numbers of corresponding author.

MANUSCRIPT FORMAT

The *Journal* complies with "Uniform Requirements for Manuscripts Submitted to Biomedical Journals" (*Annals of Internal Medicine* 1982;96:766–70), except place reference numbers in text in square brackets (not parentheses). Type on 8.5 × 11 in. white bond paper with 1 in. margins. Type all material, including tables, references, and legends, double-spaced. Do not use italic or boldface type. Indicate italic by underlining (do not underline in vivo, in

vitro, in situ, or et al.). See a recent journal issue for style. Submit original typescript and three copies with four sets of illustrations. Put each set of illustrations in an envelope. Number all pages, including title page. Order material in major articles and concise communications as: title page, footnotes to title page, abstract, text, acknowledgment, references, figure legends, tables, and figures.

Title Page: Supply a running head (40 characters and spaces), a title (160 characters and spaces), and names and affiliations of authors.

Footnote Page: Supply a separate page with footnotes to article's first page, including name, date (month and year), and location (city, state, country if not U.S.) of meeting at which information was presented (include an abstract number); state that informed consent was obtained from patients or parents or guardians and that human experimentation guidelines of the U.S. Department of Health and Human Services and/or those of the authors' institution(s) were followed or that animal experimentation guidelines were followed in animal studies; state commercial or other associations that might pose a conflict of interest (e.g., pharmaceutical stock ownership); sources of financial support (including grant numbers); name, address, and telephone and fax numbers of corresponding and reprint request author(s); and current affiliations and addresses for authors who have moved since completion of study.

Abstract: 150 words, one paragraph. State purpose of research, methods used, results (with specific data given, if possible), and conclusions. Write in third person; do not cite references.

References: Type double-spaced. Include only published works or those accepted for publication. Put unpublished observations by authors, personal communications, and manuscripts submitted for publication in parenthetical expressions in text, e.g., (unpublished data). Number references in order of appearance; number those cited only or first in tables or figures according to order in which table or figure is cited in text.

Follow format of National Library of Medicine as used in *Index Medicus* and "Uniform Requirements." Provide all authors' (or editors') names when fewer than 7; when 7 or more, list first 3 and add et al. Spell out titles of journals not listed in *Index Medicus*. To reference a doctoral dissertation include author, title, institution, location, year, and publication information, if published. Verify accuracy of references.

1. Moscicki AB, Palefsky JM, Gonzales J, Smith G, Schoolnik GK. Colposcopic and histologic findings and human papillomavirus (HPV) DNA test variability in young women positive for HPV DNA. J Infect Dis 1992;166:951–7.

2. McIntosh K. Diagnostic virology. In: Fields BN, Knipe DM, Chanock RM, et al. Fields virology. 2nd ed. Vol 1. New York: Raven Press, 1990:411–40.

3. Lee LA, Puhr ND, Bean NH, Tauxe RV. Antimicrobial resistance of *Salmonella* isolated from patients in the United States. 1989–1990 (abstract 523). In: Pro-

gram and abstracts of the 31st Interscience Conference on Antimicrobial Agents and Chemotherapy (Chicago). Washington, DC: American Society for Microbiology, 1991:186.

Acknowledgment: Optional. Place on page preceding references statements thanking those who assisted substantially with work relevant to study.

Tables and Figures: Make tables concise; organize to show results, do not merely compile raw data. Do not repeat data in both a table and a figure. Explain abbreviations and acronyms in table footnotes and figure legends.

Number tables in order of appearance in text. Do not use vertical and internal rules. Keep footnotes and explanatory material to a minimum. Place footnotes below table; designate by symbols (in order of location when table is read horizontally): *, †, ‡, §, ‖, ¶, **, ††, etc. Give columns appropriate headings; indicate units of measure.

Submit line drawings (black-and-white charts and graphs) as glossy photographs or high-quality laser prints on bond paper. Incorporate keys to symbols used. Submit half-tone images (e.g., gels photomicrographs) as glossy photographs. Label prints with first author's name and figure number. Type legends double-spaced on a separate sheet. Authors pay full cost for color figures. On photomicrographs show only most pertinent areas or mark for cropping. Include a micron bar or appropriate scale on figure.

OTHER FORMATS

CONCISE COMMUNICATIONS—Nine double-spaced pages (2000 words) of text, including a short abstract, two inserts (tables or figures), and 15 references. Present a complete study that is narrower in scope than one in a major article or that represents a new development.

CORRESPONDENCE (letters)—Four double-spaced pages (750 words) of text, one insert (table or figure), and 10 references, prepared in manuscript format, including a title page; no acknowledgments. Address an observation of unusual or scientific value or a previous publication in the Journal. Submit correspondence related to observations of unusual teaching or clinical value to Clinical Infectious Diseases.

EDITORIALS—May be on any aspect of infectious diseases. Generally invited; unsolicited editorials will be considered.

INSTITUTIONAL REPORTS—Summaries of workshops and reports from National Institutes of Health or Centers for Disease Control. Page charges may be required.

NEWS FROM THE INFECTIOUS DISEASES SOCIETY OF AMERICA—Announcements of interest to the members of the IDSA.

PERSPECTIVES—Invited overviews of articles in the *Journal* or of other research in infectious diseases.

REVIEWS—Of infectious diseases that are primarily basic in nature. Submit manuscripts that describe both basic and clinical aspects of disease to either JID or Clinical Infectious Diseases.

SUPPLEMENTS—Inquire in writing about the suitability of topic, program organization, and production to Editor.

OTHER

Manufacturers: For commercially obtained products, list full names of manufacturers and locations (city, state, and country, if not U.S.). Use generic names of drugs and other chemical compounds.

Nomenclature: Use latest widely accepted nomenclature. See *Approved Lists of Bacterial Names* (amended ed., American Society of Microbiology, 1989) or *Bergey's Manual of Determinative Bacteriology* (Williams & Wilkins, 1993) and *Enzyme Nomenclature, 1992: Recommendations of the Nomenclature Committee of the International Union of Biochemistry* and *Molecular Biology on the Nomenclature and Classification of Enzymes* (Academic Press, 1992). For names and abbreviation of drugs and chemical compounds, refer to *Merck Index* (11th ed, Merck, 1989).

Reviewers: Submit names, addresses, and telephone numbers of 4 persons who potentially could serve as unbiased and expert reviewers.

Statistical Analyses: Identify analyses used in text and in all tables and figures where results of statistical comparison are shown. For guidance on the application of statistical methods for analysis of data, see *JID* (1984;149:349–54), *JAMA* (1966;195:1123–8) and *British Medical Journal* (1977;1:85–7 and 1983;286:1489–93).

Style: Follow *CBE Style Manual* (5th ed. Bethesda, MD: Council of Biology Editors, 1983) and *The Chicago Manual of Style* (14th ed. Chicago: University of Chicago Press, 1993).

Units: Use degrees Celsius for temperature. Use SI units (see *Annals of Internal Medicine* 1987;106:114–29). Express data in metric units.

SOURCE

July 1994, 170(1): back matter.

Journal of Internal Medicine
Blackwell Scientific Publications

Journal of Internal Medicine
PO Box 22 114
S-104 22 Stockholm, Sweden
(Location: Hantverkargatan 7)
Telephone: +46 8 650 9688; Fax: +46 8 650 8234

AIM AND SCOPE

The *Journal of Internal Medicine (JIM)* publishes original research papers in the field of internal medicine. *JIM* is a clinical journal and will primarily accept papers covering clinical science, but is willing to include articles on experimental work and basic science, provided there are clear links with clinical medicine. All articles should be presented in a concise form. Great care should be taken in the length which should be short!

MANUSCRIPT SUBMISSION

Submit one original and three copies of all manuscripts (except Supplements). Follow uniform requirements for manuscripts submitted to biomedical journals (Vancouver style), found in *Br Med J* 1991; 302: 338-41, and *N Engl J Med* 1991; 324: 424-8. A booklet, *Uniform Requirements for Manuscripts Submitted to Biomedical Journals* (Philadelphia: International Committee of Medical Journal Editors, 1993) is available (Secretariat Office, *Annals of Internal Medicine,* American College of Physicians, Independence Mall West, Sixth Street at Race, Philadelphia, PA 19106-1572).

Give signed consent to publication in a cover letter. State that manuscript has not been published elsewhere, and that, if accepted, it will not be republished in any other journal in same or similar form without written consent of editor of *JIM* and Blackwell Scientific Publications Ltd. If article is accepted, a formal referral of copyright will be necessary. Disclose financial or other relationships that might lead to a conflict of interest. State that manuscript has been read and approved by all authors.

MANUSCRIPT FORMAT

If a word processor is used, text must be letter quality, with fully formed characters (no smaller than 12 pt type). Type all pages (including references, tables and captions, and figure captions) double-spaced with a wide left margin (>5 cm). Number all pages, beginning with title page = 1, abstract page = 2, etc. Begin each manuscript component on a new page: title page; abstract (consult a current issue for correct layout), six keywords in alphabetical order; main text; references; tables, each on a separate page, complete with title and footnotes (if any); captions for figures.

Title Page: First page should contain: title of paper; suggested running headline (30 characters, including letters and spaces); and author's name, department, institution, city and country (give address in English).

Abstract: Case reports and Reviews have short, conventional 100 word abstracts. Give 250 word structured abstracts for Original articles starting on page 2. Minisymposia have one abstract which covers all included articles. Frontiers in medicine abstracts can be structured or unstructured depending on style of article. Editorials do not require abstracts. Use subtitles whenever applicable to give abstract desired structure.

Objectives: Give a clear statement of main aim of study and major hypothesis tested, if any.

Design: Describe design of study and mention randomization, blinding, placebo control, case control, cross-over, criterion standards for diagnostic tests, and so on.

Setting: Describe setting of study, including level of clinical care (primary care, tertiary referral centre, private practice, and so on) and number of participating centres.

Subjects: State entry requirements, selection procedures, and number of subjects approached, entering, and completing study.

Interventions: Describe main features of interventions, including method and duration of use.

Main Outcome Measures: State primary outcome measures as planned before data were collected. If paper does not emphasize planned main outcome measures; explain.

Results: Give main results of study with 95% confidence intervals and exact level of statistical significance. Do not include data that do not appear in text.

Conclusions: State primary conclusions of study and their clinical implications; do not speculate or over-generalize. Suggest areas for further research if appropriate.

Key Words: List six keywords below abstract. Use terms from *Medical Subject Headings* list from *Index Medicus.*

Text: Consult a recent journal issue and follow all points of style (units, abbreviations, headings, etc.). Start main text with an introduction (one typed page). Indicate purpose of the paper, do not review subject extensively. Text follows, subdivided by headings such as: Study population, Methods, Results, and Discussion. Do not use footnotes in text. Acknowledge only financial support.

References: 30-40 for Original articles, 10-15 for Case reports. Number consecutively in order mentioned in text. Identify references in text, tables and captions by Arabic numerals in square brackets. Use references form of U.S. National Library of Medicine used in *Index Medicus.* Abbreviate titles of journals according to *Index Medicus.* List only works accepted for publication. Do not refer to textbooks and other standard works. List all authors when six or less: when seven or more, give six followed by *et al.* Double-space.

JOURNALS

Standard journal article

You CH, Lee KY, Chey WY, Menguy R. Electrogastrographic study of patients with unexplained nausea, bloating and vomiting. *Gastroenterology* 1980; **79:**311–14.

Corporate author

The Royal Marsden Hospital Bone-Marrow Transplantation Team. Failure of syngeneic bone-marrow graft without preconditioning in post-hepatitis marrow aplasia. *Lancet* 1977; **ii:** 242–4.

No author given

Anonymous. Coffee drinking and cancer of the pancreas (Editorial). *Br Med J* 1981; **283:** 628.

Journal supplement

Mastri AR. Neuropathy of diabetic neurogenic bladder. *Ann Intern Med* 1980; **92** (2 Part 2): 316-18.

Frumin AM. Nussbaum J, Esposito M. Functional asplenia: demonstration of splenic activity by bone marrow scan (Abstract). *Blood* 1979; **54** (Suppl. 1): 26a.

Journal paginated by issue

Seaman WB. The case of the pancreatic pseudocyst. *Hosp Pract* 1981; **16** (Sept.): 24–5.

BOOKS AND OTHER MONOGRAPHS

Personal author(s)

Eisen HN. *Immunology: an Introduction to Molecular and Cellular Principles of the Immune Response,* 5th edn. New York: Harper and Row, 1974.

Editor, compiler, chairman as author

Dausset J, Colombani J, eds. *Histocompatibility Testing, 1972.* Copenhagen: Munksgaard, 1973.

Chapter in a book

Weinstein L, Swartz MN. Pathogenic properties of invading microorganisms. In: Sodeman WA Jr., Sodeman WA, eds. *Pathologic Physiology: Mechanisms of Disease.* Philadelphia: WB Saunders, 1974; 457–72.

Published proceedings paper

DuPont B. Bone marrow transplantation in severe combined immunodeficiency with an unrelated MLC compatible donor. In: White HJ, Smith R, eds. *Proceedings of the Third Annual Meeting of the International Society for Experimental Hematology.* Houston: International Society for Experimental Hematology, 1974; 44–6.

Monograph in a series

Hunninghake GW, Gadek JE, Szapiel SV *et al.* The human alveolar macrophage. In: Harris CC, ed. *Cultured Human Cells and Tissues in Biomedical Research.* New York: Academic Press, 1980; 54–6. (Stoner GD, ed. Methods and Perspectives in Cell Biology; Vol. 1.)

Agency publication

Ranofsky AL. *Surgical Operations in Short-Stay Hospitals: United States—1975.* Hyattsville, Maryland: National Centre for Health Statistics, 1978; DHEW publication no. (PHS) 78–1785. (Vital and Health Statistics; Series 13; No. 34.)

Dissertation or thesis

Cairns RB. *Infrared spectroscopic studies of solid oxygen* (Dissertation). Berkeley, California: University of California, 1965.

OTHER PUBLISHED MATERIAL

Computer file

Renal system (computer program). MS-DOS version. Edwardsville, Kansas: MediSim, 1988.

Audiovisual

AIDS epidemic: the physicians role (video recording). Cleveland, Ohio: Academy of Medicine of Cleveland, 1987.

Last page of references carries correspondence address, i.e., name and address, telephone and fax numbers of corresponding author. Give name and address of 'second' author.

Tables: Number consecutively (Arabic numerals). Type double-spaced on separate sheets, with an appropriate caption; make table self-explanatory and understandable without reference to text. All columns and headings within tables should be ranged left. Present numerical results as means with relevant standard errors and/or statistically significant differences, quoting probability levels. Three significant figures are sufficient for mean values; quote standard errors to two more places of decimals than the mean. Cite all tables in text. Use only horizontal lines in table. Align decimals and numbers in columns.

Figures: Make line drawings in black indelible ink and twice required size for publication. Put lettering on an overlay or photocopy; make no less than 4 mm high for a 50% reduction. Make all lines thick enough to be clear when reduced 50%. Use symbols for experimental points:

○ △ ▢ ● ▲ ■. Do not use + and ×. Originals may not be larger than A4. Do not mount photographs; indicate labelling on overlay. Give magnification in caption or show by a scale or bar. Where diagrams or photographs are mounted together on one page, provide a sketch of layout. Label figures consecutively (Arabic numerals) with author(s) name and title of paper marked on back in pencil.

Type figure captions double-spaced on a separate sheet; make comprehensive so that figure is understandable without reference to text. Cite all figures in text.

OTHER FORMATS

EDITORIALS—Two printed pages (six double-spaced typed pages with wide margins), 10–12 references.

LETTERS TO THE EDITOR—One-half to one printed page used as a form of correspondence with short critical comments in response to published articles, journal policy, etc.

SUPPLEMENTS—Publication is decided by special rules obtained from Editorial Office. Do not submit manuscripts to Editorial Office until Supplement has been discussed with Editor and Publisher.

OTHER

Questions to contributors: 1 Is the reason for undertaking study scientifically sound, and is the aim of study clearly formulated? 2 Are the ethical standards of the project stated explicitly with reference to the Second Declaration of Helsinki? 3 Are materials and methods adequately described and are they suitable for elucidating the problems studied? 4 Was inclusion or exclusion of patients in clinical studies based on well-defined and described criteria, and has proper attention been paid to representativeness of the sample examined? 5 If the study makes use of controls, were patients randomly allocated to the groups compared? 6 Is choice of statistical methods well founded, and are they adequately described and followed? 7 Is essence of study presented with emphasis on its originality or innovative character? 8 Are results presented to allow reader to verify conclusions? 9 Are conclusions in harmony with presented results? 10 Is it obvious that reader will consider author's initial questions being answered by publication?

Abbreviations: Use only approved abbreviations; use as few as possible. The full term for which an abbreviation stands precedes its first use in text and abstract.

Electronic Submission: Submit final manuscripts on disks of any type. Detailed instructions for submitting a disk will be sent when article is accepted.

Spelling and Terminology: Use British English. Spelling follows *Oxford Concise Dictionary* (7th edn, Oxford University Press, 1982), and *Oxford Dictionary for Writers and Editors* (Oxford: Clarendon Press, 1988). For medical spellings and terminology use *Butterworth's Medical Dictionary* (Guildford: Butterworth & Co., 1978), and *Oxford Dictionary for Scientific Writers and Editors* (Oxford: Clarendon Press, 1991).

Units: Use SI system; for guidance see *Units, Symbols and Abbreviations.* (4th edn, London: Royal Society of Medicine Services, 1988). Traditional units may be added in parentheses. Use negative indices; use $g\,L^{-1}$, $mol\,L^{-1}\,s^{-1}$, not g/L, mol/L/s.

SOURCE

July 1994, 236(1): 105-109.

Journal of Investigative Dermatology
The Society for Investigative Dermatology, Inc.
European Society for Dermatological Research
Elsevier Science Publishing

Edward J. O'Keefe, M.D.
121 West Rosemary Street, Suite 3
Chapel Hill, NC 27516-2512

AIM AND SCOPE

The Journal of Investigative Dermatology publishes papers describing original research relevant to all aspects of cutaneous biology and skin disease. The spectrum of interest is indicated by the breadth of the editorial staff and includes biochemistry, biophysics, carcinogenesis, cellular growth and regulation, clinical research, development, epidemiology, extracellular matrix, genetics, immunology and percutaneous absorption, photobiology, physiology and structure. Review articles and letters to the editor are a standard feature. Methods papers are eligible for publication if they are judged to represent important techniques of general interest to the readers.

MANUSCRIPT SUBMISSION

All data, text, figures, and tables must be original and must not be published or submitted for publication elsewhere. Submission implies that the same or very similar data have not been and will not be submitted or published elsewhere. In cover letter indicate address and telephone and fax numbers of corresponding author. Have copyright agreement signed by all authors. Indicate in cover letter whether authors have financial, equity, patenting, or other relevant relationships or arrangements with a product or sponsor of research that might constitute a conflict of interest. Indicate sources of support in Acknowledgments section.

Submit four complete copies of both manuscript and all illustrative material. Type entire manuscript double-spaced, including face page, abstract, footnotes, references, legends, and tables. Keep 1.5 in. margins. Submission of final manuscript on disk is encouraged after acceptance.

MANUSCRIPT FORMAT

See a recent issue for appropriate format. Organize investigative studies as: Title page, Abstract, Introduction, Materials and Methods, Results, Discussion, Acknowledgments, References, Tables, Legends for Figures. Shorter papers may combine Results and Discussion. Number pages consecutively starting with title page as page one. Do not use headings for one or a few sentences; consolidate paragraphs and headings under Methods and Results. Design text, tables, and figures to save space. Include a word count in cover letter. *Journal* pages have about 1400 words; try to stay under 8 pages.

Title Page: Provide a brief, informative, not descriptive title. List title of manuscript, authors (including full first names), department(s), institution(s), city, state, and nation where work was done. Do not include academic degrees of authors. Indicate authors' affiliations with symbols after authors' names and before institutions. Acknowledge grant or other financial support under "Acknowledgments." Include name and address and fax and phone numbers of corresponding author. Give short title (45 spaces including blanks). Include up to four keywords not in title; separate by slashes.

Abstract: 250 word summary of purpose, findings, and conclusions. Make concise and self-explanatory; do not include nonstandard abbreviations, footnotes, acknowledgments of support, or references. Do not introduce abbreviations or use nonstandard abbreviations.

Footnotes: List consecutively at foot of page on which they fall, designated by superscript symbols (*, †, ‡, §, ¶, **, ††, etc.). Give abstracts used as references as footnotes or in parentheses in text.

Materials and Methods, Results, and Discussion: Readers should be able to reproduce experiments from information supplied in methods, legends and references. Discussion should not repeat Results.

References: Do not include abstracts or unpublished manuscripts in reference list. Type on a separate sheet. Number in order of appearance. Include papers in press with title of publication and journal name; give unpublished citations such as theses, "personal communication," or "in preparation" in parentheses in text. Enclose reference numbers in text within brackets. List all authors' names; do not use et al. Follow style of references and abbreviations of journal names of *Index Medicus*; show inclusive page numbers. Spell out names of unlisted journals. Cite only full-length published papers or manuscripts in press in references; list abstracts as footnotes in standard reference format or cite in text in parentheses by authors, journal, page, and date, followed by designation "(abstr.)."

Figures and Legends: One page of illustrations and legends is allowed for every three pages of text. Submit four sets of glossy prints, unmounted and sorted with each complete set of figures in a separate envelope. Submit artwork of exact column measurements and crop out unnecessary areas. On back indicate: TOP; figure number; and name of first author. Number in order of appearance. Label figure parts on figure in small lowercase letters; give title in legends, not on figure. Do not use 3-dimensional graphs unless third dimension is used for data. Label axes parallel to axis, not at top of graph. Do not use borders. In legends explain how experiment was done, do not interpret figure. Put legends on a separate sheet, double-spaced as:

Figure 1. A short title identifying point or result of figure. Use informative, not descriptive titles. Give information on how experiment was performed not included in Methods. Explain meanings of symbols, keys, and abbreviations; do not put in figure. Except for photomicrographs, place interpretive text in manuscript text, not figure legends. Put scale bars, not magnification, on micrographs; indicate scale in legend, e.g., "scale bar, 5μm." Define error bars in legend as "mean ± SD," or "mean ± SEM."

Submit reviewers' copies of artwork at initial submission and original artwork after acceptance; oversize original artwork will be reduced.

Tables: Provide informative, not descriptive titles. Do not use table legends; supply information in footnotes to title and table contents. Number (Roman numerals) in sequence of citation in text. Supply information regarding how experiment was performed; define columns or abbreviations, etc. as footnotes to tables designated by superscript italic letters. Do not submit as photographs; type double-spaced on sheets separate from text. Make self-contained and self-explanatory. Define errors in table by a footnote; e.g., "mean ± SD" or "mean ± SEM."

OTHER FORMATS

ANNOUNCEMENTS—Brief announcements of scientific meetings, availability of fellowship grants, and awards for research relevant to readership; must reach editorial office 8 weeks before anticipated publication.

COMMUNICATIONS—Four printed pages (1200 words per page) including all written material, figures, tables, and references (30). Concise reports of work of unusual interest or importance with limited but definitive data.

HIGH IMPACT MANUSCRIPTS—Of special interest or importance or to apply innovative techniques or approaches. Suggest that papers be placed in this category.

LETTERS TO THE EDITOR—Two pages typed double-spaced on appropriate subjects. Comments may be solicited from original authors. Subject to editing and abridgment.

REVIEW ARTICLES—Minireviews of three to four journal pages describing recent developments of interest to readers. Should not be extensive in nature or contain exhaustive references and should be accessible to readers not familiar with subject to increase their interest and to point out important and exciting recent developments rather than to serve as an encyclopedic reference. Longer reviews will also be considered.

OTHER

Abbreviations: Do not use for common language or for phrases used less than five times; conform to those listed in the *Council of Biology Editors Style Manual*, 5th ed. Terms used five or more times may be abbreviated; list on title page. Define commonly abbreviated terms at first use in text; do not include in list of abbreviations. Capitalize trade names; use generic names if possible. For names of chemicals, use trivial name if formal chemical name is given after first use of trivial name.

Color Charges: $1000 for first figure on a page; $200 for each additional part or figure on same page; and $400 for each additional part or figure on an additional page. If all figures can be reproduced satisfactorily as one illustration, cost is $1000. Enclose a letter agreeing to pay color charge upon submission.

Data Submission: Sequences must have an EMBL or GenBank database accession number; give number in legend of figure showing sequence.

Materials Sharing: Make nonproprietary unique reagents described (cells, antibodies, DNA, etc.) freely available to qualified scientists.

Page Charges: $35 per printed page; assessment may be waived upon appeal.

Reviewers: Supply full names, addresses, and phone and fax number of suggested referees.

Statistical Analyses: When data and error are provided in text, give number followed by "(mean ± SD)" or "(mean ± SEM)." Do not submit tables or figures showing data not bearing on main thrust of manuscript or data without statistically significant differences; state pertinent negative or tangentially related results in text. Statistical treatment of multiple experiments is preferred to representative experiments; use standard deviation (SD) unless sets of data from experiments or groups are pooled, then use standard error of the mean (SEM). If SEM is used, give number of observations in each data point. Do not use abbreviation SE.

SOURCE

July 1994, 103(1): back matter.

Journal of Investigative Medicine
American Federation for Clinical Research

Jonathan C. Weissler, MD
6900 Grove Road
Thorofare, NJ 08086-9447

AIM AND SCOPE

The *Journal of Investigative Medicine* (formerly *Clinical Research*) is the official publication of the American Federation for Clinical Research (AFCR). The Journal seeks to publish original articles and reviews of broad interest to academic physicians and scientists.

MANUSCRIPT SUBMISSION

Authors of reviews should contact Editor prior to submission. Original manuscripts must be submitted solely to *Journal of Investigative Medicine*; no part of material may be submitted to any other publication.

Include a cover letter signed by all authors. Indicate corresponding author (with address, fax, and phone number). Credit for authorship requires substantial contributions to conception and design, or analysis and interpretation of data; and drafting article or revising it critically for important intellectual content. Address situations regarding possible conflict of interest (see Guidelines for Conflict of Interest in *AFCR Newsletter* March 1990) in cover letter.

MANUSCRIPT FORMAT

Submit original and three copies of complete manuscript. including numbered pages, in order: title page. abstract, text, acknowledgments, references, legends, tables, and glossy prints of figures. Review articles do not require an abstract.

Title Page: Include title of manuscript, author(s) laboratory or institution with city and state, acknowledgment of grant support, name, address, and telephone number of corresponding author, and an abbreviated title (40 characters) to be used as a running head. Indicate whether paper was presented at a scientific meeting.

Abstract and Key Words: For original investigations prepare an abstract (250 words) labeled Background, Methods, Results, and Conclusions. Below abstract give 5 key words or short phrases for indexing. Use terms from *Medical Subject Headings* from *Index Medicus*.

References: Conform to "Uniform Requirements for Manuscripts Submitted to Biomedical Journals (Vancouver Style)" (*JAMA* May 5, 1993; 269(17):2282-86) or obtain a copy from editorial office.

Tables: Type tables on separate sheets, double-spaced, number in Arabic numerals, and give titles. Avoid excessive use of horizontal lines; do not use vertical lines.

Illustrations: Submit professionally-designed figures or high quality computer-generated graphics. Submit glossy black-and-white photographs or laser prints on hard-coated paper and greater than 300 dpi. For color illustrations provide both transparencies and prints. Photocopies or low resolution computer-generated figures are not acceptable. On back of each figure, indicate TOP, sequence number, and corresponding author's name. Print legends for illustrations double-spaced, on a separate sheet.

OTHER

Electronic Submission: Submit accepted manuscripts on diskette. Diskettes must be IBM PC compatible and either 5.25 in. double-sided, double density or 3.5 in. high density (convert MS/DOS files to ASCII before submission).

SOURCE

April 1994, 42(1): inside back cover.

Journal of Laboratory and Clinical Medicine
Central Society for Clinical Research
Mosby

Dr. Harry S. Jacob
Box 105 UMHC
University of Minnesota
500 Southeast Harvard Street
Minneapolis, MN 55455
Telephone:(612) 626-2640; Fax: (612) 626-2642
        Postal delivery
Dr. Harry S. Jacob
Room 124 K-E Building
Dwan CV Research Center
425 East River Road
Minneapolis, MN 55455
        UPS/DHL/FedEx, etc.

AIM AND SCOPE

The *Journal* encourages the submission of manuscripts describing primarily original investigations in the broad fields of laboratory and clinical medicine. Papers submitted by nonmembers of the Central Society for Clinical Research will be given equal consideration with those of members. Reports of purely laboratory or animal investi-

gations should have a clear bearing on the problems of human disease.

The *Journal* will consider manuscripts that are primarily methodologic if the method is novel, if its development poses or answers important biologic questions, or if its description includes data applying it to a study of potential interest to our readership. Papers describing refinements of routine clinical laboratory tests are not encouraged.

The *Journal* publishes reviews and welcomes submission of review manuscripts. Both state-of-the-art comprehensive reviews, directed at research scientists in specific fields, and more general informative reviews, directed at the broader community of clinical investigators, are welcome. Authors of reviews should realize that the *Journal* is multidisciplinary and that review articles for such a journal require appropriate interpretive material. Clarity of presentation is a major criterion for acceptance.

MANUSCRIPT SUBMISSION

In cover letter senior corresponding author must warrant that manuscript, as submitted, has been reviewed by and approved by all named authors; that s/he is empowered by all authors to act on their behalf with respect to the manuscript; that article is original and that neither text nor data have been published previously (abstracts excepted); and that article or a substantially similar article is not under consideration by another journal.

If manuscript is accepted, a form transferring copyright and confirming authorship will be sent to corresponding author to be signed by all authors. If U.S. Government jurisdiction precludes copyright transfer, provide a specific statement of exemption.

MANUSCRIPT FORMAT

Submit original and three copies of manuscript, each with a complete set of tables. Original must include glossy prints or prime laser prints on high-quality paper; submit clear, high-quality photocopies for review copies. All halftone figures must be glossy prints. Type double-spaced on one side of 8.5 × 11 in. paper with 1- to 1.5 in. margins.

Submit a diskette of manuscript. Most common word-processing programs in IBM or Apple formats can be accommodated.

Indicate main sections of all manuscripts with capitalized head set flush with left margin. Organize review articles appropriately for content. For manuscripts describing original investigations use:

Title Page: Include affiliations of author(s) and sources of support for investigation, when appropriate. Indicate address, business and home telephone numbers of Corresponding and reprint request author(s).

Abstract: 250 words; orient reader to problem; describe major observations; and state principal conclusions. Make easily understood without reference to text.

Running Head and Abbreviations: Include an abbreviated title (45 characters), to be used as a running head, and define abbreviations used in manuscript. Keep abbreviations to a minimum. Use only universally understood abbreviations. Use standard chemical or nonproprietary pharmaceutical nomenclature.

Introduction: Introduce and orient general scientific reader to topic.

Methods: Describe statistical methods used.

Results: Present in tables or figures; do not duplicate text.

Tables: Number tables and figures in order of mention in text. Type double-spaced on separate pages. Do not use ditto marks. Center table number at top of page; put title beneath it.

Illustrations: A reasonable number of black-and-white illustrations are permitted without extra charge. Where possible, size illustrations no smaller than 3 × 4 in. and no larger than 5 × 7 in.; larger prints may be necessary to hold detail. Consistency in size is strongly preferred. Note special sizing instructions. Indicate on back TOP, figure number, and name of first author. Make arrangements for use of figures requiring special handling with Editor at an additional charge. Type figure legends double-spaced on a separate page. Place tables and figures after references.

Avoid duplicating previously published material. If using copyrighted table, figure, or data, give full credit to original source and state that material is reprinted with permission in figure legend or table footnote.

Discussion: Set results in context and set forth major conclusions. Do not repeat information from Introduction or Results unless necessary for clarity. Present speculations concerning implications of findings clearly separated from direct inferences. Include a short concluding paragraph, under subheading "Speculations," to point out and clearly denote broader possibilities for general readership.

References: Cite all references in text. Number serially in text; list in order cited after personal acknowledgments. Follow reference format outlined in "Uniform Requirements for Manuscripts Submitted to Biomedical Journals" (Vancouver style) (*Ann Intern Med* 1988;108:258–65). Use journal abbreviations from *Index Medicus*. If not listed, do not abbreviate. Verify accuracy of references. Do not include unpublished results and personal communications in reference list; cite parenthetically in text. If six or few authors, list all; if seven or more, list first three and add *et al*.

JOURNAL ARTICLES

You CH, Lee KY, Chey WY, Menguy R. Electrogastrographic study of patients with unexplained nausea, bloating and vomiting. Gastroenterology 1980;79:311–4.

BOOKS

Langer M, Chiandussi L, Chopra IJ, Martini L, eds. The endocrines and the liver. London: Academic Press, 1984:9–34.

CHAPTERS IN BOOKS

Gustafsson JA, Eneroth P, Hokfelt T, Mode A, Norstedt G. Studies on the pituitary: A novel concept in regulation of steroid and drug metabolism. In: Langer M, Chiandussi L, Chapra IJ, Martini L, eds. The endocrines and the liver. London: Academic Press, 1984:9–34.

OTHER FORMATS

LETTERS TO THE EDITOR—500 words, 5 references. Scientific commentary about published articles directed at confirming results (from a different approach), extending original report, or refuting results or authors' interpretation.

OTHER

Experimentation, Human: State in letter of submission that research was carried out according to principles of the Declaration of Helsinki, that informed consent was obtained, and that author's institutional review board approved study. For photographs of identifiable persons, submit signed releases showing informed consent

Page Charges: $45 per page. A $35 handling fee must accompany each submission. Remit draft or check drawn on a U.S. bank only or by U.S. postal money order made payable to *The Journal of Laboratory and Clinical Medicine.*

Permission: For direct quotations, tables, or illustrations from copyrighted material, submit written permission for use from copyright owner and original author, with source complete information.

SOURCE

July 1994, 124(1): 8A-10A.

Journal of Lipid Research
Lipid Research, Inc.

Dr. Scott M. Grundy
University of Texas Southwestern Medical Center, Y-3206
5323 Harry Hines Boulevard
Dallas, TX 75235-9052
Telephone: (214) 688-8729

AIM AND SCOPE

The *Journal of Lipid Research (JLR)* publishes original articles and invited reviews on subjects involving lipids in any scientific discipline, e.g., chemistry, biophysics, biology, clinical medicine, ecology, geology, pharmacology, physiology, pathology, etc. Sound primary experimental data are major criteria for acceptance of articles. Interpretation of the data will be the authors' responsibility and speculation must be labeled as such. The editors will consider significant contributions in the field of methodology and encourage publication of sufficient details so that the methods can be reproduced. Brief "Notes on Methodology" also are accepted and should be so marked.

MANUSCRIPT SUBMISSION

The *JLR* publishes original and review articles. Except for review articles, articles may not contain a significant portion of data, in form of figures and tables, that has been published or submitted for publication elsewhere. State in cover letter, that no significant amount of data reported has been published elsewhere or is under consideration for publication elsewhere. If some data have previously been published, or is contemplated, include a reprint or a copy of manuscript. If, in editor's judgment, prior publication renders submitted article nonoriginal, manuscript will be returned.

In cover letter include a statement regarding prior publication and exact mailing address and telephone number of corresponding author.

MANUSCRIPT FORMAT

Type on 8.5 × 11 in. paper with 1 in. margins on each side, double-spaced (at least .25 in. between lines, including references, legends, footnotes).

Organize as follows; begin each section on a separate page: title page; abstract and supplementary key words; text; acknowledgments and notice of grant support (if appropriate); references; footnotes to text, if any; tables; figure legends; and figures.

Title Page: Title; author(s); institution from which manuscript is submitted; abbreviated title for a running footline (60 characters and spaces); addresses to which proofs and reprint requests should be sent. Keep footnotes to title page and text to a minimum; number consecutively; and type on a separate page.

Title: Make short, but clear, two printed lines (90 characters and spaces). Do not include chemical formulas or arbitrary abbreviations. Use chemical symbols to indicate structure of isotopically labeled compounds.

Abstract: 200 words, stating objectives, newly observed findings, and conclusions drawn from observations. Make self-explanatory. Do not use abbreviations.

Supplementary Key Words: Add up to ten supplementary key words or phrases to abstract or summary referring to subjects not included in title.

Tables: Type on separate sheets, number with Arabic figures, and provide with descriptive headings which, together with column headings and experimental details in legends (double-spaced), will make tables intelligible without reference to text. Indicate footnotes by superscript lowercase italic letters.

Figures: Keep to a minimum. Include sufficient information in legend so figure is intelligible without reference to text. Attach photocopies of line diagrams to each copy of manuscript. Submit one set of glossy prints with original manuscript. Submit halftone engravings (light or electron microscopy photographs, thin-layer plates, X-ray diffraction patterns, immunochemical and electrophoretic patterns, etc.) as glossy prints. Put each set in a separate envelope marked with first author's name. Cost of color prints is borne by authors. Submit unmounted glossy prints no smaller than 5 × 7 in. and no larger than 8 × 10 in. To letter original drawings, use upper- and lowercase letters. Charts with simple curves will be reduced to fit a single column; make ratio of original figure width to uppercase letter height 45:1 and ratio to lowercase letter height 55:1. For double-column figures, respective ratios should be 90:1 to 110:1.

References: Verify accuracy of references. Reference citations appear in numerical order in parentheses in text. Type reference list double-spaced on a new page at end of text; number consecutively in order cited. List all authors and inclusive pages. Do not include unpublished experiments, personal communications, or papers submitted for publication in reference list; place parenthetically in text. Submit written approval by person(s) cited in personal communications. Include papers accepted but not yet published in reference list with name of journal followed by "In press"; submit four copies. Abbreviate journals according to *Index Medicus*.

JOURNAL ARTICLES

First author's last name, initials, second author's initials, last name, etc. Year. Title of article, *Name of journal*. **Volume:** inclusive pages. (Burstein, M., H. R. Sholnick, and R. Morfin. 1970. Isolation of lipoproteins from human serum by precipitation. *J. Lipid Res.* **11:** 583–595.)

BOOKS

Authors' names as above. Title of Book, Publisher, City, State. Inclusive pages cited. (Snedecor, G. W., and W. G. Cochran. 1967. Statistical Methods, 6th ed. Iowa State University Press, Ames, IA. 342–343.

ARTICLES IN BOOKS

Authors' names as above. Year. Title of article. *In* Title of Book. Initials and last name(s) of editor(s). Publisher, City; State. Inclusive pages of article. (Sweeley, C. C., and G. Dawson. 1969. Lipids of the erythrocyte. *In* Red Cell Membrane. G. A. Jamieson and T. W. Greenwalt, editors. J. B. Lippincott, Philadelphia, PA, 172–228.)

OTHER

Abbreviations and Terminology: List and explain unusual abbreviations in an unnumbered footnote on title page. For common abbreviations and other points of style, follow *CBE Style Manual* (1983, 5th ed, Council of Biology Editors, Inc., Bethesda, MD); for lipid nomenclature, The Nomenclature of Lipids (Recommendations 1976), IUPAC-IUB Commission of Biochemical Nomenclature. *J. Lipid Res.* 1978. **19:** 114–128; for chemical nomenclature, *Chemical Abstracts*; and for enzyme terminology, IUB Recommendations (*Enzyme Nomenclature*, 1978, Academic Press, Inc., New York).

Monoclonal Antibodies: Include a footnote indicating how qualified investigators can obtain antibody or a sample of parent hybridoma cells.

Page Charges: Publication cost is included in reprint charges. $30.00 mandatory article charge for each paper published.

Reviewers: Submit names, addresses, and phone numbers of four or five potential reviewers.

SOURCE

January 1994, 35(1): inside back cover.

Journal of Medical Genetics
BMJ Publishing Group

Editor, *Journal of Medical Genetics*
BMA House - Tavistock Square
London WC1H 9JR

Dr. P. M. Conneally
Department of Medical Genetics

James Whitcomb Riley Hospital for Children RR129
Indiana University Medical Center
Indianapolis, IN 46223
Papers from U.S.

AIM AND SCOPE

The readership of *Journal of Medical Genetics* is world-wide and covers a broad range of workers, including clinical geneticists, scientists in the different fields of medical genetics, clinicians in other specialties, and basic research workers in a variety of disciplines. It publishes original research on all areas of medical genetics, along with reviews, annotations, and editorials on important and topical subjects. It also acts as a forum for discussion, debate, and information exchange through its Letters to the Editor columns, conference reports, and notices.

MANUSCRIPT SUBMISSION

Original papers may be on any aspect of medical and human genetics and may involve clinical or laboratory based and theoretical genetic studies. Authors may be requested to give data upon which manuscript is based to Editor. Obtain guidance on length from studying *Journal*.

Submission implies that it contains original work which has not been previously published. Submitting author must ensure that all coauthors are agreeable that their names appear on manuscript. Provide a fax number.

MANUSCRIPT FORMAT

Submit papers in triplicate and in Vancouver Style (*BMJ* 1988; **296**:401-5).

Where a patient(s) with a structural chromosome abnormality is described, state availability of a cell line(s) in text with identifying number, cell bank, and, where appropriate, contact person.

Type double-spaced with wide margins.

Include an abstract (preferably structured) giving main results and conclusions.

Figures: Keep figures to a minimum. Number consecutively in Arabic numerals. Type legends on separate sheet.

Tables: Do not include tables in body of text, type on separate pages numbered with Arabic numerals. Provide a legend.

References: Conform precisely to current style current in journal. Verify accuracy and completeness of references.

Illustrations: Submit high quality black-and-white photographs. Color illustrations can be accepted; authors pay part of cost, discuss in advance of submission. Identifiable photographs of patients must be accompanied by written permission for use.

OTHER FORMATS

ANNOTATIONS AND EDITORIALS—Written or commissioned by editors, send suggestions regarding possible topics and authors.

BOOK REVIEWS—The Journal aims to review as wide a range of relevant books as possible. Computer programs and databases, official reports, and other material relevant to the field may all be appropriate for review. Enquiries about such items are welcome.

BRIEF REPORTS—Case and family reports.

CONFERENCE REPORTS—Reports from small to medium sized meetings, especially international workshops on specific topics, will be appreciated. Authors intending to submit conference reports should liaise with the Reviews Editor to avoid duplication.

LETTERS—On any relevant topic. Those relating to or responding to previously published items in the *Journal* will be shown to those authors, where appropriate. Papers submitted as an original report may sometimes be published in shortened form as a letter, submit as a short report, unless directly related to a previous journal article.

REVIEWS—Include all aspects of medical genetics, discuss first with Reviews Editor. Contributions on historical topics, or which could form part of specific series, are particularly acceptable.

SHORT REPORTS—500 words, one or two illustrations. Continuous text with no headings. An abstract should be provided for all papers. Contributions may also be submitted as Hypotheses or Technical notes.

SPECIAL ISSUES AND SUPPLEMENTS—These are published at intervals on topics of particular relevance. Enquiries are welcome from those organizing workshops or symposia who may have material suitable for such an issue.

OTHER

Nomenclature: For chromosomes, see: *ISCN 1985. An international system for human cytogenetic nomenclature.* Basel: Karger, 1985. For genes, see: Shows TB, *et al.* In: Human Gene Mapping 5 and 7. *Cytogenet Cell Genet* 1979;**25**:96–116, 1984;**37**:340–3. For loci, use conventional nomenclature with lowercase lettering as appropriate (see: Race RR, Sanger R. *Blood Groups in Man.* 6th ed. Oxford, London: Blackwell, 1975; and Giblett ER. *Genetic Markers in Human Blood.* Oxford, London: Blackwell, 1969). For blood coagulation, see: International Committee of Haemostasis and Thrombosis (Graham JB, *et al*). A genetic nomenclature for human blood coagulation. *Thromb Haemostas* 1973;**30**:2–11. For enzymes, see: *Enzyme Nomenclature: Recommendations of the Nomenclature Committee of the International Union of Biochemistry* (New York: Academic Press, 1984).

SOURCE

January 1994, 31(1): inside back cover.

Journal of Medicinal Chemistry
American Chemical Society

Professor Philip S. Portoghese
Department of Medicinal Chemistry, Rm 8-114 HS Unit F
College of Pharmacy, University of Minnesota
308 Harvard St. S.E.
Minneapolis, Minnesota 55455
Telephone: (612) 624-6184; Fax: (612) 626-6891

AIM AND SCOPE

The *Journal of Medicinal Chemistry* invites original research contributions dealing with chemical-biological relationships. The primary objective of the *Journal* is to publish studies that contribute to an understanding of the relationship between molecular structure and biological activity or mode of action. It is not intended merely to serve as a compendium for compounds that have been prepared and tested for biological activity. In this regard, routine extensions of existing series, which utilize no novel chemical or biological approach, or which do not add significantly to a basic understanding of the structure activity relationship of the series, will not normally be accepted for publication. Some of the specific areas that are appropriate include design, synthesis, and evaluation of novel therapeutic and diagnostic agents; improvement of existing drugs by molecular modification; isolation, structural elucidation, and synthesis of naturally occurring molecules that have significant biological activity; molecular modeling and QSAR; physicochemical studies (NMR, ORD, X-ray, etc.) of biologically active compounds which furnish some insight into their interaction with recognition sites; the effect of molecular structure on the distribution, pharmacokinetics, and metabolic transformation of drugs; and studies of macromolecular targets that provide insight into the design of biologically active compounds. Manuscripts whose focus is novel methodology with broad application to medicinal chemistry are considered for publication if the methods have been tested on representative, relevant molecules. Retrospective analyses of published series are considered only if the results provide fresh insight and if the series is of current general interest.

MANUSCRIPT SUBMISSION

Submit manuscripts in quadruplicate with a signed copyright transfer form (see January 7, 1994, issue for a copy of form) and a transmittal letter that includes: telephone and fax numbers; e-mail address; manuscript category; and names and addresses of five reviewers.

See outline ethical obligations of authors in *J. Med. Chem.* **1990**, *33*, 2–3.

MANUSCRIPT FORMAT

Print manuscript double-spaced, on one side of 8.5×11 in. paper. If paper length is greater, keep print within above dimensions. Minimum acceptable type size is 12 characters per inch. Double-space tables, footnotes, and references. Number all pages consecutively, starting with title page and continue in order: abstract, test, experimental, references, tables, figure legends, figures, and elemental analysis. Attach supplementary material at end of manuscript; number pages separately. Be brief. State rationale and objectives in introductory sentences. Avoid detailed and lengthy methods presentations in introductory and discussion. State conclusions and significance of findings after discussion and results. Summary of conclusions should place research findings in an appropriate perpective.

Title: Title should reflect purposes and findings of work so that it can be used in computer-assisted search. Follow with authors' full first names, middle initials, and last names and addresses of all contributing laboratories. Mark name of corresponding author with an asterisk.

Abstract: 100–150 words presented in a findings oriented format giving critical results and summary of conclusions. Type on a separate page after title page.

Tables: Tabulation of experimental results is encouraged if presentation is more effective or less space is used. Number consecutively in order of citation in text with Arabic numerals. Give footnotes lowercase letter designations; cite as superscripts. Sequence letters by line not by column. If a reference is in both table and text, insert a lettered footnote in table to refer to numbered reference in text. Provide a descriptive heading which, together with column headings, should make table self-explanatory.

Linear formulas (e.g., $3,4\text{-}Cl_2C_6H_3CHOHCOOH$) may be used in headings. Use structural formulas only when absolutely necessary.

Figures: Submit as original drawings or as high-contrast photographic copies or laser prints. Submit color prints with each copy of manuscript. Use lettering symbols in unrelated figures of the same or comparable size in a series of figures. Ideal reduced size for letters is 2 mm. Recommended symbols are: ●■▲×+○□△◐◑. Number consecutively with Arabic numerals. Place on separate pages. Place legends together as a block. Do not designate blocks of structural formulas as figures; represent as schemes or charts.

Color Illustration: Submit only if essential for clarity. Justify need for color upon submission. If Editor approves use, authors are charged part of cost.

References and Notes: Number literature references and notes in one consecutive series by order of mention in text, with numbers as unparenthesized superscripts. Verify accuracy of references. Type complete list double-spaced on separate page(s) at end of manuscript.

JOURNALS

Rich, D. H.; Green, J.; Toth, M. V.; Marshall, G. R.; Kent, S. B. H. Hydroxyethylamine Analogues of the pl7/p24 Substrate Cleavage Site Are Tight-Binding Inhibitors of HIV Protease. *J. Med. Chem.* **1990**, *33*, 1285–1288.

MONOGRAPHS

Casy, A. F.; Parfitt, R. T. *Opioid Analgesics*; Plenum Press: New York, 1986; pp 333–384.

EDITED BOOKS

Rall, T. W.; Schleifer, L. S. Drugs Effective in the Therapy of the Epilepsies. In *The Pharmacological Basis of Therapeutics,* 7th ed.; Gilman, A. G., Goodman, L. S., Rall, T.W., Murad, F., Eds.; Macmillan Publishing Co.: New York, 1985; pp 446–472.

List submitted manuscripts as "in press" only if formally accepted for publication; otherwise, use "unpublished results" after names of authors. Incorporate footnotes to text in correct numerical sequence with eferences. Cite financial support under "Acknowledgment."

OTHER FORMATS

ARTICLES—Definitive, full accounts of significant studies.

BOOK REVIEWS—At invitation of Book Review Editor.

COMMUNICATIONS TO THE EDITOR—1600 words. For rapid publication of preliminary reports on important findings of novel and timely significance. Paper's major concept should not have been published previously. Detailed experimental procedures need not appear, but include sufficient information or supplementary materials to facilitate rapid handling of manuscript. Submit supporting data (e.g., detailed experimental conditions, CHN analyses, spectral data) for use by reviewers and editors. A detailed, expanded version should be published as soon as possible after publication.

EXPEDITED ARTICLES—Six printed journal pages (26 double-spaced manuscript pages in acceptable type size, including tables and figures). Same format as articles. Should be of extraordinary importance, novelty, and timeliness. Criteria for acceptance are identical to those for communications to the Editor. In transmittal letter justify why manuscript warrants accelerated publication. Prior publication as a communication or other form of rapid publication precludes consideration as an expedited article.

NOTES—Definitive reports with more limited scope than an article. Three journal pages (15 manuscript pages including structures, tables, and figures); format is identical to articles.

PERSPECTIVE ARTICLES—Interpretive accounts on subjects of current interest to medicinal chemistry; a forum for experts to present their perspectives of emerging or active areas of research that affect practice of medicinal chemistry. Manuscripts usually are invited; contributions can be made; contact Editor to ensure that a suggested topic is suitable and not already in process.

OTHER

Abbreviations: Use without periods. Use standard abbreviations throughout Experimental Section. Preferred forms for commonly used abbreviations: mp, bp, °C, K, min, h, mL, μL, g, mg, μg, cm, mm, nm, mol, μmol, mmol, ppm, TLC, GC, NMR, UV, and IR. Abbreviate units in table columns and with numbers only. See *ACS Style Guide*.

Analyses: Include to verify purity. Routine analyses that agree with values within ± 0.4% are not printed; indicate in text as: anal. $(C_{23}H_{31}N_5O_2)$ C,H,N. In tabular listings of new compounds, give symbols of elements or functions for which satisfactory analyses were obtained in column headed "Analyses." Natural or degradation products, novel or abnormal reactions are printed in full. If necessary to rely on normally unaccepted analysis, figures are printed in full. e.g., Anal. $(C_{16}H_{20}N_2O_8S)$ C,H,S; N; calcd, 6.99; found,7.55. Report analyses in tables as footnotes. Attach combustion analyses for new compounds as last page, labeled "Analytical Data:"

Formula	Analyses
$C_{16}H_{20}N_2O_8S$	C,H,S;N[a]

[a]N: calcd, 6.99; found, 7.55

If analytical data is not available, submit high-resolution mass spectral and chromatographic evidence of purity. Elemental analyses may not be sole criterion of purity for medium or larger peptides. Submit additional evidence for homogeneity by separation methods and amino acid analysis.

Biological Data: Submit quantitative biological data for all "final" or "target" compounds unless unusual circumstances (e.g., compound instability) prohibit. Qualitative biological data is not usually accepted. Reference biological test methods or describe in sufficient detail to permit experiments to be repeated. Place detailed descriptions of biological methods in Experimental Section. Test standard compounds or established drugs in the same system for comparison. Present data as numerical expressions or in graphical form; present biological data for extensive series of compounds in tabular form. Do not submit tables consisting primarily of negative data; submit as supplementary material for microfilm edition.

Drugs: Use chemical names. If not practical, use generic names, or names approved by U.S. Adopted Names Council, or World Health Organization. Mention registered trademark names or code numbers for drugs once in parentheses; subsequently, refer to drugs by their chemical names or compound numbers. Designate compounds widely employed a research tools and recognized primarily by their code numbers by code numbers. If a generic name is employed, give chemical name or structural formula at point of first citation.

Electronic Submission: Manuscripts prepared with either WordPerfect or Microsoft Word can be used for production. Submit disk with final accepted version of manuscript. Version on disk must exactly match final version accepted in hard copy.

Use document mode or equivalent in wordprocessing program. Do not include page-layout instructions. Left justify text. Do not insert spaces before punctuation. Conform to format of Journal *(vide infra)*. Ensure that all characters are correctly represented:, 1 (ones) and l (ells), 0 (zeros) and O (ohs), x (exs) and × (times sign). Check final copy for consistent notation and correct spelling. Check disk with a virus detection program. Disks with viruses will not be processed. Label disk with manuscript number and corresponding author name. Provide details of disk on Diskette Description form.

Put text and tabular material in one file; do not integrate graphic material. For tables create columns with tabs or spaces, not both.

Microfilm Edition and Supplementary Material: For extensive tabulations of data of interest only to readers who need more complete data, including biological data for a series with low activity, full spectra, other spectral data, molecular modeling coordinates, modeling programs, X-ray crystallographic data, etc. "Dumping" of materials is discouraged; Editors reserve the right to make downward adjustments in numbers of material presented.

Supplementary material is reproduced as submitted. Identify material for Microfilm Edition with a separate title page indicating "Supplementary Material with author names and manuscript title. Include figure captions, table titles, and other identifying captions on same page as figures or tables and not on a separate sheet. Preferable page size is 22 × 28 cm with readable material aligned parallel with 22 dimension. Type size or letter size should be large enough for easy reading. Number pages separately and sequentially. Refer to *The ACS Style Guide* for more specific information. Place a statement of availability of supplementary data at end of Experimental Section: "Supplementary Material Available: Give description of material (*x* pages). Ordering information is given on any current masthead page."

Nomenclature: Provide correct nomenclature. Conform with current American usage. Use systematic names similar to those used by Chemical Abstracts Service or IUPAC. Use "semisynthetic names" for certain specialized classes of compounds, such as steroids, peptides, carbohydrates, etc. Name should conform to generally accepted nomenclature conventions for compound class. See nomenclature rules in Appendix IV of current *Chemical Abstracts Index Guide*. For a list of ring systems, including names and numbering systems, see *Ring Systems Handbook* (American Chemical Society, Columbus, OH, 1988). For IUPAC Rules, see *Nomenclature of Organic Chemistry,* Sections A-F, H; (Pergamon: Elmsford, NY, 1979). For rules of carbohydrate nomenclature, see *Biochemistry* 1971,*10*, 3983; for steroid nomenclature, see *Pure Appl. Chem.* 1989, *61* (No. 10), 1783-1822. Other recommendations are *Enzyme Nomenclature* (Academic Press: New York, 1984); *Compendium of Biochemical Nomenclature and Related Documents* (Portland Press, Ltd.: England, 1992); prostaglandins, *J. Med. Chem.* **1974,***17*, 911–918. For CA nomenclature advice consult Manager of Nomenclature Services, Chemical Abstracts Service, P.O. Box 3012, Columbus, OH 43210. A name generation service is available for a fee through Registry Services Department, Chemical Abstracts Service, P.O. Box 3343, Columbus, OH 43210.

Page Charges: $25 per page; not a condition for publication.

Safety: Draw special attention to safety considerations such as explosive tendencies, precautionary handling procedures, and toxicity.

Software: Use software readily available from reliable sources for computer-aided drug design (e.g., molecular modeling, QSAR, etc.); specify where software can be obtained. When conformational calculations are included, give parameters employed for relevant potential functions. Indicate all details needed to reproduce numbers in paper or as supplemental material for microfilm edition. This includes coordinates of hypothetical computer-generated receptor motels. Refer to *J. Med. Chem.* 1988, *31*, 2230-2234 for publication guidelines.

Spectral Data: Include such data for: representative compounds in a series, key synthetic intermediates, novel classes of compound, in structural determinations, and as criteria of purity. Do not include routine spectral data for every compound. Papers where interpretations of spectra are critical to structural elucidation and those in which band shape or fine structure needs to be illustrated are published with spectra included. When such presentations are essential, only pertinent sections are reproduced.

Statistical Analyses: Statistical limits (statistical significance) for biological data are required. If statistical limits cannot be provided, give number of determinations and some indication of variability and reliability of results. Include references to statistical methods of calculation. Express doses and concentrations as molar quantities (e.g., mol/kg, μmol/kg, M, mM, etc.) when comparisons of potencies are made on compounds having large differences in molecular weights. Indicate routes of administration of test compounds and vehicles. Note salt forms used (hydrochlorides, sulfates, etc.). For inactive compounds, indicate highest concentration (in vitro) or dose level (in vivo) tested.

Structural Formulas: Draw structural formulas clearly. Make economical use of space. Combine structures with same skeleton but different functional groups into a single formula with an appropriate legend to specify different chain lengths, substituents, etc. Use line formulas wherever possible. Number structures with Arabic numerals.

Carefully letter (e.g., with a computer drawing program such as ChemDraw). Equations, schemes, and blocks of structural formulas are presented in either one- or in two-column format (e.g., 8.4 cm or 17.7 cm width); submit in actual size. One-column format is preferred.

If using ChemDraw use: fixed length, 18 point; line width, 0.8 point; bold width, 2.5 point; hash spacing, 3.0 point; margin width, 2.0 point; bond spacing, 18% of length. Use single-width bold and dashed lines for stereochemical notations; use 12-point Helvetica font , both for atom labels and test materials. Prepare with page setup at 80% and print in this manner with a laser printer on good quality white paper. Mark each drawing "to be reduced to 75% for publication".

X-ray Data: Results and analysis include: unit cell parameters and standard errors; formula, formula weight, and number of formula unit in unit cell; measured and calculated densities; space group; wavelength used, number of reflections observed, (for diffractometer data) number of unobserved weak reflections; method of collection of intensity data and methods of structure solution and refinement (references may suffice); final R value; statement on presence or absence of significant features on a final difference Fourier map; noteworthy bond lengths and angles; a clear representation of structure reducible to one-column width; and tables (with standard deviations) of final atomic positional parameters, atomic thermal parameters, and bond distances and angles. Submit tables as "Supplementary Material for the Microfilm Edition."

SOURCE

January 7, 1994, 37(1): 7A-10A.

Journal of Membrane Biology
Springer-Verlag

The Journal of Membrane Biology
Marine Biological Laboratory
Woods Hole, MA 02543
Fax: (508)548-2003

AIM AND SCOPE

The *Journal of Membrane Biology* will publish papers on the nature, the structure, genesis and functions of biological membranes, and on the physics and chemistry of artificial membranes with a bearing on biomembranes. It will publish articles dealing with plasma membranes (including cell surfaces and accessory surface structures, such as bacterial or plant cell walls); intracellular membranes (e.g., membranes of organelles, nuclei, mitochondria, chloroplasts, vesicles, intracellular reticula), membrane organs (e.g., skin, urinary bladder, gut, kidney, lung); biomembrane models (e.g., monolayers, bilayers, synthetic membranes); and general considerations of two-dimensional biological structures. Suitable topics are chemical and physical structure; immunochemical properties and fingerprinting; colloid and surface chemistry; transport and secretory functions, including natural and artificial transport carrier systems, membrane channels, diffusion and pinocytosis; metabolic functions, membrane-bound enzyme systems, transduction; electrical phenomena in excitable membranes, nerve, muscle, receptors, etc; membrane—membrane interaction, intercellular communication; membrane synthesis and replication; cell regulatory functions of membranes and cybernetic aspects; membrane genetics; evolution and comparative aspects of biomembranes.

The *Journal* will not publish preliminary notes. It will publish brief letters to the Editor commenting on papers in the *Journal*. The Pivotal ideas series will survey the history of ideas which have proven seminal to the biomembrane field. They are personal accounts which recall the climate and setting in which the ideas arose. Consult Editorial "Theory and Experimentalism in Biological Sciences," Vol. 129:1. Topical Reviews and articles in Pivotal ideas series are generally by invitation only, but editors welcome suggestions.

MANUSCRIPT SUBMISSION

Submission implies: work is original; that it has not been published before (except as an abstract or as part of a published lecture, review, or thesis); that it is not under consideration for publication elsewhere; that its publication has been approved by all coauthors, if any, as well as by responsible authorities at institute where work done; that, if and when accepted, authors agree to automatic transfer of copyright to publisher; that manuscript will not be published elsewhere in any language without consent of copyright holders; and that written permission of copyright holder is obtained by authors for material used from other copyrighted sources.

Submit four complete copies (original and three copies). Submit original illustration material (original drawings, etc.) and three sets of copies.

Avoid technical jargon; define specialized terminology.

MANUSCRIPT FORMAT

Type using standard size type, double-spaced on one side of letter size paper 8.5 × 11 3/4 in. and with a 2 in. margin on left. Number all pages serially. Type references, tables, footnotes and legends for illustrations double-spaced, also using standard size type, on separate pages. Manuscripts typed with a dot-matrix printer are not acceptable.

On first page: title; name(s) of author(s); name of laboratory where work was done; a running title (40 characters including spaces); and footnotes to title.

Precede paper with a 225 word summary, intelligible to general reader without reference to text; avoid abbreviations.

Follow with 6 key words.

Bibliography: Refer only to work cited in text. List references alphabetically at end of paper including full titles. Use abbreviations according to the Bibliographic Guide for Editors & Authors. Use standard size type.

Eigen, G.S. 1960. The Kinetics of cation transport. *Proc. R. Soc. London* **138:** 182–191

Gibbs, T., Charles, R.T. 1961. Enzymes in membranes. *In*: The Molecular Structure of Membranes. B.A. Selkirk, editor. pp. 53–60. Springer-Verlag, New York

Use inclusive pagination, give first and last page numbers of articles.

Give citations in text in parentheses; e.g., (Huntley & Briarly, 1967), or (Carson, 1940; Hopkins, 1943), except when author's name is part of a sentence; e.g., "Harding (1968) reported that . . .." When three authors are cited, name all authors in first citation, but subsequently name only the first author; e.g., (Miller et al., 1963). When more than three authors, name only the first author; e.g., (Smith et al., 1987). Another form of citation uses numbers in square brackets referring to an alphabetically ordered bibliography list. Either form is acceptable when used consistently.

Footnotes: Keep to a minimum; number consecutively. Mark footnotes to titles or authors by asterisks; place on title page. Also mark by asterisks footnotes to formula. Use standard size type.

Figures: Separate from text and number. Use India ink on white bristol board for original drawings and graphs. Submit clean glossy prints of these, shot in sharp focus to final size (50%). Prepare figures to fit one column (8.1 cm) or, on rare occasions, two columns (16.9 cm) after a reduction to 50%. Make lettering, thickness of lines, size of inscriptions, size of measuring points, adequate spacing of shaded and dotted areas large enough to be legible after reduction. Make final size of letters: 2 mm for capital letters and numbers; 1.6 mm for lowercase letters. If size of lettering is inadequate for reduction to 50%, figures will be returned.

Publisher reserves the right to reduce or enlarge illustrations.

Keep to a minimum illustrations requiring reproduction as half-tone plates. Photographs should be clean glossy prints in sharp focus and as rich in contrast as possible; trim at precise right angles; give scales.

Use standard size type for figure legends.

OTHER

Electronic Submission: Submit electronically prepared manuscripts in WordPerfect 5.1, Microsoft Word (Macintosh), or Microsoft Word (PC). Follow instructions exactly. Submit two double-spaced hard copies with software.

Mathematical Equations: Style so as to avoid misinterpretation by printer.

All letters in formula are set in italics as well as single letters in text. Do not underline. Underline in yellow abbreviations in formula to be set in Roman type.

Underline Greek characters in red and Script in green. Underline small letters once and capital letters twice; this applies also to Latin letters in formula (in pencil). Mark boldface type (heavy type) by wavy underlining.

Clarify subscripts and superscripts by caret and inverted caret marking: 1^2_v, 1^v_2; a subscript to a subscript, by $1^v_{2_v}$

Call obscure primes and dots to printer's attention. Mark very clearly: number 1 and letter l; zero and letter O, o; e and l; n, u, and v; primes and apostrophes. Use fraction exponents instead of root signs and solidus (/) for fractions wherever use will save vertical space, exp () notation when exponent is complicated.

Number all equations sequentially in Arabic numerals in parentheses on right side of page.

Reviewers: Suggest four to six names, including addresses and specific fields of interest, of possible referees.

Symbols and Abbreviations: Follow the *CBE Style Manual* (5th ed. rev., Council of Biology Editors, Inc., Bethesda, MD).

SOURCE

December 1994, 142(3): 413.

Journal of Molecular and Cellular Cardiology
International Society for Heart Research
Academic Press (London)

Dr. N.R. Alper
Department of Physiology and Biophysics
University of Vermont College of Medicine
Burlington, Vermont 05405-0068
Telephone: (802) 656-2540; Fax: (802) 656-0747

AIM AND SCOPE

The *Journal of Molecular and Cellular Cardiology* publishes original full papers and short communications dealing with cellular, sub-cellular, and molecular studies of the heart and cardiovascular system. Papers are published on molecular biology and genetic aspects of the heart and circulatory system; cardiac metabolism and metabolic regulation; physiology, biochemistry and biophysics of cardiac and vascular smooth muscle; cardiovascular morphology, pathology and pathophysiology, cardiac and smooth muscle electrophysiology, basic and clinical aspects of cardiovascular pharmacology; neural regulation of the heart and circulation; experimental aspects of clinical heart disease that reflect altered molecular and cellular properties of the heart and cardiovascular system. Review articles, editorials and letters are also accepted.

MANUSCRIPT SUBMISSION

Article must not have been submitted or published elsewhere. If accepted, exclusive copyright to paper is assigned to Publisher.

Submit three complete copies of manuscripts. Submit one set of unmounted glossy prints and two photocopies. Submit original electron microscopic data in triplicate.

MANUSCRIPT FORMAT

Submit manuscripts no longer than 25 typewritten pages with 6–8 Figures and/or Tables. Select the best way to present data—stated in text, plotted in Figures, or listed in Tables. Avoid repeating the same data in different presentations. Overly long or overdocumented manuscripts are subject to condensation and delay. Type papers double-spaced throughout text, legends and tables; triple-space references. Number all pages, including tables and figures, in sequence. Allow 3 cm margins. For full length papers begin following sections on separate pages:

Title Page: Include title of paper, authors' names and affiliation and name and address of corresponding author. Include a Running Head (50 characters).

Abstract: 250 words (avoid abbreviations) and 5–15 Key Words for indexing (avoid general or self-evident words).

Introduction: 1–2 pages stating problem addressed in paper with 10–15 references. End with a one sentence statement of study's major finding.

Materials and Methods: Include sources of chemicals and special apparatus. Provide descriptions of general procedures in sufficient detail to allow others to reproduce experiments. Include methodological details of individual experiments in Figure or Table legends; include general procedures in "Materials and Methods." Describe statistical methods in full; distinguish standard error (± S.E.) from standard deviation (± S.D.).

Results and Discussion: Best presented as separate sections but may be combined. Keep Discussions brief.

References: Cite by name and year of publication in parentheses. If two authors, use both names. If more than two, use only first name followed by "et al." If author's name is part of sentence, put date only in parentheses. If two or more papers by the same author(s) in a single year, place lowercase letters following year.

Cite accepted manuscripts in the same way as published work; use expected year of publication. Include in bibliography; designate as "in press." Cite unpublished material or personal communications in text by giving, in parentheses, authors' names and initials and nature of material, i.e., "personal communication," or "unpublished observations." Do not include in bibliography. Provide permission letters for citations of personal communications or unpublished work of anyone except authors upon acceptance of manuscript.

Type bibliography triple-spaced; list references alphabetically by author. If authorship and year are identical for two or more references, include lowercase letters following year. Abbreviate names of journals longer than one word as proposed in "Uniform Requirements for Manuscripts Submitted to Biomedical Journals" (1982, *Ann Intern Med* **96**: 766–771 and 1982, *Br Med J* **284**: 1766–1770). Follow bibliographic form proposed there, except year of publication (and lowercase letter, if needed) immediately follows list of authors, and give inclusive page numbers in full.

You CH, Lee KY, Chey RY, Menguy R, 1980. Electrogastrographic study. Gastroenterology **79**:311–314.

Refer to abstracts only when essential, and only if abstract is from a readily accessible periodical. In bibliography, include word (Abstract) after title.

Acknowledgments

Tables: Give titles, number in series, and give appropriate headings and explanations of data. Refer to Materials and Methods for explanations of experimental procedures.

Figure Legends: Include brief titles followed by explanations that, where possible, refer to Materials and Methods for explanation of experimental procedures.

Illustrations: Number glossy prints of illustrations (5×7 in.) sequentially; label with first author's name, abbreviated title, and TOP. Odd sized prints are discouraged. Make line drawings in black indelible ink; draw 1.5 to 2 times required size for publication. Make symbols large enough to be legible after reduction. Half tone photographs should be no smaller than required size. Indicate labeling in pencil on line drawings and on photographs and half-tones. Half tones are limited to four pages. Color may be used; additional cost is borne by authors.

For previously published illustrations, submit written permission of original author and copyright holder, with precise reference to original source.

OTHER FORMATS

EDITORIALS, REVIEW AND SPECIAL ARTICLES—Consult Editor. Subject to normal review. Editorials 8, typed pages, review articles, 20 typed pages.

LETTERS—Comments regarding articles published in the *Journal*. Subject to editorial review. Authors of article commented on are invited to reply.

RAPID COMMUNICATIONS—1500 words plus 15 references, 2 Figures and/or Tables following general format. To provide for rapid communication of data of special interest. Results and Discussion may be combined. Submit in triplicate with a brief letter explaining general importance of results and why rapid communication is desired.

OTHER

Abbreviations: Avoid unfamiliar abbreviations. A limited number of clearly defined abbreviations may be used. In a footnote list and define all abbreviations; place footnote at point where first abbreviation is used. Express metric units and physical and chemical quantities abbreviated as recommended in Instructions to Authors of current volume of *Journal of Biological Chemistry*.

Experimentation, Animal and Human: The International Society for Heart Research endorses Recommendation from the Declaration of Helsinki and the Guiding Principles in the Care and Use of Animals. All investigations involving human and animal experimentation must be conducted in conformity with these principles. Document adherence to principles. Papers in which evidence of adherence is not apparent will be rejected.

Units: Express measurements in metric units.

SOURCE

December 1994, 26(12): back matter.

Journal of Molecular Biology
Academic Press (London)

Journal of Molecular Biology
10d St Edward's Passage
Cambridge CB2 3PJ, England

AIM AND SCOPE

The *Journal of Molecular Biology* will publish studies of living organisms or their components at the molecular level. Suitable subject areas are: Proteins, Nucleic Acids and other biologically important macromolecules: Molecular structure, physical chemistry, *in vivo* modification and processing, molecular engineering, macromolecular assemblies, enzymology. Genes: Expression, replication and recombination, sequence, structure, genetics of eukaryotes and prokaryotes. Viruses and Bacteriophages: Genetics, structure, growth cycle. Cells: Organelle structure and function, motility, transport and sorting of macromolecules, energy transfer, growth control, genetics of development.

MANUSCRIPT SUBMISSION

Submission implies that material is original, and that no similar paper has been or will be submitted for publication elsewhere. Submission implies that all named authors have seen and agreed to submitted version of paper and that all included in acknowledgments or as providers of personal communications have agreed to inclusion.

Exclusive copyright in paper is assigned to Publisher who will not put any limitation on the personal freedom of the author to use material in paper in other works.

MANUSCRIPT FORMAT

Type double-spaced throughout (including tables, legends and footnotes), on one side of A4 or 28 cm $\times$ 22.5 cm paper. Number all pages serially. Submit original and three copies of each manuscript, including figures

Title and Key Words: Make title brief and informative; do not include nonstandard abbreviations. Include a short running title (50 characters including spaces), if full title is longer. Supply 5 key words and a subject classification. Give name of laboratory where work was done on title page. Indicate current addresses of all authors (if different from laboratory of origin).

Summary: 300 words intelligible without reference to text; do not include literature references.

Text: Follow conventions used in current issues of *Journal* for headings, references, etc. Divide articles into sections: 1. Introduction, 2. Materials and Methods, 3. Results, 4. Discussion. Other section headings (e.g. Experimental Methods, Theory, Results and Discussion) may be used if clarity of presentation is improved. Give descriptive topic subheadings letters in italics in text.

Personal Communications: Quote only with agreement of author cited. Copies of written agreement may be requested.

Illustrations: Provide a set of photographs, high quality computer graphics or original drawings. Figures are reduced to a width of 80 mm and a depth not exceeding 240 mm; label accordingly. Write name(s) of author(s) and number on back, indicate TOP.

Color illustrations are published without charge if Editor and Publisher decide color is essential for clarity.

Make stereos the size they are to appear in *Journal*. Mount so that corresponding points in each half are 60 mm apart.

Special illustrations such as transparent overlays and foldout sheets must be judged essential by Editor. Publisher may charge for cost of special illustrations.

References: List alphabetically at end of paper. Cite title of article, volume number and first and last pages. Abbreviate journal titles according to *Chemical Abstracts Service Source Index* (American Chemical Society).

Sanger, F. & Coulson, A. R. (1975). A rapid method for determining sequences in DNA by primed synthesis with DNA polymerase. *J. Mol. Biol.* **94**, 441–448.

Maaloe, O. (1979). Regulation of the protein-synthsizing machinery-ribosomes, tRNA, factors and so on. In *Biological Regulation and Development* (Goldberger, R. F., ed.), pp. 487–542 Plenum Press, New York.

Citations in text should read Smith & Brown (1957) or (Smith & Brown, 1957); when three or more authors, use Smith *et al.* (1957). Use (Brown, 1957*a, b*) when more than one paper by the same author(s) published in the same year is cited.

OTHER FORMATS

COMMUNICATIONS—Brief papers that make a specific well-documented point. No page limit. Include no more than 4 Figures and Tables. Text is continuous. Technical or methodological detail is printed in Table and Figure legends. Do not divide into sections; topic headings may be included.

CRYSTALLIZATION NOTES AND SEQUENCE NOTES—Discontinued (see *JMB* 241: i).

TOPICAL REVIEW ARTICLES—Deal with a more restricted field than those in review Journals. Consult Editor before submitting.

OTHER

Abbreviations and Symbols: Use SI units and system of abbreviations and symbols of the IUPAC-IUB Combined Commission on Biochemical Nomenclature, endorsed by the Commission of Editors of Biochemical Journals of the International Union of Biochemistry. Nonstandard abbreviations may be used.

In genetics papers follow proposals of Demerec *et al.* (*Genetics*, 54, 61 (1966).

Data Submission: For papers dealing with amino acid sequences of proteins or with nucleotide sequences include a statement that data have been deposited with an appropriate data bank, e.g., European Molecular Biology Laboratory or GenBank Data Libraries. Give data base accession number in text. Lengthy sequences will only be published if results are of general interest and importance.

Electronic Submission: *Journal* will endeavour to typeset manuscripts directly from disc. All word-processing packages are acceptable; any version of Microsoft Word, Wordperfect or Wordstar are preferred. Save text as a word-processing file, use "save as" in an ASCII or RTF file.

Ensure that document on disc is in order: running title; title; authors; addresses; abstract; key words; introduction; materials & methods; results; discussion; acknowledgments; references; figure legends; tables; footnotes; abbreviations. Section headings are not required for Communications.

Use two carriage returns to end headings and paragraphs—not to rearrange lines or to introduce space between headings and paragraphs. Enter references in *Journal* style. Type text without hyphenation, except for compound words. Do not use lowercase "l" (ell) for 1 (one) or "o" for 0 (zero). Do not include any copyrighted material on disc. In case of mis-match between disc and hard copy, hard copy is definitive. Label disc with *Journal* reference number, author(s), machine used, and program used to generate file. submit two hard copies, including all figures, typed double-spaced. Read proofs carefully.

Low-Resolution Structure Studies: Studies must have clear biological implications.

X-ray Analyses: Deposit structural data, including both X-ray amplitudes and phases, and derived atomic coordinates, with an appropriate protein data bank. State and give data bank reference number in paper with specified release date. Maximum delay for release of data should be 2 years; if no date, immediate availability is assumed. Communications describing preliminary crystallographic data for biological macromolecules is acceptable if these exhibit special features of interest.

SOURCE
November 18, 1994, 244(1): back matter.

Journal of Molecular Evolution
Springer International

Dr. Emile Zuckerkandl
Institute of Molecular Medical Sciences
460 Page Mill Road
Palo Alto, CA 94306

AIM AND SCOPE

The journal publishes articles in the following research fields: Biogenetic evolution (prebiotic molecules and their interaction); Evolution of informational macromolecules (primary through quaternary structure); Evolution of genetic control mechanisms; Evolution of enzyme systems and their products; Evolution of macromolecular systems (chromosomes, mitochondria, membranes, etc.); Molecular bases for organismal evolution; Evolutionary aspects of molecular population genetics.

Considered for publication are: Papers stressing experimental results; papers of theoretical content leading to a higher level of understanding or inducing new experiments; papers of experimental content, not dealing directly with evolutionary problems but having evolutionary implications that are clearly expressed.

Review articles may occasionally be accepted. In all other cases, papers should present new and previously unpublished material. They should be suited to the character of the journal and should describe the work done and methods used in a clear and reproducible manner. The treatment should be as concise as possible. Papers, that "tell a whole story" will be favored against papers that are too fragmentary.

MANUSCRIPT SUBMISSION

Submit original and 2 copies. Avoid technical jargon. Define specialized terminology. Paper must be in final version. Check form and content to minimize corrections in proof. A charge is made for changes introduced after manuscript has been set in type. If information is missing, manuscript will be returned.

Submission implies: that work described has not been published before (except as an abstract or as part of a published lecture, review, or thesis); that it is not under consideration for publication elsewhere; that its publication has been approved by all coauthors and the responsible authorities at institute where work was done; that, if and when manuscript is accepted for publication, authors agree to automatic transfer of copyright to publisher; that manuscript will not be published elsewhere in any language without consent of copyright holders; that written permission of copyright holder is obtained for material used from other copyrighted sources; and that authors accept responsibility for any costs associated with obtaining this permission.

MANUSCRIPT FORMAT

Submit electronically prepared manuscripts in WordPerfect 5.1, Microsoft Word (Macintosh) or Microsoft Word (PC). Submit a double-spaced hard copy of manuscript with software.

Type papers double-spaced on one side of letter-size paper with a wide left margin. Number all pages consecutively. Number tables and figures serially. Type captions to illustrations double-spaced on separate sheets. Mark in manuscript margin where figures and tables are to be inserted.

Do not use term *homology* for structural similarity but only in its biological sense of derivation of structures from a common ancestral structure.

Identify all organisms by scientific binomen.

First Page: title; initial(s) and name(s) of author(s); name of laboratory where work was done; footnotes to title; address of corresponding author.

Summary: Summarize main points so reader can determine quickly whether subject is of interest. Include facts and conclusions; do not cite areas and subjects that have been treated or discussed. Relate problem, methods, results, and conclusions reported by article. Use complete sentences, active verbs, third person, and past tense. Use standard nomenclature. Define unfamiliar terms, abbreviations, and symbols at first mention.

Follow with 10 key words.

References: In text literature references are by author and year; if two authors, name both; if three or more give first author's name plus "et al." List at end of paper only works mentioned in text. Arrange alphabetically by name of first author.

JOURNAL PAPERS

Name(s) and initial(s) of author(s), year in brackets, full title, name of the journal as abbreviated in *Chemical Abstracts*, volume number, first and last page numbers:

Brendel M, Khan NA, Haynes RH (1970) Interstellar molecules. Mol Gen Genet 106:289–310

BOOKS

Name(s) of author(s), year in brackets, full title, edition, publishers, place of publication, page number:

Gibbs T, Charles RT (1961) Enzymes in membranes. In: Selkirk BA (ed) The molecular structure of membranes. Springr, Berlin Heidelberg New York, p 53

Verify accuracy of references. Use as few "in press" citations as possible; include name of journal that has accepted paper.

Footnotes: Number consecutively in text . Mark footnotes to title or authors of article by asterisks and place on title page.

Figures: Use with discretion to clarify or reduce text. Do not repeat information in captions in text. Do not submit colored or previously published illustrations. Mention all figures and graphs in text. Number with Arabic numerals. Provide brief descriptive legends for each figure. Legends are part of text; append legends to text. Submit figures suitable for reproduction separately from the text. Mark TOP of each figure, author, and figure number lightly on back in soft pencil.

Line Drawings—Submit good quality glossy prints with clearly legible inscriptions; may be reduced 85%. Make letters 2 mm high.

Half-tone Illustrations—Submit well-constructed photographic prints (not photocopies), trimmed at right angles and in desired final size. Make inscriptions about 3 mm high. Group several figures into a plate on one page.

Type mathematical equations clearly so they can be interpreted properly.

Call obscure primes and dots to attention of printer. Mark very clearly number 1 and letter l. Use fractional exponents instead of root sign. Use solidus (/) for fractions when use will save vertical space. Use exp () notation when exponent is complicated.

Number equations sequentially in Arabic numerals in parentheses on right side of column.

OTHER

Abbreviations, Symbols, Units, etc.: Follow internationally agreed-upon rules, especially those adopted by Commission on Biochemical Nomenclature: see *Eur J Biochem* 1 (1967) 259, 375 and 379; 2 (1967) 1 and 127;

3 (1967); 5 (1968). Reprints available from Waldo E. Cohn, Director, NAS-NCR Office of Biochemical Nomenclature, Biology Division, Oak Ridge National Laboratory, P.O. Box Y, Oak Ridge, TN 37830.

Follow rules defined in "Suggestions and Instructions to Authors" (*Biochimica et Biophysica Acta*, 1965).

Use *ad hoc* abbreviations for names of chemical substances only when advantageous to reader (not author). Do not abbreviate names of enzymes.

Print a list of less common abbreviations as a footnote on title page.

Materials Sharing: Computer software referred to in methods must be obtainable either commercially or upon request. Furnish upon request any information necessary for reproducing methodologies on which results are based.

SOURCE

July 1994, 39(1): iv.

Journal of Morphology
Wiley-Liss, Inc.

Dr. Frederick W. Harrison
Department of Biology
Western Carolina University
Cullowhee, NC 28723
Telephone: (704) 293-5566; Fax: (704) 293-7029;
E-mail: JMORPH@wcu.edu

AIM AND SCOPE

The *Journal of Morphology* publishes original papers in cytology, protozoology, embryology, and general morphology. Preliminary notices or articles of a purely taxonomic or ecological nature are not included.

MANUSCRIPT SUBMISSION

Submit manuscript complete with author's abstract, original drawings and photographs. Do not submit papers which have already appeared. Simultaneous publication elsewhere is not allowed.

MANUSCRIPT FORMAT

Conform to *Journal of Morphology* style with respect to use of capital and lowercase letters in headings. Submit manuscript exactly as it is to appear in print. Use subdivisions, prepare each as a unit on separate sheets. Introduction through Acknowledgments constitute one unit. They follow one another consequently, utilizing all available space.

Title page (p. 1); Abstract (p. 2); Text (Introduction (p. 3), Materials and Methods, Results, Discussion, Acknowledgments); Literature Cited; Footnotes; Tables; Figure legends.

Type manuscript, including Literature Cited and other sections, double-spaced on bond or heavy-bodied paper 8.5 × 11 in. with a 1 in. margin on one side. Number pages consecutively beginning with title page. Use standard typeface (size Elite or larger), a new black ribbon and solid white paper. Make computer-generated documents letter quality; dot matrix printing and light green paper are unacceptable. Submit original manuscript and original prints of all illustrations, and two review copies.

Do not divide hyphenate words at ends of lines. Do not justify right margin. Type corrections or print legibly in ink. Do not begin sentences with abbreviations. Do not abbreviate word Figure in text, except in parentheses. Indicate italics and boldface; typing text in italics or boldface is insufficient. Spell out numbers when first word in a sentence; do not follow with abbreviations. For numbers indicating time, weight, and measurements use Arabic numerals when followed by abbreviations.

Title Page: Author's name(s); Institution from which work emanated, with city, state, and zip code; Number of text pages, figures, graphs, and charts, each on a separate line; Abbreviated title (running headline, 48 characters and spaces); Name, address, and telephone number of corresponding author; Special instructions regarding joint publication with other articles, return of artwork, and color plates.

Abstract: 250 words. Disclose findings rather than aims. Indicate techniques. Write in complete sentences; make intelligible without reference to text. Do not repeat information in title.

Introduction

Materials and Methods: Repeat name of organism (author of Latin name may be included; spell it out). Include source of material, sex, weight and, if appropriate, conditions of laboratory acclimation, and one or two introductory sentences. Indicate how, by whom and where material was collected and identified. Follow with a terse description of techniques. Even if published elsewhere, include enough information so that remaining material can be placed into context. Provide enough information to let observation and experiments be repeated.

Results: Group under appropriate subheadings. Italicize and center primary subheadings, marginalize secondary headings, and italicize and indent tertiary headings. Describe observations; do not discuss. Use present tense.

Discussion: Briefly review significant aspects of results. Review past work in terms of topics and organisms rather than an historical treatment of previous studies. Do not make authors of papers the subjects of sentences. Organize ideas so that each issue is discussed only once. Place most important points in first and last paragraphs.

Literature Cited: Type double-spaced. In text, cite references by author's surname followed by abbreviated year of publication ('94). Do not abbreviate publication dates before 1901: (1784), (1889), (1900). Do not cite items as "submitted" or "in preparation." Refer to unpublished data as personal communication. Mention full address in acknowledgments, not in literature.

When references are made to more than one paper by the same author in the same year, designate in text as (Condon et al, '90a,b) and in literature list as:

Condon, K. L. Silberstein, H.M. Blau, and W.J. Thompson (1990a) Development of muscle fiber types in the prenatal rat hindlimb. Dev. Biol. *138*:256-274 .

Condon, K. L. M. Blau, and W.J. Thompson (1990b) Differentiation of fiber types in aneural musculatrue of the prenatal rat hindlimb. Dev. Biol. *138*:275-295.

Arrange literature list alphabetically by author's surname: Author's name (or names), year of publication, complete title, volume, and inclusive pages.

Hermanson, J.W., and J.S. Altenbach (1985) Functional anatomy of the shoulder and arm of the fruit-eating bat *Artibeus jamaicensis*. J. Zool. Lond. A *205*:157-177.

Hermanson, J.W., and M.A. Cobb (1992) Four forearm flexor muscles of the horse, *Equus caballus*. Anatomy and histochemistry. J. Morphol. *212*:269-28.

Hermanson, J.Q., W. LaFramboise, and M.J. Daood (1991) Uniform myosin isoforms in the flight muscles of the little brown bats, *Myotis lucifugus*. J. Exp. Zool. *259*:174-180.

References to papers by two authors follow those by senior author. If more than one with a second author, list alphabetically. References by three or more authors follow those by senior author (including two-author references); arranged chronologically independent of order of second author.

Abbreviate journal titles according to *Index Medicus*. For non-English titles, use original language unless a different alphabet. Follow appropriate spelling and capitalization. Include accents and umlauts. In Literature Cited section never abbreviate year.

Footnotes: Limit as much as possible. Number consecutively. Clearly indicate corresponding reference numbers in text. Number additional references to identical footnotes with next following consecutive number.

Type table footnotes directly beneath table, numbered 1, 2, 3, etc. Do not number in sequence with text footnotes.

Tables: Cite in text. Make simple and uncomplicated, with as few vertical and horizontal rules as possible. Indicate in margin where to appear in text. Make titles complete but brief. Present information other than that defining data in footnotes.

Figures: Cite in text. Give photographs or drawings mounted together separate figure numbers, in lower left corner. If group-mounted illustrations are closely related, assign them a single figure number and letter individual prints as A, B, C (a, b, c), etc. in lower left corner. Put code to abbreviations for each figure in figure caption, summed at end. Order abbreviations alphabetically, followed by numerical codes. List codes consecutively, with each code entry followed by a comma and full term and a semicolon or terminal period. Do not provide a separate list of abbreviations at end of article. Give abbreviations in parentheses after first mention of term in text.

When possible, integrate figures into text. Group figures to fit a single page with legends. Refer to relevant text passages to reduce length of legends and avoid redundancy.

Photographs—Submit original drawings or high-quality photographic prints. Submit illustrations in complete and finished form with adequate labeling.

Photographic prints must have adequate contrast. Multiple prints in a single figure must have uniform tone.

Submit black-and-white prints on white nonmatte paper. Indicate reduction desired. Illustrations cannot be reduced to less than 20% of submitted size. Submitted line drawings cannot exceed 11 × 14 in. Lettering and labels must be readable after reduction. When reduced, minimum height of a capital letter cannot be less than 2.5 mm for a photomicrograph and 1 mm for a graph or chart. When printed, an individual figure or group of figures cannot exceed page dimensions of 5.5 in, wide × 7.75 in. long, or 2.62 in. wide × 7.75 in.long for single-column.

Trim photomicrographs and illustrations straight on all sides and "square." Mount on strong bristol board, about 15 points (0.4 mm) thickness, with a 1 in. margin around figure or grouping.

Attach figures to bristol board with dry mounting materials, or white or colorless cement.

When two or more figures are assembled, mount close together, separated by no more than 1/8 in. Use illustrations of similar density and tone to prevent loss of detail.

Use printed paste-on or transfer labels that are large enough to allow for suitable reduction and sturdy enough to withstand mailing and handling. Spray labeling with clear adhesive. Label directly on drawing or photographic print, not on an overlay. Place labeling at least .25 in. in from edges. Place white labels over dark backgrounds and black labels over light backgrounds, or shadow labels with an appropriately light or dark highlight.

Number figures consecutively including charts and graphs.

Submit original and two review copies of illustrations with manuscript. Copies may be photographs of originals; photocopies are not acceptable. If original drawings are too large, submit photographic prints. On reverse side indicate: author's name, figure number, TOP, and reduction requested. Do not fasten with paper clips, staples, etc. Ship flat and protected by heavy cardboard.

Cost of color illustrations is charged to author; price depends on sophistication of work. After paper is accepted, a price will be quoted for author approval. Submit color transparencies or prints. Mark transparency to indicate areas to be cropped to arrive at the critical image area.

Line Drawings—Draw figures with black ink on medium-weight white paper or light-weight artboard. Submit photographic prints in lieu of original drawings. Artwork should be sharp and black.

Use stippling and hatching to achieve tonal quality. Avoid shading (pencil, wash, or airbrush) for tonal effect unless drawing is to be reproduced as a halftone. Draw original graphs on blue-ruled paper; other colors will reproduce.

OTHER

Nomenclature: Spell out genus names the first time; abbreviate the second time in a paragraph.

Spelling: For nontechnical terms use current *Webster's International Dictionary*.

Symbols: When preceded by a digit, use following symbols: % for percent; ° for degrees.

Units: Use metric system for all measurements, weight, etc. Express temperatures in degrees Celsius (centigrade). Express metric abbreviations (*J Morphol* 1995; 223(1): 122) in lowercase without periods.

SOURCE
January 1995, 223(1): 119-123.

Journal of Muscle Research and Cell Motility
Chapman and Hall

Dr. C.C. Ashley
University Laboratory of Physiology
Oxford OX1 3PS

Dr. C.R. Bagshaw
Department of Biochemistry
University of Leicester
Leicester LE1 7RH

Professor G. Goldspink
Department of Anatomy and Developmental Biology
The Royal Free Hospital Medical School
London NW3 2PS

Dr. D.J. Manstein
National Institute for Medical Research
The Ridgeway, Mill Hill
London, NW7 1AA

Dr. J.R. Sellers
National Heart, Lung and Blood Institute
National Institutes of Health
Bethesda, MD 20892

Professor A.P. Somlyo
Dept. of Molecular Physiology and Biological Physics
The University of Virginia School of Medicine
Charlottesville, VA 22908.

AIM AND SCOPE

The *Journal* has as its main aim the publication of original research which bears on either the excitation and contraction of muscle, the analysis of any one of the processes involved therein, or the processes underlying contractility and motility of animal and plant cells. The policy of the *Journal* is to encourage any form of practical study whatever its specialist interest, as long as it falls within this broad field. Theoretical essays are welcome provided that they are concise and suggest practical ways in which they may be tested.

Full-length Reviews and short News and Views items will also be published; these will normally be commissioned, but may also be submitted. Book Reviews, Abstracts from and reports of meetings and other items will be published at the discretion of the Editors.

MANUSCRIPT SUBMISSION

Submit original copy of manuscript (with illustrations for printer) and two additional copies for referees (both accompanied by a complete set of illustrations).

Submission implies that paper presents original, unpublished work, not under consideration for publication elsewhere. By submitting a manuscript, Authors agree that copyright for their article is transferred to Publisher if and when article is accepted for publication. Copyright covers exclusive rights to reproduce and distribute article, including reprints, photographic reproductions, microfilm or any other reproductions of similar nature, and translations.

MANUSCRIPT FORMAT

Research Papers may be as long as findings justify; be complete; do not submit preliminary findings. Group halftone illustrations into plates, total number should relate to extent and significance of observations. Only five free plates are allowed; plates in excess are charged to authors, as are color plates.

Type manuscripts double-spaced throughout on one side of A4 paper with a 3 cm left margin. Check text and references thoroughly for errors. Ensure that typescript is correct in style. Underline generic and specific names and words for emphasis to be set in italics. Avoid footnotes if possible.

Divide research papers into headed sections in order: Summary, Introduction, Materials and methods, Results, Discussion, Acknowledgments, References, Figure captions.

Title Page: On first page include: full title, affiliation and address of author(s); a running title (75 letters and spaces); name, full postal address, telephone and fax numbers of corresponding author; number of plates or halftones; and an estimate of total number of words.

Summary: 300 words; reflect length of text. Make clearly written, factual and comprehensible without reference to text.

Methods: Avoid excessively detailed description or unnecessary literature citations of widely used techniques. Describe novel techniques in full detail.

Tables: Number and head with short titles. Type on separate sheets; indicate preferred point of insertion in text margin.

References: Use Harvard system in text, thus: . . . Smith (1972) . . . (Brown & Jones, 1972) . . . Smith *et al.* (1972a). List references in alphabetical order:

JONES, A.B., PRICE, C & EVANS, D. (1973) Fine structure of the engram. *J. Neurocytol.* **2,** 1–25.

SMITH, E.F. (1972) Color electron microscopy. In *Advances in Cytooptics* (edited by BROWN, G.), pp. 23–51. London: Chapman & Hall.

Abbreviate journal titles following latest edition of *World List of Scientific Periodicals.*

Illustrations: Submit line drawings, graphs, chemical formula as clear high quality photographic prints or laser printed originals. Make labelling clear and of sufficient size to be legible after reduction. Prepare lettering on line figures with a reduction of 2:1 in mind.

Submit original photographic prints with all copies of manuscript. Group photographs where possible. Submit at final publication size. Maximum dimensions for a single photograph or group of photographs are 190 × 280 mm. Maximum width for single column illustrations is 85 mm.

Indicate scale of object or magnification used. Make numbers, lettering and scale markings using Letraset or other professional aids; be consistent in style and lettering.

Label back of each illustration with first author's name and figure number; indicate orientation. In figure legends indicate content of figure; identify all symbols used.

OTHER FORMATS

CORRESPONDENCE—800 words. Submit comment on papers published in the *Journal* in the form of scientific correspondence. Published with original author's response to comment (same length restriction).

FAST STREAM PUBLICATION—3500 words including references, tables and figures (allow 250 words for a simple figure). Request for papers that authors consider to merit it, and which they expect to be accepted without amendment. Must meet conditions stated in *J Muscle Cell Res Motil* 1992;12.

NEWS AND VIEWS—3500 words including bibliography and figures (allow 250 words for a simple figure). Brief reviews intended to highlight and discuss new data and ideas, not necessarily of author but discussed from his/her point of view. Treated as Fast Stream publications.

REVIEWS—8000 words (exclusive of bibliography) plus two figures or tables. Summarize and analyze topics of current interest and progress; a personal viewpoint is encouraged while a historical one is not.

SEQUENCE PAPERS—Four printed pages and two figures (may include predicted protein, single letter code, but not nucleic acid, sequence data). Submit sequence papers which discuss unique or comparative data on proteins of interest within the broad field of the *Journal*. Must meet conditions stated in *J Muscle Res Cell Motil* 1992; 12.

Head sequence papers by a title, authors and affiliations, followed by database accession number. Divide text into headed sections in order: Summary, Discussion, Method notes (brief notes on novel features of identification and sequencing approach), Bibliography. Bibliography should allow direct or indirect access to all sequence information for this class of proteins. Submit sequence to appropriate database; obtain accession number.

OTHER

Abbreviations: Use without explanation only when in very common usage.

Color Charges: $600 per page. State in writing willingness to meet cost upon submission of manuscript.

Materials Sharing: Make cell lines, hybridomas and DNA clones available to any qualified investigator. Manuscript submission constitutes agreement to abide by this.

Units: Use SI system (Système International d'Unités). See Royal Society of Medicine booklet *Units, Symbols and Abbreviations* (1971, London).

SOURCE

February 1994, 15(1): inside front and back covers.

Journal of Nervous and Mental Disease
Williams & Wilkins

Eugene B. Brody, M.D
The Sheppard and Enoch Pratt Hospital

6501 N. Charles St.
Baltimore, MD 21285-6815

AIM AND SCOPE

The *Journal* publishes articles containing new data or ways of reorganizing established knowledge relevant to understanding and modifying human behavior, especially that called "sick" or "deviant." Our policy is summarized by the slogan, "Behavioral science for clinical practice." Articles should include at least one behavioral variable, clear definition of study populations, and replicable research designs.

MANUSCRIPT SUBMISSION

Neither a submitted article nor its data may have been previously published or be currently under review for publication elsewhere. Submit reprint permission for all materials printed in or adapted from other publications immediately after formal acceptance. Listed authors may include only primary researchers and writers; acknowledge other contributors in a footnote. Limit research reports to 15–18 typed pages; for longer evaluative review papers query Editor in advance. Brief Reports and case studies (3.25 manuscript pages) must have heuristic value.

Authors assign copyright to Williams and Wilkins upon acceptance.

Submit three clear copies. In cover letter state complete title of paper, and name, mailing address, and telephone number of corresponding author. Notify Editorial Offices of change of address change for corresponding author.

Type all sections double-spaced, including footnotes, references, and tables, on 8.5 × 11 in. paper. Use 1 in. margins on all sides. Number pages consecutively. Do not indicate authors' names on manuscript pages. Do not submit glossy photos of figures or computer diskettes. If paper is accepted, corresponding author will be notified of additional requirements, such as camera-ready prints of all figures and final accepted version of text on 5.25 in. floppy disk, preferably in IBM-PC compatible format using ASCII or DOS programming.

MANUSCRIPT FORMAT

Include sections and materials, in order:

Running Title Page: A short, identifying title (45 characters, including spaces) and name and address of corresponding author.

Complete Title Page: A brief, informative title (two lines) and names and highest degrees of all authors.

Abstract: Title of article and a concise description (150 words) of general purpose, methodology, results, and conclusions of research.

Introduction: State purpose of study, briefly survey salient literature, describe research setting if relevant, and give rationale for methodology chosen.

Methods: Precisely describe subjects, procedures, apparatus, and methods of data analysis, in sufficient detail to allow others to evaluate or replicate study.

Results: Succinctly present significant data. Use tables or figures to supplement, not repeat, text.

Discussion: Extend, do not reiterate Results. Emphasize significant principles, relationships, generalizations and implications, relevance to previous studies, limitations, and suggestions for further research.

Conclusions: State conclusions and briefly summarize evidence for each.

References: List unnumbered cited sources in alphabetical order. List all author's names. Use et al. only in text. Verify accuracy of references. If a manuscript has been accepted for publication, list as "in press," give journal name. Do not include unpublished or privately published materials and personal communications in references, cite as footnotes.

In text, cite author's name and year of publication (e.g., Mills, 1956; Mills and Smith, 1956). If more than two authors, give only first author, followed by "et al." and date (e.g., Mills et al., 1956). If more than one publication by same author in same year, add suffixes (a, b, c, etc.) to year in both text and list citations. In text show page numbers from original source for quoted material (e.g., Mills, 1956, p. 12). Do not cite more than four references in support of any point.

Lewis SW, Reveley A, Reveley M, Murray RM (1987) The familial/sporadic distinction as a strategy in schizophrenia research. *Br J Psychiatry* 151:306–313.

Gottlieb BH (Ed) (1981) *Social networks and social support.* Beverly Hills, CA: Sage.

Weissman MM, Boyd JH (1985) Affective disorders: Epidemiology. In HI Kaplan, BJ Sadock (Eds), *Comprehensive textbook of psychiatry/IV* (4th ed, Vol 1, pp 764–769). Baltimore: Williams & Wilkins

Footnotes: Give a single listing of all footnotes in order of appearance in text. Footnote 1: identify primary institutional affiliation of first author (and others) with name and address of reprint request author. In subsequent footnotes identify affiliations of authors at other institutions, followed by an unnumbered footnote describing grant support and other acknowledgments. Final numbered notes provide information on citations in text which do not qualify as references.

Legends for Figures: Include a consecutively numbered (Arabic) listing of all figure legends. Make each sufficiently explanatory to make reference to text unnecessary.

Figures: Send photocopies of professionally prepared figures. Submit camera-ready glossy or laser prints only upon acceptance. On a typed label on back of each figure give Arabic figure number, name of lead author, and title of manuscript.

Tables: Number consecutively. Make self-explanatory. Type each double-spaced throughout on separate pages, headed by brief, descriptive titles.

OTHER

Experimentation, Human: Indicate social context from which subjects were drawn and relationship to investigator. State that informed consent was obtained after procedure(s) were fully explained. Protect patient anonymity.

Style: Use active voice and first person.

SOURCE

January 1994, 182(1): inside back cover.

Journal of Neurochemistry
International Society for Neurochemistry
Raven Press, Ltd., New York

Prof. G.G. Lunt
Department of Biochemistry—4 West
University of Bath, Claverton Down
Bath BA2 7AY, U.K.
          Europe, Asia, Africa, and Australasia
Dr. A.A. Boulton
Neuropsychiatric Research Unit
A114 Medical Research Building
University of Saskatchewan
Saskatoon, Saskatchewan, Canada S7N 0W0
          Western Hemisphere

AIM AND SCOPE

The *Journal of Neurochemistry* is devoted to the prompt publication of original findings in areas relevant to molecular, chemical, and cell biological aspects of the nervous system. Papers which are wholly pharmacological, histochemical, or immunological, and methods papers which do not advance knowledge in neurochemistry will not normally be considered. Although papers in any area of neurochemistry will be considered, the *Journal* particularly encourages submissions in the areas of molecular and cellular biology.

MANUSCRIPT SUBMISSION

Submission of a paper implies that it represents original research not previously published (except as an abstract or preliminary report), that it is not being considered for publication elsewhere, and that, if accepted, it will not be published elsewhere in similar form, in any language, without consent of the International Society for Neurochemistry. Each person listed as an author must have participated in the study to a significant extent. Copyright will be held by the International Society for Neurochemistry, Ltd. Submit a signed statement expressly transferring copyright to Society in the event of publication. A copyright transfer form will be sent to senior author upon receipt of manuscript.

Submit original plus four copies of any paper including four originals of any photographs. Laboratory location where work was done determines submission.

Submit Rapid Communications without figures by facsimile to Eastern Hemisphere Office (Fax: 44-225-826828) or to Western Hemisphere Office (Fax: 306-966-8830). Send original plus three copies by regular mail.

Submit Short Reviews in quadruplicate to Prof. H. Soreq, Department of Biological Chemistry, The Institute of Life Sciences, The Hebrew University of Jerusalem, Jerusalem 91904, Israel.

MANUSCRIPT FORMAT

Manuscripts longer than ten printed pages (33 double-spaced typed pages, 10,000 words not counting figures) will be returned for shortening without review. Graphs occupy about one-fourth of a page; halftone plates one-half to a full page.

Type manuscripts double-spaced throughout with 2.5 cm margins. Submit in quintuplicate. Do not number papers in series; subtitles are acceptable.

On title page include title, author's names, laboratory of origin, name, address, and telephone and fax numbers of corresponding and reprint request author(s), and any necessary footnotes, including one defining abbreviations used in text. Indicate affiliations of each author. Put information about sources of financial support in Acknowledgment section, between Discussion and References.

On following page include a brief abstract (200 words, use only abbreviations allowed without definition), six key words, and a running title (45 letters and spaces).

Number pages in succession; title page is page 1.

Put each table on a separate page at end of manuscript, followed by a page of figure legends, typed double-spaced. Indicate approximate location of figures and tables in margin of text.

Use footnotes to text sparingly, indicate by superscript numbers, and type with corresponding numbers on a separate sheet. In Tables, reference footnotes with superscript lowercase letters.

Indicate Greek characters clearly. Identify ambiguous characters such as 1 (one or el), 0 (zero or oh), and, when handwritten, such letters as C, P, and K (upper or lowercase) in margin.

Write concisely in a readily understandable style. Do not use technical neologisms, "laboratory slang," or words not defined in dictionaries. Do not use redundant words, phrases, and sentences. Do not repeat captions of Tables and Figures, with or without paraphrasing, in text.

In an introductory statement give reasons for undertaking investigation and summarize experimental plan. Avoid exhaustive literature reviews.

Experimental Procedures or Materials and Methods: Group special chemicals, etc., with sources under a separate subheading "Materials." Give procedures in sufficient detail to permit repetition. Summarize published procedures; do not describe in detail unless substantially modified. If procedures used are described in detail in articles "in press," submit copies with manuscript.

Results: Describe findings without discussing significance. Base experimental conclusions on an adequate number of observations with statistical analysis of variance and significance of differences.

Discussion: Assess significance of findings in relation to status of field.

Tables and Figures: Construct so that with captions and legends, they will be intelligible with minimal reference to text. Do not publish the same data in a Table and a Figure.

For each Figure include an explanatory legend, typed double-spaced on a single sheet. Explain all symbols in Figure.

Give each Table column an appropriate heading.

Indicate units of measure in column headings of Tables or coordinates of Figures. If unit is the same for all data, give in caption or in a line immediately below Table.

Present data as clearly defined mean values with some measure of dispersion (e.g., standard deviation or standard error, and range). Indicate, with appropriate symbols, statistical significance of differences from control values (e.g., $*p < 0.05$; $**p < 0.01$; $***p < 0.001$). Do not include more significant figures than are warranted by accuracy of determinations.

Write figure number, authors' names and "J. Neurochem." on back. Submit line drawings camera-ready. Electron micrographs will be reproduced to author's satisfaction. Request special attention for halftones in writing. Provide an overlay showing which area requires greatest definition. Submit halftone photographs as original photographic prints. Charges for reproducing color photographs will be given on request. Submit figures about twice final size; do not exceed 20×30 cm. Make lettering large enough to be legible when reduced to single-column size.

Insert symbols on graphs etc. using following standard characters: ○ ● △ ▲ ❑ ■ × +.

Present data in bar graphs where appropriate. Use three-dimensional presentation only for data with three variants.

Plan figures to avoid wasted "white" space. Scales on coordinates do not have to start at 0. Use scale bars on micrographs.

References: In text cite by names of authors and year of publication; for papers by three or more authors use "et al." At end of paper, list references, typed double-spaced, in alphabetical order, except for papers by three or more authors (group in chronological order after any other papers by first author). Give first and last page numbers, full titles of papers, and names of all co-authors.

JOURNAL ARTICLE

Molenaar P. C., Newsom-Davis J., Polak R. L., and Vincent A. (1981) Choline acetyltransferase in skeletal muscle from patients with myasthenia gravis. *J. Neurochem.* **37,** 1081–1088.

CHAPTER IN BOOK

Yavin E. and Yavin Z. (1980) Sources of choline for developing cerebral cells, in *Tissue Culture in Neurobiology* (Giacobini E., Vernadakis A., and Shahar A., eds), pp. 277–289. Raven Pres, New York.

BOOK

Fuster, J. M. (1980) *The Prefrontal Cortex: Anatomy, Physiology, and Neuropsychology of the Frontal Lobe,* pp. 5–59, Raven Press, New York.

BOOK IN SERIES

Di Chiara G. and Gessa G. L., eds (1981) *Advances in Biochemical Psychopharmacology, Vol. 27: Glutamate as a Neurotransmitter.* Raven Press, New York.

Abbreviate journal titles according to *List of Journals Indexed in Index Medicus.*

Mention unpublished experiments, papers in preparation, etc. only in text; do not include in list of References.

Cite papers accepted for publication in References with abbreviated name of journal followed by "in press." Indicate date of acceptance. Submit two copies of "in press" and "submitted" papers. Submit written authorization for personal communications from communicator with original manuscript; mention only in text, with name of communicator.

OTHER FORMATS

MATTERS ARISING—Restrict to scientific issues of fact or interpretation relevant to interests of *Journal's* readership. Article stimulating comment must have appeared in *Journal* within six months. Interested or affected parties are offered right to reply in the same issue. Submission and reply are one printed page each.

RAPID COMMUNICATIONS—Four printed pages (26000 characters, counting spaces), including illustrations, tables, and references. Each table and figure decreases space available for text. Submit in quadruplicate. Give number of characters. In cover letter explain why paper warrants rapid publication. Include an abstract, methods, results, and discussion. Results and discussion may be combined. Too long manuscripts will be returned without review.

Brief papers which present new ideas or data of particular originality and timeliness.

SHORT REVIEWS—Ten printed pages. Follow same rules of style and presentation as other papers. Include section headings. Usually invited.

OTHER

Abbreviations and Symbols: Restrict use of abbreviations to Systeme Internationale (SI) Units [see *Biochem. J.* (1978) **169**, 1–27]. Excessive use is discouraged. See *J Neurochem* 1995, 64(1): iii-iv for abbreviations that may be used without definition.

Use other abbreviations sparingly; define in a footnote on title page and at first mention in text. Recommended forms of abbreviation to be used with definition:

ACh	acetylcholine
AChE	acetylcholinesterase
AMPA	α-amino-3-hydroxy-5-methylisoxazole-4-propionate
ChAT	choline acetyltransferase
COMT	catechol-*O*-methyltransferase
DA	3,4-dihydroxyphenylethylamine or dopamine
5-HT	5-hydroxytryptamine or serotonin
MAO	monoamine oxidase
NA	noradrenaline, norepinephrine
NeuNAc	*N*-acetylneuraminic acid (not NANA)

Do not use in titles or running titles.

Experimentation, Animal and Human: Abide by accepted ethical standards. Provide an explicit statement in Experimental Procedures or Materials and Methods that experimental protocols were approved by appropriate institutional review committee and meet guidelines of responsible governmental agencies. State steps taken to avoid unjustified procedures and discomfort. For human subjects, informed consent is essential.

Materials Sharing: Supply samples of new recombinant nucleic acid or monoclonal antibody preparations in response to reasonable scientific requests.

Nomenclature: Follow recommendations of IUPAC-IUB Joint Commission on Biochemical Nomenclature. See *Biochemical Nomenclature and Related Documents* (2nd ed., 1992, Portland Press Ltd., London).

Name simple chemical compounds in text of Experimental and Results sections by formulas if printed in a single horizontal line of type.

For isotopic specifications, conform to IUPAC system [*Biochem. J.* (1978) **169**, 1–27]. Give full details of specific activities of isotopes used, with corrections and efficiency of counting. Express results in absolute terms, e.g., Ci, dps, dpm.

Use official or approved drug names; give trade names or common names in brackets at first mention.

Quote IUB Enzyme Commission (EC) number with full name of enzyme at first mention in text. Subsequently use accepted trivial name, e.g., *Full name:* Acetyl-CoA: choline *O*-acetyltransferase (EC 2.3.1.6), *Trivial name:* Choline acetyltransferase not choline acetylase. Refer to *Enzyme Nomenclature* (1984, Academic Press, San Diego). Define units of enzyme activity in terms of rate of reaction catalyzed under specified conditions. The official Systeme Internationale (SI) Unit is the katal, i.e., mol of substrate transformed (or product formed). It may be expressed in other terms, if clear definitions are given, e.g., μmol/s or μmol/min.

Solutions: Express concentrations as mol/L rather than *M*. Give '% (wt/vol)' and '% (vol/vol)' as 'g/L' and 'ml/L', respectively. Use *M*, % (wt/vol), and % (vol/vol). *N* is not permitted. Express fractional concentrations in decimal form.

Express centrifugation conditions in terms of *g* and time; not only in revolutions per minute.

SOURCE

January 1995, 64(1): i-iv.

Journal of Neurocytology
Chapman and Hall

Professor A. R. Lieberman
Department of Anatomy
University College London, Gower Street
London WC1E 6BT, U.K.

Dr. M. W. Brightman
Building 36, Room 2A-29
National Institutes of Health
Bethesda MD 20892 or
8908 Mohawk Lane
Bethesda MD 20817
Telephone: (301) 496-5091; Fax: (301) 496-42276

AIM AND SCOPE

The *Journal of Neurocytology* will publish research papers (and occasionally reviews and letters) dealing with the structure and function of nervous tissue. The scope of the *Journal* is wide: it embraces the organization and fine structure, and the associated molecular biology, biochemistry, cytochemistry, biophysics, physiology and pharmacology of neurons, receptor and glial cells, synapses and other intercellular specializations, in vertebrates and invertebrates under normal, experimental and pathological conditions. Papers reporting multi-disciplinary studies and providing direct correlations between structure and function are especially welcome. Anyone interested in contributing a review paper should first contact the Editor.

MANUSCRIPT SUBMISSION

Submit original copy of manuscript (with illustrations for printer) and two additional copies (preferably double-sided Xerox copies) for referees.

Submission of a paper implies that it contains original, unpublished work, not under consideration for publication elsewhere, and that Authors agree to transfer copyright to publisher if and when article is accepted for publication.

MANUSCRIPT FORMAT

Type manuscripts double-spaced throughout (preferably on A4 paper) with a 3 cm left margin. Check text and references for errors. Ensure that typescript is correct in style, syntax and spelling (*Shorter Oxford English Dictionary*). Underline generic and specific names and words for italics. Avoid footnotes.

Papers may be as long as findings justify. There are no text-page charges. Group half-tone illustrations into plates. Five free plates are allowed; excess is charged to Authors.

Divide research papers into headed sections in order: Summary, Introduction, Methods, Results (or Observations), Discussion (and Conclusions) Acknowledgments, References, Figure captions. Letters contain no sections.

Title Page: First page: full title; affiliation and address of Author(s); running title (75 letters and spaces); name, postal address, telephone and fax numbers of corresponding Author; number of halftone plates; and an estimated word count. Provide five Key Words.

Summary: 300 words, clearly written, factual and comprehensible without reference to text. Avoid abbreviations.

Methods: Avoid excessively detailed description of, or unnecessary literature citations to, widely used techniques.

Table: Number and head with short titles. Type small tables where they fall in text. Indicate preferred point of insertion of large tables (typed on separate sheets) in text margin. Tables must not exceed one printed page.

References: Use Harvard System in text, thus: . . . Smith (1972) . . . (Brown & Jones, 1972) . . . Smith *et al*. (1972a). List references alphabetically after text. Write out journal and book titles in full, e.g.:

JONES, A.B., PRICE, C. & EVANS, D. (1973) Fine structure of the engram. *Journal of Neurocytology* 2, 1–25.

SMITH, E. F. (1972) Color electronmicroscopy. In *Advances in Cytooptics* (edited by BROWN, G.), pp. 23–51. London: Chapman and Hall.

Illustrations: Draw line drawings in India ink on thin white card, drawing paper or tracing paper. They will be reduced by 50%; bear in mind when labelling figure. Draw graphs in black ink on paper ruled in blue. Over-rule grid lines to be shown in black.

Group photographs for printer into plates, with narrow gutters (~1 mm) between individual figures. Mount photographs on white paper or very thin card with protective overlays. Trim mounting paper or card to A4 size (297 × 210 mm). Photographic plates are reproduced without reduction. Keep number of plates and halftones and space wastage to a minimum. Maximum display area for halftone plates is 11 × 7.5 in. Make plates that size. Use 'landscape' only when unavoidable.

Color plates are charged to Author.

Number all types of illustrations in a single sequence; refer to in text and caption as Fig. 1, Fig. 2, etc. Show Author's name and figure number on front or back. Provide figures with brief captions; do not repeat information in text. If figures do not bear scale lines, indicate magnifications in captions. Indicate preferred position for insertion of illustrations in margin of typescript.

Use Letraset or other professional aids for numbering, lettering and scale marking to ensure a uniformly high standard, consistent in style and letter size.

Illustrations for referees must be original photographs or good quality photocopies. Xerox copies of halftone illustrations are not acceptable. Group original photographs. Label exactly as printer's set. Mount on paper, not card. If photocopies of plates prepared for printer are submitted, do not mount and trim to match typescript page size.

OTHER

Electronic Submission: IBM-compatible disks in Word-Perfect file format, ASCII, Roff, Troff, T^EX or L^AT_EX may be submitted with hardcopy.

Reviewers: Provide names, addresses and telephone numbers of appropriate specialist referees, particularly if papers contain subject matter outside areas of expertise of editorial board.

Units, Symbols and Abbreviations: Use SI system (Systeme International d'Unités); Å is permitted. See Royal Society of Medicine booklet *Units, Symbols and Abbreviations*, 1971. The following abbreviations may be used without explanation: ACh, AChE, ADP, AMP, ATP, etc., ADPase, etc., CNS, CSF, DNA, DOPA, EEG, EM, GABA, 5-HT, LM, PNS, PTA, RNA, tris. In all other cases write word(s) to be abbreviated in full at first mention followed by abbreviation in parentheses.

SOURCE

January 1994, 23(1): inside front and back covers.

Journal of Neurology Neurosurgery & Psychiatry

BMJ Publishing Group

Professor R A C Hughes
Medical School Building, UMDS
Guy's Hospital, St Thomas Street
London SE1 9RT, U.K.
Telephone and Fax: 071-378 6758

MANUSCRIPT SUBMISSION

Submit four copies of manuscript and figures. Follow format of articles in current issue. Type double-spaced, on one side of paper.

All authors must have participated sufficiently in work to take public responsibility for content (see *British Medical Journal* 1991;**302**:309).

MANUSCRIPT FORMAT

Figures: Prepare to a high standard suitable for publication. Submit photographs on glossy paper, unmounted, with magnification bars when appropriate. There is a charge for color figures. Do not insert figures in text; mark back with figure number and name of first author. Submit figure legends on a separate sheet.

Tables: Put each on a separate sheet with a title at top. Vertical lines are not printed; put only three horizontal lines in each table.

References: Use Vancouver style. References appear in text by number in order of occurrence. List on a separate sheet in order. Use correct punctuation. Give journal titles in full or abbreviate in accordance with *Index Medicus*.

Millikan CH, Eaton LH, Clinical evaluation of ACTH in myasthenia gravis. *Neurology* 1951;**1**:145–52.

Penn AS. Immunological features of myasthenia gravis. In: Aguayo AJ, Karpati G, eds. *Topics in Nerve and Muscle Research*. Amsterdam: *Excerpta Medica* 1975:123–32.

Coers C, Woolf AL. *The innervation of muscle. A biopsy study:* Oxford: Blackwell, 1951:16–24.

Do not include references to unpublished work in list; include works "in press" with name of journal. Verify accuracy of references.

OTHER FORMATS

LETTERS—1000 words, five references and one illustration or table. Short letters concerning papers published in *Journal* are printed under Matters Arising.

NEUROLOGICAL PICTURES—One journal page, with a maximum of two authors.

OCCASIONAL REVIEWS, CLINICOPATHOLOGICAL CASE CONFERENCES AND EDITORIAL—Solicited by Editor and are subject to review. Seek advice of Editor in advance.

SHORT REPORTS—1500 words with a minimum of references and one figure and one table. Topics include single case reports which illustrate important new phenomena, or reports of short, original research studies. Short case reports may be selected for Lesson of the month series.

OTHER

Abbreviations: Express measurements in SI units. See *British Medical Journal* 1991;**302**:388–41; *SI Unit Conversion Guide* (1992; Boston: New England Journal of Medicine). For recognized abbreviations see *Units, Symbols and Abbreviations* (4th Ed 1988, DN Baron, ed., Royal Society of Medicine: London).

Experimentation, Human: Ethical considerations are taken into account in assessing papers. See Medical Research Council's publications on ethics of human experi-

mentation, and World Medical Association's code of ethics, the Declaration of Helsinki (*British Medical Journal* 1964; **2**:177).

Obtain consent from patient (or if patient has died, from relatives) to publish any information that might alone or in combination identify a patient, whether living or dead, adult or child. Do not falsify details.

SOURCE

January 1994, 57(1): inside front cover.

Journal of Neuropathology and Experimental Neurology
American Association of Neuropathologists, Inc.

Michael Noel Hart, M.D.
Division of Neuropathology
Department of Pathology
University of Iowa College of Medicine
148A Medical Laboratories
Iowa City, IA 52242
Telephone: (319) 335-8273; Fax: (319) 335-6510

MANUSCRIPT SUBMISSION

Submission of a manuscript implies that it is sent exclusively to this *Journal*, and that its contents have not been published previously except in abstract form. Submit transfer of copyright to AANP signed by each author.

In cover letter state that all authors have seen and approved manuscript. Disclose any conflicts of interest.

MANUSCRIPT FORMAT

Double-space all parts of manuscript on white bond paper. Do not justify right margin, do not hyphenate at ends of lines, and do not use bold or italic type (indicate italics by an underline). Submit four copies with three sets of illustrations. Type senior author's name in upper right corner of each page. Number pages consecutively. Use footnotes only in Tables. Begin each division on a new page.

Title Page: List: Short title (50 characters including spaces); complete title; authors' names, degrees, and affiliations. Indicate corresponding author with complete address, telephone and fax numbers. List sources of grant support. Supply written justification of role of each author if seven or more authors.

Abstract: 200 words, one paragraph. State purpose, findings and conclusions. Do not use references.

Key Words: List seven in alphabetical order below abstract. Use *Index Medicus* terms from *Medical Subject Headings*.

Introduction: State purpose of work, previous contributions of authors and others, and hypotheses.

Materials and Methods: Describe subjects, methods, reagents, and procedures with appropriate references.

Results: Present in logical sequence in text, tables, and illustrations. Avoid excessive repetition of data.

Discussion: Stress original aspects of study and conclusions. Cite and analyze important work of others. Do not repeat data given in "Results." Avoid claims of priority and allusions to uncompleted work. State new hypotheses if warranted, clearly label as such.

Acknowledgments: List only persons who have made substantive contributions. Acknowledge material from other publications; supply written permission from copyright owner to reprint any figure or portion.

References: Number entries consecutively in order mentioned in text. Give text citation by number in parentheses. List only references cited in manuscript. Verify accuracy and completeness of references. Reference only published works and papers in press; designate journal followed by (in press). Submit preprints of references listed as (in press). Do not include work in progress, unpublished experiments, and personal communications in reference list; acknowledge in text. Obtain written permission from appropriate person to cite unpublished data and personal communications. Abbreviate journal titles as in *List of Journals Indexed in Index Medicus.*

JOURNALS

Author(s) (list all when six or less; when seven or more, list only first three and add et al), full title of paper, abbreviated name of journal, year, volume number, inclusive pagination.

Carpenter S, Karpati G. Segmental necrosis and its demarcation in experimental micropuncture injury of skeletal muscle fibers. J Neuropathol Exp Neurol 1989;48:154–70

BOOKS

Author(s), title of book, edition, city, publisher, year.

Adams JH, Graham DI, Harriman DGF. An introduction to neuropathology. Edinburgh: Churchill Livingstone, 1988

CHAPTER IN A BOOK

Author(s), chapter title, editor(s) of book, book title, city, publisher, date, volume (when necessary), section (when necessary), inclusive pagination.

Kirkpatrick JB. Gunshots and other penetrating wounds of the central nervous system. In: Leestma JE, ed. Forensic neuropathology. New York: Raven Press, 1988:276–99.

PAPERS IN CONFERENCE PROCEEDINGS OR ABSTRACTS

Author(s), title of paper, title of conference, dates (when available), city, publisher, date, inclusive pagination.

Moossy J. Anatomy and pathology of the vertbrobasilar system. In: Berguer R, Bauer RB, eds. Vertebrobasilar arterial occusive disease. Medical and surgical management. Papers from the First International Conference on Vertebrobasilar Occusive Vascular Disease, held in Detroit, MI, November 8 & 9, 1982. New York: Raven Press, 1984:1–13

Sung JH, Manivel JC. Macrophages in ischemic infarcts . (Abstract) J Neuropathol Exp Neurol 1989;48:342

Use abstract form when reference is a letter.

Tables: Type each on a separate page; double-space all lines. Number consecutively with Arabic numbers. Supply brief titles. Give each column a short heading. Explain all abbreviations and symbols in a footnote. Identify statistical measures of variations such as SD and SEM. Omit internal horizontal and vertical lines. Cite in text in consecutive order.

Legends: Type legends for illustrations, in order cited, with Arabic numbers corresponding to each figure. Note staining methods and magnification of photomicrographs and electron micrographs. Explain arrows, numbers, letters, or symbols.

Illustrations: Send three sets of unmounted black and white glossy prints numbered on front lower left corner. If a large number of figures or parts to each figure, reviewer sets may be mounted. Do not mount printer set. Paste a label on back of each with figure number, an arrow indicating TOP and author's name. Maximum acceptable size of a single figure or array is 7 × 9 in. for full page or 3.25 × 9 in. for single column; minimum is 2 × 2 in. Apply special markings (arrows, numbers, etc.); make sufficiently large to withstand reduction. Original artwork may be requested if drawings, diagrams and similar illustrations are not well reproduced. Photocopies are not accepted. Cite in text in order. Color plates are printed at author's expense. If more than one color figure, group together to fit size requirement above. Submit letters of consent for photographs of patients where a possibility of identification exists. Author(s) may be asked to submit additional photographs, histological slides or related research data to aid in evaluating paper.

OTHER FORMATS

LETTERS—Two double-spaced typed pages, four references, no illustrations. Not a regular department, but may be submitted to comment on articles published in the *Journal* or on special aspects of neuropathology.

OTHER

Color Charges: $500 per page.

Experimentation, Human: Indicate adherence to standards of institutional committee on human experimentation, or National Research Council's *Guide for the Care and Use of Laboratory Animals.*

Page Charges: Six pages (any combination of text and illustrations) are free of charge. Excess pages will be charged at the prevailing rate.

Reviewers: In letter of submission, suggest names (include addresses, phone and Fax numbers) of three potential referees and request that certain persons not be used as referees. Editor-in-Chief will try to honor requests.

Spelling and Abbreviations: Use *Webster's Third New International Dictionary.* Scientific abbreviations and symbols should conform to those listed in current edition of *Handbook of Chemistry and Physics.* For nomenclature for dyes see *H. J. Conn's Biological Stains* (9th ed., 1977). Refer to International Steering Committee "Uniform requirements for manuscripts submitted to biomedical journals," *Ann Intern Med* 1988;108:258–65 and *Council of Biology Editors Style Manual* (5th ed., 1983) for additional information on preparation of copy.

SOURCE
January 1995, 54(1): inside front and back covers.

Journal of Neurophysiology
American Physiological Society

Journal of Neurophysiology
9650 Rockville Pike
Bethesda, MD 20814-3991

AIM AND SCOPE

The *Journal* publishes original articles on the function of the nervous system. All levels of function are included, from the membrane and cell to systems and behavior. Experimental approaches include molecular neurobiology, cell culture and slice preparations, membrane physiology, developmental neurobiology, functional neuroanatomy, neurochemistry, neuropharmacology, systems electrophysiology, imaging and mapping techniques, and behavioral analysis. Experimental preparations may be invertebrate or vertebrate species, including humans. Theoretical studies are acceptable if they are tied closely to the interpretation of experimental data and elucidate principles of broad interest.

MANUSCRIPT SUBMISSION

Research articles must be original. Do not submit articles in which, other than in abstracts, a significant portion of data in the form of figures and tables has been published or submitted for publication elsewhere. If some data could be construed this way, provide a copy of publication in question. If Editor believes prior publication significantly compromises originality of submission, manuscript will be returned. If manuscript depends critically on another paper that is as yet unpublished, submit three copies of that paper.

In cover letter, include a statement that manuscript will not be submitted for publication elsewhere until a decision has been made at the *Journal of Neurophysiology*. Each author must sign cover letter. Each author should have made an important contribution to study and should be thoroughly familiar with primary data.

A copyright transfer form will be sent to corresponding author. It must be completed and returned before work will be set in type.

MANUSCRIPT FORMAT

Type double-spaced with wide margins on 8.5 × 11 in. paper. Submit four copies, including figures. Submit glossies for photomicrographs, gels and other halftones. Photocopies of line drawings are acceptable; provide glossies when revised version is returned to Editor. Number pages in upper right corner (beginning with first text page). Arrange in order: title page, summary and conclusions, text, text footnotes, acknowledgments, references, tables, figure legends, illustrations. On title page include title of article, author(s), department and institution where work was done with city, state and zip code, or country; an abbreviated title for running head (55 characters including spaces); and name, address and telephone number of corresponding author. Type summary and conclusions on separate sheet, double-spaced. Double-space text footnotes, acknowledgments, references, and figure legends each on a separate sheet. Type each table on separate sheet double-spaced. Identify illustrations on reverse (lightly with a soft pencil) with figure number and name of first author; when necessary, mark TOP. Make text clear and concise; conform to accepted standards of English style and usage. Define unfamiliar or new terms when first used. Do not use jargon, clichés or laboratory slang.

Title: Make title informative with no unnecessary words. Be as brief as possible.

Summary and Conclusions: At beginning of text, arrange in short numbered paragraphs and overall not more than 5% of text. Summarize methods and results very briefly and clearly identify the main conclusions.

Introduction: Orient general reader to problem and its significance. Focus on the specific study; do not give a general overview of the entire field.

Methods: Include enough information about techniques and instrumentation to permit evaluation of appropriateness of these aspects. Include enough information about handling of experimental animals to indicate that policy of The American Physiological Society regarding the use and care of animals has been adhered to.

Results: Present clearly in a logical order.

Discussion: Confine to the significant implications of the results of this specific study.

Footnotes: Number text footnotes consecutively throughout. Double-space and assemble on one sheet.

References: Limit to directly pertinent published works or papers that have been accepted for publication. An abstract properly identified (Abstract) may be cited only when it is the sole source. Verify accuracy.

Type references separately, double-spaced, arranged alphabetically by author. Include appropriate author name and year for each reference in parentheses at proper point in text: one author (Brown 1982); two authors (Brown and Smith 1982); three or more authors (Brown et al. 1982). If more than two references are cited by different authors, separate entries with a semicolon (Brown 1982; Smith 1983). If more than two references are cited by the same first author (or single author) use et al. where appropriate plus the date, even if subsequent authors are not the same (Brown et al. 1982, 1983, 1986–1988). Note use of commas between two consecutive years or nonconsecutive years and dashes for ranges (Brown et al. 1982, 1983, 1986–1988). If more than two references with the same year and author(s) are cited, use lowercase letters after year (Brown 1982a, b). Insert lowercase letters in same year references in reference list.

JOURNAL ARTICLES

Last names of authors, followed by initials; title of article (first word only capitalized); name of journal (abbreviated as in *Serial Sources for the BIOSIS Data Base*, BioSciences Information Service), volume, inclusive pages, and year.

O'Mara, S., and Tarvin, R. P. C. Neural mechanisms in vibrotactile adaptation. *J. Neurophysiol.* 59:607–622, 1988.

BOOK REFERENCES

Author(s) as above; title of book (main words capitalized); city of publication; publisher; year and pages.

Include references to government technical documents only when availability is assured. For style of citation of these documents, congress proceedings, chapters in books, etc., consult recent issues of *Journal*.

Do not include citations such as "unpublished observations" or "personal communication" in reference list; add in parentheses in text. Secure permission of person cited in "personal communications."

Illustrations: Furnish as unmounted, sharp photographic prints no larger than 8.5 × 11 in. Do not submit original artwork. Laser-prints are acceptable provided image is clear, sharp black and on highest gloss paper possible. Submit in triplicate and refer to as figures. On reverse in light pencil indicate figure number (with Arabic numeral); principal author's last name; short title; and TOP.

Prepare with consideration of reduction to single column width (3.5 in.) whenever possible, otherwise for double-column width (7 3/8 in.). Text height of *Journal* page is 9.25 in.; prepare figures so as not to exceed this height when reduced.

Make recordings such as electroencephalograms and oscillograms originally on nonphoto blue-ruled recording paper and photographed so that crosshatched background is eliminated.

Have all drawings and lettering done in India ink, with graphic arts transfer lettering, or with a graphic arts lettering system, or with a high-quality computer software program with output on a high-resolution laser printer. Handwritten, typewritten, or computer-generated dot-matrix lettering is not acceptable. Make lettering proportionate to size of figure. Make all lettering and symbols on a figure the same size so that none becomes illegible with reduction. Make scale of lettering such that it will not be less than 2 mm high after reduction. Keep lettering within the framework of the illustration, including keys to symbols. When necessary to explain symbols in legend, use standard characters: ❑ ■ ○ ● ◐ △ ▲ ×.

Photographs of animals are not acceptable; instead use a good line drawing. Likewise use good line drawings of equipment.

Include actual magnification of photomicrographs in legend. A length scale on print is preferable.

Designate special features on photomicrographs with letters, numerals, arrows, and other symbols that contrast with background.

Figures in color are accepted if author assumes all printing costs. Supply three glossy positive color prints.

Indicate approximate position of each figure in margin of manuscript.

Group figure legends in numerical order and type double-spaced on one or more sheets. Each figure must have a legend. Write legends in complete sentences that present necessary details clearly.

Tables: Keep number to a minimum. Do not duplicate material in text or illustrations. Type double-spaced on separate sheet(s) with a brief title; put explanatory matter in footnotes, not in title. Submit statistical summary tables rather than tables with many lines of individual values. Number consecutively with Arabic numerals. Prepare with size of *Journal* page in mind: 3.5 in.wide, single column; 7 3/8 in. wide, double column.

Omit horizontal and vertical rules when possible. Omit nonsignificant decimal places in tabular data. Use short or abbreviated column heads; explain if necessary in footnotes.

Identify statistical measures of variations, SD, SE, etc.

List table footnotes in order of appearance; identify by standard symbols *, †, ‡, § for four or fewer; for five or more, use consecutive superior letters throughout.

Indicate approximate position of each table in margin of manuscript.

Mathematical Formulas and Equations: Address mathematical aspects of articles to readers who are not mathematicians. State mathematical strategy; summarize meaning of final mathematical statement, assumptions and limitations. Submit an informal letter or note for referees, filling in details of mathematics not explicitly stated in paper.

Simplify structural chemical formulas and complicated mathematical equations as much as possible and carefully check. Clearly identify all subscripts, superscripts, Greek letters, and other unusual characters in penciled marginal notes where they first appear. Distinguish between 1 (one) and letter l (el), zero and letter O, × (mathematical sign) and letter x. Use slant line (/) for simple fractions $(a + b)/(x + b)$ rather than the built-up fraction $\dfrac{a + b}{x + y}$.

OTHER FORMATS

RAPID PUBLICATIONS—2000 words in text, figure legends, and references (7 double-spaced manuscript pages in 10-point font and three figures). Write clearly, make understandable to general reader of journal. Label figures adequately so they are easily interpretable. Report timely, new experimental results of unusual interest to a broad readership. Use format of regular articles.

In cover letter indicate, in addition to usual assurances of originality, wish for consideration for rapid publication and that if it is not judged appropriate, whether wish paper returned or reviewed as a general article. Provide author's fax and telephone numbers or electronic mail address.

Submit three copies to an Associate Editor most appropriate for field of study, and one copy to office in Bethesda:

Wayne E. Crill, Dept. Physiology, SJ40, Univ. of Washington School of Medicine, Seattle, WA 98195, Fax: 206/685-0619

Nigel W. Daw, Dept. Physiology, Washington Univ., St. Louis, MO 63110, Fax: 314/362-9862

Charles Liberman, Dept. of Otolaryngology, Harvard Medical School, Boston, MA 02114, Fax: 617/573-3350

Mark L. Mayer, LCMN, NICHD, NIH, Bldg. 49, Rm. 5A78, Bethesda, MD 20892, Fax: 301/593/9680

Allen I. Selverston, Dept. Biology, B-022, Univ. of California, LaJolla, CA 92093, Fax: 619/534-7309

Peter L. Strick, Research Service (151), VA Medical Center, Syracuse, NY 13210, Fax: 315/423-5729.

OTHER

Abbreviations, Symbols, and Terminology: Include a list of new or special abbreviations used in paper with spelled-out form or definition. For commonly accepted abbreviations, word usage, symbols, etc. see *CBE Style Manual* (5th ed., 1983). For chemical and biochemical terms and abbreviations follow recommendations of the IUPAC-IUB Combined Commission on Biochemical Nomenclature. Isotope specification should conform to IUPAC system.

Drugs, Chemicals, and Trade Names: Capitalize proprietary (trademarked) names; check spelling carefully. Chemical or generic name precedes trade name or abbreviation of a drug at first use. Capitalize trade names of chemicals or equipment; check spelling carefully.

Electronic Submission: After revision of manuscript, provide final version both in hard copy and on disk, preferably in ASCII format on 3.5 or 5.25 in. disks created with any standard word-processing program.

Experimentation, Animal: Conduct investigations in conformity with "Guiding Principles for Research Involving Animals and Human Beings." In describing surgical procedures on animals, specify type and dosage of anesthetic agent. Curarizing agents are not anesthetics; if used, provide evidence that anesthesia of suitable grade and duration was employed. Editors/Associate Editors will refuse papers in which evidence of adherence to these principles is not apparent.

Experimentation, Human: Investigations must follow principles embodied in the Declaration of Helsinki.

Mathematical Formulas and Equations: State mathematical strategy and summarize meaning of final mathematical statement, the assumptions and limitations. Submit an informal letter or note for referees, with details of mathematics not explicitly stated in paper.

Simplify structural chemical formulas and complicated mathematical equations as much as possible; check carefully. Clearly identify subscripts, superscripts, Greek letters, and other unusual characters in penciled marginal notes at first appearance. Distinguish between 1 (one) and letter l (el), zero and letter O, × (mathematical sign) and letter x. Use slant line (/) for simple fractions

$(a + b)/(x + y)$ rather than built-up fraction $\dfrac{a + b}{x + y}$.

Page Charges: $50 a printed page. $100 charge for accepted manuscripts that do not provide a computer disk for final revised, accepted version of paper.

Spelling: Follow *Webster Third New International Dictionary* for spelling, compounding, and word division.

SOURCE

December 1994, 72(6): back matter.

Journal of Neuroscience

Society for Neuroscience
Oxford University Press

Diane M. Sullenberger
Society for Neuroscience
11 Dupont Circle N.W., Suite 500
Washington, D.C. 20036
Telephone: (202) 462-6688; Fax: (202) 462-1547
E-mail: jn@sfn.org

Dr. Thomas J. Carew
Department of Psychology, Yale University
Yale Station, P.O. Box 11A, 2 Hill House Avenue
New Haven, CT 06520
                Behavioral neuroscience

Dr. John E. Dowling
Biological Laboratories
Harvard University
16 Divinity Avenue
Cambridge, MA 02138

Dr. Charles F. Stevens
The Salk Institute
10010 North Torrey Pines Boulevard
La Jolla, CA 92037 USA
                Cellular neuroscience

Dr. Mary E. Hatten
Laboratory of Developmental Neurobiology
The Rockefeller University
1230 York Ave.
New York, New York 10021-6399

Dr. Dennis D.M. O'Leary
The Salk Institute (MNL)
10010 North Torrey Pines Road
LaJolla, CA 92037-1099
                Developmental neuroscience

Dr. Stephen F. Heinemann
The Salk Institute (MNL)
10010 North Torrey Pines Road
La Jolla, CA 92037-1099

Dr. Richard H. Scheller
Department of Molecular and Cellular Physiology, B-155
Beckman Center
Stanford University Medical Center
Stanford, CA 94305
                Molecular neuroscience

Dr. Allan I. Basbaum
Department of Anatomy, Box 0452
University of California San Francisco
513 Parnassus Ave.
San Francisco, CA 94143-0542

Dr. Stephen G. Lisberger
Department of Physiology, Box 0444
University of California, San Francisco
513 Parnassus Ave., Rm. S-762
San Francisco, CA 94143-0542
                Systems neuroscience

AIM AND SCOPE

The Journal of Neuroscience is the official journal of the Society for Neuroscience. Its purpose is to publish papers on a broad range of topics of general interest to those working on the nervous system.

MANUSCRIPT SUBMISSION

Submission of a manuscript involves tacit assurance that no similar paper, other than an abstract, has been, or will be, simultaneously submitted for publication elsewhere; submission also implies thorough understanding of and concurrence with Society's statements on use of animals in research and on ethics. Do not duplicate publication of research results. Submit five reprints of papers describing any potentially overlapping earlier work (including papers in press). For manuscripts with multiple authorship it is understood that all listed authors concur in submission and that final manuscript has been approved by all authors. Multiple part papers are discouraged. Collapse multiple papers into a single manuscript. Copyright, which must be signed by each author, is vested in the Society for Neuroscience.

Submit five copies of each manuscript (including illustrations). Submit four copies to the Society for Neuroscience Central Office; send one copy to the appropriate Section Editor: Behavioral Neuroscience, Cellular Neuroscience, Developmental Neuroscience, Molecular Neuroscience, or Systems Neuroscience. Choice of section is author's.

MANUSCRIPT FORMAT

Type double-spaced throughout, including references, tables, and figure legends; do not use footnotes.

Manuscripts must include a title page, an abstract, key words, an introductory statement, a description of experimental procedures or methods, a description of results, discussion of experimental findings (1500 words), and, on separate pages, list of references, figure legends, and tables.

Title Page: Put name of appropriate Section Editor at top right corner; complete title (do not include chemical formulas or arbitrary abbreviations); names of all authors; complete affiliations of all authors; an abbreviated title (60 characters and spaces); numbers of text pages (from first page of introduction through last page of discussion), figures, and tables; name, complete address, telephone and fax numbers, and electronic mail address of corresponding author; and acknowledgments (list of grant support and personal acknowledgments).

Abstract: 250 words; state research objective, procedures, results, and significance of the data; write in complete sentences and in a form acceptable for abstracting services. Follow by six key words or phrases; words may appear in title.

Introduction: 500 words, No separate heading; indicate objectives of study and provide background information to clarify why study was undertaken and what hypotheses are tested.

Materials and Method: Brief, but adequate to allow a qualified investigator to repeat research; make reference to published procedures, both original description and perti-

nent modifications. List all companies from which materials were obtained (with city and state or province or country). If source was an individual, list affiliation.

Results: Present experimental finds of study. Include only results essential to establish main points.

Discussion: 1500 words; briefly state principal findings, discuss validity of observations, discuss findings in light of other published work and state possible significance. Do not discuss literature extensively.

References: List only published and "in press" references in reference list at end of paper. Provide latest information on "in press" references. Cite "submitted" references only in text: (A. B. Smith, C. D. Johnson, and E. Greene, unpublished observations); for personal communications: (F. G. Jackson, personal communication).

Cite in text as follows: "The procedure used has been described elsewhere (Green, 1978)," or "Our observations are in agreement with those of Brown and Black (1979) and of White et al. (1980)," or with multiple references, in chronological order: "Earlier reports (Brown and Black, 1979, 1981; White et al., 1980; Smith, 1982, 1984) . . ." Cite each listed reference in text; list each text citation in reference section.

Type list of references double-spaced in alphabetical order according to name of first author. In two-author papers with the same first author, order is alphabetical by second author's name. In three-or-more-author papers with same first author, order is chronological. Follow name of author(s) by date in parentheses, full title of paper as it appeared in original with source of reference, volume number, and first and last pages.

JOURNAL ARTICLE

Hamill OP, Marty A, Neher E, Sakmann B, Sigworth F (1981) Improved patch-clamp techniques for high resolution current recordings from cells and cell free membrane patches. Pfluegers Arch 391:85.

Hodgkin AL, Huxley AF (1952a) The components of membrane conductance in the giant axon of *Loligo*. J Physiol (Lond) 116:473–496.

Hodgkin AL, Huxley AF (1952b) The dual effect of membrane potential on sodium conductance in the giant axon of *Loligo*. J Physiol (Lond) 116:497–506.

BOOK

Hille B (1984) Ionic channels of excitable membranes. Sunderland, MA: Sinauer.

CHAPTER IN A BOOK

Stent GS (1981) Strength and weakness of the genetic approach to the development of the nervous system. In: Studies in developmental neurobiology: essays in honor of Viktor Hamburger (Cowan WM, ed), pp 288–321. New York: Oxford UP.

Abbreviate journal titles as in *Index Medicus*. Verify correctness of reference. After manuscript revisions, double check that all in-text citations are in reference list and that all references on reference list have at least one corresponding in-text citation.

Tables: Include only essential data; type double-spaced on a separate sheet with an explanatory title and sufficient experimental detail in legend, to be intelligible without reference to text. Give each column an appropriate heading and set out material to be readily understood by compositor; do not duplicate data in both in text and table or figure.

Refer to all tables in text. Indicate approximate position in margin of manuscript.

Illustrations: Prepare to be the smallest size that will convey essential scientific information. Check carefully, author alterations in proof are not allowed. Prepare review copies at size to appear in journal. Specify on original whether figure should be reproduced as one column (8.8 cm), 1.75 columns (11.4 cm to 13.1 cm) [use is discouraged], or 2 columns (15.7 cm to 18.3 cm). Vertical dimensions can vary, but must not exceed 24 cm. Do not submit oversized figures.

Refer to all figures in text. Number consecutively. Indicate preferred location in margin. Letter and label figures appropriately with characters that will be between 2 and 6 mm high in final reproduction. If reduced lettering would be unreadable, figures may be returned for relabeling. Identify each figure on reverse side with author's name, figure number, orientation of figure (top, bottom, etc.) and desired final column width.

Original color figures must be flexible enough to be rolled around a cylinder. Composite color figures are acceptable if parts are mounted on a flexible backing and do not overlap. Do not use tape to mount figures. Rub-on lettering is permissible; avoid patch-type labels that add another layer of paper above image. For thermographic images, submit a photographic print rather than fragile wax version.

Submit original black-and-white figures as line drawings, photographic prints, or high-quality laser output (make edges of type and shapes crisp and clean, not ragged). Put photographs on white, glossy paper.

Attach a complete set of figures, at final size on sheets the same size as text, to each copy of manuscript as unmounted prints or legible photocopies. Original prints or good photocopies of photomicrographs are essential. Copies considered unsuitable for review will be returned.

Send one set of top quality figures for printer (labeled "printer's copy"), with a set of photocopies, with top copy of typescript to Society for Neuroscience Central Office. Original figures should be final size.

Figure Legends: Give each figure a title and an explanatory legend. Include title in legend; do not letter on figure. In legend include sufficient detail to make intelligible without reference to text; define all labels and symbols in figure art; and provide other essential information such as magnification or scale bar dimension. Do not state "See text," be more specific; for example, "See Results."

OTHER FORMATS

FEATURE ARTICLES—Invited contributions; submit suggestions and proposals to Features Editor and Editor-in-Chief.

OTHER

Abbreviations: Spell out at first occurrence, then introduce by placing abbreviation in parentheses after term being abbreviated. Over-defining is preferable to under-defining; copyeditor can delete definitions, but cannot always add a definition for an unfamiliar abbreviation. Use metric system for all volumes, lengths, weights, etc. Express temperatures in degrees celsius (centigrade). Units should conform to the International System of Units (SI) (see *Handbook of Chemistry and Physics*).

Color Charges: $1000 for one color figure; each additional figure is $400.

DNA Sequences: Submit only complete genomic and cDNA sequences. Confirm accuracy of sequences by analyses of both strands. Deposit sequences in an accessible data base; provide sequence accession number.

Electronic Submission: Submit a disk containing an electronic version of manuscript with hardcopy. Label disk with date, manuscript number (if revision), first author's name, file name, and software and hardware used. In cover letter, signed by any author, state that electronic file matches hard copy.

The 3.5 in. disk is preferred,. Do not write-protect disks. Include all sections of paper except tables in a single file. Tables will be typeset from hard copy. Most word processing software may be used: WordPerfect (versions 4.2 through 5.1), Ami Pro (1.2, 1.2a, 1.2b), Displaywrite (4.0, 4.2, 5.0), Microsoft Word (1.0, 1.1, 1.2 for Windows; 5.0, 5.5), MultiMate (3.3, 3.6, 4.0; Advantage II) Officewriter (6.0, 6.1, 6.11, 6.2), Rich Text Format (RTF), WordStar (3.3, 3.31, 3.4, 4.0, 5.0, 5.5, 6.0), and Xywrite III Plus (3.55, 3.56). Format may be either DOS or Macintosh; DOS is preferred. Macintosh users: save manuscript file as MS-DOS on a DOS-formatted disk. If word processing software is not listed, save manuscript file in ASCII form on an DOS-formatted disk.

Prepare electronic file accurately, consistently, and simply; avoid use of special fonts or elaborate formatting. Format paragraphs the same way throughout. Use lowercase "ell" (l) and numeral one (1), and capital "oh" (O) and numeral zero (0) correctly. Type Greeks symbols, diacritical marks, italics, superscripts, and subscripts in electronic file using software features. When a special character cannot be typed, represent it by an available character not otherwise used; provide a translation key in cover letter. If accents or other unusual characters must be drawn on manuscript, highlight and list in an accompanying note.

Ethics: See Instructions for Authors, January 1993 edition.

Experimentation, Animal: See Instructions for Authors, January 1993 edition.

Materials Sharing: Make freely available to colleagues in academic research any clones of cells, nucleic acids, antibodies, etc. that were used in research that are not available from commercial suppliers.

Page Charges: $40 per published page. Price for all work requiring special printing, including figures in color, will be determined by publisher. Page charges may be waived for work from countries that do not allow research funds to be spent for this purpose. Request waiver of charges in writing when accepted paper is submitted in final form. Other charges cannot be waived and author must pay.

Reviewers: In cover letter suggest five reviewers who are especially qualified to review work; Give complete address, phone and fax numbers and e-mail addresses.

SOURCE
February 1994, 14(2): end of issue.

Journal of Neuroscience Research
Wiley-Liss

Dr. Bernard Haber
Chief, Neurochemistry Section
Marine Biomedical Institute
University of Texas Medical Branch
200 University Blvd. #519
Galveston, TX 77555-0843
FAX: (409) 772-2920 (9382); e-mail: haber@mbian.ut-mb.edu

AIM AND SCOPE

The *Journal of Neuroscience Research* publishes basic reports on molecular, cellular, and subcellular areas of the neurosciences. The journal also publishes clinical studies that emphasize fundamental and molecular aspects of nervous system dysfunction. The journal features full-length papers, rapid communications, and mini-reviews on selected areas.

The journal features a section devoted to the rapid communication of full-length, critically reviewed papers reporting new and important advances in neuroscience research.

MANUSCRIPT SUBMISSION

Manuscripts must be submitted solely to this journal, may not have been published in any part or form in another publication of any type, professional or lay, and become the property of the publisher upon acceptance for publication. Author must sign agreement transferring copyright to publisher, who reserves copyright. No published material may be reproduced or published elsewhere without written permission of publisher and author.

MANUSCRIPT FORMAT

Present in order: Title Page, Abstract (200-250 words), Key words (3-5 for indexing; terms not in title are preferred), Introduction (1500 words), Materials and Methods, Results, Discussion, Acknowledgments (including supporting research grants), References, Tables, Figure Legends, and Figures. Type double-spaced.

On title page include full title, author's name(s), with initials in place of given names, and complete addresses in English (including zip code, and indicate which author is at which address), phone and Fax numbers, e-mail addresses, and a short running title (40 characters).

Divide material into logically organized sections with all headings and subheadings clearly delineated. Do not exceed 40 pages including tables and figures.

References: In text, cite references by name and date. For more than two authors, use first surname and et al. In final list, put in alphabetical order; include complete title of article cited, and names of all authors. Journal abbreviations follow *Index Medicus* style. Notice punctuation; do not use all capitals, do not underline.

JOURNAL ARTICLES

Sinoway MP, Kitagawa K, Timsit S, Hashim GA, Colman DR (1994): Proteolipid protein interactins in transfectants: Implications for myelin assembly. J Neurosci Res 37:551–562.

BOOKS

Hayes TE, McKayt RDG (1993): Differentiation of cell types in the central nervous system. In Levine AJ, Schmidek HH (eds): Molecular Genetics of Nervous System Tumors. New York: Alan R. Liss, Inc., pp 37–48.

Tables: Keep tables as few and as simple as possible. Type each table on a separate sheet; include at end of text. Put title above and detailed explanation of contents (if necessary) below body of table. Use horizontal lines only above and below column headings and at bottom of table: do not use vertical rules or horizontal rules within table body. Number consecutively with Roman numerals. Make sure all tables are cited in text, and that all abbreviations used are clearly spelled out. Designate footnotes by superscript, lowercase letters (a, b, etc.).

Illustrations: Submit one set of original figures and two sets of high quality copies. Have all line art professionally drawn and lettered (typed lettering is not acceptable). Make lettering on line drawings and halftones large enough to insure legibility after reduction. High contrast is necessary for reduction and reproduction.

Design all illustrations for reproduction at full-text width (6.75 in.) or single-column width (3 5/16 in.) with a maximum length of 9 in.

Place a circled number ("transfer lettering") in bottom left corner.

Indicate magnification by a micron bar or in legend. If more than one illustration is mounted as a plate, number and letter left to right, top to bottom.Type names of authors, figure number (in single Arabic numeral sequence), and an arrow indicating TOP on a gummed label and affix to back of each illustration. Do not write directly on figures. Do not use ink or ball point pen. Be sure all figures are cited in text, in order.

Type all figure legends double-spaced on separate sheets at end of text. Explain all abbreviations and symbols in legend. Include a credit line for material reproduced from a prior publication; obtain written permission from previous publisher.

Color Illustrations: Cost is charged to author. If available, supply a color transparency or negative, in addition to color prints. Author has opportunity to approve both costs and proofs prior to printing.

OTHER FORMATS

MINI-REVIEWS—12 manuscript pages, including figures and references. Focussed, critical mini-reviews of forefront areas of neuroscience. Submit to: Dr. Steve Pfeiffer, Department of Microbiology and Program in Neurological Sciences, University of Connecticut Medical School Farmington, CT 06030; FAX: + 1 (203) 679-1239; e-mail: pfeiffer@mbcg.uchc.edu

RAPID COMMUNICATIONS—Two manuscript pages, including tables and figures. In cover letter indicate reasons manuscript is deserving of high publication priority. Submit to: Dr. Jean de Vellis, Neurobiochemistry Group, Mental Retardation Research Center, #68–177, University of California, Los Angeles, 760 Westwood Plaza, Los Angeles, CA 90024, FAX: + 1 (310) 206-5061; e-mail: jdevellis%neuro2.dnet@loni.ucla.edu

OTHER

Abbreviations: Express magnitudes of variables in numerals. Abbreviations are used without punctuation. Use abbreviations for commonly used substances in *Council of Biology Editors Style Manual: A Guide for Authors, Editors, and Publishers in the Biological Science*, (5th Ed., The Council of Biology Editors, Inc., Chicago.)

Use generic names for drugs; give trade names in parentheses at first mention, use generic name thereafter.

Color Charges: $950 for one page of color. Second, and subsequent pages, up to four, are $500 each.

Experimentation, Animals: Acquire and care for animals in accordance with guidelines in NIH *Guide for the Care and Use of Laboratory Animals* (National Institutes of Health Publication NO. 85-23, revised 1985) and principles in "Guidelines for the Use of Animals in Neuroscience Research" by the Society of the Neuroscience (Washington, D.C.). In cover letter indicate that all animal protocols were approved by authors' institutional animal experimentation committee. Manuscripts that do not comply with acceptable standards for humane treatment of vertebrate animals will not be considered.

SOURCE

May 1, 1994, 38(1): 122-125.

Journal of Neurosurgery

American Association of Neurological Surgeons

John A. Jane
University of Virginia
1224 West Main St., Suite 450
Charlottesville, VA 22903
Telephone: (804) 924-5503; Fax: (804) 924-2702

AIM AND SCOPE

The *Journal of Neurosurgery* is devoted to the publication of work relating primarily to neurosurgery, including studies in clinical neurophysiology, organic neurology, ophthalmology, radiology, and pathology.

Articles on unusual cases and technical notes on special instruments or equipment that might be useful to others in the field of the neurosciences are also acceptable. Case reports should be short and not include an extensive review of the literature. The Editorial Board reserves the right to judge the appropriateness of studies involving human subjects.

MANUSCRIPT SUBMISSION

Submit a statement signed by all authors certifying that work has not been published previously and has not been submitted elsewhere for review.

A Copyright Release & Author Agreement will be sent to corresponding author on receipt of manuscript; all authors must sign. Manuscript cannot be accepted if signed agreement is not returned.

For technical reports include a statement as to whether authors have a financial interest in instrumentation or methodology used at conclusion of article.

Submit three copies of each manuscript typed double-spaced with three sets of illustrations. Mail flat in heavyweight wrappers.

MANUSCRIPT FORMAT

Title Page: Include title of article (concise yet informative); first name, middle initial, and last name of each author with highest academic degrees; name of department, department affiliation, and institution from which work should be attributed; name, address, telephone and fax numbers of corresponding and reprint request author(s); and sources of support in the form of grants, supplied drugs or equipment, etc.

Abstract: 250 words. State in first sentence purpose and rationale of investigation. State methods and results concisely; omit complicated statistical information if possible. Emphasize major and important contributions. In final sentence state principal conclusion of study.

Key Words: Provide and identify three to six key words or phrases for indexing.

Text: Divide into Introduction, Materials and Methods, Results (or Summary of Cases), and Discussion. Use subheadings to improve readability and organization. Number pages.

Introduction: State purpose of article and summarize rationale for undertaking. Reference major background reports; do not review in detail pertinent literature (belongs in discussion).

Materials and Methods/Case Material: Include enough details so methodology is clearly understood. Refer to prior work if methodology has previously been reported in detail; include enough information for reader to understand methodology without referring to previous reports.

Results: Concisely summarize study findings. Follow the train of thought in Methods section. Include only findings of current report; do not refer to previous investigations.

Discussion: Concisely emphasize major findings of study or investigation and its significance. Do not repeat information in Methods and Results sections. Use subheadings so reader can follow author's train of thought.

Illustrations: Do not submit too many illustrations. Approved black/white figures are free of charge. Authors pay cost of color reproductions. Obtain permission from subject for use of identifiable photographs. Submit three sets of illustrations of a quality suitable for reproduction between 3×5 and 8.5×11 in. in size. Do not mount; number lightly on back; indicate TOP. Make lettering or arrows large enough to permit necessary reduction; apply directly to print. Do not submit computer graphics in dot matrixes; laser-printed graphs should be bold enough for reduction.

References: Alphabetize by author(s), type double-spaced and cite by superior numbers in text. Use abbreviations in *Index Medicus*. Give names of all authors unless more than three; then give names of first three authors, followed by "et al."

ARTICLE

White NP, Kim D, Fos K, et al: Posterior fossa aneurysms. Case Report. **J Neurosurg 72:**345–349, 1990

CHAPTER IN BOOK

Jane JA, Persing JA: Neurosurgical treatment of craniosynostosis, in Cohen MM Jr (ed): **Craniosynostosis: Diagnosis and Management.** New York: Raven Press, 1986, pp 249–320.

Include only published materials in reference list. List papers accepted but not yet published with journal name and year, followed by "(In press)." Supply a full citation as soon as available. Do not include personal communication or unpublished data in reference list; included in text in parentheses. Cite a date for "personal communication."

OTHER

Experimentation, Animal: Provide enough information to assure Editorial Board that animals were handled in a humane fashion and met U.S. Public Health Service standards or equivalent.

Experimentation, Human: Indicate approval by Institutional Review Board for study.

Materials Sharing: Contributors must share any materials and methodology necessary to verify conclusions of experiments and distribute freely to interested academic researchers any material needed to reproduce reported experiments.

SOURCE

October 1994, 81(4): 650.

Journal of Nuclear Medicine
Society of Nuclear Medicine

Stanley J. Goldsmith, MD
The Journal of Nuclear Medicine
136 Madison Ave.
New York, NY 10016-6760
Telephone: (212) 889-1905; Fax: (212) 889-6582

AIM AND SCOPE

The Journal of Nuclear Medicine publishes material of interest to the practitioners and scientists in the broad field of nuclear medicine. Proffered articles describing original laboratory or clinical investigations, case reports, techni-cal notes and letters to the editor will be considered for publication.

MANUSCRIPT SUBMISSION

Manuscripts, including illustrations and tables, must be original and not under consideration by another publication.

The *Journal* follows the "Uniform Requirements for Manuscripts Submitted to Biomedical Journals" (*Ann Intern Med* (1988; 108:258–265 and *Br Med J* 1988; 296:401–405). See also *Medical Style & Format: An International Manual for Authors, Editors, and Publishers* by Edward J. Huth, MD (Philadelphia: ISI Press; 1987).

Submit one original and three copies of manuscript and four copies of figures.

In a cover letter from corresponding author include following copyright disclosure statement: "Upon acceptance by The Journal of Nuclear Medicine, all copyright ownership for the article (title) is transferred to The Society of Nuclear Medicine. We, the undersigned co-authors of this article, have contributed significantly to and share in the responsibility for the release of any part or all of the material contained within the article noted above. The undersigned stipulate that the material submitted to The Journal of Nuclear Medicine is new, original and has not been submitted to another publication for concurrent consideration.

We also attest that any human and/or animal studies undertaken as part of the research from which this manuscript was derived are in compliance with regulations of our institution(s) and with generally accepted guidelines governing such work.

We further attest that we have herein disclosed any and all financial or other relationships which could be construed as a conflict of interest, and that all sources of financial support for this study have been disclosed and are indicated in the acknowledgment."

Statement must be signed by all co-authors. Copyright requirement does not apply to work prepared by United States government employees as part of their official duties.

In cover letter also state that manuscript has been seen and approved by all authors and give any additional information which may be helpful to the Editor. Acknowledge and include appropriate permission for any prior publication of any part of work. If color illustrations are included, state that author(s) is/are willing to assume cost of color separation and reproduction.

MANUSCRIPT FORMAT

Type manuscript on white bond paper, 8.5×11 in. with 1.5 in. margins. Type on one side of paper only, double-space every page. Begin each section on separate pages and in order: title page, abstract, text, acknowledgments, references, tables (each on a separate page), and legends. Number pages consecutively, beginning with title page. Type name of senior author and page number in upper right corner of each page.

Title Page: Include: concise and informative title (200 characters); short running head or footline (40 characters letters and spaces) identified at bottom of title page; complete byline, with first name, middle initial, and last name of each author and highest academic degree(s), up to ten authors may be cited; complete affiliation for each author, with department name(s) and institution(s) to which work should be attributed; disclaimers, if any; name, address, and telephone number of corresponding and reprint request author(s) or statement that reprints are not available from author. Note financial support.

Abstract: 350 words divided into: rationale (goals of investigation), methods (description of study subjects or experiments, animals and observational and analytical techniques), results (major findings), and principle conclusions. Except for rationale, use headings Methods, Results, Conclusions. Submit 3-5 key words with abstract.

Text: Divide text of Original Scientific and Methodology Articles into sections: Introduction, Materials and Methods, Results, Discussion, and Summary or Conclusion. Text of original, scientific papers, exclusive of abstract, legends, tables and references, should not exceed 5000 words.

Use generic names throughout text. Identify instruments and radiopharmaceuticals by manufacturer name and address in parentheses and describe procedures in sufficient detail to allow other investigators to reproduce results.

References: Cite in consecutive numerical order at first mention in text. Designate by underlined reference number in parentheses. Number references in a table or figure sequentially with those in text.

Type Reference list double-spaced, numbered consecutively as in text. Follow *Index Medicus* style for references and abbreviate journal names according to the *List of Journals Indexed in Index Medicus*. Do not use "unpublished observations" and "personal communications" as references; note written not verbal communications as such in text. References cited as "in press" must have been accepted. Verify accuracy of references against original documents.

JOURNALS

List all authors when six or less; for seven or more, list first three and et al:

Baumier PL, Krohn KA, Carrasquillo JA, et al. Melanoma localization in nude mice with monoclonal Fab against p97. *J Nucl Med* 1985; 26:1172–1179.

Weissmann HS, Badia J, Sugarman LA, Kluger L, Rosenblatt R, Freeman LM. Spectrum of ^{99m}Tc-IDA cholescintigraphic patterns in acute cholecystitis. *Radiology* 1981; 138:167–175.

BOOKS AND BOOK CHAPTERS

DeGroot LJ. Evaluation of thyroid function and thyroid disease. In: DeGroot LJ, Stanbury JB, eds. *The thyroid and its diseases*. 4th ed. New York: Wiley; 1975:196–248.

Dupont B. Bone marrow transplantation in severe combined immunodeficiency with an unrelated MLC compatible donor. In: White HJ, Smith R, eds. *Proceedings of the third annual meeting of the International Society of Experimental Hematology*: Houston: International Society for Experimental Hematology; 1971:44–46.

Tables: Type double-spaced on separate pages. Do not submit as photographs.

Make self-explanatory to supplement, not duplicate, text. Cite in consecutive numerical order in text. Number consecutively with an Arabic number following word TABLE. Give descriptive, brief titles typed centered in upper and lowercase letters. Place horizontal rules below title, column headings, and at end of table. Do not use vertical lines. Give each column a short heading.

Place explanatory matter in footnotes, not in heading. Use symbols, in sequence: *, †, ‡, §, ¶, **. In footnote explain all nonstandard abbreviations used. Identify statistical measures of variations, such as standard deviation and standard error of the mean. If data from another published source are used, obtain written permission from original publisher and acknowledge fully. If data from an unpublished source are used, obtain permission from principal investigator, and acknowledge fully.

Illustrations: Use to clarify and augment text. Inferior figures will be returned for correction or replacement.

Submit four complete sets of glossy illustrations, no smaller than 3.5 × 5 in. no larger than 8 × 10 in. Do not send original artwork. Submit glossy photographs of line drawings rendered professionally on white drawing paper in black India ink, with template or typeset lettering. Do not submit hand-drawn or typewritten art. Letters, numbers, and symbols (typeset or template) must be clear and of sufficient size to be legible after reduction.

Number each illustration and cite in consecutive order in text. Identify on a gummed label affix to back with figure number, part of figure (if more than one), senior author's name, and TOP.

Color illustrations are considered, author is responsible for charges. An estimate will be sent at production. Author approval is required. Submit four complete sets of glossy color photographs, not transparencies or polaroid prints.

Legends for Illustrations: Be concise; do not repeat text. Type double-spaced on a separate page. Cite each figure in consecutive numerical order in text. Number figures with an Arabic number following word FIGURE. Use letters to designate parts of illustrations (e.g., A, B, C); describe each part in legend. Identify and describe letter designations or arrows on illustrations.

Original (not previously published) illustrations are preferred; if illustrations have been published previously, obtain and submit written permission from publisher to reprint. Cite source of original material in references; include credit line in legend: (Reprinted by permission of Ref. X.)

Acknowledgments: Acknowledge persons or agencies contributing substantially to work, including grant support.

OTHER FORMATS

CASE REPORTS—1250 words, exclusive of abstract, legends, tables and references, and 3 figures, emphasizing nuclear medicine aspects and including methodology, data, and correlative studies.

LETTERS—Concern previously published material or matters of general interest. Be brief and to the point. Submit a diskette (3.5 or 5.25 in. only) with a word processer file, labelled with name of file, word processing software, operating environment (DOS, Windows), and platform (IBM, Macintosh). Include copyright disclose statement. Submit letters commenting on previously published articles within one year of publication. Include no more than 10 references.

OTHER

Abbreviations and Symbols: Except for units of measurement, use is discouraged. For information on proper medical abbreviations, consult *CBE Style Manual* (5th Ed, Bethesda, MD: Council of Biology Editors, 1983). Precede abbreviations by full word or name of item being abbreviated.

Reviewers: In cover letter, suggest individuals who could serve as reviewers.

Style: Do no use hyperbolic terms or phrases in title, abstract or body of text. Do not make qualitative claims as to superiority or performance of an idea or an instrument.

Units: List measurements in SI units. Older conventions may be used after SI units, in parentheses.

SOURCE

January 1994, 354(1): 193-194.

Journal of Nutrition
American Institute of Nutrition

Editor, *The Journal of Nutrition*
University of Illinois, College of Medicine
190 Medical Sciences Building
506 South Mathews Avenue
Urbana, IL 61801

MANUSCRIPT SUBMISSION

Base manuscripts on original, unpublished research. Although data may have been reported in part or in abstract form, full reports of research may not have been accepted or be under consideration by another journal. Reports of preliminary research are not acceptable. Do not release data accepted for publication before publication.

Submit five complete copies of manuscript with a letter of submission; put labeled original figures in a protective envelope; and a manuscript submission and copyright release form.

Letter of submission includes: information about previous or concurrent publication of any part of work; statement of financial or other contractual agreements that may cause conflicts of interest; statement that paper has been read and approved by all authors; assurance of written permission from any person mentioned in a personal communication or an acknowledgment; and name, address, telephone and fax numbers of corresponding author.

Photocopy transfer of copyright form from January issue of the *Journal*. An original signature is required from each author. Copies of completed form may be signed by each author independently. Include signed transfer of copyright forms with submitted manuscripts

MANUSCRIPT FORMAT

Type on white paper. Double-space everything, including abstract, footnotes, tables, literature cited, and figure legends. Leave 30 mm for left and right margins. Do not justify right margin. Do not hyphenate words at end of lines. Insert line numbers in left margin. For computer-generated typescripts, print letter-quality. Number each page at top.

Arrange manuscript in order beginning each section on an new page: Title page, Abstract and key words, Text and acknowledgments, Literature cited, Text footnotes, Tables with titles and footnotes, one table per page, Figure legends consolidated on one page, and labeled Figures and illustrations without legends or page numbers.

Title Page: Type double-spaced: complete title of manuscript in upper-and lowercase letters; first names, middle initials and last names of all authors in upper- and lowercase letters; institutions, including departments, where research was performed (Give address of each, do not include street address, include zip or postal code; indicate which authors are affiliated with which institutions using symbols: *, †, **, ‡); name, complete mailing address, telephone and fax numbers of corresponding author; and shortened form of title (48 characters, including spaces).

Title should be a descriptive and concise declarative sentence naming species studied. do not begin with "Effects of ...." Do not submit numbered titles for series of papers.

Abstract: 200 words, one paragraph. Give objective of study, a concise description of plan or design and state results and conclusions. Summarize paper. Avoid abbreviations.

List five key words for subject classification in annual index. Include species studied.

Text: Make clear, concise and understandable to readers from other fields. Use grammatically correct language without euphemisms or laboratory jargon. Use declarative sentences; avoid passive voice. Use first person when appropriate. Use past tense to refer to current study. Use present tense to refer to existing knowledge or prevailing concepts and to state conclusions. Differentiate previous knowledge from new contributions.

Do not italicize or underline such commonly used phrases as "et al.," "in vivo" or "in vitro." Italicize or underline foreign words and phrases not commonly found in an English-language dictionary. When appropriate, italicize symbols for variables and constants.

Divide research articles into sections in order; center uppercase headings: Introduction (no heading), Materials and Methods, Results, and Discussion. Variation is acceptable; Materials and Methods section may be separated; Results and Discussion sections may be combined.

Materials and Methods: Document methods and materials used sufficiently to permit replication of research. Describe control and experimental subjects; give age, weight, sex, race and, for animals, breed or strain. Give name, city and state or country of suppliers of experimental animals. State source of specialized materials, diets, chemicals and instruments and other equipment, with model or catalog numbers. Include name, city and state or country of the supplier parenthetically. State purity of chemicals used.

Acknowledgments: Acknowledge technical assistance and advice in a section at end of text. Obtain written permission to mention each person named; include assurance that permission has been given in letter of submission. List financial support in a footnote to title, not in acknowledgments.

Literature Cited: Verify accuracy of literature citations. Limit number to about 30 for a full-length research article. Do not include personal communications, articles not yet accepted and unpublished data in Literature Cited section; include parenthetically in text. Personal communications must be written; obtain permission to use in writing. Include affiliation of person providing communication in text. Avoid using abstracts as references; identify as abstracts when used.

Cite articles accepted for publication but not published as "in press." Include references to oral reports, bulletins, theses and other materials in Literature Cited section.

Abbreviate journal names according *Serial Sources for the Biosis Database* (BIOSIS, Philadelphia, PA). If journal is not listed, abbreviate according to rules of the American National Standards Institute (*American National Standard for Information Sciences-Abbreviation of Titles for Publication*. ANSI Z39.5. ANSI, New York) or spell journal name in full. Do not italicize journal or book titles.

In text, cite references by author and year of publication. When citing an article with more than two authors, use "et al." When citing multiple references together, keep in alphabetical rather than chronological order. Use a comma between multiple citations in text and a space only between authors and dates. For multiple citations in one year from the same authors, use 1990a, 1990b, etc, including them in order of appearance in Literature Cited section. If two or more references from one year have different authors with same last name, use initials. In Literature Cited section, cite references alphabetically, without numbering. Do not use "et al." Include all authors' names for every article.

JOURNALS

Jones, A. C., Brown, B.C.D. & Little, C. (1992) Title of the article. Journal 11: 111–113.

BOOKS

Brown, B.C.D. & Jones, A. (1992) Article title. In: Book Title (Little, C., ed.), vol. 2, pp 1–20. Publisher, city and state or country.

ABSTRACTS

Jones, A. C., Brown, B.C.D. & Little, C. (1992) Title of the abstract. Journal 11: 111 (abs.).

THESES

Jones, A. C. (1992) Title of the Thesis. Doctoral thesis, university name, city and state or country.

IN PRESS ARTICLES

Jones, A. C., Brown, B.C.D. & Little, C. (probable publication year) Title of the article. Journal (in press).

Text Footnotes: Number consecutively, including those on title page. Number table footnotes separately for each table. Compile text footnotes on a separate page

Use title footnotes for: earlier presentation at symposia or meetings (give name, date and location of presentation; if abstract was published, cite all authors, year, title, journal, volume and page number); acknowledgments of financial support of research; institutional approval of paper for publication; and identification in an institutional series.

Use footnotes to an author's name to: indicate author(s) for correspondence and reprint requests; and give current address of an author if different from affiliation given.

Use footnotes to text for an abbreviation note or to give composition of diets and vitamin and mineral mixtures not presented in a table.

Tables: Use only if best way to present material. Design to be interpretable without reference to text. See current issues for style.

Type tables and footnotes double-spaced. If more than one page, indicate "continued" at bottom of page and "continued, Table *n*" at top of next. Use concise titles that fully describe contents. Give methods information in footnotes. Include explanatory material in a footnote rather than title.

Show statistics of variability (e.g., SD, pooled SEM) and significance of differences among data. Indicate units of measure. Place two lines above first value in each column or center over all columns to which unit applies. Do not put units in title or footnote. Cite tables sequentially in text. Boldface first reference to each table. Include references cited in tables in Literature Cited section.

Acknowledge tables adapted or reproduced verbatim from another source in a footnote; submit written proof that copyright bearer has granted permission for use.

Figures: Use if best way to present material. See current issues for style. All figures are reproduced at one-column width (9 cm) unless complexity demands two-columns (18.5 cm). Figures not complex enough for two-columns but that would be illegible after reduction to one column will be returned for revision. Compile legends for all figures and type double-spaced on a separate page (not on figures themselves). Put titles in legends not on figures. Legends should contain enough detail, including statistics, to ensure that figure is interpretable without reference to text. Affix a label to back of each figure with manuscript title, author's name and figure number, and clearly indicate TOP. Do not write on front or back of original figures. Figures without appropriate labels will be returned. Abbreviations, units and symbols must conform to *Journal* style. Do not apply transfer lettering to surface of a glossy print. Freehand and typewritten lettering are not acceptable.

Submit original, high quality line drawings (laser printed or glossy print) in a protective envelope. Include photocopies of each line drawing with manuscript copies. Unacceptable line drawings will be returned.

Submit six glossy prints of each black-and-white photograph, attach one to each manuscript copy; put one in a protective envelope. Give photomicrographs internal scale markers.

Authors bear cost of publication of color photographs.

Obtain cost information from technical editors. Submit six glossy prints of each figure

OTHER FORMATS

BIOGRAPHIES AND HISTORICAL PERSPECTIVES—Biographies are invited by biographical and historical editors. Suggest subjects and authors. Submit articles on history of nutrition without invitation. Papers will be reviewed.

BOOK REVIEWS—By invitation of editor.

CRITICAL REVIEWS AND COMMENTARIES—Critical reviews and commentaries on significant developments in nutritional science. Contact editor with suggestions of topics and possible authors.

ISSUES AND OPINIONS—Short essays presenting scientific viewpoints on issues in nutrition are invited by editor. Suggest topics and possible authors.

LETTERS TO THE EDITOR—Three double-spaced pages, including references. Constructive comments on recently published *Journal of Nutrition* articles and other issues. Author is given an opportunity for rebuttal.

OTHER—Extensive reports of research, monographs, compendia, proceedings of symposia, etc., are considered. Some may be published as supplements. Contact American Institute of Nutrition (AIN) headquarters, 9650 Rockville Pike, Bethesda, MD 20814, (301) 530-7050.

RAPID COMMUNICATIONS—Must contain timely information and meet all standard requirements.

RESEARCH ARTICLES—Concise reports of original research and manuscripts describing concepts about nutrition developed from published literature. Manuscripts describing new methods will normally not be considered unless they also contain data obtained using methods.

OTHER

Abbreviations: Avoid unless absolutely necessary. See Table 2 (*J. Nutr.* 1994, 124(1): 6) for an abridged list of abbreviations that may be used without definition. Other standard abbreviations are listed in *CBE Style Manual: A Guide for Authors, Editors, and Publishers in the Biological Sciences*, (CBE Style Manual Committee, 1983, 5th ed, Council of Biology Editors, Chicago, IL).

Abbreviate cumbersome terms when necessary after first being defined in text.

Avoid all but the most standard abbreviations in abstract. Use as few as possible (<5). Do not abbreviate simple terms such as cholesterol, dry matter and triglycerides. Do not begin sentences with abbreviations or use in titles or subtitles. Give abbreviations in parentheses after full term at first use in text. If three or more abbreviations in text, include an abbreviation footnote associated with first abbreviation defined in text. Include all abbreviations used in text section in alphabetical order, followed by definitions. Define abbreviations used in tables and figures in legend or a footnote for each table or figure.

Use standard abbreviations for SI prefixes (see Table 3, *J. Nutr.* 1994, 124(1): 7 and units of measure (see Table 4, *J. Nutr.* 1994, 124(1): 8). Do not use period or pluralize abbreviations and unit symbols.

Auxiliary Publication: To conserve space, deposit unusually lengthy descriptions of experimental procedures, extensive data, extra figures and other important supplementary information with American Society for Information Science, National Auxiliary Publications Service through editorial office. Submit material to be deposited with each copy of manuscript, clearly labeled as "material for deposition with NAPS." A footnote will be included in paper to give information on availability of photoprint or microfiche copies.

Diets: See Nelson, Margetts and Black, "Checklist for the methods section of dietary investigations" (*Br. J. Nutr.* 1993; 69:935-940 and *Metabolism* 42:258-266) for guidelines and a checklist for investigations of human dietary intake. Publish questionaires used in dietary intake studies, cite as reference, or submit with manuscript.

Detail composition of control and experimental diets, means of validation, and primary and interacting nutrients pertinent to study. When composition of a diet is published for the first time in *The Journal of Nutrition*, provide complete information on all components. Present in tabular form. Summarize composition of diets previously described in *Journal*; give literature citation. Give proximate composition of closed formula diets as amounts of protein, energy, fat and fiber.

Express components in mass concentrations (g/kg diet) or substance concentrations (μmol/kg diet) rather than percentages.

Refer to "Guidelines for Describing Diets for Experimental Animals" (*J. Nutr.* 1987; 117: 16–17), "Report of the American Institute of Nutrition ad hoc committee on standards for nutritional studies." (*J. Nutr.* 1977; 107: 1340–1348) and "Second report of the ad hoc committee on standards for nutritional studies." (*J. Nutr.* 1980; 110: 1726) for information concerning experimental diets. Refer to Baker D. H. "Construction of assay diets for sulphur-containing amino acids." (*Methods Enzymol.* 1987; 143: 297–307) for a discussion of purified diet formulation.

Electronic Submission: Submit computer diskette after all requested revisions have been made. Requirements and acceptable formats will be sent after review.

Experimentation, Animal: Include a statement that protocol was reviewed and approved by appropriate committee or complied with *Guide for the Care and Use of Laboratory Animals* (National Research Council, 1985, NIH publication no. 85-23 (rev.). Washington, D.C.). If animals were killed, state method. Do not use such euphemisms as "sacrificed" or "euthanized."

Experimentation, Human: Include a statement that protocol was reviewed and approved by appropriate institutional committee or complied with Helsinki Declaration of 1975 as revised in 1983.

Nomenclature: Chemical and biochemical terms and abbreviations conform to recommended usage of the International Union of Pure and Applied Chemistry (IUPAC) and the International Union of Biochemistry and Molecular Biology (IUBMB). (4th ed, 1989, Biochemical Society Book Depot, Essex U.K.).

Identify enzymes by their EC numbers and systematic names conforming to IUBMB enzyme nomenclature policy (IUBMB 1992, *Enzyme Nomenclature, Recommendations 1992*, Academic Press, Orlando, FL).

For names for vitamins and related compounds, follow AIN nomenclature policy "Nomenclature policy: generic descriptions and trivial names for vitamins and related compounds" (*J. Nutr.* 1990;120: 12–19).

Abbreviations for amino acids follow AIN nomenclature policy "Nomenclature policy: abbreviated designations of amino acids" (*J. Nutr.* 1987;117: 15). Use in tables and figures, not in text.

For designations for unsaturated fatty acids comply with IUPAC-IUBMB nomenclature. Use *(n-x)* system, where *x* represents number of carbon atoms from the end of the acyl chain to the nearest double bond. Include parentheses, even when referring to "*(n-x)* fatty acids" in text.

Use complete names of chemicals with complex formulas in text. Use formulas in tables and figures.

Numbers: Use Arabic rather than Roman numerals and cardinal rather than ordinal numbers throughout, including references to tables and figures. For values less than 1, include a zero before decimal point.

Page Charges: $60 per page. Upon receipt of proofs, authors lacking funding may request a waiver signed by an appropriate institutional official verifying that no funds are available for paying page charges.

Statistical Analyses: In Materials and Methods section, describe statistical tests and indicate probability level (*P*) at which differences were considered significant. If several tests were used, indicate specifically which groups or treatments were analyzed with each. Indicate whether data were transformed before analysis. Cite references for all analyses other than ANOVA, Student's *t* test or chi-square test and specify any statistical computer programs used. Give version of software and list supplier parenthetically in text with city and state or country.

In tables and figures, present results of analysis in the body or use superscripts to indicate significant differences and define superscripts in a table footnote or figure legend. Provide appropriate statistics of variability. Standard ANOVA methodology assumes a homogeneous variance, and, except in instances of unequal numbers (*n*) per treatment, use a pooled SEM rather than individual SEM (i.e., per treatment) to describe variability for response criteria. If error variance is tested and found to be heterogeneous, log transform data before ANOVA. When experimental treatment design involves built-in comparisons, use orthogonal single df (with assessment of interactions if appropriate) in preference to a range test. When quantitatively expressing a dependent variable (*y*) as a function of an independent variable (*x*), describe regression equation used and provide an indication of fit (e.g., r^2). For a discussion of variability calculations and curve-fitting procedures, see Baker, D. H. "Problems and pitfalls in animal experiments designed to establish dietary requirements for essential nutrients." (*J. Nutr.* 1986; 116: 2339–2349).

Style: Do not use phrases "fed ad libitum" or "ad libitum-fed." It is not feeding but consumption that is ad libitum. Use phrases such as "with free access to" or "were given free access to."

Do not use "fast," "fasted" or "fasting" to indicate that food was withheld. Restrict use of "fast" to studies involving freely complying human subjects. Use phrases such as "unfed," "food-deprived" and "food was withheld."

Use no more than one solidus (slash or slanting line) for "per" in a symbol expression. To abbreviate "0.6 g Ca per kg per day" use either 0.6 g Ca/(kg·d) or 0.6 g Ca·kg-1·d-1. Use solidus only for "per" except in equations. Do not use to indicate groupings; use a comma or a dash. Use a colon for ratios. Avoid phrase "and/or," replace by either "and" or "or;" solidus may be used for phrase when unavoidable.

Other points of usage for scientific writing are covered in the *CBE Style Manual* (CBE, 1983).

Units: Conform to *le Système Internationale d'Unités* (SI): see American Society for Testing and Materials "ASTM Standard for Metric Practice: E380-88." (1988. ASTM, Philadelphia, PA), Page, C. H. & Vigoureaux, P., eds. (1986) "The International System of Units (SI)." (1986, National Bureau of Standards Special Publication no. 330 (rev.), Washington, D.C.), and Young, D. S. "Implementation of SI units for clinical laboratory data, style specifications and conversion tables" (*Ann. Intern. Med.* 1987; 106: 114–129. Reprinted, *J. Nutr.* 120: 20–35).

Use metric system and Celsius scale (C). Express concentrations on a molar basis. Express energy values in joules. Report radioactivity as becquerels, not curies. Reporting enzyme activity as katals (kat, mol/ s) is optional. State previously reported values (e.g., in discussion) in SI units, even if originally reported differently.

Do not use such non-SI terms as calorie, cubic centimeter, degrees Fahrenheit, disintegrations per minute, foot, inch, micron, micromicrogram, ounce, quart and parts per million.

Do not use prefixes other than milli-, micro-, nano- or pico- with liter, e.g., 100 mL not 1 dL. Except for diet composition, avoid mass concentrations, e.g., g/L, mg%. Convert to substance concentration, e.g., mol/L. Make denominator L, not mL, 100 mL, etc. Do not use *M*, m*M*, *N*, etc.

SOURCE

January 1994, 124(1): 2-12.
"Guide for Authors" published in each January issue.

Journal of Oral and Maxillofacial Surgery
American Society of Oral and Maxillofacial Surgeons
W. B. Saunders

Dr. Daniel M. Laskin
Department of Oral and Maxillofacial Surgery
Medical College of Virginia
Box 566, MCV Station
Richmond, VA 23298-0566
Telephone: (804) 786-0602; Fax: (804) 786-0753

AIM AND SCOPE

The *Journal* publishes articles reflecting a wide range of opinions and techniques, provided they are original, contribute new information, and meet ordinary standards of scientific thought, rational procedure, and literary presentation.

MANUSCRIPT SUBMISSION

Submit two copies, original and a duplicate, with all material (text, references, tables, legends) double-spaced on 8.5 × 11 in. paper, one side only, with a 1 in. margin all around. Original articles are considered on the condition that they have not been published, or submitted for publication, elsewhere. Provide a strong justification for more than four authors.

Disclose any commercial associations that might create a conflict of interest. Indicate sources of external funds supporting work in a footnote, with corporate affiliations of authors. In cover letter, inform Editor about pertinent consultancies, stock ownership or other equity interests, or patent licensing arrangements. Information will remain confidential during review and will not influence editorial decision. If manuscript is accepted, Editor will discuss with authors how best to disclose relevant information.

Include statement in cover letter: "In consideration of the *Journal of Oral and Maxillofacial Surgery* taking action in reviewing and editing my (our) submission, the author(s) undersigned hereby transfer(s), assign(s), or otherwise convey(s) all copyright ownership to the American Association of Oral and Maxillofacial Surgeons in the event that such work is published in the *Journal of Oral and Maxillofacial Surgery*." Obtain permission of author and publisher for direct use of material (text, photos, drawings) under copyright.

Obtain waivers for use of full-face photographs unless eyes are masked to prevent identification. Waiver forms are available from the Editor-in-Chief.

MANUSCRIPT FORMAT

On a separate title page list title, authors, and their affiliations. Make titles descriptive and concise. Include a structured abstract (introductory summary) (250 words) with clinical or scientific articles. Divide into sections labeled Objective, Materials (or Patients) and Methods, Results, and Conclusions. No abstracts for case reports.

References: Type double-spaced on a separate sheet. Cite references in numerical order in text. Bibliographies and reading lists are not used. Do not exceed 30 in major contributions; use fewer in shorter articles.

JOURNALS

Give author's name, article title, journal name as abbreviated in *Index Medicus*, volume, pagination, and year.

1. Regezi JA, Batsakis JG, Courtney RM: Granular cell tumors of the head and neck. J Oral Surg 37: 402,1979

BOOKS

Give author's name, book title, location and name of publisher, and year of publication (give page numbers for direct quotations).

1. Skinner EW, Phillips RW: Science of Dental Materials (ed 5). Philadelphia, PA, Saunders, 1960, p 246

Illustrations: Includes all material that cannot be set in type, such as photographs, line drawings, graphs, and charts. Submit two copies. Number and cite in text. X-ray films are not acceptable; submit radiographs as glossy prints.

Photographs—Handle carefully, do not bend, fold, or use paper clips. Submit photographs at least 4 × 6 in., as unmounted and untrimmed, high quality, sharp, black-and-white glossy prints. Submit colored illustrations if their use contributes significantly to value of article. A charge may be made. On back of each photograph, place a label with author's name, figure number, and TOP indicated lightly in pencil.

Drawings—Have figures, charts, and graphs drawn professionally. Make lettering large and clear. Submit glossy black and white prints of drawings, not original artwork.

Legends: Type double-spaced as a group on a separate sheet, not on illustrations or on individual sheets attached to illustrations. For photomicrographs, specify magnification and stain.

Tables: Organize logically to supplement article. Type on separate sheets numbered consecutively. Put title and footnotes on same page. Do not draw vertical rules. Do not submit as glossy prints.

OTHER

Page Charges: Make special arrangements with Editor for publishing extensive illustrative or tabular material, formulas, or four-color illustrations.

SOURCE

January 1994, 52(7): front matter.
Notice to contributors appears in January and July issues.

Journal of Parasitology
American Society of Parasitologists

Gerald W. Esch
Department of Biology
Wake Forest University
P.O. Box 7629
Winston-Salem, NC 27109

AIM AND SCOPE

Manuscripts are accepted from investigators in any country whether or not they are members of the Society. The *Journal* publishes official business of the ASP and results of new, original research, primarily on parasitic animals.

MANUSCRIPT SUBMISSION

Type on one side only of good quality, white paper. Thin onion skin or rice parchment papers are not acceptable. Type all parts double-spaced (no more than 3 lines/25 mm) with 25 mm margins. Use 10 point (elite) type; do not photoreduce. Do not use proportional spacing or hyphenation; do not justify right margin. Use a single font; underline words to be set in italic type. Type authors' names in literature cited section with capitals for initials and first letter and lowercase for all other letters (these names are printed in large and small capital letters). Submit original typescript and 3 copies, plus 4 sets of figures.

Manuscripts are received with the understanding that: all authors have approved submission; results or ideas are original; work has not been published previously; paper is not under consideration for publication elsewhere and will not be submitted elsewhere unless rejected or withdrawn by written notification; if accepted for publication or published, article, or portions thereof, will not be published elsewhere unless consent is obtained in writing from the editor; reproduction and fair use of articles in *The Journal of Parasitology* are permitted in accordance with the United States Copyright Revision Law (PL94-533), provided intended use is for nonprofit educational purposes (other use requires consent and fees where appropriate); and obligation for page charges and redactory fees is accepted by authors.

MANUSCRIPT FORMAT

Organize in following format and sequence; number all pages, beginning with running head, consecutively.

Running Head: Provide last names of authors (use et al. for more than 2) and a shortened title on a separate page. Entire running head may not exceed 60 characters and spaces, e.g., RH: JONES ET AL.—LIFE CYCLE OF H. DIMINUTA

Title Page: On a separate page, give title of article, names of authors, and address of first author. At bottom of page, give name, address, and telephone number of corresponding author. Make title short and descriptive. Avoid "empty words" such as preliminary studies on ... and biology or ecology of.... Do not use author and date citations with scientific names. Spell out numbers less than 11; numbers indicating papers in a series are not accepted. Give current addresses and addresses for authors, if different from first author, as footnotes.

Abstract: 200 words on a separate page. Be factual, as opposed to indicative. Outline objective, methods used, conclusions, and significance of study. Head with word abstract paragraph indented; type in all capitals; end with a colon. Text is run in after colon, is not subdivided, and does not contain literature citations.

Index Descriptors: On a separate page, provide 3-5 key words or short phrases for annual index. Include principal species and topic studied.

Introduction: Start on a separate, unheaded, page. Establish context of paper by stating general field of interest, presenting findings of others that will be challenged or developed, and specifying specific question addressed. Limit accounts of previous work to minimum necessary for an appropriate perspective. Do not subdivide.

Materials and Methods: Give sufficient information to permit repetition of study. Indicate methods and apparatus used, mention specific brand names and models only if significant. Reference, but do not detail previously published or standard techniques. Give generic descriptions of unusual compounds.

Type primary heading in all capital letters centered on page. Do not number or punctuate heading. Put second-level headings on a separate line at left margin. Capitalize only initial letter of first word except for proper nouns. Do not number or punctuate. Indent third-level headings for a paragraph, underline, and end with a colon. Capitalize only initial letter of first word except for proper nouns. Run text in following head. Do not subdivide further. If materials and methods section is short, do not subdivide; do not provide headings for single paragraph subsections.

Results: A concise account of new information. Use tables and figures as appropriate; do not repeat information in text. Avoid detailing methods and interpreting results. Subdivide and head as in materials and methods section.

In taxonomic papers, replace with section headed description, redescription, revision, etc. Primary heading is followed by scientific name of taxon studied, synonyms, and reference to figures, each as a separate line. Text of description follows as a new paragraph, followed with a taxonomic summary section, headed as a second-level heading. Taxonomic summary section is a listing of the type host, other hosts, site, locality, specimens deposited (museum names and accession numbers are required). and other relevant data. Each topic is headed as a third-level heading. Follow taxonomic summary with a remarks section, headed as a second-level heading. Remarks replace discussion section; give comparisons to similar taxa. Sequence of subsections is repeated for each taxon. If description section does not comprise all results and discussion, incorporate format outlined into usual results section. Museum accession numbers for appropriate type material (new taxa) and for voucher specimens (surveys) ordinarily are required.

Discussion: Interpret and explain relationship of results to existing knowledge. Emphasize important new findings; identify new hypotheses. Support conclusions by fact or data. Use primary heading and subdivisions, if needed.

Acknowledgments: Be concise. Consult colleagues before acknowledging their assistance. Use primary heading as described for materials and methods section. Do not use subdivisions.

Literature Cited: Section has a primary heading as described for materials and methods.

Arrange citations alphabetically. All references cited in text must appear in literature cited section, and vice versa. Do not cite unpublished studies or reports, i.e., a volume and page number must be available for serials and a publisher, city, and page numbers for books. Cite work as "in press" if proof has been produced. If necessary, document a statement by "personal communication"; provide a copy of page signed by person cited. Cite as: (Smith, X. Y., 1989, pers. comm.); do not include in literature cited section.

Style in Text:

(Allen, 1989)

(Allen and Smith. 1989)

(Allen et al., 1989)

(Jones, 1987; Allen, 1989)—chronological

(Jones, 1987; Allen, 1989; Smith, 1989)—chronological and alphabetical within year (Jones, 1987, 1988a, 1988b, 1989)

Et al. is used for more than 2 authors providing reference can be determined from literature cited section.

Style in the Literature Cited Section:

JOURNAL

Nollen, P. M. 1990. Chemosensitivity of Philophthalmus megalurus (Trematoda) miracidia. Journal of Parasitology **76**: 439–440.

Edwards, D. D., and A. O. Bush. 1989. Helminth communities in avocets: Importance of the compound community. Journal of Parasitology **75**: 225–238.

BOOK

Schmidt, G. D., and L. S. Roberts. 1989. Foundations of parasitology, 4th ed. Times Mirror/Mosby College Publishing Company, St. Louis, Missouri, 750 p.

CHAPTER IN EDITED BOOK

Nesheim, M. C. 1989. Ascariasis and human nutrition. In Ascariasis and its prevention and control, D. W. T. Crompton, M. C. Nesheim, and Z. S. Pawlowski (eds.). Taylor and Francis, London, U.K., p. 87-100.

THESIS OR DISSERTATION

Monks, W. S. 1987. Relationship between the density of Moniliformis moniliformis and distribution within the definitive host population. M.S. Thesis. University of Nebraska-Lincoln, Lincoln, Nebraska, 64 p.

Do not abbreviate titles of serial publications. Put spaces between initials.

Footnotes: Use only for title page to indicate authors' addresses. Place on a separate page with a primary heading. Type table footnotes directly under table.

Tables: Use only to present data that cannot be incorporated into text. Do not present values from statistical tests as tables; state tests employed and probability accepted for significance in materials and methods section; indicate significant differences in tables by footnotes or in text by a statement.

Design to fit 1 or 2 columns, only rarely to fit height of a printed page. If width does not fit height of a typed page, it is too wide. Continue on following pages to accommodate length; pages may not be taped together, photoreduced, single spaced, oversized. or otherwise modified to contain more material.

Number with Roman numerals in a continuous series and reference, in sequence, in text. Type captions in Roman on page with table; underline to indicate italics. Spell out species at first use in each caption. Give all columns headings, capitalize first letter of first word and proper noun.

Avoid horizontal lines in body; do not use vertical lines or vertical symbols spanning more than 1 line of type. If such symbols are necessary, prepare table as a line drawing and treat as a figure. Do not use letters and numbers as superscripts or subscripts. Use in sequence: *, †, ‡, §, ||, #, ¶, **, ‡‡.

Figure Captions: Give each figure or plate of figures a caption. Write in paragraph style, beginning with word "Figure." Use Roman type, underline words requiring italic type. For plates, a summary statement should precede specific explanation of each figure. Do not repeat information in summary statement for each figure. Spell out species names in full at first use in each caption. Explain all abbreviations; indicate value of lines or bars used to show size (unless value is shown directly on figure). Do not indicate size by magnification in caption because figure might not be printed at size calculated.

Figures: Number consecutively in sequence mentioned in text. Do not abbreviate nonparenthetical references to figures in text; abbreviate references to figures in parentheses in text. Define all symbols, when possible by a key within body of figure. Style, including form of abbreviation, must be that used in *Journal*. Use the following symbols: ○ ❑ △ ▽ ◇ ⊖ ⊕ ⊗ ◑ ◐ ● ■ ▲ ▼ ◀ ▶ ◆. Others require artwork and additional expense billed to author. Do not label figures freehand.

Use figures singly or grouped in a plate. Mount originals on illustration board with a 25 mm margin on all sides. Do not combine photographs and line drawings in a single plate. If necessary, additional expense is billed to author. Identify figures on back by author name and figure number; indicate TOP. Do not number single figures on front; each figure in a plate must include a number, applied directly on figure without an added background. Tightly butt figures in a plate without space or masking between.

Figures and plates are printed in 1 (67 mm wide) or 2 (138 mm wide) columns. Length may be up to 204 mm; make shorter to allow room beneath for caption. Prepare figures and plates in a space proportional to printed dimensions and to withstand reduction to them.

Do not submit original line drawings, unless requested. Provide a clear, sharp, offset print or photograph for printing and good photocopies (3) or photographs for review. Photocopies of photographs are not acceptable. If figures are not mounted, print or photocopy on 8.5 × 11 in. paper.

OTHER FORMATS

BOOK REVIEWS—Books having a broad interest to Society membership are reviewed by invitation.

Number all pages, beginning with title page, consecutively. On a separate title page, give title of book being reviewed in style:

Toxoplasmosis of Animals and Man, by J. P. Dubey and C. P. Beattie. CRC Press. Boca Raton, Florida. 1988. 220 p. Hardcover $124.95.

Capitalize first letter of principal words; do not use a comma between authors when only 2. Substitute "edited by" for "by" as appropriate. Do not subdivide text. If literature must be cited, a headed literature cited section follows text in style described for articles. Do not use figures and tables. Name and address of author follow text or, if present, literature cited section. Style is as described for research notes.

CRITICAL COMMENTS—Correct errors of published fact, provide alternative interpretations of published data, or present new theories based on published information.

Number all pages, beginning with title page, consecutively. On a separate page, give title of article. At bottom of page, give name and address of corresponding author. Write text without subdivision. Literature citations are as for articles. Include acknowledgments as an unheaded final paragraph. If literature citations are used in text, literature cited section is as described for articles. Names and addresses of authors follow text or, if present, literature cited section. Style is as described for research notes. When present, Tables, Figure Captions, and Figures are as described for articles.

RESEARCH NOTES—Discrete, definitive information (as opposed to preliminary results) that does not lend itself to inclusion in a typical, more comprehensive article. Present a new or modified technique if it is not being used in ongoing studies. Ordinarily techniques are incorporated into materials and methods section of regular articles.

Number all pages, including title page, consecutively. On a separate title page, give title of article. On a separate line, give names of authors. On a third line (and following lines, if necessary) give addresses of authors, joined by semicolons, matched to authors other than first by symbols in sequence used for tables. At bottom of page, give name and address of corresponding author. Titles should be as described for articles. Provide an abstract as described for articles. Provide index descriptors as for articles. Write text without sections. Give acknowledgments, without heading, as last paragraph. Cite literature in text as described for articles. Do Literature Cited, Tables, Figure Captions, and Figures in form and sequence described for articles.

REVIEW ARTICLES—Invited only; do not submit unsolicited reviews. Suggest topics to editor or editorial board.

Use form described for article; other sections may be used in place of materials and methods, results, and discussion sections. Restrict headings to major headings and no more than 2 sublevels. Obtain and submit appropriate permissions for use of tabular data or figures from work of others.

OTHER

Acronyms and Abbreviations: At first use, place acronyms in parentheses following name written out in full. At subsequent use, use acronym alone. Acronyms may begin sentences. Do not begin sentences with abbreviations. Use abbreviations recommended in *Council of Biology Editors Style Manual*. Use all International System of Measurement (SI) metric unit abbreviations. Common CBE and SI abbreviations include the following (same abbreviation is used for plural form):

wk	week	p.	page
hr	hour; use 0–2400 hr for time	ad lib.	ad libitum
sec	second	U.S.A.	as a noun
min	minute	U.S.	as an adjective
mo	month	sp. gr.	specific gravity
day	not abbreviated	t	test
n. sp.	new species	U	test
n. gen.	new genus	P	probability
L	liter; but ml for milliliter	$\bar{x}$	arithmetic mean

g	gravity; not $\times g$	r	correlation coefficient
RH	relative humidity	n	sample size
p.o.	per os	SD	standard deviation of the mean
s.c.	subcutaneous	SE	standard error of the mean
i.pl.	intrapleural	df	degrees of freedom
i.p.	intraperitoneal	NS	not significant
PI	postinoculation		

BASIC SI UNITS

Meter	m	Ampere	A
Kilogram	kg	Volt	V
Second	sec	Mole	mole

PREFIXES FOR SI UNITS

10^{-1}	deci	d	10^{1}	deca	da
10^{-2}	centi	c	10^{2}	hecto	h
10^{-3}	milli	m	10^{3}	kilo	k
10^{-6}	micro	μm	10^{6}	mega	M
10^{-9}	nano	n	10^{9}	giga	G
10^{-12}	pico	p	10^{12}	tera	T

Experimentation, Animal: The ASP conforms to the "U.S. Government Principles for the Utilization and Care of Vertebrate Animals Used in Testing, Research and Training." Work involving vertebrate animals must have been conducted within guidelines adapted from a statement by The American Association for Laboratory Animal Science (1989, *Laboratory Animal Science* **39:** 267).

Transportation, care, and use of animals for research and teaching must conform with appropriate national guidelines (in the U.S.A., the Animal Welfare Act) and other applicable laws, guidelines, and policies. Refer to *Guide for the Care and Use of Laboratory Animals* (U.S. DHEW Publication Number [NIH] 86-23, as revised in 1985 or subsequently). Design and conduct experiments with full consideration given for animal's relevance to human or animal health, acquisition of knowledge, and/or welfare of society. Select animal species appropriate for results expected; use minimum number justified by sound statistical analysis. Experimental and maintenance procedures must avoid creating conditions that would lead to animal discomfort, distress, or pain, consistent with sound scientific practices. Use appropriate anesthesia if animals are to be subjected to momentary distress or pain. Do not conduct painful experiments on unanesthetized animals that have been paralyzed by chemicals or other procedures. Kill animals used in experiments that cause chronic pain or distress soon as experiments are concluded or, if possible, during experiment. Veterinary care for laboratory animals is essential. Keep laboratory animals in conditions appropriate for species and under conditions that contribute to their health and comfort. Staff must be well trained for the conduct of experiments on living animals.When exceptions to these principles are required, have decisions regarding animal use made by appropriate institutional animal care and use committee.

Obtain animals from natural populations in accordance with regulations and policies of appropriate national or state agencies.

Mathematical and Chemical Notations: Write mathematical equations so that they can be set in 1 line of type. When 1 unit appears in a denominator use solidus; for 2 or more units in a denominator, use negative exponents. If equation cannot fit 1 line, draft and reproduce as a figure. Handle chemical structures the same way.

Nomenclature: Write out full binomen at first use of a species name. At subsequent use, abbreviate generic component to first letter, except at beginning of sentence. Do not use author and date citations for scientific names in nonsystematic papers. In systematic papers use author and date citations the first time a taxon is mentioned in abstract and text but not subsequently except as described for tables and figures. Follow International Code of Zoological Nomenclature; be consistent. Author and date citations used only as authorities for scientific names do not appear in literature cited section.

Use terms prevalence, incidence, intensity, mean intensity, density, relative density, abundance, infrapopulation, suprapopulation, site, niche, and habitat as recommended by the ASP Ad Hoc Committee on the Use of Ecological Terms in Parasitology (1982, *Journal of Parasitology* **68:** 131–133).

Numbers: In text use Arabic numerals except when beginning a sentence. Do not use naked decimals in text, tables, legends, or on figures (i.e., 0.1, not .1). For numbers greater than 999, use commas. Use metric units; use 24-hour system to indicate time.

Page Charges: First 5 pages are published without charge. ASP members are charged $40 for each additional page; nonmembers $70. Nonmembers are encouraged to become members of the Society. Annual dues are $55. Corrections of authors' errors or revisions on proofs are $0.25 each.

SOURCE

February 1994, 80(1): 163–171.

Journal of Pathology
Pathological Society of Great Britain and Ireland
Wiley

Professor P.G. Toner
Institute of Pathology
Queen's University of Belfast
Grosvenor Road
Belfast BT12 6BL U.K.

AIM AND SCOPE

The aim of the *Journal* is to publish high quality papers in the fields of pathology and clinico-pathological correlation. Papers dealing with new aspects of histopathology and significant collections of cases are welcome. Single case reports are accepted when they are of particular interest and can be used to convey important messages of new information. They should include a full, but concise, review of the literature. Emphasis will be placed on the quality of the illustrations in all publications. Articles in the field of experimental pathology, relevant to the understanding of human disease, are welcome, as are papers on the use of techniques such as immunology and molecular biology to elucidate disease mechanisms.

Review articles, dealing with general and experimental pathology, as well as diagnostic histopathology, are published regularly. Unsolicited reviews are accepted if they are clearly written and topical. The Editorial Board wishes to encourage articles on the teaching of pathology at the undergraduate and postgraduate level, including articles on curricula and teaching methods. Editorials, a letter column and reviews of recently published books form a regular feature of the journal. Symposia proceedings may be published if the editors consider them appropriate and relevant.

It is the wish of the Editorial Board that the contents of the Journal will be of interest and value to all engaged in the fields of diagnostic and experimental pathology and also to other workers, outside the immediate field of pathology, interested in disease mechanisms and clinico-pathological correlation.

MANUSCRIPT SUBMISSION

Prepare manuscripts prepared in accordance with the Vancouver style (International Committee of Medical Journal Editors. "Uniform requirements for manuscripts submitted to biomedical journals." *Br Med J* 1982; 284: 1766–1770). Type papers on A4 (21 × 30 cm) bond paper, double-spaced with a 4 cm left margin. Submit three copies, including 3 copies of illustrations. In cover letter indicate name and full postal address, telephone and fax numbers of corresponding and reprint request author(s). Accepted manuscripts become property of *Journal* and may not be published elsewhere without written permission of both editor and publisher.

Copyright from author to publisher must be explicitly transferred. A copy of the publishing agreement is printed in each volume. Submit signed copy of agreement with article.

MANUSCRIPT FORMAT

Limit full articles to 6 journal pages (4000 words, 650/page), including tables, figures, and references (30). Explain why longer in cover letter. Short articles are 4 journal pages (2500 words), including tables, figures, and references.

Title Page: Give a concise and specific title. Identify a short running title (40 characters) at foot of page. Indicate first name, middle initial and last name of each author. Name departments and institutions to which work should be attributed. Give name and address of corresponding author.

Summary and Key Words: 200 word abstract on second page stating purpose of study or investigation, basic procedures, main findings and principal conclusions. Emphasize new and important aspects of study or observations.

Below abstract provide, and identify as such, 3-10 key words for cross indexing.

Text: Divide text of observational and experimental articles into sections with headings: Introduction, Method, Results and Discussion. Number each page, after title page, including references, tables and legends, consecutively.

References: Number consecutively in order mentioned in text. Identify in text, tables, and legends by Arabic numerals (in parentheses). Number references cited only in tables or legends in accordance with sequence established by first identification in text of table or illustration. At end of article, give full list of references with name and initials of all authors (if more than six, give first three, followed by *et al.*). Authors names are followed by title of article; title of journal abbreviated according to *List of Journals Indexed, in* January issue of *Index Medicus*; year of publication; volume number and first and last page numbers. Follow titles of books by place of publication, publisher, year and relevant page(s).

JOURNALS

Soter NA, Wasserman SI. Austen KF. Cold Urticaria: release into the circulation of histamine and eosinophil chemotactic factor of anaphylaxis during cold challenge. *N Engl J Med* 1976; **294**: 687-690

BOOKS

Elsen H N. Immunology: An introduction to Molecular and Cellular Principles of the Immune Response. 5th ed. New York: Harper & Row, 1974: 406

CHAPTER IN BOOK

Weinstein L, Swartz MN. Pathogenic properties of invading microorganisms. In: Sodeman WA Jr, Sodeman WA, eds. Pathologic Physiology: Mechanisms of Disease. Philadelphia: W B Saunders, 1974: 457–472.

Tables: Type on separate numbered pages following reference list. Number consecutively with Roman numerals. Give each a brief description heading. Do not repeat data in text or vice versa.

Illustrations: Line diagrams, graphs and half tones are figures. Submit top quality photomicrographs and electron micrographs cropped and unmounted.

Photographs should fit within print area: 16 × 20 cm (full page), 16 × 9 cm (half page), or 7.5 × 9 cm (quarter page). Orient quarter page illustrations with long axis vertical. Light photomicrographs should not exceed quarter plate size. Consider insets to illustrate higher magnifications and special features. On back indicate figure number, TOP and last name of first author.

Cost of color illustrations are borne by author(s). Current charge is £1000. Submit as prints or transparencies. Ensure that illustrations fit on a single color plate and that full plate is utilized.

Make legends brief and specific. Type on a separate sheet. Do not indicate magnification for light micrographs. Use internal scale markers for electron micrographs. Indicate type of stain used, if not haematoxylin and eosin.

SOURCE

September 1994, 174(1): inside back cover.

Journal of Pediatric Surgery
Surgical Section of the American Academy of Pediatrics
British Association of Pediatric Surgeons
American Pediatric Surgical Association
Canadian Association of Pediatric Surgeons
Pacific Association of Pediatric Surgeons
W. B. Saunders Company

Stephen L. Gans, MD
717 N Rexford Drive
Beverly Hills, CA 90210
J. C. Molenaar, MD
Kastanjeplein 105
3053 CB Rotterdam, The Netherlands
Europe
D. G. Young, FRCS
Royal Hospital for Sick Children
Yorkhill, Glasgow, G3 8SJ, Scotland
British Isles

MANUSCRIPT SUBMISSION

Submit manuscripts in duplicate. Send correspondence concerning abstracts, book reviews, notices, reports of meetings, and other announcements to Dr. Gans.

Articles must be contributed solely to this journal. Papers should be as brief as possible, consistent with subject.

If article is accepted for publication, copyright, including right to reproduce it in all forms and media, shall be assigned exclusively to publisher, who will not refuse any reasonable request by the author for permission to reproduce any of his or her contributions to the journal.

The *Journal of Pediatric Surgery* subscribes to "Uniform Requirements for Manuscripts Submitted to Biomedical Journals" (*N Engl J Med* 324:424–428, 1991).

MANUSCRIPT FORMAT

Type on 8.5 × 11 in. good bond paper, on one side only, with double- or triple-spacing and liberal margins. Do not use erasable bond.

Include an accurate address (include zip or postal codes) for editorial communications and print requests.

A brief abstract of material of paper precedes body of paper, (500 words), do not include a summary section at end. Follow with several words for indexing and computer programming, titled: Index Words.

Body of paper begins with 2 to 5 sentences, setting the general train of thought, before headings.

Illustrations and Tables: Cite figures and tables in order in text; mark their position in margin of manuscript. Use Arabic numbering for both figures and tables. Submit all line drawings in duplicate as clear, glossy, black and white, 5 × 7 in. photographs. Submit photomicrographs in duplicate; make allowance for effects of reduction. Type legends for each illustration double-spaced, together on a separate sheet; include at end of manuscript.

Type each table on a separate sheet, appropriately numbered. Type legend on sheet with tables.

Contributor pays all costs for printing color illustrations.

References: Compile at end of article in order of citation in text, not alphabetically. Type double-spaced, under heading REFERENCES. Verify all reference information. Use abbreviations for titles of medical periodicals from *Index Medicus*. Give inclusive page numbers.

JOURNAL ARTICLE, ONE AUTHOR

1. Lloyd JR: The etiology of gastrointestinal perforations. J Pediatr Surg 4:77–85, 1969

JOURNAL ARTICLE, TWO OR THREE AUTHORS

2. Kilpatrick ZM, Aseron CA: Radioisotope detection of Meckel's diverticulum causing intestinal bleeding. Z Kinderchir 13:210–217, 1973

JOURNAL ARTICLE, MORE THAN THREE AUTHORS

3. Filler RM, Eraklis AJ, Das JB, et al: Total intravenous nutrition. Am J Surg 121:454–458, 1971

JOURNAL ARTICLE, IN PRESS

4. Coran AG: The hyperalimentation of infants. Biol Neonat (in press)

COMPLETE BOOK

5. Gallagher JR: Medical Care of the Adolescent (ed 2). New York, NY, Appleton, 1966, pp 208–215

CHAPTER OF BOOK

6. Nixon HH: Intestinal obstruction in the newborn, in Rob C, Smith R (eds): Clinical Surgery, chap 16. London, England, Butterworth, 1966, pp 168–172

CHAPTER OF BOOK THAT IS PART OF PUBLISHED MEETING

7. Natvig JB, Kunkel HG, Gedde-Dahl T Jr: Chain subgroups of G globulin, in Killander J (ed): Gamma Globulins, Proceedings of the Third Nobel Symposium. New York, NY, Wiley, 1967, pp 37–54

8. Okamatsu T, Takayama H, Nakata K, et al: Omphalocele surgery. Presented at the meeting of the Pacific Association of Pediatric Surgeons, San Diego, CA, April 1973

OTHER FORMATS

SHORT CASE REPORTS—Two full pages (five typewritten pages). Estimate space occupied by title, authors, illustrations, and references to keep within two-page limit.

OTHER

Units: Use metric system for measurements.

SOURCE

January 1994, 29(1): front matter.

Journal of Pediatrics
Mosby

Joseph M. Garfunkel, M.D.
CB No. 7230, Medical School Wing C
University of North Carolina
Chapel Hill, NC 27599-7230
Telephone: (919) 966-5215

AIM AND SCOPE

The *Journal of Pediatrics* publishes articles on original research, clinical observations, and reviews of pediatric subjects and related fields.

MANUSCRIPT SUBMISSION

Have manuscript reviewed and approved for submission by department chairman or editorial committee. Articles must be submitted solely to The *Journal of Pediatrics*, and are subject to editorial revision. Papers that have been published in other journals, even if in another language, or that are being considered by another journal or are in press will not be considered. If a paper by the same author(s) contains any data previously published, in press, or under consideration by another journal, submit a reprint of previous article or a copy of other manuscript, with an explanation of overlap or duplication. If explanation is inadequate, editors of other general pediatric journals will be notified (see *J Pediatr* 1990; 117:903-4), as will an appropriate official of primary author's academic institution.

Submit following written statement, signed by all authors: "The undersigned authors transfer all copyright ownership of the manuscript entitled [title of article] to Mosby in the event the work is published. The undersigned authors warrant that the article is original, is not under consideration by another journal, and has not been previously published. The final manuscript has been read and each author's contribution has been approved by the appropriate author."

Disclose any conflict of interest, especially any financial arrangement with a company whose product is discussed in manuscript. Information is confidential during review process. If accepted for publication, an appropriate disclosure statement will be required.

For direct quotations, tables, or illustrations that have appeared in copyrighted material, submit written permission for use from copyright owner and original author with complete source information.

MANUSCRIPT FORMAT

Submit three copies (letter-quality computer printout or clean, sharp photocopy) typed on one side of white 8.5 × 11 in. paper, sequentially numbered, double-spaced (including references), with liberal margins, approximately 25 lines to a page.

Title Page: Include authors' names and academic degrees; departmental and institutional affiliations of each author; and sources of financial assistance, if any. Listed authors should include only those individuals who have made a significant, creative contribution; justify more than six. Designate one author as correspondent, and provide address, business and home telephone, and fax numbers.

Abstract: 200 words summation after title page. Use general outline described by Ad Hoc Working Group for Critical Appraisal of the Medical Literature (*Ann Intern Med* 1987;106:598-604 and 1990;113: 69-76).

References: Number in order of appearance in text, following format in "Uniform Requirements for Manuscripts Submitted to Biomedical Journals" (Vancouver style) (*N Engl J Med* 1991;324:424-8), with journal abbreviations according to *Cumulated Index Medicus*. If reference is to abstract, letter, or editorial, place appropriate term in brackets after title.

If six or fewer authors, list all; if seven or more, list first three and add *et al.*:

JOURNAL ARTICLES

Oleske J, Minnefor A, Cooper R Jr, et al. Immune deficiency syndrome in children. JAMA 1983;249:2345–9.

BOOKS

Bradley EL. Medical and surgical management. Philadelphia: WB Saunders, 1982:72–95.

CHAPTERS IN BOOKS

Bohl I, Wallenfang T, Bothe H, et al. The effect of glucocorticoids in the combined treatment of experimental brain abscess in cats. In: Schiefer W, Klinger M, Brock M, eds. Brain abscess and meningitis: subarachnoid hemorrhage—timing problems. Berlin: Springer Verlag. 1981:125–33.

Tables: Type doubled-spaced each on a separate sheet with a concise title. Make self-explanatory to supplement, not duplicate, text. If a table or any data therein have been previously published, give full credit to original source in a footnote.

Figure Legends: Give each illustration a legend. Type legends double-spaced on a separate sheet. If illustration has been previously published, give full credit to original source in legend.

Illustrations: Prepare original drawings of graphs in black India ink or typographic (press-apply) lettering. Typewritten or freehand lettering is unacceptable. Have all lettering done professionally and in proportion to drawing, graph, or photograph. Do not send original artwork, x-ray films, or ECG strips.

Computer-generated black-and-white illustrations must be legible and clearly printed in jet-black ink on heavy coated paper with either a glossy or dull finish. Patterns or shadings must be dark enough for reproduction and distinguishable from each other. Make lines, symbols, and letters smooth and complete. Do not put legend on print. Submit unmounted original individual laser or plotter prints. Laser prints should be full size at 300 dots per inch or greater full-page resolution; do not put multiple illustrations on a page. Dot matrix prints and photographic halftones are not acceptable. Computer-generated color illustrations are acceptable; make arrangements with editor. Colors must be dark enough and of sufficient contrast for reproduction. Do not use fluorescent colors. Follow preceding guidelines for black-and-white computer-generated illustrations. Authors pay extra cost associated with reproduction of color illustrations.

Submit glossy print photographs, 3 × l in. (minimum) to 5 × 7 in. (maximum), with good black-and-white contrast or color balance. Consistency in size is preferred. Note special instructions regarding sizing. Submit one set of glossy prints, unmounted, numbered, marked lightly on back with author name and TOP—and two sets of clean, sharp photocopies. If clean, sharp photocopies of photomicro-graphs, radiographs, karyotypes, etc. are unobtainable, submit three glossy sets of illustrations.

A reasonable number of black-and-white illustrations will be reproduced at no cost; obtain Editor's approval for color plates, elaborate tables, or a large number of illustrations.

OTHER FORMATS

BOOKS FOR REVIEW—Books and monographs, domestic and foreign, are reviewed depending on their interest and value to subscribers. Send to Editor of The Book Shelf section, André D. Lascari, MD, Department of Pediatrics, Albany Medical Center, 47 New Scotland Ave., Albany, NY 12208.

CLINICAL AND LABORATORY OBSERVATIONS—Three pages, 1000 words with a brief abstract (50 words), a total of two illustrations and tables, 10 references.

CLINICAL CONFERENCE—Prepare in traditional clinico-pathologic conference (CPC) style. Submit to David S. Smith, MD, St. Christopher's Hospital for Children, Erie Ave. at Front St., Philadelphia, PA 19134, or to editor.

EDITORIAL CORRESPONDENCE—300 words and three references, prepared in the same style as other manuscripts. Pertain to articles published in The *Journal* or related topics.

MEDICAL PROGRESS—Consult with Editor or Associate Editors.

NEWS ITEMS—Announce scheduled meetings, symposia, or postgraduate courses. Send to publisher at least 5 months in advance. News items of general interest to pediatricians and related specialists is considered.

OTHER

Abbreviations: Abbreviate frequently used, complex terms. List all abbreviations used in text on a separate sheet; do not put in parentheses at first use in text, except in abstract.

Experimentation, Human: Indicate that informed consent was obtained from parents or guardians of children who served as subjects of investigation and, when appropriate, from subjects themselves. If Editor or referees question propriety of human investigation with respect to risk to subjects or to means of obtaining informed consent, The *Journal* may request more detailed information about safeguards employed and procedures used to obtain informed consent. Copies of minutes of committees that reviewed and approved research may be requested.

Obtain permission from patient, or parent or guardian of a minor child, for publication of recognizable likenesses. Do not use patient initials.

Units: Describe laboratory values in both International System of Units (SI units) and metric mass units. Give SI units first and metric units in parentheses immediately thereafter. For conversion tables see *JAMA* 1986;255:2329-39 or *Ann Intern Med* 1987;106:114-29.

SOURCE

July 1994, 125(1): 13A–15A.

Journal of Periodontology
American Academy of Periodontology

Managing Editor, *Journal of Periodontology*
Suite 800, 737 North Michigan Avenue
Chicago, IL 60611-2690

AIM AND SCOPE

The *Journal of Periodontology* publishes articles relevant to the science and practice dealing with the treatment of periodontal diseases and related areas.

MANUSCRIPT SUBMISSION

Manuscripts are accepted for consideration with the understanding that text, figures, photographs, and tables have not appeared in any other publication, except as an abstract prepared and published in conjunction with a presentation by author(s) at a scientific meeting, and that material has been submitted only to this *Journal*.

In letter of submission state the above. Also, give full name, address, telephone and fax numbers and electronic mail address of corresponding author.

Accepted manuscripts become property of the American Academy of Periodontology. Letter of acceptance includes a copyright form to be signed by all authors and returned before a publication date is scheduled.

Submit original and two copies of manuscript and three original sets of illustrations. Mail flat in a heavy weight envelope.

List only those individuals who have made a substantial contribution to work and who are willing to take public responsibility for content of manuscript as authors. All authors must sign submission letter.

MANUSCRIPT FORMAT

Type manuscripts on one side of white bond paper (8.5 × 11 in.), double-spaced (including illustration legends and references) throughout. Keep 1 in. margins on both sides, top and bottom. Present materials in order: Title page, Abstract and Key Words, Text, Acknowledgments, References, Tables, and Legends.

Begin each section on a separate page. Number pages (including tables and legends) consecutively beginning with Title Page as page 1.

Title Page: Include a concise but informative title; first name, middle initial, and last name of each author with highest academic degrees and institutional affiliations; name of department(s) and institution(s) to which work should be attributed; disclaimers, if any; name and address of corresponding and reprint request author(s); sources of support in form of grants, equipment, drugs, or other significant sources of support; and a short running title (50 characters).

Abstract and Key Words: 250 word abstract on second page stating purpose of study; basic methodologies; main findings, with specific data and their statistical significance; and principal conclusions including clinical relevance, if appropriate. Do not include references.

List six key words or short phrases from MeSH documentation below abstract.

Text: Divide original articles into sections: Introduction, Materials and Methods, Results, and Discussion. Longer articles may require subheadings. For case reports, merge sections as appropriate.

Introduction: Concise review of subject area and rationale for study. Put more detailed comparisons to previous work and conclusions of study in Discussion section.

Materials and Methods: List methods used in sufficient detail so that other investigators can reproduce research. When established methods are used, refer to previously published reports; provide brief descriptions of methods that are not well known or have been modified. Identify all drugs and chemicals used with both generic and proprietary names and doses. Define populations for research involving human subjects.

Results: Present in logical sequence with reference to tables, figures, and illustrations as appropriate.

Discussion: Emphasize new and possibly important finding and conclusions that can be drawn from them. Compare present data to previous findings. Indicate limitations of experimental methods and implications for future research. Include new hypotheses and clinical recommendations and identify as such. Include recommendations, particularly clinical ones, when appropriate.

Acknowledgments: At end of discussion, acknowledge: individuals who contributed to research, including technical help or participation in a clinical study (obtain written permission from persons mentioned by name); sources of financial and material support; and financial relationships which may pose a conflict of interest. (including employment, acting as an officer, director, or owner of a company whose products or products of a competitor are being tested, owning substantial stock or having other financial interest in such companies).

References: Number consecutively in order of appearance in text. Give each item only one number; give books a different number each time, if page numbers differ.

Identify all references, whether in text, table, or legend, by Arabic numbers in parentheses. Use journal title abbreviations from U.S. National Library of Medicine in *Index Medicus* and *Index to Dental Literature*.

Use of abstracts as references is discouraged. Manuscripts accepted for publication may be cited. Cite material submitted, but not yet accepted, in text as "unpublished observations"; refer to written, not oral, personal communications in text, do not cite as references. Do not cite presented papers, unless subsequently published in a proceedings or peer-reviewed journal, as references. Limit references to materials published in peer-reviewed professional journals. Type references double-spaced.

JOURNALS

1. Standard Journal Reference (List all authors if six or less; if seven or more, list first three and add et al.).

Glass DA, Mellonig JT, Towle HJ. Histologic evaluation of bone inductive proteins complexed with coralline hydroxyapatite in an extraskeletal site of the rat. J Periodontol 1989;60:121–26.

2. Corporate Author

Federation Dentaire Internationale. Technical report No. 28. Guidelines for antibiotic prophylaxis of infective endocarditis for dental patients with cardiovascular disease. Int Dent J 1987;37:235.

3. Journal Supplement

American Academy of Periodontology. Glossary of periodontic terms. J Periodontol 1986;56(suppl):3.

4. Journal Paginated by Issue

Card SJ, Caffesse RG, Smith BA, Nasjleti CE. New attachment following the use of a resorbable membrane in the treatment of periodontitis in dogs. Int J Periodontics Restorative Dent 1989;9(1):59–69.

BOOKS AND OTHER MONOGRAPHS

5. Personal Author(s)

Tullman JJ, Redding SW. Systemic Disease ital Treatment. St. Louis: The CV Mosby Company; 1983:1–5.

6. Chapter in a Book

Rees TD. Dental management of the medically compromised patient. In: McDonald RE, Hurt WC, Gilmore HW, Middleton RA, eds. Current Therapy in Dentistry, vol. 7. St. Louis: The CV Mosby Company; 1980:3.

7. Agency Publication

Miller AJ, Brunelle JA, Carlos JP, Brown LJ, Loe H. Oral health of United States adults. Bethesda, MD: National Institute of Dental Research, 1987; NIH publication no. 87-2868.

8. Dissertation or Thesis

Teerakapong A. Langerhans cells in human periodontally healthy and diseases gingiva. [Thesis]. Houston, Texas: University of Texas, 1987. 92 p.

Footnotes: Use only to identify authors affiliation, explain symbols in tables and illustrations, and identify manufacturers' of equipment, medications, materials, and devices mentioned in Materials and Methods. Use symbols in sequence: *, †, ‡, §, ||, ¶,#, **, ††, etc.

Tables: Double-space on a separate pieces of paper. Number consecutively in Arabic numbers in order of appearance in text. Give brief descriptive titles for each. Include explanations, including abbreviations, in footnotes, not in heading. Include statistical measures of variations such as standard deviation or standard error of the mean in footnotes. Do not use internal horizontal or vertical rules.

Illustration Legends: Typed double-spaced with Arabic numbers corresponding to illustration. Explain use of arrows, symbols, numbers, or letters; explain internal scale, original magnification, and method of staining. Do not put on same page as illustration.

Illustrations: Submit three complete and original sets of illustrations. Photograph each professional drawing, roentgenogram, or other original material; submit sharp, glossy 5 × 7 in. (no larger than 8 × 10 in.) photographic prints. Do not mount. Paste a label on back indicating figure number, author's name, and TOP.

Give photomicrographs internal scale markings. Human subjects must not be identifiable in photographs, unless written permission is obtained and submitted. Make lettering, arrows, or other identifying symbols large enough to permit reduction. Apply directly to illustration.

Color photographs used at Editor's discretion at no charge.

Computer Graphics: Provide high quality, reproducible copies of all graphics. Computer-generated graphics are acceptable, when printed on a laser-quality printer and photographed; dot-matrix prints cannot be accepted. Prepare basic, simple computer designs which can be clearly understood when reproduced; do not use "three-dimensional" graphics. Unnecessarily complex designs may be returned for simplification.

OTHER FORMATS

CASE REPORTS—On unusual cases indicating observed lesions, symptoms, treatment, and outcome. Report new combinations of disease, or previously reported adverse effects of treatment which are unique and provocative. Document unusual findings or new concepts. Be concise; include quantitative data as appropriate; clearly illustrate and include pertinent literature review. Include longitudinal data if single cases are reported or documentation on several cases illustrate unusual features of condition.

GUEST EDITORIALS—Invited from authorities offering their perspective on articles published in the *Journal*, or on other items of interest to readership.

LETTERS TO THE EDITOR—Comment on articles published in *Journal*. Offer constructive criticism or address other relevant matters of concern to readership. Be brief, focus on a few specific points, and sign. If a letter comments on an article published in *Journal*, author will be provided an opportunity to respond. Incorporate citations into body of letter as parenthetical statements; do not handle as standard references.

REVIEW ARTICLES—5000 words, 100 references, and 6 tables or figures. Present a critical appraisal of knowledge in a specific area; provide an overview of knowledge in an area and the opportunity to become familiar with areas of emerging research. Often reviews will be invited by Editor from authorities in a discipline. Authors proposing to develop reviews should contact Editor.

OTHER

Experimentation, Animal: Indicate that protocol was approved by institutional experimentation committee or was in accordance with guidelines approved by Council of the American Psychological Society (1980) for the use of animal experiments.

Experimentation, Human: State that protocol was approved by institutional review committee for human subjects or that study was in accordance with the Helsinki Declaration of 1975, as revised in 1983. Do not use any designation in tables, figures, or photographs which would identify a patient, without written consent.

Manufacturers: For products, medication, devices, equipment, or material identified by brand name in text, include name and address of manufacturer as a footnote. Use generic or nonbrand name terminology. Mention brand names only in Materials and Methods to identify specific product or device used in study.

Statistical Analyses: Describe methods such that a knowledgeable reader with access to original data could verify results. Quantify result. Give appropriate indicators of measurement error or uncertainty. Avoid sole reliance on statistical hypothesis testing or normalization of data. Present data in as close to original form as reasonable for critical analysis. Include details about eligibility criteria for experimental subjects, randomization, methods for blinding of observations, treatment complications, and numbers of observations. Indicate losses to observations such as dropouts from clinical trial. List general use computer programs. Define statistical terms, abbreviations, and symbols.

Style: Follow "The Uniform Requirements for Manuscripts Submitted to Biomedical Journals" (*N Engl J Med* 1991;324:424–428 or *Brit Med J*, 1991; Feb 9; 302: 6772), or obtain from Managing Editor.

Units: Report measurements of length, height, weight, and volume in metric units or their decimal multiples. Give temperatures in degrees Celsius and blood pressure in millimeters of mercury. Report hematologic and clinical chemistry measurements in metric system in terms of International System of Units (SI).

For describing teeth, use Federation Dentaire Internationale convention (*International Dental Journal* 21: 104–106).

SOURCE

January 1994, 65(1): 98–101.
Instructions to Authors are published in January and July issues.

Journal of Pharmaceutical Sciences
American Chemical Society
American Pharmaceutical Association

William I. Higuchi, Ph.D.
Dept. of Pharmaceutics and Pharmaceutical Chemistry
301 Skaggs Hall
University of Utah
Salt Lake City, UT 84112

AIM AND SCOPE

The *Journal* will publish original research papers, original research notes, invited topical reviews, and editorial commentary and news. Area of focus shall be concepts in basic pharmaceutical science and such topics as chemical processing of pharmaceuticals, including crystallization, lyophilization, chemical stability of drugs, pharmacokinetics, biopharmaceutics, pharmacodynamics, prodrug developments, metabolic disposition of bioactive agents, dosage form design, protein-peptide chemistry and biotechnology specifically as these relate to pharmaceutical technology, and targeted drug delivery.

MANUSCRIPT FORMAT

Type double-spaced on 22 × 28 cm (or A4) paper. One copy must be a good, clear typescript, stencil, or photoduplicating machine product. Include original inked drawings or photographs of structural formulas for direct use. High-quality output from laser printers is acceptable.

Use separate sheets for title page, with authors' names and affiliations, and for abstract. Group references and notes, tables, and figure legends on separate sheets, in that order. Number every page, beginning with title page. Be consistent in author designation; supply given name, initial of second name, and last name. Give complete mailing address of place where work was done and include telephone and fax numbers and e-mail address of corresponding author. Add current address of each author, if different, on title page with a numerical superscript and footnote. Indicate corresponding author by a superscript x.

Title: Carefully choose wording to provide information on contents and to function as "points of entry" for information retrieval. Do not include symbols, formulas, or arbitrary abbreviations, except chemical symbols to indicate structure of isotopically labeled compounds.

Abstract: 80–200 words in one paragraph, presenting problem and experimental approach; state major findings and conclusions. Make self-explanatory and suitable for reproduction without rewriting. Do not use footnotes or undefined abbreviations. If a reference must be cited, give complete publication data.

Experimental Section: Describe experimental procedures in sufficient detail to enable others to repeat. Include names of products and manufacturers (with city, state address) only if alternate sources are unsatisfactory. Describe novel experimental procedures in detail; refer to published procedures by literature citation of both original and published modifications. Include purity of key compounds; describe method(s) used to determine purity. For buffers, use terminology such as "20 mM potassium phosphate buffer (pH 7.7) containing ..." Also, state w/v or v/v when appropriate.

Place identification of and precautions for handling hazardous chemicals and dangerous procedures at beginning of section. Example: "*Caution: The following chemicals are hazardous and should be handled carefully: (list of chemicals and handling procedures or references).*"

Results: Present concisely. Design tables and figures to maximize presentation and comprehension of experimental data. Do not present same data in more than one figure or in both a figure and a table. Reserve interpretation of results for discussion section; results and discussion may be combined in a single section.

Discussion: Interpret results and relate them to existing knowledge in field. Make clear and brief. Do not repeat information elsewhere in manuscript. Avoid extensive reviews of literature.

References and Notes: Number in one consecutive series by order of mention in text, with numbers as unparenthesized superscripts. Verify accuracy. Type complete list double-spaced on separate page(s) at end of manuscript.

JOURNALS

Feld, K. M.; Higuchi, W. I. *J. Pharm. Sci.* **1981**, *70*, 723–727.

BOOKS (EDITED)

Dervan, P. B.; Dougherty, D. A. In *Diradicals*; Borden, W. T., Ed.; Wiley: New York, 1982; pp 107–149.

List submitted manuscripts as "in press" only if formally accepted for publication; otherwise, use "unpublished results" after names of authors. Incorporate footnotes to text in correct numerical sequence with references.

Acknowledgments: Acknowledge financial support, technical assistance, advice from colleagues, gifts, etc. Obtain permission from persons whose contribution is acknowledged.

Tables: Tabulate experimental results if a more effective presentation is made or space is more economically used. Number consecutively with Arabic numerals. Provide a brief title with each table and a brief heading for each column. Indicate units of measure (preferably SI). Round data to nearest significant figure. Include explanatory material referring to whole table as a footnote to title. Give footnotes lowercase letter designations; cite in tables as italicized superscripts. Submit as camera-ready copy tables that require special treatment, such as insertion of arrows or other special symbols under or over alphanumeric characters, or contain many structures. Cite all tables in text.

Figures: Submit two sets of figures as original drawings or high-quality reproductions on single sheets of paper no larger than 22 × 28 cm. Use photocopies for duplicates. Make lettering sufficiently large that it can be clearly read following reduction to single-column width (8.9 cm). Ideal reduced size is 2 mm. For graphical presentation of data, make symbols clearly discernible; use brackets to indicate magnitude of statistical variation. Submit stereodiagrams in single units with left and right images correctly positioned. Authors are charged for incremental costs for color reproduction. An estimate is given upon request. With revised manuscript send letter acknowledging willingness to pay. Designate blocks of structural formulas as schemes or charts rather than figures. Make figure legends sufficiently descriptive so figure can be understood without reference to text. Write number of figure and name of first author on back of each figure at top.

OTHER FORMATS

COMMENTARY—500 words; no figures or tables. Technical or scientific subject with scope of *Journal*. Submit for *Open Forum* page. If commentary criticizes an article published in *Journal,* authors are usually asked for an opinion and are given an opportunity to reply in same issue.

COMMUNICATIONS—1000 word preliminary reports of sufficient importance and general interest to justify accelerated publication.

OTHER

Abbreviations: Use without periods. Use standard abbreviations. Keep nonstandard abbreviations to a minimum; define in text following first use.

Analyses: Provide adequate evidence to establish identity and purity for new compounds. Include elemental analysis. State purity of compounds used for biological testing; describe method used to evaluate it.

Biological Data: Reference biological test methods or describe in sufficient detail to permit repetition. Place detailed descriptions of biological methods in experimental procedures section. Present data as numerical expressions or in graphical form. Give statistical limits (statistical significance); if cannot be provided give number of determinations and indicate variability and reliability of results. Include references to statistical methods of calculation. Express doses and concentrations as molar quantities (e.g., µmol/kg, mM) when comparisons of potencies are made on compounds having large differences in molecular weights. Indicate routes of administration of test compounds and vehicles used.

Electronic Submission: Manuscripts prepared with WordPerfect or Microsoft Word can be used for production. Hardcopy is required for review. Submit disk with final accepted version of manuscript. Version on disk must exactly match final version accepted in hardcopy.

Use document mode or equivalent in word processing program. Do not include page-formatting instructions. Do not justify text. Do not insert spaces before punctuation. Follow reference format printed in *Journal*. Ensure that all characters are correctly represented: 1 (ones) and l (ells), 0 (zeros) and O (ohs), x (exs) and × (times sign). Check final copy for consistent notation and correct spelling. Check disk with a virus-detection program. Label disk with manuscript number and first author's name. Provide name and version of software used, platform used, and file names on diskette description form.

Place complete text (first) and tabular material (following) in one file. Do not integrate graphic material into file. If graphics are available, put in a separate file. For tabular material, set column alignment with either tabs or spaces, not a mixture of both.

Nomenclature: Provide correct nomenclature. Make consistent and unambiguous; conform with Rules of Nomenclature established by International Union of Pure and Applied Chemistry (IUPAC), International Union of Biochemistry, Chemical Abstracts Service (CAS), Nomenclature Committee of the American Chemical Society, and other appropriate bodies. Recommended references: *IUPAC Nomenclature of Organic Chemistry* (Pergamon: Elmsford, NY, 1979); *Nomenclature of Inorganic Chemistry, Recommendations* (Blackwell Scientific Publications: Oxford, 1990); *Enzyme Nomenclature* (Academic Press: New York, 1979); and *Compendium of Biochemical Nomenclature and Related Documents* (Portland Press, Ltd., England, 1992).

Use chemical names for drugs. If terminology is unwieldy, use nonproprietary names of drugs after first mention and identification. Use formally adopted nonproprietary names listed in *United States Adopted Names* (*USAN*). If a name has not been assigned, use *International Nonproprietary Names* (*INN*), approved by the World Health Organization. Do not use trade names and laboratory codes except as additional information.

For assistance with nomenclature problems contact Director of Nomenclature, Chemical Abstracts Service, P.O. Box 3012, Columbus, OH 43210.

Page Charges: $45.00 per printed page. Payment is not a condition for publication; articles are accepted or rejected on merit alone.

Reviewers: Recommend several qualified reviewers other than members of Editorial Advisory Board.

Spectral Data: Include such data for representative compounds in a series, for novel classes of compounds, and in structural determinations. Do not include routine spectral data for every compound. Papers where interpretations of spectra are critical to structural elucidation and those in which band shape or fine structure needs to be illustrated are published with spectra. When presentations are essential, reproduce only pertinent sections.

Structures: Clearly draw structural formulas. Construct for maximum clarity in minimum space. Group original drawings or high-quality reproductions at end of manuscript with other illustrations. Combine structures with same skeleton but containing different functional groups into a single formula with an appropriate legend to specify different chain lengths, substituents, etc. Use line formulas whenever possible.

Prepare structural formulas for direct photoreproduction. Equations, schemes, and blocks of structural formulas are presented in either one-column (3.5 in.) or two-column (7.0 in.) format. One-column format is preferred. Use when possible. In print, one-column equations and schemes will be 3.5 in. (8.9 cm) wide and two-column equations and schemes will be 7.0 in. (17.7 cm) wide. Prepare to fit one of these widths. Submit actual size.

For authors using ChemDraw program: fixed length, 18 pt; line width, 0.8 pt; bold width, 2.5 pt; hash spacing, 3.0 pt; margin width, 2.0 pt; bond spacing, 18% of width. Single-width bold and dashed lines are preferred to wedges for stereochemical notation; use 12 pt Helvetica font both for atom labels and for text material. Prepare drawings with page setup at 80% and print with a laser printer on good quality white paper. Mark each drawing "to be reduced to 75% for publication."

Style: See ACS journals and *The ACS Style Guide* (1986, ACS, Washington, D.C.).

SOURCE

January 1994 83(1): vii-ix.

Journal of Pharmacology and Experimental Therapeutics
Williams and Wilkins

Dr. John A. Harvey
Department of Pharmacology, Suite 129
The Medical College of Pennsylvania at EPPI
3200 Henry Avenue
Philadelphia, PA 19129

AIM AND SCOPE

The *Journal* invites for review original papers dealing with interactions of chemicals with biological systems. Any aspect of pharmacology and therapeutics is included, but descriptive case reports without dose response or mechanism studies are not accepted. Manuscripts dealing primarily with new methods will be reviewed only if data are presented showing new or more reliable pharmacological information has been obtained by use of the methods.

MANUSCRIPT SUBMISSION

Submit four copies of manuscript. Type double-spaced on one side only of approximately 21.6 by 28 cm paper. Submit one original drawing or photograph of each line figure and four clear copies and/or 4 original drawings or photographs of each halftone figure. Do not staple pages.

If accepted for publication, copyright in article, including right to reproduce it in all forms and media, shall be assigned exclusively to the Society for Pharmacology and Experimental Therapeutics. Reasonable requests by authors for permission to reproduce their contributions will not be refused. All authors must sign an assignment of copyright form (mailed from Field Editor's Office on provisional acceptance of manuscript). Work done by employees of U.S. Federal Government is excepted.

Submit a written statement signed by one author: "The undersigned author affirms that the manuscript (title of article) and the data it contains are original, are not under consideration by another journal and have not been published previously. The studies reported have been carried out in accordance with the Declaration of Helsinki and/or with the Guide for the Care and Use of Laboratory Animals as adopted and promulgated by the National Institute of Health." Indicate in cover letter the field of pharmacology to which research applies.

MANUSCRIPT FORMAT

Include sections in order; begin each section on a new page; number all pages consecutively.

Title Page: Give complete title of manuscript, names of all authors and laboratory of origin. Indicate support for research as a numbered footnote to title; include with other footnotes on a separate page after reference section.

Running Title Page: Include a running title (35 spaces, including punctuation) at top of page that conveys the sense of full title. Commonly used abbreviations may be used. Give name, address and telephone number of person to whom page proof are sent. List abbreviations used in paper. Define each abbreviation at first use in text.

Abstract: 200-250 words on purpose, general methods, findings and conclusions of manuscript.

Introduction: Do not label section; begin on a new page. Give a clear statement of aims of reported work or of hypotheses tested. Give a brief account of background of reported work.

Methods: Explicitly and concisely describe all new methods or procedures employed. Describe modifications of previously published methods. Cite original source of commonly used methods. Descriptions of methods should enable reader to judge accuracy, reproducibility and reliability and to repeat experiment. Submit very extended descriptions of methods with the American Society for Information Science National Auxiliary Publications Service (NAPS) c/o Microfiche Publications, P.O. Box 3513, Grand Central Station, New York, NY 10017. A text reference or footnote states that full description is available from NAPS. There is a nominal charge to authors and to those who request complete information.

Results: Include experimental data; do not discuss significance. Present results in graphic or tabular form, rather than discursively. Graphic presentation is preferred; do not duplicate in text, tables and figures. Give sufficient data to permit judgement of variability and reliability of results. Explain statistical tests used and data analyzed so findings can be interpreted and evaluated.

Discussion: Present conclusions drawn from results. Base speculative discussion on data presented; identify as such. Be concise. Adequately discuss related work by others.

Acknowledgments: Include at end of text on a separate page and a separate heading. Acknowledge gifts and assistance. Include financial support as a footnote to title.

References: In text citations, give author's name and year of publication (e.g., McCarthy, 1952; Ruth and Gehrig, 1929). With more than two authors, give name of first author, followed by "*et al.*," and date (e.g., Kennedy *et al.*, 1960). In list of references, arrange entries alphabetically by author; do not number. Give all authors' names. If reference is made to more than one publication by the same authors in the same year, add suffixes (a, b, c, etc.) to year in text citation and list of references.

List references includes in sequence: authors' names and initials, title of cited article, abbreviated title of journal in which article appeared, volume number of journal, inclusive pagination and year of publication. Abbreviate journals according to *Biosis List of Serials* (BioSciences Information Services of Biological Abstracts, Philadelphia, PA). Cite personal communications, unpublished observations and papers submitted for publication in parentheses at appropriate place in text not in list of references. Do not cite papers as "in press" unless accepted for publication; give name of journal.

Bruce, M. S.: The anxiogenic effects of caffeine. Postgrad. Med. J. 66: 18–24, 1990.

Griffiths, R. R., Bigelow, G. E. and Liebson, I. A.: Human Coffee drinking: Reinforcing and physical dependence producing effects of caffeine. J. Pharmacol. Exp. Ther. 239: 416–425, 1986.

Tallarida, R. J. and Murray, R. B.: Manual of Pharmacologic Calculations with Computer Programs. Springer-Verlag, New York, 1987.

Young, A. M. and Sannerud, C. A.: Tolerance to drug discriminative stimuli. In Psychoactive Drugs: Tolerance and Sensitization, ed. by A. J. Goudie and M. W. Emmett-Oglesby, pp. 221–270, Humana Press, Clifton, NJ, 1989.

Do not cite more than four references in support of any given point. Verify accuracy of citations; proofread citations and reference list.

Footnotes: Indicate footnotes to title, authors' names or text by Arabic numeral superscripts. Type footnotes in sequence on a separate page.

Tables: Put each on a separate page. Number consecutively with Arabic numerals. Design to fit *Journal* column or page size. Give a brief descriptive title at top of each table. Give table number in capital letters, title in boldface upper and lowercase letters (*e.g.*, TABLE 3. **Uptake of labeled drug by rat liver slices**). Reference footnotes by italicized lowercase superscript letters; place on a separate page of manuscript. Extensive tabular material not absolutely essential will be accepted for deposit with the ASIS/NAPS. A footnote will refer interested readers to NAPS from which copies may be obtained.

Legends for Figures: Number consecutively in Arabic numerals. Put on a separate page. Explain figures in sufficient detail so frequent referral to text is unnecessary; avoid unnecessary length. Use same abbreviations as in text.

Figures: Submit illustrations as unmounted glossy prints, preferably reduced to single column size (3.5 in. wide) to permit reproduction without reduction. Prints should be no larger than 8×10 in. and clear enough to permit reproduction without retouching and legibility after reduction. Place author's name and address and figure number on back of each print. Use a typed label, not a pen. Use uniform symbolism for all illustrations. Symbols, numbers and letters in illustrations must be at least 1.5 mm in height after reduction to single column size. Put figure title in legend, not on figure. Label photomicrographs and electron micrographs with a magnification calibration in micro or Angstrom units. State magnification in figure legend.

OTHER

Abbreviations: Express magnitudes of variables in numerals. Most abbreviations are without punctuation, with no distinction between singular and plural forms (*e.g.*, 1 mg, 25 mg; 1 sec, 25 sec). Use usual prefixes for multiples or submultiples of basic units (*e.g.*, kg, g, mg, μg, ng, pg). See *J Pharmacol Exp Ther* 1995, 272(1): front matter for abbreviations, or their prefixed multiples or submultiples, that may be used without definition in text, tables and figures.

For other abbreviations, define in text or in a footnote at first use. Use abbreviations for commonly used substances from *Journal of Biological Chemistry*. Define abbreviations for drugs on first use; give chemical or generic name of drug. Abbreviations may be used in running title but not in full title.

Drugs: Use generic names in text, tables and figures. Mention trade names in parentheses at first reference; do not use in titles, figures or tables. Capitalize trade names; do not capitalize generic or chemical names. Give chemical nature of new drugs when known. Indicate form of drug used in calculations of doses (*e.g.*, base or salt). If several drugs, include a separate paragraph in "Methods," or a separate table or footnote, with relevant information about all drugs employed.

Experimentation, Human: Affirm that original studies were carried out in accordance with the Declaration of Helsinki and with the *Guide for the Care and Use of Laboratory Animals* as adopted and promulgated by the National Institutes of Health.

Index Terms: List terms for annual index on last page.

Page Charges: $30 per printed page. Charges will be assessed only if they involve no personal expense to authors. Request waiver to the ASPET Executive Officer at time of billing.

The cost of excessive numbers of illustrations and of colored illustrations are defrayed by authors.

Statistical Analyses: Express statistical probability (P) in tables, figures, and figure legends as * $P < .05$, ** $P < .01$ and *** $P < .001$. For second comparisons, use symbols †. Avoid unnecessary complexity; be consistent. For multiple comparisons within a table, use footnotes, *a*, *b*, *c*, etc.

Submission Fee: $40.00 payable to ASPET. If fee entails personal financial hardship to author, fee will be waived. Submit request for waiver at submission.

SOURCE

January 1995, 272(1): front matter.
Instructions to authors appear in first issue of each volume (January, April, July, October).

Journal of Pharmacy and Pharmacology
Royal Pharmaceutical Society of Great Britain

The Editor
Journal of Pharmacy and Pharmacology
1 Lambeth High Street, London SE1 7JN

AIM AND SCOPE

The Journal of Pharmacy and Pharmacology reviews and reports all aspects of the sciences concerned with the discovery, detection, evaluation, safety, development and formulation of drugs and medicines, and welcomes original contributions relating to such research.

MANUSCRIPT SUBMISSION

Original research papers or review articles are subject to editorial revision, must not have been published elsewhere and that in consideration of publication copyright of text is assigned to the *Journal*. Research involving human subjects requires approval of an appropriate ethical committee.

Short papers and topics of immediate interest requiring rapid publication are published as Communications or Letters to the Editor.

MANUSCRIPT FORMAT

Text: Consult a current issue of *Journal*. Conform to typographical conventions, use of headings, lay-out of tables, and citation of references. Type text double-spaced on sheets not larger than 9×13 in. with a 1.25 in. margin. Send typescript and one copy with name(s) of contributor(s), name and address of laboratory where work was done and a shortened title (50 letters and spaces). Use presentation best suited to clear exposition of subject matter. Include an abstract, giving results and conclusions.

References: Include all citations in text in reference list, and vice versa. Do not include unpublished work in list unless accepted for publication.

In text, reference should appear as either: "Brown (1987) described" or "as described previously (Brown & Green 1986)." When three or more authors, use Brown et al. If year is the same for several references, identify with a, b, c, etc. [e.g., Brown (1985a, b, c)], both in text and in reference list.

In reference list, arrange references in order: name(s) (all), date, title of article, serial title, volume, first and last page numbers, with no underlining: Brown, B., Green, G. (1986) The effects of dyes on flexible films. J. Pharm. Pharmacol. 38: 990–995. List references alphabetically under first author's surname, if several references to three or more authors with the same first author, arrange chronologically. Place these "alphabetical-chronological" references after references to Brown alone or to Brown & X. Alphabetize names beginning with de, De, Van, von, etc., according to prefix, e.g., McLoud is alphabetized under MCLO . . ., not under MacL or Loud. Abbreviate serial titles in accordance with ISO 833, ANSI Z39.5 (1969), and BS 4148 (1970, 1975) or with lists (such as *Index Medicus* or *BIOSIS List of Serials*). Do not abbreviate one-word titles.

BOOKS

Flower, R. J., Moncada, S., Vane, J. R. (1985) Analgesic-antipyretics and anti-inflammatory agents; drugs employed in the treatment of gout. In: Gilman, A. G., Goodman, L. S. , Rall, T. W., Murad, F. (eds) The Pharmacological Basis of Therapeutics. 7th edn, Macmillan, New York, pp 674–715.

Tables: (for each copy of text) Type on separate sheets. Headings briefly describe content; make understandable without reference to text. Do not rule.

Illustrations: Keep to minimum number necessary for understanding of subject matter. Submit two of each, one prepared as described below, the other suitable for review. Draw line illustrations such as graphs or apparatus diagrams clearly and boldly in Indian ink on tracing paper, white paper, faintly blue-lined graph paper or Bristol board. Chose initial dimensions to allow for a reduction of at least one-half and not more than one-quarter of original size. Most illustrations are 2.0–3.0 in.wide. Select kymograph records and photographs to allow for similar reduction. Insert lettering and numbering lightly and clearly in pencil. For curves based on experimental data, use clear and bold indications of experimental points; use open or closed circles, diamonds, triangles or squares.

Put legends for illustrations together on a separate sheet appended to typescript. Write author's name, title of paper and figure number lightly in pencil on back of each. Mark approximate position of each illustration in text.

SOURCE

January 1994, 46(1): inside back cover.

Journal of Physiology
Cambridge University Press

The Publications Office, *The Journal of Physiology*
Printing House
Shaftesbury Road
Cambridge CB2 2BS, U.K.
Telephone: (01223) 68713; Fax: (01223) 312849
International: +44 1223 68713;
E-mail: physio@cup.cam.ac.uk; Web: http: //physiology.cup.cam.ac.uk/

AIM AND SCOPE

The Journal of Physiology publishes papers which present the authors' own original experimental work and which illustrate new physiological principles or mechanisms. As well as human and mammalian physiology this includes work at level of cell membrane, single cells, tissues or organs. Work on invertebrate or lower vertebrate preparations is suitable if it contributes results and further understanding applicable in general to functioning of organisms including mammals. Work on systems physiology of lower vertebrates and invertebrates, specific to that type of organism, is not suitable.

MANUSCRIPT SUBMISSION

Results submitted for publication in the journal may not repeat findings already published (or intended to be published). Refer to previous findings as if work had come from a different group. Policy applies to results in the wider sense and not simply to figures or parts of figures. Manuscripts which are merely an expanded version of work published elsewhere are not acceptable. An abstract of about 400 words is an exception.

Submit any of authors' material that overlaps content of manuscript, including preliminary notes, communications, abstracts, chapters or reviews, published in the last year, in press or submitted by any of authors.

Include following declaration: 'The attached paper entitled ...does not contain material, other than that submitted with the manuscript, which has been, or is intended to be, published anywhere except in *The Journal of Physiology*. It is understood that upon acceptance for publication the entire copyright in this paper shall pass to The Physiological Society. I/We confirm that any material of mine/ours, published in the last year, in press or submitted, which overlaps in content with this manuscript or which is cited in this manuscript, accompanies this submission and is listed below. I/We also confirm that any acknowledgments of scientific assistance or advice and any references to unpublished material or personal communications have been seen and approved by those concerned.' List overlapping material. Declaration must be signed by each author.

Submit three complete copies of paper; high quality photocopies are acceptable.

MANUSCRIPT FORMAT

Type on one side of page only on good-quality paper of uniform size. This applies to all sections including Methods, References, figure legends, table headings and footnotes. Number pages. Make dot-matrix printer output at least near-letter quality. Make insertions legible and few; type or write in ink. Consult recent volumes of *Journal* for general style of presentation.

Arrangement of Papers: Title page; Summary; Introduction; Methods; Results; Discussion; References; Acknowledgments; Tables; Figures and legends.

Papers may deviate from usual format for obvious and compelling reasons. Indicate desired position of figures and tables in text.

Title Page: Type on separate sheet with authors' names, name and address of laboratory where work was carried out, running title, key words and name, address, telephone and fax numbers and e-mail address of corresponding author. Address of origin should be minimal adequate postal address with post or zip code.

Give a running title (45 letters and spaces); may use above abbreviations. Make title informative. Do not exceed 120 letters and spaces. Include animal species. Do not submit a series of papers with same main title.

Following abbreviations may be used in title: ATP, ADP, AMP, cAMP, GTP, GDP, GMP, cGMP, GTPγS, GDPβS, DNA, RNA, mRNA, GABA, NMDA, 5-HT, PGE, ACh, InsP_3, IP$_3$, AMPA; Chemical symbols, e.g., K$^+$, CA^{2+}; Certain common peptides, e.g., 'the peptide VIP', 'the peptide ACTH'; Common cell lines, e.g., MDCK, 3T3, HeLa; and Fluorescent indicators, e.g., 'the indicator SBFI'; EPSC, IPSC, EPSP, IPSP, NMR, pH, P_{O2}, P_{CO2}.

For Subject Index, provide three key words chosen from most recent Cumulative Index.

Summary: 200 words arranged in short numbered paragraphs. Include a succinct account of problem, methods and results and conclusions. Do not cite references.

Introduction: Clarify object of research. Reference own previous work only if it has a direct bearing on subject; extensive historical review is not appropriate.

Methods: Describe only once. Do not put in figure and table legends. Give sufficient details to allow work to be repeated; give maker's name and address, including country, for nonstandard chemicals, apparatus and equipment. For materials known by a trade name, capitalize initial letter; if material is not in common use, give maker's name. Give Latin names of nonmammalian species.

Results: Present quantitative observations graphically, not in tables. For numerical results, relate number of significant figures to accuracy of results. Do not present individual results of a large number of repeated tests if number of measurements is stated. Give standard deviations and standard errors when pertinent with no more than two significant figures. Clearly distinguish theory and inference from what was observed; do not elaborate.

Discussion: Follows Results and is separate from it. State assumptions involved in making inferences from experimental results. Do not recapitulate Results. Provide a succinct conclusion to work.

An Appendix or a Theory section may be added where, for example, it is necessary to derive mathematical results required in paper.

References: Conclude paper with list of papers and books cited in text. Cite no more than 40. Give journal titles in full, capitalize both nouns and adjectives. Alphabetize references regardless of chronology. Ascertain style of citations from papers in *Journal*; punctuate correctly.

In text make references by giving author and year of publication in parentheses, e.g., (Hill, 1938), except when author's name is part of sentence, e.g., 'Hill (1938) showed that ...' Where several references are given together put in chronological order, separated by semicolons.

If two authors, give both names, for three to six authors give all names first time cited, but thereafter first name only, followed by '*et al.*' Refer to unpublished material sparingly in text by giving author's name and initials followed by 'unpublished observation' or 'personal communication'; do not put in list of references. Obtain approval for such references from persons concerned.

Acknowledgments: Keep to minimum consistent with courtesy. Obtain approval from persons concerned.

Tables: Use sparingly. Type double-spaced on separate numbered sheets. Use Arabic numerals. Avoid expression 'the following table'. Give each table its own self-explanatory title. Do not present the same information in both tabular and graphical forms. Give units of results in parentheses at top of each column; do not repeat on each line of table. Avoid long descriptive headings to columns. Do not use 'ditto' signs.

Figures and Legends: Give titles and legends to make comprehensible without reference to text; avoid undue repetition. Provide three sets of labelled copies of figures with legends.

For clarity, choose symbols on graphs etc. from: ○ ❑ △ ●
■ ▲ ◇ ◆ ☆ ★ ⊗ ⊖ ⊠ ⬦ ▼ ▽

Distinguish bars in histograms etc., using following patterns: ▨ ▰ ▨ ▨ ■ ▢

Avoid reversed hatching (◺ ◪) and vertical or horizontal, and greys. Use other symbols and patterns if more are required

Submit one original (high standard) copy of each figure as large as possible up to a maximum of 297×210 mm (A4). Do not submit poor-quality photocopies, recordings on chart paper, low-resolution computer plots, or white-on-black figures. Avoid grey shadings. Do not letter originals; lettering will be applied by Publications Office. Remove boxes surrounding figures. For half-tones and anatomical drawings (e.g. tracings of sections), unletter originals but set scale bars (defined in legend) and arrows.

Submit colored illustrations when scientific value is enhanced. Author pays a small proportion of actual costs.

OTHER FORMATS

SHORT PAPERS—3500 words, including figure legends (twelve double-spaced manuscript pages at 12 c.p.i.), four figures or tables and 20 references. Present papers in usual format with Summary, Introduction, Methods, Results and Discussion. Make papers complete in themselves. Pairs or sequences of papers are not allowed.

Send a copy of manuscript on disk. Give telephone and fax numbers and e-mail address with corresponding author's address. Send usual copyright declaration and copies of other relevant documents.

In a statement at bottom of title page and in cover letter request that manuscript be considered as a Short Paper for rapid publication.

OTHER

Abbreviations: Avoid unless easily understood and help in reading paper; keep to a minimum. Define when first in-troduced in text or figures labels. Print in upper-case letters without stops (e.g., EPSP, EMG, AHP), with certain well-established exceptions, e.g., S.E.M., S.D., p.p.m.

Experimentation, Animal: Experimental procedures must not cause unnecessary pain or discomfort. Experiments on living vertebrates or *Octopus vulgaris* must conform to U.K. legal requirements. When appropriate, include a statement indicating that experiments were performed in accordance with local/national guidelines. Provide a full description of anesthetic and surgical procedures. Provide evidence that adequate steps were taken to ensure that animals did not suffer unnecessarily at any stage of experiment.

On papers describing experiments on isolated tissues, indicate whether donor animal was killed or anesthetized, and how this was accomplished.

Experimentation, Human: Include evidence that experiments were performed with understanding and consent of each individual, if procedures were not therapeutic or carried a risk of harm. Show that study was approved by local ethics committee. For minors, include evidence that experiments were performed with understanding and consent of legal guardian. See *Code of Ethics of the World Medical Association* (Declaration of Helsinki) and to Medical Research Council's pamphlet *Responsibility in Investigations on Human Subjects* (reprinted *British Medical Journal* July 18, 1964).

Muscle Relaxants: Describe precautions taken to ensure adequacy of anesthesia. Provide sufficient detail to enable reader to determine that no unnecessary suffering occurred. Follow Physiological Society advice: When muscle relaxants are used with anesthetic agents during physiological experiments, difficulties arise in ensuring that animals do not experience unnecessary pain or distress. Person conducting experiment must ensure that anesthesia is adequate. Never use muscle relaxants without anesthesia.

Establish that the proposed anesthetic regime is adequate, in absence of relaxants, to provide analgesia for any proposed surgical procedure or noxious stimulus. When light levels of anesthesia are considered appropriate for experimental purposes, establish that deeper levels of anesthesia would interfere with purpose of experiment. Perform preparatory major surgery under full surgical anesthesia. Conduct subsequent procedures under light anesthesia in the presence of relaxants so that any residual pain from initial surgery is blocked by local anesthetics or analgesia and no further noxious stimuli are delivered. Some methods of head holding using ear bars and zygomatic bars are a potential source of pain: use other, atraumatic, methods of head restraint in lightly anesthetized animals.

There must be a protocol for continuous or regular assessment of adequacy of anesthesia. Make methods of assessment appropriate to particular anesthetic and to particular experiment.

Nitrous oxide (N_2O) has effective mood changing (tranquillizing) and analgesic properties and is useful because of its ease and consistency of delivery; even at maximum concentration feasible at normal pressures, it is not an adequate anesthetic for surgery in cats. Use caution in relying upon it for maintenance, even when precautions have been taken to avoid noxious stimuli. Assess animal's state periodically and use supplementary agents as required.

Nomenclature, Chemical and Biological: Follow conventions in chemical nomenclature adopted by the Biochemical Society (*Biochemical Journal* (1995) **305**, 1–15). Italicize or underline names of species and genera. Use Roman type for names of muscles, bones, etc., not italics).

Style: Write in English (spelling as in *The Chambers Dictionary* or the *Shorter Oxford English Dictionary*). Use up to three levels of subheading in addition to section headings (such as RESULTS). Use small print sparingly. Do not use footnotes.

Symbols, Units and Mathematical Notation: See *Quantities, Units, and Symbols* (1975, Royal Society, London). Use SI system of units. Certain traditional units which are still in common usage are also acceptable.

pressure, mmHg (SI unit kPa)

radioactivity, Ci = Curie (SI unit Bq)

O_2 uptake and CO_2 elimination, 1 min^{-1} (SI unit mol s^{-1}).

Use indexed notation, e.g., ml s^{-1}, rather than ml/s.

Always define symbols. Make symbols representing physical quantities single letters printed in italics; suffixes as inferiors, e.g., V_{max}, not Vmax. When two or more suffixes, set as inferiors; separate by commas, for example, $V_{O_2 max}$ (with certain exceptions,such as, a'_{Na}, $I_{K(Ca)}$). Ensure consistency throughout manuscript, including illustrations and tables.

SOURCE

(1995) 482 (1): iii-vii.

Journal of Physiology (Paris)
Editions Scientifiques Elsevier

Professor Michel Imbert
Centre de Recherche Cerveau & Cognition
Faculté de Médecine de Rangueil
Université Paul Sabatier
133 route de Narbonne
31062 Toulouse Cedex, France
Telephone: (33) 62 17 28 00; Fax: (33) 62 17 28 09

AIM AND SCOPE

The *Journal of Physiology (Paris)*, an Integrative Neuroscience Journal, aims at covering all aspects of neurobiology deemed relevant for a better understanding of behavior and cognition. Molecular and cellular neurobiology will be included only if the results presented have a direct impact on the understanding of the integrative functions of the brain. Along with primary contributions, the journal will publish review articles and position papers requested by the Editorial Board, conference proceedings and invited lectures and book review in functional imaging, development and plasticity, cellular neurobiology, systems physiology including neuromuscular physiology, endocrinology and cellular communications, and behavioral physiology and cognition.

MANUSCRIPT SUBMISSION

Manuscripts or data must not have been previously published or submitted elsewhere for publication. Molecular and cellular neurobiology will be included only if results have a direct impact on understanding of the integrative functions of the brain. Manuscripts may full papers or rapid communications.

Type on one side of paper double-spaced with wide margins. Mark desired positions of figures or tables in margin. Submit one original drawing or photograph of each line figure with three clear copies and four original drawings or photographs for each halftone figure. Submit four copies of manuscript.

MANUSCRIPT FORMAT

Title Page: Complete title of manuscript, names of all authors and laboratory of origin (with full addresses). In an insert at bottom of page give name and address of reprint request author.

Summary: 300 words, stating main points.

Key Words: 4-5 key words, defining field of work, for volume index and abstracting services.

Introduction: Express aims of work or subject matter in a clear and concise way without a detailed account of historical background.

Materials and Methods: Describe all new methods or procedures employed in detail so that reader can repeat experiments. Describe modifications of previously published methods; reference commonly accepted methods.

Results: Include experimental data; do not extend to a discussion of significance. Do not include very extensive data (e.g., computer programs); use graphic or tabular format for results. Include sufficient data account for variability in results and justify reliability of method(s). Use statistical tests so results can be interpreted and evaluated. Do not submit results obtained with unpublished methods.

Discussion and Conclusion: Present relevant conclusions to be interpreted from results; avoid speculative conclusions. Discuss results related to similar work by other scientists to allow readers to draw their own conclusion.

Acknowledgments: Include at end of manuscript, separated from main text.

References: Cite in text according to "Harvard" system (i.e., names and date). In list of references include only articles cited in text arranged alphabetically by author (not numbered); give all authors' names. If reference is made to more than one publication by the same author in one year, add suffixes (a, b, c, etc.) to year in text citation and list of references. If more than two authors for a cited reference, give name of first author followed by *et al.* in text citation. In reference list, cite all authors.

Provide following information in list of references: Authors' names and initials, year of publication in parentheses, full title of cited article, title of journal in which article appeared (abbreviated and underlined), volume of journal and page numbers (first and last pages).

Abeles M, Gerstein GL (1988) Detecting spatio-temporal firing patterns among simultaneously recorded single neurons. *J Neurophysiol* 60, 909–924.

References from books: Authors' names and initials, year of publication in parentheses, full title of cited article; In: full title of book (underlined), name(s) of editor(s) and initials in parentheses, publisher, address (city) and page numbers (first and last pages).

Sparks DL, Hartwich-Young R (1989) The deep layers of the superior colliculus. *In: The Neurobiology of Saccadic Eye Movements*, Elsevier Science Publishers, Amsterdam, 213–255.

Limit citations to personal communications and unpublished observations; give in parentheses in text, do not include in list of references.

Do not cite a paper as "in press" unless it has been accepted for publication; give name of journal.

Tables: Put each table on a separate page. Number consecutively with a brief descriptive title at top. Reference footnotes to tables by subscript letters beneath table.

Figure Legends: Place on a separate page. Explain figures in sufficient detail so referral to text is unnecessary.

Figures: Illustrate results. Restrict to minimum needed to clarify results. Do not submit previously published figures. Do not submit same data in graphic and tabular form.

Illustrations: Submit as unmounted glossy photographic prints on separate sheets with author's name and address and number of figure on back on a typed label. Put figure title in legend, not on figure itself. Additional cost for color reproduction is charged to author.

OTHER FORMATS

RAPID COMMUNICATIONS—1000 words (including 6 references), one table or figure, no abstract or subheadings. Indicate desired position of table or figure in margin. Put name, postal address, telephone and fax numbers of corresponding author on title page.

Allows rapid publication of exciting new results.

OTHER

Drugs: Use generic names; if commercially available drugs are used, state proprietary name, chemical composition and manufacturer in full in Methods. Indicate form of drug used (e.g., base or salt).

SOURCE

1994, 88(1): inside back cover .

Journal of Prosthetic Dentistry

Federation of Prosthodontic Organizations
The Academy of Prosthodontics
The American Prosthodontic Society
The Canadian Academy of Restorative Dentistry and Prosthodontics
Mexican Prosthodontic Society
Academy of Prosthodontics of South Africa
Indian Prosthodontic Society

International College of Prosthodontists
Israel Society of Oral Rehabilitation
The Swiss Prosthodontic Society
British Society for the Study of Prosthetic Dentistry
European Prosthodontic Association
Brazilian Society of Oral Rehabilitation
Japan Prosthodontic Society and others
Mosby

Dr. Glen P. McGivney
State University of New York at Buffalo
School of Dental Medicine - 345 Squire Hall
Buffalo, New York 14214
Telephone: (716) 829-2984; Fax: (716) 829-2985

MANUSCRIPT SUBMISSION

Submit all materials in triplicate (one original and two copies). Double-space all materials and use 1 in. margins on all borders. Number all pages. If manuscript was prepared on a Macintosh or an IBM compatible computer, also supply floppy disk (indicate operating system and word processing program used). Do not exceed 10 to 12 pages (excluding references, legends, and tables). Limit to no more than four authors. List additional contributing authors as an addendum after references.

Submit written permission for direct quotations, tables, and illustrations that have appeared in copyrighted material from copyright owner and original author along with complete source information.

Authors may not directly or indirectly advertise equipment, instruments, or products with which they have personal identity. Disclose any financial interest in products mentioned in articles.

All manuscripts must be accompanied by following statement signed by one author: "The undersigned author transfers all copyright ownership of the manuscript (title of the article) to the Editorial Council of The Journal of Prosthetic Dentistry in the event the work is published. The undersigned author warrants that the article is original, is not under consideration for publication by any other journal, and has not been previously published. I sign for and accept responsibility for releasing this material on behalf of any and all coauthors."

MANUSCRIPT FORMAT

Articles in the *Journal* can be classified as: clinical reports, research reports, descriptions of technical procedures, historical (literature) reviews, essays and articles on other professional subjects, tips from readers, and book reviews. All manuscripts contain a brief statement on the clinical significance of material presented.

Title Page: Include full name and title of each author, academic degrees, and institutional affiliations and locations. If manuscript was read before a group, give name of organization and place where manuscript was read on title page. If research was supported by a grant, give name of supporting organization and grant number.

Include an abbreviated title for use as a running head. Give mailing address, business and home telephone, and fax numbers of corresponding and reprint request author(s).

Do not use abbreviations in title. If possible, identify clinical significance of manuscript in title.

Abstract: 100 words typed double-spaced on a separate page. Do not use abbreviations. Include all important points mentioned in article. State clinical significance of manuscript.

References: Do not cite manuscripts in preparation, personal communications, and other unpublished information in reference list; mention in text in parentheses. Keep references to foreign language publications to a minimum; use only when original article has been translated into English. Indicate English translation in brackets following original title. Avoid references to abstracts. Identify references in text by superscript Arabic numerals; number in consecutive order as mentioned in text. Type reference list double-spaced at end of article in numeric sequence.

Include only references cited in text in list. Format should conform to "Uniform Requirements for Manuscripts Submitted to Biomedical Journals" (International Committee of Medical Journal Editors) (*JAMA* 1993;269:2282-6). Use journal title abbreviations from *Cumulated Index Medicus*. List all authors when six or fewer; when seven or more, list first three and add et al.):

JOURNAL ARTICLES

Jones ER, Smith IM, Doe JQ, et al. Occlusion. J Prosthet Dent 1985;53:120-9.

BOOKS

Hickey JC, Zarb GA, Bolender CL. Boucher's prosthodontic treatment for edentulous patients. 9th ed. St Louis: CV Mosby, 1985:312-23.

Tables: Double-space throughout, including column heads, footnotes, and data. Number according to order of mention in text. Submit each on a separate sheet. Omit internal horizontal and vertical lines.

Supply a concise heading describing table's content as a title. Make self-explanatory to supplement, not duplicate, text. Footnotes immediately follow table; define all abbreviations used. If table or any data therein have been previously published, give full credit to original source in a footnote.

Illustrations: Number in order of mention in text. Label all three sets of prints on back with figure number and an arrow to indicate TOP; typed labels are preferred.

Submit glossy prints of all illustrations, including slides, original artwork, EMG strips, and graphs. Typewritten or handwritten lettering is unacceptable. Have all lettering done professionally; make visible against background, and of legible proportion if reduced. If a key to an illustration requires artwork (screen lines, dots, unusual symbols) not available on a typesetting keyboard, incorporate into drawing, do not put in legend. Photographs should have good black and white contrast or color balance. Note special instructions regarding sizing, placement, or color.

Color illustrations must contribute significantly to value of manuscript. Submit two sets of color glossy prints and one set in black and white. A maximum of two *Journal* pages of color, each page containing a maximum of eight photographs and each photograph measuring 4 in. wide × 3 in. high, is reproduced free for articles other than Clinical Reports. Clinical Reports are limited to one page of color, or eight color photographs. Group photographs to fit eight to a page. Submitted glossy color prints must be 4 × 3 in. (minimum) to 7 × 5 in. (maximum). Do not mix horizontal and vertical illustrations. Do not send slides.

Figure Legends: Type double-spaced on a separate sheet; number to correspond with figures. If illustration is from previously published material, give full credit to original source.

OTHER FORMATS

CLINICAL REPORT—Three double-spaced, typed pages, eight good quality descriptive illustrations. Describe methods for meeting a treatment challenge of interest in providing care for an individual patient.

DESCRIPTION OF TECHNICAL PROCEDURES—State objective of technique. Give an orderly description of procedures. Make appropriate reference to alternate techniques. Summarize advantages and disadvantages of technique. Can be written in a step-by-step "cookbook" manner.

ESSAYS AND ARTICLES ON OTHER PROFESSIONAL SUBJECTS—Include such topics as education, communication, ethics, medicolegal problems, theoretical explanations, and analytic comparisons. Organizational pattern may vary; make presentation logical, effective, and keyed to *Journal* audience.

RESEARCH REPORT—State problem clearly so that there is no doubt about objective of research. Describe method so it can be duplicated and its validity judged. Report results accurately and brief. List conclusions drawn from research. Provide, under a separate heading, a brief statement of clinical implications of research. Provide a summary.

THE LITERATURE REVIEW—Record sequence of development of a particular phase of dentistry; be as brief as possible, but also complete. Provide documentation by references.

TIPS FROM OUR READERS—Two authors, 250 words, and two illustrations. Brief reports of helpful or timesaving procedures.

OTHER

Color Charges: $500 for each additional page of color over allowed one or two pages (eight or sixteen photographs).

Experimentation, Human: Photographs of identifiable persons must be accompanied by signed and dated releases showing informed consent.

Nomenclature, Dental: See Sixth Edition of the Glossary of Prosthodontic Terms for accepted terminology (*J Prosthet Dent* 1994;71:41-112).

Style: "Guide to Preparing Articles for the Journal of Prosthetic Dentistry," "Guidelines for Reporting Statistical Results," and "An Author's Guide to Controlling the Photograph" are available from Editor's office request.

If unusual abbreviations cannot be avoided, use expanded form when first mentioned and abbreviate thereafter. Use generic drug names (list trade names in parentheses at first mention). Product trade names must be accurate; give manufacturer, city, and state in parentheses.

SOURCE

January 1995, 73 (1): 8A–9A.

Journal of Protozoology
see Journal of Eukaryotic Microbiology

Journal of Reproduction and Fertility
Society for the Study of Fertility

Managing Editor, *Journal of Reproduction and Fertility*
22 Newmarket Road
Cambridge CB5 8DT, U.K.
Telephone: 0223 351809; Fax: 0223 359754

AIM AND SCOPE

The *Journal of Reproduction and Fertility* publishes original papers and reviews on the molecular biology, biochemistry, physiology and pathology of reproduction and early embryogenesis in man and other animals, and on the biological, medical and veterinary problems of fertility, pregnancy and lactation.

MANUSCRIPT SUBMISSION

Papers are considered for publication on the understanding that they have not been, and will not be, published elsewhere in whole or in part. Exceptions may be made if prior publication is limited to a brief abstract or is in a relatively inaccessible medium.

Submit original + 3 copies of manuscript and illustrations. In cover letter state that work has not been and will not be submitted for publication elsewhere until *Journal* decides whether to publish paper. Provide three copies of papers quoted as 'in press' in References.

MANUSCRIPT FORMAT

Type manuscripts double-spaced throughout (including Reference list and figure legends) on one side of paper only, with 2 cm margins on all sides. Number pages. Arrange manuscripts in order: Title page, Summary, Introduction, Materials and Methods, Results, Discussion, Acknowledgments, References, Tables, Figure legends and Figures. Number lines down left edge of each page.

Title Page: Give title and author(s) names (use surnames and initials) and addresses. Use superscript numbers after authors' name for addresses. Use * † ‡ § . . . for footnotes. Provide a short title (50 characters) for a running head. Indicate corresponding and reprint request author(s) with telephone and fax numbers.

Summary: In a single 300 word paragraph, state objectives of study and methods used and summarize results and conclusions. Do not use abbreviations or references.

Introduction: Set study in context by briefly reviewing relevant knowledge of subject. Concisely state study objectives.

Methods: Provide sufficient information so other workers can repeat study. Reference well established methods; provide full details of modifications. State sources of chemicals, reagents and hormones and give manufacturer's name and location in parentheses. Give generic name, dose and route of administration for drugs. Specify composition of buffers, solutions and culture media.

Results: Present in text, tables and figures as appropriate. Do not repeat data given in tables and figures in text.

Discussion: Interpret results, relate observations to other relevant studies and outline implications of results for future research. Do not repeat results.

Acknowledgment: Acknowledge technical help and financial and material support.

References: Ensure that all references cited in text are included in Reference list and vice versa. Reference list should contain only accessible articles. Refer to unpublished work, including personal communications, manuscripts in preparation and manuscripts submitted but not yet accepted for publication, in text as: (A. Stone, unpublished) (J. Brown, personal communication). List articles accepted for publication but not yet published as 'in press" in Reference list.

Cite references in text in chronological order; use *et al.* for more than two authors: Davies and Smith (1989); Williams (1990); Frost *et al.* (1992).

Cite references in list in alphabetical order. Give papers by the same author in order: single author; two authors alphabetically according to name of second author; three or more authors chronologically with (a) (b) and (c), etc. for papers published in the same year. References to articles in journals include authors' names with initials, year of publication, full title of article, journal title in full and first and last page numbers.

Larson RC, Ignotz GG and Currie WB (1992) Effect of fibronectin on early embryo development in cows *Journal of Reproduction and Fertility* **96** 289–297

References to books include authors' names, year of publication, chapter title, book title date, edition number, page numbers, names of editors, name of publisher and city of publication.

Aitken RJ (1983) The zona free hamster egg penetration test *Male Fertility* pp 75–86 Ed. TB Hargreave. Springer-Verlag, Berlin

Tables: Be concise and informative. Title is a single sentence typed at head of table; include name of organism studied. Put additional explanatory material in alphabetically ordered footnotes cross-referenced [a][b] to column entries. Make tables self contained, so they do not require further explanation; number in Arabic numbers and cite in text. Submit each on a separate sheet. Explain abbreviations in footnotes. Give each column a short heading. Do not use internal horizontal and vertical rules.

Figures: Submit each figure on a separate sheet; put authors' names and figure number on a label on back of figure. Figures are relettered by printer. Provide originals without numbering and lettering; put numbering and lettering on photocopies. Label sections of figures as (a) (b) etc. in top left corner. Color illustrations may be accepted.

Line Drawings—Draw originals in black on white paper at double size of eventual reproduction. Preferred symbols are ○ ● □ ■ △ ▲. Make symbols large enough to be clearly visible when reduced. Do not enclose in boxes.

Half-tones—Provide four sets as unmounted glossy prints with a guide to their arrangement. Halftones should be high resolution; prints from original half-tones cannot be accepted. Provide one set of originals without lettering. Indicate magnification by a scale bar on half-tone and give measurement in legend.

Figure Legends: Put legends describing figures with keys to symbols on a separate sheet. Include name of organism studied; explain all abbreviations.

OTHER

Abbreviations: Define all at first mention, except:

ACTH	EDTA	hCG	PAGE
ADP	EGTA	Hepes	PBS
ATP	ELISA	HPLC	PCR
BSA	FSH	IGG	RNA
cAMP	GDP	LH	SDS
DEAE-cellulose	GLC	LHRH	TLC
DNA	GnRH	NAD	Tris
DNAase	GTP	NADP	

Color Charges: £250 for each page on which color is printed.

Experimentation, Animal: Perform experiments in accordance with U.K. legal requirements. Provide details of procedures and anesthetics used and doses given.

Give full binomial Latin names for all experimental animals other than common laboratory animals. State breed or strain and source of animals and details of age, weight, sex and housing.

Experimentation, Human: Indicate that investigations were approved by local ethical committee and that consent was obtained from patients.

Spelling: Use *Oxford English Dictionary.*

Statistical Analysis: Give details of statistical methods at end of Materials and Methods section.

Symbols: Use SI symbols. See *Units, Symbols and Abbreviations* (1988, Royal Society of Medicine Series, London). Give concentrations in mol l^{-1}. For international units, use iu. Use U for enzyme activity.

SOURCE

January 1995, 103(1): back matter.
Guidelines for authors are printed in full in January issue. More detailed guidelines are available from Managing Editor.

Journal of Rheumatology
Journal of Rheumatology Publishing Co., Ltd.

Editor, *The Journal of Rheumatology*
920 Yonge St., Suite 115
Toronto, Canada M4W 3C7

AIM AND SCOPE

Manuscripts on clinical subjects and original research are invited from scientists working in rheumatology and related fields. Subject matter may relate to the broad field of rheumatology, rehabilitation medicine, immunology, infectious diseases or orthopedic subjects pertaining to the rheumatic diseases. The *Journal* will also consider the publication of symposia on topics of interest and new developments in the field of rheumatology, either in its regular issues or as supplements.

MANUSCRIPT SUBMISSION

Manuscripts must contain original material; neither the article nor any part of its essential substance, tables, or figures may have been or will be published or submitted for publication elsewhere. This restriction does not apply to abstracts or press reports of scientific meetings. Submit copies of related manuscripts that have been or will be published by or submitted to this or another journal. If this requirement is violated or author engages in other misconduct, Editor may reject manuscript and impose a moratorium on the acceptance of new manuscripts. If misconduct is sufficiently serious, Editor will refer matter to author's institution and/or appropriate disciplinary body.

Disclose financial arrangements with companies whose products figure in manuscript or with a company making a competing product. Information will be held in confidence during review and will not influence editorial decision. If article is accepted for publication, editors will discuss with authors how to communicate information to readers.

MANUSCRIPT FORMAT

Manuscripts should be concise and typed double-spaced on one side of good quality paper, with liberal margins. Journal titles cited in references follow abbreviations used in *Index Medicus.*

Provide an original and 3 copies of manuscript and tables. Submit 4 glossy prints of each figure. Xerographic copies are not acceptable. Number each page, including tables and figure legends, in sequence. Organize manuscript to include: Title Page, Introduction, Materials and Methods, Results, Discussion, Acknowledgment, References, Tables, Figures and Figure Legends. Indicate main sections Materials and Methods, Results, Discussion, Acknowledgment, References as headings. Use Arabic numbers except at beginnings of sentences.

Include statistical worksheet (not for publication), if applicable.

Be concise. For Discussion do not exceed half the length of whole article (preferably a third). Descriptions of clinical findings and the course of cases may be included under the heading Case Report in place of Materials and Methods and Results sections.

Title Page: On a separate page and in order: full title of manuscript (concise but informative; given surname of author(s) without degrees; structured abstract (250 words) describing Objectives, Methods, Results, Conclusion; key Indexing Terms (6 MeSH terms); name of department(s) and institution(s) to which work should be attributed; source(s) of support as grants or industrial support; initials, surnames, appointments and highest academic degrees of all authors; name and address of reprint request and corresponding author(s); and a short running head or footline (4 words).

Acknowledgment: At end of discussion and before references. Do not include grant or industrial support for fellowship awards; put on title page.

Tables: Type each on a separate sheet, double-spaced. Number consecutively and supply brief titles. Number references cited only in tables in sequence established by identification in text.

Figures: Number with author's name on back and TOP clearly marked. List legends on a separate sheet. Have figures professionally drawn and photographed; indicate critical area of radiographs or photomicrographs. Freehand or typewritten lettering is unacceptable. Send sharp glossy black and white photographic prints no larger than 5 × 7 in. Cite each figure in text in consecutive order. Contributor bears all costs of color printing.

References: Number consecutively in order mentioned in text. Identify references in text, tables and legends by Arabic numerals (in parentheses). Number references cited only in tables or in figure legends in accordance with a sequence established by first identification in text of table or illustration. Use reference form of U.S. National Library of Medicine's *Index Medicus*. Avoid using abstracts as references; do not use "unpublished observations" and "personal communications" as references; insert references to written, not verbal, communications (in parentheses) in text. Include among references manuscripts accepted but not yet published; designate journal followed by "in press." Cite information from manuscripts submitted but not yet accepted in text as "unpublished observations" (in parentheses). Verify accuracy of references against original sources.

JOURNAL

Standard Journal Article (List all authors when 6 or less; when 7 or more, list only first 3 and add et al)

Soter NA, Wasserman SSI, Austen KF: Cold urticaria: release into the circulation of histamine and eosinophil chemotactic factor of anaphylaxis during cold challenge. *N Engl J Med 1976;294:687-90.*

JOURNAL SUPPLEMENT

Dawkins RL, Garlepp MJ, McDonald BL, *et al:* Myasthenia gravis and D-penicillamine. *J Rheumatol 1981;* (suppl 7)8:169-72.

CORPORATE AUTHOR

The Committee on Enzymes of the Scandinavian Society for Clinical Chemistry and Clinical Physiology: Recommended method for the determination of gamma-glutamyl-transferase in blood. *Scand J Clin Lab Invest 1976;36:119-25.*

EDITORIAL

Coffee drinking and cancer of the pancreas (editorial). *BMJ 1981;* 283:628.

LETTER TO THE EDITOR

Gardner GC, Lawrence MK: Polyarteritis nodosa confined to calf muscles (letter) *J Rheumatol 1993;20:908-9.*

Pedrol E, Garcia F, Casademont J: Polyarteritis nodosa confined to calf muscles (reply to letter) *J Rheumatol 1993;20:90-9.*

CHAPTER IN BOOK

Weinstein L, Swartz MN: Pathogenic properties of invading microorganisms. In: Sodeman WA Jr, Sodeman WA, eds. *Pathologic Physiology: Mechanisms of Disease.* Philadelphia: WB Saunders, 1974:457-72.

PERSONAL AUTHOR

Osler AG: *Complement: Mechanisms and Functions.* Englewood Cliffs: Prentice-Hall, 1976.

CORPORATE AUTHOR

American Medical Association Department of Drugs. *AMA Drug Evaluations.* 3rd ed. Littleton: Publishing Sciences Group, 1977.

EDITOR, COMPILER, CHAIRMAN AS AUTHOR

Rhodes AJ, Van Rooyen CE, Comps: *Text of Virology: For Students and Practitioners of Medicine and the Other Health Sciences.* 5th ed. Baltimore: Williams & Wilkins, 1968.

PUBLISHED PROCEEDINGS PAPER

DuPont B: Bone marrow transplantation in severe combined immunodeficiency with an unrelated MLC compatible donor. In: White HJ, Smith R, eds. *Proceedings of the Third Annual Meeting of the International Society for Experimental Hematology.* Houston: International Society for Experimental Hematology, 1974:44-6.

AGENCY PUBLICATION

National Center for Health Statistics: *Acute Conditions: Incidence and Associated Disability:* United States. July 1968-June 1969. Rockville, MD: National Center for Health Statistics, 1972; DHEW publication no. (HSM)72-1036. (Vital and health statistics; series 10, no. 69).

DISSERTATION OR THESIS

Cairns RB: Infrared spectroscopic studies of solid oxygen (Dissertation). Berkeley, California: University of California, 1965, 156 p.

NEWSPAPER ARTICLE

Shaffer RA: Advances in chemistry are starting to unlock mysteries of the brain: discoveries could help cure alcoholism and insomnia, explain mental illness. How the messengers work. *Wall Street Journal 1977* Aug 13:1(col.1), 10(col . 1).

MAGAZINE ARTICLE

Roueche B: Annals of medicine: the Santa Claus culture. *The New Yorker 1971* Sep 4:66-81.

OTHER FORMATS

ANNOUNCEMENTS—Submit 3 months prior to publication.

CASE REPORTS—1000 words, 100 word abstract, 3 figures or tables and 15 references; no title page.

LETTER TO THE EDITOR—800 words, 10 references, 2 figures or tables and no subdivision for Abstract, Methods or Results. Editorial comments. Include full name(s) and academic appointment(s) and MeSH indexing terms.

SUPPLEMENTS—Proceedings of significant symposia with original work not previously published.

Symposium organizers should contact editor well in advance of the symposium date with a draft program. Papers must conform to *Journal* style.

OTHER

Charges: $100 allowance for tables, figures and author's additions or alterations; contributors bear costs in excess of this amount.

Reviewers: Suggest names of 3 or 4 persons who might be considered suitable reviewers.

SOURCE

January 1994, 21(1): viii-ix.

Journal of Steroid Biochemistry and Molecular Biology
Pergamon Press

Dr J. R. Pasqualini or Dr. R. Scholler
Editorial Office
Foundation for Hormone Research
26 Boulevard Brune
75014 Paris, France

AIM AND SCOPE

The *Journal of Steroid Biochemistry and Molecular Biology* is devoted to new experimental or theoretical developments in areas related to steroids. It publishes original papers as well as general and mini-reviews, proceedings of selected meetings and rapid communications (brief highly topical articles of particular interest. The journal publishes a wide range of topics related to steroids: molecular, cellular, and physiological actions of steroid hormones, steroid anti-hormones and their analogs via regulation of gene expression or other mechanisms; structure, function, regulation, and cell biology of receptors of the steroid-thyroid-retinoid superfamily; steroids and cancer: functional relationships in normal and neoplastic tissues between steroid hormones and cytokines, growth factors, growth inhibitors, and oncogenes; biosynthesis, secretion, and metabolism of steroids and of biologically-related compounds, and their regulation by peptide hormones, prostaglandins and other substances; molecular aspects of enzymes involved in steroidogenesis; novel therapeutic applications of new steroid agonists or antagonists; steroids and neuroendocrinology or neuroimmunology; steroid hormones and hypertension; steroids and their actions on bone and on the cardiovascular systems; steroid hormones during pregnancy, development, and differentiation; structure, physicochemical, chemical, and pharmacological characteristics of natural and synthetic steroids and related compounds of biological interest; innovative techniques relating to steroids and biologically-related substances; clinical and physiological studies of inborn or acquired changes of biosynthesis, metabolism or action of steroids and their receptors; comparative endocrinology and evolution of steroids; and environmental effects on steroid metabolism and action.

MANUSCRIPT SUBMISSION

Papers must be original contributions. Do not present for publication elsewhere in the same form or in another language.

Type manuscripts double-spaced. Submit in quadruplicate (1 original and 3 photocopies). Number pages, beginning with title page. Write clearly and concisely; do not exceed 20 typewritten pages of A4 size (21.0 × 29.7 cm). Submit a computer disk (5.25 or 3.5 in. HD/DD disk) containing final accepted revised version of manuscript. Send hard copy only at initial submission.

MANUSCRIPT FORMAT

On first page give title of paper, name(s) and address(es) of author(s); provide a short title (45 letters and spaces) for use in page headlines; and give name and address of corresponding author with fax number. On next page: Summary (200 words). Begin article on third page; divide into sections: Introduction, Materials and Methods, Results, Discussion (including Conclusions), Acknowledgments, References.

References: Indicate by numbers in square brackets in text. List under "References" at end of Article. For abbreviations of periodicals, use *World List of Scientific Periodicals* (4th ed, Butterworths, London).

ARTICLE IN A PERIODICAL

Name(s) of author(s), title of article, abbreviated title of journal, volume, year, first and last pages of article.

Shut D. A., Smith I. D. and Shearman R. P.: Oestrone, oestradiol-17β and oestriol levels in human foetal plasma during gestation and at term. *J. Endocr.* **60** (1974) 331–341.

BOOK

Name(s) of author(s), full title of book, publisher, town, volume (if appropriate), year and page (if a specific reference is required).

Djerassi C.: *Steroid Reactions.* Holden-Day, San Francisco (1963) p. 37.

ARTICLE IN A BOOK

Name(s) of author(s) of article, title of article, full title of book, name of editor, publisher and town, volume (if appropriate), year, first page of article, specific page reference if required.

Barraclough C. A.: Modifications in reproductive function after exposure to hormones during the prenatal and early postnatal period. In *Neuroendocrinology* (Edited by L. Martini and W. F. Ganong). Academic Press, New York, Vol. II (1967) p. 61, pp. 63–69.

Tables: Give titles that make meaning clear without reference to text. Type on separate sheets; identify by Arabic numerals.

Illustrations: Keep to a minimum. Identify by consecutive Arabic numerals. Put each on a separate sheet with author's name, title of paper, and figure number marked lightly on back. Submit line drawings in black ink on plain white drawing paper or tracing cloth. Make about twice final size. Make lettering sufficiently large and clear to permit necessary reduction. Submit photographs as glossy prints. Dye-line copies or similar reproductions may be used for copies for referees.

Number Tables and Figures on both original and copies.

OTHER FORMATS

GENERAL REVIEW—30 pages. Choice of headings is left to author.

PRELIMINARY NOTES—3 pages.

RAPID COMMUNICATIONS—Follow same format sequence as original articles.

SHORT COMMUNICATIONS—3 pages.

OTHER

Nomenclature: Follow IUPAC/IUB 1967 revised rules for nomenclature of steroids. See *J. Steroid Biochem.* **1** (1970) 143. See also *J. Steroid Biochem. Molec. Biol.* 1994; 51 (1/2): ix-xi.

Use SI symbols for units. See *Symbols, Signs and Abbreviations, Recommended for British Scientific Publications* (1969, The Royal Society, London).

Reviewers: Suggest names of 4-6 potential referees specializing in subject of paper. Provide full names and addresses.

SOURCE

October 1994, 51(1/2): v-vi.

Journal of Surgical Research: Clinical and Laboratory Investigation
Association for Academic Surgery
Association of Veterans Administration Surgeons
Academic Press

Journal of Surgical Research, Editorial Office
525 B Street, Suite 1900
San Diego, CA 92101-4495
Telephone: (619) 699-6415; Fax: (619) 699-6800

AIM AND SCOPE

The *Journal of Surgical Research* publishes original manuscripts dealing with clinical and laboratory investigations pertinent to the practice and teaching of surgery. Priority will be given to reports of clinical investigations or basic research bearing directly on surgical management, and of general interest to a wide range of surgeons and surgical investigators. Manuscripts relating to surgical specialty interests will be judged on the basis of general interest. Research need not have been done by surgeons or in surgical laboratories. The *Journal* will publish review articles and special articles relating to educational, research, or social issues pertinent to the academic surgical community. Such manuscripts should be designated as *Research Review* or *Special Article* in the cover letter, as well as on the title page. Preliminary reports of 1000 words or less which are accepted by the editorial board will be given priority for the earliest possible publication.

MANUSCRIPT SUBMISSION

In cover letter state that the same work has not been published, that it is not under consideration for publication elsewhere, and that its submission has been approved by all authors and institution where work was carried out. Include a statement that any person cited as a source of personal communication has approved such citation. Written authorization may be required.

If accepted for publication, copyright in article, including the right to reproduce article in all forms and media, shall be assigned exclusively to the Publisher. Publisher will not refuse any reasonable request by author(s) for permission to reproduce any of his or her contributions to the *Journal*.

MANUSCRIPT FORMAT

Submit manuscripts in quintuplicate (one original and four photocopies), including five sets of original figures or good quality glossy prints. Submit halftone artwork as glossy prints or originals; photocopies of halftones do not show sufficient detail for review process. Use standard manuscript form: double-spaced typing, ample margins. Number pages consecutively in order: title page, abstract, text, references, tables, legends.

Page 1: title of manuscript (70 characters); a running title (50 characters); key words; name(s) of author(s), highest degree, institutional affiliations (maximum of 2), and location; and an author's address for correspondence and mailing proofs. Page 2: abstract (250 words) stating purpose of study, methods used, results (summarize important numerical data), and conclusions. Emphasize new and important aspects of work. Organize text with internal headings designating introduction, methods, results, discussion, and references. Refer in text to tables and illustrations by number. Type footnotes together on a separate page.

References: Type double-spaced on pages separate from text. Cite references in text by Arabic numerals in square brackets and list at end of paper in consecutive order. Abbreviate journal titles in conformity with *Index Medicus,* 1981. Give complete publication data:

1. Wylds, A. C., Richard, M. T., and Karow, A. M. A model for thermal gradients during renal vascular anastomoses. *J. Surg. Res.* **43:** 532, 1988.

2. Svanes, K., Critchlow, J., Takeuchi, K., Magee, D., Ito, S., and Silen, W. Factors influencing reconstitution of frog gastric mucosa. In A. Allen, G. Flemstrom. A. Garner, W. Silen, and L. A. Turnberg (Eds.), *Mechanisms of Mucosal Protection in the Upper Gastrointestinal Tract.* New York: Raven Press, 1984. Pp. 33–39.

3. Sommer, A. *Nutritional Blindness: Xerophthalmia and Keratomalacia.* New York: Oxford Univ. Press, 1982.

Tables: Type double-spaced on pages separate from text providing a brief title for each table; number in sequence with Arabic numerals. Indicate source of nonoriginal material in a footnote. Indicate table footnotes by superscript lowercase italic letters.

Illustrations: Type legends double-spaced on pages separate from text and illustrations. Provide one legend for each illustration or illustration group; number in sequence with Arabic numerals. If an illustration is not original, indicate source at end of legend.

Illustrations must be of good quality and suitable for reproduction and for reduction as necessary to page size. On back indicate (in soft pencil) number of illustration, author's last name, and TOP.

For halftones, provide clear, sharp photographs with adequate contrast. Match contrast and density of all figures on a single plate. If not possible, provide unmounted original art. Provide an overlay indicating critical areas of micrographs. Color illustrations are accepted if authors defray cost.

Credits and Permissions: Obtain and submit written permission to use nonoriginal material (quotations exceeding 50 words, tables, or illustrations) from author and publisher of original.

OTHER

Drugs: Identify precisely all drugs and chemicals used, including generic name(s), dosage(s), and route(s) of administration.

Electronic Submission: Manuscripts may be submitted on personal computer disks after manuscript has been accepted and all revisions have been incorporated onto disk. Label disk with type of computer used, type of software and version number, and disk format. Tabular material will be typeset. Prepare art as camera-ready copy.

Send a hard copy printout that exactly matches disk file. File names must clearly indicate content of each file.

For further information on preparing disks, contact Editorial Supervisor, Journal Division, Academic Press, Inc., 1250 Sixth Avenue, San Diego, California 92101; PH: (619) 699-6415; FAX: (619) 699-6800).

Experimentation, Animal: Indicate whether institution's or National Research Council's *Guide for the Care and Use of Laboratory Animals* was followed.

Experimentation, Human: Indicate whether procedures followed were in accordance with ethical standards of committee on human experimentation of institution in which experiments were done or in accordance with Helsinki Declaration of 1975. Do not use patients' names, initials, or hospital numbers.

Statistical Analyses: Describe statistical methods employed in evaluation of data, either in methods or with results.

SOURCE

July 1994, 57(1): back matter.

Journal of the American Academy of Child & Adolescent Psychiatry
American Academy of Child & Adolescent Psychiatry
Williams & Wilkins

John F. McDermott, Jr.
University of Hawaii
School of Medicine at Kapiolani Medical Center
1319 Punahou Street, Rm. 633
Honolulu, Hl 96826-1032

AIM AND SCOPE

The *Journal's* purpose is to advance theory, research, and clinical practice in child and adolescent psychiatry. It is interested in manuscripts from a variety of viewpoints, including genetic, epidemiological, neurobiological, cognitive, behavioral, and psychodynamic. Studies of diagnostic reliability and validity as well as psychotherapeutic and psychopharmacological treatment efficacy are encouraged.

MANUSCRIPT SUBMISSION

Major manuscript categories are Regular Articles (research reports) and Case Studies. Special Articles (theoretical or critical analyses of literature) are invited. Submit suggestions for Special Sections (a group of related articles), Debates, and Clinical Perspectives directly to appropriate Assistant Editor.

Manuscripts must represent original material and may not have been submitted or accepted elsewhere, either as a whole or any substantial part. Piecemeal publication of small amounts of data from one study is not acceptable. Each publication must report enough new data to make a significant and meaningful contribution to the development of new knowledge or understanding. When data from the same study are reported in more than one publication, inform editor, either in body of manuscript or in cover letter about other manuscripts that have been published, are in press, have been submitted elsewhere, or are in preparation. Inform editor how manuscript is different from other manuscripts. Submit copies of closely related manuscripts that report data from the same study and that have been published, are in press, or have been submitted for publication.

Base authorship credit only on substantial contributions to conception and design or analysis and interpretation of data; drafting article or revising it critically for important intellectual content; and final approval of version to be published. Participation solely in acquisition of funding, collection of data, or supervision of research group does not justify authorship. Each author should have participated sufficiently in work to take public responsibility for content.

Manuscripts are subject to peer review and judged by four criteria: Is material new, true, important, and comprehensible?

When a paper is accepted, Editor sends author an agreement authorizing American Academy of Child and Adolescent Psychiatry to publish article and to own copyright.

Submit written approval of manuscript submission by all authors and expressly transfer copyright.

Include either statements of following in cover letter:

For papers submitted by all authors except those whose work is part of employment with federal government:

In consideration of the *Journal's* taking action in reviewing and editing my (our) submission [title of article], the author(s) undersigned hereby transfer(s), assign(s), or otherwise convey(s) all copyright ownership to the American Academy of Child and Adolescent Psychiatry in the event that such work is published in the *Journal* I (we) warrant that the material contained in the manuscript represents original work, has not been published elsewhere, and is not under consideration for publication elsewhere.

For papers prepared as part of an author's employment with the federal government:

The work described in [title of article] was done as part of my (our) employment with the federal government and is therefore in the public domain. The author(s) undersigned warrant(s) that the material contained in the manuscript represents original work, has not been published elsewhere, and is not under consideration for publication elsewhere.

Manuscript Format

Research reports follow IMRAD format, with separate sections titled Introduction, Method, Results, and Discussion that describe the problem, how it was studied, findings, and what they mean. In Introduction state purpose of study and a priori hypotheses and recent and relevant literature. In Method section describe design, with information on sample selection, inclusion/exclusion criteria, method of randomization (if any), determination of sample size (include power calculation), and whether or not study was "blind" in any way. State response and outcome variables. Data collection information includes response rates or follow-up rates, and possible sampling bias. Discuss representativeness of sample selected (controls and patients). Describe all analyses, with names of specific statistical tests used. Justify and reference use of unusual statistical techniques. If multiple comparisons are unavoidable, use an appropriate adjustment to control Type I error. State whether tests were one- or two-tailed.

In Results section present summary statistics (such as means and standard deviations) so readers can verify results. When reporting significant results, include statistical test used, test value, degree(s) of freedom, and probability level (p value). Report confidence intervals on main findings. Keep number of tables to a minimum, generally not more than 4 manuscript pages.

In Discussion section consider both statistical and clinical significance. Integrate findings into what is known and how these findings advance theory or practice. Point out and discuss any weaknesses in study design or execution.

Include a subsection entitled Clinical Implications in which relevance for clinical practice or developmental theory is specifically considered.

Print manuscripts on 8.5×11 in. bond paper; 10 characters per inch (preferred font: Courier); make all four margins 1 in. wide. Double-space all copy, including title page, abstract, list of references, tables, and figure captions. Number pages consecutively throughout. Blinding

beyond the title page is author's responsibility. Each manuscript should contain the following elements in order; begin each on a separate sheet.

Cover Sheet: On an unnumbered page, provide an abbreviated form of main title for a running head (40 characters and spaces). Give name, address, telephone number, and fax number of corresponding author. Include word count, including references and tables.

Title Page: On first numbered page state title. Use 10-15 words to state major variables rather than details of study. Full names of authors and their academic degrees follow on a separate line. Include a paragraph giving authors' affiliations, credit lines, and name and address for reprint requests. Paragraph should not exceed 120 words.

Abstract: 200 word structured abstract that can stand on its own. Do not include general statements that refer to text. Do not cite references. Follow with 3 to 5 key words for indexing.

Abstracts for research articles: Objective: primary purpose of study; Method: design of study and main outcome measures; Results: key findings; and Conclusions: including clinical significance.

Abstracts for special review articles: Objective: primary purpose of review; Method: data sources, study selection (number of articles reviewed and how selected); Results: methods of data synthesis and key findings; and Conclusions: summary statement of what is known including potential applications and research needs.

Abstracts for case studies are 100 words, unstructured.

Text: Begin on third numbered page. Spell out all abbreviations (other than units of measure) at first use. Do not use footnotes to text. If more than two authors in a citation, use "et al." after first author's name. If more than one citation appear together, arrange in alphabetical order. Use diagnostic classifications from *DSM-IV.*

References: Arrange reference list in alphabetical order by author names; do not number. Verify accuracy of references. Use initials and surnames of authors. List all authors' names. If several papers by one author are cited, list in chronological order. When an author has published several papers in the same year, follow date by a, b, c, d, etc. Refer to *Index Medicus* for appropriate journal abbreviations. Do not include unpublished manuscripts, submitted manuscripts, or personal communications in reference list; note in text. Cite "in press" manuscripts in reference list.

Achenbach TM, Edelbrock C (1983), *Manual for the Child Behavior Checklist and Revised Child Behavior Profile.* Burlington: University of Vermont Department of Psychiatry

Achenbach TM Verhulst RC, Baron GD, Akkerhuis GW (1987), Epidemiological comparisons of American and Dutch children I. Behavioral/emotional problems and competencies reported by parents for ages 4 to 16. *J Am Acad Child Adolesc Psychiatry* 26:317–325

Bell AC (1930), *The Child and His World.* New York Macmillan, 1975 (If year of original publication does not coincide with the edition referred to, add year of publication of edition used after publisher's name.

Terr LC (1985), Children traumatized in small groups I: *Post-Traumatic Stress Disorder in Children.* Erh S, Pynoos RS, eds. Washington, DC American Psychiatric Press, Inc., pp 47–70

Terr LC (in press), *Too Scared to Cry.* New York: Harper & Row

US Department of Health and Human Services (1987), *Report of the Surgeon General's Workshop on Children with HIV Infection and Their Families.* (DHHS Publication HRS-D-MC 87-1). Washington, DC: US Government Printing Office

Tables: Type each on a separate page with a title and legend. Number in Arabic numerals, consecutively in order of appearance. Double-space and omit all underlining. Properly align numbers, both horizontally and vertically. Use brief headings for columns and, if necessary, use abbreviations and explain them in a key to abbreviations in a footnote. Use superscript letters, ^{a,b} for footnotes. Keep footnotes to a minimum. Avoid repeating information in text and in tables. Cite each table in text in order as Table 1, Table 2, etc. Do not submit tables or figures that have appeared in other publications. Cite previously published materials as references only.

Figures: Submit camera-ready, suitable for reproduction, preferably as Xerox or glossy prints or laser copy on high quality paper. Dot-matrix printing is not acceptable. Do not mount artwork. Place a pressure-sensitive label on back of each piece with figure number, last name(s) of author(s), and TOP. Prepare with size of *Journal* page (one column or page width) in mind. List captions together on a separate sheet. Cite each figure in text in order as Figure 1, Figure 2, etc.

OTHER FORMATS

RESEARCH REPORTS—7000 words, including title page, abstract, references, and tables (100 words for a double-spaced table that fills one-half of a vertical page).

CASE STUDIES—3000 words.

Longer manuscripts will not be accepted and may be returned unreviewed. Submit in quadruplicate (original and three clear copies).

OTHER

Drugs: Use generic terms. When necessary to refer to proprietary name, list name in parentheses, with a registered mark ® attached, after generic term.

Experimentation, Human: Conduct research ethically with due regard to informed consent. Protect patient's anonymity in case studies; avoid any identifying information.

SOURCE

January 1995, 34(1): back matter.

Journal of the American Academy of Dermatology

American Academy of Dermatology
Mosby Yearbook

Richard L. Dobson, MD
Department of Dermatology
Medical University of South Carolina
171 Ashley Ave.
Charleston, SC 29425-2215
Telephone: (803) 792-9155; Fax: (803) 792-9157

AIM AND SCOPE

The *Journal of the American Academy of Dermatology* is a refereed journal designed to meet the continuing education needs of the Academy members and the international dermatologic community.

MANUSCRIPT SUBMISSION

Acknowledge, on title page, all funding sources that supported work and all institutional or corporate affiliations of authors. Disclose in a cover letter, any commercial associations that might pose a conflict of interest, including consultant arrangements, stock or other equity ownership, patent licensing arrangements, or payments for conducting or publicizing study. Disclosure will be held in confidence during review and will not influence editorial decision. If paper is accepted, Editor will determine how to disclose any conflict of interest.

For direct quotations, tables, or illustrations that have appeared in copyrighted material, submit written permission for use from copyright owner and original author with complete information as to source.

Include following statement signed by senior or corresponding author with manuscript: "The undersigned author transfers all copyright ownership of the manuscript referenced above to the American Academy of Dermatology in the event the work is published. The undersigned author warrants that the article is original, is not under consideration by another journal, and has not been published previously. I certify that all authors listed have seen this paper (and any subsequent revisions), and have approved the final version. I sign for and accept responsibility for releasing this material on behalf of any and all coauthors."

MANUSCRIPT FORMAT

Submit original numbered copy of manuscript and all supporting material plus two photocopies (not carbon copies). Type double-spaced on one side of 8.5×11 in. paper with adequate margins.

Use generic names. If a trade name is used, it should follow generic name in parentheses at first mention. Thereafter, use generic names only.

Title Page: Include title, authors full names, highest earned academic degrees and institutional affiliations and locations. Designate a corresponding author (provide address and telephone numbers).

Abstract: 150 words typed double-spaced on a separate sheet. Organize abstracts of Clinical and Laboratory Studies, Therapy, Dermatopathology, and Dermatologic Surgery papers as: Background (what major problem prompted the study?); Objective (what is the study's purpose?); Methods (how was study done?); Results (what are the most important findings?); Conclusion (what is the single most important conclusion?)

Use headings before each description. Do not state that results or other data will be presented or discussed.

References: Do not cite personal communications in reference list; state parenthetically in text. Identify references in text by superscript Arabic numerals in order of mention. Type list double-spaced at end of text in numeric sequence. Follow format of "Uniform Requirements for Manuscripts Submitted to Biomedical Journals" (Vancouver style) (*JAMA* 1993;269:2282-6). Journal title abbreviations follow *Cumulated Index Medicus*. If three or fewer authors, list all; if four or more, list first three and add *et al.*).

JOURNAL ARTICLES

Moore TJ, Dluhy RG, Williams GA, et al. Nelson's syndrome: frequency, prognosis, and effect of prior pituitary irradiation. Ann Intern Med 1976;85:731-4.

BOOKS

Hunt TK, ed. Wound healing and wound infection: theory and surgical practice. New York: Appleton-Century-Crofts, 1980:99-104.

CHAPTERS IN BOOKS

Freinkel RK, Freinkel N. Dermatologic manifestations of endocrine disorders. In: Fitzpatrick TB, Eisen AZ, Wolff K, et al, eds. Dermatology in general medicine. New York: McGraw-Hill, 1986:2063-81.

Illustrations: Tables, figures, and legends should supplement, not duplicate, text. A reasonable number or halftone photographs and line drawings are published at no charge.

For color photographs, submit original transparencies and two sets of unmounted prints on glossy (smooth surface) paper. Polaroid prints are not acceptable. Thirty-five mm transparencies are normally enlarged to twice original size. If deviating from standard, indicate upon submission. Indicate TOP for each print (or transparency). For black-and-white illustrations, send three sets of glossy prints. Do not make back-and-white photographs from color slides. Prepare original drawings or graphs in black India ink or typographic (press-apply) lettering. Typewritten or freehand lettering is unacceptable. Have lettering done professionally; keep in proportion to drawing, graph, or photograph. Do not send original artwork, x-ray films, or ECG strips. Submit glossy print photographs, 3 × 4 in. (minimum) to 5 × 7 in. (maximum), with good black-and-white contrast or color balance. Be consistent in size. Note special sizing instructions. Number illustrations in Arabic numbers according to mention in text. Indicate (lightly in pencil) first author's last name, figure number, and top or bottom on back of each illustration.

Legends: Type double-spaced on a separate sheet; insert after references. For previously published illustrations give full credit to original source.

Tables: Make self-explanatory. Number in Roman numerals according to mention in text. Provide a brief title for each. If a table or any data has been published previously, give full credit to original source in a footnote.

OTHER FORMATS

BOOK REVIEWS—Books and monographs (domestic and foreign) are reviewed depending on their interest and values. Send books to Editor, Richard L. Dobson, MD. No books will be returned.

BRIEF COMMUNICATIONS—Four double-spaced typed pages, two illustrations. No abstract. Case reports or clinical observations of unusual interest.

CASE REPORTS—Brief individual case reports of unusual interest.

CLINICAL AND LABORATORY STUDIES—Original, in-depth clinical and investigative laboratory research articles.

CLINICAL REVIEW—A review of several clinical studies or a pattern appearing in several cases.

CONTINUING MEDICAL EDUCATION—In-depth, substantiated, educational articles presenting core information for continuing medical education of practicing dermatologists. Submit answers to accompanying questions to American Academy of Dermatology office for CME credit. Submit three numbered photocopies, as well as original.

CORRESPONDENCE—Brief letters or notes to the Editor that comment on articles that have appeared in the *Journal*. Submit individual case reports as Brief Communications.

CURRENT ISSUES—Brief, provocative, opinionated communications, not necessarily documented, on one limited subject.

DERMATOLOGIC SURGERY—Articles emphasizing surgical aspect of dermatology.

DERMATOPATHOLOGY—Articles emphasizing microscopic changes in skin disease.

EDITORIALS—Brief, substantiated commentary on limited subjects.

MEETING REPORTS—Concise statements summarizing important material for dermatologists presented at major dermatology meetings and seminars.

PEARLS OF WISDOM—Three double-spaced typed pages, 3 illustrations. Special approaches useful in diagnosis and treatment or surgical management of skin diseases.

SPECIAL REPORTS—Items of special interest.

THERAPY—In-depth critical reviews of a therapeutic modality or treatment procedure.

OTHER

Color Charges: $375 per page (one side).

Electronic Submission: Final version of accepted manuscript may be submitted on diskette with three copies of printout. Content of disk must match exactly final version of printout. Guidelines for submission will be sent by editorial office.

Experimentation, Human: Submit patient consent forms for publication of recognizable photographs. Identify patients by numbers and/or letters, not by name, initials, or hospital record number. Institutional consent must also be available.

Style: Conform to acceptable English usage. Consult latest edition of *The Chicago Manual of Style* (The University of Chicago Press) or the *Manual of Style* (AMA) for current usage. Limit abbreviations to those in general usage.

Units: Express weights and measurements in metric units. Express temperatures in degrees centigrade.

SOURCE

January 1995, 32(1): 24A-26A.

Journal of the American Chemical Society

American Chemical Society

Allen J. Bard
Department of Chemistry and Biochemistry
University of Texas
Austin, TX 78712
Telephone: (512) 471-1838; Fax: (512) 471-0088

AIM AND SCOPE

Journal of the American Chemical Society is devoted to the publication of research papers in all fields of chemistry. Articles, Communications to the Editor, Book Reviews, and Computer Software Reviews are published.

MANUSCRIPT SUBMISSION

Submission of a manuscript to the *Journal* implies that work has not received prior publication and is not under consideration for publication elsewhere in any medium including electronic journals and public computer data bases.

Articles most appropriate for publication are those that deal with some phase of pure chemistry as distinguished from "applied" chemistry. Articles of high scientific quality and originality that are of interest to the wide and diverse readership of the *Journal* are given priority. Articles that mainly repeat findings previously published as Communications and only incorporate experimental data will be referred to specialized journals.

Estimating Length: Title and authors' names and affiliation(s) equal 150 words. To calculate word count: (number of lines of text × inches per average line × characters per inch) /8.4 characters per word. Use the same equation, but with a divisor of 10 characters per word, to estimate words in footnotes, references, and captions. Estimate tables at 7 words per line (including title and column headings) for a single journal column a table and at 14 words per line for a full page width table. Estimate figures, schemes, structures, and tables that contain structural components at 50 words per inch for single-column width and at 100 words per inch for double-column width. Reduction to single- or double-column width depends on amount of detail and size of lettering used (design illustrations so smallest letter will not be less than 2 mm after reduction). Do not include abstract in word count estimate.

Submit four copies of manuscript. Submit three copies of work cited as "in press" or submitted for publication. Supply three copies of supplementary material. Do not fax. Provide postal address, telephone and fax numbers and e-mail address of submitting author. Include a signed ACS copyright status form (in first issue each year).

MANUSCRIPT FORMAT

Present materials with the utmost conciseness consistent with clarity. Do not submit extensive literature reviews. Designate well-known procedures either by name or literature references. Make notes acknowledging financial assistance, citing theses, or indicating presentation at a meeting brief and preferably combined under one number. Mark name of corresponding author with an asterisk; include footnote with address if not obvious from manuscript heading. Do not include addresses and official titles or connections of other authors.

In introduction give only enough background material to show why work was done. Do not intermingle Results and Discussions. Include a Conclusion section at end of paper, briefly summarizing principal conclusions of work.

Type manuscripts double-spaced (3 lines per inch), on one side only of 8.5 × 11 in. or A4 paper in a type size ≤ 88 characters/6 in. If printer is used, output must be high quality. Number all pages consecutively, including references and notes, figure captions, and tables; group in sequence following text. Include engraver's copies of graphs, drawings of apparatus, and structural formulas; place at end; indicate in typescript where to insert.

Consult recent issues of the *Journal* and *The ACS Style Guide*; (American Chemical Society: Washington, D.C., 1986) for style, arrangement, and orthography of manuscripts.

In main heading, give address of contributing authors and laboratory with city, state or country, and postal code. Make title as briefly informative as possible; do not include formulas or other symbols. Use subheadings sparingly; paragraph indent side headings and run into following text with a period. Capitalize registered trademark names; do not capitalize trade and trivial names; give; give chemical name or composition in parentheses or in a footnote at first occurrence. Use metric units (SI) for all quantities of length, area, and volume. Number figures, tables, schemes, and charts with Arabic numerals.

Abstracts: State briefly reason for work, significant results, and conclusions.

References: Verify accuracy. Number references to literature and all notes regardless of nature (except those in tables) in one consecutive series. Type reference numbers in text as unparenthesized superscripts; enclose in parentheses in references and notes section following text. Do not use bibliographic references to classified documents and reports or to unpublished material not generally available.

   (1) Doe, J. S.; Smith, J.; Roe, P. *J. Am. Chem. Soc.* **1968**, *90*, 8234-8265.

Include initials of authors and journal abbreviation used in *Chemical Abstracts Service Source Index (CASSI) 1907-1989 Cumulative* and its supplements.Use inclusive pagination.

References to books include author, title, volume, publisher, address, date of publications, and pages.

(2) Smith, A. B. *Textbook of Organic Chemistry*; D. C`. Jones: New York, 1961; pp 123–126.

Tables: Use only to present information more effectively than text. Design to occupy fully a single column (50–75 characters and spaces), or full width of a page (110–150). Avoid any arrangement which unduly increases a table's depth. Make column heads brief; use abbreviations. Do not number lines of data. Do not give run numbers unless needed for reference in text. Do not use columns for only one or two entries. Do not repeat the same entry numerous times consecutively. Do not include data which are deducible easily by simple arithmetic from given data. Group tables at end of manuscript. Tables need not be placed alone on separate sheets.

Figures and Graphs: Submit camera ready. Prepare diagrams, graphs, charts, and other artwork using dark black ink on high quality white, smooth opaque bond paper. Avoid thin transparent or textured papers. Submit original artwork or photographic prints of originals, not photocopies. Use a high-quality graphics plotter rather than a dot-matrix or laser printer. If dot matrix or laser produced, use high quality paper and choose highest resolution output option. Put figures and reproductions on 22×28 cm paper. Design illustrations to fit width of one journal column (~3.25 in.). Make width of original drawings twice publication size; do not vary size of lettering and plotted points; make large enough that smallest letter will not be less than 2 mm after reduction. For symbols for plotted points use open or closed circles, triangles, or squares; do not use unusual symbols. Mark each piece of illustration copy on margin or back with name of author and brief article title. Do not write on front or back of image area of photograph. Number figures in series. Type all captions together on separate sheet. Group both figures and captions together at end of manuscript. Author bears incremental charges for color reproduction; charges given upon request. Include a letter acknowledging willingness to pay cost with revised manuscript.

Formulas and Equations: Arrange empirical and structural formulas and mathematical and chemical equations to fill width of a single- or double-column. Write subscripts and especially superscripts with care; set up exponents in a single line, as $e^{-60/RT}$. Space all signs such as $+$ $-$ $=$ $<$ $>$; do not space components of mathematical products. Submit structural formulas as copy suitable for direct photographic reproduction, either as artwork or as prints produced using drawing packages such as ChemDraw; fill space economically, using bonds and arrows vertically, horizontally, or at 45° angles. Do not use structures when a simple formula will suffice. Do not use multiple lines unnecessarily. Arrange all formula matter carefully; execute (preferably typewritten) with special attention to correctness of symbols, location of subscripts, superscripts, and electric charges, and placing and close join-up of single and multiple bond lines. Show phenyl groups as C_6H_5 or Ph, not as ϕ. Place small numbers showing ring positions outside rings, and if electric charges are shown, use $+$ and

–. Use a copy of structure in text at point of proper citation, but, when originals are provided, group at end of manuscript.

All art must be complete. Letter compound numbers and other material to appear in copy; it will not be added.

Analyses: Provide adequate evidence to establish both identity and purity for new compounds. Include elemental analyses. Data collected in tables may either be printed, deposited as supplementary data, or substituted by a footnote which states, e.g., "Satisfactory analytical data ($\pm 0.4\%$ for C, H, halogen, etc.) were reported for all new compounds listed in the table." Include data for examination by reviewers and editor. When necessary to rely on a normally unacceptable analysis, give figures in full, e.g., "Anal. ($C_{12}H_{20}N_2O_8S$) C, H, S; N: calcd, 6.99; found, 7.55." In tables, report such analyses as footnotes.

Produce high-resolution mass spectrometric molecular weights in lieu of elemental analyses in some cases. Include independent evidence of compound purity. Not required, for instance, when compound is too unstable at room temperature or when it is only a minor byproduct of secondary importance. For high-resolution mass spectral data, for materials with molecular weights below 1000, measured mass should generally agree to 5 parts per million or better with calculated mass. For molecular weights above 1000, give measurements at unit mass resolution. Report calculated and measured relative intensities of each significant signal in the molecular envelope.

OTHER FORMATS

BOOK REVIEWS AND COMPUTER SOFTWARE REVIEW—At invitation of the editor. Submit suggestions for reviewers; unsolicited reviews are not accepted.

COMMUNICATIONS TO THE EDITOR—Two journal page (2000 words) reports of unusual urgency, significance, and interest. State how manuscript meets criteria of urgency and significance. State principal conclusions in opening sentences. Major concepts cannot have appeared previously as a report or publication. If a previous Communication by the same authors has already appeared and present manuscript describes a technical improvement or increase in scope of work, it is not acceptable unless there is also a novel conceptual advance. If manuscript describes work of general synthetic utility, deposit sufficient experimental information as supplementary material to enable readers to employ technique. Additional supporting information may be requested even if space will not permit its publication. Do not submit multiple Communications on the same topic within a short time. Describe such work in a regular Article.

OTHER

Computations: When details are essential, give sufficient detail either within paper or in supplementary material to enable readers to reproduce calculations. This includes, e.g., force field parameters and equations defining models or references to material available in open literature.

Crystallographic Data: For determinations of molecular structure by X-ray crystallography, submit supplementary material containing tables of positional and thermal parameters. Tables of bond lengths and angles are published.

If publication of more complete crystallographic details is planned, state in a footnote, include authors' names and journal of publication. X-ray data relating to stereochemical confirmation of synthetic intermediates will not be published except for a statement (in text or footnote) that such determinations have been performed. This pertains to organic structures without any unusual or novel structural features or without any intrinsic structural interest except for stereochemistry. ORTEP figures or stereochemical structural representations and related tabulated data will not be published. Include these data and figures in submitted supplementary materials. In manuscripts describing novel structural features confirmed by X-ray analyses, these are an integral part of the research findings and will be published.

Electronic Submission: Manuscripts accepted for publication that have been prepared with WordPerfect or Microsoft Word can be used for production. Submit disk with only with final accepted version of manuscript. Disk must exactly match final version accepted in hardcopy. Use document mode. Do not include any page formatting instructions, justify margins or insert spaces before punctuation. Check disk for viruses. Label disk with manuscript number and first author's name. Provide name and version of software used, platform and file names on disk. Put text and tabular material in one file. Do not integrate graphic material.

Mathematical Expressions: Indicate italic or boldface type. Plainly write Greek or unusual characters or explain in annotation. Write simple fractional expressions with a slant line, so that only a single line of type is required. Carefully make and place subscripts and superscripts. Do not use degree sign ° or a small o to represent zero.

Nomenclature: Conform with current American usage. Use systematic names of Chemical Abstracts Service or IUPAC. For *Chemical Abstracts* nomenclature rules, see Appendix IV of current *Chemical Abstracts Index Guide*. For a list of ring systems, including names and numbering systems, see *Ring Systems Handbook* (American Chemical Society: Columbus, OH, 1988). For IUPAC rules, see *Nomenclature of Inorganic Chemistry, Recommendations, 1990;* (Blackwell Scientific Publications: Oxford, 1990); *Nomenclature of Organic Chemistry,* Sect. A-F and H (Pergamon Press: Elmsford, NY, 1979). Another important reference is *Compendium of Biochemical Nomenclature and Related Documents* (Portland Press, Ltd.: London, 1992).

Rigid and consistent conformance to these rules is not required, but give approved names of compounds at least once (in parentheses or in notes) and in proper relation to compound name. Minimize use of code or letter designations for compounds. For CA nomenclature advice consult Manager of Nomenclature Services, Chemical Abstracts Service, P.O. Box 3012, Columbus, OH 43210. A name generation service is available through the Registry Services Department, Chemical Abstracts Service, P.O. Box 3343, Columbus, OH 43210.

Page Charges: $25 per page. Payment does not affect acceptance or scheduling of papers.

Reviewers: Suggest persons competent to referee manuscript. Also request that certain persons not be used.

Spectra: Reproduction of spectra, or relevant segments, will be published only if concise numerical summaries are inadequate for paper's purpose. Papers dealing primarily with interpretation of spectra and those in which band shape or fine structure need to be illustrated may include such spectra. Furnish accurate drawings or clear glossy photographs of material. Spectra will not be published as adjuncts to characterization of compounds. Numerically summarize routine infrared, electronic, NMR, and mass spectra of new compounds in Experimental Section.

Supplementary Material: Extensive tabular, graphical, or spectral data of interest only to those readers who may need more complete data for comparison purposes are "supplementary material" and should be relegated to the microfilm edition of the *Journal*. Use to save journal space and make clearer and more readable presentations. Separately identify material for microfilm edition with authors' names and manuscript title. Put in a form easily handled for photoreproduction. Put figure captions, titles to tables, and other identifying captions on the same page as figures or tables and not on a separate sheet. Preferable page size is 22 × 28 cm with readable material aligned parallel to 22-cm side. Submit figures and illustrative material as original India ink drawings or black and white matte prints (not glossies) of originals. Make type size large enough for easy reading. Place a statement of availability of supplementary data at end of paper:

Supplementary Material Available: A listing (describe such) (no. of pages). Ordering information is given on any current masthead page.

SOURCE
January 12, 1994, 116(1): 7A-10A.
Instructions for authors and copyright transfer form are in first issue of each volume.

Journal of the American College of Cardiology
American College of Cardiology
Elsevier Science Publishing

William W. Parmley
415 Judah Street
San Francisco, CA 94122
Telephone: (415) 759-4185; Fax: (415) 759-0251

AIM AND SCOPE

The *Journal of the American College of Cardiology* (JACC) publishes peer-reviewed articles on all aspects of cardiovascular disease, including original clinical studies, review articles and experimental investigations with clear clinical relevance. JACC regularly features articles on coronary artery disease, congenital heart defects, surgery, results of new drug trials, invasive and noninvasive diagnostic and therapeutic techniques, findings from the laboratory and large-scale multicenter studies of drugs and new therapies.

In general, case reports will not be considered. As the official journal of the American College of Cardiology,

JACC publishes abstracts of reports to be presented at the Annual Scientific Sessions of the College as well as reports of the College-sponsored Bethesda Conferences.

MANUSCRIPT SUBMISSION

In cover letter state that: paper is not under consideration elsewhere, none of paper's contents have been previously published (including symposia, proceedings, transactions, books, articles, etc., except as abstracts not exceeding 400 words), and all authors have read and approved manuscript. Explain exceptions. Each author must have contributed significantly to work. If more than four authors, substantiate each contribution. Authorship includes all of the following: conception and design or analysis and interpretation of data or both; drafting or critically revising manuscript for intellectual content; and final approval. Specify corresponding author.

Disclose any commercial associations that might pose a conflict of interest. Acknowledge all funding sources supporting work in a footnote on title page, with institutional affiliations of authors, including corporate appointments. Disclose other kinds of associations, such as consultancies, stock ownership or other equity interests, or patent-licensing arrangements, to Editor in cover letter. Direct questions about this policy to Editor-in-Chief.

Upon acceptance written transfer of copyright to American College of Cardiology signed by all authors will be required. Follow guidelines in "Uniform Requirements for Manuscripts Submitted to Biomedical Journals," 3rd edition, International Committee of Medical Journal Editors (*Ann Intern Med* 1988:108:258–65 and *Br Med J* 1988:296:401–5).

Submit four complete sets of manuscript, including four sets of figures in separate envelopes. Laser prints or clean photocopies are adequate for line drawings, charts, and graphs. For halftone illustrations, including angiograms, photomicrographs and color figures, submit two sets of glossy prints, with laser prints or photocopies. Additional sets may be requested upon acceptance. Provide three copies of supplementary materials, such as "in press" references.

MANUSCRIPT FORMAT

Type double-spaced throughout on one side only of 8.5×11 in. opaque white bond paper with 3 cm margins all around (8 cm at bottom of title page). Use a standard 10 cpi font or laser printer font no smaller than 12 points. Arrange manuscript as: title page, structured abstract, condensed abstract, text, acknowledgments (if any), references, legends, tables. Number pages consecutively, beginning with title page 1.

Title Page: Include title, authors' names (including full first or middle names, degrees and, where applicable, FACC) and a short running title (45 characters). List departments and institutions with which authors are affiliated. Indicate specific affiliations if work was generated from more than one institution (use footnote symbols in Tables). Provide information on grants, contracts, or other forms of financial support and list city and state of all foundations, funds, and institutions.

Under heading "Address for correspondence: . . ." give full name, complete postal address with zip code, telephone and fax numbers of corresponding and reprint request author(s).

Structured Abstract: 200 to 300 words presenting essential data in five paragraphs headed: Objectives, Background, Methods, Results, and Conclusions. Use complete sentences; do not use abbreviations (other than units of measurement). All data must appear in text, tables or figures. For information on preparing structured abstracts, see Haynes RB, Mulrow CD, Huth EJ, Altman DG, Gardner MJ. "More informative abstracts revisited" (*Ann Intern Med* 1990;113:69–76). Review articles have nonstructured abstracts.

Condensed Abstract (for Table of Contents): 100 words, one-paragraph stressing clinical implications. Do not include data not in text to tables.

Text: Use headings and subheadings in Methods, Results and Discussion sections. Cite references, figures and tables in text in numerical order according to order of mention.

Acknowledgment: Provide letters of permission from persons listed.

References: Identify in text by Arabic numerals in parentheses on the line. Type double-spaced on sheets separate from text and numbered consecutively in order of mention in text. Do not cite personal communications, manuscripts in preparation and other unpublished data in reference list; include in text in parentheses. Cite full papers, rather than abstracts. Do not cite abstracts older than two years. Identify abstracts by abbreviation "abstr" in parentheses; letter to the editor, by "letter" in parentheses.

Journal references contain inclusive page numbers; book references contain specific page numbers.

Abbreviate journal titles as in *Index Medicus,* National Library of Medicine.

PERIODICAL

(List all authors if 6 or less; otherwise list first 3 and add et al.; do not use periods after authors' initials)

> 5. Glantz SA. It is all in the numbers. J Am Coll Cardiol 1993;21:835-7.

CHAPTER IN BOOK

> 27. Meidell RS, Gerard RD, Dambrook JF. Thrombolytic agents. In: Molecular Cardiology. Cambridge, MA: Blackwell Scientific Publications, 1993;295-324.

BOOK (PERSONAL AUTHOR OR AUTHORS)

> 23. Cohn PF. Silent Myocardial Ischemia and Infarction. 3rd ed. New York: Marcel Dekker, 1993;33.

Figure Legends: Type double-spaced on sheets separate from text. Figure numbers correspond with order of presentation in text.

Identify all abbreviations after first mention or in alphabetical order at end of each legend. Explain all symbols.

Submit written permission from publisher and author to reproduce previously published figures.

Figures: Submit four sets of laser prints or clean photocopies in separate envelopes. Submit two sets of glossy prints for all half-tone or color illustrations.

Design figures, particularly graphs, to take up as little space as possible. Make lettering of sufficient size to withstand reduction. Optimal type size, after reduction, is 8 points. Keep type and symbols similar in size. Maximal width for one column figures is 3.25 in.; for two column figures, 6 7/8 in.

Have graphs and drawings professionally prepared or done by computer and reproduced as high quality laser prints. Use only black and white, not gray in charts and graphs. Make decimals, lines, etc. strong enough for reproduction.

Indicate crop marks on photomicrographs to show only essential field. Designate special features by arrows. Symbols, arrows or letters in photomicrographs must contrast with background.

Indicate first author's last name, figure number and TOP on back of each illustration in light black pencil, preferably on a gummed label.

Put figure title and caption material in legend not on figure.

Estimates for color work are provided on acceptance All costs are charged to author(s).

Tables: Type double-spaced on separate sheets with table number and title centered above table and explanatory notes below. Table numbers are Arabic and correspond with order of presentation in text.

Provide a footnote to each table identifying in alphabetical order all abbreviations. Footnote symbols appear in order: *, †, ‡, §, ||, ¶, **, †† . . . .

Make tables self-explanatory; do not duplicate data in text or figures.

Provide written permission from publisher and author to reproduce previously published tables.

OTHER FORMATS

EDITORIAL COMMENTS—Invited only.

EDITORIALS AND EDITORIAL REVIEWS—Succinct opinion pieces.

LETTERS TO THE EDITOR—500 words, focussing on a specific article from JACC. Do not include original data. Type double-spaced and include cited article as a reference. Provide a brief title and sign letter with name and institutional affiliation. Replies will be solicited by Editors.

REVIEW ARTICLES—Adhere to preferred length guidelines. Detail in cover letter how submission differs from existing reviews of subject.

OTHER

Abbreviations: Do not use in text except for units of measure. Consult "Uniform Requirements for Manuscripts Submitted to Biomedical Journals, 3nd edition," (*Ann Intern Med* 1988;108:258-65 and *Br Med J* 1988;296:401-405) for appropriate use of units of measure.

Experimentation, Animal: Conform to "Position of the American Heart Association on Research Animal Use" adopted November 11, 1984 by AHA. If equivalent guidelines are used, so indicate.

AHA position: Animal care and use must be by qualified individuals supervised by veterinarians; all facilities and transportation must comply with current legal requirements and guidelines. Research involving animals is only done when alternative methods to yield needed information are not possible. Anesthesia must be used in all surgi-

cal interventions, all unnecessary suffering should be avoided and research must be terminated if unnecessary pain or fear results. Animal facilities must meet standards of the American Association for Accreditation of Laboratory Animal Care (AAALAC).

Experimentation, Human: Obtain informed written consent. Provide details of procedure and indicate that institutional committee on human research has approved study protocol. If radiation is used, specify radiation exposure in Methods.

Statistical Analyses: In Methods section include a subsection detailing statistical methods, including specific methods used to summarize data, methods used for hypothesis testing (if any), and level of significance used for hypothesis testing. When using more sophisticated statistical methods (beyond t tests, chi-square, simple linear regression), specify statistical package, version number, and non-default options. For information on statistical review, see Glantz SA "It's all in the numbers" (*J Am Coll Cardiol* 1993;21:835-7).

SOURCE

January 1995, 25(1): front matter.

Journal of the American College of Surgeons
American College of Surgeons

Samuel A. Wells, Jr., M.D
Department of Surgery
Campus Box 8109
Washington University School of Medicine
Wohl Hospital-Room 9902
660 South Euclid Avenue
St. Louis, Missouri 63110

AIM AND SCOPE

The *Journal of the American College of Surgeons* is a monthly journal that considers for publication original articles in all fields of surgery.

MANUSCRIPT SUBMISSION

Work, or any part, may not have been, nor may not be considered for publication by another periodical. Submit materials in question to editorial office for review if there is concern about duplicity of closely related material. In cover letter say that each author has reviewed manuscript and approves of its contents. Reveal author affiliations with organizations that have a direct or indirect financial interest in content or materials that are discussed.

Type manuscript on white bond paper, 8.5 × 11 in. with 1 in. margins. Use double-spacing throughout. Submit an original and four copies of manuscript, tables and figures.

Include a computer diskette containing complete manuscript. Include complete name, address, telephone and fax number and e-mail address of communicating author.

MANUSCRIPT FORMAT

Type manuscript double-spaced and divide into four sections: introduction, materials and methods, results and discussion. Number pages consecutively, beginning with title page. Use abbreviations sparingly, except for units of measurement. Express measurements in conventional units. When first used, abbreviations immediately follow words for which they stand. Use generic names for drugs.

Title Page: Title of article and full name of each author (with middle initial), highest academic degree and institutional affiliation. List name and address of reprint request author. Mention institutional support of described research.

Abstract: 250 words on a separate page. Divide into four sections: background, study design, results and conclusions. Emphasize new and important aspects of study. At bottom of page, list three to ten key words or short phrases to help in cross-indexing article.

References: Type triple spaced. Number consecutively as cited. Follow style in *Index Medicus*. If there are four names in reference, list all; if more than four, list first three followed by "and others."

1. Smith, R. E., Jones, J., Paulson, G. H., and Lewis, J. F. The management of patients with rectal carcinoma: the results of a randomized controlled trial. J. Am. Coll. Surg., 1994, 184: 120–138.

For references to books list authors, title of chapter, title of book, editor of book, inclusive pages, volume or edition, or both, city of publisher, name of publisher, and year.

1. Richerson, H. B., Metzger, W. J., and Hunninghake, G. W. Experimental models of bronchial asthma in man and the rabbit. In: Asthma: Clinical Pharmacology and Therapeutic Progress. Edited by A. B. Kay. Pp.23–32. Oxford, England: Blackwell Scientific, 1986.

Tables: Type each on a separate sheet double-spaced. Give each a title. Number as they appear in text. Write out all abbreviations as footnotes in legend.

Illustrations: Submit professionally designed 5 × 7 in. glossy prints of original. Letters and numerical symbols must be clear to remain legible when reduced. Computer-generated illustrations are not acceptable. Label backs with last name of author, figure number and TOP orientation. Patient's written permission must accompany photographs of patients. For color illustrations, submit both transparencies and prints. Cost of color reproduction is borne by author.

Type legends triple-spaced on a separate sheet. Obtain written permission from author and publisher of borrowed previously published material.

OTHER FORMATS

LETTERS TO THE EDITOR—750 words typed double-spaced on white bond paper, 8.5 × 11 in. with 1 in. margin. Illustrations must be a professionally designed 5 × 7 in. glossy prints of original.

THE SURGEON AT WORK—Six pages, including references with artwork as line drawings of high quality with proper legends, providing surgeons with information of a practical nature and presenting certain aspects of an operation or an invasive procedure.

OTHER

Electronic Submission: Submit a computer diskette containing complete manuscript. Label diskette with: author's last name and manuscript title type of computer (e.g., PC or Macintosh) and, name and version of wordprocessing software program.

SOURCE

January 1995, 180(1): 96.

Journal of the American Geriatrics Society

American Geriatrics Society
Williams and & Wilkins

Cynthia J.T. Clendenin
Department of Preventive Medicine
The University of Tennessee, Memphis
66 North Pauline, Suite 232
Memphis, TN 38105

MANUSCRIPT SUBMISSION

Guidelines follow "Uniform Requirements for Manuscripts Submitted to Biomedical Journals" (see *Annals of Internal Medicine* 1988;108:258–265).

Do not send by certified or registered mail. If sent by Express Mail, waive requirement for recipient signature.

Manuscripts purporting to contain original material are considered for publication with the understanding that neither the article nor any of its essentials (including tables and figures) has been or will be published or submitted for publication elsewhere before appearing in *Journal*. This rule does not apply to abstracts or press reports published as a result of a scientific meeting. Submit copies of papers or manuscripts closely related to paper being submitted. Avoid double or overlapping publication.

Submit an original and two copies. Use double-spacing for all elements of manuscript, including title and author listing, abstract, text, references, tables and footnotes, and figure legends. Provide 1 in. margins at top and bottom and either side of each page. Type page number in upper right corner of each page. Start each element of manuscript on a new page. Arrange material in order.

In a single-spaced cover letter give name, address, phone and fax numbers of corresponding author. Cite an alternate correspondent who can edit page proofs. Include any information not on title page that might be relevant to a possible conflict of interest, e. g., consultancies, stock ownership or patent-licensing arrangements. Such information will not go to reviewers and will not affect initial editorial decision. If manuscript is accepted, Editor and correspondent will discuss degree of conflict, how information will be disclosed, and, rarely, the possibility of a decision not to publish paper.

Sign and submit statements of agreement to transfer copyright and verification of participation by all authors printed at end of Information for Authors in every issue.

MANUSCRIPT FORMAT

Title Page: Make title brief and useful for indexing. List all authors' names, with highest academic degree and all relevant institutional and corporate affiliations and titles of each author. Specify where work was done, all funding sources (grants or institutional and corporate support), and meeting, if any, at which paper was presented. Give an abbreviated title (45 characters) with essence of title, for a running title.

Abstract: Submit abstracts for Clinical Investigations in structured form as used in *Annals of Internal Medicine* (modified from Hayes RB et al. More informative abstracts revisited. *Ann Intern Med* 1990,113:69–76). Include headings: Objective (do not give full background of study; state precise objective or research question addressed); Design (do not describe full method of study; briefly summarize experimental design; e.g., a crossover, double-blinded, randomized trial with two years' follow-up after intervention); Setting; Patients (or Participants); Interventions (include intervention(s) for experimental group(s) and "treatment" control group(s) received); Measurements (include a list of measurement methods, tools, and instruments for independent and dependent variables and identify which are independent and dependent variables); Main Results; and Conclusions.

Text: Clinical investigations include sections: introduction, materials and methods, results, discussion, acknowledgments and references. Start each section on a new page. Include statistical methodology.

References: Cite all listed references listed in paper by superscript numbers placed after commas, semi-colons or periods. Number in sequence of first appearance in text, list only once even if cited repeatedly. In reference list, include surnames and initials of authors up to three followed by et al. Abbreviate journal titles as in *Index Medicus*. Include only references accessible to all readers. Do not use abstracts unless published in established sources within preceding 4 years. Do not cite by number (or list as a Reference) personal communications or manuscripts in preparation or submitted for publication; include in text. Reference manuscripts for publication with page numbers indicated as 000-000; send a copy of accepted manuscript marked "in press."

JOURNAL

1. Mulrow CD, Aguilar C, Endicott JE et al. Quality-of-life changes and hearing impairment: A randomized trial. Ann Intern Med 1990;113:188–194.

BOOK CHAPTER

1. Davidson JM. Sexuality and aging. In: Hazzard WR, Andres R, Bierman EL, Blass JP, eds. Principles of Geriatric Medicine and Gerontology, 2nd Ed. New York: McGraw Hill, 1990, pp 108–114

BOOK

1. Kane RL, Ouslander JG, Abrass IIB. Essentials of Clinical Geriatrics, 2nd Ed. New York: McGraw-Hill, 1990.

Verify references against original sources for accuracy and recheck them in proof.

Tables: Number with Arabic numbers consecutively in order of appearance. Type each on a separate page. Give each a caption typed above tabular material. Use symbols for units only in column headings. Do not use internal horizontal or vertical lines; place horizontal lines between table caption and column headings, under column headings, and at bottom of table (above footnotes, if any). Do not submit as photographs.

Illustrations: Submit two unmounted, untrimmed glossy or computer-generated black-and-white prints of each figure and three photocopies of each. Make symbols, lettering and numbering large and clear enough to be legible after reduction to one column width. Have line drawings, graphs, charts, and lettering on illustrations done professionally. Protect illustrations for mailing. Do not staple, clip, or write heavily on backs. Paste a label the back of each illustration listing its number in order of appearance, author's name, and TOP. If photomicrographs are not one-column or two-column width, they may be rotated or cropped to fit; indicate if this should not be done. Indicate stain and magnification of photomicrographs. For photographs of recognizable subjects, submit signed consent of subject for publication. For previously published illustrations, submit publishers' permission. Cost of publishing color illustrations is borne by author.

Present legends for illustrations in numerical order on a separate page or pages, not on or under illustration. Legends consist of titles followed by explanatory material.

Footnotes: Use footnote symbols in tables only, in order: *, **, ***, #, ##, ###, †, ††, †††. Place parenthetical statements appropriate for footnotes to text in text within parenthesis signs.

OTHER FORMATS

CLINICAL EXPERIENCE—1000-2000 words (4-8 double-spaced pages) for single or multiple case reports and brief descriptions of new techniques, equipment, or devices. Case reports describe a previously unreported circumstance or contain at least one novel feature. No abstract; final sentence or paragraph is a summary.

CLINICAL INVESTIGATIONS—10 to 20 double-spaced text pages. Reports of typical investigation-initiated research, presenting information new and relevant to geriatrics. Subject matter may range from strictly clinical research to health sciences research to basic laboratory research, as long as latter is relevant to a geriatric condition in man.

ETHICS AND HUMANITIES—Subject matter may include any ethical issues that arise in the course of geriatric practice. Papers in the humanities relevant to geriatrics are particularly sought. Often solicited but may be submitted spontaneously.

GERIATRIC BIOSCIENCE—Solicited single-authored reviews of basic scientific information relevant to a geriatric condition or problem.

LETTERS-TO-THE-EDITOR—250-750 words of objective, constructive, and educational criticism of published material; submit within two months of publication of original paper. Letters may also discuss matters of general interest to physicians involved with geriatric patients. A few references, a small table, or an illustration may be used. Letters on papers published in this *Journal* will be submitted to authors of paper. Letter and reply will be published together.

PROGRESS IN GERIATRICS—In-depth reviews, generally solicited, of problems relevant to clinical geriatrics. Subject is addressed pathophysiologically as well as from the clinical point of view.

PUBLIC POLICY—Papers concern economic, political, environmental, and other issues of public policy relevant to geriatrics. They are frequently solicited but spontaneous submissions will be reviewed.

SPECIAL ARTICLES—Papers that do not fit into any other section. Examples: papers on history, geriatrics education and training, research design, recommendations for preventive strategies in geriatrics, and reports of meetings, committee activities, or task force.

OTHER

Abbreviations: Use of too many is discouraged. Use for units of measurement. Follow first appearance of an abbreviation text by words for which it stands.

Drugs: Use generic names whenever possible. Include brand names in parentheses after generic name at first use.

Units: Use Système International (SI) units with conventional units in parentheses. Conventional units alone are accepted.

SOURCE

January 1994, 42(1): front matter.
In every issue.

Journal of the American Medical Association
American Medical Association

Editor, George D. Lundberg, MD
JAMA
515 N State Street
Chicago, IL 60610

MANUSCRIPT SUBMISSION

Manuscripts are considered with the understanding that they have not been published previously in print or electronic format and are not under consideration by another publication or electronic medium. A complete report following presentation or publication of preliminary findings elsewhere (e.g., in an abstract) can be considered. Include copies of possibly duplicative material that has been previously published or is currently being considered elsewhere.

Designate one author as correspondent; provide a complete address and telephone and fax numbers. Limit authors to six; more requires justification. A publishable footnote may be added explaining order of authorship. See International Committee of Medical Journal Editors "Statement from the International Committee of Medical Journal Editors" (*JAMA* 1991;265:2697-2698) and Glass RM "New information for Authors and Readers: Group Authorship, Acknowledgments and Rejected Manuscripts' (*JAMA* 1992;268: 99, Correction. 1993;269:48). If authorship is attributed to a group (either solely or in addition to one or more individual authors), all members must meet full criteria and requirements for authorship. One or more authors may take responsibility "for" a group, in which case the others are not authors, but are listed in an acknowledgment.

In cover letter include following statements on authorship responsibility and financial disclosure and one of the two statements on copyright or federal employment. Each of these three statements must be read and signed by all authors. See Lundberg GD and A. Flanagan "New Requirements for Authors: Signed Statements of Authorship Responsibility and Financial Disclosure" (*JAMA* 1989;262:2003-2004).

Authorship Responsibility: "I certify that I have participated sufficiently in the conception and design of this work and the analysis of the data (when applicable), as well as the writing of the manuscript, to take public responsibility for it. I believe the manuscript represents valid work. I have reviewed the final version of the submitted manuscript and approve it for publication. Neither this manuscript nor one with substantially similar content under my authorship has been published or is being considered for publication elsewhere, except as described in an attachment. If requested, I shall produce the data upon which the manuscript is based for examination by the editors or their assignees."

Financial Disclosure: "I certify that any affiliations with or involvement in any organization or entity with a direct financial interest in the subject matter or materials discussed in the manuscript (e.g., employment, consultancies, stock ownership, honoraria, expert testimony) are disclosed below."

List research or project support in an acknowledgment.

Copyright Transfer: "In consideration of the action of the American Medical Association (AMA) in reviewing and editing this submission, the author(s) undersigned hereby transfers, assigns, or otherwise conveys all copyright ownership to the AMA in the event that such work is published by the AMA."

Federal Employment: "I was an employee of the U.S. federal government when this work was investigated and prepared for publication; therefore, it is not protected by the Copyright Act and there is no copyright of which the ownership can be transferred."

Obtain written permission from all persons named in an acknowledgment See Glass RM "New information for Authors and Readers: Group Authorship, Acknowledgments and Rejected Manuscripts' (*JAMA* 1992;268: 99, Correction. 1993;269:48). In cover letter state: "I have obtained written permission from all persons named in the Acknowledgment."

All accepted manuscripts become the permanent property of the AMA and may not be published elsewhere without written permission from both author(s) and AMA.

MANUSCRIPT FORMAT

Prepare manuscripts in accordance with *American Medical Association Manual of Style* (Iverson CL, Dan BB, Giltman P et al., 8th ed. Baltimore, MD: Williams & Wilkins; 1988) and/or "Uniform Requirements for Manuscripts Submitted to Biomedical Journal." (*N Engl J Med* 1991;324:424-428).

Submit original manuscript and three photocopies. Type on one side of standard-sized bond paper. Use 1 in. margins. Double-space throughout, including title page, abstract, text, acknowledgments, references, legends for illustrations, and tables. Start each section on a new page, number consecutively in upper right corner; begin with title page.

Provide copy that can be scanned by an optical character reader with no smudges or pencil or pen marks. Use standard 10- or 12-pitch type and spacing. Do not use 10-pitch type with 12-pitch spacing. If prepared on a word processor, do not use proportional spacing; do not justify right margins; and use letter-quality printing.

On title page type full names, highest academic degrees, and affiliations of all authors. If an author's affiliation has changed since work was done, list new affiliation.

Abstract: 250 word structured abstract for reports of original data from clinical investigations and reviews (including meta-analyses). See "Instructions for Preparing Structured Abstracts" (*JAMA*1994:272(1): 20-22). For other major manuscripts, include a conventional, unstructured abstract (150 words). Abstracts are not required for Editorials, Commentaries, and special features.

References: Number in order of mention in text; do not alphabetize. In text, tables, and legends, identify with superscript Arabic numerals. For list, follow AMA style, abbreviate journal names according to *Index Medicus*. List all authors and/or editors up to six; if more, list first three and "et al."

1. Lyketsos CG, Hoover DR, Guccione M, et al., for the MulticenterAIDS Cohort Study. Depressive symptoms as predictors of medical outcomes in HIV infection. *JAMA* 1993;270:2563-2567.

2. Marcus R, Couston AM. Water-soluble vitamins: the vitamin B complex and ascorbic acid. In: Gilman AG, Rall TW, Nies AS, Taylor P, eds. *Goodman and Gilman's The Pharmacological Basis of Therapeutics*, 8th ed. New York, NY: Pergamon Press; 1990:1530–1552.

Verify accuracy and completeness of references and correct text citation.

Tables: Double-space on separate sheets of standard-sized white bond paper. Title all tables and number in order of citation in text. If a table must be continued, repeat title on a second sheet, followed by "(cont)."

Illustrations: Submit four sets of all illustrations: 5 × 7 in. matte or glossy photographs for all graphs and black-and-white photographs; high-contrast prints for roentgenograms; color slides (and corresponding color prints) for color illustrations. Number according to order in text. Affix a label with figure number, name of first author, short form of title, and an arrow indicating TOP to back of print. Never mark on print or transparency itself.

Double-space legends (40 words) on separate pages. Indicate magnification and stain used for photomicrographs.

Acknowledge all illustrations and tables taken from other publications. Submit written permission from original publishers.

OTHER

Abbreviations: Do not use in title or abstract; limit use in text.

Drugs: Use generic name, unless specific trade name is directly relevant to discussion.

Experimentation, Human: State in "Methods" section that an appropriate institutional review board approved project. For those investigators who do not have formal ethics review committees (institutional or regional), follow principles outlined in the Declaration of Helsinki. State manner in which informed consent was obtained from subjects.

Include a signed statement of consent to publish all case descriptions and photographs from all patients (parents or legal guardians for minors) who can be identified in written descriptions and photographs.

Units: Use Système International (SI) measurements only, except when "Dual report" is indicated in SI unit conversion table in Lundberg GD "SI Unit Implementation-the Next Step" (*JAMA* 1988;260:73-76).

SOURCE

July 6, 1994, 272(1): 19-26.

Journal of the American Veterinary Medical Association
American Veterinary Medical Association

American Veterinary Medical Association
1931 N Meacham Rd.
Schaumburg, IL 60173-4360
Telephone: (800) 248-AMVA; Fax: (708) 925-1329

AIM AND SCOPE

The *JAVMA* welcomes manuscripts dealing with any subject germane to veterinary medicine, but preference will be accorded to manuscripts that have clinical or practical value. Manuscripts may involve prospective studies, clinical reports, or retrospective studies, or they may be reviews or special commentaries. The first author of a report dealing with clinical interpretations or treatments should be a veterinarian.

MANUSCRIPT SUBMISSION

A manuscript is received with the understanding that it and all revisions have been approved by each author and that neither the article nor any of its parts is under concurrent consideration by any other Publication. Consult an AVMA editor before submitting a report on information previously published (except for short descriptive abstracts) in any compiled format to ensure that report will not be disqualified. For multiple authorships, each author must have generated part of the manuscript's intellectual content and should agree with all interpretations and conclusions. Do not include more than ten authors. In cover letter, designate a corresponding author, and provide a complete address and telephone number.

Accepted manuscripts become property of the AVMA.

MANUSCRIPT FORMAT

Type manuscripts double-spaced. If computer printed, type must be letter quality. Submit three copies of manuscript and figures.

Limit acknowledgments to persons who have contributed substantially to technical content.

In text, limit references to published works that are relevant and necessary. Refer to works cited by superscript numbers. Refer to abstracts, personal communications, and theses by footnotes; cite footnotes by superscript, lowercase letters, arranged alphabetically on a separate page.

Type references double-spaced and list in order of citation in text. Send copies of items in reference list entered as in press (i.e., accepted and in final form).

Identify products and equipment by chemical or generic names or descriptions. Include trade name in a lettered footnote with name and location (city and state) of manufacturer when product or equipment is essential to outcome of experiment or treatment. If study involved evaluation of efficacy or safety of a pharmaceutical, biologic, or other product, product must be commercially and legally available.

Furnish photographs as glossy prints. Identify illustrations on back with first author's name, figure number and TOP. Color illustrations may be used, cost is billed to author.

On front border of each illustration, indicate (with wax pencil marks) cropping. If cropping cannot be done because of loss of important features, indicate whether illustration can be reduced to single column width. Give photomicrographs and electron micrographs internal scale markers. To express magnification by use of an internal scale marker, divide length of marker by original magnification.

Type legends for illustrations double-spaced on a separate sheet. Include sufficient information to allow illustration to be understood without reference to text; include stain where applicable.

OTHER FORMATS

CLINICAL REPORT—Deal with any number of clinical cases. If cases are similar, describe in detail one typical case, rather than detailed descriptions of all cases; address important differences among cases separately. For 4 or more cases, include one table of pertinent abnormal findings, provided findings are not repeated in text. If 3 or fewer cases, summarize pertinent abnormal findings in text. Use illustrations sparingly (3).

RETROSPECTIVE STUDY—Base on retrieval of case records accumulated over a period of years, include a statement of purpose, clinically relevant data, and clinically useful conclusions or interpretations derived directly from evaluation of cases described.

Prospective and retrospective studies should include an informative summary, stating what was done and what was learned or concluded.

REVIEW ARTICLES—Basic criteria for consideration: subject area is one in which important advances have been made in past 5 years; author has considerable experience in subject area; manuscript focuses on information that has or will have clinical application; manuscript is concise.

OTHER

Experimentation, Animal: Reports suggesting that animals have been subjected to adverse, stressful, or harsh conditions or treatment will not be processed unless it is convincingly demonstrated that the knowledge gained was of sufficient value to justify adverse conditions or treatment.

Submission Fee: $50 (nonrefundable) payable by check, money order, VISA/Mastercard, or institutional purchase order, required at submission.

Units: For weights and measures, use metric units. Express dosages in metric units with specific time intervals.

SOURCE

July 1, 1994, 205(1): 6.

Journal of the National Cancer Institute
National Cancer Institute
National Institutes of Health

Editor-in-Chief, *Journal of the National Cancer Institute*
R. A. Bloch International Cancer Information Center
Rm. 213, 9030 Old Georgetown Rd.
Bethesda, MD 20814
Telephone: (301) 496-6975

AIM AND SCOPE

The *Journal of the National Cancer Institute* publishes manuscripts that describe new findings of particular significance in any area relating to cancer, as well as associated news items, reviews, and opinion pieces.

MANUSCRIPT SUBMISSION

In cover letter state that contents of manuscript have not been previously published (abstracts and papers presented at scientific meetings and published as part of the proceedings are excepted) and have not been and will not be submitted or published elsewhere while under consideration by the *Journal*. Do not fragment reporting of aspects of a single investigation. If paper is one of a number of existing or planned manuscripts related to a single study, in cover letter identify paper as such and justify use of fragmented approach. Describe scope of each planned related paper. Submitted paper must clearly explain and justify fragmented approach and reveal full extent of investigation.

In cover letter state that all authors have directly participated in planning, executing, or analyzing study and have approved final submitted version. Submit written permission from: those named in acknowledgments who have been credited with substantive scientific contributions and authors whose work is cited in text as a personal communication, unpublished data, manuscript in preparation, manuscript submitted for publication, or manuscript in press.

If tables or illustrations have been published previously, obtain written permission from copyright holder. If contents are modified, permission to use adapted version is required. Notify author of original manuscript.

State any financial arrangement with a company whose product was used in a study or a competing company. If manuscript is accepted, author(s) may be asked to communicate such information in manuscript. If submitting a review or an editorial discussing a product that is made by a company in which author(s) has a substantial financial interest or that is made by a competing company, state financial interest in a footnote.

Acceptance is contingent on submission of complete and consistent data, accurate reference list, and conclusions consistent with results demonstrated in study.

Publication is contingent on acknowledgment of copyright release to the National Cancer Institute. Forms are sent to authors when manuscript has been accepted. All material in the *Journal* is in public domain.

Submit an original and three copies of manuscript.

MANUSCRIPT FORMAT

Articles describe new fundings of major importance in 4000 words (not counting abstract, methods section, reference list, or legends for tables and figures).

Title Page: Make title brief (14 words) and, except for Editorials, not in sentence form. Give full name and affiliation of each author and name and address of corresponding and reprint request author(s).

Key Words: List up to four for indexing.

Abstract: 400 words written in complete sentences for Articles and Reports (250 words for Reviews and Brief Communications). Clearly state unique elements of work. Make readable by nonspecialists as well as experts. Concisely state all important findings of study. Include headings in abstracts for Articles and Reports only: Background (note relevant previous findings, designated as those of the author(s) or other investigators), Purpose (relate to Background), Methods (state unique materials or techniques used), Results (clearly relate to Methods), Conclusion(s) (state any important Conclusion(s) demonstrated or suggested by results), and Implications (suggest possible clinical applications and/or areas for experimental research and note a particular need for confirmatory studies or offer any cautions regarding interpretation). Abstracts of Brief Communications are not structured; address each aspect in above order.

All information in abstract must be explicitly stated in text. Use "significant" only if a result is statistically significant; cite a *P* value or confidence interval in text for any statistically significant finding. For each result, in Articles and Reports, provide corresponding method in Methods sections of abstract and text.

Text: Type all material double-spaced. Write in standard grammatical English. (Refer to AMA *Manual of Style* or a recent *Journal* issue on questions of style. Number pages consecutively. Do not submit dot-matrix or italicized text. Define all symbols and abbreviations. Number references, tables, and figures in order in cited in text. References in tables continue sequence of numbers in text at point where table is first mentioned.

Methods: Describe methods used succinctly but with sufficiently detail to allow replication. Methods sections of Articles and Reports are exempted from word count. Specify full names, types, amounts, and sources of reagents (give complete names and locations of suppliers); full names as well as standard abbreviations of labeled compounds and isotopes used for labeling; concentrations of solutions; and reaction conditions (e.g., incubation times and temperatures). Accurately name, clearly and thoroughly explain, reference when appropriate, and organize under appropriate subheadings techniques and procedures used. Methods section must be complete; include methodology corresponding to each end point presented in Results. Specify experimental conditions that limit generalizability of Results.

For clinical trials, define and explain purpose of study, study design, numbers of patients, clinical staging of disease, type and sequence of treatments given before and during study, time points for evaluation of response, duration of follow-up, end points used (e.g., overall survival, disease-free survival), specific outcomes assessed, and methods of assessment. Meet requirement even if focus of a study is not the clinical trial to which it is related (e.g., association of gene, messenger RNA, or protein expression with disease therapy or prognosis). Describe methods of statistical analysis in sufficient detail that a knowledgeable reader could reproduce analysis if data were available.

Reference List: Type list double-spaced. Follow standard form of *Index Medicus*. For periodicals, provide author(s), title, journal, volume, first and last page numbers, and year. For books, give author(s), pertinent chapter, or section title, book title, name(s) of editor(s), location and name of publisher, year, and page numbers. List papers "in press." Note citations of papers in preparation or submitted for publication, unpublished observations, and personal communications parenthetically in text, not in reference list; cite names of all investigators.

Tables and Figures: Include a limited number of tables and figures. Each panel of a figure is counted as a separate illustration unless panels represent a logical sequence of closely related similar forms. Dissimilar panels (e.g., a graph and a photo, or two graphs with different axes) are counted separately. The maximum number of tables plus figures are: Articles and Reviews (eight); Reports and Commentary (four); Brief Communications (two); Correspondence (one). Type all tables double-spaced, each on a separate page. Computer printouts are acceptable. Provide titles, legends, and headings for columns. Indicate units of measurement used and define all abbreviations.

Photographs and other figures should be clear, glossy prints; good-quality laser prints of computer-generated graphs and artwork are acceptable. Send two copies of each figure. Draw or original artwork in black ink. Color photos and illustrations are considered only if essential and black-and-white will not suffice.

Type figure legends double-spaced on one separate sheet. Identify lettering or symbols in legend. On back of print indicate figure number and TOP edge. Label separate figure parts with capital letters. Do not group dissimilar illustrations (e.g., a photo and a graph, or two graphs with different categories of plots) together as parts of the same figure.

OTHER FORMATS

BOOK REVIEWS—1000 word reviews solicited by book review editors.

BRIEF COMMUNICATIONS—750 word (not counting abstract, reference list, or legends) descriptions of new findings of general interest. Do not divide text by heads and subheads.

COMMENTARY—2000–4000 word topical summaries in relatively nontechnical and informal style, of current activities or events that bear on some aspect of cancer, e.g., highlights of scientific meetings, intersection of science and public policy, or news items of general interest in relation to cancer. Pieces are usually solicited, unsolicited contributions are welcome.

CORRESPONDENCE—500 word (not including legend for a table or figure) letters to the editor express opinion about material previously published in *Journal* or views on topics of current relevance to some aspect of cancer. Letter about work published in the *Journal* are referred to author of original paper for response, which may be published with letter. Give complete references for work of others.

EDITORIALS—Usually solicited opinions (1000 words) on any subject relevant to *Journal's* concerns, e.g., significance of a paper in the issue, a recent finding published elsewhere, or a particular topic of significance. Unsolicited submissions are considered.

REPORTS—2000 words (not counting abstract, methods section, reference list, or legends) describing significant findings that do not require a more extensive format.

REVIEWS—6000 words (not counting abstract, reference list, or legends) providing a comprehensive and scholarly overview of an area or issue of current interest. Make contents of Reviews comprehensible to knowledgeable readers outside particular subject area. Most are solicited, unsolicited submissions are welcome.

OTHER

Abbreviations: Avoid whenever possible. If essential, list and define. Do not use author-created abbreviations.

Clinical Trials: See "Methodologic Guidelines for Reports of Clinical Trials" (*Cancer Treat Rep* 69:1-3, 1985).

Electronic Submission: Submit letters to the editor initially both as hard copy and on disk (3.5 or 5.25 in. disk formatted for MS-DOS. Provide disks for other manuscripts after acceptance. Use WordPerfect 5.0 or 5.1. If not, submit documents in ASCII format. Write first author's name and name of program used (or ASCII) on disk label.

Experimentation, Animal: State that care was in accord with institution guidelines. Where applicable, report dose and schedule of anesthetics and analgesics.

Experimentation, Human: Include explicit assurance that written informed consent was obtained from each subject or from his or her guardian. Include a statement that human investigations were performed after approval by a local institutional review board and in accord with an assurance filed with and approved by the U.S. Department of Health and Human Services, where appropriate.

Nomenclature, Genes: Do not italicize symbols designating genes, alleles, or loci. Follow such symbols with wording that clarifies meaning of symbol.

Use specific gene symbol of choice. For human genes, if symbol is other than that listed in Human Gene Mapping, identify parenthetically, at first mention, appropriate gene symbol listed in Human Gene Mapping report from the most recent international workshop on human gene mapping. Capitalize all Human Gene Mapping designations for human genes; for other species, use lowercase and identify species of origin at first mention. Distinguish proteins from genes by encoding them by an initial capital letter (e.g., Nm23).

SOURCE
January 5, 1994, 86(1): 74-76.

Journal of the Neurological Sciences
Elsevier Science Publishers

James F Toole, M D.
Department of Neurology
Wake Forest University Medical Center
Box 1068, Medical Center Boulevard
Winston-Salem, NC 27157-1068
Telephone: (919) 777-3979; Fax: (919) 716-5477;
Telex: 806 449 BGSM WSL.

AIM AND SCOPE

The *Journal of the Neurological Sciences* is designed for the prompt publication of studies which bridge clinical neurology and the basic sciences. Emphasis is placed on sound scientific developments which are or will soon become relevant for patient care. Its scope includes demyelination, neuromuscular diseases, dementia, infections, and disturbances of consciousness, stroke and cerebral circulation, growth and development, plasticity, metabolism and molecular neurobiology and genetic.

The mission of the *Journal* is to inform our readership of progress in clinical medicine and research, history of medicine, and social interfaces of medicine.

MANUSCRIPT SUBMISSION

Authors are those persons who accept intellectual and public responsibility for statements made and results reported. Do not attribute authorship to anyone who was not involved in study.

Papers which integrate, elucidate and educate in a succinct manner are particularly welcome. Reviews, Research Reports and Short Reports are published, as are Book Reviews and News and Notes from the World Federation of Neurology and its various Research Groups.

Submission of a paper implies that it has not previously been published (except in abstract form) and that it is not being published elsewhere. Submit a statement signed by all listed authors that they concur with submission and that manuscript has been approved by responsible authorities where work was carried out. If accepted, manuscript shall not be published elsewhere in the same form in either the same or any other language, without consent of Editor and Publisher.

Criteria for Articles:

For studies of prevention or treatment—Random allocation of participants to comparison groups, follow-up of at least 80% of those entering investigation, outcome measure of known or probable clinical importance.

For studies of prognosis—Inception cohort of individuals, all initially free of the outcome of interest, follow-up of at least 80% of participants until occurrence of a major study end point or end of study.

For studies of causation—Clearly identify comparison group for those at risk for, or having, the outcome of interest (e.g., randomized controlled trial, quasi-randomized controlled trial, nonrandomized controlled trial, cohort analytic study with case-by-case matching or statistical adjustment to create comparable groups, case-control study); blinding observers of outcome to exposure (criterion assumed to be met if outcome is objective, e.g., all-cause mortality, objective test), blinding of observers of exposure for case control studies or blinding of subjects to exposure for all other study designs.

For studies of economics of health care programs or interventions—Economic question addressed must be based on comparison of alternatives, alternative diagnostic or therapeutic services or quality assurance activities must be compared on the basis of both the outcomes produced (effectiveness) and resources consumed (costs); evidence of effectiveness must be from a study that meets the above noted criteria for diagnosis, treatment, quality assurance or a review article; results should be presented in terms of the incremental or additional costs and outcomes of one intervention over another; where there is uncertainty, in the estimates or imprecision in the measurement, do a sensitivity analysis.

MANUSCRIPT FORMAT

Submit manuscripts, in quadruplicate, typed double-spaced on pages of uniform size (preferably A4 or 8 × 11 in.). Divide papers into sections, headed by a caption (e.g., Summary, Introduction, Materials and Methods, Results, Discussion, References).

Title Page: Include authors full names, academic or professional affiliations, and complete addresses. Give name and full postal address of corresponding author with telephone and fax numbers.

Summary: 200 words at beginning of each article.

Key Words: 6-8 items included on title page. Use *Medical Subject Headings* of *Index Medicus* as a guide.

References: Cite text by author(s) and year in chronological not alphabetical order. When more than two authors, use abbreviation "et al" following name of leading author. List all references cited in text at end of paper, typed double-spaced and arranged in alphabetical order of first author's name (Harvard system). References must be complete, include initial(s) of author(s) cited, year of publication, title of paper, journal, volume, and page numbers. Abbreviate journal titles according to *List of Journals Indexed in Index Medicus*, latest edition. For citations of books use uniform sequence: author(s), year of publication, editor(s), complete title of book, publisher, place of publication, and page numbers. All references cited in text must be in reference list and vice versa. Do not include manuscripts in preparation and submitted but not accepted

as well as "personal communications" in reference list; cite at appropriate place in text. For examples of forms of reference citations see any recent *Journal* issue.

Illustrations: Submit figures suitable for immediate reproduction (i.e. originals, no photocopies). Make line drawings in black ink on drawing or tracing paper. Lettering should be of professional standard and large enough to withstand reduction. Alternatively, do lettering in fine pencil. Submit photographs as clear black and white prints on glossy paper (no smaller than 4 × 5 in. and no larger than 8 × 10 in.). Reproduction in color is possible; authors bear extra costs involved.

Type legends for figures double-spaced, on separate pages. If illustrations from other articles or books are used, submit written permission of both author and publisher.

Tables: Type double-spaced on separate pages. Give short descriptive headings and, if applicable, a legend.

OTHER FORMATS

ARTICLES—4000 words. Original research concerning the causes, mechanisms, diagnoses, course, treatment, and prevention of disease.

HISTORICAL REVIEWS—4000 word essays, reports, or biographic sketches on evolution of medicine.

LETTERS TO THE EDITOR AND BOOK REVIEWS

RAPID COMMUNICATIONS—4 printed pages (12 double-spaced typed pages, including tables and figures and adequate references). Brief but complete account of important new observations which merit urgent publication. Submit as electronic manuscript on floppy disk.

REVIEW ARTICLES—5000 words, 100 references. Detailed, critical surveys and meta-analyses of published research relevant to clinical problems.

SHORT REPORTS—Clinical association, case reports, and reports on adverse effects. 10 references, 1 figure or table.

OTHER

Electronic Submission: Preferred storage medium is a 5.25 or 3.5 in. disk in MS-DOS format; other systems are welcome, e.g., Macintosh (save file in usual manner, do not use option save in MS-DOS format). Do not split article into separate files.

Ensure that letter "I" and digit "1" (also letter "O" and digit "0") have been used properly; format article (tabs, indents, etc.) consistently. Do not leave characters not available on word processor (Greek letters, mathematical symbols, etc.) open; indicate by a unique code used consistently throughout text. List codes and provide a key. Do not allow word splits; do not use a "justified" lay-out. Adhere strictly to general instructions on styles/arrangement and, in particular, reference style.

Save file in word processor format. If word processor features option to save files in flat ASCII, do not use. Format disk correctly, Put only manuscript file on disk. Specify type of computer and word processing package used. Label disk with author name and name of file. After final acceptance submit disk plus one final printed and exactly matching printout to accepting editor. Further information may be obtained from publisher.

Experimentation, Animal: Indicate steps taken to eliminate pain and suffering. Show clearly in writing a recognition of moral issues involved.

Experimentation, Human: Specify that research received prior approval by appropriate institutional review body and that informed consent was obtained from each subject or patient.

SOURCE

March 1995, 129(1): inside back cover.

Journal of Theoretical Biology
Academic Press (London)

The Editorial Office
Journal of Theoretical Biology
525 B Street, Suite 1900
San Diego CA 92101-4495
Att. Dr. S. Kauffman

Professor L. Wolpert
Department of Anatomy and Developmental Biology
University College and Middlesex School of Medicine
Windeyer Building, Cleveland Street
London W1P 6DB, U.K.
From Western Europe

AIM AND SCOPE

The aim of the *Journal of Theoretical Biology* is to publish theoretical papers which give insight into biological processes. The biological significance should be clearly stated. Highly speculative papers not based on current biological knowledge will not be acceptable. Papers may include new experimental results which bear on the theory being presented.

It is essential that papers be accessible to as wide a readership as possible. Every effort should be made to make the main points of the paper intelligible to biologists as a whole. Authors should thus make it clear how any mathematical models relate to the biological problems they address; detailed mathematical technicalities and experimental procedures may usually be best presented in appendices so as not to impede the exposition of the central ideas.

Brief notes may be submitted in the form of letters to the Editor. We also welcome comment on current theoretical issues or papers published in the Journal.

MANUSCRIPT SUBMISSION

Submit manuscript in triplicate. Supply a contact telephone and fax number.

Submission of a manuscript implies that material is original and has not been and will not be submitted in equivalent form for publication elsewhere.

If is accepted for publication, exclusive copyright in the paper shall be assigned to Publisher, who will not put any limitation on the personal freedom of author to use material contained in paper in other works.

MANUSCRIPT FORMAT

Type double-spaced throughout (including summary, footnotes, tables and legends) on one side of A4 paper, with wide margins. Type footnotes, tables and legends for illustrations separately at end of manuscript.

Submit manuscript and illustrations fully ready for press. Check typescript (including references) for errors before submission. If mathematical equations or formulae are used, check that these are clearly written, are entirely correct and that notation is used consistently throughout. Only corrections and minor amendments in proof can be made.

Running Headline: 40 characters, including spaces, suitable for page headings. Give only if full title is long.

Summary: 300 words at beginning of all submissions other than a Letter to the Editor. Make intelligible to general readers without reference to main text. Do not use abbreviations.

References: List alphabetically at end of paper with titles of papers in full and journal titles abbreviated according to *World List of Scientific Periodicals*. Citations in text should read: Brown & Robinson (1987), or (Brown & Robinson, 1987); when more than two authors: (Smith *et al.* 1987). Use conventions (Brown, 1987*a*) (Brown, 1987*b*) when more than one paper by the same author(s) has appeared in one year.

CABANC, M. & LEBLANC, J. (1983). Physiological conflict in humans: fatigue vs. cold discomfort. Am. J. Physiol. **224**, R621–R628.

COLLIER, G. H. & ROVEE-COLLIER, C. K. (1981). A comparative analysis of optimal foraging behavior: laboratory stimulations. In: *Foraging Behavior* (Kamil, A. C. & Sargent, T. D., eds) pp. 39–76. New York: Garland.

or:

CABANC, M. & LEBLANC, J. (1983). *Am. J. Physiol.* **244**, R621–R628.

COLLIER, G. H. & ROVEE-COLLIER, C. K (1981). In: *Foraging Behavior*. New York: Garland.

Illustrations: Make drawing two or three times larger than it will appear in *Journal*; avoid very large drawings. Submit drawings and illustrations on A4 sheets. Submit originals with manuscript. Place explanations in legend; drawing should contain minimal lettering, inserted lightly in blue pencil, in lowercase except for proper names.

Refer to illustrations as Fig. 1, Fig. 2, etc. Write name(s) of author(s) and number on all illustrations; indicate TOP. Submit one duplicate set to accompany second and third copies of manuscript; these may be clear and legible photocopies. If oversize original drawings are submitted (bigger than A4 size), include three sets of small photocopies.

OTHER

Abbreviations: Except for those used for units of measurement, do not use abbreviations without spelling out when introduced. Write abbreviations consisting of groups of initials in capitals without full stops.

Reviewers: Submit names of five individuals who may be used as referees.

Spelling: Use American or British forms of spelling, but be consistent.

Units: Express all measurements in accordance with Système Internationale d'Unites (S.I. units, International Metric System).

SOURCE

November 7, 1994, 171(1): back matter.

Journal of Thoracic and Cardiovascular Surgery
American Association for Thoracic Surgery
Western Thoracic Surgical Association
Mosby

John W. Kirklin, MD
University of Alabama at Birmingham
Department of Surgery
University Station, Birmingham, AL 35294
Telephone: (205) 934-5160

AIM AND SCOPE

The *Journal* will consider for publication suitable articles on topics pertaining to thoracic and cardiovascular surgery. Acceptance is based upon significance, originality, and validity of the material presented.

MANUSCRIPT SUBMISSION

Include the following statement signed by one author: "The undersigned author transfers all copyright ownership of the manuscript (title of article) to Mosby-Year Book, Inc., in the event the work is published. The undersigned author warrants that the article is original, is not under consideration by another journal, and has not been previously published. I sign for and accept responsibility for releasing this material on behalf of any and all co-authors." Authors will be consulted, when possible, regarding republication of their material.

When proposed publication concerns a device, include a statement indicating absence financial or other interest in manufacturer or distributor of device or explaining nature of any relation with manufacturer or distributor of device.

MANUSCRIPT FORMAT

Submit an original and two copies of all material, including illustrations. Place author's name and page number in upper right corner of each page. Use double-spacing throughout. In cover letter state that material has not been previously published or submitted elsewhere for publication. Identify letter name, address, and business and home telephone numbers of corresponding author.

Title Page: Make title concise. List affiliation and academic degrees of author(s). Restrict number of authors to those making material contributions. Include where work was done, sources of support (if any), and name, address, and business and home telephone numbers of corresponding and reprint request author(s).

Abstract: 150 to 200 words on first page of manuscript. Summarize data and present inferences. Do not include tables or use acronyms.

Include one to three sentences (50 words) as an ultramini-abstract, containing essence of manuscript.

References: Number consecutively in text. List, on a separate sheet, double-spaced, at end of paper in that order. Follow format of "Uniform Requirements for Manuscripts Submitted to Biomedical Journals" (*N Engl J Med* 1991; 324:424-8), also known as "Vancouver" style for biomedical journals. For journal abbreviations see *Cumulated Index Medicus*.

JOURNALS

Authors' names and initials, title of article, journal name, date, volume number, and inclusive pages (list all authors when six or less; when more, list only three and add *et al.*):

Josa M, Khuri SF, Braunwald NS, et al. Delayed sternal closure: dealing with complications after cardiopulmonary bypass. J Thorac Cardiovasc Surg 1986;90:598-603.

Okies JE, Page UL, Bigelow JC, Kranse AH, Salomon NW. The left internal mammary artery: the graft of choice. Circulation 1984;70(Suppl):I213-21.

BOOKS

Authors' names, chapter title, editor's name, book title, edition, city, publisher, date, and pages:

Berger HJ, Zaret BL, Cohen LS. Cardiovascular nuclear medicine. In: Goldberger E, ed. Textbook of clinical cardiology. 1st ed. St. Louis: CV Mosby, 1982:326-45.

Do not list unpublished data and personal communications as references. Verify accuracy of references.

Illustrations: Number illustrations (one set of glossy prints, unmounted and three photocopies, except for halftone drawings, histologic sections, electron microscopic data, echocardiograms, and similar figures each of which requires three glossy prints) in order of mention in text. Mark lightly on back with author's name and an arrow indicating TOP. A reasonable number of halftone illustrations are reproduced free of cost; make special arrangements with Editor for color plates, elaborate tables, or extra illustrations. Prepare original drawings or graphs in black India ink or typographic (press-apply) lettering. Typewritten or freehand lettering is unacceptable. Have lettering professionally done; make in proportion to drawing, graph, or photograph. Do not send original art work, x-ray films, or ECG strips. Glossy print photographs, 3×4 in. (minimum) to 5×7 in. (maximum), with good black-and-white contrast or color balance are preferred. Make consistent in size. Note special instructions regarding sizing. Type legends double-spaced. List on a separate sheet and include at end of manuscript.

Tables: Make self-explanatory. Supplement, do not duplicate, text. Type on pages separate from text. Provide a brief title for each. Define abbreviations at bottom.

OTHER FORMATS

EDITOR'S INITIATIVE PAPERS—May be submitted to or requested by Editor. Usually pertain to an area of controversy or complexity; describe and compare different methods of treatment so that inferences can be drawn about optimal methods in specific circumstances. Base comparisons on detailed analyses of one or preferably several institutional experiences. Papers serve as reviews of subject being considered. LETTERS TO THE EDITOR—In cover letter explain circumstances relevant to "letter." Give letter title, double-space, and submit in triplicate.

The *Journal* encourages submission of substantive "letters to the Editor." Submit brief "letters to the Editor" containing substantive information concerning experimental or clinical studies, accompanied by a brief bibliography (5) and one or two tables and/or figures.

OTHER

Experimentation, Animal: Include a statement in Methods section giving assurance that all animals have received humane care in compliance with "Principles of Laboratory Animal Care" formulated by the National Society for Medical Research and "Guide for the Care and Use of Laboratory Animals" prepared by the Institute of Laboratory Animal Resources (National Institutes of Health, NIH Publication No. 86-23, revised 1985).

Experimentation, Human: Include a statement in Methods section indicating approval by institutional review board and affirmation that informed consent was obtained from each patient.

Statistical Analyses: For papers prepared with a statistician who is not a coauthor, include a footnote or an acknowledgment with name and degree(s) of consultant.

SOURCE

July 1994, 108(1): 23A-24A.

Journal of Trauma

American Association for the Surgery of Trauma
Eastern Association for the Surgery of Trauma
Trauma Association of Canada
Western Trauma Association
Williams & Wilkins

John H. Davis, MD
Department of Surgery
D319 Given Building
University of Vermont College of Medicine
Burlington, VT 05405
Telephone: (802) 656-3364; Fax: (802) 656 8837

AIM AND SCOPE

The Journal of Trauma welcomes submissions from all sources and all countries of original articles that contribute to the scientific knowledge of the surgical management of trauma.

MANUSCRIPT SUBMISSION

Submit four high-quality copies of manuscript with four sets of all illustrations and tables. In transmittal letter include name, address, and telephone number of corresponding author.

MANUSCRIPT FORMAT

Type manuscript double-spaced on 8.5 × 11 in. paper. Include: a short title (running head) page (45 characters); a complete title page, including names of all authors and their degrees; a page listing affiliations of all authors (include acknowledgments, where applicable, for financial or other support received for study; organization, date, and place of meeting for any paper previously presented in whole or in part; and reprint request author with exact address (including ZIP or country code).

Abstract: 150 words (50 words for case reports) giving factual essence of article.

Text: Divide scientific articles into sections: Introduction, Material and Methods, Results, and Discussion. Acknowledge persons who assisted in study or helped in other ways after Discussion section. Number manuscript pages consecutively.

References: Cite all references in Reference list, figures, and tables in text; number in order of appearance. Verify accuracy. Head Reference list "REFERENCES"; type double-spaced. Arrange in order of appearance in text (i.e., numerical). Journal abbreviations follow *Index Medicus* style. If a journal is not listed, give full name.

JOURNAL ARTICLE

Author names (for articles with more than three, give first three followed by et al), title of article, Journal abbreviation, volume number, beginning page number, year.

1. Koons TS, Boyden GM: Gas gangrene for parenteral injection. *JAMA* 175:46, 1961

MONOGRAPH

Author names (if more than three, give first three then et al), title of book, edition or volume number (if appropriate), city of publication, publisher name, year, page numbers of section of book cited.

CHAPTER IN BOOK

Author names (if more than three, give first three then et al), title of chapter, editor names (if more than three, give three then et al), title of book, edition or volume number (if appropriate), city of publication, publisher's name, year, inclusive page numbers.

1. Herget CH: Wound ballistics. In Bowers WB (ed): *Surgery of Trauma*. Philadelphia, JB Lippincott, 1953, pp 494–510

Illustrations: Submit four complete sets of all illustrations. Make line drawings and photographic prints suitable for reproduction in black and white. Submit figures in 3 × 5 in. or 5 × 7 in. format with a minimum of white space Type used to label figures should be large enough to be readable if reduced, including arrows and other symbols. Have line drawings professionally drawn. Convert X-ray films to high contrast glossy prints in negative mode (i.e., black areas of print should correspond to black areas of x-ray film). Graphs and charts prepared with a computer and laser printer or a plotter must be very clear, toner must have good density, and no smudges. Do not submit xerographic copies. Do not mount prints or use paper or binder clips. Place authors' names, article title, figure number, and TOP on back of each figure on a gummed label. If pen is used, make sure it will not smudge or offset to front of next figure.

Cite all figures in text; number in order of appearance. Give each figure a descriptive legend. Type legends double-spaced on a separate sheet. In legends for photomicrograph indicate original magnification unless a scale is included in photograph. For legends for slide preparations include stain used.

If an illustration is from another published source, submit a letter of permission from copyright holder. For photographs of patients, submit release signed by patient or guardian or obscure subject's face.

Color illustrations will be published if necessary to illustration's clarity. Authors bear costs of color separations.

Tables: Type each on a separate sheet double-spaced. Keep abbreviations to a minimum; define in a note at bottom of table. Do not use vertical or horizontal rules. Cite in text in numerical order. Construct tables so that like information reads down.

OTHER

Drugs: Use generic names with brand name in parentheses if appropriate.

Electronic Submission: Include a computer diskette containing word processing files of article along with hard (paper) copies of manuscript. Files in IBM-compatible WordPerfect 4.2, 5.0, or 5.1 or ASCII format are preferred; files from most popular word processing programs can be converted; Apple-compatible diskettes are accepted. Double-density (DD) or high-density (HD) 5.25 in. or 3.5 in. diskettes are acceptable. Send a copy of original files (not original files); label diskette with authors' names, title of article, type of computer, DOS version, and word processing software used.

Equipment: Give manufacturer or source, city, and state or country in parentheses for any equipment, devices, or supplies mentioned critical to replication of experiment or study.

Experimentation, Animal and Human: Include a statement, preferably in Materials and Methods section, about institutional approval of study and adherence to established guidelines on treatment of subjects.

Style: Follow guidelines in *American Medical Association Manual of Style* (8th ed, Baltimore, Williams & Wilkins, 1989).

SOURCE

July 1994, 37(1): front matter.

Journal of Urology
American Urological Association
Williams & Wilkins

Dr. Jay Y. Gillenwater
1120 North Charles Street
Baltimore, MD 21201
Telephone: (410) 539-8138
Clinical section
Dr. Stuart S. Howards
Box 495
University of Virginia School of Medicine

Charlottesville, VA 22908
Telephone: (804) 924-9554
Investigative section

AIM AND SCOPE

The *Journal of Urology* contains 3 sections: Clinical Urology, Investigative Urology and Urological Survey. The *Clinical Section* usually does not publish laboratory animal studies. The *Investigative Section* does not publish letters to the editor, case reports or clinically oriented articles. Unsolicited material is not accepted by *Urological Survey*.

MANUSCRIPT SUBMISSION

Submit a cover letter signed by all authors stating: that all authors have made a substantial contribution to information or material submitted for publication; that all have read and approved final manuscript; that they have no substantial direct or indirect commercial financial incentive associated with publishing the article; that manuscript or portions thereof are not under consideration by another journal and have not been previously published. Limit number of authors to six. If more, justify inclusion of each. Designate a corresponding author and give complete address and telephone number.

In cover letter include paragraph: "In consideration of the Editors of the *Journal of Urology* taking action in reviewing and editing my submission, the author(s) undersigned hereby transfers, assigns or otherwise conveys all copyright ownership to American Urological Association Inc., Copyright owner of the *Journal of Urology*, in the event that such work is published in that *Journal*." All authors must read and comply with requirements in Information for Authors.

If manuscript includes text and/or tables or illustrations borrowed from another journal or book, submit permission from original author and publisher and acknowledge in manuscript. Patients must not be recognizable in photographs unless written consent of subject is supplied.

MANUSCRIPT FORMAT

Submit manuscripts, tables and illustrations (glossy prints) in quadruplicate. Type double-spaced with wide margins of 8.5 × 11 in. paper. Do not exceed 2500 words of text for regular manuscripts; keep text as concise as possible. Define standard abbreviations when first used; be consistent throughout text. Do not use jargon and uncommon abbreviations. Use generic names for all drugs. Follow trivial names for compounds in parentheses by correct chemical names when first used; avoid trade names. Provide normal laboratory values in parentheses when first used.

Indicate source of extra-institutional funding, in particular that provided by commercial sources. Verify accuracy of references and all statements made in work, including changes made by copy editor. Submit complete and correct manuscripts.

Arrange original manuscripts as: Title Page, Abstract (150 words), Introduction, Materials and Methods, Results, Discussion, References, Tables, Legends. Number pages; use only one side of page. Begin each section on a separate page.

Title Page: Concise, descriptive title, names and affiliations of all authors and a brief, descriptive running head

(80 characters). Type one to five key words at bottom of page. Use terms from *Medical Subject Headings* of *Index Medicus* (National Library of Medicine, annual).

References: 20 readily available citations (except in Review Articles). Cite by superscript numerals; do not alphabetize. References include names and initials of all authors, complete title, abbreviated journal name (according to *Index Medicus*), volume, beginning page number and year.

JOURNAL REFERENCE

Jones, A. W., Smith, K. P. and Black, J. M.: Chemotherapy for bladder cancer. J. Urol., **149:** 752,1993.

BOOK REFERENCE

Smith, J. A.: Urology Review. Baltimore: Williams &Wilkins, vol. 2, p. 1. 1993.

CHAPTER IN BOOK

Robins, J. A. and Jones, K. A.: Bladder cancer in children. In: Pediatric Malignancies. Edited by J. K. Early and C. W. Diner. Baltimore: Williams & Wilkins, vol. 1 chapt. 2, pp. 35–46. 1993.

Illustrations and Tables: Keep to a necessary minimum (10); do not duplicate information in text. Do not mount. Do not attach by clips. Submit as glossy prints no larger than 5×7 in. Indicate names of all authors and TOP of illustration on reverse side.

Type legends double-spaced on a separate page. Supply magnifications for photomicrographs. Label graphs clearly. Provide reference to illustrations, numbered with Arabic numerals, in text. Blurry or unrecognizable illustrations are not acceptable.

Type tables double-spaced on separate pages, numbered and referred to in text. Present summarized rather than individual raw data.

OTHER FORMATS

CASE REPORTS—1000 words, minimal references. Should be informative and devoid of irrelevant details.

LETTERS TO THE EDITOR—500 words. Should be useful to urological practitioners.

REVIEW AND STATE OF THE ART ARTICLES, AND CLINICOPATHOLOGICAL CONFERENCES—Do not submit without prior approval. Include a detailed outline of proposed article, an abstract (750 words) and an estimate of manuscript length with inquiry.

UROLOGISTS AT WORK—Brief communication on a technique of a practical nature; include data documenting experience with technique advocated.

OTHER

Color Charges: $550 per page. Color work or an excessive number of illustrations is published at author's expense.

Statistical Analyses: Indicate and reference statistical methods used. Present enough information to allow an independent critical assessment of data.

SOURCE

October 1994, 152(4): 1354.

Journal of Vascular Surgery
The Society for Vascular Surgery
International Society for Cardiovascular Surgery,
  North American Chapter
Mosby

Calvin B. Ernst, MD
James C. Stanley, MD
Editors, *Journal of Vascular Surgery*
Henry Ford Hospital
2799 W. Grand Blvd.
Detroit, MI 48202
Telephone: (313) 876-3147; Fax: (313) 876-3364

AIM AND SCOPE

The *Journal of Vascular Surgery* is devoted to the publication of clinical and laboratory studies in vascular surgery. As the official publication of The Society for Vascular Surgery and the North American Chapter, International Society for Cardiovascular Surgery, the *Journal* will publish, after careful peer review, selected papers presented at the annual meetings of those organizations, as well as original articles from members and nonmembers.

MANUSCRIPT SUBMISSION

Manuscripts accepted for publication become property of the *Journal of Vascular Surgery* and may not be published in whole or in part without written permission of author(s) and *Journal.*

In cover letter identify and give name, address, business and home telephone and fax numbers of corresponding author.

Manuscripts from a particular institution are assumed to be submitted with approval of requisite authority.

Cover letter must include these statements, signed: "The undersigned author(s) transfer(s), assign(s), or otherwise convey(s) all copyright ownership of the manuscript [title of article] to The Society for Vascular Surgery and the North American Chapter, International Society for Cardiovascular Surgery, in the event the work is published in the *Journal of Vascular Surgery.* The undersigned author(s) warrant(s) that the article is original in form and substance, is not under consideration by another journal and has not been previously published.

The undersigned author(s) certifies (certify) that I (we) have participated to a sufficient degree in the conception and design of this work, in the analysis of data, and in the writing of the manuscript to take public responsibility for it. I (we) believe the manuscript describes truthful facts. I (we) have reviewed the final version of the manuscript and approve it for publication. Furthermore, I (we) attest that I (we) shall produce the data on which the manuscript is based for examination by the editors or their assignees, should they request it."

MANUSCRIPT FORMAT

Submit an original and five clean, sharp photocopies (not carbon copies) of all material (including abstract, tables, legends, and references), and five sets of glossy prints of illustrations. Type double-spaced on one side only on 8.5 $\times$ 11 inch paper of good quality, with 1 in. margins on all sides. Number pages consecutively in upper right corner beginning with title page, then abstract, text, references, legends, and tables.

Do not exceed 20 typed pages, including tables, illustrations, and reference list.

Title Page: Include, in order: title (and subtitle) of manuscript; full name(s) of author(s); highest earned academic degree; departmental and institutional affiliations; a brief acknowledgment of grant support provided; meeting, date, and place where paper was presented, if applicable; and reprint requests author. At bottom of page, supply a shortened title (52 characters).

Abstract: Except for case reports, special articles, and presidential addresses. In 200 words state main factual points of article in four paragraphs labeled Purpose, Methods, Results, and Conclusions. Make informative, not descriptive.

Text: Conform to guidelines promulgated by the Ad Hoc Committee on Reporting Standards, The Society for Vascular Surgery/North American Chapter, International Society for Cardiovascular Surgery, published in the *Journal*: "Suggested standards for reports dealing with lower extremity ischemia" (*J Vasc Surg* 1986;4:80–94); "Reporting standards in venous disease" (*J Vasc Surg* 1988; 8:172–81); "Suggested standards for reports dealing with cerebrovascular disease" (*J Vasc Surg* 1988;8:721–9); "Suggested standards for reporting on arterial aneurysms"(*J Vasc Surg* 1991;13:452–8); and "Standards in noninvasive cerebrovascular testing" (*J Vasc Surg* 1992;15:495–503).

References: Cite selectively. Do not cite personal communications and unpublished data as references; indicate these sources in text at appropriate place, and append to manuscript a note of approval from source for statement. References to articles in press include authors' names, title of article, and name of journal. Check spelling of references in foreign languages. Verify all references; cite consecutively in text by superscript numbers. Follow format of "Uniform Requirements for Manuscripts Submitted to Biomedical Journals" (Vancouver style) (*N Engl J Med* 1991;324:424–8). Type reference list double-spaced on a separate page in numeric order as citations appear in text. Abbreviate journal names according to *Cumulated Index Medicus.* If six or fewer authors, list all; if seven or more, list first three then et al.

JOURNAL ARTICLES

Josa M, Khuri SF, Braunwald NS, et al. Delayed sternal closure: an improved method of dealing with complications after cardiopulmonary bypass. J Thorac Cardiovasc Surg 1986;90:598–603.

BOOKS

Berger HJ, Zaret BL, Cohen LS. Cardiovascular nuclear medicine. In: Goldberger E, ed. Textbook of clinical cardiology. 1st ed. St. Louis: CV Mosby, 1982:326–45.

Illustrations: Limit to those that amplify, not duplicate, text. Prepare original drawings or graphs in black India ink or typographic (press-apply) lettering. Typewritten or freehand lettering is unacceptable; have all labeling done

professionally. Do not send original artwork, x-ray films, or ECG strips. Number figures consecutively in Arabic numerals in order of appearance in text. A reasonable number of halftone or line illustrations are reproduced without charge. Make special arrangements for color illustrations. Half of the expense of color reproduction is paid by author. Submit five sets of 5 × 7 in. (maximum size) glossy prints. Submit illustrations consistent in size. Do not mount illustrations; do not write on reverse side. On adhesive-back labels, type figure number, title of article, and first author's name and an arrow to indicate TOP.

Legends: Type consecutively, double-spaced, on a separate page. Indicate original magnification and stain for photomicrographs. If figure was previously published, obtain and submit written permission for use from copyright owner and original author and give full credit to original source.

Tables: Make self-explanatory, to supplement, not duplicate, text. Number consecutively in Roman numerals according to citation in text; provide a brief title for each. Type double-spaced on a separate page. If a table or any data therein have been previously published, obtain and submit written permission for use from copyright owner and original author and in a footnote to table give full credit to original source.

OTHER

Abbreviations: Use standard abbreviations. Use unusual or coined abbreviations sparingly; spell out at first appearance in text, followed in parentheses by abbreviation.

Drugs: Use generic names whenever possible. Cite proprietary drug names in parentheses after first mention, if desired.

Equipment: Use generic names whenever possible. For proprietary names of equipment, cite manufacturer and its city/state parenthetically immediately after proprietary name.

Experimentation, Animal: Include a statement in Methods section that care complied with "Principles of Laboratory Animal Care" (National Society for Medical Research) and the Guide for the Care and Use of Laboratory Animals (NIH Publication No. 86-23, revised 1985).

Experimentation, Human: Include a statement in Methods section indicating approval by institutional review board and noting that informed consent was obtained from each patient. For photographs of identifiable persons, submit signed releases from patients or from both living parents or guardian of minors.

Statistical Analyses: Identify, in a statement not part of text, coauthor or consultant responsible for statistical analysis.

Style: Conform to standard English usage. Consult *The Scientific Journal: Editorial Policies and Practices* (St. Louis, The C.V. Mosby Co.), *The Britannica Book of English Usage* (Garden City, N.Y., Doubleday), *A Dictionary of Modern English Usage* (New York, Oxford University Press), *A Grammar of the English Language* (Essex, Conn., Verbatim), and *Manual of Style* by the American Medical Association.

Units: State all measurements in metric units. Add English units parenthetically if measurements were originally made in that form.

SOURCE

July 1994, 20(1): 12A–14A.

Journal of Virology
American Society for Microbiology

Journals Division
American Society for Microbiology
1325 Massachusetts Ave., N.W.
Washington, DC 20005-4171

AIM AND SCOPE

The *Journal of Virology* is devoted to the timely dissemination of significant knowledge concerning the viruses of bacteria, plants, fungi, protozoa, and animals. Investigators in all areas of basic virology are invited to submit reports of original research that uses the approaches of biochemistry, biophysics, cell biology, genetics, immunology, molecular biology, morphology, physiology and pathogenesis and immunity. The original articles should contain experimental observations that address a hypothesis, lead to new concepts, and indicate new directions in research. The Journal will not publish papers that simply provide a new restriction map or nucleotide sequence or report the isolation or characterization of monoclonal antibodies, a viral variant, or a new strain or type. Such information or reagents must instead be used in further experimentation to test an idea or relate a clear set of novel conclusions that derive from these data.

The *Journal of Virology* specifically encourages publications relating the viruses under study to their host cells or organisms. In recognition of this emphasis, the sections of the journal relating to viral pathogenesis and immunity and to virus-cell interactions have been specifically set aside and identified in the table of contents. The editors wish to promote the publication of research done at the cell biology-virology-organismic biology interface.

ASM publishes a number of different journals covering various aspects of microbiology. Each journal has a prescribed scope that must be considered in determining the most appropriate journal for each manuscript.

The *Journal of Virology* will consider papers that describe the use of antiviral agents in elucidating the basic biological processes of viruses and host cells and manuscripts on the basic biology of bacterial viruses. Manuscripts describing new or novel methods or improvements in media and culture conditions will not be considered unless they are applied to the study of basic problems in virology or cell biology. By the same token, manuscripts dealing with methods for the production of monoclonal antibodies will not be considered unless the methods have been used to address fundamental questions.

Questions about these guidelines may be directed to the editor in chief of the journal being considered.

A manuscript rejected by one ASM journal on scientific grounds or on the basis of its general suitability for publication is considered rejected by all other ASM journals.

MANUSCRIPT SUBMISSION

In a cover letter state: journal to which manuscript is being submitted, most appropriate journal section, complete mailing address (including street), telephone and fax numbers of corresponding author, electronic mail address, and former ASM manuscript number and year if a resubmission. Include written assurance that permission to cite personal communications and preprints has been granted.

Submit three complete copies of each manuscript, including figures and tables. Type every portion double-spaced (6 mm between lines), including figure legends, table footnotes, and References. Use 1 in. margins on all sides. Number all pages in sequence, including abstract, figure legends, and tables. Place last two items after References section. Make characters easily distinguishable: numeral zero (0) and letter "oh" (O); numeral one (1), letter "el" (1), and letter "eye" (I); and multiplication sign (×) and letter "ex" (x). If distinctions cannot be made, mark items at first occurrence for cell lines, strain and genetic designations, viruses, etc., on modified manuscript.

To facilitate review submit three copies of "in-press" and "submitted" manuscripts.

Manuscripts must represent reports of original research. All authors must agree to submission and are responsible for its content, including appropriate citations and acknowledgments, and must also agree that corresponding author has authority to act on their behalf on matters pertaining to publication of manuscript. Submission guarantees that manuscript, or one substantially the same, was not published previously, is not being considered or published elsewhere, and was not rejected on scientific grounds by another ASM journal.

The American Society for Microbiology accepts the definition of primary publication in *How to Write and Publish a Scientific Paper* (third ed., Robert A. Day), to wit: ". . . (i) the first publication of original research results, (ii) in a form whereby peers of the author can repeat the experiments and test the conclusions, and (iii) in a journal or other source document readily available within the scientific community."

A scientific paper published in a conference report, symposium proceeding, technical bulletin, or other retrievable source is unacceptable for submission on grounds of prior publication. Preliminary disclosure of research findings published in abstract form as an adjunct to a meeting, is not considered "prior publication."

Acknowledge prior publication of data and submit copies of relevant work.

Obtain and submit permissions from both original publisher and original author, copyright owner(s), to reproduce figures, tables, or text from previous publications. Identify each as to relevant item in manuscript (e.g., "permissions for Fig. 1 in JVI 123-95").

An author is one who made a substantial contribution to the "overall design and execution of the experiments;" all authors are responsible for entire paper. Do not list individuals who provided assistance as authors; recognize in Acknowledgment section.

Authors must agree to order in which names are listed. Footnotes regarding attribution of work are not permitted. If necessary, include statement in Acknowledgment section.

Corresponding author must sign copyright transfer agreement on behalf of all authors upon acceptance of manuscript.

If all authors were employed by U.S. Government when work was performed, corresponding author should attach a statement attesting that manuscript was prepared as a part of their official duties and, as such, is a work of the U.S. government not subject to copyright. If some authors were employed by the U.S. government, but others were not, corresponding author should sign copyright transfer agreement as it applies to that portion performed by the nongovernment employee authors.

MANUSCRIPT FORMAT

Title, Running Title, and Byline: Present results of an independent, cohesive study; numbered series titles are not allowed. Avoid main title/subtitle arrangement, complete sentences, and unnecessary articles. On title page, include title, running title (54 characters and spaces), name of each author, address(es) of institution(s) at which work was performed, each author's affiliation, and a footnote indicating present address of any author no longer there. Place asterisk after name of corresponding author with telephone and fax numbers.

If desired, complete mailing address, telephone and fax numbers, and e-mail address of corresponding author will be published as a footnote. Include information in lower left corner of title page labeled "Correspondent Footnote."

Abstract: 250 words concisely summarizing basic content of paper without presenting extensive experimental details. Avoid abbreviations; do not include diagrams. When essential to include a reference, use References citation but omit article title. Make complete and understandable without reference to text.

Introduction: Supply sufficient background information to allow reader to understand and evaluate results of present study without referring to previous publications. Provide rationale for present study. Use only those references required to provide the most salient background rather than an exhaustive review of topic.

Materials and Methods: Include sufficient technical information to allow experiments to be repeated. When centrifugation conditions are critical, give enough information to enable repetition of procedure: make of centrifuge, model of rotor, temperature, time at maximum speed, and centrifugal force (x g rather than revolutions per minute). For commonly used materials and methods, a simple reference is sufficient. If several alternative methods are commonly used, identify method briefly and cite reference. Allow reader to assess method without reference to previous publications. Describe new methods completely and give sources of unusual chemicals, equipment, or microbial strains. When large numbers of microbial strains or mutants are used, include tables identifying sources and properties of strains, mutants, bacteriophages, plasmids, etc.

Describe a method, strain, etc., used in only one of several experiments in Results section or very briefly (one or two sentences) in a table footnote or figure legend.

Results: Include rationale or design of experiments. Reserve extensive interpretation of results for Discussion. Present as concisely as possible in either text, table(s), or figure(s). Limit photographs (particularly photomicrographs and electron micrographs) to those absolutely necessary to show experimental findings. Number figures and tables in order cited in text; cite all figures and tables.

Discussion: Interpret results in relation to previously published work and to experimental system at hand. Do not repeat Results or reiterate Introduction. In short papers, Results and Discussion may be combined.

Acknowledgments: Acknowledge financial assistance and personal assistance in separate paragraphs. Format: "This work was supported by Public Health Service grant CA01234 from the National Cancer Institute."

Appendixes: Appendixes contain supplementary material to aid reader. Do not use titles, authors, and References sections distinct from those of primary article. If author(s) of appendix cannot be listed in byline or Acknowledgment of primary article, rewrite appendix so that it can be considered as an independent article. Label equations, tables, and figures with letter "A" preceding numeral.

References: Include all relevant sources; cite all listed references in text. Arrange citations in alphabetical order, by first author, and number consecutively. Abbreviate journal names according to *Serial Sources for the BIOSIS Data Base* (BioSciences Information Service, 1994). Cite each listed reference in text by number.

1. **Arens, M., and T. Yamashita.** 1978. In vitro termination of adenovirus DNA synthesis by a soluble replication complex. J. Virol. **25:**698–702

2. **Barton, B., G. Harding, and A. Zuccarelli.** 1994. A method for sizing large plasmids, abstr. H-249, p. 244. *In* Abstracts of the 94th General Meeting of the American Society for Microbiology 1994. American Society for Microbiology, Washington, D.C.

3. **Berry, L. J., R. N. Moore, K. J. Goodrum, and R. E. Couch, Jr.** 1977. Cellular requirements for enzyme inhibition by endotoxin in mice, p. 321–325. *In* D. Schlessinger (ed.), Microbiology—1977. American Society for Microbiology, Washington, D.C.

4. **Cox, C. S., B. R. Brown, and J. C. Smith.** J. Gen. Genet., in press.*

5. **Fitzgerald, G., and D. Shaw.** *In* A. E. Waters (ed.) Clinical microbiology, in press. EFH Publishing Co., Boston .

6. **Gill, T. J., III.** 1976. Principles of radioimmunoassay, p. 169–171. *In* N. R. Rose and H. Friedman (ed.), Manual of clinical immunology. American Society for Microbiology, Washington, D.C.

7. **Gustlethwaite, F. P.** 1985. Letter. Lancet ii:327.

8. **Jacoby, J., R. Grimm, J. Bostic, V. Dean, and G. Starke.** Submitted for publication. [Article title is optional.]

9. **Jensen, C., and D. S. Schumacher.** Unpublished data. [Date is optional.]

10. **Jones, A.** Personal communication. [Date is optional.]

11. **Leadbetter, E. R.** 1974. Order II. *Cytophagales* nomen novum, p. 99. *In* R. E. Buchanan and N. E. Gibbons (ed.), Bergey's manual of determinative bacteriology, 8th ed. The Williams & Wilkins Co., Baltimore.

12. **Miller, J. H.** 1972. Experiments in molecular genetics, p. 352–355. Cold Spring Harbor Laboratory, Cold Spring Harbor, N.Y.

13. **Sigma Chemical Co.** 1989. Sigma manual. Sigma Chemical Co., St. Louis, Mo.

14. **Smith, J. C.** April 1970. U.S. patent 484,363,770.

15. **Smyth, D. R.** 1972. Ph.D. thesis. University of California, Los Angeles.

16. **Yagupsky, P., and M. A. Menegus.** 1989. Intraluminal colonization as a source of catheter-related infection. Antimicrob. Agents Chemother. **33:**2025. (Letter.)

"In press" references to ASM publications should state control number or name of publication if a book.

Illustrations: Put figure number and authors' names on all figures, either in margin or on back (marked lightly with a soft pencil). For micrographs especially, indicate TOP.

Do not use paper clips. Insert small figures in an envelope. Do not submit illustrations larger than 8.5 × 11 in. Data should not contain more significant figures than precision of measurement allows.

Continuous-Tone and Composite Photographs:—Keep in mind journal page width: 3 5/16 in. (single column) and 6 7/8 in. (double-column, maximum). Include only significant portion of an illustration. Photos must be of sufficient contrast to withstand loss of contrast and detail inherent in printing. Submit one photograph of each continuous-tone figure for each manuscript copy; photocopies are not acceptable. Submit figures in final size. If must be reduced, make sure that all elements, including labeling, can withstand reduction and remain legible.

If a figure is a composite of a continuous-tone photograph and a drawing or labeling, provide original composite for printer. Send original, labeled "printer's copy," with modified manuscript to editor.

Electron and light micrographs must be direct copies of original negative. Indicate magnification with a scale marker on micrograph.

Computer-Generated Images—Provide a floppy disk on which figures are stored in either tagged image file format (TIFF) or Encapsulated Postscript (EPS) from either an MS-DOS or Macintosh system.

Use Adobe Photoshop or Aldus Freehand. For Aldus, one and two column art cannot exceed 20 picas (3 5/16 in.) and 41.5 picas (6 7/8 in.) respectively. Use Helvetica (medium or bold) or Times Roman font. Adobe users: check image densities on-line. If image shadow density reads below 1.25, enter density as 1.40; if 1.25–1.6, enter as 1.65; enter readings >1.65 as actual reading. Provide computer file with prints on floppy disk (Macintosh) with accepted manuscript.

Canvas software does not allow output in TIFF or EPS. When preparing to output, select "Publish" option and then "Illustrator 88" option. This will generate two files, copy both and send to ASM. (Do not use rotated type for labeling figures.)

Images produced with other types of software may not be acceptable.

Supply a disk and an extra printed copy of each figure (as it appears on disk) when manuscript is submitted. Indicate type of software used and number of images stored on each disk. Each figure must contain all labeling (including panel identification letters). Write figure number and author names lightly in pencil on back.

For large images, use 40- or 80-megabyte Syquest cartridges or magneto-optical cartridges. For transfer from UNIX systems, submit either 9-track or 8-mm "tar" archives. Incorporate all final lettering, labeling, tooling, etc., in final supplied material. Do not include figure numbers on image.

Electronic image telephone hotline: Cadmus Journal Services, (800) 257-5331, ext 3361 (non-U.S.: (804) 261-3000, ext 3361); E-mail: image@cadmus.com.

Include a description of software/hardware used in figure legend(s).

Color Photographs—Use is discouraged. Include an extra copy at submission for a cost estimate. Cost is borne by author.

Drawings—Submit graphs, charts, sequences, complicated clinical or mathematical formulas, diagrams, and other drawings as glossy photographs made from finished drawings not requiring additional artwork or typesetting. Computer-generated graphics produced on high-quality laser printers are also acceptable. No part of graph or drawing should be handwritten. Label both axes of graphs. Graphs will be reduced to one-column width (3 5/16 in.); make elements large enough to withstand reduction. Avoid heavy letters and unusual symbols.

In figure ordinate and abscissa scales and table column headings, avoid ambiguous use of numbers with exponents. Use appropriate SI symbols (μ for 10^{-6}, m for 10^{-3}, k for 10^3, M for 10^6, etc.). For a complete listing of SI symbols see IUPAC "Manual of Symbols and Terminology for Physicochemical Quantities and Units" (*Pure Appl. Chem.* **21:**3–44, 1970).

When powers of 10 is used, associate exponent power with number shown. In representing 20,000 cells per ml, numeral on ordinate is "2" and label is "10^4 cells per ml" (not "cells per ml × 10^{-4}").

Figure Legends: Provide enough information so that figure is understandable without reference to text. Describe detailed experimental methods in Materials and Methods section, not in legend. Report a method unique to an experiment in one or two sentences. Define all symbols and abbreviations that have not been defined elsewhere.

Tables: Type each on a separate page. Arrange data so columns of like material read down, not across. Make headings clear so meaning of data will be understandable without reference to text. See Abbreviations for those that can be used in tables. Explanatory footnotes are acceptable, extensive table "legends" are not. Footnotes should not include detailed descriptions of experiment. Tables must include enough information to warrant format; incorporate fewer than six pieces of data into text. Tables that can be photographically reproduced for publication without further typesetting or artwork are "camera ready." Do not hand letter.

OTHER FORMATS

AUTHOR'S CORRECTION—Provides a means of correcting errors of omission (e.g., authors names or citations) and scientific errors that do not alter basic results or conclusions.

For omission of an author, submit a letter from both authors of article and omitted author agreeing to publication of correction. Send corrections of a scientific nature to editor who handled article. Addition of new information is not permitted.

ERRATA—Provides a means of correcting errors that occurred during writing, typing, editing or printing of a published articles. Changes in data and addition of new material are not permitted. Send directly to Journals Division.

LETTERS TO THE EDITOR—500 words (typed double-space). Comments on articles previously published in *Journal*. Include data to support argument. Send three copies to Journals Department. Letter will be sent to editor who handled article in question. If publication is warranted, a reply from corresponding author of article will be solicited.

MINIREVIEWS—6 printed pages exclusive of references. Brief summaries of developments in fast-moving areas; based on published articles; address any subject within scope. Provide three double-spaced copies.

NOTES—1000 words. Brief observations that do not warrant full-length papers. Submit the same way as full-length papers. Include a 50 word abstract. Do not use section headings in body; report methods, results, and discussion in a single section. Paragraph lead-ins are permissible. Keep figures and tables to a minimum. Describe materials and methods in text, not in figure legends or table footnotes. Present acknowledgments as in full-length papers; do not use a heading. References section is identical.

RETRACTIONS—Reserved for major errors or breaches of ethics that, for example, may call into question source of the data or validity of results and conclusions of an article. Send retraction and an accompanying explanatory letter signed by all authors directly to editor-in-chief of *Journal*. The editor who handled paper and the chairman of ASM Publications Board will be consulted. If all parties agree to the publication and content of the retraction, it will be sent to Journals Department for publication.

OTHER

Abbreviations: Use as an aid to reader, not a convenience to author; limit use. Use abbreviations other than those recommended by the IUPAC-IUB *(Biochemical Nomenclature and Related Documents*, 1978) only as a necessity, such as in tables and figures.

Use pronouns or paraphrase a long word after first use. Use standard chemical symbols and trivial names or their symbols for terms that appear in full in neighboring text.

Introduce all abbreviations except those listed below in first paragraph of Materials and Methods. Or, define each abbreviation and introduce it in parentheses at first use. Eliminate abbreviations not used at least five times in text (including tables and figure legends).

In addition to abbreviations for Système International d'Unités (SI) units of measurement, other common units (e.g., bp, kb, and Da), and chemical symbols for elements, use the following without definition in title, abstract, text, figure legends, and tables: DNA (deoxyribonucleic acid); cDNA (complementary DNA); RNA (ribonucleic acid); cRNA (complementary RNA); RNase (ribonuclease); DNase (deoxyribonuclease); rRNA (ribosomal RNA); mRNA (messenger RNA); tRNA (transfer RNA); AMP, ADP, ATP, dAMP, ddATP, GTP, etc. (for the respective 5′ phosphates of adenosine and other nucleosides) (add 2′-, 3′-, or 5′- when needed for contrast); ATPase, dGTPase, etc. (adenosine triphosphatase, deoxyguanosine triphosphatase, etc.); NAD (nicotinamide adenine dinucleotide); NAD (nicotinamide adenine dinucleotide, oxidized); NADH (nicotinamide adenine dinucleotide, reduced); NADP (nicotinamide adenine dinucleotide phosphate); NADPH (nicotinamide adenine dinucleotide phosphate, reduced); NADP$^+$ (nicotinamide adenine dinucleotide phosphate, oxidized); poly(A), poly(dT), etc. (polyadenylic acid, polydeoxythymidylic acid, etc.); oligo(dT), etc. (oligodeoxythymidylic acid, etc.); P$_i$ (orthophosphate); PP$_i$ (pyrophosphate); UV (ultraviolet); PFU (plaque-forming units); CFU (colony-forming units); MIC (minimal inhibitory concentration); MBC (minimal bactericidal concentration); Tris [tris(hydroxymethyl)aminomethane]; DEAE (diethylaminoethyl); A_{260} (absorbance at 260 nm); EDTA (ethylenediaminetetraacetic acid); and AIDS (acquired immunodeficiency [or immune deficiency] syndrome. Abbreviations for cell lines (e.g., HeLa) also need not be defined.

Use the following without definition in tables:

amt	amount	SD	standard deviation
approx	approximately	SE	standard error
avg	average	SEM	standard error of the mean
concn	concentration	sp act	specific activity
diam	diameter	sp gr	specific gravity
expt	experiment	temp	temperature
exptl	experimental	tr	trace
ht	height	vol	volume
mo	month	vs	versus
mol wt	molecular weight	wk	week
no.	number	wt	weight
prepn	preparation	yr	year

Experimentation, Human: When isolates are derived from patients in clinical studies, do not identify with patients' initials, even as part of a strain designation. Change initials to numerals or use randomly chosen letters. Do not give hospital unit numbers; if a designation is needed, use only last two digits of unit. (Note: established designations of some viruses and cell lines although they consist of initials, are acceptable [e.g., JC virus, BK virus, HeLA cells].)

Isotopically Labeled Compounds: For simple molecules, indicate isotopic labeling in chemical formula. Do not use brackets when isotopic symbol is attached to the name of a compound that in its natural state does not contain the element or to a word that is not a specific chemical name. For specific chemicals, place symbol for the isotope introduced in square brackets directly preceding the part of the name that describes labeled entity. Configuration symbols and modifiers precede isotopic symbol.

Follow conventions for isotopic labeling of the *Journal of Biological Chemistry*; for more detailed information see instructions to authors of that journal (first issue each year).

Materials Sharing: Make plasmids, viruses, and living materials such as microbial strains and cell lines newly described in the article available from a national collection or in a timely fashion and at reasonable cost to members of the scientific community for noncommercial purposes.

Nomenclature, Bacteria: Use binary names, consisting of a generic name and a specific epithet (e.g., *Escherichia coli*), for all bacteria. Names of categories above genus level may be used alone, but specific and subspecific epithets may not. A specific epithet must be preceded by a generic name at first use in paper. Thereafter, abbreviate generic name to the initial capital letter (e.g., *E. coli*), provided there can be no confusion with other genera used in paper. Names of all taxa (phyla, classes, orders, families, genera, species, subspecies) are in italics; strain designations and numbers are not.

Nomenclature, Chemical and Biochemical: *Chemical Abstracts* (Chemical Abstracts Service, Ohio State University, Columbus) and its indexes and *The Merck Index* (11th ed., 1989; Merck & Co., Inc., Rahway, N.J.) are the recognized authorities for names of chemical compounds. For guidelines on use of biochemical terminology, consult *Biochemical Nomenclature and Related Documents* (1978; The Biochemical Society, London) and instructions to authors of *Journal of Biological Chemistry* and *Archives of Biochemistry* and *Biophysics* (first issues of each year).

Do not express molecular weight in daltons; molecular weight is a unitless ratio. Express molecular mass in daltons. For enzymes, use recommended (trivial) name assigned by Nomenclature Committee of the International Union of Biochemistry (see *Enzyme Nomenclature,* Academic Press, Inc., 1992). If a nonrecommended name is used, place proper (trivial) name in parentheses at first use in abstract and text. Use EC number when one has been assigned; express enzyme activity either in katals (preferred) or in micromoles per minute.

Nomenclature, Genetic: Phenotype describes observable properties of an organism. Genotype refers to genetic constitution of an organism, usually in reference to some standard wild type. Use recommendations of Demerec et al. (*Genetics* 54:61–76,1966) as a guide to terms.

Use phenotypic designations when mutant loci have not been identified or mapped. Use to identify protein products of genes. Phenotypic designations consist of three-letter symbols; do not italicize, capitalize first letter. Use Roman or Arabic numerals to identify a series of related phenotypes. Designate wild-type characteristics with a superscript plus (Pol⁺). Use negative superscripts (Pol⁻) to designate mutant characteristics. Use lowercase superscript letters to further delineate phenotypes. Define phenotypic designations.

Indicate genotypic designations by three-letter lowercase italic locus symbols. If several loci govern related functions, distinguish by italicized capital letters following locus symbol. Indicate promoter, terminator, and operator sites as described by Bachmann and Low (*Microbiol. Rev.* 44: 1–56, 1980).

Indicate wild-type alleles with a superscript plus. Refer to an *ara* mutant rather than an *ara⁻* strain.

Italicize entire description of a virus, including designations *am* or *sus* (amber suppressible) and *ts* (temperature sensitive). Employ superscripts to indicate hybrid genomes. Genetic symbols may be one, two, or three letters. Delineate host DNA insertions into viruses by square brackets and genetic symbols; designations for inserted DNA conform to those used for host genome. Genetic symbols for phage λ can be found in Echols and Murialdo (*Microbiol. Rev.* **42:**577–591, 1978) and Szybalski and Szybalski (*Gene* **7:**217–270, 1979).

For nomenclature of restriction endonucleases follow recommendations of Roberts (p. 757–768, *in* A. I. Bukhari, J. A. Shapiro, and S. L. Adhya, ed., *DNA Insertion Elements, Plasmids, and Episomes,* Cold Spring Harbor Laboratory, Cold Spring Harbor, N.Y., 1977).

A mutation is an alteration of the primary sequence of genetic material. A mutant is a strain carrying a mutation.

Nomenclature, Viruses: Names used for viruses should be those approved by the International Committee on Taxonomy of Viruses (ICTV) and published in the 4th Report of the ICTV, Classification and Nomenclature of Viruses (*Intervirology* **17:**23–199, 1982), with modifications contained in the 5th Report of the ICTV (*Arch. Virol., Suppl.* 2, 1991. If desired, add synonyms parenthetically when name is first mentioned. Approved generic (or group) and family names may also be used.

Nucleic Acid Sequences: Present limited length nucleic acid sequences which are the primary subject of a study freestyle in the most effective format. Use the following format for longer sequences to save space. Submit as 8.5 × 11 in. camera-ready copy in standard (portrait) orientation. Print in lines of 100 to 120 nucleotides in a nonproportional (monospace), legible font to be published in 6 in. length lines. If possible, subdivide lines of nucleic acid sequence into blocks of 10 or 20 nucleotides by spaces within sequences or by marks above it. Use uppercase and lowercase letters to designate exon/intron structure, transcribed regions, etc., if lowercase letters remain legible at a 6 in. line length. Number sequence line by line; place numerals, representing first base of each line, to left of lines. Minimize spacing between adjacent lines, leave room only for sequence annotation. Annotation may include boldface, underlining, brackets, boxes, etc. Present encoded amino acid sequences, if necessary, immediately above first nucleotide of each codon, using single-letter amino acid symbols. Use same format for comparisons of multiple nucleic acid sequences.

Nucleotide Sequences: Include GenBank/EMBL accession numbers for primary nucleotide and/or amino acid sequence data in original manuscript or insert when manuscript is modified. Include accession number as a separate paragraph at end of Materials and Methods section for full-length papers or at end of text of Notes.

GenBank Submissions, National Center for Biotechnology Information; Bldg. 38A, Rm. 8N-803, 8600 Rockville Pike, Bethesda, MD 20894; e-mail (submissions): gbsub@ncbi.nlm.nih.gov, e-mail updates: update@ncbi.nlm.nih.gov.

EMBL Data Library Submissions, Postfach 10.2209, Meyerhofstrasse 1, 6900 Heidelberg, Germany; telephone: 011 49 (6221) 387258; fax: 011 49 (6221) 387306; electronic mail (data submissions): datasubs@embl.bitnet.

Page Charges: $55 per printed page if research was supported by special funds, grants (departmental, governmental, institutional, etc.) or if research was done as part of official duties. If not, request a waiver of charges to Journals Division, American Society for Microbiology with manuscript. Separate request from cover letter; indicate how work was supported and include a copy of Acknowledgment section.

Minireviews and Letters to the Editor are not subject to page charges.

Reviewers: Suggest an appropriate editor for new submissions. Recommend two or three reviewers who are not members of your institution(s) and have never been associated with you or your laboratory(ies). Provide name, address, phone and fax numbers, and area of expertise for each.

Style: Follow *CBE Style Manual* (5th ed., 1983; Council of Biology Editors, Inc., Bethesda, Md.), *ASM Style Manual for Journals and Books* (American Society for Microbiology, 1991), and Robert A. Day's *How to Write and Publish a Scientific Paper* (3rd ed., 1988; Oryx Press).

Use past tense to narrate events in the past, including procedures, observations, and data of study. Use present tense for conclusions, conclusions of previous researchers, and generally accepted facts. Most of abstract, Materials and Methods, and Results sections will be in past tense, and most of introduction and some of Discussion will be in present tense. It may be necessary to vary tense in a single sentence.

For an in-depth discussion of tense in scientific writing, see p. 158–160 in *How to Write and Publish a Scientific Paper,* 3rd ed.

Units: Use standard metric units for reporting length, weight, and volume. For these units and molarity, use prefixes m, µ, n, and p for 10^{-3}, 10^{-6}, 10^{-9}, and 10^{-12}, respectively. Use prefix k for 10^3. Avoid compound prefixes. Use µg/ml or µg/g in place of ppm. Units of temperature are: 37°C or 324 K.

When fractions are used to express units such as enzymatic activities, use whole units in denominator instead of fractional or multiple units. Use unambiguous forms such as exponential notation.

See *CBE Style Manual,* 5th ed., for more detailed information about reporting numbers and information on SI units for reporting illumination, energy, frequency, pressure, and other physical terms. Report numerical data in appropriate SI unit.

SOURCE

January 1995, 69 (1): i-x.

Journal of Zoology
The Zoological Society of London

The Editor, *Journal of Zoology*
The Zoological Society of London
Regent's Park
London NW1 4RY

AIM AND SCOPE

The *Journal of Zoology* contains original papers within the general field of experimental and descriptive zoology, and notices of the business transacted at the Scientific Meetings of the Society.

MANUSCRIPT SUBMISSION

Submission implies that material is original and that no similar paper is being or will be submitted elsewhere. Do not submit serialized studies; titles should not contain part numbers and should stand alone.

If accepted for publication, exclusive copyright in the paper shall be assigned to The Zoological Society of London. The Society will not put any limitation on personal freedom of author to use material contained in the paper in other works.

MANUSCRIPT FORMAT

Provide two copies of typescript, double-spaced throughout on one side of paper (preferably A4), with a wide margin all round. Number all pages consecutively and fasten together. Type tables and captions for illustrations separately at end of manuscript; indicate required positions in margins of text. Number text lines.

Title Page: Give a concise, specific title with name(s) of author(s) and institution(s) where work was carried out. Provide a short title for page headings.

Synopsis: 300 words, intelligible without reference to main text.

List of Contents: Except for papers too short to be divided into sections.

Summary: Not obligatory; if included, give a succinct account of subject with results obtained and conclusions.

Illustrations: Illustrations are reduced to a maximum size of 19 × 14 cm. Do not submit outsize artwork (i.e., needing more than 50% reduction).

Include a metric scale on each illustration. Indicate required reduction for both line drawings and photographs.

Mark name(s) of author(s) and number of figure or plate on back of all illustrations; give orientation if necessary.

Do not submit original illustrations until paper has been accepted.

Line Drawings: Must be good enough for direct reproduction. Prepare in black (India) ink on white card or tracing paper, preferably A4 size (30 × 21 cm). Insert necessary lettering in blue pencil on original, or write on photocopy.

Indicate graph curves by solid —— pecked - - - - or dotted ······· lines; use symbols to determine points and to key captions: ○ ● ❑ ■ △ ▽ ▲ ▼

Shade areas by hatching or cross-hatching; do not use dots. Refer to line drawings as Fig. 1, Fig. 2, etc., and sub-sections as (1), (b) etc.

Photocopies of line figures cannot be reproduced but may be submitted for reviewing purposes.

Photographs (for half-tone illustrations): Submit high quality glossy prints of maximum contrast, preferably in final size. Indicate labels and scale-lines accurately on a duplicate set. Mount on white backing board, with rubber solution for ease of removal for reproduction. Refer to Photographs as Plate I, Plate II, etc., and any sub-sections as (a), (b), etc.

Tables: Present to fit page size (19 × 14 cm) without undue reduction. Oversize tables will not be accepted.

References: Check against text to ensure that spelling of authors' names and dates given are consistent and that all authors quoted in text (in date order if more than one) are given in reference list and *vice versa*.

Give full title of paper with first and last pages. Abbreviate journal titles according to *World List of Scientific Periodicals* (4th ed).

Follow book titles by place of publication and publisher. Give name of editor(s) if different from author cited.

Arrange references first alphabetically under author(s) name(s) and then in chronological order if several papers by the same author(s) are cited. Use *a*, *b*, etc. after year to distinguish papers published by the same author(s) in the same year.

In text follow surname(s) of author(s) by date, to which *a*, *b*, etc. may be added to distinguish papers published by the same author(s) in the same year. For articles with two authors use both names and year. Do not use *et al.* For articles with three or more authors, on first citation use all authors' names (to a maximum of three) and year. Thereafter, give name of first author followed by *et al.* and date.

Currey, J. D. (1984). Effects of differences in mineralization on the mechanical properties of bone. *Phil. Trans. R. Soc. (B)* **304**: 509–518.

Pianka, E. R. (1978). *Evolutionary ecology.* (2nd edn). New York: Harper & Row.

Whitcar, M. (1992). Solitary chemosensory cell. In *Fish chemoreception:*. 103-125. Hara, T. J. (Ed.). London: Chapman & Hall

OTHER

Conventions: Use metric system and SI Units where appropriate. For further details see British Standards Institute 5775, *Quantities, Units and Abbreviations.*

Spell out whole numbers one to nine (except in Methods section) and give number 10 onwards in figures but this is not essential if many numbers appear together.

If a new taxon is described, give institution in which type material is deposited, together with details of registration.

Experimentation, Animal: Comply with standards and procedures laid down by the Home Office. The Publications Committee will not accept papers based on work involving cruelty to animals or endangering species populations.

SOURCE

September 1994, 234(1): back matter.

Kidney International
International Society of Nephrology
Blackwell Science, Inc.

Thomas E. Andreoli
Department of Medicine
University of Arkansas, College of Medicine
4301 West Markham, Slot #712
Little Rock, Arkansas 77205
University of Arkansas, College of Medicine
Hendrix Hall, Room 279
Corner of 7th and Cottage Drives
Little Rock, Arkansas 72205
                For express mail

AIM AND SCOPE

Kidney International will consider publication of any original manuscript that deals with either clinical or laboratory investigation of relevance to the broad field of nephrology. Animal and in vitro research will be considered if kidney-oriented and pertinent to normal or abnormal human biology. Research manuscripts of interest to qualified clinical nephrologist will receive strong consideration, and submission of research manuscripts dealing with diverse topics and problems (normal and abnormal renal physiology, immunology, morphology, experimental renal disease, clinical nephrology, etc.) is encouraged.

MANUSCRIPT SUBMISSION

All manuscripts must be contributed solely to *Kidney International*. Document all studies completely by reference to literature; thoroughly describe experimental method. Substantive portions of any manuscript must not have been published elsewhere except in abstract form. Published papers become property of *Kidney International*.

Submission implies that work has not been published before (except as an abstract or as part of a published lecture, review or thesis); that it is not under consideration for publication elsewhere; that its publication has been approved by all coauthors, as well as by responsible authorities at the institute where work was done; that if accepted, authors agree to automatic transfer of copyright to the society; that manuscript will not be published elsewhere in any language without consent of copyright holder; and that written permission of copyright holder was obtained for use of material from other copyrighted sources.

MANUSCRIPT FORMAT

Type manuscripts on 8.5×11 in. bond paper. Double-space and use wide margins—including abstract, footnotes, references, figure legends, and tables. Submit original typescript and four additional copies (photocopies must be clearly legible) with five sets of illustrations (a single glossy print and 4 photocopies). For illustrations such as photomicrographs, electron micrographs, etc., submit 5 original sets of glossy prints.

Arrange manuscripts in order: complete title page, short title page, abstract (summary), introduction, methods, results, discussion, acknowledgments (including gifts of material, grants and sources of support), references, tables, figure legends, and clearly labeled photocopies of all figures. Number pages consecutively beginning with title page.

Submit manuscripts on diskette, with five accurate copies of manuscript on bond paper. Use any major word processing software, either DOS-based or Macintosh. Label diskettes must be labeled with author's name, article title, and software used.

Title Page: Include complete title (85 characters including spaces); first name(s), middle initial(s), and surname of author(s); laboratory or institution of origin, including city, state, and country; and complete address for corresponding author with telephone and telex or fax numbers typed on separate sheet.

Short Title Page: A running head (45 characters, including spaces); typed on separate sheet.

Abstract: 200 words on separate page. Include on first line of narrative complete title of paper (omit names of authors and institutional affiliation). Describe precisely what was done, results obtained, and conclusion(s) drawn.

Footnotes: Number consecutively in superscripts as they appear throughout article. Type at bottom of page where used and separate from main body of text by a centered horizontal line. Footnotes to tables follow order: a b c d e.

References: Type on separate page; number in order of appearance in text. Give each reference its own number. "1A" is unacceptable. Enclose reference numbers in brackets on line of writing within narrative. Number references cited in a caption or table in sequence with text. List papers "in press" in references; enclose citations such as "to be published," "unpublished observations," or "personal communication," in parentheses on line of writing within narrative. "In press" references must have received final acceptance. Submit written permission when citing a personal communication. Reference form includes end pagination; omit number, day, or month of issue. List all authors. Use *The List of Journals Indexed in Index Medicus* for abbreviations for journals.

JOURNAL ARTICLES

Surname and initials of author(s), title of article, name of journal, volume number, first and last page, year.

CLAPP JR, ROBINSON RR: Distal sites of action of diuretic drugs in the dog nephron. *Am J Physiol* 215:228–235, 1968

BOOKS

Surname and initials of author(s), title and subtitle, edition (other than first), city, publishing house, year, volume (where applicable), page as specific reference.

GRINDLEY MF: *Manual of Histologic and Special Staining Techniques* (2nd ed). New York, Blakiston, 1968, p. 47

ARTICLES IN BOOKS

Surname and initials of author(s), title of article, chapter number (when present), title of book, editor, edition (other than first), city, publishing house, year, volume (where applicable), page.

TRUMP BF, BULGER RE: The morphology of the kidney, in *The Structural Basis of Renal Disease*, edited by BECKER EL, ELLIS J, New York, Hoeber Medical Division of Harper and Row, 1968, p. 1

ABSTRACT AND SUPPLEMENT

Surname and initials of author(s), title of abstract, abstract in parentheses, name of journal, volume number, page, year.

FLOWER RJ: The mediators of steroid action. (abstract) *Nature* 320:20, 1986

Surname and initials of author(s) title of article, name of journal, volume number, supplement number in parentheses, first and last page, year.

KAMM DE, GENIN M: Diuretic-induced azotemia. *Kidney Int* 24 (Suppl 16): S58–S60, 1983

Acknowledgments: Include in order of appearance: acknowledgment of presentation at recognized meeting or previous publication in abstract form; sources of financial support for study; awards to individual authors; technical or collaborative assistance.

Tables: Type double-spaced on separate sheets with number and title flush left at top. Number consecutively in body of manuscript. Do not use vertical lines. Double-space superscript footnotes in alphabetical order; do not use symbols or numbers. Do not submit photographs.

Figures and Illustrations: Submit one original glossy print and four photocopies. For photographs, photomicrographs, electrocardiographic tracings, and electron micrographs, submit five original glossy prints. Number and cite consecutively in order of appearance in text.

Submit clear glossy photographs of figures. Original line drawings are acceptable. Make lettering sufficiently large to be clearly legible after reduction. Make patterns or shadings in bar graphs clearly distinguishable; horizontal and diagonal lines are preferable to shading. Line drawings will be copy edited and relettered to conform to *Journal* style.

Submit photomicrographs, electron micrographs, etc., as sharp well-contrasted glossy prints, trimmed at right angles. Use cardboard less than 1/16 in. thick for multiple part figures mounted on a plate. Mounted plates must fit 7 1/8 × 9 in. If special grouping of illustrations is desired, include instructions. Submit remaining three sets unmounted. Label photographs if possible. If not, indicate in ink on transparent overlay (tissue paper) on actual photograph desired inscription(s) and exact placement. Designate margin trim on overlay. Indicate end points of desired marker lines by a fine needle prick in original photograph. Plan illustrations carefully so a minimum number are required and fit either single (3.5 in.) or double (7 1/8 in.) column width. Figures must not exceed 9 in. in length or they will be reduced. If illustration is reduced, magnification will be adjusted accordingly. Print electronically reproduced halftones using a scanner and a laser printer at highest possible resolution (1200 dpi or higher). Computer file of graphic may be submitted on a separate disk labelled with figure number and software, Put each figure in a separate file.

Captions: Type on a separate sheet. Include final magnifications. Be brief; do not duplicate information in text.

OTHER FORMATS

TECHNICAL NOTES—3000 words, including references (four journal pages, including illustrations). pertaining to clinical or laboratory topics. Appropriate subject matter includes descriptions of new laboratory or clinical methods, new apparatus, or critical modifications of established techniques. Abstract and short title page are not required, otherwise prepare exactly as for full-length manuscripts: introduction, methods, results, and discussion. Omit section headings and designate by a new paragraph.

OTHER

Abbreviations, Nomenclature, and Symbols: Follow *The Council of Biology Editors Style Manual* (1983, 5th ed, American Institute of Biological Sciences, Arlington, VA) for general usage. Limit use of abbreviations; define nonstandard abbreviations.

Color Charges: $1500 per page.

Drugs: Indicate generic names; use proprietary name only when generic name has been used first.

Electronic Submission: Prepare manuscripts using a personal computer. Submit five paper copies, together with a diskette. Data on diskette is used for typesetting. If authors do not have access to a computer, *Journal* will consider manuscripts not accompanied by a diskette.

Do not attempt to make output approximate or match typeset page. Be consistent in style (i.e., units, abbreviations). End paragraphs in a uniform manner, in a different way, e.g., two carriage returns, from line endings within paragraphs. Use code for paragraph endings to indicate required line ending. Never type letter "el" for numeral "one" or vice versa. Do not interchange numeral "zero" and capital letter "oh." Do not add extra line spacing (except as a normal paragraph ending indication) above or below titles, subheads, or between paragraphs. For double-spaced output, use printer commands in word processor to

double-space printout. Avoid using multiple spaces (horizontal). End sentences with one space. Never use multiple spaces for horizontal positioning of text. Prepare tables and figure captions in separate files. Do not divide words by hyphenating at line endings. Let text wrap. Turn off automatic hyphenation. Style references according to *Journal* guidelines. Enclose diskette in protective cover for mailing.

Illustration (halftone) Charges: $50 for each individual halftone or full plate exceeding two.

Page Charges: $75 for each page exceeding four.

Spelling: Use *Webster's Third New International Dictionary* as standard reference for spelling, hyphenating, and compounding.

SOURCE

January 1995, 47(1): xxx-xxxiii.

Laboratory Investigation
United States and Canadian Academy of Pathology, Inc.
Williams & Wilkins, Inc.

Emanuel Rubin, M.D.
Thomas Jefferson University
225 Jefferson Alumni Hall, 1020 Locust St.
Philadelphia, PA 19107
Telephone: (215) 955-4847; Fax: (215) 955-8703

AIM AND SCOPE

The goal of *Laboratory Investigation* is to provide a medium for prompt publication of significant advances in research in pathology, both human and experimental. Manuscripts dealing with normal structure and function, and those describing technical procedures, when relevant to an understanding of disease, are also accepted. Reports of original research, brief communications, and special review articles will be considered for publication.

MANUSCRIPT SUBMISSION

Submit manuscripts in triplicate. Original papers must be contributed solely to this *Journal*. Copyright of published manuscripts is held by United States and Canadian Academy of Pathology, Inc. who must receive in writing assignment of copyright from authors of accepted manuscripts. Forms for assignment will be mailed to authors from Editorial office. Include Fax number on cover letter.

MANUSCRIPT FORMAT

Double-space typescript; keep ample margins. Number all pages consecutively starting with title page as page 1. Place first author's name in upper right corner of each page. Do not use footnotes.

Title Page: Include full title of paper—short, clear and specific; authors' names and institutional affiliation.

On a separate page give name and address of corresponding and reprint requests author(s) and running title (3-4 words).

Abstract: 300 words on a separate page describing exactly what was done, results obtained, and conclusions drawn. Divide into four paragraphs headed: Background, Experimental Design, Results, and Conclusions. After abstract, on same page, supply 7 key words.

Text: Arrange text: Introduction; Experimental Design; Results and Discussion; and Methods. In Methods include detailed methodology, include new methods or those particularly pertinent to a study in Experimental Design.

Grants and Other Acknowledgments

References

Charts, Graphs, Tables, and Photographs

Figure Legends

References: Number consecutively (in parentheses) in text as cited. List all authors; if number exceeds six give six followed by et al. Cite papers quoted as "in press" in references. Do not reference citations such as "personal communication" or "unpublished data"; enclose in parentheses in appropriate place in text. Type on a separate page and number in order of citation. For abstracts add abstr in parentheses at end of reference. Abbreviate journal titles according to *Index Medicus*.

ARTICLE

Vracko R. Thorning D. Contractile cells in rat myocardial scar tissue. Lab Invest 1991:65:2114–27.

BOOK

Jones AD. Pathology of the lung. Montreal; John Green Co, 1970:42.

CHAPTER IN EDITED BOOK

Brown AB, Green XY. Jejunal pathology. Inack CD, White EF, editors. The gastrointestinal tract. New York; Sam Jones Co, 1972:30.

Illustrations: Plan graphs and photographs to suit format and page size of *Journal*. Illustrations, or groups of illustrations, may be 3.5 or 7 in. wide; maximum page length, including legends, is 9 in. Illustrations not conforming to these size limitations will be returned for replacement even when manuscript is otherwise acceptable.

Submit clear, black and white, glossy photographs unmounted in triplicate; write number, author's name, and indicate TOP margin lightly on back of each using a soft pencil. Type legends on a separate page; number consecutively. Refer to all illustrations in text. Designate one set of illustrations "set for publication.."

Tables: Type on separate pages, number, and give brief descriptive titles.

OTHER FORMATS

BRIEF COMMUNICATIONS—Three printed pages, including illustrations, tables, and references. (a full page of text approximates 1000 words). Reports of experimental data of immediate importance to other scientists is considered for accelerated publication after usual editorial review and acceptance. Paper must be of significant, immediate importance to other investigators. Adequately document reported findings. Format is the same as that for other manuscripts.

LETTERS TO THE EDITOR—Invited brief letters commenting on papers appearing in recent issues. Letters are referred to senior author of paper in question, whose response may also be published.

OTHER

Abbreviations, Nomenclature, and Symbols: Follow *Council of Biology Editors Style Manual*, (5th ed, Council of Biology Editors, Inc., Bethesda, MD).

Color Charges: $500 subsidy is provided at discretion of Editor. Other color printing costs are responsibility of author.

Page Charges: $40, including one page of black-and-white photographs with legends. Illustrations in excess $105 per page. Inability to meet charges will not affect publication.

$105 per page for Brief Communications.

SOURCE

July 1994, 71(1): front matter.
Instructions to authors appear in every issue.

Lancet

Lancet
42 Bedford Square
London WC1B 3SL U.K.
Telephone: +44 71 436 4891); Fax: +44 71 436 7550

AIM AND SCOPE

Lancet welcomes any contribution that advances or illuminates medical science and practice; territory extends to all aspects of human health.

MANUSCRIPT SUBMISSION

Follow "Uniform Requirements for Manuscripts Submitted to Biomedical Journals" (*BMJ* 1991; **302:** 338–41 or *N Engl J Med* 1991; **324:** 424–28). Submit three copies, double-spaced.

Most readers are in other specialties: before submitting a specialized paper, try it on colleagues. Assume that readers are starting from near-ignorance; Summary and Introduction deserve close attention. In cover letter include a few words about why it belongs in *Lancet* rather than another journal.

Short items are more appealing and chance of acceptance declines beyond following lengths. One *Lancet* column (half page) is 500 words, Articles: 8 columns including headings, summary (200 words), tables, figures, and references (30). Short Reports: 2.5 columns, including summary (100 words); 10 references (see *Lancet* 1990; 336: 1226). Hypotheses—3 columns. Viewpoint—2 columns.

Submission of paper or letter requires all authors signatures. Follow "Uniform Requirements" criteria of authorship: "Authorship credit should be based only on substantial contribution to (a) conception and design, or analysis and interpretation of data; and to (b) drafting the article or revising it critically for important intellectual content; and on (c) final approval of the version to be published. Conditions (a), (b), and (c) must all be met.

Participation solely in the acquisition of funding or the collection of data does not justify authorship."

For each author, indicate a single qualification (such as MD or PhD) and full professorships; information is printed with addresses in a footnote.

Reveal any financial or other conflict of interest that might have biased work. Even if bias was avoided, provide information on possible conflicts. Specify all relevant sources of financial support. Mention grants, or business interests, and consultancies. Let Editor decide what should be declared in print.

Indicate concessions that may be made, such as omitting a a table or figure. *Lancet* will store rather than print some of the fine detail, after peer review, for despatch to interested readers. Enclose copies of work that is unpublished, submitted, or in press, that is important to argument. Inform Editor if material has been published before or is to appear in part or whole elsewhere in any form.

Authors, except certain government employees, will transfer copyright to *Lancet*. At proof stage submit written agreement covering words and illustrations, photographs and other material. Obtain permission to publish material for which authors do not hold copyright. All literary matter in the form in which journal publishes it is *Lancet's* copyright.

MANUSCRIPT FORMAT

Illustrations: Submit figures as prints. Figures will be reduced to 80 mm wide (one column) or even less: choose lettering size accordingly. Trim, do not reduce histological illustrations or radiographs. Four color and single-color (other than black) illustrations must enhance a paper scientifically or educationally. In cover letter indicate if able to contribute to additional costs.

OTHER FORMATS

LETTERS TO THE EDITOR—One column (500 words or less to allow for references (5), tables, or figures). Submit one copy. Letters commenting on previous items in journal are not peer reviewed; those recording original observations may be.

SOURCE

December 24/31, 1994, 344(8939): back of list of Letters to the Editor in every issue.

Laryngoscope
Triological Foundation

The *Laryngoscope Journal*
10 S. Broadway, 14th Floor
St. Louis, MO 63102-1741

AIM AND SCOPE

The *Laryngoscope* publishes original research papers presented at annual and section meetings of the Triological Society, as well as "How I Do It," independent basic science and clinical research papers, letters, and commentary.

MANUSCRIPT SUBMISSION

The *Laryngoscope* reserves the right to exclusive publication of all accepted manuscripts. Do not submit manuscripts previously published or concurrently submitted to any other publication. Transfer of copyright to *Laryngoscope* is a prerequisite of publication. All authors must sign transfer. (This does not preclude publication of abstracts in transactions or proceedings of various societies.)

Submit one original and three copies of manuscript, bibliography, legends, tables, charts; and four sets of original illustrations.

In cover letter state title, author(s), and name and address of corresponding author.

MANUSCRIPT FORMAT

Type double-spaced, on 8.5 × 11 in. white bond paper. Provide ample margins. Number pages consecutively and put a running head with first author's name and an abbreviated title (40 characters and spaces) in upper right corner of each page. Do not exceed 30 typewritten pages.

Title Page: Include full title of paper (be succinct); names of authors and highest degrees attained; name and address of institution(s) where work was done; name and address for reprint requests; source of financial support or funding; and, if presented at a meeting, give name of society, city, and date.

Abstract: 135 word summary. State problem, method of study, results, conclusion, and significance of work.

Text: Outline for main body of paper: Introduction, Materials and Methods, Results, Discussion, Conclusion; include bibliography; figure legends, and tables following text.

Bibliography: Type references double-spaced; number consecutively in order of appearance in text; list in bibliography in same order. $3 charge for each reference over 15. Verify accuracy of references. Do not list unpublished data and personal communications in references; cite parenthetically in text. Use *Index Medicus* for journal abbreviations.

JOURNAL

Veldman, J.E., Roord, J.J., O'Connor, A.F., *et al.*: Autoimmunity and Inner Ear Disorder: An Immune Complex Mediated Sensorineural Hearing Loss. Laryngoscope, 94:501–507, 1984.

BOOK

Batsakis, J.G.: *Tumors of the Head and Neck: Clinical and Pathological Considerations* (2nd ed.) Williams & Wilkins, Baltimore, pp. 76–99,1979.

CHAPTER

Fairbanks, D.N.F.: Embryology and Anatomy. In: *Pediatric Otolaryngology*. (Chap. 23). C.D. Bluestone and S.E. Stool (Eds.). W.B. Saunders Co, Philadelphia, pp. 647–678, 1983

Legends: Provide a typed, double-spaced legend, numbered to match each illustration. For photomicrographic material, indicate stain and magnification or use an internal scale marker.

Tables: Type double-spaced; number consecutively. Use Roman numerals; provide titles. Limit size; use only for information not already in text or illustrations.

Illustrations: Submit four complete sets of unmounted illustrations. Identify each on back with number, first author's name, and an arrow indicating TOP. Place each set in a sealed envelope labeled with author's last name and title of paper. Do not use paper clips or staples. Have line drawings, graphs, and charts professionally drawn, photographed; send as prints. Do not send original artwork. Dot-matrix or other computer-generated illustrations will not be accepted without editor's consent.

Illustrations must be 5×7 in. or smaller glossy prints. Four black-and-white illustrations are published without charge. Authors are charged for additional ones.

For photographs of recognizable persons submit a signed release from patient or legal guardian that authorizes publication. Obtain permission from publisher for any illustrative material that has previously appeared in print.

Cost of color illustrations are borne by authors. Estimates are available from office.

OTHER FORMATS

"HOW I DO IT"—Six typewritten pages, double-spaced; no abstract; include a few pertinent illustrations.

OTHER

Style: Use generic names for drugs. List supplier or manufacturer for products and instruments; include city and state.

Plot audiograms according to ISO standards in black-and-white.

For commonly accepted abbreviations, consult *Logan's Medical and Scientific Abbreviations*. See also *Dorland's Illustrated Medical Dictionary* (26th Ed) and *AMA Manual of Style*.

SOURCE

January 1994, 104(1): 22a.

Life Sciences
Pergamon Press

Life Sciences, c/o Pergamon Press
655 N. Alvernon Way, Suite 212
Tucson, AZ 85711
Telephone: (602) 321-7778; Fax: (602) 321-7781
Life Sciences c/o Dr. S.Z. Langer
Synthelabo Recherche (L.E.R.S.)
B.P. 110-31 avenue Paul Vaillant Couturier
92225 Bagneux Cedex, France
Telephone: 45 36 24 11; Fax: 45 36 20 12; Telex: 634818

AIM AND SCOPE

Life Sciences is an international weekly journal publishing full length reports on research in a range of areas in the life sciences. These areas include molecular and cellular aspects of: Cardiovascular & Autonomic Mechanisms, Endocrinology, Immunology, Toxicology, Drug Metabolism, Growth Factors & Neoplasia Neuroscience.

The *Journal* publishes original research rapidly. Full-length manuscripts are favored. exciting new results requiring rapid dissemination in shorter form are also considered for either *Life Sciences* or the ultra-rapid communications section *Pharmacology Letters*.

The *Journal* favors publication of papers on seminal areas of science utilizing modern scientific technologies and exploring molecular and/or cellular mechanisms involved in explaining reported observations.

Contained in journal: New scientific information; Minireviews on selected aspects of a scientific field undergoing rapid change; Brief conceptual papers based on original and/or literature data; Relevant clinical discoveries; Collections of selected papers on current concepts in a particular field; Symposium papers.

MANUSCRIPT SUBMISSION

Submit original and two copies of manuscript. Sign "Waiver of signature" on express mail packages. Do not submit by fax.

Submit a Transfer of Copyright Agreement for every article accepted for publication. Upon acceptance, corresponding author will be sent an agreement to be completed prior to preparation for publication.

At top right of each manuscript, in pen or pencil, indicate under which section heading manuscript should be considered: Cardiovascular and Autonomic Mechanisms; Drug Metabolism; Endocrinology; Neuroscience; Toxicology; Growth Factors and Neoplasia; Immunology. *Pharmacology Letters* articles are considered separately.

MANUSCRIPT FORMAT

Following author paragraph, quadruple-space, then type a black line across entire width of text. On first line below black line, at left margin, type: Abstract. Continue on same line with a statement summarizing important points of text. Extend abstract paragraph across entire width of text. Under abstract, type another black line across width of text. Double-space below black line, and type Introduction (underlined and centered over text). Continue with Instructions beginning with "Major Headings."

Use either Times Roman, Helvetica, or Courier typeface, in 12 point size. Make table and figure legends easily discernible from text. Do not use small italic typeface. Type with a black carbon ribbon, or ink-jet or laser printer. Dot matrix print is unacceptable. No erasure marks, smudges, spots, pencil or ink corrections or creases are allowed.

Use 8.5×11 in. paper. Typing area must be exactly 6 5/8 $\times$ 9 7/8 in. Justify margins left and right (block format). Fill entire typing area of each page, leave no wasted space. Type single-spaced on good quality white bond paper; double-space between paragraphs. Ensure that symbols, superscripts and subscripts are legible. Make sure text lines are equidistant.

Title: On first page, start title 1 in. down from top text margin. Type in all capital letters, centered on width of typing area; single-space if more than one line. Spell out all words; be brief and descriptive. Double-space, then type author(s) name(s), single-space if more than one line is required. Double-space, then type author(s) address(es), also single-spaced, capitalizing first letter of main words. Quadruple-space before Summary (*Life Sciences*) or Abstract (*Pharmacology Letters*).

Summary: (for *Life Sciences*) Center, type and underline summary heading, capitalize first letter. Double-space to separate heading from summary text. Indent approximately 1/2 in. from both left and right margins. Make intelligible to reader without reference to body of paper. Begin Introduction to text (without a heading) two spaces below summary, using full margins.

Major Headings: Arrange to increase clarity: Methods, Results, Discussion, Acknowledgments and References. Capitalize first letter, underline, and center headings on width of typing area. Arrange other headings to indicate their relative importance.

Tables: Type within main body of manuscript, set apart to avoid confusion with text. Center word TABLE, all letters capitalized, with table number (Roman numerals) above table. Insert caption above table, capitalize first letters of all main words. Place legend beneath table. Single table must not overlap onto following page.

Figures: Submit glossy prints or laser-printed originals. Incorporate figures into main body of text using as little space as possible without hindering legibility after a 25% reduction. Identify figures in pencil on back. Place in a protective envelope and attach to manuscript with a paper clip. Center abbreviation "Fig." with figure number (Arabic numerals) beneath figure and above legend. Indent legend on both sides to distinguish from text.

Half-Tones (Photographs): Submit black and white prints large enough to be legible after 25% reduction.

Footnotes: Type single-spaced at bottom of appropriate page. Separate from text by at least a double-space and a short line immediately above.

References: Indicate in text by consecutive numbers in parentheses, e.g., (1,2). Cite full reference at end of manuscript. Single-space between as well as within references. Give names of all authors in CAPITAL letters (initials typed first, surnames following), title of journal (abbreviated in accordance with *World List of Scientific Periodicals),* volume number underlined, first and last page numbers of article and year of publication (in parentheses). References to books contain author(s) names, in same format as above, title (underlined and first letter of main words capitalized), editor(s) names, if any, page number(s), publisher's name and location and year of publication (in parentheses). Items "submitted for publication," are not references; place in parentheses at appropriate place in text.

1. A.B. NORMAN, G. BATTAGLIA, A.L. MORROW and I. CREESE, Eur. J. Pharmacol. 106 461–462 (1985).

2. I. CREESE, *Neurotransmitter Receptor Binding,* H.I. Yamamura, S.J. Enna, and M.J. Kuhar (Eds), 189–233, Raven Press, New York (1985).

Running Title: 40 spaces on a separate sheet.

Key Words: Provide one or more for indexing. Follow by enough explanatory information to make them useful to readers in a subject index. Do not use abbreviated forms of chemical compounds. Spell out entirely, using official nomenclature. Example: cannabidiol, effect on testicular esterase; L-dihydroxyphenylalanine (L-DOPA), effect on protein synthesis

Pagination: Number each page in light blue pencil in lower right corner.

Corresponding Author: Place at bottom of first page, under a short black line. Extend name and address across entire width of page.

OTHER FORMATS

PHARMACOLOGY LETTERS—4 to 6 pages, a figure-plus-table combination of 4, and 20 references. Provides rapid review and publication of original research involving rigorous experimental studies of immediate interest on any aspect of basic pharmacology. Clinical studies are included if they have a direct bearing on mechanism of drug action.

Following author paragraph, quadruple space, type a black line across width of text; on first line below at left margin type: Abstract. Continue on same line with statement summarizing important points of text. Make paragraph entire width of text. Under abstract, type another black line, double space and type Introduction (underlined and centered over text).

Commentaries, new methodologies and manuscripts on pharmacologic theory (either conceptual or mathematical) are also invited.

OTHER

Electronic Submission: Submit a computer disk with final version of paper. Use WordPerfect, specify which version. Specify computer used. Include both original and ASCII file on disk. Single space. Begin all text elements flush left. Keep a backup disk.

Experimentation, Animal and Human: Adhere to recommendations from the Declaration of Helsinki or NIH guidelines for care and use of laboratory animals.

Reviewers: Submit names of four qualified reviewers, including addresses and fax numbers. Each should be an expert in field of manuscript. Do not choose referees from Editorial Advisory Board unless they are familiar with topic.

Style: Use Information to contributors, especially nomenclature and abbreviations, contained in the *Proceedings of the National Academy of Sciences* as a guide for preparation of manuscripts for any points not covered.

SOURCE

1994, 55(1): last page.

Limnology and Oceanography
The American Society of Limnology and Oceanography

Managing Editor, *Limnology and Oceanography*
School of Oceanography
P.O. Box 357940
University of Washington, Seattle, WA
Telephone: (206) 543-8655

AIM AND SCOPE

Limnology and Oceanography accepts contributions that present the results of original research in any aspect of limnology and oceanography. Applied work will not be published unless it has a clear focus on understanding natural aquatic systems or if the results have exceptional general applicability. Papers are judged based upon originality and the degree to which the findings can be generalized beyond the particular aquatic system.

MANUSCRIPT SUBMISSION

Submit five copies of manuscript. Provide five copies of materials crucial to review not readily available to reviewers. Indicate if manuscript is intended as an article, note, or comment.

MANUSCRIPT FORMAT

Double-space throughout. Footnotes, summaries and appendices will not be printed.

Title Page: Include the title, authors' names, addresses. and a condensed running head (less than 28 characters).

Abstract: Limit length to 3% of text; make complete in itself.

Text: Within text cite referrences by author name and year.

Acknowledgments: Type on a separate page.

Literature Cited: Begin on new page. List alphabetically according to author names. Double space. Abbreviations of journal titles should follow *Chemical Abstracts Service Source Index* or *Biossis*.

JOURNAL ARTICLE

Fenchel, T. 1986. Protozoan filter feeding. Prog. Protistol. 1: 65–113.

BOOK

Stumm, W., and J. Morgan. 1981. Aquatic chemistry, 2nd ed. Wiley.

CHAPTER IN BOOK

Codispoti, L. A. 1983. Nitrogen in upwwelling systems, p. 513–564. In E. J. Carppenter and D. C. Capone [eds.], Nitrogen in the marine environment. Academic.

Tables: One column table is 60 characters wide; a 1.5-column table is 90 characters wide; 2-column table is 130 characters wide. Type legends double-spaced on same page as table.

Figure Legends: Type double-spaced together on a new page.

Figures: Number consecutively using Arabic numerals. Do not include mounted photographs or oversize figures. Laser prints are generally acceptable. Include a complete set of figures with each copy of the manuscript.

OTHER FORMATS

NOTE—Notes communicate brief research results.

COMMENTS—Discussion of papers previously published in *Limnology and Oceanography* or other issues of interest.

REVIEWS—Consult the editor before submission.

OTHER

Abbreviations: Use metric and Celsius units. SI units are preferred.

Page Charge: No page charge for first 10 printed pages. Reference lists that exceed 20% of length of text will be charged as will costs for color illustrations.

Reviewers: Names and addresses of potential reviewers are welcome.

SOURCE

January 1995, 40(1): front matter.

Magnetic Resonance in Medicine
Society of Magnetic Resonance in Medicine
Williams & Wilkins

Felix W. Wehrli, Ph.D.,
Department of Radiology
University of Pennsylvania Medical Center
3400 Spruce Street
Philadelphia, Pennsylvania 19104
Telephone: (215) 662–7951; Fax: (215) 349–5925

AIM AND SCOPE

Magnetic Resonance in Medicine is an international journal devoted to the publication of original investigations concerned with all aspects of the development and the use of nuclear magnetic resonance and electron paramagnetic resonance techniques for medical applications.

Reports of original investigations in the areas of mathematics, computing, engineering, physics, biophysics, chemistry, biochemistry, and physiology directly relevant to magnetic resonance will be accepted, as well as methodology-oriented clinical studies.

MANUSCRIPT SUBMISSION

The chief criteria for acceptance of papers are significance, originality, clarity, and quality of work reported. The same work must have not been published, or be under consideration for publication elsewhere. Submission must have been approved by all authors and by appropriate authority at institution where work was carried out. Persons cited as a sources of personal communication must have approved such citation. Written authorization may be required.

If accepted for publication, copyright, including right to reproduce article in all forms and media, shall be assigned exclusively to the Society of Magnetic Resonance in Medicine. The Society will not refuse reasonable requests by author for permission to reproduce any of his or her contributions.

Submit manuscripts in triplicate, including three sets of original figures. Submit copyright assignment form signed by all authors; form is printed in each journal issue.

MANUSCRIPT FORMAT

Type manuscripts double-spaced throughout (including tables, footnotes, references, and figure captions); type on one side of a good grade 8.5 × 11 in. white paper, with 1 in. margins on all sides. Mimeographed or duplicated manuscripts are accepted only if indistinguishable from good typed copies. Number pages consecutively, including references, tables, and figure legends. Indicate organization of paper by appropriate headings and subheadings.

Page 1: Include article title, author(s), and affiliation(s); at bottom, type a short running head (abbreviated title, 60 characters including spaces), and name, complete mailing address, telephone and fax numbers of corresponding author. Page 2: Abstract (150 words, 100 for *Communications* and *Notes*) describing substantive content and conclusions reached. Make self-contained; do not refer to formulas, equations, or bibliographic citations in body of manuscript.

Key Words: Four for indexing at end of abstract.

Footnotes: Use only when absolutely necessary; type consecutively, double-spaced, on a separate sheet in order of appearance in text, and identified by Arabic numbers. In text refer to footnotes by superscript Arabic numerals.

Tables: Number consecutively with Arabic numerals in order of appearance in text. Type double-spaced throughout, on a separate sheet; avoid vertical lines. Type a short descriptive title above table. Type footnotes, lettered a, b, etc., directly below appropriate table, not on a separate sheet.

Figures: Number with Arabic numerals in order of mention in text. Write author's name and number on back in pencil. If ambiguous, indicate TOP. Make lettering professional quality with consistent spacing and alignment and uniform in size and type face, Make figures consistent in format and size. Symbols, letters and numbers must be legible after reduction; make smallest data points not less than 1.5-1.7 mm high. Make data 3 mm high on 5×7 in. print and 6 mm high on 8.5×11 in. print. If too small, figure will be returned. If figure consists of several illustrations, label a,b,c... Do not place labels in figure. Crop images; remove irrelevant parts.

Type legends together on a separate sheet. Paste figures on separate numbered pages on top of appropriate legends at end of manuscript. When necessary, color illustrations are acceptable at author's expense.

References: Cite in text by a number in parentheses and list at end of paper in numerical order. Cite private communication and unpublished material in text; do not treat as a reference. "Submitted" and "in preparation" are unacceptable in reference list. Cite unpublished observations and personal communications in text; do not include in reference list. Use *Index Medicus* for journal abbreviations.

1. J. Hennig, A. Naureth, H. Friedburg, RARE imaging: a fast imaging method for clinical MR. *Magn. Reson. Med.* **3,** 823–833 (1986).

2. A. Young, L. Axel, Three-dimensional motion and deformation of the heart wall: estimation with spatial modulation of magnetization—a model-based approach. *Radiology* **185,** 241–247 (1993).

3. J. M. S. Hutchison, F. W. Smith, *in* "Nuclear Magnetic Resonance Imaging" (C. L. Partain, A. E. James, F. D. Rollo, and R. R. Price, Eds.), p. 231, Saunders, Philadelphia, 1983.

4. P. Mansfield, P. G. Morris, "NMR Imaging in Biomedicine," Academic Press, New York, 1982.

5. M. J. Lizak, B. Pfleiderer, J. Moore, E. Sun, J. Ackerman. Fluoride as a marker for bone mineral dynamics, *in* "Proc., SMRM, 12th Annual Meeting, New York, 1992," p. 1108.

OTHER FORMATS

COMMUNICATIONS—Four to five printed pages (12 to 15 manuscript pages). Preliminary accounts of work of special topicality and importance.

LETTERS TO THE EDITOR—One printed page. Provides a forum to critique papers in *Journal*. Paper's authors will have an opportunity to respond. Letters will be published consecutively in same issue.

NOTES—Four to five printed pages (12 to 15 manuscript pages). Complete accounts of work of limited scope.

OTHER

Abbreviations: Do not use final periods after abbreviations of units of measure in text or in tables, except for "in." (inch). Use *American Chemical Society Style Guide* or *Style Manual of the American Institute of Physics* as a reference. Standard abbreviations, which need no definition, include MRI, CT, RF, ECG, NMR, EPR and ESR. Define all other abbreviations before first use.

Electronic Submission: Submit diskettes of final version of manuscripts with typed revised manuscript. Diskettes produced on IBM or IBM-compatible computers are preferred; those produced on most Apple/Macintosh or Wang computers can be converted. Following word processing programs are preferred: XyWrite III Plus, WordPerfect 4.2. 5.0, or 5.1 (IBM or Macintosh), Microsoft Word (IBM or Macintosh), Word for Windows, Wang OIS (WPS), and Wordstar (IBM). Other convertible word processing systems are CTP 8000, MacWrite 2.2 or 4.5, MacWrite II, Display Write 3 or 4, Multimate, PC Write, Volkswriter, and Write Now. Do not use Fast Save option on Macintosh computers. Identify diskette with journal name, manuscript number, senior author's name, manuscript title, name of computer file, type of hardware, operating system and version number, and software program and version number. Make electronic copy of manuscript exactly identical to accepted final version of hard copy. Mathematical and tabular material are processed in traditional manner.

Equations: Type mathematical equations or symbols. Number consecutively, place number in brackets against right margin. Refer to as Eq. [3]; except at beginning of sentences spell out word Equation. Punctuate to conform to place in syntax of sentence.

Experimentation, Animal and Human: Conform to requirements of institution and country in which investigations were done. Investigate at least ten subjects in order to establish significant biological or medical conclusions. In preliminary work, investigate at least four subjects.

Symbols: Attach a complete typed list of symbols, identified typographically, not mathematically. List will not appear in print. If equations are hand-written in text, handwrite lists. Distinguish between "oh," "zero;" "ell," "one;"

"kappa," "kay;" upper- and lowercase "kay;" etc. Indicate when special type is required (German, Greek, vector, scalar, script, etc.); all other letters in mathematical expressions are set in italic type.

SOURCE

July 1994, 32(1): front matter.

Mayo Clinic Proceedings
Mayo Foundation for Medical Education and Research

Mayo Clinic Proceedings
Manuscript Office
Room 660 Siebens Building
Rochester, MN 55905

AIM AND SCOPE

The *Mayo Clinic Proceedings,* a peer-reviewed journal, publishes original articles of general interest in medicine, surgery, research, and basic science.

MANUSCRIPT SUBMISSION

State that article has not been published previously or has not been concurrently submitted for publication elsewhere. Acknowledge any potential conflict of interest. Include in cover letter name, complete mailing address, telephone and fax numbers, of corresponding author. All authors must sign letter. Indicate category of article: (e.g., original article, subject review, case report).

On acceptance of manuscript, all authors (except employees of U.S. government whose work was part of official duties) must sign copyright transfer form.

Use of previously published material is discouraged. Obtain and submit permission for reuse of material (illustrations, tables, or lengthy quotes) from copyright holder and author of original source (along with complete bibliographic information).

Prepare manuscripts in accordance with "Uniform Requirements for Manuscripts Submitted to Biomedical Journals," of the International Committee for Medical Journal Editors (*N Engl J Med* 1991; 324: 424–428).

Submit three complete copies of manuscript (including three glossy prints of each illustration).

MANUSCRIPT FORMAT

Type double-spaced throughout; begin each segment on a new page arranged as: title page, abstract, text, references, legends, and tables. Number pages consecutively and label with last name of first author.

Title Page: Include: Title; Authors (include first names, academic degrees, departmental affiliations, and institution); Running head: ≤ 50 characters; Financial support information: include grant number and granting agency; name and address of corresponding and reprint request author(s).

Abstract: 150-200 words. Organize in structured format: use headings: Background; Objective, Methods, Results, and Conclusion (see *Ann Intern Med* 1987;106:598-604).

Provide 3-5 key words or phrases for indexing.

Text: Provide an outline of headings used in text.

Do not use footnotes. Avoid specialized "jargon" and abbreviations; define abbreviations used at first mention. Use generic names for drugs. Cite references, figures, and tables consecutively as they appear in text.

References: Type completely double-spaced. Number consecutively as cited in text; use superscript numerals for text citations. Cite personal communications and other unpublished data parenthetically in text (do not include in reference list). In reference list, include names and initials of all authors (if more than six, list six followed by et al), title, source (for journal abbreviations follow *Index Medicus*), year, volume, and inclusive paging.

PERIODICAL

1. Comroe JH, Long TV, Sort AJ. The lung: clinical physiology and pulmonary function tests. Chest 1989; 65:20–22

CHAPTER IN BOOK

2. Schiebler GL, Van Mierop LHS, Krovetz LJ. Diseases of the tricuspid valve. In: Moss AJ, Adams F, editors. Heart Disease in Infants, Children, and Adolescents. 2nd ed. Baltimore: Williams & Wilkins, 1988: 134–139

BOOK

3. Guyton AC. Textbook of Medical Physiology. 8th ed. Philadelphia: Saunders, 1991: 255–262

Illustrations: Submit three glossy prints of each, clearly marked on back with figure number, author's name, and orientation. Cite all illustrations in text, and number with Arabic numerals in order of appearance.

Provide a legend for each; include definitions of abbreviations on figure.

For photomicrographs, specify stain and magnification. For illustrations with recognizable patients, submit a release form signed by patient. Do not trim or assemble component parts; indicate cropping and format on photocopy. If photograph has been altered by computer enhancement or any other means, submit original with altered version. In cover letter state which illustrations were adjusted and why.

Tables: Number consecutively with Arabic numerals in order of citation in text. Type double-spaced; place each on a separate page. Provide a title for each; define all abbreviations in a footnote. Use footnote symbols in order: *, †, ‡, §, //, ¶, #, **, ††, etc. Do not submit as glossy prints.

OTHER

Reviewers: Suggest names of potential reviewers.

Units: Express measurements in conventional units; include SI units parenthetically.

SOURCE

January 1994, 69(1): 95.

Medical Care

American Public Health Association—Medical Care Section

J.B. Lippincott Company

Medical Care
P.O. Box 38249
Cleveland, OH 44138-0249

AIM AND SCOPE

Medical Care is designed to serve as an international medium for publication of worthy articles in the broad field of medical care, and thereby to encourage progress in the research, planning, organization, financing, provision, and evaluation of health services.

MANUSCRIPT SUBMISSION

Submit authoritative original papers describing current developments in the field and findings of pertinent investigations. Selection is based on timeliness, originality, soundness of method, significance of findings to the medical care field, appropriateness of conclusions, and quality of presentation. Authors' guidelines are available from editor. Consult guidelines before final preparation of manuscript.

Submit four copies. Give full names of authors, degrees, academic or professional titles, affiliations, and complete addresses. Specify corresponding author. Notify Editor of changes of address.

Manuscripts must be submitted solely to *Medical Care,* and must not be under review nor have been published or committed for publication elsewhere. If a paper has been presented at a meeting, note on manuscript.

MANUSCRIPT FORMAT

Type double-spaced, with liberal margins, on one side of 8.5 × 11 in. white paper. Number pages consecutively. Write 3000–5000 words (12-20 pages), exclusive of tabular material. Longer articles are accepted, when subject matter requires more extensive treatment. An abstract (200 words) should accompany each article.

References: Submit in order of citation on a separate page at end of text. Use references from journals from the "Uniform Requirements for Manuscripts Submitted to Biomedical Journals;" omit last page number, so each reference will list, in order, author's name, title of article, name of journal, year of publication, volume number, and first page number. Cite references in text by number only.

Illustrations: Black-and-white illustrations are reproduced free of charge; publisher reserves the right to establish a reasonable limit to number. Colored illustrations are published at author's expense.

Submit black-and-white photographs as glossy prints. Indicate TOP of each photo. Do line and wash drawings (including graphs and charts) in black on white paper, with lettering large enough to be legible after reduction. Supply a separate typewritten sheet with legends for illustrations.

Tables: Submit on separate sheets at end of manuscript; do not exceed 8.5 × 11 in. sheet; double-space.

OTHER

Style: Use *Dorland's Medical Dictionary* and *Webster's International Dictionary* for reference. Use numerals; do not spell out numbers (not even 1 to 9) except at beginning of a sentence and where sense requires it. Use metric system for dosages. Express temperature as degrees centigrade. Avoid use of copyright or trade names of drugs; if necessary, capitalize.

SOURCE

January 1994, 32(1): front matter.

Medical Journal of Australia

Australian Medical Association

Australasian Medical Publishing Company

The Editor, *The Medical Journal of Australia*
1-5 Commercial Road, PO Box 410
Kingsgrove, NSW 2208, Australia

MANUSCRIPT SUBMISSION

The Medical Journal of Australia welcomes original contributions on all aspects of medicine. Offer manuscripts exclusively to *Journal.* Copyright is vested in the Australasian Medical Publishing Company Ltd. Authors must sign cover letter stating whether any material has appeared elsewhere. Reports of work done before 1991 are not published, unless there is a compelling reason to do so.

Type, double-spaced (including references), on one side paper. Submit two copies. All authors must have contributed significantly to paper and be prepared to take public responsibility for everything in it. Justify inclusion of more than six authors.

MANUSCRIPT FORMAT

On title page identify one author as correspondent and include complete address, telephone and fax numbers. Give first (or usual) given name of authors in full, other given name(s) as initial(s), surname, principal qualifications, and present position (plus past position, if held when research being reported was conducted).

Original articles include Introduction, Methods, Results and Discussion. Original articles, reviews and case reports have structured abstracts with headings (see *Journal,* September 3,1990: 249 and May 20, 1991: 649). Original Articles are 2500 words; Short papers 1000 words; Occasional Reviews 3000. Keep figures and tables to a minimum. Other illustrative material is welcome. Number all pages, starting with title page. Include a word count.

References: Verify accuracy. List 50 references for review articles, 25 for original articles and 10 for short papers or case reports. Cite first and last page numbers. Personal communications and unpublished references are not acceptable. Follow "Vancouver" style (*Med J Aust* 1991; 155: 197–202). Abbreviate journal names according to *Index Medicus.* Specify if reference is a letter to the editor, an editorial or an abstract.

Tables: Type double-spaced, on separate sheets. Simplify information as much as possible. Keep number of columns to a minimum; headings short. Do not duplicate information in tables in text.

Illustrations: Supply three copies of all figures labelled with figure number, principal author's name, and orientation. Professionally draw and photograph all figures; do not use freehand or typewritten lettering. Laser printed illustrations on good quality paper or glossy photographic prints are acceptable. Do not use abbreviations in illustrations, or in keys to illustrations. For numbers less than one, zero precedes decimal point. Include data in tabular form with graphs. Photographs are glossy black-and-white or color prints; place in separate heavy envelopes. A charge may be made for use of color.

Submit a disk copy with accepted computer-generated figures. Save disk copies in Adobe Illustrator, Adobe Photoshop, EPS, TIFF, or Harvard Graphics. Label disk with principal author's name, manuscript number, and names and formats of graphic files.

Obtain and submit letters of permission from copyright holders for illustrations from other publications. Do not submit original x-ray films. Photomicrographs should have internal scale markers.

Legends: 40 words. Indicate magnification and stain for photomicrographs.

Acknowledgments: Keep as brief as possible. Obtain written permission from those mentioned in this section.

OTHER FORMATS

LETTERS TO THE EDITOR—400 words, 5 references. Include signature, principal qualifications, professional affiliation, and address of each author. Do not duplicate manuscripts or letters submitted to other journals.

OBITUARIES—Brief summary of person's career, including all qualifications. Main part of obituary is an appreciation of his/her contribution—no longer than 300 words. Include a photograph. Include birth and death dates.

OTHER

Experimentation, Human: Include a statement that subjects gave informed consent and that approval of an ethics committee was obtained.

Electronic Submission: To speed editing, provide a computer disk copy of accepted manuscripts. WordPerfect text files (Macintosh or IBM/DOS) are ideal, other formats can be translated. Label disks (3.5 in. preferred) with title of article, authors' names, our manuscript reference number and wordprocessing file format.

Units: Give all measurements in SI units, except blood pressure (mmHg). Supply normal ranges for each measurement.

SOURCE

July 1994, 161(2): 168.

Medicine
Williams & Wilkins

Victor A. McKusick, M.D.
The Johns Hopkins Hospital
Baltimore, MD 21287-4922

MANUSCRIPT SUBMISSION

Articles that review original data from experience of authors as well as survey the literature are encouraged. Make subject broad enough to be of interest to *Medicine's* multispecialty readership.

In cover letter state that manuscript has been seen and approved by all authors, that it has not been published elsewhere, and is not under consideration for publication elsewhere. If accepted for publication, Williams & Wilkins must receive in writing assignment of copyright from all authors. Forms will be mailed from Editorial Office.

Inform editors of any financial connection to, or interest in, a company whose product is subject of manuscript.

MANUSCRIPT FORMAT

Double-space with wide margins on 8.5 × 11 in. bond paper. Submit original and two copies. Number pages.

On separate title page give: title, authors, institutions, city and state of origin, complete address for mailing proofs, running head (50 characters, including spaces). Type each table on a separate sheet. Type references, footnotes and legends for illustrations double-spaced on separate sheets. Submit three sets of illustrations. Identify each photograph on reverse side with figure number and author's name; mark TOP. Summary should precede references.

Follow guidelines of "Uniform Requirements for Manuscripts Submitted to Biomedical Journals" (see February 7, 1991, issues of *British Medical Journal* and *New England Journal of Medicine*). List references in alphabetical order. Follow guidelines relating to ethics, authorship, and prior or duplicate publication.

References: List in alphabetical order, according to last name of first author. Number consecutively according to alphabetical listing. Cite in text with reference number in parentheses. Include all authors.

5. Billings FT Jr, DePree HE. Diagnosis of portal vein obstruction: Studies of intestinal absorption of glucose using abdominal collateral veins. Johns Hopkins Med J 85: 183-99, 1949.

79. Lawson HC. The volume of blood. In: Hamilton WF, ed. Handbook of physiology. Washington, DC: American Physiological Society, p. 23, 1962.

183. Peters JP, Van Slyke DD. Quantitative clinical chemistry. Interpretations. Vol. I. 2nd ed. Baltimore: Williams & Wilkins, p. 425, 1946.

For references to articles in press, state name of journal and if possible, volume and year. Do not enter citations in reference list such as "Unpublished observations," "Manuscript in preparation," "Personal communication," and "Submitted for publication;" note parenthetically in text. Obtain and submit written permission to cite someone else's unpublished observations.

Check accuracy of names, initials, volume number, year, pages, and article title; also, any attributions included in article. Authors are charged for extensive reference checking and corrections.

Illustrations: Refer to all illustrations as figures. Number in Arabic numerals. Make lettering proportional to size and scale so smallest elements (subscripts, for example) are readable when reduced. Give actual magnification of photomicrographs. Submit sharp, high-contrast, unmounted photographs on glossy paper.

Obtain permission to use previously published illustrative material from copyright holder and author.

Tables: Prepare with page size of *Medicine* in mind. Number in Arabic numerals; give brief headings; and do not duplicate data in text or figures.

OTHER

Abbreviations, Nomenclature and Symbols: Follow *Style Manual for Biological Journals*. Symbols of units of measurements are not abbreviations; do not use periods. Define abbreviations and contractions at first use.

Identify any medications, materials and devices by full nonproprietary and brand name and manufacturer's name, city, state and country. Chemical name precedes trade, popular name, or abbreviation at first occurrence. Capitalize trademarked names.

Electronic Submission: Submit final version of manuscript on 3.5 or 5.25 in. diskette with typescript of final revision. Diskettes produced on IBM or IBM-compatible computers are preferred; those produced on most Apple/Macintosh or Wang computers can be converted. ASCII files can also be used. Do not use Fast Save option on Macintosh computers.

Identify diskette with name of first author and manuscript title; name of computer file; type of hardware, operating system, and version number; and name of software program and version number. Tables are processed in the traditional manner.

SOURCE

January 1994, 73(1): back of table of contents.

Medicine and Science in Sports and Exercise
American College of Sports Medicine
Williams & Wilkins

Peter B. Raven, Ph.D.
University of North Texas Health Science Center at Fort Worth
3500 Camp Bowie Blvd.
Fort Worth, TX 76107

AIM AND SCOPE

Manuscripts dealing with original investigations, clinical studies, special communications or comprehensive reviews on topics relevant to the areas of interest of the College will be considered for publication. Membership in the American College of Sports Medicine is not a requisite for publication in the Journal, nor does it influence editorial decisions.

MANUSCRIPT SUBMISSION

Manuscripts must be contributed solely to this Journal and have not been and will not be published elsewhere. The Journal is owned by the American College of Sports Medicine and is copyrighted for the protection of authors and the College. College must receive in writing exclusive assignment of copyright from all authors of all accepted manuscripts; forms are mailed from Editorial Office.

If a blind review is desired, so indicate in cover letter.

MANUSCRIPT FORMAT

Submit four copies of manuscript. Type on one side of bond paper, double-space throughout, and with a 2.5 cm wide margin on all sides. Submit an original and three copies complete with tables and figures. Number pages in upper right corner beginning with a title page as number one. Arrange in order: title pages, abstract page including index terms, text, page for acknowledgments, references, texts to figures, tables, and illustrations.

Title Page: Title of article (85 characters), concise, informative; short running head (60 characters); names of authors, including first name and middle initial; department(s) and institution(s) of origin of study; and name, address, zip code, and telephone and fax numbers of corresponding author. List only those persons having substantive input as authors. If a blind review is requested, include a second title page with title only. Include information concerning grant support on acknowledgment page; also note institutional identification for co-authors not located at the institution of origin on acknowledgment page.

Abstract and Key Words: Page 2; informative abstract (200 words). State question under investigation, brief description of methodology, main findings with specific data and statistical significance, essential conclusions, and clinical or scientific relevance. Use only approved abbreviations (see Units and fifth edition of *The Council of Biology Editors Style Manual*). At end of abstract, supply list of key words or terms (3-12) not used in title for indexing (e.g., important variables, methods, treatments, conditions). Include species of animals if not in title.

Text: Observational and experimental articles include separate sections for rationale for study, methodology employed, results obtained, and discussion of results. In introductory section, state purpose(s) of study. Provide only relevant references; do not exhaustively review subject. In methodology section, describe experimental subjects and controls.

Identify methods, apparatus, and procedures employed with sufficient detail to allow others to reproduce results. Provide references for established methods and statistical procedures. If methods utilized are not well known or accepted, provide a rationale for using them and include a description of possible limitations.

Denote statistical significance when appropriate. Include detailed statistical analyses, mathematical derivation, or computer programs within an appendix. Present findings logically in text, tables, or figures. Do not include the same data in tables or figures. In discussion emphasize original and important features of study. Avoid repeating data presented in results section. Incorporate within discussion significance of findings and relationship(s) and relevance to published observations. Provide conclusions supported by data. A summary section is not necessary. Make manuscript brief, clear, and grammatically correct.

Footnotes: Not permitted in text. Put unnumbered notes at end of article to present special information about apparatus, technique, or present addresses of authors. Type double-spaced on acknowledgment page.

Acknowledgments: Give only to persons who have made substantive contributions to manuscript. Obtain permission from persons cited. Include as acknowledgments special notes on apparatus, techniques, funding sources, disclaimers, and present addresses of authors. (Addresses as, "Present address for A. B. Smith: . . .").

References: Select 30 publications or studies already accepted for publication. Cite abstracts only when sole source of information. Number notation of references in parentheses, one reference to a number, and list alphabetically by last name of first author. Type double-spaced. For journal articles include last name of first author, followed by initials, initials and last names of each co-author, title of article (first word only capitalized), name of journal (abbreviated as in *Index Medicus*), volume, inclusive pages, and year.

Brava, E. L. and R. C. Tarazi. Converting enzyme inhibition with an orally active compound in hypertensive man. *Hypertension* 1:39–46, 1979.

Mention all, if fewer than seven authors are listed. If seven or more, list only first three. If author is unknown (corporate author):

Anonymous. Epidemiology for primary health care. *Int,. J. Epidemiol.* 5:224–225, 1976

American College of Sports Medicine position paper on blood doping as an ergogenic aid. *Med. Sci. Sports Exerc.* 19:540–543, 1987.

Use format of U.S. National Library of Medicine and in *Index Medicus*. For those not included in *Index Medicus*, adhere to form established by American National Standard for Bibliographic References.

Do not include unpublished manuscripts, M.S. theses, or personal communications as references. Include personal communication within text after securing permission of person cited. Cite unpublished work in text and identify in parentheses as "unpublished observation."

Reference books: author's last name and initials, title of book (with main words capitalized), city of publication, publisher, year, and pages. For examples for corporate authors, chapters, editors, center publication, etc., see *British Medical Journal* 1:1334–1336, 1978. Use technical documents only when availability is assured. If a publication in press is essential to substance of manuscript, include with submission, with a citation of journal in which article will appear. Designate "in press."

Figures: Submit unmounted, glossy, high-contrast black-and-white photographic prints not larger than 8.5 × 11 in.

nor smaller than 5 × 7 in. Submit two sets, each in a separate envelope protected by cardboard. Do not use paper clips or staples. Identify on back with a typed label or mark lightly with a soft pencil. Include figure number and abbreviated title of article; indicate TOP, especially if a radiological or histological photograph. For blind reviews, do not include author's name. If original drawings are submitted, prepare with black India ink. Use clean laser black-and-white print reproductions on bond in lieu of glossy prints. Number consecutively with Arabic numerals. Give each figure a legend; group in numerical order and type double-spaced on a separate sheet. Include copies of each figure with original and three copies of submission for review purposes. Make lettering of professional quality with consistent spacing and alignment. Freehand and typewritten printing are not acceptable. Make lettering uniform with respect to size and type face. Make figures consistent in format and size. Make height and thickness of data (symbols, letters, and numbers) legible after reduction (smallest data points should not be less than 1.5–1.7 mm high). To ensure adequate size, make data 3 mm high on a 5 × 7 print and 6 mm high on an 8.5 × 11 print. If data is too small after reduction, figure will be returned for modification. Keep lettering within framework of figure, including key to symbols. Use abbreviations, symbols, and terminology identical to those used in corresponding section of text. Use following symbols: ❑ ■ ○ ● △ ▲.

Photographs of equipment and animals are discouraged; good line drawings are more informative and effective. Draw and letter drawings for reduction to a given size and letter to same scale. Halftone photographs must be in focus, crisp, high-contrast, suitable for reproduction; eliminate all extraneous background. When photographing graphs, eliminate grid lines unless essential. Label each axis with quantity measured and units of measurement.

Color figures are accepted if author assumes all printing costs. Supply three positive color prints. Indicate approximate position of each figure in margin of manuscript.

Tables: Submit figures rather than tables. Type double-spaced on separate sheets, and design to fit one or two-column widths (8.5 and 18 cm, respectively). Give each a brief title; place explanatory matter in text or in caption below table. Include means and units of variation (S.D., S.E., etc.); exclude nonsignificant decimal places. Use same abbreviations as in text and figure legends; define in caption. Define symbols in order of appearance, reading horizontally across table; use standard symbols *, †, §, and ¶. When more columns of tabulated statistics need to be presented, separate into two tables. Indicate approximate position of each table in left margin of manuscript. Do not submit camera copy.

If tables are too extensive, Editor-In-Chief may recommend deposition of portions of data with National Auxiliary Publications Service of the American Society for Information Science (ASIS).

Formulas and Equations: Keep to a minimum. Present in text as a single line: (a + b)/(x + y). Presentation as $\frac{a+b}{x+y}$ requires hand composition and wastes additional space. Define or explain all unusual characters. Write names of Greek letters in margin of text. Spell out numbers below 10 in text; express numbers 10 and above numerically.

OTHER FORMATS

BOOKS FOR REVIEW—Send two copies of book to be reviewed to National Center, Book Review Editor, ACSM National Center, 401 West Michigan Street, Indianapolis, IN 46202-3233.

If chosen for review, book is forwarded to book review editor, who in turn sends it out to be reviewed. Published reviews appear monthly.

BRIEF REVIEWS—To educate nonexperts on a particular issue, topic, or problem in sports medicine or exercise science.

CASE STUDIES—Describe specific clinical cases that provide relevant information on diagnosis and therapy of a particular case that proves unique to clinical sports medicine. Make manuscripts current, concise, accurate, and understandable. Include: a brief abstract (describe case and its clinical implications); introduction (comment on clinical problem, explain using case as an example; document patients' agreement to use of clinical data); brief case report (include history, physical examination, laboratory findings, treatment and outcome); discussion (explain in detail clinical implications of course of case and key aspects of case which may be unique or may differ from similar reported cases in medical literature.

LETTERS TO THE EDITOR—500 words. Must promote intellectual discussion of an article recently published. Include an informative title; type double-spaced. If acceptable, a copy is sent to original author who can submit a rebuttal that will be considered for publication.

SPECIAL COMMUNICATIONS—New methods, important modifications of existing ones, or applications of new equipment.

SYMPOSIUM PROCEEDINGS—Organizers of symposia concerned with new developments or "state of the art" in sports medicine and exercise science should contact Editor-In-Chief about possibility of publication by submitting an outline, titles, names of participants, and a brief statement of symposium's importance.

Submit four copies of edited manuscripts with salient and unique features of presentation; do not exceed 16–20 typewritten, double-spaced pages. Manuscripts will contain data already published or accepted for publication. New data is subject to evaluation as a separate item. Do not cite abstracts as supporting evidence. Symposia are subject to review.

OTHER

Drugs: Identify drugs and chemicals by generic name(s). Explain dosage(s), concentration(s), and routes of administration.

Electronic Submission: Submit disks of accepted manuscripts for computerized editing with hard copy. Most word processing programs are acceptable. Macintosh users-do not use fast save option.

Experimentation, Animal: Conduct in conformance with policy statement of the American College of Sports Medicine (*Medicine and Science in Sports and Exercise* January and July issues). Failure to conform with guide-

lines and to guarantee conformance by a statement in manuscript will result in rejection of manuscript.

Experimentation, Human: Conform to policy statement regarding use of human subjects and informed consent as published by *Medicine and Science in Sports and Exercise* (January and July issues). Indicate that experiments were in accordance with policy statements of the American College of Sports Medicine (*Medicine and Science in Sports* 10:ix-x, 1978 and 21:vi, 1989). Do not include information that will identify human subjects.

Language: Use nonsexist language (see *American Psychologist* 30:682–684, 1975). Be sensitive to semantic description of persons with chronic diseases and disabilities (see editorial in *Medicine and Science in Sports and Exercise*, 23 (11), 1991).

Use only standardized abbreviations and symbols. If unfamiliar abbreviations are employed, define at first appearance in text. Follow *Webster's Third New International Dictionary* for spelling, compounding, and division of words.

Capitalize trademark names and verify spelling. Chemical or generic names precede trade name or abbreviation of a drug at first use in text.

Page Charges: $40 per printed page.

Units: Use standard terms generally acceptable in the field of exercise science and sports medicine. Units of measurement are Système International d'Unités (SI).

Exceptions to SI are heart rate beats per min: blood pressure—mm Hg; gas pressure—mm Hg. See *British Medical Journal* (1:1334–1336, 1978) and *Annals of Internal Medicine* (106:114–129, 1987) for the proper method to express other units or abbreviations. When expressing units, locate multiplication factor midway between lines to avoid confusion with periods, e.g., $ml \cdot min^{-1} kg^{-1}$.

The basic and derived units most commonly used in this *Journal* include: mass—gram (g) or kilogram (kg); force—newton (N); distance—meter(m), kilometer (km); temperature—degree Celsius (°C); energy, heat, work—joule (J) or kilojoule (kJ); power—watt (W); torque—newton·meter (N·m); frequency—hertz (Hz); pressure—pascal (Pa); time—second (s), minute (min), hour (h) volume—liter (l), milliliter (ml); and amount of a particular substance—mole (mol), millimole (mmol).

Selected conversion factors: 1 N = 0.102 kg (force); 1 J = 1 N·m = 0.000239 kcal = 0.102 kg·m; 1 kJ = 1000 N·m = 0.239 kcal = 102 kg·m; 1 W = 1 $J \cdot s^{-1}$ = 60 $J \cdot min^{-1}$ = 60 $N \cdot m \cdot min^{-1}$ = 6.118 $kg \cdot m \cdot min^{-1}$.

SOURCE

January 1994, 26(1): back matter.
Information for authors are in January and July issues.

Metabolism: Clinical and Experimental
W.B. Saunders Company

James B. Field, MD
Eight Windmill Point Lane
Hampton, VA 23664-2129

MANUSCRIPT SUBMISSION

Articles are accepted for publication on the condition that they are contributed solely to *Metabolism*. Articles must have a definite metabolic application. Submit case reports only if they include controlled observations of an exceptionally revealing nature. Submission implies that if manuscript is accepted for publication, copyright in article, including the right to reproduce article in all forms and media, shall be assigned exclusively to Publisher. Publisher will not refuse any reasonable request by author for permission to reproduce any contributions to journal. Make papers as brief as possible.

MANUSCRIPT FORMAT

Type on good quality paper, one side of page only, Double- or triple-space and keep 1 in. margins on all sides. Include an abstract or summary. Handwritten manuscripts, single-spaced material, and excessive interlinear corrections are unacceptable. Submit original and two complete copies.

Title Page: Include name of institution where work was done ("From the Fels Research Institute . . ."); acknowledgments for research support; and complete mailing address for corresponding and reprint request author(s).

Illustrations and Tables: Cite in order in text with Arabic numerals. Put numbers and author's name on back of illustrations; indicate TOP. Compile figure legends in a separate list at end of paper. Submit diagrams, charts, and other line drawings as glossy, black-and-white photo prints; for half-tone work, supply good photographic prints. Photocopies are not acceptable for reproduction; legible copies may be sent with duplicate manuscripts. Make tables and illustrations a size easily handled. Consider effect of reduction in size to fit Journal page. Contributors bear full cost of printing color plates.

References: Compile numerically in order of citation in text; type double-spaced; give inclusive pages:

JOURNAL ARTICLE

1. Chick WL, Lauris V, Soeldner JS, et al: Monolayer culture of a human pancreatic beta-cell adenoma. Metabolism 22:1217–1224,1973

2. Katz A, Bogardus C: Insulin-mediated increase in glucose 1,6-bisphosphate is attenuated in skeletal muscle of insulin-resistant man. Metabolism (in press)

COMPLETE BOOK

3. Wesson LG: Physiology of the Human Kidney. Philadelphia, PA, Grune & Stratton, 1969

CHAPTER IN BOOK

4. Young VR: The role of skeletal and cardiac muscle in the regulation of protein metabolism, in Munro HN (ed): Mammalian Protein Metabolism, vol 4. San Diego, CA, Academic, 1970, pp 585–674

OTHER FORMATS

PRELIMINARY REPORTS—1000 words including bibliography, but exclusive of illustrative material, simple table or small figure is allowed, but cut word allowance by an amount sufficient to allow for space taken. Include data to support conclusions stated. It can serve as a preliminary report on work just completed, or may be a final report or an observation that does not require a lengthy write-up.

SOURCE

January 1994, 43(1): front matter.

Microbiological Reviews
American Society for Microbiology

Dr. W. K. Joklik
Department of Microbiology and Immunology
P.O. Box 3020
Duke University Medical Center
Durham, NC 27710

AIM AND SCOPE

Microbiological Reviews publishes reviews dealing with all aspects of microbiology and other fields of concern to microbiologists, such as immunology. Authoritative and critical reviews of current state of knowledge are preferred, although historical analyses will be accepted if importance of subject justifies this approach. Unevaluated compilations of literature and annotated bibliographies do not fall within the scope of *Microbiological Reviews*. Manuscripts of lectures delivered at symposia and round tables are likewise unacceptable; however, their authors are encouraged to discuss with the editor the possibility of using such material as basis for preparation of a review when publication in this form seems appropriate. Because a distinctive goal of the journal is to appeal to interests of its diverse group of subscribers, authors are asked to address themselves to both specialists and generalists; this demands conscious concern for this goal in writing.

MANUSCRIPT SUBMISSION

Reviews are solicited by editors from leading investigators around the world. Unsolicited reviews may be submitted in completed form; include a curriculum vitae and bibliography for each author, with names, addresses, and telephone/fax numbers of six experts who may be asked to review the manuscript. Also, submit suggestions for reviews of subjects that have not been reviewed recently to editors; include an annotated topical outline; a one- or two-paragraph statement describing aim, scope, and relevance of review; list of key references showing author's contributions to field, with other investigators' findings; and a curriculum vitae and bibliography for each author.

Submit four complete copies of each manuscript, including figures and tables, either in original typescript or as clear, clean photocopies. Include four copies of summary (250 words) for table of contents. Indicate name, mailing address, and telephone and fax numbers of corresponding author on title page. Double or triple space all text, including quotations, tables, figure legends, and references. Number pages sequentially.

Complete mailing address, telephone and fax numbers and e-mail address of corresponding author will be printed as a footnote if desired. Include information in lower left corner of title page labeled "Correspondent Footnote."

ASM requires corresponding author to sign a copyright transfer agreement on behalf of all authors. Agreement is sent when manuscript is accepted and scheduled for publication. Unless agreement is executed (without changes and/ or addenda), ASM will not publish manuscript.

If all authors were employed by U.S. government when work was performed, corresponding author should not sign copyright transfer agreement., instead, attach to agreement a statement attesting that manuscript was prepared as a part of official duties and, as such, is a work of U.S. government not subject to copyright.

If some authors were employed by U.S. government when work was performed but others were not, sign copyright transfer agreement as it applies to that portion performed by nongovernment employee authors.

An author is one who made a substantial contribution to article; all authors are responsible for entire paper. Do not list individuals who provided assistance, e.g., supplied strains or critiqued paper, as authors; recognize in Acknowledgment section.

Authors must agree to order in which names are listed in byline. Do not footnote attribution of work (e.g., X. Jones and Y. Smith contributed equally to. . .). If necessary, include in Acknowledgment section.

Submit written assurance that permission to cite personal communications and preprints has been granted.

By submission of a manuscript, authors guarantee that manuscript, or one substantially the same, was not published previously, is not being considered or published elsewhere, and will not be published elsewhere. All authors must have agreed to its submission and are responsible for its content, including appropriate citations and acknowledgments, and must also have agreed that corresponding author has authority to act on their behalf on all matters pertaining to publication of manuscript.

The American Society for Microbiology defines primary publication as in *How to Write and Publish a Scientific Paper* (3rd ed, Robert A. Day), : ". . . (i) the first publication of original research results, (ii) in a form whereby peers of author can repeat experiments and test conclusions, and (iii) in a journal or other source document readily available within the scientific community."

Scientific papers published in a conference report, symposium proceeding, technical bulletin, or any other retrievable source are unacceptable for submission on grounds of prior publication. Preliminary disclosure of research findings published in abstract form as an adjunct to a meeting is not considered "prior publication" because it does not meet criteria for a scientific paper.

Acknowledge any prior publication of data contained in a manuscript. Submit copies of relevant work.

Obtain and submit permissions from both original publisher and original author [i.e., copyright owner(s)] to reproduce figures, tables, or text (in whole or in part) from previous publications. Identify each as to relevant item in manuscript (e.g., "permissions for MR Fig. 1").

MANUSCRIPT FORMAT

Ensure that sets of characters are easily distinguishable in manuscript: numeral zero (0) and letter "oh" (O); numeral one (1), letter "el" (l), and letter "eye" (I); and a multiplication sign (×) and letter "ex" (x). If distinctions cannot be made, mark items at first occurrence for cell lines, strain and genetic designations, viruses, etc., on modified manuscript to be identified properly for printer by copy editor.

Submit original drawings or glossy prints. Indicate magnification by a suitable scale on photograph. Number each figure and include name of author, either in margin or on back (marked lightly with a soft pencil). Do not submit illustrations larger than 8.5 × 11 in. Do not exceed 100 manuscript text pages; excess will require discussion with editor.

References: Cite all listed references in text. Arrange references in alphabetical order, by first author; number consecutively. Abbreviate journal names according to *Serial Sources for the BIOSIS Data Base* (BioSciences Information Service, 1992). Cite each reference in text by number. State control number for ASM publications "in press" or name of publication if it is a book.

1. **Anagnostopoulos, C., and J. Spizizen.** 1961. Requirements for transformation in *Bacillus subtilis.* J. Bacteriol. **81:**741-746.

2. **Berry, L. J., R. N. Moore, K. J. Goodrum, and R. E. Couch, Jr.** 1977. Cellular requirements for enzyme inhibition by endotoxin in mice, p. 321-325. *In* D. Schlessinger (ed.), Microbiology—1977. American Society for Microbiology, Washington, D.C.

3. **Cox, C. S., B. R. Brown, and J. C. Smith.** J. Gen. Genet., in press.*

4. **Dhople, A., I. Ortega, and C. Berauer.** 1989. Effect of oxygen on in vitro growth of *Mycobacterium leprae,* abstr. U-82, p. 168. Abstr. 89th Annu. Meet. Am. Soc. Microbiol. 1989.

5. **Finegold, S. M., W. E. Shepherd, and E. H. Spaulding.** 1977. Cumitech 5, Practical anaerobic bacteriology. Coordinating ed., W. E. Shepherd. American Society for Microbiology, Washington, D.C.

6. **Fitzgerald, G., and D. Shaw.** *In* A. E. Waters (ed.), Clinical microbiology, in press. ASM Press , Boston.

7. **Gustlethwaite, F. P.** 1985. Letter. Lancet **ii:**327.

8. **Jacoby, J., R. Grimm, J. Bostic, V. Dean, and G. Starke.** Submitted for publication.

9. **Jensen, C., and D. S. Schumacher.** Unpublished data.

10. **Jones, A. (Yale University).** 1990. Personal communication.

11. **Leadbetter, E. R.** 1974. Order II. *Cytophagales* nomen novum, p. 99. *In* R. E. Buchanan and N. E. Gibbons (ed.), Bergey's manual of determinative bacteriology, 8th ed. The Williams & Wilkins Co., Baltimore.

12. **Miller, J. H.** 1972. Experiments in molecular genetics, p. 352-355. Cold Spring Harbor Laboratory, Cold Spring Harbor, N.Y.

13. **Powers, R. D., W. M. Dotson, Jr., and F. G. Hayden.** 1982. Program Abstr. 22nd Intersci. Conf. Antimicrob. Agents Chemother., abstr. 448.

14. **Sigma Chemical Co.** 1989. Sigma manual. Sigma Chemical Co., St. Louis, Mo.

15. **Smith, J. C.** April 1990. U.S. patent 484,363,770.

16. **Smyth, D. R.** 1972. Ph.D. thesis. University of California, Los Angeles.

17. **Yagupsky, P., and M. A. Menegus.** 1989. Intraluminal colonization as a source of catheter-related infection. Antimicrob. Agents Chemother. **33**:2025. (Letter.)

OTHER FORMATS

LETTERS TO THE EDITOR—500 words. Include data to support argument. Comment on articles published previously in journal. Send letters to Journals Division. A reply may be solicited from author of article. Type double spaced.

OTHER

Abbreviations: Limit use. Use abbreviations other than those recommended by IUPAC-IUB *(Biochemical Nomenclature and Related Documents,* 1978) only when necessary, such as in tables and figures.

Use pronouns or paraphrase a long word after first use (e.g., "the drug," "the substrate"). Use standard chemical symbols and trivial names or their symbols for terms that appear in full in neighboring text.

Define each abbreviation and introduce in parentheses at first time use. Eliminate abbreviations that are not used at least five times in text, tables and figure legends.

Not requiring Introduction—In addition to abbreviations for Système International d'Unités (SI) units of measurement, other common units (e.g., bp, kb, and Da), and chemical symbols for elements, use the following without definition in title, abstract, text, figure legends, and tables: DNA (deoxyribonucleic acid); cDNA (complementary DNA); RNA (ribonucleic acid); cRNA (complementary RNA); RNase (ribonuclease); DNase (deoxyribonuclease); rRNA (ribosomal RNA); mRNA (messenger RNA); tRNA (transfer RNA); AMP, ADP, ATP, dAMP, ddATP, GTP, etc. (for the respective 5′ phosphates of adenosine and other nucleosides) (add 2′-, 3′-, or 5′- when needed for contrast); ATPase, dGTPase, etc. (adenosine triphosphatase, deoxyguanosine triphosphatase, etc.); NAD (nicotinamide adenine dinucleotide); NAD+ (nicotinamide adenine dinucleotide, oxidized); NADH (nicotinamide adenine dinucleotide, reduced); NADP (nicotinamide adenine dinucleotide phosphate); NADPH (nicotinamide adenine dinucleotide phosphate, reduced); NADP+ (nicotinamide adenine dinucleotide phosphate, oxidized); poly(A), poly(dT), etc. (polyadenylic acid, polydeoxythymidylic acid, etc.); oligo(dT), etc. (oligodeoxythymidylic acid, etc.); P_i (orthophosphate); PP_i (pyrophosphate); UV (ultraviolet); PFU (plaque-forming units); CFU (colony-forming units); MIC (minimal inhibitory concentration); MBC (minimal bactericidal concentration); Tris [tris(hydroxymethyl)aminomethane]; DEAE (diethylaminoethyl); A_{260} (absorbance at 260 nm); EDTA (ethylenediaminetetraacetic acid); PCR (polymerase chain reaction); and AIDS (acquired immunodeficiency [or immune deficiency] syndrome). Abbreviations for cell lines also need not be defined.

Use following abbreviations without definition in tables:

amt (amount)	SD (standard deviation)
approx (approximately	SE (standard error)
avg (average)	SEM (standard error of the mean)
concn (concentration)	sp act (specific activity)
diam (diameter)	sp gr (specific gravity)
expt (experiment)	temp (temperature)
exptl (experimental)	tr (trace)
ht (height)	vol (volume)
mo (month)	vs (versus)
mol wt (molecular weight	wk (week)
no. (number)	wt (weight)
prepn (preparation)	yr (year)

Nomenclature: Spelling of names follows *Approved Lists of Bacterial Names* (amended edition, V. B. D. Skerman, V. McGowan, and P. H. A. Sneath, ed.) and *Index of the Bacterial and Yeast Nomenclatural Changes Published in the International Journal of Systematic Bacteriology since the 1980 Approved Lists of Bacterial Names (1 January 1980 to 1 January 1989)* (W. E. C. Moore and L. V. H. Moore, ed., American Society for Microbiology, 1989), and validation lists and articles published in *International Journal of Systematic Bacteriology* since 1 January 1989. If there is reason to use a name that does not have standing in nomenclature, enclose name in quotation marks and make a statement concerning nomenclatural status of name in text (see *Int. J. Syst. Bacteriol.* **30**:547–556,1980).

Use names for viruses approved by International Committee on Taxonomy of Viruses published in 4th Report of ICTV, Classification and Nomenclature of Viruses *(Intervirology* **17**:23–199,1982), with modifications contained in 5th Report of ICTV *(Arch. Virol.,* Suppl. 2, 1991). If desired, add synonyms parenthetically when name is first mentioned. Approved generic (or group) and family names may also be used.

For enzymes, use recommended (trivial) name assigned by Nomenclature Committee of the International Union of Biochemistry as described in *Enzyme Nomenclature* (Academic Press, Inc., 1992).

For genetic nomenclature follow recommendations of Demerec et al. *(Genetics* **54**:61–76,1966) and those in instructions to authors of *Journal of Bacteriology* and *Molecular and Cellular Biology* (January issues).

Style: Conform to *CBE Style Manual* (5th ed., 1983; Council of Biology Editors, Inc. Bethesda, MD.), *ASM Style Manual for Journals and Books* (American Society for Microbiology, 1991), and Robert A. Day' s *How to Write and Publish a Scientific Paper* (3rd ed., 1988; Oryx Press), as interpreted and modified by editorial board and ASM Journals Division.

Include a table of contents showing major headings and subheadings of text. Consult a recent issue of *Microbiological Reviews* for format and style.

Summarize supporting evidence whenever possible.

When original data are presented, fully describe methods or reference previously published methods.

Units: Use standard metric units for reporting length, weight, and volume. For these units and for molarity, use prefixes m, μ, n, and p for 10^{-3}, 10^{-6}, 10^{-9}, and 10^{-12}, respectively. Use prefix k for 10^3. Avoid compound prefixes such as mμ or $\mu\mu$. Use μg/ml or μg/g in place of ambiguous ppm. Present units of temperature as: 37°C or 324 K.

When fractions are used to express units such as enzymatic activities, use whole units, such as "g" or "min," in denominator instead of fractional or multiple units, such as μg or 10 min. For example, "pmol/min" is preferable to "nmol/10 min," and "μmol/g" is preferable to "nmol/μg." Use unambiguous forms such as exponential notation."

See *CBE Style Manual*, 5th ed., for more detailed information about reporting numbers and information on SI units for reporting illumination, energy, frequency, pressure, and other physical factors. Report numerical data in applicable SI units.

SOURCE

March 1994, 58(1): i-iv.

Microbiology
Society for General Microbiology

Microbiology Editorial Office
Marlborough House, Basingstoke Road
Spencers Wood, Reading RG7 1 AE, U.K.

AIM AND SCOPE

The *Journal* publishes original work on microorganisms in the laboratory and in their natural environments and is particularly concerned with fundamental studies. Papers are published in the following subject areas: Development and Structure, Physiology and Growth, Biochemistry, Genetics and Molecular Biology, Biotechnology, Environmental Microbiology, Plant-Microbe Interactions, Pathogenicity and Medical Microbiology, Immunology, and Systematics.

MANUSCRIPT SUBMISSION

Submit: original typescript (typed double-spaced throughout) plus two copies; one printers' set (preferably unlettered) of line drawings, plus three lettered reviewers' sets; one set of unlettered prints of any halftone photographs, plus three sets (not photocopies) with lettering and labels; and two copies of papers cited as 'accepted for publication' for 'in the Press.'

Manuscripts must report unpublished work that is not under consideration for publication elsewhere. All named authors must have agreed to submission. If accepted, authors must transfer copyright to Society of General Microbiology. Work must not be published again in the same form in any language, without consent of the Society.

Wait until a piece of work is rounded off and a comprehensive paper can be written; do not write a series of papers on the same subject as results come to hand.

For guidelines on the presentation of Genome Analysis papers, refer to *J Gen Microbiol* July 1993 or contact Editorial Office. Obtain guidelines on presentation of systematics papers from Editorial Office.

MANUSCRIPT FORMAT

Submit manuscripts in triplicate (original typescript plus two copies), typed double-spaced with wide margins throughout. Do not use low-quality dot-matrix printers. Include three lettered sets of each figure for reviewers (for halftone figures, submit photographic prints, not photocopies), and an additional set of figures (unlettered in the case of halftones) for use by printers.

Write in clear and concise English, in past tense. Include: Title page, including key words; Summary; Introduction; Methods; Results; Discussion with Conclusions if appropriate; Acknowledgments, if any; References. Results and Discussion may be combined. Select figures and tables to illustrate points that cannot easily be described in text.

Title Page: Include: title of paper (include topical keywords, allude to interesting conclusions, emphasize main conclusions or ask a question, do not describe nature of study); a shortened 'running title' (50 characters and spaces); 5 keywords or short phrases; subject category for contents list (categories are listed in Aim & Scope); names of authors (one given name in full), indicate corresponding author; name and address of laboratory or laboratories where work was done, and present addresses of authors who have moved; telephone and fax numbers and electronic mail address for corresponding author (this information is a footnote on first page) and a footnote defining nonstandard abbreviations. Obtain list of abbreviations not requiring definition from Editorial Office.

Summary: Make clear and comprehensible. Do not cite References. Define abbreviations used. Introduce subject in first sentence. Present main conclusion in last sentence.

Introduction: State objectives of work when it was undertaken; do not summarize results. Make clear why area is worth studying.

Methods: Provide sufficient detail to allow work to be repeated. Indicate suppliers of chemicals and equipment if this may affect results. Omit suppliers' addresses unless essential for a particular reason.

Results: Use sufficient subheadings to make clear how work was organized, what key questions were addressed, how one experiment led to another, and what conclusions were reached. Indicate reliability and reproducibility of results of experiments.

Discussion: Do not recapitulate results. Be brief. Use subheadings where appropriate to highlight point under discussion. List main conclusions at end. Combine Results and Discussion when appropriate.

References: Cite in text: two authors, Smith & Jones (1991) or (Smith & Jones, 1991); three or more authors, Smith *et al.* (1991) or (Smith *et al.*, 1991). Distinguish references to papers by same author(s) in same year in text and reference list by letters *a*, *b*, etc. (e.g., 1991*a*, or 1991*a*, *b*). Give references at end of paper in alphabetical order. For three or more authors, list in chronological order after any other papers by first author. Include title of paper with initial and final page numbers. Abbreviate journal titles according to *Index Medicus* and *Biological Abstracts* (prior to 1994). Do not use stops after abbreviated words. Include in references to books: year of publication, title, edition, editor(s) (if any), town of publication and publisher, in order. Give inclusive page numbers and, if appropriate, chapter title when reference is to part of a book.

JOURNAL

Towbin, H., Staehelin, T. & Gordon, J. (1979). Electrophoretic transfer of proteins from polyacrylamide gels to nitrocellulose sheets; procedure and some applications. *Proc Natl Acad Sci USA* **76**, 4350–4354.

BOOK CHAPTER

Taylor, R. K. (1989). Genetic studies of enterotoxin and other potential virulence factors of *Vibrio cholerae*. In *Genetics of Bacterial Diversity*, p. 309–329. Edited by D. A. Hopwood & K. F. Chater. London: Academic Press.

Cite only those papers accepted for publication but not yet published as 'in press' in reference list; include name of journal; enclose two copies. Cite references to papers not yet accepted in text as unpublished results; give surname(s) and initials of author(s). Do not include in list of references.

Obtain and submit written permission for all personal communications cited in text.

Tables: Make comprehensible without reference to text. Do not repeat detailed descriptions of methods, or give in full abbreviations previously defined in text. Use symbols *, †, ‡, §, ||, ¶ for footnotes. For results expressed as percentages, state absolute value(s) corresponding to 100%. Include statements of reproducibility.

Figures: Select to illustrate specific points. Type legends double-spaced on a separate sheet. Points regarding comprehensibility, relative values and reproducibility for tables also apply to figures and legends. Color charges are borne by author.

Line Drawings—Submit suitable for direct reproduction approximately twice final size. Maximum printed size, including lettering and legends is 176 × 225 mm. Submit line drawings as original drawings in India ink on tracing paper or white card; glossy photographs; or high-quality computer-generated figures. Make lines 0.4 mm thick; symbols on graphs should be 3 mm in diameter. Preferred symbols are: ○ ● □ ■ △ ▲ ▼ ▽. Where possible, use same symbol for same quantity in different figures. Place scale-marks on graphs inside axes. Supply three lettered sets of each drawing. Provide a top set of figures for printers; preferably unlettered, but not essential.

Photographs—Submit well-contrasted prints on glossy paper; normally unmounted and approximately final size (maximum width 176 mm). Provide one unlettered set of prints, with three additional reviewers' copies (not photocopies) of each photograph (lettered or with a lettering guide-a photocopy or tracing-paper overlay with accurately marked outlines of features indicated). Size of photomicrographs may be altered in printing; show magnification by a bar marker on lettering guide. Crop nonessential areas of photographs. Group photographs to form a composite picture (maximum final size 176 × 225 mm, including space for legend); to avoid loss of definition, submit component parts as separate prints with a layout guide.

State if photographs of gels, autoradiograms, etc., have been digitally processed to enhance quality.

Color photographs are not published unless essential. Additional cost is borne by authors. Printers can produce black-and-white photographs from color prints if necessary; this is not recommended.

Reproducibility of Results: Give an indication of this. State how many times an experiment was repeated and whether means or representative results are shown. Indicate variability statistically; when error terms are given, state measure of dispersion and number of observations. Specify statistical techniques used; describe fully or reference. If results are expressed as percentages, state absolute value corresponding to 100%.

OTHER FORMATS

INVITED REVIEW ARTICLES—Write to Editor-in-Chief at Editorial Office.

MICROBIOLOGY COMMENT—Correspondence section for reader responses to issues raised in journal.

OTHER

Data Submission: Deposit detailed information supplementing papers (e.g., sets of spectra, extensive tables of taxonomic data, computer programs) in Supplementary Publication Scheme of the British Library Document Supply Centre. Details are available from Editorial Office.

Electronic Submission: Submit manuscript on floppy disk for typesetting, along with three paper copies for review. Disks should be 3.5, 5.25 or 8 in. Any format is acceptable. Disk should contain a single word-processor document file containing no graphics; any word-processor is acceptable. Label disk with paper number, name(s) of author(s), disk format, software used and filename(s). Send disk when requested. In the event of a discrepancy between disk and final hard-copy version, latter is taken as correct.

Experimentation, Animal: Conduct experiments in accordance with legal requirements of relevant local or national authority. Animals should not suffer unnecessarily. Give details of procedures and of anesthetics used. Papers where ethical aspects are open to doubt are not accepted.

Experimentation, Human: State approval by Ethical Committee of institution in which work was done. State that subjects gave informed consent to work.

Microbial Strains: Give source (name and brief address) or reference for each strain used. Deposit important strains in a recognized culture collection and refer to collection and strain number in paper.

Nomenclature, Chemical and Biochemical: Follow recommendations of IUPAC for chemical nomenclature, and Nomenclature Committee of IUBMB and IUPAC-IUBMB Joint Committee on Biochemical Nomenclature for biochemical nomenclature. See *Biochemical Journal* "Instructions to Authors" (first issue each year) for a summary of nomenclatural recommendations with references, See *Compendium of Biochemical Nomenclature and Related Documents* (2d ed, 1992, London: Portland Press) for full recommendations.

Nomenclature, Enzymes: Use system in *Enzyme Nomenclature* (1992, London & New York: Academic Press). Where appropriate give Enzyme Commission numbers.

Nomenclature, Genetic: Follow, for bacteria: Demerec, M. *et al.* (1966) *Genetics* **54**, 61–76 [also *Journal of General Microbiology* (1968), **50**, 1–14]. Plasmids: Novick, R. P. *et al.* (1976) *Bacteriological Reviews* **40**, 168–189. Yeasts: Sherman, F. (1981) In *The Molecular Biology of the Yeast Saccharomyces. I. Life Cycle and Inheritance*, pp. 639–640 (ed. J. N. Strathern *et al.* New York: Cold Spring Harbor Laboratory). *Aspergillus nidulans*: Clutterbuck, A. J. (1973) *Genetical Research* **21**, 291–296. *Neurospora crassa*: *Neurospora Newsletter* (1978), **25**, 29.

Nomenclature, Microorganisms: Use correct name of organism, follow international rules of nomenclature; if desired, add synonyms in parentheses when name is first mentioned. Names of bacteria must conform with current Bacteriological Code and opinions issued by the International Committee on Systematic Bacteriology. Names of algae and fungi must conform with the current International Code of Botanical Nomenclature. Names of protozoa must conform with current International Code of Zoological Nomenclature. Do not submit descriptions of new species unless a specimen has been deposited in a recognized culture collection and is designated as a type strain in a paper. Consult *Bergey's Manual of Systematic Bacteriology* (Baltimore: Williams & Wilkins). Volume 1 (1984, N. R. Krieg & J. G. Holt, eds) covers Gram-negative bacteria of general, medical or industrial importance; Volume 2 (1986, P. H. A. Sneath, N. S. Mair, M. E. Sharpe & J. G. Holt, eds) covers Gram-positive bacteria other than actinomycetes. Volume 3 (1989, J. T. Staley, M. P. Bryant, N. Pfennig & J. G. Holt, eds) covers archaeobacteria,, cyanobacteria and remaining Gram-negative bacteria. Volume 4 (S. T. Williams, M. E. Sharpe & J. G. Holt, eds) covers actinomycetes.

See also *Approved Lists of Bacterial Names* (1980, V. B. D. Skerman, V. McGowan & P. H. A. Sneath, eds., Washington, DC: American Society for Microbiology) (reprinted from *International Journal of Systematic Bacteriology* **30**, 225–420). An amended version, with minor corrections, was published in 1989. Supplements are published in *IJSB*. An *Index of the Bacterial and Yeast Nomenclatural Changes* published in the *IJSB* since the 1980 *Approved Lists* (1 January 1980 to 1 January 1989) (W. E. C. Moore & L. V. H. Moore, eds, American Society for Microbiology, 1989).

A computerized version of the Approved Lists, including nomenclatural changes validly published since January 1980 is published by the Information Centre for European Culture Collections (Braunschweig, Germany) as Bacterial Nomenclature Update.

The Yeasts, a Taxonomic Study (3rd edn, 1984, N. J. W. Kreger-van Rij, ed, Amsterdam: Elsevier).

Yeasts: Characteristics and Identification, J.A. Barnett, R.W. Payne & D. Yarrow. (2nd edn, 1990, Cambridge: Cambridge University Press).

Ainsworth and Bisby's Dictionary of the Fungi. D. L. Hawksworth, B. C. Sutton & G. C. Ainsworth, (7th edn, 1983, Slough: Commonwealth Agricultural Bureaux).

Nucleotide or Amino Acid Sequence Data: Submit substantial additional experimentation to characterize gene(s) and product(s) concerned, and/or substantial computer analysis. A sequence alone is unacceptable. For DNA sequences from double-stranded genomes, sequence both strands independently.

New sequence data must be submitted to a nucleotide database (or PIR in case of animo acid sequences) and an accession number obtained. Obtain instructions on how to submit data to GenBank, EMBL or DDBJ from addresses below. For a free program for submitting sequence data on PC or Macintosh computers write: Authorin, NCBI, Bldg 38A, Room 8N-803, 8600 Rockville Pike, MD 20894 (e-mail: authorin@ncbi.nlm.nih.gov).

Submit sequence data in computer readable form. Paper submission is acceptable. Submitted manuscripts containing sequence data should include on title page, footnote 'The Genbank [or EMBL or DDBJ] accession number for the sequence reported in this paper is X00000'. Give mnemonic in parentheses after accession number.

GenBank: GenBank Submission, National Center for Biotechnology Information, Bldg 38A, Room 8N-803, 8600 Rockville Pike, MD 20894; (e-mail: gb-sub@ncbi.nlm.nih.gov); Telephone: (301) 496-2475; Fax: (301) 480-9241.

EMBL: EMBL Data Library, Postfach 10.2209, D-6900 Heidelberg, Germany; e-mail: datasubs@embl-heidelberg.de; Telephone: +49 6221 387258; Fax: +49 6221 387519.

DDBJ: DDBJ, National Institute of Genetics, Mishima, Shizuoka 411, Japan; e-mail: ddbjsub@ddbj.nig.ac.jp

Submit figures representing nucleotide or amino acid sequences suitable for direct photographic reproduction. Print sequence with a letter-quality printer; number nucleotides or amino acid residues at appropriate intervals. Design layout to fit either full width of journal page (176 mm) or single column (84 mm). Make characters 1·5–2 mm (6-8 point) high. For single column layout, use 50-60 nucleotides per line; for full page 80-100 (60-70 if spaces between codons). Make spacing between lines as close as consistent with clarity. Sequence data must be submitted to GenBank.

Percentage Concentrations: Use '%' to express concentrations in its correct sense, i.e., g per 100 g solution: otherwise use '% (v/v)' or '% (w/v)' for solutions of concentration ≤ 1%.

Quantities, Units and Symbols: Use recommended SI units. See *Quantities, Units and Symbols* (Royal Society, London) and *Units, Symbols and Abbreviations* (Royal Society of Medicine, London). Use superscripts rather than solidus, e.g., mg ml^{-1} and not mg/ml.

Style/Layout: Consult a recent issue for guidance on the layout of headings, tables, etc.

SOURCE
January 1994, 140(1): v-ix.

Molecular and Cellular Biology
American Society for Microbiology

Journals Division
American Society for Microbiology
1325 Massachusetts Ave., N.W.
Washington, D.C. 20005-4171

AIM AND SCOPE

Molecular and Cellular Biology (MCB) is devoted to the advancement and dissemination of fundamental knowledge concerning molecular biology of eukaryotic cells, of both microbial and higher organisms. Significant papers on cellular morphology and function, genome organization, regulation of genetic expression, morphogenesis, and somatic cell genetics (not simple linkage analyses) are invited for submission. In most cases, reports that emphasize methods and nucleotide sequence data alone (without experimental documentation of functional significance of sequence) will not be considered.

ASM publishes a number of different journals covering various aspects of microbiology. Each journal has a prescribed scope that must be considered in determining most appropriate journal for each manuscript.

Most manuscripts concerning virus-infected cells should be submitted to the *Journal of Virology*. Those in which emphasis is clearly on the cell, with the virus being incidental, are appropriate for MCB.

The *Journal of Bacteriology* is a journal of basic microbiology, including that of eukaryotic microbes. The scope statements for MCB and JB are complementary. They provide authors with appropriate journals for the publication of research covering all aspects of eukaryotic microbiology.

Direct questions to editor in chief of journal being considered. Note that a manuscript rejected by one ASM journal on scientific grounds or on the basis of its general suitability for publication is considered rejected by all other ASM journals.

MANUSCRIPT SUBMISSION

Submit manuscript to ASM office. Include a cover letter stating: journal to which manuscript is being submitted; most appropriate section of journal; complete mailing address (including street), telephone and fax numbers of corresponding author; an electronic mail address; and former ASM manuscript number and year if a resubmission. Include written assurance that permission to cite personal communications and preprints has been granted.

Submit three complete copies of each manuscript, including figures and tables. Enclose three copies of each "in press" and "submitted" manuscripts.

Manuscripts must represent reports of original research. All authors must have agreed to its submission and are responsible for its content, including appropriate citations and acknowledgments, and must also have agreed that corresponding author has authority to act on their behalf on all matters pertaining to publication of manuscript. By submission, authors guarantee that manuscript, or one substantially same, was not published previously, is not being considered or published elsewhere, and was not rejected on scientific grounds by another ASM journal.

Corresponding author must sign copyright transfer agreement on behalf of all authors. Agreement is sent when manuscript is accepted and scheduled for publication.

If all authors were employed by U.S. government when work was performed, corresponding author should not sign copyright transfer agreement but should, instead, attach to agreement statement attesting that manuscript was prepared as part of their official duties and, as such, is a work of U.S. government not subject to copyright.

If some authors were employed by the U.S. government when work was performed but others were not, the corresponding author should sign copyright transfer agreement as it applies to that portion performed by nongovernment employee authors.

The American Society for Microbiology defines primary publication as in *How to Write and Publish a Scientific Paper* (3rd ed, R. A. Day): ". . . (i) the first publication of original research results, (ii) in a form whereby peers of the author can repeat experiments and test conclusions, and (iii) in a journal or other *source document* [emphasis added] readily available within the scientific community."

A scientific paper published in a conference report, symposium proceeding, technical bulletin, or any other retrievable source is unacceptable for submission on grounds of prior publication. Preliminary disclosure of research findings published in abstract form as an adjunct to a meeting is not considered "prior publication."

Acknowledge prior publication of data contained in manuscript even if such publication is not considered to be in violation of ASM policy. Submit copies of relevant work.

An author is one who made a substantial contribution to "overall design and execution of experiments;" all authors are responsible for entire paper. Do not list individuals who provided assistance, e.g., supplied strains or reagents or critiqued paper, as authors; recognize in Acknowledgments. Authors must agree to order in of names in byline. Do not use footnotes to attribute work (e.g., X. Jones and Y. Smith contributed equally to. . .). If necessary, include in Acknowledgment section.

MANUSCRIPT FORMAT

Type double-spaced (6 mm between lines), including figure legends, table footnotes, and References. Number all pages in sequence, including abstract, figure legends, and tables. Place last two items after References section. Make margins at least 1 in. on all four sides. Make these sets of characters easily distinguishable in manuscript: numeral zero (0) and letter "oh" (O); numeral one (1), letter "el" (l), and letter "eye" (I); and multiplication sign ($\times$) and letter "ex" (x). If distinctions cannot be made, mark items at first occurrence for cell lines, strain and genetic designations, viruses, etc., on modified manuscript to be identified for printer by copy editor.

MANUSCRIPT FORMAT

Title, Running Title, and Byline: Present results of an independent, cohesive study; numbered series titles are not allowed. Avoid main title/subtitle arrangement, complete sentences, and unnecessary articles. On title page include: title; running title (54 characters and spaces); name of each author; address(es) of institution(s) at which work was done; each author's affiliation; and a footnote indicating present address of any author no longer at that institution. Place asterisk after name of corresponding author, and give telephone and fax numbers. If electronic mail address of corresponding author is supplied on title page, it will be included as a footnote in published article.

Abstract. 250 words concisely summarizing basic content of paper without presenting extensive experimental details. Avoid abbreviations; do not include diagrams. When essential to include a reference, use References citation but omit article title. Make complete and understandable without reference to text.

Introduction: Supply sufficient background information to allow reader to understand and evaluate results of present study without referring to previous publications. Provide rationale for present study. Use only those references required to provide most salient background rather than an exhaustive review of topic.

Materials and Methods: Include sufficient technical information to allow experiments to be repeated. For centrifugation conditions, give enough information to enable repetition of procedure: make of centrifuge, model of rotor, temperature, time at maximum speed, and centrifugal force ($\times g$ rather than revolutions per minute). For commonly used materials and methods, a simple reference is sufficient. If several alternative methods are commonly used, identify method briefly and cite reference. Allow readers to assess method without constant reference to previous publications. Describe new methods completely; give sources of unusual chemicals, equipment, or microbial strains. When large numbers of microbial strains or mutants are used, include tables identifying sources and properties of strains, mutants, bacteriophages, plasmids, etc.

Describe a method, strain, etc., used in only one of several experiments reported in paper in Results section or in one or two sentences in table footnote or figure legend.

Results: Reserve interpretation of results for Discussion section. Present as concisely as possible in text, table(s), or figure(s). Avoid extensive use of graphs to present data more concisely presented in text or tables. For example, do not present double-reciprocal plots used to determine apparent K_m values as graphs; state values in text. Do not show graphs illustrating other methods commonly used to derive kinetic or physical constants (e.g., reduced viscosity plots, plots used to determine sedimentation velocity). Limit photographs (particularly photomicrographs and electron micrographs) to those absolutely necessary to show experimental findings. Number figures and tables in order cited in text; be sure to cite all figures and tables.

Discussion: Interpret results in relation to previously published work and to experimental system at hand. Do not repeat Results section or reiterate introduction. In short papers, combine Results and Discussion sections.

Acknowledgments: Indicate sources of financial support. Format for acknowledgment of grant support: "This work was supported by Public Health Service grant CA01234 from the National Cancer Institute." Recognize personal assistance in a separate paragraph.

Appendixes: Contain supplementary material to aid reader. Titles, authors, and References sections may not be distinct from those of primary article. List author(s) of Appendix in byline or Acknowledgment section of primary article. Label equations, tables, and figures with letter "A" preceding numeral to distinguish them from those in main body of text.

References: Include all relevant sources; cite all listed references in text. Arrange citations in alphabetical order by first author; number consecutively. Abbreviate journal names according to *Serial Sources for the BIOSIS Data Base* (BioSciences Information Service, 1992). Cite each listed reference in text by number.

1. **Clutterbuck, A. J., and D. J. Cove.** 1974. Linkage map of *Aspergillus nidulans,* p. 665–676. *In* A. I. Laskin and H. A. Lechevalier (ed.), Handbook of microbiology, vol. 4. CRC Press, Cleveland, Ohio.

2. **Cox, C. S., B. R. Brown, and J. C. Smith.** J. Gen. Genet., in press.*

3. **Dhople, A., I. Ortega, and C. Berauer.** 1989. Effect of oxygen on in vitro growth of *Mycobacterium leprae,* abstr. U-82, p. 168. Abstr. 89th Annu. Meet. Am. Soc. Microbiol. 1989.

4. **Fitzgerald, G., and D. Shaw.** *In* A. E. Waters (ed.), Clinical microbiology, in press. EFH Publishing Co., Boston.

5. **Gustlethwaite, F. P.** 1985. Letter. Lancet **ii:**327.

6. **Jacoby, J., R. Grimm, J. Bostic, V. Dean, and G. Starke.** Submitted for publication.

7. **Jensen, C., and D. S. Schumacher.** Unpublished data.

8. **Jones, A. (Yale University).** 1990. Personal communication.

9. **Liebman, S. W., J. W. Stewart, J. H. Parker, and F. Sherman.** 1977. Leucine insertion caused by a yeast amber suppressor. J. Mol. Biol. **109:**13–22.

10. **Powers, R. D., W. M. Dotson, Jr., and F. G. Hayden.** 1982. Program Abstr. 22nd Intersci. Conf. Antimicrob. Agents Chemother., abstr. 448.

11. **Sigma Chemical Co.** 1989. Sigma manual. Sigma Chemical Co., St. Louis, Mo.

12. **Smith, B. A., and D. D. Burke.** 1980. Protein synthesis during germination of *Allomyces macrogynus* mitospores. J. Bacteriol. **143:**1498–1500.

13. **Smith, J. C.** April 1970. U.S. patent 484,363,770.

14. **Smyth, D. R.** 1972. Ph.D. thesis. University of California, Los Angeles.

15. **Yagupsky, P., and M. A. Menegus.** 1989. Intraluminal colonization as a source of infection. Antimicrob. Agents Chemother. **33:**2025. (Letter.)

Include a control number with "in press" references to ASM publications or name of publication if a book.

Illustrations: Write figure number and authors' names on all figures, either in margin or on back (marked lightly with a soft pencil). For micrographs especially, indicate TOP. Do not use paper clips. Put small figures in an envelope. Do not submit illustrations larger than 8.5×11 in.

Continuous-Tone and Composite Photographs—Journal page width: 3 5/16 in.(single column) and 6 7/8 in. (double column, maximum). Include only significant portions of illustrations. Photos must be of sufficient contrast to withstand loss of contrast and detail in printing. Submit one photograph of each continuous-tone figure for each copy of manuscript; photocopies are not acceptable. If possible, submit figures in final published size, so no reduction is necessary. If must be reduced, make sure all elements, including labeling, can withstand reduction and remain legible.

If a figure is a composite of a continuous-tone photograph and a drawing or labeling, provide original composite for printer, labeled "printer's copy."

Electron and light micrographs must be direct copies of original negative. Indicate magnification with a scale marker on each.

Computer-Generated Images—Authors may provide a floppy disk with stored figures. Submit in either tagged image file format (TIFF) or Encapsulated PostScript (EPS) in either MS-DOS or Macintosh format. Use Adobe PhotoShop or Aldus Freehand. Adobe users check densities of images online. If image's shadow density is below 1.25, enter density as 1.4; if 1.25-1.6, enter as 1.65; and if >1.65, enter as actual reading. For Aldus, one and two-column art cannot exceed 20 picas (3 5/16 in.) and 41.5 picas (6 7/8 in.), respectively. Use Helvetica (medium or bold) or Times Roman text font. Canvas software does not allow output in TIFF or EPS. Select "Publish" option and then "Illustrator 88" option to prepare output. Two files are generated. Copy and submit both. Do not use rotated type for labeling.

Supply disk and extra printed copies of figures as they appear on disk upon submission. Indicate type of software used and number of images stored. Include all labeling, including panel identification letters. Indicate figure number and author names in pencil on back

For large images, use 40- or 80-megabyte Syquest cartridges or magneto-optical cartridges. For transfer from UNIX systems, submit either 9-track or 8-mm "tar" archives. Incorporate all final lettering, labeling, tooling, etc., in final supplied material. It cannot be added later. Do not include figure numbers on images.

Electronic image telephone hotline: ASM printer Cadmus Journal Service 800-257-5531, ext. 3361; non-U.S. (804)261-3000, ext. 3361; e-mail: image@cadmus.com.

Since contents of computer-generated images can be manipulated, describe software/hardware used in figure legend(s).

Color Photographs—If necessary, include an extra copy at submission so that a cost estimate may be obtained. Cost of printing is borne by author.

Drawings—Submit graphs, charts, sequences, complicated chemical or mathematical formulas, diagrams, and other drawings as glossy photographs made from finished drawings not requiring additional artwork or typesetting. Computer-generated graphics produced on high-quality laser printers are also usually acceptable. No part of graph or drawing should be handwritten. Label both axes of graphs. Most graphs will be reduced to one-column width (3 5/16 in.); make all elements large enough to withstand reduction. Avoid heavy letters and unusual symbol.

In figure ordinate and abscissa scales (and table column headings), avoid ambiguous use of numbers with exponents. Use appropriate SI symbols (μ for 10^{-6}, m for 10^{-3}, k for 10^3, M for 10^6, etc.). For a complete listing of SI symbols, see IUPAC "Manual of Symbols and Terminology for Physicochemical Quantities and Units" (*Pure Appl. Chem.* **21:**3–44, 1970).

Where powers of 10 must be used, show exponent power associated with number. In representing 20,000 cells per ml, numeral on ordinate is "2" and label is "10^4 cells per ml" (not "cells per ml $\times 10^{-4}$"). Show an enzyme activity of 0.06 U/ml as 6, with label 10^{-2} U/ml. Preferred designation is 60 mU/ml (milliunits per milliliter).

Figure Legends: Provide enough information so that figure is understandable without reference to text. Describe detailed experimental methods in Materials and Methods section, not in figure legend. Report a method unique to one of several experiments in a legend only if discussion is only one or two sentences. Define all symbols and abbreviations used in figure not defined elsewhere.

Tables: Type each on a separate page. Arrange data so columns of like material read down, not across. Make headings clear so meaning of data will be understandable without reference to text. Explanatory footnotes are acceptable, extensive table "legends" are not. Do not include in Footnotes detailed descriptions of experiment. Include enough information to warrant table format; incorporate those with fewer than six pieces of data into text. Submit camera-ready if long and complicated and when division of material and spacing are important. Do not hand letter; carefully prepare to conform with the style of the journal.

OTHER FORMATS

ERRATA—Provides a means of correcting errors in published articles. Changes in data and addition of new material are not permitted. Send to Journals Division.

LETTERS TO THE EDITOR—500 words, Include data to support writer's argument. Comment on articles published previously in *Journal*. Send to Journals Division. A reply from corresponding author of article may be solicited. Type double-spaced.

OTHER

Abbreviations: Use as an aid to reader, rather than as a convenience for author; limit use. Use abbreviations other than those recommended by the IUPAC-IUB (*Biochemi-*

cal Nomenclature and Related Documents, 1978) only when necessary, such as in tables and figures.

Use pronouns or paraphrase a long word after its first use (e.g., "the drug," "the substrate"). Use standard chemical symbols and trivial names or their symbols (folate, Ala, Leu, etc.) for terms that appear in full in neighboring text.

Define each abbreviation and introduce in parentheses at first time use. Eliminate abbreviations not used at least five times in text (including tables and figure legends).

Not requiring Introduction—In addition to abbreviations for Système International d'Unités (SI) units of measurement, other common units (e.g., bp, kb, and Da), and chemical symbols for the elements, use the following without definition in title, abstract, text, figure legends, and tables: DNA (deoxyribonucleic acid); cDNA (complementary DNA); RNA (ribonucleic acid); cRNA (complementary RNA); RNase (ribonuclease); DNase (deoxyribonuclease); rRNA (ribosomal RNA); mRNA (messenger RNA); tRNA (transfer RNA); AMP, ADP, ATP, dAMP, ddATP, GTP, etc. (for the respective 5′ phosphates of adenosine and other nucleosides) (add 2′-, 3′-, or 5′- when needed for contrast); ATPase, dGTPase, etc. (adenosine triphosphatase, deoxyguanosine triphosphatase, etc.); NAD (nicotinamide adenine dinucleotide); NAD^+ (nicotinamide adenine dinucleotide, oxidized); NADH (nicotinamide adenine dinucleotide, reduced); NADP (nicotinamide adenine dinucleotide phosphate); NADPH (nicotinamide adenine dinucleotide phosphate, reduced); $NADP^+$ (nicotinamide adenine dinucleotide phosphate, oxidized); poly(A), poly(dT), etc. (polyadenylic acid, polydeoxythymidylic acid, etc.); oligo(dT), etc. (oligodeoxythymidylic acid, etc.); P_i (orthophosphate); PP_i (pyrophosphate); UV (ultraviolet); PFU (plaque-forming units), CFU (colony-forming units); MIC (minimal inhibitory concentration); MBC (minimal bactericidal concentration); Tris [tris(hydroxymethyl)aminomethane]; DEAE (diethylaminoethyl); A_{260} (absorbance at 260 nm), EDTA (ethylenediaminetetraacetic acid); and AIDS (acquired immunodeficiency [or immune deficiency] syndrome). Abbreviations for cell lines (e.g., HeLa) need not be defined.

Use following abbreviations without definition in tables:

amt (amount)	SD (standard deviation)
approx (approximately)	SE (standard error)
avg (average)	SEM (standard error of the mean)
concn (concentration)	sp act (specific activity)
diam (diameter)	sp gr (specific gravity)
expt (experiment)	temp (temperature)
exptl (experimental)	tr (trace)
ht (height)	vol (volume)
mo (month)	vs (versus)
mol wt (molecular weight)	wk (week)
no. (number)	wt (weight)
prepn (preparation)	yr (year)

Experimentation, Human: Do not identify isolates derived from patients in clinical studies with patients' initials, even as part of a strain designation. Change initials to numerals or use randomly chosen letters. Do not give hospital unit numbers; if a designation is needed, use only last two digits of unit. Established designations of some viruses and cell lines, although they consist of initials, are acceptable (e.g., JC virus, BK virus, HeLa cells).

Isotopically Labeled Compounds: For simple molecules, indicate isotopic labeling in chemical formula (e.g., $^{14}CO_2$, ^3H_2O, $H_2^{35}SO_4$). Do not use brackets when isotopic symbol is attached to name of a compound that in its natural state does not contain element (e.g., ^{32}S-ATP) or to a word which is not a specific chemical name (e.g., ^{131}I-labeled protein, ^{14}C-amino acids, ^3H-ligands, etc.).

For specific chemicals, place symbol for isotope introduced in square brackets directly preceding part of name that describes labeled entity. Configuration symbols and modifiers precede isotopic symbol. Correct usage:

[^{14}C]Urea	[γ-32P]ATP
L-[*methyl*-^{14}C]methionine	UDP-[U-^{14}C]glucose
[2,3-^3H]serine	*E. coli* [^{32}P]DNA
[α-^{14}C]lysine	fructose 1,6-[1-^{32}P] bisphosphate

Follow conventions for isotopic labeling of the *Journal of Biological Chemists*. More detailed information can be found in instructions to authors of that journal (first issue of each year).

Materials Sharing: Make DNAs, viruses, microbial strains, cell lines, antibodies, and similar materials newly described in article available from a national collection or in a timely fashion, at reasonable cost, and in limited quantities directly to members of the scientific community for noncommercial purposes. Failure to comply may result in a suspension of publishing privileges in ASM journals for up to 5 years.

Nomenclature, Chemical and Biochemical: For names of chemical compounds see *Chemical Abstracts* (Chemical Abstracts Service, Ohio State University, Columbus) and its indexes. *The Merck Index* (11th ed., 1989; Merck & Co., Inc., Rahway, N.J.) is also an excellent source. For use of biochemical terminology, consult *Biochemical Nomenclature and Related Documents* (1978; reprinted for The Biochemical Society, London) and instructions to authors of *Journal of Biological Chemistry* and *Archives of Biochemistry and Biophysics* (first issues of each year).

Do not express molecular weights in daltons; molecular weight is a unitless ratio. Express molecular mass in daltons.

For enzymes, use recommended (trivial) name assigned by Nomenclature Committee of the International Union of Biochemistry (*Enzyme Nomenclature*, Academic Press, Inc., 1992). If a nonrecommended name is used, place proper (trivial) name in parentheses at first use in abstract and text. Use EC number when assigned. Express enzyme activity either in katals (preferred) or in micromoles per minute.

Nomenclature, Genetic:

Prokaryotes—Describe genetic properties of prokaryotes in terms of phenotypes and genotypes. Phenotype describes observable properties of an organism. Genotype refers to genetic constitution of an organism, usually in reference to some standard wild type. Follow recommendations of Demerec et al. (*Genetics* **54**:61–76, 1966) and current practices of the *Journal of Bacteriology*.

Employ phenotype designations when mutant loci have not been identified or mapped. Use to identify the protein product of a gene. Phenotype designations generally consist of three-letter symbols; do not italicize; capitalize first letter. Designate wild-type characteristics with a superscript plus. When necessary for clarity, use negative superscripts to designate mutant characteristics. Use lowercase superscript letters to further delineate phenotypes. Define phenotype designations.

Indicate genotype designations by three-letter symbol in lowercase italic. If several loci govern related functions, distinguish by an italicized capital letter following locus symbol. Distinguish mutation sites by placing serial isolation numbers (allele numbers) after locus symbol. Indicate promoter, terminator, and operator sites as described by Bachmann and Low (*Microbiol. Rev.* **44**: 1–56, 1980): e.g., *lacZp, lacAt,* and *lacZo.* In papers reporting isolation of new mutants, give allele numbers to mutations. *Escherichia coli* number registry: *E. coli* Genetic Stock Center, Department of Human Genetics, Yale University School of Medicine, P.O. Box 3333, New Haven, CT 06510. *Salmonella* registry: *Salmonella* Genetic Stock Center, Department of Biology, University of Calgary, Calgary, Alberta, Canada T2N 1N4.

Indicate wild-type alleles with a superscript plus. Do not use superscript minus to indicate a mutant locus; refer to an *ara* mutant rather than an *ara⁻* strain.

Avoid use of superscripts with genotypes (other than + to indicate wild-type alleles). Designations indicating amber mutations (Am), temperature-sensitive mutations (Ts), constitutive mutations (Con), cold-sensitive mutations (Cs), and production of a hybrid protein (Hyb) follow allele number. Define all other designations of phenotype at first occurrence. If superscripts must be used, approval by editor is needed; define at first occurrence.

Use subscripts to distinguish between genes (with the same name) from different organisms or strains, e.g., $his_{E. coli}$ or his_{K-12} for the *his* gene of *E. coli* or strain K-12 in another species or strain, respectively. Use an abbreviation if it is explained. Use a subscript to distinguish between genetic elements with the same name.

Avoid use of a genotype as a name. If no strain designation, select an appropriate word combination.

Viruses—Viruses usually have no phenotype; they have no metabolism outside host cells. Distinctions between phenotype and genotype are not made. Use superscripts to indicate hybrid genomes. Genetic symbols may be one, two, or three letters. Delineate host DNA insertions into viruses by square brackets. Genetic symbols and designations for such inserted DNA conform to those used for host genome.

Eukaryotes—Conform to current practices in identifying mutants and their genotypes. Consult *Handbook of Micro-*

biology (A. I. Laskin and H. A. Lechevalier, ed., CRC Press, 1974) or the *Handbook of Genetics* (vol. 1, *Bacteria, Bacteriophages, and Fungi,* R. C. King, ed., Plenum Publishing Corp., 1974) for designations currently in use for *Aspergillus nidulans, Schizosaccharomyces pombe, Podospora anserina, Ustilago sp., Schizophylum commune, Coprinus* sp., and *Chlamydomonas reinhardtii.*

For genetic designations for *Saccharomyces cerevisiae* follow recommendations of Sherman and Lawrence (*Handbook of Genetics,* vol. 1). Designate the two mating-type alleles by a boldface Roman "**a**" and a Greek "α." Designate gene symbols with three italicized letters. Identify genetic locus by an Arabic numeral following gene symbol; designate alleles by an Arabic numeral separated from locus number by a hyphen. Identify complementation groups of a gene by capital letters following locus number. Denote dominant and recessive genes by upper- and lowercase letters, respectively. When no confusion, designate wild-type genes as +; the + may follow locus number to designate a specific wild-type gene. Avoid superscript, but sometimes distinguish genes conferring resistance and sensitivity with superscripts r and s, respectively. Distinguish mitochondrial and non-Mendelian genotypes from chromosomal genotypes by enclosure in square brackets. When applicable, use rules for designating non-Mendelian genes. Avoid use of Greek letters and use transliterations.

For current linkage map of *S. cerevisiae* consult Mortimer and Schild, (*Microbiol. Rev.* **49**:181–213, 1985). For current list of *N. crassa* chromosomal loci consult Perkins et al. (*Microbiol. Rev.* **46**:426–570, 1982).

As indicated in *CBE Style Manual,* italicize symbols for *Drosophila* mutations and chromosome aberrations; do not include Greek letters, subscripts, or spaces. Do not italicize spelled-out names of mutants. Symbols for mutant types are abbreviations of their characterizing names. Symbol begins with first letter of its name; convention designates an initial capital letter for a dominant (*R* for roughened) and an initial lowercase letter for a recessive (*r* for rudimentary or *ry* for rosy).

Strain Designations—Do not use a genotype as a name. If a strain designation has not been chosen, select an appropriate word combination.

Transposable Elements, Plasmids, and Restriction Enzymes—For nomenclature of transposable elements (insertion sequences, transposons, phage Mu, etc.) follow recommendations of Campbell et al. (*Gene* **5**:197–206,1979), with modifications in instructions to authors in *Journal of Bacteriology.* Use system of designating transposon insertions at sites where there are no known loci, e.g., *zef123*:TnS, described by Chumley et al. (*Genetics* **91**:639–655, 1979). Use nomenclature recommendations of Novick et al. (*Bacteriol. Rev.* **40**: 168–189, 1976) for plasmids and plasmid-specified activities, of Low (*Bacteriol. Rev.* **36**: 587–607, 1972) for F-prime factors, and of Roberts (*Nucleic Acids Res.* **17**: r347–r387, 1989) for restriction enzymes and their isoschizomers. Recombinant DNA molecules constructed in vitro follow nomenclature for insertions in general. Describe DNA inserted into recombinant DNA molecules using gene symbols and conventions for organism from which DNA was obtained.

The Plasmid Reference Center (E. Lederberg, Plasmid Reference Center Department of Microbiology and Immunology, 5402, Stanford University School of Medicine, Stanford, CA 94305-2499) assigns Tn and IS numbers to avoid conflicting and repetitive use and also clears nonconflicting plasmid prefix designations.

Nomenclature, Microorganisms: Use binary names, consisting of a generic name and a specific epithet, for all microorganisms. Names of higher categories may be used alone, specific and subspecific epithets may not. Precede a specific epithet by a generic name at first use. Thereafter, abbreviate generic name to initial capital letter, provided there is no confusion with other genera in paper. Print names of all taxa (phyla, classes, orders, families, genera, species, subspecies) in italics and underline (or italicize) in manuscript; do not italicize or underline strain designations and numbers. For spelling of names follow *Approved Lists of Bacterial Names* (amended edition, V. B. D. Skerman, V. McGowan, and P. H. A. Sneath, ed.) and *Index of the Bacterial and Yeast Nomenclatural Changes Published in the International Journal of Systematic Bacteriology since the 1980 Approved Lists of Bacterial Names (I January 1980 to I January 1989)* (W. E. C. Moore and L. V. H. Moore, ed., American Society for Microbiology, 1989), and validation lists and articles published in *International Journal of Systematic Bacteriology* since 1 January 1989. If there is reason to use a name that does not have standing in nomenclature, enclose name in quotation marks and make an appropriate statement concerning nomenclatural status of name in text (for an example, see *Int. J. Syst. Bacteriol.* 30:547–556,1980).

Since classification of fungi is incomplete, determine accepted binomial for a given organism. Sources for names include *The Yeasts: a Taxonomic Study* (3rd ed., N. J. W. Kreger-van Rij, ed., Elsevier Science Publishers B.V., Amsterdam, 1984) and *Ainsworth and Bisby's Dictionary of the Fungi, Including the Lichens* (7th ed., Commonwealth Mycological Institute, Kew, Surrey, England, 1983). Use names for viruses approved by International Committee on Taxonomy of Viruses (ICTV), published in 4th Report of the ICTV, Classification and Nomenclature of Viruses (*Intervirology* 17:23–199,1982), with modifications in the 5th Report of the ICTV (*Arch. Virol.*, Suppl. 2, 1991). If desired, add synonyms parenthetically when name is first mentioned. Use approved generic (or group) and family names. Give microorganisms, viruses, and plasmids designations consisting of letters and serial numbers. Include a worker's initials or a descriptive symbol of locale, laboratory, etc., in designation. Give each new strain, mutant, isolate, or derivative a new (serial) designation. Designation should be distinct from those of genotype and phenotype; do not include genotypic and phenotypic symbols.

Nucleic Acid Sequences: Present nucleic acid sequences of limited length which are primary subject of a study freestyle in most effective format. Present longer nucleic acid sequences in following format to conserve space. Submit sequence as camera-ready copy of dimensions 8.5 × 11 in. (or slightly less) in standard (portrait) orientation. Print sequence in lines of approximately 100–120 nucleotides in

a nonproportional (monospace) font that is easily legible when published with a line length of 6 inches. If possible, further subdivide lines of nucleic acid sequence into blocks of 10 or 20 nucleotides by spaces within the sequences or by marks above it. Use uppercase and lowercase letters to designate exon/intron structure, transcribed regions, etc., if lowercase letters remain legible at a 6 in. line length. Number sequence line by line; place numerals, representing first base of each line, to left of lines. Minimize spacing between adjacent lines of sequence, leaving room only for annotation of sequence. Annotation may include boldface, underlining, brackets, boxes, etc. Present encoded amino acid sequences, if necessary, immediately above first nucleotide of each codon, using single-letter amino acid symbols. Comparisons of multiple nucleic acid sequences should conform as nearly as possible to same format.

Nucleotide Sequences: Include GenBank/EMBL accession numbers for primary nucleotide and/or amino acid sequence data as a separate paragraph at end of Materials and Methods section in original manuscript or insert when manuscript is modified.

GenBank may be contacted at: GenBank Submissions, Mail Stop K710, Los Alamos National Laboratory, Los Alamos, NM 87545; telephone: (505) 665-2177; electronic mail (submissions): gb-sub%life@lanl.gov, gb-sub@genome.lanl.gov, or gb-sub@life.lanl.gov. The EMBL Data Library may be contacted at: EMBL Data Library Submissions, Postfach 10.2209, Meyerhofstrasse 1, 6900 Heidelberg, Germany; telephone: 011 49 (6221) 387258; fax: 011 49 (6221) 387306; electronic mail (data submissions): datasubs@embl.bitnet. See nucleic acid sequence for formatting instructions.

Numerical Data: Use standard metric units for reporting length, weight, and volume. For these units and for molarity, use prefixes m, μ, n, and p for 10^{-3}, 10^{-6}, 10^{-9}, and 10^{-12}, respectively. Use prefix k for 10^3. Avoid compound prefixes such as mμ or μμ. Use μg/ml or μg/g in place of the ambiguous ppm. Present units of temperature as 37°C or 324 K.

When fractions are used to express units such as enzymatic activities, use whole units, such as "g" or "min," in denominator instead of fractional or multiple units, such as μg or 10 min. For example, "pmol/min" would be preferable to "nmol/10 min," and "μmol/g" would be preferable to "nmol/μg." It is also preferable that an unambiguous form such as exponential notation be used; for example, "μmol g^{-1} min^{-1}" is preferable to "μmol/g/min."

See *CBE Style Manual,* 5th ed., for more detailed information about reporting numbers. Also contained in this source is information on SI units for reporting illumination, energy, frequency, pressure, and other physical factors. Always report numerical data in applicable SI units.

Page Charges: $55 per page paid by authors whose research was supported by special funds, grants (departmental, governmental, institutional, etc.), or contracts or whose research was done as part of their official duties.

If research was not supported, submit a request to waive charges to Journals Division, American Society for Microbiology, 1325 Massachusetts Ave., N.W., Washington, D.C. 20005-4171, with submitted manuscript. This request, separate from cover letter, must indicate how work was supported and should be accompanied by a copy of Acknowledgment section. Letters to the Editor are not subject to page charges.

Permissions: Corresponding author must obtain permissions from both original publisher and original author [i.e., copyright owner(s)] to reproduce figures, tables, or text (in whole or in part) from previous publications. Submit signed permissions to ASM, and identify each as to the relevant item in the ASM manuscript (e.g., "permissions for Fig. 1 in MCB 123-93").

Reviewers: Suggest an appropriate editor for new submissions. If we are unable to comply with such a request, corresponding author will be notified.

Style: Conform to *CBE Style Manual* (5th ed., 1983; Council of Biology Editors, Inc., Bethesda, MD), *ASM Style Manual for Journals and Books* (American Society for Microbiology, 1991), and Robert A. Day' s *How to Write and Publish a Scientific Paper* (3rd ed., 1988; Oryx Press), as interpreted and modified by the editors and the ASM Journals Division.

Use past tense to narrate events in the past, including procedures, observations, and data of study reported. Use present tense for personal general conclusions, conclusions of previous researchers, and generally accepted facts. Most of abstract, Materials and Methods, and Results sections will be in past tense, and most of introduction and some of Discussion will be in present tense. It may be necessary to vary tense in a single sentence.

For an in-depth discussion of tense in scientific writing, see *How to Write and Publish a Scientific Paper* (3rd ed., p. 158-160).

SOURCE

January 1995, 15(1): i-xi.

Molecular and Cellular Endocrinology
Elsevier

Professer B.A. Cooke
Department of Biochemistry
Royal Free Hospital School of Medicine
Rowland Hill Street
London NW3 2PF, U.K.
Telephone: (+44-71) 7940500 ext. 4202/4989;
Fax: (+44-7 1) 7945029
Dr. E.R. Simpson
Cecil H. and Ida Green Center for Reproductive
   Biology Sciences
The University of Texas Southwestern Medical Center
   at Dallas, 5323 Harry Hines Boulevard
Dallas, TX 75235-9051
Telephone: (+1-214) 688-3260; Fax: (+1-214) 688-8683
Professor J.W. Funder
Baker Medical Research Institute
P.O. Box 348
Prahran, Victoria 3181, Australia
Telephone: (+61-3) 5224333; Fax: (+61-3) 5104368
                At the Cutting Edge and Review articles

AIM AND SCOPE

Molecular and Cellular Endocrinology was established in 1974 to meet the demand for integrated publication on all aspects related to the biochemical effects, synthesis and secretions of extracellular signals (hormones, neurotransmitters, etc.) and to the understanding of cellular regulatory mechanisms involved in hormonal control.

The journal is fulfilling this aim by publishing full-length original research papers, rapid papers, invited reviews, At the Cutting Edge essays, and book reviews.

The scope encompasses all subjects related to biochemical and molecular aspects of endocrine research and cell regulation. These include (1) mechanisms of action of extracellular signals (hormones, neurotransmitters, etc.), (2) interaction of these factors with receptors, (3) generation, action and role of intracellular signals such as cyclic nucleotides and calcium, (4) hormone-regulated gene expression, (5) structure and inactivation of hormones, neurotransmitters, etc., (7) hormonal control of differentiation, (8) related control mechanisms in non-mammalian systems, (9) methodological and theoretical aspects related to hormonal control processes, (10) clinical studies as far as they throw new light on basic research in this field, (11) control of intermediary metabolism at the cellular level, (12) ultrastructural aspects related to hormone secretion and action.

MANUSCRIPT SUBMISSION

Manuscirpts must be contributed solely to this journal, and may not have been and will not be published in whole or in part in any other journal or book. Preferred medium of final submission is on disk with accompanying reviewed and revised manuscript. Submit manuscripts in triplicate.

MANUSCRIPT FORMAT

Type text, double- or triple-spaced with liberal margins on all four sides. If a dot-matrix printer is used, typeface must be near-letter quality. Follow rules defined in "Information for Contributors" for *Biochimica et Biophysica Acta* (available from BBA Editorial Office, P.O. Box 1345, 1000 BH Amsterdam). Define unfamiliar terms, arbitrary abbreviations, and trade names when first used. Avoid unnecessary abbreviations and symbols. For drugs, use generic names; mention trade names in parentheses the first time name appears in text. Identify all items that should appear in italics (e.g., species and gene names) by underlining. Place super- and subscripts clearly above or below typeline.

Title Page: Include title of article (85 characters, including spaces), author's initials, names, hospital and academic affiliations. As a footnote, give corresponding author's full address, telephone and fax numbers.

Summary: Include a summary (150 words) at beginning of article. Mention essential information contained in article, plus authors' conclusions.

Key Words: List 3-6 key (indexing) words above abstract. Use American spelling; avoid plural terms. Use terms indicating origin of material studied as sub-keywords, i.e., within parentheses, e.g., (Rat testis), (Sertoli cell).

Tables: Give short informative titles. Number consecutively in Arabic numerals. Submit each table plus its caption on a separate sheet.

Figures: Submit as sharp original drawings, or as well-contrasting, unmounted photographs on glossy paper. Make lettering proportional to figure size, to ensure legibility after reduction. Use following symbols : △ ▽ ◇ ○ ⊙ ⊖ ⊕ □ ▲ ▼ ◆ ● ■ * × +.

Prepare figures for either one column (84 mm) or entire page width (177 mm). Maximum height is 232 mm. Write name of (first) author on back in pencil, and an arrow indicating TOP. Number consecutively in Arabic numerals.

References: Use Harvard system; give names and data in body of text and an alphabetical list of references at end of manuscript. References in text give author's surname with year of publication in parentheses. When reference is to a work by two authors, give both names; reference work by more than two authors with name of first author, followed by et al., e.g., Goustin et al. (1986). If several papers by the same author(s) and from the same year are cited, place a, b, c, etc. after year of publication.

Abbreviate journal titles according to *List of Serial Title Word Abbreviations* (International Serials Data System, Paris). If a publication is in press, make reference as complete as possible. stating name of journal and adding 'in press.' Do not include personal communications and submitted manuscripts not yet in press. Type list of references triple-spaced in alphabetical order of first author's names.

JOURNAL ARTICLES

References to articles from journals mention, in order: author(s) name(s), year of publication in parentheses. name of journal, volume number, first and last page numbers.

> Beisbroeck, R., Oram, J.F., Albers, J.l. and Bierman, E.L. (1983) J. Clin. Invest. 71, 525–539.

BOOKS

References to articles from books (or whole books) mention, in order: author(s) name(s), year of publication in parentheses, title of book (volume number if applicable), name(s) of editor(s) in brackets, first and last relevant page numbers, name and location of publisher.

> Appenzeller, O. (1986) Clinical Autonomic Failure, pp. 217-230, Elsevier, Amsterdam. Batenburg, J.J. (1984) in Pulmonary Surfactant (Robertson, B., Van Golde, L.M.G. and Batenburg, J.J., eds.), pp. 237–270, Elsevier, Amsterdam.

Acknowledgments: Place at end of text, on a separate sheet.

OTHER FORMATS

AT THE CUTTING EDGE—2000 word essays on rapidly developing areas in contemporary endocrinology. Inferences from data presented in abstract form, or as personal communications, are welcome as pointers to directions in which a particular field in evolving. Provide information on reprints and address requested. Provide a statement that copyright of article is transferred to publisher in event of acceptance. If transfer is not possible, state reasons.

CRITICAL, SHORT REVIEW ARTICLES—Sponsored by a member of the Editorial Board.

RAPID PAPERS—Four printed pages, including figures, tables, and references. Short, complete and essentially final reports. Paper must be of sufficient immediate importance to the work of other investigators to justify urgent publication. Provide information on reprints and address for requests. Provide a statement that copyright of article is transferred to publisher in event of acceptance. If transfer is not possible, state reasons.

OTHER

Electronic Submission: For initial submission, hardcopies are sufficient. For processing accepted papers electronic versions are preferred. After final acceptance, submit disk plus two, final and exactly matching printed versions. Use double density (DD) or high (HD) diskettes (3.5 or 5.25 in.). Save file saved in native format of wordprocessor program used. Label disk with name of computer and wordprocessor package used, your name, and name of file on disk. Obtain further information obtained from Publisher.

SOURCE

January 1995, 107(1): back matter.

Molecular and General Genetics
Springer International

W. Arber
Biozentrum der Universität Basel
Klingelbergstrasse 70
CH-4000 Basel, Switzerland
Fax: 41/61/2 6721 18
    Transposition in prokaryotes, restriction and modification systems
M. Ashburner
Department of Genetics
University of Cambridge
Downing Street, Cambridge CB2 3EH, U.K.
Fax: 44/2 23/33 3992
    Drosophila genetics
E. Bautz
Lehrstuhl Molekulare Genetik der Universität
Neuenheimer Feld 230
D-69120 Heidelberg, Germany
Fax: 49/62 21/56 5678
    Gene structure and function in insects and mammals
H. Böhme
Institut für Genetik und Kulturpflanzenforschung
Correusstrasse 3, D-06466 Gatersleben
Krs. Aschersleben, Germany
Fax: 49/39 48 22 80
    Plant genetics, bacterial genetics
J. Campos-Ortega, Managing Editor
P. Hardy, Assistant Editor
Universität zu Köln, Institut für Entwicklungsbiologie
D-50923 Köhn, Germany
Fax: 49/2 21/4 70 51 64
    Developmental genetics, *Drosophila* and other invertebrates

R. Devoret
Groupe d'Etude "Mutagenese et Cancerogenese"
Institut Curie-Biologie, Batiment 110
Universite Paris-Sud, F-91405 Orsay Cedex, France
Fax: 33/(1)6907 28 48
Genetic and molecular bases of repair, replication and mutagenesis

D. J. Finnegan
Department of Molecular Biology
King's Buildings, Mayfield Road
Edinburgh EH93JR, U.K.
Fax: 44/31/668 38 70
Transposable elements, genetics of *Drosophila* and other invertebrates

G. P. Georgiev
Engelhardt Institute of Molecular Biology
Russian Academy of Sciences, Vavilovstr. 32
117984, Moscow B-334, Russia
Fax: 7/09 51 35 14 05
Eukaryotic mobile elements, chromatin organization, transgenic animals

W. Goebel
Theodor-Boveri-Institut für Biowissenschaften
Lehrstuhl für Mikrobiologie (Biozentrum der
Universität Würzburg, Am Hubland, D-97074
Würzburg, Germany
Fax: 49/9 31/8 88 44 02
Genetics of archaebacteria, bacterial plasmids, molecular pathogenicity of microorganism

R. Hagemann
Department of Genetics, Martin-Luther-University
Domplatz 1, D-06108 Halle (Saale), Germany
Fax: 49/34 52 95 15
Plastid genetics, molecular genetics in plant breeding

R. G. Herrmann
Botanisches Institut der Ludwig-Maximilian Universität
Menzinger Strasse 67
D-80638 München, Germany
Fax: 49/89/17 1683
Plastid and mitochondrial gene structure, gene functions and genetics, plant somatic cell genetics

C. P. Hollenberg
Institut für Mikrobiologie, Universität Düsseldorf
Universitätsstrasse 1, D-40225 Düsseldorf, Germany
Fax: 49/2 11/3 11 53 70
Gene expression in yeast

C.A.M.J.J. van den Hondel
TNO Nutrition and Food Research
P.O. Box 5815
NL-2280 HV Rijswijk, The Netherlands
Fax: 31/15 84 39 89
Molecular genetics, gene expression and gene regulation in filamentous fungi

K. Illmensee
Institut für Genetik und Entwicklungsphysiologie

der Universität Salzburg, Hellbrunner Str. 34
A-5020 Salzburg, Austria
Fax: 43/6 22/80 44 5795
Developmental genetics—mammals, insects

K. Isono
Kobe University, Faculty of Science,
Department of Biology, Rokkodai, Kobe 657, Japan
Fax: 81/78/8 02 43 36
Bacterial genetics, genetic regulation in lower eukaryotes

B. H. Judd
Laboratory of Genetics,
NIEHS, P.O. Box 12233
Research Triangle Park, NC 27709
Fax: (919) 541-7593
Drosophila genetics, gene structure and regulation

B. J. Kilbey
Institute of Cell and Molecular Biology
University of Edinburgh
Darwin Building, Mayfield Road
Edinburgh EH93JR, U.K.
Fax: 44/31/6 68 38 70
Mechanisms of mutagenesis and DNA repair (both mainly in microbial organisms), genetics and enzymology of DNA replication

A. Kondorosi
Centre National de la Recherche Scientifique
Institut des Sciences Vegetales
Avenue de la Terasse F-91198
Gif-sur-Yvette Cedex, France
Fax: 33/(1)6 98/2 36 95
Bacterial genetics, genetics of plant-bacterium interactions, bacterial and plant gene regulation

J. Lengeler
Fachbereich Biologie/Chemie der Universität Osnabrück
Barbarastrasse 11, D-49076 Osnabrück, Germany
Fax: 49/5 41/9 69 28 70
Bacterial genetics, transport and physiology

D. M. Lonsdale
Cambridge Laboratory
John Innes Center for Plant Science Research
Colney Lane, Norwich NR4 7UJ, U.K.
Fax: 44/6 03/50 50 57 25
Mitochondrial genetics, mitochondrial DNA organization and expression

G. Melchers
Max-Planck-Institut für Biologie, Spemannstrasse 37 II
Postfach 21 09, D-72076 Tübingen, Germany
Plant cell genetics

H. Saedler
Max-Planck-Institut für Züchtungsforschung
Carl-von-Linne-Weg 10
D-50829 Köln, Germany
Fax: 49/2 21/5 06 21 13
Plant classical and molecular genetics; transposable elements

J. Schell
Max-Planck-Institut für Züchtungsforschung

Carl-von-linne-Weg 10, D-50829 Köln, Germany
Fax: 49/2 21/5 06 22 13
  Plant transformation—molecular biology, soil
  bacteria
M. Sekiguchi
ent of Biochemistry
Kyushu University 60
School of Medicine, Fukuoka, 812, Japan
Fax: 81/92/6 33 68 01
  DNA repair, replication and transcription in
  pro- and eukaryotic systems
D. Y. Thomas
Biotechnology Research Institute
National Research Council 6100, Avenue Royalmount
Montreal, Quebec, Canada H4P2R2
Fax: 1/5 14/4 96 62 13
  Gene expression in eukaryotes

AIM AND SCOPE

MGG provides publication in all areas of general and molecular genetics—developmental genetics, somatic cell genetics, and genetic engineering—irrespective of the organism. Articles on animal or plant breeding or human genetics will be accepted only if results described have significance for fundamental genetics.

Reports that are exclusively dealing with nucleic acid sequences are published only if they are judged to be of wide interest and biological significance.

MANUSCRIPT SUBMISSION

Submission implies: that work described has not been published before (except as an abstract or as part of a published lecture, review, or thesis); that it is not under consideration for publication elsewhere; that its publication has been approved by all coauthors, as well as by responsible authorities at institute where work was done; that, if manuscript is accepted for publication, authors agree to automatic transfer of copyright to publisher; and that manuscript will not be published elsewhere in any language without consent of copyright holders. Presentation should be intelligible to as wide an audience as possible.

Submit manuscripts in triplicate to an editor representing discipline involved. Submit original manuscript typed on one side of sheet. Photocopy additional copies on both sides.

MANUSCRIPT FORMAT

Submit manuscripts in final form. Insufficiently prepared manuscripts will not be considered. Type double-spaced with wide margins. If manuscript was written using a PC, also provide file on diskette. A charge will be made for changes introduced after manuscript has been set in type. Write in standard grammatical English.

Title Page: Include: title of paper, first name(s) and surname(s) of author(s), institute, footnotes referring to title (indicated by asterisks), and address to which proofs should be sent. Front page must contain telephone and fax numbers, as well as number of figures and tables.

Abstract: Include important experimental results shown in figures.

Key Words: Up to five.

Footnotes: Number consecutively, other than those referring to title.

References: List only works cited in text and those accepted for publication at end of paper in alphabetical order under first author's name. List alphabetically works by two authors according to second author's name. List chronologically works by three or more authors. If there is more than one work by the same author or team of authors in same year, add a, b, c, etc. to year both in text and in list of references. Use internationally accepted abbreviations of journal titles.

Röder MS, Sorrells ME, Tanksley SD (1992) 5S ribosomal gene clusters in wheat: Pulsed field gel electrophoresis reveals a high degree of polymorphism. Mol Gen Genet 232:215–220

Game JC (1983) Radiation-sensitive mutants and repair in yeast. In: Spencer JFT, Spencer DM, Smith ARW (eds) Yeast genetics: Fundamental and applied aspects. Springer-Verlag, New York, pp 109–137

Cite in text by author and year. If two authors, name both; if more than two, give only first author's name plus 'et al.'

Figures: Restrict to minimum needed to clarify text. Give each a short legend. Submit legends on a separate sheet. Do not repeat information in legends in text. Do not present same data in both graph and table form. Previously published illustrations are not usually accepted. Color illustrations are accepted.

Number consecutively all figures, whether photographs, graphs, or diagrams, throughout and submit on separate sheets. Write lightly in soft pencil on back of each TOP of figure, author, and figure number.

Make figures match size of column width (8.6 cm) or printing area (17.8 × 24.1 cm). Group several figures into a plate on one page (17.8 × 24.1 cm). If figures are mounted on flexible white drawing paper, paper should not exceed 0.4 mm in thickness and about 300 g/m² in weight.

Submit good quality glossy prints of line drawings in desired final size. Make inscriptions clearly legible. Letters 2 mm high are recommended.

Submit well-contrasted photographic prints of half-tone illustrations, trimmed at right angles and in desired final size. Make inscriptions about 3 mm high.

Tables: Submit on separate sheets. Give each a short title.

OTHER FORMATS

SHORT COMMUNICATIONS—3 or 4 printed pages, including a table or figure, giving brief accounts of particularly interesting results.

OTHER

Abbreviations: Follow rules and recommendations of IU-PAC-IUB Commission on Biochemical Nomenclature (CBN) and of the IUB Commission of Editors of Biochemical Journals (CEBJ).

Color Charges: Approx. DM 1200.- for first and DM 600.- for each additional page.

Experimentation, Animal: State that "Principles of Laboratory Animal Care" (NIH Publication No. 86-23, revised 1985) were followed, as well as specific national laws. Manuscripts that do not comply may be rejected.

Experimentation, Human: State that studies were reviewed by appropriate ethics committee and were performed in accordance with ethical standards of 1964 Declaration of Helsinki. State that all persons gave informed consent. Omit details that might identify subjects.

Materials Sharing: Make available to colleagues in academic research any of the genetic lines (organisms or cells), nucleic acids, antibodies, etc., used in research reported and are not available from commercial suppliers.

Nucleic Acid Sequence Data: Deposit novel nucleic acid sequence data with EMBL Data Library. Make deposition immediately after paper has been accepted for publication; add relevant accession number from Data Library at end of text at page proof stage.

Page Charges: DM 150.00 for each page exceeding six printed pages (18 manuscript pages, including figures and tables).

Style: Underline genus or species names, gene symbols, mathematical formula and any words to be given special emphasis for italics. Clearly distinguish similar signs, e.g., 1 (one) and l (letter), or 0 (zero) and O (letter).

SOURCE

October 17, 1994, 245(1): A4.

Molecular Biology and Evolution

Society for Molecular Biology and Evolution
University of Chicago Press

Molecular Biology and Evolution, Editorial Office
Biology Department, Hutchinson Hall
University of Rochester
Rochester, NY 14627

AIM AND SCOPE

Molecular Biology and Evolution is a bimonthly journal devoted to the interdisciplinary science between molecular biology and evolutionary biology. The journal will emphasize experimental papers, but theoretical papers are also published if they have a solid biological basis. Although this journal is primarily for original papers, review articles and book reviews are also published.

MANUSCRIPT SUBMISSION

Do not submit manuscripts or parts thereof that have been published or submitted for publication elsewhere. Upon acceptance, an agreement transferring copyright to publisher must be signed. No published material may be reproduced or published elsewhere without written permission of copyright owner. Reviewers may request sequences to verify claimed results; sequences may be requested on diskette.

Submit one original and two high-quality copies of manuscripts.

MANUSCRIPT FORMAT

Organize paper in sequence. Type each section, including tables and figure legends, double-spaced on heavyweight, nonerasable bond paper with 1.25 in. margin on left and 1 in. margin on right. Make type no more than 15 characters per inch (12 point Times Roman type) nor more than three double-spaced lines per inch (24 point spacing). Do not justify right margin. Dot matrix printing is acceptable only if it is standard typewriter quality. Identify handwritten items (e.g., Greek letters) in margin. Use correct diacritics for nonEnglish words. Begin each major part of paper on a new page, number pages consecutively throughout, beginning with title page and continuing through abstract, text, Appendix, Literature Cited, footnotes, tables, and figure legends. Figures appear last.

Title Page: Include paper's title, names of all authors, institution(s) at which research was done, current affiliations of all authors, name and address, telephone and fax numbers, and e-mail address for correspondence, list of keywords, a list of nonstandard abbreviations used, and a running head (50 characters and spaces).

Key Words: 3–8 to index subject matter of article on title page. Include name of organism studied.

Abstract: 350 word factual condensation of entire paper, including statement of purpose, clear description of observations and findings, and concise presentation of conclusions. Do not assert that findings are discussed. Minimize references to cited literature.

Text: Include sections: Introduction, Material and Methods, Results, Discussion, and Acknowledgments (if any). Results and Discussion may be combined. Conclusion section may be added. Be concise.

Introduction: Provide sufficient background to understand what follows. Do not do a literature review. Do not summarize conclusions. State questions to be addressed.

Materials and Methods: Provide brief descriptions of important methods, even if previously published.

Results and Discussion: Separation of sections is sometimes awkward and arbitrary. Results may contain interpretations if necessary.

Conclusions and Speculations: Include all alternative conclusions; explain why yours is preferred. Label speculation as such.

References: Cite by author and year and, where citation is to a book, relevant pages thereof. Give text citations of two or more works in chronological order; if two or more works have the same publication year, give in alphabetical order. If a paper has three or more authors, give name of first author plus "et al."

Arrange Literature Cited section at end of paper alphabetically by author(s) and then chronologically. Show only works specifically cited in text. Cite references to unpublished papers the same as for articles, except that "submitted" or "accepted" (with journal name) replaces volume and page numbers. Submit copies of those papers. Do not reference material that has not been submitted in Literature Cited section; note "unpublished data" at appropriate place in text.

JOURNAL ARTICLES

BARRY, P. A., A. J. JEFFREYS, and A. F. SCOTT. 1981. Evolution of the β-globin gene cluster in man and the primates. J. Mol. Biol. 149:319–336.

BOOKS

INGRAM, V. M. 1963. The hemoglobins in genetics and evolution. Columbia University Press, New York. (The pages of the book that are appropriate to the item cited must be given in the text where the book is cited.)

BOOK CHAPTERS

HALL, B. G. 1983. Evolution of new metabolic functions in laboratory organisms. Pp. 234–257 *in* M. NEI and R. K. KOEHN, eds. Evolution of genes and proteins. Sinauer Associates, Sunderland, Mass.

Use abbreviations of periodicals in *Chemical Abstracts.*

Include "Reviewed under the auspices of (name of editor or associate editor with whom author has corresponded)" at end of Literature Cited section.

Footnotes: Do not use in text. Reference footnotes to tables by superscript letters except for significance levels (use asterisks); type table footnotes on page with table.

Tables: Type each on a separate page, double-space. Give titles that describe content; number with Arabic numerals.

Legends: Type on pages at end of manuscript, after tables. Make descriptive so that illustration can be understood apart from text; define abbreviations.

Illustrations: Submit originals, not photocopies. Keep figures separate; use uniform lettering. Number consecutively, following sequence of mention in text. Write names of authors, figure number and an arrow indicating TOP away from figure itself or lightly in pencil on back. Make line drawings high quality; typewritten or hand lettering is unacceptable. Make text and lines of a size and thickness that will withstand reduction to *MBE* page format. Photographs should be high-contrast, glossy prints. Indicate magnifications by a micron bar or in legend. Sequence alignments must not exceed two published pages Submit alignments to EMBL and obtain an alignment number, which will be published in article. Dot matrices and hydropathy plots are rarely accepted. Color illustrations incur a moderate charge.

OTHER FORMATS

LETTERS TO THE EDITOR—8 manuscript pages, including literature with 1 figure, 1 table and no abstract. Do not divide into sections. Short communications containing new and unpublished information. Do not submit opinion, commentary, or discussion.

OTHER

Abbreviations: Those used by the *Journal of Biological Chemistry* are regarded as standard; define nonstandard abbreviations collectively in a footnote.

DNA Sequences: Sequence both strands fully; minimize length of each primer read; control for PCR misincorporation when sequencing cloned PCR products; and institute rigorous quality control procedures to minimize clerical errors in reading, transcribing, and communicating nucleotide sequences. Determination of 80-90% of any given sequence on both strands is standard.

If only one strand is sequenced, so state in Material and Methods and in entry submitted to sequence database. It may be appropriate to sequence one strand when a number of sequences from within a species or from a set of closely related taxa are being compared. Employ methods such as in Nachman et al. (1994) *Genetics* 136: 1105-1120, fig 3. to ensure that all differences are detected. When a single strand is sequenced, quality of data must be adequate; justify purpose and nature of analysis. Indicate ambiguous positions in a DNA sequence with appropriate IUB-single letter code. State undefined ambiguities.

Electronic Submission: Include electronic versions of final revised versions of manuscripts on 3.5 in. disk formatted for either Macintosh or DOS, and with files in any word processing format. Provide hardcopy with disk. Do not send disk with initial submission.

Equations: Type or print mathematical equations carefully; ensure spacing between characters is correct as typed. All characters will be italicized unless otherwise specified at first appearance. Number sequentially in Arabic numerals in parentheses on right side of page.

Materials Sharing: Make available any clone of cells, DNA, or antibodies used in reported experiments and any computer programs or nucleic acid or protein sequences on which paper is based. Provide nucleic acid sequences to GenBank in computer-readable format with appropriate associated data. GenBank accession number will be printed in article.

Nomenclature: Identify all organisms mentioned by scientific binomens; underline. For abbreviations for taxa, use form Gsp (Genus, species) as, for example, "Hsa" for *Homo sapiens*. Underline symbols for genetic loci; follow established rules of genetic nomenclature for organisms. Include formal IUB name and number of enzymes.

Page Charges: $35 per page or fraction of page. A waiver of charges may be requested of editor. Articles are selected without regard to ability to pay charges.When an institution is responsible for charges, submit an official purchase order. List vendor as The University of Chicago Press.

Style: Keep jargon to a minimum. Limit excessive verbiage and excessive referencing. Cite only important literature, not every possible publication on topic.

Follow guidelines in the *Council of Biology Editors Style Manual* (4th ed., 1978). All material should conform to *CBE* format. See also recent journal issues.

Terminology: Use any word provided that its meaning is clear, consistent, and serves to increase paper's comprehensibility. Following preferred usages are assumed unless defined otherwise.

Where alignments disagree, they are differences rather than change. "Invariant" means, invariable and unvaried; be sure meaning is clear. Mutations are changes before selection has operated. If two molecules are alike in some degree, they are similar. If you infer from their similarity that they have a common ancestor, they are homologous, but, if similarity was acquired by convergence, they are analogous. When homology arises via gene duplication, it is paralogy; when it arises via speciation, it is orthology; when it arises by horizontal gene transfer, it is xenology. "Insertions and/or deletions" may be reduced to indels. Gaps are introduced into sequences to increase similarity rather than to optimize similarity, unless an algorithm is employed that guarantees an optimized result according to the way similarity is defined. Do not assert similarity to be significant unless patently obvious or accompanied by a probability statement and its method of determination.

For trees determined by methods that can produce branch lengths, place those lengths upon them. Use bootstrapping; place those numbers on tree. Alternatively, provide estimated standard error of each branch's length on tree. If more than one tree is obtained for same taxa, perform rotations about ancestral nodes to make order of taxa agree with each other. If there is a table of distances or a set of aligned sequences, order of taxa should also match that of one of the trees obtained from them.

IUB single-letter code for nucleotide bases including ambiguity is: A = adenine, C = cytosine, G = guanine, T = thymine, U = uracil, R = A/G (purine), Y = C/T (pyrimidine), M = A/C, W = A/T, S = C/G, K = G/T, B = C/G/T (not A), D = A/G/T (not C), H = A/C/T (not G), V = A/C/G (not T), N = X = A/C/G/T (any or unknown). For ambiguous nucleotides, T and U are equivalent.

SOURCE

January 1995, 12(1): 183-187.

Molecular Biology of the Cell
American Society for Cell Biology

Ms. Rosalba A. Kampman
American Society for Cell Biology
9650 Rockville Pike
Bethesda, Maryland 20814-3992
Telephone: (301) 530-7153; Fax: (301) 571-8304

AIM AND SCOPE

Molecular Biology of the Cell will publish papers that describe and interpret results of original research concerning molecular aspects of cell structure and function. Studies whose scope bridges several areas of biology are particularly encouraged, for example cell biology and genetics. The aim of the *Journal* is to publish papers describing substantial research progress in full: papers should include all previously unpublished data and methods essential to support conclusions drawn. Journal will not publish papers that are merely confirmatory, preliminary reports of partially completed or incompletely documented research, findings of uncertain significance or reports documenting well-known processes in new organisms or cell types. Methodological papers are considered only when some new result of biological significance has been achieved with method.

MANUSCRIPT SUBMISSION

Submit one original and three copies of each paper. Type double-spaced on bond paper, in a final form requiring minimal editing. Manuscripts printed on a nonletter quality dot matrix printer will be returned. Accepted manuscripts may be submitted on diskette; include operating system and word processing program used. Include high quality glossy prints of each figure with original manuscript and each copy. Submit color figures as both glossy prints and slides.

Submission of a manuscript implies that it has not been submitted for publication elsewhere and that it contains unpublished, new information. Submit copies of closely related papers that are in press or have been submitted elsewhere. Quote personal communications only with agreement of person cited. Obtain and submit letter of permission, stating that person involved has seen text of quotation and gives permission for its use. Send all authors copies of submitted manuscript.

Submit fully documented, original research papers. Provide all data and methods essential to conclusions. Reviewers must certify that central conclusions of each paper do not depend on unpublished work, data not shown, or preliminary summaries of data intended for publication elsewhere. Readers should be able to reproduce experiments relying solely on paper describing them and published work cited by paper. Cite published precedents for results or conclusions; reviewers will reject papers with grossly insufficient or inappropriate citation of previous work.

MANUSCRIPT FORMAT

Write in concise, logical, and grammatically correct English. Organize into: Abstract, Introduction, Methods, Results, Discussion, Acknowledgments, References, Tables, and Figure Legends. Be brief. Do not repeat in one section material described in another.

Title Page: Include authors' full names and affiliations, running title (40 characters), and telephone and fax numbers of corresponding author.

Abstract: 200 words. Briefly summarize background of research reported. Elaborate theoretical background to experimental design; do not summarize data.

Materials and Methods: Describe experimental protocols and origin of unusual or special materials, tissue, cell lines or organisms; give genotypes in full. Provide data to support identity or purity of reagents (e.g., specificity of an antibody preparation), reliability of methods (e.g., linearity of assays), sensitivity of instruments, or essential features of genotypes. Include most experimental detail into section; use Results section for exposition of experimental design and results. Reference all standard procedures. Be complete enough so results can be verified by others.

Results: Present experiments that support conclusions to be drawn later in Discussion in a logical order. State results exactly; do not interpret, extend lines of inference, arguments or speculations. Conform to a high standard of rigor.

Discussion: Propose interpretation of results, and suggest what they might mean in a larger context. May be imaginative. Subdivide Results and Discussion further if subheadings give manuscript more clarity.

Acknowledge all sources of financial support for work.

Figures: *MBC* will notify authors of substandard figures, so figures can be brought up to standard before final acceptance of paper. Cite in numerical order in text.

Type legends double-spaced consecutively on a separate sheet. Include a general figure title followed by explanation of specific parts. Use Arabic numerals for figures and uppercase letters for multiple parts of a single figure. For complete figure preparation guidelines, see Instructions to Authors, January 1993 issue.

Tables: Type double-spaced on sheets separate from text. Make self-contained and self-explanatory. Do not use vertical rules. Label each at top with a Roman numeral followed by title. Insert explanatory material and footnotes below table. Supply units of measure at heads of columns.

Key Words: List 5 on title page. Do not duplicate words in title. Also use phrases no longer than three words.

References: Cite in text by name and date, not by number (Beckerle *et al.*, 1987 or Nagafuchi and Takeichi, 1989). List only articles published or in press. Include complete titles and inclusive page numbers. Abbreviate names of journals as in *Serial Sources for the Biosis Data Base*. Cite unpublished results, including personal communications and submitted manuscripts, as such in text. Submit permission letters with personal communications unless from authors' laboratory.

OTHER

Abbreviations: Follow abbreviations of *Council of Biology Editors Style Manual*. For chemical nomenclature, use Subject Index of *Chemical Abstracts*. Capitalize trade names and give manufacturers' names and addresses.

Terms not on *CBE Style Manual* abbreviations list must be used five times to qualify as an abbreviation. Spell out on first mention and follow with abbreviated form in parentheses. Thereafter, use abbreviated form. Supply a footnote of nonstandard abbreviations, in alphabetical order; give each abbreviation followed by spelled-out version. Keep to a minimum.

Crystallographic Data: Submit crystallographic studies of proteins and other biopolymers relevant structural data to Protein Data Bank (Chemistry Department, Brookhaven National Laboratory, Upton, NY 11973) [see Commission on Biological Macromolecules (1989) *Acta Crystallogr.* Sect. A45, 65]; specify submission in a footnote to paper.

Experimentation, Animal and Human: Research involving recombinant DNA, humans, and animals must have been carried out in accordance with recommendations from Declaration of Helsinki and appropriate NIH guidelines; research protocols must be approved where necessary by appropriate institutional committees.

Materials Sharing: Make available all propagative materials (e.g., mutant organisms, cell lines, recombinant plasmids, vectors, viruses, and monoclonal antibodies) used to obtain results. Interested scientists must provide authors with a written statement that they will be used for non-commercial research purposes only.

Nucleotide Sequences: Nucleotide sequences must have an EMBL database accession number. Accepted manuscript that do not have such a number by page proof stage will be held until number is provided.

Page and Plate Charges: $25 per page. Inability to pay will not affect publication. Surcharge for halftone illustrations is $15 per halftone; $1000 for printing color figures.

SOURCE

January 1994, 5(1): back matter.

Molecular Brain Research
see Brain Research: Molecular

Molecular Endocrinology
Endocrine Society

Anthony R. Means, Ph.D.
2200 West Main St., Suite B-210, Room 26
Durham, NC 27705
Telephone: (919) 286-6430; Fax: (919) 286-7239

AIM AND SCOPE

The Endocrine Society publishes papers describing results of original research in field of endocrinology and metabolism in the *Journal of Clinical Endocrinology and Metabolism*, *Endocrinology*, *Molecular Endocrinology*, and *Endocrine Reviews*.

The *Journal of Clinical Endocrinology and Metabolism* publishes endocrine and metabolic studies related to human and primate physiology and disease. *Endocrinology* publishes nonprimate biochemical and physiological studies, although work on material of primate origin is not excluded. *Molecular Endocrinology* publishes mechanistic studies of effects of hormones and related substances on molecular biology and genetic regulation of nonprimate and primate cells. *Endocrine Reviews* publishes bimonthly and features in-depth review articles on both experimental and clinical endocrinology. Authors are invited to consult with Editors concerning most appropriate journal to which to submit manuscripts.

Molecular Endocrinology is a journal devoted to rapid publication of papers pertaining to the molecular mechanisms by which hormones and related compounds regulate function. The journal will promote the interaction of disciplines by accepting papers in fields studying processes with overlapping mechanisms, including hormones, growth factors, oncogenes and their products, growth inhibitory factors, ions, lymphokines, etc. Two types of articles will be entertained for publication. The bulk of the journal will be comprised of regular research papers based on new data, previously unpublished in reviewed journals or unreviewed publications. These papers may be of any length; however, brevity is no contraindication for submission. Secondly, minireviews and overviews may be published from time to time. If an author wishes to suggest a topic for a minireview, he first should contact an editor of *Molecular Endocrinology* to ascertain its appropriateness.

MANUSCRIPT SUBMISSION

Provide assurance, on form provided in every issue or copy, that no substantial part of work has been published or is being submitted or considered for publication elsewhere. When a portion of results is to appear in another journal, or in publications of congresses, symposia, workshops, etc., supply details to Editor, and a copy of other paper(s) submitted, with expected date of publication.

The Journals are copyrighted by The Endocrine Society. The Society grants to authors royalty free right of republication in any work of which they are author or editor, subject only to giving proper credit in work to original publication of article by The Endocrine Society.

Submit affidavit (in every issue) affirming originality and assigning copyright to The Endocrine Society. Manuscripts will not be considered until affidavit and processing fee are received.

Present a clear, honest, accurate, and complete account of research performed. Describe complete study or completed phase of an extended study. Avoid fragmenting reports. When some results are to appear in another journal, publications of congresses, symposia, workshops, etc., supply details to editor, and submit copies of paper(s).

List as authors only those individuals who made significant contributions to intellectual and procedural aspects of study. An author should have participated in conception and planning of work, interpretation of results, and writing of paper. Acknowledge others who contributed to a lesser extent. Authors signatures on Affirmation of Originality and Copyright Release form (in every issue) indicates approval of final version of manuscript and readiness to take public responsibility for work. Failure to notify editor that some results are being or have been previously published will result in placement of a notice in journal that authors have violated the Ethical Guidelines for Publication of Research in Endocrine Society journals.

MANUSCRIPT FORMAT

Submit original and three copies of each article, including data, figures, tables, etc. Type double-spaced, including references, tables, and figure legends. Provide three sets of glossy prints of figures.

Describe work in sufficient detail to allow others to repeat work. Include all relevant data, including those which may not support hypothesis being tested. Cite those publications which have a direct bearing on novelty and interpretation of results.

Title Page: Title, name of authors, university affiliations, running title, and corresponding author's address and telephone and fax numbers. Provide a list of key words.

Abstract: 250 words, no references.

Introduction: Begin with a brief introductory statement that places work to follow in historical prospective and explains its intent and significance.

Results and Discussion: Present separately, or combined into a single section. Results present experimental data, including tables and figures. Discussion explains interpretation and significance of findings with concise objective comments which describe relation to other work in area. Results and Discussion comprise bulk of article,

Materials and Methods: Describe and reference completed experiments so they can be repeated by others. Give source of hormones, unusual chemicals and reagents, and special pieces of equipment.

Acknowledgments: If desired for example for assistance in preparation of article and grant support for research.

References: List in numerical order of first citation in text. Include articles listed as "in press" or "submitted" as reference material and submit with manuscript. Abbreviate journal titles as in *Index Medicus*.

Sakaue Y, Thompson EB 1977 Characterization of two forms of glucocorticoid hormone-receptor complex separated by DEAE-cellulose column chromatography. Biochem Biophys Res Commun 77:533–541

CHAPTERS IN BOOKS

Thompson EB, Simons SS, Hammon JM 1982 Deacylcortivazol, a potent glucocorticoid with unusual structure and unusual antileukemic cell activity. In: Leavitt WW (ed) Hormones and Cancer. Plenum, New York, pp 315–324

BOOKS

Turner CD 1960 General Endocrinology, ed. 3. WB Saunders Co, Philadelphia, p 426

PUBLISHED ABSTRACTS

Kaptein EM, Grieb D, Wheeler W, Peripheral thyroid hormone kinetics in critical illness. Program of the 62nd Annual Meeting of The Endocrine Society, Washington, DC, 1980, p 189 (Abstract)

Figure Legends: Type on separate page.

Figures: Usually single-column; plan content and format accordingly. If information cannot be presented clearly in that configuration, indicate by a note on back at submission. Submit as glossy photographs. Supply copies of sufficient quality to impart full information or additional prints. On back of each in light pencil write authors names, condensed article title, and an arrow indicating TOP. Do not mount figures.

Color figures are printed at authors' expense. A cost estimate is provided when manuscript is submitted to printer; provide written or verbal indication of acceptance of charges.

Tables: Submit as typewritten copy on separate page.

OTHER

Abbreviations: Use standard abbreviations that appear in January and July issues of *Endocrinology* each year. For abbreviations not defined, follow form in January issue of *Journal of Biological Chemistry*. In case of different definitions, *Endocrinology* definition takes precedence. Keep nonstandard abbreviations to a minimum; define each at first use.

Electronic Submission: Submit electronic diskettes of final version of manuscripts along with typed revised manuscript. Send most recent version of manuscript, and matched print copy that was accepted for publication. File should contain all parts of manuscript in one file. Mathematics and tabular material is processed in traditional manner and may be excluded from diskette file.

Label diskette with journal name, manuscript number, senior author's name, and name of computer file. Provide: name used to access paper on diskette; computer used; operating system and version; word processing program and version; special characters used; manuscript number; corresponding author; telephone and fax numbers.

Experimentation, Animal: Conduct in accordance with highest standards of humane care. Consider appropriateness of experimental procedures, species and number of animals used, in study design. Acquire and use all research animals in compliance with federal, state, and local laws and institutional regulations. Maintain animals in accordance with NIH *Guide for Care and Use of Laboratory Animals.* Animals must receive appropriate tranquilizers, analgesics, anesthetics, and care to minimize pain and discomfort during preoperative, operative, and postoperative procedures. Make choice and use of drugs in accordance with NIH Guide. If use of anesthetics would negate results, justify and show approval by Committee on Animal Care and Use of local institution and follow accepted veterinary medical practice. Monitor health of animals. If either study or condition of animals requires that they be killed, do it in a humane manner.

Materials Sharing: Make any clones, whether of cells or genes, published in journal available to other researchers upon proper request. Provide propagative materials such as DNA clones, cell lines, and hybridomas developed by authors and used in studies to qualified investigators.

Page Charges: $45 per printed page. No paper will be rejected because of authors' inability to pay charge; contact Editor-in-Chief to make special arrangements, after completion of scientific review process.

Reviewers: Provide a list of possible reviewers; request that specific reviewer(s) not be used.

Style: Follow *CBE Style Manual,* Fifth Edition.

Submission Fee: Submit $70 processing fee and a computer diskette of complete manuscript (most word processing languages are acceptable) or $170 wtihout a diskette. Submit diskettes with final, revised version of manuscript (i.e. after it has been reviewed and returned). Pay fee in U.S. currency. Manuscripts will not be considered for publication until fee is received.

SOURCE

January 1994, 8(1): front matter.

Molecular Immunology
Pergammon Press

Dr. Steve Dower
Immunex Corporation, 51 University Street
Seattle, WA 98101
Dr. David Holowka
Department of Chemistry, Cornell University
Ithaca, NY 14853
          The Americas

Prof. Michel Fougereau
Centre d'Immunologie de Marseille-Luminy, Case 906
13288 Marseille Cedex 9 France
Dr. Ed Palmer
Basel Institute for Immunology, Grenzacherstrasse 487
CH-4005 Basel, Switzerland
Dr. Janine Borst
Division of Cellular Biochemistry, the Netherlands Cancer Institute, Plesmaniaan 121
1066 CX Amsterdam, The Netherlands
Dr. Hidde Ploegy
Massachusetts Institute of Technology
Center for Cancer Research, 40 Ames Street, E17-322
Cambridge, MA 02139
          Europe, Middle East and Africa
Dr. Steve Gerondakis
The Walter and Eliza Hall Institute of Medical Research
Royal Melbourne Hospital
Victoria 3050, Australia
          Far East, Japan and Australasia

AIM AND SCOPE

Molecular Immunology is primarily devoted to the publication of immunological knowledge which can be delineated at the molecular level. Within this framework are included contributions concerned with the molecular analysis of the cellular components and processes which underlie the physiological behavior of cells involved in immune phenomena. Papers which merely describe new monoclonal antibodies and/or epitope mapping will no longer be accepted unless they have a direct relevance to the understanding of the immune system.

MANUSCRIPT SUBMISSION

Submit all manuscripts and material to a Regional Editor with a cover letter stating category or section preferred in *Journal.* Give name and address of corresponding author.

MANUSCRIPT FORMAT

For Research Reports, submit detailed and complete descriptions of research problem, experimental approach and findings, providing new and significant knowledge and containing information essential for verification of conclusions by other laboratories. Manuscripts are subject to detailed scrutiny by expert reviewers whose recommendations are the primary determinants of acceptability for publication. Groups of research reports considered as 'mini symposia' may also be organized by a Guest Editor, in agreement with the board of Regional Editors. Such reports will also be subject to critical review.

Submit all contributions typed double-spaced, with margins which provide sufficient space for notations. Submit three copies. Supply two originals of each photograph. Include an English abstract. Divide manuscripts dealing with experimental work into sections: Introduction (optional), Experimental (with Materials, Methods and Results) followed by Discussion, Conclusions and References. If a large number of abbreviations are used, include a glossary defining terms.

Place title, authors and affiliations, on a separate page. Give an abbreviated title for use as a running headline.

References: Cite in text by authors' names and date in parentheses, e.g., Cohen and Eisen (1977). Cite full references in an alphabetical list at end of paper.

Cohen T. J. and Eisen H. N. (1977) Interactions of macromolecules on cell membranes and restrictions of T-cell specificity by products of the major histocompatibility complex. *Cell Immun.* **32,** 1–9.

Haber E., Margolies M. N. and Cannon L. E. (1977) Structure of the framework and complementarity regions of elicited antibodies. In *Antibodies in Human Diagnosis and Therapy* (Edited by Haber E. and Krause R. M.), p. 45. Raven Press, New York.

Footnotes: Use symbols for statements or references which do not fit into text: *, †, ‡, §, ||, ¶; start afresh on each page. Type at bottom of page on which they appear.

Tables: Reproduced by photo-offset directly from typed manuscripts. Type on separate manuscript pages; make intelligible without reference to text. Make headings concise, clearly present subject matter, and make part of table. Place footnotes to tables on same page; designate by alphabetical superscripts.

Graphs and Illustrations: Keep figures and illustrations to a minimum. Do not present data in duplicate forms. Include separately and not as part of text. Make about twice published size; clearly mark on back TOP, number and author's name. Avoid unduly small letters in published version. Use either original or glossy prints. Provide immunoelectrophoretic patterns and other optical patterns as photographic reproductions of actual pattern, not line drawings. Submit a complete set of labeled figures and illustrations with each copy of manuscript. Label optical patterns (gel diffusion, immunoelectrophoresis, Ouchterlony) with symbols descriptive of materials employed and which do not require reference to legend for identification. Indicate position of figures or illustrations on manuscript. Give legends on a separate page; give enough information to make figure understandable. Use standard symbols on line diagrams: ○ ● + × □ ■ △ ▲ ▼ ▽.

OTHER FORMATS

ANNOUNCEMENTS—Submit announcements of forthcoming meetings and conferences of immunological interest in duplicate to a Regional Editor.

BRIEF STRUCTURAL DATA REPORTS—Provides an opportunity to publish new protein or gene sequences or other definitive structural data of obvious immunological interest. Submit camera-ready (2 pages). Include a brief outline of sequencing strategy. Whenever suited, include comparison with relevant sequence(s). Provide an accession number, to be obtained from one of the classical data banks.

REVIEW ARTICLES—Invited by Regional Editors. Anyone interested in submitting a scholarly and critical review of a timely subject in molecular immunology should contact the appropriate Regional Editor.

SHORT COMMUNICATIONS—Two to four pages providing rapid dissemination of timely and significant observations

in molecular immunology. These manuscripts will also be evaluated by experts.

SUMMARIES OF MEETINGS AND CONFERENCES—Reports prepared by selected individuals in attendance.

VIEWPOINTS—Brief communications devoted to formulation of challenging new ideas, which possess the potential of suggesting novel experiments which meet the criterion of falsifiability (Karl Popper, *The Logic of Scientific Discovery*, 1959).

OTHER

Electronic Submission: Submit a computer disk (5.25 or 3.5 in. HD/DD) containing final version of paper with manuscript. Specify software used and release. Specify computer used (either IBM compatible PC or Apple Macintosh). Include text file and separate table and illustration files. Make file follow general instructions on style/arrangement and especially reference style of journal. Single space; let text wrap (do not use returns at ends of lines). Begin all textual elements flush left, do not indent paragraphs. Place two returns after elements such as title, headings, paragraphs, figure and table callouts.

SOURCE
January 1994, 31(1): iii-iv.

Molecular Microbiology
Blackwell Scientific Publications

Chris Higgins
Imperial Cancer Research Fund
University of Oxford
Institute of Molecular Medicine
John Radcliffe Hospital
Oxford OX3 9DU, U.K.
Telephone: 44 (0) 865 222423; Fax: 44 (0) 865 222431;
E-mail: C_HIGGINS@ICRG.AC.UK
Virginia Miller
Department of Microbiology and Molecular Genetics
UCLA, 1602 Molecular Sciences Bldg.
405 Hilgard Ave.
Los Angeles, CA 90024-1489
Telephone: (310) 206-3077; Fax: (310) 206-5231;
E-mail: virginia@microbio.lifesci.ucla.edu
          Papers on bacterial pathogenicity and virulence
          from North America
Bob Simons
Department of Microbiology and Molecular Genetics
UCLA, 1602 Molecular Sciences Building
Los Angeles, CA 90024-1489
Telephone: (310) 825-8890; Fax: (310) 206-5231;
E-mail: bobs@microbio.lifesci.ucla.edu
          Papers on basic molecular biology from North
          America

AIM AND SCOPE

Molecular Microbiology publishes high quality, original research papers addressing any microbiological question at a molecular level. The journal invites papers describing the molecular biology, genetics, biochemistry, pathogenicity and cell biology of any microorganism, prokaryotic or eukaryotic. Papers in the field of biotechnology will be considered, but only where they address fundamental biological questions.

MANUSCRIPT SUBMISSION

Be as concise as possible, compatible with clarity and completeness. Do not exceed 8 printed pages (17 pages of double-spaced typescript with 3 figures/tables, 15 pages and 6 figures/tables or 14 pages and 8 figures/tables.). Only complete reports will be published; notes or preliminary communications are not considered.

Submit four copies of manuscript. Submit papers to any Editor. Submit papers exclusively to *Molecular Microbiology*. Papers must not have been, and will not be, published elsewhere. If accepted, papers become copyright of journal. Authors must give signed consent to publication in cover letter, but permission to use material elsewhere will normally be granted on request.

MANUSCRIPT FORMAT

Manuscripts are published from disk; provide in that form. Submit four paper copies of each manuscript. Double-space text and, where appropriate, provide original photographs with each copy of manuscript.

Title page includes author's name(s), affiliations and address, telephone and fax numbers and e-mail address of corresponding author. Indicate present addresses of authors as a footnote. Provide a running title (50 characters) with six key words for indexing.

Include a summary (200 words). Subdivide main text into Introduction, Results, Discussion and Experimental Procedures. Results and Discussion may be combined. Include additional subheadings. Sufficiently detail Experimental procedures to enable experiments to be reproduced.

Number all pages consecutively. Submit tables, figure legends and acknowledgments on separate sheets following main text. Indicate preferred position of tables and figures in margin of text. Avoid footnotes.

References: Include only full articles which have been published or are 'in press' in reference list. In text, refer to unpublished studies as such (e.g., J. M. Smith, unpublished), or as a personal communication. Obtain permission to include work as a personal communication. In text, insert references in parentheses: (Ames, 1974; Ames *et al.,* 1977; Ames and Nikaido, 1978). Put reference list in alphabetical order according to first named author. Papers with two authors follow those of first named author, arranged in alphabetical order according to name of second author. Articles with more than two authors follow those of first named author in chronological order. Include all authors' names and title of article. Use standard abbreviations of journal titles, as in *Index Medicus*.

Ames, G.F.-L. (1974) Resolution of bacterial proteins by polyacrylamide gel electrophoresis. *J Biol Chem* **249:** 634–644.

Ames, G.F.-L., and Nikaido, K. (1978) Identification of a membrane protein as a histidine transport component in *Salmonella typhimurium*. *Proc Natl Acad Sci USA* **75:** 5447–5451.

Ames, G.F.-L., Noel, K.D., Taber, H., Spudich, E.N., Nikaido, K., and Ardeshir, F. (1977) Fine-structure map of the histidine transport genes in *Salmonella typhimurium*. *J Bacteriol* **129:** 1289–1297.

Ames, G.F.-L. (1985) The histidine transport system of *Salmonella typhimurium*. In *Current Topics in Membranes and Transport*. Adelberg, E.A., and Slayman, C.W. (eds). London: Academic Press, pp. 103–119.

Reviews: Abbreviated references are used, e.g. Ames. G.F.-L. *et al.* (1977) *J Bacteriol* **129:** 1289–1297.

Tables and Figures: Supply original drawings or photographs. Figures will be reduced to single column (80 mm), two-thirds page (110 mm) or full page (169 mm).

Provide photographs as glossy prints. Cost of color plates is borne by authors. If authors cannot meet charges, and editors feel color figures are necessary, part or all of charges can be waived.

OTHER FORMATS

MICROCORRESPONDENCE—Forum for discussion of scientific issues raised by papers in this or other journals.

MICROREVIEWS—Short topical reviews in areas of particular interest and current importance. Many are invited; contact Review Editor, Tony Pugsley, (Unite de Genetique Molecularie, Institut Pasteur, 25 rue du Dr Roux, 75724 Paris Cedex 15, France; fax: (33)1 45 68 89 60; e-mail: Tony.Pugsley@Pasteur.fr) for further details of format and content.

OTHER

Abbreviations: Use standard abbreviations as recommended in *Quantities, Units, and Symbols* (The Royal Society, 1988). Abbreviations of nonstandard terms should follow, in parentheses, first full usage.

Electronic Submission: Use any software; specify type of computer/word processor used and type of software package used. Supply titles in both word processor and ASCII formats. Define keyboard characters used to represent characters not on keyboard (e.g., Greek). Use separate files for: main text, reference list, figure legends, and tables. Disk information form is supplied on request.

Materials Sharing: Distribute strains, clones or antibodies described in report for use in academic research.

Nomenclature, Genetic: Use standard genetic nomenclature. Consult Bachmann *(Microbiol Rev* **47:** 180–230, 1983) for *E. coli* K-12; Sanderson and Roth *(Microbiol Rev* **47:** 310–453, 1983) for *Salmonella typhimurium*; Holloway *et al. (Microbiol Rev* **43:** 73–102, 1979) for *Bacillus subtilis*; Perkins *et al (Microbiol Rev* **46:** 426–570, 1982) for *Neurospora crassa*; and the *Handbook of Genetics* vol. 1 (R. C. King, ed., Plenum Press, 1974) for *Saccharomyces cerevisiae*.

Nucleotide Sequences: Include nucleotide data as part of a larger study. Primary sequence data is not published as figures. Reference data in text citing database accession number for reference. Deposit data in EMBL/GenBank/DDBJ Nucleotide Sequence Data Libraries and cross reference accession number in manuscript. Obtain forms for submitting data to library from editors. Journal is unlikely to accept a manuscript if the only data are nucleotide sequences, especially if sequence of same gene has been reported in another species, unless sequence data themselves provide new and important biological insights.

SOURCE

October 1994, 14(1): back matter.

Molecular Pharmacology
American Society for Pharmacology and Experimental Therapeutics
Williams & Wilkins

Dr. Raymond J. Dingledine
Emory University School of Medicine
5001 Rollins Research Center, Clifton Road
Atlanta, GA 30322-3090

AIM AND SCOPE

Molecular Pharmacology will publish results of investigations that contribute significant new information on drug action or selective toxicity at the molecular level. The term "drug" is defined broadly to include chemicals that selectively modify biological function.

Suitable papers are those that describe applications of methods of biochemistry, biophysics, genetics, and molecular biology to problems in pharmacology or toxicology. Also suitable are reports of fundamental investigations which, although not concerned directly with drugs, nevertheless provide an immediate basis for further study of molecular mechanism of drug action. Observations of phenomena that shed no light upon underlying molecular interactions are not appropriate for publication. Comparative studies, such as those involving drug-receptor or drug-enzyme interactions that already have been well characterized in other types of cells or tissues, also are inappropriate for publication unless they contribute significant new insight into mechanisms.

Specific areas of interest include: identification and characterization of receptors for hormones, growth factors, neurotransmitters, toxins, and other drugs; analysis of receptor response pathways; drug effects on metabolic pathways, biosynthesis and degradation of macromolecules, and cellular regulatory mechanisms; analysis of drug-receptor and drug-enzyme interactions; effects of drugs on structure and properties of macromolecules and membranes; relationships between drug structure and activity; molecular mechanisms of drug metabolism; distribution and transport between biological compartments; molecular mechanisms of chemical mutagenesis, carcinogenesis, and teratogenesis; and molecular mechanisms of selective toxicity, drug allergy, and pharmacogenetics.

Molecular Pharmacology has adopted a uniform policy for evaluation of manuscripts utilizing molecular modeling. Key aspects of content that determine suitability and eventual acceptance include: use of modeling technology to generate predictions concerning new molecules, modeling studies that offer significant new insights into the mechanism of actions of drugs, and inclusion of experimental data that support predictions of molecular modeling. It is not necessary that each aspect be reflected in every manuscript; however, manuscripts that are purely theoretical in nature or that simply generate structural predictions without correlating these to drug action or new biological data will be returned as unsuitable for publication.

MANUSCRIPT SUBMISSION

Manuscripts and results they contain may not have been published previously and may not be submitted elsewhere. If submitted manuscript utilizes or makes bibliographic reference to articles in press, include copies. All persons listed as authors must approve submission of paper; each person cited as a source of personal communications must approve citation. Written authorization may be required. If and when a manuscript is published, it becomes the sole property of Journal. Copyright in article, including right to reproduce article in all forms and media, is assigned exclusively to the Society for Pharmacology and Experimental Therapeutics. No reasonable request by author for permission to reproduce any of his or her contributions will be refused.

MANUSCRIPT FORMAT

Type double-spaced with ample margins on one side of 8.5 × 11 in. paper. Submit four complete copies of manuscript and four copies of each figure, plus one original drawing or photograph of each figure. Each half-tone figure requires four original drawings or photographs. Number all pages consecutively beginning with title page.

Include in order: Title (avoid numbered footnotes; acknowledge financial support in an unnumbered footnote to title). Names of authors, their laboratory and institution. A running title (60 characters and spaces). Summary. Text (refer to footnotes by superscript numbers and references by numbers in parentheses). References. Footnotes (number according to order of appearance in text). Tables. Figures. Legends to figures. Name and address of person to receive galley proof.

Authors are allowed maximum freedom in organizing and presenting material, and in expressing their ideas, provided that clarity and conciseness are achieved. For most manuscripts, suitable format is: Summary, Introduction, Materials and Methods, Results, and Discussion.

Summary: 250 words.

References: Limit to 40. Cite references to papers that have been accepted for publication, but have not appeared, like other references with abbreviated name of journal followed by "in press." Submit three copies if their findings have a direct bearing on submitted paper. Cite "Personal Communications" and "Unpublished Observations" in footnotes to text; do not include in reference list.

Number according to order of citation in text, including title and complete pagination.

1. Goren, J. H., L. G. Bauce, and W. Vale. Forces and structural limitations of binding of thyrotropin-releasing receptor: the pyroglutamic acid moiety. *Mol. Pharmacol.* **13**:606–614 (1977).

2. Chernow, B., and J. T. O'Brian. Overview of catecholamines in selected endocrine systems, in *Norepinephrine*

(M. G. Ziegler and C. R. Lake, eds.). Williams and Wilkins, Baltimore, 439–449 (1984).

3. Snedecor, G. W., and W. G. Cochran. *Statistical Methods.* Iowa State University Press, Ames (1967).

Tables: Number with Arabic numerals; design to fit single-column width. Give explanatory titles and sufficient experimental detail in a paragraph following title to be intelligible without reference to text (unless procedure is given in Methods section, or under another table or figure). Place footnotes beneath tables themselves; designate by lowercase italic superscript letters.

Figures: Number with Arabic numerals. Each manuscript copy must contain all figures. Only original set must be suitable for reproduction; submit four sets of photographs or original drawings of halftones. Submit unmounted glossy photographs (or original India-ink drawings). Figures will be reduced to one column width (85 mm); numbers after reduction should be at least 1.5 mm high. Figures must be ready for direct reproduction: no lettering or other art work will be done by publisher. If symbols are not explained on figure, use only standard characters ($\times \bigcirc \bullet \square \blacksquare \triangle \blacktriangle$). On back of photographs put its number, and TOP at appropriate edge. Give list of legends for figures captions with sufficient experimental detail, as required for tables.

OTHER FORMATS

ACCELERATED COMMUNICATIONS—Provides a mechanism for rapid publication of novel experimental findings of unusual and timely significance. Not intended for publication of preliminary results. Present novel results that are clearly documented and make a conceptual advance in their field. In transmittal letter outline significance of work; list three appropriate reviewers.

Manuscripts that require major revisions or that do not fit criteria will be returned for revision and further consideration as a regular paper.

Submit in same style as regular manuscripts. Results and Discussion may be combined. Do not exceed five printed pages, 25 double-spaced typewritten pages (1 in. margins), including all components of manuscript and counting each figure as a page of text. Manuscripts that are too long will be considered as regular papers.

OTHER

Abbreviations: Define all essential abbreviations in a single footnote when first introduced. Abbreviate journal names in style of *Biological Abstracts.*

Electronic Submission: Submit electronic diskettes of final version of manuscripts with typed revised manuscript. Diskettes produced on IBM or IBM-compatible computers are preferred; Apple/Macintosh or Wang computer disks can be converted. Preferred word processing programs are: XyWrite III Plus, Word Perfect 4.2, 5.0, or 5.1 (IBM or Macintosh), Microsoft Word (IBM or Macintosh), Wang OIS (WPS), and Wordstar (IBM). Other word processing systems can be converted: CPT 8000, MacWrite 2.2 or 4.5, DisplayWrite 3 or 4, Multimate, PCWrite, Volkswriter, and WriteNow. If using Macintosh computers, do not use Fast Save option. ASCII files can be used, but are not preferred. Identify diskette with journal name, manuscript number, senior author's name, manuscript title, name of computer file, type of hardware, operating system and version number, and software program and version number.

Page Charges: $40.00 per page. Payment is not a condition for publication. In case of personal financial hardship, charges will be waived.

Style: For chemical and mathematical formulas and abbreviations follow *Instructions to Authors of the Journal of Biological Chemistry* (261: 1-11, January 10, 1986). Refer to drugs by generic or chemical names; identify by trade name in parentheses or a footnote. Include systematic name and number given by Commission on Enzymes of the International Union of Biochemistry for each enzyme of importance, at point in Summary or Introduction where enzyme is first mentioned. Minimize use of abbreviations; avoid use in Summary.

Submission Fee: $40 (in U. S. funds drawn on a U S. bank payable to ASPET) or a validated purchase order from authors' institution. If fee entails a personal financial hardship to author(s), fee will be waived. Submit a request for waiver of fee when manuscript is submitted.

SOURCE

July 1994, 46(1): front matter.
Instructions to Authors appears in every issue.

Muscle and Nerve
American Association of Electrodiagnostic Medicine
John Wiley & Sons

Jun Kimura, MD
Department of Neurology
Kyoto University Hospital
Shogoin, Sakyoku Kyoto, 606 Japan
Telephone: 81 75 751 3770; Fax: 81 75 761 9780

AIM AND SCOPE

Muscle & Nerve is an international and interdisciplinary publication of original contributions, in both health and disease, concerning studies of muscle, neuromuscular junction, peripheral motor and sensory neurons, and central nervous system where behavior of peripheral nervous system is clarified. Appearing monthly, *Muscle & Nerve* publishes clinical studies and clinically relevant research reports in fields of anatomy, biochemistry, cell biology, electrophysiology and electrodiagnosis, epidemiology, genetics, immunology, pathology, pharmacology, physiology, toxicology, and virology. The Journal welcomes articles and reports on clinical electrophysiology and electrodiagnosis including those related to motor control and all forms of evoked potential studies, and basic research papers with actual or potential implications for clinical diagnosis or management of neuromuscular disorder.

MANUSCRIPT SUBMISSION

Submit four high-quality copies of manuscript (1 original and 3 copies) and 4 complete sets of figures, accompanied by cover letter including name, address, phone and fax numbers of corresponding author; copies of any published reports that may duplicate material in submitted manuscript; written permission of author(s) and publisher(s) to use published material (figures, tables, or quotations of more than 100 words); releases signed by patient(s) or guardian(s) for any recognizable patient photographs; and optionally, names of potential referees and addresses as well as any requests that specific individuals not be invited to review.

Manuscripts must not have appeared elsewhere and may not be concurrently under review elsewhere.

Publication agreement printed in journal must be signed by all authors and accompany manuscript or letter to the Editor at submission.

MANUSCRIPT FORMAT

Type on white bond 8.5 × 11 in. paper with broad margins. Word processor printers must be of letter quality. Double-space throughout, including reference section.

Organize in order; begin each component on a separate page; type running title and page number in upper right corner of each page.

Title Page: Page 1: title of article (80 spaces); authors full names (first name, middle initial, surname) with degrees; affiliations (name of department and institution, city, and state or country where work was done) indicating which authors are associated with which affiliations; acknowledgments of grant support and of individuals who were of help in preparation of study; name and address of reprint request author; and running title (30 spaces). If part or all of material was presented at a national meeting, include organization, city, and date of presentation as a footnote.

Abstract: Page 2: title of article followed by abstract (150 words for a main or review article, 75 words for a case of the month). State purpose of study, basic procedures, most important findings, principal conclusions, and their clinical relevance, with an emphasis on new aspects of study. Spell out all nonstandard abbreviations at first use.

Key Words: 5 below abstract page; cover all major points.

Text: Organize: Introduction, Materials and Methods, Results, and Discussion. Use other descriptive headings and subheadings if appropriate. Avoid jargon, spell out all nonstandard abbreviations at first mention, and present contents of study as clearly and as concisely as possible.

Identify methods, apparatus (including manufacturer's name and address), and procedures in sufficient detail to allow other investigators to reproduce results. Give references for discussions of previous studies and for nonstandard methods used.

Cite references, tables and figures within text. Number tables and figures in order of appearance. Summarize, do not duplicate data in tables of figures in text. Check all data cited in text against corresponding data in tables to ensure that they correspond; check names cited in text against references to ensure correct spelling. Identify ambiguous symbols (e.g., letter "O" versus numeral "0," letter l versus numeral 1").

References: Type double-spaced in alphabetical order; number accordingly. Identify references in text, tables, and legends by Arabic numerals typed as superscripts.

Ensure accuracy and completeness of references. For journal articles include: author names (surnames followed by initials); title of article with spellings and accent marks as in original; journal title abbreviated as in *Index Medicus* or spelled out if not listed there; date of publication; volume number; and inclusive page numbers. For books include chapter title, chapter authors, editors of book, title of book (including volume or edition number), publisher's name and location, date of publication, and page numbers.

1. Dawkins RL, O'Reilly C, Grimsley G, Ziko PJ: Myasthenia gravis: the role of immunodeficiency. *Ann NY Acad Sci* 1976; 274:461–467.

2. Siegel IM: Orthopedic correction of muscolusketal deformity in muscular dystrophy, in Griggs RC, Moxley RT III(eds): *Advances in Neurology*. New York, Raven Press, 1973, vol 17, pp 343–364

Do not include "unpublished observations," "personal communications," and information from manuscripts "submitted for publication" but not yet accepted in references. Cite in parentheses in text. "Unpublished observations" include authors and year; submit letters of permission from individuals cited. For quotations from manuscripts submitted for publication include: authors, title of manuscript, and date. Manuscripts that have been accepted for publication but have not yet been published may appear in references. Include authors, manuscript title, and journal name, followed by "to be published" in parentheses.

Tables: Type double-spaced on a separate sheet; do not submit as photographs. If exceeds one typewritten page, duplicate all headings on second sheet. Avoid wide tables; break up into smaller tables. Number in order of citation in text. Give each a title, and every column, including left (stub) column, a heading. Define all abbreviations and indicate units of measurement for all values. Use commas for numbers exceeding 999; use zeros before decimals for numbers less than 1. Organize tables so like data read vertically, not horizontally. Do not use internal horizontal or vertical lines. Explain all empty spaces or dashes. Indicate footnotes with symbols, in order: *, †, ‡, §, ||, ¶, #. Use lowercase italic letters if more than seven footnotes. Put symbols (or letters) after commas and periods, before colons and semicolons; make superscript. Obtain permission and cite source in legend if data from any other source, published or unpublished, are used.

Figures: Submit professionally drawn and photographed; glossy, high-contrast (black-and-white) prints 8 – 16 cm in width. Make letters, numbers, and symbols clear and large enough to remain legible when reduced. Use correct spelling with no broken letters or uneven type. Use abbreviations consistent with those in text.

Label back with article's running title and TOP. Do not write directly on backs of photographs. Do not trim, mount, clip, or staple illustrations.

Submit photomicrographs in final desired size (preferably 8 cm wide). Photomicrographs must include a calibration bar of appropriate length. Contrast symbols used in micrographs with background.

For photographs of persons, obtain written permission from subject. Mask eyes to prevent identification.

Four color illustrations are considered for publication; authors bear costs of publication. Supply color transparency or negative, in addition to color prints.

Figure Legends: Type double-spaced; number with Arabic numerals corresponding to illustrations. Explain symbols, arrows, numbers, or letters used to identify parts of illustrations in legend. For photomicrographs, define internal scale markers and give methods of staining. If figure has been previously published, include a credit line.

OTHER FORMATS

AAEM NEWS AND COMMENTS—Items of interest to MEM members as prepared by AAEM office and its officers. Direct submissions to MEM office in Rochester, Minnesota.

BOOK REVIEWS—Solicited by Editor. Publishers should send one copy to: Robert L. Rodnitzky, MD, Assistant Editor, Muscle & Nerve, Department of Neurology, University of Iowa Hospitals, Iowa City, IA 52242.

CALENDAR OF EVENTS—Notices of forthcoming meetings in field and current events. Send notices to Editor three months prior to desired publication date.

CASE OF THE MONTH—2000–3000 words; 1 table and 4 figures. Unique or illustrative studies of neuromuscular disorders.

ISSUES AND OPINIONS—2000–3000 words; 1 table and 4 figures. Current topics related to etiology, pathogenesis, electrodiagnosis or therapy of neuromuscular disorders. Need not be data based. A hypothesis and a review of controversial subjects are welcome. When appropriate, opposing views will be presented.

LETTERS TO THE EDITOR—500 words; 1 figure or table. Comment on papers published in this journal or other relevant matters.

MAIN ARTICLES—6000 words. Present original clinical and laboratory research and related topics.

REVIEW ARTICLES—Current topics of importance, primarily by invitation. All papers undergo review process.

SHORT REPORTS—1000 words, 1 figure or table. Preliminary communications of new data, research methods, brief case studies, new ideas and techniques.

OTHER

Abbreviations: Use only standard abbreviations, as listed in *CBE Style Manual* and *AMA Stylebook and Editorial Manual* without definition. Terms appearing frequently may be abbreviated; spell out at first citation, with abbreviation following parentheses.

Electronic Submission: Submit final, accepted version of manuscript on diskette and harpcopy printout. If different, paper copy is considered definitive. Submit 5.25 or 3.5 in. diskette in Macintosh, IBM MS-DOS or Windows format. WordPerfect®is preferred software; manuscripts prepared on any microcomputer word processor are acceptable. Use of Aldus Pagemaker, Quark Xpress and other desktop publishing software is discouraged. Keep document as simple as possible. Refrain from complex formatting. Do not use footnote function.

Submit each article as a single file on one diskette. Name file with last name (8 letters, truncate if too long) followed by a period, and 3-letter descriptive extension MUS. If using Macintosh, maintain MS-DOS file-naming convention of eight letters, period, and three-letter extension.

Label disk with name, title of manuscript, file name, and word processing program used.

Submission of electronic illustrations is encouraged, not required. Submit on separate diskette from text. Use TIFF and EPS files or native application files. Files may be submitted on SyQuest 44 or 88 megabyte cartridges. For grey scale, color submissions, and more information, contact: Gerry Grenier, internet: GGrenier@Jwiley.com; telephone: (212) 850-8860; fax: (212) 850-8888.

Experimentation, Animal: Indicate whether institution's or National Research Council's Guide for Care and Use of Laboratory Animals was followed. For drugs and chemicals, use generic name at first mention and, preferably, thereafter. Put trade name in parentheses; capitalize.

Experimentation, Human: Indicate whether procedures were in accord with standards of Committee on Human Experimentation of institution in which experiments were done or in accord with the Helsinki Declaration of 1975. Do not use patients' names, initials, or hospital numbers.

Style: Use *Webster's Third New International* or *New Collegiate* dictionaries (G. & C. Merriam Co., Springfield, MA) for spelling and hyphenation of nonmedical terms, and *Dorland's Illustrated Medical Dictionary* (WB Saunders, Philadelphia) for medical terms. Good sources for general style (grammar, punctuation, capitalization, etc.) are: *A Manual of Style* (The University of Chicago Press, Chicago) and *The Elements of Style* (Strunk and White, Macmillan Publishing Co., New York). For units of measure, symbols, and nomenclature for biochemistry and biology, use *CBE Style Manual* (American Institute of Biological Sciences, Arlington, VA) and for medicine, use AMA *Stylebook and Editorial Manual* (American Medical Association, Chicago).

Units: Use numerals for all units of measure and time, and for all enumerations). Place SI unit conversions in parentheses following units of measure. Spell out numbers one through nine only for general usage. Spell out numbers beginning a sentence.

SOURCE

January 1995, 18(1): front and back matter.

Mutation Research
Elsevier

AIM AND SCOPE

Mutation Research is an international journal on mutagenesis, chromosome breakage and related subjects. The Journal consists of seven different sections, each with a different, but overlapping character, authorship, and readership.

Mutation Research contains three types of publications: papers reporting results of original fundamental research concerning mutagenesis, chromosome breakage and related subjects; review articles; and short communications (to be published in the section *Mutations Research Letters*).

MANUSCRIPT SUBMISSION

These instructions apply to all sections. Specific section "Aims and Scopes" and submission addresses are listed below.

Submission of a manuscript implies that it contains original work and that it has not been published or submitted for publication elsewhere. It also implies transfer of Copyright from author to publisher. Submit in triplicate, one original plus two copies, with three sets of original illustrations.

MANUSCRIPT FORMAT

Type double-spaced (final version preferably accompanied by a diskette). Include key words (3-6 words or short phrases), abstract (300 words). Introduction summarizes research problem and pertinent findings. Divide text into sections such as Materials and methods, Experimental, Results, Discussion. Number main sections. In papers mentioning chemicals, include *CAS Registry Numbers*.

Title Page: Include title, authors' full names and complete addresses of academic or professional affiliations, indicate corresponding author with an asterisk. Give telephone and fax numbers and e-mail addresses in a footnote.

Figures: Submit in triplicate as unmounted glossy prints, suitable for reproduction across a single column (76 mm) or across a whole page (160 mm); maximum height is 206 mm. Make figures and lettering in proportion and large enough to allow for reduction. Type legends for all figures on a separate page. Extra cost of color reproduction is borne by author.

Tables: Type double-spaced. Give headings.

References: Use Harvard System; give names and dates in body of text and an alphabetical list of references at end of manuscript. Abbreviate journal titles according to *Chemical Abstracts Bibliographic Guide for Authors and Editors 1974.*

Ames, B.N., J. McCann and E. Yamaski (1975) Methods for detecting carcinogens and mutagens, Mutation Res., 31, 347-363.

Eisenberg, L. and C.A. Wachtmeister (1977) Safety precautions in work with mutagenic and carcinogenic chemicals, in: B.J. Kibley, M.S. Legator, W. Nichols and C. Ramel (Eds.), Handbook of Mutagenicity Test Procedures, Elsevier, Amsterdam, pp. 401-410.

OTHER

Electronic Submission: Preferred storage medium is 5.25 or 3.5 in. disks in MS-DOS format; other systems are welcome, e.g., NEC and Macintosh (save file in usual manner, do not use option 'save in MS-DOS format'). Do not split article into separate files.

Ensure that letter 'l' and digit '1' (also letter 'O' and digit '0') have been used properly, and structure article (tabs, indents, etc.) consistently. Do not leave characters not available on word processor (Greek letters, mathematical symbols, etc.) open, indicate by a unique code (e.g., gralpha, @, #, etc., for the Greek letter α). Use codes consistently throughout text. Make a list of codes and provide a key. Do not allow word processor to introduce word splits; do not use a 'justified' layout. Adhere strictly to general instructions on style/arrangement and, in particular, the reference style of the journal.

Save file in wordprocessor format; do not use option to save files in flat ASCII. Format disk correctly. Ensure that only the relevant file is on disk. Specify type of computer and word processing package used. Label disk with author name and name of file on disk. After final acceptance, submit disk plus one final, printed and exactly matching version (as a printout) to accepting Editor. File on disk and printout must be identical.

SOURCE

January 1995, 336(1): inside back cover.

Mutation Research, DNA Repair
Elsevier

Prof. E.C. Friedberg
Department of Pathology
University of Texas Southwestern Medical Center
5323 Harry Hines Blvd,
Dallas, TX 75235-9072
Telephone: (214) 648-4025; Fax: (214) 648-4067;
E-mail: IN%"SIEDE@UTSW.SWMED.EDU"

Prof. A.A. van Zeeland
Dept. of Radiation Genetics and Chemical Mutagenesis
State University of Leiden, Sylvius Laboratories
Wassenaarseweg 72, P.O. Box 9503
2300 RA Leiden The Netherlands
Telephone: +31(71)276150/276151; Fax: +31(71)221615

Prof. H. Takebe
Department of Experimental Radiology
Faculty of Medicine
Kyoto University, Kyoto 606 Japan
Telephone: +81(75)7534410; Fax: +81(75)7534419

AIM AND SCOPE

Mutation Research, DNA Repair provides a forum for the comprehensive coverage of cellular responses to DNA damage in pro- and eukaryotes. It publishes papers on original observations on cellular, biochemical and molecular aspects of DNA damage and repair, and their relationship to human hereditary diseases, cancer, and aging, and to cell cycle progression. The journal also welcomes contributions which incorporate other aspects of the pathogenesis of cancer especially on the role of activated oncogenes and tumor suppressor genes and of genomic instability. Original papers and short communications are published at regular intervals. In addition, invited mini-reviews on selected topics that provide a 'state of the art' synopsis are regularly featured.

SOURCE

January 1995, 336(1): inside back cover.

Mutation Research, DNAging: Genetic Instability and Aging
Elsevier

Richard B. Setlow
Biology Department, Brookhaven National Laboratory
Upton, Long Island, NY 11973

Takashi Sugimura
National Cancer Center
1-1, Tsukiji 5-chome, Tokyo 104 Japan

Jan Vijg
Molecular Genetics Section, Gerontology Division
Harvard Medical School of Beth Israel
330 Brookline Avenue
Boston, MA 02215

AIM AND SCOPE

Mutation Research, DNAaging: Genetic Instability and Aging focuses on papers dealing with: age-related changes in cellular macromolecules, especially alterations in DNA sequence and structure (e.g., mutations, gene rearrangements, changes in methylation, chromosomal aberrations, changes in chromatin conformation), accuracy of transcription and translation; molecular mechanisms of genomic destabilization, studies on endogenous processes and exogenous factors contributing to the maintenance and/or chronic disturbance of genomic integrity (e.g., antioxidant defenses, DNA repair, cellular stress responses, genomic instability disorders, transposons, free radical scavengers); age-related macromolecular damage, studies on endogenous or exogenous agents which can affect DNA, RNA and protein structure and function, and their long-term consequences; age-related changes in information retrieval, studies on the accuracy of transcription and translocation, and on the control of gene expression; and molecular mechanisms of age-related diseases.

SOURCE

February 1994, 316(1): inside front cover.

Mutation Research, Environmental Mutagenesis and Related Subjects
Elsevier

Prof. J.M. Gentile
Biology Department, Hope College
Holland, MI 49423

Prof. K. Sankaranarayanan
Dept. of Radiation Genetics and Chemical Mutagenesis
State University of Leiden
Sylvius Laboratories, Wassenaarseweg 72
P.O. Box 9503
2300 RA Leiden The Netherlands

Prof. B.W. Glickman, Director
Centre for Environmental Health
Department of Biology, University of Victoria
P.O. Box 1700
Victoria, British Columbia V8W 2Y2 Canada
                    Molecular genetics

AIM AND SCOPE

Mutation Research, Environmental Mutagenesis and Related Subjects publishes complete papers that address research topics on the effects of environmental genotoxins on humans and on species composition in ecosystems. Papers dealing with human population monitoring/surveillance for genotoxic effects, new methodologies, and validation of methodologies for assessing genetic damage to humans and other organisms within an ecosystem are particularly encouraged. This section will also publish a selected number of meeting abstracts, summaries of society meetings, and society news that require dissemination to the world community and will thus serve as the Journal's link to scientific societies in the field.

SOURCE

February 1995, 334(1): inside front cover.

Mutation Research, Fundamental and Molecular Mechanisms of Mutagenesis
Elsevier

Prof. J.M. Gentile
Biology Department, Hope College
Holland, MI 49423

Prof. K. Sankaranarayanan
Dept. of Radiation Genetics and Chemical Mutagenesis
State University of Leiden
Sylvius Laboratories, Wassenaarseweg 72
P.O. Box 9503
2300 RA Leiden The Netherlands

Prof. B.W. Glickman, Director
Centre for Environmental Health
Department of Biology, University of Victoria
P.O. Box 1700
Victoria, British Columbia V8W 2Y2 Canada
                    Molecular genetics

AIM AND SCOPE

Mutation Research, Fundamental and Molecular Mechanisms of Mutagenesis publishes complete research papers in all areas of mutation research which focus on fundamental mechanisms underlying phenotypic and genotypic expression of genetic damage, molecular mechanisms of mutagenesis including the relationship between genetic damage and its manifestation as hereditary diseases and cancers. Additional 'special issues' which bring together research papers in specific themes of topical interest will also appear in this section.

SOURCE

January 1995, 336(1): inside front cover.

Mutation Research, Genetic Toxicology
Elsevier

Dr. M.D. Shelby
NIEHS. P.O. Box 12233
Research Triangle Park, NC 27709
                    Non-European papers

Prof. P.H.-M. Lohman
Dept. of Radiation Genetics and Chemical Mutagenesis
State University of Leiden, Sylvius Laboratories
Wassenaarseweg 72, P.O. Box 9503
2300 RA Leiden The Netherlands
Telephone: +31(71)276150/276151; Fax: +31(71)221615

AIM AND SCOPE

Mutation Research, Genetic Toxicology publishes papers on the development and evaluation of testing methods; the testing of agents for genetic toxicity; the detection and monitoring of genetic effects in humans; and the assessment of genetic risks in the overall safety evaluation of chemical exposures and their overall toxicity. Health risks of concern range from genetic diseases and cancer through abnormal reproductive outcomes. Papers will not be limited to any research technique or level of organism.

SOURCE

November 1994, 341(1):back matter.

Mutation Research, Reviews in Genetic Toxicology
Elsevier

Dr. F.J. de Serres
Toxicology Branch, Environmental Toxicology Program
National Institute of Environmental Health Sciences
P.O. Box 12233, MD 19-02
Research Triangle Park, NC 27709

Dr. J.S. Wassom
Human Genome and Toxicology Group
Oak Ridge National Laboratory
1060 Commerce Park, MS-6480
Oak Ridge, TN 37831-6480

AIM AND SCOPE

Mutation Research, Reviews in Genetic Toxicology publishes timely, comprehensive and critical reviews on the potential genetic hazards of environmental agents and periodically, on methodologies and intercomparison of text results with different batteries of assays for mutagenicity. Additionally, this section will consider papers on new technologies in genetic toxicology and their application to mutation research.

SOURCE

Novermber 1994, 341(1): back matter.

Mutation Research Letters
Elsevier

Dr. S.M. Galloway
Merck Research Laboratories, W 45
West Point, PA 19486
Fax: (215) 652-7758

Dr. M.D. Shelby
NIEHS. P.O. Box 12233
Research Triangle Park, NC 27709
Fax: (919) 541-4634
                    U.S.A.

Dr. M. Hayashi
Division of Genetics and Mutagenesis
National Institute of Hygienic Sciences
1-18-1 Kamiyoga, Setagaya-ku
Tokyo 158 Japan
Fax: 3 3703 6950
                    Japan

Dr. L.-D. Adler
GSF-Institut fur Saugetiergenetik
Neuherberg, Ingolstadter Landstr 1
D-85758 Oberschleissheim, Germany
Fax: 49 89 3187 3099
                    Europe

AIM AND SCOPE

Mutation Research Letters aims at rapid publication of short, complete research papers in all areas of mutation research such as fundamental mechanisms of mutagenesis, DNA damage and repair, and genotoxicity testing. In addition, brief preliminary reports will be considered if they describe novel/important observations that warrant rapid publication.

SOURCE

January 1995, 346(1): inside back cover.

Nature
Macmillan

Editor
4 Little Essex Street
London WC2R 3LF, U.K.

1234 National Press Building
Washington, D.C. 20045

AIM AND SCOPE

Nature is an international journal covering all the sciences.

MANUSCRIPT SUBMISSION

Include telephone and fax numbers. Declare manuscripts or proofs sent by air courier to London as 'manuscripts' and 'value $5.' Brevity is highly valued. One printed page, without display items, contains about 1300 words.

MANUSCRIPT FORMAT

Type manuscripts, double-spaced, on one side of paper only. Submit original and four copies, each with artwork. Submit five sets of original photographs; for line drawings, send one set of originals and four good quality photocopies. Put reference lists, figure legends and tables on separate sheets, double-spaced and numbered. Include three copies of relevant manuscripts in press or submitted for publication elsewhere, clearly marked as such.

Limit articles to 3000 words of text (excluding figure legends), 50 references, and six display items to occupy five pages. Use a few short subheadings. Begin with a heading (50–80 words) advertising content in general terms. Do not use numbers, abbreviations or measurements. Introduce study in first two or three paragraphs, briefly summarizing results and implications.

Title: Make brief and simple. Avoid active verbs, numerical values, abbreviations and punctuation. Include one or two key words for indexing.

Artwork: Mark individually and clearly with author's name and, when known, manuscript number. Make no larger than 28 by 22 cm. Avoid figures with parts, use only if parts are closely related, either experimentally or logically. Provide unlettered originals of photographs.

Use three-letter codes for amino acids in protein/nucleotide sequences. One column width can accommodate 20 amino acids or 60 base pairs.

Figure Legends: 300 words; describe figure first, then, method. Refer to a published method rather than giving a full description. Do not describe methods in text.

References: Number sequentially as appear in text, followed by those in tables and then those in figure legends. Number and include only papers published or in press in reference list. Cite other forms of reference in text as a personal communication, manuscript submitted or in preparation. Abbreviate journal titles according to *World List of Scientific Periodicals* (Butterworths, London, 1963–65). Give first and last page numbers. References to books include publisher, place and date.

Acknowledgments: Make brief. Place after reference list; do not include grant and contribution numbers.

OTHER FORMATS

COMMENTARY ARTICLES—Deal with issues in, or arising from, research that are also of interest to general readers.

LETTERS TO NATURE—1000 words, 30 references, and four display items; outstanding novel findings whose implications are general and important enough to be of interest to those outside the field. In first paragraph (150 words) describe, without abbreviations, background, rationale and chief conclusions of study. Do not use subheadings.

NEWS AND VIEWS ARTICLES—Inform nonspecialists about new scientific advances, sometimes as a conference report. Most are commissioned; make proposals to Editor.

PROGRESS ARTICLES—4 pages reviewing particularly topical developments for a nonspecialist readership. Make suggestions to Reviews Coordinator in a brief synopsis.

REVIEW ARTICLES—Survey recent developments in a field. Most are commissioned; suggestions are welcome. Submit a one page synopsis to Reviews Coordinator.

SCIENTIFIC CORRESPONDENCE—500 words; discussions of topical scientific matters, and for miscellaneous contributions.

SUPPLEMENTARY INFORMATION—Material relevant to Articles or Letters which cannot, for lack of space, be published in full. It is available from *Nature* on request.

OTHER

Abbreviations, Symbols, Units and Greek Letters: Identify at first use. Avoid acronyms; if used, define. Do not use footnotes in text.

Color Charges: £500 per page. Inability to pay will not prevent publication; explain circumstances.

Data Submission: Deposit sequence and crystallographic data in databases for this purpose.

Reviewers: Suggest reviewers; limited requests for exclusion of specific reviewers are usually heeded.

SOURCE

October 20, 1994, 371: 720.

Nature Genetics
Nature America

Editor, *Nature Genetics*
1234 National Press Building
529 14th St. NW
Washington D.C. 20045
Telephone: (202) 626-2513; Fax: (202) 628-1609;
E-mail: natgen@linneus.naturedc.com

AIM AND SCOPE

Nature Genetics is an international monthly journal publishing important advances in all fields affecting human and mammalian genetics and the Genome Project, including the identification and characterization of genes and gene products involved in hereditary disease, genetic diagnosis and therapy, molecular analysis and mapping of the genome (in all organisms), animal models, cancer and developmental genetics.

MANUSCRIPT SUBMISSION

Submit an original and three copies, each accompanied by artwork, together with a computer diskette. Put reference lists, figure legends and tables on separate sheets, double-spaced. Include four copies of any relevant manuscript in press or submitted for publication. In cover letter suggest potential reviewers. Inform Editor of potential conflicts of interest. Provide current fax and phone numbers of corresponding author.

MANUSCRIPT FORMAT

Type manuscripts double-spaced, on one side of paper only. Color prints are partly paid for by authors unless otherwise agreed. Follow style and format of articles in *Nature,* with a Summary, Introduction, Results and Discussion. There is a separate Methodology section following main text. Include full titles of papers in reference list. Text should be between 1000 and 3000 words.

Title: Keep simple and concise.

Summary: 100 words explaining goals, results and chief conclusions of work. Do not include references.

Results: Include short cross-headings to define main aspects of study. Deposit sequence data in databases, and provide accession number in paper.

Methodology: At end of text, before references; do not include in figure legends.

References: Number sequentially as they appear in text, followed by those in figure legends and tables. Do not include any annotation. Include full titles of paper. List all authors unless six or more, then substitute *'et al.'* Include first and last page numbers; references to books include publisher, place and date. Include in list only papers published or in press; cite abstracts, papers submitted or in preparation and personal communications in text. Where possible, use a reference to a review; keep references to a minimum.

1. Grantham, J. J. & Burrow, C. R. The nucleotide sequence of genes coding for Brutons disease. Mol. Cell. Biol. **15**, 110–116 (1990).

2. Miliew, J. *et al.* Introns of the *ras* genes. Science **1124**, 774–778 (1989).

3. Jay, R. A. *Manual of Genetic Transformations.* (Benjamin, Menlo Park, California, 1995).

4. Wake, R. E. The basis for tumor formation. in *Tumor Suppressor Genes* (ed. Klein, G.) 217–243 (Marcel Dekker, Inc., New York, 1990).

Figures: Submit original artwork. Avoid oversized art.

Acknowledgments: Keep as brief as possible.

OTHER

Electronic Submission: Page proofs are set directly from computer discs provided by authors. Any common Macintosh or PC word-processing package is compatible and preferable to a text/ASCII file. Manuscripts written in WordStar cannot be processed.

SOURCE

January 1995, 9(1): back matter.

Nature Medicine
Nature America

Editor, *Nature Medicine*
1234 National Press Building
529 14th St. NW
Washington D.C. 20045
Telephone: (202) 626-2513; Fax: (202) 628-1609;
E-mail: natgen@linneus.naturedc.com

AIM AND SCOPE

Nature Medicine is an international monthly journal publishing important advances in all fields of biomedical science, especially those relating to molecular medicine that represent a conceptual advance or an original approach to understanding the molecular basis of pathogenesis and therapy. Original research articles in diverse fields of biomedical science are welcomed, including gene therapy, neuroscience, pharmcology, and advanced medical techniques.

MANUSCRIPT SUBMISSION

In cover letter suggest potential reviewers. Inform Editor of potential conflicts of interest. Submit original and three copies, each accompanied by artwork, together with a computer diskette. Place reference lists, figure legends and tables on separate sheets, double-spaced. Include four copies of any relevant manuscript in press or submitted for publication. Provide current fax and phone numbers of corresponding author on all submissions.

Manuscripts previously rejected by *Nature Genetics* on editorial grounds or due to space limitations may be resubmitted to *Nature Medicine*. Such papers may be published without peer review if work is technically sound and meets editorial requirements of *Nature Medicine*.

MANUSCRIPT FORMAT

Type manuscripts, double-spaced, on one side of paper only. Color prints are partly paid for by authors unless otherwise agreed. Generally follow style and format of an article in *Nature,* with an Abstract, Introduction, Results and Discussion. A separate Methodology section follows main text. Include full titles of papers in reference list. Text should be between 1000 and 3000 words in length.

Title: Keep simple and concise.

Abstract: 100 words explaining goals, results and chief conclusions of work; do not include references.

Introduction: Provide sufficient background to clarify recent history of field, present aims of current work and advance presented. Write for a non-specialist so work can be placed in context.

Results: Include short cross-headings to define main aspects of study. Deposit sequence data in databases, and provide an accession number in paper.

Methodology: At end of text, before references; do not include in figure legends. Present all methods and details of protocols in detail, using references where appropriate.

References: Number references sequentially as they appear in text, followed by those in figure legends and tables. Do not include any annotation. Include full titles of papers. List all authors unless six or more, then substitute '*et al.*' Include first and last page numbers in full; references to books include publisher, place and date. Include in list only papers published or in press; cite abstracts, papers submitted or in preparation and personal communications in text. Where possible, reference a review; keep references to a minimum.

1. Levi, J. J. & Brown, C. R. Movement of HIV is mediated by CD120. Mol. Cell. Biol. **15**, 110–116 (1990).

2. Moss, J. *et al.* Nucleotide repeats at the end of chromosome 4 as the basis for Alzheimers disease. Science **1124**, 774–778 (1995).

3. Jay, R. A. *Manual of Genetic Transformations.* (Benjamin, Menlo Park, California, 1995).

4. Wake, R. E. The basis for tumor formation. in *Tumor Suppressor Genes* (ed. Klein, G.) 217–243 (Marcel Dekker, Inc., New York, 1990).

Figures: Submit original artwork. Avoid oversized art.

Acknowledgments: Keep as brief as possible. Include grant numbers.

OTHER

Electronic Submission: Page proofs are set directly from computer discs provided by authors. Any common Macintosh or PC word-processing package is compatible and preferable to a text/ASCII file. Manuscripts written in WordStar cannot be processed.

SOURCE
January 1995, 1(1): 38.

Nature Structural Biology
Nature America

Editor, *Nature Structural Biology*
1234 National Press Building
529 14th St. NW
Washington DC 20045
Telephone: (202) 626-2513; Fax: (202) 628-1609;
E-mail: natgen@linneus.naturedc.com

AIM AND SCOPE

Nature Structural Biology is an international monthly journal publishing important advances in all fields relating to the structures of biological macromolecules as determined by X-ray crystallography, NMR, or electron diffraction, including improvements in the available methods for the determination of such structures, studies of biological mechanism or function in which the methods used are primarily structural, studies of macromolecular structure using molecular biological techniques, and theoretical and applied methods for structure predictions. Preference is given to structures of special interest and structural studies elucidating biological function.

MANUSCRIPT SUBMISSION

Submit original and three copies of manuscript, each accompanied by artwork, together with a computer diskette. Put reference lists, figure legends and tables on separate sheets, also double-spaced. Include four copies of any relevant manuscript in press or submitted for publication. Provide current fax and phone numbers of corresponding authors. In cover letter suggest potential reviewers. Inform Editor of potential conflicts of interest.

Manuscripts previously rejected by *Nature* on editorial grounds or due to space limitations may be resubmitted to *Nature Structural Biology*. Such papers may be published without peer review if work is technically sound and meet editorial requirements of *Nature Structural Biology*.

MANUSCRIPT FORMAT

Type manuscripts double-spaced, on one side of paper only. Color prints are partly paid for by authors unless otherwise agreed. Follow style and format of articles in *Nature,* with a Summary, Introduction, Results and Discussion. A separate Methodology section follows main text. Include full titles of papers in reference list. Text should be less than 5000 words.

Title: Keep simple and concise.

Summary: 100 words explaining goals, results and chief conclusions of work. Do not include references.

Results: Include short cross-headings to define main aspects of study. Deposit sequence data in databases; provide accession number in paper.

Methodology: At end of text, before references; do not include in figure legends.

References: Number sequentially as they appear in text, followed by those in figure legends and tables. Do not include any annotation. Include full titles of paper. List all authors unless six or more, then substitute *'et al.'* Include first and last page numbers; references to books include publisher, place and date. Include in list only papers published or in press; cite abstracts, papers submitted or in preparation and personal communications in text. Where possible, use a reference to a review; keep references to a minimum.

1. Gram, J. J. & Brown, C. R. Protein folding across the nuclear membrane as a basis for ribosomal transport. Mol. Cell. Biol. **15**, 110–116 (1990).

2. Miller, J. *et al.* Three dimensional analysis of a DNA binding protein. Science **1124**, 774–778 (1995).

3. Jay, R. A. *Manual of X-ray Crystallography.* (Benjamin, Menlo Park, California, 1995).

4. Wake, R. E. The basis for tumor formation. in *Tumor Development* (ed. Klein, G.) 217–243 (Marcel Dekker, Inc., New York, 1990).

Figures: Submit original artwork. Avoid oversized art.

Acknowledgments: Keep as brief as possible.

OTHER

Electronic Submission: Page proofs are set directly from computer discs provided by authors. Any common Macintosh or PC word-processing package is compatible and preferable to a text/ASCII file. Manuscripts written in WordStar cannot be processed.

SOURCE
January 1995, 2(1): back matter.

Naunyn-Schmiedeberg's Archives of Pharmacology
Springer International

Dr. M. Göthert
Institut für Pharmakologie und Toxikologie der Universität Bonn
Reuterstrasse 2b
D-53113 Bonn, Germany
Fax: (0)2 28 73 54 04

Dr. K.H. Jakobs
Institut für Pharmakologie
Universitätsklinikum Essen
Hufelandstrasse 55
D-45122 Essen Germany
Fax: (0)20 17 23 59 68

AIM AND SCOPE

Naunyn-Schmiedeberg's Archives of Pharmacology will consider manuscripts in all fields of pharmacology for publication as full papers or Short communications. The publication must make a significant contribution to pharmacological knowledge.

MANUSCRIPT SUBMISSION

Submission of a manuscript implies that it has not been published before, and is not being submitted for publication elsewhere in whole or in part. This restriction does not apply to results published as an abstract, as part of a lecture or a review, or symposium contribution, provided submission adds significantly to previously published information. Submission also implies that publication has approval of all authors.

Submission implies that if manuscript is accepted, authors agree to automatic transfer of copyright to publisher, and that manuscript will not be published elsewhere in any language without consent of copyright holders.

Submit in triplicate. Submit one set of illustrations of a quality suitable for reproduction.

MANUSCRIPT FORMAT

Type double-spaced on one side of paper only with wide margins. Number pages. Do not divide words at end of lines. Use hyphens, capital letters, abbreviations, units, etc. consistently. Mark desired position of figures and tables in margin. Consult recent issues of journal for general layout and details. Arrange manuscript in order: title page; Summary and Key words; Introduction; Methods; Results; Discussion; Acknowledgments; References; Tables; Figures and figure legends.

Title Page: Title of paper; names of authors; full address(es) of institution(s) where work was performed; if work was carried out at more than one institution, follow names of authors by superscript numbers preceding names of corresponding institutions; "Correspondence to" followed by name of author and either "at the above address" or another address as appropriate.

Summary and Key Words: Begin with Summary (three unnumbered paragraphs), concisely presenting purpose of study, general methods used, results and conclusions. If unavoidable, give full references without title of paper. Below, give eight key words for indexing.

Introduction: Avoid detailed historical introductions. Briefly define subject matter and clearly state aim of study.

Methods: Describe in enough detail to enable repetition of experiments.

Results: Include experimental results but not conclusions or theoretical considerations unless reason for performing an experiment is unclear. Do not repeat data in text, tables and figures. In short papers, Results and Discussion may be combined (Results and discussion).

Discussion: Interpret results against background of existing knowledge. Avoid recapitulation of results.

References: Include only works cited in text. Abbreviate journal titles according to *Index Medicus*.

Meyer W, Nose M, Schmitz W, Scholz H (1984) Adenosine and adenosine analogs inhibit phosphodiesterase activity in the heart. Naunyn-Schmiedeberg's Arch Pharmacol 328:207–209

Trendelenburg U (1984) Metabolizing systems. In: Fleming WW, Graefe KH, Langer SZ, Weiner N (eds) Neuronal and extraneuronal events in autonomic pharmacology. Raven Press, New York, pp 93–109

List at end of paper in alphabetical order under first author's name. If papers by several authors with the same first author, order is: papers with one coauthor (list alphabetically by coauthor, and then chronologically); papers with more than one coauthor (list chronologically and as only first author and "et al.," followed by year of publication in text). If more than one work by same author(s) in same year, add a, b, c, etc. to year both in text and in list.

Include in reference list papers accepted for publication with name of journal and "in press;" submit copies. Do not include submitted papers, not yet accepted. Mention unpublished observations or personal communications in text with names of authors or communicators.

Tables: Type each on a separate page. Number consecutively followed by a brief descriptive title. Give explanations at bottom of table. Make comprehensible with minimal reference to text.

Figures: Restrict to minimum needed to clarify text. Present clear illustrations to be legible after reduction to a width of 86 mm (column width), 178 mm (full print area), or 118 mm (with legend at side). Make symbols, numbers and letters about 2 mm in height after reduction. Color illustrations are accepted.

Submit one set of illustrations suitable for reproduction. Submit original drawings in black ink, high-quality graphs generated by computer, or sharp, glossy photographs no larger than 210×297 mm (A4). Indicate authors' names, number, and TOP on back.

Illustrations for referees may be clear photocopies of line drawings; halftones must be photographs. Place each figure on a separate sheet with its legend typed below, as it will appear in print. Legends make figures comprehensible with minimal reference to text. Explain abbreviations and symbols in figure in legend. Make figure legends and explanations in tables consistent in wording. Submit three sets of figures with legends typed below.

OTHER FORMATS

SHORT COMMUNICATIONS—2-3 printed pages, including two display items (figures, tables). Brief, but complete series of experiments with results of unusual interest; accelerated review process. Explain in cover letter why data merit publication in this form.

OTHER

Abbreviations: Use International System of Units (SI units). See *Naunyn-Schmiedeberg's Arch Pharmacol* 1994; 350(1): 348(1):A4-A5 for abbreviations for units of physico-chemical quantities, including space, time, mechanical, and thermodynamic quantities, prefix abbreviations, and abbreviations that may be used without definition.

Color Charges: DM 1200.- for first and DM 600.- for each additional page.

Electronic Submission: Submit diskettes after completion of review process. Use diskettes formatted for DOS or Macintosh systems. Store text in two versions: standard data file format offered by word processing system and either (in declining order of preference) RTF (Microsoft Rich Text Format), DCA/RFT (Document Containment Architecture/Revisable Form Text), DCA/FFT (Document Containment Architecture/Final Form Text), ASCII or "text only"

Keep a copy of diskette. Make sure diskette is adequately packed for mailing. Enclose a printout of final text with diskette. Text file and printout must correspond exactly. Deviations may delay processing. Do not incorporate special page layout in text. Delete annotations or comments from final text file. Send only final updated version.

Input text continuously; insert hard returns at ends of paragraphs or headings, subheadings, lists, etc. Do not use space bar to make indents; use tabulator or indent command. Use automatic pagination function. Indicate words or phrases in text to be emphasized in italic script or, by underlining. Use boldface type in text for certain mathematical symbols, e.g., vectors. In table titles, boldface word "Table" and table number. Boldface figure legends, abbreviation "Fig.," figure number, and letters referring to figure parts (a, b, etc.). Boldface headings for emphasis.

Hyphen/dash coding

Hyphen	high-resolution
En-dash	1990–1992, Diabetologia 25:345–352
Em-dash	Bacteria—in high numbers—were found
Minus sign	at a temperature of −75°C

Place tables at end of file. Separate columns using tabulators, not space bar.

Experimentation, Human and Animal: Studies must be in accordance with recommendations in the Declaration of Helsinki and with internationally accepted principles concerning the care and use of laboratory animals.

Footnotes: Title footnotes do not have symbols; indicate footnotes to authors' names with asterisks; those to text by consecutive numbering.

Nomenclature: Use International Nonproprietary Names of drugs. Indicate proprietary equivalent in round brackets. If a drug has no International Nonproprietary Name, use chemical name or suitable abbreviation. Identify drugs referred to by code numbers chemically at first mention.

For stereoisomers, state whether racemate or an enantiomer was used. Use prefixes (+)-, (–)-, and (±)- to indicate optical rotation. Small capital D and L refer to absolute configuration.

Use following names of receptors: Acetylcholine receptors or cholinoceptors; subtypes are muscarine (muscarinic) receptors, with M_1 etc. subtypes, and nicotine (nicotinic) receptors. Adrenoceptors: use names noradrenaline and adrenaline for their endogenous agonists. Dopamine receptors with D_1, etc. subtypes: there is no need to use DA_1 and DA_2 for peripheral receptors. 5-Hydroxytryptamine receptors or serotonin receptors with 5-HT_1, etc. subtypes. Opioid μ-, δ-, and K-receptors. Use suffix "ergic" only for nerve fibres or transmission process.

SOURCE

July 1994, 350(1): 348(1): A3–A6.

Nephron
Karger Medical and Scientific Publishers

S. Karger AG
Editorial Dept. *Nephron*
Allschwilerstrasse 10
P.O. Box CH-4009 Basel, Switzerland

MANUSCRIPT SUBMISSION

Manuscripts must not be under simultaneous consideration by any other publication. Submission implies transfer of copyright from author to publisher upon acceptance. Accepted papers become permanent property of *Nephron* and may not be reproduced by any means, in whole or in part, without written consent of publisher. Obtain permission to reproduce illustrations, tables, etc. from other publications.

Submit in triplicate (with one original set of illustrations) typed double-spaced on one side of paper, with wide margin.

MANUSCRIPT FORMAT

Consult leaflet 'Rules for the Preparation of Manuscripts', available on request from the Publisher, S. Karger AG, P.O. Box, CH-4009 Basel (Switzerland).

Number each page consecutively; put author's name in upper right corner.

Title Page: On first page list title (main title underlined), authors' names, and institute where work was conducted; a short title for a running head; and full address of corresponding author. Give exact postal address complete with postal code at bottom of title page. Supply phone and fax numbers.

Key Words: For indexing, 3-10.

Abstract: 10 lines. Last sentence of abstract should summarize work without mathematics.

Small Type: Indicate paragraphs to be set in smaller type (case histories, test methods, etc.) with a 'p' (petit) in margin on left side.

Footnotes: Avoid footnotes. When essential, number consecutively and type at foot of appropriate page.

Tables and Illustrations: Number in Arabic numerals and prepare on separate sheets. Give tables a heading and figures a legend, also on a separate sheet. Submit good drawings and original photographs; do not submit negatives or photocopies. When possible, group illustrations on one block for reproduction (max. size 181×223 mm) or provide crop marks. On back of each illustration, indicate number, author's name, and TOP. Color illustrations are reproduced at author's expense.

References: In text identify by Arabic numerals [in square brackets]. Note material submitted for publication but not yet accepted as 'unpublished data;' do not include in reference list. Include only those publications which are cited in text in list. Do not alphabetize; number references in order of mention in text. Give surnames of authors followed by initials. Use a comma to separate authors. Cite all authors, 'et al' is not sufficient. Abbreviate journal names according to *Index Medicus*. (Also see International Committee of Medical Journal Editors: Uniform requirements for manuscripts submitted to biomedical journals. *Br Med J* 1988;296:401–405.)

PAPERS PUBLISHED IN PERIODICALS

Kauffman HF, van der Heide S. Beaumont F, Blok H, de Vries K: Class-specific antibody determination against *Aspergillus fumigatus* by means of the enzyme-linked immunosorbent assay. III. Comparative study: IgG, IgA, IgM ELISA titers, precipitating antibodies and IgE binding after fractionation of the antigen. Int Arch Allergy Appl Immunol 1986;80:300–306.

MONOGRAPHS

Matthews DE, Farewell VT: Using and Understanding Medical Statistics. Basel, Karger. 1985.

EDITED BOOKS

Hardy WD Jr, Essex M: FeLV-induced feline acquired immune deficiency syndrome: A model for human AIDS; in Klein E (ed): Acquired Immunodeficiency Syndrome. Prog Allergy. Basel, Karger, 1986. vol 37, pp 353–376.

OTHER

Color Charges: SFr. 600.-/U.S. $400 per page for up to 6 color illustrations per page.

Electronic Submission: Store on 3.5 or 5.25 in. disk. Preferred word processing package is Microsoft Word. Other commonly used PC text programs are also accepted, as well as ASCII (144 byte HD) and Macintosh formats. Submit double-spaced printout that exactly matches text. Mark subsequent revised versions as such on disk and printout. Check galley proofs against copyedited manuscript. Label disk with title of journal, short paper title, author's name, software and hardware used. and file names contained (one article per file).

Use same type style for headings of same ranking throughout text. Be consistent, follow same patterns throughout text. Do not use automatic page numbering; number pages by hand. Text and headings should be ranged left throughout without word breaks. Do not center, space or use block letters (except for abbreviations). Do not split words. Do not use indentations. Leave a blank line between paragraphs. Do not use hyphenating program. Use word processing capabilities to boldface, italicize, make super- and subscripts, etc. Insert one space after each punctuation mark. For special characters not on keyboard, use the same readily identifiable code throughout. Supply a list of codes used at beginning of text or include a 'Read-me-file.' Do not use spaces in tables; use tabulator. Format operation data within tables in one column; do not split into two. Illustrations and mathematical formulae are handled conventionally.

Page Charges: None for papers of 4 (Letters to the Editor: 1) or fewer printed pages (including tables, illustrations and references). Each additional complete or partial page is SFr. 245.- / US $196. Allocation is equal to approx. 12 (Letters to the Editor - 2) manuscript pages (including tables, illustrations and references).

Units: Submit data in both SI Units and in mks/cgs units, with exception of blood pressure (millimeters of mercury).

SOURCE
January 1995, 69(1): front matter.

Neuroendocrinology
International Society of Neuroendocrinology

Dr. Claude Kordon
INSERM U. 159
2ter, rue d'Alésia
F-75014 Paris, France
Dr. Richard Weiner
Department of Obstetrics, Gynecology
   and Reproductive Sciences
Reproductive Endocrinology Center
University of California
San Francisco, CA 94143-0556
            From North America
Dr. Hiroo Imura
Department of Medicine
Kyoto University
Sakyo-ky, Kyoto 606-01 Japan
            From Southeast Asia

AIM AND SCOPE

The Editorial Board of *Neuroendocrinology* welcomes articles for review that report new observations on interactions between the brain and the endocrine system. Papers that are published deal with both basic and clinical subjects. Editorials, summaries of meetings, and reviews of timely topics in basic and clinical neuroendocrinology are also published, but these items are accepted only at invitation of Editor-in-Chief.

MANUSCRIPT SUBMISSION

Submit unpublished results of research work; be concise. Type or letter-quality print on one side of paper; double-space with wide margin. Send original and three first class copies of manuscript. One set of original illustrations plus three copies is adequate, except for photomicrographs or other illustrations that do not reproduce well, submit four clear prints of each illustrations. Present as: Abstract, Introductory statement on purpose of studies, Materials and methods, Results, Discussion, and References.

MANUSCRIPT FORMAT

Title Page: On first page list title, authors full first and last names, and name of institute or department where work was done. Include: short title for use as a running head; 3-9 key words for indexing; name and complete mailing address, phone and fax numbers of corresponding and reprint request author(s).

Abstract: Keep short. Describe procedures, observations and conclusions.

Footnotes: Avoid if possible. When essential, number consecutively and type at foot of appropriate page.

Acknowledgments: Include credit to sources of grant support.

Tables and Illustrations: Number in Arabic numerals. Prepare on separate sheets; give each a suitable heading.

Number illustrations in Arabic numerals and submit legends for figures on a separate page. Submit only good drawings and original photographs; do not submit negatives or photocopies. When possible, group several illustrations on one block for reproduction (max. size 181 × 223 mm) or provide crop marks. On back of each illustration, indicate number, author's name, and TOP. Color illustrations are reproduced at author's expense.

References: Identify by Arabic numerals [in square brackets]. Note material submitted for publication but not yet accepted as 'unpublished data;' do not include in reference list. Include only those publications cited in text. Do not alphabetize; number references in order of mention in text. Give surnames of all authors followed by initials. Use a comma to separate authors. Cite all authors; 'et al' is not sufficient. Abbreviate journal names according to *Index Medicus*. (Also see International Committee of Medical Journal Editors: Uniform requirements for manuscripts submitted to biomedical journals. *Br Med J* 1988;296:401–405.)

PAPERS PUBLISHED IN PERIODICALS

Kauffman HF, van der Heide S, Beaumont F, Blok H, de Vries K: Class-specific antibody determination against *Aspergillus fumigatus* by means of the enzyme-linked immunosorbent assay. III. Comparative study: IgG, IgA, IgM ELISA titers, precipitating antibodies and IgE binding after fractionation of the antigen. Int Arch Allergy Appl Immunol 1986;80:300–306.

MONOGRAPHS

Matthews DE, Farewell VT: Using and Understanding Medical Statistics. Basel, Karger. 1985.

EDITED BOOKS

Hardy WD Jr, Essex M: FeLV-induced feline acquired immune deficiency syndrome: A model for human AIDS; in Klein E (ed): Acquired Immunodeficiency Syndrome. Prog Allergy. Basel, Karger, 1986. vol 37, pp 353–376.

Do not list papers that have been submitted to a journal but are not yet accepted; refer to data in them in text as unpublished observations. If details of methods are only available in papers that are in press, submit copies with manuscript.

Consult leaflet 'Rules for the Preparation of Manuscripts', from Publisher at above address.

OTHER FORMATS

RAPID COMMUNICATIONS—7 manuscript pages (double-spaced, including figures, tables and references) presenting new findings of sufficient importance to justify accelerated appearance.

OTHER

Abbreviations: Use commonly accepted abbreviations (e.g., ACTH, TSH, DNA) throughout text and tables; define the first time each abbreviation is used. Limit to those in common use; avoid other abbreviations and acronyms.

Color Charges: 6 per page; SFr. 600.- /U.S. $400 per page.

Experimentation, Animal: A criterion considered in reviewing manuscripts is the humane and proper treatment of animals. While the use of anesthetics, analgesics, and tranquilizers may defeat the purpose of some experiments, the use of painful or otherwise noxious stimuli must be carefully and thoroughly justified in paper. Papers that do not meet these criteria will not be accepted for publication.

SOURCE

July 1994, 60(1): front matter.

Neurology
American Academy of Neurology

Robert B. Daroff, M.D.
University Hospitals of Cleveland
2074 Abington Road
Cleveland, Ohio 44106

AIM AND SCOPE

Neurology, the official journal of the American Academy of Neurology, publishes clinical and research articles in neurology, neuroscience, and related fields. Research studies on animals must have clear applicability to humans or diseases of the nervous system in humans. Papers and posters presented at the Academy meetings are the property of the Academy, and *Neurology* has the right of first acceptance of the full manuscript. Authors who wish to publish the manuscript elsewhere must request written release from the Editor-in-Chief. We will not reconsider revisions of manuscripts previously rejected by *Neurology*.

MANUSCRIPT SUBMISSION

Manuscripts, or their content. must be previously unpublished, except in abstract form, and may not be under simultaneous consideration by another journal. Do not submit data from an accepted manuscript to another journal until after actual publication in *Neurology;* and cite it appropriately so Editorial Board and readers are aware of the duplication. If there is a possibility of repetitive, duplicated, or redundant publication explain, in cover letter, reason for simultaneous or sequential publication of same or overlapping material.

Submission implies that all coauthors have seen and approved final version of manuscript.

Comply with International Committee of Medical Journal Editors statement on "Conflict of Interest" *(Ann Intern .Med* 1993;118:646-647). In cover letter, advise Editor-in Chief of pertinent financial interest (ownership, equity position, stock options, patent-licensing arrangements), consulting fees. honoraria, or expert testimony associated with manufacturer of drug or product, or commercial laboratory, within past 5 years. This applies to all authors and their immediate families. Editor-in-Chief and authors will agree on necessity and form of disclosure in journal. Failure to reveal information constitutes fraudulent submission and may cause a published paper to be retracted, authors to be prohibited from further submission to *Neurology*, or both.

MANUSCRIPT FORMAT

Submit four copies of article, including illustrations. Text and tables may be photocopies. When referencing a paper "in press," include four copies.

Double-space manuscripts, including text, tables, figure legends, and references. Type on only one side of page. Indent paragraphs five spaces. Place page number and lead author's last name in upper right corner of each page including reference pages, tables, and figure legends.

Type manuscript with a good quality ribbon on 8.5 × 11 in. or 8.25 × 11.75 in. bond paper; left margin = 1 .5 in., and right margin .5 in. or more; do not justify right margin. White out and retype corrections. Manuscripts that are not of letter quality will be returned.

Title Page: Include names of authors followed by highest academic degrees (MD, PhD) and institutional affiliations. Use full given names or initials except when more than four authors; then use only initials. Provide three to five key words for indexing. List name, address, telephone, and fax numbers of corresponding author. List acknowledgments, including financial support from manufacturers of drugs or products in manuscript.

Abstract: Type on a separate page. Required for full-length articles, Brief Communications, and Issues of Neurological Practice, but not for Clinical/Scientific Notes; optional for Views & Reviews and Historical Neurology. Use narrative form (six typewritten lines) for Brief Communications and may be narrative (150 words) or structured (260 words) for other categories of articles. Use phrases rather than complete sentences. See "Guidelines for the preparation of structured abstracts" in *Ann Intern Med* 1990;113:69-76 (Appendixes 1 and 2).

Figures and Tables: Figures must be first-generation glossy prints or original artwork. Do not mount glossy prints or make larger than 8 × 10 in. Composites that are photographs of photographs may be used for review purposes, but are not acceptable for reproduction. Retain printers copies until manuscnpt is finally accepted. Indicate cropping on photocopy, not on original photograph. Single-column photograph width should not exceed 8 cm; two columns (full-page width) should not exceed 17 cm. Make symbols, lettermg, and numbering sufficiently large to remain legible after reduction to fit single column width. If a figure is prepared for one column (8 cm), make letters, numbering and symbols 1.5 -3 mm high; for two columns (17 cm), 3- 6 mm high. In graphs, use standard symbols for data points in order: ● ■ ▲ ◆ ○ ▢ △ ◊. Not acceptable are symbols like ⊕ ⊗. To prevent wasted white space, axes should end not more than one increment beyond final data points. Explanatory lettering should not extend beyond ends of axes. Submit permission statements for photographs of recognizable patients.

On back of each figure, indicate name of lead author and TOP. Include internal scale markers on all microscopic photographs; those without scale markers are returned for placement.

Place legends for illustrations on a separate page. Author must bear part of expense for color reproduction. Provide both transparencies (slides) and prints.

Figures of analog data (e.g., from strip chart records) should be of raw data. When a figure is traced or redrawn, identify it as such in caption. Editor-in-Chief may request original raw data for reviewers.With figures reflecting tabular data (e.g., bar graphs of relative frequencies, life tables), submit tables of actual data unless data are already included in, or derivable from, manuscript. Type each table, with a title, on a separate sheet. Number tables and figures in order mentioned in text.

Lengthy, complex tables are required for renewers, but need not be published; file with National Auxiliary Publications Service (NAPS), which keeps information on file and provides a file number to be used as a reference citation. Information is available for a small fee to any reader from ASIS/NAPS, New York.

References: Cite references in numerical order in text. List all authors when six or fewer; when seven or more, list first three and add et al. Use *Index Medicus* abbreviations for journal names; eliminate Amencan cities in parentheses after name of journal. Do not reference submitted papers; mention in body of text and provide four copies. Mention personal communications in body of text. References follow Vancouver style described in "Uniform Requirements for Manuscripts Submitted to Biomedical Journals"; pagination must be complete.

JOURNAL ARTICLE

1. Kurtzke JF, Hyllested K. Multiple sclerosis in the Faroe Islands. II. Clinical update, transmission, and the nature of MS Neurology 1986;36:307–328.

PUBLISHED ABSTRACT

2. Olney RK, Aminoff MJ. Diagnostic sensitivity of different electrophysiologic techniques in Guillain-Barr syndrome [abstract]. Neurology 1989;39(suppl 1):354.

UNPUBLISHED MATERIAL

3. Mark MH, Dickson DW, Schwarz KO, et al. Familial diffuse Lewy body disease. Presented at the 10th International Symposium on Parkinson's Disease; October 19,1991; Tokyo.

LETTER

4. McCrank E. PSP risk factors [letter]. Neurology 1990;40:1637.

BOOK

5. Caplan LR, Stein RW. Stroke. A clinical approach. Boston: Butterworths, 1986.

BOOK CHAPTER

6 Munsat TL. Spinal muscular atrophies. In: Rowland LP, ed. Merritt's textbook of neurology. 8th ed Philadelphia: Lea & Febiger, 1939:678–682.

OTHER FORMATS

BRIEF COMMUNICATIONS–1250 words, three tables or figures. and 10 references.

CLINICAL/SCIENTIFIC NOTES—750 words, one table or figure, seven references, and no abstract.

CORRESPONDENCE—500 word comments about papers published previously in *Neurology.*

HISTORICAL NEUROLOGY—Articles, except under unusual circumstances, follow same size guidelines as Brief Communications.

ISSUES OF NEUROLOGICAL PRACTICE—Contemporary practice of neurology in the United States.

VIEWS & REVIEWS—Either review articles or opinion statements on timely clinical or scientific subjects. Manuscripts proposing a"Hypothesis" are considered only if wntten by an acknowledged contributor to field.

OTHER

Color Charges: $910 minimum per-page. Publisher will individually estimate cost.

Materials Sharing: Submission implies agreement to share material used in experiments such as clones of cells, antibodies, viruses, DNA and protein sequences, and software programs in response to reasonable requests.

Numbers: Use Arabic numerals for all numbers above nine, for designators (e.g., case 5, day 2, etc.) and for units of measure; spell out numbers if below 10, at begining and end of sentences, and for fractions below one.

Style: Refer to "Uniform Requirements for Manuscripts Submitted to Biomedical Journals" (*N Engl J Med* 1991:324:424-428) and "Suggestions to Authors" (*Neurology* 1993:43:231-232) for style and publication guidelines. Follow recommendations of International Committee of Medical Journal Editors concerning order of authors and protection of patient anonymity (*BMJ* 1991:302:1194).

Avoid use of passive voice in Abstract, Introduction, and Discussion. Passive voice is acceptable in Methods and Results. Express clmlcal laboratory data in conventional rather than SI units. Provide SI units in parentheses.

SOURCE

January 1994, 44(1): 6A.

Neuron

Cell Press

Editorial Office, *Neuron*
Cell Press
50 Church Street
Cambridge, Massachusetts 02138
Telephone: (617) 661-7063; Fax: (617) 661-7061

AIM AND SCOPE

Neuron publishes reports of novel results in any area of experimental neuroscience. Papers will be considered for publication if they report results of unusual significance or general interest.

MANUSCRIPT SUBMISSION

Make papers as concise as possible. Prncipal text of manuscripts (Summary through Discussion) is 17 double-spaced pages; this corresponds toabout 60K, about 35K of principal text with about 25K of all other text. Submit fewer than 8 figures; all figures should fit into no more than 2 pages.

Submit four copies of each manuscript. Text should be letter quality and double spaced. If a manuscript is closely related to papers in press or submitted elsewhere, provide copies of papers. Paper may not contains any data that have been or will be submitted for publication elsewhere (including symposium volumes).

MANUSCRIPT FORMAT

Summary: 120 word single paragraph.

Running Title: 50 characters.

Introduction: Succinct, no subheadings.

Results and Discussion: May be divided by subheadings or combined. Do not use footnotes; transfer material to text.

Experimental Procedures: Include sufficient detail so all procedures can be repeated, in conjunction with cited references.

References: Include only articles that are published or in press. Cite unpublished data, submitted manuscripts, or personal communications within text. Document personal communication by a letter of permission. Do not cite abstracts of work presented at meetings.

Miller, C. (1989). Genetic manipulation of ion channels: a new approach to structure and mechanism. Neuron 2, 1195–1205.

Fallon, J. H., and Loughlin, S. E. (1993). Implications of the anatomical localization of neurotrophic factors. In Neurotrophic Factors, S. E. Loughlin and J. H. Fallon, eds. (San Diego, California: Academic Press), pp. 1–24.

Figures: Provide each copy of manuscript with a set of figures of sufficient quality for reviewers to judge data. Indicate magnification by a bar scale. Put lettering on halftone figures generated by computer outside area of halftone. Color figures are included if reviewers believe they are essential.

Figure Legends and Tables: Include as separate sections, follow style of journal. Double-space tables.

OTHER FORMATS

MINIREVIEWS—Briefly discuss a sharply focused topic of recent experimental research making it accessible to researchers in other areas. Provide a critical but balanced view of field. Submit proposals to Editor.

OTHER

Abbreviations: Define nonstandard abbreviations when first used in text.

Color Charges: $1500 for first page, $1000 for second, and $250 for each additional page.

Electronic Submission: Provide a disk copy of each manuscript. Format disks preferably under MC-DOS (PC-DOS); however, we can handle 3.5 in. disks from other operating systems. Indicate word processor program, type of computer, and operating system used.

Materials Sharing: Publication implies that authors are prepared to distribute freely to academic researchers for their own use any materials (e.g., cells, DNA, antibodies, genetically engineered mice) used in published experiments. In cases of dispute, authors may be required to make primary data available to Editor. Deposit nucleic acid and protein sequences as well as X-ray crystallographic coordinates in appropriate data base.

Page Charges: $35 per page. Inability to pay will not influence acceptance; authors unable to meet this charge should make reason known upon publication.

Style: Italicize genetic loci; protein products of loci are not italicized, nor are journal names, foreign phrases, or species names.

Units: Conform with International System of Units (SI). If special circumstances necessitate use of other units, define on first appearance in terms of SI units.

SOURCE

July 1994, 13(1): back matter.

Neuropharmacology
Elsevier

Neuropharmacology
Cellular and Molecular Neuroscience Group
The Medical School, The University of Birmingham
Edgbaston, Birmingham B15 2TT, U.K.
Telephone: +44 (0) 21 414 4497;
Fax: +44 (0) 21 414 6637].

AIM AND SCOPE

Neuropharmacology publishes monthly articles concerned with the actions of biologically active substances on the central and peripheral nervous systems. Since the field of neuropharmacology is interdisciplinary in nature, the journal welcomes papers on all aspects of the subject. However, papers within the area of cellular and molecular neuroscience are particularly encouraged. Papers will only be considered for publication providing they make an original and important contribution to the field of neuropharmacology.

MANUSCRIPT SUBMISSION

Submission implies that manuscript has not been, and will not be submitted elsewhere until a decision has been made by *Neuropharmacology*. Manuscripts may not have been published and may not be simultaneously submitted or published elsewhere. Submission implies that authors agree that copyright for article is transferred to publisher if and when article is accepted for publication. Assignment of copyright is not required from authors who work for organizations which do not permit assignment.

Submit original and three high quality copies of manuscript. In cover letter indicate number of pages, figures and tables; suggest an appropriate Executive Editor for handling paper.

MANUSCRIPT FORMAT

Type manuscripts, including references and figure legends, double-spaced on one side of A4 paper (206 × 294 cm) or equivalent with 2.5 cm margins. Type should be no smaller than 12 point. Number pages consecutively, starting with title page. Either U.K. or U.S. spelling may be used, be consistent.

Use following headings. An additional level of subheadings may be used.

Title Page: Include a brief and informative title, a running title (40 characters), names and addresses of authors and a list of key words or phrases. Indicate corresponding author, with telephone and xax numbers.

Summary: Second page. 200 words readily accessible to nonspecialists; include important points of paper.

Introduction: Third page. Succinct account of why work was done. Do not give long historical background. Include a statement accessible to a lay audience indicating potential benefit of work to man or animals. If any work in manuscript has been published previously as an abstract, this must be referenced.

Methods: Give sufficient detail to enable others to repeat experiments. Give full descriptions of all analgesic, anaesthetic and surgical procedures. Where abbreviations are used in place of long chemical names, provide full chemical name. Give details of statistical analyses performed. Use Sl units.

Results: Fully illustrate. Note negative findings.

Discussion: Be as concise as possible. Discuss results in context with current state of field.

Acknowledgments: Be as brief as courtesy allows.

References: 40. Abbreviate according to 4th edition of *World List of Scientific Periodicals.*

Davies J., Francis A. A., Jones A. W. and Watkins J. C. (1981) 2-Amino-5-phosphonovalerate (2APV), a potent and selective antagonist of amino acid-induced and synaptic excitation. *Neurosci. Lett.* 21: 77–81.

Ascher P. and Johnson J. W. (1989) The NMDA receptor, its channel, and its modulation by glycine. In: *The NMDA Receptor* (Watkins J. C. and Collingridge G. L., Eds), pp. 109–121. IRL Press at Oxford University Press, Oxford.

Cite papers accepted for publication as In press; provide photocopy of manuscript. Refer to papers submitted or in preparation as unpublished observations/personal communications within text, if absolutely necessary. In text give references as: Smith (1964) or (Smith. 1964). For multiple authorship use *et al.* throughout, i.e., Smith *et al* (1964). If works published by same author(s) in the same year are cited, distinguish by letters a b, c, etc. Arrange reference list alphabetically according to surname of first author, and chronologically if several papers by same author(s) are referenced.

Illustrations: Submit two sets of figures as high quality photographs, line drawings or laser prints with figure legends typed double-spaced on separate sheets. Letter figures; identify on reverse with authors, figure numbers and orientation. Include additional sets of figures in copies of manuscript. These may be high quality photocopies; include legend with each figure (single-spaced if necessary).

Tables: Keep to a minimum; make self explanatory without reference to text. Do not reproduce data presented in figures or text. Indicate suitable location for placement of tables and figures in text.

OTHER FORMATS

RAPID COMMUNICATIONS—2-3 printed pages (6-7 A4 pages including title page, figure legends and references), one illustration and 10 references. Work of a particularly novel or timely nature, that is accepted without modification or rejected. Follow same format as full papers without headings; precede by a 50 word summary. If submitted material does not conform exactly to these requirements it will be returned without review.

REVIEWS—Short, timely reviews by invitation. Suggestions for reviews are welcome; send to Chief Editor.

OTHER

Abbreviations: Keep to a minimum. Write out in full when first used and put in parentheses.

Electronic Submission: Submit computer disk (5.25 or 3.5 in. HD/DD disk) containing final version of paper along with final manuscript. Send only hard copy when first submitting paper. Make sure that disk and hard copy match exactly. Specify what software was used, including release. Specify what computer was used (either IBM-compatible PC or Apple Macintosh). Include text file and separate table and illustration files. Follow general instructions on style/arrangement and, in particular, reference style of journal. Single-space file; use wraparound end-of-line feature, i.e., no returns at the end of each line. Begin all textual elements flush left; no paragraph indents. Place two returns after every element such as title, headings, paragraphs, figure and table call-outs. Keep a backup disk.

Experimentation, Animal: Carry out all experiments in accordance with the U.K. Animals (Scientific Procedures) Act, 1986 and associated guidelines, of European Communities Council Directive of 24 November 1986 (86/609/EEC) or National Institutes of Health guide for e care and use of laboratory animals (NIH Publications No. 80-23, revised 1978). Submit a statement that all efforts were made to minimize animal suffering, to reduce number of animals used, and to utilize alternatives to in vivo techniques, if available. Consult "A fair press for animals" [*New Scientist* (1992) 1816: 18-30] before preparing manuscript. Editors reserve the right to reject papers if there is doubt about use of suitable procedures.

Experimentation, Human: Conduct experiments in accordance with Declaration of Helsinki. Include a statement that all procedures were carried out with adequate understanding and written consent of subjects.

SOURCE

January 1994, 33(1): back matter.

Neuropsychiatric Genetics

see American Journal of Medical Genetics

Neuropsychologia
Pergamon

Prof. G. Berlucchi
Dipartimento di Scienze
Neurologiche e della Visione
Strada Le Grazie 8, 1-37134 Verona, Italy
Fax: 39 45 580881

AIM AND SCOPE

Neuropsychologia considers for publication empirical rather than purely theoretical papers, using experimental methods and dealing with the relationship between brain functions and behaviour. This includes research in the various fields of behavioural neurosciences both in humans (normal and brain lesioned) and in animals.

MANUSCRIPT SUBMISSION

Submission implies that it is an original paper which has not previously been published (except as an abstract or preliminary report), and that it is not being considered for publication elsewhere.

MANUSCRIPT FORMAT

Submit three double-spaced copies typed only one side of each page. Allow generous margins. Mark passages which may be printed in small type.

Abstract: 200 words that should supplement title in giving essentials of paper.

Key Words: Six key words (which do not appear in title) follow abstract, for indexing purposes.

Illustrations: Do not insert in text. Refer to all figures, charts and diagrams as "Figures" (abbreviated to "Fig."). Number consecutively in order referred to in text. Submit figures in form suitable for direct reproduction; provide original figures or glossy prints. Do not submit prints with weak lines. Make illustrations twice final size required. Use standard symbols (○ ● △ ▲ ❑ ■ + × ▽ ▼).

Tables: Construct so as to be intelligible without reference to text. Provide every table and column with a heading, and make suitable for direct reproduction. Clearly indicate units of measurement. Summarize results by an accepted method of expression, e.g., standard deviation (S.D.). Do not reoproduce the same information in both tables and figures. Tables are reproduced by photographic means directly from typed manuscripts.

References: Indicate in text by numbers, referring to an alphabetical listing. Give full references in a list at end of paper in alphabetical order.

Barlow, H. B., and Penigrew, J. D. Mechanisms of binocular stereopsis. *J. opt. Soc. Am.* **57**, 572–573,1967.

Yockey, H. P. In *Symposium on Information Theory* , H. P. Yockey, and H. Quastler (Editors), pp. 50–51. Pergamon Press, New York, 1958.

References contain names of all authors with initials, title of paper, name of Journal abbreviated in accordance with *World List of Scientific Periodicals* (4th Edn, Butterworths, London, 1963–1965, 3 vols), volume number, inclusive page numbers, and year of publication.

Footnotes: Indicate by a consistent series of symbols commencing anew on each page. Do not include in numbered reference system.

SOURCE

January 1995, 33(1): inside back cover.

Neuroscience
International Brain Research Organization
Pergamon Press

Prof. P.G. Kostyuk
Department of General Physiology
A. A. Bogomoletz Institute of Physiology
Academy of Sciences
Ukrainian SSr, 4 Bogomoletz Street
Kiev 24, 252601 GSP, Ukraine

Dr. R. Llinas
Department of Physiology and Biophysics
New York University School of Medicine
550 First Avenue
New York 10016 U.S.A.

Prof. A.D. Smith
Department of Pharmacology
University of Oxford
Mansfield Road, Oxford OX1 3QT U.K.

AIM AND SCOPE

Neuroscience is devoted to the prompt publication of results of original research on any aspect of the scientific study of the nervous system. Since one of the chief aims is to promote communication between neuroscientists, the *Journal* will also include occasional commentaries on specific areas of neuroscience.

MANUSCRIPT SUBMISSION

Report new observations clearly and succinctly; make study meaningful to scientists of other disciplines.

Original research reports must describe significant, new and confirmed findings and must give adequate experimental detail. Preliminary communications are not accepted.

Submission implies that manuscript represents original research not previously published (except as an abstract or preliminary report), and that it is not being considered for publication elsewhere in similar form, in any language.

MANUSCRIPT FORMAT

Type double- or triple-spaced throughout with 2.5 cm margins. Use A4 size (206 × 294 mm) paper. Submit three copies. Send photocopies of diagrams, or rough prints of photographs, with second and third copies; copies of photographs are not acceptable for referees. Number pages in succession; title page is page 1.

Title Page: Make title short and consistent with clarity. Do not number papers in series; subtitles are accepted. List authors' names, laboratory of origin, name and address of corresponding author, and any footnotes. List abbreviations used in text at bottom of page.

Running Title: 56 letter and spaces on a separate sheet.

Tables/Figures: Place tables and figures on separate pages at end of manuscript. Indicate desired approximate locations in margin of text.

Footnotes: Use sparingly; indicate locations with symbols *, †, ‡, §, ||, ¶, in order.

Abstract: Make brief, not exceeding 5% of length of paper. In first paragraph summarize results obtained; in final paragraph summarize major conclusions so that a reader not familiar with techniques used can see implications for his area of neuroscience. Do not use abbreviations.

Text: Organize remaining text, in four main sections:

Introduction: 'Set the scene' for a non-specialist; continue with specific reasons for undertaking investigation. Avoid exhaustive reviews of literature; do not indicate results obtained. Omit heading "Introduction."

Experimental Procedures: Give in sufficient detail to permit repetition of work by others. Briefly summarize published procedures; describe in detail only if they have been substantially modified. Group special chemicals, drugs, etc. with their sources of supply under a separate subheading, Materials.

Results: Describe findings without discussing their significance. Use subsections to clarify expression of results.

Discussion: Interpret findings and assess their significance in relation to previous work. Do not repeat material in 'Results.' Use subsections wherever possible; separate subsections dealing with technical or highly specialized matter from result of text.

Presentation of Data in Tables or Figures: Construct tables and figures so they, with captions and legends, will be intelligible with minimal reference to text. Give each figure a caption and explanatory legend typed on a separate sheet. Present data in a precise manner. Do not use simple histograms when data can equally well be given in a table.

Figures: Put number, authors' name and 'Neuroscience' on back. Put line drawings on white card, faintly blue- or green-lined graph paper or on tracing cloth or paper. Unglazed photographs of line drawings are acceptable. Submit illustrations in a form suitable for direct reproduction. Make line drawings above twice final size; do not exceed 20 × 30 cm. Line diagrams will be reduced to fit within a single journal column.

Line drawings in two or more colors will be charged to authors. Cost is given on request. Prepare drawings in black, with registration marks in diagonally opposite corners.

Insert symbols on graphs, etc.; use standard characters: ○ ● △ ▲ □ ■ × +.

Include original mounted halftones only when essential. Prepare micrographs as follows: For halftones (including color plates), maximum area is 170 × 250 mm. Make use of as much of space as possible; make allowance for figure legends if desired. Label prints, but not within 10 mm of edges of half-tones.

References: In text, quote by superscript numbers; correspond numbers to those in alphabetical reference list.

If a reference has to be inserted at a late stage, do not renumber references, call new reference, e.g., [21a].

Type reference list on a separate sheet(s) at end of manuscript, in alphabetical order and arranged as: number (in alphabetical sequence), authors' names and initials, year, title of article, abbreviated title of journal, volume, first and last page numbers. Abbreviate journal titles according to 4th edition of *World List of Scientific Periodicals* (Butterworths, 1965). Give first and last pages and that "and" is used in text and list of References.

12. Vogt M. and Wilson G. (1972) Concentration of 5-hydroxytryptamine and its acid metabolite in ventricle-near regions of the rat brain. *J. Neurochem.* **19,** 1599–1600.

References to books include authors' names and initials, year, title of book, volume, page numbers, publisher and place of publication. Where relevant, give title of a paper within a book, and editor's name.

Baker P. F. (1972) The sodium pump in animal tissues and its role in the control or cellular metabolism and function. In *Metabolic Pathways* (ed. Hokin L. E.), Vol. 6. pp. 243–268, Academic Press, New York.

Mention unpublished experiments only in text; do not include in list of References; give initials and surnames for authors whose unpublished experiments are quoted.

Cite papers accepted for publication but which have not appeared in References with abbreviated name of journal followed by "in press." Indicate date of acceptance.

If work by authors that is submitted or in press elsewhere is referred to, enclose a copy.

Use personal communications only when written authorization from communicator is submitted with original manuscript; mention only in text.

OTHER FORMATS

COMMENTARIES—3000–10000 words. Commentaries on a specific topic in neuroscience are normally published only on invitation from a Chief Editor. Obtain agreement of a Chief Editor before submitting a manuscript. Commentaries are not exhaustive reviews but short articles intended either to draw attention to developments in a specific field for workers in other scientific disciplines, or to bring together observations over a wide area that seem to point in a new direction, or to give author's personal views on a controversial topic, or to direct soundly based criticism at some widely held dogma or widely used technique in neuroscience. Make commentary understandable to neuroscientists of other disciplines

LETTERS TO NEUROSCIENCE—1000 words of text (excluding references and figure legends; 4 display items (tables or figures). Short reports which describe important discoveries of general interest to all neuroscientists. Describe significant and substantial new findings of a particularly novel kind. Write with nonspecialist readers in mind; use minimal abbreviations and technical jargon. Submit in triplicate with a title page giving: title, authors names and affiliations, telephone and fax numbers and number of words in full text.

Provide line figures (approx. twice final size) in a form suitable for reproduction without modification, either pho-

tographic prints or laser printed. Do not use simple histograms when data can be presented more concisely numerically. Label photographs intended for halftone or color reproduction with high-quality commercial lettering; prepare mounted on card exactly (same size) as they will appear in Journal (maximum size: 170×250 mm).

Do not use subheadings. Begin with an introductory paragraph (200 words, no abbreviations); explain rationale for work; summarize main findings with possible implications for nonspecialists. Describe remaining text results in a clear and succinct manner, including interpretations and conclusions drawn by authors. Confine technical details of methods used to legends of appropriate figures or tables; wherever possible, reference published experimental procedures. List abbreviations used in text. Put references in Journal style; no more than 35.

Submit a signed cover letter from corresponding author stating: "The work described has not been submitted for publication elsewhere and all authors listed have approved the manuscript that is enclosed."

MATTERS ARISING—1500 words, excluding references. Submit to any Chief Editor. Important scientific points that arise out of papers previously published in *Neuroscience*. Replies by authors of original papers may be invited.

OTHER

Abbreviations: Do not use excessively. Awkward and unfamiliar abbreviations and those intended to express concepts or experimental techniques are not permitted. Define when first used by placing in parenthesis after full term; use consistently thereafter. List abbreviations used in text on bottom of title page. Do not use abbreviations as short form for experimental procedures or for concepts.

Color Charges: £300 (U.S. $600) for one full page of photographs. Exact cost is given on request.

Drugs: Use official or approved names; give trade or common names in brackets at first mention. Give manufacturer's name. Give doses of drugs as unit weight/unit body weight, e.g., mmol/kg or mg/kg. Give concentrations in terms of molarity, e.g., nM or μM, or as unit weight/unit volume solution; state whether weight refers to salt or active drug component. State molecular weight, inclusive of water of crystallization, if doses are given as unit weight.

Electronic Submission: Submit final revised versions of manuscripts and illustrations on 3.5 or 5.25 in. IBM or compatible diskettes with hardcopy from current version of disk; mark on outside with content, software package and hardware used. For illustrations, preferred software package is *Chemdraw*; for manuscripts, preferred software packages are MS-Word, WordPerfect, LocoScript, PC Write and WordStar. Supply manuscripts from word processors other than these as ASCII files. ASCII files can usually be produced by saving document as a text only file or by saving file without formatting information.

Isotopes: Specifications conform to IUPAC system [*Biochem. J.* (1975) **145,** 1–20].

Nomenclature, Chemical and Biochemical: Follow conventions used in *Biochemical Journal* (1975; **145,** 1–20).

Style: Write concisely in English in a readily understandable style. Do not use technical jargon, "laboratory slang" or words not defined in dictionaries.

Do not use redundant words, phrases, and sentences. Do not repeat captions of tables and figures in text, with or without paraphrasing.

Units: Restrict symbols for physical units to Système Internationale (S.I.) Units. For examples of commonly used symbols see *Biochem. J.* (1975) **145,** 1–20 and *Quantities, Units and Symbols* (1971, The Royal Society, London).

SOURCE
January 1994, 58(1): v-vii,

Neuroscience Letters
Elsevier

M. Zimmermann
Universität Heidelberg
II Physiologisches Institut der Universität
Im Neuenheimer Feld 326
D-69120 Heidelberg, F.R.G.

AIM AND SCOPE
The journal provides a rapid publication of short, complete reports, but not preliminary communications, in all areas in the fields of neuroanatomy, neurochemistry, neuroendocrinology, neuropharmacology, neurophysiology, neurotoxicology, molecular neurobiology, behavioral sciences, biocybernetics and clinical neurobiology.

MANUSCRIPT SUBMISSION
Criteria for publication are novelty and interest to a multidisciplinary audience. Papers not sufficiently substantiated by experimental detail will not be published. Manuscripts must be complete in all respects and must be original material not previously published. Submission of a paper implies that it is not being submitted for publication elsewhere. Submission under multiple authorship implies that manuscript has been approved by all authors and by responsible authorities at laboratories where word was done. If accepted, manuscript will not be published elsewhere in the same form, without consent of editors and publisher.

Submit manuscripts to Editor-in-Chief or the Associate Editors located in your country or continental region (see inside front cover for addresses). In cover letter, state that manuscript or parts of it have not been and will not be submitted elsewhere for publication.

MANUSCRIPT FORMAT
Submit three copies of each manuscript typed double-spaced with at least a 4 cm margin on pages of uniform size. No section headings. No revisions or up-dating will be incorporated after manuscript has been accepted and sent to Publisher (unless approved and instructed to do so by Editor.

Include on a separate title page authors' full names, academic and professional affiliations, and complete addresses. Clearly specify name and address, plus telephone, fax and Bitnet E-mail numbers of corresponding author.

Length of manuscripts should be adequate to yield 3 printed pages, and in no case more than 4 printed pages. An approximate guide for judging length of paper: heading + abstract = 0.5–0.6 pages; 3 typewritten (double-spaced) pages = 1 printed page; (when using a word-processor) 900 words or 5600 characters = 1 printed page; 3 single-column wide or 2 double-column wide figures plus legends = 1 printed page; 3 single-column wide or 2 double-column wide tables = 1 printed page; 17 references = 0.5 printed page.

Provide more precise length calculations by exactly determining the final size of figures and tables including captions.

Preferred medium of final submission is on disk with accompanying reviewed and revised manuscript (see Electronic Submission below).

Manuscripts should be written in English and be accompanied by: Title page, Key Words and Abstract.

Title Page: Include title, authors' names and affiliations, corresponding author's name, address, fax and phone numbers. Make informative; include animal species, brain part or preparation used. Do not give abbreviations except those widely known and used. If abbreviations are unavoidable in title, explain by a circumscription, e.g., 'MPTP, a dopaminergic neurotoxin'.

Key Words: 6–8.

Abstract: 100–130 words on a separate page. Make informative, containing a brief description of methods used and results obtained.

References: Give citation of literature references in text at appropriate places by numbers in square brackets. List all references cited in text at end of manuscript on a separate page (also double-spaced), arranged in alphabetical order of first author and numbered consecutively. Cite all items in Reference list in text; all references cited in text must be present in list. Make complete, including name and initials of all authors, title of paper, abbreviated title of periodical, volume, year, and first and last page numbers of article. Abbreviate journal titles according to *List of Serial Title Word Abbreviations*, (CIEPS/ISDS, Paris, 1985), Example 2. The form of literature references to books should be: author, initials, title of book, publisher and city, year and page numbers referred to (Example 1). References to authors contributing to multi-author books or to proceedings printed in book-form should be similar to those for books (Example 3).

1 Swanson, L.W., Brain Maps: Structure of the Rat Brain, Elsevier, Amsterdam, 1992, 244 pp.

2 Stroemer, R.P., Kent, T.A. and Hulsebosch, C.E., Increase in synaptophysin immunoreactivity following cortical infarction, Neurosci. Lett., 147 (1992) 21–24.

3 Kolb, B., Animal models for human PFC-related disorders. In H.B.M. Uylings, C.G. van Eden, J.P.C. de Bruin, M.A. Corner and M.G.P. Feenstra (Eds.), The Prefrontal Cortex: its Structure, Function and Pathology, Progress in Brain Research, Vol. 85, Elsevier, Amsterdam, 1990, pp. 501–519.

Illustrations: Submit in triplicate in a form and condition suitable for reproduction either across a single column (= 8.3 cm) or a whole page (= 17.6 cm). Give authors name and number in Arabic numerals according to sequence of appearance in text, where referred to as Fig. 1, Fig. 2, etc. Provide a figure caption that explains the technical details of figure and abbreviations and symbols used. In the case of graphs, label coordinate scales with descriptors and units of measurement. Put line drawings in black ink on drawing or tracing paper or submit glossy sharp photographs. Make lettering clear and of adequate size to be legible after reduction. Professional labelling is preferable, if not possible, present in fine pencil. Supply photographs, including roentgenograms, electroencephalograms and electron micrographs as clear black-and-white prints on glossy paper, not copies, usually larger than final size but not more than 21 by 29 cm (A4 format). Give micrographs a scale bar, rather than a magnification factor in legend. Assume that the same degree of reduction will be applied to all figures in paper.

Calculation of the Printed Size of an Illustration: Arrange figure material to fit into a printed format either single column (8.3 cm) or a page (17.5 cm). Column width is preferred. Sets of multiple micrographs are better reproduced at page width. Original figure is usually larger than final figure. Using the original width (OW, in cm) and original length (OL, in cm), calculate.

A figure at column width requires: $[OL \times 8.3 \text{ cm}/OW] + 5$ cm.

A figure at page width requires: $[OL \times 17.5 \text{ cm}/OW] + 3.5$ cm.

Reproduction in color must be approved by Editor. Submit color figures as separate prints, not mounted on cardboard. Avoid press-on lettering. Submit slides of labelled prints or electronic files of figure(s) in all standard graphics formats (e.g., Adobe Illustrator, Adobe Photoshop, Quark XPress, Corel Draw, Aldus Freehand, TIFF and EPS. Give each illustration a legend, typed double-spaced on a separate page; begin with number of illustration. If illustrations or other small parts of articles or books already published elsewhere are used, include written permission of original author and publisher. Indicate original source in legend of illustration.

Tables: Type each double-spaced on a separate page and number in sequence in Arabic numerals (Table 1, 2, etc.). Provide a heading and refer to in text as Table 1, Table 2, etc.

Calculation of the Printed Size of a Table: Tables are printed at column width or page width format. The number of characters per line (including 5 blanks to separate columns) is the determining factor: if 65 characters or less per line, table will fit a single column; otherwise, full page width is required.

To estimate printed length, count number of horizontal rows (NHR) in table, including blank space or drawn lines to subdivide table material vertically and multiply by 0.4 cm to obtain length of printed column or page. Add 5.0 cm of column length for a column width table and 3.5 cm of page length for a page width column to allow for heading, caption, and blank space above and below table.

A table at column width requires $[NHR \times 0.4 \text{ cm}] + 5$ cm. A table at page width requires $[NHR \times 0.4 \text{ cm}] + 3.5$ cm.

OTHER

Color Charges: $970 for first page of color and approximately $810 per page for all additional pages in color.

Electronic Manuscripts: The preferred storage medium is a 5.25 or 3.5 in. disk in MS-DOS format although other systems are welcome, e.g., Macintosh (save file in usual manner, do not use option 'save in MS-DOS format') and NEC. Submit disk plus one final, printed and exactly matching version (as a printout) to accepting editor. File on disk and printout must be identical. Specify type of computer and word-processing package used (do not convert textfile to plain ASCII). Ensure that letter 'l' and digit '1' (also letter 'O' and digit '0') have been used properly, and format article (tabs, indents, etc.) consistently. Do not leave characters not available on word processor (Greek letters, mathematical symbols, etc.) open; indicate by a unique code (e.g., gralpha, @, #, etc., for the Greek letter α) or write out in full. Use codes consistently throughout text. List codes and provide a key. Do not allow word processor to introduce word splits. Adhere strictly to general instructions on style/arrangement and in particular, reference style of journal. Obtain further information from Publisher.

Experimentation, Animals: Show that attention has been paid to guidelines in "Editorial: ethical principles for the maintenance and use of animals in neuroscience research." (*Neurosci. Lett.* 1987; 73:1).

SOURCE

November 21, 1994, 182(1): v-vi.

Neurosurgery
Congress of Neurological Surgeons
Williams & Wilkins

Michael L. J. Apuzzo, M.D.
1200 North State Street, Suite 5046
Los Angeles, CA 90033

AIM AND SCOPE

The goal of *Neurosurgery* is to provide a medium for prompt publication of scientific papers dealing with clinical or experimental neurosurgery, solicited manuscripts on specific subjects from experts, case reports, and other information of interest to neurosurgeons.

MANUSCRIPT SUBMISSION

Submit manuscripts (including all charts, illustrations, and references) in triplicate, doubled-spaced with ample margins. The Congress of Neurological Surgeons requires written exclusive assignment of copyright from all authors of all accepted manuscripts. Copyright forms are mailed from Editorial Office. If any form of preliminary publication (other than an abstract or 400 word condensation) is contemplated or has appeared, submit a reprint or photocopy of publication. Mail all submitted material flat.

MANUSCRIPT FORMAT

Submit title page with full title of paper—short, clear, and specific—and authors' full names with highest academic degrees and institutional affiliations. Include separate page with running title (3–4 words), and name, address (including zip code), telephone and fax numbers of corresponding author.

On a separate page include an abstract (300 words) describing what was done, results obtained, and conclusions drawn. In alphabetical order, list 7 key words for coding and indexing. Consult *Index Medicus* for appropriate terms.

Follow with Text; Grants/other acknowledgments; References cited (not footnotes); Figure legends; Tables (typewritten, not photographed); and Charts, graphs, and photos.

Number all pages consecutively, starting with title page as page 1. Place first author's name in upper right corner of each page.

References: Verify accuracy and completeness of references. Enclose reference numbers in parentheses in text on line of type, preferably at ends of sentences. Type list on a separate page(s), double-spaced. Arrange alphabetically with all authors listed. Number consecutively according to alphabetical order of authors. If more than one title for an author (or same group of authors), order entries according to date of publication. Cite references by number in text. Papers quoted as "in press" (accepted for publication elsewhere) appear in references. Do not reference papers "submitted for publication" but not yet accepted and citations such as "personal communication" or "unpublished data;" enclose in parentheses in appropriate place in text. Identify abstracts with "abstr" in parentheses at end of reference. Use abbreviations for journal titles from *Index Medicus*. Note punctuation and spacing.

1. Alper MG, Blazina VJ: Clinical neurophysiology of orbital tumors, in Laws ER Jr (ed): *The Diagnosis and Management of Orbital Tumors*. Mount Kisco, Futura Publishing Co., 1988, pp 29–92.

2. Barrow DL, Tindall GT: Visual loss following transsphenoidal surgery. Neurosurgery (in press).

3. Giannotta SL, Oppenheimer MD, Levy ML: Management of intra-operative aneurysm rupture without hypotension. Proceedings of the Western Neurosurgical Society, in Neurosurgery 26:713, 1990 (abstr).

4. Laws ER, Ots M: Fibrin tissue adhesive: A role in transsphenoidal neurosurgery. Presented at the 42nd Annual Meeting of the Neurosurgical Society of America, Tuckertown, Bermuda. May 10–13, 1989

5. Russell DS, Rubinstein LJ: *Pathology of Tumors of the Nervous System*. Baltimore, Williams & Wilkins, 1977, ed 4, pp 127–141.

6. Watts C: Chemonucleolysis: An appeal for objectivity. J Neurosurg 42:488, 1975 (letter).

7. Wilkins RH: The natural history of intracranial vascular malformations: A review. Neurosurgery 16:421–430, 1985.

Illustrations: Use Arabic number to designate figures and tables. Refer to all figures and tables in numerical order in text. Graphs and photographs must be 1, 2, or 3 columns wide. Submit clear glossy photographs unmounted in triplicate; write lightly on back with a soft pencil indicating figure number, first author's name, and TOP. Authors are responsible for costs of color reproductions. Designate one set of illustrations "set for publication."

Figure Legends: Type double-spaced on a separate page and number consecutively. Identify all legends. Include keys for symbols in illustrations. Give credit for previously published illustration in corresponding legend:

(From Watson CB, Reierson N, Norfelect EA: Clinically significant muscle weakness induced by oral dantrolene sodium prophylaxis for malignant hyperthermia. Anesthesiology 65:312–314, 1986.)

Tables: Type on separate pages, number consecutively, and add a brief, descriptive title.

Permissions: Obtain written permission from author and original publisher (copyright holder) for use of previously published material and for direct quotations (more than 50 words). Give credit in legends and/or text for borrowed materials. Submit written permission from appropriate investigator to cite unpublished data. Obtain and submit written permission from persons shown in photographs, unless faces are masked to prevent identification.

OTHER FORMATS

RAPID COMMUNICATIONS—Submit camera-ready. Contact editorial office for forms and detailed instructions.

OTHER

Abbreviations, Nomenclature, and Symbols: Conform to those found in *CBE Style Manual* (5th Edition, 1983, Council of Biology Editors, Inc., Bethesda, MD). Identify medications, materials, and devices by full nonproprietary name as well as brand name and manufacturer's name, city, state, and country. Place information in parentheses in text, not in a footnote. State whether they have any personal or institutional financial interest in drugs, materials, or devices described in submissions. Use standard international units (SIU) accompanied by former notation.

Experimentation, Human: Submit statement to Editor, indicating mechanism used for reviewing ethics of research conducted, such as a photocopy of the IRB's statement of approval.

Statistical Analyses: When statistical significance is attributed, cite specific method of analysis and use upper case P ($P < 0.01$).

SOURCE

July 1994, 35(1): front matter.

Neurotoxicology and Teratology

Behavioral Toxicology Society
Neurobehavioral Teratology Society
Pergamon

Donald E. Hutchings
New York State Psychiatric Institute
722 W. 168th St.
New York, NY 10032
Telephone: (212) 960-5717

AIM AND SCOPE

Neurotoxicology and Teratology will publish original reports of systemic studies in the areas of neurotoxicology and developmental toxicology in which the primary emphasis and theoretical context are on the nervous system and behavior. Brief Communications describe a new method, technique, or apparatus and results of experiments which can be reported briefly with limited figures and tables. Minireviews provide a timely update on selected aspects of a scientific field undergoing rapid change or that present special methodological or interpretive problems. Behavioral studies as such will not be published unless they are obviously pertinent and make a significant contribution to neurotoxicology and developmental toxicology. A limited number of relevant theoretical articles, results of symposia, and more comprehensive studies as monograph supplements will also be published.

MANUSCRIPT SUBMISSION

Suggest appropriate journal section. Include office telephone and fax numbers and E-mail address. Transfer of Copyright Agreement will be sent to submitting author. Form must be completed and returned to publisher before article can be published.

MANUSCRIPT FORMAT

Submit manuscript in triplicate. Type, double-spaced with wide margins on good quality paper. If a word processor is used, a letter quality printer must be used and computer generated illustrations must be of the same quality as professional line drawings.

Title Page: Include title of paper; author(s); laboratory or institution of origin with city, state zip code, and country; complete address for mailing proofs including telephone and FAX numbers, and E-mail address; a running head (40 characters including spaces).

Type references, footnotes, and legends for illustrations on separate sheets, double spaced. Identify illustrations (unmounted photographs) on reverse with figure number and author(s) name; when necessary mark TOP. Type each table on a separate sheet double-spaced.

Do not use italics for emphasis. Follow style of *Council of Biology Editors* (CBE). Make clear, concise statements of facts and conclusions. Do not fragment material into numerous short reports.

Title: 85 characters, including spaces between words.

Abstract: 170 words suitable for use by abstracting journals. Prepare as follows:

MYERS, R. D., C. MELCHIOR AND C. GISOLFI. *Feeding and body temperature: Changes produced by excess calcium ions* . . . NEUROTOXICOL TERATOL **15**(X) 000-000,1993-Marked differences in extent of diffusion have been . . .

Key Words: Type 3-6 words or short phrases at bottom of abstract page to be printed with paper at end of abstract.

Footnotes: Number title page footnotes consecutively. If senior author is not to receive reprint requests, use a footnote to designate to whom requests should be sent. Do not use text footnotes; incorporate material into text.

References: Prepare literature cited according to Numbered/Alphabetized style of CBE. Cite references by number, in parentheses, within text (only one reference to a number) and listed in alphabetical order (double-spaced) on a separate sheet at end of manuscript. Do not recite names of authors within text.

Journal citations in reference list contain: surnames and initials of all authors (surname precedes initials); title of article; journal title abbreviated as in *List of Journals Indexed in Index Medicus;* volume, inclusive pages, and year.

1. Banks, W. A.; Kastin, A. J. Peptides and the blood-brain barrier: Lipophilicity as a predictor of permeability. Brain Res. Bull. 15: 287–292; 1985.

For book references give in order: author, title, city of publication, publisher, year, and pages.

1. Mello. N. K. Behavioral studies of alcoholism. In: Kissin B.; Begleiter, H., eds. The biology of alcoholism, vol. 2. Physiology and behavior. New York: Plenum Press; 1972:219–291.

2. Myers. R. D. Handbook of drug and chemical stimulation of the brain New York: Van Nostrand Reinhold Company; 1974.

Illustrations: Prepare for use in single column width. Draw and letter all drawings for reduction to a given size to the same scale. Refer to all illustrations as figures and number in Arabic numerals. Do lettering in India ink or other suitable material; make proportionate to size of illustrations so it is legible after reduction. Size lettering so that smallest elements (subscripts or superscripts) will be readable when reduced. Place lettering within framework of illustration; place key to symbols on face of chart. Use standard symbols: ○ ● △ ▲ □ ■ +. Give actual magnification of all photomicrographs. Indicate dimension scale. Submit sharply contrasting unmounted photographs of figures on glossy paper. Submit illustrations in black-and-white unless color reproduction is requested. Submit color prints in actual size; authors are responsible for additional costs.

Tables: Give each table sa brief heading; put explanatory matter in footnotes, not in title. Indicate table footnotes in body of table in order of appearance with symbols: *, †, ‡, §, ‖, ¶, #, **, etc. Do not duplicate material in text or illustrations. Omit vertical rules. Use short or abbreviated column heads. Identify statistical measures of variation, SD, SE, etc. Do not submit analysis of variance tables; incorporate significant F's where appropriate within text. The appropriate form for reporting F value is: $F(ll, 20) = 3.05$, $p < 0.01$.

Formulas and Equations: Keep structural chemical formulas, process flow-diagrams, and complicated mathematical expressions to a minimum. Draw chemical formulas and flow-diagrams in India ink for reproduction as line cuts. Identify subscripts, superscripts, Greek letters, and unusual characters.

OTHER

Drugs: Capitalize proprietary (trademarked) names. Chemical name precedes trade, popular name, or abbreviation of a drug at first occurrence.

Electronic Submission: Submit a computer disk (5.25 or 3.5 in. HD/DD) with final version of paper with final manuscript. Send only hard copy when first submitting paper. When paper has been refereed, revised and accepted, send a disk containing final version with final hard copy. Make sure disk and hard copy match exactly. Specify software, including release and computer (either IBM compatible PC or Apple Macintosh) used. Include text file and separate table and illustration files. Follow general instructions on style/arrangements and, in particular, reference style of journal. Single-space and use wrap around end-of-line feature, i.e., no returns at end of each line. Begin all textual elements flush left; no paragraph indents. Place two returns after every element, e.g., title, headings, paragraphs, figure and table callouts. Keep a back-up disk.

Experimentation, Animal: In describing surgical procedures on animals, specify type and dosage of anesthetic agent. Curarizing agents are not anesthetics; if used, provide evidence that anesthesia of suitable grade and duration was employed.

Units: Specify dimensions and measurements in metric system. Use standard nomenclature, abbreviations and symbols, as specified by Royal Society Conference of Editors, "Metrication in Scientific Journals," *Am. Scient.* 56:159-164;1968.

SOURCE

January/February 1995, 17(1): inside back cover.

New England Journal of Medicine
Massachusetts Medical Society

Editor, *New England Journal of Medicine*
10 Shattuck Street
Boston, MA 02115-6094

MANUSCRIPT SUBMISSION

Guidelines are in accordance with "Uniform Requirements for Manuscripts Submitted to Biomedical Journals" (*British Medical Journal,* February 9, 1991, and *New England Journal of Medicine,* February 7, 1991).

Manuscripts containing original material, nor any part of its essential substance, tables, or figures, may not have been nor will not be published or submitted for publication elsewhere before appearing in *Journal.* This does not apply to abstracts or press reports published in connection with scientific meetings. Submit copies of closely related manuscripts. Submission of more than one article dealing with related aspects of same study is discouraged.

Submit original manuscript and one set of original figures and two copies of complete manuscript. Include a word count (not including abstract or references). Type on standard-sized typewriter paper; triple-space throughout. In a cover letter signed by all authors, identify corresponding author (with address and telephone number); clearly state that final manuscript has been seen and approved by all authors and that they have taken care to ensure integrity of work. At least one person's name must accompany a group name (Thelma J. Smith, for the Boston Porphyria Group). As stated in Uniform Requirements, credit for authorship requires substantial contributions to: conception and design, or analysis and interpretation of data; and drafting article or revising it critically for important intellectual content. If more than 12 authors are listed for a multicenter trial, or more than 8 for a study from a single institution, each must sign a statement attesting that s/he fulfills authorship criteria of Uniform Requirements. Acknowledgments are limited to a column of *Journal* space and will be listed only once (see editorial, Nov. 21, 1991).

Disclose at time of submission any financial arrangement with a company whose product figures prominently in submitted manuscript or with a company making a competing product. Information is held in confidence during review and will not influence editorial decision, but if article is accepted for publication, editors will discuss the manner in which information is communicated to reader.

Authors of reviews and editorials may not have any financial interest in a company (or its competitor) that makes a product discussed in article. Potential authors who have questions about these issues should contact Editor.

MANUSCRIPT FORMAT

Titles and Authors' Names: Provide a page giving title of paper (make title concise and descriptive, not declarative); a running head (40 letter spaces); name(s) of author(s), including first name(s) and no more than two graduate degrees; name of department and institution in which work was done; institutional affiliation of each author; and name and address of reprint request author. Mention grant support that requires acknowledgment.

Abstracts: 250 words on separate page. Divide into four paragraphs: Background, Methods, Results, and Conclusions. Briefly describe, respectively, problem being addressed in study, how study was performed, salient results, and what authors conclude from results.

Key Words: Three to 10 key words or short phrases at bottom of abstract page, for indexing article. Use terms from *Medical Subject Headings* from *Index Medicus.*

References: Type triple-spaced; number consecutively as cited. Number references first cited in tables or figure legends in sequence with references cited in text. Style of references is from *Index Medicus.* List all authors when six or fewer; when seven or more, list first three, then "et al."

1. Lahita R, Kluger J, Drayer DE, Koffler D, Reidenberg MM. Antibodies to nuclear antigens. N Engl J Med 1979;301:1382-5.

Do not number references to personal communications, unpublished data, and manuscripts either "in preparation" or "submitted for publication." If essential, incorporate material in appropriate place in text.

Tables: Type double-spaced on separate sheets, and provide titles for each. If article is accepted, *Journal* will arrange to deposit extensive tables of important data with National Auxiliary Publications Service (NAPS); we will pay for the deposit and add an appropriate footnote to text. This service makes available microfiche or photocopies of tables at moderate charges upon request.

Illustrations: Have professionally designed. Send glossy black-and-white photographs. Do not send photocopied or computer-generated figures. Make symbols, lettering, and numbering clear and large enough to remain legible after figure has been reduced to fit width of a single column.

Indicate on back, sequence number, name of author, and TOP. Do not mount on cardboard. Crop photomicrographs to a width of 8 cm; give electron photomicrographs internal scale markers.

If photographs of patients are used, either subjects are not identifiable or pictures are accompanied by written permission for use. Forms are available from Editor.

Type legends for illustrations (triple-spaced) on a separate sheet; information should not appear on illustrations.

Color illustrations are encouraged. Send both transparencies and prints.

OTHER

Abbreviations: Except for units of measurement, abbreviations are discouraged. Consult *CBE Style Manual* (5th ed, Bethesda, MD: Council of Biology Editors, 1983) for lists of standard abbreviations. At first use, precede abbreviation by words for which it stands.

Drugs: Use generic names. When proprietary brands are used, include brand name in parentheses in Methods.

Permissions: Material from other sources must be accompanied by a written permission from both author and publisher for reproduction.

Obtain written permission in writing from at least one author of papers still in press. unpublished data, and personal communications.

Units: Express all measurements in conventional units; give Système International (SI) units in parentheses. For figures and tables use conventional units, with conversion factors in legends or footnotes.

SOURCE

July 7, 1994, 331(1): 68.

Nucleic Acids Research
IRL Press

The Editor, *Nucleic Acids Research*
Oxford University Press
PO Box Q
McLean VA 22101, USA
Telephone: (703) 356-4301; Fax: (703) 356-4303
    From the Americas (Full-length papers, Short
    Reports and Computational Biology)
The Editor, *Nucleic Acids Research*
6819 Elm Street, McLean, VA 22101
    Courier mail

Dr R.T.Walker
School of Chemistry, The University of Birmingham
Edgbaston, Birmingham B15 2TT U.K..
Telephone: (+4421 or 021) 476 1688;
Fax: (+4421 or 021) 477 5376
    From the Rest of the World (Full-length papers
    and Short Reports)
The Editor, *Nucleic Acids Research*
Oxford University Press, Journals Department
Walton Street, Oxford OX2 6DP U.K.
Telephone: (+44865 or 0865) 56767;
Fax: (+44865 or 0865) 267773
    From the Rest of the World
    (For Short Methods)

AIM AND SCOPE

Nucleic Acids Research provides rapid publication for papers on physical, chemical, biochemical and biological aspects of nucleic acids, and proteins involved in nucleic acid metabolism interactions.

MANUSCRIPT SUBMISSION

Submit three hard copies of manuscript, including one original set of line copy figures and two others suitable for referees, and three original sets of halftones, accompanied by a disk if appropriate, and a completed manuscript submittal form (end of instructions in each issue), including telephone and fax numbers and e-mail address. Submit disk at time of revision.

Submit all computational biology to McLean. If a customs declaration is required, list contents as OF NO COMMERCIAL VALUE. Bills for Customs or VAT will be returned to author.

Papers are listed under: Surveys & Summaries, RNA, Molecular Biology, Enzymology, Chemistry, Genome Structure and Mapping, Computational Biology, Structural Biology, Methods and Short Reports.

Present some novel development; meet criteria of originality, timeliness, significance and scientific excellence.

Technical progress has greatly increased ease of obtaining new sequence information. Manuscripts with new sequences must include complementary data with relevance to genomic organization, transcription, RNA processing, expression and genetic analysis; reported sequence must shed new light on basic questions of structural or functional interest. Sequences of cDNAs or genes whose products are not relevant to nucleic acids are not acceptable. Comparative sequence analysis is considered if genes are of relevance to nucleic acid metabolism or interactions; some new important findings must also emerge. Sequences of genes for well-studied RNAs (as well as RNA sequences themselves) are not acceptable unless of unusual interest. It is unnecessary and undesirable to print all the sequence. Provided that sequence is available for referees and has been deposited in databank, only those parts relevant to discussion of results are printed.

Manuscripts dealing with characterization of promoters or other regulatory elements must demonstrate careful characterization of the element and significant new insights into regulatory mechanisms.

Manuscripts presenting construction of vectors or clone banks, preparation of monoclonal antibodies, the isolation of nucleic acids or routine synthesis of oligonucleotides or oligonucleotide analogues must be based on completely new principles or accompanied by novel biological applications. Manuscripts dealing with small molecule-nucleic acid interactions must present substantial new information relevant to nucleic acid structure.

Submit only complete chromosome maps for prokaryotic genomes. For higher organisms, genome structure and mapping papers must introduce new technical advances or describe a complete intron-exon structure for a gene relevant to nucleic acid metabolism and interactions.

Chromosome assignment papers where only cytological data are presented are not acceptable. Simple NMR assignment papers are not acceptable. Submit only those that present new information not previously available from other methods.

Computational Biology papers must describe new applications or novel algorithms relevant to nucleic acid sequence determination or analysis. Small improvements in existing programs or variations of well established algorithms are not suitable. Manuscripts dealing with computer analysis of already published sequences are considered if new and interesting findings emerge.

Significant new methods are published in three forms. In methods section of a paper presenting new data; paper is listed in Table of Contents under both normal subject category and in Methods category. As full-length Methods papers if an outstanding and potentially very useful advance is clearly documented. As Short Methods: useful improvements on existing methods. Describe methods in enough detail for it to be reproduced. Give evidence that method is reproducible and reliable.

New motifs that represent signals in DNA or RNA are considered. Modifications to previously described motifs must include a demonstration of increased utility. Protein sequence motifs are limited to those present in proteins relevant to nucleic acid metabolism and interactions.

New restriction enzymes and methylases, including enzymes with novel specificities are considered. Usually isoschizomers will not be suitable, unless they differ significantly from other known enzymes. If any other subject, consult Executive Editor before submission.

Submission of a paper implies that it reports unpublished work and that it is not under consideration elsewhere. Signature of corresponding author is on behalf of all authors and assumes that they are in complete agreement with contents of paper and are prepared to abide by policies of the Journal.

If previously published tables, illustrations, or more than 200 words of text are included, obtain and submit copyright holder's permission with paper. Obtain and submit authorization for all personal communications. Notes stating that two authors have contributed equally are not allowed. Requests for back-to-back publication are honored if papers are submitted together with a joint signed undertaking indicating consent.

MANUSCRIPT FORMAT

Full-length papers should not exceed 8 printed pages; 1240 words, allow 1/3-page for each figure or table. Includes papers placed in Molecular Biology, Enzymology, Chemistry, Genome Structure and Mapping, Computational Biology, Structural Biology and Methods headings.

Abstract: 200 words, one paragraph.

References: Number in order of appearance; list numerically. Abbreviate titles as in *Chemical Abstracts*.

1. Holdsworth, M.J., Bird, C.R., Ray, J., Schuch, W, and Grierson, D. (1987 *Nucleic Acids Res.*, **15**, 7312-739.

2. Huynh, T.V., Young, R.A. and Davies, R.W. (1988) In Glover, D.M. (ed.). DNA cloning—A Practical Approach. IRL Press, Oxford, Vol. I, pp. 49–78.

3. Maniatis, T., Fritsch, E.F. and Sambrook, J. (1982) Molecular Cloning: A Laboratory Manual. Cold Spring Harbor University Press, Cold Spring Harbor.

4. Muller, S.J. and Caradonna. S. (unpublished). X52486.

Do not use references of the type Smith *et al.* (1989).

Figures: Figures are handled conventionally. Do not supply on disk. Refer to each figure in text. On back of each figure mark number, name of first author, and indicate TOP. Submit figures in desired final size. The type area of a page is 240 mm (height) × 184 mm (width); do not exceed. A single column is 88 mm wide and a double column is 184 mm wide; figures should fit either a single or double column. Type sequence figures full width of page. Make lettering about 2 mm in height. If figures require reduction, ensure that lettering will be clearly legible after reduction to final size.

Provide photographs as high quality glossy prints to withstand loss of contrast and detail inherent in printing process. Provide line drawings as good quality hard copies suitable for reproduction as submitted. No additional art work, redrawing or typesetting is done by Publishers. Color figures are subject to a charge, which may be waived for academic authors if color is essential for clarity of figure. Include figure legends on disk.

Submit two copies of an executable version of program and instructions for use by referees with computer papers. Specify costs associated with acquiring program in text.

OTHER FORMATS

SHORT METHODS—2 printed pages. Short methods that do not merit a full-length paper but which are complete, original and useful.

SHORT REPORTS—1 printed page. Contains data not appropriate for publication as a full paper but which are of use to other scientists. Not for preliminary publication of data which will eventually be published elsewhere or for information that would better be included in a full length paper.

SURVEYS & SUMMARIES—8-10 printed pages. Provides a format for brief reviews. Present material not normally acceptable in a formal review article.

Obstetrics and Gynecology 587

OTHER

Atomic Coordinates: Deposit atomic coordinates for crystal structures with a database prior to manuscript submission. Appropriate databases are: Nucleosides, nucleotides and other small molecules: Cambridge Crystallographic Data Center (CCDC). Proteins, polypeptides, etc.: Protein Data Bank (PDB). Oligonucleotides: CCDC or PDB.

Electronic Submission: Papers will be typeset from disk. In the case of a mismatch between disk and copy, hardcopy is taken as definitive.

After manuscript has gone through review and editing, copy onto a clean newly formatted disk. Do not copy irrelevant files and/or back-up files onto disk. For Apple Mac users, ensure wastebasket is empty. Use first-named author or manuscript number for disk label and file name. Protect disk for mailing. Keep a back-up.

Most computer and word processor disks are acceptable; preferred combinations are: PC MS-DOS, PC Windows or Apple Mac, and either Microsoft Word or WordPerfect.

Enter text in style and order of Journal. Insert figure captions and Tables at end of file. Type references in correct order and style of Journal.

Type text unjustified, without hyphenation (except for compound words). Type headings in Journal style. Press tab key once for paragraph indents. Use Times for text font and Symbol for Greek and special characters. Use word processing features to indicate bold, italic, Greek and maths, super- and subscript characters. Indicate special characters drawn by hand.

Do not enter carriage returns for spacing between lines, paragraphs, references, etc. Space is generated automatically by typesetters. Use only one space between sentences in a paragraph. Do not use automatic page numbering, running titles, or footnote features. Number hardcopy by hand at bottom of page. Do not include any copyright material (e.g., word processor software or operating systems files) on disk.

Materials Sharing: Make all strains, clones, cell lines, hybridomas, X-ray and NMR coordinates, and computer programs described in the Journal immediately available to any qualified investigator upon request. Editors are prepared to deny further publication rights in the Journal to authors unwilling to abide by this principle.

Nucleotide Sequences: Submit all new sequence information, including that which extends a previously determined sequence already in the database (and which already has an accession number) to Data Library for a new accession number. Provide number when submitting manuscript. For details see *Nucl Acids Res* 1994; 21(1): i-vi.

Nucleotide sequence data reported in *Nucleic Acids Research* with an EMBL database accession number can be retrieved electronically from EMBL File Server, a free service to anyone who has access to electronic mail networks. To get the sequence with the accession number X12399, send a message to the Internet address netserv@embl-heidelberg.de including the following line: GET NUC:X12399. The requested data will be returned automatically via electronic mail. More introductory documentation can be obtained by including the line: HELP.

SOURCE

January 11, 1994, 21(1): i-vi.

Obstetrics and Gynecology
American College of Obstetrics and Gynecology
Elsevier Science Publishing Co.

The Editor, *Obstetrics and Gynecology*
1100 Glendon Avenue, Suite 1655
Los Angeles, CA 90024-3520

MANUSCRIPT SUBMISSION

Original submissions must be contributed solely to *Obstetrics and Gynecology*. If any form of publication elsewhere of any of the material, other than an abstract of not more than 300 words, has occurred or is planned, identify in cover letter; include a copy of other publication. Failure to comply may lead to a judgment of redundant publication; authors found responsible for redundant publication without attribution of other sources may be barred for up to 3 years from submitting manuscripts; and a statement identifying nature and source of redundant publication will be printed prominently in the journal.

Identify potential conflicts of interest, both financial and personal. List all sources of financial support for study on title page with institutional and corporate affiliations. Disclose any commercial affiliations, whether or not it is a source of funding. Identify specifically any financial involvement (e.g., employment, stock holdings, consultantships, honoraria) within the past 3 years with a commercial organization that might have any potential interest in subject or materials discussed in manuscript. If uncertain as to what constitutes potential conflict of interest, err on side of full disclosure. Financial information will be held in confidence during review process. If paper is accepted, Editor, after consultation with author, will determine if financial disclosure is important for readership.

A submitted manuscript must be accompanied by an agreement with original signature of each author. A copy of this agreement is in January and July Journal issues. By signing form, each author agrees that: paper and work contained constitute original research contributed solely to *Obstetrics and Gynecology*; each author has participated meaningfully in conception and design of research, analysis of data, and writing and approval of manuscript, to a degree sufficient that he or she can take public responsibility for work described; any financial interests, commercial affiliations, or any possible conflicts of interest are disclosed in an accompanying letter; and if paper is accepted, copyright will transfer to The American College of Obstetricians and Gynecologists.

Articles generated from work done while author is a United States government employee are in public domain and copyright considerations do not apply; identify such employment by author's signature in bottom section of agreement. Agreement should indicate name, address, and telephone and Fax numbers for corresponding author.

MANUSCRIPT FORMAT

Type entire manuscript (including bibliography, tables, and figure legends) in 10-pitch (12-point) type double-spaced on nonerasable bond paper with 1 in. margins on top, bottom, and each side. Do not use bold or italic type. Type first author's name in upper right corner of each page. Number each page consecutively, beginning with title page. Mail two complete sets of manuscript.

Use of subheadings is discouraged in all but the most complex papers.

Title Page: First page: title of article, name(s), affiliations, and major degree(s) of author(s), and source of work or study. Depending on nature of study, up to six authors may be listed; more than six requires justification. If authorship is attributed to a group, one or more members must take responsibility for group; list other members in a footnote. Allow titles no more than 100 characters (counting letters and spaces). Do not use abbreviations.

Give acknowledgments on title page. Acknowledge financial and other substantive support. Other acknowledgments, such as advice, secretarial services, or contribution of study subjects, are not permitted. If all or part of the paper was presented at the Annual Clinical Meeting of The American College of Obstetricians and Gynecologists, note that presentation; acknowledge no other organizational presentations.

List name, address, telephone number, and fax number of corresponding author on title page. If another author will receive reprint requests, identify on title page.

Provide a short title (30 characters and spaces) at bottom of title page for use as a running foot.

Précis and Abstract: On second page, provide a *Précis* for table of contents (single sentence, 25 words); state concisely and in simple declarative, and nonquantitative terms conclusion(s) of report.

On third page, provide an Abstract. For original research reports, case reports, or review articles, submit structured abstracts. For all other articles, make abstract a single 250 word paragraph that states what was done, what was found, and what findings mean.

For original research reports, structured abstract is 250 words with headings: Objective (main question, objective, or hypothesis), Methods (study design, participants, outcome measures), Results (measurements, including confidence intervals and level of statistical significance where appropriate), and Conclusions (those directly supported by data, along with any clinical implications).

For case reports, structured abstract (150 words) uses headings: Background (importance of subject matter and specific purpose of report), Case (summary of pertinent features of the clinical findings, important laboratory abnormalities, treatment, and outcome), and Conclusion (summary of principal finding and why it is unique or worthy of mention, indicating relevance to clinical practice).

Use structured abstracts (250 words) for review articles, with headings: Objective (statement of purpose), Data Sources (sources searched, including dates, terms, and constraints), Methods of Study Selection (number of studies reviewed and selection criteria), Data Extraction and

Synthesis (guidelines used for abstracting data, main results of review, and methods used to obtain these results), and Conclusions (primary conclusions and their clinical applications).

For structured abstracts, each heading begins a separate paragraph. All information in abstract must be found in text, tables, or figures. For more information, see Haynes RB, Mulrow CD, Huth EJ, Altman DG, Gardner MJ. "More informative abstracts revisited" *Ann Intern Med* 1990;113: 69-76).

Introduction: Orient reader to problem(s) addressed by report and state purpose or objective of research. Avoid a detailed literature review; cite important published works.

Materials and Methods: Describe methodology used in research in sufficient detail so others can duplicate work. If methodology has been published, cite appropriate publication(s); do not repeat description. Identify statistical analysis methods used and, state basis (including alpha and beta error estimates) for their selection.

Results: Present findings in appropriate detail. Use tables and/or figures; avoid duplication between text and tables or figures.

Discussion: Raise implications of study and compare findings with those of similar reports. Recapitulate some findings, but avoid needless repetition of results. Some degree of speculation is permissible; avoid unfounded conclusions.

Figures and Legends: Submit two sets of glossy photographs for graphs and line drawings and five sets of glossy photographs for photomicrographs, radiographs, sonographs, color illustrations, and figures that do not photocopy well.

Journal column size of 3.25 in. Make lettering and identifying marks (arrows) clear, sharp, and large enough to accommodate reduction in size. Make related figures the same size. Identify critical area(s) of x-rays and photomicrographs. Check for typographical errors. Do not write on back of figures or attach paper clips, glue, or tape to any part of figure; firmly affix adhesive label designating TOP, (first) author's name, and (Arabic) number of figure on back. Do not mount on cardboard. Place figures in a labeled envelope. For color illustrations, submit five sets of glossy prints and original transparencies.

Type legends on a single separate page, not on figure itself. Define abbreviations in figures. Use Arabic numbers for figures. Because of production costs, do not substitute figures for tables unless understanding is enhanced. Author is responsible for cost of color illustrations. Estimates are available from publisher.

Tables: Type each on a separate sheet, number with Arabic numerals, and include a clear and concise title. Use explanatory footnotes; define abbreviations used. Include sufficient information so table can be understood by itself. Do not use a table for data that can be described adequately in two or three sentences in text. Observe ordinary rules of capitalization in tables. Do not use vertical lines.

Permissions: For lengthy direct quotations, tables, or figures from previously published sources, submit written permission from author and copyright holder. Provide complete information on source of material.

References: Verify accuracy. Follow specifications of "Uniform Requirements for Manuscripts Submitted to Biomedical Journals" of the International Committee of Medical Journal Editors (*N Engl J Med* 1991;324:424–8).

Number consecutively in order of appearance in text. Cite either by superscript without parentheses or on line with parentheses. Cite each reference. Cite references appearing for the first time in a table or figure in text where table or figure is mentioned.

Cite references from generally accessible peer-review publications. Do not list unpublished data, personal communications, papers presented at meetings and symposia, abstracts, letters, and manuscripts "submitted for publication" in bibliography. Cite, if absolutely necessary, in text with sources in parentheses. List papers accepted by peer-reviewed publications but not yet published in bibliography with words "in press" substituting for year, volume, and pages. Books include publisher and year of publication.

Type list of references double-spaced. Abbreviate journal names according to *Index Medicus*.

JOURNAL ARTICLE, UP TO SIX AUTHORS

List all authors when six or fewer; with seven or more, list first three and add "et al."

1. Jonnavithula S, Warren MP, Fox RP, Lazaro MI. Bone density is compromised in amenorrheic women despite return of menses: A 2-year study. Obstet Gynecol 1993;81: 669–74.

JOURNAL ARTICLE, MORE THAN SIX AUTHORS

2. Evans MI, Dommergues M, Wapner RJ, et al. Efficacy of transabdominal multifetal pregnancy reduction: Collaborative experience among the world's largest centers. Obstet Gynecol 1993;82:61–6.

BOOK

3. Cunningham FG, MacDonald PC, Gant NF, Leveno KJ, Gilstrap LC III. Williams obstetrics. 19th ed. Norwalk, Connecticut: Appleton & Lange, 1993.

CHAPTER IN A BOOK

4. Thomason JL. Perinatal infections. In: Shaver DC, Phelan ST, Beckmann CRB, Ling FW, eds. Clinical manual of obstetrics. 2nd ed. New York: McGraw-Hill, 1993:264–78.

EDITED BOOK

5. Folb PI, Dukes MNG, eds. Drug safety in pregnancy. Amsterdam: Elsevier, 1990.

Do not use periods after authors' initials or after journal abbreviations. Use "et al" when six authors. Place a period at end of each reference.

OTHER FORMATS

AFTER OFFICE HOURS—An article for light reading addressing a pertinent topic. No data, tables, or figures are needed. Humorous or satirical material may be included.

CASE REPORT—Describe up to three cases (with 8 references) of a particular condition that is unusual and provides new insight into diagnosis or management.

CLINICAL COMMENTARY— 12 double-spaced, typed pages; 12 references. A short essay expressing opinions, experiences, or perspectives of clinical relevance to obstetrics and gynecology.

EDITORIAL—A signed statement regarding a topic of current interest and importance involving broad aspects of the discipline or a matter of concern specifically to the Journal. Usually written by an editor or member of the Editorial Board; authors may be invited.

GRADUATE EDUCATION—A report involving educational techniques or programs of interest that apply to medical students, residents, fellows, and practicing obstetrician-gynecologists.

INSTRUMENTS AND METHODS—A description of a new operation, instrument, technology, or method of diagnosis or treatment. Brief Introduction outlines need for new development. Method or Technique section describes innovation, usually with illustrations. Experience section reports experience with it and outcomes. In Comment section, describe implications of findings.

LETTER TO THE EDITOR—Three double-spaced typed pages, four references. Questions or challenges to articles published in journal must be received within 6 weeks of publication. Do not use to report additional experience. Put all authors' full names, and corresponding author's address and telephone and fax numbers at end of letter. Disclose any financial associations and potential conflicts of interest.

ORIGINAL RESEARCH REPORT—A full-length report of basic or clinical investigation that provides sufficient information to permit critical and rigorous evaluation. Include tables and figures as appropriate and a bibliography of selected pertinent references.

REVIEW ARTICLE—30 double-spaced, typed pages. A comprehensive scholarly review of prior publications relating to an important clinical subject, accompanied by critical analysis and leading to rational conclusions.

OTHER

Abbreviations: Use without first spelling out term.

ACOG	American College of Obstetricians and Gynecologists
ACTH	adrenocorticotropic hormone
ADH	antidiuretic hormone
AIDS	acquired immunodeficiency syndrome
D&C	dilation and curettage
DNA	deoxyribonucleic acid
FSH	follicle-stimulating hormone
GnRH	gonadotropin-releasing hormone
hCG	human chorionic gonadotropin
HLA	human leukocyte antigen
hPL	human placental lactogen
LH	luteinizing hormone
RNA	ribonucleic acid
TSH	thyroid-stimulating hormone

VDRL   Venereal Disease Research Laboratory

For abbreviations that may be used after spelling term fully at first used in Abstract and again in body of paper, see Appendix B, last page of Instructions to Authors.

Units: Abbreviations for units of measurement are given in *AMA Manual of Style*, (8th edition, Williams & Wilkins Co.). Other abbreviations may not be used.

SOURCE

January 1995, 85(1): front matter.

Oncogene
Macmillan Press, Limited

John Jenkins
Marie Curie Research Institute
The Chart, Oxford, Surrey, RH8 0TL U.K.
Telephone: 0883 717273; Fax: 0883 717448

E. Premkumar Reddy
The Fels Institute for Cancer Research
   & Molelcular Biology
Medical Research Bldg., 3420 North Broad St.
Philadelphia, PA 19140

AIM AND SCOPE

The prime objective of the journal will be the advancement and dissemination of knowledge concerning the molecular basis of malignant change. Overriding criteria for publication will be originality, exemplary scientific merit and general interest. *Oncogene* will publish full and detailed papers as well as short communications relevant to all aspects of oncogene research including the following topics: cellular oncogenes and their mechanisms of activation, structure and functional aspects of their encoded proteins oncogenes in RNA and DNA tumor viruses, presence of oncogenes in human tumors, relevance and biology, cell cycle control, immortalization, cellular senescence, regulatory genes and 'antioncogenes,' growth factors and receptors.

MANUSCRIPT SUBMISSION

Submit manuscript and two good quality copies (including figures) either directly to US or British Editorial Office, or submit two copies directly to an editorial board member (see journal front matter) who will arrange refereeing. Submit top copy directly to British Editorial Office with a copy of cover letter sent to editorial board member.

Before publication authors must assign world copyright of manuscript to Macmillan Press Ltd. Submit a signed statement that article is original, is not under consideration or has not been previously published elsewhere and its content has not been anticipated by any previous publication. Authors will be entitled to publish any part of their paper elsewhere without permission, provided usual acknowledgments are given. Submission of a manuscript implies that authors have obtained permission from employers or institution to publish.

MANUSCRIPT FORMAT

Submit well-presented manuscripts; check grammar, spelling and punctuation. Subdivide into sections in order: Title page, Abstract, Introduction, Results, Discussion, Materials and methods, Acknowledgments, References, Tables, Legends to figures.

Type original and two copies on A4 or American quarto paper, double-spaced with generous margins. Number each page (title page is 1). Underline only words or letters to appear in italics. Indicate position of each figure and table in margin.

Title Page: Keep title short and to the point. Follow surname and initials of each author by his or her department, institution, city with postal code and country. Give changes of address in numbered footnotes. Indicate corresponding author(s). Provide a running title (50 characters). Include a contact telephone number.

Abstract: On second page, a single 200 word paragraph. Make comprehensible to readers before they have read paper. Avoid abbreviations and reference citations.

Acknowledgments: Include at end of text and not in footnotes. Personal acknowledgments precede those of institutions or agencies.

References: Verify accuracy. Include articles in press (state accepting journal). In text cite references by author and date; do not cite more than two authors per reference; if more, use *et al.* At end of manuscript type citations in alphabetical order, with authors' surnames and initials inverted. References include, in order: authors' names, year, journal title, volume number, inclusive page numbers, and name and address of publisher (for books only). Abbreviate name of journal according to *International List of Periodical Title Word Abbreviations*. Underlined to indicate italics.

Hamming, J., Arnberg, A. & Gruber, M. (1981). *Nucleic Acids Res.* **9**, 1339–1350.

Koster, H. (ed.) (1980). *Nucleic Acids Synthesis: Applications to Molecular Biology and Generic Engineering*. IRL Press: London.

Norris, K.E., Norris, F. & Brunfeldt, K. (1980). *Nucleic Acids Synthesis: Applications to Molecular Biology and Generic Engineering*. Koster, H. (ed.). IRL Press: London. pp 233–241.

Avoid personal communications. Cite manuscripts in preparation or submitted, but not yet accepted, in text; do not include in list of references.

Tables: Type on separate sheets. Number consecutively with Arabic numerals. Make self-explanatory. Include a brief descriptive title. Indicate footnotes by lowercase letters; do not include extensive experimental detail.

Illustrations: Refer to all illustrations in text as Figure 1, etc. Write title of paper, name of first author and figure number lightly in pencil on back of each. On manuscript indicate with an arrow in margin most appropriate position for figure. Provide line drawings as clear, sharp prints, suitable for reproduction as submitted.

Photographs: Present with sufficient high quality with respect to detail, contrast and fineness of grain to withstand reduction and loss of contrast and detail inherent in printing process. Indicate magnification by a rule on photograph. If several prints of varying quality of same figure are provided, indicate which print should be used.

Color Plates: Subject to a special charge.

Figure Legends: Put on a separate, numbered manuscript sheet. Define all symbols and abbreviations used in figure.

OTHER

Abbreviations: Restrict use to SI symbols and those recommended by IUPAC. Define in brackets after first mention in text. Use standard units of measurements and chemical symbols of elements without definition in text.

Materials Sharing: Authors agree to make freely available to colleagues in academic research any of the cells, nucleic acids, antibodies. etc. used in research reported that are not available from commercial suppliers.

Style: Follow conventions of *Committee of Biological Editors Style Manual* (5th edn. 1983). Use *Chemical Abstracts* and its indexes for chemical names. For use of biochemical terminology follow recommendations of IUPAC-IUB Commission on Biochemical Nomenclature, in *Biochemical Nomenclature and Related Documents* (Biochemical Society, UK). For enzymes, use recommended name assigned by IUPAC-IUB Commission on Biochemical Nomenclature, 1978, as given in *Enzyme Nomenclature* (Academic Press, New York, 1980). Use recommended SI (Système International) units.

Italicize genotypes (underline in typed copy); do not italicize phenotypes. For bacterial genetics nomenclature follow Demerec *et al.* (1966). *Genetics* **54,** 61–76.

SOURCE
January 1994, 9(1): inside back cover.

Ophthalmology
American Academy of Ophthalmology
J.B. Lippincott Co.

Paul R. Lichter, MD, Editor in Chief
Ophthalmology
WK Kellogg Eye Center
1000 Wall Street
Ann Arbor, MI 48105-1994

AIM AND SCOPE

The objective of the American Academy of Ophthalmology in publishing its journal, *Ophthalmology*, is to provide opportunities for the free exchange of ideas and information.

MANUSCRIPT SUBMISSION

Ophthalmology adheres to the policies in the "Uniform Requirements for Manuscripts Submitted to Biomedical Journals" of the International Committee of Medical Journal Editors (*Ann Intern Med* 1988; 108:258–65).

Each author must have made a significant intellectual contribution to research project and/or to writing manuscript and take full responsibility for contributions. Collaborative efforts (e.g., committees, study groups) must list at least one individual as an author who will take responsibility for content of manuscript.

In cover letter disclose individual or family investments, stock ownership, consultancies, retainers, patents, or other commercial interests that could cause or be perceived as a conflict of interest. Describe greatest degree of involvement, not smallest or least complicated. Disclosure will not affect review. Where appropriate, make a positive statement that no such relationship exists.

Manuscripts may not have appeared in other publications; differently written reports of previously published studies, updates of previously published studies that add small amounts of data or numbers of patients, or slightly different studies of a previously described patient pool will not be considered. If unsure whether specific printed material comprises prior or repetitive publication, alert Editor in cover letter and include copies of publications in question.

Submit one original and two copies of text and figures. Annual Meeting free papers require three copies of text. Submit one set of slides with any color illustrations.

In a cover letter to Editor-in-Chief, identify corresponding author. List meetings where material was presented, and disclose any area that could be construed as a conflict of interest. Disclose any material that could be construed as a prior publication. Include a copyright transfer.

Material (in whole or in part) may not currently be under consideration by another publication, in press in any other format, and or previously published.

Each author must sign a statement transferring copyright ownership to the American Academy of Ophthalmology. Request a copyright transfer form from Journal office, or include following statement, signed by each author, in cover letter: "In consideration of the journal, *Ophthalmology*, taking action in reviewing and editing my (our) submission, the author(s) undersigned hereby transfers, assigns, or otherwise conveys all copyright ownership to the American Academy of Ophthalmology in the event that such work is published in the journal *Ophthalmology*."

Federal employees should sign the following statement: "The above-named article is a work of authorship prepared as part of the undersigned author's (authors) official duties as an officer or employee of the U.S. government and is therefore in the public domain. Should, however, the article ever be determined to be copyrightable, I (we) hereby transfer, assign, or otherwise convey all copyright ownership in the above-named article to the American Academy of Ophthalmology."

Inform journal office if material in manuscript has been or is going to be submitted as an Annual Meeting poster or paper. Do not jeopardize your poster or oral presentation by printing the same findings in advance of Annual Meeting. This is contrary to Academy policy. If you submit a manuscript to Journal and later decide to submit an abstract to Program Advisory Committee, inform us.

MANUSCRIPT FORMAT

Title Page:

Title—meaningful and brief as possible (135 characters); do not use declarative titles.

Names of authors—give full first name, middle initial, advanced degrees, and professional certification.

Institutional affiliation—identify each author's affiliation during course of performing the study. Use superscript numbers, not symbols.

Meeting presentation—if the material has been presented previously, supply names, places, and dates of meetings.

Financial support—identify all sources, public and private. Provide agency name and city, company name and city, fellowship name, and grant number.

Proprietary interest statement—Disclose any type of financial interest. Disclose greatest extent of interest, not smallest or easiest to describe. Sample statements include:

"This study was supported in part by (name of company)."

"(Author name) is a consultant to (name of company) or to a competing company."

"(Author name) has a financial interest in this or a competing (instrument/drug/piece of equipment)."

"(Author name) holds patent rights to this or a competing (instrument/drug/piece of equipment)."

"(Author name and/or) author's family owns or has warrant or other potential rights to more than 1% of stock in (name of company) or a competing company."

"Each author states that s/he has no proprietary interest in the development or marketing of this or a competing (instrument/drug/piece of equipment)."

Running head—no longer than 60 characters; and Address for reprints if reprints are going to be available.

Structured Abstract: 250 words on a separate page immediately following title page; four separate paragraphs: Purpose or Background, Methods, Results, and Conclusion. Manuscripts without a structured abstract will be returned. For useful information see: Ad Hoc Working Group for Critical Appraisal of the Medical Literature. A proposal for more informative abstracts of clinical articles. (*Ann Intern Med* 1987;106:598–604); Huth EJ. Structured abstracts for papers reporting clinical trials [editorial]. (*Ann Intern Med* 1987;106:626–7); Haynes RB, Mulrow CD, Huth EJ, et al. More informative abstracts revisited. (*Ann Intern Med* 1990;113:69–76); Lichter PR. Structured abstracts now required for all submissions to the journal [editorial]. (*Ophthalmology* 1991;98: 1611–12).

Text: Follow recognized standards for presenting scientific material. See *CBE Style Manual* (5th ed, Council of Biology Editors; 1983). Do not give an extensive review of literature in Introduction, include only literature pertinent to purpose of study and its relationship to work in field. Write Materials and methods in enough detail so that others can duplicate. Results must be concise. Restrict Discussion to significant findings presented. Wide digressions and theorizing cannot be published.

Acknowledgments: Do not acknowledge those who reviewed, discussed, edited, or typed a manuscript, or gave "technical," "helpful," "crucial," or "moral" support or similar collegial efforts. Acknowledge those who referred patients, translated references, provided extensive statistical assistance, provided essential tissue, equipment, or other materials without which the study could not have been accomplished. See *Ophthalmology's* March 1988 editorial, "The Author Wishes to Thank."

References: Follow text. Number consecutively by order of appearance in text. In text, designate numbers either as superscripts or on line in parentheses.

Do not list a reference unless you have read it. Do not cite unpublished studies as references. Place "unpublished data" or "personal communication" in parentheses in text. Do not cite abstracts as references; this is considered unpublished data; note in text. Do not cite oral or poster presentations as references; cite in parentheses in text as "presented as a poster at the American Academy of Ophthalmology Annual Meeting, Anaheim, 1991."

Cite articles, books, and chapters in press (not simply submitted) as such. Use journal abbreviations from *Index Medicus*. If correct abbreviation is in doubt, cite complete journal name. Follow format and punctuation precisely. Do not underline journal titles. Do not use periods in abbreviations of journal titles or in author initials.

JOURNAL ARTICLE—four or fewer authors

Richards JE, Kuo C-Y, Boehnke M, Sieving PA. Rhodopsin Thr58Arg mutation in a family with autosomal dominant retinitis pigmentosa. Ophthalmology 1991;98:1797–1805.

JOURNAL ARTICLE—five or more authors (list first three and add et al.)

Culbertson WW, Brod RD, Flynn HW Jr, et al. Chickenpox-associated acute retinal necrosis syndrome. Ophthalmology 1991;98:1641–6.

CHAPTER IN A BOOK

Parks MM, Mitchell PR. Cranial nerve palsies. In: Tasman W, Jaeger EA, eds. Duane's Clinical Ophthalmology, rev ed. Philadelphia: Harper & Row, 1991; v. 1, chap. 19.

BOOK

Miller NR. Walsh and Hoyt's Clinical Neuro-Ophthalmology, 4th ed. Vol. 4. Baltimore: Williams & Wilkins, 1991;2102–14.

LETTER TO THE EDITOR

Sneed SR, Blodi CF, Berger BB, et al. *Pneumocystis carinii* choroiditis in patients receiving inhaled pentamidine [letter]. N Engl J Med 1990;322:936–7.

Tables: Follow references. Number consecutively in order mentioned in text. Give each table a title and each column a heading. Explain abbreviations in legend. Type only one table on a page.

Legends: Follow tables. Number figures consecutively in text. For histologic figures, note stains and magnifications. Acknowledge source and obtain proof of permission (from copyright holder) to use figures published elsewhere. Identify symbols or letters on prints.

Double-space all material on white bond, with 1 in. margins on all sides.

Illustrations: Submit three identical complete sets of prints in three separate envelopes (note on envelope which set is printer's copy). On back of each print, affix a label noting figure number, last name of first author, and orientation (top/left/right). Do not write on print. Use transfer sets to affix symbols, letters, numbers. Submit color prints with slides. Place slides in jackets; number identically to prints. Write "front" and "top" on front of slide. Orientation must match prints. There may be a charge for printing color. Call Journal Office for information.

For clinical photographs where patient could be recognized, submit a statement signed by patient or guardian granting permission to publish photograph for educational purposes. If permission was not obtained, photograph will be cropped to assure that patient's identity is not disclosed.

OTHER

Abbreviations: Restrict to those that are widely used and understood. Avoid using abbreviations that have meaning only in this manuscript. At first use, write out term in full with abbreviation following in parentheses.

Drugs: Use generic names only. Cite trade name in parentheses first time generic name appears.

Experimentation, Animal: State if animals were used in study. Describe protocol followed briefly; name any institutional body granting approval.

Experimentation, Human: State that informed consent was obtained and that appropriate institutional committee or review board approved protocol.

SOURCE

January 1994, 101(1): 211.

Oral Surgery, Oral Medicine, and Oral Pathology

Mosby-Year Book, Inc.

Dr. Larry J. Peterson, Editor in Chief
Department of Oral and Maxillofacial Surgery
College of Dentistry, The Ohio State University
2131 Postle Hall, 305 West 12th Avenue
Columbus, OH 43210-1241
            Oral and Maxillofacial Surgery
Dr. Jed J. Jacobson
University of Michigan School of Dentistry, Room G306
Ann Arbor, MI 48109
            Oral Medicine
Dr. Carl M. Allen
Oral and Maxillofacial Surgery and Pathology
College of Dentistry, Ohio State University
305 W. 12th Ave.
Columbus, OH43210
            Oral and Maxillofacial Pathology
Dr. Richard E. Walton
Department of Endodontics
The University of Iowa, College of Dentistry
435 Dental Science Bldg. S
Iowa City, IO 52242-1001
            Endodontics
Dr. Allan G. Farman
Division of Radiology and Imaging Sciences
Department of Biology and Physical Sciences
School of Dentistry, University of Louisville
Louisville, KY 40292
            Oral and Maxillofacial Radiology
Dr. Peter G. Fotos
Department of Oral Pathology, Radiology, and Medicine
University of Iowa College of Dentistry, DSB
Iowa City, IA 52242
            Oral Diagnosis

MANUSCRIPT SUBMISSION

Send manuscripts to Editor-in-Chief or appropriate section Editor. Complete and have all authors sign copyright statement that appears in each issue of the Journal. Submit with manuscript.

Good case reports will be published if they: are of rare or unusual lessons that need documentation; are well-documented cases showing unusual or "atypical" clinical or microscopic features or behavior; are cases showing good long-term follow-up information, particularly in areas where good statistics on results of treatment are needed. Abbreviated case reports are published from time to time; three to four paragraphs with one or two illustrations.

MANUSCRIPT FORMAT

Type double-spaced on one side of paper only, with liberal margins. Submit original and three copies.

Title Page: Include: title of article, full name of author(s), academic degrees, positions, and institutional affiliations. Give corresponding author's address, business and home telephone and fax numbers. For title page of three copies list article title only; no authors' names or affiliations.

Abstract: 50–150 words, double-spaced, before introduction.

Structured Abstract: 150 words; may be used for data-based research articles. Use the following major headings: Objective(s)—reflect purpose of study, that is, hypothesis being tested; Study Design—include setting for study, subjects (number and type), treatment or intervention, and type of statistical analysis; Results—include outcome of study and statistical significance if appropriate; and Conclusion(s)—state significance of results.

References: Cite selectively. Do not cite personal communications and unpublished data as references; cite in parentheses at appropriate place in text. Verify all reference. Cite consecutively in text by superscript numbers. Reference list format conforms to "Uniform Requirements for Manuscripts Submitted to Biomedical Journals" (Vancouver style) (*JAMA* 1993;269:2282-6). For references to articles in press include authors' surnames and initials, title of article, and name of journal.

Type reference list double-spaced on a separate page and number in order as reference citations appear in text. For journal citations, include surnames and initials of authors, complete title of article, name of journal (abbreviate according to *Cumulated Index Medicus*), year of publication, volume number, and inclusive page numbers. For book citations, given surnames and initials of authors, chapter title, editors' surnames and initials, book title, volume number, edition number, city and full name of publisher, year of publication, and inclusive page numbers of citation. If six or fewer authors, list all; if seven or more, list first three and add *et al.*):

FORMAT FOR PERIODICAL REFERENCES

Pullon PA, McGivney J. Computer utilization in an oral biopsy service. Int J Oral Surg 1977;6:251–5.

FORMAT FOR BOOK REFERENCES

Seakins J, Saunders R, eds. Treatment of inborn errors of metabolism. London: Churchill Livingstone, 1973:51–6.

FORMAT FOR CHAPTER REFERENCES

Hudson FB, Hawcroft J. Duration of treatment in phenylketonuria. In: Seakins J, Saunders R, eds. Treatment of inborn errors of metabolism. London: Churchill Livingstone, 1973:51–6.

Illustrations: Submit three sets of glossy prints, numbered, with suitable legends, and marked lightly on back with article title and an arrow to indicate TOP.

A reasonable number of halftone illustrations or line drawings are reproduced at no charge; make special arrangements with Editor in Chief for color plates, elaborate tables, or extra illustrations. Prepare drawings or graphs in black India ink. Typewritten or freehand lettering is not acceptable; all lettering must be done professionally. Letters should be in proportion to drawings or photographs. Do not send original artwork or x-ray films. Good black-and-white contrast is essential for reproduction, submit only glossy print photographs (no smaller than 3×4 inches and no larger than 5×7 inches). Consistency in size of illustrations is preferred; note special instructions concerning sizing. Supply roentgenograms as prints showing bone areas as white.

If color illustrations are deemed essential, there is no charge. Supply either color transparencies or prints. Make size standard size of journal illustration.

Legends to Illustrations: Type double-spaced on a separate sheet. If an illustration is from published material, give full credit to original source.

Tables: Type double-spaced, including column heads, data, and footnotes; submit on separate sheets. Make self-explanatory and supplement, not duplicate, text. Supply a concise title for each. Give all columns concise headings describing data therein. Type footnotes immediately below table; define abbreviations. If a table or any data therein have been previously published, give full credit to original source in a footnote.

OTHER FORMATS

ANNOUNCEMENTS—Submit to the Editor, Dr. Peterson, 8 weeks before desired month of publication. Items published at no charge include those received from a sponsoring society of the Journal; courses and conferences sponsored by state, regional, or national dental organizations; and programs for the dental profession sponsored by government agencies. All other announcements carry a charge of $60 U.S.; fee must accompany request to publish.

CLINICOPATHOLOGIC CONFERENCE—Present interesting, challenging, or unusual cases. Presentation should stimulate clinical work-up, including differential diagnosis. Include complete diagnostic evaluation, management, and follow-up. Organize CPC articles into five parts: *Clinical presentation*—describe clinical and imaging characteristics of lesion. Use clinical photographs and radiographs as appropriate. *Differential diagnosis*—list and discuss lesions to be considered as reasonable diagnostic possibilities. *Diagnosis*—histopathologic findings illustrated with photomicrographs. *Management*—describe treatment of patient and response to treatment. *Discussion*—concentrate on most interesting aspect of case. Submit articles to: Lewis R. Eversole, DDS, MSD, MA, CPC Editor, Oral Surgery, Oral Medicine, Oral Pathology, Section of Diagnostic Sciences, UCLA School of Dentistry, 10833 Le Conte Avenue, Los Angeles, CA 90024.

REVIEWS—Manuscripts that review the current status of a given topic, diagnosis, or treatment. Do not present an exhaustive review of literature but rather a review of contemporary thought with respect to topic. Bibliography should include only seminal, pertinent and contemporary references deemed by author to be most important.

OTHER

Electronic Submission: After a manuscript has been accepted, revised version may be submitted on disk, along with four printed copies exactly matching disk version. Most word-processing programs are acceptable (IBM-WordPerfect or Macintosh-Microsoft Word are preferred. Label disks with computer type (IBM or Macintosh), word-processing program, title of paper, and first author's name. Use word processor's default typeface and type size. Do not use text fomatting capabilities. Number pages consecutively beginning with title page.

Permissions: For direct quotations, tables, or illustrations that have appeared in copyrighted material, obtain and submit written permission for use from copyright owner and original author along with complete information as to source. Obtain and submit signed releases showing informed consent for photographs of identifiable persons.

SOURCE

January 1994, 79(1): 12A-14A.

Pacing and Clinical Electrophysiology (PACE)

North American Society of Pacing and Electrophysiology

International Cardiac Pacing and Electrophysiology Society

Asian-Pacific Working Group or Cardiac Pacing Futura Publishing Company

Seymour Furman, M.D.
Montefiore Medical Center
111 East 210 Street, Bronx, New York 10467
Fax: (718) 547-1795

AIM AND SCOPE

The *Journal* welcomes original and review communications in laboratory and clinical cardiac pacing, electrophysiology, and the electrostimulation of other organs and cardiac assist.

MANUSCRIPT SUBMISSION

Manuscripts may not be under simultaneous consideration by any other publication and should not have been published elsewhere in substantially similar form. Affirm in cover letter. No part of a paper published by *PACE* may be reproduced or published elsewhere without written permission of author(s) and Publisher.

Disclose associations that may pose a conflict of interest. for employees of a commercial or industrial organization; identification will suffice. Acknowledge commercial/industrial author's individual affiliation if part of a noncommercial group. Industrial employees may not evaluate or comment on products of a competitor. Do not include commercial names in manuscript titles. Do not make claims of priority.

MANUSCRIPT FORMAT

Submit in triplicate (one original and two copies); type all sections double-spaced (text, references, legends, etc.) on 8.5 × 11 in. white bond paper with 25 lines per page and all margins 1 in. Submit in order: Title Page, Abstract, Text, References, Legends, Tables, and Figures. Number all pages; abstract is page 1. On each page, beginning with summary, include senior author's surname typed on upper, left corner. Manuscripts not prepared according to instructions may be returned.

Title Page: Include full name(s), degree(s), and affiliation(s) of author(s); list under title; affiliations of commercial or industrial employees; running title of 3 to 6 words. At bottom of page, include information about financial support, i.e., grants, if applicable. Give address for reprints with full name, address, telephone and fax numbers.

Abstract: 250 words on Page 1, after title page. At end of abstract, provide 6 key words for indexing. Use standard abbreviations for all measurements.

Text: Follow abstract; begin on a new page (References, Tables, and Legends also). Minimize abbreviations; define those used at first use. Cite references in numerical order; also tables and figures. Use written personal communications as references; unpublished data may not be used.

References: Number in order of appearance in text. Abbreviate journal titles according to *Index Medicus.*

PERIODICALS

1. Fisher JD, Kim SG, Waspe LE, et al. Amiodarone: Value of programmed electrical stimulation and Holter monitoring. PACE 1986; 9:422–435. (If more than three authors, please use "et al.").

BOOKS (EDITED BY OTHER THAN AUTHOR OF ARTICLE)

2. Griffin JC. Pacemaker programmability: Its role in the maintenance of pacing function. In GA Feruglio (ed.): Electrophysiology and Pacemaker Technology. Padova, Italy, Piccin Medical Books, 1982. pp. 759–760.

BOOKS (IDENTICAL AUTHOR AND EDITOR)

3. Chung EK. Principles of Cardiac Arrhythmias. Baltimore, MD, Williams & Wilkins, 1977, pp. 97–188.

PERSONAL COMMUNICATIONS

4. Smith I. December 10, 1986, personal communication.

ABSTRACTS

5. Same as periodicals and followed by "(abstract)."

Tables: Supplement, do not duplicate, text. Number consecutively in order of appearance in text. Give Roman numerals and titles at top of page. Footnote abbreviations; explain in alphabetical order. Footnote material not self-explanatory. Submit written permission from publishers and authors for previously published material.

Figures: Submit 3 black-and-white glossy prints and two photocopies (5 × 7 in.). On back, write lightly with soft, black pencil or on a gummed label, number, senior author's surname, and TOP. Submit written permission from publisher(s) for use of any previously published figure. Submit in an envelope backed by cardboard. Make lettering or scale of measurement large enough to be legible after half-size reduction. Do not send original art-work, X-rays, or ECGs.

Submit written permission from person in photographs if identifiable. State specifically what is being consented to and what restrictions are placed upon publication. Observe all restrictions. Color illustrations are charged to author. Ask about cost from publisher before submitting.

Legends: Must correspond with figures. Do not duplicate text. Identify abbreviations at end of legend, if not used in text. Use standard abbreviations for measurements.

OTHER FORMATS

CASE REPORTS—6 pages, including all typed material, plus a title page and three figures all in standard *PACE* format; 100 word abstract. Report one or two cases.

LETTERS TO THE EDITOR—500 words, usual *PACE* format, usually about an article in *PACE*; a response from those authors will be solicited.

OTHER

Electronic Submission: Submit computer diskettes with final revised versions of manuscripts. IBM or IBM-compatible diskettes are preferred; others can be converted. Enter all material in a single column run-in style. Delete all print commands from disk, but commands should appear on hardcopy. Label diskette with journal name, authors names, and name and version of word processing software. Computer-generated artwork supplied on diskette will be output electronically. EPS, PIC, and TIFF files capable of being sized are preferred. Submit figures on a separate diskette; specify hardware and software (name and version) used.

Experimentation, Animal and Human: Authors are responsible for ethical and accurate human and/or animal data in manuscript. Manuscripts that involve patients must have a medical practitioner as an author. Animal studies must have a medical or veterinary practitioner as an author.

Manufacturers: Give name(s), address(es) of manufacturer(s) or supplier(s) of trademarked or registered items. Supply generic names.

Style: Use *CBE Style Manual* (Committee on Form and Style of the Council of Biology Editors, American Institute of Biological Sciences, Arlington, Virginia, 1978) as a guide in writing scientific papers and for standard abbreviations for measurements and other scientific terms.

Also follow *Uniform Requirements for Manuscripts Submitted to Biomedical Journals* (International Committee of Medical Journal Editors, *Ann Intern Med* 1988; 108:258-265).

SOURCE

January 1994, 17(1): xxxix.

Pain
International Association for the Study of Pain
Elsevier

Prof. P.D. Wall
The Royal College of Anesthetists
48/49 Russell Square
London WC1B 4JY, U.K.
          Outside the Americas

Dr. R. Dubner
Neurobiology and Anesthesiology Branch
National Institute of Dental Research
NIH, Bldg. 49, Room 1A-11
Bethesda, MD 20892
          Within the Americas

AIM AND SCOPE

This journal publishes original research on nature, mechanisms and treatment of pain. The journal provides a forum for dissemination of research in basic and clinical sciences of multidisciplinary interest.

MANUSCRIPT SUBMISSION

Submit manuscripts in quadruplicate, complete in all respects. Preferred medium of final submission is on disk with accompanying reviewed and revised manuscript. Submission implies that it has not previously been published (except in abstract form) and that it is not being considered for publication elsewhere. Submission of manuscripts under multiple authorship implies that: all authors listed concur with submitted version of manuscript and with listing of authors; authorship credit is based on important contributions in one or more of these areas: conception and design, analysis and interpretation of data, drafting of manuscript or making intellectual contributions to its content; and final manuscript has been tacitly or explicitly approved by responsible authorities in laboratory or institution where work was carried out. For use of illustrations or other parts of articles or books already published, include written permission of author and publisher. Indicate original source in legend. In cover letter include a statement of financial or other relationships that might lead to a conflict of interest.

MANUSCRIPT FORMAT

Type double-spaced with at least 4 cm margin on pages of uniform size. Include a brief summary on a separate page followed by Introduction, Methods, Results, Discussion, Acknowledgments, Reference. List 6 key words.

Include authors' full names, academic or professional affiliations and complete addresses on a separate title page. Give name, address, telephone and fax numbers of corresponding author.

References: Cite at proper points in text. Include authors and year in chronological order in parentheses. When papers written by more than two authors are cited in text, use "et al.," following name of lead author, even if subsequent authors are not same in all references. List all references cited in text at end of paper, typed double-spaced in alphabetical order by author. Include initial(s) of author(s) cited, year of publication, title of paper referred to, journal, volume, and page numbers. If more than two references with same year and author(s) are cited, use lowercase letters after year. Abbreviate journal titles according to *Index Medicus, List of Journals Indexed*, latest edition. For citations of books author(s), title of article, editor(s), complete title of book, publisher, place of publication, year and page numbers. Do not include in manuscripts in preparation and submitted but not accepted, as well as "personal communications."

Adams, C.W.M., Neurohistochemistry, Elsevier, Amsterdam, 1965, 67 pp.

Goldenberg, D.L., Psychiatric and psychological aspects of fibromyalgia syndrome, Rheum. Dis. Clin. N. Am., 15 (1989a) 105–115.

Goldenberg, D.L., Fibromyalgia and its relation to chronic fatigue syndrome, viral illness and immune abnormalities, J. Rheumatol., 16 (1989b) 91–93.

Turner, J.A., Coping and chronic pain. In: M.R. Bond, J.E. Charlton and C.J. Woolf (Eds.) Pain Research and Clinical Management, Vol. 4, Proc. VIth World Congress on Pain, Elsevier, Amsterdam, 1991, pp. 219–227.

Figures: Submit in quadruplicate as unmounted glossy photographs in a form and condition suitable for reproduction either across a single column (= 8.4 cm) or a whole page (=17.6 cm). Make lettering clear and of adequate size to be legible after reduction. Supply photographs, including roentgenograms, electroencephalograms, and electron micrographs as clear black-and-white prints on glossy paper, usually larger than final size of reproduction, but not more than 20×25 cm. Same degree of reduction will be applied to all figures in paper. Reproduction in colors must be approved by Editors; cost is borne by author(s). Place authors' names and number of figure on back of each. Number legends for Figures consecutively in Arabic numerals. Type on a separate page at end of manuscript. Explain all symbols and abbreviations in Figure. Present data in Tables and Figures as clearly defined mean values with some measure of dispersion (standard deviation, standard error, range) and an appropriate indication of statistical significance of differences from control values. Indicate number of individual values represented by a mean.

Tables: Type double-spaced on a separate page, numbered in sequence in Roman numerals; provide with a heading, and refer to in text as Table I, Table II, etc.

Acknowledgments: Place at end of text before References. List and specify: contributions that need acknowledging but do not justify authorship: acknowledgments of technical help; acknowledgments of financial and material support, specifying nature of support; and financial arrangements that may represent a conflict of interest.

OTHER FORMATS

BASIC SCIENCE RESEARCH REPORTS—200-300 word original research reports.

BASIC SCIENCE REVIEWS—Send reviews or plans for such reviews to Prof. G.F. Gebhart, Department of Pharmacology, University of Iowa, Iowa City, IA 52242.

CLINICAL NOTES—Submit brief reports on clinical cases to Dr. J.E. Charlton, Department of Anesthesia, Pain Management Clinic, Royal Victoria Infirmary, Queen Victoria Road, Newcastle upon Tyne, NEW 4LP, U.K.

CLINICAL RESEARCH REPORTS—2000-3000 word original research, reports.

CLINICAL REVIEWS—Submit Clinical Reviews or plans for reviews to Prof. H. Merskey, Department of Psychiatry, London Psychiatric Hospital, 850 Highbury Ave., London, Ont. N6A 4H1, Canada.

LETTERS TO THE EDITOR—Invited.

OTHER

Electronic Submission: Preferred storage medium is a 5.25 or 3.5 in. disk in MS-DOS format; other systems are welcome, e.g., NEC and Macintosh (save file in usual manner, do not use option 'save in MS-DOS format'). Do not split article into separate files. Ensure that letter 'l' and digit '1' (also letter 'O' and digit '0') have been used properly; format article (tabs, indents, etc.) consistently. Do not leave characters not available on word processor (Greek letters, mathematical symbols, etc.) open. Indicate by a unique code (e.g., gralpha, @, #, etc., for Greek letter α); use codes consistently throughout entire text. Make a list of codes and provide a key. Do not allow word processor to introduce word splits; do not use a 'justified' layout. Adhere strictly to general instructions on style/arrangement and, in particular, reference style of journal. Save file in word processor format; do not save in "flat ASCII." Format disk correctly; ensure that only relevant file is on disk. Specify type of computer and word-processing package used; label disk with your name and file on disk. After final acceptance, submit disk plus one final, printed and exactly matching version (as a printout) to accepting editor. File on disk and printout must be identical.

Experimentation, Animal: Show that attention was given to proposals of the Committee for Research and Ethical Issues of IASP published in *Pain*, 16 (1983) 109–110. Indicate if experimental work was reviewed by an institutional animal care and use committee or its equivalent.

Experimentation, Human: Submit evidence that work has been approved by an institutional clinical research panel or its equivalent.

SOURCE
July 1994, 58(1): inside front and back covers.

Parasitology
Cambridge University Press

Professor F. E. G. Cox
Immunology Section
King's College London, Campden Hill Road
London W8 7AH U.K.
        Protozoology and/or Immunology

Professor C. Arme
Parasitology Research Laboratory
University of Keele
Keele, Staffs ST5 5BG, U.K.
        All other manuscripts

AIM AND SCOPE

Parasitology publishes definitive papers on all aspects of pure and applied parasitology including biochemistry, molecular biology, immunology, genetics, ecology and physiology and also the application of new techniques, long term epidemiology studies, advances in the understanding of life-cycles, chemotherapy and major systematic revisions.

MANUSCRIPT SUBMISSION

Papers should be full length with an explanatory introduction and detailed discussion of findings reported. Shorter reports of important findings are also considered. Invited reviews of topics not usually covered elsewhere and of particular interest to those teaching or studying parasitology are also published; contact Editor before writing.

There is no maximum or minimum length; excessively long papers are unlikely to be accepted. Short papers reporting definitive findings are acceptable; prepare in approved journal form.

Submit original and one copy of both text and figures, no larger than A4 size in a C4 size envelope (325 × 330 mm).

Submission of a manuscript implies that it has been approved by all named authors, that it reports unpublished work and that it is not being considered in whole or in part for publication elsewhere. Manuscripts not prepared in correct style will be returned for revision.

MANUSCRIPT FORMAT

Type on A4 paper (210 × 295 mm) double-spaced throughout with a left-hand margin of about 40 mm. Make all headings flush left and number all pages.

Title Page: A concise but informative full title plus a running title (40 letters and spaces), name(s) of author(s) and address, including post, zip or other code, of institute where work was done. Include footnotes for other addresses. Nothing else should appear on title page.

Summary: 150– 200-word informative précis of contents and conclusions of paper in a form suitable to be used as an abstract, plus 3–6 key words suitable for indexing.

Introduction: Keep short; give background and reasons for work.

Materials and Methods: Give sufficient experimental details to enable others to repeat work.

Give full binomial name for all organisms and all animals except those commonly used in laboratories. Give generic names in full at first mention and subsequently if confusion is likely. Follow *International Rules for Nomenclature* and, if new names are introduced, *International Code for Zoological Nomenclature*. State all strains and sources of hosts and parasites.

Results: Be concise; do not include methods or discussion. Text, tables and figures should not duplicate information.

Tables: Make self-explanatory, with title at top; organize to fit into one or two column widths. Avoid rules, particularly vertical ones. Type double-spaced, on a separate page; number consecutively; indicate position in text.

Figures: Refer to line drawings or halftones consecutively as figures, e.g., Fig. 1; indicate positions in text. Maximum size for any figure or group of figures is a single page. Identify on back with author's name, short title of paper and figure number. Type captions on a separate page.

Line Drawings—Make no larger than twice final size to fit into either one (80 mm) or two columns (166 mm). Make lines bold enough to stand reduction to 0·25–0·35 mm. Preferred symbols are ○ ● □ ■ △ ▲; use consistently. Keep lettering to a minimum (make self-explanatory and unambiguous). Lettering will be inserted by printer. Indicate in soft pencil or on a transparent overlay. If using own high-quality labelling, use following typeface and size, 2 mm for numbers and uppercase and 1 mm for lowercase, when reduced. If in doubt, submit unlabeled figures with labelled overlays or labelled Xerox copies. The best reproduction is obtained from original drawings on drawing or tracing paper; high-quality glossy prints are acceptable. Prepare suitably reduced figures to judge final appearance.

Halftone Illustrations—Submit in size that will appear in journal to fit one column (80 mm) or two columns (166 mm). Mount composite illustrations on thin card; do not exceed final page size (166 × 258 mm). Identify component parts of a composite illustration as A, B, C, etc. Insert explanatory labelling or leave for printer. Indicate lettering for printer on a transparent overlay. Lettering must be in following typeface and size, 2 mm for numbers and uppercase and 1 mm for lowercase. Give scale bars and units. Statements of magnification are not acceptable.

Discussion: Do not be excessive, repeat results, nor give new information; emphasize significance and relevance of results reported.

References: Verify accuracy of references. Use Harvard System for citation of references in text, e.g., (Brown & Green, 1989) or Brown & Green (1989). Include names of all authors up to three when first cited in text, e.g., Brown, Green & White (1989). For all subsequent citations use Brown *et al.*; use this form for all citations, including first, for four or more authors. Distinguish different papers published in same year by a letter after date. Avoid confusion of multiauthor papers with same senior author. Cite references

to unpublished observations, abstracts or papers 'in preparation' only in exceptional circumstances. Cite papers 'in Press' with full title and journal. List references in alphabetical order; and give both title and name of journal in full.

CHAPPELL, L. H. (1988). The interactions between drugs and the parasite surface. *Parasitology* **96** (Suppl.), S167–S193.

GARDNER, R. A. & MOLYNEUX, D. H. (1988). *Schizotrypanum* in British bats. *Parasitology* **97**, 43–50.

MEIS, J. F. G. M. & VERHAVE, J. P. (1988). Exoerythrocytic development of malaria parasites. In *Advances in Parasitology*, Vol. 27 (ed. Baker, J. R. & Muller, R.), pp. 1–61. London: Academic Press.

SMYTH, J. D. & HALTON, D. W. (1971). *The Physiology of Trematodes*, 2nd edn. Cambridge: Cambridge University Press.

OTHER

Abbreviations: Use sparingly and unambiguously. These are commonly used and need not be spelled out: ADP, AMP, ATP, bp, kDa, cpm, d (day), D.F., DNA, ED_{50}, Fig., g, h (hour), i.m., i.p., M_r min, NAD, NADP, No., pH, p.i. (post-infection), ppm, %, rpm, RNA, sp., spp., S.C., S.D., S.E., WHO, [^{3}H]alanine, [6-^{14}C]glucose. Use standard chemical, biochemical and molecular abbreviations. In case of doubt, spell out term in full on first usage followed by abbreviation in parentheses.

Units: Use S.I. units wherever appropriate.

SOURCE

April 1995, 109(3): i-ii.

Pediatric Infections Disease Journal
Williams & Wilkins

Chief Editors, *The Pediatric Infectious Disease Journal*
The University of Texas Southwestern Medical
Center at Dallas
5323 Harry Hines Blvd.
Dallas, TX 75235-9063

AIM AND SCOPE

The Pediatric Infectious Disease Journal® publishes original articles, reviews, and instructive cases on all aspects of infectious diseases in children which have relevance to clinical practice.

MANUSCRIPT SUBMISSION

Articles must be submitted solely to *Journal* and should not have been previously published. This does not apply to published abstracts or to press reports resulting from scientific meetings.

Submission under multiple authorship implies that all listed authors concur in submission, and that a copy of final manuscript has been seen and approved by all authors.

Assemble manuscript in order: Letter of transmittal; Title page; Abstract; Body of manuscript; Acknowledgments; References; Tables; Legends; Illustrations. Submit three copies. Include a fourth copy if there is statistical analysis.

Williams & Wilkins must receive in writing assignment of copyright from all authors. Forms will be mailed from publisher.

In cover letter, indicate all affiliations with or financial involvement in any organization or entity with a direct financial interest in subject matter or materials of research discussed in manuscript (e.g., employment, consultancies, stock ownership).Information is held in confidence during review process. If accepted, Editors will discuss extent of disclosure appropriate.

MANUSCRIPT FORMAT

Type on one side of standard-size white bond paper, double-spaced throughout, including Title Page, Tables and References. Use ample margins. If a word processor is used, do not justify right hand margins. Use a letter-quality printer.

Title Page: Make concise and clear; avoid subtitles. Supply a condensed title (75 spaces). List authors with given and family names, highest earned degree, mailing address, and telephone and fax numbers of corresponding and reprint request author(s). Provide a list of key words for indexing.

Abstract: 150 words; state reason for study, major findings and conclusions. Provide specific data rather than generalizations. Abstract not required for Reviews, Instructive Cases or Brief Reports.

Body: Divide into Introduction, Methods, Results and Discussion unless material does not lend itself to that format. Do not duplicate in text material in tables. Do not repeat results in Discussion.

Acknowledgments: Acknowledge grant support, technical assistance, advice, referral of patients, etc. Do not acknowledge secretarial work.

References: Type double-spaced and number consecutively as appear in text. Use superscript numbers in text. Abbreviate journal names using *Index Medicus* style. For a standard journal article, list all authors when six or less; when seven or more, list first three and add et al. Insert references to personal communications, unpublished data and manuscripts either "in preparation" or "submitted for publication" in text within parentheses.

Tables: Type double-spaced on a separate sheets. Number consecutively as cited in text with Arabic numerals and provide a concise, descriptive title.

Illustrations: Put on glossy black-and-white photographs 5 × 7 in. paper. Make lettering and symbols easily recognizable when reduced to one-column width (9 cm). If photographs of patients could result in their recognition as individuals, submit written permission of person, parent or legal guardian. On back, indicate TOP, figure number and name of first author. Supply three originals of each, unless photocopies of original have sufficient detail to allow evaluation by reviewers. Authors are charged for color photographs. Type legends double-spaced on a separate sheet headed "Legends for Illustrations."

OTHER FORMATS

BRIEF REPORTS—Five double-spaced pages, one table or figure and ten references. Report limited clinical or laboratory observations.

LETTERS—Type double-spaced; provide a title. Letters relating to material previously published in the Journal and other brief comments on matters of general interest.

OTHER

Abbreviations: Standard abbreviations do not require explanation. Avoid overuse of contrived abbreviations. At first use put uncommon abbreviations in parentheses following full word or words. Use generic names for drugs.

Electronic Submission: Submit electronic diskettes of final version of manuscript along with typescript of final revision. Diskettes produced on IBM or IBM-compatible computers are preferred; those produced on most Apple/Macintosh or Wang computers can be converted. ASCII files can also be used. Macintosh users: do not use FastSave option. Identify diskette with name of first author and manuscript title; name of computer file; type of hardware, operating system, and version number; and name of software program and version number. Tables are processed in traditional manner.

Experimentation, Human: Indicate that informed consent was obtained from parents or guardians of children who served as subjects of investigation and, when appropriate, from subjects themselves.

SOURCE

January 1994, 13(1): front matter.

Pediatric Research

International Pediatric Research Foundation
American Pediatric Society
European Society for Pediatric Research
Society for Pediatric Research
Williams & Wilkins

George Lister
Department of Pediatrics
Yale University School of Medicine
333 Cedar Street
PO Box 208064
New Haven CT 06520–8064
Peter J. J. Sauer
Department of Pediatrics
Erasmus University Hospital
Sophia Children's Hospital
Dr Nolewaterplein 60
3015 GJ Rotterdam, The Netherlands

AIM AND SCOPE

Pediatric Research publishes original papers on research pertinent to normal and abnormal human development and pathogenesis of disease in the fetus, infant, child, and adolescent. Use of theoretical models, animals, or *in vitro* techniques relevant to developmental biology or medicine is as acceptable as human studies. Research describing basic physical, chemical, molecular, or genetic mechanisms of developmental processes and their pathophysiologic alterations is particularly encouraged.

The journal wishes to publish the highest quality original articles related to, or reflecting on, the span of human development from the fetus to the adolescent. Research that enhances our understanding of physiologic and pathologic processes is of prime interest. The range of acceptable research is as broad as that found at international, national, or regional meetings of our sponsoring societies, and no particular subspecialty discipline has priority. Acceptable studies may use any of a variety of systems including, but not limited to, theoretic models, cell culture, isolated organs, intact animals, and humans. What is most important is that the research address a clear hypothesis or question, or elucidate a mechanism, and that the chosen system and methods be appropriate for the issue under study.

MANUSCRIPT SUBMISSION

Include completed Manuscript Submission Form (in every issue), with each submission. Submit four clear copies to either American or European editorial office. With each copy submit clear copies of figures.

By signing manuscript submission form, corresponding author verifies that all authors approve of material submitted, that material has not been reported previously and is not under consideration for publication elsewhere, and that human and animal studies have been approved by the authors' Institutional Review Board. Conduct human investigations according to principles of the Declaration of Helsinki. If any related preliminary report other than an abstract has been published or submitted, send copies. All authors must have read final version of submitted manuscript and share full responsibility for what is written.

Articles must be original reports; data must not have been submitted or published elsewhere without full disclosure and submission of the other publication (except data reported in abstracts). Deviations from these policies is considered potentially fraudulent. Under such circumstances, editorial board will first ask corresponding author for a formal explanation. If a satisfactory response is not received, a more extensive inquiry will be requested from institution from which manuscript originated. A decision will then be made by editorial board based on that response. Such ethical breeches are unusual. A more common issue is the submission of a portion of a project that has been divided into small reports. Authors of short reports that address different aspects of the same study often submit them simultaneously to different journals without disclosure. The editors of seven major pediatric journals published a policy statement regarding this undesirable practice (Bier DM, Fulginiti, VA, Garfunkel JM, Lucey JF, Spranger J, Valman B, Zetterström R 1990 Duplicate publication and related problems. *Pediatr Res* 28:561); they will inform each other whenever such instances occur. Read this statement and abide by its spirit.

MANUSCRIPT FORMAT

Submit results of original research investigations as a regular article. Limit to 20 double-spaced typed pages (5000 words) for text, references, tables, figures, and figure legends, excluding running title page, title page, and abstract. Begin each section on a new page; assemble in order: running title page, title page, abstract, abbreviations, text, references, tables, figure legends, and illustrations.

Running Title Page: Running title (40 letters and spaces), complete address, telephone and fax numbers for corresponding author and reprint request author(s).

Title page: Make title concise and descriptive. Mention if observations were made on animals; all other titles refer to studies on humans or human tissues. Include complete name and institutional affiliation of each author. Credit for authorship requires substantial contribution to concept and design, analysis and interpretation of data, and important intellectual input in drafting or revising manuscript.

Abstract: 250 words. Describe problem addressed, hypothesis, how study was performed, important findings, and interpretation of results.

Introduction: Orient reader to state of knowledge in specific area under investigation; delineate questions and hypothesis of research.

Methods: Describe and reference in sufficient detail so others can repeat work. Subdivide methods into sections (subjects, measurements, protocol, and data analysis). Have complex data analyses reviewed by a statistician. Give manufacturer's name and location for unusual chemicals, reagents, and special pieces of apparatus. Give measurements in SI units.

Results: Present experimental data in most appropriate form; do not submit the same data in tabular and graphic form. When possible, present data in graphic form.

Key Words: Five key words or brief phrases at bottom of abstract page, using standard MeSH-MEDLINE major subject headings. Do not use words or phrases in title.

Discussion: Focus on implications of results and limitations of study. Relate findings to other relevant studies.

Acknowledgments: Acknowledge support from extramural sources, technical assistance, critical advice, or other assistance after discussion.

References: Type double-spaced beginning on a separate sheet. Number in order of appearance in text. Include all authors' names (do not use "*et al.*"), complete article titles, and inclusive page numbers. Abbreviate titles of journals according to *Index Medicus*. If an article is *in press,* submit four copies of manuscript. Cite as such in text manuscripts in preparation, unpublished observations, and personal communications; do not include in reference list.

ARTICLES IN JOURNALS

Give in order: names of all authors with initials, year, title of paper (capitalize only initial letter of initial word and proper nouns), abbreviated journal name without punctuation, volume number and first and last page of article.

1. Gomez RA, Meernik JG. Kuehl WD, Robillard JE 1984 Developmental aspects of renal response to hemorrhage during fetal life. Pediatr Res 18:10–46.

BOOKS

Give authors' last names with initial(s), year, title of book, publisher, city of publication, and inclusive pages.

2. Berne RW, Levy MN 1981 Cardiovascular Physiology. CV Mosby Co, St. Louis, pp 52-260.

CHAPTERS IN BOOKS

3. Seegmillere JE 1983 Disorders of purine metabolism. In: Emery AH, Rimoin DL (eds) Principles of Medical Genetics. Churchill-Livingstone, Edinburgh. pp 1286–1305.

Tables: Type double-spaced, one per page. Number and give titles. Make intelligible without reference to text. Minimize redundant or repetitious entries. For instance, rather than stating "$p < 0.05$" for each of 20 comparisons in a table, use an asterisk or other symbol to denote specified probability levels defined in a footnote.

Figure Legends: Type double-spaced on a separate page and not on figure. Number and give titles. Include sufficient detail to make intelligible without reference to text.

Figures: Prepare line drawings to minimize space required. Make lettering as large as possible to permit maximal reduction in size. Tabular material may be inserted into "blank" areas. Five by 7 in. prints are preferred. Submit professional quality line drawings or graphic material as clear laser prints or sharp photographs on glossy paper. Do not mount photographs, trim to exclude all but essential areas; number on back, with TOP identified. Submit only one set of unmounted glossy figures. Submit clear, numbered photocopies, with one figure on a page for other review copies.

For gels or halftones such as photomicrographs or electron micrographs, submit an original with three sharp sets of glossy prints (not photocopies) no larger than 8.5×11 in. Show only most pertinent areas; mark for cropping to avoid reduction in size. Include a micron bar of appropriate scale marking. Submit each set in a separate envelope. Authors pay cost for color figures; cost quotations are sent before publication.

OTHER FORMATS

RAPID PUBLICATIONS—10 double-spaced typed pages of unusual scientific importance; indicate request for rapid publication on checklist. If manuscript is rejected for rapid publication, further evaluation through normal review process for regular articles may be requested.

REVIEW PAPERS—15 double-spaced typed pages that describe important new developments not widely appreciated by readership, findings of interest to investigators in multiple disciplines related to pediatrics, or general themes that have not been critically summarized elsewhere. Usually invited, but suggestions are welcome. Provide a one to two page proposed outline.

OTHER

Abbreviations: Some standard abbreviations are listed in "Instructions to Authors" (*Pediatr Res* 1994; 35(1): front matter). Define unusual abbreviations used at least three times immediately after first use and in list of abbreviations. Define abbreviations used only in tables or figures in legends. For enzymes, include EC number if assigned.

Electronic Submission: After a manuscript has been reviewed and the final version accepted, submit manuscript on a disk. Disk should contain only and exactly the final, accepted version. Submit a printout of disk. On disk's label, specify file name, MS-DOS or Macintosh, word processor software used, and version numbers of operating system and software.

Experimentation, Animal and Human: All studies that use biologic materials, animals, or human subjects must have been approved by institutional review committees for animal or human research, to conform to Declaration of Helsinki: Recommendations for Conduct of Clinical Research (Bankowski Z, Levine RJ 1993 Ethics and Research on Human Subjects: International Guidelines. Council for International Organizations of Medical Sciences, Geneva, Switzerland. Annex I) or Guiding Principles in the Care and Use of Animals (Guide for the Care and Use of Laboratory Animals, DHEW Publication   No. (NIH) 85-23, Revised 1985, Office of Science and Health Reports, DRR/NIH, Bethesda, MD). Acknowledge such approval.

Page Charge: $50 per printed page. Editors may waive charges upon appeal.

Source

January 1994, 35(1): front matter.

Pediatrics
American Academy of Pediatrics

Jerold F Lucey, MD
Medical Center Hospital
Burlington, VT 05401
Telephone: (802) 862-8778

MANUSCRIPT SUBMISSION

In cover letter signed by all authors, state: that manuscript is being submitted only to *Pediatrics*, that it will not be submitted elsewhere while under consideration, that it has not been published elsewhere, and, should it be published in *Pediatrics,* that it will not be published elsewhere—either in similar form or verbatim—without permission of editors (restrictions do not apply to abstracts or press reports of presentations at scientific meetings). Also state that authors are responsible for reported research; that they have participated in the concept and design; analysis and interpretation of data; drafting or revising of manuscript, and that they have approved manuscript as submitted. Disclose any affiliation, financial agreement, or other involvement of any author with any company whose product figures in the manuscript (editors will discuss with affected authors whether to print information and in what manner).

Upon acceptance, authors are sent a standard Copyright Agreement which must be signed by all authors; return to Editor. Accepted manuscripts become permanent property of the American Academy of Pediatrics and may not be published elsewhere, in whole or in part, without written permission from the Academy. Authors who were employees of the United States Government at time work was done should so state on Copyright Agreement.

Relevance to readers is of major importance in manuscript selection. *Pediatrics* generally accepts manuscripts in: reports of original research, particularly clinical research; special articles; and experience and reason.

Reports of original research are judged on importance and originality of research, scientific strength, clinical relevance, clarity of presentation, and number of submissions on same topic. Decision to publish is not based on direction of results.

Unsolicited commentaries or editorials are considered, although most are solicited by Editors. Review articles generally are not appropriate. Case reports are of interest only when they present a new entity or illustrate a major new aspect of a previously reported entity.

MANUSCRIPT FORMAT

Prepare manuscripts, including tables, illustrations, and references, according to "Uniform requirements for manuscripts submitted to biomedical journals" (*Br Med J.* 1991; 302: 338–341 or *N Engl J Med*. 1991; 324:424–428). Only information not included is stated here.

Prepare abstracts with a structured format. Include four elements: why did you start (objective), what did you do (methodology, including design, setting, patients or other participants, interventions, and outcome measures), what did you find (results), and what does it mean (conclusions). See Ad Hoc Working Group for Critical Appraisal of the Medical Literature, "A proposal for more informative abstracts of clinical articles" (*Ann Intern Med*. 1987;106: 598–604). Label each section with subheading. Experience and Reason and Commentaries do not require abstracts.

Acknowledge research or project support as a footnote to title page; identify technical and other assistance in an appendix to text.

All cited references must have been read by authors. Appropriately note citation of review articles. Otherwise, do not cite secondary sources.

Submit four complete copies, including tables (in type no smaller than text of article) and glossy prints of illustrations. Do not send original artwork or printed forms. A reasonable number of black-and-white illustrations are printed without charge. Payment for color illustrations and other special processing is responsibility of authors. Arrange before manuscript is processed.

OTHER

Abbreviations: Limit to those listed in Chapter 11 of *AMA Manual of Style* (Iverson C, Dan BB, Glitman P, et al. *American Medical Association Manual of Style*, 8th ed. Baltimore, MD: Williams & Wilkins; 1988). List any uncommon abbreviations at beginning of article.

Style: Grammar punctuation, and scientific writing style follow *American Medical Association Manual of Style* (8th edition).

Units: Use conventional system measurements followed in parentheses by equivalent Système International (SI) values. See Lundberg GD, "SI unit implementation: the next step" (*JAMA*. 1988;260:73–76) and Système International conversion factors for frequently used laboratory components (*JAMA*. 1991;266:45–47).

SOURCE
January 1994, 93(1): A5.

Perception & Psychophysics
Psychonomic Society, Inc.

Myron L. Braunstein
Department of Cognitive Sciences
School of Social Sciences
University of California
Irvine, CA 92717-5100

AIM AND SCOPE

Perception & Psychophysics publishes articles that deal with sensory processes, perception, and psychophysics. While the majority of published articles are reports of experimental investigations in these content areas, articles that are primarily theoretical or that present integrative and evaluative reviews are accepted. Studies employing either human or animal subjects are welcome. There are no explicit length restrictions on acceptable articles other than the requirement of clear and concise writing and relevance of the various contents.

MANUSCRIPT SUBMISSION

Submit manuscripts in quintuplicate. Package with care to avoid damage or loss in mail; use a padded envelope.

MANUSCRIPT FORMAT

On title page, include address, telephone number, and e-mail address. Include an abstract (100–150 words) in form used by American Psychological Association. Adhere to conventions concerning references, manuscript format, etc., described in *Publication Manual of the American Psychological Association* (3rd ed.); abbreviations of physical units follow style of American Institute of Physics. See *Percept Psychophys* 48: 205 or consult a recent journal issue.

Figures and Illustrations: Make graphs, shading, and lettering clear and sharp enough to accommodate reduction for publication. Give special attention to halftones; both detail and figure-ground contrast. See *Percept Psychophys* 48:207.

Tables: Adhere to conventions in *Publication Manual of the American Psychological Association* (3rd ed.). For further information, see *Percept Psychophys* 48: 206.

OTHER

Electronic Submission: Submit copy of manuscript on computer disk in ASCII (text only) format with final accepted version of manuscript. Also include exact version in hard copy. Text in hard copy and disk should match exactly. For details, see *Percept Psychophys* 48: 208 or call Publications Office (512-462-2442).

Experimentation, Animal and Human: Human and animal subjects must be treated ethnically. Mistreatment of subjects is grounds for rejection.

SOURCE
January 1995, 57(1): inside front cover.

Pharmacological Reviews

American Society for Pharmacology and
Experimental Therapeutics
Williams & Wilkins

Dr. Richard Stitzel
Department of Pharmacology and Toxicology
West Virginia University Health Sciences Center
Morgantown, WV 66506

MANUSCRIPT SUBMISSION

Papers are, for the most part, invited by the Editor after consideration of recommendations from Associate Editors and other consultants and evaluation of outlines and statements of central theme submitted by prospective authors. Others interested in writing for *Pharmacological Reviews* should send proposals to the Editor.

Articles deal mainly with current status of subject under review. Write clearly and concisely and intelligible to non-specialists; define unfamiliar technical terms and explain difficult or controversial points. At same time, present review as precise and detailed to command attention and respect of experts in the field. Be selective rather than exhaustive in covering literature. Cite previous reviews of subject and related fields. Be critical of methods, results, and conclusions and challenge accepted concepts where warranted. Present conflicting points of view objectively in good perspective. Point out deficiencies in field; indicate avenues for further work.

Provide an estimate of length of review article, usually 12–50 printed pages (40–150 manuscript pages). Include diagrams, tables, and, occasionally, illustrations to bring out new concepts and important relationships, or when access to original sources would be difficult.

MANUSCRIPT FORMAT

Type double-spaced on good quality white paper. Standard 8.5 × 11 in. paper is preferred; longer paper is also acceptable. Type each table double-spaced, on a separate sheet. Begin legends for figures, footnotes in text, and references on separate sheets, all double-spaced.

Illustrations: Submit in duplicate as unmounted glossy photographic prints, no larger than 8 × 10 in. Place authors' name and address, and number of figure on back. Present clear enough to permit reproduction without retouching, and legible after reduction to single-column size (approximately 3.5 in. wide). Use uniform symbolism for all illustrations. Make symbols, numbers, and letters at least 1.5 mm in height after reduction to single column size. Place title in legend, not on figure itself. Label photomicrographs and electron micrographs with a magnification calibration in micro or Angstrom units. State magnification in figure legend.

References: Group at end of manuscript under heading References. Arrange in alphabetical order. Determine order by the following rules, applied in sequence: 1. Arrange alphabetically according to last name of first author. 2. If two or more first authors with identical last names, arrange alphabetically according to first author's initials. 3. References with coauthors follow those by first author alone.

Arrange alphabetically according to first coauthors' name. If two or more papers with the same first two authors, arrange alphabetically according to second coauthor's name, etc. 4. If two or more references by same author, or by same sequence of authors, arrange according to date of publication, earliest paper first; if date of publication is same, choice is arbitrary. References to "Unpublished observations" or to a "Personal communication" may follow other references by same author or authors or enter in text in parentheses and omit from list of references.

Place authors' names in text in parentheses. Where two or more references are cited within parentheses, place in chronological order. Include in bibliographic citations names and initials of all authors, full title of article in original language, abbreviated name of journal, full pagination, and year. Abbreviations for journals conform to those of *Biosis List of Serials* (Biosciences Information Service); use capital and lowercase letters, accents, umlauts, and other diacritical marks.

BACANER, M. B.: Treatment of ventricular fibrillation and other acute arrhythmias with bretylium tosylate. Am. J. Cardiol. **21:** 530–543, 1968.

BOJAR, H.: Quality control requirements in estrogen receptor determination. Cancer Res. (suppl.) **46:** 4249s–4255s, 1986.

BROWN, R. H., JR., AND NOBLE, D.: Effect of pH on ionic currents underlying pacemaker activity in cardiac Purkinje fibers. J. Physiol. (Lond.) **224:** 38P–39P, 1972.

BULBRING, E., AND TOMITA, T.: Catecholamine action on smooth muscle. Pharmacol. Rev. **39:** in press, 1987.

CRANEFIELD, P. F.: The Conduction of the Cardiac Impulse, 404 pp., Futura Publishing Co., Mt. Kisco, NY, 1975.

DEGEEST, H., KESTELOOT, H., AND PIESSENS, J. L.: Ischemic heart disease: a long term controlled lidoflazine study. Acta Cardiol. Suppl. **24:**1,1979.

FLECKENSTEIN, A.: Experimental heart failure due to disturbances in high-energy phosphate metabolism. *In* Proceedings of the Fifth European Congress of Cardiology, Athens, Sept. 1986, pp. 255-269, 1968.

GODFRAIND, T., ECLEME, C., AND WIBO, M.: Effects of dihydropyridines on vessels. *In* Proceedings of "Bayer Symposia," ed. by A. Fleckenstein and C. Van Breeman, vol. IX, pp. 309-325, Springer Verlag, Berlin, 1985.

STARKE, K., AND SCHUMANN, H. J.: Wirkung von nifedipine auf die funktion der sympathischen nerven des herzens. Arzneim. Forsch. **23:**193-197, 1973.

WATANABE, Y.: Effects of electrolytes on atrioventricular conduction. In Symposium on Cardiac Arrhythmias, ed. by E. Sandae, E. Flensted-Jensen, and K. H. Olesen, pp. 535-558, AB Astra, Sodertalje, Sweden, 1970.

OTHER

Abbreviations: For standard abbreviations see *CBE Style Manual* or *J. Biol. Chem.* **262:** 1-11, 1987.

Drugs: On first mention of an unfamiliar drug, give official or generic (nonproprietary) name by which drug is known in U.S. or country of origin of manuscript. Followed by parentheses with official or generic names of drug in other countries, selected familiar trade names, and/or chemical name. (A diagram of chemical structure in a figure is often preferable to spelled-out chemical name.) Thereafter, use whichever nonproprietary name is most suitable without giving synonyms.

Electronic Submission: After an article has been accepted for publication, authors with access to Apple Macintosh, IBM-PC, or compatible personal computers may submit final manuscript on diskette. Software preference is XyWrite III Plus or Word Perfect(4.2 or 5.0) on 5.25 in. diskettes. Other commonly used software programs are accepted. Submit two double-spaced hard copies with diskette. Give name of software program and version used.

Style: For definitions, usage, spelling, and punctuation use: *Webster's Ninth New Collegiate Dictionary* (G. & C. Merriam Co., Springfield, 1985); *Webster's Third New International Dictionary* (G. & C. Merriam Co., Springfield, 1986); *The Concise Oxford Dictionary* (7th ed., J. B. Sykes (ed.), Oxford University Press, Oxford, 1982); *Dorland's Illustrated Medical Dictionary* (26th ed., J. P. Friel (ed.), W. B. Saunders, Philadelphia, 1981); *CBE Style Manual: A Guide for Authors, Editors, and Publishers in the Biological Sciences* (5th ed., CBE Style Manual Committee, Council of Biology Editors, Inc., Bethesda, MD, 1983; and Strunk, W., Jr., and White, E. B., *The Elements of Style* (2nd ed., Macmillan, New York, 1979).
For words spelled differently in American and British usage, follow either usage but do so consistently.

Units: Use Arabic numerals for definite weights and degrees of temperature. Give all weights, volumes, doses, etc., in metric units. Relate doses to unit weight, surface area, or other standard. Indicate concentrations of solutions as normal, molar, or %. If percentages are used, indicate whether strict percentage (w/w), % by volume (v/v), or a given weight of solute in 100 ml of solution (w/v). Express quantities in units that give the closest approximation to unity, e.g., 0.5 mg and 1.5 g rather than 0.0005 g and 1500 mg. Abbreviations conform to internationally accepted usage, viz., kg, g, mg, μg, l, ml, μl, m, cm, mm, μ, μm, Å, cm^2, cm^3, etc. For per cent, use symbol %.

SOURCE

March 1994, 46(1): i-ii.
Suggestions to contributors appear in every issue.

Pharmacology & Therapeutics
International Encyclopedia of Pharmacology and Therapeutics
Pergamon Press

A.C. Sartorelli
Yale University School of Medicine
Department of Pharmacology
333 Cedar Street, P.O. Box 208066
New Haven, CT 06510
    Chemotherapy, Toxicology and
    Metabolic Inhibitors

W. C. Bowman
University of Strathclyde
Department of Physiology & Pharmacology
Royal College, 204 George Street
Glasgow, G1 1XW Scotland
    General and Systematic Pharmacology

A. M. Breckenridge
University of Liverpool
Department of Pharmacology & Therapeutics
P. O. Box 147
Liverpool L69 3BX England
    Clinical Pharmacology and Therapeutics

AIM AND SCOPE

Pharmacology and Therapeutics presents lucid, critical and authoritative reviews of currently important topics in pharmacology. Articles are normally specially commissioned, although uninvited review papers are occasionally published. When all manuscripts covering a certain topic have been published, they are reappraised and, if necessary, updated, and then published as a definitive hard bound volume of the *International Encyclopedia of Pharmacology and Therapeutics*.

MANUSCRIPT SUBMISSION

Uninvited reviews are welcome for consideration; write to relevant Executive Editor to obtain guidance on how to prepare manuscripts prior to submission.

Manuscripts must have not been published and may not be simultaneously submitted or published elsewhere. By submitting a manuscript, authors agree that copyright for article is transferred to publisher if and when article is accepted. Assignment of copyright is not required from authors who work for organizations which do not permit such assignment.

OTHER

Drug and Dosage Selection: Ensure accuracy of information, particularly with regard to drug selection and dose. Consult appropriate information sources, especially for new or unfamiliar drugs or procedures. Practitioners must evaluate the appropriateness of a particular opinion in context of actual clinical situations and with due consideration to new developments.

SOURCE

1994, 64(1): front matter.

Pharmacology Biochemistry and Behavior
Pergamon Press

Professor Sandra E. File
Psychopharmacology Research Unit
UMDS Division of Pharmacology
Guy's Hospital
P.O. Box 3448
London SE1 9QH, U.K.
Fax: 44 (0)71 955-4627
    From Europe

Professor Matthew J. Wayner
Division of Life Sciences
The University of Texas at San Antonio
San Antonio, TX 78249-0662
Fax: (210) 691-4510
          From elsewhere

AIM AND SCOPE

Pharmacology Biochemistry and Behavior will publish original reports of systematic studies in the areas of pharmacology, biochemistry, toxicology, and behavior in which primary emphasis and theoretical context are behavioral; and brief communications that describe a new method, technique, or apparatus and results of experiments that can be reported briefly with limited figures and tables. Behavioral studies as such will not be published unless they are obviously pertinent and make a significant contribution to pharmacology, biochemistry, or toxicology. Accelerated publication will be available for original and high-quality manuscripts not to exceed four printed pages in length. These will be published as Rapid Communications. A limited number of relevant reviews and theoretical articles, results of symposia, and more comprehensive studies as monograph supplements will also be published.

MANUSCRIPT SUBMISSION

Manuscripts may have not been published nor will not be simultaneously submitted or published elsewhere. Do not fragment material into numerous short reports.

By submitting a manuscript, authors agree that copyright for article is transferred to publisher if and when article is accepted for publication. A Transfer of Copyright Agreement will be sent to submitting author. Complete form and returned to publisher.

MANUSCRIPT FORMAT

Submit in triplicate, double-spaced with wide margins on good quality paper. If a word processor is used, use a letter quality printer; computer-generated illustrations must be of same quality as professional line drawings. Clearly and concisely state conclusions. Do not use italics for emphasis.

On title page: title of paper (85 characters, including spaces); author(s); laboratory or institution of origin with city, state, zip code, and country; complete address for mailing proofs; a running head (40 characters including spaces).

Type references, footnotes, and legends for illustrations on separate sheets, double-spaced. Identify illustrations (unmounted photographs) on reverse with figure number and author(s) name; clearly mark TOP. Type each table on a separate sheet, double-spaced.

Specify dimensions and measurements in metric system. Use standard nomenclature, abbreviations and symbols; see Royal Society Conference of Editors' "Metrification in Scientific Journals" (*Am. Scient.* 56:159-164; 1968.

Abstract: 170 words.
MYERS, R. D., C. MELCHIOR AND C. GISOLFI. *Feeding and body temperature: Changes produced by excess calcium ions. . .* PHARMACOL BIOCHEM BEHAV. Marked differences in extent of diffusion have been . . .

List 3-12 (or more) words or short phrases for indexing terms at bottom of abstract page.

Footnotes: Number title page footnotes consecutively. If senior author is not to receive reprint requests, use a footnote to designate who will receive requests. Do not use text footnotes; incorporate the material into text.

References: Prepare according to Numbered/Alphabetized style of the Council of Biology Editors. Cite references by number, in parentheses, within text, one reference to a number and list in alphabetical order (double-spaced) on a separate sheet at end of manuscript. Do not recite names of authors within text. Journal citations in reference list contain: surnames and initials of all authors (surname precedes initials); title of article; abbreviated journal title as in *List of Journals Indexed in Index Medicus*; and volume, inclusive pages, and year.

1. Banks, W. A.; Kastin, A. J. Peptides and the blood-brain barrier: Lipophilicity as a predictor of permeability. Brain Res. Bull. 15: 287–292; 1985.

Put book references in order: author, title, city of publication, publisher, year, and pages.

1. Mello, N. K. Behavioral studies of alcoholism. In: Kissin, B.; Begleiter, H., eds. The biology of alcoholism, vol. 2. Physiology and behavior. New York: Plenum Press; 1972:219–291.

2. Myers, R D. Handbook of drugs. New York: Van Nostrand Reinhold Company; 1974.

Illustrations: Prepare for use in a single column width. Draw and letter all drawings for reduction to a given size to same scale. Refer to all illustrations as figures and number in Arabic numerals. Do lettering in India ink or other suitable material and make proportionate to size of illustrations to be legible after reduction. Size lettering so that smallest elements (subscripts or superscripts) will be readable when reduced. Put lettering within framework of illustration; put key to symbols on face of chart. Use standard symbols: ○ ● △ ▲ ☐ ■ +. Give actual magnification of all photomicrographs. Indicate dimension scale. Submit sharply contrasting unmounted photographs of figures on glossy paper. Submit illustrations in black-and-white unless color reproduction is requested. Submit color prints in actual size. Authors are responsible for additional costs.

Tables: Brief heading; footnote explanatory matter, do not put in title. Indicate table footnotes in body of table in order of appearance with symbols: * † ‡ § ¶ # **, etc. Tables must not duplicate material in text or illustrations. Omit vertical rules. Use short or abbreviated column heads. Identify statistical measures of variation, SD, SE, etc. Do not submit analysis of variance tables, incorporate significant F where appropriate within text. The appropriate form for reporting F value is: $F(11, 20) = 3.05, p < 0.01$.

Formulas and Equations: Keep structural chemical formulas, process flow-diagrams, and complicated mathematical expressions to a minimum. Clearly identify all subscripts, superscripts, Greek letters, and unusual characters.

OTHER

Drugs: Capitalize proprietary (trademarked) names. Chemical names precede trade, popular name, or abbreviation of a drug at first occurrence.

Electronic Submission: Submit a computer disk containing final version of paper along with final manuscript. Specify software used, including release. Specify what computer was used. Include both text file and ASCII file on disk. Single-space file; use wrap-around end-of-line feature (i.e., no return at end of line). Begin all textual elements flush left, no paragraph indents. Place two returns after every element, such as title, headings, paragraphs, figure and table callouts, etc. Keep a backup disk.

Experimentation, Animal and Human: In cover letter include a statement that experimental protocol was approved by an Institutional Review Committee for use of Human or Animal Subjects or that procedures are in compliance with the Declaration of Helsinki for human subjects, or National Institutes of Health Guide for Care and Use of Laboratory Animals (Publication No. 85-23, revised 1985), U.K. Animals Scientific Procedures Act 1986 or European Communities Council Directive of 24 November 1986 (86/609/EEC). Manuscripts will be returned without sufficient evidence that these accepted procedures and good ethical standards were followed.

In describing surgical procedures on animals, specify type and dosage of anesthetic agent. Curarizing agents are not anesthetics; if used, provide evidence that anesthesia of suitable grade and duration was employed.

SOURCE

January 1994, 47(1): inside back cover.

Physiology and Behavior
Pergamon Press

Susan S. Schiffman
Department of Psychiatry
Box 3259
Duke University Medical School
Durham, NC 27710

Barry J. Everitt
Department of Experimental Psychology
University of Cambridge
Downing Street
Cambridge CB2 3EB England

AIM AND SCOPE

Physiology and Behavior will publish original reports of systematic studies in areas of physiology and behavior, in which at least one variable is physiological and primary emphasis and theoretical context are behavioral, and brief communications that describe a new method, technique or apparatus, and results of experiments that can be reported briefly, with limited figures and tables. Accelerated publication will be available for original and high quality manuscripts not to exceed four printed pages in length. These will be published as Rapid Communications. A limited number of pertinent review and theoretical articles, results of symposia, and more comprehensive studies as monograph supplements will also be published.

MANUSCRIPT SUBMISSION

Submit all manuscripts to Editors-in-Chief for review and processing. Submit in triplicate, typed double-spaced with wide margins on good quality paper. If a word processor is used, use a letter quality printer. Computer-generated illustrations must be professional quality line drawings.

A Transfer of Copyright Agreement will be sent to submitting author. Complete form and return to publisher.

MANUSCRIPT FORMAT

Do not fragment material into numerous short reports. Make clear, concise statement of facts and conclusions. Do not use italics for emphasis.

On title page include: title of paper (85 characters, including spaces); author(s); laboratory or institution of origin with city, state, zip code, and country; complete address for mailing proofs; a running head (40 characters including spaces).

Type references, footnotes, and legends for illustrations on separate sheets, double-spaced. Identify illustrations (unmounted photographs) on reverse with figure number and author(s) name; clearly mark TOP. Type each table double-spaced on a separate sheet.

Specify dimensions and measurements in metric system. Use standard nomenclature, abbreviations and symbols, as specified by Royal Society Conference of Editors ("Metrication in Scientific Journal" *Am. Scient.* 56:159–164; 1968.

Abstract: 170 words, suitable for abstracting journals.

MYERS, R. D., C. MELCHIOR AND C. GISOLFI. *Feeding and body temperature: Changes produced by excess calcium ions.* . . PHARMACOL BIOCHEM BEHAV. Marked differences in extent of diffusion have been . . .

List 3-12 (or more) words or short phrases suitable for indexing terms at bottom of abstract page.

Footnotes: Number title page footnotes consecutively. If senior author is not to receive reprint requests, designate reprint request author in a footnote. Do not use text footnotes; incorporate material into text.

References: Prepare according to Numbered/Alphabetized style of Council of Biology Editors. Cite references by number, in parentheses, within text (one reference to a number); list in alphabetical order (double-spaced) on a separate sheet at end of manuscript. Do not recite names of authors within text. Journal citations in reference list contain: surnames and initials of all authors (surname precedes initials); title of article; journal title abbreviated as in *List of Journals Indexed in Index Medicus;* volume, inclusive pages, and year.

1. Banks, W. A.; Kastin, A. J. Peptides and the blood-brain barrier: Lipophilicity as a predictor of permeability. Brain Res. Bull. 15: 287–292; 1985.

Book references: author, title, city of publication, publisher, year, and pages:

1. Mello, N. K. Behavioral studies of alcoholism. In: Kissin, B.; Begleiter, H., eds. The biology of alcoholism, vol. 2. Physiology and behavior. New York: Plenum Press; 1972:219–291.

2. Myers, R D. Handbook of drug and chemical stimulation of the brain. New York: Van Nostrand Reinhold Company; 1974.

Illustrations: Prepare for use in single column width. Draw and letter all drawings for reduction to a given size to same scale. Refer to all illustrations as figures; number in Arabic numerals. Do lettering in India ink or other suitable material; make proportionate to size of illustrations to be legible after reduction. Size lettering so that smallest elements (subscripts or superscripts) will be readable when reduced. Place lettering within framework of illustration; place keys to symbols on face of charts. Use standard symbols: $\bigcirc$ $\bullet$ $\triangle$ $\blacktriangle$ $\square$ $\blacksquare$ +. Give actual magnification of photomicrographs. Indicate dimension scale. Submit sharply contrasting unmounted photographs of figures on glossy paper. Submit illustrations in black-and-white unless color reproduction is requested. Submit color prints in actual size. Authors pay additional costs.

Tables: Brief heading; footnote explanatory matter, do not include in title. Indicate footnotes in body in order of appearance with symbols: * † ‡ § ¶ # **, etc. Do not duplicate material in text or illustrations. Omit vertical rules. Use short or abbreviated column heads. Identify statistical measures of variation, SD, SE, etc. Do not submit analysis of variance tables, incorporate significant F where appropriate within text. The appropriate form for reporting F value is: $F(11, 20) = 3.05, p < 0.01$.

Formulas and Equations: Keep structural chemical formulas, process flow-diagrams, and complicated mathematical expressions to a minimum. Draw chemical formulas and flow-diagrams in India ink for reproduction as line cuts. Clearly identify all subscripts, superscripts, Greek letters, and unusual characters.

OTHER

Drugs: Capitalize proprietary (trademarked) names. Chemical name precedes trade, popular name, or abbreviation of a drug at first time use.

Electronic Submission: Submit a computer disk containing final version of paper along with final manuscript. Specify software used, including release. Specify computer used. Include both text and ASCII file on disk; single space file; use wrap-around end-of-line feature (i.e., no returns at ends of lines). Begin all textual elements flush left, no paragraph indents. Place two returns after every element, such as title, headings, paragraphs, figure and table callouts, etc. Keep a backup disk.

Experimentation, Animal: In describing surgical procedures on animals, specify type and dosage of anesthetic agent. Curarizing agents are not anesthetics; if used, provide evidence that anesthesia of suitable grade and duration was employed.

SOURCE
January 1994, 55(1): inside back cover.

Plant Physiology
American Society of Plant Physiologists

Deborah Weiner, Managing Editor
American Society of Plant Physiologists
15501 Monona Drive
Rockville, MD 20855-2768

AIM AND SCOPE
Plant Physiology is an international journal open to papers of merit dealing with all phases of experimental plant biology.

MANUSCRIPT SUBMISSION
Manuscripts must be original research reports that have not been submitted elsewhere, other than as an abstract of an oral or poster presentation, Three types of manuscripts can be submitted: full-length articles, rapid communications, and plant gene register articles (electronic submission only). Assign manuscript to a category: Biochemistry and Enzymology; Whole Plant, Environmental, and Stress Physiology; Development and Growth Regulation; Gene Regulation and Molecular Genetics; Cell Biology and Signal Transduction; Bioenergetics; Plant-Microbe and Plant-Insect Interactions.

Authors are presumed to have read paper and agreed to authorship. Mention persons who only made minor contributions to work in Acknowledgments; do not list as authors.

A copynght assignment/acknowledge form is sent to corresponding author when manuscript is received. Sign and return form to ASPP headquarters promptly.

MANUSCRIPT FORMAT
Make length (3-10 journal pages) consistent with data presented; include an abstract (200 words).

Submit four copies of typescript and original figures. Articles with faint or illegible type or with substandard illustrations will be returned unreviewed. Type original manuscripts on nonerasable bond paper, 8.5 × 11 in. or A4 with a 1 in. margin all around. Double-space all typed material (blank space between lines ≤ 6 mm), including "Literature Cited," tables, table titles and legends, figure legends, and footnotes. Submit manuscript with elements arranged in order, number all pages consecutively.

Page 1: Running head (60 characters and spaces); name, address, and telephone and fax numbers and e-mail address of corresponding author.

Page 2: Title of article; all authors' full names; institution address(es).

Page 3: Footnotes in order: financial source and experiment station or institution paper number; present address(es) of authors if different from heading; corresponding author with fax number and e-mail address; abbreviations (unnumbered footnote).

Page 4: Abstract (include binomial); 200 words.

Page 5 and subsequent pages: Text; Acknowledgments. Enter dates of manuscript receipt and acceptance after Acknowledgments. Number pages sequentially beginning with title page, p. 1.

Literature Cited

Figure Captions and Legends (grouped, double spaced)

Tables with brief and concise titles and legends (one table per page, double spaced).
Original figures

Literature Cited: Cite references in text by last names and year of publication. Arrange text citations from earliest to most recent year, alphabetized by name within same year. For entries in 'Literature Cited,' alphabetize by authors' last names and follow styles exactly for capitalizabon, punctuation, and order of elements.

JOURNAL ARTICLES

Author AB, Author BB (1977) Title of article. Plant Physiol **59**: 121–125

BOOK ARTICLES

Author AB, Author BB, Author CC (1974) Title of article. *In* A Smith, B Jones, eds, Title of Book, Ed 2 Vol 3. Publisher City, pp 14–19

THESES

Author BC (1974) Title of thesis. PhD thesis. University, City

ABSTRACT IN SUPPLEMENT TO *PLANT PHSYIOLOGY*

Author DD (1980) Title of abstract (abstract No. xx). Plant Physiol 65: S-page number

NO AUTHORS OR EDITORS

Title of Booklet, Pamphlet, etc. (1975). Publisher (or Company), City

Write out in full all one-word journal titles. Use *BIOSIS List of Serials* for other journal title abbreviations; write out in full names of unlisted journals.

Designate articles 'in press' as such in 'Literature Cited.' Do not refer to an article as 'in press' unless it has been accepted for publication; cite journal in which article will appear.

Do not include unpublished data, submitted articles, articles in preparation, and personal communications as literature citations; refer to them parenthetically in text. Verify personal communications,with author of information and obtain approval for use; include letter of permission with manuscript.

Tables: Present data either in tables or figures, not both. Number tables consecutively with Roman numerals. Mention in text in sequential order: indicate first mention of each in margin of text. Provide with short, concise titles followed by legends that make general meaning comprehensible without reference to text. Provide descriptive headings for columns. Place tables with legends, double-spaced throughout, on a separate page; write authors' names on back of each page. Submit complex or large tables as camera-ready figures produced on a laser printer. Do not use double spaces in camera-ready tables except where necessary for legibility.

Illustrations: Make figures self-explanatory without much reference to text. Mark treatments or variables on figure itself with words so that reader can understand illustrated experiment. Line drawings must be sharp, black on white (laser prints are acceptable), and photographs must have good contrast and focus. Submit complicated formulas, flow diagrams, and pathways as figures. When designing figures, use reducing mode of a copy machine to visualize what it will look like when printed. Type should not be smaller than 6 points (2 mm) after reduction.

Submit one set of original, unmounted figures in a separate envelope. Mark each figure back with authors' names, figure number, and an arrow to indicate TOP. Submit three sets of high-quality reproductions for reviewers, i.e., photocopies of line drawings and actual photographs of gels and micrographs. Photocopies of halftone gels are unacceptable. Do not use a writing utensil that will smear or damage figures.

Number figures with Arabic numerals and mention sequentially in text. Indicate first mention of each figure in margin of text.

Group composite figures with different parts (A, B, C, or plates of micrographs) together and mount on lightweight, flexible cardboard. If a figure is a composite with several parts, label as A, B, C, etc. and not as separate figures grouped together.

Provide a caption and an explanatory legend for each figure. Type captions and legends double-spaced on a separate manuscript sheet. Explain all symbols or abbreviations in figure.

Use journal's accepted abbreviations (*Plant Physiol.* 1995; 107(1): xiii-xv) for units of measurement. Use powers of 10 with units of measurement.

Extend abscissa and ordinate only as far as contents of graph demand. Enclose graphs in a ruled-in box of same weight as abscissa and ordinate.

Create only as many figures as are necessary to accompany and clarify research.

For two-dimensional gels (e.g., combined IEF and SDS separations), present photographs with basic side to right. Label maximum and minimum pls of IEF gels at top of photograph. Label positions of M_r markers to left of photograph. Photographs may be mounted on lightweight, flexible cardboard to facilitate labeling.

Attach a tissue overlay to photographs, especially electron micrographs, and indicate important areas.

OTHER FORMATS

PLANT GENE REGISTER ARTICLES—Submit via electronic mail only to pgr@crcvms.unl.edu. Send a separate file for each manuscript.

Plant Gene Registers are concise descriptions of full-length sequences of transcribed regions of DNA. Determine sequences for both DNA strands. A gene description must not have been published or be in press elsewhere; sequence may not be simply an allele of a previously published sequence unless allele causes an extraordinary change in phenotype or biochemistry. State whether characteristics such as expression or function are attributed to gene based on experimental evidence or inferred on the basis of homology. Provide a GenBank/EMBL accession number upon submission.

Article should include: A brief discussion of rationale that led to sequencing and of importance of finding. A table of information on organism from which sequence was obtained, techniques used, and characteristics of sequence. Do not repeat information in text. Use previously published articles as a guide. A "Literature Cited" section (10 references). Do not include abstracts or figures; actual nucleotide or amino acid sequences will not be published. Edition will determine whether comparison sequences will be published.

Text is limited to 75 characters/line; put left margin at first column of screen.

Begin all lines at left margin except for paragraph indentations in text. Title—Include in parentheses database accession number for sequence being reported (blank line), Authors (blank line), Author affiliations: See Journal for format (blank line), Corresponding author e-mail address and fax number (blank line), Footnotes (blank line), Text: Limit to 75 lines. Indent paragraphs. Cite literature by author and date (blank line). Acknowledgments (blank line), Table: Not necessary, include to summarize basic information not included in text. e g., cloning and sequencing strategy (blank line), Literature Cited: Limit of 10 references. Use journal format.

Convert files prepared with word processing software to ASCII files before submission. Set left, top, and bottom margins and page offset to 0.

Spell out special characters (e.g., alpha, beta, micro); do not use underlining, italics, etc.

Before submitting file, view contents in mail system to assure that it is in correct format. Verify that there are no extraneous characters, extra blank lines, displaced margins, etc. Be sure that text begins at first column of screen.

Author must assure accuracy of files at time of submission. What is submitted will be directly deposited in database if accepted.

Do not include primary nucleotide or amino acid sequence in submission. Those interested can access information using database accession number in title. A comparison between different sequences may be necessary to show particularly unique features of sequence being described. Alignments of sequences are not necessary for most submissions.

RAPID COMMUNICATIONS—Three journal pages, including a 50 word abstract, text, tables, figures, and literature cited, representing either a preliminary report or a complete accounting of a significant research contribution. In cover letter explain why findings deserve rapid communication.

OTHER

Abbreviations: Do not abbreviate words or measures in title other than those standard for international usage. Chemical symbols can be used in titles. Units of measure can be abbreviated in abstract. In remainder of text and running head, abbreviations listed in *Plant Physiol* 1995;107: xiii–xv may be used without definition. Define all other abbreviations alphabetically in a single, unnumbered footnote.

Color Charges: $500 for first figure, $500 for second, and $250 for third and all subsequent color figures in one article. Include a letter indicating willingness to pay charges.

Gases: To indicate volume,, use microliters per liter (μL L^{-1}) or nanoliters per liter (nL L^{-1}) rather than ppm or ppb.

Growth Room Conditions: Describe according to guidelines in *CBE Style Manual* (1983, 5th ed.), pp. 169–170, or *HortScience,* 1983, 18: 662–664, or *Physiologia Plantarum,* 1982, 56: 231–235.

Electronic Submission: Submit a diskette with accepted manuscript with final version of text, references, and figure legends (tabular and equation material will not be captured from diskette but will be rekeyed). Send diskette to monitoring editor with revised manuscript. For manuscripts accepted without revision, send diskette directly to Managing Editor upon notification of acceptance.

Most word processing software progams can be used. If using Macintosh computers do not use Fast Save option. ASCII files can be submitted, they are not a preferred format. Do not submit diskettes created on desktop publishing systems or created on proprietary typesetting systems. Diskette must contain only files pertinent to manuscript. Identify diskette with journal name, manuscript number, corresponding author's name, manuscript title, type of computer, operating system and version, and software program and version number. Contact managing editor at (301) 251-0560, extension 18, if there are any questions.

Fees: $250 per accepted manuscript. Fee may be waived provided a written application explaining reason for request is made to managing editor. Fee for electronic Plant Gene Register article is $100.

Ions: Represent as: Na$^+$, Mn^{3+}, Br$^-$, PO$_4^{3-}$.

Isotopically Labeled Compounds: For simple molecules, indicate labeling by writing chemical formulae, for example ^{14}CO$_2$, H$_2^{18}$O. For other molecules, place isotopic symbol in square brackets attached to name or formula without hyphen or space: [^{14}C]glucose, [^{32}P]ATP. In case of generic names, write isotope without brackets and follow with a hyphen: ^{131}I-albumin, ^{14}C-amino acids. Place letter and symbols indicating configuration, etc. before square bracket: D-[^{14}C]glucose, L-[^{14}C]-alanine, α-[^{14}C]napthaleneacetic acid. Indicate positions of isotopic labeling by Arabic numerals, Greek letters, or prefixes placed in square bracket and before symbol of element concerned to which they are attached by hyphen: D-[3-^{14}C]lactate, L-[2-^{14}C]leucine. Use U to indicate isotope is uniformly distributed among all six carbons: [U-^{14}C]glucose.

Molecular Weight and Mass: 'Molecular weight' (M_r), is the ratio of mass of a molecule to one-twelfth of the mass of carbon 12 and is dimensionless. 'Molecular mass' (mass of one molecule of a substance) is not a ratio and can be expressed in daltons (D). Say molecular mass of X is 20000 daltons (20 kD) or molecular weight (M_r) is 20000, but do not express M_r in daltons. Expressions such as 20-kD peptide and mass of a band on a gel is 240 kD' are acceptable for an entity that is not a definable molecule.

Nomenclature: In title, abstract, at first mention in text, and in Materials and Methods, include complete botanical names (genus, species, authority for binomial, and, when appropriate, cultivar) for all experimental plants. Following first mentions, generic names may be abbreviated to initial, except when confusion could arise. Identify algae and microorganisms by a collection number or that of a comparable listing.

Numbers and Fractions: Write out numerals one through nine, except with units of measure. Write out numbers or fractions that begin a sentence or rephrase sentence. Use preposition 'to' between numerals (do not use a hyphen). Exceptions: in tables, figures, graphs, legends, and within parenthesis in text,use hyphens. Decimals are preferred over fractions; when simple fractions are used, write out as a hyphenated unit: 'two-thirds.'

Numerals: Check tabular data and numerical values for proper number of significant figures For decimals smaller than one, insert a zero before decimal point.

Powers: To avoid numbers with many digits, express numbers as powers of 10. Change unit with prefixes such as "m" or "μ ." For example: enter 5 to express a g value of 0.005 under heading $g \times 10^3$ or a g value of 5000 under heading $g \times 10^{-3}$; express a concentration of 0.0015 M as 1.5 under heading concn (mM), as 1500 under heading concn (μM), or as 15 under heading $10^4 \times$ concn (M).

Ratios: In describing mixtures, use 'to' if a ratio is stated in words: 'the chloroform to methanol ratio; use a colon if numerical ratio is provided: chloroform:methanol (2:1, v/v); use a hyphen if numerical value is not given: "used in chloroform-methanol."

Reviewers: Suggest appropriate reviewers in their field.

Solutions: Describe solutions of common acids and bases in terms of normality (N), e.g., 1 N NaOH, and those of salts in terms of molarity (M). Express fractional concentrahons by decimals: 0.1 N acetic acid (not N/10 acetic acid). Define % as (w/w), (w/v), or (v/v); 10% (w/v) signifies 10 g/100 mL. Express concentrations as micrograms per gram (μg g^{-1}) or micrograms per milliliter (μg mL^{-1}) rather than as parts per million (ppm).

Statistical Analyses: Include statistical analysis. Define all statistical measures clearly. Identify number of replications of experimental treatments and number of times individual expenments were duplicated.

Style: Write in simple declarative sentences. Conform to accepted standards of English style and usage. Consult issues since January 1993 for style and placement of main headings, subheadings, and paragraph headings and for other details of format. See *Council of Biology Editors Style Manual* (6th ed., 1994, Council of Biology Editors, Bethesda, MD).

Trade Names: Provide names and addresses of manufacturers or suppliers of special material. Capitalize trade names. Avoid use of trade names and code numbers of experimental chemical compounds used in research; identify such compounds by common name (American Standards Association) or by chemical name and structural formula.

Units: Metric system is standard; use Sl units as much as possible. Use negative exponents to indicate units in denominator when three or more units are used (e.g., μmol m^{-2} s^{-1} rather than μmol/m^2/s).

SOURCE

January 1995, 107(1): front matter.

Planta
Springer International

Professor Russell L. Jones
University of California
Department of Plant Biology
Berkeley, CA 94720
Fax: (510) 642 4995

North, Central and South America, Australia and New Zealand

Professor Andreas Sievers
Botanisches Institut Universität Bonn
Venusbergweg 22
D-53115 Bonn, Germany
Fax: 49 (228) 73 26 77

Continental Europe

Professor Malcolm B. Wilkins
Botany Department
Glasgow University
Glasgow G12 8QQ U.K.
FAX: 44 (141) 330 4447

U.K., Africa and Asia

AIM AND SCOPE

Planta publishes original articles in all aspects of plant biology, particularly in molecular and cell biology, ultrastructure, biochemistry, metabolism, growth, development and morphogenesis, ecological and environmental physiology (including crop physiology), biotechnology, plant-microorganism interactions. Preference is given to experimental articles and articles serving as basis for experimental work.

MANUSCRIPT SUBMISSION

Copyright of article is transferred to Springer-Verlag when article is accepted for publication. Copyright covers exclusive and unlimited rights to reproduce and distribute article in any form (printing, electronic media or any other form); it also covers translation rights for all languages and countries, For U.S. authors copyright is transferred to the extent transferable.

Submission of a manuscript implies that work has not been published before (except as an abstract or as part of a published lecture, review of thesis); that it is not under consideration for publication elsewhere; that its publication has been approved by all coauthors and by responsible authorities where work was done; that if and when manuscript is accepted for publication, authors agree to automatic transfer of copyright to publisher; and that manuscript will not be published elsewhere in any language without consent of copyright holders.

Submit manuscnpts in final form ready for pnnting. Mark approximate position of figures and tables in margins.

Provide telephone and fax numbers and/or E-mail address of corresponding author

MANUSCRIPT FORMAT

Type manuscripts double-spaced with wide margins. Submit in triplicate (original typed on one side of paper only plus two copies photocopied on both sides). Place main headings (Introduction, Materials and methods, etc.) on separate lines. Mark footnotes, Material and methods sections, Acknowledgements, References, Tables and figure legends for small print.

Abstract: Give plant material (binomial including authority) and EC numbers.

Key Words: Suggest six key words in alphabetical order characterizing scope of paper, principal plant materials (scientific generic names only) and main topics treated.

Running Title: 75 letters and intervals.

Footnotes: Number consecutively.

Literature Citations: Give author(s) and year. If more than two author(s), name only first, followed by "et al." Examples: Manning (1994) has shown ... Experiments in this laboratory (Radetzky and Langheinrich 1994) have shown that .. Liu et al. 1994) have shown that...

References: Include only publications cited in text. Put in alphabetical order by name of first author with all authors and complete title of each work cited. Second and subsequent lines are indented.

ARTICLES FROM JOURNALS AND OTHER SERIAL PUBLICATIONS (E.G., ANNUAL REVIEWS)

Webb MC, Gunning BES (1994) Embryo sac development in *Arabidopsis thaliana*. II. The cytoskeleton during megagametogenesis. Sex Plant Reprod 7: 153–162

ARTICLES FROM NON-SERIAL COLLECTIVE PUBLICATIONS (SYMPOSIA VOLUMES, ENCYCLOPEDIAS, ETC. AND BOOKS

Rincón M, Boss WF (1990) Second-messenger role of phosphoinositides. In: Morre DJ, Boss WF, Loewus FA (eds) Inositol metabolism in plants. Wiley, New York. pp 173–200

Tables: Number consecutively with Arabic numbers. Indicate footnotes by lowercase superscript letters, beginning with [a] in each table.

Illustrations: Limit number to minimum needed to clarify text. Do not double document the same points in Figures and Tables.

Number all figures (photographs, graphs, diagrams) consecutively. Submit halftone illustrations as sharp, glossy, high-quality photographic prints. Supply line drawings (graphs and diagrams) as black-and-white drawings suitable for reproduction. Figures that are to appear together should be either photographed as a group or mounted together on flexible white drawing paper (0.4 mm thick, 300 g/m^2). Trim figures at right angles and submit as a size permitting direct printing no more than 8.6 cm across for column width, no more than 17.8 cm for page width, no higher than 24 cm.

Color illustrations can be published if authors agree to bear some of extra costs

Line Drawings: Make lines uniformly thick, letters and numbers should be professional quality and proper dimensions (2 mm high after reproduction). Computer drawings are acceptable if of comparable quality to line drawings

Photographs: Should exhibit high contrast. Insert arrows, letters and numbers with template rub-on letters. If not possible, make inscriptions on a transparent overlay (not on photograph). Micrographs should have an internal magnification marker; state magnification in legend.

Legends: Provide each illustration with a concise, descriptive legend. List after references and Tables at end of typescript.

OTHER FORMATS

Rapid Communications–Three printed pages (nine typescrlpt pages) including tables, etc. Short papers making a single point. Declsion as to whether a paper qualifies rests with editors

OTHER

Abbreviations: For correct usage of abbreviations, consult list of Units, symbols and abbreviations in first issue of each volume and instructions in *Eur J Biochem* (1993) 213: 1–3.

Color Charges: No charge for first page and DM 1.500 for each additional page.

Electronic Submission: Submission of diskettes is encouraged; do not send until changes requested following reviewing process have been entered and authors have been notified of acceptance for publication.

Genus and Species Names: Type in italic or underline genus and species names, along with words to be emphasized.

Units and Symbols: Express temperatures expressed in degrees Celsius; time in seconds (s), minutes (min). hours (h). days (d) etc. Use International System of Units (SI), wherever possible. Consult, e.g., U.S. Department of Commerce. Natlonal Bureau of Standards, Special Publication 330. *The International System of Units*. latest edition. or Rotter F (1979) Das internationale Einheitensystem in der Praxis (*Physik in unserer Zeit* 10: 23-51).

SOURCE

1995, 196(1): A2, A5.

Plastic and Reconstructive Surgery

American Society of Plastic and Reconstructive Surgeons, Inc.
American Association of Plastic Surgeons
American Society for Aesthetic Plastic Surgery, Inc.
American Society of Maxillofacial Surgeons
Williams & Wilkins

Robert M. Goldwyn, M.D.
1101 Beacon Street
Brookline, Massachusetts 02146

AIM AND SCOPE

The goal of *Plastic and Reconstructive Surgery* is to inform its readers of significant developments in all areas related to plastic and reconstructive surgery. This journal provides a forum for responsible discussion among identified individuals.

Significant papers on any aspect of plastic surgery—operative procedures, clinical or laboratory research, case reports, and special topics are invited for publication.

MANUSCRIPT SUBMISSION

Use acceptable English usage and syntax; make contents clear, accurate, coherent, and logical. Originality, teaching value, and validity are considered.

Do not exceed 4000 words; be frugal with illustrations.

Submit original manuscript and two copies, with three sets of illustrations; retain one complete copy.

Articles may not have been published elsewhere (in part or in full, in other words, or the same) and may not be submitted elsewhere unless rejected. If an author violates requirement or engages in similar misconduct, Editorial Board may reject manuscript and impose a moratorium on acceptance of new manuscripts from author; if misconduct is sufficiently serious, will refer matter to author's academic institution or hospital, to appropriate state or local disciplinary body, and/or to Ethics Committee of the American Society of Plastic and Reconstructive Surgeons, Inc. Published manuscripts become sole property of the journal and will be copyrighted by the American Society of Plastic and Reconstructive Surgeons, Inc. Submission implies that each author agrees to each of above conditions. Author(s) explicitly assigns copyright ownership to Society if article is published.

Disclose any commercial associations that might create a conflict of interest. Disclose all sources of funds supporting work in a footnote, and all institutional or corporate affiliations of authors. In cover letter inform Editor about consultancies, stock ownership or other equity interests, or patent licensing arrangements. Information will remain confidential during review and will not influence editorial decision. If manuscript is accepted, Editor will discuss with authors how best to disclose relevant information.

MANUSCRIPT FORMAT

Type double-spaced, on one side only, on 8.5 × 11 in. (22 × 28 cm) white bond paper, with 1.5 in. (4 cm) margins at left, top, and bottom and a 1 in. (2.5 cm) margin at right. Double-space all copy including text, footnotes, bibliographies, legends, tables, and headings. Supply text references in sequence for all tables and figures. At end of text, supply name and complete address of principal author.

Title Page: State full title of article, followed by authors' names, degrees, and city. Place footnotes giving principal affiliation of each author and where and when paper was presented at bottom of page.

Abstract: 100-150 words for every original article. Case Reports and Ideas and Innovations have a summary. Type on a separate page; give factual information, not generalities (problem investigated, method used, results obtained, and conclusions reached).

References: Verify accuracy and completeness of references. Type on separate pages and cite in text in numerical order, not alphabetically.

JOURNAL ARTICLES

Include: author(s), title, journal name (as abbreviated in *Index Medicus),* volume number, first page number, and year, in that order.

BOOKS

Include author(s), chapter title (if any), editor (if any), title of book, city of publication, publisher, and year. Include volume and edition numbers, specific pages, and name of translator when appropriate.

Legends: Required for illustrations; type on separate pages. Keep brief and pertinent. Identify stain and state magnification in legends for photomicrographs.

Illustrations and Tables: Consider size and shape of journal page when planning illustrations; arrange to conserve vertical space. Photographs must be in sharp focus, have good contrast, and be neither retouched nor altered in any way. Glossy prints are preferable. Before-and-after photographs of patients must be identical in terms of size, position, and lighting. Backgrounds should be clean and uncluttered. Color photographs that significantly enhance the presentation will be considered at author's expense; obtain estimates from publisher. Computer-generated graphs and line drawings are acceptable if prepared on high quality paper. Photostats are preferred.

Acknowledgments: Acknowledge illustrations from other publications by a complete credit line in legend. Submit publisher's letter of permission.

OTHER FORMATS

CORRESPONDENCE AND BRIEF COMMUNICATIONS—500 words typed double-spaced. Published at editors' discretion as space permits.

OTHER

Statistical Analyses: Explain how performed at end of section preceding results. Reference unusual or complex methods of analysis.

SOURCE

January 1994, 93(1): front matter.

Postgraduate Medical Journal
Fellowship of Postgraduate Medicine
Macmillan Press, Ltd., Scientific & Medical Division

Editor, *Postgraduate Medical Journal*
12 Chandos Street
LondonW1M 9DE England

AIM AND SCOPE

Postgraduate Medical Journal publishes original papers on subjects of current clinical importance. Each issue contains commissioned review and leading articles but unsolicited submissions are also welcome. Each issue also includes 'Letters to the Editor', book reviews and an international postgraduate diary. Many issues contain papers or abstracts of symposia devoted to a single subject, and the full proceedings of meetings are published as supplements to the *Journal*. The *Journal*, as the organ of the Fellowship of Postgraduate Medicine, is dedicated to advancing the understanding and practice of postgraduate medical education and training.

MANUSCRIPT SUBMISSION

Prepare manuscripts in accordance with guidelines of the International Committee of Medical Journal Editors (*Br Med J* 1988, **296**: 401–405). All material is assumed to be submitted exclusively to *Journal* unless otherwise stated. Principal author must ensure that all authors agree to submission of typescript.

MANUSCRIPT FORMAT

Submit two copies. Type double-spaced, on one side of A4 paper (297 mm × 210 mm) with a 5 cm margin. On first page type names of author(s) and name and address of laboratory or institution where work was done and title of paper. Give full address and fax number of principal author to whom proofs will be sent as a footnote. Footnote permanent changes of address or appointment. Give a short (running) title (45 characters). Number all pages including title page.

Acknowledge written or illustrative material which has been or will be published elsewhere; send written consent of authors and publishers concerned.

Divide papers into: Title page, Summary (250 words stating what was done, main findings and how work was interpreted), Introduction, Materials and methods, Results, Discussion, Acknowledgments, References, Tables, Figures and captions. Avoid numbered paragraphs.

References: Follow Vancouver format. In text, show as numbers. At end of paper, list (double-spaced) in numerical order corresponding to order of citation in text. List all authors for papers with up to six authors; for papers with more, list first three followed by *et al.* Abbreviate titles of medical periodicals according to latest edition of *Index Medicus*. Provide first and last page numbers for each reference. Identify abstracts and letters as such.

1. Clements R. & Gravelle I.H. Radiological appearances of hydatid disease. *Postgrad Med J* 1986 **62**: 167–173.

2. Greenberger J.S. Long-term hematopoietic cultures. In: Golde, W. (ed) *Hematopoiesis*. Churchill Livingstone, New York 1984, pp 203–242.

Figures: In text, use Arabic numbers; specifically refer to all illustrations in text, e.g., (Figure 2). Submit all illustrations at about 1.5 times intended final size. Number as figures whether photographs, representational drawings or line diagrams and graphs.

Photographs and Photomicrographs: Submit unmounted glossy prints; do not retouch; exclude technical artifacts. Indicate magnification by a line representing a defined length within photographs. Indicate areas of key interest and or critical reproduction on an attached flimsy overlay or on a photocopy. Similarly indicate annotations

and lettering, preferably not on original print. Clearly contrasted and focused prints are essential.

Line Diagrams and Graphs: Put on separate sheets. Draw with black India ink on white paper, or supply as photographic prints of originals. Keep lettering to a minimum; do not duplicate legend. Use symbols consistently within papers; explain symbols in caption, not on figure. Submit photocopies of all illustrations.

Tables: Supply as few as possible; present only essential data. Type on separate sheets with a title or caption; give Roman numbers.

OTHER

Drugs: For manuscripts reporting adverse drug reactions, submit evidence showing that reaction has been reported on a 'yellow card' or to appropriate drug licensing authority, and to drug manufacturer. Give approved generic names of drugs. If a proprietary (brand) name must be used, begin with a capital letter.

Style: For abbreviations and symbols use standard and SI units throughout. Use acronyms sparingly; explain fully when first used. Statistical analyses must explain methods used. Do not use footnotes. Use single quotation marks. Underline words to be italicized. Use *Concise Oxford English Dictionary* as a reference for spelling and hyphenation.

SOURCE

January 1994,70(819): inside back cover.
A notice to contributors is usually published on inside back cover of each issue.

Proceedings of the National Academy of Sciences of the United States of America
National Academy of Sciences

Proceedings of The National Academy of Sciences USA
2101 Constitution Avenue, N.W.
Washington, D.C. 20418
Telephone: (202) 625-4725; Fax: (202) 625-4747
U.S. Postal Service
1010 Wisconsin Avenue N.W., Suite 530
Washington, D.C. 20007
Express and courier services

AIM AND SCOPE

Proceedings of the National Academy of Sciences USA publishes reports, commentaries, reviews, and colloquium papers.

MANUSCRIPT SUBMISSION

Journal reports describe results of original theoretical or experimental research of exceptional importance and broad interest to diverse groups of scientists. Reports must be contributed by a member or foreign associate of the Academy or communicated by an Academy member or foreign associate on behalf of a nonmember. Papers should be of the highest scientific quality and should be intelligible to a broad scientific audience. Do not exceed five printed pages.

Members or foreign associates of the National Academy of Sciences who submit manuscripts assume responsibility for their propriety and scientific standards. Each member may submit (contribute and/or communicate) five papers per calendar year.

Reports may not have been published previously or submitted for publication elsewhere and may not be published elsewhere without attribution to *Proceedings*.

Limit authorship to those persons who have contributed substantially to major aspects of planning, execution, and interpretation of work and preparation of manuscript. Persons providing materials or resources or giving general advice are not authors. All authors must concur in submission, having seen and approved final copy of manuscript; all authors agree to share responsibility for paper.

Return appropriately completed copyright assignment form to editorial office with proofs.

Do not include statements of priority or novelty or descriptions of work in progress or planned. Failure to provide necessary documentation upon submission may delay publication.

Consider conflicts of financial interest in submitting papers. Conflicts include officers, directors, or full-time employees of a company; having a substantial personal or family financial interest in a company for which publication could result in a change in value of equity; having a close, sustained, and lucrative consulting arrangement with a company whose paper is being communicate; or reporting clinical trials of a product on which contributor or communicator holds a patent or with whose manufacturer member or foreign associate has a financial arrangement from which s/he could benefit. Be alert to conflicts of interest of authors for whom papers are communicated. Discuss matter with the Chairman of the Editorial Board, a Contributing or Consulting Editor, or Managing Editor. Alternatively, another member or foreign associate or one of the Editors can participate in review process.

For their own manuscripts, members must provide assurance that the manuscript has been reviewed by a knowledgeable colleague who is not a coauthor.

Give name, address, and telephone and fax numbers of corresponding author(s) in transmittal letter.

MANUSCRIPT FORMAT

Manuscripts that do not meet specifications will be returned for correction or, for illustrations, may be corrected by *Proceedings* staff. Adhere to length limits and format specifications. Failure to comply will delay publication. Articles estimated to exceed length limit will be returned to be shortened. Submit three complete copies of manuscript; two extra copies of both title page and abstract for Editorial Board review; three complete sets of illustrations; and three complete sets of photocopies of figures

Type legibly on one side of each page, on standard paper. Double or triple space entire manuscript, including title page, references, legends, and tables; keep generous margins (at least 4 cm, sides, top, and bottom). Number pages in sequence; title page is p. 1.

Be as brief as full documentation allows. Do not exceed five printed pages (6000 words, if no figures or tables). Review articles may be eight pages. Title, key terms, names of authors and their affiliation(s), and statement of communication or contribution occupy about 300-400 words. Make allowance for space for footnotes, references, tables, and figures with legends.

Title Page: Provide classification, title, key terms, author affiliation, abbreviations footnote, and data deposition.

Classification: Based on titles of Academy Sections. Papers are generally listed under one classification only. Listing under more than one classification is acceptable only when the two fall under different major headings).

Physical Sciences: Applied Mathematics, Applied Physical Sciences, Astronomy, Chemistry, Computer Sciences, Engineering, Geology, Geophysics, Mathematics, Physics, and Statistics.

Biological Sciences: Agricultural Sciences, Applied Biological Sciences, Biochemistry, Biophysics, Cell Biology, Developmental Biology, Ecology, Evolution, Genetics, Immunology, Medical Sciences, Microbiology, Neurobiology, Pharmacology, Physiology, Plant Biology, Population Biology, and Psychology.

Social Sciences: Anthropology, Economic Sciences, Psychology, Political Sciences, and Social Sciences.

Title: Be brief, specific, and use informative words; do not include nonstandard abbreviations. Do not use serial titles; a serial title followed by a colon and specific title may be used. Do not use footnotes. If paper is part of a series, footnote on an appropriate location in Introduction: "This is paper no. 19 in a series. Paper no. 18 is ref..." Include preceding paper in list of references.

Key Terms: 5, optional. Do not repeat title. Give below title, enclose within parentheses, separate by slashes (/).

Author Affiliation: Furnish department, institution, city, state, and ZIP code or country for each author. Use complete, spelled-out names of institutions not acronyms. If several authors with different affiliations, match authors to respective institutions by superscript symbols after names: *, †, ‡, §, ¶, ‖, **, ††, ‡‡.

Primary affiliation is where work was done. If author has moved, give information as a "Present address" footnote.

Indicate reprint request author by a footnote stating "To whom reprint requests should be addressed."

Abbreviations Footnote: List nonstandard abbreviations used five or more time. Keep such abbreviations to a minimum; do not use in title or key terms; define where first mentioned in text, after which use abbreviation only.

Data Deposition: If sequence or crystallographic data have been deposited in appropriate database, so state.

Abstract: Second page, 250 words. State subject and general conclusions of article. Avoid abbreviations and symbols unless used five times in abstract. Make understandable to reader before paper is read, suitable for reproduction in abstract service publications, and unambiguous without recourse to manuscript. Use only essential references (for example, those that acknowledge work of others); use complete citations.

Text: Write to be understandable to scientists in many disciplines. Avoid laboratory slang and minimize jargon.

Describe procedures in sufficient detail so work can be repeated. Reference other papers describing techniques. Give correct chemical names and specify strains of organisms. Identify trade names by an initial capital letter with remainder lowercase (do not use superscript "TM" or "R"). Provide names and addresses of suppliers of uncommon reagents or instruments.

Use units and symbols of the Système International (SI). When not used, give factor for conversion to SI units in parentheses where first mentioned. Note that "kilo" is a prefix to names of units and its abbreviation is always lowercase. The expression "273 K" indicates temperature at which water freezes, not molecular weight 273,000. Use prefixes (e.g., "m," "μ," "n"), not power-of-ten notation.

Footnotes: Keep to a minimum; indicate in text by symbols in order *, †, ‡, §, ¶, ‖,**, ††, ‡‡. Use each only once, including title page (symbols used in tables or figures do not affect symbols used in text). If more are needed, use superscript lowercase letters.

Acknowledgments: Include dedications and acknowledgments, and statements regarding equivalence of contributions by various authors. Acknowledgments to individuals precede those for financial support. Spell out names of grant sources.

References: Check for accuracy. Cite only published or in-press papers and books in reference list. Do not include abstracts of papers presented at meetings (unless published in a recognized serial publication) in list. Limit such references to those necessary to give credit to others; cite parenthetically in text or as footnotes. Cite references in numerical order as they appear in text. Give each a separate number; multiple citations under one numeral are not used. Number references in tables and figures according to appearance in text. Use Arabic numerals separated by commas (except for sequences of three or more, when a dash is used between first and last numeral) and enclose reference number in parentheses and set on-line. Give full citation at end of manuscript. Name all authors and give inclusive pagination. If reference is an abstract, note in citation, after page number(s).

Data bases are ephemeral references. Cite in text or as footnote. Other citations are either "unpublished data" or "personal communication; " include directly in text where first mentioned. Former implies that data are from one or more of the authors; attribution is by initials of author(s) and names of others who have contributed to the work. "Personal communication" means a source other than an author; give a name or names. When reference is made to a personal communication, unpublished work, or a paper in press that does not involve a submitting author, furnish a statement that authorizes citation of such material signed by one of the persons cited.

JOURNAL ARTICLES

10. Neuhaus, J.-M., Sticher, L., Meins, F., Jr., & Boller, T. (1991) *Proc. Natl. Acad. Sci.* USA **88**, 10362–10366.

For abbreviations of journal titles, use *Chemical Abstracts Service Source Index* (CASSI) (American Chemical Society, Washington, D.C.).

ARTICLES OR CHAPTERS IN BOOKS

11. Jones. C. D. & Shapiro, L. M. (1984) in *Enzyme Reactions in Protozoa*, eds. Smith. T. G. & Williams, H. I. (Universal, New Brunswick, NJ), Vol. 2. pp, 646–672.

12. Green. J. (1985). Probability Theory (Van Nostrand, Princeton), p. 474.

Figures: Submit original drawings, laser prints, or high-quality photographs in triplicate. Provide three photocopies of each. Identify on reverse side with a soft pencil. Indicate orientation for by an arrow and TOP. Type legends double-spaced, in numerical order, on a separate page. See *Proc. Natl.Acad. Sci. USA* 1994; 91(1): vi for examples of style and standards for figures.

Line Drawings—Mark with index lines and label with scales on x and y axes. For scales that involve large or small numbers use numbers multiplied by powers of 10. Follow convention: 3000 cpm is represented as 3 and axis is labeled cpm × 10^{-3}. Use both separate symbols and various lines (solid, broken, etc.). Nomenclature, abbreviations, and units used in figures must agree with those in text. Do not mount illustrations unless a composite figure is required. Identify a composite figure by a single figure number; identify individual parts by letters. Indicate magnifications of photomicrographs in legend or by scale bars (or both).

Submit illustrations in final size or in larger size for reduction by printer. Lettering must be such that in printed version (after reduction) it is 8–10 points (in 8-point type, uppercase letters are 2 mm tall; legends are set in 8-point type). Except for single-letter locants or markers, make lettering no larger than 10 points. Do not widely vary type sizes within a single figure; do not boldface. Serif type is preferred to sans serif. Use uppercase and lowercase letters just as in text. Illustrations with hand-drawn lettering, heavy lines, or uneven lettering or lines are not acceptable.

Placement and orientation of lettering are important. Do not obscure or detract from figure components with lettering. Center axis labels along axis and orient to be read from bottom or from left. For numbering scales on graphs, consider omitting alternate numbers (but keep marker lines) to permit large-enough lettering.

Choose symbols, shading, and lines to survive reduction. At severe reduction, hexagons are indistinguishable from circles and light gray fades to white. In crowded figures, identify circles, squares, and triangles in legend. Identify unusual symbols and patterns of hatching, cross-hatching, stippling, etc. in a key that can be photoreproduced as part of figure. Use different kinds of lines if differences do not disappear on reduction.

Identify figures that have been published elsewhere and provide written permission from the copyright holder.

Color Art:—Submit as prints or slides. Submit prints in duplicate, either unmounted or mounted on a lightweight, flexible backing. Mark slides for orientation with arrow and TOP; include a print or color photocopy to indicate desired magnification and cropping limits. Publication is subject to a surcharge. Obtain an estimate by telephone; a quotation will be sent on receipt of work.

Tables: Prepare so that they are self-explanatory. Do not use vertical rules; include horizontal rules between title and column heads, between heads and body, and between body and legend or footnote only. Give each a brief title. Number (Arabic numerals) in order cited in text. Place each on a separate page; type double-spaced throughout. Use nonstandard abbreviations sparingly; define in legend at bottom of table, if not in abbreviations footnote on title page. Give each data column a heading. Group headings by use of straddle (spanner) rules. Refer to footnotes with symbols *, †, ‡, §, ¶, ‖,**, ††, ‡‡, in order. If more are required use superscript lowercase letters.

Do not waste space. Large amounts of empty space in body of a table indicates poor design. Other poor features are vertical columns in which all entries are the same (incorporate information in title or legend and delete column); a change in column heading part way down column (make into two tables or change format); only a few entries in first column or line spaces between groups of data (use an interior center heading).

Take into account relationship between size of heading and size of entries in a column. If bulky headings appear over small numbers, consider redesigning table so that these headings become entries in first column.

Simplify tables by using powers of 10 in place of multiple zeros. Make heading show what was done to original data to obtain numbers shown.

A table can have a legend or footnotes or both below it. Legend provides information relevant to entire table. Footnotes provide information relevant to parts or single entries, as shown by placement of symbol; place symbol to include all entries affected.

OTHER FORMATS

COLLOQUIUM—Based on presentations held under Academy auspices. The Convenor of the Colloquium serves as the Communicating Member.

COMMENTARIES—Three pages calling attention of readers to papers that are appearing or have appeared in the *Proceedings* or elsewhere in scientific literature. Guide readership to an appreciation of the importance of a particular work and to describe the scientific context into which it fits. Commentaries are written by or at the invitation of a Contributing Editor.

REVIEWS—Comparatively short (8 pages) statements of the status of a scientific problem or idea. Ordinarily solicited by a Contributing Editor; unsolicited reviews are accepted only if contributed by a member or foreign associate or communicated by an Academy member or foreign associate on behalf of a nonmember and are subject to review by the editorial board.

OTHER

Abbreviations and Symbols: Standard abbreviations includes symbols for chemical elements; three-letter codes for amino acids, carbohydrates, lipids, and nucleotides; two-letter codes for chemical radicals; and units of Système International. See *Proc. Natl. Acad. Sci. USA* 1994;91(1): ix-x Tables 2 and 3 for some standard abbreviations and symbols and some standard abbreviations for units of measurement and physical and chemical quanti-

ties. Most other abbreviations are nonstandard; keep to a minimum; spell out on first usage in both abstract and text. Use only for terms mentioned five or more times. Nonstandard abbreviations should be unambiguous. Refer to *IUPAC Manual of Symbols and Terminology for Physico-chemical Quantities and Units* (1970, Butterworths, London); *National Bureau of Standards Special Publication 330* (1981, United States Government Printing Office. Washington, D.C.); or *Pure and Applied Chemistry* (1970) **21**, 3–44.

Too many abbreviations, symbols, and acronyms make a paper difficult to read. Abbreviate units of measure after numerals to emphasize the numeral. Common terms sometimes are more easily recognized in short form—e.g., DNA, ATP; these are preferred. Short forms common in one field may not be recognized by all readers.

Crystallographic Data: Submit relevant structural data of proteins and other biopolymers, to Protein Data Bank (Chemistry Department, Brookhaven National Laboratory, Upton, NY 11973). See Commission on Biological Macromolecules (1989) *Acta Crystallogr. Sect. A* **45**, 658. Supply data bank accession; specify embargo periods, which may not exceed those recommended by International Union of Crystallography.

Experimentation, Human: For research involving human subjects or use of materials of human origin submit copy of document authorizing proposed research, issued and signed by appropriate official(s) of institution where work was conducted.

Materials Sharing: Submission of manuscripts based on unique materials (e.g., cloned DNAs; antibodies; bacterial, animal, or plant cells; viruses; and computer programs) implies that these materials will be made available for noncommercial purpose to all qualified investigators.

Nomenclature, Chemistry: *The ACS Style Guide: A Manual for Authors and Editor* (ed. Dodd, J. S., 1986, American Chemical Society Publications, Washington, D.C.).

Nomenclature, Genetics: Bacterial—Demerec, M., Adelberg, E. A., Clark, A. J. & Hartman, P. E. (1966) *Genetics* **54**, 61–76. Human—Klinger, H. P., ed. (1992) *Human Gene Mapping 12: Twelfth International Workshop on Human Gene Mapping* (Karger, Basel). *Other.* O'Brien, S. J., ed. (1993) *Genetic Maps: Locus Maps of Complex Genomes* (Cold Spring Harbor Lab. Press, Plainview, NY, 6th Ed).

Nomenclature, Immunology: For human immunoglobulins and their genetic factors, use rules of World Health Organization or first reference book listed for Life Sciences.

Nomenclature, Life Sciences: *Council of Biology Editors Style Manual* (1983, 5th Ed. Council of Biology Editors, Chicago, IL). *Biochemical Nomenclature and Related Documents* (1992, Portland Press Ltd., London, or Chapel Hill, NC). This compendium contains the International Union of Biochemistry rules of nomenclature for amino acids, peptides, nucleic acids, polynucleotides, vitamins, coenzymes, quinones, folic acid and related compounds, corrinoids, lipids, enzymes, proteins, cyclitols, steroids, carbohydrates, carotenoids, peptide hormones, and human immunoglobulins.

Identify enzymes by recommended name followed in parentheses by systematic name and Enzyme Commission (EC) number on first mention, in both abstract and text. Refer to: *Enzyme Nomenclature: Recommendations (1992) of the Nomenclature Committee of the International Union of Biochemistry* (1992, Academic Press, New York).

Nomenclature, Mathematics: *A Manual for Authors of Mathematical Papers* (1970, reprinted with corrections 1980, American Mathematical Society, Providence, RI)

Nomenclature, Physics: *AIP Style Manual* (1990, American Institute of Physics, New York, NY)

Nomenclature, Psychology: *Publication Manual* (1983, 3rd Ed. American Psychological Association, Washington, D.C.).

Nucleic Acid Sequences: Submit relevant data to GenBank/EMBL/ DNA Data Bank of Japan (National Center for Biotechnology Information, Bldg. 38A, Room 8N-803, 8600 Rockville Pike, Bethesda, MD 20894); supply accession number of sequence.

Page Charges: Charge is adjusted as costs change; ascertain current charge from *Proceedings* Office. Authors are also billed for costs of extensive changes made in proof, for color reproductions, and for other special items.

Recombinant DNA Research: Physical and biological containment levels used must conform to guidelines of the National Institutes of Health or corresponding agency of country where research was conducted (see, e.g., *Federal Register,* May 7, 1986); Editorial Board may request proof.

Spelling: Follow American spelling and usage as in *Webster's Third New International Dictionary* (G. & C. Merriam, Springfield, MA) or *Random House Dictionary of the English Language* (Random House, New York).

SOURCE

January 4, 1994, 91(1): iii-x (Revised 1994).

Proceedings of the Royal Society: Biological Sciences Series B
The Royal Society

Proceedings B Editorial Office
The Royal Society
6 Carlton House Terrace
London SW1 5AG, U.K.

AIM AND SCOPE

Proceedings series B welcomes papers on any aspect of biological science. It publishes, rapidly, announcements of important results, normally not exceeding 4000 words, including the abstract and references (plus four figures and/or tables; equivalent to five printed pages). With the same restriction on length, reviews containing original and interesting ideas, and extensions to, or criticisms of, papers already published (subject to the criteria of interest, originality and good manners) will also be acceptable. The Editor will also consider short reviews, but only if they contain original and interesting new ideas. Short additions to, or criticisms of, papers that have already been published (subject to their originality and interest, and to the rules of good manners) will also be acceptable. Preliminary reports ('letters') are not encouraged.

Proceedings: Biological Sciences contains announcements of important new developments in biology. Papers crossing the boundaries of subjects are particularly welcome.

MANUSCRIPT SUBMISSION

Submitted papers must not have been published previously, nor be under consideration for publication elsewhere. Include telephone and fax numbers, and/or electronic mail addresses in correspondence about paper.

Submit four copies of typescript and figures (with one set of original drawings and prints). Include a word count. Submit extra copies of photographs as prints rather than photocopies.

Authors must assign copyright in article to the Society. In assigning copyright, authors will not forfeit the right to use their original material elsewhere subsequently without seeking permission and subject only to normal acknowledgment to the journal. Inform Society.

MANUSCRIPT FORMAT

Type double-spacing throughout, on one side of paper with a margin of at least 3 cm all round; number sheets serially and securely clip together. Spelling should conform to preferred spelling of *Shorter Oxford English Dictionary*. Avoid footnotes.

Title and Summary: Make title and summary comprehensible and interesting to nonspecialists. Make titles as short and general as possible. On a separate cover sheet include names of authors and laboratory or other place where work was done. Give addresses for correspondence, if different from place of work, indicating corresponding author with telephone and fax numbers. Give a very short title (50 letters and spaces) for page headings. Give a 200 word Summary; be precise and informative.

Sections: Divide papers into sections, described by short headings. Do not use subsections. Mark Materials and methods sections in margin for small type.

Illustrations: Supply duplicate figures (e.g., Xerox or photographic copies as appropriate) with each copy. Write author's name and number of figure on back. Number figures in one sequence throughout paper.

Color illustrations are included only if scientifically necessary and if cost is met by author (unless an acceptable case is made why funds are not obtainable). Clearly mark position of each illustration in typescript thus:

> Figure 2 near here

Line Drawings: Apply labelling necessary for understanding figure directly on original drawings before making duplicate copies. Make lettering lowercase except for initial capital letters of proper names or where capitals are essential e.g. for chemical abbreviations. Use Times or a close equivalent. Make height of capital letters after reduction 2 mm. When in doubt use smaller rather than larger lettering. For assistance call Editorial Office (telephone 071-839 5561, extension 229).

Legends: Type double-spaced on a separate sheet at end of paper. Follow style:

Figure 7. Time-course of changes in fibre type composition during post-stimulation recovery. (*a*) Type 1 fibres. (*b*) Type 2A fibres, including the transitional fibres (asterisks) referred to in the text. (*c*) Type 2B fibres. Bands indicate the range (mean ± s.d.) for the corresponding fibre type in control muscles.

Photographs: When essential to include photographs make efficient use of space. Restrict area covered by photographs to subject in question or to a minimum representative area in photomicrographs, etc. This enables photograph to be reproduced at largest possible scale. Text area available is 55 mm × 16 mm. Photographs are printed with text, not on plates.

Supply unlettered, unmounted glossy prints marked on back with authors' names, number of figure; indicate top and bottom. Provide a rough set with required lettering clearly marked. For micrographs include a scale bar, either applied directly to original or marked on rough set, with an indication of exact length.

Tables: Number in Arabic numerals. Refer to in text by numbers. Show position of each table as follows:

Table 3 near here

Table headings should be a brief title; describe experimental detail starting on a new line in parentheses. Present column headings in lowercase lettering except for capital initial letters of proper names. Place units of measurement and any numerical factors at head of column.

References: Type references to literature cited double-spaced in alphabetical order at end of paper. Cite references in text are by name and year method; do not number.

OTHER

Electronic Submission: Submission of final version on computer disk is welcomed. Hard copy is required for refereeing. A definitive copy should accompany disk. MS-DOS and Macintosh disk formats are acceptable; preferred word-processor format is Word-Perfect; documents prepared in Microsoft Word and Wordstar can be used.

Statistical Analyses: Follow guidelines published each year in July issue of the *Proceedings*. When referring to computer programs, specify procedures used, and quote publications that will allow reader to ascertain how they are carried out.

Units, Symbols and Abbreviations: Follow recommendations in *Quantities, Units, and Symbols* (1975, The Royal Society); use International System of Units (SI) whenever practicable.

Clearly differentiate between handwritten symbols of comparable shape. Use marginal indications and differential underlinings where necessary. Follow normal conventions where applicable e.g., use wavy underline to signify bold characters. Underline mathematical variables.

Use internationally agreed abbreviations; see, for example, list of accepted abbreviations in *Biochemical Journal*.

SOURCE

October 22, 1994, 258(1351): back matter.

Proceedings of the Society for Experimental Biology and Medicine
Society for Experimental Biology and Medicine
Blackwell Science, Inc.

Dr. Gregory W. Siskind, *PSEBM*
162 W. 56th St., Suite 203
New York, NY 10019
Telephone: (212) 541-7855

MANUSCRIPT SUBMISSION

The Journal is organized into a series of sections: Biochemistry/Nutrition; Cellular and Organ Physiology; Development/Growth/Aging; Endocrinology/Metabolism; Experimental Medicine/Pathology; Genetics/Molecular Biology/Molecular Medicine; Hematology/Oncology/Radiobiology; Host-Parasite Interaction/Immunology/Microbiology/Virology; Pharmacology/Toxicology; Minireviews; Comments. Indicate which section is most appropriate for paper.

Write in clear, concise and grammatical English; conform to general style of Journal. Manuscripts that are not adequately prepared will be returned. If accepted for publication, copyright in article, including right to reproduce in all forms and media, shall be assigned exclusively to the Society. Society will not refuse any reasonable request by an author for permission to reproduce their contribution.

Submit manuscripts in quadruplicate (one original and three copies). Send only original prints of photomicrographs. Cover letter must be signed by all authors. If author(s) cannot sign due to distance, state that non-signing authors have read and approved manuscript.

Submit signed approval from person(s) cited in manuscript as a sources of "personal communication" or "unpublished data."

Disclose commercial relationships that might be viewed as representing a conflict of interest in a footnote to applicable author's name. Make disclosure on a separate page that will not be sent to reviewers or influence editorial decisions; information is included in published paper.

Work may not have been and, if accepted by PSEBM, may not be published or submitted for publication elsewhere. Submission must have been approved by all authors (as confirmed by signing covering letter) and by institution where work was performed. Unnecessary subdivision of a study into several manuscripts is not acceptable. If any material published previously is included indicate with appropriate citation in manuscript. Submit a copy of any material that might be regarded as "duplicate publication," such as preliminary reports (including reviews, symposia and proceedings) or other publications (submitted, in press or published) containing data or other material included in submitted manuscript.. Clearly label such material to avoid confusion with submitted manuscript. It is acceptable to include detailed journal material that has been presented previously in summary form in reviews, symposia, or conferences, provided appropriate citation is included.

MANUSCRIPT FORMAT

Be as concise as possible; give sufficient detail to permit critical appraisal. Do not exceed 40 typed pages (including tables, charts and references).

Double- or triple-space, including tables, legends and footnotes). Number pages consecutively.

Research articles should be divided into an Abstract (300 words), Introduction, Materials and Methods, Results, and Discussion sections. First page: Complete title of paper, category for "Table of Contents" (select from "Section Headings" list), names of authors (without degrees), affiliations (including zip codes), and a running title (40 characters, including spaces). Second page: give name, complete address, zip code and telephone number of corresponding author.

Title: Limit to 15 words. Do not use abbreviations. Acknowledge all grant, contract and industrial support of work in a footnote to title.

Figures: Cite consecutively by Arabic numerals in text. type figure legends on a separate sheet. Legends should contain sufficient experimental detail to permit figures to be interpreted without reference to text. Indicate units in figures. Submit figures and illustrations so as to permit photographic reproduction without retouching or redrawing. This includes lettering, which is reproduced as part of figure, not set in type.

Send original prints of photomicrographs. Draft line drawings with black India ink on white drawing paper or blue drafting cloth, no larger than 8.5×11.5 in. overall. Make lettering large enough to allow a reduction of two thirds. High quality glossy prints are acceptable. Colored illustrations are published if authors cover costs.

Tables: Number with Roman numerals; cite consecutively in text. Title each table. Double-space on separate sheets. Refer to current journal issues for acceptable style. Title should clearly indicate nature of contents; include sufficient experimental detail in footnotes to permit interpretation of results without reference to text. Indicate units for each entry in table.

Footnotes: In text identify by superscript Arabic numerals. Type on a separate sheet; identify table footnotes by superscript lowercase letters a, b, c, etc., and placed at bottom of table.

Acknowledgments: Type on a separate page and place at end of text pages.

References: Provide appropriate attribution and credit to previous investigators. Avoid extensive citation; adhere to applicable conventions of scientific community. Obtain permission to reproduce from appropriate.

Arrange references numerically at end of manuscript ; cite in text with Arabic numerals in parentheses, set on text line. Number consecutively in order of citation. Abbreviate journal titles as in *Index Medicus*. Follow style of capitalization and punctuation.

1. Wang BC, Bettice JA, Brown EB Jr. Effect of body temperature on salicylate-induced hyperventilation. Proc Soc Exp Biol Med 174:102–106,1983.

2. Abramson DI. Circulation in the Extremities. New York: Academic Press, p000, 1967.

3. Langford MP, Weigent DA, Stanton GJ, Baron S. Virus plaque-reduction assay for interferon: Mircoplaque and regular macroplaque reduction assay. In: Pestka S, Ed. Methods in Enzymology. New York: Academic Press, Vol Part A 78: p000, 1981.

4. Hylden JL, Wilcox GL. Intrathecal morphine in mice: A new technique. Eur J Pharmacol (in press).

Do not include "personal communication," "unpublished," and "submitted" items in reference list; cite in a footnote. If a cited manuscript has been accepted for publication, include in reference list, giving journal, year, etc., as available.Citation of numerous abstracts is discouraged, although acceptable if necessary to provide proper attribution or credit.

OTHER FORMATS

COMMENTS—Letters or essays discuss articles that have recently appeared in Journal or other timely topics of interest to readership. Letters discussing a published article can include supporting information, clarifications, criticisms, corrections, alternate interpretations or perspectives, etc. Letters are sent to corresponding author of article prior to publication. If corresponding or other author wishes, her/his response will be published together with original letter. Essays or letters discussing controversial issues may be sent to persons representing alternative viewpoints for a response.

CORRECTIONS—Notice of errors introduced inadvertently into a manuscript during publication process are published as "Errata." Corrections of errors author's introduced into manuscript or into data analysis or interpretation is published under "Comments," with responses by dissenting authors.

MINIREVIEWS—No page limit, reviews are generally short, reflecting the state-of-the-art of area being reviewed and often include a short history of field; do not include an exhaustive literature survey. Emphasize a particular hypothesis or point of view or personal opinions or interpretation; clearly indicate this focus in paper. Most reviews are solicited by Editor-in-Chief and are not subject to review process. Unsolicited reviews are welcome but will be reviewed.

OTHER

Experimentation, Animal: Conduct studies in compliance with applicable laws and regulations as well as principles expressed in National Institutes of Health, USPHS, Guide for the Care and Use of Laboratory Animals. Use only lawfully acquired animals.

Experimentation, Human: Conform to ethical standards set by Belmont Report, Ethical Principles and Guidelines for the Protection of Human Subjects of Research. Study must have been reviewed and approved by appropriate institutional Review Board (IRB). Present so as to assure preservation of anonymity of subjects.

Experimentation, Recombinant DNA: Must be reviewed and approved as required by applicable laws and regulations.

Page Charges: $30 per page. Payment is not a condition for publication. SEMB members are exempt from page charges.

Reviewers: In cover letter give names and addresses of at least three U.S. scientists with specific expertise for potential use as referees; Editor-in-Chief may not use suggested referees.

Sequence Submission: Submit original nucleotide or amino acid sequences to GenBank and obtain an accession number before publication, Include accession number in manuscript as a footnote or in appropriate figure legend.

Trade Name, Popular Name or Abbreviations: When first used, precede by chemical, scientific or technical name; thereafter, use any name or abbreviation. Capitalize trade names. Use structural formulas of chemicals only when absolutely necessary. Use abbreviations listed at the end of Notice of Contributors. Units of weights, measures, etc., when used in conjunction with numerals, are abbreviated and unpunctuated, e.g., 6 R, 3 9, 5 ml, 8%.

SOURCE
January 1995, 208(1): back matter.

Progress in Cardiovascular Diseases
W. B. Saunders Co.

Edmund H. Sonnenblick, MD
Professor of Medicine and Chief
Division of Cardiology
Albert Einstein College of Medicine
1300 Morris Park Ave, Bronx, NY 10461

Michael Lesch, MD
Department of Medicine
Henry Ford Hospital and Medical Group
E-263, 2799 W. Grand Blvd, Detroit, MI 48202-2689

MANUSCRIPT SUBMISSION

If accepted for publication, copyright in the article, including the right to reproduce the article in all forms and media, shall be assigned exclusively to Publisher.

Submit photographic prints rather than original electrocardiographic tracings or original x-ray films as illustrations.

SOURCE
July/August 1994, xxxvii(1): front matter.

Prostaglandins
Butterworth Heinemann

Dr. Peter Rarnwell
Department of Physiology and Biophysics
Georgetown University
3900 Reservoir Road
Washington, D.C. 20007

AIM AND SCOPE

Prostaglandins invites concise reports of original research in the experimental and clinical aspects of all areas of prostaglandins research. The purpose of this journal is to provide a medium for the rapid communication of advances and new knowledge in this important field. The editor anticipates receiving manuscripts from workers in the following areas of research: chemistry, biochemistry physiology, endocrinology, biology, the medical sciences, and demography. Minireviews on areas of general interest are invited and should consist of not more than 8 manuscript pages. Notice or announcements of future meetings and scientific courses are also encouraged and should be submitted as far in advance of the event as possible.

MANUSCRIPT SUBMISSION

Submit two reprints of article or two copies of manuscript of any report by same author(s) that deals in any respect whatever with same patients, same animals, same laboratory experiments, or same data, in part or in full, as those in submitted manuscripts. Inform Editor of circumstances, similarities, and differences. This includes papers in which a few different patients, animals, laboratory experiments, or data were added to those previously reported. Articles published in another language are not considered.

For manuscripts with two or more authors, each author must qualify by having participated actively and sufficiently in study. Inclusion in authorship list is based on (1) substantial contributions to (a) concept and design, or analysis and interpretation of data and (b) drafting manuscript or revising it critically for important intellectual content; and (2) on final approval by each author of version of manuscript. Conditions 1 (a and b) and 2 must both be met. Recognize contributions of others in an Acknowledgment. In cover letter, confirm that all authors fulfilled both conditions.

MANUSCRIPT FORMAT

Submit original and two copies of text, tables, and figures. Type double-spaced, on good quality bond paper.

Title Page: Center near top of first page with name(s) and affiliation(s) of author(s) just below. For multiple authors and multiple affiliations, use symbols to key affiliations with authors. Indicate name and complete address of corresponding and reprint request author(s) in Acknowledgments/Footnotes section, preceding references.

Text: Include: Abstract (150 words), Introduction, Materials and Methods, Results, Discussion, and Acknowledgment/Footnotes.

Figures: Make figures (photographs, drawings, diagrams, charts) clear, easily legible and cited consecutively by Arabic numerals in text. Type figure legends on one separate page. Legends should contain sufficient detail to permit figure interpretation without reference to text. Indicate units of measure in figures. Submit glossy prints of figures and original prints of photomicrographs. Label on back, in pencil, to indicate figure number, TOP margin and authors. Cost for color plates are charged to authors.

Tables: Make concise and as simple as possible. Cite consecutively by Arabic numerals in text. Type each on a separate sheet. Titles should clearly indicate nature of contents. Include sufficient detail in table footnote to facilitate interpretation.

References: Adhere to specifications of "Uniform Requirements for Manuscripts Submitted to Biomedical Journals" of the International Committee of Medical Journal Editors.

Number consecutively in order of appearance in text. Make citations superscripts without parentheses; on line with parentheses is also permissible. Cite each reference. Reference articles from peer-reviewed, accessible publications. Do not cite unpublished data, personal communications, papers presented at meetings and symposia, abstracts, and manuscripts "submitted for publication" as references. Cite information from such sources in text with sources in parentheses. Cite papers "in press" with journal title and year (if known). Books must include publisher and year of publication (if known).

List references in numerical order in Reference Section, immediately following Acknowledgments section. Abbreviate journal names according to *Index Medicus*, National Library of Medicine. List all authors when six or fewer. When more than six, list first three and add "et al."

(1) Nugteren. D.H., and Hamberg, M. Absolute Configuration of the Prostaglandins. Nature *212*:38.1966.

(2) Ramwell, P.W., ed., *Prostaglandins* Vol. III. Plenum, London, 1969, p. 65.

(3) Karim, S.M.M. In: Prostaglandins, Peptides, and Amines. (P. Mantegazza and E. W. Horton, eds.) Academic Press, London, 1969, p. 65.

OTHER

Abbreviations, Symbols and Terminology: Refer to the *Council of Biology Editors Style Manual* (5th ed., Council of Biology Editors, 1983). Define new or special abbreviations in text of in a footnote at first use. Prostaglandin nomenclature need not be defined. Abbreviate chemical and biochemical terms and abbreviations in accordance with recommendations of IUPAC-IUB Combined Commission on Biochemical Nomenclature. Isotope specifications conform to IUPAC system, with mass number placed as a superscript preceding chemical symbol. For spelling, compounding, and word division, follow either *Webster's Third International* or the *Oxford Dictionary*.

Drugs: Do not use trade names of drugs and chemicals in titles, figures or tables. A trademarked name may be mentioned in parentheses in first text reference to drug.

Electronic Submission: Most MS-DOS software formats are acceptable. Return final edited version on diskette; indicate: name and version number of word processing program used; full name and extension of file containing manuscript; and type of file (ASCII or NON-ASCII). Editorial office will query authors via mail as to whether or not final manuscript will be made available on diskette.

Experimentation, Animal and Human: It is assumed that manuscripts emanating from a particular institution are submitted with approval of the requisite authority. Human experimentation that requires local institutional approval must have approval before experiment is started; indicate approval in Methods section. For reports of experiments on animals, state in Methods section that guidelines for the care and use of the animals approved by the local institution were followed. Name species of nonhuman animals in title, abstract, and key words.

SOURCE

January 1994, 47(1): inside back cover and preceding page.

Protein Engineering
Oxford University Press

Prof. G.A. Petsko
Rosenstiel Center, Brandeis University
415 South Street, Waltham, MA 02254-9110 *or*
Dr. C. Craik
Dept. of Pharmaceutical Chemistry and Biochemistry/Biophysics, University of California, 513 Parnassus Ave.
San Francisco, CA 94143-0446
    For the Americas
Dr T. Imoto
Faculty of Pharmaceutical Sciences
Kyushu University 62, Maidashi, Higashi-ku
Fukuoka 812, Japan *or*
Dr H. Nakamura
Protein Engineering Research Institute
6-2-3 Furuedai, Suita, Osaka 565, Japan
    For Japan
Prof. O. B. Ptitsyn
Institute of Protein Research
Russian Academy of Sciences, 142292 Puschino
Moscow Region, Russia
    For the former Soviet Union
Prof. A. R. Rees
c/o Journals Department, Oxford University Press
Walton Street, Oxford OX2 6DP, U.K.
    For Europe and the rest of the world

AIM AND SCOPE

The field of protein engineering is a new but fast-moving area of biology characterized by the coordinated application of a number of specialized theoretical and experimental disciplines. The objectives of those engaged in this area of research are to investigate the principles by which particular structural features in proteins relate to the mechanisms through which the biological function is expressed, and to test these principles in an empirical fashion by introduction of specific changes followed by evaluation of any altered structural and/or functional properties. The specialized areas alluded to include determination and prediction of protein conformation (secondary and tertiary), experimental studies of protein folding, chemical modification, mutagenesis (generalized and site-specific), and physical and/or biochemical methods for establishing structure-activity relationships.

Protein Engineering provides for rapid publication of full-length papers describing original research of specific relevance to advancing our understanding of the structural and biochemical basis of protein function, employing some or all of the approaches described above. Preference is given to those papers that both satisfy this criterion and report novel findings that significantly advance the field. While the editors encourage submission of papers that employ a multidisciplinary approach, papers describing the application of a single technique are accepted if they are specifically relevant to the field. Papers should be as intelligible as possible to the broad audience likely to be interested in protein engineering and particular attention should be paid to indicating the general conclusions of the study and how it has advanced our understanding of protein structure and function. In this context, the Introduction and Discussion should be used to indicate the novel aspects of the research reported and how it has added to existing knowledge.

In addition to main papers, *Protein Engineering* publishes Review articles, Commentaries, by which readers are kept up to date with results and developments in techniques reported in other journals, and Protocols—short communications giving details of improvements in experimental methods in any field relevant to the study of protein structure and function.

MANUSCRIPT SUBMISSION

Submit manuscript (original and two copies) to appropriate executive editor. Submission of a paper implies that it reports unpublished work and that it is not under consideration for publication elsewhere. If previously published tables, illustrations or more than 200 words of text are included, obtain and submit copyright holder's written permission. Manuscripts may be submitted on floppy disk.

MANUSCRIPT FORMAT

Submit manuscripts in final form so proofs will require only correction of typographical errors. Subdivide manuscripts into: Title page, Abstract, Introduction, Materials and methods, Results, Discussion, Acknowledgments, References, Tables, Legends to figures.

Give each manuscript a brief Introduction to a major section consisting of Materials and methods. Indicate applications of methodology described. Include references only where essential.

Main papers should be six printed pages; to estimate length, allow 1400 words for each full page of printed text, and allow sufficient space for tables/illustrations.

Type on A4 or American quarto paper double-spaced (space between lines not less than 6 mm). Leave 1 in. margins at sides, top and bottom of each page. Number each page top right (Title page is 1). Avoid footnotes; use parenthesis within brackets. Underline words or letters to appear in italics. Identify unusual or handwritten symbols and Greek letters. Differentiate between letter O and zero, and letters I and l and number 1. Mark position of each figure and table in margin.

Title Page: Make title short, specific and informative. Follow surname and initials of each author by his/her department, institution, city with postal code and country. Give changes of address in numbered footnotes. Indicate author(s) to whom proofs are to be addressed. Provide a running title (50 characters).

Key Words: Up to five in alphabetical order, below title, separated by a slash (/).

Abstract: (main papers only) Second page must contain Abstract only; a single 200 word paragraph. Make comprehensible to readers before they have read paper. Avoid abbreviations and reference citations.

Acknowledgments: Include at end of text and not in footnotes. Personal acknowledgments precede those of institutions or agencies.

References: Verify accuracy. Include published articles and those in press (state journal which has accepted them and enclose a copy). In text cite references by author and date; do not cite more than two authors per reference; if more than two, use *et al*. At end of manuscript type citations in alphabetical order, with authors' surnames and initials inverted. References include, in order: authors' names, year, journal title, volume number, inclusive page numbers, and name and address of publisher (for books only). Abbreviate journal names according to *World List of Scientific Periodicals;* underline to indicate italics.

Hamming, J., Arnberg, A. and Gruber, M. (1981) *Nucleic Acids Res,* **9,** 1339–1350.

Koster, H. (ed.) (1980) *Nucleic Acids Synthesis: Applications to Molecular Biology and Genetic Engineering.* IRL Press, London.

Norris, K.E., Norris, F. and Brunfeldt, K. (1980) In Köster, H. (ed.), *Nucleic Acids Synthesis: Applications to Molecular Biology and Genetic Engineering.* IRL Press, London, pp. 233–241.

Have personal communications (J. Smith, personal communication) authorized in writing by those involved; cite unpublished data as (unpublished data). Use both as sparingly as possible and only when data is peripheral rather than central to discussion. Cite references to manuscripts in preparation, or submitted, but not yet accepted, in text as (B. Smith and N. Jones, in preparation); do not include in list of references.

Tables: Type on separate sheets and number consecutively with Roman numerals. Make self-explanatory; include a brief descriptive title. Indicate footnotes by lowercase letters; do not include extensive experimental detail. Use an arrow in text margin to indicate insertion in text.

Illustrations: Refer to all illustrations (line drawings and photographs) in text as Figure 1, etc; abbreviate to "Fig. 1" only in figure legend. Write title of paper, name of first author, and figure number lightly in blue pencil on back of each; indicate TOP. Indicate with an arrow in margin most appropriate position for the figure.

Photographs—Submit in desired final size to avoid reduction. Type area of a page is 244 (height) × 183 mm (width); photographs, including legends, must not exceed this area. Single column is 88 mm wide; double column, 183 mm wide; photographs should fit either. Photographs should be of sufficiently high quality with respect to detail, contrast and fineness of grain to withstand loss of contrast and detail inherent in printing. Indicate magnification by a rule on photograph. When several prints of varying quality of same figure are provided, indicate which should be used for reproduction.

Color Plates—Inclusion is encouraged, particularly for reproduction of computer graphics displays or protein structures.

Line Drawings: Provide as clear, sharp prints, suitable for reproduction as submitted. No additional artwork, redrawing or typesetting is done. Make labelling with a lettering set. Ensure that size of lettering is in proportion with overall dimensions of drawing. Submit line drawings in desired final size to avoid reduction (maximum size 244 × 183 mm including legends) to fit either a single (88 mm) or a double column (183 mm). If line drawings require reduction, check that lettering will be clearly legible after reduction; letters may not be smaller than 1.5 mm high.

Figure Legends—Type on a separate, numbered sheet. Define all symbols and abbreviations. Do not redefine common abbreviations and others in preceding text.

OTHER

Abbreviations: Restrict to use of SI symbols and those recommended by IUPAC; use Angström units (Å) in place of nanometers (nm). Define in brackets after first mention in text. Use standard units of measurements and chemical symbols of elements without definition in body of paper.

Color Charges: £650 maximum for each color page. If more than one page per article or per issue, there may be savings. Total cost of color reproduction in each issue is divided proportionally between all relevant authors.

Conventions: Follow conventions of *CBE Style Manual* (Council of Biology Editors, Bethesda MD, 1983, 5th edn).

Follow *Chemical Abstracts* and its indexes for chemical names. For use of biochemical terminology follow recommendations of IUPAC-IUB Commission on Biochemical Nomenclature, in *Biochemical Nomenclature and Related Documents* (Biochemical Society, U.K.). For enzymes, use recommended name assigned by IUPAC-IUB Commission on Biochemical Nomenclature, 1978, in *Enzyme Nomenclature* (Academic Press, New York, 1980). Use recommended SI (Système International) units.

For bacterial genetics nomenclature follow Demerec *et al.* (1966) *Genetics,* **54,** 61–76. For details of nomenclature to be used for mutant proteins, see *Protein Eng* 1986; 1(1).

Materials Sharing: Authors agree, where feasible, to make freely available to colleagues in academic research any cells, nucleic acids, proteins, etc. used in reported research that are not available from commercial suppliers.

Stereoscopic Images: A 3-picture system enables readers with normal and cross-over stereo vision to visualize stereoscopic images. Submit figures as line drawings or photographs (including color figures). The center image should be a right image flanked by two left images, with a spacing of 65 mm between image centers (maximum overall width 183 mm).

SOURCE

January 1994, 7(1): back matter.

Proteins: Structure, Function, and Genetics
Wiley-Liss

Eaton E. Lattman
Department of Biophysics and Biophysical Chemistry
Johns Hopkins Medical School
Baltimore, MD 21205-2185
Telephone: (410) 955-9275; Fax: (410) 955-0637

AIM AND SCOPE

Proteins: Structure, Function, and Genetics publishes original reports of significant experimental and analytic research in all areas of protein research: structure, function, computation, genetics, and design. The journal encourages reports that present new experimental or computational approaches for interpreting and understanding data from biophysical chemistry, structural studies of proteins and macromolecular assemblies, alterations of protein structure and function engineered through techniques of molecular biology and genetics, functional analyses under physiologic conditions, as well as the interactions of proteins with receptors, nucleic acids, or other specific ligands or substrates. Research in protein and peptide biochemistry directed toward synthesizing or characterizing molecules that simulate aspects of the activity of proteins, or that act as inhibitors of protein function, is also within the scope of *Proteins*. In addition to full-length reports, short communications (usually not more than 4 printed pages) and preliminary crystallographic reports are welcome. Reviews are typically by invitation; authors are encouraged to submit proposed topics for consideration.

MANUSCRIPT SUBMISSION

Send four copies of manuscript to Editor, with three sets of original color or black-and-white illustrations and tables. Suggest an Associate Editor (see inside front cover) who would be appropriate to coordinate review. Include a fax number in cover letter, and indicate whether you are willing to receive reviews and other editorial correspondence at that number.

Manuscripts must be submitted solely to this journal. Manuscripts may not have been published in another publication of any type, professional or lay. Upon acceptance, author will be requested to sign an agreement transferring copyright to publisher, who reserves copyright. No published material may be reproduced or published elsewhere without written permission of publisher and author.

MANUSCRIPT FORMAT

Type double-spaced. Submit original and two copies (including tables and illustrations) on one side of good quality 8.5 × 11 in. paper with 1 in. margins. Number all pages in sequence, beginning with title page. Follow guidelines in CBE Style Manual Committee (*CBE Style Manual: A Guide for Authors, Editors, and Publishers in the Biological Sciences*, 5th ed. rev. and expanded, Bethesda. MD: Council of Biology Editors, Inc.; 1983).

Title Page: Complete title of manuscript, names and affiliations of all authors, institution at which work was performed, name, address and telephone and fax numbers of corresponding author, a short title (45 characters, including spaces), and 5–10 key words not in title to index subject matter of article.

Abstract: 200 words summarizing purpose, methods, results, and major conclusions of work.

Introduction, Materials and Methods, Results, Discussion, and Conclusion: Conform to standard scientific reporting style. Give sufficient data so study can be replicated. Results and Discussion sections may be combined.

References: Cite in text numerically. Provide full titles and complete page numbers of all works cited in reference section. Refer to *CBE Style Manual* for style of reference.

JOURNAL

1. Craik, C., Largman, C., Fletcher, T., Roczniak, S., Barr, P.J., Fletterick, R.J., Rutter, W.J. Redesigning pancreatic trypsin: Alteration of substrate specificity. Science 228:291–297, 1985.

BOOK

2. Nelson, H.C.M., Hecht, M.H., Sauer, R.T. (eds.). Mutations defining the operator-binding sites of Bacteriophage λ repressor. In: "Structures of DNA." Cold Spring Harbor Symposia on Quantitative Biology. Vol. XLVII-Part I. Cold Spring Harbor, New York: The Cold Spring Harbor Laboratory. 1983: 441–449.

Tables and Illustrations: Indicate placement of all tables and illustrations in text. Number tables in order of appearance with Roman numerals; illustrations with Arabic numerals. Give each illustration a legend, and each table a title. Define all abbreviations. Lettering must meet professional standards and be legible after reduction in size. Identify all illustrations with a gummed label on back listing number of illustration, first author's name, short title, and an arrow indicating TOP. Do not submit original recordings, radiographic plates, or artwork. Color is printed at author's expense. Inquire in advance if cost is of concern; upon acceptance of color illustrations, Publisher will provide cost quotes. Either prints or transparencies are acceptable, provide both if possible. Unless otherwise arranged, stereo pairs will be printed to a scale that yields a separation of 55-60 mm between corresponding points in left and right images. Provide each stereo pair in a premounted or composite form; do not submit "cross-eyed" pairs.

OTHER

Crystallographic Studies: *Proteins* has developed a set of guidelines to promote uniform standards for the evaluation of manuscripts that describe structures at the atomic level. Obtain a copy from Editorial Office before preparing manuscript. Deposit atomic coordinates prior to publication; instructions for data deposition are contained in guidelines for crystallographic studies and reports.

Electronic Submission: Submit final, accepted version of manuscript on diskette; submit hard copies for review. Storage medium is 5.25 or 3.5 in. IBM MS-DOS, Windows, or Macintosh format. Preferred software is Word-Perfect; any microcomputer wordprocessor is acceptable. Use of desktop publishing is discouraged; if used, export text to a wordprocessing format. Keep document as simple as possible. Refrain from complex formatting. Submission

of electronic illustrations is encouraged, not required. Submit on a separate disk. EPS, TIFF or native application files are acceptable. For grey scale and color figures contact Wiley-Liss production department. Submit files on SyQuest 44 or 88 megabyte cartridges.

Submit each article as a single file. Name each file with your name (8 letters), followed by a period, plus three-letter extension indicating software used. For Macintosh, follow MS-DOS file naming convention, 8 letters, a period, and a three letter extension. Label diskette with your name, file name and wordprocessing program used. Submit hardcopy; if disk and paper copy differ, paper copy is considered definitive.

SOURCE

1995, 21(1): back matter and inside back cover.

Protoplasma
Springer-Verlag

Dr. R. D. Allen
Pacific Biomedical Research Center
University of Hawaii at Manoa
2538 The Mall, Honolulu, HI 96822

Dr. Y. Anraku
Department of Plant Sciences
Graduate School of Science
University of Tokyo
Hongo, Tokyo 113 Japan

Dr. F.-W. Bentrup
Institut für Pflanzenphysiologie
Universität Salzburg
Heilbrunner Strasse 34
A-5020 Salzburg Austria

Dr. R. A. Bloodgood
Deparment of Anatomy and Cell Biology
School of Medicine
University of Virginia
Box 439
Charlottesville, VA 22908

Dr. R. M. Brown, Jr.
The Department of Botany
The University of Texas at Austin
Austin, TX 78713-7640

Dr. B. E. S. Gunning
Plant Cell Biology Group
Research School of Biological Sciences
The Australian National University
P.O. Box 475
Canberra City, A.C.T. 2601, Australia

Dr. P. K. Hepler
Biology Department
Morrill Science Center
Universityof Massachusetts
Amherst, MA 01003

Dr. R, Kuriyama
Department of Cell Biology and Neruoanatomy
Medical School, University of Minnesota
321 Church Street SE
Minneapolis, MN 55455

Dr. T. Kuroiwa
Department of Plant Sciences,
Graduate School of Science
University of Tokyo, Hongo, Tokyo 113, Japan

Dr. D. J. Morré
Arthur G. Hansen Life Sciences Research Building
Purdue University
West Lafayette, IN 47907

Dr. P. Navas
Departamento de Biologia Celular
Universidad de Cordoba
Avenida San Alberto Magno s/n
E-14004 Cordoba Spain

Dr. J. Pickett-Heaps
School of Botany, University of Melbourne
Parkville, Victoria 3052 Australia

Dr. E. Schnepf
Lehrstuhl für Zellenlehre der Universität Heidelberg
Im Neuenheimer Feld 230, D-W-6900 Heidelberg
Federal Republic of Germany

Dr. F. Wieland
Institut für Biochemie I
Universität Heidelberg
Im Neuenheimer Feld 328
D-69120 Heidelberg, Federal Republic of Germany

AIM AND SCOPE

Protoplasma, an international journal of cell biology, is specializing in the areas between microscopic anatomy and molecular biology where new approaches integrate structural and functional studies at molecular, biochemical and physiological levels. The journal is pleased to consider original manuscripts dealing with cell biology of both single and multicellular organisms, plant, animal or fungal. Appropriate topics include molecular cytology; experimental and quantitative ultrastructure, membrane biology, including biogenesis, dynamics, energetics and electrophysiology; cellular and intracellular transport and trafficking phenomena; the cytoskeleton; organelles, cyto- and histochemistry; and the cell cycle. Also considered are papers concerned with development, differentiation, cellular dynamics, pathology, or pathogenesis and environmental aspects of cell biology. In addition to regular research articles and reviews, *Protoplasma* includes essays of topical interest under the heading "New Ideas in Cell Biology," items of significance to the historical development of cell biology as "Classics Revisited" and provides expedited publications as "Rapid Communications."

MANUSCRIPT SUBMISSION

The journal publishes original papers, rapid communications, and critical review articles. Submission of a manuscript implies that it has not been published before (except as abstract or as part of a published lecture, review, or thesis), that it is not under consideration for publication elsewhere, that its publication has been approved by all authors and by the responsible authorities, tacitly or explicitly, in the institutes where work was done, and that if accepted it will not be published elsewhere without consent of copyright holders. By submitting a authors agree that copyright of article is transferred to published if and when it is accepted for publication.

MANUSCRIPT FORMAT

Submit original and two copies of manuscript. On first page give full name(s) of author(s), institute where paper originated, and exact address to which proofs should be sent marked with an asterisk. Provide original photographs mounted as plates in separate sheets, labeled and numbered, with each copy. Type double-spaced with broad margins on one side of page only. Underline Latin terms and words to be emphasized. Original papers must have a brief summary.

Key Words: List alphabetically 3–6 key words which characterize scope of paper, principal research material(s) and main subjects of work.

Illustrations: Submit on separate sheets, mounted as plates, labeled and numbered, 8 or 17 cm, height 6 to 22 cm. Clear and exact line drawings are required. Supply halftone illustrations as sharp, well-contrasted glossy prints, trimmed at right angles. Provide multi-panel illustrations or other composites as such, not rephotographed, for reproduction. Provide descriptive legends to illustrations. Do not submit more than 4 plates; if editor agrees up to 6 plates may be included. For more than 6, author will be charged DM 268.—for each additional page required.

References: Give at end of paper in alphabetical order:

BOOKS

Family name(s) and initials of author(s), year of publication in parentheses, title of article, editors, book title, volume, publisher, place of publication, and first and last page numbers:

Pavelka M (1987) Functional morphology of the Golgi apparatus. In: Beck F, Hild W, Ortmann R, Pauly JE, Schiebler TH (eds) Advances in anatomy, embryology and cell biology, vol 106. Springer, Berlin Heidelberg New York Tokyo, pp 1–94

JOURNALS

Family name(s) and initials of author(s), year of publication in parenthesis, title of paper, abbreviated name of journal, volume, and first and last page numbers:

Porter KR, Kenyon K, Badenhausen S (1967) Specialization of the unit membrane. Protoplasma 63: 262–274

In text, cite literature by author's name and year of publication. Cite up to two authors in text; if more, give first author followed by "et al." and year.

OTHER FORMATS

RAPID COMMUNICATIONS—Four printed pages and one plate of illustrations covering a single point. Explain in cover letter why publication is urgent.

OTHER

Abbreviations: Submit list of generally understandable abbreviations used in paper.

Color Charges: One color illustration (single figure or multipanel plate) is free; additional figures or plates are DM 1275-.

SOURCE

1994, 183(1–2): front matter.

Psychological Bulletin
American Psychological Association

Robert J. Sternberg
Department of Psychology, Yale University
P.O. Box 208205
New Haven, CT 06520-8205

AIM AND SCOPE

The *Psychological Bulletin* publishes evaluative and integrative reviews and interpretations of substantive and methodological issues in scientific psychology. Original research is reported only for illustrative purposes. Manuscripts dealing with issues of contemporary social relevance; minority, cultural, or underrepresented groups; or other topics at the interface of psychological science and society are welcomed. Original theoretical articles should be submitted to the *Psychological Reviews*, even when they include reviews of research literature. Literature reviews should be submitted to the *Bulletin*, even when they develop an integrated theoretical statement.

MANUSCRIPT SUBMISSION

APA policy prohibits authors from submitting the same manuscript for concurrent consideration by two or more publications. It is a violation of APA Ethical Principles to publish "as original data, data that have been previously published" (Standard 6.24). APA policy prohibits publication of any manuscript that has already been published in whole or substantial part elsewhere. Consult journal editors concerning prior publication of data upon which article depends.

MANUSCRIPT FORMAT

Submit five copies of each manuscript. Present all copies as clear and readable; use good quality paper. A dot matrix or unusual typeface is acceptable if it is clear and legible. Supply addresses, phone and fax numbers, and electronic mail addresses. Prepare manuscripts according to *Publication Manual of the Psychological Association* (3rd ed.). Include an abstract (960 characters and spaces, approximately 120 words) typed double-spaced on a separate sheet. Typing instructions and instructions on preparing tables, figures, references, metrics, and abstracts appear in the *Manual*.

Masked review is an option. If not requested in cover letter, it is the prerogative of the processing editor. If requesting masked review, include with each copy of manuscript a cover sheet, with title of manuscript, authors' names and institutional affiliations, and date of submission. Omit author's name and affiliation on first page, but include title and submittal date. Place footnotes containing author information on separate pages. Ensure that manuscript itself contains no clues to authors' identity.

OTHER FORMATS

METHODOLOGICAL CONTRIBUTIONS—Descriptions of quantitative methods and research designs, whether expository or critical. Articles on broadly applicable methods are encouraged; spell out range of application and limitations. A detailed description of policies on methodological contributions appears in the January 1987 issue of *Psychological Bulletin* (101(1): 3–4).

SUBSTANTIVE CONTRIBUTIONS—Integrative reviews that summarize a literature may set forth major developments within a particular research area or provide a bridge between related specialized fields within psychology or between psychology and related fields. Reviews that develop connections between areas of research are particularly valuable. Expository articles may be published if they are deemed accurate, broad, clear, and pertinent.

OTHER

Experimentation, Animal and Human: State in writing compliance with APA ethical standards in treatment of sample, human or animal, or describe details of treatment. Obtain a copy of APA Ethical Principles from APA Ethics Office, 750 First Street, NE, Washington, D.C. 20002-4242.

Materials Sharing: After publication, authors may not withhold data on which conclusions were based from other competent professionals who seek to verify substantive claims through reanalysis and who intend to use data only for that purpose, provided that participant confidentiality can be protected and unless legal rights concerning proprietary data preclude their release (Standard 6.25). Authors are expected to have available data throughout editorial review process and for at least 5 years after publication.

SOURCE

January 1994, 115(1): front matter.

Psychological Medicine
Cambridge University Press

Professor Eugene Paykel
Department of Psychiatry
University of Cambridge, Addenbrooke's Hospital
Cambridge CB2 2QQ U.K.

AIM AND SCOPE

Psychological Medicine is a journal primarily for the publication of original research in clinical psychiatry and the basic sciences related to it. These comprise not only the several fields of biological enquiry traditionally associated with medicine but also the various psychological and social sciences, the relevance of which to medicine has become increasingly apparent.

MANUSCRIPT SUBMISSION

Submit three copies of text, tables and figures. Copies other than first may be Xeroxed. In addition to longer articles, preliminary and brief communications of 1500–2500 words are accepted.

Submission of a paper implies that it contains original work that has not been previously published and that it is not being submitted for publication elsewhere.

MANUSCRIPT FORMAT

Type manuscripts double-spaced on one side of paper with wide margins throughout, including references and notes; number consecutively. On first page (title sheet), give: title and short title for running head (60 characters); authors' names; and department in which work was done. Provide a short synopsis of about half an A4 page double-spaced at the beginning. Indicate name of corresponding author with full postal address in footnote. Place acknowledgments at end of text (before Reference section).

Indicate italic type by underlining and bold type by wavy underlining. Follow foreign quotations and phrases by a translation.

References: Use Harvard (author-date) system in text. Give a complete list of References cited at end of article. In a text citation of a work by more than two authors cite first author's name followed by *et al.*; give names of all authors in References section. Where several references are cited together, list in rising date order. Type References section in alphabetical order on a separate sheet. Give Journal titles in full.

Brown, G.W. (1974). Meaning, measurement and stress of life events. In *Stressful Life Events: Their Nature and Effects* (ed. B. S. Dohrenwend and B. P. Dohrenwend), pp. 217–244. John Wiley: New York.

Brown, J. (1970). *Psychiatric Research*. Smith: Glasgow.

Brown, J., Williams, E. & Wright, H. (1970). Treatment of heroin addiction. *Psychological Medicine* **1**, 134–136.

Figures and Tables: Include only if essential. Provide un-mounted photographs on glossy paper. Letter magnification on these. Trim prints to column width (70 mm).

Diagrams: Do not include in text. Submit in a form suitable for direct reproduction (printed version is normally reduced to 70 mm wide). Letter in either Letraset or stencil; make lettering and symbols of comparable size. Refer to all photographs, graphs, and diagrams as figures; number consecutively in Arabic numerals. Mark figure number on back of the photograph or artwork with name of author and paper title. Type captions for figures double-spaced on separate sheets.

Tables: Number consecutively in text in Arabic numerals. Type each on a separate sheet after Reference section. Type titles above table.

OTHER

Abbreviations: Spell out in full any abbreviations used.

Statistical Analyses: Check guidelines for presentation in: Altman, D.G., Gore, S.M., Gardner, M.J. & Pocock, S.J. (1983). Statistical guidelines for contributors to medical journals. *British Medical Journal* **286**, 1489–1493.

Units: Use S.I. units throughout in text, figures and tables.

SOURCE

February 1994, 24(1): inside back cover.

Psychological Review
American Psychological Association

Robert A. Bjork
Department of Psychology, University of California
Los Angeles, CA 90024-1563

AIM AND SCOPE

Psychological Review publishes articles that make important theoretical contributions to any area of scientific psychology. Preference is given to papers that advance theory rather than review it and to statements that are specifically theoretical rather than programmatic. Papers that point up critical flaws in existing theory or demonstrate the superiority of one theory over another will also be considered. Papers devoted primarily to surveys of the literature, problems of method and design, or reports of empirical findings are ordinarily not appropriate. Discussions of previously published articles will be considered for publication as Theoretical Notes on the basis of the scientific contribution represented.

MANUSCRIPT SUBMISSION

Submit manuscript in quadruplicate to Editor.

Prepare manuscripts according to the *Publication Manual of the American Psychological Association* (4th ed.).

APA policy prohibits submitting the same manuscript for concurrent consideration by two or more publications. It is a violation of APA Ethical Principles to publish "as original data, data that have been previously published" (Standard 6.24). This is a primary journal that publishes original material only. Do not submit material that has already been published in whole or substantial part elsewhere. Consult journal editors concerning prior publication of any data upon which article depends. APA Ethical Principles specify that "after research results are published, psychologists do not withhold the data on which their conclusions are based from other competent professionals who seek to verify the substantive claims through reanalysis and who intend to use such data only for that purpose, provided that the confidentiality of the participants can be protected and unless legal rights concerning proprietary data preclude their release" (Standard 6.25). Make data available throughout the editorial review process and for at least 5 years after date of publication.

MANUSCRIPT FORMAT

Include an abstract (960 characters and spaces, approximately 120 words) typed double-spaced on a separate sheet. For typing instruction on preparation of tables, figures, references, metrics and abstracts, see *Publication Manual of the American Psychological Association* (4th ed.).

All manuscripts are subject to editing for sexist language.

All copies should be clear, readable, and on paper of good quality. A dot matrix or unusual typeface is acceptable only if clear and legible. Do not submit dittoed or mimeographed copies. In addition to addresses and phone numbers, supply electronic mail addresses and fax numbers.

Masked Reviews are optional; request when submitting manuscript. Include a separate title page with author(s) name and affiliation; these should not appear anywhere else on manuscript. Type footnotes that identify author(s) on a separate page. Make sure manuscript contains no clues to identifies.

OTHER

Experimentation, Animal and Human: State in writing compliance with APA ethical standards in treatment of sample, human or animal, or describe details of treatment. Obtain a copy of APA Ethical Principles from APA Ethics Office, 750 First Street, NE, Washington, D.C. 20002-4242.

Materials Sharing: After publication, authors may not withhold data on which conclusions were based from other competent professionals who seek to verify substantive claims through reanalysis and who intend to use data only for that purpose, provided that participant confidentiality can be protected and unless legal rights concerning proprietary data preclude their release (Standard 6.25). Authors are expected to have available data throughout editorial review process and for at least 5 years after publication.

SOURCE

April 1995 102(2): p. 210.

Psychopharmacology
Springer International

Daniel E. Casey, M.D.
Psychiatry Service
P.O. Box 1034, V.A. Medical Center
Portland, OR 97207
    Clinical psychopharmacology/North America and South America

J. Gerlach, M.D.
Department P, St. Hans Hospital
DK-4000 Roskilde, Denmark
    Clinical psychopharmacology/The rest of the world

David R. Sibley, Ph.D.
Experimental Therapeutics Branch
National Institute of Neurological Disorders & Stroke
National Institutes of Health
Building 10, Room 5C-108
Bethesda, MD 20892 USA
    Biochemical neuropharmacology, drug metabolism and pharmacokinetics/North America and South America

J. Leysen, Ph.D.
Janssen-Pharmaceutica
Department of Biochemical-Pharmacology
Turnhourtseebaan 30, B-2340 Beerse, Belgium
    Biochemical neuropharmacology, drug metabolism and pharmacokinetics/The rest of the world

Klaus A. Miczek, Ph.D.
Research Building, Tufts University
490 Boston Ave., Medford, MA 02155
    Behavioral pharmacology in laboratory animals/North America and South America

T.W. Robbins, Ph.D.
Department of Experimental Psychology
University of Cambridge
Downing Street, Cambridge CB2 3EB, U.K.
    Behavioral pharmacology in laboratory animals/The rest of the world

Mark A. Geyer, Ph.D.
Department of Psychiatry, 0804
School of Medicine, University of California San Diego
La Jolla, CA 92093-0804
    Preclinical psychopharmacology/North America and South America

A.R. Green, Ph.D.
Astra Neuroscience Research Unit
1 Wakefield Street
London WC1N 1PJ, U.K.
    Preclinical psychopharmacology/The rest of the world

H. de Wit, Ph.D
Department of Psychiatry
University of Chicago MC3077
5841 S. Maryland Ave, Chicago, IL 60637
    Alcohol and substance abuse in humans/Human cognitive psychopharmacology/North America and South America

D.M. Warburton
Department of Psychology
University of Reading
Building 3, Early Gate
Whiteknights, Reading RD6 2AL, U.K.
    Human cognitive psychopharmacology/Alcohol and substance abuse in humans/The rest of the world

AIM AND SCOPE

Psychopharmacology is an international journal of research and scholarship, the aims of which are to cover the general area of elucidating mechanisms by which drugs affect behavior, both terms intended in the broadest sense. The scope of the Journal extends from clinical psychopharmacology (including trials), to experimental studies on the effects of drugs on cognition and behavior in humans and laboratory studies in experimental animals. The research methodologies (techniques covered) may range from neurochemical assays and electrophysiological recording to studies of functional neuro-imaging, and the methods of experimental psychology and ethology, as well as clinical neurology and psychiatry. The Journal is particularly interested in articles that integrate these levels of analysis.

MANUSCRIPT SUBMISSION

Follow APA policy regarding duplication of submission and publication. Do not submit the same manuscript for concurrent consideration by two or more journals. Do not submit a manuscript that has already been published in whole or in substantial part in another journal.

Author(s) transfer(s) copyright to his/their article to Springer-Verlag effective if and when article is accepted for publication. Copyright covers exclusive and unlimited rights to reproduce and distribute article in any form and any language.

Submit manuscripts directly to Managing Editor of category for which manuscript seems most appropriate.

Clinical psychopharmacology encompasses studies of therapeutic drug effects on patients. *Human experimental psychopharmacology* refers to experiments on normal human subjects. *Behavior pharmacology* presents experimental analysis of drug effects on behavioral variables in animals, including interaction of drugs with learned modification of behavior and multivariate analysis of unconditioned behavior; for example, with schedule-controlled responding or operant behavior, learning, memory, discrimination, and the microstructure of unconditioned behavior. *Preclinical psychopharmacology* includes studies of the pharmacological actions of drugs using unconditioned or spontaneous behavior as an assay or screening test in animals; including, for example, hotplate or tail-flick tests or analgesia, sleep, spontaneous motor activity, neuropharmacological or physiological measurements, such as receptor binding and EEC recording, and observations in the natural environment.

Manuscripts which fall into more than one category or about which the author has some doubt should be sent to Miczek or Robbins, who will take appropriate action.

Disclose any commercial or other association that might pose a conflict of interest. Acknowledge all funding sources supporting work and institutional or corporate affiliations of authors on title page.

MANUSCRIPT FORMAT

Submit manuscripts and figures in quadruplicate (i.e., original plus three copies). Present text as clear and concise as possible. Do not exceed 6 printed pages (15 typed pages). Carefully check both form and content of paper to exclude need for excessive corrections in proof.

Type double-spaced, on one side of letter-size paper with a wider margin on left. Mark margin for insertion of figures and tables. Number all pages consecutively. Type (double-spaced) tables, footnotes, and legends for illustrations on separate pages. Use metric system for text and figures.

First page: Title; name(s) of author(s); name of laboratory where work was done; footnotes to title (marked by asterisks); and a telex or fax number. Begin paper with a brief Abstract (200 words). For original investigations state problem concisely and arrange material under Materials and Methods (marked for small print), Results, Discussion, and References.

Key Words: Immediately following abstract, supply some relevant key words for subject indexing.

Footnotes: Keep to a minimum; number consecutively throughout paper.

References: Only include work cited in text; list sources alphabetically at end of paper.

JOURNALS

K.A. Perkins, A. Marco, R.L. Stille (1994) Nicotine discrimination in smokers. Psychopharmacology 116: 407-413

BOOKS

Friede RL (1989) Developmental neuropathology, 2nd edn. Springer, Berlin Heidelberg New York

SINGLE CONTRIBUTION IN A BOOK

Sjöqvist F (1989) Pharmacogenetics of antidepressants. In: Dahl SG, Gram LF (eds) Clinical pharmacology in psychiatry (Psychopharmacology series, vol 7). Springer, Berlin Heidelberg New York, pp 181–191

When several publications by the same author or group of authors, list in chronological order; distinguish those that appear in same year by a, b, c, etc. Give citations in text in parentheses, e.g., (Beckman 1958, Brown 1963), except when author's name is part of a sentence, e.g., "Mann (1966) reported that . . ." Where two authors, name both, with three or more give first author's name plus "et al."

Figures: Use with discretion only to clarify or reduce text. Do not repeat information in captions in text. Previously published illustrations are not accepted. Color illustrations are accepted; authors are expected to make a contribution. Mention figures and graphs in text; number with Arabic numerals. Provide a brief descriptive legend for each figure; legends are part of text; append to it. Submit figures in a form suitable for reproduction separately from text. Using a soft pencil, indicate TOP of figure on back with author and figure number.

Line Drawings—Submit good-quality glossy prints in desired final size. Make inscriptions clearly legible. Letters (capitals) 2 mm high are recommended.

Half-tone Illustrations—Provide well-contrasted photographic prints (not photocopies) trimmed at right angles in desired final size. Inscriptions should be about 3 mm high.

Figure, including legend, should not exceed print area (17.8 × 24 cm, 8.6 cm column width). Group several figures into a plate on one page. If line drawings or halftones are reduced, state alternative scale desired. Publisher may reduce or enlarge illustrations.

OTHER FORMATS

LETTERS TO THE EDITORS—3 typed pages, technical comments or points of controversy.

RAPID COMMUNICATIONS— 5 typed pages, one figure or table, plus a few references and 100 word abstract. Important new results worthy of immediate publication. Present current results of special interest with minimal delay.

REVIEW ARTICLES—35 typed pages including references.

OTHER

Color Charges: Approx. DM 1200 for the first and DM 600 for each additional page.

Experimentation, Animal: State that "Principles of Laboratory Animal Care" (NIH Publication No. 85-23, revised 1985) and specific national laws were followed.

Experimentation, Human: State in text that studies were approved by appropriate ethics committee and were performed in accordance with ethical standards of 1964 Declaration of Helsinki. State that all persons gave their informed consent. Omit details that might identify subjects.

SOURCE

January 1995, 117(1): A11, inside back cover.

Psychophysiology
Society for Psychophysiological Research

John T. Cacioppo
Department of Psychology
Ohio State University
1885 Neil Avenue
Columbus, OH 43210-1222
Telephone: (614) 292-1916; Fax: (614) -292-5326;
E-mail: cacioppo.1@osu.edu

AIM AND SCOPE

Psychophysiology publishes original full-length articles in any area of psychophysiological research: experimental studies, theoretical papers, evaluative reviews of literature, and methodological developments (e.g., experimental procedures, instrumentation, and computer techniques). Archival documents of the Society for Psychophysiological Research are also published in the journal. We also provide a special publication option—Special Reports—for short empirical articles of special interest. Book reviews and Letters to the Editor are also considered, again subject to space availability. Consult recent editorials in the journal for further information about editorial policies and review procedures (volume 32, pp. 1-3).

MANUSCRIPT SUBMISSION

Submit an original and four copies of each manuscript. Manuscript cannot have been published previously, nor be under review for publication elsewhere. For papers with multiple authors, all authors must approve submitted manuscript and concur with its submission. Senior authors must sign and return Copyright Transfer Agreement, with certain specified rights reserved by author, prior to publication.

In a cover letter, indicate whether manuscript is submitted for consideration as a regular, full-length article Brief Report or Special Report. Provide telephone and fax number, and address for electronic mail.

MANUSCRIPT FORMAT

Do not exceed 30 pages of text (12 printed pages), except in cases of reports of multi-experiment studies or comprehensive literature reviews.

To qualify as a Special Report, do not exceed 3500 words including references, tables, figures, figure legends, and abstract. Each page of tables and each figure corresponds to 250 words.

Type entire manuscript double-spaced on 8.5 × 11 in. or A4 paper. Follow guidelines in *Publication Manual of the American Psychological Association* (3rd edition, American Psychological Association, Washington, D.C.). Arrange pages of manuscript as follows:

Title Page: (First Page) Title, names and addresses of authors, including academic or other affiliations, acknowledgments and support, and name and address for requests for reprints.

Running Title: (Second Page) 45 characters and spaces.

Abstract: (Third Page) 150 words, state problem, method, results, and conclusions of experimental and methodological articles or a summary of major issues, source of observations, and conclusions of theoretical articles. Follow by six descriptor terms.

(Fourth Page) Begin first page of text proper with full title and names of authors followed by beginning of introductory section.

Unless compelling reasons for variation, continue with a Methods section, Results section, and Discussion section, with subsections as needed. Further sections begin on a new page: References, Footnotes, Tables, Figure Legends, and Figures. Put each table and figure on a separate page.

Figures: Good quality laser-printed copies or photocopies are acceptable. Identify by figure number, first author's name, and TOP on back of both originals and copies. Construct figures with notations and data points of sufficient size to permit legible photo reduction to one column of a two-column page. No character should be smaller than 1 mm wide following reduction. Color figures are accepted. Extra costs are paid by author.

OTHER

Abbreviations: Except for widely used abbreviations, use is discouraged, even if defined. Use only if they contribute to better comprehension, which is rare. Use metric system (SI) units.

Effect Size: When describing results, report measures of effect size in addition to probability values.

Electronic Submission: For Special Reports provide disk versions (IBM PC compatible or Macintosh) of manuscript at first submission. For regular articles submit disk version following acceptance. Include original (native) word-processing file and ASCII file.

Experimentation, Human: In cover letter state that informed consent was obtained and that the rights of subjects were protected. If infrahuman subjects were used, indicate that they were treated in accordance with institutional guidelines.

Repeated Measures: Consult Methodology section of *Psychophysiology*, July 1987 pp. 474–478, for journal policy concerning repeated measures designs. Correct format for reporting results of ANOVAs with repeated measures and more than 2 degrees of freedom is: $F(29,522) = 2.89$, $p < .05$, $_e = .0998$. Uncorrected degrees of freedom are provided along with correction factor (epsilon).

SOURCE

January 1995, 32(1): inside front cover.

Psychosomatic Medicine
American Psychosomatic Society
Williams & Wilkins

Joel E. Dimsdale, M.D
University of California, San Diego
9500 Gilman Drive
La Jolla, CA 92093-0804

AIM AND SCOPE

The journal welcomes original research articles, reviews, and case reports.

MANUSCRIPT SUBMISSION

Manuscripts must be original, may not have been published other than as an abstract and may not be under simultaneous review elsewhere. Describe in a few words paper's objectives and significance. All co-authors should sign letter of submission. Upon acceptance, author(s) will be asked to transfer copyright to American Psychosomatic Society.

MANUSCRIPT FORMAT

Submit original and 4 copies (5 copies all together) typed double-spaced, including references, on 8.5×11 in. paper. Editor welcomes, but is not bound by, suggestions of referees.

On cover page, include title, full names of authors with degrees and academic or professional affiliations, and complete address, telephone and fax numbers of corresponding authors. Number pages consecutively beginning with abstract page. Do not exceed 35 pages including references, tables, and figures. If title exceeds 45 characters, supply an abbreviated running title on cover page.

Abstract: 250 words in outline format, using underscored headings: Objective, Method, Results, and Conclusions; follow by 6 key words for indexing.

Tables: Double-space, including all headings. Give descriptive titles. Number sequentially in Arabic numerals. Use only horizontal rules.

Illustrations: For line artwork, submit black ink drawings of professional quality, or high-contrast glossy photographs of original drawings. On back of each, give its number, author's name, and indication of top. Include a separate sheet of legends, typed double-spaced.

Footnotes: Indicate footnotes to by Arabic numeral superscripts numbered consecutively throughout paper and placed at foot of each page on which they are cited.

References: In text, cite by full sized numbers in parentheses. List in order cited in text. Number consecutively, using Arabic numerals. Type double-spaced and place at end of text on separate pages.

JOURNALS

Mei-Tal V, Meyerowitz S, Engel GL: The role of psychological process in a somatic disorder: Multiple sclerosis. Psychosom Med 32:67–86, 1970

BOOK

Kao F: Introduction to Respiratory Physiology. New York, Elsevier, 1973, 50–61

EDITED BOOK

Lewi PJ: A model of coronary collateral circulation in dogs. In Schaper W (ed), The Collateral Circulation of the Heart, Vol 1. New York, Elsevier, 1971, 181–211

Periodical abbreviations follow those in *Index Medicus*.

OTHER FORMATS

RAPID COMMUNICATION—Text, including references and 1 table, is less than 8 pages and manuscript does not require major revision. If major revision is required, manuscript will not be processed as a Rapid Communication.

SOURCE

January/February 1995, 57(1): front matter.

Quarterly Journal of Medicine

Association of Physicians of Great Britain and Ireland

Oxford University Press

Dr. Julian Hopkin, Executive Editor
Quarterly Journal of Medicine
Churchill Hospital, Oxford OX3 7LJ
Telephone: 0865 225225; Fax: 0865 67659

AIM AND SCOPE

The journal welcomes for publication contributions that promote medical science and practice: (1) original articles, describing either clinical research or basic scientific work relevant to medicine; (2) review articles on significant advances or controversies in clinical medicine or medical science; and (3) correspondence commenting on articles published in the journal.

MANUSCRIPT SUBMISSION

Present Manuscripts as 2 hard copies and one copy on computer disk, together with fax and telephone numbers.

Articles published become the property of the Journal and can be republished in full or part only with Editor's permission.

MANUSCRIPT FORMAT

Type double-spaced. Original manuscripts include a 200 word summary and sections of Introduction; Subjects or Materials and methods; Results; and Discussion.

Most popular wordprocessor formats are acceptable; if possible supply 3.5 in. or 5.25 in. MS-Dos or AppleMac disks using Wordperfect or Microsoft Word. Type copy in style and order of journal.

References: Number consecutively by superscript Arabic numerals in order of mention in text. Where references are cited in a table or figure legend, correspond numbering to first mention of table or figure in text. Follow Vancouver system for references. Include all authors, title of article, title of journal (as abbreviated in *Index Medicus*), year of publication, volume number, first and last pages. For book references give author, title, edition, volume and page number (where appropriate), with town of publication, name of publisher and year of publication.

Evans DGK and Harris R. Heterogeneity and genetic conditions. *Q J Med* 1992; **84**:563–5.

West JB. *Blood Flow and Gas Exchange* (2nd edn). Oxford, Blackwell, 1970.

Illustrations: Present line figures in black ink with symbols large enough to be legible when reduced to text size. Do not mount photographs and other illustrations; do not mark front surface. Mark in soft pencil on back authors' names and figure number and TOP.

Where a special point is indicated by arrows, stippling, etc. show these on a transparent overlay or by Letraset arrows. If color plates, or large numbers of half-tones are required, authors may be charged with part of cost. Type legends, clearly identified for appropriate figures, on a separate sheet.

SOURCE

January 1995, 88(1): back matter.

Quarterly Review of Biology
State University of New York at Stony Brook
University of Chicago Press

The Editors, *The Quarterly Review of Biology*
Division of Biological Sciences
State University of New York
Stony Brook, NY 11794-5275
Telephone: (516)632-6977; Fax: (516)632-9282;
E-mail: QRB@sbbiovm.bitnet

AIM AND SCOPE

The review articles published by *The Quarterly Review of Biology* are intended especially for the informed general biologist, and should present not only a synthesis of recent investigations but also a critical evaluation of them. Theoretical papers should include a critical synthesis of the literature bearing on the theory and promote further research. Both invited and unsolicited articles are considered by *The Quarterly*, and the Editors encourage the submission of unsolicited articles that conform to the objectives of the journal.

MANUSCRIPT SUBMISSION

Submit an original and two copies of all articles. Write in concise and sufficiently nontechnical language to be intelligible both to specialists in other fields and to general biologists. Interpretive diagrams are desirable.

MANUSCRIPT FORMAT

See recent issues as a model for style with respect to: section headings in text; literature citations in text (note that name-and-date citation form is used); and list of references. Give article a brief abstract that summarizes its principal conclusions.

Send glossy prints of illustrations; do not send original artwork unless requested.

Footnotes to textual material are not permitted; explanatory material of secondary importance may be set off in the text in small print.

Limit reference citations to those essential for documenting specific statements. Abbreviate journal titles according to *BIOSIS List of Serials* (BioSciences Information Service of Biological Abstracts, Philadelphia, PA). Give all entries full title and inclusive page numbers. Provide photocopies of all quoted passages, including page numbers.

OTHER

Electronic Submission: Submission of manuscripts on computer disks or via BITNET is encouraged. File must be IBM compatible or transformable into an ASCII file. Inquire for details.

Page Charges: $30 per journal page for first 20 journal pages from authors who attribute support to a funding agency with a policy of publication support; $60 for each journal page in excess of 20. Authors not supported by funding agencies and authors of invited articles are eligible for 20 free journal pages (approximately 50 double-spaced typed manuscript pages or equivalent in bibliographic or illustrative matter). When such articles are longer than 20 journal pages, authors will be charged $60 for each extra page.

Style: Follow recommendations of *The Chicago Manual of Style* (13th Edition, The University of Chicago Press, Chicago,); in respect to form of references cited, see a recent issue of the journal, or request from the office of *The Quarterly* a detailed set of instructions for preparation of reference list.

SOURCE

March 1995, 70 (1): inside back cover.

Radiation Research
Radiation Research Society

Dr. R. J. M. Fry
Biology Division
Oak Ridge National Laboratory
P. O. Box 2009, Oak Ridge, TN 37831-8077
Telephone: (615) 574-5874; Fax: (615) 576-4149
Biology Division
Y-12 Plant, Bear Creek Road, Building 9207
Oak Ridge, TN 37830
          Courier service

AIM AND SCOPE

Radiation Research will publish original articles dealing with radiation effects and related subjects in areas of physics, chemistry, biology and medicine. The term radiation is used in its broadest sense and includes specifically ionizing radiation and ultraviolet, visible and infrared light as well as microwaves, ultrasound and heat. Effects may be physical, chemical or biological. Related subjects include (but are not limited to) dosimetry methods and instrumentation, isotope techniques and studies with chemical agents contributing to the understanding of radiation effects.

MANUSCRIPT SUBMISSION

Manuscripts are accepted for review with the understanding that the same work has not been published, that it is not presently submitted elsewhere, and that its submission for publication has been approved by all authors and by institution where work was carried out; that persons cited as a source of personal communications have approved such citation. Include statement in cover letter: "Submission of this manuscript has been approved by all authors."

Include copies of unpublished articles (those submitted or in press) containing information crucial to contents of manuscript under review.

If accepted for publication, copyright in article, including right to reproduce article in all forms and media, shall be assigned exclusively to Radiation Research Society. Radiation Research Society will not refuse any reasonable request by author for permission to reproduce any of his or her contributions to journal.

MANUSCRIPT FORMAT

Do not exceed 30 double-spaced typed pages including references, footnotes, tables, figures and legends. Type on one side only of 8.5×11 in. white bond paper with 1 in. margins on all sides. Include original and 3 copies, complete with figures, tables, etc.

Page 1: Article title, author name(s) and complete affiliation(s), number of copies submitted, and number of figures and tables. Page 2: Running head (abbreviated form of title, 50 characters including letters and spaces); list name, complete mailing address, and telephone and fax numbers and e-mail address of corresponding author. Page 3: Abstract (200 words); do not use proprietary terms and abbreviations. Put authors' names and article title at top of abstract:

Bedford J. S. and Mitchell, J. B. Dose-Rate Effects in Synchronous Mammalian Cells in Culture *Radiat. Res.*

Tables: Number consecutively with Roman numerals. Type double-spaced on individual sheets; consider proportions of printed page in designing table. Identify footnotes to tables with superscript lowercase italic letters; place footnotes at bottom of page containing table.

Footnotes: Indicate footnote material by superscript Arabic numerals; cite consecutively throughout article starting with title. Type double-spaced on a separate page.

Figures: Consider all illustrations as figures. Number each graph, drawing or photograph in sequence with Arabic numerals. Plan to fit proportions of printed page (7 1/8 $\times$ 9 in; column width 3.5 in.), with originals no larger than 8.5×11 in. Identify each figure on back with author's name and figure number. Type figure legends double-spaced on a separate sheet.

Artwork: Line drawings must be of a sufficient quality for reproduction. Lines should not be less than 1 point in weight. Submit high-quality high-contrast computer graphics or glossy prints of original line drawings. Present lettering on drawings of professional quality or generated by high-resolution computer graphics and large enough (10–12 points) to take a reduction of 50 to 60%. Freehand, penciled or typewritten lettering is not acceptable. Make symbols used to identify points within a graph large enough to be distinguishable after reduction. Submit halftone photographs as single glossy prints with strong contrast. Do not use paper clips. Match contrast and density of all figures on a single plate. Indicate magnification scales on photographs by means of bars (⊢). Color plates will be paid for by author.

References: Cite in text by Arabic numerals in parentheses (in order of appearance). Type double-spaced on a separate page in numerical order. Limit to material in open literature. Give reports, private communications, etc. as footnotes with adequate information as to source and availability. References should be appropriate and not unnecessarily numerous. Identify unpublished results and private communications directly in text, in parentheses.

1. T. Stamato and N. Denko, Asymmetric field inversion gel electrophoresis: A new method for detecting DNA double-strand breaks in mammalian cells. *Radiat. Res.* **121,** 196–205 (1990).

2. D. M. Bates and D. G. Watts, *Nonlinear Regression Analysis and Its Applications.* Wiley, New York, 1988.

3. F. J. Burns and R. E. Albert, Dose-response for radiation-induced cancer in rat skin. In *Radiation Carcinogenesis and DNA Alterations* (F. J. Burns, A. C. Upton and G. Silini, Eds), pp 51–70. Plenum, New York, 1986.

Abbreviate journal names according to *Chemical Abstracts Service Source Index,* 1985. Give inclusive pagination.

OTHER FORMATS

BRIEF RESEARCH NOTE—Original data that do not merit treatment as a full-length paper, or new research that author does not expect to pursue in foreseeable future.

COMMENTARIES—Deal with subjects of interest that do not fall into other categories. Commentary should be of interest, important and factually correct; it can express unaccepted views that have not been shown to be unequivocally incorrect. Commentary can be inflammatory but not defamatory, Commentaries are reviewed to ensure that facts are correct but not to eliminate hypotheses or reasonable speculation.

LETTERS TO THE EDITOR—Substantive comments on papers and Editorials published in journal. Select most appropriate format for presenting communication.

RAPID COMMUNICATIONS—Short articles that report a new method or technique and new concepts that are testable. The aim is to provide a more rapid publication for those papers that provide new approaches to technical and conceptual problems and that may be useful and timely.

SHORT COMMUNICATIONS—Brief reports and other manuscripts that do not fit overall requirements for general research papers; including pertinent discussion of current problems in the field of radiation research as well as contributions of a didactic nature. Topicality and general significance of subject are of primary importance; original research data is not a prerequisite.

Do not exceed 12 double-spaced typewritten pages including references, footnotes, tables, figures and legends. Adhere to format outlined for general research papers.

OTHER

Electronic Submission: Submit final version of paper, including text, tables and figures if possible, on diskette. Preferred text format is Microsoft Word; other formats are acceptable; a list of text and graphics formats that can be submitted is included with letter of acceptance.

Experimentation, Animal: Indicate that guidelines for humane treatment of animals that have been followed; for example, approval by institution's animal use committee.

Units: Use International System of Units (SI). Do not use centigray or centisievert.

SOURCE

January 1994, 137(1): back matter.

Radiology
Radiological Society of North America

Stanley S. Siegelman, MD
Johns Hopkins Medical Institutions
550 N Broadway, Suite 206, Baltimore, MD 21205
Telephone: (410) 327-0124; Fax: (410) 276-0353

MANUSCRIPT SUBMISSION

Radiology is published under the supervision of the Board of Directors of the Radiological Society of North America. Inc, which selects all material submitted for publications. Instructions are in accord with *Uniform Requirements for Manuscripts Submitted to Biomedical Journals (JAMA* 1993; 269:2282–2286).

Send copyright agreement with initial submission of all manuscripts for regular articles, editorials, letters, and technical notes. Use a copy of form that appears in January issue. All authors must sign agreement. In the case of officers or employees of U.S. government, the Society recognizes that works prepared as part of official government duties are in public domain, but they still must sign copyright agreement.

Describe any direct or indirect financial interest author may have in subject matter of a submitted manuscript, and authorize Society to publish financial disclosure with article if appropriate. The Society's Financial Disclosure policy: "Each author shall describe (a) any direct financial interest which that author has in subject matter discussed in submitted manuscript, and (b) any affiliation or financial involvement which that author has with or in any organization with a direct financial interest in subject matter discussed in submitted manuscript. Such information will be held in confidence of journal editor during review process. If necessary, in the editor's view, this information may be shared with reviewers after discussion between editor and author. Author shall grant permission for RSNA to publish financial information described above, or an appropriate summary thereof, with manuscript if accepted for publication."

No manuscript on same or similar material may have been or will be submitted to another journal before work appears in *Radiology.*

Obtain written permission from publisher and author to reproduce previously published figures; note such material and give source in manuscript. Submit letters of permission with recognizable photographs of patients or technologists; otherwise, block out eyes to prevent identification.

MANUSCRIPT FORMAT

Type double- or triple-spaced (all pages) on one side of paper with 3 cm margins. If a dot matrix printer is used, ensure legibility. Begin each manuscript component on a new page in order: Title page, Abstract, Text, Acknowledgments, References, Tables (each on a separate page), and Captions for illustrations. Do not number title page and abstract page. Begin sequential numbering with introduction. Submit three copies of manuscript and three complete sets of mounted figures.

Arrange observational and experimental studies in sections with headings Introduction, Materials and Methods, Results, and Discussion. Long papers may need subheadings to clarify content, especially in Results and Discussion sections.

Title Page: Include first names, middle initials, and degrees of all authors, name and street address (not a postal box) of institution from which work originated, telephone number, fax or telex numbers of corresponding author, and information concerning grants. Reprint address should include complete name, street address (not a postal box), and zip code. Indicate on title page whether paper has been presented at an RSNA meeting (give year) or has been accepted for presentation at a future meeting.

Abstract and Key Words: 60 words for case report or Technical Developments and Instrumentation paper. State what was done, what was found, and what was concluded. Submit 150 word abstracts for major papers. Divide into four paragraphs with headings: Purpose—State hypothesis being tested or procedure being evaluated; Materials and Methods—Briefly state what was done and what materials were used, include number of subject and methods used to assess data; Results—Provide findings of study, including indicators of statistical significance and actual numbers, as well as percentages; Conclusions—Summarize (one or two sentences) conclusion made on basis of findings. Below abstract, list three to six key words from terms used in most recent *RSNA Index to Imaging Literature.*

References: Number consecutively in order mentioned. Check all documentation for accuracy. Abbreviate periodical titles according to *Index Medicus.* For journal articles, list surnames and initials of all authors when six or less.

1. Stuart MJ, Elrad H, Graeber JE, Hakanson DO, Sunderji SG, Barvinchak MK. Increased synthesis of prostaglandin endoperoxides and platelet hyperfunction in infants of mothers with diabetes mellitus. J Lab Clin Med 1979; 94:12–26.

When seven or more authors are listed, identify only first three names, followed by "et al,":

1. Hibbard CT, Campbell S, Sabbagha RE, et al. Demonstration of tissue interfaces within the body by ultrasonic echo sounding. Br J Radiol 1961; 34:539-549.

Note abstracts, editorials, and letters to the editor as such. For books, provide authors of a chapter, title of chapter, editor(s), title of book, edition, city, publisher, year, and specific pages.

1. Brown M, Gray L. Indications for hematology. In: Wintrobe MM, ed. Clinical hematology. 3rd ed. Philadelphia, Pa: Lea & Febiger, 1975;1146-1167.

Illustrations: Limit to those required to show essential features in paper. Submit unretouched glossy prints no larger than 5×7 in. Prints to be combined into one cut, such as anteroposterior and lateral views, should be same height or same width and should correspond in appearance to tonal relations of original radiograph (i.e,, showing bones white on dark background, with patient's right to the observer's left; computed tomograms should employ "view from below"). Submit professional drawings and charts in India ink on white paper. For computer-generated artwork, do shading in continuous tones (i.e,, solid black, solid gray, solid white): Do not use dotted shadings.

Number all photographs and drawings and indicate TOP on back. Attach each figure to paper (one illustration per page), and reindicate number and TOP on this page; do not use glue, staples, or corner mount; we recommend Scotch Removable Magic Tape (3M, St. Paul, Minn). Do not add letters or numbers to face of illustration to identify figure.

Place arrows and other keys only on final print; they should be professional quality, and removable. Professional artist service for such items is available without charge. Indicate desired additions on tissue or transparent overlay attached securely to print.

Captions for Illustrations: Supply one for each illustration; do not duplicate text material. Place collectively on one or more pages separate from text, and also below corresponding illustration.

Tables: Number in Arabic numerals and give titles. Explain abbreviations used in a footnote. Present in Journal style. Do not use vertical lines or shading. Avoid excessive use of horizontal lines.

Introduction: Clearly state purpose of study or observation. Give only strictly pertinent references; do not review subject extensively.

Materials and Methods: Describe selection of observational or experimental subjects (patients or experimental animals, including controls). Identify methods, instrumentation (manufacturer's name and address in parentheses), and procedures in sufficient detail to allow other workers to reproduce results. Do not include name of institution where work was performed, either within text or at top of manuscript pages. Reference established methods, including published statistical methods that are not well known; describe new or substantially modified methods, give reasons for using these techniques, and evaluate limitations. Include numbers of observations and statistical significance of findings when appropriate. Detailed statistical analyses, mathematical derivations, etc. may be suitably presented in an appendix.

Results: Present in logical sequence in text, tables, and illustrations. Do not repeat in text data in tables and/or illustrations; summarize only important observations.

Discussion: Emphasize new and important aspects of study and conclusions that follow. Do not repeat in detail data in Results section. Include implications of findings and limitations, and relate observations to other relevant studies. Link conclusions with goals of study, but avoid unqualified statements and conclusions not supported by data. Avoid claiming priority and alluding to work that has not been completed. State new hypotheses when warranted, clearly labelled as such. Recommendations, when appropriate. may be included.

OTHER

Experimentation, Animal: Maintenance and care of experimental animals to provide humane treatment and to ensure reliable results are described in National Institutes of Health guidelines for use of laboratory animals. Comply with these guidelines; acknowledge compliance in manuscript.

Experimentation, Human: Include statement that informed consent was obtained after nature of procedure(s) had been fully explained.

Units and Abbreviations: Give radiation measurements and laboratory values in *International System of Units (SI)* See "SI Units in Radiation Protection and Measurements," NCRP Report no. 82 [August 1985]; "Now Read This: The SI Units Are Here " (*JAMA* 1986; 255:2329-2339). Spell out abbreviations at first use in text. Avoid laboratory slang, clinical jargon, and uncommon abbreviations. Restrict discussion of previous literature and material presented to significant findings.

SOURCE
January 1994, 190(1): 79A–81A.

Regulatory Peptides
Elsevier

M. Ian Phillips
Department of Physiology
Box J274
College of Medicine, University of Florida
Gainesville, FL 32610
Rolf Häkanson
Department of Pharmacology
University of Lund
Sölvegatan 10
S-223 62 Lund Sweden

AIM AND SCOPE
Regulatory Peptides provides a medium for the rapid publication of interdisciplinary studies on the physiology and pathology of peptides of the gut, endocrine and nervous systems which regulate cell or tissue function. Articles emphasizing these objectives may be based on either fundamental or clinical observations obtained through the disciplines of morphology, cytochemistry, biochemistry, physiology, pathology, pharmacology or psychology.

MANUSCRIPT SUBMISSION
Regulatory Peptides publishes high quality, novel, original papers on laboratory and clinical experimental studies. Reviews of current areas of research are by invitation. Comments on reports and reviews will be considered for publication in Letters.

Submission of a paper implies that it has not previously been published (except in abstract form), and that it is not being considered for publication elsewhere. Manuscripts submitted under multiple authorship are reviewed on the assumption that all listed authors concur with submission and that a copy of final manuscript has been approved by all authors and tacitly or explicitly by responsible author ities in laboratories where work was carried out. If accepted, manuscript shall not be published elsewhere in same form, in either the same or another language, without consent of Editors and Publisher.

Submit in quadruplicate complete in all respects (including 4 copies of all illustrations). Preferred medium of final submission is on disk with accompanying reviewed and revised manuscript.

Do not exceed 12 printed pages. Important new and exciting data can be published as a Rapid communication.

MANUSCRIPT FORMAT

Type double-spaced on pages of uniform size. For papers 21.5 × 28 cm, with 4 cm margins: 2 typed pages or 3 'average' figures with legends or 25 references = 1 printed page.

Divide papers longer than 5 printed pages into sections, headed by captions (Introduction; Materials and Methods; Results; Discussion; Acknowledgments). On first page include title and a single-spaced abstract (200 words) describing experimental objectives, results and conclusions.

Include authors' full names, academic or professional affiliations, and complete postal addresses on a separate title page. Include name and address of corresponding author, including telephone and fax numbers.

Key Words: List no more than 6, not included in title.

Literature References: In text give at appropriate places as numbers in square brackets. List all references cited at end of paper on a separate page (double-spaced), arranged in numerical order of citation in text, not in alphabetical order. Make literature references complete, including initials of authors cited, title of paper referred to, abbreviated journal title, volume, year, and first and last page numbers of article in a periodical. For abbreviations of journal titles use *List of Serial Title Word Abbreviations* (International Serials Data System, Paris, France.)

For references to books include: author(s), initials, title of book, publisher and city, year and page number(s).

1 Lovrecich, M. and Rubessa, F., Effect of loading parameters on theophylline release from polystyrene beads, Int. J. Pharm., 40 (1987) 63–72.

2 Deming, S. N. and Morgan, S. L., Experiment Design: a Theory Approach, Elsevier, Amsterdam, 1987, 286 pp.

3 Mendell, L. M., Munson, J. B. and Collins III, W. F., Changes in peripheral and central axonal projections of sensory fibers following muscle nerve transection. In F. J. Seil, E. Herbert and B. M. Carlson (eds.), Neural Regeneration, Progress in Brain Research, Vol. 71, Elsevier, Amsterdam, 1987, pp. 231–238.

Do not include numbered references to personal communications and unpublished data. If essential, incorporate such material at appropriate place in text.

Illustrations: Submit in quadruplicate, in a form and condition suitable for reproduction. Illustrations should bear author's name. Number in Arabic numerals according to sequence of appearance in text; refer to as Fig. 1, Fig. 2, etc. Present line drawings in black ink on drawing or tracing paper. Make lettering clear and of adequate size to be legible after reduction. Professional labelling is preferable; or do lettering in fine pencil.

Supply photographs, including radiographs and electron micrographs as clear black-and-white prints on glossy paper. Figures will usually be reduced to fit 12.5 × 19 cm area. Reproduction in color must be approved by Editors and Publisher. Extra cost is charged to author(s). Prices are available from the Publisher.

Give each illustration a legend, typed double-spaced on a separate page; begin with number of illustration.

Tables: Type double-spaced, each on a separate page, number in sequence in Roman numbers (Table I, II, etc.), provide with a concise heading, and refer to in text as Table I, Table II, etc.

OTHER

Electronic Submission: Preferred storage medium is a 5.25 or 3.5 in. disk in MS-DOS format; other systems are welcome, e.g., NEC and Macintosh (save file in usual manner, do not use option 'save in MS-DOS format'). Do not split article into separate files. Ensure that letter 'l' and digit '1' (also letter 'O' and digit '0') have been used properly. Format article (tabs, indents, etc.) consistently. Do not leave characters not available on word processor (Greek letters, mathematical symbols, etc.) open. Indicate by a unique code (e.g., gralpha, @, #, etc., for the Greek letter α). Use codes consistently throughout text. List codes and provide a key. Do not allow word processor to introduce word splits; do not use a 'justified' layout. Adhere strictly to general instructions on style arrangement and, in particular, reference style of journal. Save file in wordprocessor format; do not save 'in flat ASCII.' Format disk correctly; include only relevant file on disk. Specify type of computer and word-processing package used; label disk with your name and name of file on disk. After final acceptance, submit disk plus one final, printed and exactly matching version (as a printout) together to the accepting editor. File on disk and printout must be identical.

Units: Use S.I. units throughout.

SOURCE

March 7, 1995, 56(1): v-vi .

Respiration Physiology
Elsevier

Dr. Peter Scheid
Institut für Physiologie
Ruhr-Universität Bochum
D-44780 Bochum, Germany
Telephone: +49-(0) 234700.4852; Fax: -49-(0)234-709.4191

AIMS AND SCOPE

Respiration Physiology publishes original articles or notes concerning the field of respiration in its broadest sense: all mechanisms responsible for, and connected with, the supply of oxygen to the mitochondria of animals and man. Accordingly the journal covers all aspects, from the molecular to the organismic level, of: gas exchange in lungs, gills, skin, and tissues: gas transport by diffusion and convection; acid-base balance: mechanics of breathing: control of breathing respiration at rest and exercise respiration in normal and unusual conditions, like high or low pressure or changes of temperature, low ambient oxygen embryonic and adult respiration comparative respiratory physiology non-respiratory lung function. Papers on clinical aspects and articles on original methods, as well as theoretical papers in respiratory physiology are also considered as long as they foster the understanding of respiratory physiology and pathophysiology. In addition, state-of-the-art reviews, prepared on invitation, will appear regularly under the title of *Frontiers review*.

MANUSCRIPT SUBMISSION

Articles must deal with original research that has not been published previously or is not being considered for publication elsewhere. Submission of a manuscript implies transfer of copyright from author(s) to publishers.

MANUSCRIPT FORMAT

Preferred medium of final submission is on disk with accompanying reviewed and revised manuscript. Type manuscript double-spaced with wide margins. Submit three complete copies, each with tables and copies of figures, plus one set of high-quality prints of figures. Number pages in order. Include a title page, abstract page, text pages, acknowledgments, and references followed by figure legends and tables.

Title Page: Include full title (85 characters and spaces); list of authors; laboratory of origin (full postal address if more than one, indicate each author's full postal affiliation); corresponding author with phone and fax numbers; present address of authors.

Abstract Page: 160 words stating what was done, what was found, and what was concluded. References give authors, journal, volume, inclusive pages, and year, e.g., Butler et al., Respir. Physiol. 93: 5156, 1993. List keywords for indexing.

References: Start on a new page; type double spaced. Limit to 25. In text, indicate references by last name of author(s), followed by year of publication in brackets. For three or more authors, follow first author et al.; to cite several papers in the same year, place small a, b, etc. after year of publication (also in reference list). Arrange references in reference list alphabetically; list each author. List papers by the same first author in chronological sequence.

Kuwahira, I., N. Heisler, J. Piiper and N.C. Gonzalez (1993). Effect of chronic hypoxia on hemodynamics, organ blood flow and O_2 supply in rats. Respir. Physiol. 92: 227–238.

Dejours, P. (1988). Respiration in Water and Air. Amsterdam, New York, Oxford: Elsevier, 179 p.

Fencl, V. (1986). Acid-base balance in cerebral fluids. In: Handbook of Physiology, Section 3: The Respiratory System, Vol. II: Control of Breathing, part 1, edited by N.S. Cherniack and J.G. Widdicombe. Washington, DC: American Physiological Society, pp. 115–140.

Figures and Tables: Number in Arabic numerals. Refer to in text as Fig. 1, Table 2, etc. Abbreviate Fig(s). at beginning of sentence. Prints of figures should be well contrasted and unmounted. If a figure has panels, submit a single print of final arrangement of all panels. Make lettering large enough to be legible after reduction to final size: maximal print width is 16 cm for full-width figures 7.5 cm for single-column figures. Include a calibration bar on micrographs. Indicate figure number and orientation on back. Type figure legends double-spaced on a separate sheet. Type each table double-spaced on a separate sheet. Indicate approximate position of figures and tables in margin of text.

OTHER

Electronic Submission: For initial submission of manuscripts, hardcopies are sufficient. For processing accepted papers, electronic versions are preferred. After final acceptance, submit disk plus two final and exactly matching printed versions. Use double density (DD) or high density (HD) diskettes (3.5 or 5.25 in.). Save file in native format of wordprocessor program used. Label disk with name of computer and wordprocessing package used, your name, and name of file on disk. Obtain further information from Publisher.

Nomenclature: Use standard nomenclature. Define unfamiliar or new terms, arbitrary abbreviations and trade names at first use, in Abstract and again in main text. Avoid unnecessary abbreviations and symbols.

Symbols and Units: Make meaning of symbols clearly understood from context. Define all symbols that are not commonly used on first appearance in Abstract and in main text. If typewriter cannot produce symbols, include a list that shows each symbol as it appears in manuscript and how it should appear in print. Make symbols conform with glossary of terms and symbols in respiratory physiology proposed by International Union of Physiological Sciences. For a complete description of units, symbols and abbreviations see *Cumulative Author & Subject Index Vols 51-75.*

F Fractional concentration in dry gas phase; P Gas pressure in general; C Concentration; D Diffusing capacity; f Frequency; $\dot{V}$ Volume; V Volume per unit time, e.g., flow, ventilation; S Saturation; R Respiratory exchange ratio.

Modifiers (small capitals and ordinary small letters, on the same line as main symbol): I inspired; E expired; ET end-expired, end-tidal; A alveolar; T tidal; D dead space; B barometric; H heart, cardiac; L pulmonary; R respiratory; b blood in general; a arterial; c capillary; c´ end-capillary; v venous; $\bar{v}$ mixed venous blood; STPD Standard temperature and pressure, dry (0°C, 760 mmHg); BTPS Body temperature and pressure saturated with water vapor; ATPS Ambient temperature and pressure, saturated with water. Symbols referring to gas species are in subscript. Dash above a symbol designates a mean value.

Units: Use basic quantities and units of Système International d'Unités (SI); modify abbreviations for clarity, e.g., sec instead of s, L instead of l (but ml). Express amounts of gases as mole rather than gas volume (see Piper et al.

Respir. Physiol. 13: 292304, 1971). For pressure, use mmHg (=Torr), cmH_2O and atm and SI unit kPa (=7.5 mmHg ≈ 10 cmH_2O).

SOURCE

January 1995, 99(1): inside front and back covers.

Scandinavian Journal of Clinical & Laboratory Investigation
Scandinavian Society for Clinical Chemistry
Medisinsk Fysiologisk Forenings Forlag
Blackwell Scientific Publications

Editorial Secretary, *Scandinavian Journal of Clinical & Laboratory Investigation*
Institute of Clinical Biochemistry, Rikshospitalet
N-0027 Oslo 1, Norway

MANUSCRIPT SUBMISSION

The journal publishes original articles, editorials, invited reviews, and abstracts from Nordic meetings. The publication of short technical notes (improvements of methods, etc.), is encouraged. Follow Vancouver requirements for manuscripts submitted to biomedical journals. Detailed instructions for preparation of manuscripts, tables, figures, and reference lists can be found in *British Medical Journal* 1982; 284: 1766–9 or in *Nordisk Medicin* 1981; 96:209–12. When preparing manuscripts, refer to copies of journal printed after 1 January 1985 and follow style of the journal as closely as possible.

For requirements and regulations concerning authorship and acknowledgments, follow Vancouver style as outlined in *British Medical Journal* 1991; 302: 338–41.

MANUSCRIPT FORMAT

Type double-spaced throughout on one side of paper with 5 cm margin. Submit original with two copies. Use separate sheets for: title page, with author's name, institution, and, if title is longer than forty letters and spaces, a short title not exceeding this limit for use in running heads; an abstract (250 words); tables with headings; legends to figures; and references. Insert name and full postal address of author to whom proofs and reprints are to be sent under key words. Acknowledge grants and other assistance at end of text.

Key Words: 10; append to abstract in alphabetical order. Do not include words in title. Use words from *Medical Subject Headings* of *Index Medicus*.

Tables: Type on separate pages; number with Roman numerals. Make intelligible without reference to text. Give each a short descriptive heading. Place units in which results are expressed at top of column. Define abbreviations in a footnote at first use.

Author may be charged for expensive typesetting of complicated tables or formula. Submit formula as original drawings or photographs for presentation as figures.

Figures: Submit three copies of each; one must be suitable for reproduction, other two may be photostats. Number in sequence with Arabic numerals. Identify each by its number, name of author, and first words of title, written in pencil on back. Indicate TOP. Do not use paper clips.

Give each figure a legend containing sufficient information to make figure intelligible without reference to text. Type all legends together, double-spaced on a separate sheet(s). Size of letters, number and symbols in figures should be 1.3–1.6 mm when printed. Do line drawings with India ink and lettering in black India ink.

Plot graphs on blue graph or plain white paper; ink in black grid lines that are to show in engraving.

Submit photographs as unmounted glossy enlargements showing good details.

References: Number consecutively in order of mention in text. Identify references in text, tables, and legends by Arabic numerals (in parentheses). Use form of references of the U.S. National Library of Medicine and *Index Medicus*. Cite in text information from manuscripts submitted but not yet accepted as 'unpublished observations' (in parentheses). Include among references manuscripts accepted but not yet published; designate journal followed by 'in press' (in parentheses).

OTHER FORMATS

SUPPLEMENTS—Papers exceeding 24 printed pages may be published as supplements; cost of publication is paid by author. Obtain permission from Managing Editor. Full responsibility for contents and presentation of supplement rests with author. Obtain conditions for printing supplements from Blackwell Scientific Publications Ltd, Osney Mead, Oxford OX2 0EL, U.K.

OTHER

Abbreviations and Terminology: Properly define any nonstandard abbreviations (avoid if possible) in text at first use. For chemical nomenclature, follow *Chemical Abstracts*, and for enzyme terminology, the IUB Recommendations in *Enzyme Nomenclature* (Elsevier Publishing Company, 1965). Express quantities and units according to Système International (SI). Express hydrostatic pressures in body fluids in mmHg.

SOURCE

February 1994, 54(1): inside back cover.

Scandinavian Journal of Gastroenterology
Gastroenterological Societies of Denmark, Finland, Iceland, Norway, and Sweden
Scandinavian University Press

Editors, *Scandinavian Journal of Gastroenterology*
Medical Dept. A. Rikshospitalet
N-0027 Oslo, Norway

MANUSCRIPT SUBMISSION

Submit original manuscript with two additional copies. Manuscript must not have been published, simultaneously submitted, or already accepted for publication elsewhere. Acceptance for publication implies transfer of exclusive copyright for article to Scandinavian University Press. Clearly type on one side of the paper only, double-spaced throughout with margins of at least 3 cm. Begin each section on separate pages: title page, abstract and key words, text, acknowledgments, references, individual tables, and figure legends. Number pages consecutively, beginning with title page. Divide text into introduction, methods, results, and discussion. Use a clear system of headings to divide and clarify text; do not use more than three grades of headings. Indicate desired position of figures and tables in margin. Submit manuscripts in final, fully corrected form. Articles should not exceed 5 printed pages.

MANUSCRIPT FORMAT

Title Page: Title of article, concise but informative. Use subtitles if necessary, but keep short. A short title (40 letters and spaces) for use as a running headline. Full name of each author. Departments and institutions (addresses) to which work should be attributed.

Abstracts and Key Words: Present abstract (150 words) on second page. Structure abstract using sections: Background, Methods, Results, and Conclusions. Use only standard abbreviations. Above abstract type authors' names, article title, etc., in accordance with style of journal. Type ten key words in alphabetical order below. Use terms from *Medical Subject Heading*s list from *Index Medicus*. Type name and current address of author(s) to whom correspondence, proofs, and offprints are to be addressed below key words.

Acknowledgments: Acknowledge only persons who have made substantive contributions to study. Obtain written permission from those acknowledged by name. Acknowledge sources of support in the form of grants, etc.,

Tables: Type each double-spaced, on a separate sheet. Do not submit as photographs. Number consecutively with Roman numerals and give each a short descriptive heading. Give each column a short or abbreviated heading. Place explanatory matter in footnotes, not in heading. Explain in footnotes all nonstandard abbreviations. For footnotes, use symbols in sequence: *, †, ‡, §, ||, ¶, **, ††, etc. Omit internal horizontal and vertical rules. If data from another published or unpublished source are used, obtain permission and acknowledge without reference to text.

Figures: All illustrations are considered figures. Submit in triplicate; number consecutively with Arabic numerals. Letters, numbers, and symbols must be clear, in proportion to each other, and large enough to be legible after reduction. Decide whether a figure is to cover one, one-and-a-half, or two columns when printed; plan accordingly.

Submit photographs as near to printed size as possible. If magnification is significant, indicate by a bar on print, not by a magnification factor in figure legend. Submit line drawings in India ink two or three times larger than they will appear in print. Draw in proportion to each other, so letters, numbers, and symbols in different figures will be same size after reduction. Stencil lettering in black India ink or use dry-transfer or pressure-sensitive lettering sets. Do not hand-letter line drawings and photographs. Plot graphs on plain white or blue squared paper; ink in black the grid lines that are to show on engraving.

Affix labels on backs of photographs with figure number,
name(s) of author(s), and TOP. Do not write on backs or damage with paper clips. Pack so that they will not be bent or damaged. If photographs of persons are used, either subjects must not be identifiable or submit their written permission to use photographs.

Give each figure a legend containing sufficient information to make intelligible without reference to text. Type legends together, double-spaced, on a separate sheet(s). If a figure has been published previously, acknowledge original source and submit written permission from copyright holder to reproduce.

Submit color figures as color negatives or positive transparencies and, when necessary, drawings marked to indicate region to be reproduced. Authors pay for extra cost of color illustrations.

References: Type double-spaced, on a separate sheet(s); number consecutively in order of mention in text. Reference list is not in alphabetical order. Include in list all authors referred to in text, and in tables or legends. Verify accuracy of reference list.

Identify references in text, tables, and legends by Arabic numerals in parentheses. In text, make references made by giving name of author followed by appropriate number in parentheses, e.g., Hansen (1) has reported . . . , Hansen & Olsen (2) have reported . . . , or by just the appropriate number in parentheses, e.g., As has recently been reported (1,2) . . . In text, when more than two authors, give only name of first author, followed by "et al." Cite in text information from manuscripts submitted but not yet accepted as 'Unpublished observations' (in parentheses).

Follow style in *Uniform Requirements for Manuscripts Submitted to Biomedical Journals* (Vancouver style). Abbreviate journal titles according to *Index Medicus*.

JOURNALS

1. Hoofnagle JH, Bergasa NV. Methotrexate therapy of primary biliary cirrhosis: promising but worrisome [editorial]. Gastroenterology 1991; 101:1440–2.

2. Sarosick J, Peura DA, Guerrant RL, et al. Myocolytic effects of *Helicobacter pylori*. Scand J Gastroenterol 1991; 26 Suppl 187:47–55.

BOOKS

3. Pope CE II. Anatomy and developmental anomalies. In: Sleisenger MH, Fordtran JS, editors. Gastrointestinal disease. Pathophysiology, diagnosis, management. Section A. The Esophagus, 3rd ed. Philadelphia: Saunders, 1983:407–14.

OTHER

Abbreviations: Use standard abbreviations. Consult *CBE Style Manual* (5th ed., Council of Biology Editors, Inc., Bethesda, MD, 1983). Explain nonstandard abbreviations (avoid if possible) in text at first use. Avoid abbreviations in title. Express quantities and units in accordance with recommendations of the Système International d'Unités.

Experimentation, Animal: Include a statement verifying that care of laboratory animals followed accepted standards.

Experimentation, Human: Indicate whether procedures followed were in accordance with ethical standards of committee on human experimentation of the institution in which experiments were done and in accordance with the Declaration of Helsinki.

Page Charges: $144 (NOK 930) per printed page or part in excess of 5.

Statistical Analyses: See Special Article in June 1990 issue (Vol. 25, No. 6, pp. 545–547).

SOURCE

January 1994, 29(1): inside back cover.

Scandinavian Journal of Immunology
Blackwell Scientific Publications

P. Brandtzaeg
J.B. Natvig
Institute of Immunology and Rheumatology
Fr. Qvamsgt. 1
N-0172 Oslo 1, Norway

G. Möller
Department of Immunology
Biology Building F, Stockholm University
S-10691 Stockholm Sweden

AIM AND SCOPE

Scandinavian Journal of Immunology publishes papers within the various fields of cellular and molecular immunology. The epithet 'Scandinavian' does not imply a journal restricted to Scandinavian research. We very much welcome international papers and hope that both Scandinavian and other investigators will find journal suitable for their publications.

MANUSCRIPT SUBMISSION

Scientific standard is the only criterion for accepting a manuscript for publication. Send editorials directly to A. Coutinho, The Editorials Editor, Department of Immunology, Institut Pasteur, 28 rue du Dr Roux, F-75724 Paris Cedex 15, France. Papers are accepted for publication on condition that they are submitted only to this journal, and on acceptance become copyright of journal. All authors must sign a statement that they agree to submitting manuscript to journal for publication.

MANUSCRIPT FORMAT

Type double-spaced throughout on one side of paper with 5 cm wide margin. Submit original with three additional copies. Use separate sheets for: title page, with author's name, institution, and, if title is longer than forty letters and spaces, a short title not exceeding this limit for use in running heads; abstract (250 words); tables with headings; legends to figures; and references. Type name and full postal address of author to whom proofs and reprints are to be sent under key words. Acknowledge grants and other assistance at end of text.

Tables: Type on separate pages; number with Arabic numerals. Make understandable without reference to text. Give each a descriptive heading. State conditions specific to particular experiment(s) below table. Put units in which results are expressed at top of each column. Keep footnotes to a minimum.

Figures: Submit three copies of each; one must be original and suitable for reproduction. Number in sequence with Arabic numerals. Identify by name of author and number of figure, written in pencil on back. Indicate TOP.

Give each figure a legend containing sufficient information to make intelligible without reference to text. Type legends together, double-spaced, on a separate sheet(s). Explain on figure itself symbols used to identify results. If not possible, use symbols: ○ △ ❏ ● △ ▲ ■ ×, and identify in figure legend. Do line drawings and lettering in India ink. Plot graphs on blue graph or plain white paper; ink in black grid lines that are to show in engraving. Submit photographs as unmounted glossy enlargements showing good details.

References: Use Vancouver Reference System. Number references consecutively in order of mention. Identify references in text, tables, and legends by Arabic numerals [in brackets]. Number references cited only in tables or in legends to figures in accordance with a sequence established by first identification in text of table or illustration.

Abbreviate titles of journals according to *Index Medicus*.

Avoid using abstracts as references. Do not use 'unpublished observations' or 'personal communications' as references; insert references to written, not verbal, communications (in parentheses) in text. Include among references manuscripts accepted but not yet published; designate journal followed by 'in press' (in parentheses). Cite information from manuscripts submitted but not yet accepted in text as 'unpublished observations' (in parentheses).

STANDARD JOURNAL ARTICLE

List all authors when six or less; when seven or more, list only first three and add *et al.*

Soter NA, Wasserman SI, Austen KF. Cold urticaria: release into the circulation of histamine and eosinophil chemotactic factor of anaphylaxis during cold challenge. N Engl J Med 1976;294:687–90.

CORPORATE AUTHOR

The Committee on Enzymes of the Scandinavn Society for Clinical Chemistry and Clinical Physiology. Recommended method for the determination of gamma-glutamyltransferase in blood. Scand J Clin Lab Invest 1976;36:119–25.

Anonymous. Epidemiology for primary health care. Int J Epidemiol 1976;5:224–5.

BOOKS AND OTHER MONOGRAPHS
Personal Author(s)

Osler AG. Complement: Mechanisms and Functions. Englewood Cliffs: Prentice-Hall, 1976.

Corporate Author

American Medical Association Department of Drugs. AMA drug evaluations. 3rd edn. Littleton: Publishing Sciences Group. 1977.

Editor, Compiler, Chairman as Author

Rhodes AJ, Van Rooyen CE, comps. Textbook of Virology: for Students and Practitioners of Medicine and the other Health Sciences. 5th edn. Baltimore: Williams & Wilkins, 1968.

CHAPTER IN BOOK

Weinstein L, Swartz MN. Pathogenic properties of invading microorganisms. In: Sodeman WA Jr, Sodeman WA, eds. Pathologic Physiology: Mechanisms of Disease. Philadelphia: WB Saunders, 1974:457–72.

PUBLISHED PROCEEDINGS PAPER

DuPont B. Bone marrow transplantation in severe combined immunodeficiency with an unrelated MLC compatible donor. In: White HJ, Smith R, eds. Proceedings of the Third Annual Meeting of the International Society for Experimental Hematology. Houston: International Society for Experimental Hematology, 1974:44–6.

OTHER FORMATS

SHORT PAPERS—Four printed pages, three figures and tables.

SUPPLEMENTS—Papers exceeding 16 printed pages may be published as supplements, cost is paid by author. Manuscripts are subject to editorial acceptance and revision. Follow typographical style of journal. Obtain further information from publishers.

OTHER

Abbreviations: Properly define any nonstandard abbreviations (avoid is possible) in text at first use.

Electronic Submission: Supply revised manuscript on floppy disk with printouts. Submit original and three copies in typewritten form.

Final version of hard copy and floppy disk must be the same. Disk must contain all last minute changes. Give file name(s) with disk, with separate file names for: text, references, tables, figure captions. Supply files in both word processor and ASCII formats. Be consistent. Use same presentation for all headings, etc. that are to appear the same in finished printing. Advise of use of a keyboard character to represent a character that is not on your keyboard, e.g. Greek. Tables may not be set from disk; supply adequate hard copy. File Description Forms are available from Editorial office.

SOURCE

January 1994, 39(1): back matter.

Science

American Association for the Advancement of Science

Editor-in-Chief, *Science*
1333 H Street NW
Washington D.C. 20005

Senior Editor, European Office, *Science*
Thomas House
George IV St.
Cambridge CB2 1NH U.K.

AIM AND SCOPE

Science is a weekly, peer-reviewed journal, that publishes research in every field of scientific endeavor.

MANUSCRIPT SUBMISSION

Make submitted manuscripts intelligible to readers in a variety of disciplines; be brief and write clearly.

To reprint illustrations or tables from other publications, obtain written permission. Include complete citation from copyright owner (usually publisher) to reprint illustrations in *Science*.

Law requires copyright transfer from authors of each paper published in *Science*. Copyright forms are sent to all authors prior to acceptance. U.S. government employees sign section of form stating exemption from copyright laws. Do not alter or make substitutions on form.

MANUSCRIPT FORMAT

Double-space throughout text, tables, figure legends, and references and notes, and leave margins of at least 2.5 cm. Put name on each page; number pages starting with title page.

Titles and Subheadings: Use descriptive clauses, not complete sentences or questions. Titles are 102 characters and spaces for general articles, and 98 characters and spaces for research articles and reports. Abstracts should explain to general reader why research was undertaken and why results are important; convey paper's main point and outline results or conclusions.

Text: In a brief introduction describe paper's significance intelligibly to readers in different disciplines. Define technical terms. Cite all tables and figures in numerical order.

Figures and Tables: Submit on separate pages from text. For each figure submit four high-quality prints, laser prints, or original drawings no larger than 8.5×11 in. On back of figures write first author's name and figure number and indicate correct orientation.

Photocopies of figures are not acceptable; do not send transparencies, slides, or negatives. Papers that include a large number of figures or tables and a small amount of text present layout problems. Maintain sufficient text to wrap around figures.

Uncertainties and Reproducibility: Explicitly state evidence that results are reproducible and conditions under which this reproducibility (replication) was obtained. Discuss effect of limitations in experimental conditions on generalizability of results. State uncertainties in terms of variation expected in independent repetitions of experiments; include an allowance for possible systematic error arising from inadequacies in assumed model and other known sources of possible bias. Probabilities from statistical tests of significance should not replace reporting of results and associated uncertainties.

Acknowledgments: Make a brief statement at end of references and notes; include funding information.

Equations and Formulas: Type quadruple-spaced if to be set off from text. Define all symbols; number all equations.

Illustrations: Most illustrations are printed 2.3 in. (1 column) or 4.8 in. (2 columns) wide. Some illustrations (for example, bar graphs, simple line graphs, and gels) are reduced to a smaller width. Make symbols and lettering large enough to be legible after reduction. Label composite figures A, B, C, ... If mounting is necessary, use cardboard.

Type legends double-spaced in numerical order on a separate page. No legend should be longer than one page. Match nomenclature, abbreviations, symbols, and units used in a figure to those in text. Give figure title as first line of legend.

Line Drawings—Label on ordinate and abscissa with parameter or variable being measured, units of measure, and scale. Present scales with large or small numbers as powers of 10. Define symbols in figure legend, not in figure. Simple symbols (circles, squares, triangles, and diamonds, solid or open) survive reduction best. Recommended symbols: ● ○ ❑ ▲ △.

Avoid use of light lines, shading, and stippling. Use heavy lines or boxes for emphasizing or marking off areas of figure; use black, white, hatched, and cross-hatched designs in place of stippling in bar graphs and ball-and-stick molecular models. If using computer graphics, choose screens between 20 and 60%.

Halftones—Submit as high-quality prints or originals (do not send irreplaceable artwork). Use scale bars in place of, or in addition to, magnifications. In gels, number lanes and identify by number in figure legend.

Color Art—Provide a positive slide and a print or laser proof. Indicate positioning, lettering, and cropping limits on print. For composite figures, send original composite board rather than a print if quality of original is much better. Do not send irreplaceable artwork.

Lettering—Helvetica font is preferable. Use boldface type for axis labels and for labels in composite figures; use italic type only as used in text (e.g., for variables and genes). Make first letter of each entry uppercase; otherwise, use uppercase letters as they would be used in text (e.g., for acronyms). Avoid wide variation in type size within single figures. In printed version of figure, make letters about 7 point (2 mm) high.

Sequences: May be reduced considerably, so make sure typeface in original is clear. Include 130 characters (including spaces) per line for a sequence occupying full width of printed page and about 84 characters per line for a sequence occupying two columns.

References and Notes: Number in order cited, first through text and then through table and figure legends. List a reference only one time. Group references always cited together under a single number. Reference unpublished data with a number in text and place, in correct sequence, in references and notes. Use conventional abbreviations for well-known journals; provide complete titles for other journals. Do not use op. cit.

JOURNALS

1. N. Tang, *Atmos. Environ.* **14**, 819 (1980). [one author]

2. J. C. Smith and M. Field, *Proc. Natl. Acad. Sci. U.S.A.* **51**, 930 (1964). [Two authors.].

3. J. C. Cheeseborough III, S. Trajmar, J. -T. Yang, *EMBO J.*, in press. [Three to five authors.]

4. G. Sunshine *et al, Lancet* i,711(1975). [More than five authors.]

5. M. Schmidt, *Sci. Am.* **251**, 58 (November 1984). [Journal paginated by issue.]

6. J. Brown, *ibid.*, p. 67.

TECHNICAL REPORTS

1. D. E. Shaw, *Technical Report No. CUCS-29-82* (Columbia University, New York, 1982).

2. F. Press, "A report on the computational needs for physics" (National Science Foundation, Washington, DC, 1981). [Unpublished or access by title.]

3. "Assessment of the carcinogenicity and mutagenicity of chemicals," *WHO Tech. Rep. Ser. No. 546* (1974).

PROCEEDINGS

1. *Proceedings of the Fifth IEEE Pulsed Power Conference,* Arlington, VA, inclusive dates of meeting (publisher, publisher's location, year).

2. *Proc. IEEE* **88**, 452 (1968)

3. *Title of symposium published as a book*, sponsoring organization, location of meeting, dates (publisher, location, year).

PAPER PRESENTED AT A MEETING (NOT PUBLISHED)

1. M. Konishi, paper presented at the 14th Annual Meeting of the Society for Neuroscience, Anaheim, CA, 10 October 1984. [Sponsoring organization should be mentioned if it is not part of the meeting name.]

THESES AND UNPUBLISHED MATERIAL

1. B. Smith, thesis, Georgetown University(1973).

2. J. A. Norton, unpublished material.

BOOKS

1. A. M. Lister, *Fundamentals of Operating Systems* (Springer-Verlag, New York, ed. 3,1984), pp 7–11. [Third edition.]

2. J. B. Carroll Ed., *Language, Thought and Reality: Selected Writings of Benjamin Lee Whorf* (MIT Press, Cambridge, MA, 1956). 3. R. Davis and J. King, in *Machine Intelligence,* E. Acock and D. Michie, Eds (Wiley, New York, 1976), vol 8, chap 3.

4. D. Curtis *et al., in Clinical Neurology of Development.* B Walters, Ed. (Oxford Univ. Press, New York, 1983), pp 60–73. [*Et al.* = more than five authors.]

5. F. R. Sabier, *Contributions to Embryology* (Publ. 18, Carnegie Institution of Washington, Washington, DC, 1917), p. 61.

6. *Principles and Procedures for Evaluating the Toxicity of Household Substances* (National Academy of Sciences, Washington, DC, 1977). [Organization as author and publisher.]

Tables: Supplement, do not duplicate, text. Number in order cited in text. Place each on a separate page with legend double-spaced above. Make first sentence of legend a brief descriptive title. Insert horizontal lines at top and bottom of table and between column headings and table body. Do not use vertical lines between columns.

Give every vertical column a heading consisting of a title with unit of measure in parentheses. Do not change units within a column. Use centered headings to break entries into groups.

Footnotes contain information relevant to specific entries or parts of table. Use symbols in sequence: *, †, ‡, §, ||, ¶, #, **, ††, ‡‡, ...

OTHER FORMATS

BOOK REVIEW—Selections are made by editors. Instructions and specifications are sent to reviewers by editors.

GENERAL ARTICLES—3000–5000 words (three to five printed pages) reviewing new developments in a field that will be of interest to readers in other fields; describe a current research problem or a technique of interdisciplinary significance; or discuss some aspect of the history, logic, policy, or administration of science. Readers should learn what has been firmly established and what are unresolved questions or future directions. Many are solicited by editor-in-chief, but unsolicited articles are welcome. Include a note with authors' names, titles, addresses; an abstract (50 to 100 words); an introduction that outlines the main point of article; and brief subheadings to indicate main ideas. Reference list (40) should not be exhaustive. Figures and tables with their legends occupy about one printed page.

LETTERS—250-500 words. Selected for pertinence to material published in *Science* or because they discuss problems of general interest to scientists. Letters about material published in *Science* may correct errors, provide support or agreement, or offer different points of view, clarifications, or additional information. Personal remarks about an author are inappropriate. Author of the paper in question is usually given an opportunity to reply.

PERSPECTIVES—One to two printed pages, briefly analyzing recent research and impact of new developments on future investigations, rather than new results and hypotheses; do not primarily discuss the author's own work.

POLICY FORUM—2000 words (two printed pages) provides a platform to present discussions of policy issues relevant to science.

REPORTS—7700 words (three printed pages) containing important research results. List addresses for authors on title page; indicate corresponding author by an asterisk. Include an abstract (100 words), an introductory paragraph, and 30 references. Figures and tables with their legends occupy no more than three quarters of a printed page.

RESEARCH ARTICLES—4000 words (four printed pages) about new data representing a major breakthrough in a field. Include an author note, abstract, introduction, and sections with brief subheadings, and 40 references. Figures and tables with their legends occupy about one printed page.

TECHNICAL COMMENTS—500 words; may criticize articles or reports published in *Science* within previous 6 months or offer useful additional information. Authors of

original paper are asked for an opinion of comment and are given an opportunity to reply in same issue. Comments and replies are subject to review and editing.

OTHER

Color Charges: $600 for first color figure or figure part and $300 for each additional figure or figure part. There is an additional charge for color figures in reprints.

Data Submission: Deposit archival data sets (such as sequence and crystallographic data) in appropriate data banks and send identifier code to *Science* for inclusion in published manuscript. Coordinates should be released no later than 1 year after publication.

Electronic Submission: Authors may provide a copy of manuscript on disk upon acceptance. Specific instructions are provided when manuscript is returned for revision.

Experimentation, Animal: Include a statement that care was in accordance with institutional guidelines. For animals subjected to invasive procedures, include anesthetic, analgesic, and tranquilizing agents used, as well as amounts and frequency of administration.

Experimentation, Human: Include a statement indicating that informed consent was obtained after the nature and possible consequences of studies were explained.

Materials Sharing: Make materials and methods necessary to verify conclusions of experiments reported available to other investigators under appropriate conditions.

Symbols, Abbreviations, and Acronyms: Define at first use.

Units: Give in metric. If measurements were made in English units, give metric equivalents.

SOURCE

January 7, 1994, 263 (5143): 37–39.

Southern Medical Journal
Southern Medical Society

Editor, *Southern Medical Journal*
35 Lakeshore Drive, PO Box 190088
Birmingham, AL 35219-0088
Telephone: (800) 423-4992; Fax: (205) 945-1548

MANUSCRIPT SUBMISSION

Neither manuscript nor its essential substance may have been published nor simultaneously submitted for publication elsewhere. Indicate in cover letter name, address, telephone and fax numbers of corresponding author. Submit a double-spaced typed original and two copies of complete manuscript (text, references, tables, glossy prints of figures, and legends).

Complete and send *Journal's* Manuscript Submission Agreement, (in every issue). If form is not received, a copy will be sent to corresponding author. Accepted articles become permanent property of *Journal,* and may not be published elsewhere without its permission. After a manuscript is accepted, supply a computer disk containing final, revised manuscript with hard copy.

Acknowledge all financial support or provision of supplies used in study. On title page include a footnote with grant or contract numbers and complete name and location of funding agency or institution.

Submit a confidential statement to Editor specifying any commercial or proprietary interest in any drug, device, or equipment mentioned in article. Identify any financial interest any author might have (as a consultant, stock owner, employee, evaluator, etc.) in any item mentioned in article.

Submit a written statement from both author and publisher giving *Journal* permission to reproduce material from other sources.

For further details on manuscript preparation see "Uniform Requirements for Manuscripts Submitted to Biomedical Journals," prepared by the International Committee of Medical Journal Editors.

MANUSCRIPT FORMAT

Title Page: Contains: concise title, name(s) of author(s), including first name and academic degrees, department and institution in which work was done, and name and address of reprint request author.

Abstracts: 150 words, giving only essential features of report (i.e., problem, method, results, conclusions). Do not duplicate introduction; it replaces a summary. All articles except Editorials and Correspondence have an abstract.

Acknowledgments: Reserve for those who have made a special contribution to an article beyond their regular duties. Do not acknowledge secretarial services.

References: Double-space and number consecutively as cited. List names and initials of first three authors, followed by et al if more than three. Do not include personal communications and unpublished data as numbered references; footnote in text.

1. Jones JJ, Smith AW, Nelson EC, et al: Carcinoma of the parathyroid. *South Med J* 1971; 63:510–515

2. Avery ME: *The Lung and Its Disorders in the Newborn Infant.* Philadelphia, WB Saunders Co, 1964

3. Jones JJ, Smith AW, Wilson WW: Diseases of connective tissue. *Textbook of Pediatrics* Nelson WE, Vaughn VA III, McKay RJ (eds). Philadelphia, WB Saunders Co. 9th Ed, 1969, p 995

Tables: Type double-spaced on separate sheets; number consecutively; give adequately descriptive titles. Explain abbreviations and symbols in footnotes. Do not duplicate material in text; excessive tabular data are discouraged.

Illustrations: Number and cite consecutively. Have professionally drawn and photographed; submit as high-contrast black and white glossy prints (5 × 7 in.) in triplicate. Affix label on back of each figure with name of first author, figure number, and TOP. Do not mount, staple, or clip prints.

Mask identifying features of patients in photograph or submit copies of signed permission for their use. For photomicrographs, state magnifications and stains used in legends. Submit both transparencies and prints for color photographs. Type figure legends consecutively, double-spaced, on a separate sheet.

OTHER FORMATS

CASE REPORTS—Brief individual case reports of unusual interest demonstrating a clinical point or new finding.

CLINICAL AND LABORATORY STUDIES—Original in-depth clinical investigative laboratory research articles. Manuscripts must have direct relevance to clinical practice.

CLINICAL REVIEWS—In-depth, appropriately referenced educational articles presenting core information for continuing medical education of physicians.

CONTROVERSIES IN MEDICINE—Appropriately referenced reviews of topics on which more than one clinical approach has been suggested.

CORRESPONDENCE—Brief letters or notes to the Editor. Individual case reports, unless of unusual interest, are discouraged.

EDITORIALS—Brief, appropriately referenced commentary on limited subjects.

MEDICAL INTELLIGENCE—Brief state of the art reviews on new technologies or approaches to disease.

SPECIAL REPORTS—Items of special interest, not necessarily related to clinical medicine.

OTHER

Abbreviations and Nomenclature: Keep to a minimum; spell out terms at first mention, with abbreviation immediately following in parentheses. Use generic names of drugs. If proprietary names are included, capitalize and enclose in parentheses after generic name. Express measurements in metric system.

Color Charges: $750 per page.

Electronic Submission: Submit accepted manuscript on computer disk. Macintosh format is preferred; DOS-based disks are acceptable. ASCII or RTF files are preferred.

Reviewers: Suggest three reviewers; supply reviewers' addresses, telephone, and fax numbers. Reviewers may not be from same institution or locality as authors. If anonymous evaluation is desired, exclude all identifying data except for title page.

SOURCE

January 1994, 87(1): 105.

Stroke
American Heart Association

Marie-Germaine Bousser
Hopital Saint Antoine; 184, rue du Faubourg
Saint-Antoine; 75571 Paris Cedex 12 France
         Clinical studies from Europe

Mark L. Dyken, MD
Dept. of Neurology
Indiana University School of Medicine
Bryce Bldg., BA402, 1001 West 10th Street
Indianapolis, IN 46202-2879

         All other manuscripts, including clinical, basic
         science, and animal studies

MANUSCRIPT SUBMISSION

Submit articles for publication that describe Original Research; Short Communications; Case Reports; Comments, Opinions, and Reviews; Letters to the Editor; and Special Reports. Submit all articles solely to *Stroke*. Submit four copies of a separate page with author and funding information. Manuscripts and figures should not include author identification. Identify figures with abbreviated titles.

Indicate in cover letter that work submitted has not been published and is not under consideration for publication elsewhere, in whole or in part, with exception of abstracts of not more than 400 words. Letter must be signed by all coauthors and affirm that each has read and approved manuscript. Provide six address labels providing name and address of corresponding.

Authorship means those persons, and only those persons, who accept intellectual and public responsibility for statements made and results reported. Authorship should not be attributed to anyone not directly involved in study, to physicians or technicians who provided routine services, or to technical advisors. Include justification in cover letter for inclusion of more than five authors. A group study should use group name and reference contributing authors in a footnote on title page. Recognize all other contributors in Acknowledgments section.

Do not submit original contributions that constitute preliminary findings intended to be expanded upon later unless study is described as such. Disclose in cover letter any relevant information, and include copies of relevant articles, pertaining to possible prior publication of any part of manuscript. This includes reworked data already described in whole or in part in another article; patients previously described in another published study; or content already published in another format or different language. Editor will consider submissions without bias if authors provide full disclosure and four copies of each relevant article.

Acknowledge in cover letter any financial support from commercial sources or pecuniary interest in such enterprises that could pose a conflict of interest. Disclosure will remain confidential during review process and will not prejudice process. If article is accepted, Editor will determine extent of disclosure in consultation with authors.

Do not submit manuscripts as sequential articles to be published together. Make every effort to submit a single, internally consistent manuscript that does not depend on any other study for acceptability.

Authors must release first and subsidiary rights to manuscript at submission. The Editorial Office will send a Copyright Transfer Agreement to corresponding author with acknowledgment letter for manuscript. Complete, sign by all authors, and return to Editorial Office. Provide original signatures. Imprints, facsimiles, or photocopies are not acceptable. The American Heart Association grants authors the right to use portions of their manuscripts, without charge, in books or articles of which they are authors or editors.

MANUSCRIPT FORMAT

Conform to specifications in "Uniform Requirements for Manuscripts Submitted to Biomedical Journals" (*Ann Intern Med* 1988;108:258–265 and *Br Med J* 1982;284:1766–1770). Submit manuscript, tables, and illustrations in quadruplicate. Type double-spaced, including references, on good-quality, nonerasable paper. Begin title page, abstract, text, acknowledgments, references, legends, tables, and figures on a separate sheet and follow in order. Define abbreviations at first appearance and do not use at all in title or abstract. Use generic names of drugs. Number references by first appearance in text, and cite by that superscripted number.

For paragraph headings use only: Introduction, Materials (or Subjects) and Methods, Results, and Discussion. Indent new paragraphs five spaces; make top, bottom, and left margins 2.5 cm; right margins 2.5 cm and unjustified. Do not begin or end sentences with an abbreviation or Arabic numeral.

Include four copies of a separate author information page. Include name, address, academic affiliation, and highest degree(s) for each author; acknowledgment of all sources of support from both profit and nonprofit organizations; and address of corresponding author. Include telephone and fax numbers.

Title Page: Brief (80 characters); no abbreviations. For animal studies include name of animal in title. Include a short title (40 characters or less) for a running head and 3–5 key words. Use words listed in "Subject Index" published in *Stroke* each December.

Abstracts: 250 words. Do not use abbreviations. Structure in four paragraphs with headings: Background and Purpose; Methods; Results; and Conclusions. In each paragraph briefly describe, respectively, problem or question that study addresses; how study was carried out; important results (including probability values for statistical significance); and what authors conclude from results.

Introduction: Briefly state what was studied and why, including its relationship to previous work. Include only major references and pertinent recent reviews.

Methods: Describe all methods concisely, but in sufficient detail that other investigators can replicate study. For apparatus used in research, give name, city, and state or country of manufacturer(s).

Results: Describe positive and relevant negative findings in text, supported by tables and figures when these summarize or enhance data. Do not repeat information in text and table or figure; summarize what graphic shows.

Discussion: Interpret results, describe implications, and discuss importance and limitations of findings. Show a clear relationship between results and original hypothesis; relate this to previous studies.

References: Type double-spaced in numerical sequence.

1. Kelm M, Schrader J. Control of coronary vascular tone by nitric oxide. *Cirr. Res.* 1990;6:1561–1575

Consult *AMA Manual of Style* for formatting special publications. Provide all authors' names; do not use "et al."

Give inclusive page numbers. Book references contain specific page numbers. Cite personal communications and unpublished data in parentheses in text, without numerical reference at end of article. If citation is from someone other than authors, submit a letter with direct quotation and author's signature.

Illustrations: Enclose four sets of figures in four separate envelopes; do not use clips. Write figure number lightly in pencil on back margin of each figure, and indicate TOP.

Type legends double-spaced on a separate sheet. Provide sufficient description to interpret figure; number to correspond with numbers in text. Use abbreviations in legends only if they appear in figure; use American spelling.

Submit all illustrations as 5×7 in. glossy prints unmounted and untrimmed except for graphs, which may be printed using a high-quality laser printer. If laser prints are submitted, provide originals, not photocopies.

Prepare drawings and graphs (if not prepared with a laser printer) with black India ink on white background. Make letters, numbers, and symbols clear and even throughout and large enough to remain legible after reduction. Submit photographs of original drawings and provide internal scale markers for photomicrographs. On all multipart figures, including photographs, label each panel (A, B, C, D, etc.). Attach arrows and arrowheads to indicate specific areas in photographs. Specific permission for facial photographs of patients is required.

Request permission from publisher and author for figures or tables reproduced or adapted from a previously published article. Request world rights.

Tables: Type double-spaced, on separate pages; number with Arabic numerals in order cited in text. Make title concise; describe content of table so reader may understand without referring to text. Omit vertical rules; use extra space to delineate sections. Use abbreviations here that are not permitted in text, explain each in footnote. For footnotes, use symbols in sequence, *, †, ‡, §, ||, ¶, #; double as necessary.

OTHER FORMATS

CASE REPORTS—1500 words presenting important and unique clinical experience. Limit description of negative and normal findings. Length depends on number of patients; include only the most relevant references. Limit figures to those which truly enhance study.

COMMENTS, OPINIONS, AND REVIEWS—Summarize present state of knowledge pertaining to some aspect of cerebrovascular disease without objectivity required in a Progress Review. Length and format vary.

EXPEDITED PUBLICATION—10 pages (double-spaced) that are acceptable with only minor revision. Article must be unique and contain information that might make a significant difference in medical practice or constitute an important advance in basic knowledge.

State reasons for request for expedited publication in cover letter. Include a diskette containing manuscript in a word processing file, with four paper copies of manuscript and figures. Indicate which computer software and hardware was used to prepare file. List five or more possible referees. Editors may choose other referees.

LETTERS TO THE EDITOR—1000 words, excluding references, expressing views about a specific article published in *Stroke* or ideas or findings of scientific interest that do not constitute original research. Editor invites responses as appropriate. Double-space text and references, provide brief title, and obtain signatures from all authors.

SHORT COMMUNICATIONS—1500 words, three tables and/or figures, 100 word abstract. Case studies with more than three patients.

OTHER

Experimentation, Animal: For experimental animals, state species, strain, number used, and other pertinent descriptive characteristics. When describing surgical procedures on animals, identify preanesthetic and anesthetic agents used; state amount or concentration and route and frequency of administration for each. Use of paralytic agents, such as curare or succinylcholine, is not an acceptable substitute for anesthetics. For other invasive procedures on animals, report analgesic or tranquilizing drugs used; if none were used, justify exclusion. Indicate that procedures followed were in accordance with institutional guidelines.

Experimentation, Human: Indicate that study was approved by an institutional review committee and that subjects gave informed consent. Describe characteristics of human subjects or patients. Indicate that procedures followed were in accordance with institutional guidelines.

Page Charges: $50 per printed page. $50 per printed page for excessive author alterations.

Style: For guidance on grammar, punctuation, and scientific writing, consult *American Medical Association Manual of Style* (8th Ed, American Medical Association, Chicago). Avoid use of passive voice in Abstract, Introduction, and Discussion; limit its use in Methods and Results. Be as concise as possible: do not say in ten words what could be said in four.

Units: Use Système International units of measurement. Change molar (M) to mol/L, mg/dL to mmol/L, cm to mm, and percentages to decimal fractions. Make conversions before submitting manuscript.

SOURCE

January 1994, 25(1): A7–A8.

Surgery

Society of University Surgeons
Central Surgical Association
American Association of Endocrine Surgeons
Mosby–Year Book, Inc.

Walter F. Ballinger, M.D.
Washington University School of Medicine
Campus Box 8221
St. Louis, MO 63110
Telephone: (314) 362-7407
          Author surnames A through L
George D. Zuidema, M.D.
University of Michigan
M4101 Medical Science I Bldg., C-Wing

1301 Catherine Rd.
Ann Arbor, MI 48109-0608
Telephone: (313) 936-1481
        Author surnames M through Z

AIM AND SCOPE

The *Journal* invites concise, original articles of new matter in the broad field of clinical and experimental surgery as well as surgical organization, education, and history. Emphasis for acceptance includes conciseness and clarity of presentation as well as appropriateness of English usage.

MANUSCRIPT SUBMISSION

Observe rules against dual publication. State financial association between one or more authors and a commercial company that makes a product that figures prominently in article. Editors reserve the right to reject article if a significant conflict of interest is present.

Submit a letter signed by all authors giving approval of work submitted. Transfer ownership by following written statement, which must accompany manuscript and be signed by one author: "The undersigned author transfers all copyright ownership of manuscript [title of article] to Mosby–Year Book, Inc., in event work is published. The undersigned author warrants that article is original, is not under consideration by another journal, and has not been published previously. "I sign for and accept responsibility for releasing this material on behalf of any and all coauthors."

For direct Quotations, Tables, or Illustrations that have appeared in copyrighted material, obtain and submit written permission for use from copyright owner and original author with complete information as to source. For photographs of identifiable persons submit signed releases showing informed consent.

MANUSCRIPT FORMAT

Submit four complete sets of manuscript typed double-spaced on one side of 8.5 × 11 in. paper, with liberal margins on all sides. Provide photocopies; carbon copies are unacceptable.

Title Page: Include name and highest achieved degree of each author, institution from which work originated, sources of financial support, and exact and complete address, business and home phone numbers, and fax number of corresponding author.

Synopsis or Abstract: 200 words; four paragraphs, labelled: Background (stating purpose of study), Methods, Results, and Conclusions. Number abstract page two. Include with Original Communications, not Clinical Reviews, Brief Clinical Reports, Surgical Techniques, or Editorials.

Illustrations: Submit with each copy of manuscript as glossy prints, 3 × 4 in. (minimum) to 5 × 7 in. (maximum); number in order of appearance in text. Present with good black-and-white contrast or good color balance. Make illustrations consistent in size. Note special sizing instructions. Mark each print lightly on back in pencil with figure number and first author's surname; indicate TOP. Indicate approximate position of figures in text in margins. Six illustrations are produced free of charge; make special arrangements with Editors and publisher for color plates or more than six illustrations. Prepare original drawings or graphs in black India ink or typographic (press-apply) lettering. Typewritten or freehand lettering is unacceptable; have all lettering done professionally. Do not send original artwork, x-ray films, or electrocardiographic tracings; submit glossy prints; good black-and-white contrast is essential.

Legends: Type double-spaced on a separate sheet. If a figure has been previously published, legend must give full credit to original source.

Tables: Type double-spaced on separate pages. Number in order of mention in text; give brief, descriptive titles. Omit horizontal or vertical rules from body. Do not submit glossy prints or reduced versions of typed tables. Explain in a footnote all acronyms, abbreviations, and unusual units of measurement in title, headings, or body of table. For footnotes, use symbols in sequence: *, †, ‡, §, ||, ¶, **, ††. If a table or any data therein have been previously published, give full credit to original source in a footnote.

References: Include only references cited in text in reference list; cite references in text by superscript numbers. Number reference list according to order of mention in text. Type list double-spaced, and conform format in "Uniform Requirements for Manuscripts Submitted to Biomedical Journals" (Vancouver style) (*N Engl J Med* 1991;324:424–8). For journal abbreviations follow style in *Cumulated Index Medicus*. For periodical references, give surnames of authors and their initials, title of article, publication name, year, volume and inclusive page numbers. For books, give surnames of authors and their initials, chapter title (if applicable), editors' surnames and initials, book title, volume number (if applicable), edition number (if applicable), city of publisher, full name of publisher, year of publication, and inclusive page numbers of citation. If six or few authors, list all; if seven or more, list first three and add *et al.*):

JOURNALS

Blaisdell WF, Clauss RH, Galbraith JG, Smith JA Jr. Joint study of extracranial arterial occlusion, IV. A review of surgical considerations. JAMA 1969;209:1899–95.

BOOKS

Bergel DH, ed. Cardiovascular dynamics, 2nd ed. London: Academic Press, Inc, 1974:115–9.

CHAPTERS

Sagawa K. The use of control theory and systems analysis. In: Bergel DH, ed. Cardiovascular dynamics, 2nd ed. London: Academic Press, Inc, 1974:115–9.

OTHER FORMATS

BRIEF CLINICAL REPORTS—Four double-spaced manuscript pages, five references. Include one or two pertinent illustrations; no abstract. Send four complete sets of manuscript.

EDITORIALS—1000 words; minimum references, expressing personal opinion of author which may include not only timely subjects of clinical interest, but also material of general interest to surgical community, including topics of social significance. Send four copies.

LETTERS TO THE EDITORS—500 words, including complete references, comments in the form of letters that express differences of opinion or supporting views of previously published editorials or recently published papers in *Surgery*. Type double-spaced; submit four sets.

OTHER

Abbreviations: Use of unusual abbreviations is discouraged, if necessary, spell out at first use, followed by abbreviation in parentheses. Use consistently. For currently accepted usage, consult *Manual for Authors & Editors* (American Medical Association), *Council of Biology Editors (CBE) Style Manual,* and *The Chicago Manual of Style* (University of Chicago Press).

Experimentation, Human: Indicate that informed consent was obtained from patients who served as subjects. If either Editors or referees question propriety of human investigation with respect to risk to subjects or to means of obtaining informed consent, *Surgery* may request more detailed information about safeguards employed and procedures used to obtain consent. Minutes of local human experimentation committees that reviewed and approved the research may be requested.

SOURCE

January 1994, 115(1): 7A-8A.

Systematic and Applied Microbiology
Gustav Fischer

Professor Dr. K. H. Schleifer
Lehrstuhl für Mikrobiologie
Technische Universität Munchen
Arcisstr. 21, D-8000 Munchen 19, Germany

AIM AND SCOPE

The journal publishes the following types of papers: Full length papers; Short communications of up to 2 printed pages without detailed references to methods and comprehensive survey of literature.

Papers will be accepted from the following branches of microbiology: 1. Systematics: for instance new descriptions and revisions of taxa, methods for determination of taxonomical and genealogical relationships. 2. Morphology and physiology: comparative studies, particularly concerning classification or phylogenetic assignment of organisms, mode of life and role in environment, importance for agriculture, food processing and biotechnology. 3. Applied microbiology: all aspects of agricultural, industrial, food and sewage microbiology, inasmuch as main emphasis concerns role or characteristics of microorganisms; studies on composition and dynamics of microorganism populations associated with manufacture and decay of all types of products; qualitative and quantitative determination procedures, etc. 4. Ecology: all aspects of soil, water and air microbiology including analysis of populations occurring at various locations, role in material cycle and effect of human activity upon them.

MANUSCRIPT SUBMISSION

Submit manuscripts in duplicate to managing editor or any member of the editorial board.

MANUSCRIPT FORMAT

Full length papers do not exceed 8 printed pages (20 pages of one-and-one-half-spaced typescript).

Paper should consist of an Introduction, Materials and Methods, Results and Discussion. Underline Latin names of organisms. Write names of authors in text in capitals. Mark descriptions of methods and less important parts of text at margin for small print.

Title Page: Include: title of paper (avoid subtitles); given name(s) and surname(s) of author(s); laboratory or institution; running title (60 letters); address to which proofs are to be sent; list of nonstandard abbreviations (standard abbreviations are those used in biochemical literature, e. g. , in *European J. Biochem. 1,* 259–266 (1967), and need not be defined).

Summary: Follow on second page.

Key Words: 10 after summary, indicating scope of paper.

References: In text, cite by author's name and year (where more than two authors, use first author's name plus et al.). In list at end of paper, cite journal papers by names and initials of all authors, full title, abbreviated journal title, volume number, first and last page numbers and year (in brackets). For books, quote names and initials of all authors, full title, edition, place of publication, publisher and year. Arrange list alphabetically by name of first author.

JOURNAL PAPER CITATION

Mitchell, T. G., Handrie, M. S., Shewan, J. M.: The taxonomy of Cytophage. J. Appl. Bact. 32, 40–50 (1969).

BOOK CITATION

Baird-Parker, A. C.: Micrococcaceae, pp. 478–490. In: Bergey's manual of determinative bacteriology (*R. E. Buchanan, N. E. Gibbons,* eds.) 8th ed., Baltimore, Williams & Wilkins 1974.

Figures: Submit good quality copies without flaws.

Line Drawings—Present schematic representations as original ink drawings or photographs. On back of copy indicate desired size in print, taking type-area into consideration. Make drawing and lettering large enough to allow a 50% reduction in size.

Halftone Photographs—Submit glossy prints, not pasted on paper. On back of print, note author's name, figure number and size reduction in soft pencil. If only certain areas of print are to be reproduced, clearly demarcate. Submit figures consisting of several component parts ready assembled (do not exceed type-area 17.2 × 22.5 cm).

Sequences, drawings, tables, chemical formulas, complicated mathematical or physical formulas, portions of genetic maps, diagrams, and flow schemes can be submitted camera-ready. Prepare carefully to conform to Journal style; submit as glossy prints.

Lettering—Use Letraset (Instant Lettering) technique for lettering figures. Choose script size so that inscriptions are at least 2 mm high following reduction.

Legends: Provide on a separate sheet of paper.

OTHER FORMATS

SHORT COMMUNICATIONS—2 printed pages, including illustrations. Precede paper by a brief summary. Division of text into sections may be dispensed with.

OTHER

Color Charges: Approx. 1100.- DM per page.

Units: Quote weights and measures data in SI units.

SOURCE

December 1992, 15(4): back matter.

Tetrahedron
Pergamon

Professor W. Steglich
Institut für Organische Chemie der Universität München
Kalistr. 23
D-80333 München 2, Germany
Fax: 49 89 5469225
Austria, Germany, Switzerland and all papers in German
Professor L. Ghosez
Laboratoire de Chemie Organique de Synthese
Universite Catholique de Louvain
Place L.Pasteur1
B-1348 Louvain-la-Neuve, Belgium
Fax: 3210 47 41 68
all papers in French
Professor Lin Guo-Qiang
Shanghai Institute of Organic Chemistry
Academia Sinica, 345 Lingling Lu
200032 Shanghai, China
Fax: 86 214376263
China, Taiwan
Professor N. K. Kochetkov
N. D. Zelinsky Institute of Organic Chemistry, Academy of Sciences, Leninsky Prospekt 47
Moscow, B-334, Russia
Fax: 7 095 135 5328
all papers in Russian
Professor T. Shioiri
Faculty of Pharmaceutical Sciences
Nagoya City University
Tanabe-dori, Mizuho-ku
Nagoya 467, Japan
Fax: 81 52 834 4172
Japan, North and South Korea, Taiwan, Hong Kong and Macao
Professor A. R. Katritzky FRS
Department of Chemistry
University of Florida
Gainesville, FL 32611-2046
Fax: (904) 392-9199

The Americas: Physical Organic Chemistry, Computer Oriented Chemistry and Heterocyclic Chemistry
Professor S. F. Martin
Department of Chemistry and Biochemistry
University of Texas
Austin, TX 78712-1167
Fax: 800 435 6857
The Americas: Natural Products: Structure and Synthesis, Synthetic Methods and Synthesis other than Heterocyclic Compounds
Professor A. McKillop
University of East Anglia
School of Chemical Sciences
University Plain
Norwich, NR4 7TJ, U.K.
Fax: 44 1603 507773
All other regions
Professor A. Nickon
Department of Chemistry
The Johns Hopkins University
Baltimore, MD 21218
Fax: (410) 516-8420
Tetrahedron Report The Americas
Professor W. B. Motherwell
Department of Chemistry
University College London
20 Gordon Street
London, WC1H 0AJ, U.K.
Fax: 44 171 380 7463
Tetrahedron Report All other regions

AIM AND SCOPE

Tetrahedron seeks to publish experimental or theoretical research results of outstanding significance and timeliness in the field of Organic Chemistry. Publications may be in the form of Articles, Reports and Symposia-in-Print. In order to achieve rapid publication, Articles and Symposia-in-Print, as well as camera-ready Reports, will be reproduced directly from the author's typescript.

Articles for *Tetrahedron* should describe original research in organic chemistry of high quality and timeliness. Because of restrictions on the size of the journal, priority will be given to those contributions describing scientific work having as broad appeal as possible to the diverse readership. We ask referees to help in the selection of articles which have this breadth and suggest that papers covering narrower aspects of the field be sent to journals specializing in those areas. Papers in French or German will be considered.

MANUSCRIPT SUBMISSION

Many word processing programs are available. Use a system that allows text and graphics to be integrated in printout. Use a laser printer (proportional Times fonts). Use any computer printer except band printers and dot matrix printers with fewer than 21 pins or typewrite (Prestige Elite typeface).

Guidelines for typefaces, styles (e.g., bold, italic) or sizes are preferences. Strictly adhere to rules regarding image area.

Submit camera-ready manuscript and graphical abstract (three copies of each) directly to Regional Editor. Determine appropriate Editor by address of main author.

MANUSCRIPT FORMAT

Submit manuscripts as camera-ready copy on good quality white paper either U.S. Letter size paper (8.5 ×11 in.) or A4 size paper (8.3 × 11.6 in.). Number manuscript pages in a front corner of each page in blue pencil and include first words of title.

Provide a graphical abstract in triplicate on separate pages for publication at front of issue. Present essence of paper in a concise, pictorial form. Start with title of paper in 10 pt Times Roman bold capitals, followed by name(s) and address(es) of author(s) in 10 pt Times Roman, and a very brief verbal description of work (25 words), also 10 pt Times Roman, followed by graphical abstract presented linearly across the page. Abstract should fit horizontally within image area (7.75 in) and should occupy a 2 in. vertical space. Leave top right corner left free. Abstract will be linearly reduced 25% during reproduction.

First page of text occupies 7.38 in. wide by 9.45 in. high. Graphics and text for other pages occupy an area of 17.38 in. wide by 10.34 in. high. For U.S .letter size paper, create an image area by setting a top margin of 0.5 in., a bottom margin of 0.25 in., left margin of 0.5 in., and right margin of 0.6 in. For A4 paper, setting should be top margin 0.8 in., bottom margin 0.5 in., left and right margins 0.4 in.

Center title of paper in bold 14-16 pt Times Roman with initial letter of each major word capitalized. Center author names in 12 pt Times Roman, giving first name, middle initial, last name and indicating corresponding author with a superscripted asterisk. Center address of authors in 10 pt Times Roman. If authors are at more than one location leave one blank line between each set. Abstract (4-6 lines) follows single-spaced in 10 pt Times Roman beginning with emboldened word Abstract: indented at both left and right margins by 0.8 in.

Print manuscript using 12 pt Times Roman font and 1.25 or 1.50 line spacing. Indent paragraphs by 0.4 in., do not separate by a blank line. Directly enter all italics, emboldening, super- and subscripts, special symbols, equations, graphs and drawings into text. Denote compounds with bold Arabic numerals where appropriate. Center major heading, bold, all capitalized, with one blank line before and after. Italicize regular headings, flush left, with one blank line before and no blank lines between it and following paragraph. Italicize and indent subheadings followed by a full stop (period); run on with body of following text. Separate subheadings from preceding text by one line.

Adjust spacing and content of paper to make page count an even number (not including graphical abstract page).

References and Notes: Present citations of references and notes in text by superscripted 8 or 10pt Arabic numerals. List of references and notes at end of text under major heading REFERENCES or REFERENCES AND NOTES. Follow reference number by a full stop (period) and tabulate to 0.4 in. from left margin. Indent following lines of reference by 0.4 in.

Abbreviate journal titles as in *Chemical Abstracts Service Source Index* (CASSI) 1907-1984 cumulative and later supplements. Give inclusive pagination. Book references include author(s), book title, editor(s), publisher, address, volume, date and pages.

1. Barton, D. H. R.; Yadav-Bhatnagar, N.; Finet, J.-P.; Khamsi, J. *Tetrahedron Lett.* **1987,** *28,* 3111–3114.

2. Katritzky, A. R. *Handbook of Heterocyclic Chemistry;* Pergamon Press: Oxford, 1985; pp. 53–86.

3. Smith, D. H.; Masinter, L. M.; Sridharan, N. S. Heuristic DENDRAL: Analysis of Molecular Structure. In *Computer Representation and Manipulation of Chemical Information*; Wipke, W. T.; Heller, S. R.; Feldmann, R. J.; Hyde, E. Eds.; John Wiley and Sons, Inc.: New York, 1974; pp. 287–298.

4. Cato, S. J. *Studies in Novel Heterocyclic Polymer,* University of Florida 1987.

Figures, Schemes and Graphs: Prepare to actual size and insert in text appropriately close to first reference. For text in figures use Times Roman or Helvetica between 10 and 12 pt. Make figure captions 12 pt Times Roman, centered 0.2 in. below figure.Treat schemes similarly. Number figures and schemes in Arabic numerals. Glue, do not tape, materials that cannot be directly printed in appropriate place in text above corresponding caption.

Tables: Give captions with initial letters of major words capitalized. Below caption is a blank line then column headings, a blank line, body of table, a blank line, followed by any notes to table. Align decimal points. Place short notes to table below last line in 8 or 10 pt type.

OTHER FORMATS

REPORTS—Specially commissioned critical reports reviewing research results of topical importance in appropriate fields. Submit either in camera-ready form (for expedited publication) or to be typeset by printer. Submit all tables, figures and drawings as camera-ready copy.

SYMPOSIA-IN-PRINT—Collections of original research papers (including experimental sections) covering timely areas of Organic Chemistry. Guest Editor will invite authors to submit papers which are reviewed and processed for publication by Guest Editor under the usual refereeing system. Opportunity is also provided for other active investigators to submit contributions.

OTHER

Chemical Structures: Draw using standard bond lengths so that all structures are to same scale. Make final size of lettering 12 pt, i.e., the same size as text of article.

Equations: Number with Arabic numerals in parentheses against right margin. Separate from body text by one blank line above and below.

New Compounds: Fully characterize all new compounds with relevant spectroscopic data. Include microanalyses whenever possible. High resolution mass spectra may serve in lieu of microanalysis, if accompanied by suitable NMR criteria for sample homogeneity.

Nomenclature, Organic Chemical: Consult *IUPAC Nomenclature of Organic Chemistry,* (The Blue Book), edited by J. Rigaudy and S. P. Klesney, Pergamon Press, Oxford, 1979.

Reviewers: Suggest names of suitable referees.

Safety: Draw attention to hazardous compounds by adding word CAUTION followed by a brief descriptive phrase and literature reference if appropriate.

Trademarks: Acknowledge trademark protection in standard fashion, e.g., K-Selectride ®. Do not use for words which have entered general usage: pyrex, etc. Use letters TM or R in superscripted 8 or 10 pt type for trademarks and registered trademarks, respectively.

X-Ray Crystallographic Data: Submit lists of refined coordinates and e.s.d's for deposition by Editor at Cambridge Crystallographic Data Center.

SOURCE

January 2, 1995, 51(1): I-VII.

Tetrahedron Letters
Pergamon

Professor W. Steglich
Institut für Organische Chemie der Universität München
Kalistr. 23
D-80333 München 2, Germany
Fax: 49 89 5469225
    Austria, Germany, Switzerland and all papers in German

Professor G. Ourisson
Centre National de la Recherche Scientifique
Centre de Neurochimie
5 rue Blaise Pascal
67084 Strasbourg, France
Fax: 33 88 607620
    all papers in French

Professor Lin Guo-Qiang
Shanghai Institute of Organic Chemistry
Academia Sinica, 345 Lingling Lu
200032 Shanghai, China
Fax: 86 214376263
    China, Taiwan

Professor S. Ito
Faculty of Pharmaceutical Sciences
Tokushima Bunri University
Yamashiro-cho, Tokushima 770, Japan
Fax: 81 886 55 8774
    Japan, North and South Korea, Taiwan, Hong Kong and Macao

Professor H. H. Wasserman
Department of Chemistry, Yale University
P. O. Box 208107
New Haven, CT 06520-8107
Fax: (203) 432-9990
    The Americas

Professor N. K. Kochetkov
N. D. Zelinsky Institute of Organic Chemistry, Academy of Sciences, Leninsky Prospekt 47
Moscow, B-334, Russia
Fax: 7 095 135 5328
    all papers in Russian

Professor A. McKillop
University of East Anglia
School of Chemical Sciences
University Plain
Norwich, NR4 7TJ, U.K.
Fax: 44 1603 507773
    All other regions

AIM AND SCOPE

Tetrahedron Letters is a medium for the rapid publication of short communications in English (papers in French or German are also accepted) and offers maximum dissemination of outstanding developments in Organic Chemistry. The Letters attempts to provide rapid publication of important new contributions, and in order to do this, authors' typescripts will be directly reproduced. Communications may announce either experimental or theoretical results of special interest, and will be evaluated for their novelty, timeliness and general interest.

MANUSCRIPT SUBMISSION

Submit camera-ready manuscript and graphical abstract (three copies of each) to Regional Editor. Fax a copy (supplying their own telephone and fax numbers) to Regional Editor and mail camera-ready manuscript. Appropriate Editor is determined by permanent address of main author.

Many word processing programs are available. Use a system that allows text and graphics to be integrated in printout. Use a laser printer (proportional Times fonts). Use any computer printer except band printers and dot matrix printers with fewer than 21 pins or typewrite (Prestige Elite typeface).

Guidelines for typefaces, styles (e.g., bold, italic) or sizes are preferences. Strictly adhere to rules regarding image area.

MANUSCRIPT FORMAT

Submit manuscripts as camera-ready copy on good quality white paper either U.S. letter size paper (8.5 ×11 in.) or A4 size paper (8.3 × 11.6 in.). Number manuscript pages in a front corner of each page in blue pencil and include first words of title.

Provide a graphical abstract in triplicate on separate pages for publication at front of issue. Present essence of paper in a concise, pictorial form. Start with title of paper in 10 pt Times Roman bold capitals, followed by name(s) and address(es) of author(s) in 10 pt Times Roman, and a very brief verbal description of work (25 words), also 10 pt Times Roman, followed by graphical abstract presented linearly across the page. Abstract should fit horizontally within image area (7.75 in) and should occupy a 2 in. vertical space. Leave top right corner left free. Abstract will be linearly reduced 25% during reproduction.

First page of text occupies 7.38 in. wide by 9.45 in. high. Graphics and text for other pages occupy an area of 17.38 in. wide by 10.34 in. high. For U. S letter size paper, create an image area by setting a top margin of 0.5 in., a bottom margin of 0.25 in., left margin of 0.5 in., and right margin of 0.6 in. For A4 paper, setting should be top margin 0.8 in., bottom margin 0.5 in., left and right margins 0.4 in.

Center title of paper in bold 14-16 pt Times Roman with initial letter of each major word capitalized. Center author names in 12 pt Times Roman, giving first name, middle initial, last name and indicating corresponding author with a superscripted asterisk. Center address of authors in l0 pt Times Roman. If authors are at more than one location leave one blank line between each set. Abstract (4-6 lines) follows single-spaced in l0 pt Times Roman beginning with emboldened word Abstract: indented at both left and right margins by 0.8 in.

Print manuscript using 12pt Times Roman font and 1.25 or 1.50 line spacing. Indent paragraphs by 0.4 in. Do not separate by a blank line. Directly enter all italics, emboldening, super- and subscripts, special symbols, equations, graphs and drawings into text. Denote compounds with bold Arabic numerals where appropriate. Center, boldface, use all capitals, and insert one blank line before and after major headings. Italicize, indent and follow subheadings a full stop (period), but run on with body of following text. Separate subheadings from preceding text by one line. Adjust spacing and content of paper to make page count 2 or 4 pages (not including page with graphical abstract).

References and Notes: Present citations of references and notes in text by superscripted 8 or 10 pt Arabic numerals. List of references and notes at end of text under major heading REFERENCES or REFERENCES AND NOTES. Follow reference number by a full stop (period) and tabulate to 0.4 in. from left margin. Indent following lines of reference by 0.4 in.

Abbreviate journal titles as in *Chemical Abstracts Service Source Index* (CASSI) 1907-1984 cumulative and later supplements. Give inclusive pagination. Book references include author(s), book title, editor(s), publisher, address, volume, date and pages.

1. Barton, D. H. R.; Yadav-Bhatnagar, N.; Finet, J.-P.; Khamsi, J. *Tetrahedron Lett.* **1987,** *28,* 3111–3114.

2. Katritzky, A. R. *Handbook of Heterocyclic Chemistry;* Pergamon Press: Oxford, 1985; pp. 53–86.

3. Smith, D. H.; Masinter, L. M.; Sridharan, N. S. Heuristic DENDRAL: Analysis of Molecular Structure. In *Computer Representation and Manipulation of Chemical Information;* Wipke, W. T.; Heller, S. R.; Feldmann, R. J.; Hyde, E. Eds.; John Wiley and Sons, Inc.: New York, 1974; pp. 287–298.

4. Cato, S. J. *Studies in Novel Heterocyclic Polymer,* University of Florida 1987.

Figures, Schemes and Graphs: Prepare to actual size and insert in text appropriately close to first reference. For text in figures use Times Roman or Helvetica between 10 and 12 pt. Make figure captions 12 pt Times Roman, centered 0.2 in. below figure.Treat schemes similarly. Number figures and schemes in Arabic numerals. Glue, do not

tape, materials that cannot be directly printed in appropriate place in text above corresponding caption.

Tables: Give captions with initial letters of major words capitalized. Below caption is a blank line then column headings, a blank line, body of table, a blank line, followed by any notes to table. Align decimal points. Place short notes to table below last line in 8 or 10 pt type.

OTHER FORMATS

REPORTS—Specially commissioned critical reports reviewing research results of topical importance in appropriate fields. Submit either in camera-ready fonn (for expedited publication) or to be typeset by printer. Submit all tables, figures and drawings as camera-ready copy.

SYMPOSIA-IN-PRINT—Collections of original research papers (including experimental sections) covering timely areas of Organic Chemistry. Guest Editor will invite authors to submit papers which are reviewed under the usual refereeing system. Opportunity is also provided for other active investigators to submit contributions.

OTHER

Chemical Structures: Draw using standard bond lengths so that all structures are to same scale. Make final size of lettering 12 pt, i.e., the same size as text of article.

Equations: Number with Arabic numerals in parentheses against right margin. Separate from body text by one blank line above and below.

New Compounds: Fully characterize all new compounds with relevant spectroscopic data. Include microanalyses whenever possible. High resolution mass spectra may serve in lieu of microanalysis, if accompanied by suitable NMR criteria for sample homogeneity.

Nomenclature, Organic Chemical: Consult *IUPAC Nomenclature of Organic Chemistry,* , edited by J. Rigaudy and S. P. Klesney, Pergamon Press, Oxford, 1979.

Reviewers: Suggest names of suitable referees.

Safety: Draw attention to hazardous compounds by adding word CAUTION followed by a brief descriptive phrase and literature reference if appropriate.

Trademarks: Acknowledge trademark protection in standard fashion, e.g., K-Selectride®. Do not use for words which have entered general usage: pyrex, etc. Use letters TM or R in superscripted 8 or l0 pt type for trademarks and registered trademarks respectively.

X-Ray Crystallographic Data: Submit lists of refined coordinates and e.s.d's for deposition by Editor at Cambridge Crystallographic Data Centre.

SOURCE
January 3, 1994, 35(1): I-VI.

Theoretical and Applied Genetics
Springer International

L. Alföldi
Magyar Tudomanyos Akademia
Szegodi Biologiai Központ

Genetikai Intezete, Szeged, Pl. 521 Hungary
          Somatic cell genetics, microbial genetics

J. S. F. Barker
Department of Animal Science
The University of New England
Armidale, N. S. W. 2351 Australia
          Animal breeding, experimental quantitative ge-
          netics (animals), population genetics

J. S. Beckmann
Centre d'Etude du Polymorphisme Humain (C.F.P.H.)
27 rue Juliciic Dodu.
F-Paris, 75010 France
          Molecular genetics, RFLPs, structure and func-
          tion of genes

H. K. Dooner
Waksman Institute
P. O. Box 0759
Piscataway, NJ 08855
          Plant classical and molecular genetics

E. J. Eisen
North Carolina State University
School of Agriculture and Life Sciences
Department of Animal Science
Box 7621
Raleigh, NC 27695-7621
          Animal breeding and quantitative genetics

Yu. Gleba
American Cyanamid Company
Agricultural Research Division
P.O. Box 400
Princeton, NJ 08543-0400
Fax: (609) 799-1842
          Plant cell genetics, somatic cell hybridization,
          plant cell and protoplast culture

R. Hagemann
Institut für Genetik
Fachbereich Biologie
Martin-Luther-Universität
Domplatz J, D-06108 Halle, Saale, Germany
          Plasmatic and plastid inheritance, genetic insta-
          bility, paramutation, tomato genetics

G. E. Hart
Department of Soil and Crop Sciences
Texas A & M University
College Station, TX 77843-2474
          Genetics of crop plants (especially cereals), genome
          mapping, biochemical genetics, cytogenetics

A. L. Kahler
Biogenetics Services, Inc.
2308 6th Street E., P. O. Box 710
Brookings, SD 57006
          Plant genetics, population genetic, plant breed-
          ing and production

M. Koormneef
Agricultural University
Department of Genetics
Dreijenlaan 2, 6703 Wageningen, The Netherlands
          Plant genetics, especially *Ardridopsis* and tomato

H. F. Linskens
c/o PRODUserv Springer Produktions-Gesellschaft
Heidelberger Platz 3, D-14197 Berlin, Germany
          Biochemical genetics, incompatibility, incongruity

J. MacKey
Swedish University of Agricultural Sciences
Department of Plant Breeding
P.O. Box 7003, S-750 07 Uppsala, Sweden
          Autogamous plant breeding, taxonomy of crop
          plants, yield structure, disease resistance

P. Maliga
Waksman Institute
Rutgers, the State University of New Jersey
Piscataway, NJ 08855-0759
          Plant cell genetics

F. Mechelke
Universität Hohenheim
Institut für Genetik (240)
D-70593 Stuttgart, Germany
          Structure, function and evolution of chromo-
          somes

G. Melchers
Max-Planck-Institut für Biologie
Spemannstr. 37, Postfach 2109
D-72076 Tübingen, Germany
          Combination of cell and conventional genetics

P. L. Plahter
Agronomy Department
Institute of Food and Agricultural Sciences
University of Florida
Gainesville, FL 32611
          Genetics and breeding of maize, rye and oats

I. Potrykus
Institut für Pflanzenzwissenschaften
ETII Zentrum, Universitätsstrasse 2
CII-8092 Zürich, Switzerland
          Plant cell genetics

D. R. Pring
USDA-ARS, Plant Pathology Department
1453 Fifield Hall, University of Florida
Gainesville, FL 32611
          Cytoplasmic male sterility, cytoplasmic inherit-
          ance, organelle molecular biology

F. Salamini
Max-Planck-Institut für Züchtungsforschung
D-50829 Köln-Vogelsang, Germany
          Genetics and breeding of cereals, biochemical
          and developmental genetics of plants

C. Smith
University of Guelph
Ontario Agricultural College
Department of Animal and Poultry Science
N1G 2W1, Canada
          Animal breeding and quantitative genetics

J. W. Snape
Cambridge Laboratory
JI Centre for Plant Science Research
Colney Lane, Norwich, NR4 7UJ Great Britain
          Classical and biotechnological breeding of crop
          plants

P.M.A. Tigerstedt
Department of Plant Biology
Faculty of Agriculture and Forestry
University of Helsinki
SI-00710 Helsinki 71 Finland
>   Population and ecological genetics of plants,
>   plant breeding theory, genetics of forest eco-
>   systems

K. Tsunewaki
Laboratory of Genetics
Faculty of Agriculture
Kyoto University, Kyoto 606 Japan
>   Plant genetics (especially of wheat), cytogenet-
>   ics, cytoplasma inheritance

L. D. Van Vieck
A-218 Animal Sciences
University of Nebraska
Lincoln, NE 68583-0908
>   Quantitative genetics especially as applied to
>   animals

G. Wenzel, Managing Editor
Lehrstuhl für Pflanzenbau und Pflanzenzüchtung
TU München-Welhenstephan
D-85350 Freising (Germany)
>   Cell genetics and haploids in plant breeding

AIM AND SCOPE

TAG will publish original articles in the following areas:
Genetic fundamentals of plant and animal breeding; Physiological fundamentals of plant and animal breeding; Application of cell genetics to breeding.

MANUSCRIPT SUBMISSION

Limit papers to six printed pages (18 manuscript pages including figures and tables).

Submission of a manuscript implies: that work described has not been published before (except as an abstract or as part of a published lecture, review, or thesis); that it is not under consideration for publication elsewhere; that its publication has been approved by all coauthors, and by responsible authorities at institute where work was done; that, if and when manuscript is accepted for publication, authors agree to automatic transfer of copyright to publisher; and that manuscript will not be published elsewhere in any language without consent of copyright holders.

MANUSCRIPT FORMAT

Write in standard grammatical English. Submit in duplicate with one set of original illustrations to one of the Editors according to his field of competence or to the Managing Editor.

Type double-spaced on one side of paper, clearly legible and completely ready for press. Photocopy additional copies on both sides.

Arrange material under appropriate headings and present in a good, readable style. Precede each contribution with a summary. Immediately following, supply up to five key words, indicating scope of paper.

Mark passages to be printed in small print with a vertical line in left margin. Number footnotes, other than those referring to title heading, consecutively. Underline genus and species names in blue. Do not mark anything else. Check that symbols are used consistently in figures, tables and text (lowercase or capital letters, Arabic or Roman numerals).

Title Page: Title of paper, initial(s) of first name(s) and surname(s) of author(s), complete address of author's institute, footnotes referring to title (indicate by asterisks). Indicate to whom proofs should be sent.

Tables: Number with Arabic numerals, separately from figures. Give each a heading. Do not include in text but submit together at end of paper. Give footnotes in tables in small letters consecutively.

References: Place list of references after text. Arrange as:

Hutchinson (1971 a)
Hutchinson (1971 b)
Hutchinson (1973)
Hutchinson and Hamilton (1969)
Hutchinson and Smith (1968)
Hutchinson, Smith, and Hamilton (1953)
Hutchinson, Hamilton, and Smith (1963).

For journal title abbreviations, use *Bibliographic Guide for Editors and Authors* (American Chemical Society).

Falk DE, Kasha KJ (1983) Genetic studies of the cross-ability of hexaploid wheat with rye and *Hordeum bulbosum*. Theor Appl Genet 64: 303–307.

Michelsen A (1974) Hearing in invertebrates. In: Keidel WD, Neff WD (eds) Auditory system. Springer, Berlin Heidelberg New York (Handbook of sensory physiology, vol V/1, pp 389–422)

Paigen K (1980 Temporal geness. In: Leighton T, Loomis WF (eds The molecular genetics of development. Academic Press, London New York, pp 419–470

To cite in text: Solari and Favret (1967) or (Michelsen 1974). If more than two authors, name only the first, followed by "et al."

Illustrations: No more than 10. Originals must be reproducible. If desired size is not stated, publisher may reduce or enlarge. Prepare for reduction to column width (8.6 cm) or smaller and up to full page size (17.6 cm). Originals should have maximum width of 17.2 or 35.2 cm for 50% reduction.

Submit clear, glossy photos of excellent quality for halftones and line drawings. Make inscriptions and symbols clear and large enough to be legible (2-3 cm high after reduction). Assemble illustrations with several parts (marked a,b,c...) in correct order; observe maximum width and allow 2 mm space between parts. Block cannot exceed 45 cm in height. Mark author's name and figure number on reverse.

Submit figures in plates mounted on regular paper, not cardboard.

Check that figures and tables are referred to in text. Mark desired position of illustrations and tables in margin of manuscript.

Figure Legends: Make as brief as possible. Attach to manuscript on a separate sheet.

OTHER

Color Charges: DM 1200.- for first and DM 600.- for each additional page.

Electronic Submission: Submit diskette only after copyediting. Submit text on diskettes formatted for DOS or Macintosh systems. Store text in two versions: standard data file format of word processing system; and a data interchange format (RFT, DCA, DCA/FFT, or ASCII or text only). Diskette must match final printout exactly. Do not incorporate special page layout in text. Input text continuously; insert hard returns only at ends of paragraphs, headings, lists, etc. Do not use spacebar to make indents. Use tabulator or indent command. Use automatic paginating function; do not insert page numbers manually. Italicize or underline words or phrases to be emphasized. Use boldface only in tables (word "Table" and table numbers) and figure legends (abbreviation "Fig." and figure number). Main headings can be boldfaced. Place tables at end of file. Separate columns with tabulators, not spacebar. Delete annotations and comments from file. To code hyphens/dashes: hyphen=high-resolution; en-dash=1990--1992; em-dash=Bacteria -- often in high; minus sign=temperature of --75 C.

Page Charges: DM 200.00 for each additional printed page over six.

SOURCE

March 1995, 90(3-4i): inside back cover.

Thorax
British Thoracic Society
BMJ Publishing Group

Dr. S. G. Spiro
Private Patients' Wing, University College Hospital
25 Grafton Way
London WC1E 6DB England

AIM AND SCOPE

Thorax is the journal of the British Thoracic Society. It is intended primarily for the publication of original work relevant to diseases of the thorax.

MANUSCRIPT SUBMISSION

Contributors do not have to be members of the society. Papers are returned if presented in an inappropriate form.

Submit original typescript and three copies of all papers. Editorial and historical articles are normally commissioned; Editor may accept uncommissioned articles. Submit declaration, signed by all authors, that paper is not under consideration by any other journal at the same time and that it has not been accepted for publication elsewhere. Include name, address and telephone and fax numbers of corresponding author. Authors may be asked to supply copies of similar material they have published previously. Papers may undergo editorial revision.

MANUSCRIPT FORMAT

Follow requirements of International Steering Committee of Medical Editors (*BMJ* 1979;i:532–5). Type double-spaced with wide margins for correction on one side of paper only. Include a structured abstract on a separate sheet. Include adequate reference to previous work on subject.

Abstract: 250 words. State clearly why study was done, how it was carried out (including number and brief details of subjects drug doses, and experimental design), results, and main conclusions. Structure under headings "Background," "Methods," "Results," and "Conclusions."

Illustrations: Prepare line drawings, graphs, and diagrams to professional standards. Submit as originals or as unmounted glossy photographic prints. Take particular care with photomicrographs, detail is easily lost. It is often more informative to show a small area at a high magnification than a large area. Use scale bars to indicate magnification. For size of symbols and lettering (upper and lowercase, not all capitals) and thickness of lines take account of reduction usually to a width of 65 mm. Submit three copies of each. Affix to each a label on back marked in pencil with names of authors and number of figure; indicate TOP. Type legends on a separate sheet. Authors pay for color illustrations.

References: Check for accuracy and completeness. Reference to work published in abstract form is allowed only in exceptional circumstances, for example, to acknowledge priority or indebtedness for ideas. Number in order of mention. Identify in text, tables, and legends to figures by Arabic numerals above the line. Number references cited only (or first) in tables or legends s according to where table or figure is first mentioned in text. Type list of references double-spaced in numerical order on separate sheets. Include reference number, authors' names and initials (all authors if less than six, if more, follow first six names by *et al*), title of article, and for journal articles name of journal (abbreviated according to *Index Medicus*), year of publication, volume, and first and last page numbers. Order and punctuation are important.

1 Anderson HR. Chronic lung disease in the Papua New Guinea Highlands. *Thorax* 1979;34: 647–53.

2 Green AB, Brown CD. *Textbook of pulmonary disease.* 2nd ed. London: Silver Books, 1982:49.

3 Grey EF. Cystic fibrosis. In: Green AB, Brown CD, eds. *Textbook of pulmonary disease*. London: Silver Book, 1982:349–62.

OTHER FORMATS

CASE RESPORTS—850 words, one table or illustration, a short unstructured abstract, and 10 references. A single case study.

CORRESPONDENCE—300 words and three references (listed at end). Letters related to articles published in *Thorax*. Type double-spaced with wide margins; must be signed by all authors.

SHORT REPORTS—Two pages, 1400 words, inclusive of structured abstract, tables, illustrations and references. Experimental work, new methods, or a preliminary report.

OTHER

Experimentation, Human: Descriptions of experimental procedures on patients not essential for the investigation or treatment of their condition must include written assurance that they were carried out with informed consent of subjects and with agreement of local ethics committee.

Statistical Analyses: Refer to Altman DG, Gore SM, Gardner MJ, Pocock SJ. Statistical guidelines for contributors to medical journals, *BMJ* 1983;286:1489–93. Name statistical methods used and give details of randomization procedures. For large numbers of observations, give mean values and an estimate of the scatter (usually 95% confidence intervals) with a footnote stating from whom full data may be obtained. Give the power of the study to detect a significant difference when appropriate; may be requested by referees. Give standard deviation (SD) and standard error (SE) in parenthesis (not preceded by ±) and identify by SD or SE at first mention.

Units: Cite units in which measurements were made. If not SI units, give factors for conversion as a footnote.

SOURCE

January 1994, 49(1): inside front cover.

Thrombosis and Haemostasis
International Society on Thrombosis and Haemostasis
Schattauer

Prof. Dr. J. Vermylen
University of Leuven, Campus Gasthuisberg
O & N, Herestraat 49, B-3000 Leuven, Belgium
Telephone: (32) 16 34 6015; Fax: (32) 16 34 6016

AIM AND SCOPE

Thrombosis and Haemostasis publishes original papers in English contributing new information about any aspect of thrombosis and hemostasis. Papers with laboratory or clinical orientation, original articles, letters to the editor and review articles will be accepted.

MANUSCRIPT SUBMISSION

Do not submit material which has already been printed elsewhere or is planned for submission to another journal. Submit a statement from authors that manuscript or essential parts of it have not been published. Scientific meeting abstracts do not qualify for publication. In cover letter, specify Table of Contents category to which manuscript should be assigned: Atherosclerosis, Clinical Studies, Coagulation, Fibrinolysis, Molecular Biology, Platelets and Vessel Wall. Submit original typescript and 2 copies including tables, plus 2 prints of figures and original figures.

MANUSCRIPT FORMAT

Submit in a form suitable for sending directly to printer. Type double-spaced, on one side of good quality paper with wide margins; number all pages. Collate typescript in order: title page; summary (150 words); introduction; materials and methods; results; discussion; acknowledgments; references; tables; legends to figures and tables.

Title Page: Title of paper, name(s) of author(s) and institution where work was done. Include name, address and fax number of author responsible for reading proofs.

References: Identify by Arabic numerals; number consecutively in order of first mention in text. Arrange reference list accordingly with names of all authors, entire title of article cited, abbreviated name of journal according to *Index Medicus*, year of publication, volume number and first and last page numbers. Refer to books by giving all authors, full title of chapter followed by In: full title of book, editors (eds), publisher, place and year of publication and number of pages.

JOURNALS

Quick AJ, Stanely-Brown M, Bancroft FW. A study of the coagulation defect in hemophilia and in jaundice. Am J Med Sci 1935; 190: 501–11.

CHAPTER IN A BOOK

Hawiger J. Adhesive Interaction of Blood Cells and Vessel Wall. In: Hemostasis and Thrombosis. Basic Principles and Clinical Practice. Colman RW, Hirsh J, Marder VJ, Salzman EW, eds. J.B. Lippincott Company, Philadelphia, PA 1987; pp 3–17.

Tables: Type each on a separate sheet; number appropriately. Cite in numerical order in text using Arabic numbers. Give each an explicit descriptive title. Include experimental detail or explanatory comments in legend below table.

Illustrations: Refer to in text as: *Fig. 1, Figs. 1–5*. Submit original diagrams or good quality glossy prints. Draw original diagrams at least twice the size of final reproduction with lines sufficiently thick and letters and symbols large enough to stand reduction. Use * symbol only for statistical analysis. Submit photographs and photomicrographs as unmounted glossy prints. Color illustrations are charged to author. Identify original drawings and photographs on back in pencil; indicate figure number, author(s), title of paper and TOP (indicate by an arrow).

Legends to Figures and Tables: Type on a separate sheet.

Footnotes: Do not use.

OTHER FORMATS

ANNOUNCEMENTS—Scientific meetings, conferences and job opportunities that are of interest to readership. Submit at least 4 months before intended date of publication.

LETTERS TO THE EDITOR—Two typed pages including no more than 1 table or 1 figure.

RAPID COMMUNICATIONS—10 double-spaced typed pages including tables, figures and references. Brief papers of original work of scientific importance. Does not include case studies, methods papers or preliminary reports. Short papers that do not meet criteria may be reviewed as regular manuscripts upon approval of authors.

REVIEW ARTICLES—Survey recent developments on: Atherosclerosis, Clinical Studies, Coagulation Fibrinolysis, Molecular Biology, Platelets and Vessel Wall. Solicited by Editors, but individual suggestions are welcome. Contact Editor before preparing a review paper.

OTHER

Nomenclature and Abbreviations: Follow internationally approved rules.

Statistical Analyses: Describe statistical techniques used in analysis and presentation of results. When a measure of variation is reported (e.g., 14.2 ± 3.7), state which measure has been calculated (e.g., standard deviation, standard error of mean, etc.) and from how many observations; this also applies to error bars in figures. When a level of statistical significance is quoted, indicate test used. For common tests the name (e.g., paired t-test) is sufficient; otherwise, give a reference or more detailed description.

Submission Fees: $50. Pay by check in U.S. currency payable to "ISTH" and drawn on a U.S. bank of U.S. branch; an institutional purchase order; or by Visa or MasterCard/EuroCard. Mark payment with first author's name. When paying by institutional purchase order or using Visa or MasterCard/EuroCard, corresponding author shall include in letter of submission a request for a "Manuscript Processing Fee Invoice." If Visa or MasterCard/EuroCard is used, processing fee is $55.

ISTH address: International Headquarters of ISTH, 416 Burnett-Womack CB #7035, UNC Medical School, Chapel Hill, NC 27599-7035. Telephone: (919) 929-3807; Fax: (919) 929-3935.

SOURCE

January 1994, 71(1): vii.

Thrombosis Research
Pergamon

Birger Blomback
Executive Editorial Office, *Thrombosis Research*
Karolinska Institutet
Doktorsringen 4
S-171 77 Stockholm, Sweden
Telephone and Fax: +46 (8) 30 74 20

Colvin M. Redman
The New York Blood Center
310 East 67th Street
New York, N.Y. 10021
Telephone: (212)570-3059; Fax: (212) 879-0243

Editors:

Gerardo Casillias
Sección Hemostasia y Thrombosis
Instituto de Investigaciones Hematológicas
Academia Nacional de Medicina
J.A. Pacheco de Melo 3081
(1425) Buenos Aires, Argentina

Klaus Lechner
Department of Hematology and Blood Coagulation
University Hospital
Währinger Gürtel 18-20
A-10,0 Vlenna, Austria

R. L. Kinlough-Rathbone
Department of Pathology, Room 2E5C, HSC
McMaster University
1200 Main Street West
Hamilton, Ontario, L8N 3Z5 Canada

Chen Wen-Chieh
Chinese Academy of Medical Sciences

Institute of Hematology
288 Nanjins Road, Tianjin 300 020 China

Josef Hladovec
Institute for Clinical and Experimental Medicine
4-Krc. Videnska 800, 14622 Praha, Czech Republic

Ulla Hedner
Novo Industri A/S
Niels Steensensvej
DK-2880 Gentofte, Denmark

Vesa Rasi
Finnish Red Cross
Blood Transfusion Service
Kivihaantie 7
SF-003 10 Helsinki, Finland

Hugues Chap
Inserm Unité 326, Hôpital Purpan
Place du Docteur Baylac
F-310 59 Toulouse Cedex, France

Sylviane Levy-Toledano
Inserm Unité 348
Hôpital Lariboisère
8, rue Guy Patin
F-75475 Paris Cedex 10, France

Claudine Soria
Laboratoire d'Hématologie
Hôpital Lariboisière
2 rue Ambroise Paré
75475 Paris Cedex 10, France

Wolfram Bode
Max-Planck-Institut für Biochemie
Am Klopferspitz 18 a
D-82152 Martinsried, Germany

Klaus T. Preissner
Max-Planck-Gesellschaft
Kerckhof-klinik, Sprudelhof 11
D-61231 Bad Nauheim, Germany

Rüdiger E. Scharf
Institute of Experimental Hematology and Transfusion Medicine
University of Bonn
Sigmund-Freud-Str. 25
D-53105 Bonn, Germany

Jörg Stürzebecher
Zentrum für Vaskuläre Biologie und Medizin
Institut für Biochemie
Nordhäuser Straße 78
D-99089 Erfurt, Germany

T. Mandalaki
Haemophilia Treatment Center
2nd Regional Blood Transfusion Center
Laiko General Hospital
Agiou Thoma 17,11527
Athens, Greece

Laszlo Muszbek
Department of Clinical Chemistry
University School of Medicine - PF 40
4012 Debrecen, Hungary

Marco Cattaneo
Hemophilia and Thrombosis Center
Maggiore Hospital

Via Pace, 9
1-20122 Milano, Italy

Giovanni De Gaetano
Istituto "Mario Negri"
Consorzio "Mario Negri Sud"
Centro di Ricerche Farmacologiche e Biomediche
Via Nazionale
66030 S. Maria Imbaro (Chieti), Italy

Giovanni Di Minno
Clinica Medica II
Nuovo Policlinico
Via S. Pansini 5
I-80131 Napoli, Italy

Yasuo Ikeda
Division of Hematology
School of Medicine, Keio University
35 Shinanomachi, Shinjuku-ku
Tokyo 160, Japan

Sadaaki Iwanaga
Department of Biology, Faculty of Science
Kyushu University 33
Fukuoka 812, Japan

Hidehiko Saito
First Department of Internal Medicine
Nagoya School of Medicine
65 Tsurumal-cho, Showa-ku
Nagoya 466 Japan

Koji Suzuki
Department of Molecular Biology on Genetic Disease
Mie University School of Medicine
Tsu-city, Mie 514, Japan

Yumiko Takada
Department of Physiology
Hamamatsu University, School of Medicine
3600 Handa-cho, Hamamatsu-shi
Shizuioka-Ken 431-31, Japan

Willem Nieuwenhuizen
Ivvo Tno
Gaubius Laboratory
P.O. Box 430, 2300 AK Leiden, Netherlands

Peter Kierulf
Clinical Chemistry Department
Ulleval University Hospital
N-0407 Oslo 4, Norway

Nils Olav Solum
Research Institute for Internal Medicine
Section on Thrombosis and Hemostasis
Rikshospitalet, N-0027 Oslo 1, Norway

Czeslaw S. Cierniewski
Department of Biophysics
Medical University
3 Lindley Street, 90-131 Lodz, Poland

Mark A. Rosenfeld
Institute of Chemical Physics
Academy of Science
4 Kosygin Street, Moscow 117 977, Russia

Eva Bastida
Servicio de Hemoterapia y Hemostasia
Hospital Clinic i Provincial

Villarroel, 170
08036 Barcelona, Spain

Björn Dahlbäck
University of Lund, Department of Clinical Chemistry
Malmö General Hospital
S-214 01 Malmö, Sweden

Jesper Swedenborg
Department of Surgery
Karolinska Hospital
S-171 76 Stockholm, Sweden

Michael Furlan
Haematologisches Zentrallabor Inselspital
CH-3010 Bern, Switzerland

Volodymir Kibirev
Institute of Bioorganic Chemistry and Oil Chemistry
Academy of Sciences of Ukraine
Murmanskaya St. 1
Kiev 252660, Ukraine

Edzard Ernst
Center for Complementary Health Studies
Streatham Court, Rennes Drive
Exeter EX4 4PU, United Kingdom

Patrick J. Gaffney
National Institute of Biological Standards and Control
Potters Bar
Hertfordshire EN6 3QG, United Kingdom

David A. Lane
Department of Haematology
Charing Cross Hospital
Fulham Road
London W6 8RF, United Kingdom

Jerard Seghatchian
North London Blood Transfusion Centre
Colindale Avenue
London NW9 5BG, United Kindgom

Elspeth B. Smith
University of Aberdeen
Medical School Buildings
Department of Clinical Biochemistry
Aberdeen Royal Infirmary
Foresterhill, Aberdeen AB9 2ZD, United Kingdom

Alessandra Bini
The Lindsley Kimball Research Institute
The New York Blood Center
310 East 67th Street
New York, NY 10021

Randal J. Kaufman
Howard Hughes Medical Institute
University of Michigan Medical Center
MSSR II, Room 4554
1150 West Medical Center Drive
Ann Arbor, MI 48109-0650

Ariel G. Loewy
Department of Biology
Haverford College
370 Lancaster Avenue
Haverford, PA 19041-1392

Susan T. Lord
Department of Pathology
The School of Medicine, CB#7525

University of North Carolina
Chapel Hill, NC 27599-7525
Joel L. Moake
Biomedical Engineering Laboratory
Rice University
6565 Fannin, Mail Station 902
Houston, TX 77030
Roman Procyk
Bristol-Myers Squibb, H.3119
P.O. Box 4000
Route 206 and Province Line Road
Princeton, NJ 08543-4000
John R. Shainoff
Research Division, The Cleveland Clinic
Cleveland, OH 44195
Joan Sobel
Columbia University
Department of Medicine
630 West 168th Street
New York, NY 10032
James G. White
Depts. of Laboratory Medicine and Pathology, Pediatrics
    Medical School
University of Minnesota, Box 490
Minneapolis, MN 55455

AIM AND SCOPE

Thrombosis Research serves as a world forum for the rapid dissemination of original research on hemostasis, bleeding disorders, thrombosis-atherogenesis and thrombolysis. The Editors welcome scientific contributions which give new directions to research in these areas. The Journal's policy encourages the publication of research, including clinical investigations, which will lead to new approaches in diagnosis, therapy, prognosis and prevention of thrombotic and hemorrhagic diseases. The policy is also to publish selected State of the Art Articles pertaining to thrombosis and hemostasis or other rapidly expanding fields of research.

Topics include: Cell biology, biochemistry and physiology of blood and vessel wall; aggregation and disaggregation of blood cellular elements; molecular biology and molecular genetics of hemostatic and fibrinolytic factors; structure and function of factors involved in hemostasis, thrombosis and fibrinolysis; rheological and other physical properties of blood, fibrin gels and thrombi, blood vessels and constituent tissues; atherogenesis; thrombogenesis; thromboembolization; thrombolysis; epidemiology of thrombosis.

Thrombosis Research presents peer-reviewed original research on thrombosis, hemostasis and fibrinolysis to the international scientific community. Both basic and clinical studies are published. The Journal's policy encourages the publication of research, including clinical investigations, which will lead to novel approaches in diagnosis, therapy, prognosis and prevention of thrombotic and hemorrhagic diseases.

MANUSCRIPT SUBMISSION

Submit manuscript to Editor of choice, preferably within country or region. Mail manuscripts well packed, preferably strengthened by cardboard. Do not fold Figures and Diagrams, Use First Class Air Mail. Supply corresponding author's telephone and fax number, and e-mail address.

A Transfer of Copyright Agreement for articles accepted for publication will be sent to author by Editor. Complete Form and return to Editor immediately.

MANUSCRIPT FORMAT

Instructions allow direct reproduction of manuscript for rapid publishing. If not followed, manuscript may be returned for retyping before final acceptance. Preferred typefaces, in order of desirability: Times Roman, Helvetica and Courier. Small typefaces are unsuitable. If using software that allows for a variety of type sizes, use 12 pt. type for text and 14 pt. type for title. Use NY4/A4 lay-sheets, obtainable from Editor upon request. Typing area must be 16.5 cm × 24 cm. Fill entire typing area. Type text single-spaced on good quality white paper; double-space between paragraphs. Insure that symbols, superscripts and subscripts are legible and do not overlap onto lines above or below. Make table and figure legends easily discernible from text. Erasure marks, smudges, spots, pencil or ink corrections or creases are not allowed. Type with a black carbon ribbon, inkjet or laser printer. Ink in black material that cannot be typed or produced on software, such as symbols and formulae. Do not use dot matrix printers.

Type name and address of correspondiing author single-spaced using full margins at bottom of first page, one line space under key words or under abbreviations.

Main Text Headings and Subheadings: Begin introduction to main text, without heading, three single-line spaces below Abstract using full margins. Single-space paragraphs in text; double-space between paragraphs. Italicize or underline generic names. Arrange headings to increase clarity. In Papers, Brief Communications, and Mini-Reports center major headings, e.g., MATERIALS AND METHODS, RESULTS, DISCUSSION and REFERENCES; do not underline; use all capitals. Double-space between text and major headings. Put subheadings in lowercase letters; begin at left margin, single-spaced from text that follows. Italicize or underline subheadings. Avoid using bold types in headings. At end of Discussion, double-space and in lowercase letters type Acknowledgments, centered on width of typing area; do not underline.

Title, Authors and Affiliation: On first page, start title 3 cm down from top margin of typing area. Type title in capital letters (except formulae), centered on width of typing area; single-space if more than one line. Make title brief, descriptive; spell out all words. Double-space between title and name(s) of Author(s). Type name(s) single-spaced and centered on width of typing area. Use format: First name(s), Surname, e.g., Angela M.V. Silveira. Single-space and type author(s) affiliation(s), also single- spaced. Capitalize first letter of main words. For Papers, make six line spaces between author affiliation and abstract. For Brief Communications, Mini-Reports, and Mini-Reviews, use same spacing before introduction (without heading) to text. Never type page numbers; number each sheet lightly near lower right corner, with a light blue pencil.

Abstract: Type 'Abstract' in lowercase letters, beginning at left margin. Two spaces to right, type text of abstract in lowercase lettering, single-spaced in block style. Make intelligible to reader without reference to body of paper, and suitable for reproduction by abstracting services. Never use more space than Title Page provides.

Symbols and Formula: Type materials. Prepare complex equations separately and paste onto manuscript.

Footnotes: Type single-spaced, two line spaces below text at bottom of appropriate pages and separated from text by a short line. Place within allowed typing space. Reference in text by superscript Arabic numerals.

Tables: Type within main body of manuscript, but set apart, to avoid confusion with text. Do not overlap onto following page. Center word 'TABLE' in capitals followed by table number (Roman numerals) above table. Leave two line spaces between main text and TABLE and two line spaces between TABLE and caption. Capitalize first letters of all main words in caption; do not make wider than width of table. Type legends beneath table; make easily discernible from main text. Make 5 space indentations on both margins for table, caption and legend. Position tables and figures at very top or bottom of typing area.

Figures: Leave appropriate space for each Figure and line diagram in text above or next to a descriptive caption. Capitalize abbreviation FIG., with figure number (Arabic numerals), centered above caption. Single-space caption. Allow two line spaces between FIG. and caption and two line spaces between end of caption or following text. Make indentations as described for Tables when figure is above caption, and on one side of margin if caption is next to (left) of figure.

Supply halftone pictures of figures in triplicate as glossy prints in final actual size (or slightly larger). Indicate on all photographs important details essential for comprehension. Details in figure must be large enough to allow for 15% reduction, figures should not occupy more space than necessary. Do not paste in figures; supply separately with number and author's name indicated in pencil on reverse side. Do not use ink. If number of Figures and/or Tables are excessive, Editor may ask Author(s) to eliminate some. Figures should be black-and-white, clear enough to permit reproduction without retouching, and legible after reduction to journal page size. If publication of figure(s) in color is requested, authors must share expense of reproduction.

References: Use consecutive numbers in parenthesis to indicate references in text, e.g., (1,2); do not superscript numbers. Cite full reference in a numbered list without parenthesis at end of manuscript. Single-space references. References contain names of all authors in capital letters (surnames first, followed by initials), title of communication in lowercase letters, title of Journal (abbreviated, in accordance with *World List of Scientific Periodicals*), volume number in italics or underlined, first and last page number and year of publication. References to books contain names in the same format, title (underlined or in italics and first letter of main words capitalized), publisher's name and location and year of publication (in parentheses). References to books with editor(s) contain author(s) names in the same format as above, title of contribution, title of book (in italics or underlined), editors name(s), page number(s), publisher's name and location and year of publication (in parentheses). Material "submitted for publication" is not a reference; place in parentheses at appropriate place in text, e.g., (Hessel, B, submitted for publication, 1991).

1. COPLEY, A.L. Fibrin(ogen), platelets and a new theory of atherogenesis. Thrcmb Res *14*, 249–263, 1979.

2. DAVIES, J.T. and RIDEAL, E.K. *Interfacial Phenomena*. Academic Press, New York-London (1961).

3. BLOMBÄCK, B. Fibrinogen to fibrin transformation. In: *Blood Clotting Enzymology*. W.H. Seegers (ed.), pp. 143-215, Academic Press, New York-London (1967).

4. COPLEY, A.L., KING, R.G., KUDRYK, B. and BLOMBÄCK, B. Viscoelasticity studies of surface layers of highly purified fibrinogen (Fg) systems and the role of pH on the rigidity of such layers. Thromb Res *18*, 213–230, 1980.

Key Words: Type single-spaced using full margins, two line spaces below text at bottom of page and separated from text by a short line. Take six Key Words for subject indexing from *Medical Subject Headings* of *Index Medicus*.

Running Title: 32 characters, including spaces, on a separate sheet.

OTHER FORMATS

ANNOUNCEMENTS—1 page. Submit to Executive Editorial Office.

BRIEF COMMUNICATIONS—6 pages. Short reports of new work; no abstracts.

COMMUNICATIONS—Submit one original plus two copies to an Editor of choice; mail or fax copy of title page (with Editor's name) to Executive Editorial Office. All communications are photographically reproduced from original typescripts; manuscript must conform accurately to specifications.

EDITORIALS—Submit comments on matters significant to readers to Executive Editorial Office.

INVITED STATE OF THE ART ARTICLES—Reviews solicited by Editors-in-Chief are published in first issue of each volume. Submit on 3.5 or 5.25 in. disks as ASCII or text files. Mark disk with author's name, article and journal title, and type of software used. Submit printed version of manuscript with disk.

LETTERS TO THE EDITORS— 2 pages. Matters of opinion and criticism on contributions published in journal and other matters of interest to researchers in field.

MINI-REPORTS—4 pages. Address practical problems in laboratory and in clinical investigation; no abstract or summary. Introduction is 8 lines. End with a brief half-page comment.

MINI-REVIEWS—12 pages. Brief outline of scientific issues of general interest; no abstract.

NEWS ITEMS—Submit to Executive Editorial Office.

REPORTS OF SCIENTIFIC MEETINGS—12 pages. By invitation. Follow same format and layout requirements.

SUPPLEMENT ISSUES—Cover various topics in field of thrombosis and hemostasis.

OTHER

Abbreviations: Use standard abbreviations in text and illustrations. When not defined in text, type single-spaced using full margins on title page one line space below Key Words.

Page Charges: $50 for each additional 2 pages exceeding listed page limits.

SOURCE

October 1, 1994, 76(1): back matter.

Toxicology and Applied Pharmacology
Society of Toxicology
Academic Press

Toxicology and Applied Pharmacology Editorial Office
525 B Street, Suite 1900
San Diego, CA 92101-4495
Telephone: (619) 699-6469; Fax: (619) 699-6859

AIM AND SCOPE

Toxicology and Applied Pharmacology publishes original scientific research pertaining to the action of chemicals, drugs, or natural products on the structure or function of animal (including human) cells and/or tissues. Manuscripts should address mechanistic approaches to physiological, biochemical, cellular, or molecular understanding of toxicologic/pathologic lesions and to methods used to describe these responses. Papers dealing with alternatives to use of experimental animals are encouraged.

MANUSCRIPT SUBMISSION

The same work may not have been published and may not be under consideration for publication elsewhere. Submission must have been approved by all authors and by institution where work was done. All persons cited as sources of personal communications must have approved such citation.

If accepted for publication, copyright in article, including right to reproduce article in all forms and media, shall be assigned exclusively to the Publisher, who will not refuse any reasonable request by author for permission to reproduce any of his or her contributions to the journal.

MANUSCRIPT FORMAT

Submit in quadruplicate (one original and three photocopies), including four sets of figures. Submit halftone photographs and electron micrographs as glossy prints or originals; do not send photocopies. Papers in French or German can be considered; contact Editor prior to submission. Type double-spaced (including references) on one side of 8.5 × 11 in. white paper with 1 in. margins on all sides. Do not use dot matrix printers.

Page 1: Article title, names and affiliations of all authors, abbreviated form of title (55 characters including letters and spaces) and name, telephone number, and complete mailing address of corresponding author. Make article title comprehensive and descriptive. Do not use proprietary names in titles; identify in footnotes.

Page 2: Abstract (250 words starting with title of paper), authors' names and reference to *Toxicology and Applied Pharmacology*; leave blanks for volume and page numbers. Abstract is a concise summary of what was done, results obtained, and valid conclusions. Mention compounds or families of compounds studied, their actions, and species of animals. Use important words which are used as index terms, but not proprietary names.

A Biologically Based Toxicokinetic Model for Pyrene in Rainbow Trout. LAW, F. C. P., ABEDINI, S., and KENNEDY, C. J. (1991). *Toxicol. Appl. Pharmacol.* **110**, 390–402.

Index Terms: List terms suitable for subject index on a separate sheet; include chemical name, generic name, common name, and proprietary name of each important substance used in experiments. Include biological activity of each. Terms should elucidate important relationships.

Introduction: State why investigation was carried out, note relevant published work, and delineate objective of investigation.

Methods: Describe new methods or significant improvements of methods or changes in old methods. Do not describe methods for which adequate reference can be cited. Draw attention to chemical or biological hazards involved in carrying out experiments described. Describe relevant safety precautions. If an accepted code of practice has been followed, reference relevant standards. Give details regarding animal housing conditions.

Results: Do not duplicate material presented in tables and figures. Avoid tabular presentation of masses of negative data; replace with a statement in text. Include what was done, how it was done, how data were analyzed, a measure of variability, and significance of result.

Discussion: Relate significance of work to existing knowledge in field and indicate importance of contribution of study. Avoid needless detailed recapitulation of results. Omit unsupported hypotheses and speculation.

References: Cite names of authors in text with year of publication in parentheses. When more than two authors, use first author's name followed by *et al.* List all papers mentioned in text in reference list and vice versa. List references alphabetically, typed double-spaced on a separate sheet at end of paper; include all authors. Abbreviate journal titles according to *Chemical Abstracts Service Source Index*, 1985.

Cole, R. J. and Cox, R. H. (1981). *Handbook of Toxic fungal Metabolites.* Academic Press, New York.

Randall, C. (1977). Teratogenic effects of *in utero* ethanol exposure. In *Alcohol and Opiates* (K. Blum, Ed.), pp 91–107 Academic Press, New York.

Shain, W., Bush, B., and Seegal, R. (1991). Neurotoxicity of polychlorinated biphenyls: Structure-activity relationship of individual congeners. *Toxicol. Appl. Pharmacol.* **111**, 33–42.

Footnotes: Identify in text by superscript Arabic numerals; cite consecutively throughout paper. List in order typed double-spaced on a separate sheet. Use to identify proprietary names of substances and names and addresses of suppliers. If paper has been presented orally in whole or in part, include date and occasion in a footnote.

Tables: Type double-spaced on a separate sheet with title typed directly above. Do not use abbreviations in title. Title and footnotes must contain all information necessary to understand and interpret table without reference to text. Number consecutively with Arabic numerals. Identify footnotes by superscript lowercase italic letters; place at bottom of table page. Assign letters alphabetically in order of appearance as table is read horizontally, not vertically.

Figures: Number consecutively with Arabic numerals. Type legends consecutively on a separate sheet. Plan to fit proportions of printed page (7 1/8 × 9 in.; column width, 3.5 in.); wherever possible, figures will be reduced to single-column width. Make lettering professional quality or generated by high-resolution computer graphics and large enough (10–12 points) to be legible after reduction to single-column width (approximately 50 to 60%). Differences in type size within a single figure should be no more than approximately 15%. Freehand, penciled, or typewritten lettering is not acceptable. Present drawings with equivalent of black India ink on tracing linen, smooth-surface white paper, or Bristol board. High-quality computer graphics are accepable. Make symbols used to identify points within a graph large enough to be easily distinguishable after reduction. Plot graphs with black India ink on blue coordinate or white paper no larger than 8.5 × 11 in. Ink grid lines in black.

Submit all copies of halftone photographs and electron micrographs as glossy prints or originals; do not send photocopies. Keep photographs to a minimum; submit as glossy prints (5 × 7 or 8 × 10 in.). Use overlays indicating parts of electron micrographs to be emphasized. Submit electron micrographs no larger than 6.5 × 4.5 in. Indicate magnification by a bar on photograph. Illustrations in color are accepted if authors defray cost.

Identify all figures on back, in pencil, with authors' names and figure number, and indicate TOP.

OTHER FORMATS

ANNOUNCEMENTS—Announcements of interest to toxicologists such as notices of society meetings, symposia, or other matters.

HIGHLIGHTS—12 double-spaced, typewritten manuscript pages, 2 tables and/or 2 figures. Limited number of highly significant, complete (not preliminary), and timely reports that have progressed to the stage at which the science of Toxicology would be advanced were the results made available as soon as possible. Submit manuscript in quadruplicate and conform strictly to journal form and style.

LETTERS TO THE EDITOR—250 words dealing with papers published in *Toxicology and Applied Pharmacology* with comments of scientific value. Submit within two months of mailing of journal. Do not introduce new data. Letter will be submitted to author of original paper in order that any reply may be published simultaneously with letter.

OTHER

Abbreviations: Use International Système (SI) units as adopted by the 11th General Conference on Weights and Measures. Do not use periods after abbreviations (note exceptions, e.g,. in. for inches).
Common abbreviations to be used in this journal are:

m	meter	ppm	parts per million
cm	centimeter	cpm	counts per minute
mm	millimeter	dpm	disintegrations per minute
μm	micrometer	sc	subcutaneous
nm	nanometer	ic	intracutaneous
kg	kilogram	im	intramuscular
g	gram	ip	intraperitoneal
mg	milligram	iv	intravenous
μg	microgram	po	oral
ng	nanogram	LD_{50}	medial lethal dose
ml	milliliter	LC_{50}	medial lethal concentration
μl	microliter	Hz	hertz
mol	mole	°C	centigrade
M	molar	sec	seconds
mM	millimolar	min	minutes
μM	micromolar	hr	hours
n	normal	SD	standard deviation
Ci	Curie	SE	standard error
$\bar{x}$	mean	TLV	threshold limit value

Electronic Submission: Submit manuscripts on personal computer disks after manuscript has been accepted and after all revisions have been incorporated onto disk. Label disk with computer used, software and version number, and disk format. All tabular material is typeset conventionally. Prepare art as camera-ready copy.

Supply an exactly matching hard copy printout of manuscript. File names must indicate content of file.

For further information on preparing disks for typesetting conversion, contact Editorial Supervisor, Journal Division, Academic Press, Inc., 525 B Street, Suite 1900, San Diego, California 92101-4495; telephone: (619) 699-6415; fax: (619) 699-6800).

Experimentation, Animal: Describe procedures employed for animal care and handling. Conduct experiments in accordance with the *Guiding Principles in the Use of Animals in Toxicology,* adopted by the Society of Toxicology in 1989. A statement of these principles is published in the January issue.

Nomenclature: For styling of isotope, enzyme, and biochemical nomenclature, consult extended Instructions to Authors published in January issue of *Archives of Biochemistry and Biophysics*.

Submission Fee: $100 payable to Society of Toxicology in United States currency, must accompany original submission of manuscript. Include name(s) of author(s) and title of manuscript in check identification information. If payment poses an extraordinary hardship to authors, send an explanatory letter indicating circumstances to Editor with manuscript.

SOURCE
January 1994, 124(1): back matter.

Toxicon

International Society on Toxicology
Pergamon

Prof. Alan L. Harvey
Dept. of Physiology and Pharmacology
University of Strathelyde
Royal College 204 George Street
Glasgow Gl IXW U.K
Dr. Gerhard Habermehl
Department of Chemistry
School of Veterinary Medicine
Bisehofsholer Damm 15
3000 Hannover 1, Germany
              Review editor

MANUSCRIPT SUBMISSION

Submit manuscripts of regular papers, short communications, letters to the editor and announcements in triplicate. Submission implies that contents of manuscript have not previously been published (except as an abstract or preliminary report), is not presently under consideration elsewhere, and will not be submitted for publication elsewhere, in any language, if accepted in *Toxicon*. All authors in a multiauthored manuscript must be aware of and in agreement with contents of manuscript. It is assumed that papers submitted have received appropriate clearances and approvals from institution of origin.

Manuscripts in French, German, or Spanish are acceptable if accompanied by a 200-400 word English summary. Do not use technical jargon, laboratory slang and words not defined in dictionaries.

MANUSCRIPT FORMAT

There is no set standard length for papers. Present data concisely, consistent with good reporting. Do not fragment report into several short papers. Type manuscripts double-spaced on one side of 8.5 × 11 in. bond paper with 25 mm margins on all sides. Submit an original and two clear carbon or photostat copies. Number all pages consecutively starting with title page as number 1. Following title page, divide manuscript into major headings, typed in capitals: Abstract, Introduction, Materials and Methods, Results, Discussion, Acknowledgments (if any), References. Indicate subheadings by italics without numerical or alphabetical designations.

Title Page: Give title of paper; author(s); laboratory or institute of origin including city, state and country; a running title (40 letters and spaces); an address to which proof is to be sent: and telephone and fax numbers of corresponding author. If manuscript is based upon work carried out at more than one institution, use superscript Arabic numerals by authors' names and institutions to correlate authors to institutions. If a species is mentioned in title, give its common name and complete Latin name in parentheses.

Abstract: Page 2 : Author(s). Title, *Toxicon* ● ●, ● ● ● – ● ● ●, *19* ● ●.— An informative abstract (100 to 300 words). State concisely what was done, what was observed and what was concluded. Be as specific as possible. Avoid abbreviations.

Introduction: Summarize reasons for carrying out study and why it is of interest. Adequately reference earlier work with direct bearing on study. Do not present details of data obtained by other workers or discuss relationship to results.

Materials and Methods: Indicate sources of venoms, toxins and chemicals, including address (city and country). Acknowledge gifts of materials. Indicate purity of agents used wherever appropriate. Do not describe standard methods, for which adequate, readily available literature citations can be provided, in detail. In a sentence or two, summarize principle of analytical method, physiological procedure, etc. Describe all original methods or modifications of procedures in sufficient detail to be duplicated by other workers.

Results: Include clear concise presentations of findings; reserve extensive discussions of data for Discussion. Use Tables and Figures for data which do not lend themselves to presentation in text. Refer to all Tables and Figures in text; indicate approximate positions in margin.

Present data as means with an index of variation (standard deviation, standard error of the mean, range). Indicate number of values upon which each mean is based. Provide statistical significance (P-values) of differences from control values and method of calculation (e.g., Student's t-test). Present lethality data as LD_{50} values. There is theoretically and statistically no such thing as a 100% lethal dose, an LD_{50} dose, a minimum lethal dose, a maximum tolerated dose, etc. Do not use such terms.

Tables: Refer to in text consecutively by Arabic numerals. Do not duplicate material in text or figures. Put each Table on a separate page. Place all Tables together at end of manuscript. Prepare for printing horizontal. Above body, place Table number and a clear concise descriptive title. Below place information and footnotes (*, †, ‡, §, ||, ¶) needed for making Tables understandable without referring to text. Give each column a heading; indicate units of measurement. Present data as mean values with some measure of dispersion (standard deviation, standard error of the mean, range). Present an indication of statistical significance of differences from control values either in body of Table or in footnotes below Table. Present number of individual values upon which each mean is based. Avoid tabular presentation of masses of negative data; replace with a statement in text listing what was done, results obtained (including a measure of variability), how data were analysed and significance of analysis. Do not list nonsignificant figures in Tables. Tables are reproduced by photooffset means directly from typed manuscript.

Figures: Refer to in text consecutively by Arabic numerals; do not duplicate material in text or Tables. Put each Figure on a separate page. Place all Figures together at end of manuscript. Place legends for Figures together on a page headed "Legends for Figures" immediately preceding Figures. In Legends give sufficient experimental detail so Figures are intelligible without reference to text.

Submit Figures in triplicate; keep to a minimum. Only one original set must suitable for reproduction, i.e., high quality India ink drawings on thick white paper or glossy photographic prints. Two other sets must be adequate for referees. Photocopies of light or electron micrographs are not suitable; submit three original sets. Group several related Figures together to form a plate on one page with a single number with alphabetical subdivisions such as Fig. 1a, 1b, 1c and 1d. Submit Figures of an adequate size to ensure clarity, avoid excessive white space. Make approximately twice final size and no smaller than 10×13 cm nor larger than 21×28 cm. Show points of actual observations on line graphs; make large enough to be visible when reduced. Make letters and numbers at least 4 mm in height. Label both x and y axes.

In line and bar graphs show means and indicate measure of dispersion (standard error of the mean, standard deviation or range) by vertical lines if more than two replicates were performed for each point. Indicate numbers of experiments in legend. Include statistical significance of differences from control values either in Figure or in legend beneath Figure. Show actual points of observation if data are based upon one or two determinations. Use standard symbols on line drawings: ▲ △ ⊡ □ ● ○ △ ⊙ ⊕ ⊖ ■ ⊗.

For micrographs (optical or electron) or other photographs that need a unit of length, include a bar scale directly on Figure. Do not use terms in legend such as $\times 20,000$ since published Figure size may be different. On back of each Figure lightly write number, author(s) names, "Toxicon" and along appropriate edge, TOP.

Discussion: Discuss data in detail including: sources of error, significance, relationship to data obtained by other workers and possible reasons for or significance of differences. Avoid needless detailed recapitulation of Results. Omit unsupported hypotheses and speculation.

Acknowledgments: Brief statement of thanks to individuals for services or advice. Note grants or fellowships.

References: Listed references must have been read; verify correctness of citation. In text give references in name and date form; use *et al.* if more than two authors. Examples: (Jones, 1976), (Jones and Smith, 1975), (Jones and Weinberg, 1960: Jones *el al.*, 1976), . . . as reported by Jones and Smith (1975), . . . as described by Jones *et al.* (1976). List references alphabetically on a separate page at end of manuscript. If more than one reference to the same author in one author papers, arrange references chronologically. If more than one reference to same first author in papers with two authors, arrange references first alphabetically by second author and then chronologically if there are references to publications by the same two authors in different years. Arrange references chronologically if three or more authors, regardless of second and other authors' names. If more than one reference with the same author(s) and the same year of publication, differentiate reference by *a, b,* etc. (e.g., 1964*a,b*). Abbreviate journal titles according to *World List of Scientific Periodicals* (4th Ed, 1965, Butterworths), except for Japan and Japanese which are abbreviated to *Jpn* and *Jpn.*, respectively.

List references to articles which have been accepted but not yet published in reference list; add 'in press' added after journal or book citation. Cite unpublished observations, personal communications and references to graduate theses in text where referred to, do not include in reference list.

JOURNAL REFERENCE

Minton, S. A., Jr. (1968) Preliminary observations on the venom of Wagler's pit viper *(Trimeresurus wgleri). Toxicon* **6,** 93-97.

TEXTBOOK REFERENCE

Smith, D. S., Cayer. M. L., Russell, F. E. and Rubin, R. W. (l978) Fine structure of stingray spine epidermis with special reference to a unique microtubular component of venom secreting cells. In: *Toxins: Animals. Plant. and Microbial.* pp. 565 582 (Rosenberg, P., Ed.). Oxford: Pergamon Press.

OTHER FORMATS

ANNOUNCEMENTS—Announcements of great interest to toxicologists, such as notices of appropriate meetings and symposia and activities of the International Society on Toxicology.

LETTERS TO THE EDITOR—Of interest to the broad field of toxicology or of special significance to a smaller group of workers in a specialized field of toxicology. Head "Letter to the Editor:" followed by a title. Include names of authors and affiliation at end of letter.

REVIEWS—Abstracts of articles of interest to toxicologists published in other journals.

SHORT COMMUNICATIONS—Three double-spaced typewritten pages (not including references, tables and figures), two tables or two figures or one of each. Complete, coherent and self-contained research studies that do not lend themselves to extended presentation. Quality is expected to be as good as that of full articles, and will be refereed in identical manner. Form is identical to full manuscripts except do not divide into Introduction, Materials and Methods, Results and Discussion. Provide an abstract (75 words).

OTHER

Abbreviations: Avoid abbreviations, or use very sparingly. Do not use awkward or unfamiliar abbreviations. Do not use abbreviations in Title or Abstract. Define all abbreviations either in text or in a footnote on first page where any abbreviation is used.

Use numerals for numbers from 11 up. Spell out numbers up to ten; use numerals in conjunction with symbols and units of measurement. Never begin sentences with a numeral; spell out number. Use commas to separate groups of three digits in numbers of five or more digits. Include a a zero in front of decimal point for numbers between zero and one. Place isotope mass number before the atomic symbol. Write ions: Ca^{2+}, Mg^{2+}, Fe^{3+}. Use term absorbance (A) not extinction (E) or optical density (OD). Express centrifugation conditions in terms of g and time, not as rev/min.

Give scientific name as well as common name for all species, other than common domestic and laboratory animals when first mentioned; then use common name or abbreviated Latin name if there is no ambiguity. Refer to Common Names Index (*Toxicon* **25** 799 1987*)* for correct common names. Use terms such as crotalid or clapid as adjectives; do not capitalize or underline. Use spelling of Latin species names in latest *Toxicon* Species Index *(Toxicon* **28** 751 1990). or primary taxonomic literature. Author of specific name may be added.

Italicize or underline non-English terms such as *in vitro, in vivo, et al*. and all Latin species names.

Biological Weapons: Manuscript must be in accord with the UN Convention on prohibition of development. production and stockpiling of bacteriological (biological) and toxin weapons (1975).

Chemical Names: Use trade names or abbreviations of chemical names if initially preceded by chemical or scientific name. Thereafter, trade names, common names or abbreviations may be used, chemical or scientific name is preferred. Quote International Union of Biochemistry (IUB) Enzyme Commission (EC) number with full names of enzyme when first mentioned in text. Subsequently, use accepted trivial name. See *Enzyme Nomenclature (1973). Recommendations of the International Union of Biochemistry* (Elsevier, Amsterdam).

Electronic Submission: Submit a computer disk (5.25 or 3.5 in. HD/DD disk) containing final version of paper with final manuscript. Specify software used, including release, Specify computer used (either IBM compatible PC or Apple Macintosh). Include text file and separate table and illustration files, if available. File should follow general instructions on style/arrangement and, in particular, reference style of journal. Single-space file. Use wrap-around end-of-line feature (i.e., no returns at end of each line). Begin all textual elements flush left, no paragraph indents. Place two returns after every element such as title, headings, paragraphs, figure and table callouts, etc. Keep a back-up disk.

Experimentation, Animal: In Material and Methods section describe species, strain, total number of animals, sex and weight range of experimental animals used, route of application and solvents used for dissolving venoms or toxins under investigation. If anesthetics were used, describe dose and application route. Detailed guidelines for animal experiments have been adopted by the International Society on Toxicology (*Toxicon* 1993;31, 1-12).

Spelling: Consult *Oxford English Dictionary* or *Webster's New International Dictionary* . Use either British or American spelling consistently. Do not use Latin plurals if English equivalent has become accepted form, e.g.. formulas not formulae.

Style: Follow *The Council of Biology Editors (CBE) Style Manual* (American Institute of Biological

SOURCE

January 1995, 33(1): 1-9.

Transactions of the Royal Society of Tropical Medicine and Hygiene
Royal Society of Tropical Medicine and Hygiene

The Editor, *Transactions of the Royal Society of Tropical Medicine and Hygiene*
Manson House, 26 Portland Place
London, W1N 4EY, U.K.

AIM AND SCOPE

Publication in the *Transactions*, the Society's only regular journal, is not restricted to Fellows of the Society. Subjects covered are all aspects of health and diseases in developing countries; health services, tropical medicine and hygiene, communicable diseases of the tropics and their vectors; related aspects of human and animal biology, with special emphasis on epidemiology and control, and clinical and experimental studies.

MANUSCRIPT SUBMISSION

Submit a letter signed by all authors stating that manuscript was offered to the Society exclusively. When published, copyright becomes property of the Society; authors of accepted papers must sign a copyright assignment form. Exceptions are for work deemed to be in the public domain and not subject to copyright. Submit manuscripts and figures in triplicate.

MANUSCRIPT FORMAT

Type or print with a good quality printer, on one side only of good quality A4 paper, double-spaced throughout (including references) with 3 cm margins on both sides.
Papers are 3500 words. Head with: title, name(s) of author(s) and address(es) where work was done. Include an abstract (150-200 words) which precedes main text of paper and conveys its scope. Divide into Introduction, Materials and Methods, Results and Discussion (or similar sections).

Illustrations: Keep photographs, line drawings, graphs and tables to minimum. Do not repeat details of results presented in illustrations in text. Submit photographs as well contrasted black-and-white glossy prints, unmounted. Indicate magnification by a line or bar on print. Color plates are at author's expense.
Put line drawings, maps and graphs in black ink on tracing paper, white paper or pale blue line graph paper.
Submit in a size suitable for reduction to single column (75 mm) or full page (160 mm) width. Reduction often improves final quality; note figures which must not be reduced. Make lettering, numerals etc., for figures and plates either in Letraset® (or equivalent) suitable for reproduction, or in pencil to be set by printer. Take particular care of diagrams produced by computer print-out.
Type tables and legends to figures on separate sheets.

References: Include in list at end of paper only references cited in text and only those accepted for publication in a named journal. In text, cite: "...it has been shown (MANSON, 1879)" or "...as shown by MANSON (1879)." If there are two authors, name both; if more than two, name only first, followed by "*et al*.," in text. If several references are cited consecutively in text, arrange chronologically; put those with same date in alphabetical order. In reference list arrange alphabetically by name(s) of author(s), if one or two; papers by more than 2 authors follow any by first author alone or with only one co-author arranged alphabetically by first author's name only; secondary arrangement, of all references, should be chronological, using lowercase letters after date if necessary.

Listed references include: name(s) and initial(s) of all authors (not 'et al.'), year of publication, full title of paper, full journal title, volume number, first and last page numbers.

Jones, B. (1967). Hepatitis in the vervet monkey. *Journal of Liver Diseases*, **99**, 101–109.

Citations of books include name(s) and initial(s) of author(s), year of publication, title of book or paper, editor(s) if appropriate, town of publication and publishers.

Smith, J. (1969). Immunology in onchocerciasis. In: *Research in Immunology*, Robinson, A. & Green, W. (editors), 2nd edition. London: Oliver & Boyd, pp. 99–150.

Cite Annual Reports with "Anonymous" as author. Verify accuracy of references.

OTHER FORMATS

CORRESPONDENCE—500 words on topics of general interest related to tropical medicine and hygiene, comments on papers published in the *Transactions* and, if appropriate, replies to comments. Tables and figures are rarely accepted. Give references only if essential, in abbreviated form within parentheses in text; e.g. "(Smith, 1986: *Transactions*, **80**, 1058)" or "a recent letter by Brown (1986: *British Medical Journal*, **250**, 7075)..."

SHORT REPORTS—800 words, one printed page, including references (as few as possible, in style used for full papers), illustrations and tables (one of each). Head with: title, name(s) of author(s) and address(es) where work was done; no abstract and need not be divided into sections. Suitable for brief clinical reports and preliminary results of important research.

OTHER

Experimentation, Animal: Of the Royal Society of Tropical Medicine and Hygiene the alleviation of suffering and disease in human and veterinary medicine is an objective. Referees apply the following criteria. Was experiment necessary and compatible with objectives of Society? Was experiment scientifically valid, using a sufficient but not excessive number of animals? Was experiment performed humanely with minimum possible suffering by animals and, where they exist, in accordance with local, national or institutional guidelines? Where applicable, was it justifiable to use endangered animal species or species phylogenetically close to man?

Experimentation, Human: Conform to Proposed International Guidelines for Biomedical Research involving Human Subjects, issued by the Council for International Organizations of Medical Sciences (Geneva, 1982).

Style: For spelling, follow *Oxford English Dictionaries*, use Système Internationale units (give equivalents in pa-

rentheses if required), and for symbols and abbreviations follow conventions in *Units, Symbols and Abbreviations: a Guide for Biological and Medical Editors and Authors* (4th ed, London: The Royal Society of Medicine, 1988). Write nonstandard abbreviations in full when first used.

SOURCE

January-February 1994, 88(1): 127–128, Revised February 1993.

Transfusion
American Association of Blood Banks

Jeffrey McCullough, MD
University of Minnesota Hospital and Clinic
Box 198, Room D211 Mayo
Harvard Street at East River Road
Minneapolis, MN 55455
Telephone: (612) 626-3313; Fax: (612) 624-5411

AIM AND SCOPE

Transfusion, the Journal of the American Association of Blood Banks, provides an international forum for the publication in English of communications that advance knowledge related to transfusion therapy, immunohematology, and transplantation. The scope of the Journal includes all scientific, technical, and administrative aspects of blood banking and aspects of hematology, immunology, biochemistry, and physiology that apply to blood transfusion, immunohematology, and transplantation. Acceptance of papers for publication is based on merit; equal consideration will be given to papers submitted by nonmembers and members of the Association.

MANUSCRIPT SUBMISSION

Submit manuscripts solely to *Transfusion*. Except for abstracts (500 words published for a scientific meeting), substantial parts may not have been submitted for publication elsewhere.

In cover letter include the following statements signed by each author:

"In consideration of *Transfusion* taking action in reviewing and editing my(our) submission, author(s) undersigned hereby transfer(s), assign(s), or otherwise convey(s) all copyright ownership to the American Association of Blood Banks in the event that such work is published in *Transfusion*. Author(s) represent and warrant that his/her(their) part of work, as submitted, will not infringe upon any common law or statutory copyright of any party whatsoever, and that such material does not contain matter that is libelous, in violation of any right of privacy, or otherwise contrary to law."

U.S. federal employee(s)—State that they are/were employee(s) of the U.S. federal government where this work was investigated and prepared for publication; therefore, it is not protected by Copyright Act and there is no copyright of which ownership can be transferred.

"Author(s) represent that they have contributed substantively to development of content of this paper including (1) conception or design of work; (2) analysis or interpretation of data; (3) drafting or revising article for content; and (4) approval of version to be published. Author(s) believe manuscript represents valid work and take public responsibility for it."

"Author(s) certify that they have no affiliation with or financial involvement in any organization or entity with a direct financial interest in subject matter or materials discussed in manuscript (e.g., employment, consultancies, stock ownership, honoraria), except as disclosed in an attachment. Any research or project support is identified on title page of manuscript."

Submit written permission to reproduce previously published materials or photographs that may identify individuals, or to acknowledge individuals by name.

Submit manuscripts in quadruplicate, with cover letter containing copyright release statement and $50. Provide information helpful to the Editors when reviewing manuscript. Keep a copy of all materials submitted.

MANUSCRIPT FORMAT

Prepare in accordance with instructions in "Uniform Requirements for Manuscripts Submitted to Biomedical Journals" (Declaration of Vancouver) (*New England Journal of Medicine* 1991;324:424–8). Consult most recent issue of Journal for matters of style. Manuscripts that do not conform may be returned.

Type manuscripts completely double-spaced, including title page, abstract, text, acknowledgments, references, tables, and legends. Do not justify right margins.

Begin each component on a new page in sequence: Title page; Abstract and key words; Text; Acknowledgments; References; Tables (with title and footnotes, each on a separate page); Illustrations (good-quality, unmounted, glossy prints, 5 × 7 in. - 8 × 10 in.); and Legends for illustrations.

Title Page: Contains: Title of article (concise but informative, no abbreviations); short running head (40 characters counting letters and spaces) typed at bottom of title page and identified as such; first initials and last name of each author (no titles, degrees, positions, or academic ranks); name(s) of department(s) and institution(s) to which work should be attributed, disclaimers, if any; name, address, and telephone and fax number of corresponding author; include an exact street address for overnight mail delivery, not a post office box number; either name and address of author responsible for reprint requests or a statement that reprints will not be available; and source(s) of support (grants, equipment, or drugs).

Abstract: Second page, 200 words, four sections: Background, Study Design and Methods, Results, and Conclusions. Describe, respectively, objectives of study or investigation, basic procedures (study subjects and analytic methods), main findings (specific data and statistical significance), and principal conclusions. Emphasize new and important aspects of study.

Key Words: Below abstract, provide and identify 3-10 key (indexing) terms. Use terms from *Medical Subject Headings* in *Index Medicus*.

Text: Divide observational and experimental articles into sections: Introduction, Materials and Methods, Results, and Discussion. Use subheadings in long articles to clarify content, especially in Materials and Methods, Results, and Discussion sections.

Introduction: Clearly state purpose of article. Summarize rationale for study or observations. Give pertinent references: do not review subject extensively. Do not include data or conclusions from work being reported.

Materials and Methods: Describe selection of observational or experimental subjects (patients or animals, including controls, and number in each study group). Identify methods, apparatus, equipment, reagents, and procedures in sufficient detail to allow repetition of results. Provide model name or number, manufacturer's name, and city and state address (in parentheses) at first mention of reagents, apparatus, or equipment. Give references to established methods; provide references and brief descriptions of methods that are not well-known; describe new or substantially modified methods, give reasons for using them, and their limitations. Identify all drugs and chemicals used, including generic name(s), dose(s), and route(s) of administration.

Results: Present in logical sequence in text, tables, and/or illustrations. Do not repeat in text data in tables and/or illustrations; emphasize or summarize only important observations, and avoid tables displaying data showing insignificant differences among groups.

Discussion: Emphasize new and important aspects of study and conclusions that follow from them. Do not repeat in detail data in Results or in tables or illustrations. Include implications of findings and their limitations; relate observations to other relevant studies. Link conclusions with purpose of study; avoid unqualified statements and conclusions not completely supported by data presented. Do not claim priority or allude to incomplete work. State new hypotheses when warranted; clearly label as such. Include recommendation, when appropriate.

Acknowledgments: At appropriate place(s) (e.g., title page, footnote, or appendix), specify: contributions that should be acknowledged but do not justify authorship, such as support by departmental chair; acknowledgments of technical help; acknowledgments of financial and material support, specifying nature of support; and/or financial relationships that may pose a conflict of interest.

Name persons who have contributed intellectually to paper, but whose contributions do not justify authorship; describe their function or contribution as, e.g., "scientific adviser," "critical review of study proposal," "data collection," or "participation in clinical trial." Obtain written permission from persons acknowledged by name. Acknowledge technical help in a separate paragraph from acknowledgments of other contributions.

References: Number consecutively in order of mention in text. Identify in text tables, and legends by superscript Arabic numerals (after punctuation). Number references cited in tables or legends in accordance with sequence of mention in text of table or illustration.

Use form of references in *Index Medicus*. Abbreviate names of journals, according to *List of Journals Indexed in Index Medicus*.

Avoid using abstracts as references; do not use "unpublished observations" and "personal communications;" insert references to written, not oral, communications (in parentheses) in text. Give name of person from whom and date (month and year) communication was received. Include references to manuscripts accepted but not yet published; designate journal name, followed by "in press" (in parentheses). Cite information from manuscripts submitted but not yet accepted in text as "unpublished observations" or "submitted for publication" (in parentheses).

Verify references against original documents. List inclusive page numbers of articles.

JOURNAL

(List all authors when six or less; when seven or more, list first three and add et al.)

Soter NA, Wasserman SI, Austen KF. Cold urticaria: release into the circulation of histamine and eosinophil chemotactic factor of anaphylaxis during cold challenge. N Engl J Med 1976;284:687–90.

Corporate Author

The Committee on Enzymes of the Scandinavian Society for Clinical Chemistry and Clinical Physiology. Recommended method for the determination of gammaglutamlyl-transferase in blood. Scand J Clin Lab Invest 1976;36:119-25.

No Author Given

Coffee drinking and cancer of the pancreas (editorial). Br Med J 1981;283:628.

Journal Supplement

Mastri AR. Neuropathy of diabetic neurogenic bladder. Ann Intern Med 1980;92:316-8.

Frumin AM, Nussbaum J, Esposito M. Functional asplenia: demonstration of splenic activity by bone marrow scan (abstract). Blood 1979;54(Suppl 1):26a.

Journal Paginated by Issue

Seaman WB. The case of the pancreatic pseudocyst. Hosp Pract 1981;16:24–5.

BOOKS AND OTHER MONOGRAPHS

Personal Author(s)

Osler AG. Complement: mechanisms and functions. Englewood Cliffs: Prentice-Hall, 1976.

Corporate Author

American Medical Association Department of Drugs. AMA drug evaluations. 3rd ed. Littleton, CO: Publishing Sciences Group, 1 977.

Editor, Compiler, Chairman as Author

Rhodes AJ, Van Rooyen CE, comps. Textbook of virology: for students and practitioners of medicine and the other health sciences. 5th ed. Baltimore: Williams & Wilkins, 1968.

Chapter in Book

Weinstein L, Swartz MN. Pathogenic properties of invading microorganisms. In: Sodeman WA, Jr, Sodeman WA, eds. Pathologic physiology; mechanisms of disease. Philadelphia: WB Saunders, 1974:457–72.

Agency Publication

National Center for Health Statistics. Acute conditions: incidence and associated disability, United States. July 1968–June 1969. Rockville, MD: National Center for Health Statistics, 1972. (Vital and health statistics. Series 10: Data from the National Health survey, no. 69. (DHEW Publication no. [HSM] 72–1036).

Published Proceedings Paper

DuPont B. Bone marrow transplantation in severe combined immunodeficiency with an unrelated MLC compatible donor. In: White HJ, Smith R, eds. Proceedings of the third annual meeting of the International Society for Experimental Hematology, 1974:44–6.

Monograph in a Series

Hunninghake GW, Gadek JE, Szapiel SV, et al. The human alveolar macrophage. In: Harris CC, ed. Cultured human cells and tissues in biomedical research. New York: Academic Press, 1980:546. (Stoner GD, ed. Methods and perspectives in cell biology; vol. 1).

Dissertation or Thesis

Cairns RB. Infrared spectroscopic studies of solid oxygen (Dissertation). Berkeley, CA: University of California, 1965,156 p.

Association Publications

Widmann FK, ed. Technical manual. 9th ed. Arlington: American Association of Blood Banks, 1985.

Schmidt PJ, ed and comp. Standards for blood banks and transfusion services. 11 th ed. Arlington: American Association of Blood Banks, 1985.

OTHER ARTICLES

Newspaper Article

Shaffer RA. Advances in chemistry are starting to unlock mysteries of the brain: discoveries could help cure alcoholism and insomnia, explain mental illness. How the messengers work. Wall Street Journal 1977; Aug 12:1 (col 1).

Magazine Article

Roueche B. Annals of medicine: the Santa Claus culture. The New Yorker 1971 Sep 4:66–81.

Tables: Type double-spaced on separate sheets. Do not submit as photographs. Number consecutively and provide brief titles for each on same sheet. Give each vertical row a short heading. Place explanatory information in footnotes, not in headings. For footnotes, use symbols in sequence: *, †, ‡, §, ||, ¶, **, ††, etc. Identify statistical measures of variations such as SD and SEM. Omit internal horizontal and vertical rules. Consider page width, 6.7 in.; plan accordingly. A small table covers one column (3.25 in.) with four vertical rows. For a large, 2-column table, use no more than 8 to 10 vertical rows over a 6.7 in. width.

In text, cite tables in consecutive order; mark appropriate location in margin of text.

If data from another source are used, obtain and submit written permission to publish in *Transfusion*; acknowledge source fully.

One table per 500 words of text is acceptable. Do not re-state data presented in text or figures. Editor may recommend that tables containing important backup data be deposited with National Auxiliary Publications Service or be made available by author(s); if so, an appropriate statement will be added as a footnote. Submit tables for consideration with manuscript.

Illustrations: Illustrations include figures, photomicrographs, and photographs. Submit 4 complete sets. Have figures professionally drawn and photographed. Freehand or typewritten lettering and computer-generated graphics printed with dot-matrix printers are not acceptable. Do not submit original drawings, roentgenograms, and other material, send sharp, glossy back-and-white photographic prints, preferably 5 × 7 in., but no larger than 8 × 10 in. When plotting points, use symbols: ○ ● △ ▲ ❑ and ■. Figures are reduced to 1 column (3.25 in.) in width. Draw and photograph all letters, numbers, and symbols to be about 3.32 in. after reduction (5 mm before reduction) and make clear and even throughout.

Put titles for illustrations in legends, not on illustrations. Paste a label on back of each figure indicating number of figure, name(s) of author(s), and TOP. Do not write directly on backs of figures, mount on cardboard, or mark with paper clips. Do not bend figures. Give photomicrographs internal scale markers; state magnification. Symbols, arrows, or letters in photomicrographs should contrast with background.

If persons are identifiable in photographs, submit written permission from subjects(s) to use photographs.

Cite each figure in text in consecutive order; mark approximate location in margin of text. If a figure has been published, acknowledge original source and submit written permission from copyright holder (usually publisher or journal) to reproduce material. Permission is required, even for one's own publications, except for documents in public domain.

Photographs in color, approved by Editor, are published at author's expense. Prices are quoted by Publisher. Supply color negatives or positive transparencies; mark accompanying drawings indicating region to be reproduced.

Legends for Illustrations: Type double-spaced, starting on a separate page, with Arabic numbers corresponding to illustration numbers. Identify symbols, arrows, numbers, or letters and explain each in legend. Explain internal scales; identity methods of preparation and staining.

Author(s) Information: Type double-spaced, first name, middle initial(s), and last name of each author, with highest academic degree(s), position(s) held, and department(s) and institution(s) with which each author is currently affiliated. If authors have identical departmental and institutional affiliation, list affiliation only once. Clearly indicate author responsible for correspondence, proofs, and reprints.

OTHER FORMATS

ANNOUNCEMENTS—Information about meetings of general interest to readers is published on Calendar page at Editor's discretion at no charge. Send information to Editor as far in advance of the meeting as possible. Box announcements concerning educational events (meetings, seminars, or conferences) and scholarship or achievement awards cost $75 per quarter-page or $150 per half-page.

BRIEF REPORTS, CASE REPORTS—2 typeset pages (8 typewritten pages) in same format as a manuscript. Focus on individual histories.

EDITORIALS—2 typeset pages (8 typewritten pages), including text and references. Solicited by Editors, may also be submitted without solicitation.

LETTERS TO THE EDITOR—Record observations, comment on or contend with regular articles, raise technical or medical questions, or present significant short descriptions or new data are encouraged. Include a transmittal letter containing each author's signature and copyright release statement. Provide a title for each letter. Type text, 2 pages, double-spaced on plain bond paper. Limit references to 5, typed double-spaced. Limit tables, figures, and illustrations to a total of 2. Closing salutation should include author's(s') name(s), degree(s), and address(es) but not professional appointments. Send two copies to: Peter D. Issitt, PhD, Transfusion Service, Duke University Medical Center, P.O. Box 2928, Durham, NC 27710.

REVIEWS—10 typeset pages (40 typewritten pages), including text, references, and illustrations. Detailed, critical surveys of concepts or experience in transfusion medicine. Consult with Editor before writing. Emphasize topics representing recent developments in basic science relevant to transfusion medicine.

OTHER

Abbreviations: Use only standard abbreviations. Full term for which an abbreviation stands must precede its first use in text unless it is a standard abbreviation for a unit of measurement. See list of Commonly Used Approved Abbreviations (*Transfusion* 1994; 34(1): 94). For additional standard abbreviations consult: *CBE Style Manual: A Guide for Authors, Editors, and Publishers in the Biological Sciences* (5th ed. Bethesda: Council of Biology Editors 1983); O'Conner M, Woodford FP. *Writing Scientific Papers in English: An ELSE-Ciba Foundation Guide for Authors* (Amsterdam, Oxford, New York: Elsevier-Excerpta Medica, 1975); and Day RA. *How to Write and Publish a Scientific Paper* (Philadelphia: Institute for Scientific Information Press, 1979).

Blood Group Terminology: Use conventions and style in P.D. Issitt and M.C. Crookston "Blood group terminology: current conventions" (*Transfusion* 1984;24:2-7).

For nomenclature of platelet-specific antigens follow conventions of A.E.G. von dem Borne and F. Décary in "Nomenclature of platelet-specific antigens (letter) (*Transfusion* 1990;30:477).

Electronic Submission: Send revised and accepted manuscript on disk with a hard copy. Use IBM compatible systems on 3.5 or 5.25 in. high- or low-density disks with WordPerfect, DOS Text, or text in ASCII file. If any other software, check with production office first; or use Macintosh systems with: 3.5, 1.4 meg, high density disk and run file through Apple File Exchange.

Type manuscript on white bond (not erasable) paper 8 by 10.5 in. or 8.5 by 11 in., with 1 in. margins. Do not justify right margins. Use letter-quality typewriters and printers; not dot matrix printers. Double-space throughout, including title page, abstract, text, acknowledgments, references, tables, and legends. Begin sections on a separate page: title page, abstract and key words, text, acknowledgments, references, individual tables, and legends. Number pages consecutively, beginning with title page, in upper right corner; precede each number with last name of first author.

Experimentation, Animal: Indicate whether institution's or National Research Council's guide for care and use of laboratory animals was followed.

Experimentation, Human: Indicate whether procedures followed were in accordance with ethical standards of local institution's committee(s) on human experimentation. Patients (and relatives) have a right to anonymity in published clinical documentation. Avoid details that might identify patients unless essential for scientific purposes. Do not use patients' names, initials or hospital numbers in a manner that would reveal identity. If identification of patients is unavoidable, obtain informed consent.

Recombinant DNA Research: Indicate physical and biologic containment procedures practiced, in accordance with Health Guidelines for Research Involving Recombinant DNA Molecules of the National Institutes of Health.

Statistical Analyses: Describe statistical methods in enough detail to enable a knowledgeable reader with access to original data to verify results. When possible, quantify findings and present with appropriate indicators of measurement error or uncertainty (such as confidence intervals). Avoid sole reliance on statistical hypothesis testing, such as use of p values. Discuss eligibility of experimental subjects. Give details about randomization. Describe methods for and success of blinding of observations. Report treatment complications. Give numbers of observations. Report losses to observation (such as dropouts from a clinical trial). Reference standard works (with pages) for study design and statistical methods, rather than to papers in which designs or methods were originally reported. Specify general-use computer programs used.

Put general descriptions of methods in Methods section. Avoid nontechnical uses of technical terms in statistics, such as "random" (which implies a randomizing device), "normal," "significant," "correlation," and "sample." Define statistical terms, abbreviations, and most symbols.

Submission Fee: $50 (U.S.) check, money order, or institutional purchase request made payable to American Association of Blood Banks, or major credit card (VISA, MasterCard, or American Express only) number and expiration date.

Units: Report measurements of length, height, weight, and volume in metric units (meter, kilogram, liter) or their decimal multiples. Give temperatures in degrees Celsius. Give blood pressure in torr.

Report measurements in units in which measurements were made; provide International Système of Units (SI) equivalent in parentheses following each measurement.

SOURCE
January 1994, 34(1): 90–95.

Transplantation
Transplantation Society
Williams & Wilkins

A. P. Monaco, M.D.
New England Deaconess Hospital, 185 Pilgrim Road
Boston, Massachusetts 02215
Telephone: (617)632-8549; Fax: (617)632-9929

Peter J. Morris, M.D.
Nuffield Department of Surgery, University of Oxford
John Radcliffe Hospital,
Headington, Oxford OX3 9DU U.K.
Telephone: 0865-221310; Fax: 0865-63545

MANUSCRIPT SUBMISSION

Manuscripts dealing with completed research and brief or preliminary reports are considered for publication. Submit in triplicate, including three sets of illustrations.

Manuscript must be an original contribution that has not been, and will not be submitted elsewhere while it is under consideration for publication in *Transplantation*. Photocopy and complete *Transplantation* submission form printed in each issue.

When manuscript is accepted, authors must transfer copyright to The Williams & Wilkins Company. Forms will be supplied by Editorial Offices.

MANUSCRIPT FORMAT

Abstract: Summarize objective of study, methods used, and conclusions reached.

Introduction: Include a statement of purpose of work, problem that stimulated it and a brief summary of relevant published investigations.

Materials and Methods: Present in sufficient detail to enable other investigators to repeat work.

Results: Be concise; avoid redundant tables and figures illustrating same data.

Discussion: Interpret results, with minimal recapitulation of findings.

References, Tables, Figures, and Legends: Type all parts of manuscript double-spaced with 1 in. margins on all sides.

Title Page: (Page 1) Contains title (15 words); do not use a sentence. Give full first name, middle initials, and family name of each author, as well as name(s) of department(s) and institution(s) to which work should be attributed, with address(es), including postal codes. Put mailing address for proofs in lower right corner.

Footnotes: (Page 2) Include footnotes to title, giving sources of support, and to authors' names, giving current addresses and an address for corresponding author. include all footnotes to text, designated by superscript Arabic numbers and given in numerical sequence.

Abbreviations: (Page 3) List abbreviations not likely to be familiar to reader alphabetically, with meanings. Do not abbreviate terms unless frequently used.

Abstract (Page 4) 250 words.

References: List only published works and manuscripts that have been accepted for publication in References. Refer to manuscripts in preparation, unpublished observations, and personal communications in parentheses in text. Cite as footnotes to text completed manuscripts submitted for publication.

Type double-spaced on a separate page and number in order of citation in text. In text designate by full-sized numbers in parentheses. List only first six authors; if more, include only first three and add "et al." Abbreviate journal titles according to *Index Medicus.*

1. Winearls CJ, Fabre JW, Morris PJ. Use of cyclophospamide and enhancing serum to suppress renal allograpft rejection in the rat. Transplantation 1979; 28: 271.

2. Weinstein L, Swartz MN. Pathogenic properties of invading microorganisms. In: Sodeman WA Jr, Sodeman WA, eds. Pathologic physiology: mechanisms of disease. Philadelphia: Saunders, 1974: 457.

3. National Center for Health Sciences. Acute conditions: incidence and associated disability, United States July 1968-June 1969. Rockville, MD: National Center for Health Statistics, 1972. (DHEW publication No. [HSN]72-1036).

Tables: Do not submit as photographs. Type each double-spaced throughout, including column headings, footnotes and data, on a separate page. Number in sequence in Arabic numerals; give concise, informative titles. Give each column a heading describing data in column. Use lower-case superscript letters to designate footnotes; type footnotes below table. Cite tables in text in numerical order. Make understandable without reference to text.

Figures: Supply halftone illustrations (photographs, photomicrographs, etc.), charts, and graphs as computer-generated originals or glossy prints. Make lettering on graphs large enough to be 1.5 mm high after reduction to 8.8 cm single column width. Make photomicrographs and other photographs no wider than 8.8 cm.

Cite figures in text in numerical order. Number in sequence with Arabic numbers. Label backs with a soft pencil, indicating TOP, figure number, and author's name. Mail flat with protective stiffeners.

Cost of color figures is borne by authors. Submit positive color prints.

Figure Legends: Number to correspond to figure. Type double-spaced on a separate page.

OTHER FORMATS

BRIEF COMMUNICATIONS—10 pages, including tables and legends, Brief or preliminary reports of new techniques or significant new findings. Do not divide into sections include an abstract.

LETTERS TO THE EDITOR—Comment on articles in *Transplantation,* or provide information on other pertinent subjects.

OVERVIEWS—10000 words, including references, Comprehensive reviews of topics of special interest to investigators in the field of transplantation. Do not divide into usual sections.

RAPID COMMUNICATIONS—20 pages, including tables and legends, Short articles of high originality and interest. Nonacceptance as a Rapid Communication does not preclude publication by normal route.

OTHER

Electronic Submission: Submit only final accepted revision of manuscript in electronic format. Submit electronic diskettes of final version with two hard copies, Diskettes produced on IBM or IBM-compatible computers are preferred; most Apple/Macintosh or Wang diskettes can be converted. Preferred word processing programs are: XyWrite III Plus; Word Perfect 4.2, 5.0, or 5.1 (IBM or Macintosh); Microsoft Word (IBM or Macintosh); Words for Windows; Wang OIS (WPS); and Wordstar (IBM). Word processing systems that can be converted are: CPT 8000; MacWrite 2.2, 4.5, or II; Display Write 3 or 4; Multimate; PC Write; Volkswriter; and Write Now. When preparing diskettes on Macintosh computers do not use Fast Save option. Files in ASCII can be used, but are not preferred.

Submit: diskette (3.5 or 5.25 in.), 2 hard copies of manuscript (8.5 × 11 in.) double-spaced, name of software program and version number used. Identify diskette with journal name, manuscript number, senior author's name, manuscript title, name of computer file, type of hardware, operating system and version number, and software program and version number.

Page Charges: $65 per printed page. If no source of grant or other support, appeal to Editors to waive charges.

SOURCE

July 15, 1994, 58(1): front matter.

Transplantation Proceedings
The Transplantation Society
The Japan Society for Transplantation
The Hellenic Transplantation Society
The European Society for Transplantation
The Canadian Transplantation Society
The Transplantation Society of Australia and New Zealand
The Scandinavian Transplantation Society
The Latin American Transplantation Society
The Pan-American Society for Dialysis & Transplantation
The Society for Organ Sharing
The Catalan Transplantation Society
The Asian Transplantation Society
The International Liver Transplantation Society
The Cell Transplant Society
The Middle East Society for Organ Transplantation
Société Francaise De Transplantation
Appleton & Lange

Felix T. Rapaport MD
Department of Surgery, Health Sciences Center
State University of New York at Stony Brook
Stony Brook, NY 11794

MANUSCRIPT SUBMISSION

Transplantation Proceedings publishes several different categories of manuscripts. The first type consists of sets of papers providing an in-depth expression of the current state of the art in various rapidly developing components of world transplantation biology and medicine. These manuscripts emanate from the International Congresses of the Transplantation Society, from the Affiliated Societies, from Symposia sponsored by the Societies, and special Conferences and Workshops covering related topics.

Transplantation Proceedings also publishes several special sections, including Transplantation Japonica and Clinical Transplantation Proceedings.

If accepted for publication, copyright in article is assigned to publisher. Submit diskette with three accurate double-spaced paper copies of manuscript.

MANUSCRIPT FORMAT

Type double-spaced on good quality 8.5×11 in. white paper with 1 in. margins. Do not use erasable bond paper.

First Page: title of paper; authors' names; name of institution where work was done; acknowledgment of research support; name and address of corresponding author.

Illustrations and Tables: Cite in order in text using Arabic numerals. Mark on back of each: name of first author, figure number, and designate TOP with a soft lead pencil. Avoid mounting illustrations on boards unless necessary to insure proper placement. Illustrations must be in black-and-white, and should show optimal clarity of detail. Print photographs on glossy paper.

Professionally letter all drawings with type or India ink; make no larger than 8×10 in. to allow for reduction. Do not submit hand-drawn or hand-lettered illustrations. Type a legend for each illustration double-spaced on a separate sheet at end of manuscript. Type each table on a separate sheet and number appropriately. Type table titles on sheet with table.

References: Cite by number in body of text; compile at end of article in order of citation. Type double-spaced, under heading References. Abbreviate titles of medical periodicals according to *List of Journals Indexed in Index Medicus).* Include authors' last names and initials, journal, volume, page, and year. Include city and publisher's name for books.

OTHER

Electronic Submission: Label diskette with title, author's name, article title, and software and hardware used. Use 3.5 or 5.25 in. diskettes produced on IBM, IBM-compatible, Apple (do not use "Fast Save" option), or Wang computers.

SOURCE

February 1994, XXVI(1): front matter.

Trends in Biochemical Sciences
International Union of Biochemistry and Molecular Biology
Elsevier

Elsevier Trends Journals
68 Hills Road
Cambridge CB2 1LA, U.K.

AIM AND SCOPE

Review journal in biochemistry and molecular biology that enables researchers, teachers and their students to follow new and recent developments across this broad field.

MANUSCRIPT SUBMISSION

Review articles are generally invited by Editors but ideas for *Reviews* and, in particular, *Talking Point* and *Open Question* features are welcome. Prospective authors should send a brief summary, citing key references to Staff Editor in Cambridge, who will supply guidelines on manuscript preparation if proposal is accepted. Do not submit completed articles without prior consultation. Reviews and Features articles are peer reviewed before acceptance and publication cannot be guaranteed.

MANUSCRIPT FORMAT

Cite references within text by superscript numbers. List references at end of manuscript in numerial order of citation (not alphabetically).

1 Spudich, J.L. and Long, S.R. (1992) *J. Bacteriol.* 175, 775-777

2 Fell et al. (1994) *Development* 116, 110-115

OTHER FORMATS

FRONTLINES—Includes *Journal Club* articles, which highlight recent research papers of particular interest and also contain other short articles on a wide range of subjects.

REFLECTIONS ON BIOCHEMISTRY—Provides a historical perspective.

REVIEWS— These articles, invited from leading researchers in a specific field, summarize and assess recent and important developments.

TALKING POINT—Articles contribute new perspectives to existing debates, usually based on experimental findings.

TECHNIQUES—New techniques and their application, or recent developments of established methods.

SOURCE

January 1994, 1: 3

Trends in Genetics
Elsevier Trends Journals

Elsevier Trends Journals
68 Hills Road
Cambridge CB2 1LA, U.K.

MANUSCRIPT SUBMISSION

Articles are normally commissioned, but all submitted material will be considered for publication. Submit proposals for reviews to Editor with a brief statement of scope of review, plus key references.

MANUSCRIPT FORMAT

Cite references within text by superscript numbers. List references at end of manuscript in numerial order of citation (not alphabetically).

1 Spudich, J.L. and Long, S.R. (1992) *J. Bacteriol.* 175, 775-777

2 Fell et al. (1994) *Development* 116, 110-115

OTHER FORMATS

TECHNICAL TIPS—Brief papers describing new methods in genetics or molecular biology, or significant new modifications or applications of existing techniques and 'handy hints' that other ressearchers may find useful. Be as brief as possible, but give sufficient information to enable others to repeat method.

SOURCE

October 1994 10: 381.

Trends in Neurosciences
Elsevier Trends Journals

Elsevier Trends Journals
68 Hills Road
Cambridge CB2 1LA, U.K.

AIM AND SCOPE

Trends in Neurosciences is a leading neuroscience review journal, publishing reviews and other feature articles to enable researchers and teachers to keep up to date with current developments.

MANUSCRIPT SUBMISSION

Articles are normally commissioned, but all submitted material will be considered. Submit proposals for reviews Editor, with a brief statement of scope of review, plus key references.

MANUSCRIPT FORMAT

Cite references within text by superscript numbers. List references at end of manuscript in numerial order of citation (not alphabetically).

1 Spudich, J.L. and Long, S.R. (1992) *J. Bacteriol.* 175, 775-777

2 Fell et al. (1994) *Development* 116, 110-115

OTHER FORMATS

LETTERS TO THE EDITOR—400 words. Comments concerning recently published articles are welcomed.

SOURCE

October 1994, 17:4 32.

Trends in Pharmacological Sciences
International Union of Pharmacology
International Union of Toxicology
Elsevier Trends Journals

Elsevier Trends Journals, 68 Hills Road
Cambridge CB2 1LA, U.K.

AIM AND SCOPE

The purpose of *Trends in Pharmacological Sciences* is to restore coherence to pharmacology by, for example, keeping pharmacologists in one speciality in a particular environment in a given country informed about the activities of those in other specialities, environments and countries. To this end *Trends in Pharmacological Sciences* consists of review articles written in English covering all aspects of pharmacology and toxicology. The journal also covers the toxicological sciences.

MANUSCRIPT SUBMISSION

Review articles are invited, but proposals for reviews are welcome. Submit synopses of proposed contents to the Editor. Letters and other brief items intended for publication may be sent without prior formality.

MANUSCRIPT FORMAT

Cite references within text by superscript numbers. List references at end of manuscript in numerial order of citation (not alphabetically).

1 Spudich, J.L. and Long, S.R. (1992) *J. Bacteriol.* 175, 775-777

2 Fell et al. (1994) *Development* 116, 110-115

OTHER FORMATS

LETTERS TO THE EDITOR—500 words. Comments concerning recently published articles are welcomed.

SOURCE

August 1994, 15: 272.

Urology
Cahners Publishing Co.

Joseph E. Oesterling, M.D.
Suite 250, 201 First Avenue, SW
Rochester, MN 55902
Telephone: (507) 284-8525; Fax: (507) 284-8518

MANUSCRIPT SUBMISSION

Manuscripts will be accepted for consideration with the understanding that they are contributed solely to *Urology,* have never before been published, nor submitted simultaneously elsewhere, and become property of the publisher. Manuscripts are subject to editorial modification to bring them into conformity with the journal style. Submit written permission from publisher and author for use of material reproduced from other sources.

MANUSCRIPT FORMAT

Submit as typed hardcopy or on any type of computer disk. Submit 4 copies of manuscript including references, tables, and figures. Maximum acceptable length of an article is 3000 words; for each table or figure, shorten by 250 words. Type double-spaced with ample margins on each side. Include an introduction, material and methods section, results, comment section, and conclusions. Abstracts for all manuscripts except case reports, clinical and basic science reviews, and surgeon's workshop articles—should be 250 words, divided into sections with subheadings: Objectives, Methods, Results, Conclusions.

References: Limit to those cited in text; number as they appear consecutively in text; and indicate positions in text. In any one reference with 10 authors listed, use *et al.* for remaining authors.

JOURNAL ARTICLES

Surname and initials of author(s), title of article, name of journal, volume number, first and last pages, and year.

1. Jones CJ, and Smith TH: Cysts of the kidney. J Urol 33: 102–105, 1988.

BOOKS

Surname and initials of author(s), title and subtitle, edition (other than first). City, publishing house, year and pages as specific reference.

1. Jones CJ, and Smith TH: *Kidney Diseases*, Boston, Little Brown & Company, 1973, pp 50–53.

ARTICLES OR CHAPTERS IN BOOKS

Surname and initials of author(s), title of article, chapter number (if any), surname (other than first). City, publishing house, year and pages.

1. Jones CJ, and Smith TH: Value of Cystography, in Roberts MD: *Kidney Diseases*, New York, Oxford University Press, 1973, vol. 5, pp 200–206.

References to articles in press must state name of journal and, if possible, volume and year.

Tables: Type separate pages ; supplement, do not duplicate, text. List in numerical order; give each a precise heading.

Illustrations: Black-and-white photographs should be no larger than 4 × 5 in.; color prints are welcomed, but cost will be charged to author. Send color prints only, not original transparencies, slides, or negatives. Number all illustrations, indicate TOP, and cite in text. Give each illustration a concise legend.

OTHER FORMATS

ADULT UROLOGY—Original work relating to all aspects of urology in adults.

BASIC SCIENCE REVIEW ARTICLE—5000 word comprehensive reviews of latest developments on a basic science subject, such as "physiology of erectile function," genetics, tumor markers, etc.

BRIEF COMMUNICATIONS—1500 words on a new idea that is supported by limited but significant data and should include: justification for idea, historical data, etc., description of concept, preliminary results obtained, and statement on future validation of concept.

CASE REPORTS—1500 words, an abstract (100 words), introduction, case presentation and management, discussion, and 15 or fewer references. Unique cases demonstrating concepts of diagnosis and management.

EDITORIALS—2000 words and several pertinent references) on a controversial urologic topic by 2 well-recognized authorities on a subject—one "pro" and one "con."

LETTERS TO THE EDITOR—500 words supported by several relevant references regarding previous articles or comments on timely topics in letter form. Authors of cited article will have opportunity to reply.

PEDIATRIC UROLOGY—Features original work relating to all aspects of pediatric urology.

RAPID COMMUNICATION ARTICLES—Special manuscripts that are extremely timely and of utmost importance, and as a result, warrant rapid publication. Submission fee is $300.

REVIEW ARTICLE—500 word comprehensive review-type articles covering timely urologic topics of clinical relevance; must be well referenced. The article should serve as a source for the practicing urologist and resident-in-training to glean most current information on a subject.

SOCIOECONOMIC SECTION—3000 word comprehensive reviews on a timely socioeconomic, political, or legal issue containing information of relevance.

SURGEON'S WORKSHOP—1500 word articles (50 word abstract) plus photos and/or drawings on "how I do it" techniques.

SOURCE

January 1994, 43(1): 1K

Virology
Academic Press

Editorial Office, *Virology*
525 B Street, Suite 1900
San Diego, CA 92101-4495

AIM AND SCOPE

The aim of *Virology* is to publish the results of basic research in all branches of virology, including the viruses of vertebrates and invertebrates, plants, bacteria, and yeasts/fungi. In particular, we invite articles on the nature of viruses; on the molecular biology of virus multiplication, including the nature of viral genomes and the mechanisms that regulate the expression of the information that they encode; on molecular pathogenesis, that is, the molecular and immunological mechanisms of virus-host interactions, both at the cellular and at the organismal level; and on molecular aspects of the control and prevention of viral infections. Approaches and techniques used are expected to encompass those of many disciplines, including molecular genetics, molecular biology, biochemistry, biophysics, cell biology, immunology, and morphology.

MANUSCRIPT SUBMISSION

Manuscripts must report original research results; same or similar work, wholly or in part, must not have been nor will not be published elsewhere. Manuscripts must report research that is fundamental, rather than applied. Emphasize understanding of mechanisms by which viruses function, rather than describing virological phenomena. Report pioneering research; break new conceptual ground. Papers that describe results that have already been obtained with closely related virus or cell systems will only be accepted if they add significantly to knowledge. Papers that describe the cloning and sequencing of genes must deal with complete genes; new sequence information must be used to test an idea or be related to a clear set of novel conclusions that derive from these data. The same considerations apply to articles on the construction of reagents such as vectors, monoclonal antibodies, etc. Report the use of these reagents in further experiments that provide novel conclusions or insight.

Write as concisely as possible; make length the minimum compatible with a clear presentation and interpretation of experimental data. Avoid repetition, such as between Materials and Methods section and figure legends or Results section and Discussion.

Submission of manuscript must have been approved by all authors and institution where work was carried out; persons cited as sources of personal communications must have approved such citations; and that locally and nationally prescribed safety regulations governing exposure of laboratory personnel to hazardous situations and contamination of the environment must have been observed.

Written verification of conditions may be required.

Submit manuscripts in triplicate, by First Class mail in flat form.

If accepted for publication, copyright to article shall be assigned exclusively to the Publisher, who will not refuse any reasonable request by author(s) for permission to reproduce contributions to the journal.

MANUSCRIPT FORMAT

Generate manuscripts by high-quality printers, with wide margins, on high-quality bond paper, double-spaced throughout. Manuscripts generated by dot-matrix printers will be returned. Submit three copies of each manuscript; one must be the ribbon copy or of equal quality. Submit glossy prints of all halftone figures with each copy and one set of top quality figures for printer.

Subdivide full-length articles into Abstract, Introduction, Materials and Methods, Results, Discussion, Acknowledgments, and References. Make title succinct and informative; do not imply conclusions that are not fully substantiated by results. Provide a running head (37 characters). Abstract—200 words. Introduction should contain only material directly relevant to research. State aims of investigation in light of related work. In Materials and Methods provide sufficient information to permit work to be repeated. Refer to previously published procedures. In Discussion (two pages) be constructively interpretive; do not merely restate experimental results. Check for errors in tables, figures, names, quotations, and bibliography (and text references).

Figures and Tables: Number consecutively with Arabic numbers; give descriptive legends. Present illustrations in finished form suitable for reproduction. Plan to fit proportion of printed page (7 × 9 in.; column width, 3 3/8 in.). Make lettering on drawings of professional quality or generate by high resolution computer graphics and large enough to take a reduction of 50-60%. Make drawings with India ink on tracing linen, smooth surface white paper, or Bristol board. High-quality computer graphics are acceptable. Graph paper should be ruled in blue.

Submit photographs as glossy prints with strong contrast. Indicate magnification by a scale in micrographs. Indicate groupings of photographs. Illustrations in color are accepted if authors defray cost of production. Write author's name, figure number, and an arrow to indicate TOP on all figures, either in margin or on back, using a soft pencil.

Type tables on separate pages, number consecutively with Arabic numerals, and collect at end of manuscript. Give all tables descriptive headings and make understandable without reference to text.

References: Mention names of authors in text with year of publication in parentheses. List references alphabetically at end of paper.

Dawson, W. O., and Dodds, J. A. (1982). Characterization of subgenomic double-stranded RNAs from virus infected plants. *Biochem. Biophys. Res. Commun.* **107**, 1230–1235.

Abbreviate names of Journals according to *Chemical Abstracts Service Source Index,* 1985. References to books include: Author's name, year of publication, title of book (editor, if any), edition (if not first) publisher, and place of publication. Include in reference list articles that are "in press;" incorporate into text or use footnotes for "manuscript in preparation," "unpublished observations," and "personal communication."

OTHER FORMATS

MINIREVIEWS—Discuss brief and concisely focused topics in virology. Examples of formats are discussions of extant bodies of information in light of new data and analyses of newly emerging fields based on recent developments. Make understandable and accessible to readers who work outside the specific topics being addressed. Do not be comprehensive, covering every aspect of field in question, but rather should focus on what author feels are crucial and key experiments that have defined the topic.

Length is flexible, about three printed pages, about 12 pages of double-spaced manuscript. Schematic diagrams to illustrate specific points are acceptable; keep to a minimum. Do not introduce actual experimental data.

Limit references to those necessary to support thesis; cite and list as in regular manuscripts. Cite work in press; avoid citation of unpublished data and of personal communications.

SHORT COMMUNICATIONS—Six double-spaced typed pages including tables and illustrations. Present in brief form preliminary reports of important findings, and complete accounts of defined significant pieces of research of special interest and wide applicability. Include an Abstract; do not use section headings. Keep illustrative material to a minimum (one table or figure). Cite literature by giving references serial numbers in text; list consecutively at end of text (serial number, author, name of journal, volume, first and last page, and year of publication).

OTHER

Abbreviations and Symbols: Follow usage established by *Chemical Abstracts Service Index,* 1985.

Electronic Submission: Accepted manuscripts may be submitted on personal computer disks after all revisions have been incorporated onto disk. Label disk with type of computer used, type of software and version number, and disk format. Tabular material is typeset conventionally. Prepare art as camera-ready copy. Supply a hard copy printout of manuscript that exactly matches disk file. File names must clearly indicate content of each file.

For further information on preparing disks for typesetting conversion, contact Editorial Supervisor, Journal Division, Academic Press, Inc., 525 B Street, Suite 1900, San Diego, California 92101-4495; telephone: (619) 699-6415; fax: (619) 699-6800).

Experimentation, Animal: Perform in accordance with institutional and national guidelines as specified in appropriate NIH publications.

Materials Sharing: Make novel variant virus and cell strains, gene, plasmid, and vector constructs, and monoclonal antibodies/polyvalent antisera described in them freely available for distribution upon request to all qualified members of the scientific community for research purposes. This stipulation is based on the principle that published results must be verifiable.

Nomenclature, Genetic: Use recommendations of Demerec *et al.* (*Genetics* **54,** 61–76, 1966). Indicate phenotypes (when mutant loci have not been identified or mapped) by three-letter symbols with first letter capitalized. Designate genotypes by three letters, in lowercase italics. For genetic symbols for bacteriophage lambda, consult Echols and Murialdo (*Microbiol. Rev.* **42,** 577–591 1978) and Szybalski and Szybalski (*Gene* **7,** 217–220, 1979).

Nomenclature, Restriction Endonucleases: Follow recommendations of Roberts (in Bukhari *et al.,* "DNA Insertion Elements, Plasmids and Episomes," pp. 757–768, Cold Spring Harbor Laboratory, 1979).

Nomenclature, Virus: Use nomenclature approved by the International Committee for the Taxonomy of Viruses (ICTV) in Francki, R. I. B., Fauquet, C. M., Knudson, D. L., and Brown, F. (1991). Classification and Nomenclature of Viruses. Fifth Report of the International Committee on Taxonomy of Viruses. *Arch. Virol. Suppl.* Springer-Verlag, Vienna. Commonly used synonyms or vernacular names may be used after viruses have been correctly identified and specified first.

Nucleotide Sequence Data: Describe sequencing strategy employed, include a figure. Only in exceptional circumstances will sequences be published that were not derived by analysis of both strands.

Deposit sequences with GenBank as a prerequisite for manuscript acceptance. Sequences may either be deposited with GenBank before manuscripts are submitted for review; include footnote: "The nucleotide sequence data reported in this paper have been submitted to GenBank nucleotide sequence database and have been assigned the accession number XNNNNN," or authors will be asked to deposit sequences with GenBank if manuscript is potentially acceptable and is returned for revision. In that case deposit sequences prior to resubmission and include same footnote. Submit sequence information to: GenBank Submissions, National Center for Biotechnology Information, Bldg. 38A, Room 8N-803, 8600 Rockville Pike, Bethes-da, MD 20894; E-mail submission of new sequences: gb-sub@ncbi.nlm.nih.gov.

Reviewers: Submit names of four to six distinguished scientists as potential reviewers.

Style: Make sure every word, phrase, sentence, and paragraph is essential and arranged in logical sequence. Ensure correct use of verb tense. Use past tense except for general statements, such as generally accepted facts and author's own general conclusions, and statements such as "the data indicate," "Fig 1 illustrates," and the like.

X-ray Crystallographic Protein Structure Data: Submit to the Protein Data Bank all structural data required to validate discussion, including both X-ray amplitudes and phases and derived atomic coordinates. If paper discusses a protein structure only at level of main chain alpha carbon atoms, deposit only alpha carbon coordinates. If discussion involves higher resolution data, then deposit the full set of X-ray data and coordinate list. Manuscript will not be sent to printer until confirmation has been received from author, if not initially supplied, that required information has been sent to Protein Data Bank.

If requested by authors, editors will ask Data Bank not to distribute information until a specified date. For coordinate lists, this date may not be more than one year beyond acceptance date of manuscript. For full structure amplitude and phase data, time interval before distribution may not exceed four years. Release date specified by author will appear in a footnote to paper with statement that information has been submitted to Protein Data Bank. In the absence of a specified release date, it is assumed that information is available immediately.

SOURCE

January 1994, 198(1): ix-xii.

Vision Research
Pergamon Press

Vision Research
P.O. Box 12011
1100 AA Amsterdam-Zuidoost, The Netherlands
        Central Office

Yutaka Fukuda
Department of Physiology
Osaka University Medical School
2-2, Yamadaoka, Suita, 565 Osaka Japan
        Neurobiology

Samuel M. Wu
Cullen Eye Institute
Baylor College of Medicine
Houston, TX 77030
        Neurobiology

Dennis M. Levi
University of Houston, Central Campus
College of Optometry
4901 Calhoun Boulevard
Houston, TX 77204-6052
        Psychophysics

Davida Y. Teller
Department of Psychology NI-25
University of Washington
Seattle, WA 98195 .
            Psychophysics

Han Collewijn
Department of Physiology I
Erasmus University Rotterdam
P.O. Box 1738
3000 DR Rotterdam, The Netherlands
            Behavioral Physiology and Visuomotor Control

Christa Neumeyer
Institut für Zoologie (Abt. III)
Johannes Gutenberg Universität
55099 Mainz, Germany
            Behavioral Physiology and Visuomotor Control

Ellen C. Hildreth
Department of Computer Science
Wellesley College
Wellesley, MA 02181
            Computational Vision

Michael J. Morgan
Department of Pharmacology
University of Edinburgh
1 George Square
Edinburgh EH8 9JZ, Scotland
            Computational Vision

Ivan Bodis-Wollner
Department of Neurology
University of Nebraska Medical Center
600 South 42nd Street
Omaha, NE 68198-2045
            Clinical Vision Sciences

Eberhart Zrenner
Department of Pathophysiology of Vision and Neuro-
    Ophthalmology
University of Tübingen
Schleichstrabe 12–16
72076 Tübingen, Germany
            Clinical Vision Sciences

AIM AND SCOPE

Vision Research is a journal devoted to the functional aspects of human, vertebrate and invertebrate vision and publishes experimental and observational studies, reviews, and theoretical papers firmly based upon current facts of visual science. *Vision Research* also accepts experimental studies in which clinical material has been used to address an issue of basic research interest, or where basic research methods have been used to address an issue of clinical importance, or where basic research may have, as yet unapplied, clinical relevance. The words clinical and vision sciences should be interpreted in the broadest sense. Papers reporting detailed investigations are encouraged and authors should be advised to include enough background material in the Introduction of their papers so that they are comprehensible to the non-specialist. The purpose of theoretical papers is to give a higher sense of order to the facts as they are presently known, or to point to new observations which can be verified experimentally. Papers dealing with questions in the history of visual science should lay stress upon the history of ideas in this field. *Vision Research* has always welcomed the broadest interpretation of visual science.

MANUSCRIPT SUBMISSION

Manuscripts may not have been published or simultaneously submitted for publishing elsewhere. By submitting a manuscript, authors agree that copyright is transferred to publisher if and when article is accepted for publication. Assignment of copyright is not required from authors who work for organizations which do not permit such assignment.

Submit five double-spaced typewritten copies of manuscript and illustration. Submit original plus three copies to central office. Send one copy to one of the 10 Section Editors whose subject area most closely relates to work submitted (indicate which Section Editor was selected).

Do not submit papers of monograph length or a series of numbered papers.

State that work has not been published elsewhere and is not under review with another journal, and that if published in *Vision Research,* it will not be reprinted elsewhere in any language in the same form without consent of publisher, who holds copyright.

MANUSCRIPT FORMAT

Type single-sided, double-spaced with generous margins. On title page include five key words selected from key word list published annually in first issue of *Vision Research*. If a new key word is needed, underline. Give animal classification. Provide a running head (40 characters).

Abstract: 100 words. Supplement title. Inform reader of essential points of paper.

References: Cite in text by last name of author (or authors) followed by year of publication in parentheses, e.g., Hays (1963), Van den Brink and Bouman (1957) or (Zrenner & Gouras, 1981). If three or more authors, Kuhn, Bennett, Michel-Villaz and Chabre (1981), first citation; Kuhn *et al.* (1981), second citation. If more than one work by an author(s) in a given year, label alphabetically within each year (e.g., Rushton, 1965a,b).

Type full references on a separate page at end of article. Do not give footnotes. Include names of all authors and their initials, year of publication in parentheses, full title of article or book, name of journal, volume number and pages. For books, give city of publication and publisher.

Zrenner E. & Gouras. P. (1981). Characteristics of the blue sensitive cone mechanism in primate retinal ganglion cells. *Vision Research, 21,* 1605–1609.

Semmlow, J. L. & Huny. G. K. (1983). The near responses: Theories of control. In Schor, C. M. & Ciuffreda, K. J. (Eds), *Vergence eye movements: Basic and clinical aspects* (pp. 175–195). London: Butterworths.

Cite unpublished work, work in press, or conference proceedings only exceptionally. Submit preprints if essential to argument.

Illustrations and Tables: Submit only standard size (8.5 × 12 in.). Do not affix in place; indicate proper places for insertion in text. Refer to all photographs, graphs and diagrams as "Figures" and number consecutively, also e tables. Type corresponding legends on a separate sheet at end of article. Write author's name, number, and indicate orientation lightly on back of each figure or table.

Submit halftone illustrations as glossy prints; approximately 1.5 times final size. Lines or letters to appear on photographs are inserted by publisher. Draw position on a leaf of drawing paper firmly attached to photograph, or on an additional print. Include a guide to Production Department to identify salient features to be reproduced.

Draw line drawings on stiff white paper or tracing cloth; high quality computer graphics are acceptable. Present approximately twice final size. Draw letters and symbols to be clearly legible when reduced. Use standard symbols: ○ △ ❑ ● ▲ ■ + ×. Make other special symbols (e.g., Greek Letters) particularly legible and explicit.

OTHER FORMATS

LETTERS TO THE EDITOR—If in response to a published article, letter is always sent to authors of initial article for a reply. Letter and reply are sent for review together.

MINIREVIEWS—Not a comprehensive history of the subject; a survey of recent developments in fast-growing and active areas of vision research covering the last few years.

Submit proposals to appropriate Section Editor; include: title; justification at this time on topic; a rough outline; and a firm date for submission of completed work.

SHORT COMMUNICATIONS, LANCET-STYLE LETTERS, RESEARCH REPORTS incorporating an abstract and TECHNICAL OR CLINICAL NOTES are welcomed.

SPECIAL NON-RECURRENT SYMPOSIA—Symposium organizers are requested to contact the Chairman. Decision is made by the Editorial Board of *Vision Research* at annual meeting during ARVO.

OTHER

Electronic Submission: Submit a computer disk (5.25 or 3.5 in. HD/DD disk) containing final version of paper along with final manuscript. Specify software used, including release (e.g., WordPerfect 4.0). Specify computer used (either IBM compatible PC or Apple Macintosh). Include text file and separate table and illustration files, if available. File should follow general instructions on style/ arrangement and, in particular, reference style of this journal. Single-space file and use wrap-around end-of-line feature (i.e., no returns at end of each line). Begin all textural elements flush left, no paragraph indents. Place two returns after every element such as title, headings, figure and table call-outs, etc. Keep a back-up disk.

Experimentation, Animal: Indicate in Methods section that adequate measures were taken to minimize pain or discomfort, in accordance with European Communities Council Directive of 24 November 1986 (86/609/EEC) or with U.S. NIH Guidelines regarding care and use of animals for experimental procedures. Editors reserve the right to reject papers if there is doubt whether appropriate procedures have been used.

Experimentation, Human: Submit a statement that experiments were undertaken with the understanding and written consent of each subject. Be aware of the Code of Ethics of the World Medical Association (Declaration of Helsinki), (*British Medical Journal,* 18 July 1964).

SOURCE

January 1994, 34(1): last page and inside back cover.

Yeast
Wiley

S.G. Oliver
Dept. of Biochemistry and Applied Molecular Biology
UMIST, PO Box 88, Manchester, M60 1 QD, U.K.
E-mail: sgolir@mailhost.mcc.ac.uk

R.B. Wickner
National Institutes of Health
Building 8, Room 209, Bethesda, MD, 20892
E-mail: wickner@ helix.nih.gov.

AIM AND SCOPE

Yeast is a journal published for the international community of yeast researchers. It contains original research articles, major and minor reviews and short communications on all aspects of *Saccharomyces* and other yeast genera. By focusing on the most significant developments in this research area the journal will be invaluable reading for those wishing to keep up to date with this rapidly moving field.

MANUSCRIPT SUBMISSION

Submit three copies of manuscript to closest editorial office. Major reviews will be commissioned, but minor reviews and original research papers are solicited, together with any other reports, letters or announcements of potential interest to readers of journal.

Transfer of copyright from author to publisher must be explicit. A copy of Publishing Agreement appears in each issue; additional copies are available from journal editors or publisher; agreement may be photocopied from journal. Enclose a signed copy with every article.

MANUSCRIPT FORMAT

Type double-spaced on A4 paper. Leave 1.5 in. margins at top, bottom and left side of page. Type at top of first page name and full address of corresponding author. Title of paper must describe contents fully but concisely. First page: Full title and names and affiliations of all authors. Denote with a dagger (†) corresponding author. Give full postal address for author who will check proofs, with telephone, telex and fax number.

Divide text into: Introduction; Materials and Methods; Results and Discussion. Avoid section numbering. There is no length limit.

Speedy publication will be facilitated by submission of material on diskette.

Abstract: 250 words setting out essential contents. Make self-contained and comprehensive. Type on a separate sheet.

References: For reviews, cite references by number in text. Alphabetize reference list and number in sequence. References in text will not be in numerical sequence. In original research papers, citation should be by name and date. Abbreviate journal names according to *Bibliographic Guide for Editors and Authors* (1974, American Chemical Society); include first and last page numbers.

JOURNALS

Cooper, T. G. and Sumrada, R. (1975) Urea transport in *Saccharomyces cerevisiae. J. Bacteriol.* **121,** 571-576.

BOOKS

Spencer, J. F. T., Spencer, D. M. and Smith, A. R. W. (Eds) (1984). *Yeast Genetics: Fundamental and Applied Aspects,* Springer-Verlag.

Dart, E. C., Pioli, D. and Atherton, K. T. (1981). Genetic Engineering in Norris, J. R. and Richmond, M.H. (Eds), *Essays in Microbiology,* John Wiley & Sons Ltd, pp. 3/1–3/32.

Illustrations: Submit only top quality figures; present line drawings in black ink on tracing cloth or paper or white board; halftones (particularly micrographs) as glossy bromide prints. Make each illustration twice final size and ensure that lettering will still be legible when reduced. Submit lettering on halftones as an overlay. Label figures lightly in pencil with figure number and first author's name. Color illustrations are accepted at author's expense.

OTHER FORMATS

YEAST MAPPING REPORTS—Short communications reporting determination of relative position of genes on genetic map of *Saccharomyces cerevisiae.* While evidence from pulsed field gels will be accepted for primary assignment of a cloud gene to a chromosome, genetic map position should be determined by tetrad analysis, preferably using threepoint crosses. Give full details of such analyses and of computational methods used to calculate map distances. Section Editors reserve the right to re-calculate distances following a re-appraisal of data submitted, or in light of other data. Professor Mortimer will advise authors on the naming of genes in order to avoid confusion in literature.

Submit directly to either: Professor R.K. Mortimer: Department of Molecular and Cell Biology, Division of Biophysics, University of California, California 94720 or Dr. J. R. Johnston: Department of Bioscience and Biotechnology, University of Strathclyde, Glasgow, G1 1XW, Scotland.

YEAST SEQUENCING REPORTS—Sequence data may be published rapidly with little commentary beyond precise location of the sequenced region within its chromosome. Data is accepted in electronic format on floppy discs. Minimal continuous sequence acceptable for publication is not tightly defined to permit rapid publication of short sequences of high biological interest. In the absence of such considerations, the minimal sequence shall be 3.0 kb.

Keep format simple; confine to nucleotides and amino acids. Highlight important regions by nucleotide sequence number in an accompanying glossary. The journal will produce a restriction map of sequenced fragment, to a uniform style, using computer methods. Transmit details of new genes to R. K. Mortimer for inclusion in the genetic map which will be updated regularly.

Submit to Professor S. G. Oliver at address above.

OTHER

Abbreviations: Define in full at first use.

Nomenclature: For genetic nomenclature follow *The Molecular Biology of the Yeast Saccharomyces; Life Cycle and Inheritance,* Appendix 1, pp 639–640 (Strathern J.N., Jones E.W. and Broach J.R., 1981, Cold Spring Harbor Laboratories). Send details naming a new gene or renaming an old one to Professor R. K. Mortimer, Department of Biophysics, University of California, Berkeley, CA 94720. An annual list of gene designations is published in *Yeast.*

Use yeast names consistent with *Yeasts: Characteristics and Identification* (Strathern J.N., Jones E.W. and Broach J.R., 1981, 2nd ed, Cambridge University Press).

SOURCE

February 1994, 10 (2): inside back cover.

Uniform Requirements for Manuscripts Submitted to Biomedical Journals
International Committee of Medical Journal Editors

Kathleen Case
Secretariat Office *Annals of Internal Medicine*
Independence Mall West
Sixth Street at Race
Philadelphia, PA 19106-1572
Telephone: (800) 523-1546, ext. 2631

AIM AND SCOPE

The *Uniform Requirements for Manuscripts Submitted to Biomedical Journals* was developed by the International Committee of Medical Journal Editors (also known as the Vancouver Group). Copies are available free of charge from the Secretariat Office. The first set of guidelines was issued in 1978. This is the fourth edition, revised slightly in January 1993.

The *Uniform Requirements for Manuscripts Submitted to Biomedical Journals* provides instructions to authors on how to prepare manuscripts.

Almost 500 international biomedical journals have agreed to receive manuscripts prepared in accordance with these requirements. These journals have agreed that if manuscripts are prepared in the style specified, their editors will not return manuscripts for changes in style before considering them for publication. Authors sending manuscripts to a participating journal need not try to prepare them in accordance with publication style of that journal but can prepare manuscripts in style described in the *Uniform Requirements*. Journals may later alter accepted manuscripts to conform with details of journal's publication style.

With regard to scope of an article, follow journal instructions to authors as to what topics are suitable for that journal and types of papers that may be submitted—for example, original articles, reviews, or case reports. Journal instructions contain other requirements unique to that journal, such as number of copies of manuscripts, acceptable languages, length of articles, and approved abbreviations.

MANUSCRIPT SUBMISSION

Follow journal's instructions for transfer of copyright. Keep copies of everything submitted.

Most journals will not consider a paper on work that has already been reported in a published article or is described in a paper submitted or accepted for publication elsewhere, in print or in electronic media. This policy does not usually preclude consideration of a paper that has been rejected by another journal or of a complete report that follows publication of a preliminary report, usually as an abstract, or a paper that has been presented at a scientific meeting, if not published in full in a proceedings or similar publication. Press reports of meeting should not amplify additional data or distribute copies of tables and illustrations. When submitting a paper, make a full statement to the editor about all submissions and previous reports that might be regarded as prior or duplicate publication of the same or very similar work. Include copies of such material with submitted paper.

Multiple publication, i.e., the publication more than once of the same study, irrespective of whether the wording is the same, is rarely justified. Secondary publication in another language is one possible justification, providing the following conditions are met. The editors of both journals must be fully informed; editor of secondary publication should have a photocopy, reprint, or manuscript of primary version. Respect priority of primary publication by a publication interval of at least 2 weeks. Write the paper for secondary publication for a different group of readers; do not simply translate the primary paper; an abbreviated version will often be sufficient. The secondary version must reflect faithfully the data and interpretations of the primary version. Include a footnote on the title page stating that paper was edited, and is being published, for a national audience in parallel with a primary version based on the same data and interpretations. A suitable footnote might read: "This article is based on a study first reported in the [title of journal, with full reference]."

Multiple publication other than as defined above is not acceptable. Violation of this rule may result in appropriate editorial action.

Preliminary release, usually to public media, of scientific information described in a paper accepted but not yet published is a violation of the policies of many journals. Preliminary release of data may be acceptable, with special arrangement with editor, for example, to warn the public of health hazards.

MANUSCRIPT SUBMISSION

Submit required number of copies of manuscripts and illustrations, enclosing manuscript copies and figures in cardboard, if necessary, to prevent bending of photographs during mail handling. Place photographs and transparencies in a separate heavy-paper envelope, with a cover letter, and permissions to reproduce previously published material or to use illustrations that may identify human subjects.

Cover letter must be signed by all coauthors. Include information on prior or duplicate publication or submission elsewhere of any part of work; statement of financial or other relationships that might lead to a conflict of interest; statement that manuscript has been read and approved by all authors, that requirements for authorship have been met, and that each coauthor believes manuscript represents honest work; and name, address, and telephone number of corresponding author, responsible for communicating with other authors about revisions and final approval of proofs. Give any additional information that may be helpful to editor, such as type of article the manuscript represents and whether author(s) are willing to meet cost of reproducing color illustrations.

Submit copies of permissions to reproduce published material, to use illustrations or report sensitive personal information about identifiable persons, or to name persons for their contributions.

All persons designated as authors must qualify for authorship. Order of authorship is a joint decision of coauthors. Each author should have participated sufficiently in the work to take public responsibility for its content.

Base authorship credit only on substantial contributions to conception and design, or analysis and interpretation of data; and to drafting article or revising it critically for important intellectual content; and on final approval of version to be published. All conditions must be met. Participation solely in acquisition of funding or collection of data does not justify authorship; neither does general supervision of the research group. Any part of an article critical to its main conclusions must be the responsibility of at least one author. Editors may require authors to justify assignment of authorship.

Multicenter trials are often attributed to a corporate author. All group members named as authors, either in authorship position below title or in a footnote, should fully meet the criteria for authorship. List group members who do not meet these criteria, with their permission, under acknowledgments or in an appendix.

MANUSCRIPT FORMAT

Begin each component on a new page, in sequence: title page, abstract and key words, text, acknowledgments, references, tables (put each table with title and footnotes on a separate page), and legends for illustrations.

Type or print manuscript double-spaced, including title page, abstract, text, acknowledgments, references, tables, and legends, on white bond paper, 8.5 × 11 in. or ISO A4 with margins of at least 1 in. Type or print on only one side of paper. Double-space throughout, including title page, abstract, text, acknowledgments, references, individual tables, and legends. Number pages consecutively, beginning with title page. Put page number in upper or lower right corner of each page.

Title Page: Title of article, concise but informative; first name, middle initial, and last name of each author, with highest academic degree(s) and institutional affiliations; name of department(s) and institution(s) to which work is attributed; disclaimers, if any; name and address of corresponding author; name and address of author to whom requests for reprints should be addressed or statement that reprints will not be available from author; source(s) of support as grants, equipment, drugs, etc.; and a short running head or foot line (40 characters, counting letters and spaces) placed at foot of title page and identified.

Abstract and Key Words: Second page. 150 words for unstructured abstracts or 250 words for structured abstracts. State purposes of study or investigation, basic procedures (selection of study subjects or laboratory animals; observational and analytical methods), main findings (give specific data and statistical significance), and principal conclusions. Emphasize new and important aspects of study or observations.

Below abstract provide, and identify 3 to 10 key words or short phrases for cross-indexing article. Use terms from *Medical Subject Headings* (MeSH) list of *Index Medicus*; if suitable MeSH terms are not yet available for new terms, use present terms.

Text: Text of observational and experimental articles is usually divided into sections with headings Introduction, Methods, Results, and Discussion. Long articles may need subheadings within sections to clarify content, especially Results and Discussion. Other types of articles such as case reports, reviews, and editorials need other formats. Consult individual journals for guidance.

Introduction: State purpose of article. Summarize rationale for study or observation. Give only strictly pertinent references; do not review article extensively. Do not include data or conclusions from work being reported.

Methods: Describe selection of observational or experimental subjects (patients or laboratory animals, including controls). Identify methods, apparatus (manufacturer's name and address in parentheses), and procedures in sufficient detail to allow other workers to reproduce results. Give references to established methods, including statistical methods; provide references and brief descriptions for methods that have been published but are not well known; describe new or substantially modified methods, give reasons for using them, and evaluate their limitations. Identify all drugs and chemicals used, including generic name(s), dose(s), and route(s) of administration.

Results: Present in logical sequence in text, tables, and illustrations. Do not repeat in text data in tables or illustrations; emphasize or summarize only important observations.

Discussion: Emphasize new and important aspects of study and conclusions that follow. Do not repeat in detail data or other material in Introduction or Results. Include implications of findings and their limitations, including implications for future research. Relate observations to other relevant studies. Link conclusions with goals of study but avoid unqualified statements and conclusions not completely supported by data. Do not claim priority or allude to work that has not been completed. State new hypotheses when warranted, but clearly label as such. Include recommendations, when appropriate.

Acknowledgments: At an appropriate place in article (title-page footnote or appendix to text; see journal requirements), specify: contributions that need acknowledging but do not justify authorship, such as general support by a departmental chair; acknowledgments of technical help; acknowledgments of financial and material support, specifying nature of support; and financial relationships that may pose a conflict of interest. Name persons who have contributed intellectually to paper but whose contributions do not justify authorship and describe their function or contribution, e.g., "scientific adviser," "critical review of study proposal," "data collection," or "participation in clinical trial." Obtain permission from persons so named, because readers may infer their endorsement of data and conclusions. Acknowledge technical help in a paragraph separate from those acknowledging other contributions.

References: Number consecutively in order of first mention. Identify in text, tables, and legends by Arabic numerals in parentheses. Number references cited only in tables or in legends to figures in accordance with a sequence established by first identification in text of particular table or figure.

Follow style of examples, which are based on formats used by the U.S. National Library of Medicine in *Index*

Medicus. Consult *List of Journals Indexed in Index Medicus* for journal title abbreviations. Avoid using abstracts as references. Do not use "unpublished observations" and "personal communications" as references; insert references to written, not oral, communications (in parentheses) in text. Include in references papers accepted but not yet published; designate journal and add "In press." Cite in text information from manuscripts submitted but not yet accepted as "unpublished observations" (in parentheses). Verify references against original documents.

ARTICLES IN JOURNALS

Standard journal article (list all authors, but if the number exceeds six, give six followed by et al.):

You CH, Lee KY, Chey RY, Menguy R. Electrogastrographic study of patients with unexplained nausea, bloating and vomiting. Gastroenterology 1980 Aug;79(2):311–4.

As an option, if a journal carries continuous pagination throughout a volume, the month and issue number may be omitted:

You CH, Lee KY, Chey RY, Menguy R. Electrogastrographic study of patients with unexplained nausea, bloating and vomiting. Gastroenterology 1980;79:311–4.

Goate AM, Haynes AR, Owen MJ, Farrall M, James LA, Lai LY, et al. Predisposing locus for Alzheimer's disease on chromosome 21. Lancet 1989;1:352–5.

Organization as author:

The Royal Marsden Hospital Bone-Marrow Transplantation Team. Failure of syngeneic bone-marrow graft without preconditioning in post-hepatitis marrow aplasia. Lancet 1977;2:742–4.

No author given:

Coffee drinking and cancer of the pancreas [editorial]. BMJ 1981;283:628.

Article not in English:

Massone L, Borghi S, Pestarino A, Piccini R, Gambini C. Localisations palmaires purpuriques de la dermatite herpetiforme. Ann Dermatol Venereol 1987;114:1545–7.

Volume with supplement:

Magni F, Rossoni G, Berti F. BN-52021 protects guinea-pig from heart anaphylaxis. Pharmacol Res Commun 1988;20 Suppl 5: 75–8.

Issue with supplement:

Gardos G, Cole JO, Haskell D, Marby D, Paine SS, Moore P. The natural history of tardive dyskinesia. J Clin Psychopharmacol 1988;8(4 Suppl):31S–37S.

Volume with part:

Hanly C. Metaphysics and innateness: a psycho-analytic perspective. Int J Psychoanal 1988;69(Pt 3):389–99.

Issue with part:

Edwards L, Meyskens F, Levine N. Effect of oral isotretinoin on dysplastic nevi. J Am Acad Dermatol 1989;20(2 Pt 1):257–60.

Issue with no volume:

Baumeister AA. Origins and control of stereotyped movements. Monogr Am Assoc Ment Defic 1978;(3):353–84.

No issue or volume:

Danoek K. Skiing in and through the history of medicine. Nord Medicinhist Arsb 1982:86–100.

Pagination in Roman numerals:

Ronne Y. Ansvarsfall. Blodtransfusion till fel patient. Vardfacket 1989;l3:XXVIXXVII.

Type of article indicated as needed:

Spargo PM, Manners JM. DDAVP and open heart surgery [letter]. Anaesthesia 1989;44:363–4.

Fuhrman SA. Joiner KA. Binding of the third component of complement C3 by Toxoplasma gondii [abstract]. Clin Res 1987;35:475A.

Article containing retraction:

Shishido A. Retraction notice: Effect of platinum compounds on murine lymphocyte mitogenesis [Retraction of Alsabti EA, Ghalib ON, Salem MH. In: Jpn J Med Sci Biol 1979;32:53–65]. Jpn J Med Sci Biol 1980;33:235–7.

Article retracted:

Alsabti EA, Ghalib ON, Salem MH. Effect of platinum compounds on murine lymphocyte mitogenesis [Retracted by Shishido A. In: Jpn J Med Sci Biol 1980;33:235–7]. Jpn J Med Sci Biol 1979:32:53–65.

Article containing comment:

Piccoli A, Bossatti A. Early steroid therapy in IgA neuropathy: still an open question [comment]. Nephron 1989;51:289–91. Comment on: Nephron 1988;48:12–7.

Article commented on:

Kobayashi Y, Fujii K, Hiki Y, Tateno S, Kurokawa A, Kamiyama M. Steriod therapy in IgA nephropathy: a retrospective study in heavy proteinuric cases [see comments]. Nephron 1988;48:12–7. Comment in: Nephron 1989;51:289–91.

Article with published erratum:

Schofield A. The CAGE questionnaire and psychological health [published erratum appears in Br J Addict 1989;84:701]. Br J Addict 1988;83:761-4.

BOOKS AND OTHER MONOGRAPHS

Personal author(s):

Colson JH, Armour WJ. Sports injuries and their treatment. 2nd rev. ed. London: S. Paul, 1986.

Editor(s), compiler as author:

Diener HC, Wilkinson M, editors. Drug induced headache. New York: Springer Verlag, 1988.

Organization as author and publisher:

Virginia Law Foundation. The medical and legal implications of AIDS. Charlottesville: The Foundation, 1987.

Chapters in a book:

Weinstein L, Swartz MN. Pathologic properties of invading microorganisms. In: Sodeman WA Jr, Sodeman WA, editors. Pathologic physiology: mechanisms of disease. Philadelphia: Saunders, 1974:457–72.

Conference proceedings:

Vivian VL, editor. Child abuse and neglect: a medical community response. Proceedings of the First AMA National

Conference on Child Abuse and Neglect; 1984 Mar 30-31; Chicago. Chicago: American Medical Association, 1985.

Conference paper:

Harley NH. Comparing radon daughter dosimetric and risk models. In: Gammage RB, Kaye SV, editors. Indoor air and human health. Proceedings of the Seventh Life Sciences Symposium; 1984 Oct 29-31; Knoxville (TN). Chelsea (MI): Lewis, 1985:69–78.

Scientific or technical report:

Akutsu T. Total heart replacement device. Bethesda (MD): National Institutes of Health, National Heart and Lung Institute; 1974 Apr. Report No.: NIH-NHLI-69-2185-4.

Dissertation:

Youssef NM. School adjustment of children with congenital heart disease [dissertation]. Pittsburgh (PA): Univ. of Pittsburgh, 1988.

Patent:

Harred JF, Knight AR, McIntyre JS, inventors. Dow Chemical Company, assignee. Epoxidation process. US patent 3,654,317. 1972 Apr 4.

OTHER PUBLISHED MATERIAL

Newspaper article:

Rensberger B, Specter B. CFCs may be destroyed by natural process. The Washington Post 1989 Aug 7;Sect. AP:2 (col. 5).

Audiovisual:

AIDS epidemic: the physician's role [video recording]. Cleveland (OH): Academy of Medicine of Cleveland, 1987.

Computer file:

Renal system [computer program]. MSDOS version. Edwardsville (KS): MediSim, 1988.

Legal material:

Toxic Substances Control Act: Hearing on S. 776 Before the Subcomm. on the Environment of the Senate Comm. on Commerce. 94th Cong., 1st Sess. 343 (1975).

Map:

Scotland [topographic map]. Washington: National Geographic Society (US), 1981.

Book of the Bible:

Ruth 3:1–18. The Holy Bible. Authorized King James version. New York: Oxford Univ. Press, 1972.

Dictionary and similar references:

Ectasia. Dorland's illustrated medical dictionary. 27th ed. Philadelphia: Saunders, 1988:527.

Classical material:

The Winter's Tale: act 5, scene 1, lines 13-16. The complete works of William Shakespeare. London: Rex, 1973.

UNPUBLISHED MATERIAL

In press:

Lillywhite HD, Donald JA. Pulmonary blood flow regulation in an aquatic snake. Science. In press.

Tables: Type or print double-spaced on a separate sheet. Do not submit as photographs. Cite each table in text.

Number consecutively in order of citation in text; supply a brief title for each. Give each column a short or abbreviated heading. Place explanatory matter in footnotes, not in heading. Explain in footnotes all nonstandard abbreviations. For footnotes, use symbols, in sequence: *, †, ‡, §, ||, ¶, **, ††, ‡‡, ...

Identify statistical measures of variations such as standard deviation and standard error of the mean.

Do not use internal horizontal and vertical rules.

If data is from another published or unpublished source, obtain permission and acknowledge fully. Use of too many tables in relation to length of text produces difficulties in layout of pages. Examine issues of journal to estimate how many tables can be used per 1000 words of text.

Editor may recommend that additional tables containing important backup data too extensive to publish be deposited with an archival service, such as the National Auxiliary Publications Service in the United States, or be made available by authors. In that event an appropriate statement will be added to text. Submit such tables with manuscript.

Illustrations (Figures): Submit required number of complete sets of figures professionally drawn and photographed; freehand or typewritten lettering is unacceptable. Do not submit original drawings, roentgenograms, and other material; send sharp, glossy black-and-white photographic prints, 5×7 in., (no larger than 8×10 in.). Present letters, numbers, and symbols clear and even throughout and of sufficient size that, when reduced for publication, each item will still be legible. Place titles and detailed explanations in legend for illustration, not on illustration.

Affix label on back of each figure indicating figure number, author's name, and TOP. Do not write on back or use paper clips. Do not bend figures or mount on cardboard.

Give photomicrographs internal scale markers. Symbols, arrows, or letters in photomicrographs should contrast with background.

Number figures consecutively according to order of citation in text. If a figure has been published, acknowledge original source and submit written permission from copyright holder to reproduce material. Permission is required irrespective of authorship or publisher, except for documents in public domain.

For illustrations in color, ascertain whether color negatives, positive transparencies, or color prints are required. Mark accompanying drawing to indicate region to be reproduced. Some journals publish illustrations in color only if author pays for extra cost.

Legends for Illustrations: Type or print double-spaced, starting on a separate page, with Arabic numerals corresponding to illustrations. Identify and explain symbols, arrows, numbers, or letters used to identify parts of illustrations in legend. Explain internal scale and identify method of staining in photomicrographs.

OTHER

Abbreviations and Symbols: Use only standard abbreviations. Avoid use in title and abstract. Full term for which an abbreviation stands should precede its first use in text unless it is a standard unit of measurement.

Some Standard Abbreviations and Symbols	
absorbance	A
acceleration of gravity	g
acetate	OAc
acetyl	Ac
N-acetylglucosamine	GlcNAc
acquired immunodeficiency syndrome	AIDS
activated complement	C′
adenine	Ade
adenosine	A, Ado
adenosine 5′-mono-, di-, and tri-phosphate	AMP, ADP, ATP
adenosine 3′,5′-cyclic monophosphate	cyclic AMP or cAMP
adenosine triphosphatase	ATPase
adrenocorticotropin	ACTH
alanine	Ala
alternating current	ac
amino acid(s)	aa
ampere (milliampere)	A (mA)
angstrom (0.1 nm)	Å
antigen	Ag
antigen-binding fragment	F(ab)
antigen-presenting cell	APC
apolipoprotein	apo
arginine	Arg
arginine vasopressin	AVP
asparagine	Asn
asparagine or aspartic acid	Asx
aspartic acid	Asp
atomic weight	at. wt.
atto (10^{-18})	a
audio frequency	af
barn	b
base pair(s)	bp
becquerel	Bq
before present	B.P.
benzoyl	Bz
body-centered-cubic	bcc
boiling point	b.p.
bovine serum albumin	BSA
5-bromodeoxyuridine	BrdUrd
butyl	Bu
by mouth	p.o.
calculated	Calc., calc
calorie	cal
candela	cd
carbon monoxide hemoglobin	HbCO
carbon monoxide myoglobin	MbCO
O-carboxymethylcellulose	CM-cellulose
centi (10^{-2})	c
central nervous system	CNS
chemically pure	CP
circular dichroism	CD
coenzyme A	CoA

colony-forming unit	CFU
colony-stimulating factor	CSF
complement	C
complementary DNA, complementary RNA	cDNA, cRNA
complete Freund's adjuvant	CFA
concanavalin A	Con A
concentration	Conc.
constant	const
constant region of Ig	C region
correlation coefficient	r
coulomb	C
counts per minute	cpm
counts per second	cps
curie(s)	Ci
cycles per second (hertz)	Hz
cyclic AMP	cAMP
cyclic GMP	cGMP
cysteine or half-cystine	Cys
cytidine	C, Cyd
cytomegalovirus	CMV
cytotoxic T lymphocyte	CTL
dalton	Da
day	d
deka (10^1)	da
deci (10^{-1})	d
degree Celsius	°C
degrees of freedom	df
density	p
deoxy (carbohydrates and nucleotides)	d
deoxyribonucleic acid (deoxyribonucleate)	DNA
deoxyribonuclease	DNase
diameter, inside or outside	i.d. or o.d.
diethylaminoethyl	DEAE
diethylaminoethylcellulose	DEAD-cellulose
dimethylsulfoxide	DMSO
dinitrophenyl	DNP
direct current	dc
disintegrations per minute	dpm
disintegrations per second	dps
dissociation constant	k_d
dithiothreitol	DTT
diversity region of Ig or T cell receptor for antigen	D region
double-stranded (as dsDNA)	ds
effective dose, 50%	ED_{50}
effector to target ratio	E:T ratio
electrocardiogram	ECG
electroencephalogram	EEG
electromotive force	emf
electron microscopy	EM
electron paramagnetic resonance	EPR
electron spin resonance	ESR

electron volt	eV
enzyme-linked immunosorbent assay	ELISA
Epstein-Barr virus	EBV
equilibrium constant	K
equivalent	Eq
equivalent weight	equiv. wt.
erythrocyte	E
ethidium	Etd
ethyl	Et
ethylenediaminetetraacetic acid	EDTA
ethylene glycol bis(β-aminoethyl ether)-N,N,N',N'-tetraacetic acid	EGTA
exa (10^{18})	E
experiment(al)	Exp., exp
farad	F
femto (10^{-15})	f
fetal calf serum	FCS
flavin-adenine dinucleotide	FAD
flavin mononucleotide	FMN
fluorescein isothiocyanate	FITC
fluorescence-activated cell sorter	FACS
follicle-stimulating hormone	FSH
formylmethionine	fMet
formyl-methionyl-leucyl-phenylalanine	FMLP
free fatty acids	FFA
freezing point	f.p.
fructose	Fru
fucose	Fuc
galactosamine	GalN
galactose	Gal
gas chromatography, gas/liquid chromatography	GC, GLC
gauss	G
giga (10^9)	G
glomerular filtration rate	GFR
gluconic acid	GlcA
glucosamine	GlcN
glucose	Glc
glucuronic acid	GlcA, GlcUA
glutamic acid	Glu
glutamic acid or glutamine	Glx
glutamine	Gln
glycine	Gly
gram	g
gray	Gy
guanine	Gua
guanosine	G, Guo
guanosine 5'-monophosphate (guanosine-monophosphate, guanylic acid)	GMP
half-life (half-time)	$t_{1/2}$
Hanks' balanced salt solution	R
heavy chain of Ig	H chain
hecto (10^2)	h
height	ht
hemoglobin	Hb
henry	H
herpes simplex virus	HSV
hertz	Hz
N-2-hydroxyethylpiperazine-N'-2-ethanesulfonic acid	HEPES
4-(2-hydroxyethyl)-1-piperazine ethanesulfonic acid	Hepes
high density lipoprotein	HDL
high frequency	hf
high-pressure(performance) liquid chromatography	HPLC
histidine	His
hour	h, hr
human growth hormone	hGH
human histocompatibility leukocyte antigens	HLA
human immunodeficiency virus	HIV
hyperfine structure	hfs
hypoxanthine, aminopterin, thymidine	r
idiotype	Id
immune response	Ir
immunoglobulin	Ig
immunoglobulin G, M, etc.	IgG, IgM, etc.
incomplete Freund's adjuvant	IFA
infective dose, 50%	ID_{50}
infrared	IR
inhibition constant	k_i
inhibitory concentration, 50%	IC_{50}
inosine	I, Ino
insulin-like growth factor I or II	IGF-I (II)
interferon (e.g., IFN-γ)	IFN
interleukin	IL
international unit	IU
intramuscular(ly), intraperitoneal(ly), intravenous(ly)	i.m., i.p., i.v.
intravenous	i.v.
isoelectric focusing	IEF
isoelectric point	pI
isoleucine	Ile
isotopes	^{14}C, ^3H, ^{32}P, etc.
joining region of Ig or T cell receptor for antigen	J region
joule	J
katal	kat
kelvin	K
kilo (10^3)	k
kilobase(s)	kb
kilocalorie	kcal
kilogram	kg
lethal dose, median	LD_{50}
leucine	Leu
light chain of Ig	L chain
limit	lim

lipopolysaccharide	LPS	nucleoside (unknown)	N, Nuc
logarithm, common	log	nucleotide(s)	nt
logarithm, natural	ln	number of observations	n
low density lipoprotein	LDL	observed	Obs., obs
lumen	lm	ohm	Ω
luteinizing hormone	LH	optical density	OD
lux	lx	optical rotatory dispersion	ORD
lysine	Lys	osmole	osmol
lytic unit	LU	ovalbumin	OVA
major histocompatibility complex	MHC	oxyhemoglobin	HbO_2
mannose	Man	parathyroid hormone	PTH
mass spectroscopy	MS	partial pressure of CO_2	PCO_2
maximum velocity	V_{max}	partial pressure of O_2	PO_2
mean	x	parts per million	ppm
mega (10^6)	M	pascal (newton/meter^2)	Pa
melting point	m.p.	per	/
2-mercaptoethanol	2-ME	per cent	%
messenger RNA	mRNA	peripheral blood lymphocyte	PBL
meter	m	peripheral blood mononuclear cell	PBMC
methionine	Met	peta (10^{15})	P
methyl	Me	phenyl	Ph
Michaelis constant	K_m	phenylalanine	Phe
micro (10^{-6})	μ	phenylmethylsulfonyl fluoride	PMSF
microgram	μg (not γ)	phorbol myristate acetate	PMA
microliter	μl (not λ)	phosphate-buffered saline	PBS
micrometer (not micron)	$\mu\mu$ (not μ)	phosphate (in compounds)	P or p
milli (10^{-3})	m	phosphate (inorganic)	P_i
milliliter	ml	phytohemagglutinin	PHA
minimum essential medium	MEM	pico (10^{-12})	p
minute	min	1,4-piperazinediethanesulfonic acid	Pipes
mitochondrial DNA	mtDNA	pokeweek mitogen	PWM
mixed leukocyte culture	MLC	polyacrylamide gel electrophoresis	PAGE
mixed leukocyte reaction	MLR	poly(adenylic acid), polyadenylate	poly(A)
molar (mol/liter)	M	polyethylenimine-cellulose	PEI-cellulose
mole	mol	polyethylene glycol	PEG
molecular weight	M_r	polymerase chain reaction	PCR
monoclonal antibody	MAb	probability	P
month	mo	proline	Pro
myoglobin	Mb	propyl	Pr
nano (10^{-9})	n	prostaglandin	PG
natural killer cell	NK cell	proton-motive force	pmf
negative logarithm of hydrogen ion activity	pH	quantum electrodynamics	QED
newton	N	radiation (ionizing, absorbed dose)	rad
New Zealand Black	NZB	radio frequency	rf
New Zealand White	NZW	radioimmunoassay	RIA
nicotinamide-adenine dinucleotide	NAD$^+$, NADH	rate constant	k
nicotinamide-adenine dinucleotide phosphate	NADP$^+$, NADPH	receptor (e.g., IL-2R)	R
nicotinamide mononucleotide	NMN	recombinant (e.g., rIFN-γ)	r
normal (concentration)	N	red blood cell	RBC
not determined	ND	relative molecular mass	M_r
not significant	NS		
nuclear magnetic resonance	NMR		

...ren fragment length polymorphism	RFLP
respectively	resp (mathematics)
respiratory quotient	RQ
retardation factor	R_f
revolutions per minute	rpm
ribonuclease	RNase
ribonucleic acid	RNA
ribose	Rib
ribosomal DNA, ribosomal RNA	rDNA, rRNA
ribosylthymine	T or Thd
root-mean-square	rms
second	s, sec
sedimentation coefficient	s
serine	Ser
sheet red blood cells	SRBC
siemens (mho)	S
sievert	Sv
simian virus 40	SV40
single-stranded (e.g., ssDNA)	ss
sodium dodecyl sulfate	$NaDodSO_4$, SDS
specific activity	sp act
specific gravity	sp gr
square centimeter	cm^2
standard atmosphere	atm
standard deviation of series	SD
standard error, standard error of mean	SE, SEM
standard saline citrate	SSC
standard temperature and pressure	stp
Student's t test	t test
subcutaneous(ly)	s.c.
Svedberg unit of sedimentation (coefficient (10^{13}s)	S
systemic lupus erythematosus	SLE
Tell receptor for antigen	TCR
tera (10^{12})	T
T helper cell	Th cell
thermodynamic temperature (kelvin)	K
thin-layer chromatography	TLC
threonine	Thr
thymidinedeoxyribose (also UdR, AdR)	TdR
thymidine (2′-deoxyribosylthymine)	dT or dThd
thymine	Thy
thyrotropin	TSH
transfer RNA	tRNA
trichloroacetic acid	TCA
trinitrophenyl	TNP
tris(hydroxymethyl) methylamine	Tris

tryptophan	Trp
T suppressor cell	Ts cell
tumor necrosis factor	TNF
tyrosine	Tyr
ultrahigh frequency	UHF
ultraviolet	UV
unit	U
unknown or "other" amino acid	Xaa
uracil	Ura
uridine	U, Urd
valine	Val
variable region of Ig	V region
variance ratio	F
very low density lipoprotein	VLDL
volt	V
volume	vol
volume ratio (volume per volume)	vol/vol
watt	W
week	wk
weight	wt
weight per volume	wt/vol
weight ratio (weight per weight)	wt/wt
year	yr

Electronic Submission: For papers that are close to final acceptance, some journals require authors to provide manuscripts in electronic form (on diskettes) and may accept a variety of word-processing formats or text (ASCII) files. When submitting diskettes: include a printout of manuscript version on diskette; put only latest version of manuscript on diskette; name file clearly; label diskette with file format and file name; and provide information on hardware and software used.

Consult journal's information for authors for acceptable formats, file- and diskette-naming conventions, number of copies to be submitted, and other details.

Experimentation, Animal: Indicate whether institution's or National Research Council's guide for, or any national law on, the care and use of laboratory animals was followed.

Experimentation, Human: Indicate whether procedures followed were in accordance with ethical standards of responsible committee on human experimentation (institutional or regional) or with Helsinki Declaration of 1976, as revised in 1983. Do not use patients' names, initials, or hospital numbers, especially in illustrative material.

If photographs of persons are used, either subjects must not be identifiable or obtain and submit written permission to use photographs.

Statistical Analyses: Describe statistical methods with enough detail to enable a knowledgeable reader with access to original data to verify reported results. Quantify findings and present with appropriate indicators of measurement error or uncertainty (such as confidence intervals). Avoid sole reliance on statistical hypothesis testing, such as P values, which fail to convey important quantitative information. Discuss eligibility of experimental subjects. Give details about randomization. Describe methods for and success of blinding of observations. Report treatment complications. Give numbers of observations. Report losses to observation (such as dropouts from a clinical trial). References for study design and statistical methods should be to standard works (with pages stated) when possible rather than to papers in which designs or methods were originally reported. Specify general-use computer programs used.

Put general description of methods in methods section. When data are summarized in Results section, specify statistical methods used for analysis. Restrict tables and figures to those needed to explain argument of paper and to assess its support. Use graphs as an alternative to tables with many entries; do not duplicate data in graphs and tables. Avoid nontechnical uses of technical terms in statistics, such as "random" (which implies a randomizing device), "normal," "significant," "correlation," and "sample." Define statistical terms, abbreviations, and symbols.

Units: Report measurements of length, height, weight, and volume in metric units (meter, kilogram, or liter) or their decimal multiples. Give temperatures in degrees Celsius and blood pressures in millimeters of mercury. Report hematologic and clinical chemistry measurements in metric system in terms of International System of Units (SI). Editors may request that alternative or non-SI units be added by authors before publication.

Units	Abbrev..	Interpretation
absolute	abs.	
amino acid		avoid amino-acid, amino acid except in compound words, e.g., amino aciduria
angstrom	Å	use SI unit: 1 Å = 0.1 mm
atmosphere (as unit of pressure)		give SI unit: 1 atm = 101.325 kN/m^2
atomic weight		use relative atomic mass (i.e., referred to the unit which is one-twelfth the mass of the atom ^{12}C)
atto-(10^{-18}×)	a	
average	av.	
bar		use SI unit: 1 bar = 100 kN/m^2 (100 kPa)
billion		avoid: specify whether 10^9 or 10^{12}
calorie		use joule: 1 cal = 4.1868 J
candela	cd	
centi- (10^{-2}×)	c	
centigrade		use Celsius (°C)
centimeter	cm, cm^2	

square centimeter, cubic centimeter	cm^3	
centimeters of water (pressure)	cm H$_2$O	give SI equivalent 1 cm H$_2$O = 98 N/m^2 (98 Pa) at s.t.p.
coenzyme A	CoA	
coenzyme I		use NAD
coenzyme II		use NADP
compare	cf	
concentrated	conc.	
concentration, molar (mol per liter)	c	
concentration, molecular	C	mol/l, mM = mmol/l
coulomb	C	
counts per minute	ct/min	
cubic	cu	but with units, e.g., cubic millimeter: mm^3
curie	Ci	1 Ci = 3.7 × 10^{10} disintegrations s^{-1}
cycles per second	Hz	hertz is preferred to c/s
dalton (not to be used for molecular masses)		1/12 of the mass of the pure nuclide ^{12}C, (i.e., 1.663 × 10^{-24} g
deca (10 ×)	da	
decay constant	λ	
deci (10^{-2} ×)	d	
decibel	dB	
decimal point		typed on the line: printed on the line or centrally; the comma should not be used as the decimal sign, since it is as a punctuation mark
degree Celsius (centigrade)	°C	
degree Fahrenheit	°F	avoid, give Celsius
dextrose		use glucose
diameter	diam.	
diameter, inside	i.d.	
diameter, outside	o.d.	
disintegrations (per min.)	d/min	
dissociation constant, negative logarithm of dyne	pK	use SI unit: 1 dyn = 10 μN
edition	edn	
editor(s)	ed (eds)	
einstein	Einstein	one "mol" of light (6.023 × 10^{23} quanta)
equation	eqn.	

Term	Symbol	Note
equivalent (unit of substance)		use mol
erg		use SI unit: 1 erg = 0.1 μJ
ethanol		not ethyl alcohol
extinction: $\log_{10} (I_0/I)$	E	
extinction coefficient-molar	ε	
femto ($10^{15} \times$)	f	e.g. femtoliter, fl
fluid ounce		give SI equivalent 1 fl oz = 28 ml
foot		give SI equivalent 1 ft - 0.3048 m
gallon (U.K.)		give SI equivalent 1 gal = 4.5460921 liters
gamma: γ (10^{-6} gram)		use microgram (μg)
gauss	G	use SI unit, the tesla = 1 G = 10^{-4} T
genus	gen.	
genus, new	gen.nov.	
giga ($10^9 \times$)	G	
gram	g	
gram-ion		use mol
gram-molecule		use mol
gravity, due to acceleration	g	e.g., $10,000 \times g$
hecto ($10^2 \times$)	h	
henry	H	
hertz	Hz	use for frequency of repetition of cyclic phenomena
hour(s)	h	not hr, hrs
horsepower	hp	give SI equivalent: 1 hp = 747.700 W
hundredweight		give SI equivalent 1 cwt = 51 kg
inch	in	give SI equivalent: 1 in = 25.4 mm
inhibitor constant	K_i	
joule	J	
kelvin	K	replaces °K
kilo-($10^3 \times$)	k	e.g., kilogram: kg
kilocalorie	kcal	give SI equivalent 1 kcal = 4.1868 kJ
kilogram force; kilopond		use SI unit: 1 kgf = 1 kp = 9.80665 N

Term	Symbol	Note
kilowatt hour		use SI unit, joule 1 kW h = 3.6 J
Krebs cycle		use tricarboxylic cycle
Krebs-Ringer solution		give reference or composition
lambda: λ (10^{-6} liter)		use microliter (μl) or mm^3
liter	l	do not abbreviate in typing or printing if it could be confused with 1 (numeral one)
logarithm (any base)	log	
logarithm to base e	$\log_e$ or ln	
logarithm to base 10	$\log^{10}$ or lg	
maximum	max.	
maximum velocity	V_{max}	
meta-	m-	
methanol		not methyl alcohol
meter	m	
Michaelis constant	K_m	
micro ($10^{-6} \times$)	μ	e.g., microgram: μg
microgram	μg	
micromicro- ($10^{-12} \times$)		use pico-
micron: μ (10^{-6} meter)		use micrometer (μm)
mile		give SI equivalent: 1 statute mile = 1.60934 km 1 nautical mile = 1.85318 km
milli- ($10^{-3} \times$)	m	e.g., milligram: mg
milliequivalent	mEq	the symbol is conventional, although equivalent = equiv. millimol (mmol) is preferred
millimeters of mercury (pressure)	mm Hg	give SI equivalent 1 mm Hg = 133.322 Pa
millimicro- ($10^{-9} \times$)		use nano-
millimicron		use nanometer (nm)
millimol	mmol	
milliosmol		millimol (mmol) is preferred
minimum	min.	
minute (time)	min	
molal		use mol/kg
molar (concentration)	M	better mol/l; molar strictly means

molar (with chemical formula	M-	"molecular mass divided by volume"; the symbol M never means mol (amount of substance)
mole (unit of amount of substance)	mol	replaces gram-molecule, gram-ion, gram-formula, gram-atom, equivalent, etc.
molecular mass	mol.mass	
molecular weight	mol.wt.	ratios; it is incorrect to add the word 'daltons'
month		do not abbreviate
nano- ($10^{-9} \times$)	n	e.g., nanometer: nm
newton	N	
normal (concentration)	N	avoid; mol/l is preferred
normal (with chemical formula)	N-	
normal temperature and pressure		avoid; use "standard temperature and pressure"
optical density	OD	
ortho	o-	
ounce	oz	give SI equivalent: 1 oz = 28.3495 g
oxygen consumption	Q_{O2}	measured as μl per mg dry mass per hour; Q_{O2} is acceptable and simpler to print
parts per million	parts/10^6	ppm is not allowed
pascal	Pa	name for newton per square meter
per cent	%	or in full; never use for 100 ml; avoid for concentrations of solutions
per thousand	/10^3	avoid $^0/_{00}$; preferably write in full
petroleum ether		use light petroleum; give b.p. range
pico- ($10{-12} \times$)	p	e.g., picogram, pg; not μμg
poise	P	use SI units/1P = 0.1 N s m^{-2}
pound	lb	give SI equivalent: 1 lb = 0.4536 kg
pound-force	lbf	give SI equivalent: 1 lbf/in^2 = 6.89476 kN/m^2
pressure	p	
probability	P	
retardation factor	R_f	

revolutions per minute	rpm	
roentgen	R	
second (unit of time)	s	
solidus – per		e.g., mol/min or mol · min $^{-1}$ or mol per min; avoid more than one solidus as this leads to mathematically ambiguous statements: e.g.. not mol/min/g protein, but mol/min · g protein or mol · min $^{-1}$ · (g protein)$^{-1}$
soluble	sol.	
solution	soln.	
species	sp.	plural spp.
species, new	sp.nov.	
stokes	St	use SI unit: 1 St = 10^{-4}m^2/s
strain (taxonomy)	str.	
temperature	temp.	
tension (of gases in liquids; physiology)		use partial pressure, e.g., P_{CO2}, P_{O2}
tera- ($10^{12} \times$)	T	
tesla	T	
time	t	
torr	Torr	use SI unit: 1 Torr = 1 mm Hg = 133.322 pa
volume	vol.	
vol (symbol for quantity)	V	
volume, liquid phase (physiology)	Q	
volume by volume	v/v	only for two components; otherwise use "by vol."
volt	V	
wavelength	λ	
watt	W	
weber	Wb	
weight by volume	w/v	preferably use mass/vol.
yard	yd	give SI equivalent: 1 yd = 0.9144 m
year	a	this is international symbol; yr is conventional

SOURCE
JAMA 1993; 269:2282–2286.